Fundamentals
of Ceramics

Fundamentals of Ceramics

Michel W. Barsoum
Drexel University

The McGraw-Hill Companies, Inc.
New York   St. Louis   San Francisco   Auckland   Bogotá
Caracas   Lisbon   London   Madrid   Mexico City   Milan   Montreal
New Delhi   San Juan   Singapore   Sydney   Tokyo   Toronto

McGraw-Hill

A Division of *The McGraw·Hill Companies*

FUNDAMENTALS OF CERAMICS

1 2 3 4 5 6 7 8 9 0 DOC DOC 9 0 9 8 7 6

ISBN 0-07-005521-1

This book was set in Times Roman by M&N Toscano.
The editors were B.J. Clark and John M. Morriss;
the production supervisor was Leroy A. Young.
The cover was designed by Christopher Brady.
R. R. Donnelley & Sons Company was printer and binder.

Library of Congress Cataloging-in-Publication Data

Barsoum, M. W.
     Fundamentals of Ceramics / by Michel Barsoum.
         p.   cm. — (McGraw-Hill series in materials science and engineering)
     Includes bibliographical references.
     ISBN 0-07-005521-1 — ISBN 0-07-005522-X (SM)
     1. Ceramic engineering. I. Title. II. Series.
TP807.B37   1997
666–dc20
                                            96-24295

http://www.mhcollege.com

About the Author

Professor Michel W. Barsoum received a B.Sc. degree in materials engineering from the American University in Cairo, Egypt, in 1977, an M.Sc. from the University of Missouri-Rolla, Rolla, Missouri, in the field of ceramics engineering in 1980, and a Ph.D. from the Massachusetts Institute of Technology in materials science and engineering, ceramics program, in 1985. Upon graduation from MIT he joined the faculty of the Department of Materials Engineering at Drexel University in Philadelphia, Pennsylvania. Professor Barsoum has lectured and published extensively in the field of ceramics. He is a member of the American Ceramics Society. In late 1995 Professor Barsoum and his coworkers at Drexel University synthesized two families of ternary compounds, namely, Ti_3SiC_2, Ti_3GeC_2 and the H-phases. The polycrystalline nanolaminate structure of these compounds endowed them with a set of unique and remarkable properties.

Dedicated to my parents and Patricia with love.

CONTENTS

PREFACE

It is a mystery to me why, in a field as interesting, rich, and important as ceramics, a basic fundamental text does not exist. My decision to write this text was made almost simultaneously with my having to teach my first introductory graduate class in ceramics at Drexel a decade ago. Naturally, I assigned Kingery, Bowen, and Uhlmann's *Introduction to Ceramics* as the textbook for the course. A few weeks into the quarter, however, it became apparent that KBU's book was difficult to teach from and more importantly to learn from. Looking at it from the student's point of view it was easy to appreciate why — few equations are derived from first principles. Simply writing down a relationship, in my opinion, does not constitute learning; true understanding only comes when the trail that goes back to first principles is made clear. However, to say that this book was influenced by KBU's book would be an understatement – the better word would be inspired by it, and for good reason — it remains an authoritative, albeit slightly dated, text in the field.

In writing this book I had a few guiding principles. First, nearly all equations are derived, usually from first principles, with the emphasis being on the physics of the problem, sometimes at the expense of mathematical rigor. However, whenever that trade-off is made, which is not often, it is clearly noted in the text. I have kept the math quite simple, nothing more complicated than differentiation and integration. The aim in every case was to cover enough of the fundamentals, up to a level deep enough to allow the reader to continue his or her education by delving, without too much difficulty, into the most recent literature. In todays's fast-paced world, it is more important than ever to understand the fundamentals.

Second, I wanted to write a book that more or less "stood alone" in the sense that it did not assume much prior knowledge of the subject from the reader. Basic chemistry, physics, mathematics, and an introductory course in materials science or engineering are the only prerequisites. In that respect I believe this book will appeal to, and could be used as a textbook in, other than material science and engineering departments, such as chemistry or physics.

Pedagogically I have found that students in general understand concepts and ideas best if they are given concrete examples rather than generalized treatments. Thus maybe, at the expense of elegance and brevity, I have opted for that approach. It is hoped that once the concepts are well understood, for at least one system, the reader will be able to follow more advanced and generalized treatments that can be found in many of the references that I have included at the end of every chapter.

Successive drafts of this book have been described by some reviewers as being arid, a criticism that I believe has some validity and that I have tried to address. Unfortunately, it was simply impossible to cover the range of topics, at the depth I wanted to, and be flowery and descriptive at the same time (the book is already over 650 pages long).

Another area where I think this book falls short is in its lack of what I would term a healthy skepticism (à la Feynman lectures, for instance). Nature is too complicated, and ceramics in particular, to be neatly packaged into monosize dispersed spheres and their corresponding models, for example.

I thus sincerely hope that these two gaps will be filled in by the reader and especially the instructor. First, a little bit of "fat" should make the book much more appetizing — examples from the literature or the instructor's own experience would be just what is required. Second, a dose of skepticism concerning some of the models and their limitation is required. Being an experimentalist, I facetiously tell my students that when theory and experiment converge one of them is probably wrong.

This book is aimed at junior, senior, and first-year graduate students in any materials science and engineering program. The sequence of chapters makes it easy to select material for a one-semester course. This might include much of the material in Chapters 1 to 8, with additional topics from the later chapters. The book is also ideally suited to a two-quarter sequence, and I believe there may even be enough material for a two-semester sequence.

The book can be roughly divided into two parts. The first nine chapters deal with bonding, structure, and the physical and chemical properties that are influenced mostly by the type of bonding rather than the microstructure, such as defect structure and the atomic and electronic transport in ceramics. The coverage of the second part, Chaps. 11 to 16, deals with properties that are more microstructure dependent, such as fracture toughness, optical, magnetic, and dielectric properties. In between the two parts lies Chap. 10, which deals with the science of sintering and microstructural development. The technological aspects of processing have been deliberately omitted for two reasons. The first is that there are a number of good undergraduate texts that deal with the topic. Second, it is simply not possible to discuss that topic and do it justice in a section of a chapter.

Chapter 8 on phase diagrams was deliberately pushed back until the notions of defects and nonstoichiometry (Chap. 6) and atom mobility (Chap. 7) were introduced. The chapter on glasses (Chap. 9) follows Chap. 8 since once again the notions introduced in Chaps. 6, 7, and 8 had to be developed in order to explain crystallization.

And while this is clearly not a ceramics handbook, I have included many important properties of binary and ternary ceramics collected over 10 years from numerous sources. In most chapters I also include, in addition to a number of well-tested problem sets with their numerical answers, worked examples to help the student through some of the trickier concepts. Whenever a property or phenomenon is introduced, a section clearly labeled experimental details has been included. It has been my experience that many students lacked a knowledge of how certain physical properties or phenomena are measured experimentally, which needless to say makes it rather fruitless to even try to attempt to explain them. These sections are *not* intended, by any stretch of the imagination, to be laboratory guides or procedures.

Finally, it should also be pointed out that Chaps. 2, 5, and 8 are by no means intended to be comprehensive — but are rather included for the sake of completion, and to highlight aspects that are referred to later in the book as well as to refresh the reader's memory. It is simply impossible to cover inorganic chemistry, thermodynamics, and phase equilibria in three chapters. It is in these chapters that a certain amount of prior knowledge by the reader is assumed.

I would like to thank Dr. Yoachim Maier for hosting me, and the Max-Planck Institute fur Festkorperforchung in Stuttgart for its financial support during my sabbatical year, when considerable progress was made on the text. The critical readings of some of the chapters by C. Schwandt, H. Naefe, N. Nicoloso, and G. Schaefer is also gratefully acknowledged. I would especially like to thank Dr. Rowland M. Cannon for helping me sort out, with the right spirit I may add, Chaps. 10 through 12 — his insight, as usual, was invaluable.

I would also like to thank my colleagues in the Department of Materials Engineering and Drexel University for their continual support during the many years it took to finish this work. I am especially indebted to Profs. Roger Doherty and Antonious Zavaliangos with whom I had many fruitful and illuminating discussions. Finally I would like to take this opportunity to thank all those who have, over the many years I was a student, first at the American University in Cairo, Egypt, followed by the ones at the University of Missouri-Rolla and, last but not least, MIT, taught and inspired me. One has only to leaf through the book to appreciate the influence Profs. H. Anderson, R. Coble, D. Kingery, N. Kreidl, H. Tuller, D. Uhlmann, B. Wuench, and many others had on this book.

Comments, criticisms, suggestions, and correction, from all readers, especially students, for whom this book was written, are most welcome. Please send them to me at the Department of Materials Engineering, Drexel University, Philadelphia, PA 19104, or by e-mail at Barsoumw@post.drexel.edu.

Finally, I would like to thank my friends and family, who have been a continuous source of encouragement and support.

Michel W. Barsoum

Fundamentals
of Ceramics

CHAPTER 1

Introduction

All that is, at all
Lasts ever, past recall,
Earth changes,
But thy soul and God stand sure,
Time's wheel runs back or stops:
Potter and clay endure.

Robert Browning

1.1
INTRODUCTION

The universe is made up of elements that in turn consist of neutrons, protons, and electrons. There are roughly 100 elements, each possessing a unique electronic configuration determined by its atomic number Z, and the spatial distribution and energies of their electrons. What determines the latter requires some understanding of quantum mechanics and is discussed in greater detail in the next chapter.

One of the major triumphs of quantum theory was a rational explanation of the **periodic table** (see inside cover) of the elements that had been determined from experimental observation long before the advent of quantum mechanics. The periodic table places the elements in horizontal rows of increasing atomic number and vertical columns or **groups**, so that all elements in a group display similar chemical properties. For instance, all the elements of group VII B, referred to as halides, exist as diatomic gases characterized by a very high reactivity. Conversely, the elements of group VIII, the noble gases, are monoatomic and are chemically extremely inert.

A large majority of the elements are solids at room temperature, and because they are shiny, ductile, and good electrical and thermal conductors, they are considered *metals*. A fraction of the elements — most notably, N, O, H, the halides, and the noble gases — are gases at room temperature. The remaining

1

elements are covalently bonded solids that, at room temperature, are either insulators (B, P, S, C[1]) or semiconductors (Si, Ge). These elements, for reasons that will become apparent very shortly, will be referred to as *nonmetallic elemental solids* (NMESs). Now that these distinctions have been made, it is possible to answer the not too trivial question, What is a ceramic?

1.2
DEFINITION OF CERAMICS

In the previous section, the elements were classified as being metallic, gaseous, or NMES. Very few elements, however, are used in their pure form; most often they are alloyed or reacted with other elements to form engineering materials. The latter can be broadly classified as metals, polymers, semiconductors, or ceramics, with each class having distinctive properties that reflect the differences in the nature of the bonding. In metals, the bonding is predominantly metallic, where delocalized electrons provide the "glue" that holds the positive ion cores together. This delocalization of the bonding electrons has far-reaching ramifications since it is responsible for properties most associated with metals: ductility, thermal and electrical conductivity, reflectivity, and other distinctive properties.

Polymers consist of very long, for the most part, C-based chains to which other organic atoms (for example; C, H, N, Cl, F) and molecules are attached. The bonding within the chains is strong, directional, and covalent, while the bonding between chains is relatively weak. Thus, the properties of polymers as a class are dictated by the weaker bonds, and consequently they possess lower melting points, higher thermal expansion coefficients, and lower stiffnesses than most metals or ceramics.

Semiconductors are covalently bonded solids that, in addition to Si and Ge already mentioned, include GaAs, CdTe, and InP, among others. The usually strong covalent bonds holding semiconductors together make their mechanical properties quite similar to those of ceramics (i.e.; brittle and hard).

Ceramics can be defined as *solid compounds that are formed by the application of heat, and sometimes heat and pressure, comprising at least one metal and a nonmetallic elemental solid or a nonmetal, a combination of at least two nonmetallic elemental solids, or a combination of at least two nonmetallic elemental solids and a nonmetal*. To illustrate, consider the following examples:

---

[1]   In the form of diamond. It is worth noting that although graphite is a good electrical conductor, it is not a metal since it is neither shiny nor ductile.

Magnesia,[2] or MgO, is a ceramic since it is a solid compound composed of a metal, Mg, bonded to the nonmetal, O_2. Silica is also a ceramic since it combines an NMES and a nonmetal. Similarly, TiC and ZrB_2 are ceramics since they combine the metals (Ti, Zr) and the NMESs (C, B). SiC is a ceramic because it combines two NMESs. Also note that ceramics are not limited to binary compounds: $BaTiO_3$, $YBa_2Cu_3O_7$, and Ti_3SiC_2 are all perfectly respectable class members.

It follows that the oxides, nitrides, borides, carbides, and silicides (not to be confused with silicates) of all metals and NMESs are ceramics; which, needless to say, leads to a vast number of compounds. This number becomes even more daunting when it is appreciated that the silicates are also, by definition, ceramics. Because of the abundance of oxygen and silicon in nature, silicates are ubiquitous; rocks, dust, clay, mud, mountains, sand — in short, the vast majority of the earth's crust — are composed of silicate-based minerals. When it is also appreciated that even cement, bricks, and concrete are essentially silicates, the inescapable conclusion is that we live in a ceramic world.

In addition to their ubiquitousness, silicates were singled out above for another reason, namely, as the distinguishing chemistry between traditional and advanced ceramics. Before that distinction is made clear, however, it is important to explore how atoms are arranged in three dimensions.

1.3
CRYSTALLINE VERSUS AMORPHOUS SOLIDS

The arrangement of atoms in solids, in general, and ceramics, in particular, will exhibit **long-range order**, **only short-range order**, or a combination of both.[3] Solids that exhibit long-range order[4] are referred to as **crystalline solids**, while those in which that periodicity is lacking are known as **amorphous**, **glassy**, or **noncrystalline solids**.

---

[2] A note on nomenclature: The addition of the letter a to the end of an element name usually implies that one is referring to the oxide of that element. For example, while silicon refers to the element Si, silica is SiO_2 or the oxide of silicon. Similarly, alumina is the oxide of aluminum or Al_2O_3; magnesium; magnesia; etc.

[3] Strictly speaking, only solids in which grain boundaries are absent, i.e., single crystals, can be considered to possess *only* long-range order. As discussed below, the vast majority of crystalline solids possess grain boundaries that are areas in which the long-range order breaks down, and thus should be considered as a combination of amorphous and crystalline areas. However, given that in most cases the volume fraction of the grain boundary regions is much less than 0.01, it is customary to describe polycrystalline materials as possessing only long-range order.

[4] Any solid that exhibits long-range order must also exhibit short-range order, but not vice versa.

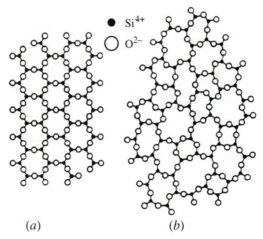

$\bullet$ Si^{4+}

$\bigcirc$ O^{2-}

(a)   (b)

Figure 1.1
(a) Long-range order; (b) short-range order.

The difference between the two is best illustrated schematically, as shown in Fig. 1.1. From the figure it is obvious that a solid possesses long-range order when the atoms repeat with a periodicity that is much greater than the bond lengths. Most metals and ceramics, with the exception of glasses and glass-ceramics (see Chap. 9), are crystalline.

Since, as discussed throughout this book, the details of the lattice patterns strongly influence the macroscopic properties of ceramics, it is imperative to understand the rudiments of crystallography.

1.4
ELEMENTARY CRYSTALLOGRAPHY

As noted above, long-range order requires that atoms be arrayed in a three-dimensional pattern that repeats. The simplest way to describe a pattern is to describe a **unit cell** within that pattern. A *unit cell* is defined as the smallest region in space that, when repeated, completely describes the three-dimensional pattern of the atoms of a crystal. Geometrically, it can be shown that there are only seven unit cell *shapes*, or **crystal systems,** that can be stacked together to fill three-dimensional space. The seven systems, shown in Fig. 1.2, are cubic, tetragonal, orthorhombic, rhombohedral, hexagonal, monoclinic, and triclinic. The various systems are distinguished from one another by the lengths of the unit cell edges and the angles between the edges, collectively known as the **lattice parameters** or **lattice constants** ($a, b, c,\ \alpha,\ \beta$, and γ in Fig. 1.2).

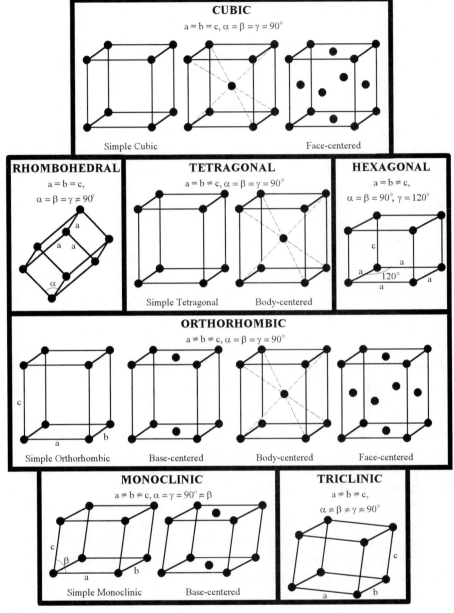

Figure 1.2
Geometric characteristics of the 7 crystal systems and 14 Bravais lattices.

It is useful to think of the crystal systems as the *shape* of the "bricks" that make up a solid. For example, the bricks can be cubes, hexagons, parallelepipeds, etc. And while the shape of the bricks is a very important descriptor of a crystal structure, it is insufficient. In addition to the shape of the brick, it is imperative to know the *symmetry* of the lattice pattern within each brick as well as the actual location of the atoms on these lattice sites. Only then would the description be complete.

It turns out that if one considers only the symmetry within each unit cell, the number of possible permutations is limited to 14. The 14 arrangements, shown in Fig. 1.2, are also known as the **Bravais lattices**. A **lattice** can be defined as an indefinitely extending arrangement of points, each of which is surrounded by an identical grouping of neighboring points. To carry the brick analogy a little further, the Bravais lattice represents the *symmetry* of the *pattern* found on the bricks

Finally, to describe the atomic arrangement, one must describe the symmetry of the **basis**, defined as the atom or grouping of atoms located at each lattice site. When the basis is added to the lattices, the total number of possibilities increases to 32 **point groups**.[5]

1.5
CERAMIC MICROSTRUCTURES

Crystalline solids exist as either single crystals or polycrystalline solids. A *single crystal* is a solid in which the periodic and repeated arrangement of atoms is perfect and extends throughout the entirety of the specimen without interruption. A *polycrystalline solid* is composed of a collection of many single crystals, termed **grains**, separated from one another by areas of disorder known as **grain boundaries** (see Chap. 6 for more details). See Fig. 1.3.

Typically, in ceramics the grains are in the range of 1 to 50μm and are visible only under a microscope. The shape and size of the grains, together with the presence of porosity, second phases, etc., and their distribution describe what is termed the **microstructure**. As discussed in later chapters, many of the properties of ceramics are microstructure-dependent.

---

[5]   For more information, see, e.g., A. Kelly and G. W. Groves, *Crystallography and Crystal Defects*, Lougmans, London, 1970.

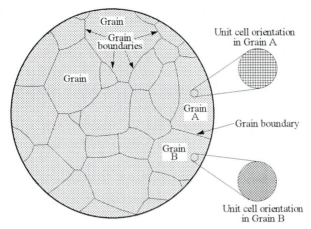

Figure 1.3
Schematic of a polycrystalline sample. A polycrystal is made up of many grains separated from one another by regions of disorder known as grain boundaries.

1.6
TRADITIONAL VERSUS ADVANCED CERAMICS

Most people associate the word *ceramics* with pottery, sculpture, sanitary ware, tiles, etc. And whereas this view is not incorrect, it is incomplete because it considers only the traditional, or silicate-based, ceramics. Today the field of ceramic science or engineering encompasses much more than silicates and can be divided into traditional and advanced ceramics. Before the distinction is made, however, it is worthwhile to trace the history of ceramics and people's association with them.

It has long been appreciated by our ancestors that some muds, when wet, were easily moldable into shapes that upon heating became rigid. The formation of useful articles from fired mud must constitute one of the oldest and more fascinating of human endeavors. Fired-clay articles have been traced to the dawn of civilization. The usefulness of this new material, however, was limited by the fact that when fired, it was porous and thus could not be used to carry liquids. Later the serendipitous discovery was made that when heated and slowly cooled, some sands tended to form a transparent, water-impervious solid, known today as glass. From that point on, it was simply a matter of time before glazes were developed that rendered clay objects not only watertight, but also quite beautiful.

With the advent of the industrial revolution, structural clay products such as bricks and heat-resistant refractory materials for the large-scale smelting of metals were developed. And with the discovery of electricity and the need to distribute it, a market was developed for electrically insulating silicate-based ceramics.

Traditional ceramics are characterized by mostly silicate-based porous microstructures that are quite coarse, nonuniform, and multiphase. They are typically formed by mixing clays and feldspars, followed by forming either by slip casting or on a potter's wheel, firing in a flame kiln to sinter them, and finally glazing.

In a much later stage of development, other ceramics that were not clay- or silicate-based depended on much more sophisticated raw materials, such as binary oxides, carbides, perovskites, and even completely synthetic materials for which there are no natural equivalents. The microstructures of these advanced ceramics were at least an order of magnitude finer and more homogeneous and much less porous than those of their traditional counterparts. It is the latter — the **advanced** or **technical ceramics** — with which this book is mainly concerned.

1.7
GENERAL CHARACTERISTICS OF CERAMICS

As a class, ceramics are hard, wear-resistant, brittle, prone to thermal shock, refractory, electrically and thermally insulative, intrinsically transparent, nonmagnetic, chemically stable, and oxidation-resistant. As with all generalizations, there will be exceptions; some ceramics are electrically and thermally quite conductive, while others are even superconducting. An entire industry is based on the fact that some ceramics are magnetic.

One of the main purposes of this book is to answer the question of why ceramics exhibit the properties they do. And while this goal will have to wait until later chapters, at this point it is worthwhile to list some of the applications for which ceramics have been or are being developed.

1.8
APPLICATIONS

Traditional ceramics are quite common, from sanitary ware to fine chinas and porcelains to glass products. Currently ceramics are being considered for uses that only two decades ago were inconceivable; applications ranging from ceramic engines to optical communications, electrooptic applications to laser materials, and substrates in electronic circuits to electrodes in photoelectrochemical devices. Some of the recent applications for which ceramics are used and/or are prime candidates are listed in Table 1.1.

Historically, ceramics were mostly exploited for their electrical insulative properties, for which electrical porcelains and aluminas are prime examples. Today, so-called electrical and electronic ceramics play a pivotal role in any modern technological society. For example, their insulative properties together with their low-loss factors and excellent thermal and environmental stability make them the materials of choice for substrate materials in electronic packages. The development of the perovskite family with exceedingly large dielectric constants holds a significant market share of capacitors produced. Similarly, the development of magnetic ceramics based on the spinel ferrites is today a mature technology. Other electronic/electrical properties of ceramics that are being commercially exploited include piezoelectric ceramics for sensors and actuators, nonlinear $I-V$ characteristics for circuit protection, and ionically conducting ceramics for use as solid electrolytes in high-temperature fuel cells and as chemical sensors.

These applications do not even include the recently discovered superconducting ceramics, currently being developed for myriad applications.

Mechanical applications of ceramics at room temperature usually exploit hardness, wear, and corrosion resistance. The applications include cutting tools, nozzles, valves, and ball bearings in aggressive environments. However, it is the refractoriness of ceramics and their ability to sustain high loads at high temperatures, together with their low densities, that has created the most interest. Applications in this area include all ceramic engines for transportation and turbines for energy production. In principle, the advantages of an all-ceramic engine are several and include lower weight, higher operating temperatures which translates to higher efficiencies, and less pollution. It is also envisioned that such engines would not require cooling and maybe not even any lubrication, which once more would simplify the design of the engine, reducing the number of moving parts and lowering the overall weight of the vehicle.

TABLE 1.1

Properties and applications of advanced ceramics

Property	Applications (examples)
Thermal	
Insulation	High-temperature furnace linings for insulation (oxide fibers such as silica, alumina, and zirconia)
Refractoriness	High-temperature furnace linings for insulation and containment of molten metals and slags
Thermal conductivity	Heat sinks for electronic packages (AlN)
Electrical and dielectric	
Conductivity	Heating elements for furnaces (SiC, ZrO_2, $MoSi_2$)
Ferroelectricity	Capacitors (Ba-titanate-based materials)
Low-voltage insulators	Ceramic insulation (porcelain, steatite, forsterite)
Insulators in electronic applications	Substrates for electronic packaging and electrical insulators in general (Al_2O_3, AlN)
Insulators in hostile environments	Spark plugs (Al_2O_3)
Ion-conducting	Sensors, fuel cells, and solid electrolytes (ZrO_2, β-alumina, etc.)
Semiconducting	Thermistors and heating elements (oxides of Fe, Co, Mn)
Nonlinear I-V characteristics	Current surge protectors (Bi-doped ZnO, SiC)
Gas-sensitive conductivity	Gas sensors (SnO_2, ZnO)
Magnetic and superconductive	
Hard magnets	Ferrite magnets[(Ba, Sr)O $\cdot$ $6Fe_2O_3$]
Soft magnets	Transformer cores [(Zn, M)Fe_2O_3, with M = Mn, Co, Mg]; magnetic tapes (rare-earth garnets)
Superconductivity	Wires and SQUID magnetometers ($YBa_2Cu_3O_7$)
Optical	
Transparency	Windows (soda-lime glasses), cables for optical communication (ultra-pure silica)
Translucency and chemical inertness	Heat- and corrosion-resistant materials, usually for Na lamps (Al_2O_3, MgO)
Nonlinearity	Switching devices for optical computing ($LiNbO_3$)
IR transparency	Infrared laser windows (CaF_2, SrF_2, NaCl)
Nuclear applications	
Fission	Nuclear fuel (UO_3, UC), fuel cladding (C, SiC), neutron moderators (C, BeO)
Fusion	Tritium breeder materials (zirconates and silicates of Li, Li_2O); fusion reactor lining (C, SiC, Si_3N_4, B_4C)
Chemical	
Catalysis	Filters (zeolites); purification of exhaust gases
Anticorrosion properties	Heat exchangers (SiC), chemical equipment in corrosive environments
Biocompatibility	Artificial joint prostheses (Al_2O_3)
Mechanical	
Hardness	Cutting tools (SiC whisker-reinforced Al_2O_3, Si_3N_4)
High-temperature strength retention	Stators and turbine blades, ceramic engines (Si_3N_4)
Wear resistance	Bearings (Si_3N_4)

1.9
THE FUTURE

Paradoxically, because interest in advanced ceramics came later than interest in metals and polymers, ceramics are simultaneously our oldest and newest solids. Consequently, working in the field of ceramics, while sometimes frustrating, can ultimately be quite rewarding and exciting. There are a multitude of compounds that have never been synthesized, let alone characterized. Amazing discoveries are always around the corner, as the following two examples illustrate.

In 1986, the highest temperature at which any material became superconducting, i.e.; the ability to conduct electricity with virtually no loss, was around –250°C, or 23 K. In that year a breakthrough came when Bednorz and Muller,[6] shattered the record by demonstrating that a layered lanthanum, strontium copper oxide became superconducting at the relatively balmy temperature of 46 K. This discovery provoked a worldwide interest in the subject, and a few months later the record was again almost doubled, to about 90 K. The record today is in excess of 120 K.

Toward the end of 1995, we discovered a family of ternary compounds with truly remarkable properties.[7, 8] These compounds combine many of the best attributes of metals and ceramics. Like metals, they are excellent electrical and thermal conductors, are superbly machinable, are *not* susceptible to thermal shock, and behave plastically at higher temperatures. Like ceramics they are oxidation-resistant and extremely refractory (melting temperatures in excess of 3000°C), and most importantly they maintain their strength to very high temperatures, when compared to even the best superalloys known today! Furthermore, as a consequence of their bonding they behave like polycrystalline nanolaminates.

Traditional ceramics have served humanity well for at least the past 10 millennia. However, the nature of modern technology, with its ever-mounting demands on materials, has prompted researchers to take a second look at these stone-age materials, and it now appears that our oldest material is shaping up to be a material of the future. It is my sincerest hope that this book will inspire a new generation of talented and dedicated researchers to embark on a voyage of discovery in this most exciting of fields.

---

[6] T. G. Bednorz and K. A. Muller, *Z. Phys. B*, **64**: 189 (1986).
[7] M. W. Barsoum and T. El-Raghy, *J. Amer. Cer. Soc.*, **79**, [7], 1953 (1996).
[8] M. W. Barsoum and T. El-Raghy, submitted for publication.

PROBLEMS

1.1. (*a*) According to the definition of a ceramic given in the text, would you consider Si_3N_4 a ceramic? How about CCl_4, $SiCl_4$, or SiF? Explain.

(*b*) Would you consider $TiAl_3$ a ceramic? How about AlC_4, BN, CN, or SiB_6? Explain.

1.2. (*a*) How many crystal systems would you expect in two dimensions? Draw them and characterize them by their lattice parameters.

Answer: 4

(*b*) How many Bravais lattices are there in two dimensions?

ADDITIONAL READING

1. W. D. Kingery, H. K. Bowen, and D. R. Uhlmann, *Introduction to Ceramics*, 2d ed., Wiley, New York, 1976.
2. A. R. West, *Solid State Chemistry and Its Applications*, Wiley, Chichester, England, 1984.
3. R. J. Brook, Ed., *Concise Encyclopedia of Advanced Ceramic Materials*, Pergamon, New York, 1991.
4. D. Richerson, *Modern Ceramic Engineering*, 2d ed., Marcel Dekker, New York, 1992.
5. K. Easterling, *Tomorrow's Materials*, Institute of Metals, London, 1988.
6. P. A. Cox, *The Electronic Structure and Chemistry of Solids*, Oxford University Press, New York, 1987.
7. J. P. Schaffer, A. Saxena, S. D. Antolovich, T. H. Sanders, and S. B. Warner, *The Science and Design of Engineering Materials*, Irwin, Chicago, 1995.
8. C. Kittel, *Introduction to Solid State Physics,* 6th ed., Wiley, New York, 1986.
9. L. Solymar and D. Walsh, *Lectures on the Electrical Properties of Materials*, 4th ed., Oxford University Press, New York, 1988.
10. N. N. Greenwood, *Ionic Crystals, Lattice Defects and Non-Stoichiometry*, Butterworth, London, 1968.
11. L. Pauling, *The Nature of the Chemical Bond*, Cornell University Press, Ithaca, New York, 1960.
12. L. Pauling and E. B. Wilson, *Introduction to Quantum Mechanics with Applications to Chemistry*, McGraw-Hill, New York, 1935.
13. L. Azaroff, *Introduction to Solids*, McGraw-Hill, New York, 1960.

CHAPTER 2

Bonding in Ceramics

All things are Atoms: Earth and Water, Air and Fire, all
Democritus foretold. Swiss Paracelsus, in alchemic lair
Saw sulfur, salt and mercury unfold Amid Millennial
hopes of faking Gold. Lavoisier dethroned
Phlogiston; then molecules analysis made bold
Forays into the gases: Hydrogen
Stood naked in the dazzled sight of Learned Men.

John Updike; *The Dance of the Solids*[†]

2.1
INTRODUCTION

The properties of any solid and the way its atoms are arranged are determined primarily by the nature and directionality of the interatomic bonds holding the solid together. Consequently, to understand variations in properties, it is imperative to appreciate how and why a solid is "glued" together.

This glue can be strong, which gives rise to *primary bonds*, which can be ionic, covalent, or metallic. Usually Van der Waals and hydrogen bonds are referred to as *secondary bonds* and are weaker. *In all cases, however, it is the attractive electrostatic interaction between the positive charges of the nuclei and the negative charges of the electrons that is responsible for the cohesion of solids.*

Very broadly speaking, ceramics can be classified as being either ionically or covalently bonded, and for the sake of simplicity, this notion is maintained throughout this chapter. However, that this simple view needs some modification will become apparent in Chap. 4; bonding in ceramics is neither purely covalent nor purely ionic, but a mixture of both.

[†]   J. Updike, **Midpoint and other Poems**, A. Knopf, Inc., New York, New York, 1969. Reprinted with permission.

Before the intricacies of bonding are described, a brief review of the shape of atomic orbitals is presented in Sec. 2.2. The concept of electronegativity and how it determines the nature of bonding in a ceramic is introduced in Sec. 2.3. In Secs. 2.4 and 2.5, respectively; the ionic bond is treated by a simple electrostatic model, and how such bonds lead to the formation of ionic solids is discussed.

The more complex covalent bond, which occurs by the overlap of electronic wave functions, is discussed in Secs. 2.6 and 2.7. In Sec. 2.8, how the interaction of wave functions of more than one atom results in the formation of energy bands in crystalline solids is elucidated.

It is important to point out, at the outset, that much of this chapter is only intended to be a review of what the reader is assumed to be familiar with from basic chemistry. Most of the material in this chapter is covered in college-level chemistry textbooks.

2.2
STRUCTURE OF ATOMS

Before bonding between atoms is discussed; it is essential to appreciate the energetics and shapes of single atoms. Furthermore, since bonding involves electrons which obey the laws of quantum mechanics, it is worthwhile to review the major conclusions, as they apply to bonding, of quantum theory.

1. The confinement of a particle results in the quantization of its energy levels. Said otherwise, whenever a particle is attracted to or confined in space to a certain region, its energy levels are necessarily quantized. As discussed shortly, this follows directly from Schrödinger's wave equation.
2. A given quantum level cannot accept more than two electrons, which is *Pauli's exclusion principle*.
3. It is impossible to simultaneously know with certainty both the momentum and the position of a moving particle, which is the *Heisenberg uncertainty principle*.

The first conclusion pertains to the shape of the orbitals; the second explains why higher energy orbitals are stable and populated; and the third elucidates, among other things, why an electron does not spiral continually and fall into the nucleus.

In principle, the procedure for determining the shape of an atomic or molecular orbital is quite simple and involves solving Schrödinger's equation with the appropriate boundary conditions, from which one obtains the all-important wave function of the electron, which leads in turn to the probability of finding the electron in a given volume. To illustrate, consider the simplest possible case, that of the hydrogen atom, which consists of a proton and an electron.

2.2.1 The Hydrogen Atom

Schrödinger's time-independent equation in one dimension is given by:

$$\frac{\partial^2 \psi}{\partial x^2} + \frac{8\pi^2 m_e}{h^2}\left(E_{tot} - E_{pot}\right)\psi = 0 \tag{2.1}$$

where m_e is the mass of the electron, 9.11×10^{-31} kg; h is Plank's constant, 6.625×10^{-34} J·s, and E_{tot} is the total (kinetic + potential) energy of the electron. The potential energy of the electron E_{pot} is nothing but the coulombic attraction between the electron and the proton,[9] given by:

$$E_{pot} = \frac{z_1 z_2 e^2}{4\pi\varepsilon_0 r} = -\frac{e^2}{4\pi\varepsilon_0 r} \tag{2.2}$$

where z_1 and z_2 are the charges on the electron and nucleus, -1 and $+1$, respectively; e is the elementary electronic charge 1.6×10^{-19} C; ε_0 is the permittivity of free space, 8.85×10^{-12} C^2/(J·m); and r is the distance between the electron and the nucleus.

Now ψ is the wave function of the electron and by itself has no physical meaning, but

$$|\psi(x,y,z;t)|^2 dx\,dy\,dz$$

gives the probability of finding an electron at any time t in a volume element $dx\,dy\,dz$. The higher ψ^2 is in some volume in space, the more likely the electron is to be found there.

For the simplest possible case of the hydrogen atom, the orbital is spherically symmetric; and so it is easier to work in spherical coordinates. Thus instead of Eq. (2.1), the differential equation to solve is

$$\frac{h^2}{8\pi^2 m_e}\left(\frac{\partial^2 \psi}{\partial r^2} + \frac{2}{r}\frac{\partial \psi}{\partial r}\right) + \left(E_{tot} + \frac{e^2}{4\pi\varepsilon_0 r}\right)\psi = 0 \tag{2.3}$$

---

[9] For the hydrogen atom z_1 and z_2 are both unity. In general, however, the attraction between an electron and a nucleus has to reflect the total charge on the nucleus, i.e., the atomic number of the element involved.

where E_{pot} was replaced by the value given in Eq. (2.2). The solution of this equation yields the functional dependence of ψ on r, and it can be easily shown that (see Prob. 2.1)

$$\psi = \exp(-c_0 r) \tag{2.4}$$

satisfies Eq. (2.3), provided the energy of the electron is given by

$$E_{tot} = -\frac{m_e e^4}{8\varepsilon_0^2 h^2} \tag{2.5}$$

and

$$c_0 = \frac{\pi m_e e^2}{\varepsilon_0 h^2} \tag{2.6}$$

As mentioned above, ψ by itself has no physical significance, but ψ^2 is the probability of finding an electron in a given volume element. It follows that the probability distribution function W of finding the electron in a thin spherical shell between r and $r + dr$ is obtained by multiplying $|\psi|^2$ by the volume of that shell (see hatched area in Fig. 2.1a), or

$$W = 4\pi r^2 |\psi|^2 dr \tag{2.7}$$

In other words, the y axis is simply a measure of the probability of finding the electron at any distance r. Figure 2.1a indicates that the probabilities of finding an electron at the nucleus or very far from the nucleus are negligible, but that somewhere in between that probability is at a maximum. This distance is known as the **Bohr radius** r_B (see Fig. 2.1a). The importance of this result lies in appreciating that fact that (1) while the electron spends most of its time at a distance r_B, its spatial extent is clearly not limited to that value and (2) the best one can hope for when discussing the location of an electron is to talk about the probability of finding it in some volume. It is worth noting here that by combining Eqs. (2.4) to (2.7) and finding the location of the maximum, it can be easily shown that $r_B = 1/c_0$.

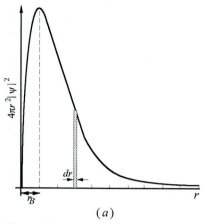

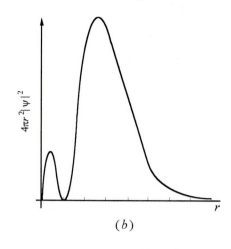

(a)   (b)

Figure 2.1

(a) Radial distribution function of 1s state electron. The crosshatched strip has a volume $4\pi r^2 dr$ which, when multiplied by $|\psi|^2$, gives the probability of finding the electron between r and $r + dr$. The probability of finding the electron very near or very far from the nucleus approaches zero. The most probable position for the electron is at a distance $r_B = 1/c_0$. (b) Radial distribution function for an electron in the 2s level. Energy of this electron is one-fourth that of the 1s state.

WORKED EXAMPLE 2.1. Calculate the ground state energy level of the electron in the hydrogen atom, and compare the result with the experimentally derived value of −13.6 eV.

Answer[10]

Using Eq. (2.5) gives

$$E_{tot} = -\frac{me^4}{8\varepsilon_0^2 h^2} = -\frac{(9.1 \times 10^{-31})(1.6 \times 10^{-19})^4}{8(8.85 \times 10^{-12})^2(6.63 \times 10^{-34})^2} \qquad (2.8)$$

$$= -2.165 \times 10^{-18} \text{J} = -13.6 \text{ eV}$$

That value is the lowest energy level of a hydrogen electron, a fact which was experimentally known well before the advent of quantum mechanics. This result was one of the first and greatest successes of quantum mechanics. It is important to note

---

[10] In all problems and throughout this book, SI units are used almost exclusively.

that the energy of the electron is a negative number, which means that the electron in the vicinity of the proton has a lower energy than it would have if it were an infinite distance away (which corresponds to zero energy).[11]

Equation (2.4) is but one of many possible solutions. For example, it can also be shown that

$$\psi(r) = A(1 + c_1 r)\exp\left(\frac{-rc_0}{2}\right) \tag{2.9}$$

is another perfectly legitimate solution to Eq. (2.3), provided that Eq. (2.5) is divided by 4. The corresponding radial distribution function is plotted in Fig. 2.1b. It follows that the energy of this electron is $-13.6/4$ and it will spend most of its time at a distance given by the second maximum.

To generalize, for a spherically symmetric wave function, the solution (given here without proof) is

$$\psi_n(r) = e^{-c_n r} L_n(r)$$

where L_n is a polynomial. The corresponding energies are given by

$$E_{\text{tot}} = \frac{-me^4}{8n^2 \varepsilon_0^2 h^2} = -\frac{13.6 \text{ eV}}{n^2} \tag{2.10}$$

where n is known as the *principal quantum number*. As n increases, the energy of the electron increases (i.e., becomes less negative) and its spatial extent increases.

2.2.2   Orbital Shape and Quantum Numbers

Equations (2.4) and (2.9) were restricted to spherical symmetry. An even more generalized solution is

---

11   An interesting question had troubled physicists for some time, as they were developing the theories of quantum mechanics: What prevented the electron from continually losing energy, spiraling into the nucleus, and releasing an infinite amount of energy? Originally the classical explanation was that the angular momentum of the electron gives rise to the apparent repulsion — this explanation is invalid in this case, however, because s electrons have no angular momentum (see Chap. 15). The actual reason is related to the Heisenberg uncertainty principle and goes something like this: If an electron is confined to a smaller and smaller volume, the uncertainty in its position Δx decreases. But since $\Delta x \Delta p = h$ is a constant, it follows that its momentum p, or, equivalently, its kinetic energy, will have to increase as Δx decreases. Given that the kinetic energy scales with r^{-2} but the potential energy scales only as r^{-1}, an energy minimum has to be established at a given equilibrium distance.

$$\psi_{n,l,m} = R_{nl}(r)Y_l^m(\theta, \pi)$$

where Y_l depends on θ and π. Consequently, the size and shape of the orbital will depend on the specific solution considered. It can be shown that each orbital will have associated with it three characteristic interrelated quantum numbers, labeled n, l, and m_l, known as the *principal*, *angular*, and *magnetic quantum numbers*, respectively.

The **principal quantum number** n determines the *spatial extent* and *energy* of the orbital. The **angular momentum quantum number**[12] l, however; determines the *shape* of the orbital for any given value of n and can only assume the values 0, 1, 2, 3 .... $n-1$. For example, for $n = 3$, the possible values of l are 0, 1, and 2.

The **magnetic quantum number** m_l is related to the *orientation* of the orbital in space. For a given value of l, m_l can take on values from $-l$ to $+l$. For example, for $l = 2$, m_l can be -2, -1, 0, $+1$, or $+2$. Thus for any value of l there are $2l+1$ values of m_l. All orbitals with $l = 0$ are called *s orbitals* and are spherically symmetric (Fig. 2.1). When $l = 1$, the orbital is called a *p orbital*, and there are three of these (Fig. 2.2a), each corresponding to a different value of m_l associated with $l = 1$, that is, $m_l = -1$, 0, $+1$. These three orbitals are labeled p_x, p_y, and p_z because their lobes of maximum probability lie along the x, y, and z axes, respectively. It is worth noting that although each of the three p orbitals is nonspherically symmetric, their sum gives a spherically symmetric distribution of ψ^2.

When $l = 2$, there are five possible values of m_l; and the *d orbitals*, shown schematically in Fig. 2.2b, result. Table 2.1 summarizes orbital notation up to $n = 3$. The physical significance of l and m_l and their relationships to the angular momenta of atoms are discussed in greater detail in Chap. 15.

One final note: The conclusions arrived at so far tend to indicate that all sublevels with the same n have exactly the same energy, when in reality they have slightly different energies. Also a fourth quantum number, the **spin quantum number** m_s, which denotes the direction of electron spin, was not mentioned. Both of these omissions are a direct result of ignoring relativistic effects which, when taken into account, are fully accounted for.

---

12   Sometimes l is referred to as the *orbital-shape quantum number*.

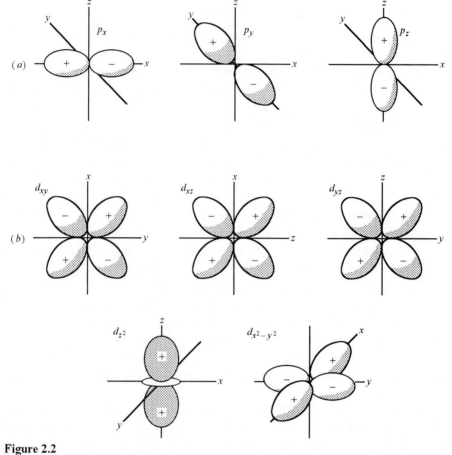

Figure 2.2
(*a*) Shape of *p* orbitals (top three) and (*b*) *d* orbitals (lower five).

TABLE 2.1
Summary of orbitals and their notation

n	l	Orbital name	No. of m_l orbitals	Full designation of orbitals
1	0	$1s$	1	$1s$
2	0	$2s$	1	$2s$
	1	$2p$	3	$2p_x, 2p_y, 2p_z$
3	0	$3s$	1	$3s$
	1	$3p$	3	$3p_x, 3p_y, 3p_z$
	2	$3d$	5	$3d_{zy}, 3d_{x^2-y^2}, 3d_{xy}, 3d_{xz}, 3d_{yz}$

2.2.3   Polyelectronic Atoms and the Periodic Table

Up to now the discussion has been limited to the simplest possible case, namely, that of the hydrogen atom — the only case for which an exact solution to the Schrödinger equation exists. The solutions for polyelectronic atoms are similar to that of the hydrogen atom except that these solutions are inexact and are much more difficult to obtain. Fortunately, the basic shapes of the orbitals do not change, the concept of quantum numbers remains useful, and, with some modifications, the hydrogen-like orbitals can account for the electronic structure of atoms having many electrons.

The major modification involves the energy of the electrons. As the nuclear charge or atomic number Z increases, the potential energy of the electron has to decrease accordingly, since a large positive nuclear charge now attracts the electron more strongly. This can be accounted for, as a first and quite crude approximation, by assuming that the electrons are noninteracting, in which case it can be shown that the energy of an electron is given by

$$E_n = -13.6 \frac{Z^2}{n^2} \text{ eV}$$

The actual situation is more complicated, however, due to electron-electron repulsions and electron screening — with both effects contributing to an increase in E_n. Conceptually this is taken into account by introducing the **effective nuclear charge** Z_{eff} which takes into account the notion that the actual nuclear charge experienced by an electron is always less than or equal to the actual charge on the nucleus. This can be easily grasped by comparing the actual first ionization energy (IE) of helium (He), that is; –24.59 eV (see Table 2.2), for which $Z = 2$ and $n = 1$, to what one would expect had there been no electron-electron interaction, or $2^2 \times (-13.6)/1^2$, or –54.4 eV. This brief example illustrates the dramatic effect of electron-electron interactions on the ionization energy of He and the importance of the concept of effective charge. Note that the measured second ionization energy for He listed in Table 2.2 is exactly –54.4 eV!

As the number of electrons increases, they are forced by virtue of Pauli's exclusion principle to occupy higher and higher energy levels, i.e., higher n values. This in turn leads to the **aufbau principle**, the periodic table (see inside cover), and a unique electronic configuration for each element as summarized in Table 2.2.

TABLE 2.2
Electronic configuration and first and second ionization energies of the elements

Z	Atom	Orbital electronic configuration	First IE, eV	Second IE, eV
1	H	$1s^1$	13.598	—
2	He	$1s^2$	24.587	54.416
3	Li	$(He)2s^1$	5.392	75.638
4	Be	$(He)2s^2$	9.322	18.211
5	B	$(He)2s^2 2p^1$	8.298	25.154
6	C	$(He)2s^2 2p^2$	11.260	24.383
7	N	$(He)2s^2 2p^3$	14.534	29.601
8	O	$(He)2s^2 2p^4$	13.618	35.116
9	F	$(He)2s^2 2p^5$	17.422	34.970
10	Ne	$(He)2s^2 2p^6$	21.564	40.962
11	Na	$(Ne)3s^1$	5.139	47.286
12	Mg	$(Ne)3s^2$	7.646	15.035
13	Al	$(Ne)3s^2 3p^1$	5.986	18.828
14	Si	$(Ne)3s^2 3p^2$	8.151	16.345
15	P	$(Ne)3s^2 3p^3$	10.486	19.725
16	S	$(Ne)3s^2 3p^4$	10.360	23.330
17	Cl	$(Ne)3s^2 3p^5$	12.967	23.810
18	Ar	$(Ne)3s^2 3p^6$	15.759	27.630
19	K	$(Ar)4s^1$	4.340	31.625
20	Ca	$(Ar)4s^2$	6.113	11.871
21	Sc	$(Ar)4s^2 3d^1$	6.540	12.800
22	Ti	$(Ar)4s^2 3d^2$	6.820	13.580
23	V	$(Ar)4s^2 3d^3$	6.740	14.650
24	Cr	$(Ar)4s^1 3d^5$	6.766	16.500
25	Mn	$(Ar)4s^2 3d^5$	7.435	15.640
26	Fe	$(Ar)4s^2 3d^6$	7.870	16.180
27	Co	$(Ar)4s^2 3d^7$	7.860	17.060
28	Ni	$(Ar)4s^2 3d^8$	7.635	18.168
29	Cu	$(Ar)4s^1 3d^{10}$	7.726	20.292
30	Zn	$(Ar)4s^2 3d^{10}$	9.394	17.964
31	Ga	$(Ar)4s^2 3d^{10} 4p^1$	5.999	20.510
32	Ge	$(Ar)4s^2 3d^{10} 4p^2$	7.899	15.934
33	As	$(Ar)4s^2 3d^{10} 4p^3$	9.810	18.633
34	Se	$(Ar)4s^2 3d^{10} 4p^4$	9.752	21.190
35	Br	$(Ar)4s^2 3d^{10} 4p^5$	11.814	21.800
36	Kr	$(Ar)4s^2 3d^{10} 4p^6$	13.999	24.359
37	Rb	$(Kr)5s^1$	4.177	27.280
38	Sr	$(Kr)5s^2$	5.695	11.030
39	Y	$(Kr)5s^2 4d^1$	6.380	12.240
40	Zr	$(Kr)5s^2 4d^2$	6.840	13.130
41	Nb	$(Kr)5s^1 4d^4$	6.880	14.320
42	Mo	$(Kr)5s^1 4d^5$	7.099	16.150

TABLE 2.2 (continued)

Electronic configuration and first and second ionization energies of the elements

Z	Atom	Orbital electronic configuration	First IE, eV	Second IE, eV
43	Tc	$(Kr)5s^2 4d^5$	7.280	15.260
44	Ru	$(Kr)5s^1 4d^7$	7.370	16.760
45	Rh	$(Kr)5s^1 4d^8$	7.460	18.080
46	Pd	$(Kr)4d^{10}$	8.340	19.430
47	Ag	$(Kr)5s^1 4d^{10}$	7.576	21.490
48	Cd	$(Kr)5s^2 4d^{10}$	8.993	16.908
49	In	$(Kr)5s^2 4d^{10} 5p^1$	5.786	18.869
50	Sn	$(Kr)5s^2 4d^{10} 5p^2$	7.344	14.632
51	Sb	$(Kr)5s^2 4d^{10} 5p^3$	8.641	16.530
52	Te	$(Kr)5s^2 4d^{10} 5p^4$	9.009	18.600
53	I	$(Kr)5s^2 4d^{10} 5p^5$	10.451	19.131
54	Xe	$(Kr)5s^2 4d^{10} 5p^6$	12.130	21.210
55	Cs	$(Xe)6s^1$	3.894	25.100
56	Ba	$(Xe)6s^2$	5.212	10.004
57	La	$(Xe)6s^2 5d^1$	5.577	11.060
58	Ce	$(Xe)6s^2 4f^1 5d^1$	5.470	10.850
59	Pr	$(Xe)6s^2 4f^3$	5.420	10.560
60	Nd	$(Xe)6s^2 4f^4$	5.490	10.720
61	Pm	$(Xe)6s^2 4f^5$	5.550	10.900
62	Sm	$(Xe)6s^2 4f^6$	5.630	11.070
63	Eu	$(Xe)6s^2 4f^7$	5.670	11.250
64	Gd	$(Xe)6s^2 4f^7 5d^1$	5.426	13.900
65	Tb	$(Xe)6s^2 4f^9$	5.850	11.520
66	Dy	$(Xe)6s^2 4f^{10}$	5.930	11.670
67	Ho	$(Xe)6s^2 4f^{11}$	6.020	11.800
68	Er	$(Xe)6s^2 4f^{12}$	6.100	11.930
69	Tm	$(Xe)6s^2 4f^{13}$	6.180	12.050
70	Yb	$(Xe)6s^2 4f^{14}$	6.254	12.170
71	Lu	$(Xe)6s^2 4f^{14} 5d^1$	5.426	13.900
72	Hf	$(Xe)6s^2 4f^{14} 5d^2$	7.000	14.900
73	Ta	$(Xe)6s^2 4f^{14} 5d^3$	7.890	—
74	W	$(Xe)6s^2 4f^{14} 5d^4$	7.980	—
75	Re	$(Xe)6s^2 4f^{14} 5d^5$	7.880	—
76	Os	$(Xe)6s^2 4f^{14} 5d^6$	8.700	—
77	Ir	$(Xe)6s^2 4f^{14} 5d^7$	9.100	—
78	Pt	$(Xe)6s^1 4f^{14} 5d^9$	9.000	—
79	Au	$(Xe)6s^1 4f^{14} 5d^{10}$	9.225	—
80	Hg	$(Xe)6s^2 4f^{14} 5d^{10}$	10.437	18.756
81	Tl	$(Xe)6s^2 4f^{14} 5d^{10} 6p^1$	6.108	20.428
82	Pb	$(Xe)6s^2 4f^{14} 5d^{10} 6p^2$	7.416	15.032
83	Bi	$(Xe)6s^2 4f^{14} 5d^{10} 6p^3$	7.289	16.600

Source: Adapted from J. Huheey, *Inorganic Chemistry*, 2d ed., Harper & Row, New York, 1978.

WORKED EXAMPLE 2.2. (*a*) What are the electronic configurations of He, Li, and F? (*b*) Identify the first transition metal series. What feature do these elements have in common?

Answer

(*a*) Helium ($Z = 2$) has two electrons, which can be accommodated in the $1s$ state as long as their spins are opposite. Hence the configuration is $1s^2$. Since this is a closed shell configuration, He is a very inert gas. Lithium ($Z = 3$) has three electrons; two are accommodated in the $1s$ shell, and the third has to occupy a higher energy state, namely, $n = 2$ and $l = 0$, giving rise to the electronic configuration of Li: $1s^2 2s^1$.   Similarly, the nine electrons of fluorine are distributed as follows: $1s^2 2s^2 2p^5$.

(*b*) The first series transition metals are Sc, Ti, V, Cr, Mn, Fe, Co, and Ni. They all have partially filled d orbitals. Note that Cu and Zn, which have completely filled d orbitals, are sometimes also considered to be transition metals, although strictly speaking, they would not be since their d orbitals are filled (see Table 2.2).

2.3
IONIC VERSUS COVALENT BONDING

In the introduction to this chapter, it was stated that ceramics, very broadly speaking, can be considered to be either ionically or covalently bonded. The next logical question that this chapter attempts to address is, What determines the nature of the bond?

Ionic compounds generally form between very active metallic elements and active nonmetals. For reasons that will become clear shortly, the requirements for an AB ionic bond to form are that A be able to lose electrons easily and B be able to accept electrons without too much energy input. This restricts ionic bonding to mostly metals from groups 1A, IIA, and part of IIIA as well as some of the transition metals and the most active nonmetals of groups VIIA and VIA (see the periodic table).

For covalent bonding to occur, however, ionic bonding must be unfavorable. This is tantamount to saying that the energies of the bonding electrons of A and B must be comparable, because if the electron energy on one of the atoms is much lower than that on the other, electron transfer from one to the other would occur and ionic bonds would tend to form instead.

These qualitative requirements for the formation of each type of bonding, while shedding some light on the problem, do not have much predictive capability as to the nature of the bond that will form. In an attempt to semiquantify the answer, Pauling[13] established a scale of relative **electronegativity** or "electron greed" of atoms and defined electronegativity to be *the power of an atom to attract electrons to itself.* Pauling's electronegativity scale is listed in Table 2.3 and was obtained by arbitrarily fixing the value of H at 2.2. With this scale it becomes relatively simple to predict the nature of a bond. If two elements forming a bond have similar electronegativities, they will tend to share the electrons between them and will form covalent bonds but if the electronegativity difference Δx between them is large (indicating that one element is much greedier than the other), the electron will be attracted to the more electronegative element, forming ions which in turn attract each other. Needless to say, the transition between ionic and covalent bonding is far from sharp, and except for homopolar bonds that are purely covalent, all bonds will have both an ionic and a covalent character (see Prob. 2.15). However, as a very rough guide, a bond is considered predominantly ionic when $\Delta x > 1.7$ and predominantly covalent if $\Delta x < 1.7$.

Each type of bond and how it leads to the formation of a solid will be discussed now separately, starting with the simpler, namely, the ionic bond.

2.4
IONIC BONDING

Ionically bonded solids are made up of charged particles — positively charged ions, called **cations**, and negatively charged ions, called **anions**. Their mutual attraction holds the solid together. As discussed at greater length throughout this book, ionic compounds tend to have high melting and boiling points because the bonds are usually quite strong and omnidirectional. Ionic compounds are also hard and brittle and are poor electrical and thermal conductors.

To illustrate the energetics of ionic bonding consider the bond formed between Na and Cl. The electronic configuration of Cl (atomic number $Z = 17$) is $[1s^2 2s^2 2p^6]3s^2 3p^5$, while that of Na ($Z = 11$) is $[1s^2 2s^2 2p^6]3s^1$. When an Na and a Cl atom are brought into close proximity, a bond will form (the reason will become evident in a moment) by the transfer of an electron from the Na atom to the Cl atom, as shown schematically in Fig. 2.3. The Na atom configuration becomes

---

13   L. Pauling, *The Nature of the Chemical Bond*, 3d ed., Cornell University Press, Ithaca, NY, 1960.

TABLE 2.3
Relative electronegativity scale of the elements

Element	Electronegativity		Element	Electronegativity
1. H	2.20		42. Mo(II)	2.16
2. He			Mo(III)	2.19
3. Li	0.98		43. Tc	1.90
4. Be	1.57		44. Ru	2.20
5. B	2.04		45. Rh	2.28
6. C	2.55		46. Pd	2.20
7. N	3.04		47. Ag	1.93
8. O	3.44		48. Cd	1.69
9. F	3.98		49. In	1.78
10. Ne			50. Sn(II)	1.80
11. Na	0.93		Sn(IV)	1.96
12. Mg	1.31		51. Sb	2.05
13. Al	1.61		52. Te	2.10
14. Si	1.90		53. I	2.66
15. P	2.19		54. Xe	2.60
16. S	2.58		55. Cs	0.79
17. Cl	3.16		56. Ba	0.89
18. Ar			57. La	1.10
19. K	0.82		58. Ce	1.12
20. Ca	1.00		59. Pr	1.13
21. Sc	1.36		60. Nd	1.14
22. Ti(II)	1.54		62. Sm	1.17
23. V(II)	1.63		64. Gd	1.20
24. Cr(II)	1.66		66. Dy	1.22
25. Mn(II)	1.55		67. Ho	1.23
26. Fe(II)	1.83		68. Er	1.24
Fe(III)	1.96		69. Tm	1.25
27. Co(II)	1.88		71. Lu	1.27
28. Ni(II)	1.91		72. Hf	1.30
29. Cu(I)	1.90		73. Ta	1.50
Cu(II)	2.00		74. W	2.36
30. Zn(II)	1.65		75. Re	1.90
31. Ga(III)	1.81		76. Os	2.20
32. Ge(IV)	2.01		77. Ir	2.20
33. As(III)	2.18		78. Pt	2.28
34. Se	2.55		79. Au	2.54
35. Br	2.96		80. Hg	2.00
36. Kr	2.90		81. Tl(I)	1.62
37. Rb	0.82		82. Pb(II)	1.87
38. Sr	0.95		83. Bi	2.02
39. Y	1.22		90. Th	1.30
40. Zr(II)	1.33		92. U	1.70
41. Nb	1.60			

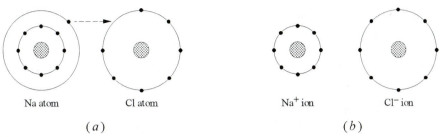

Na atom   Cl atom   Na$^+$ ion   Cl$^-$ ion

(a) (b)

Figure 2.3

(a) Transfer of an electron from Na atom to Cl atom results in the formation of (b) a cation and an anion. Note that the cation is smaller than the atom and vice versa for the anion.

$[1s^2 2s^2 2p^6]$ and is now +1 positively charged. The Cl atom, however, gains an electron, acquires a net negative charge with an electronic structure $[1s^2 2s^2 2p^6]3s^2 3p^6$, and is now an anion. Note that after this transfer of charge, the configuration of each of the ions corresponds to those of the noble gases, Ne and Ar, respectively.

The work done to bring the ions from infinity to a distance r apart is once again given by Coulomb's law [Eq. (2.2)]:

$$E_{pot} = \frac{z_1 z_2 e^2}{4 \pi \varepsilon_0 r} \qquad (2.11)$$

In this case, z_1 and z_2 are the *net* charges on the ions (+1 and −1 for NaCl, −2 and +3 for Al$_2$O$_3$, etc.). When z_1 and z_2 are of opposite signs, E_{pot} is negative, which is consistent with the fact that energy is released as the ions are brought together from infinity. A plot of Eq. (2.11) is shown in Fig. 2.4a (lower curve), from which it is clear that when the ions are infinitely separated, the interaction energy vanishes, as one would expect. Equation (2.11) also predicts that as the distance between the ions goes to zero, the ions should fuse together and release an infinite amount of energy! That this does not happen is obvious; NaCl does, and incidentally we also, exist.

It follows that for a stable lattice to result, a repulsive force must come into play at short distances. As discussed above, the attraction occurs from the *net* charges on the ions. These ions, however, are themselves made up of positive and negative entities, namely, the nuclei of each ion, but more importantly, the electron cloud surrounding each nucleus. As the ions approach each other, these like charges repel and prevent the ions from coming any closer.

The repulsive energy term is positive by definition and is usually given by the empirical expression

$$E_{rep} = \frac{B}{r^n} \tag{2.12}$$

where B and n are empirical constants that depend on the material in question. Sometimes referred to as the **Born exponent**, n usually lies between 6 and 12. Equation (2.12) is also plotted in Fig. 2.4a (top curve), from which it is clear that the repulsive component dominates at small r, but decreases very rapidly as r increases.

The net energy E_{net} of the system is the sum of the attractive and repulsive terms, or

$$E_{net} = \frac{z_1 z_2 e^2}{4 \pi \varepsilon_0 r} + \frac{B}{r^n} \tag{2.13}$$

When E_{net} is plotted as a function of r (middle curve in Fig. 2.4a), it goes through a minimum, at a distance denoted by r_0. The minimum in the curve corresponding to the equilibrium situation can be found readily from

$$\frac{dE_{net}}{dr}\bigg|_{r=r_0} = 0 = -\frac{z_1 z_2 e^2}{4 \pi \varepsilon_0 r_0^2} - \frac{nB}{r_0^{n+1}} \tag{2.14}$$

By evaluating the constant B and removing it from Eq. (2.13), it can be easily shown that (see Prob. 2.3) the depth of the energy well E_{bond} is given by

$$E_{bond} = \frac{z_1 z_2 e^2}{4 \pi \varepsilon_0 r_0} \left(1 - \frac{1}{n}\right) \tag{2.15}$$

where r_0 is the equilibrium separation between the ions. The occurrence of this minimum is of paramount importance since it defines a bond; i.e., when two ions are brought closer together from infinity, they will attract each up to an equilibrium distance r_0 and liberate an amount of energy given by Eq. (2.15). Conversely, E_{bond} can be thought of as the energy required to pull the ions apart.

It is important to note that Eq. (2.14) is also an expression for the *net* force between the ions, since by definition

$$F_{net} = \frac{dE_{net}}{dr} = -\frac{z_1 z_2 e^2}{4 \pi \varepsilon_0 r^2} - \frac{nB}{r^{n+1}} \tag{2.16}$$

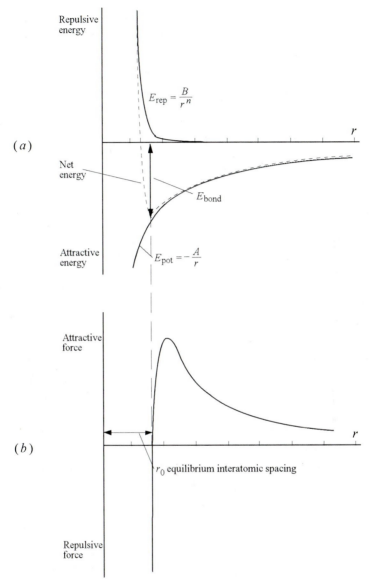

Figure 2.4

(a) Energy versus distance curves for an ionic bond. The net energy is the sum of attractive and repulsive energies, which gives rise to an energy well. (b) Corresponding force versus distance curve. This curve is the derivative of the net energy curve shown in (a). Note that when the energy is at a minimum, the net force is zero.

F_{net} is plotted in Fig. 2.4*b*. For distances greater than r_0 the net force on the ions is attractive; and for distances less than r_0; the net force is repulsive. At r_0 the net force on the ions is zero [Eq. (2.14)] which is why r_0 is the equilibrium interatomic spacing. Figure 2.4*a* and *b* illustrates a fundamental law of nature, namely, that at equilibrium the energy is minimized and the net force on a system is zero.

2.5
IONICALLY BONDED SOLIDS

The next logical question is, How do such bonds lead to the formation of a solid? After all, an ionic solid is made up of roughly 10^{23} of these bonds. The other related question of importance has to do with the energy of the lattice. This energy is related to the stability of a given ionic structure and directly or indirectly determines such critical properties as melting temperatures, thermal expansion, stiffness, and others, discussed in Chap. 4. This section addresses how the lattice energy is calculated and experimentally verified, starting with the simple electrostatic model that led to Eq. (2.15).

2.5.1   Lattice Energy Calculations

First, a structure or packing of the ions has to be assumed;[14] and the various interactions between the ions have to be taken into account. Begin with NaCl, which has one of the simplest ionic structures known (Fig. 2.5), wherein each Na ion is surrounded by 6 Cl ions and vice versa. Focusing on the central cation in Fig. 2.5 shows clearly that this cation is attracted to 6 Cl^- at distance r_0, repelled by 12 Na ions at distance $\sqrt{2}r_0$, attracted to 8 Cl^- ions at $\sqrt{3}r_0$, etc. Summing up the electrostatic interactions,[15] one obtains

---

[14]   This topic is discussed in greater detail in the next chapter and depends on the size of the ions involved, the nature of the bonding, etc.

[15]   Strictly speaking, this is not exact, since in Eq. (2.17) the repulsive component of the ions that were not nearest neighbors was neglected. If that interaction is taken into account, an exact expression for E_{sum} is given by

$$E_{sum} = \frac{-z_1 z_2 e^2 \alpha}{4\pi \varepsilon_0 r} + \frac{B\beta}{r^n}$$

where β is another infinite series. It is important to note that such a refinement does not in any way alter the final result, namely, Eq. (2.18).

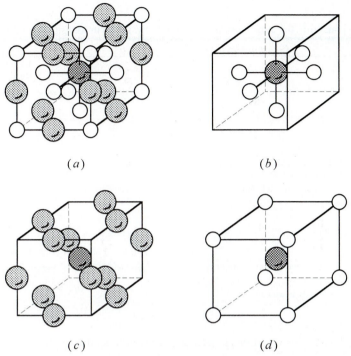

(a) (b)

(c) (d)

Figure 2.5
(a) Schematic of the NaCl structure. (b) The first 6 nearest neighbors are attracted to the central cation, (c) the second 12 nearest neighbors at a distance $\sqrt{2}r_0$ are repelled, (d) the third 8 nearest neighbors are attracted, etc.

$$E_{\text{sum}} = \frac{z_1 z_2 e^2}{4\pi\varepsilon_0 r_0}\left(1 - \frac{1}{n}\right)\left(\frac{6}{1} - \frac{12}{\sqrt{2}} + \frac{8}{\sqrt{3}} - \frac{6}{\sqrt{4}} + \frac{24}{\sqrt{5}} - \cdots\right)$$

$$= \frac{z_1 z_2 e^2}{4\pi\varepsilon_0 r_0}\left(1 - \frac{1}{n}\right)\alpha$$

(2.17)

The second term in parentheses is an alternating series that converges to some value α, known as the **Madelung constant.** Evaluation of this constant, though straightforward, is tedious because the series converges quite slowly. The Madelung constants for a number of ceramic crystal structures are listed in Table 2.4.

The total electrostatic attraction for 1 mole of NaCl in which there are twice Avogadro's number N_{Av} of ions but only N_{Av} bonds is

$$E_{\text{latt}} = \frac{N_{\text{Av}} z_1 z_2 e^2 \alpha}{4\pi\varepsilon_0 r_0}\left(1 - \frac{1}{n}\right)$$

(2.18)

TABLE 2.4
Madelung constants for some common ceramic crystal structures (see Chap. 3)

Structure	Coordination number	α [†]	α_{conv} [‡]
NaCl	6:6	1.7475	1.7475
CsCl	8:8	1.7626	1.7626
Zinc blende	4:4	1.6381	1.6381
Wurtzite	4:4	1.6410	1.6410
Fluorite	8:4	2.5190	5.0387
Rutile	6:3	2.4080[§]	4.1860[§]
Corundum	6:4	4.1719[§]	25.0312[§]

[†]   Does not include charges on ions; i.e., assumes structure is made of isocharged ions that factor out.

[‡]   The problem of structures with more than one charge, such as Al_2O_3, can be addressed by making use of the relationship

$$E_{sum} = \alpha_{conv} \frac{(Z\pm)^2 e^2}{4\pi\varepsilon_0 r_0}\left(1-\frac{1}{n}\right)$$

where Z± is the *highest common factor* of z_1 and z_2, i.e., 1 for NaCl, CaF_2, and Al_2O_3, 2 for MgO, TiO_2, ReO_3, etc.

[§]   Exact value depends on c/a ratio.

According to this equation, sometimes referred to as the *Born-Lande equation,* the information required to calculate E_{latt} is the crystal structure, which determines α, the equilibrium interionic spacing, both easily obtainable from X-ray diffraction, and n, which is obtainable from compressibility data. Note that the lattice energy is not greatly affected by small errors in n.

In deriving Eq. (2.18), a few terms were ignored. A more exact expression for the lattice energy is

$$E_{latt} = -\frac{A}{r_0} + \frac{B}{r_0^n} - \left(\frac{C}{r_0^6} + \frac{D}{r_0^8}\right) + \frac{9}{4}h\nu_{max} \qquad (2.19)$$

The first two terms, which have been discussed in detail up to this point, dominate. The term in parentheses represents dipole-dipole and dipole-quadrapole interactions between the ions. The last term represents the zero-point correction, with ν_{max} being the highest frequency of the lattice vibration mode. Finally in this section it is worth noting that this ionic model is a poor approximation for crystals containing large anions and small cations where the covalent contribution to the bonding becomes significant (see Chap. 3).

WORKED EXAMPLE 2.3. Calculate the lattice energy of NaCl given that $n = 8$.

Answer

To calculate the lattice energy, r_0, n, and the structure of NaCl all are needed. The structure of NaCl is the rock salt structure and hence its Madelung constant is 1.748 (Table 2.4).

The equilibrium interionic distance is simply the sum of the radii of the Na^+ and Cl^- ions. The values are listed at the end of Chap. 3 in Appendix 3A. Looking up the values the equilibrium interionic distance, $r_0, = 167 + 116 = 283$ pm.

$$E_{latt} = \frac{(-1)(+1)(6.02 \times 10^{23})(1.6 \times 10^{-19})^2 (1.748)}{4\pi(8.85 \times 10^{-12})(283 \times 10^{-12})}\left(1 - \frac{1}{8}\right) = -750 \text{ kJ/mol}$$

2.5.2   Born-Haber Cycle

So far, a rather simple model has been introduced in which it was assumed that an ionic solid is made up of ions attracted to each other by coulombic attractions. How can such a model be tested? The simplest thing to do would be to compare the results of the model, say E_{latt}, to experimental data. This is easier said than done, however, given that E_{latt} is the energy released when 1 mol of *cations and anions* condenses into a solid — an experiment that, needless to say, is not easy to perform.

An alternate method is to make use of the first law of thermodynamics, namely, that energy can be neither created nor destroyed. If a cycle can be devised where all the energies are known except E_{latt}, then it can be easily calculated. For such a cycle, known as the *Born-Haber cycle*, shown in Fig. 2.6, it is necessary that

$$\Delta H_{form}(\text{exo}) = E_{latt}(\text{exo}) + E_{ion}(\text{endo}) + E_{EA}(\text{endo or exo})$$
$$+ E_{diss}(\text{endo}) + E_{vap}(\text{endo})$$

Each of these terms is discussed in greater detail below with respect to NaCl.

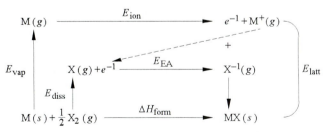

Figure 2.6
The Born-Haber cycle.

Enthalpy of formation or reaction

When the reaction

$$Na(s) + \frac{1}{2}Cl_2(g) \rightarrow NaCl(s)$$

occurs, ΔH_{form} is the thermal energy liberated. For NaCl at 298 K, this reaction is accompanied by the release of –411 kJ/mol. Enthalpies of formation of most compounds are exothermic.

Dissociation energy

Energy E_{diss} is needed to break up the stable Cl_2 molecule into two atoms, i.e., the energy change for the reaction

$$\frac{1}{2}Cl_2 \rightarrow Cl(g)$$

This energy is always endothermic and for the reaction as written equals 121 kJ/mol.

Heat of vaporization

The energy required for the reaction

$$Na(s) \rightarrow Na(g)$$

is the **latent heat of vaporization** E_{vap}, which is 107.3 kJ/mol for Na and is always endothermic.

Values of ΔH_{form}, E_{diss}, and E_{vap} can be found in various sources[16] and are well documented for most elements and compounds.

Ionization energy

The **ionization energy** E_{ion} is the energy required to completely remove an electron from an isolated atom in the gas phase. Ionization energies are always endothermic since in all cases work has to done to remove an electron from its nucleus. Table 2.2 lists the first and second ionization potentials for selected elements of the periodic table. For Na that value is 495.8 kJ/mol.

---

[16]   A reliable source for thermodynamic data is JANAF *Thermochemical Tables*, 3d ed., which lists the thermodynamic data of over 1800 substances.

Electron affinity

Electron affinity (EA) is a measure of the energy change that occurs when an electron is added to the valence shell of an atom. Some selected values of E_{EA} for nonmetals are listed in Table 2.5. The addition of the first electron is usually exothermic (e.g., oxygen, sulfur); further additions, when they occur, are by necessity negative since the second electron is now approaching a negatively charged entity. The electron affinity of Cl is -348.7 kJ/mol.

The lattice energy of NaCl was calculated (see Worked Example 2.3) to be -750 kJ/mol. If we put all the pieces together, the Born-Haber summation for NaCl yields

$$\Delta H_{form}(exo) = E_{latt}(exo) + E_{ion}(endo) + E_{EA}(exo)$$
$$+ E_{diss}(endo) + E_{vap}(endo)$$
$$= -750 + 495.8 - 348.7 + 121 + 107.3 = -374.6 \text{ kJ/mol}$$

which compares favorably with the experimentally determined value of -411 kJ/mol. If Eq. (2.19) is used, even better agreement is obtained.

This is an important result for two reasons. First; it confirms that our simple model for the interaction between ions in a solid is, for the most part, correct. Second, it supports the notion that NaCl can be considered an ionically bonded solid.

TABLE 2.5
Electron affinities[†] of selected nonmetals at 0 K

Element	EA (kJ/mol)	Element	EA (kJ/mol)
$O \rightarrow O^-$	141 (exo)	$Se \rightarrow Se^-$	195 (exo)
$O^- \rightarrow O^{2-}$	780 (endo)	$Se^- \rightarrow Se^{2-}$	420 (endo)
$F \rightarrow F^-$	322 (exo)	$Br \rightarrow Br^-$	324.5 (exo)
$S \rightarrow S^-$	200 (exo)	$I \rightarrow I^-$	295 (exo)
$S^- \rightarrow S^{2-}$	590 (endo)	$Te \rightarrow Te^-$	190.1 (exo)
$Cl \rightarrow Cl^-$	348.7 (exo)		

[†] Electron affinity is usually defined as the energy *released* when an electron is added to the valence shell of an atom. This can be quite confusing. To avoid any confusion, the values listed in this table clearly indicate whether the addition of the electron is endo- or exothermic. Adapted from J. Huheey, *Inorganic Chemistry*, 2d ed., Harper & Row, New York, 1978.

2.6
COVALENT BOND FORMATION

The second important type of primary bond is the covalent bond. Whereas ionic bonds involve electron transfer to produce oppositely charged species, covalent bonds arise as a result of electron sharing. In principle, the energetics of the covalent bond can be understood if it is recognized that electrons spend more time in the area *between* the nuclei than anywhere else. The mutual attraction between the electrons and the nuclei lowers the potential energy of the system forming a bond. Several theories and models have been proposed to explain the formation of covalent bonds. Of these the molecular orbital theory has been particularly successful and is the one discussed in some detail below. As the name implies, molecular orbital (MO) theory treats a molecule as a single entity and assigns orbitals to the *molecule as a whole*. In principle, the idea is similar to that used to determine the energy levels of isolated atoms, except that now the wave functions have to satisfy Schrödinger's equation with the appropriate expression for the potential energy, which has to include all the charges making up the molecule. The solutions in turn give rise to various *molecular orbitals*, with the number of filled orbitals determined by the number of electrons needed to balance the nuclear charge of the molecule as a whole subject to Pauli's exclusion principle.

To illustrate, consider the simplest possible molecule, namely, the H_2^+ molecule, which has one electron but two nuclei. This molecule is chosen in order to avoid the complications arising from electron-electron repulsions already alluded to earlier.

2.6.1   Hydrogen Ion Molecule

The procedure is similar to that used to solve for the electronic wave function of the H atom [i.e.; the wave functions have to satisfy Eq. (2.1)] except that the potential energy term has to account for the presence of *two* positively charged nuclei rather than one. The Schrödinger equation for the H_2^+ molecule thus reads

$$\frac{\partial^2 \psi}{\partial x^2} + \frac{8\pi^2 m_e}{h^2}\left(E_{tot} + \frac{e^2}{4\pi\varepsilon_0 r_a} + \frac{e^2}{4\pi\varepsilon_0 r_b} - \frac{e^2}{4\pi\varepsilon_0 R} \right)\psi = 0 \qquad (2.20)$$

where the distances, r_a, r_b, and R are defined in Fig. 2.7a. If it is assumed that the distance R between the two nuclei is fixed, then an exact solution exists, which is quite similar to that of the H atom, except that now *two solutions* or wave functions emerge. One solution results in an increase in the electron density between the nuclei (Fig. 2.7c) whereas the second solution decreases it (Fig. 2.7d). In the first case, both nuclei are attracted to the electron between them, which results in the lowering of the energy of the system relative to the isolated-atom case and is thus known as a **bonding orbital** (Fig. 2.7b). The second case results in an increase in energy relative to the isolated atoms, because now the unsheathed or partially bared nuclei repel one another. This is known as the **antibonding orbital,** shown in Fig. 2.7b.

The solution for the H_2 molecule is quite similar, except that now an extra potential energy term for the repulsion between the two electrons has to be included in Schrödinger's equation. This is nontrivial, but fortunately the end result is similar to that of the H_2^+ case; the individual energy levels split into a bonding and an antibonding orbital. The atomic orbital overlap results in an increased probability of finding the electron between the nuclei. Note that in the case of the H_2 molecule, the two electrons are accommodated in the bonding orbital. A third electron, i.e., H_2^-, would have to go into the antibonding orbital because of Pauli's exclusion principle.

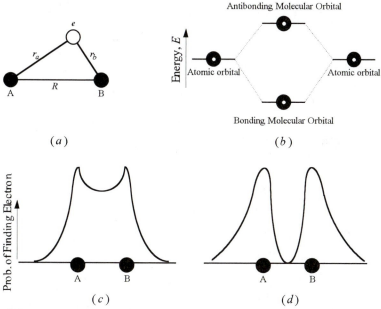

Figure 2.7
(a) Coordinates for the H_2^+ molecule used in Eq. (2.20). (b) Interaction of the two atomic orbitals results in bonding and antibonding orbitals. (c) Probability function for the bonding case in which electron density between the nuclei is enhanced. (d) Probability function for antibonding case, where the probability of finding the electron is decreased in the volume between the nuclei, resulting in a higher-energy orbital.

2.6.2   HF Molecule

In the preceding section, the electronegativities of the two atoms and the shapes (both spherical) of the interacting orbitals making up the bond were identical. The situation becomes more complicated when one considers bonding between dissimilar atoms. A good example is provided by the HF molecule. The electron configuration of H is $1s^1$, and that of F is (He) $2s^2 2p^5$. The valence orbitals of the F atom are shown in Fig. 2.8a (the inner core electrons are ignored since they are not involved in bonding). The atoms are held at the distance that separates them, which can either be calculated or obtained experimentally, and the molecular orbitals of HF are calculated. The calculations are nontrivial and beyond the scope of this book; the result, however, is shown schematically in Fig. 2.8b. The total number of electrons that have to be accommodated in the molecular orbitals is eight (seven from F and one from H). Placing two in each orbital fills the first four orbitals and results in an energy for the molecule that is lower (more negative) than that of the sum of the two noninteracting atoms, which in turn renders the HF molecule more stable relative to the isolated atoms.

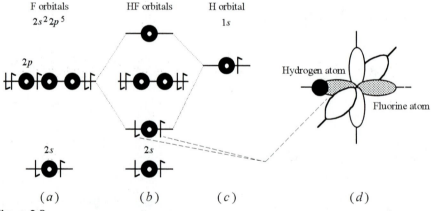

Figure 2.8
(a) The F atomic orbitals. (b) The HF molecular orbitals. (c) The H atomic orbital. (d) Interaction of H $1s$ orbital with one of the fluorine p orbitals. The overlap of these two orbitals results in a lowering of the energy of the system. The dotted lines joining (b) to (d) emphasize that it is only the fluorine p orbital which overlaps with the H orbital that has a lower energy. The two pairs of unpaired electrons (unshaded lobes) have the same energy in the molecule that they did on the F atom, since these so-called lone pairs are unperturbed by the presence of the hydrogen atom.

Figure 2.8 can also be interpreted as follows: The F $2s$ electrons, by virtue of being at a much lower energy than hydrogen (because of the higher charge on the F nucleus) remain unperturbed by the hydrogen atom.[17]   The $1s$ electron wave function of the H atom and one of the $2p$ orbitals on the fluorine will overlap to form a primary σ bond (Fig. 2.8d). The remaining electrons on the F atom (the so-called lone pairs) remain unperturbed in energy and in space.

As mentioned above, the calculation for Fig. 2.8 was made for a given interatomic distance. The same calculation can be repeated for various interatomic separations. At infinite separation, the atoms do not interact, and the energy of the system is just the sum of the energies of the electrons on the separate atoms. As the atoms are brought closer together; the attractive potential energy due to the mutual attraction between the electrons and the nuclei decreases the energy of the system up to a point, beyond which a repulsive component comes into play and the energy starts increasing again. In other words, at some interatomic distance, a minimum in the energy occurs, and a plot of energy versus interatomic distance results in an energy well that is not unlike the one shown in Fig. 2.4a.

2.7
COVALENTLY BONDED SOLIDS

Up to this point the discussion has focused on the energetics of a single covalent bond between two atoms. Such a bond, however, will not lead to the formation of a strong solid, i.e., one in which all the bonds are primary. To form such a solid, each atom has to be simultaneously bonded to at least two other atoms. For example, HF cannot form such a solid because once an HF bond is formed, both atoms attain their most stable configuration — He for H and Ne for F, which in turn implies that there are no electrons available to form covalent bonds with other atoms. It follows that HF is a gas at room temperature, despite the fact that the HF bond is quite strong.[18]

As discussed in greater detail in the next chapter, many predominantly covalently bonded ceramics, especially the Si-based ones such as silicon carbide, silicon nitride, and the silicates, are composed of Si atoms simultaneously bonded to four other atoms in a tetrahedral arrangement. Examining the ground state configuration of Si, that is, (Ne) $3s^2 3p^2$ (Fig. 2.9a), one would naturally expect only two primary bonds to form. This apparent contradiction has been explained by

---

[17]   For orbitals to overlap, they must be relatively close to each other in energy.
[18]   If sufficiently cooled, however, HF will form a solid as a result of secondary bonds.

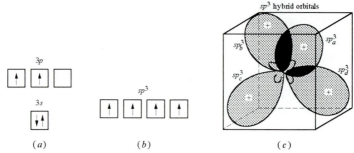

Figure 2.9
(*a*) Ground state of Si atom. (*b*) Electronic configuration after hybridization.
(*c*) Directionality of sp^3 bonds. Note that each bond lobe contains one electron, and thus the atom can form four covalent bonds with other atoms.

postulating that **hybridization** between the s and p wave functions occurs. Hybridization consists of a mixing or linear combination of s and p orbitals in an atom in such a way as to form new hybrid orbitals. This hybridization can occur between one s orbital and one p orbital (forming an sp orbital), or one s and two p orbitals (forming an sp^2 trigonal orbital). In the case of Si, the s orbital hybridizes with all three p orbitals to form what is known as sp^3 **hybrid orbitals.** The hybrid orbital possesses both s and p character and directionally reaches out in space as lobes in a tetrahedral arrangement with a bond angle of 109°, as shown in Fig. 2.9c. Each of these orbitals is populated by one electron (Fig. 2.9b); consequently each Si atom can now bond to four other Si atoms, or any other *four* atoms for that matter, which in turn leads to three-dimensional structures. Promotion of the electron from the s to the hybrid orbital requires some energy, which is more than compensated for by the formation of four primary bonds.

2.8
BAND THEORY OF SOLIDS

One of the more successful theories developed to explain a wide variety of electrical and optical properties in solids is the **band theory of solids.** In this model, the electrons are consigned to bands that are separated from each other by energy gaps. Bands that are incompletely filled (Fig. 2.10a) are termed **conduction bands**, while those that are full are called **valence bands**. The electrons occupying the highest energy in a conduction band can rapidly adjust to an applied electric or electromagnetic field and give rise to the properties characteristic of metals, such as

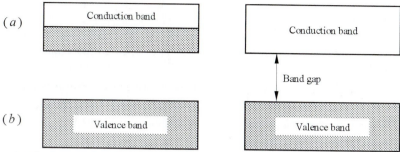

Figure 2.10

Band structure of (*a*) a metal with an incompletely filled conduction band and (*b*) an insulator or semiconductor. At 0 K such a solid is an insulator because the valence band is completely filled and the conduction band is completely empty. As the temperature is raised, some electrons are promoted into the conduction band, and the material starts to conduct.

high electrical and thermal conductivity, ductility, and reflectivity. Solids where the valence bands are completely filled (Fig. 2.10*b*), on the other hand, are poor conductors of electricity and at 0 K are perfect insulators. It follows that understanding this model of the solid state is of paramount importance if the electrical and optical properties of ceramics are to be understood.

The next three subsections address the not-so-transparent concept of how and why bands form in solids. Three approaches are discussed. The first is a simple qualitative model. The second is slightly more quantitative and sheds some light on the relationship between the properties of the atoms making up a solid and its band gap. The last model is included because it is physically the most tangible and because it relates the formation of bands to the total internal reflection of electrons by the periodically arranged atoms.

2.8.1 Introductory Band Theory

In the same way as the interaction between two hydrogen atoms gave rise to two orbitals, the interaction or overlap of the wave functions of $\approx 10^{23}$ atoms in a solid gives rise to energy bands. To illustrate, consider 10^{23} atoms of Si in their ground state (Fig. 2.11*a*). The band model is constructed as follows:

1. Assign four localized tetrahedral sp^3 hybrid orbitals to each Si atom, for a total of 4×10^{23} hybrid orbitals (Fig. 2.11*b*).

2. The overlap of each of two neighboring sp^3 lobes forms one bonding and one antibonding orbital, as shown in Fig. 2.11*d*.
3. The two electrons associated with these two lobes are accommodated in the bonding orbitals (Fig. 2.11*d*).
4. As the crystal grows, every new atom added brings *one orbital to the bonding and one to the antibonding orbital set.* As the orbitals or electron wave functions overlap; they must *broaden* as shown in Fig. 2.11c, because of the Pauli exclusion principle.

    Thus in the solid a spread of orbital energies develops within each orbital set, and the separation between the **highest occupied molecular orbital** (or HOMO), and the **lowest unoccupied molecular orbital** (or LUMO) in the molecule becomes the **band gap** (Fig. 2.11c). It is worth noting that the new orbitals are created near the original diatomic bonding σ and antibonding σ^* energies (Fig. 2.11*d*) and move toward the band edges as the size of the crystal increases.

5. In the case of Si, each atom starts with 4 valence electrons, and the total number of electrons that has to be accommodated in the valence band is 4×10^{23}. But since there are 2×10^{23} levels in that band and each level can accommodate 2 electrons, *it follows that the valence band is completely filled and the conduction band is empty.*[19]

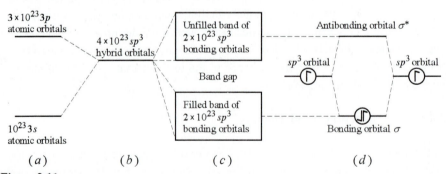

Figure 2.11

(*a*) Ground state of Si atoms. (*b*) The sp^3 hybrid orbitals. (*c*) Interaction of sp^3 orbitals to form energy bands. (*d*) Localized orbital energy levels between two Si atoms to form an Si_2 molecule. Note that the energy bands are centered on the energy of the diatomic bonds.

---

[19]   As discussed later, this is only true at 0 K. As the temperature is raised, the thermal energy will promote some of the electrons into the conduction band.

This last statement has far-reaching implications. If the band gap, usually denoted by E_g, lies somewhere between 0.02 and 3 eV, the material is considered to be a semiconductor. For higher values of E_g, the solid is considered an insulator. Said otherwise, if all the electrons are used in bonding, none is left to move freely and conduct electricity. Table 2.6, in which the band gaps of a number of binary and ternary ceramics are listed, clearly indicates that most ceramics are insulators.

TABLE 2.6
Summary of band gaps for various ceramics

Material	Band gap, eV	Material	Band gap, eV
	Halides		
AgBr	2.80	MgF_2	11.00
BaF_2	8.85	MnF_2	15.50
CaF_2	12.00	NaCl	7.30
KBr	0.18	NaF	6.70
KCl	7.00	SrF_2	9.50
LiF	12.00	TlBr	2.50
	Binary oxides, carbides, and nitrides		
AlN	6.2	Ga_2O_3	4.60
Al_2O_3 parallel	8.8	MgO (pericalse)	7.7
Al_2O_3 perpendicular	8.85	SiC (α)	$2.60 - 3.20$
BN	4.8	SiO_2 (fused silica)	8.3
C (diamond)	5.33	UO_2	5.20
CdO	2.1		

	Transition metal oxides		
Binaries		Ternaries	
CoO	4.0	$BaTiO_3$	$2.8 - 3.2$
CrO_3	2.0	$KNbO_3$	3.3
Cr_2O_3	3.3	$LiNbO_3$	3.8
CuO	1.4	$LiTaO_3$	3.8
Cu_2O	2.1	$MgTiO_3$	3.7
FeO	2.4	$NaTaO_3$	3.8
Fe_2O_3	3.1	$SrTiO_3$	3.4
MnO	3.6	$SrZrO_3$	5.4
MoO_3	3.0	$Y_3Fe_5O_{12}$	3.0
Nb_2O_5	3.9		
NiO	4.2		
Ta_2O_5	4.2		
TiO_2 (rutile)	$3.0 - 3.4$		
V_2O_5	2.2		
WO_3	2.6		
Y_2O_3	5.5		
ZnO	3.2		

Note that the degree of interaction between the orbitals depends on the interatomic distance or the spatial delocalization of the interacting electrons (the two are not unrelated). For example, the band gaps of C (diamond), Si, and Ge are, respectively, 5.33, 1.12, and 0.74 eV. In C, the interaction is between the $n = 2$ electrons, whereas for Si and Ge one is dealing with the $n = 3$ and $n = 4$ electrons, respectively. As the interacting atoms become larger, the interaction of their orbitals increases, rendering the bands wider and consequently reducing the band gap.[20]

Orbital overlap, while important, is not the only determinant of band gap width. Another important factor is how tightly the lattice binds the electron. This is dealt with in the following model.

2.8.2   Tight Binding Approximation[21]

In this approach, not unlike the one used to explain the formation of a covalent bond, Schrödinger's equation

$$\frac{\partial^2 \psi}{\partial x^2} + \frac{8\pi^2 m_e}{h^2}\left[E_{tot} - E_{pot}(x)\right]\psi = 0 \qquad (2.21)$$

is solved by assuming that the electrons are subject to a periodic potential E_{pot} which has the same periodicity as the lattice. By simplifying the problem to one dimension with interatomic spacing a and assuming that $E_{pot}(x) = 0$ for regions near the nuclei and $E_{pot} = E_0$ for regions in between, and further assuming that the width of the barrier is w (see Fig. 2.12a), Eq. (2.21) can be solved. Despite these simplifications, the mathematics of this problem is still too complex to be discussed here, and only the final results are presented.[22] It turns out that solutions are possible only if the following *restricting* conditions are satisfied:

---

[20] Interestingly enough, a semiconducting crystal can be made conductive by subjecting it to enormous pressures which increase the level of interaction of the orbitals to such a degree that the bands widen and eventually overlap.

[21] Also known as the *Kronig-Penney model.*

[22] The method of solving this problem lies in finding the solution for the case when $E = 0$, that is,

$$\psi_0 = A\exp(i\phi x) + B\exp(-i\phi x)$$

with $\phi = \sqrt{2\pi m E_{tot}}/h$. And the solution for the case where $E = E_0$, that is,

$$\psi_v = C\exp\beta x + D\exp(-\beta x)$$

where $\beta = 2\pi\sqrt{2\pi m(E_0 - E_{tot})}/h$. By using the appropriate boundary conditions, namely, continuity of the wave function at the boundaries, and ensuring that the solution is periodic, A, B, C, and D can be solved for. If further it is assumed that the barrier area, i.e., the product of

$$\cos ka = P\frac{\sin \phi a}{\phi a} + \cos \phi a \qquad (2.22)$$

where

$$P = \frac{4\pi^2 ma}{h^2} E_0 w \qquad (2.23)$$

and

$$\phi = \frac{2\pi}{h}\sqrt{2mE_{\text{tot}}} \qquad (2.24)$$

k is the wave number, defined as:

$$k = \frac{2\pi}{\lambda} \qquad (2.25)$$

where λ is the wavelength of the electron.

Since the left-hand side of Eq. (2.22) can take only values between +1 and −1, the easiest way to find possible solutions to this equation is to do it graphically by plotting the right-hand side of Eq. (2.22) as a function of ϕa, as shown in Fig. 2.12b. Whenever that function lies between +1 and −1 (shaded areas in Fig. 2.12b), that represents a solution. Given that ϕ is proportional to the energy of the electron [Eq. (2.24)], what is immediately apparent from Fig. 2.12b is that there are regions of energy that are permissible (crosshatched areas in Fig. 2.11b) and regions of forbidden energy (uncrosshatched areas). This implies that *an electron moving in a periodic potential can only move in so-called allowed energy bands that are separated from each other by forbidden energy zones.* Furthermore, the solution clearly indicates that the energy E_{tot} *of the electron is a periodic function of k.*

The advantage of using this model over others is that a semiquantitative relationship between the bonding of an electron to its lattice and the size of the band gap can be accounted for. This is reflected in the term P – for atoms that are very electronegative, V_0, and consequently P, is large. As P increases, the right-hand side of Eq. (2.22) becomes steeper, and the *bands narrow and the regions of forbidden energy widen.* It follows that if this model is correct, an empirical relationship between the electronegativities of the atoms or ions making up a solid and its band gap should exist. That such a relationship, namely,

---

wE_0, is a constant, Eqs. (2.22) and (2.23) follow. See R. Bube, *Electrons in Solids*, 2d ed.; Academic Press, New York; 1988, for more details.

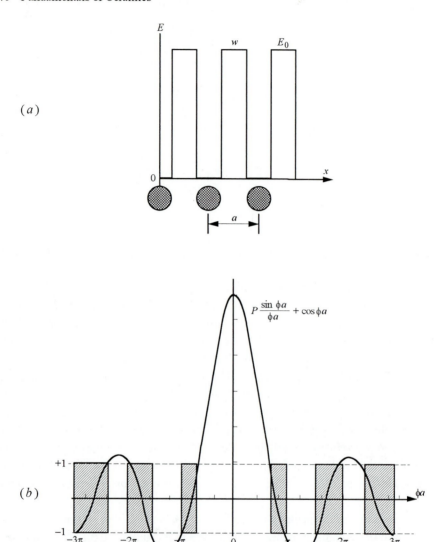

Figure 2.12

(a) Approximation of periodic potential that an electron is subjected to in a one-dimensional crystal of periodicity a. Here w is the width of the barrier, and E_0 is the depth of the energy well. (b) A plot of the right-hand side of Eq. (2.22) versus ϕa. The x axis is proportional to the energy of the electron, and the crosshatched areas denote energies that are permissible, whereas the energies between the crosshatched areas are not permissible.

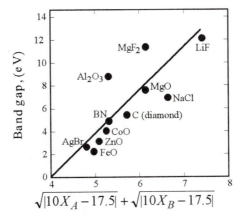

Figure 2.13
Empirical correlation between the electronegativities of the atoms making up a solid and its band gap; X_A and X_B are the electronegativities of the constituent atoms or ions.

$$E_g(\text{eV}) \approx -15 + 3.75\left(\sqrt{|10X_A - 17.5|} + \sqrt{|10X_B - 17.5|}\right)$$

does exist is illustrated nicely in Fig. 2.13. Here X_A and X_B represent the electronegativities of the atoms making up the solid.

Before moving on to the next section, it is instructive to look at two limits of the solution arrived at above:

1. When the interaction between the electrons and the lattice vanishes; i.e., as E_0 or P approaches 0. From Eq. (2.22), for $P = 0$, it follows that $\cos ka = \cos k\phi$, that is; $k = \phi$, which when substituted in Eq. (2.24) and upon rearranging yields

$$E_{tot} = \frac{h^2 k^2}{8\pi^2 m} \qquad (2.26)$$

which is nothing but the well-known relationship for the energy of a free electron (see App. 2A).

2. At the boundary of an allowed band, i.e., when $\cos ka = \pm 1$ or

$$k = \frac{n\pi}{a} \quad \text{where } n = 1,\ 2,\ 3 \qquad (2.27)$$

This implies that discontinuities in the energy occur whenever this condition is fulfilled. When this result is combined with Eq. (2.26) and the energy is plotted versus k, Fig. 2.14 results. The essence of this figure lies in appreciating that at the bottom of the bands the electron dependence on k is

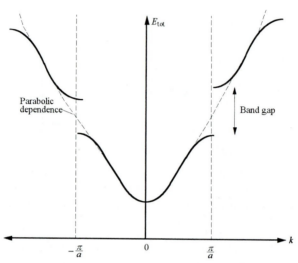

Figure 2.14

Functional dependence of E_{tot} on k. The discontinuities occur at $k = n\pi/a$, $n = 1, 2, 3, \dots$.

parabolic; in other words, the electrons are behaving as if they were free. However, as their k values increase, periodically, Eq. (2.27) will be satisfied and a band gap develops. The reason for the formation of such a gap is discussed in the next section.

2.8.3   Nearly Free Electron Approximation

The physical origin of the band gap predicated by the previous model can be understood as follows: As a totally empty band is filled with electrons, they have to populate levels of higher energies or wave numbers. Consequently, at some point the condition $k = n\pi/a$ will be fulfilled, which is another way of saying that a pattern of standing waves is set up, and the electrons can no longer propagate freely through the crystal because as the waves propagate to the right, they are reflected to the left, and vice versa.[23]

---

[23]   The condition $k = n\pi/a$ is nothing but the well-known Bragg reflection condition, $n\lambda = 2a\cos\theta$, for $\theta = 0$. See Chap. 3 for more details.

High-energy configuration – probability of finding electrons is lowest where cores are located.
Bottom of conduction band.

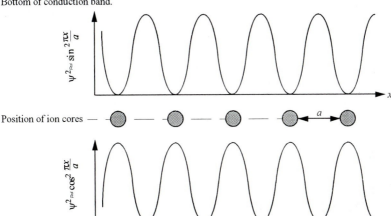

Low-energy configuration – probability of finding electrons is highest where cores are located.
Top of valence band.

Figure 2.15

Probability of finding electrons relative to location of the cores for the two standing waves that form when $k = n\pi/a$. In the bottom case, the standing waves distribute the charge over the ion cores, and the attraction between the negative electrons and the positive cores reduces the energy of the system relative to the top situation, where the electrons spend most of their time between the ion cores.

It can be shown further that[24] these standing waves occur with amplitude maxima either at the positions of the lattice points, that is $\psi^2 = (\text{const.})\cos^2(n\pi x/a)$ (bottom curve in Fig. 2.15), or in between the lattice points, that is $\psi^2 = (\text{const.})\sin^2(n\pi x/a)$, (top curve in Fig. 2.15). In the former case, the attraction of the electrons to the cores reduces the energy of the system — an energy that corresponds to the top of the valence band. In the latter case, the energy is higher and corresponds to that at the bottom of the conduction band. The difference in energy between the two constitutes the energy gap.

It is important to emphasize that the band model of solids, while extremely successful, is simply one approach among several that can be used in describing the properties of solids. It is an approach that is elegant, powerful, and amenable to quantification. However, the same conclusions can be deduced by starting from other assumptions. For instance, the band gap can be viewed simply as the energy required to break the covalent bond in a covalently bonded solid, or to ionize the

---

24   See, e.g.; L. Solymar and D. Walsh, *Lectures on the Electrical Properties of Materials*, 4th ed., Oxford University Press, New York, 1988, p. 130.

anions in an ionic solid.   At absolute zero, there are no atomic vibrations, the electrons are trapped, and the solid is an insulator.   At finite temperatures, the lattice atoms will vibrate randomly, and occasionally the amplitude of vibration can be such as to break the bond and release an electron.   The higher the temperature, the greater the probability of breaking the bond and the more likely the electron is to escape.

2.9
SUMMARY

1. The confinement of an electron to a nucleus results in the quantization of its energy. The probability of finding the position of an electron then becomes a function of its energy and the orbital it has to populate. The shapes of the orbitals differ and depend on the quantum number of the electron. The s orbitals are spherically symmetric while p orbitals are lobed and orthogonal to each other.

2. Ionic bonds are formed by the transfer of electrons from an electropositive to an electronegative atom. The long-range coulombic attraction of these charged species for each other, together with a short-range repulsive energy component, results in the formation of an ionic bond at an equilibrium interatomic distance.

3. Covalent bonds form by the overlap of atomic wave functions. For two wave functions to overlap, they must be close to each other in energy and be able to overlap in space as well.

4. The sp^3 hybridization results in the formation of four energetically degenerate bonds arranged tetrahedrally with each containing one electron. This allows an atom to bond to four other atoms simultaneously.

5. The interactions and overlap of the wave functions of many atoms or ions in a solid give rise to energy bands. If the outermost bands are not filled, the electrons are said to be *delocalized* and the solid is considered to be a metal. If the bands are separated from each other by a band gap, the solid is considered a semiconductor or insulator depending on the size of that gap.

APPENDIX 2A

Kinetic Energy of Free Electrons

The total energy of a free electron, i.e., one for which $E_{pot} = 0$, is simply its kinetic energy or

$$E_{tot} = \frac{1}{2}mv^2 = \frac{p^2}{2m} \tag{2A.1}$$

where p is its momentum, and v its velocity. p in turn is related to the de Broglie wavelength λ of the electron by

$$p = \frac{h}{\lambda} \tag{2A.2}$$

Combining this equation with Eq. (2.25), it follows that

$$k = \frac{2\pi p}{h} \tag{2A.3}$$

In other words, the wave number of an electron is directly proportional to its momentum. Combining these three equations, it follows that for a free electron

$$E_{tot} = \frac{h^2 k^2}{8\pi^2 m} \tag{2A.4}$$

Note that in the presence of a periodic field, the electron's energy can have nonzero values despite that fact that its velocity could be zero.

PROBLEMS

2.1. (a) Show that Eq. (2.4) is indeed a solution to Eq. (2.3), provided E_{tot} is given by Eq. (2.5) and c_0 is given by Eq. (2.6).

(b) Calculate the radius of the first Bohr orbit.
Answer: 0.0528 nm

(c) Consider two hydrogen atoms. The electron in the first is in the $n = 1$ state, whereas in the second the electron is in the $n = 3$ state. (i) Which atom is in the ground state configuration? Why? (ii) Which orbit has the larger radius? (iii) Which electron is moving faster? (iv) Which electron has the lower potential energy? (v) Which atom has the higher ionization energy?

2.2. (*a*) Show that

$$\psi(r) = A(1 + c_1 r)\exp\left(\frac{-rc_0}{2}\right)$$

is also a solution to the Schrödinger equation [i.e., Eq. (2.5)], and find an expression for c_1.

(*b*) Show that the energy of this level is equal to -3.4 eV.

(*c*) Determine the value of A. *Hint*: The total probability of finding an electron somewhere must be unity.

2.3. Starting with Eq. (2.13), derive Eq. (2.15).

2.4. Calculate the third ionization of Li. Explain why this calculation can be carried out exactly with no approximations required.

Answer: -122.4 eV

2.5. Given 1 mol of Na^+ and 1 mol of Cl^-, calculate the energy released when the Na and Cl ions condense as

(*a*) Noninteracting ion pairs; i.e., consider only one pairwise interactions.

Answer: -490 kJ/mol

(*b*) Noninteracting ion squares; i.e., every four ions, 2Na and 2Cl interact with each other, but not with others.

Answer: -633 kJ/mol

(*c*)   1/8 unit cell of NaCl; i.e., 8 atoms interact.

Answer: -713 kJ/mol

(*d*)   Compare with -755 kJ/mol for solid NaCl lattice.

Hint: Make sure you include all pairwise attractions and repulsions. You can assume the Born exponent $n = \infty$.

2.6. Assuming NeCl crystallizes in the NaCl structure and, using the Born-Haber cycle, show why NeCl does not exist. Make any necessary assumptions.

2.7. (*a*) Plot the attractive, repulsive, and net energy between Mg^{2+} and O^{2-} from 0.18 and 0.24 nm in increments of 0.01 nm. The following information may be useful: $n = 9$, $B = 0.4 \times 10^{-105}$ J·m^9.

(*b*) Assuming that Mg^+O^- and $Mg^{2+}O^{2-}$ both crystallize in the rock salt structure and that the ionic radii are not a strong function of ionization and taking $n = \infty$, calculate the difference in the enthalpies of formation ΔH_{form} of $Mg^{2+}O^{2-}$ and Mg^+O^-. Which is more stable?

Answer: Difference $= 1214$ kJ/mol

(*c*)   Why is MgO not written as $Mg^{3+}O^{3-}$?

2.8. Calculate the Madelung constant for an infinite chain of alternating positive and negative ions $+-+-+$ and so on.

Answer: $2\ln 2$

2.9. Write the first three terms of the Madelung constant for the NaCl and the CsCl structures. How does the sum of these terms compare to the numbers listed in Table 2.4? What are the implications, if any, if the Madulung constant comes out negative?

2.10. (*a*) He does not form He_2. Why do you think this is the case? What does this statement imply about the energies of the bonding and antibonding orbitals relative to those of the isolated ions?

(*b*) Explain in terms of molecular orbital theory why He_2 is unstable and does not occur while He_2^+ has a bond energy $\approx H_2^-$.

2.11. (*a*) Boron reacts with oxygen to form B_2O_3. (i) How many oxygens are bonded to each B, and vice versa? (ii) Given the electronic ground states of B and O, propose a hybridization scheme that would explain the resulting bonding arrangement.

(*b*) Repeat part (*a*) for BN.

2.12. The total energy (electronic) of an atom or a molecule is the sum of the energies of the individual electrons. Convince yourself that the sum of the energies of the HF molecule is indeed lower (more negative) than the sum of the energies of the two isolated atoms.

2.13. (*a*) Which has the higher ionization energy — Li or Cs, Li or F, F or I? Explain.

(*b*) Which has the higher electron affinity — Cl or Br, O or S, S or Se? Explain.

2.14. The symbol *n* has been used in this chapter to represent two completely distinct quantities. Name them and clearly differentiate between them by discussing each.

2.15. (*a*) To what inert gases do the ions Ca^{2+} and O^{2-} correspond?

(*b*) Estimate the equilibrium interionic spacing of the $Ca^{2+}-O^{2-}$ bond.

(*c*) Calculate the force of attraction between a Ca^{2+} ion and an O^{2-} ion if the ion centers are separated by 1 nm. State all assumptions.

2.16. The fraction ionic character of a bond between elements A and B can be approximated by

$$\text{Fraction ionic character} = 1 - e^{-\frac{(X_A - X_B)^2}{4}}$$

where X_A and X_B are the electronegativities of the respective elements.

(a) Using this expression, compute the fractional ionic character for the following compounds: NaCl, MgO, FeO, SiO_2, and LiF.

(b) Explain what is meant by saying that the bonding in a solid is 50 percent ionic and 50 percent covalent.

ADDITIONAL READING

1. W. D. Kingery, H. K. Bowen, and D. R. Uhlmann, *Introduction to Ceramics*, 2d ed., Wiley, New York, 1976.

2. A. R. West, *Solid State Chemistry and Its Applications*, Wiley, Chichester, England, 1984.

3. N. N. Greenwood, *Ionic Crystals, Lattice Defects and Non-Stoichiometry*, Butterworth, London, 1968.

4. P. W. Atkins, *Physical Chemistry*, 4th ed., Oxford University Press, New York, 1990.

5. M. Gerloch, *Orbitals, Terms and States*, Wiley, Chichester, England, 1986.

6. J. B. Goodenough, *Prog. Solid State Chem.*, **5:** 145, 1971.

7. P. A. Cox, *The Electronic Structure and Chemistry of Solids*, Oxford University Press, New York, 1987.

8. C. Kittel, *Introduction to Solid State Physics,* 6th ed., Wiley, New York, 1986.

9. L. Solymar and D. Walsh, *Lectures on the Electrical Properties of Materials*, 4th ed., Oxford University Press, New York, 1988.

10. J. Huheey, *Inorganic Chemistry*, 2d ed., Harper & Row, New York, 1978.

11. R. J. Borg and G. D. Dienes, *The Physical Chemistry of Solids*, Academic Press, New York, 1992.

12. A. F. Wells, *Structural Inorganic Chemistry*, 4th ed., Clarendon Press, Oxford, England, 1975.

13. J. C. Slater, *Introduction to Chemical Physics*, McGraw-Hill, New York, 1939.

14. L. Pauling, *The Nature of the Chemical Bond*, Cornell University Press, Ithaca, New York, 1960.

15. L. Pauling and E. B. Wilson, *Introduction to Quantum Mechanics with Applications to Chemistry*, McGraw-Hill, New York, 1935.

16. C. A. Coulson, *Valence*, Clarendon Press, Oxford, England, 1952.

17. L. Azaroff, *Introduction to Solids*, McGraw-Hill, New York, 1960.

CHAPTER 3

Structure of Ceramics

The Solid State, however, kept its grains
of microstructure coarsely veiled until
X-ray diffraction pierced the Crystal Planes
That roofed the giddy Dance, the taut Quadrille
Where Silicon and Carbon Atoms will
Link Valencies, four-figured, hand in hand
With common Ions and Rare Earths to fill
The lattices of Matter, Glass or Sand
With tiny Excitations, quantitatively grand.

John Updike; *The Dance of the Solids*[†]

3.1
INTRODUCTION

The previous chapter dealt with how atoms form bonds with one another. This chapter is devoted to the next level of structure, namely, the arrangement of ions and atoms in crystalline ceramics. This topic is of vital importance because many properties, including thermal, electrical, dielectric, optical, and magnetic ones, are quite sensitive to crystal structures.

Ceramics, by definition, are composed of at least two elements, and consequently their structures are, in general, more complicated than those of metals. While most metals are face-centered cubic (FCC), body-centered cubic (BCC), or hexagonal close-packed (HCP), ceramics exhibit a much wider variety of structures. Furthermore, and in contrast to metals where the structure is descriptive of the atomic arrangement, ceramic structures are named after the mineral for which the structure was first decoded. For example, compounds where the anions and

---

[†]   J. Updike, **Midpoint and other Poems**, A. Knopf, Inc., New York, New York, 1969. Reprinted with permission.

cations are arranged as they are in the rock salt structure, such as NiO and FeO, are described to have the rock salt structure. Similarly, any compound that crystallizes in the arrangement shown by corundum (the mineral name for Al_2O_3) has the corundum structure, and so forth.

Figure 3.1 illustrates a number of common ceramic crystal structures with varying anion-to-cation radius ratios. These can be further categorized into the following:

- AX-type structures which include the rock salt structure, CsCl, zinc blende, and wurtzite structures. The rock salt structure (Fig. 3.1a), named after NaCl, is the most common of the binary structures, with over one-half of the 400 compounds so far investigated having this structure. In this structure, the **coordination number** (defined as the number of nearest neighbors) for both cations and anions is 6. In the CsCl structure (Fig. 3.1b), the coordination number for both ions is 8. ZnS exists in two polymorphs, namely, the zinc blende and the wurtzite structures shown in Fig. 3.1c and d, respectively. In these structures the coordination number is 4; that is, all ions are tetrahedrally coordinated.
- AX_2-type structures. Calcium fluorite (CaF_2) and rutile (TiO_2), shown, respectively, in Fig. 3.1e and f, are two examples of this type of structure.
- $A_mB_nX_p$ structures in which more than one cation, A and B (or the same cation with differing valences), are incorporated in an anion sublattice. Spinels (Fig. 3.10) and perovskites (Fig. 3.9) are two of the more ubiquitous ones.

Note at the end of this brief introduction that the structures shown in Fig. 3.1 represent but a few of a much larger number of possible ones. Since a comprehensive survey of ceramic structures would be impossible within the scope of this book, instead some of the underlying principles which govern the way atoms and ions arrange themselves in crystals, which in turn can aid in understanding the multitude of structures that exist are outlined. This chapter is structured as follows: The next section outlines some of the more important and obvious factors that determine *local* atomic structure (i.e., the coordination number of the cations and anions) and how these factors can be used to predict the type of structure a certain compound will assume. In Sec. 3.3, the binary ionic structures are dealt with from the perspective of ion packing. In Sec. 3.4, the more complex ternary structures are briefly described. Sections 3.5 and 3.6 deal with Si-based covalently bonded ceramics such as SiC and Si_3N_4 and the silicates. The structure of glasses will be dealt with separately in Chap. 9. The last section deals with lattice parameters and density.

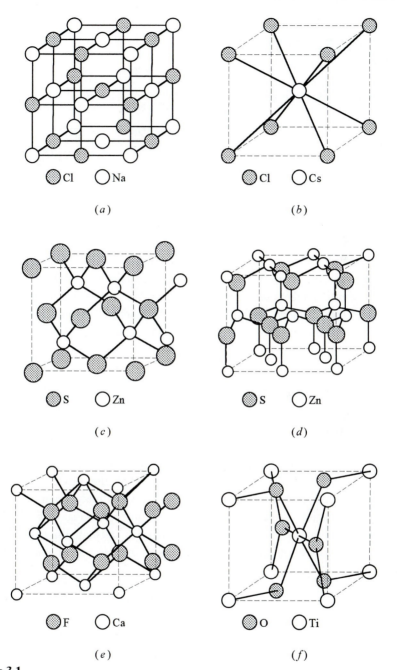

Cl ◯ Na

(a)

Cl ◯ Cs

(b)

S ◯ Zn

(c)

S ◯ Zn

(d)

F ◯ Ca

(e)

O ◯ Ti

(f)

Figure 3.1

Some common ceramic structures: (a) rock salt, (b) cesium chloride, (c) zinc blende, (d) wurtzite, (e) calcium fluorite, (f) rutile.

3.2
CERAMIC STRUCTURES

3.2.1   Factors Affecting Structure

Three factors are critical in determining the structure of ceramic compounds: crystal stoichiometry, the radius ratio, and the propensity for covalency and tetrahedral coordination.

Crystal stoichiometry

Any crystal has to be electrically neutral; i.e., the sum of the positive charges must be balanced by an equal number of negative charges, a fact that is reflected in the chemical formula of the compound. For example, in alumina, every two Al^{3+} cations have to be balanced by three O^{2-} anions, hence the chemical formula Al_2O_3. This requirement places severe limitations on the type of structure the ions can assume. For instance, an AX_2 compound cannot crystallize in the rock salt structure because the stoichiometry of the latter is AX, and vice versa.

Radius ratio[25]

To achieve the state of lowest energy, the cations and anions will tend to maximize attractions and minimize repulsions. Attractions are maximized when each cation surrounds itself with as many anions as possible, with the proviso that neither the cations nor the anions "touch." To illustrate, consider the four anions surrounding cations of increasing radii as shown in Fig. 3.2. The atomic arrangement in Fig. 3.2a is not stable because of the obvious anion-anion repulsions. Figure 3.2c, however, is stabilized by the mutual attraction of the cation and the anions. When the anions are just touching (Fig. 3.2b), the configuration is termed *critically stable* and is used to calculate the critical radii at which one structure becomes unstable with respect to another (Worked Example 3.1).

---

[25]   This radius ratio scheme was first proposed by L. Pauling. See, e.g., *The Nature of the Chemical Bond*, 3d ed., Cornell University Press, Ithaca, New York, 1960.

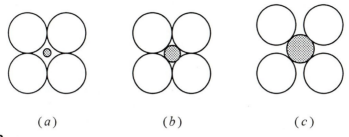

(a) (b) (c)

Figure 3.2
Stability criteria used to determine critical radius ratios.

Since cations are usually smaller than anions, the crystal structure is usually determined by the maximum number of anions that it is possible to pack around the cations, which, for a given anion size, will increase as the size of the cation increases. Geometrically, this can be expressed in terms of the radius ratio r_c/r_a, where r_c and r_a are the cation and anion radii, respectively. The critical radius ratios for various coordination numbers are shown in Fig. 3.3. Even the smallest cation can be surrounded by two anions and results in a linear arrangement (not shown in Fig. 3.3). As the size of the cation increases, i.e., as r_c/r_a increases, the number of anions that can be accommodated around a given cation increases to 3 and a triangular arrangement becomes stable (top of Fig. 3.3). For $r_c/r_a \geq 0.225$, the tetrahedral arrangement becomes stable, and so forth.

Propensity for covalency and tetrahedral coordination

In many compounds, tetrahedral coordination is observed despite the fact that the radius ratio would predict otherwise. For example, many compounds with radius ratios greater than 0.414 still crystallize with tetrahedral arrangements such as zinc blende and wurtzite. This situation typically arises when the covalent character of the bond is enhanced, such as when

- Cations with high polarizing power (for example, Cu^{2+}, Al^{3+}, Zn^{2+}, Hg^{2+}) are bonded to anions that are readily polarizable[26] (I^-, S^{2-}, Se^{2-}). As discussed in greater detail in Chap. 4, this combination tends to increase the covalent character of the bond and favor tetrahedral coordination.
- Atoms that favor sp^3 hybridization, such as Si, C, and Ge, tend to stabilize the tetrahedral coordination for obvious reasons.

---

[26] Polarizing power and polarizability are discussed in Chap. 4.

Coordination number	Arrangement of ions around central ion	Range of cation/anion ratios	Structure
3	corners of a triangle	≥ 0.155	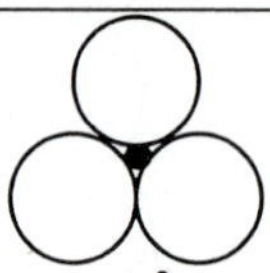
4	corners of a tetrahedron	≥ 0.225	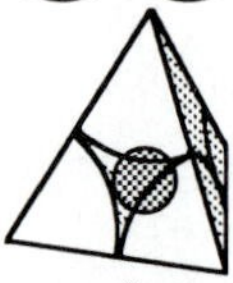
6	corners of a octahedron	≥ 0.414	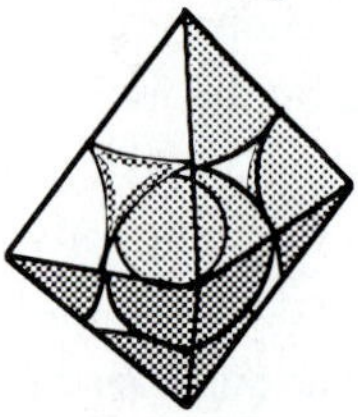
8	corners of a cube	≥ 0.732	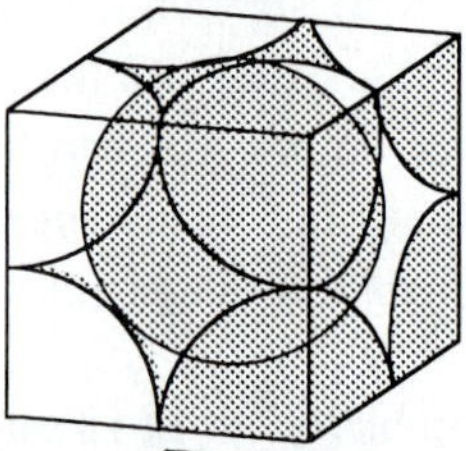
12	corners of a cuboctahedron	≈ 1.000	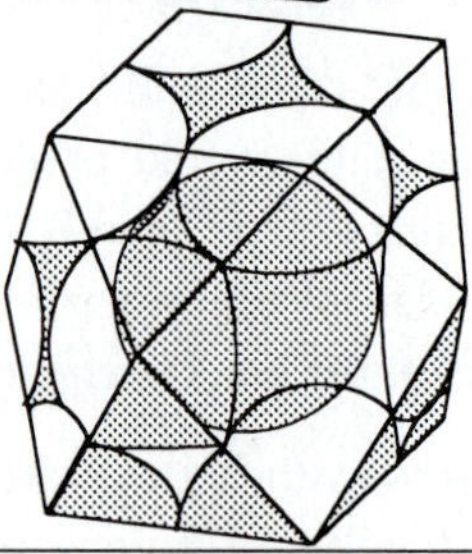

Figure 3.3

Critical radius ratios for various coordination numbers. The most stable structure is usually the one with the maximum coordination allowed by the radius ratio.

WORKED EXAMPLE 3.1. Derive the critical radius ratio for the tetrahedral arrangement (second from top in Fig. 3.3).

Answer

The easiest way to derive this ratio is to appreciate that when the radius ratio is critical, the cations just touch the anions, while the latter in turn are just touching one another (i.e., the anions are closely packed). Since the coordinates of the tetrahedral position in a close-packed arrangement (Fig. 3.4b) are 1/4, 1/4, 1/4, it follows that the distance between anion and cation centers is

$$r_{cation} + r_{anion} = \sqrt{\left(\frac{a}{4}\right)^2 + \left(\frac{a}{4}\right)^2 + \left(\frac{a}{4}\right)^2} = \sqrt{3}\,\frac{a}{4}$$

where a is the lattice parameter. Referring to Fig. 3.4b, the critical condition implies that the anions are just touching along the face diagonal, thus $4r_{anion} = \sqrt{2}a$. Combining these two equations yields $r_{cation}/r_{anion} = 0.225$.

3.2.2   Predicting Structures

It follows from the foregoing discussion that, at least in principle, it should be possible to predict the local arrangement of ions in a crystal if the ratio r_c/r_a is known. To illustrate the general validity of this statement, consider the oxides of group IV elements. The results are summarized in Table 3.1, and in all cases the observed structures are what one would predict based on the radius ratios.

TABLE 3.1
Comparison of predicted and observed structures based on radius ratio r_c/r_a

Compound	Radius ratio[†]	Prediction	Observed structure
CO_2	0.23	Linear coordination	CO_2 linear molecule
SiO_2	0.32	Tetrahedral coordination	Quartz — tetrahedral
GeO_2 [‡]	0.42	Tetrahedral coordination	Quartz — tetrahedral
SnO_2	0.55	Octahedral coordination	Rutile — octahedral
PbO_2	0.63	Octahedral coordination	Rutile — octahedral
ThO_2	0.86	Cubic arrangement	Fluorite — cubic

[†]   The radii used are the ones listed in App. 3A.

[‡]   Here the ratio is slightly greater than 0.414, but the tetrahedral coordination is still favored because of sp^3 hybridization of Ge.

This is not to say that the radius ratio should be taken absolutely; there are notable exceptions. For instance, according to the radius ratios, the Cs in CsCl should be octahedrally coordinated, when in fact it is not. Why that is the case is not entirely understood, to this date.

Clearly one of the more important parameters needed for understanding crystal structures and carrying out lattice energy calculations, etc., is the radii of the ions. Over the years there have been a number of compilations of ionic radii, probably the most notable among them being the one by Pauling.[27] More recently, however, Shannon and Prewitt[28] (SP) compiled a comprehensive set of radii that are about 14 pm larger for cations and 14 pm smaller for anions than the more traditional set of radii (see Table 3.2).

From X-ray diffraction the distance between ions (that is; $r_c + r_a$) can be measured with great precision. However, knowing where one ion ends and where the other begins is a more difficult matter. When careful X-ray diffraction measurements have been used to map out the electron density between ions and the point at which the electron density is a minimum is taken as the operational definition of the limits of the ions involved, and the results are compared with the SP radii, the match is quite good, as shown in Table 3.2. For this reason the SP radii are considered to be closer to representing the real size of ions in crystals than those of other compilations. A comprehensive set of SP radii is listed at the end of this chapter in App. 3A.

TABLE 3.2
Comparison of ionic radii with those measured from X-ray diffraction[†]

Crystal	r_{M-X}	Distance of minimum electron density from X-ray, pm	Pauling radii, pm	Shannon and Prewitt radii, pm
LiF	201	$r_{Li} = 92$	$r_{Li} = 60$	$r_{Li} = 90$
		$r_F = 109$	$r_F = 136$	$r_F = 119$
NaCl	281	$r_{Na} = 117$	$r_{Na} = 95$	$r_{Na} = 116$
		$r_{Cl} = 164$	$r_{Cl} = 181$	$r_{Cl} = 167$
KCl	314	$r_K = 144$	$r_K = 133$	$r_K = 152$
		$r_{Cl} = 170$	$r_{Cl} = 181$	$r_{Cl} = 167$
KBr	330	$r_K = 157$	$r_K = 133$	$r_K = 152$
		$r_{Br} = 173$	$r_{Br} = 195$	$r_{Br} = 182$

[†]   Source: Adapted from J. Huheey, *Inorganic Chemistry*, 2d ed., Harper & Row, New York, 1978, p. 86.

27   L. Pauling, *The Nature of the Chemical Bond*, 3d ed., Cornell University Press, Ithaca, New York, 1960, pp. 537–540.
28   R. D. Shannon and C. T. Prewitt, *Acta Crstallogr.*, **B25**:925(1969).

3.3
BINARY IONIC COMPOUNDS

The close packing of spheres occurs in one of two stacking sequences: ABABAB or ABCABC. The first stacking results in a hexagonal close-packed (HCP) arrangement, while the latter results in the a cubic close-packed or face-centered cubic (FCC) arrangement. Geometrically, and regardless of the stacking sequence, both arrangements create **two types of interstitial sites: octahedral and tetrahedral**, with coordination numbers 6 and 4, respectively (Fig. 3.4a[29]). The locations of these interstitial sites relative to the position of the atoms are shown in Fig. 3.4b for the FCC and in Fig. 3.4c and d for the HCP arrangements.

The importance of this aspect of packing lies in the fact that a majority of ceramic structures can be succinctly described by characterizing the *anion packing together with the fractional occupancy of each of the interstitial sites that are defined by that anion packing.* Table 3.3 summarizes the structure of the most prevalent ceramic materials according to that scheme. When they are grouped in that manner, it becomes immediately obvious that for most structures the *anions are in a close-packed arrangement* (second column), with the cations (fourth column) occupying varying fractions of the interstitial sites defined by the anion packing. How this results in the various ceramic structures is described in the remainder of this chapter. Before we tackle that subject, however, it is useful to examine one of the simplest ionic structures: CsCl.

3.3.1   CsCl Structure

In this structure, shown in Fig. 3.1b, the anions are in a simple cubic arrangement, and the cations occupy the centers of each unit cell.[30] Note that this is not a BCC structure because two different kinds of ions are involved.

3.3.2   Binary Structures Based on Close Packing of Anions

Cubic close-packed

The structures in which the anions are in an FCC arrangement are many and include rock salt, rutile, zinc blende, antifluorite (Fig. 3.5), perovskite (Fig. 3.9), and spinel (Fig. 3.10). To see how this scheme works, consider the rock salt structure in which, according to Table 3.1, the anions are in FCC arrangement. It

---

[29]   The sites are named for the number of faces of the shapes that form around the interstitial site.

[30]   Unit cells and lattice parameters are discussed in Chap. 1 and Sec. 3.7.

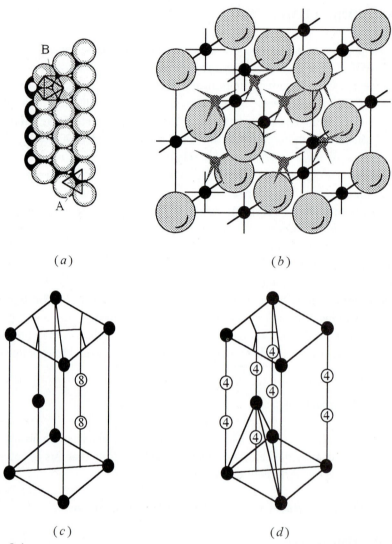

Figure 3.4

(*a*) When two close-packed planes of spheres are stacked, one on top of the other, they define the octahedral (B) and tetrahedral (A) sites between them. (*b*) Location of tetrahedral and octahedral interstitial sites within the cubic close-packed arrangement. The number of octahedral sites is always equal to the number of atoms, while the number of tetrahedral sites is always double the number of atoms. (*c*) Location of octahedral sites in the hexagonal close-packed arrangement. (*d*) Location of tetrahedral sites.

TABLE 3.3
Ionic structures grouped according to anion packing

Structure name	Anion packing	Coordination no. of M and X	Sites occupied by cations	Examples
		Binary compounds		
Rock salt	Cubic close-packed	6:6 MX	All oct.	NaCl, KCl, LiF, KBr, MgO, CaO, SrO, BaO, CdO, VO, MnO, FeO, CoO, NiO
Rutile	Distorted cubic close-packed	6:3 MX_2	1/2 oct.	TiO_2, GeO_2, SnO_2, PbO_2, VO_2, NbO_2, TeO_2, MnO_2, RuO_2, OsO_2, IrO_2
Zinc blende	Cubic close-packed	4:4 MX	1/2 tet.	ZnS, BeO, SiC
Antifluorite	Cubic close-packed	4:8 M_2X	All tet.	Li_2O, Na_2O, K_2O, Rb_2O, sulfides
Wurtzite	Hexagonal close-packed	4:4 MX	1/2 tet.	ZnS, ZnO, SiC, ZnTe
Nickel arsenide	Hexagonal close-packed	6:6 MX	All oct.	NiAs, FeS, FeSe, CoSe
Cadmium iodide	Hexagonal close-packed	6:3 MX_2	1/2 oct.	CdI_2, TiS_2, ZrS_2, MgI_2, VBr_2
Corundum	Hexagonal close-packed	6:4 M_2X_3	2/3 oct.	Al_2O_3, Fe_2O_3, Cr_2O_3, Ti_2O_3, V_2O_3, Ga_2O_3, Rh_2O_3
CsCl	Simple cubic	8:8 MX	All cubic	CsCl, CsBr, CsI
Fluorite	Simple cubic	8:4 MX_2	1/2 cubic	ThO_2, CeO_2, UO_2, ZrO_2, HfO_2, NpO_2, PuO_2, AmO_2, PrO_2
Silica types	Connected tetrahedra	4:2 MO_2	—	SiO_2, GeO_2
		Complex structures		
Perovskite	Cubic close-packed	12:6:6 ABO_3	1/4 oct. (B)	$CaTiO_3$, $SrTiO_3$, $SrSnO_3$, $SrZO_3$, $SrHfO_3$, $BaTiO_3$
Spinel (normal)	Cubic close-packed	4:6:4 AB_2O_4	1/8 tet. (A) 1/2 oct. (B)	$FeAl_2O_4$, $ZnAl_2O_4$, $MgAl_2O_4$
Spinel (inverse)	Cubic close-packed	4:6:4 $B(AB)O_4$	1/8 tet. (B) 1/2 oct. (A, B)	$FeMgFeO_4$, $MgTiMgO_4$
Illmenite	Hexagonal close-packed	6:6:4 ABO_3	2/3 oct. (A, B)	$FeTiO_3$, $NiTiO_3$, $CoTiO_3$
Olivine	Hexagonal close-packed	6:4:4 AB_2O_4	1/2 oct. (A) 1/8 tet. (B)	Mg_2SiO_4, Fe_2SiO_4

Source: Adapted from W. D. Kingery, H. K. Bowen, and D. R. Uhlmann, *Introduction to Ceramics*, 2d ed., Wiley, New York, 1976.

should be obvious at this point that placing cations on each of the octahedral sites in Fig. 3.4b results in the rock salt structure shown in Fig. 3.1a. Similarly, the zinc blende (Fig. 3.1c) structure is one in which half the tetrahedral sites are filled.

Hexagonal close-packed

Wurtzite, nickel arsenide, cadmium iodide, corundum, illmenite, and olivine are all structures in which the anion arrangement is HCP. For example, in corundum (Al_2O_3) the oxygen ions are hexagonally close-packed, and the Al ions fill two-thirds of the available octahedral sites. In contrast, if one-half the tetrahedral sites are filled, the resulting structure is wurtzite (Fig. 3.1d).

Fluorite and antifluorite structures

The antifluorite structure is best visualized by placing the anions in an FCC arrangement and filling all the tetrahedral sites with cations, as shown in Fig. 3.5. The resulting stoichiometry is M_2X with the oxides and chalcogenides of the alkali metals, for example, Li_2O, Na_2O, Li_2S, Li_2Se, crystallizing in this structure.

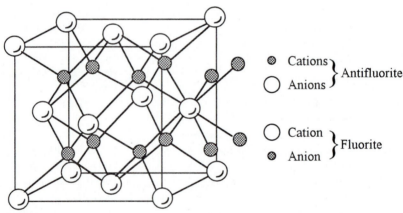

Figure 3.5
Relationship between fluorite and antifluorite structures. Note that in the fluorite structure the coordination number of the cations is 8, and the anions are in a simple cubic arrangement. Both these structures can be viewed as two interlaced structures, an FCC and a simple cubic.

In the fluorite structure, also shown in Fig. 3.5, the situation is reversed with the anions filling all the tetrahedral interstices of the close-packed cation sublattice. The resulting compound is MX_2. The oxides of large quadrivalent cations (Zr, Hf, Th) and the fluorides of large divalent cations (Ca, Sr, Ba, Cd, Hg, Pb) both crystallize in that structure. Another way to view this structure is to focus on the anions, which are in a *simple cubic* arrangement (see Fig. 3.5) with alternate cubic body centers occupied by cations. If viewed from this perspective, the eightfold coordination of the cations becomes obvious, which is not surprising since r_c/r_a now approaches 1, which according to Fig. 3.3 renders the cubic arrangement stable.

3.3.3 Rutile Structure

An idealized version of the rutile structure, shown in Fig. 3.6, can be viewed as consisting of TiO_6 octahedra that share corners and faces in such a way that each oxygen is shared by three octahedra. The structure also can be viewed as rectilinear ribbons of edge-shared TiO_6 octahedra joined together by similar ribbons, with the orientations of the adjacent ribbons differing by 90°. The relationship between the unit cell (Fig. 3.6b) and the stacking of the octahedra is shown in Fig. 3.6c. It should be noted that the actual structure comprises distorted octahedra rather than the regular ones shown here.

3.3.4 Other Structures

Table 3.3 does not include all binary oxides. However, for the most part, those not listed in the table are derivatives of the ones that are. To illustrate, consider the structure of yttria, shown in Fig. 3.7. Each cation is surrounded by six anions located at six of the eight corners of a cube. In half the cubes, the missing oxygens lie at the end of a face diagonal, and for the remaining half the missing oxygen falls on a body diagonal. The unit cell contains 48 oxygen ions and 32 yttrium ions; i.e., the full unit cell contains four layers of these minicubes, of which only the first row is shown here for clarity's sake. The structure shown in Fig. 3.7 is an idealized version; the actual positions of the oxygen atoms are shifted from the cube corners so that each yttrium atom occupies a strongly distorted octahedral site. This structure may appear quite complicated at first sight, but upon closer inspection its relationship to the fluorite structure becomes obvious.

WORKED EXAMPLE 3.2. Consider two hypothetical compounds MX and MX_2 with r_c/r_a values of 0.3 and 0.5, respectively. What possible structures can either of them adopt?

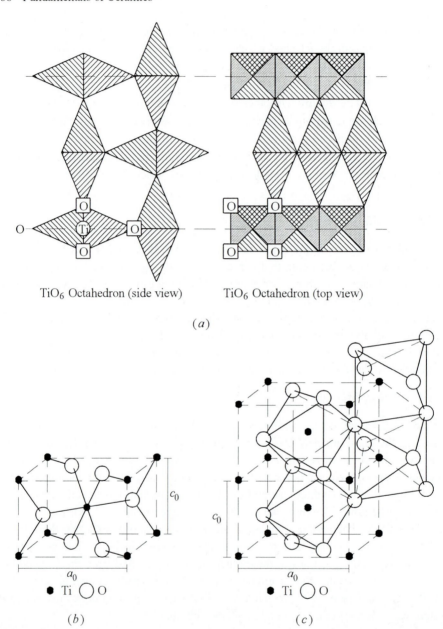

TiO$_6$ Octahedron (side view)   TiO$_6$ Octahedron (top view)

(a)

c_0

a_0

● Ti ◯ O

(b)

c_0

a_0

● Ti ◯ O

(c)

Figure 3.6

(a) Idealized stacking of TiO$_6$ octahedra in rutile. (b) Unit cell of rutile showing Ti-O bonds. (c) Stacking of TiO$_6$ octahedra and their relationship to the unit cell. Two unit cells are shown by dotted lines.

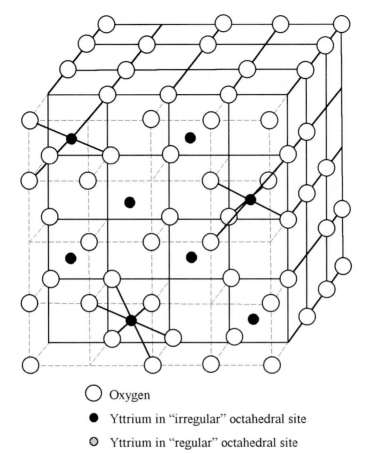

○ Oxygen

● Yttrium in "irregular" octahedral site

◉ Yttrium in "regular" octahedral site

Figure 3.7

Idealized crystal structure of Y_2O_3. Note two types of octahedral cation sites in alternating layers.

Answer

For the MX compound, the radius ratio predicts that the tetrahedral arrangement is the most stable. From Table 3.3, the only structure that would simultaneously satisfy the radius ratio requirements and the chemistry is zinc blende or wurtzite; all others would be eliminated. Which of these two structures is more stable is a more difficult question to answer and is a topic of ongoing research that depends subtly on the interactions between ions.

By using similar arguments, the case can be made that the only possible structures for the MX_2 compound are rutile and cadmium iodide.

3.4
COMPOSITE CRYSTAL STRUCTURES

In the preceding section, the structures of binary ceramics were discussed. As the number of elements in a compound increases, however, the structures naturally become more complex since the size and charge requirements of each ion differ. And while it is possible to describe the structures of ternary compounds by the scheme shown in Table 3.1, an alternative approach, which is sometimes more illustrative of the coordination number of the cations, is to imagine the structure to be made of the various building blocks shown in Fig. 3.3. In other words, the structure can be viewed as a three-dimensional jigsaw puzzle. Examples of such composite crystal structures are shown in Fig. 3.8. Two of the more important complex structures are spinels and perovskites; described below.

3.4.1   Perovskite Structure

Perovskite is a naturally occurring mineral with composition $CaTiO_3$. It was named after a 19th-century Russian mineralogist, Count Perovski. The general formula is ABX_3; and its idealized cubic structure is shown in Figs. 3.8b and 3.9, where the larger A cations, Ca in this case, are surrounded by 12 oxygens, and the smaller B (Ti^{4+}) ions are coordinated by 6 oxygens.

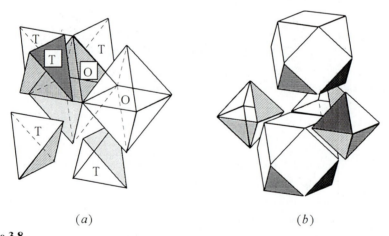

(a)   (b)

Figure 3.8

Examples of composite crystal structures. (a) Antifluorite structure, provided the octahedra are not occupied. (b) Perovskite structure ($CaTiO_3$). At the center of each cuboctahedron is a Ca ion. Each Ca cuboctahedron is surrounded by eight titania octahedra. Also see Fig. 3.9.

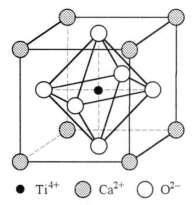

$\bullet$ Ti^{4+} $\ominus$ Ca^{2+} $\bigcirc$ O^{2-}

Figure 3.9

The perovskite structure centered on the Ti ion. See Fig. 3.8b for representation centered on a Ca ion.

Perovskites, like the spinels discussed in the next subsection, are able to accommodate a large number of cationic combinations as long as the overall crystal is neutral. For instance; $NaWO_3$, $CaSnO_3$, and $YAlO_3$ all crystallize in that structure or modified versions of it. The modified versions usually occur when the larger cation is small, which tends to tilt the axis of the B octahedra with respect to their neighbors. This results in puckered networks of linked B octahedra which are the basis for one of the unusual electrical properties of perovskites, namely, piezoelectricity, discussed in greater detail in Chap. 15.

Also note that several AB_3 structures can be easily derived from the perovskite structure (Fig. 3.9) by simply removing the atom in the body-centered position. Several oxides and fluorides, such as ReO_3, WO_3, NbO_3, NbF_3, and TaF_3, and other oxyfluorides such as $TiOF_2$ and $MoOF_2$ crystallize in that structure.

3.4.2 Spinel Structure

This structure is named after the naturally occurring mineral $MgAl_2O_4$, and its general formula is AB_2O_4, where the A and B cations are in the +2 and +3 oxidation states, respectively. The structure is shown in Fig. 3.10a, where emphasis is on the FCC stacking[31] of the oxygen ions; the cations, on the other hand, occupy one-eighth of the tetrahedral sites and one-half of the octahedral sites (see Table 3.3). The same structure, when viewed from a unit cell perspective, is shown in Fig. 3.10b.

---

31 Figure 3.10b illustrates nicely the ABCABC or FCC stacking sequence of the anions and how that stacking defines two types of interstitial sites.

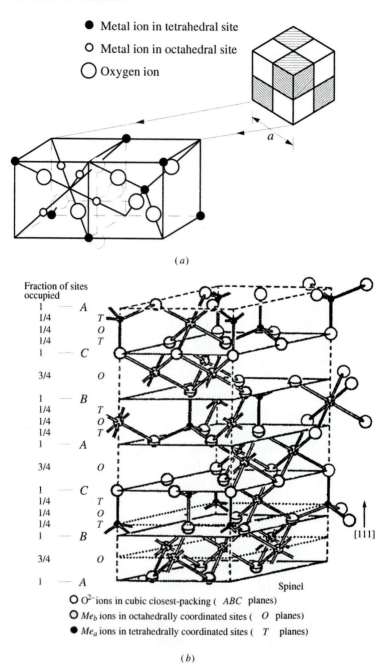

● Metal ion in tetrahedral site

○ Metal ion in octahedral site

◯ Oxygen ion

a

(*a*)

Fraction of sites
occupied

1	—	*A*	
1/4			*T*
1/4			*O*
1/4			*T*
1	—	*C*	
3/4			*O*
1	—	*B*	
1/4			*T*
1/4			*O*
1/4			*T*
1	—	*A*	
3/4			*O*
1	—	*C*	
1/4			*T*
1/4			*O*
1/4			*T*
1	—	*B*	
3/4			*O*
1	—	*A*	

[111]

Spinel

○ O^{2-} ions in cubic closest-packing (*ABC* planes)

○ Me_b ions in octahedrally coordinated sites (*O* planes)

● Me_a ions in tetrahedrally coordinated sites (*T* planes)

(*b*)

Figure 3.10

(*a*) Two octants of the spinel structure. (*b*) Spinel structure viewed by stacking the oxygens in close packing.

When the A^{2+} ions exclusively occupy the tetrahedral sites and the B^{3+} ions occupy the octahedral sites, the spinel is called a **normal spinel**. Usually the larger cations tend to populate the larger octahedral sites, and vice versa. In the **inverse spinel**, the A^{2+} ions and one-half the B^{3+} ions occupy the octahedral sites, while the other half of the B^{3+} ions occupy the tetrahedral sites.

As discussed in greater detail in Chap. 6, the oxidation states of the cations in spinel need not be restricted to +2 and +3, but may be any combination as long as the crystal remains neutral. This important class of ceramics is revisited in Chap. 15, when magnetic ceramics are dealt with.

3.5
STRUCTURE OF COVALENT CERAMICS

The building block of silicon-based covalent ceramics, which include among others the silicates (dealt with separately in the next section) SiC and Si_3N_4, is in all cases the Si tetrahedron; SiO_4 in the case of silicates, SiC_4 for SiC, and SiN_4 for Si_3N_4. The reasons Si bonds tetrahedrally were discussed in the last chapter.

Si_3N_4 exists in two polymorphs α and β. The structure of the β polymorph is shown in Fig. 3.11, where a fraction of the nitrogen atoms are linked to two silicons and others to three silicons. The structure of SiC also exists in many polymorphs, the simplest of which is cubic SiC, which has the zinc blende structure and is shown in Fig. 3.12 and 3.1c.

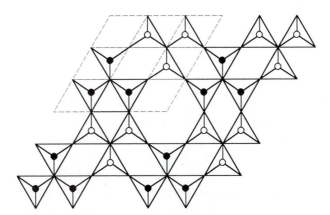

Figure 3.11
Structure of β-Si_3N_4 is hexagonal and made up of puckered six-member rings linked together at corners. The dark tetrahedra stick out of the plane of the paper while the light ones are pointed into the plane of the paper. The unit cell is dashed.

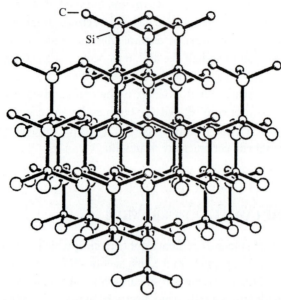

Figure 3.12
Structure of hexagonal SiC which crystallizes in the wurtzite structure.

3.6
STRUCTURE OF SILICATES

The earth's crust is about 48 percent by weight oxygen, 26 percent silicon, 8 percent aluminum, 5 percent iron, and 11 percent calcium, sodium, potassium, and magnesium combined. Thus it is not surprising that the earth's crust and mantle consist mainly of silicate minerals. The chemistry and structure of silicates can be quite complex indeed and cannot possibly be covered in detail here. Instead, a few guidelines to understanding their structure are given below.

Before proceeding much further, it is important to distinguish between two types of oxygens that exist in silicate structures, namely, **bridging** and **nonbridging oxygens.** An oxygen atom that is bonded to two Si atoms is a bridging oxygen, whereas one that is bonded to only one Si atom is nonbridging. Nonbridging oxygens (NBOs) are formed by the addition of, for the most part, either alkali or alkali–earth metal oxides to silica according to

$$-O-\overset{\overset{\displaystyle |}{O}}{\underset{\underset{\displaystyle |}{O}}{Si}}-O-\overset{\overset{\displaystyle |}{O}}{\underset{\underset{\displaystyle |}{O}}{Si}}-O+M_2O \rightarrow -O-\overset{\overset{\displaystyle |}{O}}{\underset{\underset{\displaystyle |}{O}}{Si}}\overset{M^+}{\underset{\underset{\displaystyle M^+}{O^-}}{{}^-O}}\overset{\overset{\displaystyle |}{O}}{\underset{\underset{\displaystyle |}{O}}{Si}}-O-$$

where O^- denotes a nonbridging oxygen. It is worth noting here that the NBOs are negatively charged and that local charge neutrality is maintained by having the cations end up adjacent to the NBOs. Furthermore, based on this equation, the following salient points are noteworthy:

1. The number of NBOs is proportional to the number of moles of alkali or alkali–earth metal oxide added (see Worked Example 3.3).
2. The addition of alkali or alkali–earth metal oxides to silica must increase the overall *O/Si ratio* of the silicate.
3. Increasing the number of NBOs results in the progressive breakdown of the silicate structure into smaller units.

It thus follows that *a critical parameter that determines the structure of a silicate is the number of NBOs per tetrahedron, which in turn is determined by the O/Si ratio.* How this ratio determines structure is discussed below; but before addressing this point, it is important to appreciate that in general the following principles also apply:

1. The basic building block is the SiO_4 tetrahedron. The Si-O bond is partly covalent and the tetrahedron satisfies both the bonding requirements of covalent directionality and the relative size ratio.
2. Because of the high charge on the Si^{4+} ion, the tetrahedral units are rarely joined edge to edge and never face to face, but almost always share corners, with no more than two tetrahedra sharing a corner. The reason behind this rule, first stated by Pauling, is demonstrated in Fig. 3.13, where it is obvious that the cation separation distance decreases in going from corner to edge to face sharing. This in turn results in cation-cation repulsions and a decrease in the stability of the structure.

The relationship between the O/Si ratio, which can only vary between 2 and 4, and the structure of a silicate is illustrated in Table 3.4.[32] Depending on the shape of the repeat units, these structures have been classified as three-dimensional networks, infinite sheets, chains, and isolated tetrahedra. Each of these structures is discussed in some detail below.

---

32 Implicit in Table 3.4 is that each O atom **not** shared between two Si tetrahedra, i.e., the nonbridging oxygens, is **negatively** charged.

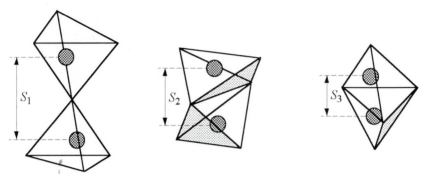

Figure 3.13
Effect of corner edge, and face sharing on cation-cation separation. The distances $S_1:S_2:S_3$ are in the ratio $1:0.58:0.33$; that is; cation-cation repulsion increases on going from left to right, which tends to destabilizes the structure.

Silica

For a ratio of 2, that is, SiO_2, each oxygen is linked to two silicons and each silicon is linked to four oxygens, resulting in a three-dimensional network, as shown at the top of Table 3.4. The resulting structures are all allotropes of silicas which, depending on the exact arrangement of the tetrahedra, include among others quartz, tridymite, and cristobalite. If, however, long-range order is lacking, the resulting solid is labeled *amorphous silica* or *fused quartz* (see Chap. 9 for more details concerning the structure of fused silica).

Sheet silicates

When three out of four silicons are shared, i.e., for an O/Si ratio of 2.5, a sheet structure results (Table 3.4). Clays, talcs $Mg_3(OH)_2(Si_2O_5)_2$, and micas $KAl_2(OH)_2(AlSi_3O_{10})$ are typical of that structure. Kaolinite clay $Al_3(OH)_4(Si_2O_5)$, shown schematically in Fig. 3.14a, is composed of $(Si_2O_5)^{2-}$ sheets that are held together by positively charged sheets of Al-O,OH octahehdra (Fig. 3.14b). This structure helps explain why clays absorb water so readily; the polar water molecule is easily absorbed between the top of the positive sheets and the bottom of the silicate sheets (Fig. 3.14c).

In mica, shown in Fig. 3.14d, aluminum ions substitute for one-fourth of the Si atoms in the sheets, requiring an alkali ion such as K^+ in order for the structure to remain electrically neutral. The alkali ions fit in the "holes" of the silicate sheets and bond the sheets together with an ionic bond that is somewhat stronger than that in clays (Fig. 3.14d). Thus whereas mica does not absorb water as readily as clays do, little effort is required to flake off a very thin chip of the material.

TABLE 3.4
Relationship between silicate structure and the O/Si ratio

Structure	O/Si ratio	No. of oxygens per Si		Structure and examples
		Bridg.	Non-bridg.	
	2.00	4.0	0.0	Three-dimensional network Quartz, tridymite, cristabolite are all polymorphs of silica
Repeat unit $(Si_4O_{10})^{4-}$	2.50	3.0	1.0	Infinite sheets $Na_2Si_2O_5$ Clays (kaolinite)
Repeat unit $(Si_4O_{11})^{6-}$	2.75	2.5	1.5	Double chains, e.g., asbestos
Repeat unit $(SiO_3)^{2-}$	3.00	2.0	2.0	Chains $(SiO_3)_n^{2n-}$, Na_2SiO_3, $MgSiO_3$
$(SiO_4)^{4-}$	4.00	0.0	4.0	Isolated SiO_4^{4-}, tetrahedra Mg_2SiO_4 olivine, Li_4SiO_4

† The simplest way to determine the number of nonbridging oxygens per Si is to divide the charge on the repeat unit by the number of Si atoms in the repeat unit.

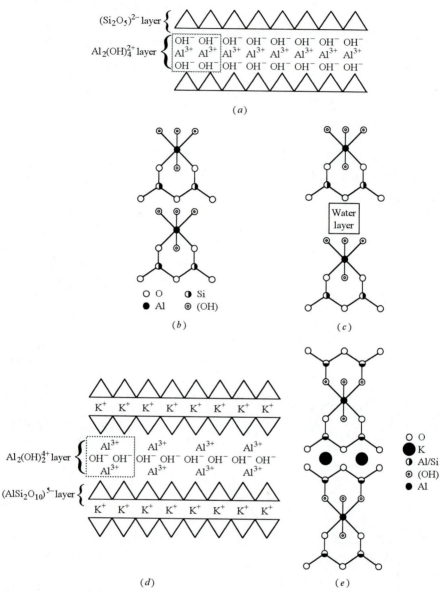

Figure 3.14

(*a*) Structure of kaolinite clay (showing layered structure). (*b*) Same structure as in (*a*) but emphasizing bonding of Al^{3+} ions. (*c*) Same as (*b*) but hydrated. Note polar water molecule easily absorbs in between the layers. (*d*) Structure of mica. (*e*) Same as (*d*) but emphasizing nature of bonding between sheets.

Chain silicates

For O/Si ratios of 3.0, infinite chains or ring structures result. The most notorious of this class is asbestos, in which the silicate chains are held together by weak electrostatic forces that are easier to pull apart than the bonds holding the chains together. This results in the stringy, fibrous structures that embed themselves in the human lung with devastating long-term consequences.

Island silicates

When the O/Si ratio is 4, the structural units are the isolated $(SiO_4)^{4-}$ tetrahedra which cannot join to each other but are connected by the positive ions in the crystal structure. The resulting structure is termed an *island silicate* for which garnets $(Mg, Fe^{2+}, Mn, Ca)_3 (Cr, Al, Fe^{3+})_2 (SiO_4)_3$ and olivines $(Mg, Fe^{2+})_2 (SiO_4)$ are examples.[33] Here the $(SiO_4)^{4-}$ tetrahedron behaves as an anion and the resulting pseudobinary structure is ionically bonded.

Aluminosilicates

Aluminum plays an interesting role in silicates. Either the Al^{3+} ions can substitute for the Si^{4+} ion in the network, in which case the charge has to be compensated by an additional cation (e.g., mica), or it can occupy octahedral and/or tetrahedral holes between the silicate network, as in the case for clays.

When Al substitutes for Si in the network, the appropriate ratio for determining the structure is the (Al + Si)/O ratio. So, e.g., for albite $(NaAlSi_3O_8)$, anorthite $(CaAl_2Si_2O_8)$, eucryptite $(LiAlSiO_4)$, orthoclase $(KAlSi_3O_8)$, and spodune $(LiAlSi_2O_6)$, the ratio O/(Al + Si) is 2; and in all cases a three-dimensional structure is expected and indeed observed. As a result of this three-dimensionality, the melting points of some of these silicates are among the highest known.

It should be obvious from the preceding discussion that with the notable exception of silica and some of the aluminosilicates, most silicates exhibit mixed bonding, with the bonding within the silicate network, i.e., the Si–O–Si bonds, being quite different from those bonds holding the units together, which can be either ionic or weak secondary bonds depending on the material.

---

[33] Separating elements by a comma denotes that these elements can be found in various proportions without changing the basic structure. For example, the end members $Mg_2(SiO_4)$ and $Fe_2(SiO_4)$ and any combination in between denoted as $(Mg,Fe)(SiO_4)$ would all exhibit the same structure.

WORKED EXAMPLE 3.3. (a) Derive a generalized expression relating the number of nonbridging oxygens per Si atom present in a silicate structure to the mole fraction of metal oxide added. (b) Calculate the number of bridging and nonbridging oxygens per Si atom for $Na_2O \cdot 2SiO_2$. What is the most likely structure for this compound?

Answer

(a) The simplest way to obtain the appropriate expression is to realize that in order to maintain charge neutrality, the number of NBOs has to equal the total cationic charge. Hence starting with a basis of y mol of SiO_2, the addition of η mol of $M_\zeta O$ results in the formation of $z(\zeta\eta)$ NBOs, where z is the charge on the modifying cation. Thus the number of nonbridging oxygens per Si atom is simply:

$$NBO = \frac{z(\zeta\eta)}{y}$$

The corresponding O/Si ratio, denoted by R, is

$$R = 2 + \frac{\eta}{y}$$

(b) For $Na_2O \cdot 2SiO_2$, $\eta = 1$, $\zeta = 2$, and $y = 2$. Consequently, $NBO = (2 \cdot 1)/2 = 1$, and so the number of bridging oxygens per Si atom is $4 - 1 = 3$. Furthermore, since $R = 2.5$, it follows that the most likely structure of this silicate is a sheet structure (Table 3.4).

3.7
LATTICE PARAMETERS AND DENSITY

Lattice parameters

As noted in Chap.1, every unit cell can be characterized by six **lattice parameters** — three edge lengths a, b, and c and three interaxial angles α, β, and γ. On this basis, there are seven possible combinations of a, b, and c and α, β, and γ that correspond to seven crystal systems (see Fig. 1.2). In order of decreasing symmetry, they are cubic, hexagonal, tetragonal, rhombohedral, orthorhombic, monoclinic, and triclinic. In the remainder of this section, for the sake of simplicity the discussion is restricted to the cubic system for which $a = b = c$ and $\alpha = \beta = \gamma = 90°$. Consequently, this system is characterized by only one parameter, usually denoted by a.

The *lattice parameter* is the length of the unit cell, which is defined to be the smallest repeat unit that satisfies the **symmetry** of the crystal. For example, the rock salt unit cell shown in Fig. 3.1*a* contains four cations and four anions, because this is the smallest repeat unit that also satisfies the requirements that the crystal possess a fourfold symmetry (in addition to the threefold symmetry along the body diagonal). It is not difficult to appreciate that if only one quadrant of the unit cell shown in Fig. 3.1*a* were chosen as the unit cell, such a unit would *not* possess the required symmetry. Similar arguments can be made as to why the unit cell of Y_2O_3 is the one depicted in Fig. 3.7, or that of spinel is the one shown in Fig. 3.10, etc.

Density

One of the major attributes of ceramics is that as a class of materials, they are less dense than metals and hence are very attractive when specific (i.e., per unit mass) properties are important. The main factors that determine density are, first, the masses of the atoms that make up the solid. Clearly, the heavier the atomic mass, the denser the solid, which is why NiO, for example, is denser than NaCl. The second factor relates to the nature of the bonding and its directionality. Covalently bonded ceramics are more "open" structures and tend to be less dense, whereas the near-close-packed ionic structures, such as NaCl, tend to be denser. For example, MgO and SiC have very similar molecular weights (≈ 40g) but the density of SiC is less than that of MgO (see Worked Example 3.4, and Table 4.3).

WORKED EXAMPLE 3.4. Starting with the radii of the ions or atoms, calculate the theoretical densities of MgO and SiC.

Answer

The density of any solid can be determined from a knowledge of the unit cell. The density can be calculated from

$$\rho = \frac{\text{weight of ions within unit cell}}{\text{volume of unit cell}} = \frac{n'(\sum M_C + \sum M_A)}{V_C N_{Av}}$$

where n' = number of formula units within the unit cell.

$\sum M_C$ = sum of atomic weights of all cations within unit cell

$\sum M_A$ = sum of atomic weights of all anions within unit cell

V_C = unit cell volume

N_{Av} = Avogadro's number

MgO has the rock salt structure which implies that the ions touch along the side of the unit cell. Refer to Fig. 3.1. The lattice parameter $= 2r_{Mg} + 2r_O = 2(126 + 86) = 424$pm.

The atomic weight of Mg is 24.31 g/mol, whereas the atomic weight of O is 16 g/mol. Since there are four magnesium and four oxygen ions within the unit cell, it follows that

$$\rho = \frac{4(16 + 24.31)}{(6.022 \times 10^{23})(424 \times 10^{-10})^3} = 3.51 \text{g}/\text{cm}^3$$

To calculate the lattice parameter a for SiC (Fig. 3.1c) is a little trickier since the atoms touch along the body diagonal with length $\sqrt{3}a$. The Si-C distance is thus equal to one-fourth the length of the body diagonal. The atomic[34] radius of Si is 118 nm, while that of C is 71 nm. It follows that

$$\frac{\sqrt{3}}{4}a = 118 + 71$$

$$a = 436 \text{ nm}$$

Given that each unit cell contains four C and four Si atoms, with molecular weights of 12 and 28.09, respectively, applying the formula for the density gives

$$\rho = \frac{4(12 + 28.09)}{(6.022 \times 10^{23})(436 \times 10^{-10})^3} = 3.214 \text{g}/\text{cm}^3$$

Note that while the weights of the atoms in the unit cell are very comparable, the lower density for SiC is a direct consequence of the larger lattice parameter that reflects the more "open" structure of covalently bonded solids.

EXPERIMENTAL DETAILS: DETERMINING CRYSTAL STRUCTURES, LATTICE PARAMETERS, AND DENSITY

Crystal Structures and Lattice Parameters

By far the most powerful technique to determine crystal structure employs X-ray or neutron diffraction. The essentials of the technique are shown in Fig. 3.15 where a collimated X-ray beam strikes a crystal. The electrons of the crystal scatter the beam through a wide angle, and for the most part the scattered rays will interfere with each other destructively and will cancel. At various directions, however, the scattered X-rays will interfere constructively and will give rise to a strong reflection.

---

[34]   The use of ionic radii listed in App. 3A is inappropriate in this case because the bonding is almost purely covalent. (See periodic table printed on inside cover.)

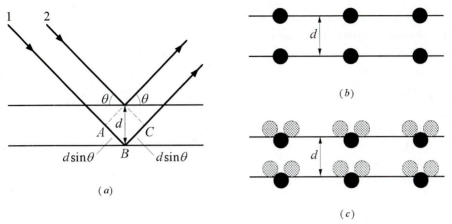

Figure 3.15

(*a*) Scattering of waves by crystal planes. While the angle of the scattered waves will depend only on *d*, the intensity will depend on the nature of the scatterers, as is clear when one compares (*b*) and (*c*). They both have the same lattice type, but quite different unit cells and crystal structures, which in turn would be reflected in the intensity of the scattered waves.

The condition for constructive interference corresponds to that when the scattered waves are in phase. In Fig. 3.15, the wavefront labeled 1 would have to travel a distance $AB + BC$ farther than the wavefront labeled 2. Thus if and when $AB + BC$ is a multiple of the wavelength of the incident X-ray λ, that is,

$$AB + BC = n\lambda$$

coherent reflection will result. It is a trivial exercise in trigonometry to show that

$$AB + BC = 2d \sin \theta$$

where θ is the angle of incidence of the X-ray on the crystal surface defined in Fig. 3.15. Combining the two equations results in the diffraction condition, also known as **Bragg's law**:

$$2d_{hkl} \sin \theta = n\lambda \quad n = 1, 2, \ldots \tag{3.1}$$

where d_{hkl} is the distance between adjacent planes in a crystal.

There are numerous X-ray diffraction techniques. The idea behind them all, however, is similar: Either the beam is moved relative to the diffracting crystals, and the intensity of the diffracted beam is measured as a function of angle θ; or the beam is fixed, the crystal is rotated, and the angles at which the diffraction occurs are recorded.

Note that the angle at which diffraction occurs is only part of the information that is needed and used to determine crystal structures — the intensity of the diffracted beam is also an indispensable clue. This can be easily grasped by comparing the two lattices in Fig. 3.15b and c. If the only information available were the angle of diffraction, then these two quite different structures would be indistinguishable. Constructive or destructive interference between the atoms within the molecules, in Fig. 3.15c, would clearly result in X-ray intensities that would be different from the ones shown in Fig. 3.15b, for example. Thus while the angle at which scattering occurs depends on the lattice type, the intensity depends on the nature of the scatterers.

Density

Measuring the density of a 100 percent dense ceramic is relatively straightforward. If the sample is uniform in shape, then the volume is calculated from the dimensions, and the weight is accurately measured by using a sensitive balance. The ratio of mass to volume is the density.

A more accurate method for measuring the volume of a sample is to make use of Archimedes' principle, where the difference between the sample weight in air w_{air} and its weight in a fluid w_{fluid}, divided by the density of the fluid ρ_{fluid}, gives the volume of the liquid displaced, which is identical to the volume of the sample. The density of the sample is then simply

$$\rho = \frac{w_{air}}{(w_{air} - w_{fluid})/\rho_{fluid}}$$

Ceramics are not always fully dense, however, and open porosities can create problems in measuring the density. Immersion of a porous body in a fluid can result in the fluid penetrating the pores, reducing the volume of fluid displaced, which consequently results in densities that appear higher than the actual ones. Several techniques can be used to overcome this problem. One is to coat the sample with a very thin layer of molten paraffin wax, to seal the pores prior to immersion in the fluid. Another is to carry out the measurement as described above, remove the sample from the fluid, wipe any excess liquid with a cloth saturated with the fluid, and then measure the weight of the fluid-saturated sample. The difference in weight $w_{sat} - w_{air}$ is a measure of the weight of the liquid trapped in the pores, which when divided by ρ_{fluid}, yields the volume of the pores. For greater detail it is best to refer to the ASTM test methods.

3.8
SUMMARY

Ceramic structures can be quite complicated and diverse, and for the most part they depend on the type of bonding present. For ionically bonded ceramics, the stoichiometry and the radius ratio of the cations to the anions are critical determinants of structure. The former narrows the possible structures, and the latter determines the local arrangement of the anions around the cations. The structures can be best visualized by focusing first on the anion arrangement which, for the vast majority of ceramics, is FCC, HCP, or simple cubic. Once the anion sublattice is established, the various structures that arise will depend on the fractional cationic occupancy of the various interstitial sites defined by the anion sublattice.

The structures of covalent ceramics that are Si-based are based on the SiX_4 tetrahedron. These tetrahedra are usually linked to each other at the corners. For silicates, the building block is the SiO_4 tetrahedron. The most important parameter in determining the structure of silicates is the O/Si ratio. The minimum ratio is 2 and results in a three-dimensional network. The addition of modifier oxides to silica increases that ratio and results in both the formation of nonbridging oxygens and the progressive breakdown of the structure. As the O/Si ratio increases; the structure changes to sheets, chains, and finally island silicates, when the ratio is 4.

APPENDIX 3A

███████████

Ionic Radii

TABLE 3A.1
Effective ionic radii of the elements

Ion	Coordination no.	pm	Ion	Coordination no.	pm	Ion	Coordination no.	pm
Ac^{3+}	4	126.0	Ag^{2+}	6	108.0	Am^{3+}	6	111.5
Ag^+	2	81.0	Ag^{3+}	4 SQ	81.0		8	123.0
	4	114.0		6	89.0	Am^{4+}	6	99.0
	4 SQ	116.0	Al^{3+}	4	53.0		8	109.0
	5	123.0		5	62.0	As^{3+}	6	72.0
	6	129.0		6	67.5	As^{5+}	4	47.5
	7	136.0	Am^{2+}	7	135.0		6	60.0
	8	142.0		8	140.0	At^{7+}	7	76.0
Ag^{2+}	4 SQ	93.0		9	145.0	Au^+	6	151.0

TABLE 3A.1 Continued
Effective ionic radii of the elements

Ion	Coordination no.	pm	Ion	Coordination no.	pm	Ion	Coordination no.	pm
Au^{3+}	4 SQ	82.0	Cd^{2+}	6	109.0	Cr^{5+}	6	63.0
	6	99.0		7	117.0		8	71.0
Au^{5+}	6	71.0		8	124.0	Cr^{6+}	4	40.0
B^{3+}	3	15.0		12	145.0		6	58.0
	4	25.0	Ce^{3+}	6	115.0	Cs^+	6	181.0
	6	41.0		7	121.0		8	188.0
Ba^{2+}	6	149.0		8	128.3		9	192.0
	7	152.0		9	133.6		10	195.0
	8	156.0		10	139.0		11	199.0
	9	161.0		12	148.0		12	202.0
	10	166.0	Ce^{4+}	6	101.0	Cu^+	2	60.0
	11	171.0		8	111.0		4	74.0
	12	175.0		10	121.0		6	91.0
Be^{2+}	3	30.0		12	128.0	Cu^{2+}	4	71.0
	4	41.0	Cf^{3+}	6	109.0		4 SQ	71.0
	6	59.0	Cf^{4+}	6	96.1		5	79.0
Bi^{3+}	5	110.0		8	106.0		6	87.0
	6	117.0	Cl^-	6	167.0	Cu^{3+}	6 LS	68.0
	8	131.0	Cl^{5+}	3PY	26.0	D^+	2	4.0
Bi^{5+}	6	90.0	Cl^{7+}	4	22.0	Dy^{2+}	6	121.0
Bk^{3+}	6	110.0		6	41.0		7	127.0
Bk^{4+}	6	97.0	Cm^{3+}	6	111.0		8	133.0
	8	107.0	Cm^{4+}	6	99.0	Dy^{3+}	6	105.2
Br^-	6	182.0		8	109.0		7	111.0
Br^{3+}	4 SQ	73.0	Co^{2+}	4 HS b	72.0		8	116.7
Br^{5+}	3 PY	45.0		5	81.0		9	123.5
Br^{7+}	4	39.0		6 LS c	79.0	Er^{3+}	6	103.0
	6	53.0		HS	88.5		7	108.5
C^{4+}	3	6.0		8	104.0		8	114.4
	4	29.0	Co^{3+}	6 LS	68.5		9	120.2
	6	30.0		HS	75.0	Eu^{2+}	6	131.0
Ca^{2+}	6	114.0	Co^{4+}	4	54.0		7	134.0
	7	120.0		6 HS	67.0		8	139.0
	8	126.0	Cr^{2+}	6 LS	87.0		9	144.0
	9	132.0		HS	94.0		10	149.0
	10	137.0	Cr^{3+}	6	75.5	Eu^{3+}	6	108.7
	12	148.0	Cr^{4+}	4	55.0		7	115.0
Cd^{2+}	4	92.0		6	69.0		8	120.6
	5	101.0	Cr^{5+}	4	48.5		9	126.0

TABLE 3A.1 Continued
Effective ionic radii of the elements

Ion	Coordination no.	pm	Ion	Coordination no.	pm	Ion	Coordination no.	pm
F^-	2	114.0	Hg^{2+}	8	128.0	Mn^{2+}	5 HS	89.0
	3	116.0	Ho^{3+}	6	104.1		6 LS	81.0
	4	117.0		8	115.5		HS	97.0
	6	119.0		9	121.2		7 HS	104.0
F^{7+}	6	22.0		10	126.0		8	110.0
Fe^{2+}	4 HS	77.0	I^-	6	206.0	Mn^{3+}	5	72.0
	4 SQ HS	78.0	I^{5+}	3PY	58.0		6 LS	72.0
	6 LS	75.0		6	109.0		HS	78.5
	HS	92.0	I^{7+}	4	56.0	Mn^{4+}	4	53.0
	8 HS	106.0		6	67.0		6	67.0
Fe^{3+}	4 HS	63.0	In^{3+}	4	76.0	Mn^{5+}	4	47.0
	5	72.0		6	94.0	Mn^{6+}	4	39.5
	6 LS	69.0		8	106.0	Mn^{7+}	4	39.0
	HS	78.5	Ir^{3+}	6	82.0		6	60.0
	8 HS	92.0	Ir^{4+}	6	76.5	Mo^{3+}	6	83.0
Fe^{4+}	6	72.5	Ir^{5+}	6	71.0	Mo^{4+}	6	79.0
Fe^{6+}	4	39.0	K^+	4	151.0	Mo^{5+}	4	60.0
Fr^+	6	194.0		6	152.0		6	75.0
Ga^{3+}	4	61.0		7	160.0	Mo^{6+}	4	55.0
	5	69.0		8	165.0		5	64.0
	6	76.0		9	169.0		6	73.0
Gd^{3+}	6	107.8		10	173.0		7	87.0
	7	114.0		12	178.0	N^{3-}	4	132.0
	8	119.3	La^{3+}	6	117.2	N^{3+}	6	30.0
	9	124.7		7	124.0	N^{5+}	3	4.4
Ge^{2+}	6	87.0		8	130.0		6	27.0
Ge^{4+}	4	53.0		9	135.6	Na^+	4	113.0
	6	67.0		10	141.0		5	114.0
H^+	1	−24.0		12	150.0		6	116.0
	2	−4.0+	Li^+	4	73.0		7	126.0
Hf^{4+}	4	4.0		6	90.0		8	132.0
	6	85.0		8	106.0		9	138.0
	7	90.0	Lu^{3+}	6	100.1		12	153.0
	8	97.0		8	111.7	Nb^{3+}	6	86.0
Hg^+	3	111.0	Mg^{2+}	4	71.0	Nb^{4+}	6	82.0
	6	133.0		5	80.0		8	93.0
Hg^{2+}	2	83.0		6	86.0	Nb^{5+}	4	62.0
	4	110.0		8	103.0		6	78.0
	6	116.0	Mn^{2+}	4 HS	80.0		7	83.0

TABLE 3A.1 Continued
Effective ionic radii of the elements

Ion	Coordination no.	pm	Ion	Coordination no.	pm	Ion	Coordination no.	pm
Nb^{5+}	8	88.0	P^{5+}	5	43.0	Pt^{5+}	6	71.0
Nd^{2+}	8	143.0		6	52.0	Pu^{3+}	6	114.0
	9	149.0	Pa^{3+}	6	118.0	Pu^{4+}	6	100.0
Nd^{3+}	6	112.3	Pa^{4+}	6	104.0		8	110.0
	8	124.9		8	115.0	Pu^{5+}	6	88.0
	9	130.3	Pa^{5+}	6	92.0	Pu^{6+}	6	85.0
	12	141.0		8	105.0	Ra^{2+}	8	162.0
Ni^{2+}	4	69.0		9	109.0		12	184.0
	4 SQ	63.0	Pb^{2+}	4 PY	112.0	Rb^{+}	6	166.0
	5	77.0		6	133.0		7	170.0
	6	83.0		7	137.0		8	175.0
Ni^{3+}	6 LS	70.0		8	143.0		9	177.0
	HS	74.0		9	149.0		10	180.0
Ni^{4+}	6 LS	62.0		10	154.0		11	183.0
No^{2+}	6	124.0		11	159.0		12	186.0
Np^{2+}	6	124.0		12	163.0		14	197.0
Np^{3+}	6	115.0	Pb^{4+}	4	79.0	Re^{4+}	6	77.0
Np^{4+}	6	101.0		5	87.0	Re^{5+}	6	72.0
	8	112.0		6	91.5	Re^{6+}	6	69.0
Np^{5+}	6	89.0		8	108.0	Re^{7+}	4	52.0
Np^{6+}	6	86.0	Pd^{+}	2	73.0		6	67.0
Np^{7+}	6	85.0	Pd^{2+}	4 SQ	78.0	Rh^{3+}	6	80.5
O^{2-}	2	121.0		6	100.0	Rh^{4+}	6	74.0
	3	122.0	Pd^{3+}	6	90.0	Rh^{5+}	6	69.0
	4	124.0	Pd^{4+}	6	75.5	Ru^{3+}	6	82.0
	6	126.0	Pm^{3+}	6	111.0	Ru^{4+}	6	76.0
	8	128.0		8	123.3	Ru^{5+}	6	70.5
OH^{-}	2	118.0		9	128.4	Ru^{7+}	4	52.0
	3	120.0	Po^{4+}	6	108.0	Ru^{8+}	4	50.0
	4	121.0		8	122.0	S^{2-}	6	170.0
	6	123.0	Po^{2+}	6	81.0	S^{4+}	6	51.0
Os^{4+}	6	77.0	Pr^{3+}	6	113.0	S^{6+}	4	26.0
Os^{5+}	6	71.5		8	126.6		6	43.0
Os^{6+}	5	63.0		9	131.9	Sb^{3+}	4 PY	90.0
	6	68.5	Pr^{4+}	6	99.0		5	94.0
Os^{7+}	6	66.5		8	110.0		6	90.0
Os^{8+}	4	53.0	Pt^{2+}	4 SQ	74.0	Sb^{5+}	6	74.0
P^{3+}	6	58.0		6	94.0	Sc^{3+}	6	88.5
P^{5+}	4	31.0	Pt^{4+}	6	76.5		8	101.0

TABLE 3A.1 Continued
Effective ionic radii of the elements

Ion	Coordination no.	pm	Ion	Coordination no.	pm	Ion	Coordination no.	pm
Se^{2-}	6	184.0	Tc^{7+}	6	70.0	U^{6+}	4	66.0
Se^{4+}	6	64.0	Te^{2-}	6	207.0		6	87.0
Se^{6+}	4	42.0	Te^{4+}	3	66.0		7	95.0
	6	56.0		4	80.0		8	100.0
Si^{4+}	4	40.0		6	111.0	V^{2+}	6	93.0
	6	54.0	Te^{6+}	4	57.0	V^{3+}	6	78.0
Sm^{2+}	7	136.0		6	70.0	V^{4+}	5	67.0
	8	141.0	Th^{4+}	6	108.0		6	72.0
	9	146.0		8	119.0		8	86.0
Sm^{3+}	6	109.8		9	123.0	V^{5+}	4	49.5
	7	116.0		10	127.0		5	60.0
	8	121.9		11	132.0		6	68.0
	9	127.2		12	135.0	W^{4+}	6	80.0
	12	138.0	Ti^{2+}	6	100.0	W^{5+}	6	76.0
Sn^{4+}	4	69.0	Ti^{3+}	6	81.0	W^{6+}	4	56.0
	5	76.0	Ti^{4+}	4	56.0		5	65.0
	6	83.0		5	65.0		6	74.0
	7	89.0		6	74.5	Xe^{8+}	4	54.0
	8	95.0		8	88.0		6	62.0
Sr^{2+}	6	132.0	Tl^{+}	6	164.0	Y^{3+}	6	104.0
	7	135.0		8	173.0		7	110.0
	8	140.0		12	184.0		8	115.9
	9	145.0	Tl^{3+}	4	89.0		9	121.5
	10	150.0		6	102.5	Yb^{2+}	6	116.0
	12	158.0		8	112.0		7	122.0
Ta^{3+}	6	86.0	Tm^{2+}	6	117.0		8	128.0
Ta^{4+}	6	82.0		7	123.0	Yb^{3+}	6	100.8
Ta^{5+}	6	78.0	Tm^{3+}	6	102.0		7	106.5
	7	83.0		8	113.4		8	112.5
	8	88.0		9	119.2		9	118.2
Tb^{3+}	6	106.3	U^{3+}	6	116.5	Zn^{2+}	4	74.0
	7	112.0	U^{4+}	6	103.0		5	82.0
	8	118.0		7	109.0		6	88.0
	9	123.5		8	114.0		8	104.0
Tb^{4+}	6	90.0		9	119.0	Zr^{4+}	4	73.0
	8	102.0		12	131.0		5	80.0
Tc^{4+}	6	78.5	U^{5+}	6	90.0		6	86.0
Tc^{5+}	6	74.0		7	98.0		7	92.0
Tc^{7+}	4	51.0	U^{6+}	2	59.0		8	98.0
							9	103.0

HS = high spin, LS = low spin; SQ = square, PY = pyramid
Source: R. D. Shannon, *Acta. Crystallogr.*, **A32**, 751, 1976.

PROBLEMS

3.1. (*a*) Show that the minimum cation/anion radius ratio for a coordination number of 6 is 0.414.

(*b*) Repeat part (*a*) for coordination number 3.

(*c*) Which interstitial site is larger; the tetrahedral or the octahedral? Calculate the ratio of the sizes of the tetrahedral and octahedral sites.

(*d*) When oxygen ions are in a hexagonal close-packed arrangement, what is the ratio of the octahedral sites to oxygen ions? What is the ratio of the tetrahedral sites to oxygen ions?

3.2. Starting with the cubic close packing of oxygen ions:

(*a*) How many tetrahedral and how many octahedral sites are there per unit cell?

(*b*) What is the ratio of octahedral sites to oxygen ions? What is the ratio of tetrahedral sites to oxygen ions?

(*c*) What oxide would you get if one-half of the octahedral sites are filled? Two-thirds? All?

(*d*) Locate all the tetrahedral sites, and fill them up with cations. What structure do you obtain? If the anions are oxygen, what must be the charge on the cation for charge neutrality to be maintained?

(*e*) Locate all the octahedral sites, fill them with cations, and repeat part (*d*). What structure results?

3.3. Given the information given in Table 3.3, draw the zinc blende structure. What, if anything, does this structure have in common with the diamond cubic structure? Explain.

3.4. The structure of lithium oxide has anions in cubic close packing with Li ions occupying all tetrahedral positions.

(*a*) Draw the structure and calculate the density of Li_2O. *Hint*: Oxygen ions do not touch, but O–Li–O ions do.

Answer: $\rho = 1.99 \, g/cm^3$

(*b*) What is the maximum radius of a cation which can be accommodated in the vacant interstice of the anion array in Li_2O?

Answer: $r_c = 0.89 \, Å$

3.5. Look up the radii of Ti^{4+}, Ba^{2+}, and O^{2-} listed in App. 3A, and making use of Pauling's size criteria, choose the most suitable cage for each cation. Based on your results, choose the appropriate composite crystal structure and draw the unit cell of $BaTiO_3$. How many atoms of each element are there in each unit cell?

3.6. Garnets are semiprecious gems with the chemical composition $Ca_3Al_2Si_3O_{12}$. The crystal structure is cubic and is made up of three building blocks: tetrahedra, octahedra, and dodecahedra (distorted cubes).

(*a*) Which ions do you think occupy which building block?

(*b*) In a given unit cell, what must the ratio of the number of blocks be?

3.7. The oxygen content y for $YBa_2Cu_3O_y$ has been found to vary between 6 and 7. The loss of oxygen also leads to a tetragonal to orthorhombic change in structure. Find and draw the unit cells of $YBa_2Cu_3O_6$ and $YBa_2Cu_3O_7$. What structure do these compounds most resemble?

3.8. Beryllium oxide (BeO) may form a structure in which the oxygen ions are in an FCC arrangement. Look up the ionic radius of Be^{2+} and determine which type of interstitial site it will occupy. What fraction of the available interstitial sites will be occupied? Does your result agree with that shown in Table 3.3? If not, explain possible reasons for the discrepancy.

3.9. Cadmium sulfide has a cubic unit cell with a density of 4.82 g/cm^3. X-ray diffraction data indicate that the lattice parameter is 0.234 nm. How many Cd^{2+} and S^- ions are there per unit cell?

3.10. The compound MX has a density of 2.1 g/cm^3 and a cubic unit cell with a lattice parameter of 0.57 nm. The atomic weights of M and X are, respectively, 28.5 and 30 g/mol. Based on this information, which of the following structures is (are) possible: NaCl, CsCl, or zinc blende? Justify your choices.

3.11. What complex anions are expected in the following compounds?

(*a*) Tremolite or $Ca_2Mg_5(OH)_2Si_8O_{22}$

(*b*) Mica or $CaAl_2(OH)_2(Si_2Al_2)O_{10}$

(*c*) Kaolinite $Al_2(OH)_4Si_2O_5$

3.12. Determine the expected crystal structure including the ion positions of the hypothetical salt AB_2, where the radius of A is 154 pm and that of B is 49 pm. Assume that A has a charge of $+2$.

3.13. (*a*) The electronic structure of N is $1s^2 2s^2 2p^3$. The structure of Si_3N_4 is based on the SiN_4 tetrahedron. Propose a way by which these tetrahedra can be joined together in three dimensions to form a solid, maintaining the 3:4 ratio of Si to N, other than the one shown in Fig. 3.11.

(*b*) Repeat part (*a*) for SiC. How many carbons are attached to each Si, and vice versa? What relationship, if any, do you think this structure has to the diamond cubic structure?

3.14. (*a*) Write an equation for the formation of a nonbridging oxygen. Explain what is meant by a nonbridging oxygen. How does one change their number? What do you expect would happen to the properties of a glass as the number of nonbridging oxygens increases?

(*b*) What happens to silicates as the O/Si ratio increases.

3.15. What would be the formulas (complete with negative charge) of the silicate units shown in Fig. 3.16?

3.16. (*a*) Derive an expression relating the mole fractions of alkali earth oxides to the number of nonbridging oxygens per Si atom present in a silicate structure.

(*b*) Repeat Worked Example 3.3*b* for the composition $Na_2O \cdot 0.5CaO \cdot 2SiO_2$.

Answer: 1.5

(*c*) Show that chains of infinite length would occur at a mole fraction of Na_2O of 0.5. What do you think the structure would be for a composition in between 0.33 and 0.5?

(*d*) Show that for any silicate structure the number of nonbridging oxygens per Si is given by $NBO = 2R - 4$ and the number of bridging oxygens is $8 - 2R$, where R is the O/Si ratio.

3.17. (*a*) Talc, $Mg_3(OH)_2(Si_2O_5)_2$, has a slippery feel that reflects its structure. Given that information, draw its structure.

(*b*) Draw a schematic representation of the structure of $Al(OH)_3$, assuming the Al^{3+} ions are octahedrally coordinated and that the aluminum octahedra are from sheets which are joined to each other by hydroxyl bonds.

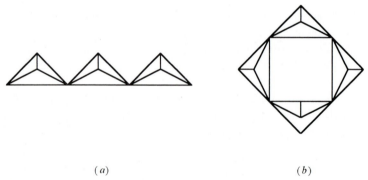

(*a*)   (*b*)

Figure 3.16
Silicate units.

ADDITIONAL READING

1. R. W. G. Wyckoff, *Crystal Structures*, vols. 1 to 6, Wiley, New York; 1971.
2. F. D. Bloss, *Crystallography and Crystal Chemistry, An Introduction*, Holt, Rinehart and Winston, New York, 1971.
3. W. D. Kingery, H. K. Bowen, and D. R. Uhlmann, *Introduction to Ceramics*, 2d ed., Wiley, New York, 1976.
4. L. Van Vlack, *Elements of Materials Science and Engineering*, 5th ed., Addison-Wesley, Reading, Massachusetts, 1985.
5. O. Muller and R. Roy, *The Major Ternary Structural Families*, Springer-Verlag, Berlin, 1974.
6. N. N. Greenwood, *Ionic Crystals, Lattice Defects and Non-Stoichiometry*, Butterworth, London, 1968.
7. R. J. Borg and G. D. Dienes, *The Physical Chemistry of Solids*, Academic Press, New York, 1992.
8. A. F. Wells, *Structural Inorganic Chemistry*, 4th ed., Clarendon Press, Oxford, England, 1975.
9. N. B. Hannay, ed., *Treatise on Solid Chemistry*, vols. 1 to 6, Plenum, New York, 1973–1976.

CHAPTER 4

Effect of Chemical Forces on Physical Properties

*Now how curiously our ideas expand by watching
these conditions of the attraction of cohesion! —
how many new phenomena it gives us beyond
those of the attraction of gravitation!
See how it gives us great strength.*

Michael Faraday, *On the Various Forces of Nature*

4.1
INTRODUCTION

The forces of attraction between the various ions or atoms in solids determine many of their properties. Intuitively, it is not difficult to appreciate that a strongly bonded material would have a high melting point and stiffness. In addition, it can be shown, as is done below, that its theoretical strength and surface energy will also increase, with a concomitant decrease in thermal expansion. In this chapter, semiquantitative relationships between these properties and the depth and shape of the energy well, described in Chap. 2, are developed.

In Sec. 4.2, the importance of the bond strength on the melting point of ceramics is elucidated. In Sec. 4.3, how strong bonds result in solids with low coefficients of thermal expansion is discussed. In Sec. 4.4, the relationship between bond strength, stiffness, and theoretical strength is developed. Sec. 4.5 relates bond strength to surface energy.

4.2
MELTING POINTS

Fusion, evaporation, and sublimation result when sufficient thermal energy is supplied to a crystal to overcome the potential energy holding its atoms together. Experience has shown that a pure substance at constant pressure will melt at a fixed temperature, with the absorption of heat. The amount of heat absorbed is known as the **heat of fusion** ΔH_f, and it is the heat required for the reaction

$$\text{Solid} \rightarrow \text{Liquid}$$

ΔH_f is a measure of the enthalpy difference between the solid and liquid states at the melting point. Similarly, the entropy difference ΔS_f between the liquid and solid is defined by

$$\Delta S_f = \frac{\Delta H_f}{T_m} \tag{4.1}$$

where T_m is the melting point in kelvins. The entropy difference ΔS_f is a direct measure of the degree of disorder that arises in the system during the melting process and is by necessity positive, since the liquid state is always more disordered than the solid. The melting points and ΔS_f values for a number of ceramics are listed in Table 4.1, which reveals that in general as a class, ceramics have higher melting temperatures than, say, metals or polymers. Inspection of Table 4.1 also reveals that there is quite a bit of variability in the melting points.[35] To understand this variability, one needs to understand the various factors that influence the melting point.

---

[35] Interestingly enough, for most solids including metals, the entropy of fusion per ion lies in the narrow range between 10 and 12 J/(mol·deg). This is quite remarkable, given the large variations in the melting points observed, and it strongly suggests that the structural changes on the atomic scale due to melting are similar for most substances. This observation is even more remarkable when the data for the noble-gas solids such as Ar are included — for Ar with a melting point of 83 K, $\Delta S_f = 14 \, \text{J}/(\text{mol} \cdot \text{K})$.

TABLE 4.1
Melting points and entropies of fusion for selected inorganic compounds

Compound	Melting point, °C	Entropy of fusion, $J/(mol \cdot °C)$	Compound	Melting point, °C	Entropy of fusion, $J/(mol \cdot °C)$
Oxides					
Al_2O_3	2054 ± 6	47.70	Mullite	1850	
BaO	2013	25.80	Na_2O (α)	1132	33.90
BeO	2780 ± 100	30.54	Nb_2O_5	1512 ± 30	58.40
Bi_2O_3	825		Sc_2O_3	2375 ± 25	
CaO	2927 ± 50	24.80	SrO	2665 ± 20	25.60
Cr_2O_3	2330 ± 15	49.80	Ta_2O_5	1875 ± 25	
Eu_2O_3	2175 ± 25		ThO_2	3275 ± 25	
Fe_2O_3	Decomposes at 1735 K to		TiO_2 (rutile)	1857 ± 20	31.50
	Fe_3O_4 and oxygen		UO_2	2825 ± 25	
Fe_3O_4	1597 ± 2	73.80	V_2O_5	2067 ± 20	
Li_2O	1570	32.00	Y_2O_3	2403	≈ 38.70
Li_2ZrO_3	1610		ZnO	1975 ± 25	
Ln_2O_3	2325 ± 25		ZrO_2	2677	29.50
MgO	2852	25.80			
Halides					
AgBr	434		LiBr	550	
AgCl	455		LiCl	610	22.60
CaF_2	1423		LiF	848	
CsCl	645	22.17	LiI	449	
KBr	730		NaCl	800	25.90
KCl	776	25.20	NaF	997	
KF	880		RbCl	722	23.85
Silicates and other glass-forming oxides					
B_2O_3	450 ± 2	33.20	$Na_2Si_2O_5$	874	31.00
$CaSiO_3$	1544	31.00	Na_2SiO_3	1088	38.50
GeO_2	1116		P_2O_5	569	
$MgSiO_3$	1577	40.70	SiO_2 (high	1423 ± 50	4.60
Mg_2SiO_4	1898	32.76	quartz)		
Carbides, nitrides, borides, and silicides					
B_4C	2470 ± 20	38.00	ThN	2820	
HfB_2	2900		TiB_2	2897	
HfC	3900		TiC	3070	
HfN	3390		TiN	2947	
HfSi	2100		$TiSi_2$	1540	
$MoSi_2$	2030		UC	2525	
NbC	3615		UN	2830	
NbN	2204		VB_2	2450	
SiC	2837		VC	2650	
Si_3N_4	At 2151 K partial pressure of		VN	2177	
	N_2 over Si_3N_4 reaches 1 atm		WC	2775	
			ZrB_2	3038	
TaB_2	3150		ZrC	3420	
TaC	3985		ZrN	2980 ± 50	
$TaSi_2$	2400		$ZrSi_2$	1700	
ThC	2625				

4.2.1   Factors Affecting Melting Points of Ceramics That Are Predominantly Ionically Bonded

Ionic charge

The most important factor determining the melting point of a ceramic is the bond strength holding the ions in place. In Eq. (2.15), the strength of an ionic bond E_{bond} was found to be proportional to the product of the ionic charges z_1 and z_2 making up the solid. It follows that the greater the ionic charges, the stronger the attraction between ions and consequently the higher the melting point. For example, both MgO and NaCl crystallize in the rock salt structure, but their melting points are, respectively, 2852 and 800°C — a difference directly attributable to the fact that MgO is made up of doubly ionized ions, whereas in NaCl the ions are singly ionized. Said otherwise, everything else being equal, the energy well of MgO is roughly 4 times deeper than that of NaCl. It is therefore not surprising that it requires more thermal energy to melt MgO than it does to melt NaCl.

Covalent character of the ionic bond

Based on Eq. (4.1), melting points are proportional to ΔH_f, and consequently whatever reduces one reduces the other. It turns out, as discussed below, that increasing the covalent character of a bond tends to reduce ΔH_f by stabilizing discrete units in the melt, which in turn reduces the number of bonds that have to be broken during melting, which is ultimately reflected in lower melting points.

It is important to note at the outset, however, that covalency per se does not necessarily favor either higher or lower melting points. The important consideration depends on the melt structure; if the strong covalent bonds have to be broken in order for melting to occur, extremely high melting temperatures can result. Conversely, if the strong bonds do not have to be broken for melting, the situation can be quite different.[36]

The effect of covalency on the structures of three MX_2 compounds is shown graphically in Fig. 4.1. In the figure, the covalent character of the bond increases in going from left to right, which results in structural changes in the structure from three-dimensional in TiO_2, to a layered structure for CdI_2, to a molecular lattice in the case of CO_2. Also shown in Fig. 4.1 are the corresponding melting points; the effect of the structural changes on the melting points is obvious.

---

[36] An extreme example of this phenomenon occurs in polymers, where the bonding is quite strong within the chains and yet the melting points are quite low, because these bonds do not have to be broken during melting.

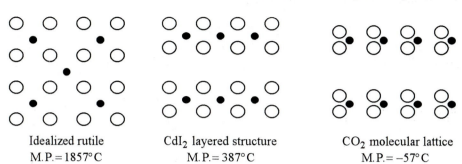

Idealized rutile	CdI$_2$ layered structure	CO$_2$ molecular lattice
M.P.= 1857°C	M.P.= 387°C	M.P.= −57°C

Figure 4.1

Effect of polarization on crystal structure and melting temperature.

It follows from this brief introduction that in order to understand the subtleties in melting point trends, one needs to somewhat quantify the extent of covalency present in an ionic bond. In Chap. 2, the bonds between ions were assumed to be either predominantly covalent or ionic. As noted then, and reiterated here, the reality of the situation is more complex — ionic bonds possess covalent character and vice versa. Historically, this complication has been addressed by means of one of two approaches. The first was to assume that the bond is purely covalent and then consider the effect of shifting the electron cloud toward the more electronegative atom. The second approach, discussed below, was to assume the bond is purely ionic and then impart a covalent character to it.

The latter approach was championed by Fajans[37] and is embodied in Fajans' rules, whose basic premise is summarized in Fig. 4.2. In Fig. 4.2*a* an idealized ion pair is shown for which the covalent character is nonexistent (i.e., the ions are assumed to be hard spheres). In Fig. 4.2*b* some covalent character is imparted by shifting the electron cloud of the more polarizable anion toward the polarizing cation. In the extreme case that the cation is totally embedded in the electron cloud of the anion (Fig. 4.2*c*) a strong covalent bond is formed. The extent to which the electron cloud is distorted and shared between the two ions is thus a measure of the covalent character of that bond. The covalent character thus defined depends on three factors:

Polarizing power of cation.   High charge and small size increase the polarizing power of cations. Over the years many functions have been proposed to quantify the effect, and one of the simplest is to define the **ionic potential** of a cation as:

---

37   K. Fajans, *Struct. Bonding*, **2**:88 (1967).

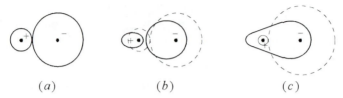

Figure 4.2

Polarization effects: (*a*) idealized ion pair with no polarization; (*b*) polarized ion pair; (*c*) polarization sufficient to form covalent bond.

$$\phi = \frac{z^+}{r}$$

where z^+ is the charge on the cation and r its radius. The ionic powers of a few selected cations are listed in Table 4.2, where it is clear that high charge and small size greatly enhance ϕ and consequently the covalent character of the bond.

To illustrate compare MgO and Al_2O_3. On the basis of ionic charge alone, one would expect the melting point of Al_2O_3 (+3, −2) to be higher than that of MgO (−2, +2), and yet the reverse is observed. However, based on the relative polarizing power of Al^{3+} and Mg^{2+}, it is reasonable to conclude that the covalent character of the Al–O bond is greater than that of the Mg–O bond. This greater covalency appears to stabilize discrete units in the liquid state and to lower the melting point. Further evidence that the Al_2O_3 melt is more "structured" than MgO is reflected in the fact that ΔS_{fusion} *per ion* for Al_2O_3 [9.54 J/(mol·K)] is smaller than that of MgO [12.9 J/(mol·K)].

Polarizability of anions. The *polarizability* of an ion is a measure of the ease with which its electron cloud can be pulled away from the nucleus, which, as discussed in greater detail in Chap. 14, scales with the cube of the radius of the ion, i.e., its volume. Increasing polarizability of the anion increases the covalent character of the bond, which once again results in lower melting points. For example, the melting points of LiCl, LiBr, and LiI are, respectively, 613, 547, and 446°C.[38]

TABLE 4.2
Ionic potential of selected cations, 1/nm

Li^+	17.0	Be^{2+}	64.0	B^{3+}	150.0
Na^+	10.5	Mg^{2+}	31.0	Al^{3+}	60.0
K^+	7.0	Ca^{2+}	20.0	Si^{4+}	100.0

[38] Another contributing factor to the lowering of the melting point that cannot be ignored is the fact that increasing the radii of the anions decreases E_{bond} by increasing r_0. This is a second-order effect, however.

Electron configuration of cation. The d electrons are less effective in shielding the nuclear charge than the s or p electrons and are thus more polarizing. Thus ions with d electrons tend to form more covalent bonds. For example, Ca^{2+} and Hg^{2+} have very similar radii (114 and 116 pm, respectively); and yet the salts of Hg have lower melting points than those of Ca. — $HgCl_2$ melts at 276°C whereas $CaCl_2$ melts at 782°C.

4.2.2   Covalent Ceramics

The discussion so far has focused on understanding the relationship between the interatomic forces holding atoms together and the melting points of mostly ionic ceramics. The melting points and general thermal stability of covalent ceramics are quite high as a result of the very strong primary bonds that form between Si and C, N, or O. Covalent ceramics are very interesting materials in that some do not melt but rather decompose at higher temperatures. For example, Si_3N_4 decomposes at temperatures in excess of 2000°C, with the partial pressure of nitrogen reaching 1 atm at those temperatures.

4.2.3   Glass Forming Liquids

These include SiO_2, many of the silicates, B_2O_3, GeO_2, and P_2O_5. What is remarkable about these oxides is that they possess anomalously low entropies of fusion. For SiO_2 ΔS_f is 4.6! J/(mol·K). This signifies that at the melting point, the solid and liquid structures are quite similar. Given that glasses can be considered supercooled liquids, it is not surprising that these oxides, called *network formers*, are the basis of many inorganic glasses (see Chap. 9 for more details).

4.3
THERMAL EXPANSION

It is well known that solids expand upon heating. The extent of the expansion is characterized by a **coefficient of linear expansion** α, defined as the fractional change in length with change in temperature at constant pressure, or

$$\alpha = \frac{1}{l_0}\left(\frac{\partial l}{\partial T}\right)_p \tag{4.2}$$

where l_0 is the original length.

The origin of thermal expansion can be traced to the anharmonicity or asymmetry of the energy distance curve described in Chap. 2 and reproduced in Fig. 4.3. The asymmetry of the curve expresses the fact that it is easier to pull two atoms apart than to push them together. At 0 K, the total energy of the atoms is potential energy, and the atoms are sitting at the bottom of the well (point a). As the temperature is raised to, say, T_1, the average energy of the system increases correspondingly. The atoms vibrate between positions x_1 and x_2, and their energy fluctuates between purely potential at x_1 and x_2 (i.e., zero kinetic energy) and speed up somewhere in between. In other words, the atoms behave if they were attached to each other by springs. The average location of the atoms at T_1 will thus be midway between x_1 and x_2, that is, at x_{T_1}. If the temperature is raised to, say, T_2, the average position of the atoms will move to x_{T_2}, etc. It follows that with increasing temperature, the average position of the atoms will move along line ab, shown in Fig. 4.3, and consequently the dimensions of a crystal will also increase.

In general, the asymmetry of the energy well increases with decreasing bond strength, and consequently the thermal expansion of a solid scales inversely with its bond strength or melting point. For example, the thermal expansion coefficient of solid Ar is on the order of $10^{-3}\,^\circ C^{-1}$, whereas for most metals and ceramics (see below) it is closer to $10^{-5}\,^\circ C^{-1}$.

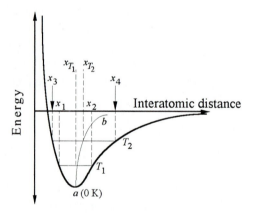

Figure 4.3

Effect of heat on interatomic distance between atoms. Note that asymmetry of well is responsible for thermal expansion. The average position of the atoms in a perfectly symmetric well would not change with temperature.

TABLE 4.3
Mean thermal expansion coefficients and theoretical densities of various ceramics

Ceramic	Theo. density, g/cm^3	$\alpha(°C^{-1}) \times 10^6$	Ceramic	Theo. density, g/cm^3	$\alpha(°C^{-1}) \times 10^6$
		Binary Oxides			
α-Al_2O_3	3.98	7.2 – 8.8	Na_2O	2.27	
BaO	5.72	17.8	SiO_2 (L.	2.32	
BeO	3.01	8.5 – 9.0	cristabolite)		
		(25 – 1000)	SiO_2 (L.	2.65	
Bi_2O_3 (α)	8.90	14.0	quartz)		
		(RT – 730°C)	ThO_2	9.86	9.2
Bi_2O_3 (δ)	8.90	24.0	TiO_2	4.25	8.5
		(650 – 825°C)	UO_2	10.96	10.0
CeO_2	7.20		WO_2	7.16	
Cr_2O_3	5.22		Y_2O_3	5.03	9.3
Dy_2O_3	7.80	8.5			(25 – 1000)
Gd_2O_3	7.41	10.5	ZnO	5.61	8.0 (c axis)
Fe_3O_4	5.24				4.0 (a axis)
Fe_2O_3	5.18		ZrO_2	5.83	7.0
HfO_2	9.70	9.4 – 12.5	(monoclinic)		
MgO	3.60	13.5	ZrO_2	6.10	12.0
			(tetragonal)		
		Mixed oxides			
$Al_2O_3 \cdot TiO_2$		9.7 (average)	Cordierite	2.51	2.1
$Al_2O_3 \cdot MgO$	3.58	7.6	$MgO \cdot SiO_2$		10.8
$5Al_2O_3 \cdot 3Y_2O_3$		8.0			(25 – 1000)
		(25 – 1400)	$2MgO \cdot SiO_2$		11.0
$BaO \cdot TiO_2$	5.80				(25 – 1000)
$BaO \cdot ZrO_2$		8.5	$MgO \cdot TiO_2$		7.9
		(25 – 1000)			(25 – 1000)
$BeO \cdot Al_2O_3$	3.69	6.2 – 6.7	$MgO \cdot ZrO_2$		12.0
$CaO \cdot HfO_2$		3.3			(25 – 1000)
		(25 – 1000)	$2SiO_2 \cdot 3Al_2O_3$	3.20	5.1
$CaO \cdot SiO_2$ (β)		5.9	(mullite)		(25 – 1000)
		(25 – 700)	$SiO_2 \cdot ZrO_2$	4.20	4.5
$CaO \cdot SiO_2$ (α)		11.2	(zircon)		(25 – 1000)
		(25 – 700)	$SrO \cdot TiO_2$		9.4
$CaO \cdot TiO_2$		14.1			(25 – 1000)
$CaO \cdot ZrO_2$		10.5	$SrO \cdot ZrO_2$		9.6
$2CaO \cdot SiO_2$ (β)		14.4	$TiO_2 \cdot ZrO_2$		7.9
		(25 – 1000)			(25 – 1000)

TABLE 4.3 continued

Mean thermal expansion coefficients and theoretical densities of various ceramics

Ceramic	Theo. density, g/cm^3	$\alpha(°C^{-1}) \times 10^6$	Ceramic	Theo. density, g/cm^3	$\alpha(°C^{-1}) \times 10^6$
			Borides, nitrides; carbides; and silicides		
AlN	3.26	5.6	SiC	3.20	4.3 – 4.8
		(25 – 1000)	TaC	14.48	6.3
B_4C	2.52	5.5	TiB_2	4.50	7.8
BN	2.27	4.4	TiC	4.95	7.7 – 9.5
Cr_3C_2	6.68	10.3	TiN	5.40	9.4
$CrSi_2$	4.40		$TiSi_2$	4.40	10.5
HfB_2	11.20	5.0	WC	15.70	
HfC	12.60	6.6	ZrB_2	6.11	5.7 – 7.0
$HfSi_2$	7.98		ZrC	6.70	6.9
$MoSi_2$	6.24	8.5			(25 – 1000)
$\beta - Mo_2C$	9.20	7.8	$ZrSi_2$	4.90	7.6
NbC	7.78	6.6			(25 – 2700)
Si_3N_4	3.20	3.1 – 3.7	ZrN	7.32	7.2
			Halides		
CaF_2	3.20	24.0	LiCl	2.07	12.2
LiF	2.63	9.2	LiI	4.08	16.7
LiBr	3.46	14.0	MgF_2		16.0
KI	3.13		NaCl	2.16	11.0
			Glasses		
Soda-lime glass		9.0	Fused silica	2.20	0.55
Pyrex		3.2			

Perusal of Table 4.3, in which the *mean* thermal expansion coefficients of a number of ceramics are listed, makes it clear that α for most ceramics lies between 3 and $10 \times 10^{-6}°C^{-1}$. The functional dependence of the fractional increase in length on temperature for a number of ceramics and metals is shown in Fig. 4.4. Given that the slope of these lines is α, one can make the following generalizations:

1. Ceramics in general have lower α values than metals.
2. The coefficient α increases with increasing temperature. This reflects the fact that the energy well becomes more asymmetric as one moves up the well, i.e., with increasing temperature. Thus it is important to specify the temperature range over which the thermal expansion coefficient is measured, since as the temperature range is expanded, the mean thermal expansion coefficient will also increase.

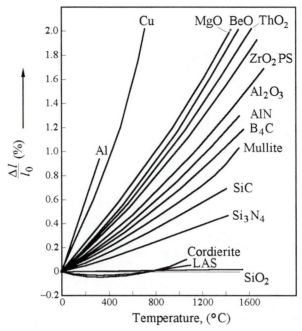

Figure 4.4

$\Delta L/L_0$ (%) versus temperature for a number of ceramics. The slopes of these lines at any temperature are α. For most ceramics, α is more or less constant with temperature. For anisotropic solids, the c axis expansion is reported.[39]

3. Covalently bonded ceramics, such as SiC and Si_3N_4, have lower α's than more close-packed ceramic structures such as NaCl and MgO. This is a reflection of the influence of atomic packing on α. In contradistinction to close-packed structures, where all vibrations result in an increase in the dimensions of the crystal, the more open structures of covalent ceramics allow for other modes of vibration that do not necessarily contribute to thermal expansion. In other words, the added thermal energy can result in a change in the bond angles without significant change in bond length. (Think of the atoms as vibrating into the "open spaces" rather than against each other.)

    One of most striking examples of the importance of atomic packing on α is silica. Vitreous silica has an extremely low α, whereas quartz and cristobalite have much higher thermal expansion coefficients, as shown in Fig. 4.5.

39   Adapted J. Chermant, *Les Ceramiques Thermomechaniques*, CNRS Presse, France, 1989.

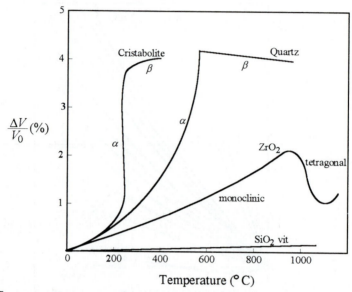

Figure 4.5

$\Delta V/V_0$ (%) versus temperature for cristobalite, quartz, zirconia, and vitreous or amorphous SiO$_2$.[40] The abrupt changes in behavior with temperature are a result of phase transformations (see Chap. 8).

4. Although not explicitly stated, the discussion so far is only strictly true for isotropic, e.g., cubic, polycrystalline materials. Crystals that are noncubic and consequently are anisotropic in their thermal expansion coefficients behave quite differently. In some cases, a crystal can actually shrink in one direction as it expands in another. When a polycrystal is made up of such crystals, the average thermal expansion can be very small, indeed. Cordierite and lithium-aluminosilicate (LAS) (see Fig. 4.4) are good examples of this class of materials. As discussed in greater detail in Chap. 13, this anisotropy in thermal expansion, which has been exploited to fabricate very low-α materials, can result in the buildup of large thermal residual stresses that can be quite detrimental to the strength and integrity of ceramic parts.

40 Adapted from W. D. Kingery, H. K. Bowen, and D. R. Uhlmann, *Introduction to Ceramics*, 2d ed., Wiley, New York, 1976.

4.4

YOUNG'S MODULUS AND THE STRENGTH OF PERFECT SOLIDS

In addition to understanding the behavior of ceramics exposed to thermal energy, it is important to understand their behavior when they are subjected to an external load or stress. The objective of this section is to interrelate the shape of the energy versus distance curve $E(r)$, discussed in Chap. 2, to the elastic modulus, which is a measure of the stiffness of a material and the theoretical strength of that material. To accomplish this goal, one needs to examine the forces $F(r)$ that develop between atoms as a result of externally applied stresses. As noted in Sec. 2.4, $F(r)$ is defined as

$$F(r) = \frac{dE(r)}{dr} \tag{4.3}$$

From the general shape of the $E(r)$ curve, one can easily sketch the shape of a typical force versus distance curve, as shown in Fig. 4.6. The following salient features are noteworthy:

- The net force between the atoms or ions is zero at equilibrium, i.e., at $r = r_0$.
- Pulling the atoms apart results in the development of an *attractive restoring force* between them that tends to pull them back together. The opposite is true if one tries to push the atoms together.
- In the region around $r = r_0$ the response can be considered, to a very good approximation, linear. In other words, the atoms act as if they are tied together by miniature springs. It is in this region that Hooke's law (see below) applies.
- The force pulling the atoms apart cannot be increased indefinitely. Beyond some separation r_{fail}, the bond will fail. The force at which this occurs represents the maximum force F_{max} that the bond can withstand before failing.

In the remainder of this section, the relationships between stiffness and theoretical strength, on one hand, and $E(r)$ and $F(r)$, on the other hand, are developed.

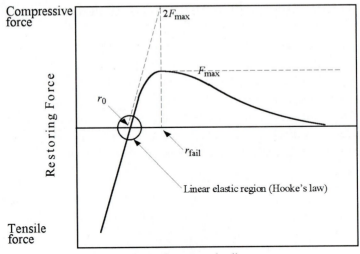

Figure 4.6

Force-distance curves showing construction used to derive Eq. (4.8) in text. Slope of line going through r_0 is the stiffness of the bond. It is assumed in this construction that the maximum force is related to the stiffness as shown. This is quite approximate but serves to illustrate the relationship between stiffness and theoretical strength.

An atomic view of Young's modulus

Experience has shown that all solids will respond to small stresses σ by stretching in proportion to the stress applied, a phenomenon that is described by **Hooke's law**:

$$\sigma = Y\varepsilon \tag{4.4}$$

where Y is Young's modulus and ε is the **strain** experienced by the material, defined as

$$\varepsilon = \frac{L - L_0}{L_0} \tag{4.5}$$

Here L is the length under the applied stress, and L_0 is the original length.

Refer once more to the force/distance curve shown in Fig. 4.6. In the vicinity of r_0, the following approximation can be made:

$$F = S_0(r - r_0) \tag{4.6}$$

where S_0 is the **stiffness** of the bond, defined as

$$S_0 = \left(\frac{dF}{dr}\right)_{r=r_0} \tag{4.7}$$

Note that Eq. (4.6) is nothing but an expression for the extension of a linear spring.

Dividing Eq. (4.6) by r_0^2 and noting that F/r_0^2 is approximately the stress on the bond, while $(r - r_0)/r_0$ is the strain on the bond, and comparing the resulting expression with Eq. (4.4), one can see immediately that

$$\boxed{Y \approx \frac{S_0}{r_0}} \tag{4.8}$$

Combining this result with Eqs. (4.3) and (4.7), it is easy to show that

$$Y = \frac{1}{r_0}\left(\frac{dF}{dr}\right)_{r=r_0} = \frac{1}{r_0}\left(\frac{d^2E}{dr^2}\right)_{r=r_0} \tag{4.9}$$

This is an important result because it says that the stiffness of a solid is directly related to the curvature of its energy/distance curve. Furthermore, it implies that strong bonds will be stiffer than weak bonds; a result that is not in the least surprising, and it explains why, in general, given their high melting temperatures, ceramics are quite stiff solids.

Theoretical strengths of solids

The next task is to estimate the theoretical strength of a solid or the stress that would be required to *simultaneously* break all the bonds across a fracture plane. It can be shown (see Prob. 4.2) that typically most bonds will fail when they are stretched by about 25%, i.e., when $r_{\text{fail}} \approx 1.25 r_0$. It follows from the geometric construction shown in Fig. 4.6 that

$$S_0 \approx \frac{2F_{\text{max}}}{r_{\text{fail}} - r_0} \approx \frac{2F_{\text{max}}}{1.25 r_0 - r_0} \tag{4.10}$$

Dividing both sides of this equation by r_0 and noting that

$$\frac{F_{\text{max}}}{r_0^2} \approx \sigma_{\text{max}} \tag{4.11}$$

i.e., the force divided by the area over which it operates, one obtains

$$\sigma_{\text{max}} \approx \frac{Y}{8} \tag{4.12}$$

For a more exact calculation, one starts with the energy/interatomic distance function in its most general form, i.e.,

$$E_{bond} = \frac{C}{r^n} - \frac{D}{r^m}$$ (4.13)

where C and D are constants and $n > m$. Assuming $\sigma_{max} \approx F_{max}/r_0^2$, one can show (see Prob. 4.2) that σ_{max} is better approximated by

$$\sigma_{max} = \frac{Y}{[(n+1)/(m+1)]^{(m+1)/(n-m)}} \frac{1}{n+1}$$ (4.14)

Substituting typical values for m and n (say, $m=1$ and $n=9$) for an ionic bond yields $\sigma_{max} \approx Y/15$.

Based on these results, one may conclude that the theoretical strength of a solid should be roughly one-tenth of its Young's modulus. Experience has shown, however, that the actual strengths of ceramics are much lower and are closer to $Y/100$ to $Y/1000$. The reason for this state of affairs is discussed in greater detail in Chap. 11, and reflects the fact that real solids are not perfect, as assumed here, but contain many flaws and defects that tend to locally concentrate the applied stress, which in turn significantly weaken the material.

4.5
SURFACE ENERGY

The **surface energy** γ of a solid is the amount of energy needed to create a unit area of new surface. The process can be pictured as shown in Fig. 4.7a, where two new surfaces are created by cutting a solid in two. Given this simple picture, the surface energy is simply the product of the number of bonds N_s broken per unit area of crystal surface and the energy per bond E_{bond}, or

$$\gamma = N_s E_{bond}$$ (4.15)

For the sake of simplicity, only first-neighbor interactions will be considered here, which implies that E_{bond} is given by Eq. (2.15). Also note that since N_s is a function of crystallography, it follows that γ is also a function of crystallography.

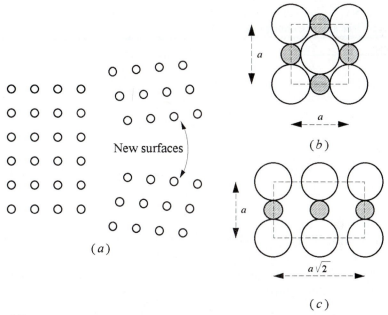

Figure 4.7
(a) The creation of new surface entails the breaking of bonds across that surface.
(b) Structure of (100) plane in the rock salt structure, (c) Structure of (110) plane in same structure. Note that the coordination number of ions in this plane is 2, which implies that to create a (110) plane, only two bonds per ion would have to be broken.

To show how to calculate surface energies by starting with Eq. (4.15), consider cleaving a rock salt crystal along its (100) plane,[41] shown in Fig. 4.7b. This plane contains two cations and two anions and has an area of $(2r_0)^2$, where r_0 is the equilibrium interionic distance. Note, however, that the total surface area created is twice that, or $2 \times (2r_0)^2$. Since four bonds have to be broken, it follows that $N_s = 4/\left[2 \times (2r_0)^2\right]$. Combining this result with Eqs. (2.15) and (4.15) yields

$$\gamma_{100} \approx -E_{bond}\left[\frac{4}{2(2r_0)^2}\right] \approx -\frac{z_1 z_2 e^2}{8\pi\varepsilon_0 r_0^3}\left(1 - \frac{1}{n}\right) \tag{4.16}$$

The minus sign is introduced because energy has to be consumed to create a surface. Calculations of surface energies based on Eq. (4.16) invariably yield values that are substantially greater than the measured ones (see Table 4.4). The reason for this discrepancy comes about because in the simple model that leads to

---

[41] It is assumed here that the reader is familiar with Miller indices, a topic that is covered in almost all introductory materials science or engineering textbooks.

TABLE 4.4
Measured free surface energies of solids

Substance	Surface	Environment	Temp., K	Surface energy, J/m^2
Mica	(0001)	Air	298	0.38
		Vacuum	298	5.00
MgO	(100)	Air	298	1.15
KCl	(100)	Air	298	0.11
Si	(111)	Liquid N_2	77	1.24
NaCl	(100)	Liquid N_2	77	0.32
CaF_2	(111)	Liquid N_2	77	0.45
LiF	(100)	Liquid N_2	77	0.34
$CaCO_3$	$(10\overline{1}0)$	Liquid N_2	77	0.23

this expression, surface relaxation and rearrangement of the atoms upon the formation of the new surface were not allowed. When the surface is allowed to relax; then much of the energy needed to form it is recovered, and the theoretical predictions do indeed approach the experimentally measured ones.

WORKED EXAMPLE 4.1. Estimate the surface energy of the (100) and (110) planes in NaCl, and compare your results with those listed in Table 4.4.

Answer

For NaCl, $r_0 = 2.83 \times 10^{-10}$ m, which when substituted in Eq. (4.16), assuming $n = 9$ yields a value for the surface energy of ≈ 4.5 J/m^2. By comparing this value with the experimentally measured value listed in Table 4.4, it is immediately obvious that it is off by more than an order of magnitude, for reasons alluded to above.

The (110) plane (Fig. 4.7c) has an area of $\sqrt{2}(2r_0)(2r_0)$ but still contains two Na and two Cl ions. However, the coordination number of each of the atoms in the plane is now 2 instead of 4, which implies that each ion is coordinated to *two* other ions above and below the plane (here, once again for simplicity, all but first-neighbor interactions are considered). In other words, to create the plane, one needs to break two bonds per ion. It follows that

$$N_s = \frac{2 \times 4}{2\sqrt{2}(2r_0)^2} \text{ bonds}/m^2$$

and the corresponding surface energy is thus 6.36 J/m^2.

EXPERIMENTAL DETAILS

Melting Points

Several methods can be used to measure the melting point of solids. One of the simplest is probably to use a **differential thermal analyzer** (DTA for short). The basic arrangement of a differential thermal analyzer is quite simple and is shown schematically in Fig. 4.8a. The sample and an inert reference (usually alumina powder) are placed side by side in a furnace, and identical thermocouples are placed below each. The temperature of the furnace is then slowly ramped, and the difference in temperature $\Delta T = T_{sample} - T_{ref}$ is measured as a function of the temperature of the furnace, which is measured by a third thermocouple (thermocouple 3 in Fig. 4.8a). Typical results are shown in Fig. 4.8b and are interpreted as follows. As long as both the sample and the reference are inert, they should have the same temperature and $\Delta T = 0$. However, if for any reason the sample absorbs (endothermic process) or gives off (exothermic process) heat, its temperature vis-à-vis the reference thermocouple will change accordingly. For example, melting, being an endothermic process, will appear as a trough upon heating. The melting point is thus the temperature at which the trough appears. In contrast, upon cooling, freezing, being an exothermic process will appear as a peak.

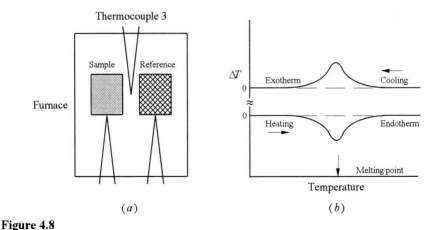

Figure 4.8
(a) Schematic of DTA setup. (b) Typical DTA traces upon heating (bottom curve) and cooling (top curve).

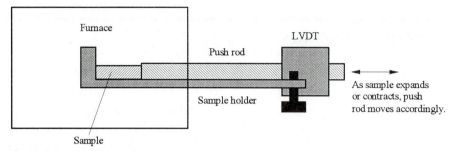

Figure 4.9
Schematic of a dilatometer.

Thermal Expansion Coefficients

Thermal expansion coefficients are measured with a dilatometer, which is essentially a high-temperature furnace from which a rod sticks out (Fig. 4.9). One side of the rod is pushed against the sample for which the thermal expansion is to be measured, and the other side is attached to a device that can measure the displacement of the rod very accurately, such as a linear variable differential transformer or LVDT. In a typical experiment, the sample is placed inside the furnace and is heated at a constant rate, while simultaneously the displacement of the push rod is measured. Typical curves for a number of ceramics and metals are shown in Fig. 4.4.

Surface energies

A variety of methods can be used to measure the surface energy of ceramics. One technique, of limited applicability (see below), is to measure the force needed to cleave a crystal by starting with an atomically sharp notch of length c. In Chap. 11, the following relationship between the surface energy, Young's modulus, and the applied stress at fracture σ_{app} is derived:

$$\gamma = \frac{A' c \sigma_{app}^2}{2Y} \qquad (4.17)$$

where A' is a geometric factor that depends on the loading conditions and the specimen geometry. Once σ_{app} is measured for a given c, γ is easily calculated from Eq. (4.17) if the modulus is known. In deriving this equation, it is implicit that all the mechanical energy supplied by the testing rig goes into creating the new surfaces. Also implicit is that there were no energy-consuming mechanisms occurring at the crack tip, such as dislocation movements; i.e., the failure was a pure brittle failure. It is important to note that this condition is only satisfied for a small number of ionic and covalent ceramics, some of which are listed in Table 4.4.

4.6
SUMMARY

1. The strengths of the bonds between atoms or ions in a solid, by and large, determine many of its properties, such as its melting and boiling points, stiffness, thermal expansion, and theoretical strength.
2. The stronger the bond, the higher the melting point. However, partial covalency to an ionic bond will tend to stabilize discrete units in the melt and lower the melting point.
3. Thermal expansion originates from the anharmonic vibrations of atoms in a solid. The asymmetry of the energy well is a measure of the thermal expansion coefficient α, with stronger bonds resulting in more symmetric energy wells and consequently lower values of α. In addition, the atomic arrangement can play an important role in determining α.
4. As a first approximation, the curvature of the energy/distance well is a measure of the stiffness or Young's modulus of a solid. In general, the stronger the bond, the stiffer the solid. Other factors such as atomic arrangement are also important, however.
5. The theoretical strength of a bond is on the order of $Y/10$. The actual strengths of ceramics, however, are much lower for reasons to be discussed in Chap. 11.
6. The surface energy of a solid not only scales with the bond energy but also depends on crystallographic orientation.

PROBLEMS

4.1. (*a*) The equilibrium interatomic spacings of the Na halides and their melting points are listed below. Explain the trend observed.

	NaF	NaCl	NaBr	NaI
Spacing, nm	0.23	0.28	0.29	0.32
Melting point, °C	988	801	740	660

(*b*) Explain the melting point trends observed for the alkali metal chlorides as one goes from HCl (–115.8°C) to CsCl.

(*c*) Which of these pairs of compounds would you expect to have the higher melting points; CaF_2 versus ZrO_2; UO_2 versus CeO_2; CaF_2 versus CaI_2? Explain.

4.2. Starting with Eq. (4.13) in text, do the following:

(a) Derive the following relationship:

$$S_0 = \frac{mD}{r_0^{m+2}}(n-m)$$

Using this equation, calculate S_0 for NaCl. Assume $n=9$.

Answer: 81 N/m

(b) Derive the following expression for Young's modulus. State all assumptions.

$$Y \approx \frac{mD}{r_0^{m+3}}(n-m)$$

(c) Show that the distance at which the bond will break r_{fail} is given by

$$r_{\text{fail}} = \left(\frac{n+1}{m+1}\right)^{\frac{1}{n-m}}(r_0)$$

For ionic bonds, $m=1$ and $n \approx 9$; for van der Waals bonds, $m=6$ and $n=12$. Calculate the strain at failure for each bond.

(d) Derive Eq. (4.14) in the text, and show that for an ionic bond $\sigma_{\text{fail}} \approx Y/15$.

4.3. (a) Show that for the rock salt structure $\gamma_{(111)}/\gamma_{(100)} = \sqrt{3}$.

(b) Calculate from first principles the surface energies of the (100) and (111) planes of MgO. How do your values compare with those shown in Table 4.4? Discuss all assumptions.

(c) It has been observed that NaCl crystals cleave more easily along the (100) planes than along the (110) planes. Show, using calculations, why you think that is the case.

4.4. Calculate the number of broken bonds per square centimeter for Ge (which has a diamond cubic structure identical to the one shown in Fig. 3.1c except that all the atoms are identical) for the (100) and (111) surfaces. Which surface do you think has the lower surface energy? Why? The lattice constant of Ge is 0.565 nm, and its density is 5.32 g/cm^3.

Answer: For (100), 1.25×10^{15} bonds/cm^2 ; for (111), 0.72×10^{15} bonds/cm^2

4.5. Take the C–C bond energy to be 376 kJ/mol. Calculate the surface energy of the (111) plane in diamond. Repeat for the (100) plane. Which plane do you think would cleave more easily? Information you may find useful: Density of diamond is 3.51 g/cm^3 and its lattice parameter is 0.356 nm.

Answer: $\gamma_{(111)} = 9.820$ J/m^2

4.6. Would you expect the surface energies of the Noble-gas solids to be greater than, about the same as, or smaller than those of ionic crystals? Explain.

4.7. Estimate the thermal expansion coefficient of alumina from Fig. 4.4. Does your answer depend on the temperature range over which you carry out the calculation? Explain.

4.8. Estimate the order of magnitude of the maximum displacement of Na and Cl ions in NaCl from their equilibrium position at 300 and at 900 K.

4.9. Prove that the linear expansion coefficient α, with very little loss in accuracy, can be assumed to be one-third that of the volume coefficient for thermal expansion α_v. You can assume that $l = l_0(1 + \alpha)$ and $v = v_0(1 + \alpha_v)$ and $v_0 = l_0^3$.

4.10. (a) "A solid for which the energy distance curve is perfectly symmetric would have a large thermal expansion coefficient." Do you agree with this statement? Explain.

(b) The potential energy $U(x)$ of a pair of atoms that are displaced by x from their equilibrium position can be written as $U(x) = \alpha x^2 - \beta x^3 - \gamma x^4$, where the last two terms represent the anharmonic part of the well. At any given temperature, the probability of displacement occurring relative to that it will not occur is given by the Boltzmann factor $e^{-U/(kT)}$, from which it follows that the average displacement at this temperature is

$$\bar{x} = \frac{\int_{-\infty}^{\infty} x e^{-U/(kT)}}{\int_{-\infty}^{\infty} e^{-U/(kT)}}$$

Show that at small displacements, the average displacement is

$$\bar{x} = \frac{3\beta kT}{4\alpha^2}$$

What does this final result imply about the effect of the strength of the bond on thermal expansion?

ADDITIONAL READING

1. L. Van Vlack, *Elements of Materials Science and Engineering*, 5th ed., Addison-Wesley, Reading, Massachusetts, 1985.
2. N. N. Greenwood, *Ionic Crystals, Lattice Defects and Non-Stoichiometry*, Butterworth, London, 1968.

3. L. Azaroff, *Introduction to Solids*, McGraw-Hill, New York, 1960.
4. J. Huheey, *Inorganic Chemistry*, 2d ed., Harper & Row, New York, 1978.
5. L. Solymar and D. Walsh, *Lectures on the Electrical Properties of Materials*, 4th ed., Oxford University Press, New York, 1988.
6. C. Kittel, *Introduction to Solid State Physics,* 6th ed., Wiley, New York, 1986.
7. B. H. Flowers and E. Mendoza, *Properties of Matter*, Wiley, New York, 1970.
8. A. H. Cottrell, *The Mechanical Properties of Matter*, Wiley, New York, 1964.
9. M. F. Ashby and R. H. Jones, *Engineering Materials*, Pergamon Press, New York, 1980.

CHAPTER 5

Thermodynamic and Kinetic Considerations

$$S = k \ln \Omega$$

Boltzmann

5.1
INTRODUCTION

Most of the changes that occur in solids in general and ceramics in particular, especially as a result of heating or cooling, come about because they lead to a reduction in the free energy of the system. For any given temperature and pressure, every system strives to attain its lowest possible free energy, kinetics permitting. The beauty of thermodynamics lies in the fact that while it will not predict what can happen, it most certainly will predict what cannot happen. In other words, if the calculations show that a certain process will increase the free energy of a system, then one can with utmost confidence dismiss that process as impossible.

Unfortunately thermodynamics, for the most part, is made confusing and very abstract. In reality, thermodynamics, while not being the easiest of subjects, is not as difficult as generally perceived. As somebody once noted, some people use thermodynamics as a drunk uses a lamppost — not so much for illumination as for support. The purpose of this chapter is to dispel some of the mystery surrounding thermodynamics and hopefully illuminate and expose some of its beauty. It should be emphasized, however, that one chapter cannot, by any stretch of the imagination, cover a subject as complex and subtle as thermodynamics. This chapter, as noted in the Preface, is included more for the sake of completion and a reminder of what the reader should already be familiar with, than an attempt to cover the subject in any but a cursory manner.

119

This chapter is structured as follows. In the next three subsections, enthalpy, entropy, and free energy are defined and explained. Sec. 5.3 deals with the conditions of equilibrium and the corresponding mass action expression. The chemical stability of ceramics is discussed in Sec. 5.4. In Sec. 5.5 the concept of electrochemical potentials is presented, which is followed by the closely related notion of charged interfaces and Debye length. In Sec. 5.7 the Gibbs-Duhem relation for binary oxides is introduced and in the final section a few remarks are made concerning the kinetics and driving forces of various processes that can occur in solids.

5.2
FREE ENERGY

If the condition for equilibrium were simply that the energy content or enthalpy of a system be minimized, one would be hard pressed to explain many commonly occurring phenomena, least of all endothermic processes. For example, during melting, the energy content of the melt is greater than that of the solid it is replacing, and yet experience has shown that when they are heated to sufficiently high temperatures, most solids will melt. Gibbs was the first to appreciate that it was another function that had to be minimized before equilibrium could be achieved. This function, called the **Gibbs free-energy function**, is dealt with in this section and comprises two terms, namely, enthalpy H and entropy S.

5.2.1   Enthalpy

When a substance absorbs a quantity of heat dq, its temperature will rise accordingly by an amount dT. The ratio of the two is the **heat capacity**, defined as

$$c = \frac{dq}{dT} \tag{5.1}$$

Since dq is not a state function, c will depend on the path. The problem can be simplified by introducing the **enthalpy function**, defined as

$$H = E + PV \tag{5.2}$$

where E, P, and V are, respectively, the internal energy, pressure, and volume of the system. By differentiating Eq. (5.2) and noting that, from the first law of thermodynamics, $dE = dq + dw$, where dw is the work done on the system, it follows that

$$dH = d(E + PV) = dq + dw + P\,dV + V\,dP \tag{5.3}$$

If the heat capacity measurement is carried out at constant pressure $dP = 0$, and since by definition $dw = -P\,dV$, it follows from Eq. (5.3) that $dH = dq|_p$. In other words, the heat absorbed or released by any substance at constant pressure is a measure of its enthalpy.

From this result it follows from Eq. (5.1) that

$$c_p = \left(\frac{dq}{dT}\right)_p = \left(\frac{dH}{dT}\right)_p \tag{5.4}$$

where c_p is the heat capacity measured at constant pressure. Integrating Eq. (5.4) shows that the enthalpy content of a crystal is given by

$$H^T - H_{\text{form}}^{298} = \int_{298}^{T} c_{p,\,\text{comp}}\,dT \tag{5.5}$$

Given that there is no absolute scale for energy, the best that can be done is to arbitrarily define a standard state and relate all other changes to that state — thermodynamics deals only with relative changes. Consequently and by convention, the formation enthalpy of the *elements* in their standard state at 298 K is assumed to be zero; i.e., for any element, $H^{298} = 0$.

Given that the heat liberated or consumed during the formation of a *compound from its elements* can be determined experimentally, it follows that the enthalpy of formation of compounds at 298 K, denoted by $\Delta H_{\text{form}}^{298}$, is known and tabulated. At any temperature other than 298 K, the heat content of a compound ΔH^T is given by

$$H^T - H_{\text{form}}^{298} = \int_{298}^{T} c_{p,\,\text{comp}}\,dT \tag{5.6}$$

Finally it is worth noting that the heat capacity data are often expressed in the empirical form

$$c_p = A + BT + \frac{C}{T^2}$$

and the heat content of a solid at any temperature is thus simply determined by substituting Eq. (5.7) in (5.5) or (5.6) and integrating.

WORKED EXAMPLE 5.1. Given that the c_p of Al is given by

$$c_p = 20.7 + 0.0124T$$

in the temperature range of 298 to 932 K and that c_p of Al_2O_3 is given by

$$c_p = 106.6 + 0.0178T - 2,850,000T^{-2}$$

in the temperature range of 298 to 1800 K, and its enthalpy of formation from its elements at 298 K is -1675.7 kJ/mol, calculate the enthalpy content of Al and Al_2O_3 at 298 and 900 K.

Answer

The enthalpy content of Al at 298 is zero by definition. The heat content of Al at 900 K is thus

$$H_{Al}^{900} - H_{Al}^{298} = \int_{298}^{900}(20.7 + 0.0124T)dT = 16.93 \text{kJ/mol}$$

At 298 K, the heat content of Al_2O_3 is simply its enthalpy of formation from its elements, or -1675.7 kJ/mol. At 900 K,

$$H_{Al_2O_3}^{900} - H_{Al_2O_3}^{298} = \int_{298}^{900}(106.6 + 0.0178T - 2,850,000T^2)dT = 70.61 \text{kJ/mol}$$

$$\therefore H_{Al_2O_3}^{900} = -1675.7 + 70.61 = -1605.0 \text{ kJ/mol}$$

5.2.2   Entropy

Disorder constitutes entropy, macroscopically defined as

$$dS \equiv \frac{dq_{rev}}{T} \tag{5.7}$$

where q_{rev} is the heat absorbed in a reversible process. Boltzmann further related entropy to the microscopic domain by the following expression:[42]

$$\boxed{S = k \ln \Omega_\beta} \tag{5.8}$$

where k is Boltzmann's constant and Ω_β is the total number of different configurations in which the system can be arranged at constant energy. There are several forms of entropy, which include:

- Configurational, where the entropy is related to the number of configurations in which the various atoms and/or defects can be arranged on a given number of lattice sites,
- Thermal, where Ω_β is the number of possible different configurations in which the particles (e.g., atoms or ions) can be arranged over existing energy levels

---

[42]   This expression is inscribed on Boltzmann's tomb in Vienna.

- Electronic
- Other forms of entropy, such as that arising from the randomization of magnetic or dielectric moments

Each will be discussed in some detail in the following sections.

Configurational entropy

This contribution refers to the entropy associated with atomic disorder. To illustrate, let's consider the entropy associated with the formation of n point defects or vacancies (see Chap. 6 for more details). Combinatorially, it can be shown that the number of ways of distributing n vacant sites and N atoms on $n + N$ sites is given by:[43]

$$\Omega_\beta = \frac{(n + N)!}{n!\, N!} \tag{5.9}$$

When Eq. (5.9) is substituted in Eq. (5.8) and Stirling's approximation[44] is applied, which for large x reduces to

$$\ln x! \approx x \ln x$$

the following expression for the configuration entropy

$$S_{\text{config}} = -k\left(N \ln \frac{N}{N + n} + n \ln \frac{n}{n + N} \right) \tag{5.10}$$

is obtained (see Prob. 5.1).

It is worth noting here, however, that a very similar expression, namely,

$$S_{\text{config}} = -R(x_A \ln x_A + x_B \ln x_B) \tag{5.11}$$

results for the mixing of two solids A and B that form an ideal solution. Here x_A and x_B are the mole fractions of A and B, respectively. R is the universal gas constant, where $R = kN_{\text{Av}}$.

WORKED EXAMPLE 5.2. (*a*) Calculate the total number of possible configurations for eight atoms and one vacancy. Draw the various configurations. (*b*) Calculate the entropy change with introducing 1×10^{18} vacancies in a mole of a perfect crystal. Does the entropy increase or decrease?

---

43 See, e.g., C. Newey and G. Weaver, eds., *Materials Principles and Practice*, Butterworth, London, 1990, p. 212.
44 $\ln x! \cong x \ln x - x + 1/2 \ln 2\pi x$

Answer

(a) Substituting $N = 8$ and $n = 1$ in Eq. (5.9) yields $\Omega_\beta = 9$. The various possible configurations are shown in Fig. 5.1.

(b) Applying Eq. (5.10) gives

$$\Delta S_{config} = \left(-1.38 \times 10^{-23}\right)\left(6.02 \times 10^{23} \ln\frac{6.02 \times 10^{23}}{6.02 \times 10^{23} + 10^{18}} + 10^{18}\frac{10^{18}}{6.02 \times 10^{23} + 10^{18}}\right)$$

$$= 0.0002 \text{ J/K}$$

Since the result is positive, it implies that the entropy of the defective crystal is higher than that of the perfect crystal.

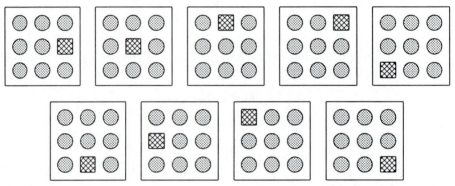

Figure 5.1
Various configurations for arranging eight atoms (circles) and one vacancy (squares). Note that the exact same picture would have emerged had the circles been A atoms and the squares B atoms.

Thermal entropy

As the atoms or ions vibrate in a solid, the uncertainty in the exact value of their energy constitutes thermal entropy, S_T. Combining 5.4 and 5.7 it follows that:

$$dS_T \equiv \frac{dq_{rev}}{T} = \frac{c_p}{T}dT$$

from which it directly follows that for any substance[45]

---

[45] In contrast to energy, one can assign an absolute value for entropy if it is postulated that the entropy of a perfect (i.e., defect-free) solid goes to zero at absolute zero (third law). One of the

$$\Delta S_T = \int_0^T \frac{c_p}{T} dT \qquad (5.12)$$

Microscopically, to understand the concept of thermal entropy, or heat capacity for that matter, one needs to appreciate that the vibrational energy levels of atoms in a crystal are quantized. If the atoms are assumed to behave as **simple harmonic oscillators,** i.e., miniature springs, it can be shown that their energy will be quantized with a spacing between energy levels given by

$$\varepsilon = \left(n + \frac{1}{2} \right) h\nu \text{ where } n = 0, 1, 2, \ldots \qquad (5.13)$$

where h, n, and ν are, respectively, Planck's constant, an integer, and the characteristic vibration frequency of the bond. The last is related to the spring constant of the bond, (see Chap. 4), S_0 [46] by

$$\omega_0 = 2\pi\nu \approx \sqrt{\frac{S_0}{M_{red}}} \qquad (5.14)$$

where ω_0 is the angular frequency in rad s^{-1} and M_{red} is the reduced mass of the oscillator system, i.e., the oscillating atoms. For a two-body problem with masses m_1 and m_2, $M_{red} = m_1 m_2 / (m_1 + m_2)$. By combining Eqs. (5.13) and (5.14), it becomes obvious that the spacing of energy levels for strongly bonded (i.e., high-S_0) solids is greater than that for weakly bonded solids, a result that has far-reaching ramifications, as discussed shortly.

At absolute zero, the atoms populate the lowest energy levels available, and only one configuration exists. Upon heating, however, the probability of exciting atoms to higher energy levels increases, which in turn increases the number of possible configurations of the system — which is another way of saying that the thermal entropy has increased.

The details of lattice vibrations will not be discussed here.[47] But for the sake of discussion, the main results of one of the simpler models, namely, the **Einstein solid**, are given below without proof. By assuming the solid to consist of Avogadro's number N_{Av} of independent harmonic oscillators, all oscillating with the same frequency ν_e, Einstein showed that the thermal entropy per mole is given by

---

implications of the third law is that every substance has a certain amount of "S" associated with it at any given temperature above absolute 0 K.

46   S_0 is not to be confused with entropy.

47   For more details see, e.g., K. Denbigh, *The Principles of Chemical Equilibrium*, 4th ed., Cambridge University Press, New York, 1981, Chap. 13.

$$S_T = 3N_{Av}k\left[\frac{h\nu_e}{kT\left(e^{h\nu_e/kT}-1\right)} - \ln\left(1-e^{-h\nu_e/kT}\right)\right] \qquad (5.15)$$

For temperatures $kT >> h\nu_e\left(e^x \approx 1+x\right)$, Eq. (5.15) simplifies to

$$\boxed{S_T = 3R\left(\ln\frac{kT}{h\nu_e}+1\right)} \qquad (5.16)$$

On the basis of this result,[48] it is possible to make the following generalizations concerning S_T:

1. Thermal entropy S_T is a monotonically increasing function of temperature; i.e., S_T increases as T increases. This comes about because as the temperature is raised, the atoms can populate higher and higher energy levels. The uncertainty of distributing these atoms among the larger number of *accessible* energy levels constitutes entropy.

2. The thermal entropy decreases with increasing characteristic frequency of vibration of the atoms, that is, ν_e. Given that ν_e scales with the strength of a bond [Eq. (5.14)], it follows that *for a given temperature, the solid with weaker bonds will have the higher thermal entropy.* The reason is simple. If the bonds are strong, that is, S_0 is large, then the spacing between energy levels will also be large, and thus for a given ΔT increase in the temperature of a system, only a few levels are accessible and the thermal entropy is low. In a weakly bound solid, on the other hand, for the same ΔT, many more levels are accessible and the uncertainty increases. As discussed in greater detail later, this conclusion is important when one is dealing with temperature-induced polymorphic transformations since they tend to occur in the direction of increased thermal entropy. In other words, polymorphic transformations will tend to occur from phases of higher cohesive energy (e.g., close-packed structures) to those of lower cohesive energies (more open structures).

Another implication of Eq. (5.16) is that if the vibrational frequency of the atoms changes from, say, a frequency ν to ν', as a result of a phase transformation or the formation of defects, e.g., the associated entropy change is

$$\Delta S_T^{\text{trans}} = 3R\ln\left(\frac{\nu}{\nu'}\right) \qquad (5.17)$$

Note that if $\nu > \nu'$, $\Delta S_T^{\text{trans}}$ will be positive.

---

[48] The more accurate Debye model, which assumes a distribution of frequencies rather than a single frequency, yields virtually the same result at higher temperatures.

WORKED EXAMPLE 5.3. (*a*) Sketch the various possible configurations for three particles distributed over three energy levels subject to the constraint that the total energy of the system is constant at 3 units. (*b*) By defining the Einstein characteristic temperature to be $\theta_e = h\nu_e/k$, it can be shown that

$$c_v = 3 N_{Av} k \left(\frac{\theta_e}{T} \right)^2 \left[\frac{e^{\theta_e/T}}{\left(e^{\theta_e/T} - 1 \right)^2} \right]$$

where c_v is the molar heat capacity at constant volume. Usually θ_e is determined by choosing its best value that fits the experimental c_v versus T data. For KCl, $\theta_e \approx 230$ K. Estimate the frequency of vibration for KCl from the heat capacity data, and compare your result with that calculated based on Eq. (5.14). Assume the Born exponent $n = 9$ for KCl. Atomic weight of Cl is 35.5 g/mol and that of K is 39.1 g/mol.

Answer

(*a*) The various configurations are shown in Fig. 5.2. They total 10.

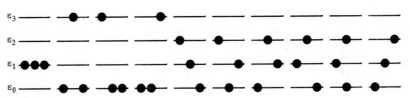

Figure 5.2
Possible configurations of arranging three particles in a system with a total energy of 3 units.[49] Here it is assumed that $\varepsilon_0 = 0$, $\varepsilon_1 = 1$ unit, $\varepsilon_2 = 2$ units, etc.

(*b*) The interatomic distance for KCl is 319 pm. If follows that for KCl (see Prob. 4.2)

$$S_0 = \frac{z_1 z_2 e^2}{4 \pi \varepsilon_0 r_0^3} (n - m) = 56.7 \frac{N}{m}$$

$$M_{red} = \frac{m_1 m_2}{(m_1 + m_2)} = \frac{18.9}{1000 N_{Av}} = 3.134 \times 10^{-26} \text{ kg}$$

Applying Eq. (5.14):

---

49 See D. Gaskell, *Introduction to Metallurgical Thermodynamics*, 2d ed., Hemisphere, New York, 1981, Chap. 4, for more details.

$$v = \frac{1}{2\pi}\sqrt{\frac{S_0}{M_{red}}} = \frac{1}{2\pi}\sqrt{\frac{56.7}{3.1 \times 10^{-26}}} = 6.8 \times 10^{12}\,s^{-1}$$

Calculating v_e from θ_e using the expression given above yields $v_e = 4.8 \times 10^{12}\,s^{-1}$. The agreement is quite good, considering the many simplifying assumptions made to arrive at Eq. (5.14). The importance of these calculations lies more in appreciating that ions in solids vibrate at a frequency on the order of $10^{13}\,s^{-1}$.

Electronic entropy

In the same manner as randomization of atoms over available energy levels constitutes entropy, the same can be said about the distribution of electrons over their energy levels. At 0 K, electrons and holes in semiconductors and insulators are in their lowest energy state, and only one configuration exists. As the temperature is raised, however, they are excited to higher energy levels, and the uncertainty of finding the electron in any number of excited energy levels constitutes a form of entropy as well. This point will be dealt with in greater detail in Chaps. 6 and 7.

Other forms of entropy

Some elements and compounds have magnetic or dielectric moments. These moments can be randomly oriented, or they may be ordered. For example, when they are ordered, the magnetic entropy is zero since there is only one configuration. As the temperature is increased, however, the entropy increases as the number of possible configurations increases. The same argument can be made for dielectric moments (see Chap. 14).

Total entropy

Since entropies are additive, it follows that the total entropy of a system is given by

$$S_{tot} = S_{config} + S_T + S_{elec} + S_{other} \tag{5.18}$$

5.2.3 Free Energy, Chemical Potentials, and Equilibrium

As noted at the outset of this section, the important function that defines equilibrium, or lack thereof, is neither enthalpy nor entropy, but rather the free-energy function G, defined by Gibbs as

$$G = H - TS \tag{5.19}$$

It follows that the free energy changes occurring during any reaction or transformation are given by

$$\Delta G = \Delta H - T\Delta S \tag{5.20}$$

where ΔS now includes all forms of entropy change.

Furthermore, it can be shown that at equilibrium $\Delta G = 0$. To illustrate, consider changes occurring in a system as a function of a given reaction variable ξ that affects its free energy as shown schematically in Fig. 5.3. The variable ξ can be the number of vacancies in a solid, the number of atoms in the gas phase, the extent of a reaction, the number of nuclei in a supercooled liquid, etc. As long as $\xi \neq \xi_0$ (Fig. 5.3), then $\Delta G \neq 0$ and the reaction will proceed. When $\xi = \xi_0$, ΔG is at a minimum and the system is said to be in equilibrium, since the driving force for change $\Delta G/\Delta\xi = \Delta G$ vanishes. Thus the condition for equilibrium can be simply stated as

$$\boxed{\Delta G|_{P,T,n_i} = \frac{\Delta G}{\Delta\xi} = 0} \tag{5.21}$$

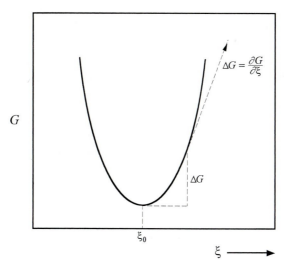

Figure 5.3

Schematic of free-energy versus reaction coordinate curve. At $\xi = \xi_0$, that is, when $\Delta G = 0$, the system is said to be in equilibrium.

Equation (5.21), despite its apparent simplicity, is an extremely powerful relationship because once the free energy of a system is formulated as a function of ξ, the state of equilibrium can simply be determined by differentiation of that function[50] (i.e., locating the minimum). In Chap. 6 how this simple and yet powerful method is used to determine the equilibrium number of vacancies in a solid is described. It is important to emphasize that this condition for equilibrium is valid only when the changes are occurring at constant temperature and pressure.

Free-energy change ΔG is an extensive property; i.e., it depends on the size of the system. If normalized, however, to a per-mole or per-atom basis, it becomes known as the **chemical potential**. The formal definition of the chemical potential of species, i, is

$$\mu_i = \frac{\partial G}{\partial n_i}\bigg|_{P,T,j} \tag{5.22}$$

The *chemical potential thus defined is the work that would be required to remove an atom from the bulk of an uncharged solid to infinity at constant pressure and temperature while keeping all other chemical components, j, in the system fixed.*

Once again, as in the case of enthalpy, since one is dealing with energy, there are no absolute values. To circumvent this problem, the **standard chemical potential** *of a pure element or compound* μ_i°, is defined, and all changes that occur in a system are then referred to that standard state.[51]

To take into account the fact that an element or compound is not in its standard state, the concept of activity has been introduced. Mathematically μ_i can be described by

$$\mu_i = \mu_i^\circ + RT \ln a_i \tag{5.23}$$

where a_i is the activity of that species, which is further described as

$$a_i = \gamma_i X_i \tag{5.24}$$

where X_i and γ_i are the mole fraction and activity coefficient, respectively. It follows directly from the definition of the standard state that a_i of a pure element in its standard state is 1.0, and $\mu_i = \mu_i^\circ$.

---

[50] Needless to say, the real difficulty does not lie in determining the location of the minimum — that is the easy part. The hard part is determining the relationship between G and ξ — therein lies the challenge.

[51] This value is unknown. This is not a major problem, however, because what is of interest is the change $\mu_i - \mu_i^\bullet$.

The activity coefficient is generally a function of composition. However, if a solution is ideal or dilute enough such that the solute atoms do not interact with each other, it is found that the activity coefficient can be assumed to be constant and

$$a_i = \gamma_i^\circ X_i \qquad (5.25)$$

where γ_i° is known as the *henrian activity coefficient*, which is not a function of composition. It also follows that for an ideal solution $\gamma_i = 1$.

EXPERIMENTAL DETAILS: MEASURING ACTIVITIES

Whereas it is possible to define activities mathematically by using Eq. (5.23), it is only when it is appreciated how a_i is measured that a better understanding of that concept becomes clear. There are several ways a_i can be measured; the most tangible, however, entails measuring the partial pressure P_i of the species for which the activity is to be determined and comparing that value to the partial pressure of the same species when it is in its pure standard state. The activity is then related to the partial pressures by[52]

$$a_i = \frac{P_i}{P^\circ} \qquad (5.26)$$

where P° is the partial pressure of the same species i in its standard state, i.e., pure. Note that for gases P° is taken to be 1 atm or 0.1 MPa.

To further illustrate, consider the following thought experiment. Take an element M, place it in an evacuated and sealed container, and heat the system at a given temperature until equilibrium is attained; then measure the pressure of gas atoms in the container. By definition, the measured quantity represents the equilibrium partial pressure P_M°, of pure M. This value, which is solely a function of temperature, is well documented and can be looked up.

By proceeding further with the thought experiment and alloying M with a second element N, such that the molar ratio is, say, 50 : 50 and repeating the aforementioned experiment, one of the following three outcomes is possible:

---

[52] This can be easily seen by noting that the work done in transferring one mol of atoms from a region where the pressure is P_i to one where the pressure is P° is simply $\Delta\mu = RT\ln(P_i/P^\circ)$. This work has to be identical to the energy change for the reaction $M_{pure} \Rightarrow M_{alloy}$ for which $\Delta\mu = RT\ln(a_i/1)$. Note that this is essentially how Eq. (5.23) is obtained.

1. The fraction of M atoms in the gas phase is equal to their fraction in the alloy, or 0.5, in which case the solution is termed *ideal* and $a_i = P_i/P° = 0.5 = X_i$, and $\gamma_i = 1$.
2. The fraction of M atoms in the gas phase is less than 0.5. So $a_i = P_i/P° < 0.5$, hence $\gamma_i < 1$. This is termed *negative deviation* from ideality and implies that the M atoms prefer being in the solid or melt to being in the gas phase relative to the ideal mixture.
3. The fraction of M atoms in the gas phase is greater than 0.5. So $a_i = P_i/P° > 0.5$ and $\gamma_i > 1$. This is termed *positive deviation* from ideality and implies that the M atoms prefer being in the gas phase relative to the ideal mixture.

Thus by measuring the partial pressure of an element or a compound in its pure state and by repeating the measurement with the element or compound combined with some other material, the activity of the former can be calculated.

5.3
CHEMICAL EQUILIBRIUM AND THE MASS ACTION EXPRESSION

Consider the reaction

$$M(s) + \frac{1}{2}X_2(g) \Rightarrow MX(s) \quad \Delta G_{rxn} \tag{I}$$

where ΔG_{rxn} represents the free-energy change associated with this reaction. Clearly ΔG_{rxn} will depend on the state of the reactants. For instance, one would expect ΔG_{rxn} to be greater if the partial pressure of X_2 were 1 atm than if it were lower, and vice versa.

Mathematically, this is taken into account by appreciating that the driving force for any reaction is composed of two terms: The first is how likely one expects the reaction to occur under standard conditions, and the second factor takes into account the fact that the reactants may or may not be in their standard states. In other words, it can be shown (App. 5A) that the driving force ΔG_{rxn} for any reaction is given by

$$\Delta G_{rxn} = \Delta G°_{rxn} + RT \ln K \tag{5.27}$$

where $\Delta G°_{rxn}$ is the *free-energy change associated with the reaction when the reactants are in their standard state*. And K is known as the **equilibrium constant** of the reaction. For reaction (I),

$$K = \frac{a_{MX}}{a_M (P_{X_2})^{1/2}} \tag{5.28}$$

where a_{MX}, a_M, and P_{X_2} are respectively, the activities of MX and M, and the partial pressure of X_2 at any time during the reaction. Equation (5.28) is also known as the **mass action expression** for reaction (I).

At equilibrium, $\Delta G_{rxn} = 0$, and Eq. (5.27) simplifies to the well-known result

$$\Delta G_{rxn}^\circ = -RT \ln K_{eq} \tag{5.29}$$

At equilibrium, $K = K_{eq} = \exp\left[\Delta G_{rxn}^\circ / (RT)\right]$.

Before one proceeds further, it is instructive to dwell briefly on the ramifications of Eq. (5.27). First, this equation says that if the reactants and products are in their standard state,[53] that is, $P_{X_2} = a_M = a_{MX} = 1$, then $K = 1$ and $\Delta G_{rxn} = \Delta G_{rxn}^\circ$, which is how ΔG_{rxn}° was defined in the first place. The other extreme occurs when the driving force for the reaction is zero, that is, $\Delta G_{rxn} = 0$, which by definition is the equilibrium state, in which case Eq. (5.29) applies.

It is worth noting that for the generalized reaction

$$aA + bB \Rightarrow cC + dD$$

the equilibrium constant is given by

$$K = \frac{a_C^c a_D^d}{a_A^a a_B^b} \tag{5.30}$$

where the a_i values represent the activities of the various species raised to their respective stoichiometric coefficients a, b, c, etc.

Armed with these important relationships, it is now possible to tackle the next important topic, namely, the delineation of the chemical stability domains of ceramic compounds.

---

[53] As noted above, the standard state of a gas is chosen to be the state of 1 mol of pure gas at *1 atm* (0.1 MPa) pressure and the temperature of interest. One should thus realize that whenever a partial pressure P_i appears in an expression such as Eq. (5.28), it is implicit that one is dealing with the *dimensionless ratio*, $P_i/1$ (atm).

5.4
CHEMICAL STABILITY DOMAINS

The chemical stability domain of a compound represents the range of activity or gaseous partial pressure over which that compound is stable. For example, experience has shown that under sufficiently reducing conditions, all oxides are unstable and are reducible to their parent metal(s). Conversely, all metals, with the notable exception of the noble ones, are unstable in air — their oxides are more stable. From a practical point of view, it is important to be able to predict the stability or lack thereof of a ceramic in a given environment. A related question, whose answer is critical for the successful reduction of ores, is this: At what oxygen partial pressure will an oxide no longer be stable?

To illustrate, it is instructive to consider an oxide MO_z, for which a higher oxide MO_y also exists (that is, $y > z$) and to calculate its stability domain. The equilibrium partial pressure of the oxide that is in equilibrium with the parent metal is determined by applying Eq. (5.29) to the following reaction:

$$\frac{z}{2}O_2 + M \leftrightarrow MO_z \qquad \Delta G_f^I \qquad\qquad \text{(II)}$$

or

$$\ln P_{O_2} = \frac{2\Delta G_f^I}{zRT} \qquad\qquad (5.31)$$

Further oxidation of MO_z to MO_y occurs by the following reaction:

$$O_2 + \frac{2}{y-z}MO_z \leftrightarrow \frac{2}{y-z}MO_y \qquad \Delta G_f^{II} \qquad\qquad \text{(III)}$$

and the corresponding equilibrium oxygen partial pressure is given by

$$\ln P_{O_2} = \frac{\Delta G_f^{II}}{RT} \qquad\qquad (5.32)$$

where

$$\Delta G_f^{II} = \frac{2}{y-z}\left[\Delta G_{f,\,MO_y} - \Delta G_f^I\right]$$

It follows that the oxygen partial pressure regime over which MO_z is stable is bounded by the values obtained from Eqs. (5.31) and (5.32). The following worked example should further clarify the concept. Needless to say, carrying out the type of calculations described above would be impossible without a knowledge of the temperature dependence of the standard free energies of formation of the oxides involved. Figure 5.4 plots such data for a number of binary oxides.

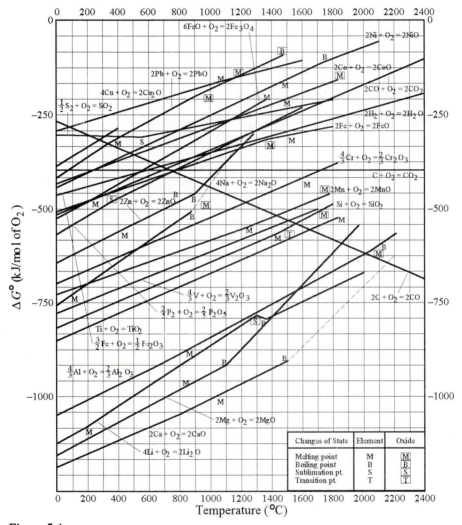

Figure 5.4

Standard free energies of formation of a number of binary oxides as a function of temperature.[54]

---

54 Adapted from L. S. Darken and R. W. Gurry, *Physical Chemistry of Metals*, McGraw-Hill, New York, 1953.

WORKED EXAMPLE 5.4. Calculate the chemical stability domains for the phases in the Fe-O system at 1000 K, given the following standard free energies of formation:[55]

$$\Delta G_{FeO} \text{ at } 1000K = -206.95 \text{ kJ/mol}$$
$$\Delta G_{Fe_3O_4} = -792.6 \text{ kJ/mol}$$
$$\Delta G_{Fe_2O_3} = -561.8 \text{ kJ/mol}$$

Answer

At equilibrium between Fe and FeO, the pertinent reaction is:

$$Fe + \frac{1}{2}O_2 \Rightarrow FeO \quad \Delta G_{FeO}$$

Applying Eq. (5.31) and solving for the equilibrium P_{O_2} at 1000 K yields 2.4×10^{-22} atm.

As the oxygen partial pressure is further increased, Fe_3O_4 becomes the stable phase according to[56]

$$3FeO + \frac{1}{2}O_2 \Rightarrow Fe_3O_4 \quad \Delta G_{r,1} = \Delta G_{Fe_3O_4} - 3\Delta G_{FeO}$$

Once again solving for P_{O_2} gives 1.14×10^{-18} atm.

Similarly Fe_3O_4 is stable up to a partial pressure given by the equilibrium between it and Fe_2O_3, or

$$\frac{2}{3}Fe_3O_4 + \frac{1}{6}O_2 \Rightarrow Fe_2O_3 \quad \Delta G_{r,2} = \Delta G_{Fe_2O_3} - \frac{2}{3}\Delta G_{Fe_3O_4}$$

with an equilibrium partial pressure of 3.4×10^{-11} atm.

To summarize at 1000 K: below a P_{O_2} 2.4×10^{-22} atm, Fe is the stable phase, between 2.4×10^{-22} and 1.14×10^{-18}, FeO is stable. Fe_3O_4 is stable between 1.14×10^{-18} and 3.4×10^{-11}. At oxygen partial pressures greater than 3.4×10^{-11}, Fe_2O_3 is the stable phase up to 1 atm (see Fig. 6.8c for a graphical representation of these results as a function of temperature).

---

[55] One of the more comprehensive and reliable sources of thermodynamic data is the JANAF thermochemical tables.

[56] The stoichiometry of the phases of interest can be read easily from the pertinent phase diagram.

5.5
ELECTROCHEMICAL POTENTIALS

In the previous section, the chemical potential of species i in a given phase was defined as the work needed to bring a mole of that species from infinity into the bulk of that phase. This concept is of limited validity for ceramics, however, since it only applies to *neutral* specie or uncharged media, where in either case the electric work is zero. Clearly, the charged nature of ceramics renders that definition invalid. Instead the pertinent function that is applicable in this case is the **electrochemical potential** η_i, defined for a particle of net charge z_i by:

$$\eta_i = \frac{\mu_i}{N_{Av}} + z_i e \phi \tag{5.33}$$

where μ_i is the chemical potential per mole and ϕ is the electric potential. On a molar basis, this expression reads

$$\eta_i^{molar} = \mu_i + z_i F \phi \tag{5.34}$$

where F is Faraday's constant ($F = N_{Av} e = 96500 \, C/\text{equivalent}$). In other words, Eq. (5.33) states that η_i is the sum of the *chemical* and *electrical* work needed to bring a particle of charge $z_i e$ from infinity to that phase. Note once again that if z_i were zero, the electrochemical and chemical potentials would be identical, or $\eta_i = \mu_i$, which is the case for metals and other electronically conducting materials. Also note that this conclusion is also valid when one is dealing with the insertion or the removal of a *neutral* species from charged media, such as ionic ceramics or liquid electrolytes.[57] The fundamental problem in dealing with ionic ceramics arises, however, if the problem involves *charged* species. In that case, one has to grapple with the electrochemical potential.

---

[57] An interesting ramification of this statement is that it is impossible to measure the activities or chemical potentials of individual *ions* in a compound, for the simple reason that it is impossible to indefinitely add or remove only one type of ion without having a charge buildup. For example, if one starts removing cations from an *MX* compound, it will very quickly acquire a net negative charge that will render removing further ions more and more difficult. In other words, because it is impossible to measure the "partial pressure" of, say, Na *ions* above an NaCl crystal, it follows that it is impossible to measure their activity. Interestingly enough, it is, in principle, possible to measure the partial pressure of Na metal, Cl_2 gas, or NaCl vapor over an NaCl crystal. In other words, it is only possible to measure the activity of **neutral** entities. This problem is by no means restricted to ionic solids. The problem was historically first looked at in liquid electrolytic solutions. For an excellent exposition of that problem, see J. Bockris and A. K. N. Reddy, *Modern Electrochemistry*, vol. 2, Plenum, New York, 1970, Chap. 7.

It can be shown (see Chap. 7) that the driving force on a charged species is the gradient in its electrochemical potential. It follows directly that the condition for equilibrium for any given species i is that the gradient vanishes, i.e., when

$$\boxed{d\eta_i = 0} \tag{5.35}$$

In other words, *at equilibrium the electrochemical potential gradient of every species everywhere must vanish*. It follows that for charged species the condition for equilibrium occurs *not* when $d\mu = 0$ but rather when $d\eta = 0$.

The astute reader may argue at this point that since the bulk of any material has to be neutral, it follows that ϕ was constant across that material and therefore the electric work was a constant that could be included in μ°, for instance. The fundamental problem with this approach, however, is that in order to insert a charged particle into a given phase, an interface has to be crossed. It follows that if a given interface is charged with respect to the bulk, then the electric work can no longer be neglected.

5.6
CHARGED INTERFACES, DOUBLE LAYERS, AND DEBYE LENGTHS

The next pertinent question is, Are interfaces charged, and if so, why? The answer to the first part is simple: Almost all interfaces and surfaces are indeed charged. The answer to the second part is more complicated; it depends on the type of interface, class of material, etc., and is clearly beyond the scope of this book. However, to illustrate the concept the following idealized and simplified thought experiment is useful. Consider the bulk of an MO oxide depicted in Fig. 5.5a. Focus on the central ion. It is obvious that this ion is being tugged at equally in all directions. Imagine, now, that the crystal is sliced in two such that an interface is created in the near vicinity of the aforementioned ion. The cutting process bares two surfaces and causes an imbalance of the forces acting on ions that are near the surface, depicted in Fig. 5.5b.

This asymmetry in force in turn induces the ion to migrate one way or another. If it is further assumed, again for the sake of simplicity, that in this case the O ions are immobile and that the driving force is such as to induce the M ions to migrate to the surface, then it follows that the interface or surface will now be positively charged with respect to the bulk — a charge that has to be balanced by a negative one in the bulk. For a pure MX compound,[58] this is accomplished automatically, because as the ions migrate to the surface, the vacancies that are left behind are negatively charged (see Chap. 6). The formation of a surface sheet of charge that is balanced by a concentration of oppositely charged bulk entities constitutes a **double layer** (Fig. 5.5c).

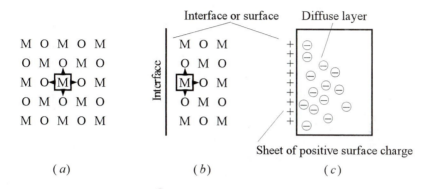

(*a*)   (*b*)   (*c*)

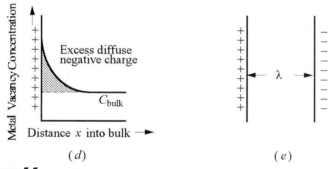

(*d*)   (*e*)

Figure 5.5
(*a*) Ion in bulk is subjected to a symmetric force. (*b*) Near an interface the forces are no longer symmetric, and the ions migrate one way or another. (*c*) Schematic of diffuse layer extending into the bulk from the interface. In this case, for simplicity it is assumed that the anions are immobile and the cations move toward the surface. The net positive charge is compensated by a distribution of negatively charged cation *vacancies* in the bulk.
(*d*) Charge distribution of cation vacancies. (*e*) A sheet of charge equal in magnitude to surface charge at a distance λ from interface.

---

[58] Most oxides contain impurities, which in an attempt to reduce the strain energy of the system tend to migrate to the interfaces, grain boundaries, and surfaces. It is usually the segregation of these impurities that is responsible for the surface charge. This charge is usually compensated, however, with bulk ionic defects.

For reasons that will become apparent in Chap. 7 (namely, diffusion), it can be shown that the compensating charges to the one at the interface, i.e., the cation vacancies in this case, are not concentrated in a plane but rather are diffusely distributed in the bulk of the solid, as shown in Fig. 5.5c and d. It can also be shown that a measure of the thickness of this so-called double layer, also known as the **Debye length** λ, is given by[59]

$$\lambda = \left(\frac{e^2 n_i z_i^2}{k' \varepsilon_0 kT} \right)^{-1/2} \tag{5.36}$$

where z_i and n_i are the charge and number density (particles per cubic meter) of the defects in the *bulk* of the material and ε_0 and k' are, respectively, the permittivity of free space and the relative dielectric constant of the solvent (see Chap. 14). All other symbols have their usual meaning; λ is the distance at which the diffuse charge can be replaced by an equivalent sheet of charge (Fig. 5.5e) that would result in the same capacitance as that of the diffuse charge. Note that Eq. (5.36) is only applicable to dilute solutions and breaks down at higher concentrations.

Finally, it is worth noting that charged interfaces are created not only at free surfaces, but whenever two dissimilar phases come into contact. Electrified interfaces are at the heart of much of today's technology; from drug manufacturing to integrated circuits. Life itself would be impossible without them. More specifically in ceramics, the electric double layer is responsible for such diverse phenomena as varistor behavior, chemical sensing, and catalysis, to name but a few.

5.7
GIBBS-DUHEM RELATION FOR BINARY OXIDES

The chemical potentials of the various components in a multicomponent system are interrelated. The relationship for binary compounds, known as the **Gibbs-Duhem equation**, is developed here. Its applicability and usefulness, however, will only become apparent later in Chap. 7.

In terms of the building blocks of a binary MO_ξ compound, one can write

$$MO_\xi \Leftrightarrow M^{\xi(2+)} + \xi O^{2-} \tag{5.37}$$

---

[59] See, for example, J. Bockris and A. K. N. Reddy, *Modern Electrochemistry*, Plenum, New York, 1970.

from which it follows that

$$\eta_{MO_\xi} = \eta_{M^{\xi(2+)}} + \xi\eta_{O^{2-}}$$

At equilibrium, by definition, $d\eta_{MO_\xi} = 0$, and consequently, $d\eta_{M^{\xi(2+)}} = -\xi d\eta_{O^{2-}}$, or

$$d\mu_{M^{\xi(2+)}} + 2\xi e\, d\phi = -\xi\left(d\mu_{O^{2-}} - 2e\, d\phi\right)$$

Furthermore, since locally, the anions and cations are subjected to the same potential ϕ, it follows that for a binary oxide

$$d\mu_{M^{\xi(2+)}} = -\xi d\mu_{O^{2-}} \tag{5.38}$$

This expression is known as the **Gibbs-Duhem relationship** and it expresses the fact that the changes in the chemical potentials of the building blocks of a binary crystal (i.e., anions and cations) are interrelated.[60]

Note that Eq. (5.37) could have been written as

$$MO_\xi \Leftrightarrow M + \frac{\xi}{2}O_2 \tag{5.39}$$

from which it follows that

$$\boxed{d\mu_M = -\frac{\xi}{2}d\mu_{O_2}} \tag{5.40}$$

As noted above, the importance and applicability of these relationships will become apparent in Chap. 7.

WORKED EXAMPLE 5.5. (*a*) Assume a crystal of MgO is placed between Mg metal on one side and pure oxygen on the other side as shown in Fig. 5.6. Calculate the chemical potential of each species at each interface at 1000 K, given that at that temperature $\Delta G^\circ_{MgO} = -492.95\,kJ/mol$. (*b*) Show that the Gibbs-Duhem relationship holds for the MgO crystal described in part (*a*).

Answer

(*a*) The pertinent reaction and its corresponding equilibrium mass action expression are, respectively,

---

60 In textbooks of solution thermodynamics the Gibbs-Duhem is derived as $x_A\, d\mu_A + x_B\, d\mu_B = 0$, where x_A and x_B denote the mole fractions of A and B, respectively.

$$Mg + \frac{1}{2}O_2 \Rightarrow MgO(s)$$

$$\Delta G° = -RT \ln K = -RT \ln \frac{a_{MgO}}{a_{Mg}P_{O_2}^{1/2}}$$

Since the MgO that forms in this case is pure, i.e., in its standard state (e.g., not in solid solution), it follows that by definition $a_{MgO} = 1$ on either side. On the metal side, $a_{Mg} = 1.0$, and solving for P_{O_2} yields 3.2×10^{-52} atm, or 3.2×10^{-53} MPa.

Conversely, on the oxygen side, $P_{O_2} = 1$ atm, and a_{Mg} is calculated to be

$$a_{Mg} = 1.7 \times 10^{-26}$$

The results are summarized in Fig. 5.6.

(b) The Gibbs-Duhem expression simply expresses the fact that the chemical potentials of the constituents of a binary compound are interrelated. Referring to Fig. 5.6, the following applies:

$$\mu_{MgO}|_{oxygen} = \mu_{MgO}|_{metal}$$

or

$$\mu_{O_2}° + \frac{RT}{2}\ln P_{O_2} + \mu_{Mg}° + RT \ln a_{Mg}|_{oxygen} = \mu_{O_2}° + \frac{RT}{2}\ln P_{O_2} + \mu_{Mg}° + RT \ln a_{Mg}|_{metal}$$

which simplifies to

$$\ln a_{Mg}|_{oxygen} = \frac{1}{2}\ln P_{O_2}|_{metal}$$

Insertion of the appropriate values for the activity of Mg and P_{O_2} at each interface shows that this identity is indeed fulfilled.

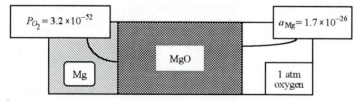

Figure 5.6
Equilibrium conditions for an MgO crystal simultaneously subjected to Mg metal on one side and pure oxygen at 1 atm on the other.

5.8
KINETIC CONSIDERATIONS

In the preceding sections, the fundamental concept of equilibrium was discussed. Given sufficient time, all systems tend to their lowest energy state. Experience has shown, however, that many systems do not exist in their most stable configuration, but rather in a metastable form. Most materials are generally neither produced nor used in their equilibrium states. For example, glasses are metastable with respect to their crystalline counterparts, yet are of great utility because at the temperatures at which they are typically used, the kinetics of the glass-crystal transformation are negligible.

In general, the kinetics or rate of any transformation is assumed to be proportional to a driving force F

$$\text{Rate} = \beta F \tag{5.41}$$

where the proportionality constant β is a system property that depends on the process involved. For instance, β can be a diffusion coefficient, a reaction rate constant, or a conductance of any sort.

The driving force is a measure of how far a system is from equilibrium. Referring to Fig. 5.2, the driving force is nothing but $\partial G/\partial \xi$, or ΔG. Thus the importance of thermodynamics lies not only in defining the state of equilibrium, but also in quantifying the driving force — it is only by knowing the final equilibrium state that the rate at which a system will approach that state can be estimated.

All changes and transformations require a driving force, the nature and magnitude of which can vary over many orders of magnitude depending on the process involved (Table 5.1). For example, the driving forces for chemical reactions, such as oxidation, are usually quite large, in the range of a few hundred kilojoules per mole. On the other hand, the driving forces for boundary migration, coarsening, and densification are much smaller, on the order of 100 J/mol or less. This, in turn, partially explains why it is much easier to oxidize a fine metal powder than it is to sinter it.

The four most important driving forces operative in materials science are those due to

1. Reduction in free energies of formation as a result of chemical reactions and phase transformations, e.g., oxidation or crystallization
2. Reduction of energy due to applied stresses, e.g., creep
3. Reduction of surface or interfacial energy, e.g., sintering and grain growth
4. Reduction of strain energy, e.g., fracture, segregation

TABLE 5.1
Typical orders of magnitude of driving forces governing various phenomena discussed in this book

Process	Driving force	Typical values, J/mol[†]	Comments
Fracture (Chap. 11)	$V_m \sigma^2 / (2Y)$	0.5	σ is stress at failure and Y is Young's modulus
Grain growth (Chap. 10)	$2\gamma_{gb}/r$	20.0	γ_{gb} is grain boundary energy, and r is radius of a particle
Sintering or coarsening (Chap. 10)	$2\gamma/r$	100.0	γ is surface energy term (Chap. 4)
Creep (Chap. 12)	σV_m	1000.0	σ is applied stress and V_m, molar volume
Crystallization (Chap. 9)	$\Delta H \, \Delta T / T_m$	3000.0	ΔH is enthalpy of transformation, ΔT is undercooling, and T_m is melting point
Interdiffusion (Chap. 7)	$RT(x_a \ln x_a + x_b \ln x_b)$	5000.0	Assuming ideal solution [see Eq. (5.11)]
Oxidation (Chap. 7)	ΔG°_{form}	50,000.0 – 500,000.0	ΔG°_{form} free energy of formation of oxide-normalized to a per-mole-of-oxygen basis

†   Assumptions: 1000 K, molar volume: $10^{-5} \, \text{m}^3/\text{mol} \, (10 \, \text{cm}^3/\text{mol})$; $r = 1\mu\text{m}$, $\gamma = 1 \, \text{J}/\text{m}^2$; $\sigma = 100 \, \text{MPa}$.

At this point the expressions and order-of-magnitude values of these driving forces are simply listed (Table 5.1). However, each will be revisited and discussed in detail in subsequent chapters. Fracture is dealt with in Chap. 11, grain growth and sintering in Chap. 10, crystallization in Chap. 9, creep in Chap. 12, and oxidation and interdiffusion in Chap. 7.

The second important parameter that determines the rate at which a given process will occur is the factor β. And since, with the notable exception of fracture, all the processes listed in Table 5.1 require the movement of atoms, β *is usually equated to the rate at which an atom or ion will make a jump.* This concept is discussed in greater detail in Chap. 7, where diffusion is elucidated.

5.9
SUMMARY

The free energy is a function made up of two terms, an enthalpy term and an entropy term. Entropy can be of various kinds, but fundamentally it is a measure of the disorder in a system.

For a system that is at constant pressure and temperature, the state of equilibrium is defined as that state of the system for which the free energy is at a minimum.

For a chemical reaction equilibrium dictates that $\Delta G_{rxn} = 0$ and consequently

$$\Delta G^\circ = -RT \ln K$$

where K is the equilibrium constant for that reaction.

In ionic ceramics it is not the chemical but the electrochemical potential that defines equilibrium.

APPENDIX 5A

Derivation of Eq. (5.27)

Reaction (I) in text reads

$$M(s) + \frac{1}{2}X_2(g) \Rightarrow MX(s) \quad \Delta G_{rxn} \tag{5A.I}$$

Applying Eq. (5.23) to reactants and products, one obtains

$$\mu_{MX} = \mu_{MX}^\circ + RT \ln a_{MX} \tag{5A.1}$$

$$\mu_M = \mu_M^\circ + RT \ln a_M \tag{5A.2}$$

$$\mu_{X_2} = \frac{1}{2}\mu_{X_2}^\circ + RT \ln P_{X_2}^{1/2} \tag{5A.3}$$

It follows that the free-energy change associated with this reaction is

$$\Delta G_{form} = \mu_{MX} - \mu_M - \frac{1}{2}\mu_{X_2} \tag{5A.4}$$

Combining Eqs. (5A.1) to (5A.4) gives

$$\Delta G_{form} = \left(\mu_{MX}^\circ - \mu_M^\circ - \frac{1}{2}\mu_{X_2}^\circ\right) + RT \ln \frac{a_{MX}}{a_M P_{X_2}^{1/2}} \tag{5A.5}$$

If one defines ΔG° as

$$\Delta G^\circ = \mu_{MX}^\circ - \mu_M^\circ - \frac{1}{2}\mu_{X_2}^\circ \tag{5A.6}$$

and K by Eq. (5.28), it is easily shown that Eqs. (5A.5) and (5.27) are identical. Furthermore, since at equilibrium $\Delta G_{form} = 0$, Eq. (5A.5) now reads

$$\Delta G^\circ = -RT \ln K_{eq} = -RT \ln \frac{a_{MX}}{a_M P_{X_2}^{1/2}} \tag{5A.7}$$

PROBLEMS

5.1. Starting with Eq. (5.9), and making use of Stirling's approximation, derive Eq. (5.10).

5.2. Pure stoichiometric ZnO is heated to 1400 K in an evacuated chamber of a vapor deposition furnace. What is the partial pressure of Zn and O_2 generated by the thermal decomposition of ZnO? Information you may find useful: ΔG_{ZnO}° at 1400 K $= -183$kJ/mol.
Answer: $\log P_{O_2} = -4.75$, $\log P_{Zn} = -4.45$

5.3. Calculate the driving force for the oxidation of pure Mg subjected to an oxygen partial pressure of 10^{-12} atm at 1000 K. Compare that value to the driving force if the oxygen partial pressure is 1 atm.
Answer: -378.1 kJ/mol

5.4. (a) Evaluate the equilibrium partial pressure of oxygen for the Si-silica system at 1000 K.
(b) If the oxidation is occurring with water vapor, calculate the equilibrium constant and the H_2/H_2O ratio in equilibrium with Si and silica at 1000 K.
(c) Compute the equilibrium partial pressure of oxygen for a gas mixture with an H_2/H_2O ratio calculated in part (b). Compare your result with the oxygen partial pressure calculated in part (a).

5.5. Given at 1623 K:

$$\Delta G_{SiO_2}^\circ = -623\text{kJ/mol}$$

$$\Delta G_{Si_2N_2O}^\circ = -446\text{kJ/mol}$$

$$\Delta G_{Si_3N_4}^\circ = -209\text{kJ/mol}$$

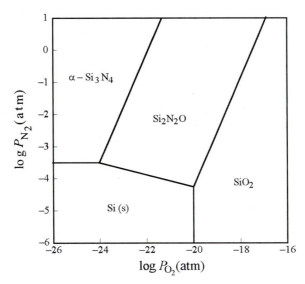

Figure 5.7
Si-N-O stability diagram at 1623 K.

Confirm that the stability diagram in the Si-N-O system at 1623 K is the one shown in Fig. 5.7.

5.6. (a) Can Al reduce Fe_2O_3 at 1200°C? Explain.
Answer: Yes
(b) Is it possible to oxidize Ni in a CO/CO_2 atmosphere with a ratio of 0.1?
(c) Will silica oxidize zinc at 700°C. Explain.
Answer: No

5.7. Calculate the stability domains of NiO and CoO at 1000 K. Compare your results with those listed in Table 6.1.

5.8. A crucible of BN is heated in a gas stream containing N_2, H_2, and H_2O at 1200 K. The partial pressure of nitrogen is kept fixed at 0.5 atm. What must the ratio P_{H_2}/P_{H_2O} have to be or exceed in order for B_2O_3 not to form? Information you may find useful: at 1200 K; $\Delta G^\circ_{BN} = -743\,\text{kJ/mol}$; $\Delta G^\circ_{B_2O_3} = -957.47\,\text{kJ/mol}$; $\Delta G^\circ_{H_2O} = -181.425\,\text{kJ/mol}$.

ADDITIONAL READING

1. R. A. Swalin, *Thermodynamics of Solids*, 2d ed., Wiley, New York, 1972.
2. D. R. Gaskell, *Introduction to Metallurgical Thermodynamics*, 2d ed., Hemisphere Publishing, New York, 1981.
3. K. Denbigh, *The Principles of Chemical Equilibrium*, 4th ed., Cambridge University Press, New York, 1981.
4. C. H. P. Lupis, *Chemical Thermodynamics of Materials*, North-Holland, Amsterdam, 1983.
5. *The Scientific Papers of J. W. Gibbs, vol. 1, Thermodynamics*, Dover, New York, 1961.
6. J. Bockris and A. K. N. Reddy, *Modern Electrochemistry*, vols. 1 and 2, Plenum, New York, 1970.
7. P. Shewmon, *Transformation in Metals*, McGraw-Hill, New York, 1969.
8. R. DeHoff, *Thermodynamics in Materials Science*, McGraw-Hill, New York, 1993.
9. L.S. Darken and R. W. Gurry, *Physical Chemistry of Metals*, McGraw-Hill, New York, 1951.

Thermodynamic Data:

10. M. W. Chase, C. A. Davies, J. R. Downey, D. J. Frurip, R. A. McDonald and A. N. Syverud, JANAF Thermodynamic Tables, 3d ed., J. Phys. Chem. Ref. Data, **14**, (1985), Supp. 1.
11. J. D. Cox, D. D. Wagman, and V. A. Medvedev, *CODATA Key Values of Thermodynamics*, Hemisphere Publishing, New York, 1989.
12. Online access to the SOLGASMIX program is available through F*A*C*T (Facility for the Analysis of Chemical Thermodynamics), Ecole Polytechnique, CRCT, Montreal, Quebec, Canada.
13. I. Barin, O. Knacke, and O. Kubaschewski, *Thermodynamic Properties of Inorganic Substances Supplement*, Springer-Verlag, New York, 1977.

CHAPTER 6

Defects in Ceramics

Textbooks and Heaven only are Ideal;
Solidity is an imperfect state.
Within the cracked and dislocated Real
Nonstoichiometric crystals *dominate.*
Stray Atoms sully and precipitate;
Strange holes, excitons, *wander loose, because*
Of Dangling Bonds, a chemical Substrate
Corrodes and catalyzes — surface Flaws
Help Epitaxial Growth to fix adsorptive claws.

John Updike, *The Dance of the Solids*[†]

6.1
INTRODUCTION

Alas, as John Updike so eloquently points out, only textbooks (present company excepted), and heaven are ideal. Real crystals, however, are not perfect but contain imperfections that are classified according to their geometry and shape into point, line, and planar defects. A *point defect* can be defined as any lattice point which is not occupied by the proper ion or atom needed to preserve the long-range periodicity of the structure. Dislocations are defects that cause lattice distortions centered on a line and are thus classified as *linear defects.* *Planar defects* are surface imperfections in polycrystalline solids that separate grains or domains of different orientations and include grain and twin boundaries. In addition, there are three-dimensional bulk defects such as pores, cracks, and inclusions; these are not treated in this chapter, however, but are considered in Chap. 11, where it is shown that these defects are critical in determining the strength of ceramics.

[†] J. Updike, **Midpoint and other Poems**, A. Knopf, Inc., New York, New York, 1969. Reprinted with permission.

The importance of defects in general and point defects in particular cannot be overemphasized. As will become apparent in subsequent chapters, many of the properties considered are strongly affected by the presence or absence of these defects. For instance, in Chap. 7, the one-to-one correlation between the concentration of point defects and atom movement or diffusion is elucidated. In metals, but less so in ceramics except at higher temperatures, it is the presence and movement of dislocations that is responsible for ductility and creep. In Chap. 11, the correlation between grain size and mechanical strength is made. As discussed in Chap. 16, the scattering of light by pores is responsible for their opacity.

Generally speaking, in ceramic systems more is known about point defects than about the structure of dislocations, grain boundaries, or free surfaces — a fact that is reflected in the coverage of this chapter in which the lion's share is devoted to point defects.

6.2
POINT DEFECTS

In contrast to pure metals and elemental crystals for which point defects are rather straightforward to describe (because only one type of atom is involved and charge neutrality is not an issue), the situation in ceramics is more complex. One overriding constraint operative during the formation of ceramic defects is the preservation of electroneutrality at all times. Consequently, the defects occur in neutral "bunches" and fall in one of three categories:

Stoichiometric defects

These are defined as ones in which the crystal chemistry, i.e., the ratio of the cations to anions, does not change, and they include, among others, Schottky and Frenkel defects (Fig. 6.3).

Nonstoichiometric defects

These defects form by the selective addition or loss of one (or more) of the constituents of the crystal and consequently lead to a change in crystal chemistry and the notion of nonstoichiometry discussed below. The basic notion that the composition of compounds is a constant with simple ratios between the numbers of constituent atoms is one that is reiterated in every first-year college chemistry course. For instance, in MgO the cation/anion ratio is unity, that for Al_2O_3 is 2/3, etc. In reality, however, it can be rigorously shown using thermodynamic arguments that the composition of *every* compound *must* vary within its existence regime.[61]

A material accommodates those changes in composition by selectively losing one of its constituents to its environment by the creation or elimination of defects (see Fig. 6.4). In so doing, a compound will adjust its composition to reflect the externally imposed thermodynamic parameters. This leads to the idea of **nonstoichiometry** where the simple ratio between the numbers of the constituent atoms of a compound breaks down. For example, if an oxide were annealed in a high oxygen partial pressure, it would be fair to assume that the number of oxygen atoms should be relatively greater than the number of cations. Conversely, if the oxygen partial pressure were very low, one would expect the cation concentration to be higher.

The importance of nonstoichiometry lies in the fact that many physical properties such as color, diffusivity, electrical conductivity, photoconductivity, and magnetic susceptibility can vary markedly with small changes in composition.

Extrinsic defects

These are defects created as a result of the presence of impurities in the host crystal.

The remainder of this section attempts to answer, among others, the following questions: Why do point defects form? What are the different types of defects that can form? And how is their concentration influenced by temperature and externally imposed thermodynamic parameters, such as oxygen partial pressure? Before we proceed, however, it is imperative to describe in greater detail the various defects that can form and to formulate a scheme by which they can be notated.

6.2.1 Point Defects and Their Notation

In a *pure* binary compound, the following lattice defects, shown schematically in Fig. 6.1, exist:

1. *Vacancies:* sites where an atom is missing. These can occur on either sublattice.
2. *Interstitial atoms:* atoms found in sites that are normally unoccupied.

---

[61] The *existence regime* of a compound defines the range of chemical potential of the constituents of that compound over which it is thermodynamically stable. For example, it was shown in Chap. 5 that MgO was stable between the oxygen partial pressures of 1 atm and 3.2×10^{-52} atm — below 3.2×10^{-52}, MgO decomposed to Mg metal and oxygen.

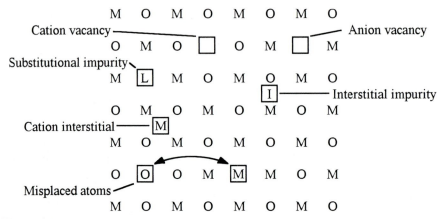

Figure 6.1
Various types of defects typically found in ceramics. Misplaced atoms can only occur in covalent ceramics due to charge considerations.

3. *Misplaced atoms:* types of atoms found at a site normally occupied by other types. This defect is only possible in covalent ceramics, however, where the atoms are not charged.

The following electronic defects also exist:

4. *Free electrons:* electrons that are in the conduction band of the crystal.
5. *Electron holes:* positive mobile electronic carriers that are present in the valence band of the crystal (see Chap. 7).

In addition to the aforementioned, an *impure* crystal will contain

6. *Interstitial and substitutional impurities:* As depicted in Fig. 6.1, these can occur on either sublattice.

Over the years, several schemes have been proposed to denote defects in ceramics. The one that is now used almost universally is the **Kroger-Vink notation** and is thus the one adopted here. In this notation, the defect is represented by a main symbol followed by a superscript and a subscript.

Main symbol. The main symbol is either the species involved, i.e., chemical symbol of an element, or the letter V for vacancy.

Subscript. The subscript is either the crystallographic position occupied by the species involved or the letter *i* for interstitial.

Superscript. The superscript denotes the **effective electric charge** on the defect, defined as the difference between the real charge of the defect species and that of the species that would have occupied that site in a perfect crystal.[62] The superscript is a prime for each negative charge, a dot for every positive charge, or an x for zero effective charge.

The best way to explain how the notation works is through a series of examples.

EXAMPLE 1. Consider the possible defects that can occur in a *pure* NaCl crystal:

(a) Vacancy on the Na^+ sublattice: $V'_{Na} \Rightarrow$ site on which vacancy resides

The symbol V is always used for a vacancy. The superscript is a prime (representing a single negative charge) because the effective charge on the vacancy is $0 - (+1) = -1$.

(b) Vacancy on Cl^- sublattice: $V^{\bullet}_{Cl}$

In this case the superscript is a small dot (which denotes a positive charge) because the effective charge on the vacancy in this case is $0 - (-1) = +1$.

(c) Interstitial position on Na sublattice: $Na^{\bullet}_i \Rightarrow$ always used for interstitials

The main symbol here is the misplaced Na ion; the subscript i denotes the interstitial position, and the effective charge is $+1 - 0 = +1$.

EXAMPLE 2. Consider the addition of $CaCl_2$ to NaCl. The Ca cation can substitute for a Na ion or go interstitial (needless to say, because of charge considerations, only cations will substitute for cations and only anions for anions). In the first case, the defect notation is $Ca^{\bullet}_{Na}$, and the effective charge $[+2 - (+1) = 1]$ is $+1$. Conversely, an interstitial Ca ion is denoted as $Ca^{\bullet\bullet}_i$.

EXAMPLE 3. Instead of adding $CaCl_2$, consider KCl. If the K ion, which has the same charge as Na, substitutes for a Na ion, the notation is K^X_{Na}, since the effective charge in this case is 0 (denoted by an x). If the K ion goes interstitial, the notation is $K^{\bullet}_i$.

EXAMPLE 4. Dope the NaCl crystal with Na_2S. Again only anions can substitute for anions, or they can go interstitial. Two possibilities are S'_{Cl} and S''_i.

---

[62] The charge is so called because it denotes not the real charge on the defect, but the effective charge *relative* to the perfect crystal. It is this effective charge that determines the direction in which the defect will move in response to an electric field. It also denotes the type of interaction between the defects, for instance, whether two defects would attract or repel each other.

EXAMPLE 5. One would expect to find the following defects in pure Al_2O_3: $Al_i^{\bullet\bullet\bullet}$, O_i'', V_{Al}''', and $V_O^{\bullet\bullet}$.

After this brief introduction to defects and their notation, it is pertinent to ask why point defects form in the first place. However, before the more complicated case of defects in ceramics is tackled in Sec. 6.2.3, the simpler situation involving vacancy formation in elemental crystals such as Si or Ge or pure metals is treated.

6.2.2 Thermodynamics of Point Defect Formation in Elemental Crystals

There are several ways by which vacancy formation can be envisioned. A particularly useful and instructive one is to remove an atom from the bulk of a crystal and place it on the surface. The enthalpy change Δh associated with such a process has to be endothermic because more bonds are broken than are re-formed. This brings up the legitimate question, If it costs energy to form defects, why do they form? The answer lies in the fact that at equilibrium, as discussed in Chap. 5, it is the free energy rather than the enthalpy that is minimized. In other words, it is only when the entropy changes associated with the formation of the defects are taken into account that it becomes clear why vacancies are thermodynamically stable and their equilibrium concentration can be calculated. It follows that if it can be shown that at any given temperature, the Gibbs free energy associated with a perfect crystal G_{perf} is *higher* than that of a crystal containing n_v defects, i.e., that $G_{def} - G_{perf} < 0$, where G_{def} is the free energy of the defective crystal, then the defective crystal has to be more stable. The procedure is as follows:

Free energy of a perfect crystal[63]

For a perfect crystal,

$$G_{perf} = H_{perf} - TS_{perf}$$

where H is the enthalpy; S, the entropy; and T, the absolute temperature of the crystal.

As noted in Chap. 5, the total entropy of a collection of atoms is the sum of a configuration term and a vibration entropy term, or

---

[63] This approach is not strictly orthodox because G_{perf} cannot be calculated on an absolute scale. However, the approach is still valid because before the final result is reached, that energy will be subtracted from G_{def}.

$$S = S_{\text{config}} + S$$

For a perfect crystal, $S_{\text{config}} = 0$ since there is only one way of arranging N atoms on N lattice sites. The vibration component, however, is given by Eq. (5.16), or

$$S = Nk\left(\ln\frac{kT}{h\nu} + 1\right)$$

where N is the number of atoms involved, k is Boltzmann's constant, and ν is the vibration frequency of atoms in the perfect crystal. Adding the various terms, one obtains

$$G_{\text{perf}} = H_{\text{perf}} - TS_{\text{perf}} = H_{\text{perf}} - NkT\left(\ln\frac{kT}{h\nu} + 1\right) \tag{6.1}$$

Free energy of a defective crystal

If one assumes that it costs h_d joules to create *one* defect, it follows that the enthalpy of the crystal upon formation of n_v vacancies increases (i.e., becomes less negative) by $n_v h_d$. Hence the enthalpy of the defective crystal is now

$$H_{\text{def}} = H_{\text{perf}} + n_v h_d \tag{6.2}$$

Furthermore, the configurational entropy is no longer zero because the n_v vacancies and N atoms can now be distributed on $N + n_v$ total atomic sites. The corresponding configuration entropy [see Eq. (5.10)] is given by

$$S_{\text{config}} = -k\left(N\ln\frac{N}{N+n_v} + n_v\ln\frac{N}{N+n_v}\right) \tag{6.3}$$

It is fair to assume that only the atoms in the near vicinity of each vacancy will vibrate at a different frequency ν' than the rest — the remainder of the atoms will be unaffected and will continue to vibrate with frequency ν. If one further assumes only nearest-neighbor interactions, then for a coordination number ζ of the vacancies, the total number of atoms affected is simply ζn_v. The vibration entropy term is then given by

$$S = k(N - \zeta n_v)\left(\ln\frac{kT}{h\nu} + 1\right) + n_v\zeta k\left(\ln\frac{kT}{h\nu'} + 1\right) \tag{6.4}$$

where the first term represents the atoms whose vibration frequencies have been unaffected by the vacancies and the second term represents those that have, and are now vibrating with, a new frequency.

Combining Eqs. (6.2) to (6.4) yields

$$G_{def} = H_{perf} + n_v h_d$$

$$-kT\left[(N - n_v\zeta)\left(\ln\frac{kT}{h\nu} + 1\right) + n_v\zeta\left(\ln\frac{kT}{h\nu'} + 1\right) - N\ln\frac{N}{n_v + N} - n_v\ln\frac{n_v}{n_v + N}\right]$$

(6.5)

Subtracting Eq. (6.1) from Eq. (6.5) yields the sought-after result

$$\Delta G = G_{def} - G_{perf} = n_v h_d + kTn_v\zeta\ln\frac{\nu'}{\nu} + kT\left(N\ln\frac{N}{n_v + N} + n_v\ln\frac{n_v}{n_v + N}\right)$$
(6.6)

This is an important result because it says that the free-energy change upon the introduction of n_v defects in an otherwise perfect crystal is a function of both n_v and T. If T is kept constant and ΔG is plotted versus n_v, as shown in Fig. 6.2a, it is immediately obvious that this function goes through a minimum.[64] In other words, the addition of vacancies to a perfect crystal will initially lower its free energy up to a point beyond which further increases in the number of vacancies is no longer energetically favorable, and the free energy increases once again.[65] The number of vacancies at which the minimum in ΔG occurs, i.e., when $\partial \Delta G/\partial n_v = 0$, is the equilibrium number of vacancies n_{eq} at that temperature and is given by (see Prob. 6.1)

$$\frac{n_{eq}}{n_{eq} + N} \approx \frac{n_{eq}}{N} \approx \exp\left(-\frac{h_d - T\Delta s_{vib}}{kT}\right) = \exp\left(-\frac{\Delta g_d}{kT}\right)$$
(6.7)

where $\Delta g = h_d - T\Delta s_{vib}$ and $\Delta s_{vib} = \zeta k\ln(\nu/\nu')$. Note that the final expression does not contain any configuration entropy terms, but depends solely on the free energy associated with the formation of a single defect.

Equation (6.7) predicts that the equilibrium number of vacancies increases exponentially with temperature. To understand why, it is instructive to compare Fig. 6.2a and b, where Eq. (6.6) is plotted, on the same scale, for two different temperatures. At higher temperatures (Fig. 6.2b), the configurational entropy term becomes more important relative to the enthalpy term, which in turn shifts n_{eq} to higher values.

---

64 For the sake of simplicity, the second term in Eq. (6.6) was omitted from Fig. 6.2.
65 Note that here n_v is the reaction variable discussed in Chap. 5 (Fig. 5.2).

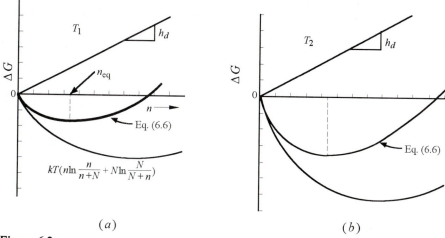

Figure 6.2

(a) Free-energy change as a function of number of defects n_v. The top line represents the energy needed to create the defects. The lower curve is the free energy gained as a result of the configuration entropy. The centerline represents the sum of the two components [i.e., a plot of Eq. (6.6)], which clearly goes through a minimum. (b) Same plot as in (a) except at a higher temperature. Note that the equilibrium number of defects increases with increasing temperature.

At this point, the slightly more complicated problem of defects in ceramics is dealt with. The complications arise because, as noted above, the charges on the defects preclude their forming separately — they always form in bunches so as to maintain charge neutrality. In the following section, defect formation in ceramics is dealt with by writing down balanced-defect reactions. Expressions for the equilibrium concentration of these defects are then calculated by using two approaches. The first uses the statistical approach used to derive Eq. (6.7). The second approach (Sec. 6.2.5) makes use of the mass action expression of the pertinent defect reactions. Needless to say, the two approaches should and do yield the same results.

6.2.3 Defect Reactions

The formation of the various point defects is best described by chemical reactions for which the following rules have to be followed:

- *Mass balance*: Mass cannot be created or destroyed. Vacancies have zero mass.
- *Electroneutrality* or charge balance: Charges cannot be created or destroyed.

- *Preservation of regular site ratio*: The ratio between the numbers of regular cation and anion sites must remain constant and equal to the ratio of the parent lattice.[66] Thus if a normal lattice site of one constituent is created or destroyed, the corresponding number of normal sites of the other constituent must be simultaneously created or destroyed so as to preserve the site ratio of the compound. This requirement recognizes that one cannot create one type of lattice site without the other and indefinitely extend the crystal. For instance, for an MO oxide, if a number of cation lattice sites are created or destroyed, then an equal number of anion lattice sites have to be created or destroyed. Conversely, for an M_2O oxide, the ratio must be maintained at 2 : 1, etc.

To generalize, for an M_aX_b compound, the following relationship has to be maintained at all times:

$$a(X_X + V_X) = b(M_M + V_M)$$

that is, the ratio of sum of the number of atoms and vacancies on each sublattice has to be maintained at the stoichiometric ratio, or

$$\frac{M_M + V_M}{X_X + V_X} = \frac{a}{b}$$

Note that this does not imply that the *number of atoms or ions has to maintain that ratio but only the number of sites*.

In the following subsections these rules are applied to the various types of defects present in ceramics.

Stoichiometric defect reactions

A stoichiometric defect reaction by definition is one where the chemistry of the crystal does not change as a result of the reaction. Said otherwise, a stoichiometric reaction is one in which no mass is transferred across the crystal boundaries. The three most common stoichiometric defects are Schottky defects, Frenkel defects, and antistructure disorder or misplaced atoms.

Schottky defects. In the Schottky defect reaction, electric-charge-equivalent numbers of vacancies are formed on each sublattice. In NaCl, for example, a

---

66 Interstitial sites are not considered to be regular sites.

Schottky defect entails the formation of Na and Cl vacancy pairs (Fig. 6.3a). In general, for an MO oxide, the reaction reads[67]

$$\text{Null (or perfect crystal)} \Rightarrow V''_M + V^{\bullet\bullet}_O \qquad \Delta g_S \qquad (6.8)$$

where Δg_S is the free-energy change associated with the formation of the Schottky defect.

Similarly, for an M_2O_3 oxide,

$$\text{Null (or perfect crystal)} \Rightarrow 2V'''_M + 3V^{\bullet\bullet}_O$$

In general for an M_aO_b oxide,

$$\text{Null (or perfect crystal)} \Rightarrow aV^{b-}_M + bV^{a+}_O$$

It is left as an exercise to the reader to ascertain that as written, these reactions satisfy the aforementioned rules.

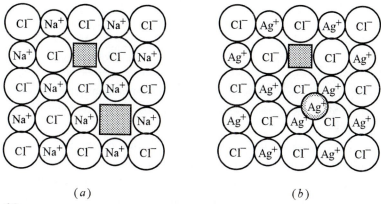

(a)　　　　　　　　　　　　(b)

Figure 6.3
(a) Schottky defect in NaCl; (b) Frenkel defect in AgCl.

---

[67] To see how that occurs, consider the formation of a defect pair in an MO oxide by the migration of a cation and an anion to the surface. In that case, one can write

$$O^x_O + M^x_M \Rightarrow O^x_{O,s} + M^x_{M,s} + V^{\bullet\bullet}_O + V''_M$$

where the subscript s refers to the surface sites. But since the ions that migrated to the surface covered ions previously located at the surface, this equation is usually abbreviated to Eq. (6.8).

Equation (6.7) was derived with the implicit assumption that only one type of vacancy forms. The thermodynamics of Schottky defect formation is slightly more complicated, however, because disorder can now occur on both sublattices. This is taken into account as follows: Assuming the number of ways of distributing the cation vacancies V_{cat} on $N_{cat} + V_{cat}$ sites is Ω_1, and the number of ways of distributing V_{an} anion vacancies on $N_{an} + V_{an}$ sites is Ω_2, it can be shown that the configuration entropy change upon the introduction of both defects is given by

$$\Delta S = k \ln \Omega = k \ln \Omega_1 \Omega_2$$

where

$$\Omega = \frac{(N_{cat} + V_{cat})! \; (N_{an} + V_{an})!}{(N_{cat})!(V_{cat})! \; (N_{an})!(V_{an})!}$$

where N_{cat} and N_{an} are the total numbers of cations and anions in the crystal, respectively. Following a derivation similar to the one shown for the elemental crystal and taking an MO oxide as an example, i.e., subject to the constraint that $(N_{cat} + n_{cat})/(N_{an} + n_{an}) = 1$, one can show that (see Prob. 6.1c) at *equilibrium*

$$\boxed{\frac{V_{an}^{eq} V_{cat}^{eq}}{\left(N_{an} + V_{an}^{eq}\right)\left(V_{cat}^{eq} + N_{cat}\right)} \approx \frac{V_{an}^{eq} V_{cat}^{eq}}{N_{an} N_{cat}} = \exp\left(-\frac{\Delta h_S - T \Delta s_S}{kT}\right)} \qquad (6.9)$$

where V_{cat}^{eq}, and V_{an}^{eq} are, respectively, the equilibrium numbers of cation and anion vacancies. And Δs_S and Δh_S are, respectively, the entropy and enthalpy associated with the formation of a Schottky pair, or $\Delta g_S = \Delta h_S - T \Delta s_S$.

This result predicts that the product of the cation and anion vacancy concentrations is a *constant* that depends on only temperature and holds true as long as equilibrium can be assumed.[68] In certain cases, discussed in greater detail below, when Schottky defects dominate, that is, $V_{an}^{eq} = V_{cat}^{eq} \gg$ the sum of all other defects, Eq. (6.9) simplifies to

$$[V_a] = [V_c] = \exp\frac{\Delta s_S}{2k} \exp\left(-\frac{\Delta h_S}{2kT}\right) \qquad (6.10)$$

where:

$$[V_c] = \frac{V_{cat}}{V_{cat} + N_{cat}} \quad \text{and} \quad [V_a] = \frac{V_{an}}{V_{an} + N_{an}} \qquad (6.11)$$

---

[68] A good analogy comes from chemistry, where it is known that for water at room temperature the product of the concentrations of H^+ and OH^- ions is a constant equal to 10^{14}, a result that is always valid. Increasing the proton concentration decreases the OH^- concentration, and vice versa.

Note that from here on, in equations in which defects are involved, *square brackets will be used exclusively to represent the mole or site fraction of defects.*

Frenkel defects. The Frenkel defect (Fig. 6.3*b*) is one in which a vacancy is created by having an ion in a regular lattice site migrate into an interstitial site. This defect can occur on either sublattice. For instance, the Frenkel reaction for a trivalent cation is

$$M_M^X \Rightarrow V_M''' + M_i^{\bullet\bullet\bullet} \tag{6.12}$$

while that on the oxygen sublattice is

$$O_O^X \Rightarrow O_i'' + V_O^{\bullet\bullet} \tag{6.13}$$

Note that the Frenkel reaction as written does not violate rule 3 since interstitial sites do not constitute regular lattice sites. FeO, NiO, CoO, and Cu_2O are examples of oxides that exhibit Frenkel defects.

Similar to the Schottky formulation, the number of ways of distributing n_i interstitials on N^* interstitial *sites* is

$$\Omega_1 = \frac{N^*!}{(N^* - n_{f,i})! n_i!}$$

Similarly, the number of configurations of distributing V_{cat} vacancies on N_T total sites is

$$\Omega_2 = \frac{N_T!}{(N_T - V_{cat})! V_{cat}!}$$

The configurational entropy is once again $\Delta S = k \ln \Omega_1 \Omega_2$. At equilibrium,

$$\boxed{\frac{V_{cat}^{eq} n_i^{eq}}{N_T N^*} \approx \exp\left(-\frac{\Delta g_F}{kT}\right)} \tag{6.14}$$

where Δg_F is the free-energy change associated with the formation of a Frenkel defect.

It is worth noting that N^* will depend on the crystal structure. For instance, for 1 mol of NaCl, if the ions migrate to tetrahedral sites, $N^* \approx 2 N_{Av}$.

WORKED EXAMPLE 6.1. Estimate the number of Frenkel defects in AgBr (NaCl structure) at 500°C. The enthalpy of formation of the defect is 110 kJ/mol, and the entropy of formation is 6.6*R*. The density and molecular weights are $6.5 g/cm^3$ and 187.8 g/mol, respectively. State all necessary assumptions.

Answer

By taking a basis of 1 mol, assuming that the Frenkel disorder occurs on the cation sublattice, and further assuming that the silver ions go into the tetrahedral sites (i.e., number of interstitial sites = double the number of lattice sites $\approx 2N_{Av}$), it follows that

$$\frac{V_{cat}^{eq} n_i^{eq}}{2(6.02 \times 10^{23})^2} = \exp\frac{6.6R}{R} \exp\left\{-\frac{110 \times 10^3}{8.314(500 + 273)}\right\} = 2.7 \times 10^{-5}$$

or

$$V_{cat}^{eq} n_i^{eq} = 1.957 \times 10^{43} \text{ defects}/\text{mol}^2$$

As long as the crystal is in equilibrium, this expression is always valid; i.e., the left-hand side of the equation will always be equal to 2.7×10^{-5}. Under certain conditions, discussed below, the Frenkel defects can dominate, in which case

$$V_{cat}^{eq} = n_i^{eq} = 4.43 \times 10^{21} \text{ defects}/\text{mol}$$

and the corresponding number of defects per cubic centimeter is $4.43 \times 10^{21} \times 6.5/187.7 = 1.5 \times 10^{20} \text{ defects}/\text{cm}^3$.

Antistructure disorder or misplaced atoms. These are sites where one type of atom is found at a site normally occupied by another. This defect does not occur in ionic ceramics, but it has been postulated to occur in covalent ceramics like SiC. The notation for such a defect would be Si_C or C_{Si}, and the corresponding defect reaction is

$$C_C + Si_{Si} \Rightarrow Si_C + C_{Si}$$

where the effective charge is assumed to be zero throughout.

Finally, note that for a stoichiometric reaction, all that is happening is the rearrangement of the atoms or ions comprising the crystal on a larger number of lattice sites, which consequently increases the configurational entropy of the crystal. In a stoichiometric reaction, the ratio of the atoms comprising the crystal does not change.

Nonstoichiometric defects

In nonstoichiometric defect reactions, the composition of the crystal changes as a result of the reaction. Said otherwise, a nonstoichiometric reaction is one in which mass is transferred across the boundaries of the crystal. The possible number of nonstoichiometric defect reactions is quite large, and covering even a fraction of them is not feasible here. The best that can be done is to touch on some of their more salient points.

One of the more common nonstoichiometric reactions that occurs at low oxygen partial pressures is shown in Fig. 6.4, where one of the components (oxygen in this case) leaves the crystal. The corresponding defect reaction is

$$O_O^x \Rightarrow \frac{1}{2}O_2(g) + V_O^x \qquad (6.15)$$

As the oxygen atom escapes, an oxygen vacancy is created. Given that the oxygen has to leave as a neutral species,[69] it has to leave two electrons (the ones that belonged to the cations in the first place!) behind (Fig. 6.4a). As long as these electrons remain localized at the vacant site, it is effectively neutral $\{-2 - (-2) = 0\}$. However, the electrons in this configuration are usually weakly bound to the defect site and are easily excited into the conduction band; i.e., V_O^x acts as a donor — see Chap. 7. The ionization reaction can be envisioned to occur in two stages:

$$V_O^x \Rightarrow V_O^\bullet + e'$$
$$V_O^\bullet \Rightarrow V_O^{\bullet\bullet} + e'$$

in which case the net reaction reads

$$O_O^x \Rightarrow \frac{1}{2}O_2(g) + V_O^{\bullet\bullet} + 2e' \qquad (6.16)$$

and the oxygen vacancy is said to be doubly ionized (Fig. 6.4c) and carries an effective charge of +2.

Another possible nonstoichiometric defect reaction is one in which oxygen is incorporated into the crystal interstitially, i.e.,

$$\frac{1}{2}O_2(g) \Rightarrow O_i^x \qquad (6.17)$$

Ionization can also occur in this case, creating holes in the valence band (i.e., the defect acts as an acceptor) such that

---

[69] The reason for this is quite simple: If charged entities were to escape, a charge would build up near the surface that would very rapidly prevent any further escape of ions. See the section on electrochemical potentials in Chap. 5.

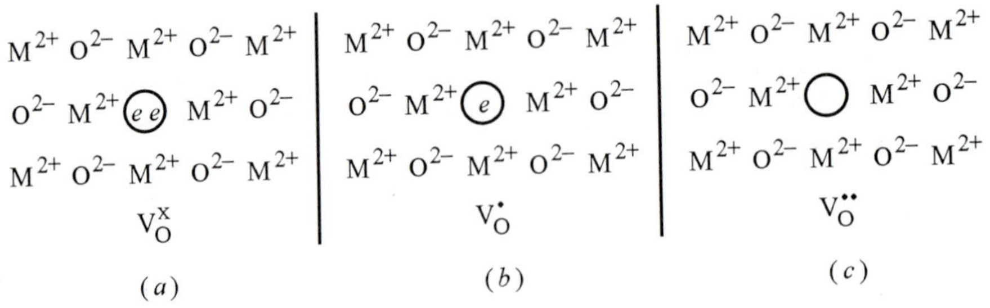

$$M^{2+} \ O^{2-} \ M^{2+} \ O^{2-} \ M^{2+}$$
$$O^{2-} \ M^{2+} \ \boxed{e \ e} \ M^{2+} \ O^{2-}$$
$$M^{2+} \ O^{2-} \ M^{2+} \ O^{2-} \ M^{2+}$$
$$V_O^X$$

(a)

$$M^{2+} \ O^{2-} \ M^{2+} \ O^{2-} \ M^{2+}$$
$$O^{2-} \ M^{2+} \ \boxed{e} \ M^{2+} \ O^{2-}$$
$$M^{2+} \ O^{2-} \ M^{2+} \ O^{2-} \ M^{2+}$$
$$V_O^{\bullet}$$

(b)

$$M^{2+} \ O^{2-} \ M^{2+} \ O^{2-} \ M^{2+}$$
$$O^{2-} \ M^{2+} \ \bigcirc \ M^{2+} \ O^{2-}$$
$$M^{2+} \ O^{2-} \ M^{2+} \ O^{2-} \ M^{2+}$$
$$V_O^{\bullet\bullet}$$

(c)

Figure 6.4

(a) The formation of an oxygen vacancy by the loss of an oxygen atom to the gas phase. This is a nonstoichiometric reaction because the crystal chemistry changes as a result. Note that as drawn, the electrons are localized at the vacancy site, rendering its effective charge zero. (b) A $V_O^{\bullet}$ site is formed when one of these electrons is excited into the conduction band. (c) The escape of the second electron creates a $V_O^{\bullet\bullet}$ site.

$$O_i^X \Rightarrow O_i' + h^{\bullet}$$
$$O_i' \Rightarrow O_i'' + h^{\bullet}$$

with the net reaction being $\quad \dfrac{1}{2}O_2(g) \Rightarrow O_i'' + 2h^{\bullet}$ \hfill (6.18)

Nonstoichiometric defect reactions, with the selective addition or removal of one of the constituents, naturally lead to the formation of nonstoichiometric compounds. The type of defect reaction that occurs will determine whether an oxide is oxygen- or metal-deficient. For example, reaction (6.16) will result in an oxygen-deficient oxide,[70] whereas reaction (6.18) will result in an oxygen-rich oxide.

When one assumes that the electrons or holes generated as a result of **redox reactions**, such as Eqs. (6.16) or (6.17), end up delocalized (i.e., in the conduction or valence bands, see Chap. 7), the implicit assumption is that the cations were only stable in one oxidation state (e.g., Al or Mg). For oxides in which the cations can exist in more than one oxidation state, such as the transition metal ions, an alternate possibility exists.

As long as the energy associated with changing the oxidation state of the cations is not too large, the electronic defects can — instead of being promoted to the conduction band — change the oxidation state of the cations. To illustrate,

---

70 Note that oxygen deficiency is also equivalent to the presence of excess metal. One possible such reaction is

$$M_M + O_O \Rightarrow M_i^X + \dfrac{1}{2}O_2(g)$$

consider magnetite, Fe_3O_4, which has a spinel structure with two-thirds of the Fe ions in the +3 state and one-third in the +2 state. One can express the oxidation of Fe_3O_4 in two steps as follows:

$$\frac{1}{2}O_2(g) \Leftrightarrow O_O^x + V_{Fe}'' + 2h^\bullet$$

$$2\,Fe^{2+} + 2h^\bullet \Rightarrow 2\,Fe^{3+}$$

for a net reaction of

$$\frac{1}{2}O_2(g) + 2\,Fe^{2+} \Rightarrow 2\,Fe^{3+} + O_O^x + V_{Fe}''$$

In other words, the holes that are created as a result of the oxidation are used to change the valence state of the cations from +2 to +3.[71]

Extrinsic defects

The discussion so far has applied to pure crystals. Most crystals are not pure, however, and their properties, especially electrical and optical, are often dominated by the presence of trace amounts of impurities (see Worked Example 6.3). These impurities cannot be avoided; and even if the starting raw materials are exceptionally pure, it is difficult to maintain the purity levels during subsequent high-temperature processing. The next task is thus to consider impurity incorporation reactions — once again, a task that very rapidly gets out of hand, what with literally thousands of compounds and reactions. What is attempted here instead is to present some simple guidelines for addressing the issue.

First and foremost, impurities usually substitute for the host ion of electronegativity nearest their own, even if the sizes of the ions differ. In other words, cations substitute for cations and anions for anions, irrespective of size differences.[72] For example, in NaCl, Ca and O would be expected to occupy the cation and anion sites, respectively. In more covalent compounds where the electronegativities may be similar, size may play a more important role. Whether an impurity will occupy an interstitial site is more difficult to predict. Most interstitial atoms are small, but even large atoms are sometimes found in interstitial sites.

---

[71] Magnetite can be considered a solid solution of FeO and Fe_2O_3. Thus upon oxidation, it makes sense that the average oxidation state should move toward Fe_2O_3, that is, more Fe^{3+} should be present.

[72] This topic is addressed again in Chap. 8, when solid solutions and phase diagrams are considered.

In writing a defect incorporation reaction, the following simple bookkeeping operation can be of help:

1. Sketch a unit or multiple units of the host (solvent) crystal, as shown in Fig. 6.5*a*.
2. Place a unit or multiple units of the dopant (solute) crystal on top of the sketch drawn in step 1, such that the cations are placed on top of the cations and the anions on top of the anions . *It is important to note that the locations of the ions so placed are not where they end up in the crystal. This is simply a bookkeeping operation.*
3. Whatever is left over is the defect that arises, with the caveat that one should try to *minimize* the number of defects formed.

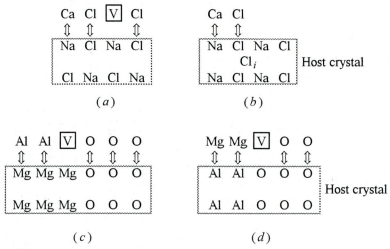

(*a*)   (*b*)

(*c*)   (*d*)

Figure 6.5

Bookkeeping technique for impurity incorporation reactions. (*a*) $CaCl_2$ in NaCl leaves a vacancy on cation sublattice. (*b*) An alternate reaction is for the extra Cl ion to go interstitial. This reaction is unlikely, however, given the large size of the Cl ion. (*c*) Al_2O_3 in MgO creates a vacancy on the cation sublattice. (*d*) MgO in Al_2O_3 creates a vacancy on the anion sublattice.

To illustrate, consider the following examples

EXAMPLE 1.   Incorporate $CaCl_2$ into NaCl. From Fig. 6.5a, it is immediately obvious that one possible incorporation reaction is

$$CaCl_2 \underset{2\,NaCl}{\Rightarrow} Ca^{\bullet}_{Na} + V'_{Na} + 2Cl^{x}_{Cl}$$

A second perfectly legitimate incorporation reaction is shown in Fig. 6.5b, for which the corresponding defect reaction is

$$CaCl_2 \underset{NaCl}{\Rightarrow} Ca^{\bullet}_{Na} + Cl'_{i} + Cl^{x}_{Cl}$$

Note that in both cases the overriding concern was the preservation of the regular site ratios of the host crystal. In the first case, two Cl lattice sites were created by the introduction of the dopant, and hence the same number of lattice sites had to be created on the cation sublattice. But since only one Ca cation was available, a vacancy on the Na sublattice had to be created. In the second case (Fig. 6.5b), there is no need to create vacancies because the number of lattice sites created does not change the *regular* site *ratios* of the host crystal (interstitial sites are not considered regular sites).

EXAMPLE 2. Doping MgO with Al_2O_3 (Fig. 6.5c):

$$Al_2O_3 \underset{3\,MgO}{\Rightarrow} 2Al^{\bullet}_{Mg} + V''_{Mg} + 3O^{x}_{O}$$

EXAMPLE 3.   Doping Al_2O_3 with MgO (Fig. 6.5d), one possible incorporation reaction is

$$2MgO \underset{Al_2O_3}{\Rightarrow} 2Mg'_{Al} + V^{\bullet\bullet}_{O} + 2O^{x}_{O}$$

It should be emphasized at this point that it is difficult to determine a priori what the actual incorporation reaction would be. For the most part, that is determined from experiments such as density measurements (see Probs. 6.8 and 6.9).

Oxides with multiple substitution of ions. In some oxides, the structure is such as to be able to simultaneously accommodate various types of cations. These multiple substitutions are allowed as long as charge neutrality is maintained. The preservation of site ratios is no longer an issue because the distinction blurs between a regular lattice site and a regular lattice site that is vacant. Good examples are clays, spinels (Fig. 3.10), and the β-alumina structure (Fig. 7.9).

Consider the clay structure shown in Fig. 3.13b. The substitution of divalent cations for the trivalent Al ions between the sheets occurs readily as long as for every Al^{3+} substituted, the additional incorporation of a singly charged cation, usually an alkali-metal ion from the surrounding, occurs to maintain charge neutrality such that at any time the reaction

$$Al_2(OH)_4(Si_2O_5) \Rightarrow (Al_{2-x}Na_xMg_x)(OH)_4(Si_2O_5)$$

holds.

The chemistry of spinels is also similar in that multiple substitutions are possible as long as the crystal remains neutral. For instance, the unit cell of normal spinel, $Mg_8Al_{16}O_{32}$, can be converted to an inverse spinel by substituting the 8 Mg ions by four Li and four Al ions to give $Li_4Al_{20}O_{32}$, where the Li ions now reside on the octahedral sites and the Al ions are distributed on the remaining octahedral and tetrahedral sites. It is worth noting here that the vast number of possible structural and chemical combinations in spinels and the corresponding changes in their magnetic, electric, and dielectric properties have rendered them indispensable for the electronics industry. In essence, spinels can be considered to be cation "garbage cans," and within reasonable size constraints, any combination of cations is possible as long as, at the end, the crystal remains neutral. In that respect, spinels can be compared to another "universal" solvent, namely, glasses (see Chap. 9).

6.2.4 Electronic Defects

In a perfect semiconductor or insulating crystal at 0 K, all the electrons are localized and are firmly in the grasp of the nuclei, and free electrons and holes do not exist. At finite temperatures, however, some of these electrons get knocked loose into the conduction band as a result of lattice vibrations. As elaborated on in Chap. 7, for an *intrinsic* semiconductor the liberation of an electron also results in the formation of an electron hole such that the intrinsic electronic defect reaction can be written as

$$\text{Null} \Leftrightarrow e' + h^\bullet \tag{6.19}$$

Given that the energy required to excite an electron from the valence to the conduction band is the band gap energy E_g (see Chap. 2), by a derivation similar to the one used to arrive at Eq. (6.14), it can be easily shown that

$$\frac{np}{N_vN_c} = \exp\left(-\frac{E_g}{kT}\right) = K_i \qquad (6.20)$$

where n and p are, respectively, the numbers of free electrons and holes per unit volume; N_c and N_v are the density of states per unit volume in the conduction and valence bands, respectively. It can be shown (App. 7A) that for an intrinsic semiconductor, N_c and N_v are given by

$$N_c = 2\left(\frac{2\pi m_e^* kT}{h^2}\right)^{3/2} \text{ and } N_v = 2\left(\frac{2\pi m_h^* kT}{h^2}\right)^{3/2} \qquad (6.21)$$

where m_e^* and m_h^* are the effective masses of the electrons and holes, respectively, h is Planck's constant, and all other terms have their usual meanings.

It is worth noting here that the mathematical treatment for the formation of a Frenkel defect pair is almost identical to that of an electron-hole pair. A Frenkel defect forms when an ion migrates to an interstitial site, leaving a hole or a vacancy behind. Similarly, an electron-hole pair forms when the electron escapes into the conduction band, leaving an electron hole or vacancy in the valence band. Conceptually, N_c and N_v (in complete analogy to N^* and N_T) can be considered to be the number of energy levels or "sites" over which the electrons and holes can be distributed. The multiplicity of configurations over which the electronic defects can populate these levels is the source of the configurational entropy necessary to lower the free energy of the system.

6.2.5 Defect Equilibria and Kroger-Vink Diagrams

One of the aims of this chapter is to relate the concentration of defects to temperature and other externally imposed thermodynamic parameters such as oxygen partial pressure, a goal that is now almost at hand. This is accomplished by considering defects to be structural elements which possess a chemical potential[73] and hence activity and expressing their equilibrium concentrations by a mass action expression similar to Eq. (5.30):

$$\frac{x_C^d x_D^c}{x_A^a x_B^b} = \exp\left(-\frac{\Delta G^\circ}{kT}\right) = K^{eq} \qquad (6.22)$$

---

[73] A distinction has to be made here between chemical and virtual potentials. As discussed at some length in Chap. 5, since the activity or chemical potential of an individual ion or charged defect cannot be defined, it follows that its chemical potential is also undefined. The distinction, however, is purely academic, because defect reactions are always written so as to preserve site ratios and electroneutrality, in which case it is legitimate to discuss their chemical potentials.

This expression is almost identical to Eq. (5.30), except that here ideality has been assumed and the activities have been replaced by the mole fractions x_i.

To illustrate, consider an MO oxide subjected to the following oxygen partial pressure regimes:

Low oxygen partial pressure

At very low oxygen partial pressures, it is plausible to assume that oxygen vacancies will form according to reaction (6.16), or

$$O_O^x \Leftrightarrow V_O^{\bullet\bullet} + 2e' + \frac{1}{2}O_2(g) \quad \Delta g_{red} \tag{I}$$

The corresponding mass action expression is:[74]

$$\frac{[V_O^{\bullet\bullet}]n^2 P_{O_2}^{1/2}}{[O_O^x]} = K_{red} \tag{6.23}$$

where $K_{red} = \exp(-\Delta g_{red}/kT)$. Note that as long as $V_{an} \ll N_{an}$, $[O_O^x] = N_{an}/(N_{an} + V_{an}) \approx 1$

Intermediate oxygen partial pressure

Here if it is assumed that Schottky equilibrium dominates, i.e.

$$M_M^x + O_O^x \Leftrightarrow V_M'' + V_O^{\bullet\bullet} \quad \Delta g_S \tag{II}$$

Applying the mass action law yields:

$$\frac{[V_M''][V_O^{\bullet\bullet}]}{[M_M^x][O_O^x]} = K_S = \exp\left(-\frac{\Delta g_S}{kT}\right) \tag{6.24}$$

which, not surprisingly, is identical to Eq. (6.9), since $[O_O^x] \approx [M_M^x] \approx 1$.

---

[74] In keeping with the notation scheme outlined above, the following applies for electronic defects:

$$[n] = \frac{n}{N_c} \text{ and } [p] = \frac{p}{N_v}$$

It is worth emphasizing once more that both $[n]$ and $[p]$ are dimensionless, whereas p and n have the units of $1/m^3$. The advantage of using site fractions, instead of actual concentrations, in the mass action expression is that the left-hand side of the mass action expression [for example, Eq. (6.23)] would be dimensionless and thus equal to $\exp\{-\Delta g/kT\}$. If other units are used for concentration, the K values have to change accordingly (see Worked Example 6.2).

High oxygen partial pressure

In this region, a possible defect reaction is:[75]

$$\frac{1}{2}O_2(g) \Leftrightarrow O_O^x + 2h^\bullet + V_M'' \qquad \Delta g_{oxid} \qquad \text{(III)}$$

for which

$$\frac{[O_O^x][V_M''][p]^2}{P_{O_2}^{1/2}} = K_{oxid} \qquad (6.25)$$

where $K_{oxid} = \exp\{-\Delta g_{oxid}/(kT)\}$. Note here that increasing the oxygen partial pressure increases the cation vacancies.[76]

In addition to Eqs. (6.23) to (6.25), the following reaction

$$\text{Null} \Leftrightarrow e' + h^\bullet \qquad \text{(IV)}$$

is relevant, and at equilibrium

$$[n][p] = K_i = \exp\left(\frac{-E_g}{kT}\right) \qquad (6.26)$$

which, not surprisingly, is identical to Eq. (6.20).

At equilibrium, the concentrations of the various defects have to *simultaneously* satisfy Eqs. (6.23) to (6.26) together with one further condition, namely, that the crystal as a whole remain electrically neutral or

$$\sum \text{positive charges } (m^{-3}) = \sum \text{negative charges } (m^{-3})$$

Note that in writing the neutrality condition, it is the *number of defects per unit volume that is important rather than their mole fractions.* For the example

---

chosen, if one assumes that the only defects present in any appreciable quantities are $h^\bullet$, e', $V_O^{\bullet\bullet}$, and V_M''; the neutrality condition reads

$$p + 2V_O^{\bullet\bullet} = 2V_M'' + n \tag{6.27}$$

At this point, all the necessary information needed to relate the concentrations of the various defects to the oxygen potential or partial pressure surrounding the crystal is available. In Eqs. (6.23) to (6.27), there are four unknowns [n, p, $V_O^{\bullet\bullet}$, V_M''] and five equations. Thus in principle, these equations can be solved simultaneously, provided, of course, that all the Δg's for the various reactions are known. Whereas this is not necessarily a trivial exercise, fortunately the problem can be greatly simplified by appreciating that under various oxygen partial pressure regimes, one defect pair will dominate at the expense of all other pairs and only two terms remain in the neutrality condition. How this **Brouwer approximation** is used to solve the problem is illustrated now:

At sufficiently low P_{O_2}, the driving force to lose oxygen to the atmosphere is quite high [i.e., reaction (I) is shifted to the right], and consequently the number of oxygen vacancies in the crystal increases. If the oxygen vacancies are doubly ionized, it follows that for every oxygen that leaves the crystal, two electrons are left behind in the conduction band (see Fig. 6.4c). In this case it is not unreasonable to assume that at sufficiently low oxygen partial pressures

$$n \approx 2V_O^{\bullet\bullet} >>> \Sigma(\text{all other defects}) \tag{6.28}$$

Combining Eq. (6.28) with (6.23) and solving for n or $V_O^{\bullet\bullet}$ yields

$$n = 2V_O^{\bullet\bullet} = [2K_{red}']^{1/3} P_{O_2}^{-1/6} = [2K_{red}N_{an}N_c^2]^{1/3} P_{O_2}^{-1/6} \tag{6.29}$$

where $K_{red}' \approx K_{red}N_{an}N_c^2$. According to this relationship, a plot of $\log n$ (or $V_O^{\bullet\bullet}$) versus $\log P_{O_2}$ should yield a straight line with a slope of $-1/6$ (Fig. 6.6a, range I), i.e., both n and $V_O^{\bullet\bullet}$ decrease with increasing P_{O_2}. The physical picture is simple: Upon reduction, the oxygen ions are being "pulled" out of the crystal, leaving electrons and oxygen vacancies behind.

By similar arguments, in the high-P_{O_2} regime, the electroneutrality condition can be assumed to be $p \approx 2V_M''$, which, when combined with Eq. (6.25), results in

$$p \approx 2V_M'' = [2K_{oxid}']^{1/3} P_{O_2}^{1/6} = [2K_{oxid}N_{cat}N_v^2]^{1/3} P_{O_2}^{1/6} \tag{6.30}$$

where $K_{oxid}' \approx K_{oxid}N_{cat}N_v^2$. In this region, a plot of the defect concentration versus $\log P_{O_2}$ yields a straight line with a *positive* slope of $1/6$ (Fig. 6.6a, range III).

In the intermediate P_{O_2} regime, two possibilities exist:

1. $K_s \gg K_i$, in which case the neutrality condition becomes

$$V_O^{\bullet\bullet} = V_M'' = \sqrt{K_s'} \qquad (6.31)$$

where $K_s' = N_{cat} N_{an} K_s$ and the point defect concentrations become independent of P_{O_2} (Fig. 6.6a, range II).

By combining the three regimes, the functional dependence of the defect concentrations over a wide range of oxygen partial pressures can be succinctly graphed in what is known as a **Kroger-Vink diagram**, shown in Fig. 6.6a.

2. $K_i \gg K_s$, in which case the neutrality condition reads

$$n = p = \sqrt{K_i'}$$

where $K_i' = N_c N_v K_i$. It is left as an exercise to the reader to show that the corresponding Kroger-Vink diagram is the one shown in Fig. 6.6b.

Up to this point, the focus has been on the effect of the oxygen partial pressure on the **majority defects**, that is, $V_O^{\bullet\bullet}$ and n under reducing conditions, V_M'' and p under oxidizing conditions, and so forth. What about the electron holes and the metal vacancies in that region, the so-called **minority defects**? To answer this question, it is important to appreciate that at equilibrium Eqs. (6.23) to (6.26) have to be satisfied at all times. For example, equilibrium dictates that at all times and under all circumstances the product $[V_O^{\bullet\bullet}][V_m'']$ has to be remain a constant equal to $\sqrt{K_s}$. And since it was just established that in the low oxygen pressure region [Eq. (6.29)]:

$$V_O^{\bullet\bullet} = (\text{const.})\left[P_{O_2}\right]^{-1/6}$$

it follows that for Schottky equilibrium to be satisfied, V_M'' has to increase by the same power law, or

$$V_M'' = (\text{const.})\left[P_{O_2}\right]^{1/6}$$

Similarly, since $n = (\text{const.})[P_{O_2}]^{1/6}$, it follows that to satisfy Eq. (6.26), $p = (\text{const.})[P_{O_2}]^{1/6}$. The behavior of the minority defects in region I is plotted in Fig. 6.6.a and b (lower lines).

In the intermediate region when $K_s \gg K_i$, Eq. (6.31) holds and $V_O^{\bullet\bullet} = \sqrt{K_s'}$ and is thus independent of the oxygen partial pressure. Substituting this result in Eq. (6.29) yields

$$n = \left[\frac{K'_{red} P_{O_2}^{-1/2}}{V_O^{\bullet\bullet}} \right]^{1/2} = \sqrt{\frac{K'_{red}}{\sqrt{K'_s}}} P_{O_2}^{-1/4} \qquad (6.32)$$

In other words, n is decreasing with 1/4 power with increasing P_{O_2}, which implies that p is increasing with the same power in that range.

Diagrams such as Fig. 6.6 are very useful when trying to understand the relationship between the externally imposed thermodynamic parameters, such as the partial pressure of one of the components of a crystal, and any property that is related to the crystal's defect concentration. For instance, as described in detail in Chap. 7, the diffusivity of oxygen is proportional to its vacancy concentration. Thus it follows that if the Kroger-Vink diagram of that given oxide is the one shown in Fig. 6.6a, then the diffusivity of oxygen will be highest at the extreme left, i.e., under reducing conditions, and will decrease with a slope of $-1/6$ with increasing P_{O_2}, will become constant over an intermediate P_{O_2} regime, and finally will start to drop again with a slope of $-1/6$ at higher P_{O_2} values. Similarly, if the oxide is an electronic conductor and its conductivity is measured over a wide enough oxygen partial pressure range, the conductivity is expected to change from n type at low oxygen partial pressures to p type at higher oxygen potentials.

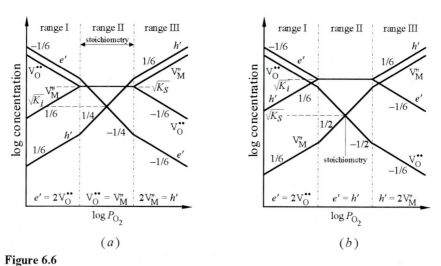

Figure 6.6
Variation in defect concentration in an MO oxide as a function of oxygen partial pressure for: (a) $K_s \gg K_i$ and (b) $K_i \gg K_s$. Note that in (a) the oxide is stoichiometric over a large range of oxygen partial pressures, but that it reduces to a point where $V_M'' = V_O^{\bullet\bullet}$ in the case where $K_i \gg K_s$.

WORKED EXAMPLE 6.2. The following information for NaCl is given: At 600 K:

$$K'_S = 3.74 \times 10^{35}\,\text{cm}^{-6} \quad \text{and} \quad K'_F \text{ (on cation sublattice)} = 5.8 \times 10^{34}\,\text{cm}^{-6}$$

At 800 K:

$$K'_S = 7.06 \times 10^{37}\,\text{cm}^{-6} \quad \text{and} \quad K'_F = 1.7 \times 10^{37}\,\text{cm}^{-6}$$

Calculate the equilibrium number of defects at 600 K and 800 K.

Answer

The three pertinent equations are

$$\left(V'_{Na}\right)\left(V^{\bullet}_{Cl}\right) = K'_S$$

$$\left(V'_{Na}\right)\left(Na^{\bullet}_i\right) = K'_F$$

$$V'_{Na} = Na^{\bullet}_i + V^{\bullet}_{Cl} \qquad \text{electroneutrality condition}$$

Note that here the concentrations are given in cm^{-3} instead of mole fractions because the K values are given in cm^{-6}. Combining the first two equations, one obtains

$$V'_{Na}\left(Na^{\bullet}_i + V^{\bullet}_{Cl}\right) = K'_S + K'_F$$

which when combined with the electroneutrality condition, yields

$$(V'_{Na})^2 = K_S + K_F$$

Solving for the various concentrations at 600 K one obtains

$$V'_{Na} = 6.6 \times 10^{17}\,\text{cm}^{-3} \quad Na^{\bullet}_i = 8.8 \times 10^{16}\,\text{cm}^{-3} \quad V^{\bullet}_{Cl} = 5.7 \times 10^{17}\,\text{cm}^{-3}$$

whereas at 800 K

$$V'_{Na} = 9.4 \times 10^{18}\,\text{cm}^{-3} \quad Na^{\bullet}_i = 1.8 \times 10^{18}\,\text{cm}^{-3} \quad V^{\bullet}_{Cl} = 7.5 \times 10^{18}\,\text{cm}^{-3}$$

6.2.6 Stoichiometric Versus Nonstoichiometric Compounds

Based on the foregoing analysis, stoichiometry (defined as the point at which the numbers of anions and cations equal a simple ratio based on the chemistry of the crystal) is a **singular** point that occurs at a very specific oxygen partial pressure. This immediately begs the question: If stoichiometry is a singular point in a partial pressure domain, then why are some oxides labeled stoichiometric and others nonstoichiometric? To answer the question, examine Table 6.1 in which a range of stoichiometries and chemical stability domains for a number of oxides are listed. The deviation from stoichiometry, defined by Δx, where Δx is the difference between the maximum and minimum values of b/a in an $MO_{b/a}$ oxide, varies from oxide to oxide. Note that FeO and MnO exhibit only positive deviations from stoichiometry, i.e., they are always oxygen-rich, whereas TiO exhibits both negative and positive deviations.

From Table 6.1, one can conclude that an oxide is labeled stoichiometric if Δx is a weak function of oxygen partial pressure. Conversely, an oxide is considered nonstoichiometric if the effect of oxygen partial pressure on the composition is significant. This concept can be better appreciated graphically as shown in Fig. 6.7.

TABLE 6.1

Range of stoichiometry and existence domains of a number of binary oxides at 1000 K[77]

| Oxides | Deviation from stoichiometry | | | Stablity or existance region[‡] $-\log P_{O_2}$ | |
	x_{min}	x_{max}	Δx [†]	Min	Max
		Nonstoichiometric oxides			
TiO	0.8	1.3	0.5	44.2[§]	41.5
Ti_2O_3	1.501	1.512	0.011	41.5	30.1
TiO_2	1.992	2.00	0.008	25.7	—
VO	0.8	1.3	0.5	35.9	33.2
MnO	1.00	1.18	0.18	34.5[§]	10.7
FeO	1.045	1.2	0.155	21.6[§]	17.9
Fe_3O_4	1.336	1.381	0.045	17.9	10.9
CoO	1.00	1.012	0.012	17.1[§]	2.5
NiO	1.00	1.001	0.001	16.5[§]	—
Cu_2O	0.500	0.5016	0.0016	9.97[§]	7.0
		Stoichiometric oxides			
Al_2O_3	1.5000	1.5000		71.3[§]	—
MgO	1.00000	1.0000		51.5[§]	—

†   For an M_aO_b oxide, $x = b/a - \delta$.

‡   See Sec. 5.4 for more details.

§   In equilibrium with the parent metal.

77   T. B. Reed, *The Chemistry of Extended Defects in Non-Metallic Solids*, L. Eyring and M. O'Keeffe, eds., North-Holland, Amsterdam, 1970.

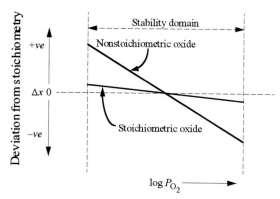

Figure 6.7

Distinction between a stoichiometric and a nonstoichiometric $MO_{b/a\pm\delta}$ oxide, where the functional dependence of the changes in stoichiometry on the oxygen partial pressure for two hypothetical compounds having the same range of chemical stability is compared. From the foregoing discussion, it follows that the oxide for which Δx varies widely over the stability domain will be labeled nonstoichiometric, and vice versa.

As a typical example of a nonstoichiometric compound, consider the variations in composition in MnO as the oxygen partial pressure is varied at 1000 K (Fig. 6.8a). MnO is stable between a P_{O_2} of $10^{-34.5}$ atm (below which Mn is the stable phase and the $O/Mn = 1.0$) and $10^{-10.7}$ atm (above which Mn_3O_4 is the stable phase and $O/M = 1.18$). The range of stoichiometry is depicted by the dotted lines normal to the x axis. Such a variation is quite large and consequently MnO is considered a nonstoichiometric oxide.

Similarly, the phase diagram of the Fe-O system is shown in Fig. 6.8b and c. Note that while FeO and Fe_3O_4 are nonstoichiometric, Fe_2O_3 is stoichiometric.

It is worth noting that as a class, the transition metal oxides are more likely to be nonstoichiometric than stoichiometric. The reason is simple: The loss of oxygen to the environment and the corresponding adjustments in the crystal are much easier when the cations can readily change their oxidation states.

WORKED EXAMPLE 6.3. (*a*) Given the expense and difficulty of obtaining powders that contain much less than 10 ppm of aliovalent impurities, estimate the formation enthalpy for intrinsic defect formation above which one would expect the properties of a material to be dominated by the impurities. State all assumptions. (*b*) Repeat part (*a*) for MnO in equilibrium with Mn_3O_4 at 1000 K.

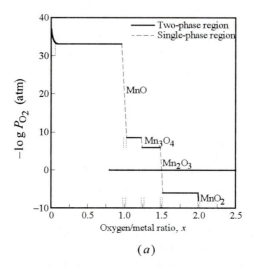

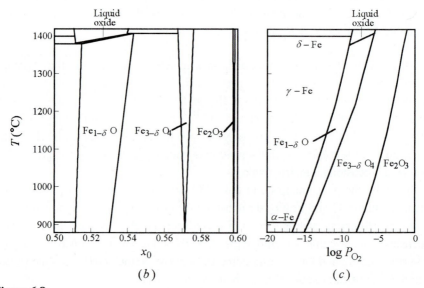

Figure 6.8

(*a*) Stability domains of various phases in the Mn-O system and the corresponding deviations in stoichiometry.[78] (*b*) Phase diagram of Fe-O system and (*c*) stability domains of the various phases in Fe-O system.[79]

78 T. B. Reed, *The Chemistry of Extended Defects in Non-Metallic Solids*, L. Eyring and M. O'Keeffe, eds., North-Holland, Amsterdam, 1970.

79 R. Dieckmann, *J. Electrochem. Soc.*, **116**, 1409 (1969).

Answer

Make the following assumptions:

- A fraction (1 ppm) of the impurities create vacancies on one of the sublattices in rough proportion to their concentration. In other words, it is assumed that the mole fraction of extrinsic vacancies on one of the sublattices is on the order of 10^{-6}.
- The stoichiometric defects are Schottky defects for which the formation enthalpy is Δh_S.
- Ignore Δs_S.
- Temperature is 1000°C.

For the solid to be dominated by the intrinsic defects their mole fraction has to exceed the mole fraction of defects created by the impurities (i.e., 10^{-6}). Thus

$$\left[V_M''\right]\left[V_O^{\bullet\bullet}\right] \approx 10^{-12} = \exp\left(-\frac{96,500\Delta h_S}{1273 \times 8.314}\right)$$

Solving for Δh_S, one obtains ≈ 3 eV.

This is an important result because it predicts that the defect concentrations in stoichiometric oxides or compounds for which the Schottky or Frenkel defect formation energies are much greater than 3 eV will most likely be dominated by impurities.

(*b*) According to Table 6.1, at 1000 K, MnO in equilibrium with Mn_3O_4 has the composition $MnO_{1.18}$. It follows that for this oxide to be dominated by impurities, the dopant would have to generate a mole fraction of vacancies in excess of 0.18!

EXPERIMENTAL DETAILS: MEASURING NONSTOICHIOMETRY

Probably the easiest and fastest method to find out whether an oxide is stoichiometric is to carry out thermogravimetric measurements as a function of temperature and oxygen partial pressure. In such experiments, a crystal is suspended from a sensitive balance into a furnace. The furnace is then heated, and the sample is allowed to equilibrate in a gas of known partial pressure. Once equilibrium is established (weight change is zero), the oxygen partial pressure in the furnace is changed suddenly, and the corresponding weight changes are recorded as a function of time.

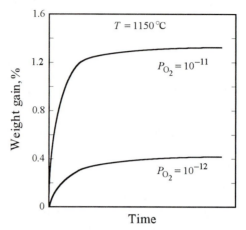

Figure 6.9
Typical thermogravimetric results for the oxidation of FeO_x.

Typical curves are shown in Fig. 6.9. From the weight gain the new stoichiometry can be easily calculated (see Prob. 6.10). If the same experiment were repeated on, say, MgO or Al_2O_3, the weight changes, over a wide range of oxygen partial pressures, would be below the detectability limit of the most sensitive balances, which is why they are considered stoichiometric oxides.

At this stage it is not a bad idea to think of a nonstoichiometric crystal as some kind of oxygen "sponge" that responds to the oxygen partial pressure in the same way as a sponge responds to water. How the oxygen is incorporated and diffuses into the crystal is discussed in the next chapter.

6.2.7 Energetics of Point Defects Formation

Clearly, a knowledge of the free-energy changes associated with the various defect reactions described above is needed to be able to calculate their equilibrium concentrations. Unfortunately, much of that information is lacking for most oxides and compounds.

As noted at the outset of this chapter, the creation of a vacancy can be visualized by removing an ion from the bulk of the solid to infinity, which costs $\approx E_{bond}$, and bringing it back to the surface of the crystal and recovering $E_{bond}/2$. Similarly, the formation of a Schottky vacancy *pair* of a binary MX compound costs $\approx 2E_{bond}/2 \cong E_{bond}$. In general, the lattice energies for the alkali halides fall in the range of 650 to 850 kJ/mol, and hence one would expect the enthalpy for the formation of a Schottky defect to be of the same order. Experimentally, however,

TABLE 6.2
Defect formation and migration energies for various halides

Crystal	Defect type	Δh_{form}, kJ/mol	Δs_{form}, in units of R	ΔH_{mig}, kJ/mol	ΔS_{mig}, in units of R
AgCl	Frenkel	140	9.4R	28 (V'_{Ag}) 1 – 10 $(Ag^{\bullet}_i)$	–1.0 (V'_{Ag}) –3.0 $(Ag^{\bullet}_i)$
AgBr	Frenkel	110	6.6R	30 (V'_{Ag}) 5 – 20 $(Ag^{\bullet}_i)$	
BaF$_2$	Frenkel	190		40 – 70 $(V^{\bullet}_F)$ 60 – 80 (F'_i)	
CaF$_2$	Frenkel	270	5.5R	40 – 70 $(V^{\bullet}_F)$ 80 – 100 (F'_i)	1 – 2 $(V^{\bullet}_F)$ 5 (F'_i)
CsCl	Schottky	180	10.0R	60 (V'_{Cs})	
KCl	Schottky	250	9.0R	70 (V'_K)	2.4 (V'_K)
LiBr	Schottky	180		40 (V'_{Li})	
LiCl	Schottky	210		40 (V'_{Li})	
LiF	Schottky	230	9.6R	70 (V'_{Li})	1 (V'_{Li})
LiI	Schottky	110		40 (V'_{Li})	
NaCl	Schottky	240	10.0R	70 (V'_{Na})	1 – 3 (V'_{Na})
SrF$_2$	Frenkel	170		50 – 100 $(V^{\bullet}_F)$	

Source: J. Maier, *Angewandte Chemie*, **32**: 313–335 (1993).

the enthalpies of formation of Schottky and Frenkel defects in alkali halides fall in the range of 100 to 250 kJ/mol (Table 6.2). The discrepancy arises from neglecting (1) the long-range polarization of the lattice as a result of the formation of a charged defect and (2) the relaxation of the ions surrounding the defect. When these effects are taken into account, better agreement between theory and experiment is usually obtained.

6.3
LINEAR DEFECTS

Dislocations were originally postulated to account for the large discrepancy between the theoretical and actual strengths observed during the plastic deformation in metals. For plastic deformation to occur, some part of the crystal has to move or shear with respect to another part. If whole planes had to move simultaneously, i.e., all the bonds in that plane had to break and move at the same time, then plastic deformation would require stresses on the order of $Y/10$ as estimated in Chap. 4. Instead, it is a well-established fact that metals deform at much lower stresses. The defect that is responsible for the ease of plastic deformation is known as a *dislocation.* There are essentially two types of dislocations: edge (shown in Fig. 6.10*a*) and screw (not shown). Every dislocation is characterized by the **Burgers vector**, **b**, which is defined as the unit slip distance for a dislocation shown in Fig. 6.10*a*. For an edge dislocation, the Burgers vector is always perpendicular to the dislocation line. For a screw dislocation it is parallel.

It is worth noting here that dislocations are thermodynamically unstable, since the entropy associated with their formation does not make up for their excess strain energy. They must thus form during solidification from the melt or as a result of thermal or mechanical stresses.

In ionic solids, the structure of dislocations can be quite complex because of the need to maintain charge neutrality. For example, for an edge dislocation to form in an NaCl crystal, it is not possible to simply insert one row of ions as one would do in a metallic crystal. Here two half-planes have to be inserted, as shown in Fig. 6.10*b*. The plane shown here is the (010) plane in NaCl and slip would occur along the $(10\bar{1})$ plane.

The structure of dislocations in diamond lattices, which, as discussed in Chap. 3, is adopted quite frequently by elements that have tetrahedral covalent bonding, has to conform to the comparatively rigid tetrahedral bonds, as shown in Fig. 6.10*c*. This, as discussed in Chap. 11, makes them highly resistant to shear, and so solids such as Si, SiC, and diamond are brittle at room temperatures.

6.4
PLANAR DEFECTS

Grain boundaries and free surfaces are considered to be planar defects. Free surfaces were discussed in Chap. 4. This section deals with grain boundary structure and grain boundary segregation.

Grain boundary structure

A grain boundary is simply the interface between two grains. The two grains can be of the same material, in which case it is known as a **homophase boundary**, or of two different materials, in which case it is referred to as a **heterophase boundary**. The situation in ceramics is further complicated because often other phases that are only a few nanometers thick can be present between the grains, in which case the grain boundary represents the three phases. These phases usually form during processing (see Chap. 10) and can be crystalline or amorphous. In general, the presence or absence of these films has important ramifications on processing, electrical properties, and creep, hence their importance.

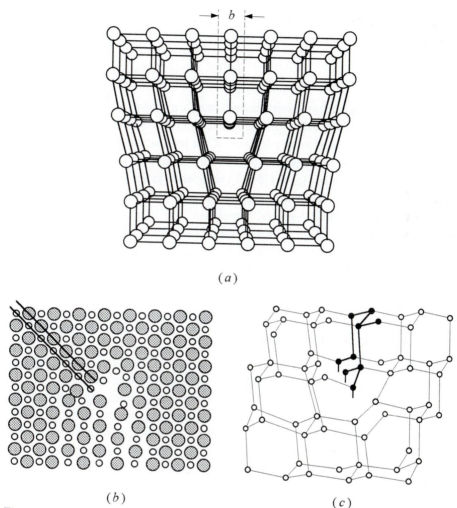

Figure 6.10

(*a*) Edge dislocation, the width of which is characterized by Burgers vector **b**. (*b*) Edge dislocation in NaCl produced by the insertion of two extra half-planes of ions (solid lines). (*c*) A 60° dislocation along <110> in the diamond cubic structure. The glide plane is (111) and the extra half-plane is shown in heavier lines.

Typically grain boundaries are distinguished according to their structure as low-angle (<15°), special, or random. The easiest to envision is the **low-angle grain boundary**, which can be described as consisting of arrays of dislocations separated by areas of strained lattice. An example of such a grain boundary is shown in Fig. 6.11*a*, where the dislocations are solid lines. The angle of the grain boundary is determined from the dislocation spacing λ_d and $\mathbf{b}$. From Fig. 6.11a it is easy to appreciate that the angle of tilt or misorientation is given by

$$\sin \theta = \frac{\mathbf{b}}{\lambda_d} \tag{6.33}$$

Special or coincident grain boundaries are those in which a special orientational relationship exists between the two grains on either side of the grain boundary. In these boundaries, a fraction of the total lattice sites between the two grains coincide. For example, at 36.87° the Ni and O ions in NiO coincide periodically, as shown in Fig. 6.11b. These special grain boundaries have lower grain boundary energy, less enhanced grain boundary diffusivity, and higher grain boundary mobility relative to general boundaries.

The vast majority of grain boundaries, however, are neither low-angle nor special, but are believed to be composed of islands of disordered material where the fit is bad separated by regions where the fit is relatively good. This so-called island model was first proposed by Mott[80] and appears to qualitatively describe the present view of grain boundary structures.

Impurity segregation at grain boundaries

The role of grain boundary chemistry on properties cannot be overemphasized. For many ceramics, the presence of small amounts of impurities in the starting material can vastly influence their mechanical, optical, electrical, and dielectric properties. The effect of impurities is further compounded since it is now appreciated that these impurities have a tendency to segregate at grain boundaries. If the concentration of solute is not too large, then the ratio of the grain boundary concentration C_{gb} to bulk concentration C_{bulk} depends on the free energy change due to segregation ΔG_{seg} and is given by

$$\frac{C_{gb}}{C_{bulk}} = \exp\frac{\Delta G_{seg}}{kT} \tag{6.34}$$

One of the contributions to the decrease in free energy comes from the decrease in strain energy resulting from solute misfit in the lattice. It can be shown that this decrease in strain energy scales as $[(r_2 - r_1)/r_1]^2$, where r_1 and r_2 are the ionic radii of the solvent and solute ions, respectively. Hence the larger the radii differences, the greater the driving force for segregation, which has been experimentally verified. Note that it is the absolute size difference that is important; i.e., both smaller and larger ions will segregate to the grain boundary. The reason is simple:

---

80 N. F. Mott, *Proc. Phys. Soc.*, **60**: 391 (1948).

The grain boundary is a region of disorder that can easily accommodate the different-sized ions as compared to the bulk. Consequently, if ΔG_{seg} is large, the grain boundary chemistry can be quite different from that of the bulk, magnifying the effect of these impurities.

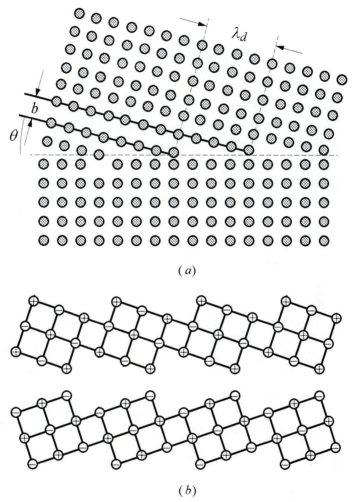

(a)

(b)

Figure 6.11

(a) Schematic representation of a low-angle tilt grain boundary made up of a series of dislocations with the Burgers vector **b** spaced λ_d apart. (b) Structure of special or coincident boundary in NiO.[81]

---

81 D. M. Duffy and P. W. Tasker, *Phil. Mag.*, **A47**: 817 (1983).

6.5
SUMMARY

Point and electronic defects reduce the free energy of a system by increasing its entropy. The concentration of defects increases exponentially with temperature and is a function of their free energy of formation.

In compound crystals, balanced-defect reactions must conserve mass, charge neutrality, and the ratio of the regular lattice sites. In *pure* compounds, the point defects that form can be classified as either stoichiometric or nonstoichiometric. By definition, stoichiometric defects do not result in a change in chemistry of the crystal. Examples are Schottky (simultaneous formation of vacancies on the cation and anion sublattices) and Frenkel (vacancy-interstitial pair).

Nonstoichiometric defects form when a compound loses one (or more) of its constituents selectively. Mass is transferred across the crystal boundary, and compensating defects have to form to maintain charge neutrality. For instance, when exposed to severe enough reducing conditions, most oxides will lose oxygen, which in turn results in the simultaneous formation of oxygen vacancies and free electrons. An oxide is labeled nonstoichiometric when its composition is susceptible to changes in its surroundings, and this is usually correlated to the ease with which the cations (or anions) can change their oxidation state.

Extrinsic defects form as a result of the introduction of impurities. The incorporation of aliovalent impurities in any host compound results in the formation of defects on one of the sublattices, in order to preserve the lattice site ratio.

To relate the concentrations of point and electronic defects to temperature and to the externally imposed thermodynamic conditions such as oxygen partial pressures, the defects are treated as chemical species and their equilibrium concentrations are calculated from mass action expressions. If the free-energy changes associated with all defect reactions were known, then in principle diagrams, known as Kroger-Vink diagrams, relating the defect concentrations to the externally imposed thermodynamic parameters, impurity levels, etc., can be constructed.

With the notable exception of transition metal oxides that generally exhibit wide deviations from stoichiometry, the concentration of intrinsic or nonstoichiometric defects in most ceramic compounds is so low that their defect concentrations are usually dominated by the presence of impurities.

In addition to point and electronic defects, ceramic crystals contain dislocations and grain boundaries.

PROBLEMS

6.1. (*a*) Starting with Eq. (6.6), derive Eq. (6.7).

(*b*) On the same graph, plot Eq. (6.6) for two different values of h_d for the same temperature, and compare the equilibrium number of vacancies. Which will have the higher number of defects at equilibrium? Why?

(*c*) Following the same steps taken to get to Eqs. (6.6) and (6.7), derive Eq. (6.9).

6.2. (*a*) A crystal of ferrous oxide Fe_yO is found to have a lattice parameter $a = 0.43$ nm and a density of 5.72 g/cm^3. What is the composition of the crystal (i.e., the value of y in Fe_yO)? Clearly state all assumptions.

Answer: $y = 0.939$

(*b*) For $Fe_{0.98}O$, the density is 5.7 g/cm^3. Calculate the site fraction of iron vacancies and the number of iron vacancies per cubic centimeter.

Answer: Site fraction $= 0.02$; $V''_{Fe} = 9.7 \times 10^{20} cm^{-3}$

6.3. (*a*) Write two possible defect reactions that would lead to the formation of a metal-deficient compound. From your knowledge of the structures and chemistry of the various oxides, cite an example of an oxide that you think would likely form each of the defect reactions you have chosen.

(*b*) Write possible defect reactions and corresponding mass action expressions for

(i) Oxygen from atmosphere going interstitial

(ii) Schottky defect in M_2O_3

(iii) Metal loss from ZnO

(iv) Frenkel defect in Al_2O_3

(v) Dissolution of MgO in Al_2O_3

(vi) Dissolution of Li_2O in NiO

6.4. Calculate the equilibrium number of Schottky defects n_{eq} in an MO oxide at 1000 K in a solid for which the enthalpy for defect formation is 2 eV. Assume that the vibrational contribution to the entropy can be neglected. Calculate ΔG as a function of the number of Schottky defects for three concentrations, namely, n_{eq}, $2n_{eq}$, and $0.5n_{eq}$. State all assumptions. Plot the resulting data.

Answer: $\Delta G_{n_{eq}} = -0.15J$, $\Delta G_{2n_{eq}} = -0.095J$, and $\Delta G_{0.5n_{eq}} = -0.1294J$

6.5. Compare the concentration of positive ion vacancies in an NaCl crystal due to the presence of 10^{-4} mol fraction of $CaCl_2$ impurity with the intrinsic concentration present in equilibrium in a pure NaCl crystal at 400°C. The formation energy Δh of a Schottky defect is 2.12 eV, and the mole fraction of Schottky defects at the melting point of 800°C is 2.8×10^{-4}.

Answer: $V''_{Na\,(extrinsic)}/V''_{Na\,(intrinsic)} = 6.02 \times 10^{19}/1.85 \times 10^{17} = 324$

6.6. (a) Using the data given in Worked Example 6.2, estimate the free energies of formation of the Schottky and Frenkel defects in NaCl.

Answer: $\Delta g_S = 104.5$ kJ/mol, $\Delta g_F = 114$ kJ/mol

(b) Repeat Worked Example 6.2, assuming Δg_F is double that calculated in part (a). What implications does it have for the final concentrations of defects?

6.7. The crystal structure of cubic yttria is shown in Fig. 3.7.
   (a) What structure does yttria resemble most?
   (b) Calculate its theoretical density. Compare your result with the actual value of 5.03 g/cm^3. Why do you think the two values are different?
   (c) What stoichiometric defect do you think such a structure would favor? Why?
   (d) The experimentally determined density changes as a function of the addition of ZrO_2 to Y_2O_3 are as follows:

Composition, mol % ZrO_2	0.0	2.5	5.2	10.0
Density, g/cm^3	5.03	5.04	5.057	5.082

Propose a defect model that would be consistent with these observations.

6.8. (a) If the lattice parameter of ZrO_2 is 0.513 nm, calculate its theoretical density.

Answer: 6.06 g/cm^3

(b) Write down two possible defect reactions for the dissolution of CaO in ZrO_2. For each of your defect models, calculate the density of a 10 mol % CaO-ZrO_2 solid solution. Assume the lattice parameter of the solid solution is the same as that of pure ZrO_2, namely, 0.513 nm.

Answer: Interstitial 6.03, vacancy 5.733 g/cm^3

6.9. The O/Fe ratio for FeO in equilibrium with Fe is quite insensitive to temperature from 750 to 1250°C and is fixed at about 1.06. When this oxide is subjected to various oxygen partial pressures in a thermobalance at 1150°C, the results obtained are as shown in Fig. 6.9.
   (a) Explain in your own words why the higher oxygen partial pressure resulted in a greater weight gain.
   (b) From these results determine the O/Fe ratio for FeO at the two oxygen partial pressures indicated.

Answer: 1.08 and 1.12

   (c) Describe atomistically what you think would happen to the crystal that was equilibrated at the higher oxygen partial pressure if the partial pressure were suddenly changed to 10^{-12} atm.

6.10. (*a*) What renders an oxide stoichiometric or nonstoichiometric? Would you expect Δg_{red} to be small or large for a stoichiometric oxide compared to a nonstoichiometric one?

(*b*) Carrying out a calculation similar to the one shown in Worked Example 6.3(*a*), show that an oxide can be considered fairly resistant (i.e., stoichiometric) to moderately reducing atmospheres ($\approx 10^{-12}$ atm) at 1000°C as long as the activation energy for reduction Δh_{red} is greater than ≈ 6 eV. What would happen if Δh_{red} were lower? State all assumptions.

(*c*) Show that for any material $\Delta h_{red} + \Delta h_{ox} = 2E_g$. Discuss the implications of this result.

6.11. Using the relevant thermodynamic data, calculate the chemical stability domain (in terms of oxygen partial pressure) of FeO and NiO. Plot to scale a figure such as Fig. 6.7 for each compound, using the data given in Table 6.1. Which of these two oxides would you consider the more stoichiometric? Why?

6.12. As depicted in Fig. 6.11*a*, the grain boundaries in polycrystalline ceramics can be considered to be made up of a large accumulation of edge dislocations. Describe a simple mechanism by which such grains can grow when the material is annealed at an elevated temperature for a long time.

ADDITIONAL READING

1. F. A. Kroger and H. J. Vink, *Solid State Physics*, vol. 3, F. Seitz and D. Turnbull, eds., Academic Press, New York, 1956, Chap. 5.
2. L. A. Girifalco, *Statistical Physics of Materials*, Wiley-Interscience, New York, 1973.
3. F. A. Kroger, *The Chemistry of Imperfect Crystals*, North-Holland, Amsterdam, 1964.
4. I. Kaur and W. Gust, *Fundamentals of Grain and Interphase Boundary Diffusion*, 2d ed., Zeigler Press, Stuttgart, 1989.
5. P. Kofstad, *Nonstoichiometry, Diffusion and Electrical Conductivity in Binary Metal Oxides*, Wiley, New York, 1972.
6. O. T. Sorensen, ed., *Nonstoichiometric Oxides*, Academic Press, New York, 1981.
7. W. D. Kingery, "Plausible Concepts Necessary and Sufficient for Interpretation of Ceramic Grain-Boundary Phenomena, I and II," *J. Amer. Cer. Soc.*, **57**: 1–8, 74–83 (1974).
8. D. Hull, *Introduction to Dislocations*, Pergamon Press, New York, 1965.
9. F. R. N. Nabarro, *Theory of Crystal Dislocations*, Clarendon Press, Oxford, England, 1967.
10. N. Tallan, ed., *Electrical Conduction in Ceramics and Glasses*, Parts A and B, Marcel Decker, New York, 1974.

CHAPTER 7

Diffusion and Electrical Conductivity

Electroconductivity depends
On Free Electrons: in Germanium
A touch of Arsenic liberates; in blends
Like Nickel Oxide, Ohms thwart Current. From
Pure Copper threads to wads of Chewing Gum
Resistance varies hugely. Cold and Light
as well as "doping" modify the sum
of Fermi levels, Ion scatter, site
Proximity, and other Factors recondite.

John Updike, *Dance of the Solids*[†]

7.1
INTRODUCTION

The solid state is far from static. Thermal energy keeps the atoms vibrating vigorously about their lattice positions and continually bumping into each other and exchanging energy with their neighbors and surroundings. Every now and then, an atom will gain sufficient energy to leave its mooring and migrate. This motion is termed **diffusion**, without which the sintering of ceramics, oxidation of metals, tempering of steels, precipitation hardening of alloys, and doping of semiconductors, just to name a few phenomena, would not be possible. Furthermore, diffusion is critical in determining the creep and grain growth rates in ceramics, hence its importance.

For reasons that will become clear shortly, a prerequisite for diffusion and electrical conductivity is the presence of point and electronic defects. Consequently,

---

[†] J. Updike, **Midpoint and other Poems**, A. Knopf, Inc., New York, New York, 1969. Reprinted with permission.

this chapter and the preceding one are intimately related, and one goal of this chapter is to make that relationship clear.

In many ceramics, diffusion and electrical conductivity are inextricably linked for two reasons. The first is that *ionic species* can be induced to migrate under the influence of a chemical potential gradient (diffusion) or an electric potential gradient (electrical conductivity). In either case, the basic atomic mechanism is the same, and one of the major conclusions of this chapter is that the diffusivity of a given species is directly related to its conductivity. The second important link is that the defects required for diffusion and electrical conductivity are often created in tandem. For example, as discussed in Chap. 6, the reduction of an oxide can result in the formation of oxygen vacancies and free electrons in the conduction band — this not only renders the oxide more electronically conductive but also increases the diffusivity of oxygen in that oxide.

This chapter is structured as follows: Sec. 7.2.2 deals with the atomistics of diffusion. The relationship between atom diffusivities and activation energies, temperature, and concentration of the defects responsible for their motion is developed. In Sec. 7.2.3, the diffusion of ions and defects subjected to a chemical potential gradient is dealt with in detail, without reference to their electrical conductivity. But since the two cannot be separated (after all, the diffusion of charged defects is nothing but a current!), Sec. 7.2.4 makes the connection between them. Section 7.2.5 goes a step further, where it is shown that in essence the true driving force that gives rise to a flux of any charged species is the gradient in the electrochemical potential.

In Sec. 7.3.1, the concept of electrical conductivity is introduced. Now in addition to ionic conductivity, the influence of mobile electronic defects has to be factored into the total conductance. How these electronic defects are introduced in a crystal was first encountered in the previous chapter and is further elaborated in Sec. 7.3.2.

In Sec. 7.4, situations where the fluxes of the different species are *coupled* are considered. This coupled or ambipolar diffusion is of paramount importance since it is responsible for such diverse phenomena as creep, sintering of ceramics, and high-temperature oxidation of metals.

In Sec. 7.5, the relationships between the various diffusion coefficients introduced throughout this chapter are made clear.

7.2
DIFFUSION

There are essentially three mechanisms by which atoms will diffuse, as shown schematically in Fig. 7.1*a* to *c*. The first, the *vacancy mechanism*, involves the jump of an atom or ion from a regular site into an adjacent vacant site (Fig. 7.1*a*). The second, *interstitial diffusion*, occurs as shown schematically in Fig. 7.1*b* and requires the presence of interstitial atoms or ions. The third, less common mechanism is the *interstitialcy mechanism*, shown in Fig. 7.1*c*, where an interstitial atom pushes an atom from a regular site into an interstitial site.

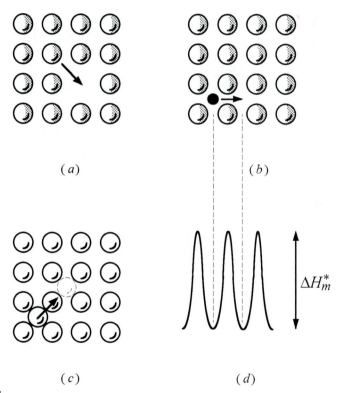

(a)

(b)

(c)

(d)

Figure 7.1
(*a*) Diffusion of atoms by vacancy mechanism. (*b*) Interstitial diffusion mechanism.
(*c*) Interstitialcy mechanism. (*d*) For the interstitial atom shown in (*b*), to make a jump, it must overcome an energy barrier ΔH_m^*.

In all cases, to make the jump, the atom has to squeeze through a narrow passage, which implies it has overcome an activation or energy barrier. This barrier is known as the *energy of migration* and is shown schematically in Fig. 7.1*d* for the diffusing interstitial ion shown in Fig. 7.1*b*.

7.2.1   Phenomenological Equations

In Chap. 5 it was noted that for many physical phenomena that entail transport — whether it is charge, mass, or momentum — the assumption is usually made that the flux J is linearly proportional to the driving force F, or

$$J = KF \qquad (7.1)$$

where K is a material property. In case of diffusion, the relationship between the flux J and the concentration gradient dc/dx is given by Fick's first law,[82] namely,

$$J_A^B \left(\frac{mol}{m^2 s} \right) = -D_A^B \left(\frac{\partial c_A}{\partial x} \right) \left(\frac{m^2}{s} \frac{mol}{m^3 \cdot m} \right) \qquad (7.2)$$

where D_A^B is the chemical diffusion coefficient of species A in matrix B. The units of D are square meters per second; c_A is the concentration, and it can be expressed in a number of units, such as moles or kilograms per cubic meter. The resulting flux is then expressed in units consistent with those chosen for c_A.

The **self-diffusivity** D of an atom or ion is a *measure of the ease and frequency with which that atom or ion jumps around in a crystal lattice in the absence of external forces*, i.e., in a totally random fashion. Experimentally, it has long been appreciated that D was thermally activated and could be expressed as

$$D = D_0 \exp\left(-\frac{Q}{kT} \right) \qquad (7.3)$$

where Q is the activation energy for diffusion which is not a function of temperature, whereas D_0 was a weak function of temperature. It also has been long appreciated that diffusivity depends critically on the stoichiometry and purity level of a ceramic. To understand how these variables affect D, the phenomenon of diffusion has to be considered at the atomic level. Before we do so, however, it is useful to briefly explore how one measures D.

---

[82]   The reason why Fick's first law does not have the same form as Eq. (7.1) — after all, a concentration gradient is not a force! — is discussed in greater detail later on.

EXPERIMENTAL DETAILS: MEASURING DIFFUSIVITIES

There are many techniques by which diffusion coefficients can be measured. The most common is to anneal a solid in an environment that well defines the activity or concentration of the diffusing species at a given location for a given time and then to measure the resulting concentration profile, that is, $c(x,t)$, of the diffusing species. The profile will clearly depend on the diffusivity (the larger D is, the faster the diffusing species will penetrate into the material), time, and temperature of the diffusion anneal and the boundary and initial conditions.   To determine the diffusivity, Fick's second law

$$\frac{\partial c}{\partial t} = \frac{\partial}{\partial x}\left(D\frac{\partial c}{\partial x}\right) \tag{7.4}$$

has to be solved by using the appropriate boundary and initial conditions. The derivation of this equation can be found in most textbooks on diffusion, and it is nothing but a conservation of mass expression. If D is not a function of position, which implies it is also not a function of concentration, then Eq. (7.4) simplifies to

$$\frac{\partial c}{\partial t} = D\frac{\partial^2 c}{\partial x^2} \tag{7.5}$$

where c is a function of x and t.

Once solved by using the appropriate initial and boundary conditions employed during the experiment, the value of D that best fits the experimental profile is taken to be the diffusivity of that species at the anneal temperature.

A convenient method to measure $c(x,t)$ is to use radioactive isotopes of the atom for which the diffusivity is to be measured. For instance, if one is interested in the diffusivity of Mn in MnO, a layer of radioactive ^{54}MnO is applied as a thin film on one end of a long rod of nonradioactive MnO. After an appropriate anneal time t at a given temperature, the rod is quenched and sectioned normal to the direction of the diffusing species, and the experimental concentration profile is evaluated by measuring the radioactivity of each section. The solution of Fick's second law for these conditions is given by[83]

---

[83]   The methods of solution will not be dealt with here. The interested reader can consult J. Crank, *Mathematics of Diffusion*, 2d ed., Clarendon Press, Oxford, 1975, or H. S. Carslaw and J. C. Jaeger, *Conduction of Heat in Solids*, Clarendon Press, Oxford, 1959. See also R. Ghez, *A Primer of Diffusion Problems*, Wiley, New York, 1988.

$$c(x,t) = \frac{\beta}{2\sqrt{\pi Dt}} \exp\left(-\frac{x^2}{4Dt}\right) \tag{7.6}$$

where β is the total quantity per unit cross-sectional area of solute present initially that has to satisfy the condition

$$\int_0^\infty c(x)\,dx = \beta$$

According to Eq. (7.6), a plot of $\ln c(x)$ versus x^2 should result in a straight line with a slope equal to $1/(4Dt)$. Given t, D can be calculated.

It is important to note that what one measures in such an experiment is known as a **tracer diffusion coefficient D_{tr}**, which is *not* the same as the self-diffusion coefficient D defined above. The two are related, however, by a **correlation coefficient f_{cor}**, the physics of which is discussed in greater detail in Sec. 7.5.

7.2.2 Atomistics of Solid State Diffusion

The fundamental relationship relating the self-diffusion coefficient D of an atom or ion to the atomistic processes occurring in a solid[84] is

$$D = \alpha \Omega \lambda^2 \tag{7.7}$$

where Ω is the frequency of *successful* jumps, i.e., number of successful jumps per second; λ is the elementary jump distance which is on the order of the atomic spacing; and α is a geometric constant that depends on the crystal structure, and its physical significance will become clearer later on. Here we only remark that for

---

[84] Eq. (7.7) can be derived from random walk theory considerations. A particle after n random jumps will, on average, have traveled a distance proportional to $\sqrt{n}$ times the elementary jump distance λ. It can be easily shown that, in general, the characteristic diffusion length is related to the diffusion coefficient D and time t through the equation

$$x^2 \approx Dt$$

from which it follows that:

$$\left(\sqrt{n}\lambda\right)^2 \propto Dt$$

Rearranging yields

$$D \propto \lambda^2 \, n/t \propto \lambda^2 \Omega$$

where Ω is defined as n/t, or the number of successful jumps per second. For further details see P. G. Shewmon, *Diffusion in Solids*, McGraw-Hill, New York, 1963, Chap. 2.

cubic lattices, where only nearest-neighbor jumps are allowed and diffusion is by a vacancy mechanism, $\alpha = 1/\zeta$, where ζ is the coordination number of the vacancy.

Frequency Ω is the product of the probability of an atom's having the requisite energy to make a jump ν and the probability θ that the site adjacent to the diffusing entity is available for the jump, or

$$\Omega = \nu\,\theta \tag{7.8}$$

From this relationship it follows that to understand diffusion and its dependence on temperature, stoichiometry, and atmosphere requires an understanding of how ν and θ vary under the same conditions. Each is dealt with separately in the following subsections.

Jump frequency ν

For an atom to jump from one site to another, it has to be able to break the bonds attaching it to its original site and to squeeze between adjacent atoms, as shown schematically in Fig. 7.1d. This process requires an energy ΔH_m^*, which is usually much higher than the average thermal energy available to the atoms. Hence at any instant only a fraction of the atoms will have sufficient energy to make the jump.

Therefore, to understand diffusion, one must first answer the question, At any given temperature, what fraction of the atoms have an energy $\geq \Delta H_m^*$ and are thus capable of making the jump? Or to ask a slightly different question, how often, or for what fraction of time, does an atom have sufficient energy to overcome the diffusion barrier?

To answer this question, the *Boltzmann distribution law* is invoked, which states that the probability P of a particle having an energy ΔH_m or greater is given by:[85]

$$P\left(E > \Delta H_m^*\right) = (\text{const.})\exp\left(-\frac{\Delta H_m^*}{kT}\right) \tag{7.9}$$

where k is Boltzmann's constant and T is the temperature in kelvin.

It follows that the frequency ν with which a particle can jump, provided that an adjacent site is vacant, is equal to the probability that it is found in a state of

---

[85] Equation (7.9) is only valid if ΔH_m^* is much larger than the average energy of the system. In most cases for solids, that is the case. For example, the average energy of the atoms in a solid is on the order of kT which at room temperature is ≈ 0.025 eV and at 1000°C is ~ 0.11 eV. Typical activation energies for diffusion, vacancy formation, etc., are on the order of a few electron volts. So all is well.

sufficient energy to cross the barrier multiplied by the frequency ν_0 at which that barrier is being approached. In other words,

$$\nu = \nu_0 \exp\left(-\frac{\Delta H_m^*}{kT}\right) \tag{7.10}$$

where ν_0 is the vibration of the atoms[86] and is on the order of $10^{13} s^{-1}$ (see Worked Example 5.3b). For low temperatures or high values of ΔH_m^*, the frequency of successful jumps becomes vanishingly small, which is why, for the most part, solid-state diffusion occurs readily only at high temperatures. Conversely, at sufficiently high temperatures, that is, $kT \gg \Delta H_m^*$, the barrier ceases to be one and every vibration would, in principle, result in a jump.

Probability θ of site adjacent to diffusing species being vacant

The probability of a site being available for the diffusing species to make the jump will depend on whether one considers the motion of the defects or of the ions themselves. Consider each separately.

Defect diffusivity. As noted above, the two major defects responsible for the mobility of atoms are vacancies and interstitials. For both, at low concentrations (which is true for a vast majority of solids) the site adjacent to the defect will almost always be available for it to make the jump and $\theta \approx 1$.

There is a slight difference between vacancies and interstitials, however. An interstitial can and will make a jump with a rate that depends solely on its frequency of successful jumps ν_{int}. By combining Eqs. (7.7), (7.8), and (7.10), with $\theta_{int} = 1$, the interstitial diffusivity D_{int} is given by

$$D_{int} = \alpha_{int} \lambda^2 \nu_0 \exp\left(-\frac{\Delta H_{m,int}^*}{kT}\right) \tag{7.11}$$

where $\Delta H_{m,int}^*$ is the activation energy needed by the interstitial to make the jump.

For a vacancy, however, the probability of a successful jump is increased ζ-fold, where ζ is the number of atoms adjacent to that vacancy (i.e., the coordination number of the atoms), since if *any* of the ζ neighboring atoms attains the requisite energy to make a jump, the vacancy will make a jump. Thus, for vacancy diffusion

$$\nu_{vac} = \zeta \nu_0 \exp\left(-\frac{\Delta H_m^*}{kT}\right)$$

---

[86] This is easily arrived at by equating the vibrational energy $h\nu$ to the thermal energy kT. At 1000 K, $\nu_0 \cong 2 \times 10^{13} s^{-1}$.

Combining this equation with Eqs. (7.7) and (7.8), again assuming $\theta_{vac} \approx 1$, yields

$$D_{vac} = \alpha \zeta \lambda^2 \nu_0 \exp\left(-\frac{\Delta H_m^*}{kT}\right) \qquad (7.12)$$

Atomic or ionic diffusivity. In contrast to the defects, for an atom or ion in a regular site, $\theta << 1$, because most of these are surrounded by other atoms. The probability of a site's being vacant in this case is simply equal to the mole or site fraction, denoted by $\measuredangle$, of vacancies in that solid. Thus the frequency of successful jumps for diffusion of atoms by a *vacancy mechanism* is given by

$$\Omega = \theta \nu_{vac} = \measuredangle \zeta \nu_0 \exp\left(-\frac{\Delta H_m^*}{kT}\right)$$

The factor ζ appears here because the probability of a site next to a diffusing atom being vacant is increased ζ-fold. The diffusion coefficient is given by

$$D_{ion} = \alpha \lambda^2 \measuredangle \zeta \nu_0 \exp\left(-\frac{\Delta H_m^*}{kT}\right) \qquad (7.13)$$

Comparing Eqs. (7.12) and (7.13) reveals an important relationship between vacancy and ion diffusivity, namely,

$$D_{ion} = \measuredangle D_{vac} \qquad (7.14)$$

Given that usually $\measuredangle << 1$, it follows that $D << D_{vac}$, a result that at first sight may appear to be paradoxical — after all one is dealing with the same species.[87] Going one step further, however, and noting that $\measuredangle \approx c_{vac}/c_{ion}$ [Eq. (6.11)], where c_{vac} and c_{ion} are the concentrations of vacancies and ions, respectively, we see that

$$\boxed{D_{ion}c_{ion} = D_{vac}c_{vac}} \qquad (7.15)$$

Now the physical picture is a little easier to grasp: The defects move often (high D) but are not that numerous — the atoms move less frequently, but there are a lot of them. The full implication of this important result will become obvious shortly.

In deriving Eq. (7.13), for the sake of simplicity, the effect of the jump on the vibration entropy was ignored. This is taken into account by postulating the

---

[87] The implication here is that the vacancies are jumping around much more frequently than the atoms, which indeed is the case. For example, in a simulation of a diffusion process, if one were to focus on a vacancy, its hopping frequency would be quite high since it does not have to wait for a vacant site to appear before making a jump. If, however, one were to focus on a given atom, its average hopping frequency would be much lower because it will hop only if and when a vacancy appears next to it.

existence of an excited equilibrium state (Fig. 7.5a), albeit of very short duration, that affects the frequency of vibration of its neighbors and is associated with an entropy change given by $\Delta S_m^* \approx kT \ln(\nu'/\nu)$, where ν and ν' are the frequencies of vibration of the ions in their ground and activated states, respectively. A more accurate expression for D_{ion} thus reads

$$D_{ion} = \alpha \lambda^2 \not\leftarrow \zeta \nu_0 \exp\left(-\frac{\Delta G_m^*}{kT}\right) \tag{7.16}$$

where ΔG_m^* is defined as

$$\Delta G_m^* = \Delta H_m^* - T\Delta S_m^* \tag{7.17}$$

Putting all the pieces together, one obtains a final expression that most resembles Eq. (7.3), namely,

$$\boxed{D_{ion} = \nu_0 \lambda^2 \alpha \zeta \not\leftarrow \exp\frac{\Delta S_m^*}{k}\exp\left(-\frac{\Delta H_m^*}{kT}\right) = D_0 \exp\left(-\frac{Q}{kT}\right)} \tag{7.18}$$

except that now the physics explaining why diffusivity takes that form should be clearer. The temperature dependence of diffusivity of some common ceramics is shown in Fig. 7.2.

The values of the activation energies Q and their variation with temperature are quite useful in deciphering the nature of the diffusion process. For instance, if $\not\leftarrow$ is thermally activated, as in the case of intrinsic point defects, then the energy needed for the defect formation will appear in the final expression for D. If, however, the vacancy or defect concentration is fixed by impurities, then $\not\leftarrow$ is no longer thermally activated but is proportional to the concentration of the dopant. The following worked examples should make that point clearer. Finally, it is worth noting here that the preexponential term D_0, calculated from first principles, i.e., from Eq. (7.18), does not, in general, agree with experimental data, the reason for which is entirely clear at this time.

WORKED EXAMPLE 7.1. For Na^+ ion migration in NaCl, ΔH_m^* is 77 kJ/mol, while the enthalpy and entropy associated with the formation of a Schottky defect are, respectively, 240 kJ/mol and 10 K (see Table 6.2).
(a) At approximately what temperature does the diffusion change from extrinsic (i.e., impurity-controlled) to intrinsic in an NaCl – CaCl$_2$ solid solution containing 0.01 percent CaCl$_2$? You can ignore ΔS_m^*.
(b) At 800 K, what mole percent of CaCl$_2$ must be dissolved in pure NaCl to increase D_{Na^+} by an order of magnitude?

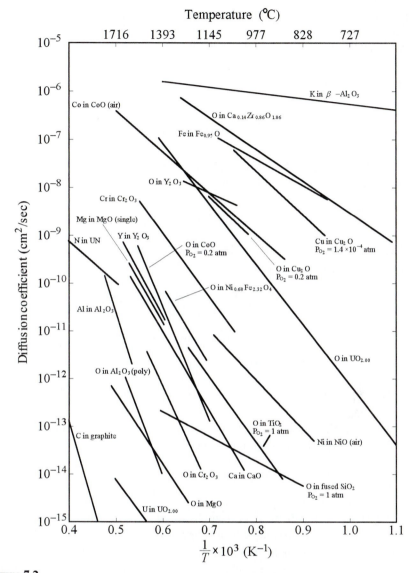

Figure 7.2

Temperature dependence of diffusion coefficients for some common ceramic oxides.[88]

---

88 Adapted from W. D. Kingery, H. K. Bowen and D. R. Uhlmann, *Introduction to Ceramics*, 2d ed., Wiley, New York, 1976. Reprinted with permission.

Answer

(a) To solve this problem, an expression for D_{Na^+} as a function of temperature in both the extrinsic and intrinsic regions has to derived.   Once derived, the two expressions are equated, and T is solved for.   In the intrinsic region, the vacancy concentration is determined by the Schottky equilibrium, or

$$[V'_{Na}][V^{\bullet}_{Cl}] = \exp\frac{\Delta S_S}{k} \exp\left(-\frac{\Delta H_S}{kT}\right)$$

where ΔH_S and ΔS_S are, respectively, the Schottky formation energy and entropy. Assuming that the Schottky defects dominate, that is, $[V'_{Na}] = [V^{\bullet}_{Cl}]$,

$$[V'_{Na}] = [V^{\bullet}_{Cl}] = \exp\frac{\Delta S_S}{2k} \exp\left(-\frac{\Delta H_S}{2kT}\right) \tag{7.19}$$

which when combined with Eq. (7.18), and noting that $\measuredangle = [V'_{Na}]$, yields the desired expression in the intrinsic regime

$$D_{Na^+} = \lambda^2 \alpha \zeta \nu_0 \exp\frac{\Delta S_S}{2k} \exp\left(-\frac{\Delta H_S}{2kT}\right) \exp\left(-\frac{\Delta H^*_m}{kT}\right) \tag{7.20}$$

If the following incorporation reaction (see Chap. 6)

$$CaCl_2 \Rightarrow V'_{Na} + Ca^{\bullet}_{Na} + 2Cl^X_{Cl}$$

is assumed, it follows that in the extrinsic region, for every 1 mol of $CaCl_2$ dissolved in NaCl, 1 mol of Na vacancies is created.   In other words, $\measuredangle = [V'_{Na}] = [Ca^{\bullet}_{Na}] = 0.0001$. Thus in the extrinsic regime, the vacancy concentration is fixed and independent of temperature. In other words,

$$D_{Na^+} = [Ca^{\bullet}_{Na}]\lambda^2 \alpha \zeta \nu_0 \exp\left(-\frac{\Delta H^*_m}{kT}\right) \tag{7.21}$$

Equating Eqs. (7.19) and (7.20) and solving for T yield a temperature of 743°C.

(b) At 800 K the intrinsic mole fraction of vacancies [Eq. (7.19)] is 2.2×10^{-6} or 1.3×10^{18} vacancies per mole. To increase D_{Na^+} by an order of magnitude, the doping must create ten times the number of intrinsic vacancies. It follows that the addition of 2.2×10^{-5} mol fraction of $CaCl_2$ to a mole of NaCl would do the trick.

Note that when the defect concentration was intrinsically controlled, the activation energy for its formation appeared in the final expression for D [i.e., Eq. (7.20)], whereas when the concentration of the defects was extrinsically controlled, the final expression included only the energy of migration.   How this fact is used to experimentally determine both ΔH^*_m and ΔH_S is discussed in the following worked example.

WORKED EXAMPLE 7.2. Calculate the migration enthalpy for Na ion migration and the enthalpy of Schottky defect formation from the data shown in Fig. 7.3. Discuss all assumptions.

Answer

The behavior shown in Fig. 7.3 is typical of many ceramics and indicates a transition from intrinsic behavior at higher temperatures to extrinsic behavior at lower temperatures. In other words, at higher temperatures Eq. (7.20) applies and the slope of the line equals $\Delta H_m^*/k + \Delta H_S/2k$. At lower temperatures Eq. (7.21) applies and the slope of the line is simply equal to $\Delta H_m^*/k$. Calculating the corresponding slopes from the figure and carrying out some simple algebra, one obtains $\Delta H_m^* \approx 77$ kJ/mol and $\Delta H_S \approx 227$ kJ/mol.

It is worth noting here that measurements such as those in Fig. 7.3 are one technique by which the data reported in Table 6.2 are obtained.

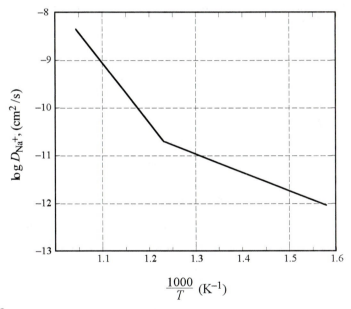

Figure 7.3
Temperature dependence of diffusivity of Na ions in NaCl.

WORKED EXAMPLE 7.3. The functional dependence of the diffusion of ^{54}Mn in MnO on the oxygen partial pressure is shown in Fig 7.4. Explain the origin of the slope. Would increasing the temperature alter the slope?

Answer

Since D_{Mn} clearly depends on the oxygen partial pressure, the first step in solving the problem is to relate the diffusivity to P_{O_2}. Assuming the diffusivity of Mn in MnO occurs by a vacancy diffusion mechanism, i.e., $\measuredangle = [V_{Mn}]$, and replacing D_0 in Eq. (7.18) by $[V_{Mn}]D_0'$, one sees that

$$D_{Mn} = [V_{Mn}]D_0'\exp\left(-\Delta H_m^*/kT\right)$$

The next step is to relate $[V_{Mn}]$ to P_{O_2}. Given that the slope of the curve is $+1/6$, that is, increasing the oxygen partial pressure increases the diffusivity of Mn, the most likely defect reaction occurring is reaction (III) in Chap. 6, for which, assuming the neutrality condition to be $p = 2[V_{Mn}'']$, the mass action expression is given by Eq. (6.30), or

$$[V_{Mn}''] = (\text{const})P_{O_2}^{+1/6}$$

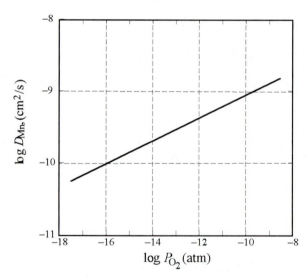

Figure 7.4
Functional dependence (on log-log plot) of the diffusion coefficient of ^{54}Mn in MnO on oxygen partial pressure.

Combining this result with the expression for D_{Mn} reveals the observed behavior. The physics of the situation can be summed up as follows: Increasing the oxygen partial pressure (i.e., going from left to right in Fig. 7.4) decreases the concentration of oxygen vacancies which, in order to maintain the Schottky equilibrium, results in an increase in $[V_{Mn}]$ and a concomitant increase in D_{Mn}.

Note that had $[V_{Mn}]$ been fixed by extrinsic impurities, then D_{Mn} would not be a function of oxygen pressure. Increasing the temperature should, in principle, only shift the lines to higher values but not alter the slope.

7.2.3   Diffusion in a Chemical Potential Gradient

In the foregoing discussion, the implicit assumption was that diffusion was totally *random*, a randomness that was assumed in defining D by Eq. (7.7). This self-diffusion, however, is of no practical use and cannot be measured. Diffusion is important inasmuch as it can be used to effect compositional and microstructural changes. In such situations, atoms diffuse from areas of higher free energy, or chemical potential, to areas of lower free energy, in which case the process is no longer random but is now biased in the direction of decreasing free energy.

Consider Fig. 7.5b, where an ion is diffusing in the presence of a chemical potential gradient $d\mu/dx$. If the chemical potential is given per mole, then the gradient or *force per atom*, f, is given by

$$f = \frac{\Xi}{\lambda} = -\frac{1}{N_{Av}}\frac{d\mu}{dx} \tag{7.22}$$

where N_{Av} is Avogadro's number.[89] Consequently (see Fig. 7.5b), the difference Ξ between the energy barrier in the forward and backward directions is

$$\Xi = \lambda f = -\lambda \frac{d[\mu/N_{Av}]}{dx} \tag{7.23}$$

The forward rate for the atom to jump is thus proportional to

$$\nu_{forward} = \alpha\zeta\nu_0 \exp\left(-\frac{\Delta G_m^*}{kT}\right) \tag{7.24}$$

while the backward jump rate is

$$\nu_{back} = \alpha\zeta\nu_0 \exp\left(-\frac{\Delta G_m^* + \Xi}{kT}\right) \tag{7.25}$$

---

[89]   Note that f is defined here, and throughout this book, as **positive**.

and $\nmid$ appears here because, as noted earlier, for a jump to be successful, the site into which it is jumping must be vacant. Also α is the same constant that appears in Eq. (7.7) and whose physical meaning becomes a little more transparent here — it is a factor that takes into account that only a fraction of the total $\zeta \nu$ hops are hops in the x direction. For instance, in cubic lattices for which $\zeta = 6$, only one-sixth of successful jumps are in the forward x direction, that is, $\alpha = 1/6$ and thus $\alpha \zeta = 1$.

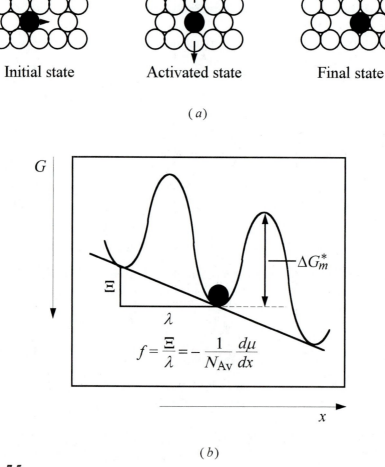

Initial state   Activated state   Final state

(a)

$$f = \frac{\Xi}{\lambda} = -\frac{1}{N_{Av}} \frac{d\mu}{dx}$$

(b)

Figure 7.5
(a) Schematic of activated state during diffusion. (b) Diffusion of an ion down a chemical potential gradient.

The existence of a chemical potential gradient will bias the jumps in the forward direction, and the net rate will be proportional to

$$v_{net} = v_{forward} - v_{back} = \mathcal{N} \alpha \zeta v_0 \left\{ \exp\left(-\frac{\Delta G_m^*}{kT}\right) - \exp\left(-\frac{\Delta G_m^* + \Xi}{kT}\right) \right\}$$

$$= \mathcal{N} \alpha \zeta v_0 \exp\left(-\frac{\Delta G_m^*}{kT}\right) \left\{ 1 - \exp\left(-\frac{\Xi}{kT}\right) \right\} \qquad (7.26)$$

In general, chemical potential gradients are small compared to the thermal energy, i.e., $\Xi/(kT) \ll 1$, and Eq. (7.26) reduces to ($e^{-x} \cong 1 - x$ for small x)

$$v_{net} = \mathcal{N} \alpha \zeta v_0 \left(\frac{\Xi}{kT}\right) \exp\left(-\frac{\Delta G_m^*}{kT}\right) \qquad (7.27)$$

The average drift velocity v_{drift} is given by λv_{net}, which, when combined with Eqs. (7.27) and (7.23), gives

$$v_{drift} = \lambda v_{net} = \frac{\alpha \lambda^2 \zeta v_0 \mathcal{N}}{kT} \exp\left(-\frac{\Delta G_m^*}{kT}\right) f \qquad (7.28)$$

and the resulting flux

$$J_i = c_i v_{drift} = \frac{c_i}{kT} \left\{ \alpha \lambda^2 \mathcal{N} \zeta v_0 \exp\left(-\frac{\Delta G_m^*}{kT}\right) \right\} f \qquad (7.29)$$

where c_i is the total concentration (atoms per cubic meter) of atoms or ions diffusing through the solid. Since the term in brackets is nothing but D_{ion} [Eq. (7.16)] it follows that

$$\boxed{J_{ion} = \frac{c_{ion} D_{ion}}{kT} f} \qquad (7.30)$$

This equation is of fundamental importance because

1. It relates the flux to the *product* $c_{ion} D_{ion}$. The full implication of Eq. (7.15), namely, that $D_{ion} c_{ion} = c_{vac} D_{vac}$, should now be obvious — when one is considering the diffusion of a given species, it is immaterial whether one considers the ions themselves or the defects responsible for their motion: *the two fluxes have to be, and are, equal.*
2. It relates the flux to the *driving force, f.* Given that f has the dimensions of force, Eq. (7.29) can be considered a true flux equation in that it is identical in form to Eq. (7.1). Note also that this relationship has general validity and is

not restricted to chemical potential gradients. For instance, as discussed in the next two sections, f can be related to gradients in electrical or electrochemical potentials as well.

3. It can be shown (see App. 7A) that for ideal and dilute solutions Eq. (7.30) is identical to Fick's first law, that is, Eq. (7.2).

7.2.4 Diffusion in an Electric Potential Gradient

The situation where the driving force was a chemical potential gradient has just been addressed. If, however, the driving force is an electric potential gradient, then the force on the ions due to an electric potential gradient is given by[90]

$$f_i = -z_i e \frac{d\phi}{dx} \tag{7.31}$$

where ϕ is the electric potential in volts and z_i is the net charge on the moving ion. The current density $I_i \left(A/m^2 \right) = C/\left(m^2 \cdot s \right)$ is related to the ionic flux J_{ion} $\left\lfloor atoms/\left(m^2 \cdot s \right) \right\rfloor$, by

$$I_i = z_i e J_{ion} \tag{7.32}$$

Substituting Eqs. (7.31) and (7.32) in Eq. (7.30) shows that

$$I_i = z_i e J_{ion} = \frac{z_i e c_{ion} D_{ion}}{kT} f = -\frac{z_i e c_{ion} D_{ion}}{kT} \left[z_i e \frac{d\phi}{dx} \right] \tag{7.33}$$

which, when compared to Ohm's law $I = -\sigma_{ion} \, d\phi/dx$ [see Eq. (7.39)], yields

$$\boxed{\sigma_{ion} = \frac{z_i^2 e^2 c_{ion} D_{ion}}{kT} = \frac{z_i^2 e^2 c_{def} D_{def}}{kT}} \tag{7.34}$$

where σ_{ion} is the ionic conductivity. This relationship is known as the **Nernst-Einstein relationship**, and it relates the self-diffusion coefficient to the ionic conductivity. The reason for the connection is obvious: In both cases, one is dealing with the jump of an ion or a defect from one site to an adjacent site. The driving forces may vary, but the basic atomic mechanism remains the same.

In applying Eq. (7.34), the following points should be kept in mind:

1. The conductivity σ refers to only the ionic component of the total conductivity (see next section for more details).

---

[90] Note that as defined here, f is positive for positive charges and negative for negative charges. This implies that positive charges will flow down the potential gradient, whereas negative charges flow "uphill," so to speak.

2. This relationship is valid only as long as θ for the defects is ≈ 1 (i.e., at high dilution).

3. The variable c_i introduced in Eq. (7.29) and now appearing in Eq. (7.34) represents the *total* concentration of the diffusing ions in the crystal.[91] For example, in calcia-stabilized zirconia which is an oxygen ion conductor, c_{ion} is the total number of oxygen ions in the crystal and not the total number of defects (see Worked Example 7.4). On the other hand, in a solid in which the diffusion or conductivity occurs by an interstitial mechanism, c_{ion} represents the total number of interstitial ions in the crystal, which is identical to the number of defects.

7.2.5   Diffusion in an Electrochemical Potential Gradient

In some situations, the driving force is neither purely chemical nor electrical but rather electrochemical, in which case the flux equation has to be modified to reflect the influence of both driving forces. This is taken into account as follows: Expressing Eq. (7.30) as a current by use of Eq. (7.32) and combining the results with Eqs. (7.34) and (7.22) one obtains

$$I'_k = -\frac{z_k e c_k D_k}{kT}\frac{d\tilde{\mu}_k}{dx} = -\frac{\sigma_k}{z_k e}\frac{d\tilde{\mu}_k}{dx} \qquad (7.35)$$

---

[91] It may be argued that since only a few ions are moving at one time, the use of c (the concentration of all the ions in the system) is not warranted. After all, most of the ions are not migrating down the chemical potential gradient simultaneously. The way out of this apparent dilemma is to appreciate that if given enough time, indeed *all* the ions would eventually migrate down the gradient. To illustrate, assume that a single crystal of a binary oxide, in which diffusion of the cations is much faster than that of the anions and occurs via a vacancy mechanism, separates two compartments of differing oxygen partial pressures. At the *high oxygen partial pressure side, oxygen atoms will adsorb on the surface*, creating cation vacancies and holes. These defects will in turn diffuse ambipolarly (see Sec. 7.4), i.e., together, toward the low oxygen partial pressure side, where they will be eliminated (i.e., combine with oxygen vacancies) in order to maintain the local Schottky equilibrium. But since the movement of cation vacancies toward the low oxygen pressure side is tantamount to cations moving toward the *high oxygen partial pressure side, the net result is that the crystal will be growing at the high oxygen partial pressure side and shrinking at the low oxygen pressure side.* Thus for all practical purposes, the solid is actually moving with respect to the laboratory frame of reference in very much the same way as a fluid flows in a pipe — the pipe in this case is an imaginary external frame.

where $d\tilde{\mu}_k/dx$ is now the driving force *per ion*,[92] that is, $\tilde{\mu}_k = \mu_k/N_{Av}$. Assuming that the total current due to an ion subjected to both a chemical and an electrical potential gradient is simply the sum of Eq. (7.35) and Ohm's law, or $I''_k = -\sigma_k \, d\phi/dx$, one obtains the following fundamental equation

$$I_k = I'_k + I''_k = -\frac{\sigma_k}{z_k e}\left(\frac{d\tilde{\mu}_k}{dx} + z_k e \frac{d\phi}{dx}\right) = -\frac{\sigma_k}{z_k e}\left(\frac{d\tilde{\eta}_k}{dx}\right) \qquad (7.36)$$

relating the gradient in electrochemical potential $\tilde{\eta}_k = \tilde{\mu}_k + z_k e\phi$ to the current density. The corresponding flux equation (in particles per square meter per second) is

$$J_k = -\frac{D_k c_k}{kT}\frac{d\tilde{\eta}_k}{dx} = -\frac{\sigma_k}{(z_k e)^2}\frac{d\tilde{\eta}_k}{dx} \qquad (7.37)$$

Equations (7.36) and (7.37) are of fundamental importance and general validity since they describe the flux of any charged species under any conditions. The following should be clear at this point:

1. The driving force acting on a charged species is the gradient in its electrochemical potential.
2. For neutral species, the electric potential does not play a role, and the driving force is simply the gradient in the chemical potential.
3. In the absence of an electric field, Eq. (7.37) reverts to Eq. (7.30).
4. If the driving force is simply an electric field, i.e., $d\mu/dx = 0$, then Eq. (7.36) degenerates to Ohm's law [Eq. (7.39)].
5. Equilibrium is achieved only when the gradient in η_k vanishes, as briefly discussed in Chap. 5.
6. In all the equations dealing with flux, the product Dc appears and once more is thus entirely equivalent whether one focuses on the defects and their diffusivity or on the ions and their diffusivity.

7.3
ELECTRICAL CONDUCTIVITY

Historically ceramics were exploited for their electric insulation properties, which together with their chemical and thermal stability rendered them ideal insulating materials in applications ranging from power lines to cores bearing wire-wound

---

[92] For the remainder of this chapter, the tilde over μ or η will denote an energy per ion or atom.

resistors. Today their use is much more ubiquitous — in addition to their traditional role as insulators, they are used as electrodes, catalysts, fuel cells, photoelectrodes, varistors, sensors, and substrates, among many other applications.

This section deals solely with the response of ceramics to the application of a constant electric field and the nature and magnitude of the steady-state current that results. As discussed below, the ratio of this current to the applied electric field is proportional to a material property known as *conductivity*, which is the the focus of this section. The displacement currents or non-steady-state response of solids which gives rise to capacitive properties is dealt with separately in Chaps. 14 and 15 which treat the linear and nonlinear dielectric properties, respectively.

In metals, free electrons are solely responsible for conduction. In semi-conductors, the conducting species are electrons and/or electron holes. In ceramics, however, because of the presence of ions, the application of an electric field can induce these ions to migrate. Therefore, when dealing with conduction in ceramics, one must consider *both* the ionic and the electronic contributions to the overall conductivity.

Before one makes that distinction, however, it is important to develop the concept of conductivity. A good starting point is *Ohm's law*, which states that

$$V = iR \qquad (7.38)$$

where V is the applied voltage (V) across a sample, R its resistance in ohms (Ω), and i the current (C/s) passing through the solid. Rearranging Eq. (7.38), dividing both sides by the cross-sectional area through which the current flows A, and multiplying the right-hand side by d/d, where d is the thickness of the sample, one gets

$$I = \frac{i}{A} = \frac{d}{RA}\frac{V}{d}$$

where $I = i/A$ is the current density passing through the sample. Given that V/d is nothing but the electric potential gradient $d\phi/dx$, *Ohm's law* can be rewritten as[93]

$$\boxed{I_i = -\sigma_i \frac{d\phi}{dx}} \qquad (7.39)$$

where

$$\sigma = \frac{d}{RA}$$

---

[93] The minus sign appears for the same reason as it appears in Fick's first law. A current of positive charges is positive when it flows *down* an electric potential gradient.

Equation (7.39) states that the flux I is proportional to $d\phi/dx$. The proportionality constant σ is the **conductivity** of the material, which is the conductance of a cube of material of unit cross section. The units of conductivity are siemens per meter or Sm^{-1}, where $S = \Omega^{-1}$.

The range of electronic conductivity (Fig. 7.6, right-hand side) in ceramics is phenomenal — it varies over 24 orders of magnitude, and that does not even include superconductivity! Few, if any, other physical properties vary over such a wide range. In addition to electronic conductivity, some ceramics are known to be ionic conductors (Fig. 7.6, left-hand side). In order to understand the reason behind the this phenomenal range and why some ceramics are ionic conductors while others are electronic conductors, it is necessary to delve into the microscopic domain and relate the macroscopically measurable σ to more fundamental parameters, such as the carrier mobilities and concentrations. This is carried out in the following subsections.

7.3.1   Generalized Equations

If one assumes there are c_m mobile carriers per cubic meter drifting with an average drift velocity v_d, it follows that their flux is given by

$$I_i = |z_i| e v_{d,i} c_{m,i} \tag{7.40}$$

The **electric mobility** $\mu_d (m^2/V \cdot s)$ is defined as the average drift velocity per electric field, or

$$\mu_{d,i} = \frac{-v_{d,i}}{d\phi/dx} \tag{7.41}$$

Combining Eqs. (7.39) to (7.41) yields the important relationship

$$\boxed{\sigma_i = c_{m,i} e |z_i| \mu_{d,i}} \tag{7.42}$$

between the macroscopically measurable quantity σ and the microscopic parameters μ_d and c_m results.

In deriving this equation, it was assumed that only one type of charge carrier was present. However, in principle, any mobile charged species can and will contribute to the overall conductivity. Thus the total conductivity is given by

$$\boxed{\sigma_{tot} = \sum_i c_{m,i} |z_i| e \mu_{d,i}} \tag{7.43}$$

The absolute value sign about z_i ensures that the conductivities are always positive and additive regardless of the sign of the carrier.

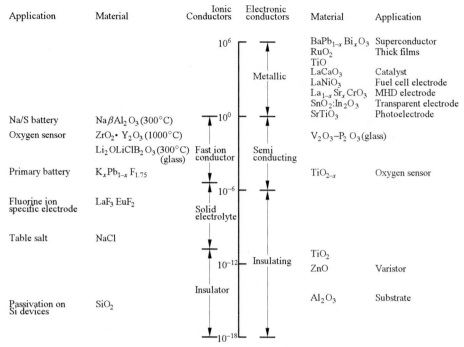

Application	Material	Ionic Conductors	Electronic conductors	Material	Application
		10^6	Metallic	$BaPb_{1-x}Bi_xO_3$	Superconductor
				RuO_2	Thick films
				TiO	
				$LaCaO_3$	Catalyst
				$LaNiO_3$	Fuel cell electrode
				$La_{1-x}Sr_xCrO_3$	MHD electrode
				$SnO_2{:}In_2O_3$	Transparent electrode
Na/S battery	$Na\beta Al_2O_3\,(300°C)$	10^0		$SrTiO_3$	Photoelectrode
Oxygen sensor	$ZrO_2 \cdot Y_2O_3\,(1000°C)$			$V_2O_3{-}P_2O_3\,(glass)$	
	$Li_2OLiClB_2O_3\,(300°C)$ (glass)	Fast ion conductor	Semi conducting		
Primary battery	$K_xPb_{1-x}F_{1.75}$			TiO_{2-x}	Oxygen sensor
		10^{-6}			
Fluorine ion specific electrode	$LaF_3\;EuF_2$	Solid electrolyte			
Table salt	NaCl				
		10^{-12}	Insulating	TiO_2	
				ZnO	Varistor
		Insulator		Al_2O_3	Substrate
Passivation on Si devices	SiO_2				
		10^{-18}			

Figure 7.6

Range of electronic (right-hand side) and ionic (left-hand side) conductivities in $\Omega^{-1}cm^{-1}$ exhibited by ceramics and some of their uses.[94]

The total conductivity is sometimes expressed in terms of the **transference** or **transport number**, defined as

$$t_i = \frac{\sigma_i}{\sigma_{tot}} \tag{7.44}$$

from which it follows that $\sigma_{tot} = t_{elec}\sigma_{tot} + t_{ion}\sigma_{tot}$, where t_{ion} is the ionic transference number and includes both anions and cations and t_{elec} is the electronic transference number which includes both electrons and electron holes. For any material $t_{ion} + t_e = 1$.

From Eq. (7.40) it follows that an understanding of the factors that affect conductivity boils down to an understanding of how both the mobility and the concentration of mobile carriers, be they ionic or electronic, vary with temperature, doping, surrounding atmosphere, etc.

---

[94] H. Tuller in *Glasses and Ceramics for Electronics*, 2d ed., Buchanan, (ed.), Marcel Dekker, New York, 1991. Reprinted with permission.

7.3.2   Ionic Conductivity

By definition, t_{ion} for an ionic conductor should be ≈ 1, that is, $\sigma_{elec} \ll \sigma_{ion} \approx \sigma_{tot}$. In these solids, the mobile carriers are the charged ionic defects, or $c_{m,i} = c_{def}$, where c_{def} represents the concentration of vacancies and/or interstitials.[95] Replacing $c_{m,i}$ by c_{def} in Eq. (7.42) and comparing the resulting expression with Eq. (7.34), one sees immediately that

$$\mu_{d,i} = \frac{|z_i|e_i D_{def}}{kT} = \frac{|z_i|e_i D_{ion}}{kT \not{s}} \tag{7.45}$$

This is an important result because it implies that the mobility of a charged species is directly related to its defect diffusivity, a not-too-surprising result since the mobility of an ion must reflect the ease by which the defects jump around in a lattice.   Note that if diffusion is occurring by a vacancy mechanism, $\not{s} \approx c_{def}/c_{ion} \ll 1.0$, whereas if diffusion is occurring by an interstitial mechanism, then $\not{s} \cong 1.0$ and $\mu_{int} = |z_i|e_i D_{int}/(kT)$.

EXPERIMENTAL DETAILS: MEASURING IONIC CONDUCTIVITY

There are several techniques by which ionic conductivity can be measured; one of the simpler setups is shown schematically in Fig. 7.7. Here two compartments of, say, molten Na are separated by a solid electrolyte or membrane that is known to be an Na^+ ion conductor (i.e., $t_e \ll t_{ion} \approx 1.0$). The application of a dc voltage V will result in the flow of an ionic current I_{ion} from the anode to the cathode. If one assumes that electrode polarization effects can be ignored, then the ratio V/I_{ion} is a measure of the ionic resistance of the solid, which is easily converted to a conductivity if the cross-sectional area through which the current is flowing and the thickness of the solid membrane are known.

---

[95]   The rationale for using the number of mobile carriers rather than the total number of ions involved is similar to the one made for Eq. (7.29). Here it can be assumed that one sublattice is the "pipe" through which the conducting ions are flowing. Referring to Fig. 7.7, the ions or defects enter the solid on one side and leave at the other. In contrast to the case where the crystal as a whole is placed in a chemical potential gradient, here the crystal itself does *not* move relative to an external frame of reference.

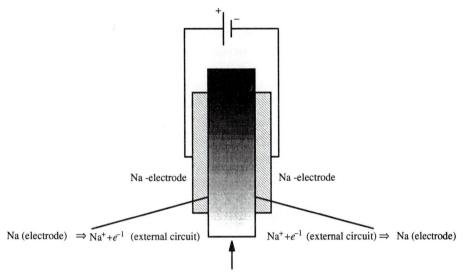

Na (electrode) $\Rightarrow$ Na$^+$+e^{-1} (external circuit)

Na$^+$+e^{-1} (external circuit) $\Rightarrow$ Na (electrode)

Na$^+$ ion conductor or solid electrolyte

Figure 7.7

Experimental setup for measuring ionic conductivity. If the electrodes are nonblocking, then the ionic conductivity is simply $R_{ion} = V/I_{ion}$.

For this experiment to work, the following reaction

$$Na(electrode) \Rightarrow Na^+(in\ solid) + e^{-1}(in\ external\ circuit)$$

has to occur at the anode. Simultaneously, the reverse reaction

$$Na^+(solid) + e^{-1}(from\ molten\ Na) \Rightarrow Na(electrode)$$

has to occur at the cathode. (Given the polarity shown in Fig. 7.7, eventually all the Na will end up on the right-hand side.) It follows that to be able to measure ionic conductivity, two requirements are critical:

1. The solid has to conduct ions rather than electrons or holes. If that were not the case, the current would simply be carried by electronic defects.
2. The electrodes have to be nonblocking to the electroactive species (Na$^+$ in this case), i.e., the Na$^+$ ions had to be able to cross unhindered from the solid electrolyte into the liquid electrode and vice versa. None of this is very surprising to anyone familiar with rudimentary electrochemistry; the only difference here is that the ions actually pass through a solid.

For the most part, ceramics, if they conduct at all, are electronic conductors. Sometimes if the band gap is large (see Worked Example 7.4) and the ceramic is exceptionally pure, it is possible to measure its ionic conductivity. In general, however, these conductivities are quite low. There is a certain class of solids, however, that exhibit ionic conductivities that are exceptionally high and can even approach those of molten salts (that is, $\sigma > 10^{-2}\,\text{Scm}^{-1}$). These solids are known as **fast ion conductors** (FICs), sometimes also referred to as **solid electrolytes**. Figure 7.8 shows the temperature dependence of a number of these solids, both crystalline and amorphous.

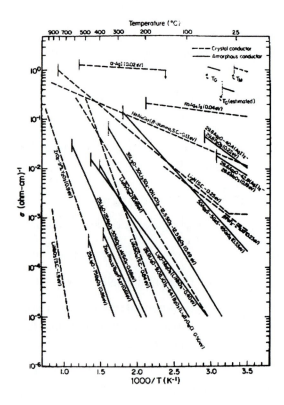

Figure 7.8

Ionic conductivities as a function of reciprocal temperature of a number of fast Ag, Na, Li, and $\overline{\text{F}}$ ion conductors.[96] T_g denotes the glass transition temperature (see Chap. 9).

---

[96] H. Tuller, D. Button, and D. Uhlmann, *J. Non-Crystalline Solids*, **40**: 93, 1980. Reprinted with permission.

FICs fall in roughly three groups. The first is based on the halides and chalcogenides of silver and copper. The most notable of these is α - AgI which is a silver ion conductor. The second is alkali metal conductors based on nonstoichiometric aluminates, the most important of which is β- Al_2O_3, with the approximate formula $Na_2O \cdot 11Al_2O_3$. The third is based on oxides with the fluorite structure that have been doped with aliovalent oxides to create a large number of vacancies on the oxygen sublattice and hence are oxygen ion conductors. In all these structures, the concentration of defects is quite large and, depending on the class of FIC, is accomplished either intrinsically or extrinsically. The halides and the β-aluminas are good examples of "intrinsic" FIC where, as a consequence of their structures (see Fig. 7.9, for example), the conduction planes contain a large number of vacant sites. Furthermore, the activation energy needed for migration ΔH_m^* between sites is quite small and is in the range of 0.01 to 0.2 eV, which results in quite large defect mobilities and hence conductivities.

Calcia-stabilized zirconia is an example where the number of defects is extrinsically controlled by aliovalent doping. For every 1 mol of CaO added to ZrO_2, 1 mol of oxygen vacancies is created according to

$$CaO \underset{ZrO_2}{\Rightarrow} Ca_{Zr}'' + V_O^{\bullet\bullet} + O_O^x$$

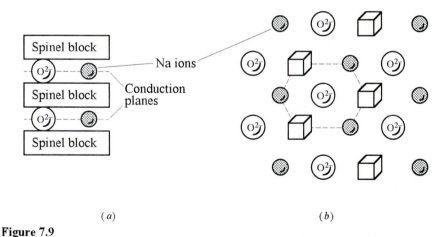

(a)   (b)

Figure 7.9

Structure of ß-alumina. (a) Plane parallel to c axis. (b) Arrangement of atoms in conduction plane (i.e., top view of conduction planes). Empty squares denote equivalent Na ion sites that are vacant.

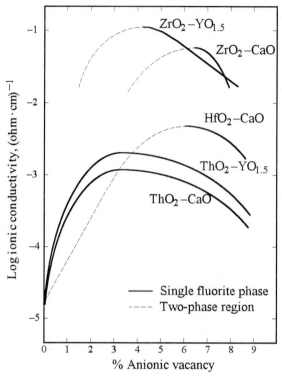

Figure 7.10

Effect of doping on the ionic conductivity of a number of oxygen ion conductors.[97]

Consequently one would expect the conductivity to be a linear function of doping, which is indeed the case, as shown in Fig. 7.10. The conductivity does not increase monotonically, however, but rather goes through a maximum at higher doping levels, a fact that has been attributed to defect-defect interactions and the breakdown of the dilute approximation.

In general, most fast ionic conductors are characterized by the following two common structural features. First, a highly ordered, immobile sublattice provides the framework and defines continuous open channels for ion transport. Second, a highly disordered complementary mobile carrier sublattice has an excess of total equipotential sites n_0, compared to the number of available mobile ions n_{mob} that fill them. Under these conditions, it can be shown (see Prob. 7.2b) that the conductivity of FICs can be expressed as[98]

97 Adapted from W.H. Flygare and R. A. Huggins, *J. Phys. Chem. Sol.*, **34**: 1199, 1973.
98 This expression is valid only if defect-defect interactions are ignored.

$$\sigma_{FIC} = (const)\frac{\beta[1-\beta]}{kT}\exp\left(-\frac{\Delta H_m^*}{kT}\right) \tag{7.46}$$

where $\beta = n_{mob}/n_0$. This function reaches a maximum when $\beta = 1/2$. Refer once again to Fig. 7.9b. The reason why this expression takes this form becomes obvious: Maximum conductivity will occur when the number of ions in the plane equals the number of vacant sites.

7.3.3 Electronic Conductivity

Electronic conductivity, like its ionic counterpart, is governed by Eq. (7.42), and is proportional to the concentration of mobile *electronic* carriers, both electrons and holes, and their mobility. In general, there are three ways by which these mobile electronic carriers are generated in ceramics, namely, (1) by excitation across the band gap (intrinsic), (2) due to impurities (extrinsic), or (3) as a result of departures from stoichiometry (nonstoichiometric). Each is considered in some detail below.

Intrinsic semiconductors

In this case the electrons and holes are generated by excitation across the band gap of the material. For every electron that is excited into the conduction band, a hole is left behind in the valence band; consequently, for an intrinsic semiconductor, $n = p$.

To predict the number of electrons which are excited across the band gap at any given temperature, both the density-of-states function and the probability of occupancy of each state must be known (see App. 7B for more details). The **density of states** is defined as the number of states per unit energy interval in the vicinity of the band edges. The probability of their occupancy is given by the Fermi energy function, namely,

$$f(E) = \frac{1}{1 + \exp[(E - E_f)/(kT)]} \tag{7.47}$$

This equation is plotted as a function of temperature in Fig. (7.11). Here E is the energy of interest, and E_f is the Fermi energy, defined as the energy for which the probability of finding an electron is 0.5. And k and T have their usual meanings. It can be shown (see App. 7B) that for this energy function, the number of electrons per cubic meter in the conduction band is given by:

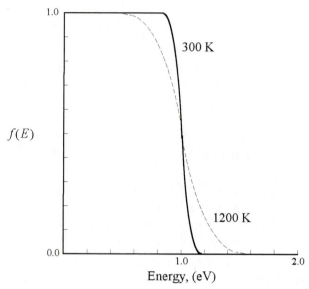

Figure 7.11
Fermi distribution function for two different temperatures; E_f was assumed to be 1 eV. Note that whereas the distribution shifts to higher energies as the temperature increases, E_f, defined as the energy at which the probability of finding an electron is 0.5, does not change, however.

$$n = N_c \exp\left(-\frac{E_c - E_f}{kT}\right) \qquad (7.48)$$

and the number of holes by

$$p = N_v \exp\left(-\frac{E_f - E_v}{kT}\right) \qquad (7.49)$$

where E_c and E_v refer to the energy of the lowest and highest levels in the conduction and valence bands, respectively (see Fig. 7.12); N_c and N_v were defined in Chap. 6 and are given by (see App. 7B)

$$N_c = 2\left(\frac{2\pi m_e^* kT}{h^2}\right)^{3/2} \text{ and } N_v = 2\left(\frac{2\pi m_h^* kT}{h^2}\right)^{3/2}$$

where m_i^* is the effective mass of the migrating species. The product of Eqs. (7.48) and (7.49) yields

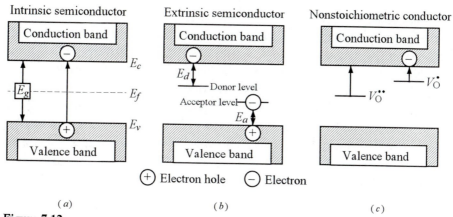

Figure 7.12

Schematic of energy levels for (*a*) intrinsic semiconductors, (*b*) extrinsic semiconductors, and (*c*) nonstoichiometric semiconductors.

$$np = N_c N_v \exp\left(-\frac{E_g}{kT}\right) = N_c N_v K_i \tag{7.50}$$

which is not a function of E_f and which, not surprisingly, is identical to Eq. (6.20). As noted above for an intrinsic semiconductor, $n = p$, that is,

$$n = p = \sqrt{N_c N_v} \exp\left(-\frac{E_g}{2kT}\right) \tag{7.51}$$

and the conductivity is given by

$$\sigma = e\mu_n n + e\mu_p p = ne(\mu_n + \mu_p) \tag{7.52}$$

Up to this point, the mobility of the electronic carriers and their temperature dependencies were not discussed. The effect of temperature on the mobility of the electrons and holes will depend on several factors, with the width of the conduction and/or valence bands being the most important. For wideband materials (not to be confused with wide-band-gap materials), the mobility of the electronic carriers decreases with increasing temperature as a result of lattice or phonon scattering, not unlike what happens in metallic conductors. In can be shown that in this case both μ_n and μ_p are proportional to $T^{-3/2}$. Thus the temperature dependence of $\sigma_{\text{el, int}}$ is given by

$$\sigma_{\text{el, int}} = K'' T^{-3/2} T^{3/2} \exp\left(-\frac{E_g}{2kT}\right) = K'' \exp\left(-\frac{E_g}{2kT}\right) \tag{7.53}$$

This result is applicable to an intrinsic semiconductor for which phonon scattering is responsible for the temperature dependence of the electronic mobility, i.e., one in which the mobility *decreases* with *increasing* temperature. Other possibilities exist, however; two of the more important ones are the **small** and **large** **polaron** mechanisms discussed below. A *polaron* is a defect in an ionic crystal that is formed when an excess of charge at a point polarizes or distorts the lattice in its immediate vicinity. For example, if an oxygen vacancy captures an electron[99] (Fig. 6.4*b*), the cations surrounding it will be attracted to the defect and move toward it, whereas the anions will move away. This polarization essentially traps or slows down the electronic defect as it moves through the lattice.

Small polaron. In this mechanism, conduction occurs by the "hopping" of the electronic defects between adjacent ions of, usually but not necessarily, the same type but with varying oxidation states. Because of the ease by which transition-metal ions can vary in oxidation states, this type of conduction is most often observed in transition metal oxides. For example, if the charge carrier is an electron, the process can be envisioned as

$$M^{+n} + e^{-1} \Rightarrow M^{n-1}$$

Polarization of the lattice results in a reduction of the energy of the system, and the carrier is then assumed to be localized in a potential energy well of effective depth E_B. It follows that for migration to occur, the carrier has to be supplied with at least that much energy, and consequently the mobility becomes thermally activated. It can be shown (see Prob. 7.2*c*) that polaron conductivity can be described by an expression very similar to that for FICs, that is,

$$\sigma_{\text{hop}} = (\text{const}) \frac{x_i[1 - x_i]}{kT} \exp\left(-\frac{E_B}{kT}\right) \qquad (7.54)$$

where x_i is the fraction of sites occupied by one charge and $[1 - x_i]$ the fraction of sites occupied by the other charge. For example, for polaron hopping between Fe cations, x and $1 - x$ represent the concentrations of Fe^{2+} and Fe^{3+} cations, respectively. Based on this simple model, one would expect a conduction maximum at $x \approx 0.5$. Experimentally, this is not always observed, however, for reasons that are not entirely clear.

Large polaron. If the distortion is not large enough to totally trap the electron but is still large enough to slow it down, the term *large polaron* is applicable. Large polarons behave as free carriers except that they have a higher effective mass

---

99 Note that the combination of a trapped electron or hole at an impurity or defect is called a *color center*, which can have very interesting optical and magnetic properties (see Chap. 16).

than a free electron. It can be shown that the large polaron mobility is proportional to $T^{-1/2}$, and consequently, the temperature dependence of an intrinsic semiconductor for which the conductivity occurs by large polarons is

$$\sigma = (\text{const})T^{3/2}T^{-1/2}\exp\left(-\frac{E_g}{2kT}\right) = (\text{const})T\exp\left(-\frac{E_g}{2kT}\right) \quad (7.55)$$

WORKED EXAMPLE 7.4. A good solid electrolyte should have an ionic conductivity of at least 0.01 $(\Omega \cdot \text{cm})^{-1}$ with an electronic transference number that should not exceed 10^{-4}. At 1000 K, show that the minimum band gap for such a solid would have to be ≈ 4 eV. Assume that the electronic and hole mobilities are equal and that each is $100\,\text{cm}^2/(\text{V}\cdot\text{s})$. State all other assumptions.

Answer

Based on the figures of merit stated, the electronic conductivity should not exceed

$$\sigma_{\text{elec}} = t_e\sigma_{\text{tot}} \approx t_e\sigma_{\text{ion}} = 0.01 \times 10^{-4} = 1 \times 10^{-6}(\Omega \cdot \text{cm})^{-1} = 1 \times 10^{-4}(\Omega \cdot \text{m})^{-1}$$

Inserting this value for σ_{elec} in Eq. (7.52), assuming $\mu_e = \mu_p = 0.01\,\text{m}^2/(\text{V}\cdot\text{s})$ at 1000 K, and solving for n yields

$$n = \frac{1 \times 10^{-4}}{1.6 \times 10^{-19}(0.01 + 0.01)} = 3.12 \times 10^{16} \text{ electrons}/\text{m}^3$$

Furthermore by assuming that $m_e = m_e^*$, it follows that

$$N_c = 2\left(\frac{2\pi m_e^* kT}{h^2}\right)^{3/2} = 2\left(\frac{2\pi \times 1.9 \times 10^{-31} \times 1.38 \times 10^{-23} \times 1000}{(6.63 \times 10^{-34})^2}\right)^{3/2} = 1.5 \times 10^{26}\,\text{m}^{-3}$$

Finally, assuming $N_c = N_v$, substituting the appropriate values in Eq. (7.51) and solving for E_g give

$$E_g = 2kT\ln\frac{n}{N_c} = 2 \times 1000 \times 8.62 \times 10^{-5}\ln\frac{3.12 \times 10^{16}}{1.5 \times 10^{26}} = 3.84 \text{ eV}$$

Extrinsic semiconductors

The conductivity of extrinsic semiconductors is mainly determined by the presence of foreign impurities. The best way to illustrate the notion of an extrinsic semiconductor is to take the specific example of elemental silicon. Consider the addition of a known amount of a group V element, such as phosphorus, such that each P atom substitutionally replaces a Si atom. Four electrons will be used up to bond with the Si, leaving behind a dangling extra electron (Fig. 7.13a). Given that this electron will not be as tightly bound as the others, it is easily promoted into the conduction band. In other words, each P atom donates one electron to the conduction band and in so doing increases the conductivity of the silicon host, which is now termed an **n-type semiconductor**. In terms of band theory, a donor is depicted by a localized level within the band gap, at an energy E_d lower than E_c (Fig. 7.12b), where E_d is a measure of the energy binding the electron to its site.

Conversely, if Si is doped with a group III atom, such as Al (Fig. 7.13b) which only has three electrons in its outer shell, a hole will be created. Upon application of an electric field, the hole acts as a vacancy into which electrons from neighboring atoms can jump. Such a material is called a **p-type semiconductor**, and the corresponding energy diagram is also shown in Fig. 7.12b (bottom of diagram).

To quantify the conductivity of an extrinsic semiconductor, consider an n-type semiconductor doped with a concentration N_D of dopant atoms. The ionization reaction of the donor can be written as

$$D \Leftrightarrow D^{\bullet} + e'$$

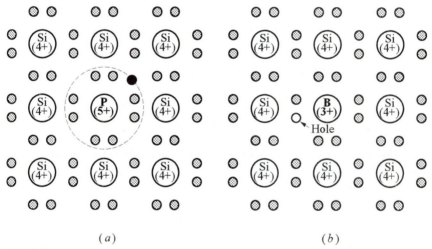

(a)   (b)

Figure 7.13

Schematic of (a) an n-type semiconductor and (b) a p-type semiconductor. The missing electron (definition of a hole) is formed when Al is added to Si. With the input of energy E_a, an electron from the valence band is promoted into the site labeled *electron hole*, and a hole is created in the valence band.

Mass balance dictates that

$$N_D = D + D^\bullet$$

where D and $D^\bullet$ denote, respectively, the concentrations of un-ionized and ionized donors. The corresponding mass action expression is

$$\frac{[D^\bullet][n]}{[D]} = \exp\left(-\frac{E_d}{kT}\right) \tag{7.56}$$

where E_d is the energy required to ionize the donor (see Fig. 7.12b). Note once again that $[D]$ and $[D^\bullet]$ denote, respectively, the mole fractions of un-ionized and ionized donors, whereas $[n] = n/N_c$ and $[D^\bullet] = D^\bullet/N_D$.

To make the problem more tractable, consider the following three temperature regimes:

Low-temperature region (region 1 in Fig. 7.14). At low temperatures, when only a few donors are ionized, the neutrality condition can be written as $n = D^\bullet \ll N_D$. In other words, $[D] \approx 1$ and $n = D^\bullet$. Making the appropriate substitutions in Eq. (7.56), one obtains

$$\sigma_{\text{elec}} = \sigma_n = e\mu_n\sqrt{N_D N_c}\,\exp\left(-\frac{E_d}{2kT}\right) \tag{7.57}$$

Thus in this region a plot of log σ_n versus reciprocal temperature should yield a straight line with slope $E_d/(2k)$.

Intermediate-temperature regime (region 2 in Fig. 7.14). In this region, $kT \approx E_d$, and one can assume that most of the donor atoms are ionized. Hence the total number of mobile carriers will simply equal N_D, in which case

$$\sigma_{\text{elec}} = \sigma_n = eN_D\mu_n \tag{7.58}$$

In this region, the conductivity will be a weak function of temperature and may even decrease with increasing temperature if phonon scattering becomes important.

High-temperature region (region 3 in Fig. 7.14). Here it is assumed that the temperature is high enough that the number of electrons excited from the valence band into the conduction band dominates, in which case the semiconductor behaves intrinsically and conductivity is given by Eq. (7.53). In this region, a plot of log σ_n versus reciprocal temperature yields a straight line with slope $E_g/(2k)$ (Fig. 7.14).

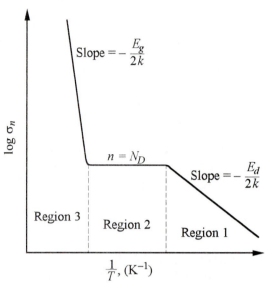

Figure 7.14

Temperature dependence of the electronic conductivity of an extrinsic semiconductor. Region 2 is sometimes referred to as the *exhaustion region*.

To reiterate: At low temperatures, the conductivity is low because of the paucity of mobile carriers — they are all trapped. As the temperature increases, the defects start to ionize and the conductivity increases with an activation energy needed to release the electrons from their traps. At intermediate temperatures, when $kT \approx E_d$, most of the impurities will have donated their electrons to the conduction band, and a saturation in the conductivity sets in. With further increases in temperature, however, it is now possible (provided the crystal does not melt beforehand) to excite electrons clear across the band gap, and the conductivity starts increasing again, but this time with a slope that is proportional to $E_g/2k$.

Nonstoichiometric semiconductors

In the preceding subsection, the number of electronic defects was fixed by the doping level, especially at lower temperatures, and the concepts of donor and acceptor localized levels were discussed. The band picture for nonstoichiometric electronic semiconductors is very similar to that of extrinsic semiconductors, except that the electronic defects form not as a result of doping, but rather by varying the stoichiometry of the crystal.

To appreciate the similarities between an extrinsic semiconductor and a nonstoichiometric oxide, compare Fig. 7.13a with Fig. 6.3a or b. In both cases the electron(s) is(are) loosely bound to its(their) mooring(s) and is(are) easily excited into the conduction band. The corresponding energy diagrams for the singly ionized and doubly ionized oxygen vacancies are shown in Fig. 7.12c. In essence, a nonstoichiometric semiconductor is one where the electrons and holes excited in the conduction and valence bands are a result of reduction or oxidation. For example, the reduction of an oxide entails the removal of oxygen atoms, which have to leave their electrons behind to maintain electroneutrality. These electrons, in turn, are responsible for conduction.

The starting point for understanding the behavior of nonstoichiometric oxides involves constructing their Kroger-Vink diagram, as discussed in Sec. 6.2.5. To illustrate, consider the following examples:

ZnO. Experimentally, it has been found that the electrical conductivity of ZnO decreases with increasing oxygen partial pressure, as shown schematically in Fig. 7.15a. When plotted on a log-log plot, the resulting slope is measured to be 0.25, which is explained as follows. The defect incorporation reaction is presumed to be

$$O_O^x + Zn_{Zn}^x \Leftrightarrow \frac{1}{2}O_2(g) + Zn_i^{\bullet} + e^{-1}$$

for which the mass action expression is

$$P_{O_2}^{1/2}[Zn_i^{\bullet}][n] = \text{const}$$

Combining this expression with the electroneutrality condition $Zn_i^{\bullet} \approx n$ results in

$$\sigma \propto n = (\text{const})P_{O_2}^{-1/4}$$

as observed. Note that had one assumed the Zn interstitials to be doubly ionized, the P_{O_2} dependence predicted would not have been consistent with the experimental results.

Cu_2O. In contradistinction to ZnO, the conductivity of Cu_2O increases with increasing oxygen partial pressure with a slope of $\approx +1/7$. To explain this result, the following incorporation reaction is assumed:

$$\frac{1}{2}O_2(g) \Leftrightarrow O_O^x + 2V_{Cu}' + 2h^{\bullet}$$

which, when combined with the neutrality condition $V_{Cu}' = p$ and the mass action expression, results in

$$\sigma \propto p = (\text{const})P_{O_2}^{1/8}$$

which is not too far off the measured value of 1/7 shown in Fig. 7.15b.

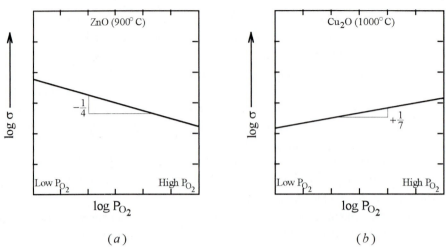

Figure 7.15
Schematic representation of changes in conductivity as a function of oxygen partial pressure for (a) ZnO and (b) Cu_2O. These curves are based on actual experimental results.

CoO. CoO is a metal-deficient oxide[100] $Co_{1-x}O$ (see Table 6.1), where the conductivity is known to be p type and thermally activated, i.e., occurs by polaron hopping such that $\mu = (const)[\exp - E_B/kT]$. At high oxygen partial pressures, the conductivity changes are steeper with a slope of $\approx +1/4$, whereas at lower partial pressure the slope changes to $\approx +1/6$. This suggests that at high oxygen partial pressures the defect reaction is given by

$$\frac{1}{2}O_2(g) \Leftrightarrow O_O^x + V_{Co}' + h^\bullet \quad \Delta G_{V_{Co}}$$

with an equilibrium constant $K_{V_{Co}'} = \exp[-\Delta G_{V_{Co}}/(kT)]$. Making use of the corresponding mass action expression and the electroneutrality condition $p = V_{Co}'$, one obtains

$$\sigma = \sigma_p = pe\mu_p = \frac{const}{T}(K_{V_{Co}})^{1/2} P_{O_2}^{1/4} \exp\left(-\frac{E_B}{kT}\right)$$

which is the pressure dependence observed. Furthermore, since

100 The actual situation for CoO is not as simple as put forth here. For the latest interpretation, see H-I Yoo, J.-H. Lee, M. Martin, J. Janek, and H. Schmelzried, *Solid State Ionics*, **67**:317–322 (1994).

$$K'_{V_{Co}} = \exp\frac{\Delta S_{V'_{Co}}}{k}\exp\left(-\frac{\Delta H_{V'_{Co}}}{kT}\right)$$

the final expression for the electrical conductivity is given by

$$\sigma = \sigma_p = pe\mu_p = \frac{\text{const}}{T}\exp\left(-\frac{\Delta H_{V'_{Co}}}{2kT}\right)\exp\left(-\frac{E_B}{kT}\right)P_{O_2}^{1/4} \qquad (7.59)$$

Similarly, at lower oxygen potentials, the data suggest that the corresponding reaction is one where doubly ionized oxygen vacancies form, namely,

$$\frac{1}{2}O_2(g) \Leftrightarrow O_O^x + V_{Co}'' + 2h^\bullet \qquad \Delta G_{V_{Co}''}$$

It is left as an exercise for the reader to show that in the low oxygen partial pressure regime

$$\sigma = \sigma_p = \frac{\text{const}}{T}\exp\left(-\frac{\Delta H_{V_{Co}''}}{3kT}\right)\exp\left(-\frac{E_B}{kT}\right)P_{O_2}^{1/6} \qquad (7.60)$$

where $\Delta H_{V_{Co}''}$ is the enthalpy of formation of the doubly ionized oxygen vacancies.

It is worth noting that when the conductivity is dominated by redox reactions such as the ones discussed here, the final expression does not depend on the band gap or on doping but rather depends on the ease with which an oxide is oxidized or reduced, i.e., on Δg_{red} and Δg_{oxid}. Generally this is directly related to the ability of the cations to exist in more than one oxidation state — and consequently it is intimately related to the range of nonstoichiometry.

Zirconia. The Kroger-Vink diagram for yttria-doped zirconia is shown in Fig. 7.16a, the construction of which is left as an exercise to the reader. In pure zirconia, the concentration of oxygen vacancies is simply $\sqrt{K_s}$. However, as noted earlier, that value can be dramatically increased by doping with aliovalent cations such as Ca^{2+} or Y^{3+}. Based on this diagram, in the range where the conductivity is ionic, the minority carriers are electronic defects: At high oxygen partial pressures, $p \gg n$; at low oxygen partial pressures, the conduction is mixed.

Actual data for yttria-doped zirconia are shown in Fig. 7.16b as a function of temperature, and generally confirm the results shown in Fig. 7.16a. At 1000°C and 10^{-10} atm pressure, $t_{ion} \approx 10^{-1}/(10^{-1}+1\times10^{-6}) \approx 1.0$, whereas at 10^{-28} atm., $t_{ion} \approx t_{elec} \approx 0.5$.

Note that, in contrast to electronic conduction, the ionic conductivity is P_{O_2} independent because the concentration of ionic defects is fixed extrinsically. Such an independence of conductivity on partial pressure is usually taken to be strong evidence that a solid is indeed an ionic conductor.

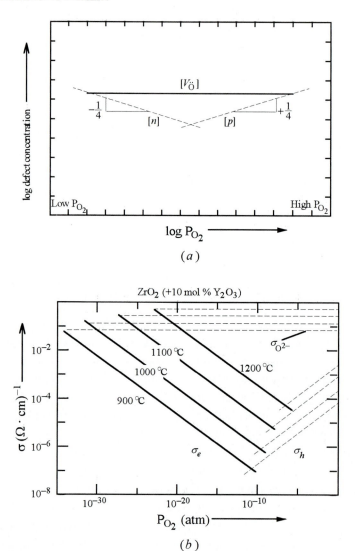

Figure 7.16

(a) Schematic of defect concentration dependence on oxygen partial pressure for yttria-doped zirconia. (b) Functional dependence of ionic and electronic conductivities on oxygen partial pressure and temperature[101] for ZrO_2 (+10 mol % Y_2O_3). Note that except at very low oxygen partial pressures where $\sigma_e \approx \sigma_{ion}$, the conductivity is ionic and independent of oxygen partial pressure.

101 L. D. Burke, H. Rickert, and R. Steiner; Z. Physik. Chem. N.F., **74**:146 (1971).

WORKED EXAMPLE 7.5. In a now classic paper,[102] Kingery et al. measured the conductivity of $Zr_{0.85}Ca_{0.15}O_{1.85}$ as a function of oxygen partial pressure and temperature. They found that the conductivity (S/m) was independent of oxygen partial pressure and obeyed the relation

$$\sigma = 1.5 \times 10^5 \exp\left(-\frac{1.26 \text{ eV}}{kT}\right)$$

The diffusion coefficient of the oxygen ions (m^2/s) was also measured in a *separate* experiment on the same material and was found to obey

$$D = 1 \times 10^{-6} \exp\left(-\frac{1.22 \text{ eV}}{kT}\right)$$

What conclusions can be reached regarding the conduction mechanisms in this oxide and its defect structure? Information you may find useful: density of zirconia $\approx 6.1 g/cm^3$ and molecular weight of Zr is $91.22 g/mol$.

Answer

Since the total conductivity was not a function of oxygen partial pressure, one can assume that the conductivity was ionic.[103] The total number of oxygen ions per cubic meter is

$$c_{ions} = \frac{6.1 \times 1.85 \times 6.02 \times 10^{23}}{(91.22 + 32) \times 10^{-6}} = 5.52 \times 10^{28} \text{ ions}/m^3$$

Using the Nernst-Einstein relationship and converting all units to SI units, one obtains the following expression for the conductivity:

$$\sigma_{O \text{ ions}} = \frac{(2^2)(1.6 \times 10^{-19})^2 \times (5.5 \times 10^{28}) \times (1 \times 10^{-6})}{1.38 \times 10^{-23} \times T} \exp\left(-\frac{1.22 \text{ eV}}{kT}\right)$$

$$= \frac{4.07 \times 10^8}{T} \exp\left(-\frac{1.22 \text{ eV}}{kT}\right) S/m$$

At 1000 K, the preexponential term yields a value of 4.07×10^5 S/m, which is in fairly close agreement with the preexponential term in the conductivity expression shown above. The fit is even better at higher temperatures. These results unambiguously prove that conduction in calcia-stabilized zirconia occurs by the movement of oxygen ions.

---

[102] W. D. Kingery, J. Pappis, M. E. Doty, and D. C. Hill, "Oxygen Mobility in $Zr_{0.85}Ca_{0.15}O_{1.85}$," *J. Amer. Cer. Soc.*, **42**:(8): 393–398 (1959).

[103] This is not always the case. There are situations where the conductivity can be electronic and yet partial-pressure-independent.

7.4
AMBIPOLAR DIFFUSION

In the discussion so far, the diffusional and electrical fluxes of the ionic and electronic carriers were treated separately. However, as will become amply clear in this section and was briefly touched upon in Sec. 5.6, in the absence of an external circuit such as the one shown in Fig. 7.7, the diffusion of a charged species by itself is very rapidly halted by the electric field it creates and thus *cannot* lead to steady-state conditions. For steady state, the fluxes of the diffusing species have to be coupled such that electroneutrality is maintained. Hence, in most situations of great practical importance such as creep, sintering, oxidation of metals, efficiency of fuel cells, and solid-state sensors, to name a few, it is the *coupled diffusion*, or *ambipolar diffusion*, of two fluxes that is critical. To illustrate, four phenomena that are reasonably well understood and that are related to this coupled diffusion are discussed in some detail in the next subsections. The first deals with the oxidation of metals, the second with ambipolar diffusion in general in a binary oxide, the third with the interdiffusion of two ionic compounds to form a solid solution. The last subsection explores the conditions for which a solid can be used as a potentiometric sensor.

7.4.1   Oxidation of Metals

To best illustrate the notion of ambipolar diffusion, the oxidation of metals will be used as an example following the elegant treatment first developed by C. Wagner.[104] Another reason to go into this model is to appreciate that it is usually the electrochemical potential, rather than the chemical or electric potential, that is responsible for the mobility of charged species in solids. It also allows a link to be made between the notions of chemical stability and nonstoichiometry. However, before one proceeds much further, it is instructive to briefly review how oxidation rates are measured and to introduce the parabolic rate constant.

EXPERIMENTAL DETAILS: MEASURING OXIDATION RATES

Oxidation rates can be measured by a variety of methods. One of the simplest is to expose the material for which the oxidation resistance is to be measured (typically metal foils) to an oxidizing atmosphere of a given oxygen partial pressure for a

---

104 C. Wagner, *Z. Physikal. Chem.*, **B21**:25 (1933).

given time, cool, and measure the thickness of the oxide layer that forms as a function of time. Long before any atomistic models were put forth, it was empirically fairly well established that for many metals the oxidation rate was parabolic. In other words, the increase in thickness Δx of the oxide layer was related to time by

$$\Delta x^2 = 2K_x t \tag{7.61}$$

with the proportionality constant $K_x \left(m^2/s \right)$, known as the **parabolic rate constant,** a function of both temperature and oxygen partial pressure.

An alternate technique is to carry out a thermogravimetric experiment and measure the weight gain of the material for which the oxidation resistance is to be measured as a function of time (Fig. 7.17). In this case, the weight change per *unit area* Δw is related to time by[105]

$$\Delta w^2 = K_w t \tag{7.62}$$

where $K_w \left(kg^2/m^4 \cdot s \right)$ is also a constant that depends on temperature and oxygen partial pressure. Needless to say, K_x and K_w are related (see Prob. 7.17). K_w is also known as a parabolic rate constant.

In the remainder of this subsection, the goal is to relate these phenomenological rate constants to more fundamental parameters of the growing oxide layer. Table 7.1 lists the parabolic rate constants for a number of metals oxidized in pure oxygen at 1000 °C.

TABLE 7.1
Parabolic rate constants K_w, for various metals oxidized in 1 atm oxygen at 1000 °C

Metal-oxide	Fe-FeO	Ni-NiO	Co-CoO	Cr-Cr$_2$O$_3$	Si-SiO$_2$
$K_w \left(kg^2/m^4 \cdot s \right)$	4.8×10^{-5}	2.9×10^{-6}	2.1×10^{-8}	1.0×10^{-8}	1.2×10^{-10}

For the sake of illustration, consider one of the simplest possible cases depicted schematically in Fig. 7.18a. Here a metal is exposed to oxygen at elevated temperatures, and an oxide layer is formed by the *outward* diffusion of *cation interstitials,* henceforth denoted by the subscript "def" for defect, together with electrons. Note that both the cations and electrons are diffusing in the same direction — ZnO is a good example of such an oxide. It follows that at the metal/oxide interface (Fig. 7.18a), the incorporation reaction is

---

[105] Sometimes Eq. (7.62) is written with a factor of 2 in analogy to Eq. (7.61). In this book K_w is defined as shown.

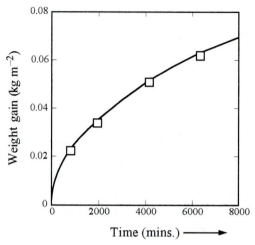

Figure 7.17
Weight gain during the oxidation of Ti_3SiC_2 in air at 1000°C.

$$M^* \leftrightarrow M_{def}^{z+} + ze^{-1} \qquad (7.63)$$

where z is the valence on the cation and M^* denotes the neutral or metallic species. From here on, the superscript asterisk* denotes neutral species.

Conversely, at the oxide/gas interface, the cation and electron will combine with oxygen according to

$$\frac{z}{4}O_2(g) + M^{z+} + ze^{-1} \leftrightarrow MO_{z/2} \qquad (7.64)$$

forming a reaction layer that grows with time with a velocity that should be proportional to the flux of the defects through that layer. The net reaction is simply the sum of Eqs. (7.63) and (7.64), or

$$\frac{z}{4}O_2(g) + M^* \leftrightarrow MO_{z/2} \quad \Delta G_{MO_{z/2}} \qquad (7.65)$$

where $\Delta G_{MO_{z/2}}$ is the free energy of formation of the growing oxide layer.

Make the following assumptions:

1. The process is diffusion-limited through a scale that is compact, fully dense, and crack-free. Such layers will occur only if the volume change upon oxidation is not too great. Otherwise, the growing oxide layer cannot accommodate the mismatch strain that develops and will tend to crack or buckle.

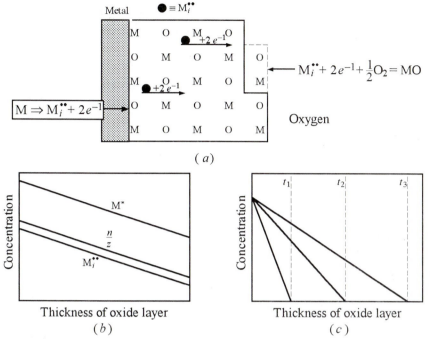

Figure 7.18

(*a*) Growth of an oxide layer by the outward diffusion of metal cations and electrons.
(*b*) Concentration profile of defects and neutral species. (*c*) Schematic of quasi-steady-state approximation – concentration profile, which is proportional to the flux at any time *t*, is not a function of *x*. Nevertheless, as the thickness of oxide layer *x* increases, the flux decreases (slope decreases). In both (*b*) and (*c*) the profile is assumed to be linear for simplicity.

A good indicator of whether an oxide layer is protective is given by the **Pilling-Bedworth ratio**

$$\text{P - B ratio} = \frac{V_{MO_{z/2}}}{V_M} = \frac{MW_{MO_{z/2}}\rho_M}{MW_M\rho_{MO_{z/2}}} \tag{7.66}$$

that compares the molar volume of the metal V_M to that of the oxide, $V_{MO_{z/2}}$. MW and ρ denote the molecular weights and densities of the metal and oxide, respectively.

For metals having a P-B ratio less than unity, the oxide tends to be porous and unprotective because the oxide that forms is insufficient to cover the underlying metal surface. For ratios greater than unity, compressive stresses develop in the film, and if the mismatch is too great, (P-B ratios > 2), the oxide coating tends to buckle and flake off, continually exposing fresh metal, and is thus nonprotective. The ideal P-B ratio is 1, but protective coatings normally form for metals having P-B ratios between 1 and 2.

2. Diffusion of charged species is by independent paths. In other words, it is assumed that the flux of species i is proportional to its electrochemical potential gradient solely and is independent of the gradient in the electrochemical potential of the other components.

3. Charge neutrality is maintained, and there is no *macroscopic* separation of charge. For instance, in the example shown above, the electronic defects, as a result of their higher mobility, will attempt to move along faster than the ions; but as they do so, they will create a *local electrical potential gradient* $d\phi/dx$ which will hold them back. That same field, however, will enhance the flux of the ionic defects. As discussed in greater detail shortly, it is this *coupling* of the two fluxes that gives rise to an effective or ambipolar diffusion coefficient.

4. There is local equilibrium both at the phase boundaries and throughout the scale. This implies that at every point, reaction (7.63) holds, or

$$\tilde{\mu}_{M^*} = z\tilde{\eta}_e + \tilde{\eta}_{def} = z(\tilde{\mu}_e - e\phi) + (\tilde{\mu}_{def} + ze\phi) \tag{7.67}$$

where ϕ is the local electric potential acting on the defects.[106] It thus follows that across the layer

$$\frac{d\tilde{\mu}_{M^*}}{dx} = \frac{d(z\tilde{\mu}_e)}{dx} + \frac{d(\tilde{\mu}_{def})}{dx} \tag{7.68}$$

Given these assumptions, now the flux of the defects can be related to the rate of growth of the layer. Assuming one-dimensional diffusion, the defect and electronic flux densities (particles per square meter per second) subject to an electrochemical potential gradient $d\tilde{\eta}_i/dx$ are given by Eq. (7.37), or

$$J_{def} = -\frac{\sigma_{def}}{(ze)^2}\frac{d\tilde{\eta}_{def}}{dx} = -\frac{\sigma_{def}}{(ze)^2}\left(\frac{d\tilde{\mu}_{def}}{dx} + ez\frac{d\phi}{dx}\right) \tag{7.69}$$

$$J_e = -\frac{\sigma_e}{e^2}\frac{d\tilde{\eta}_e}{dx} = -\frac{\sigma_e}{e^2}\left(\frac{d\tilde{\mu}_e}{dx} - e\frac{d\phi}{dx}\right) \tag{7.70}$$

To maintain electroneutrality, these two fluxes have to be equal, and mass balance dictates that they in turn must be equal to the flux of the neutral metal species J_{M^*}. *In other words,*

---

[106] As noted earlier, the tilde denotes that the quantities are expressed **per defect or per electron** rather than per mole.

$$\frac{J_e}{z} = J_{\text{def}} = J_{\text{M}^*} \tag{7.71}$$

Using this condition to solve for $d\phi/dx$ results in

$$\frac{d\phi}{dx} = \frac{t_e}{e}\frac{d\tilde{\mu}_e}{dx} - \frac{t_{\text{def}}}{ze}\frac{d\tilde{\mu}_{\text{def}}}{dx} \tag{7.72}$$

which, when substituted back in Eq. (7.69) or (7.70) while use is made of Eq. (7.68), yields

$$\boxed{J_{\text{M}^*} = \frac{-\sigma_{\text{def}}\, t_e}{(ze)^2}\frac{d\tilde{\mu}_{\text{M}^*}}{dx} = \frac{-\sigma_{\text{def}}\,\sigma_e}{(ze)^2(\sigma_{\text{def}}+\sigma_e)}\frac{d\tilde{\mu}_{\text{M}^*}}{dx}} \tag{7.73}$$

where J_{M^*} is the *number of neutral metal atoms passing through a unit area of the oxide per second* and $d\mu_{\text{M}^*}/dx$ is the chemical potential gradient of the neutral metal species.

This is an important result because it implies that for oxidation or corrosion to occur, the oxide layer must conduct *both* ions and electrons — if either σ_e *or* σ_{def} vanishes, the permeation flux also vanishes. This comes about because it was assumed early on that both the electronic and ionic defects must diffuse together to maintain charge neutrality. It also follows from Eq. (7.73) that it is the slower of the two diffusing species that is rate-limiting. If an oxide is predominantly an electronic conductor, then $t_e \approx 1$, and the permeation flux will be determined by the ionic conductivity. Conversely, if $t_{\text{ion}} \approx 1$, then the permeation rate will be determined by the rate at which the electronic defects move through the oxide layer.

Equation (7.73) also implies that the driving force for the growth of the layer is nothing but the gradient in the chemical potential of the *neutral* species, which, in turn, is nothing but the free-energy change associated with reaction (7.65). In other words, the more stable the oxide, the higher the driving force for its formation.

To relate the permeation flux to the parabolic rate constant, Wagner further assumed quasi-steady-state growth conditions. This assumption implies that the flux into the reaction layer is equal to the flux out of it and that there was no accumulation of material in the film. In other words, at any time, the flux was *not* a function of x, but was only a function of time. This condition is schematically shown in Fig. 7.18c for various times during scale growth. Mathematically it implies that the flux is inversely proportional to Δx, and hence dx in Eq. (7.73) can be replaced by Δx. Make that substitution, and note that the rate at which the oxide layer is growing is given by

$$\frac{d(\Delta x)}{dt} = J_{\text{M}^*}\Omega_{\text{MO}} = \frac{-\sigma_{\text{def}}\,\sigma_e\Omega_{\text{MO}}}{(ze)^2(\sigma_{\text{def}}+\sigma_e)}\frac{1}{\Delta x}d\tilde{\mu}_{\text{M}^*} \tag{7.74}$$

where Ω_{MO} is the atomic volume of an MO molecule.[107] Rearranging terms and integrating, one obtains

$$\Delta x^2 = 2\left[\frac{\Omega_{MO}}{(ze)^2}\int_{\mu_{M^*} \text{ at metal/oxide interface}}^{\mu_{M^*} \text{ at oxide/gas interface}}\frac{-\sigma_{def}\sigma_e}{\sigma_{def}+\sigma_e}d\tilde{\mu}_{M^*}\right]t \qquad (7.75)$$

Comparing this result to Eq. (7.61), one sees immediately that the parabolic rate constant is given by[108]

$$K_x = \frac{\Omega_{MO}}{(ze)^2}\int_{\mu_{M^*} \text{ at metal/oxide interface}}^{\mu_{M^*} \text{ at oxide/gas interface}}\frac{-\sigma_{def}\sigma_e}{\sigma_{def}+\sigma_e}d\tilde{\mu}_{M^*} \qquad (7.76)$$

By making use of the Gibbs-Duhem relation [Eq. (5.40)], that is, $d\tilde{\mu}_{M^*} = -d\tilde{\mu}_{O_2} = -(kT/2)d\ln P_{O_2}$, this relationship can be recast as

$$K_x = \frac{kT\Omega_{MO}}{2(ze)^2}\int_{P_{O_2}^I}^{P_{O_2}^{II}}\frac{\sigma_{def}\sigma_e}{\sigma_{def}+\sigma_e}d\ln P_{O_2} \qquad (7.77)$$

where $P_{O_2}^I$ and $P_{O_2}^{II}$ are the oxygen partial pressures at the metal/oxide and oxide/gas interfaces, respectively. Integrating Eq. (7.77) is nontrivial since both the electronic and ionic defect concentrations are functions of P_{O_2}. What is customarily done, however, is to assume average values for the conductivities across the layer, with the final result being

$$\boxed{K_x = \frac{kT\Omega_{MO}}{2(ze)^2}\frac{\overline{\sigma}_{def}\overline{\sigma}_e}{\overline{\sigma}_{def}+\overline{\sigma}_e}\ln\frac{P_{O_2}^{II}}{P_{O_2}^I}} \qquad (7.78)$$

At this point, it is a useful exercise to recast Eq. (7.73) in terms of Fick's first law, in order to get a different perspective on the so-called ambipolar diffusion coefficient. It can be shown (see App. 7C) that

$$\frac{d\mu_{M^*}}{dx} = kT\left(\frac{z}{n}+\frac{1}{c_{def}}\right)\frac{dc_{M^*}}{dx} \qquad (7.79)$$

---

[107] To see how this comes about, multiply the flux J_{M^*} by the area A of the layer, which gives the total number of metal atoms per second (AJ_{M^*}) reaching the surface and reacting with oxygen. If the volume of an MO molecule is Ω_{MO}, it follows that its thickness is simply Ω_{MO}/A. Hence, the rate at which the layer grows is simply $\Omega_{MO}J_{M^*}$.

[108] Wagner defined a **rational scaling rate constant** K_r as the time rate of formation of an oxide expressed as equivalents per unit scale thickness. *An equivalent is defined as the fraction of the compound that transports one positive and one negative unit charge.* In general, for an M_aX_b oxide, the number of equivalents $\phi = (b|Z^-|)^{-1} = (a|Z^+|)^{-1}$. For example, for $Ni_{0.5}O_{0.5}$, $\phi = 0.5$ while for $Al_{1/3}O_{1/2}$, $\phi = 1/6$. It can be shown that $K_r = K_x/(\phi\Omega_{MO})$ (see Prob. 7.17).

which when combined with Eq. (7.73) results in

$$J_{M^*} = J_{\text{def}} = -\frac{J_e}{z} = \frac{-\sigma_{\text{def}}\sigma_e kT}{(ze)^2(\sigma_{\text{def}}+\sigma_e)}\left(\frac{z}{n}+\frac{1}{c_{\text{def}}}\right)\frac{dc_{M^*}}{dx} \tag{7.80}$$

Further, by making use of the Nernst-Einstein equation it can be shown that

$$J_{M^*} = (t_e D_{\text{def}} + t_{\text{def}} D_e)\frac{dc_{M^*}}{dx} \tag{7.81}$$

Since this equation is in the form of Fick's first law, it follows that the chemical or ambipolar diffusion coefficient responsible for oxidation is

$$\boxed{D_{\text{ambi}}^{\text{oxid}} = t_e D_{\text{def}} + t_{\text{ion}} D_e} \tag{7.82}$$

which is an alternate way of saying that the rate at which permeation will occur depends on the diffusivities or conductivities of both the ionic and electronic carriers. If either of the two vanishes, then $D_{\text{ambi}}^{\text{oxid}}$ vanishes and the oxide layer behaves as a passivating layer, protecting the metal from further oxidation. Aluminum provides an excellent example — the aluminum oxide that grows on Al is quite insulating, water-insoluble and adherent, which is why aluminum need not be protected from the elements in the way, say, Fe is.

WORKED EXAMPLE 7.6. The self-diffusion coefficient of Ni in NiO was measured at 1000°C to be 2.8×10^{-14} cm^2/s. At the same temperature in air, K_x was measured to 2.9×10^{-13} cm^2/s. NiO is known to be a predominantly electronic conductor. What conclusions can be drawn concerning the rate-limiting step during the oxidation of Ni? The lattice parameter of NiO is 0.418 nm. The free energy of formation of NiO at 1000 °C is −126 kJ/mol.

Answer

Given that the oxide is predominantly an electronic conductor, Eq. (7.77) simplifies to

$$K_x = \frac{kT\Omega_{\text{MO}}}{2(ze)^2}\int_{P_{O_2}^{\text{I}}}^{P_{O_2}^{\text{II}}}\sigma_{\text{def}}\,d\ln P_{O_2}$$

Substituting for σ_{def}, using the Nernst-Einstein relationship [Eq. (7.34)], and integrating gives

$$K_x = \frac{\Omega_{\text{NiO}}c_{\text{Ni}}D_{\text{Ni}}}{2}\ln\frac{P_{O_2}^{\text{II}}}{P_{O_2}^{\text{I}}}$$

Note that in this case $\Omega_{NiO}c_{Ni} = 1$. The limits of integration are $P_{O_2}^{II}$ in air (0.21 atm) and $P_{O_2}^{I}$ at the Ni/NiO interface. The latter is calculated as follows: For the reaction $Ni + 1/2O_2 \Rightarrow NiO$, the equilibrium partial pressure is given by (see Worked Example 5.4 for method)

$$P_{O_2} = \exp\left(\frac{-2 \times 126,000}{8.314 \times 1273}\right) = 4.56 \times 10^{-11} \text{ atm}$$

From which

$$\ln\frac{0.21}{P_{O_2}} = 22.24$$

It follows that if the diffusion of Ni were the rate-limiting step, then the theoretically calculated K_x would be

$$K_x^{theo} = \frac{2.8 \times 10^{-14} \times 22.24}{2} = 3.1 \times 10^{-13} \text{ cm}^2/\text{s}$$

which is in excellent agreement with the experimentally determined value of $2.9 \times 10^{-13} \text{ cm}^2/\text{s}$, indicating that the oxidation of Ni is indeed rate-limited by the diffusion of Ni ions from the Ni side to the oxygen side.

7.4.2 Ambipolar Diffusion in a Binary Oxide

The problem considered here is slightly different from the one just examined. Consider, for simplicity, an MO oxide subjected to an electrochemical potential gradient $d\eta_{MO}/dx$ which in turn must result in the mass transport of MO "units" from one area to another. Typically this occurs during sintering or creep where as a result of curvature or externally imposed pressures, the oxide diffuses down its electrochemical potential gradient (see Chaps. 10 and 12). To preserve electroneutrality and mass balance, the fluxes of the M and O ions have to be *equal and in the same direction.*

Applying Eq. (7.37) to the fluxes of the M^{2+} and O^{2-} ions, one obtains

$$J_{M^{2+}} = -\frac{D_M c_{M^{2+}}}{kT}\frac{d\eta_{M^{2+}}}{dx} = -\frac{D_M c_{M^{2+}}}{kT}\left(\frac{d\tilde{\mu}_{M^{2+}}}{dx} + 2e\frac{d\phi}{dx}\right) \qquad (7.83)$$

$$J_{O^{2-}} = -\frac{D_O c_{O^{2-}}}{kT}\frac{d\eta_{O^{2-}}}{dx} = -\frac{D_O c_{O^{2-}}}{kT}\left(\frac{d\tilde{\mu}_{O^{2-}}}{dx} - 2e\frac{d\phi}{dx}\right) \qquad (7.84)$$

where c_i and D_i represent the concentration and diffusivity of species i, respectively. Making use of the following three conditions that reflect electroneutrality, local equilibrium, and mass balance, respectively, one obtains

$$J_{M^{2+}} = J_{O^{2-}}, \quad \mu_{MO} = \mu_{O^{2-}} + \mu_{M^{2+}} \quad \text{and} \quad c_{O^{2-}} = c_{M^{2+}} = c_{MO}$$

where c_{MO} is the molar concentration of MO "molecules" per unit volume (that is, $c_{MO} = 1/V_{MO}$, where V_{MO} is the molar volume). In complete analogy to how Eq. (7.73) was derived, it is a lengthy but not difficult task to show that the flux of MO units or molecules is given by

$$J_{MO} = \frac{D_M D_O}{D_M + D_O} \frac{c_{MO}}{kT} \frac{\partial \tilde{\mu}_{MO}}{\partial x} \tag{7.85}$$

Note that the driving force in this case is the chemical potential gradient in MO. By reformulating this expression in terms of Fick's first law (see App. 7A) it can be shown that[109]

$$D_{ambi} = \frac{D_M D_O}{D_M + D_O} \tag{7.86}$$

This expression is only valid for an MO oxide; for the more general case of an $M_\kappa O_\beta$ oxide, however, the appropriate expression is

$$\boxed{D_{ambi} = \frac{D_M D_O}{\kappa D_M + \beta D_O}} \tag{7.87}$$

Equation (7.87) has far-reaching ramifications and basically predicts that in binary ionic compounds, diffusion-controlled processes are determined by D_{ambi}, which is turn is a function of the individual component diffusivities. For most oxides, however, there are typically orders-of-magnitude differences between the diffusivities on the different sublattices (see, e.g., Fig. 7.2). Consequently, with little loss in accuracy, D_{ambi} can be equated to the slower diffusing species. For instance, in MgO, $D_{Mg^{2+}} \gg D_{O^{2-}}$ and $D_{ambi} \approx D_{O^{2-}}$.

Before one proceeds much further, it is important to be cognizant of the underlying assumptions made in deriving Eq. (7.87):

1. The oxide is a pure intrinsic oxide, where the dominant defects are *Schottky* defects. This was implied when it was assumed that $c_{O^{2-}} = c_{M^{2+}} = c_{MO}$.
2. The vacancy concentrations are everywhere at equilibrium.

---

[109] The term D_{ambi} is used here to differentiate it from D_{chem}. It should be emphasized, however, that in the literature, in most complex expressions for diffusivity, i.e., for any process where there is some coupling between fluxes, the term *chemical diffusion* is used almost exclusively.

3. Local electroneutrality holds everywhere.
4. The ionic transport number is unity.

7.4.3 Reaction between Solids — Interdiffusion

The reactions that can occur between solids are quite diverse and for the most part are quite complicated and not well understood. In this subsection, the focus is on one very simple case, namely, that of interdiffusion of two ionic crystals in which the cations have the same charge, e.g., AO and BO. To further simplify the problem, the following assumptions are made:

- The anion sublattice is immobile.
- Cations A and B counterdiffuse independently, with self-diffusion coefficients D_{A^+} and D_{B^+}, respectively, that are not functions of composition.
- Electroneutrality is maintained by having the counterdiffusing cation fluxes coupled. Note that for this to happen, the system must be predominantly an ionic conductor, that is, $t_e \ll t_i$ — if not, decoupling of the fluxes will occur (see below).
- Within the interdiffusion layer the system behaves ideally.

The flux equations for the two cations are given by

$$J_{A^+} = -\frac{D_{A^+} c_{A^+}}{kT} \frac{d\tilde{\eta}_{A^+}}{dx} = -\frac{D_{A^+} c_{A^+}}{kT}\left(\frac{d\tilde{\mu}_{A^+}}{dx} + ez_{A^+}\frac{d\phi}{dx}\right) \tag{7.88}$$

$$J_{B^+} = -\frac{D_{B^+} c_{B^+}}{kT} \frac{d\tilde{\eta}_{B^+}}{dx} = -\frac{D_{B^+} c_{B^+}}{kT}\left(\frac{d\tilde{\mu}_{B^+}}{dx} - ez_{B^+}\frac{d\phi}{dx}\right) \tag{7.89}$$

Mass conservation requires that the sum of the mole fractions $X_A + X_B = 1$, from which it follows that if the solution is ideal, $dc_{A^+}/dx = -dc_{B^+}/dx$. Making use of this result together with the fact that electroneutrality requires $J_{A^+} = J_{B^+}$, and following a derivation similar to the one that led to Eq. (7.86), one can show that the interdiffusion coefficient D_{AB} is given by

$$\boxed{D_{AB} = \frac{D_{A^+} D_{B^+}}{X_{AO} D_{A^+} + X_{BO} D_{B^+}}} \tag{7.90}$$

This expression is sometimes referred to as the **Nernst-Planck** expression, and given the many simplifying assumptions made in deriving it, it should be used with care.

Now D_{AB} is not to be confused with D_{ambi}; comparing Eqs. (7.90) and (7.86) or (7.87) makes it obvious that the two expressions are not equivalent. When two charged carriers are moving in opposite directions, D_{AB} results. It is the appropriate coefficient to use whenever the two constituents of the same charge migrate in opposite directions — it is the appropriate expression to use, e.g., in analyzing ion-exchange experiments.

Note that Eq. (7.90) is valid only if the system is predominantly an ionic conductor, since only under these conditions can a diffusion or Nernst potential be built up. For a predominantly electronic conductor, however, no coupling occurs between the fluxes, in which case D_{AB} is given by an equation of the type[110]

$$D_{AB} = X_{AO}D_B + X_{BO}D_A \qquad (7.91)$$

For example, MgO-NiO interdiffusion has been interpreted by using such an expression.

7.4.4   EMF of Solid-State Galvanic Cells

Technologically, one important use of ionic ceramics is for potentiometric sensors. These are solids that, by virtue of being predominantly ionic conductors, are capable of measuring the absolute thermodynamic activities of various species.

It can be shown that (see App. 7D) when a solid is placed between two electrodes with chemical potentials μ_I and μ_{II}, as shown schematically in Fig. 7.19a, the open-cell voltage V of such a cell is given by

$$V = -\frac{1}{z\chi e} \int_{\mu_I}^{\mu_{II}} t_{ion} d\mu_{M^*} \qquad (7.92)$$

where z and χ are, respectively, the charge and stoichiometry of the electroactive ion in its standard state.[111] It follows that if the solid separating the electrodes is an ionic conductor, that is, $t_{ion} \approx 1.0$, then the open cell voltage of such a cell is simply

$$V = -\frac{1}{z\chi e}(\mu_{II} - \mu_I) = -\frac{kT}{z\chi e}\ln\frac{a_{II}}{a_I} \qquad (7.93)$$

where the activities a_I and a_{II} correspond to the chemical potentials μ_I and μ_{II}, respectively. It follows directly from this expression that if one of the electrodes is in its standard state, say, $a_I = 1$, then a measure of the voltage generated by such a cell is a direct measure of the activity a_{II} in the second electrode. Needless to say, this is a very powerful and elegant technique to measure thermodynamic parameters such as activities, activity coefficients, heats of solutions, solubility limits, and extent of nonstoichiometry.

---

110 In the metallurgical literature this expression is known as a Darken-type expression.
111 For all metals, $\chi = 1$, whereas for O_2, Cl_2, etc., $\chi = 2$.

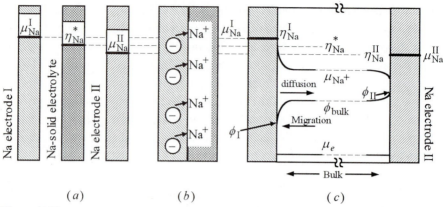

Figure 7.19

Development of a space charge and corresponding potential difference $\phi_I - \phi_{II}$ across a solid electrolyte subjected to electrodes of different chemical potentials. (*a*) Initially, before contact is made, it is assumed that $\mu_{Na}^I > \eta_{Na}^{SE} > \mu_{Na}^{II}$. (*b*) Upon contact because $\mu_{Na}^I > \eta_{Na}^*$, Na ions will jump across the interface, leaving their electrons behind and creating a space charge. (*c*) Local dynamic equilibrium is established at each electrode when the migrational and diffusional fluxes become equal.

To understand how the observed voltage develops, it is instructive to go through the following thought experiment, depicted schematically in Fig. 7.19. To simplify the problem, the following assumptions are made:

1. The solid electrolyte SE is a perfect Na ion conductor; that is, $t_i = 1.0$.
2. One of the electrodes is pure Na; that is, $a_I = 1.0$; and the other is an electrode in which $a_{II} < 1.0$.
3. Initially the electrodes are separated from the SE, and the initial conditions are such that $\eta_{Na}^I > \eta_{Na}^{SE} > \eta_{Na}^{II}$ (Fig. 7.19a). Note that this condition is identical to $\mu_{Na}^I > \eta_{Na}^{SE} > \mu_{Na}^{II}$.
4. The electrode is perfectly blocking to electrons. In other words, there are no surface states in the electrolyte that electrons can jump into.

To illustrate how a voltage develops in such a system, consider what happens when the electrodes are brought into intimate contact with the SE (Fig. 7.19b). Upon contact, at the interface where $\eta_{Na}^I > \eta_{Na}^{SE}$, the Na ions will jump across the interface from the electrode into the electrolyte, *leaving their electrons behind* (Fig. 7.19b). This has two consequences. The first is to increase the Na ion concentration in the vicinity of the interface, which will induce them to diffuse into the bulk (i.e., a diffusional flux of Na ions into the bulk results, depicted by the arrow labeled *diffusion* in Fig. 7.19c). The second consequence is the creation of a space charge or an electric field at the interface. This space charge or voltage ϕ_I creates an electric field that tends to attract the same ions back to the electrode (see arrow labeled *migration* in Fig. 7.19c). The voltage developed thus increases up to a point where the electric field responsible for the migrational flux is exactly balanced by the diffusional flux into the bulk. At this point the system is said to have reached a state of *local* dynamic equilibrium. Note that another way to look at the process is to appreciate that it will continue until $\eta_{Na}^I = \eta_{Na}^{SE}$ — which is the definition of equilibrium [Eq. (5.35)].

Conversely, at the electrode at which it was assumed that $\eta_{Na}^{SE} > \eta_{Na}^{II}$, the Na ions will jump from the SE into the electrode. Once again, the process will proceed until a voltage ϕ_{II} (opposite in polarity to the one developed at interface I) develops that is sufficient to equate the electrochemical potentials $\eta_{Na}^{II} = \eta_{Na}^{SE}$ across that interface. To summarize: The space charge that forms at the electrode/electrolyte interface gives rise to a measurable voltage difference $V = \phi_{II} - \phi_I$, which is related to the activities of the electroactive species in the electrodes — a fact that is embodied in Eq. (7.93).

7.5
RELATIONSHIPS BETWEEN SELF-, TRACER, CHEMICAL, AMBIPOLAR, AND DEFECT DIFFUSION COEFFICIENTS

Up to this point, quite a number of diffusion coefficients, listed below, have been introduced. For the reader not well versed in the field, this could lead, understandably enough, to some confusion. The purpose of this section is to shed some light on the subject.

During the course of this chapter, the following diffusion coefficients were introduced and discussed:

- *Self-diffusion D_{ion}*: This is a measure of the ease and frequency with which A atoms hop in pure A, but it applies equally to compounds $M_\kappa X_\beta$ where the species M and X form two independent sublattices.
- *Tracer diffusion D_{tr}^** is a measure of the ease and frequency with which radioactive or tagged atoms are diffusing in a matrix. It can be shown that $D_{tr}^* = f_{cor} D_{ion}$ where f_{cor} is a correlation coefficient that depends on the crystal structure and diffusion mechanism.
- *Defect diffusion D_{def}*. This is a measure of the ease and frequency with which the defects are hopping in a solid; $D_{ion} = \lambda D_{def}$, where λ is the fraction of sites available for the diffusing atom or ions to make a jump.

- *Chemical diffusion* D_{chem}. Formally this quantity is defined as

$$D_{chem} = -\frac{J_i}{dc_i/dx}$$

where J_i is the flux of species i and dc_i/dx is the gradient in its concentration.[112]   In essence, D_{chem} represents the phenomenological coefficient that describes the effective rate at which a given diffusional process is occurring.   As discussed below, to relate D_{chem} to more fundamental parameters in a given system, such as the diffusion of component ions or the diffusivities of defects, more information about the latter must be available.

- *Ambipolar diffusion coefficients* D_{ambi}^{oxid} and D_{ambi}. These diffusion coefficients are special cases of D_{chem}, and they reflect the fact that in ionic compounds the fluxes of the ions and defects are by necessity coupled, in order to maintain charge neutrality.

- *Interdiffusion diffusion* D_{AB}.   This is a measure of the rate at which a diffusional process will occur when ions are interdiffusing.

To illustrate the subtle differences and nuances between the various diffusion coefficients, it is instructive to take an example such as NiO, which was considered earlier in Worked Example 7.6. To obtain a measure of how fast Ni diffuses into NiO, one can carry out a tracer diffusion experiment, as described earlier. By analyzing the concentration profile of the radioactive tracer, it is possible to determine the so-called tracer diffusivity D_{tr} of Ni in NiO.[113]   The tracer diffusivity is then related to the self-diffusivity D_{Ni} by a correlation coefficient f_{cor}.   The coefficient f_{cor} has been calculated for many structures and can be looked up.[114]

To relate D_{ion} to defect diffusivities, however, more information about the system is needed. For starters, it is imperative to know the diffusion mechanism — if diffusion is by vacancies, their concentration or mole fraction λ has to be known to relate the two since, according to Eq. (7.14), $D_{vac} = D_s/\lambda$. Needless to say, if the number of defects is not known, the two cannot be related. If, however, the

---

112 This is why one strives to cast the flux equations in the form of Fick's first law [Eq. (7.81)]. Once in that form, the ratio of J to dc/dx is, by definition, a chemical D.

113 To measure the tracer diffusivity of oxygen, typically a crystal is exposed to a gas in which the oxygen atoms are radioactive.

114 See, e.g., J. Philibert, *Atom Movements, Diffusion and Mass Transport in Solids*, Les Editions de Physique, in English, trans. S. J. Rothman, 1991.

tracer diffuses by an interstitial mechanism, then it can be shown that $D_{int} = D_{tr} = D_s$. Note that in this case the correlation coefficient is unity.[115]

The next level of sophistication involves the determination of the nature of the rate-limiting step in a given process, i.e., relating D_{chem} to D_{tr} or D_{ion}. For example, it was concluded, in Worked Example 7.6, that the rate-limiting step during the oxidation of Ni was the diffusion of Ni ions through the oxide scale.

This conclusion was only reached, however, because the nature of the conductivity in the oxide layer was known. To generalize the argument, consider the following two limiting cases:

1. The oxide layer that forms is predominantly an electronic conductor, that is, $t_e \gg t_{ion}$. Hence, according to Eq. (7.82), $D_{ambi}^{oxid} \approx D_{def}$ and the permeation is rate-limited by the diffusivity of the ionic defects. Furthermore, it can be shown that under these conditions

$$K_x = D_{fast} \frac{\left|\Delta\tilde{G}_{MO_{z/2}}\right|}{kT} \tag{7.94}$$

where D_{fast} is the diffusion coefficient of the faster of the two *ionic* species and $\Delta\tilde{G}_{MO_{z/2}}$ is the free energy of formation of the $MO_{z/2}$ oxide according to Eq. (7.65). For most transition-metal oxides, $D_{O^{2-}} \ll D_{M^+}$ and Eq. (7.94) reads

$$K_x = D_{M^+} \frac{\left|\Delta\tilde{G}_{MO_{z/2}}\right|}{kT}$$

which, not surprisingly, is the conclusion reached in Worked Example 7.6.

Note that measuring the electrical conductivity of NiO yields no information about the conductivity of the ions, only about the electronic defects.

2. The oxide layer that forms is predominantly an ionic conductor, so, according to Eq. (7.82), $D_{ambi}^{oxid} \approx D_e$ or D_h (note that in order to determine which is the case, even more information about whether the oxide was p or n type is required). In this situation Eq. (7.78) reads

---

115 Only when diffusion occurs by uncorrelated elementary steps are D_{ion} and D_{tr} equal. A case in point is interstitial diffusion where after every successful jump the diffusing particle finds itself in the same geometric situation as before the diffusional step. However, for diffusion by a vacancy mechanism, this is no longer true. After every successful jump, the tracer ion has exchanged places with a vacancy, and thus the probability of the ions jumping back into the vacant site and canceling the effect of the original jump is much greater than the probability that the ion will move in any other direction.

$$K_x = \frac{\overline{\sigma}_{elec} \Omega_{MO}}{e^2} \left| \Delta \tilde{G}_{MO_{z/2}} \right| \tag{7.95}$$

where $\overline{\sigma}_{elec}$ is the average partial electronic conductivity across the growing layer. Note that in contrast to the aforementioned situation, measuring K_x yields information not about the diffusivity of the ions, but rather about the electronic carriers, which is not too surprising since in these circumstances the rate-limiting step is the diffusion of the electronic defects. Information about the ionic diffusivity, however, can be deduced from conductivity experiments by relating the conductivity to the diffusivity via the Nernst-Einstein relationship (see, e.g., Worked Example 7.5).

7.6
SUMMARY

1. Ions and atoms will move about in a lattice if they have the requisite energy to make the jump across an energy barrier ΔG_m^* and if the site adjacent to them is vacant.

2. In general, diffusivity in ceramics can be expressed as follows:

$$D_k = \phi D_0 \exp\left(\frac{-\Delta H_m^*}{RT} \right)$$

where ϕ is the probability that the site adjacent to the diffusing ion is vacant. For defects, $\phi \approx 1.0$; and D_0 is a temperature-independent term that includes the vibrational frequency of the diffusing ions, the jump distance, and the entropic effect of the atomic jump.

3. The temperature dependence of the diffusion coefficient will depend on the diffusion mechanism. If diffusion occurs interstitially, the temperature dependence of D will include only the migration energy term, ΔH_m^*, since the probability of the site adjacent to an interstitial atom being vacant is ≈ 1.0.

4. For diffusion by a vacancy mechanism, the temperature dependence of diffusivity will depend on both the migration enthalpy ΔH_m^* and the energy required to form the vacancies if the latter are thermally activated; i.e., the concentration of intrinsic defects is much greater than the concentration of extrinsic defects. If, however, ϕ is fixed by doping, it becomes a constant independent of temperature. The activation energy for diffusion in the latter case will only depend on ΔH_m^*.

5. The diffusivity will also depend on the chemical environment surrounding a crystal. This is especially true of nonstoichiometric oxides in which the stoichiometry and consequently the concentration of the defects are a relatively strong function of the oxygen partial pressure. In these instances, the partial pressure dependence of the diffusivity is the same as that of the defects responsible for the diffusivity.

6. The presence of a potential gradient, whether chemical or electrical, will favor jumps down that gradient and will result in an ionic flux down the potential gradient. Consequently, the ionic conductivity is directly proportional to the ionic diffusivity through the Nernst-Einstein relationship.

7. The most fundamental and general equation relating the flux of an ionic species to the gradient in its electrochemical potential is

$$J_k = -\frac{D_k c_k}{kT}\frac{d\tilde{\eta}_k}{dx} = -\frac{\sigma_k}{(z_k e)^2}\frac{d\tilde{\eta}_k}{dx}$$

This expression embodies both Fick's first law and Ohm's law.

8. The total electrical conductivity is governed by

$$\sigma = \Sigma |z_i| c_{m,i}\mu_{d,i}$$

where $c_{m,i}$ and $\mu_{d,i}$ are, respectively, the concentrations and mobilities of the mobile species. The total conductivity is the sum of the partial electronic and ionic conductivities.

9. Since the ionic conductivity and diffusivities are related by the Nernst-Einstein relationship, what governs one governs the other. Fast ionic conductors are a class of solids in which the ionic conductivity is much larger than the electronic conductivity. For a solid to exhibit fast ion conduction, the concentration and mobility of ionic defects must be quite large. The band gap of the material must also be quite high to minimize the electronic contribution to the overall conductivity.

10. The electronic conductivity depends critically on the concentration of free electrons and holes. There are essentially three mechanisms by which mobile electronic carriers can be generated in a solid:

   a. Intrinsically by having the electrons excited across the band gap of the material. In this case, the conductivity is mostly determined by the size of the band gap E_g and is a strong function of temperature.

   b. Extrinsically by doping the solid with aliovalent impurities that result in the generation of holes or electrons. In this case, if the dopant is fully ionized, the conductivity is fixed by the concentration of the dopant and is almost temperature-independent.

c. As a result of departures from stoichiometry. The oxidation or reduction of an oxide can result in the generation of electrons and holes.

11. Exposing a binary compound to a chemical potential gradient of one of its components results in a flux of that component through the binary compound as a neutral species. The process, termed *ambipolar diffusion*, is characterized by a chemical diffusion coefficient D_{chem} which is related to the defect and electronic diffusivities by

$$D_{chem} = t_i D_{elec} + t_e D_{def}$$

Since this process involves the simultaneous, coupled diffusion of ionic and electronic defects, it is the slower of the two that is rate-limiting. To maximize oxidation resistance, electrochemical sensing capabilities, and the successful use of ceramics as solid electrolytes, D_{chem} should be minimized.

12. Exposing a binary compound as a whole to a chemical potential, i.e., for $d\mu_{MX}/dx \neq 0$, results in the ambipolar migration of both constituents of that compound down that gradient. The resulting ambipolar diffusion coefficient for an MO oxide is given by

$$D_{ambi} = \frac{D_M D_O}{D_M + D_O}$$

from which it can be easily shown that D_{ambi} is determined by the slower of the two components.

13. In a quasi-binary system, interdiffusion of ions also results in a so-called interdiffusion diffusion that is also rate-limited by the diffusivity of the slower of the two ions. This process occurs, e.g., when solid-state reactions between ceramics or ion-exchange experiments are carried out.

14. Solid electrolytes can be used as sensors to measure thermodynamic data, such as activities and activity coefficients. The voltage generated across these solids is directly related to the activities of the electroactive species at each electrode.

APPENDIX 7A

Relationship Between Fick's First Law and Eq. (7.30)

The chemical potential and concentration are related [see Eq. (5.23)] by

$$\mu_i = \mu_i^\circ + RT \ln a_i = \mu^\circ + RT \ln c_i \gamma_i \tag{7A.1}$$

where γ_i is the activity coefficient.[116] It follows that Eq. (7.22) can be written as

$$\tilde{f} = -\frac{1}{N_{Av}}\frac{d\mu}{dx} = -\frac{RT}{N_{Av}}\left[\frac{d\ln c}{dx} + \frac{d\ln\gamma}{dx}\right] = -kT\left[\frac{1}{c}\frac{dc}{dx} + \frac{1}{\gamma}\frac{d\gamma}{dx}\right] \quad (7A.2)$$

For ideal or dilute solutions, γ is a constant and the second term inside the brackets drops out. Substituting Eq. (7A.2) in (7.30) yields

$$J_{ion} = -\frac{c_{ion}D_{ion}}{kT}\left\{kT\left[\frac{1}{c_{ion}}\frac{dc_{ion}}{dx}\right]\right\} = -D_{ion}\frac{dc_{ion}}{dx} \quad (7A.3)$$

which is nothing but Fick's first law. This is an important result since it indicates that whenever $\gamma \neq f(x)$, that is, it is not a function of c, the generalized flux equation [Eq. (7.30)] degenerates into Fick's first law.

If γ_i is a function of concentration, then the second term in Eq. (7A.2) cannot be ignored. Noting that

$$\frac{\partial\ln\gamma}{dx} = \frac{\partial\ln\gamma}{\partial\ln c}\frac{\partial\ln c}{dx}$$

and carrying out the same procedure to obtain to Eq. (7A.3), one obtains

$$D_{chem} = D_{atom}\left(1 + \frac{\partial\ln\gamma_{atom}}{\partial\ln c_{atom}}\right)$$

where the term in parentheses is known as the **thermodynamic factor**. In other words, the self-diffusivity of the atoms is modified by a factor that takes into account that the diffusing particles now interact with one another. Note that since one cannot define an activity coefficient for a charged species (see Chap. 5), this expression for D_{chem} is valid only for neutral species, hence the subscript "atom".

---

116 Strictly speaking, Eq. (7A.1) should read $\mu_i = \mu_i^\circ + RT\ln x_i\gamma_i$, where x_i is the mole fraction of species i. However, c_i and x_i are related by

$$x_i = \frac{c_iV_i}{\sum_i c_iV_i}$$

where V_i is the molar volume of species i. Assuming there are only two species, it follows that for *dilute* concentrations $X_1 \approx c_1V_1/(c_2V_2)$. But since $V_1/(c_2V_2)$ is approximately a constant, it can be incorporated into μ_i°.

APPENDIX 7B

Effective Mass and Density of States

For free electrons in a metal, it can be shown that the E–k relationship in three dimensions is

$$E = \frac{h^2}{8\pi^2 m_e}\left(k_x^2 + k_y^2 + k_z^2\right) \tag{7B.1}$$

where k_i is the electron wave number in the three principal directions and m_e is the rest mass of an electron. An almost identical relationship for the density of states in a semiconductor or insulator (see Chap. 2) is

$$E = \frac{h^2}{8\pi^2 m_e^*}\left(k_x^2 + k_y^2 + k_z^2\right) \tag{7B.2}$$

where the effective electron mass m_e^* replaces the rest mass. The **effective electron mass** is defined as

$$m_e^* = \frac{h}{2\pi}\left(\frac{\partial^2 E}{\partial k^2}\right)^{-1} \tag{7B.3}$$

This is another way of saying that as the electron energy approaches that of a band edge, its effective mass, or the force needed to accelerate the electron, becomes very large.[117]

To calculate the total number of electrons in the conduction band at any temperature T, the following integral

$$n = \int_{\text{bottom of cond. band}}^{\text{top of cond. band}} (\text{density of states})(\text{prob. of electron occupying given state })dE$$

or

$$n = \int_{E_g}^{\infty} f(E)Z(E)dE \tag{7B.4}$$

has to evaluated. Here $f(E)$ is the density of states, given by Eq. (7.47), and $Z(E)dE$ is the density or number of electronic states per unit volume having

---

[117] In the limit that the electron energy satisfies the Bragg diffraction condition (i.e., at the top of the valence band), the electron forms a standing wave, and even though it may be experiencing a force, it is "going nowhere." In other words, it behaves as an infinitely heavy object.

energies between E and $E + dE$. Taking zero energy to be at the top of the valence band, one can show that for the electrons[118]

$$Z(E)dE = \chi_e(E - E_g)^{1/2} dE \text{ where } \chi_e = \frac{4\pi(2m_e^*)^{3/2}}{h^3} \quad (7B.5)$$

For $E - E_F \gg kT$, the Fermi function [Eq. (7.47)] may be approximated by

$$f(E) = \exp\left(-\frac{E - E_f}{kT}\right) \quad (7B.6)$$

Substituting Eqs. (7B.5) and (7B.6) in Eq. (7B.4) and integrating lead to the final result, Eq. (7.48),

$$n = N_c \exp\left(-\frac{E_c - E_f}{kT}\right) \quad (7B.7)$$

where

$$N_c = 2\left(\frac{2\pi m_e^* kT}{h^2}\right)^{3/2} \quad (7B.8)$$

Similarly, for holes

$$p = N_v \exp\left(-\frac{E_f - E_v}{kT}\right) \quad (7B.9)$$

where

$$N_v = 2\left(\frac{2\pi m_h^* kT}{h^2}\right)^{3/2} \quad (7B.10)$$

and m_h^* is the effective mass of a hole.

It is important to note that while the density of states increases monotonically with energy [Eq. (7B.5)], as shown in Fig. 7.20a, the probability of occupancy of the higher levels drops rapidly (Fig. 7.20b), such that in the end the *filled* electron states are all clustered together near the bottom of the conduction band (Fig. 7.20c). Finally, note that for many ceramic materials, the effective masses of the electrons and holes are not known, and the assumption that $m_e = m_e^* = m_h^*$ is oftentimes made.

---

[118] L. Solymar and D. Walsh, *Lectures on the Electrical Properties of Materials*, 4th ed. Oxford University Press, New York, 1988.

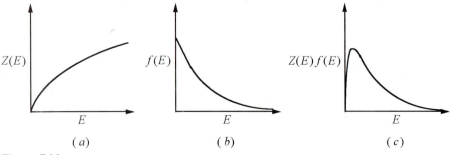

Figure 7.20
(a) Dependence of the density of states near the bottom of the conduction band on energy.
(b) Probability of finding electron at energy E, that is, $f(E)$ or Eq. (7B.6). (c) A plot of
$f(E)Z(E)$ versus E showing that most of the electrons are clustered near the bottom of the
conduction band. Conversely, holes would be clustered near the top of the valence band.

APPENDIX 7C

Derivation of Eq. (7.79)

In the dilute approximation regime where defect-defect interaction can be ignored,
it is possible to express their chemical potential as

$$\mu_{def} = \mu^{\circ}_{def} + kT \ln c_{def} \tag{7C.1}$$

and

$$\mu_e = \mu^{\circ}_e + kT \ln n \tag{7C.2}$$

Electroneutrality and mass balance dictate that $c_{def} = n/z = c_{M^*}$ from which

$$\frac{dc_{M^*}}{dx} = \frac{dn}{dx} = \frac{dc_{def}}{dx} \tag{7C.3}$$

Combining this result with Eqs. (7C.1) and (7C.2) together with Eq. (7.68), one
can easily show that

$$\frac{d\mu_{M^*}}{dx} = kT\left[\frac{z}{n} + \frac{1}{c_{def}}\right]\frac{dc_{M^*}}{dx} \tag{7C.4}$$

which is Eq. (7.79).

APPENDIX 7D

Derivation of Eq. (7.92)

The situations depicted in Figs. 7.18a and 7.19c are very similar in that in both cases a driving force exists for mass transport. The origin of this force is the chemical potential gradient $d\mu/dx$ that exists across the growing oxide layer in one case and the solid electrolyte or sensor in the other case.

The magnitudes of Na ion and electronic fluxes are given by [Eq. (7.37)]

$$J_{Na^+} = -\frac{\sigma_{Na^+}}{e^2}\frac{d\tilde{\eta}_{Na^+}}{dx} \tag{7D.1}$$

$$J_e = -\frac{\sigma_e}{e^2}\frac{d\tilde{\eta}_e}{dx} \tag{7D.2}$$

Since electroneutrality requires that $J_{Na^+} = J_e$, it follows that

$$\frac{d\tilde{\eta}_{Na^+}}{dx} = \frac{\sigma_e}{\sigma_{Na^+}}\frac{d\tilde{\eta}_e}{dx} \tag{7D.3}$$

The assumption of local equilibrium implies that

$$d\tilde{\mu}_{Na}\big|_{electrode} = \big|d\tilde{\eta}_e + d\tilde{\eta}_{Na^+}\big|_{SE} \tag{7D.4}$$

where SE refers to the solid electrolyte. Combining this equation with (7D.3), one obtains

$$\frac{d\tilde{\mu}_{Na^+}}{dx} = \frac{\sigma_e}{\sigma_{Na^+}}\frac{d\tilde{\eta}_e}{dx} + \frac{d\tilde{\eta}_e}{dx} = \frac{1}{t_{ion}}\frac{d\tilde{\eta}_e}{dx} \tag{7D.5}$$

which upon rearrangement and integration and by noting that

$$\Delta\tilde{\eta}_e = \eta_e^{II} - \eta_e^{I} = -eV$$

it follows that

$$V = -\frac{1}{e}\int_{\mu_{Na}^{I}}^{\mu_{Na}^{II}} t_{ion}\,d\tilde{\mu}_{Na} \tag{7D.6}$$

In the more generalized case, where the charge on the cation is not 1 but z and the stoichiometry of the electroactive species is not 1 but χ, the more general result given in Eq. (7.92) holds.

PROBLEMS

7.1. (a) Calculate the activation energy at 300 K given that $D_0 = 10^{-3}\,\text{m}^2/\text{s}$ and $D = 10^{-17}\,\text{m}^2/\text{s}$.

(b) Estimate the value of D_0 for the diffusion of Na ions in NaCl, and compare with the experimental value of $\approx 0.0032\,\text{m}^2/\text{s}$. State all assumptions. (Consult Table 6.2 for most of requisite information.)

Answer: $D_0 = 0.0017\,\text{m}^2/\text{s}$

(c) Calculate the mobility of oxygen ions in UO_2 at 700°C. The diffusion coefficient of the oxygen ions at that temperature is $10^{-17}\,\text{m}^2/\text{s}$. State all assumptions.

Answer: $2.4 \times 10^{-16}\,\text{m}^2/(\text{V}\cdot\text{s})$

(d) Compare this mobility with electron and hole mobilities in semiconductors.

7.2. (a) Explain why for a solid to exhibit predominantly ionic conductivity, the concentration of mobile ions must be much greater than the concentration of electronic defects.

(b) Derive Eq. (7.46) and show that the conductivity should reach a maximum when one-half the sites are occupied.

(c) Derive Eq. (7.54) and comment on the similarity of this expression to that derived in part (b).

7.3. Estimate the number of vacant sites in an ionic conductor at room temperature in which the cations are the predominant charge carriers. Assume that at room temperature the conductivity is $10^{-17}\,(\Omega\cdot\text{m})^{-1}$ and the ionic mobility is $10^{-17}\,\text{m}^2/(\text{V}\cdot\text{s})$. State all assumptions.

7.4. (a) What determines the type of conductivity in a ceramic (i.e., whether it is ionic or electronic)?

(b) It is often said that reducing a ceramic will increase its electronic conductivity. Do you agree with this statement? Explain.

(c) Distinguish between p- and n-type oxides with respect to the oxygen partial pressure dependence of their majority carriers. Describe an experiment by which you could distinguish between the two.

7.5. (a) A stoichiometric oxide M_2O_3 has a band gap of 5 eV. The enthalpy of Frenkel defect formation is 2 eV, while that for Schottky defect formation is 7 eV. Further experiments have shown that the only mobile species are cation interstitials, with a diffusion coefficient $D_{M, \text{int}}$ at 1000 K of $1.42 \times 10^{-10}\,\text{cm}^2/\text{s}$. The mobilities of the holes and electrons were found to be 2000 and 8000 $\text{cm}^2/(\text{V}\cdot\text{s})$, respectively. At 1000 K would you expect this oxide to be an ionic, electronic, or mixed conductor? Why? Assume the number of interstitial sites is twice the number of atomic sites. You may find this information useful: molecular weight of oxide $= 40$ g/mol, density $= 4 \text{g/cm}^3$. Assume the density of states for holes and electrons to be on the order of 10^{22} cm^{-3}.

Answer: $\sigma_{ion} = 3.8 \times 10^{-9}$ S/cm, $\sigma_p = 8.45 \times 10^{-7}$ S/cm, $\sigma_n = 3.4 \times 10^{-6}$ S/cm

(b) If the oxide in part (a) is doped with 5 mol % of an MbO oxide [final composition: $(MbO)_{0.05}(M_2O_3)_{0.95}$], write two possible defect reactions for the incorporation of MbO in M_2O_3 oxide. Calculate the molar fraction of each defect formed.

(c) Assume one of the defect reactions in part (b) involves the creation of Mb_i. Recalculate the ionic conductivity, given that the diffusivity of Mb interstitials D_{Mb_i} in M_2O_3 at 1000 K is 10^{-9} cm^2/s. *Hint:* Start with 1 mol of final composition, and calculate the fraction of Mb ions that go interstitially. Make sure you take the effective charge into account.

Answer: $\sigma_{Mb_i} = 7.44 \times 10^{-6}$ S/cm

7.6. (a) Construct the Kroger-Vink diagrams for pure zirconia.

(b) Repeat part (a) for calcia-doped zirconia and compare to Fig. 7.16a. State all assumptions. On the same diagram, explain what happens if the dopant concentration is increased.

7.7. (a) To increase the electron (*n*-type) conductivity of ZnO, which would you add, Al_2O_3 or Li_2O? Explain.

Answer: Al_2O_3 (not a typo)

(b) The resistivity of ZnO was found to decrease from 4.5 to 1.5 $\Omega \cdot$cm as the doping level was increased from 0.23 and 0.7 mol %. Which dopant do you think was used (Al_2O_3 or Li_2O) and why? Derive an expression for the conductivity of this oxide that takes into account the dopant concentration. Is the expression derived consistent with the changes in conductivity observed?

(c) Zinc oxide is a semiconductor. Except at very high temperatures the carrier concentration is related to the excess zinc which dissolves into the structure as interstitials. The first ionization energy of the Zn interstitial is 0.04 eV, the second is ≈ 1.5 eV. The electronic structure of the Zn atom is . . . . . $3d^{10}4s^2$, of an oxygen atom is . . . $2s^2 2p^4$. The band gap is 3.2 eV.

 (i) Sketch schematically the band structure of ZnO. Label the valence band, conduction band, and two zinc interstitial defect levels.

 (ii) What is the defect reaction for the incorporation of Zn vapor into ZnO and what kind of electronic carriers result?

 (iii) Samples of ZnO annealed at 1300°C in Zn vapor were quenched to room temperature and studied in the range −200 to +300°C. The results are shown below. What is the reason for the decrease in the carrier concentration below room temperature?

(iv) Is the mobility due to free carriers or to "hopping"? Explain.

7.8. Positive and negative vacancies are attracted to one another coulombically. Show that if the energy of attraction of such pairs is E_p, then the fraction of such pairs at equilibrium is given by

$$\frac{V_p}{(V - V_p)^2} = K = 6e^{E_p/RT}$$

where V_p is the number of pairs and $V = V_{cat} = V_{an}$. State all assumptions.

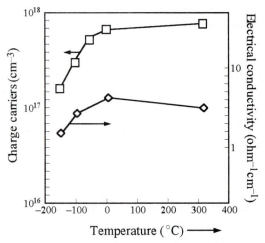

Figure 7.21
Temperature dependence of electrical conductivity (right axis) and charge carrier concentration (left axis).

7.9. Thomas and Lander[119] measured the solubility and conductivity of hydrogen in ZnO and found that they varied as $P_{H_2}^{1/4}$. Derive a model that explains their observation. *Hint:* Hydrogen dissolves interstitially and ionizes.

7.10. Undoped PbS is an n-type semiconductor and crystallizes in the NaCl structure. The predominant defects are Frenkel defect pairs on the Pb sublattice.
(a) Would you expect the diffusivity of Pb or S to be greater in PbS? Why?
(b) Explain what would happen to the diffusional flux of Pb upon the addition of Ag_2S to PbS.

119 D. G. Thomas and J. J. Lander, *J. Chem. Phys.*, **25**:1136–1142 (1956).

Answer: Pb ion flux will increase with additions (not a typo).

(c) How would Bi_2S_3 additions affect the diffusion of Pb?

7.11. Cuprous chloride is an electronic *p*-type conductor at high Cl pressures. As the chlorine pressure decreases, ionic conductivity takes over.

(a) Suggest a mechanism or combination of mechanisms to explain this behavior.

(b) Obtain a relationship between conductivity and the defect population for your proposed mechanism(s) that is consistent with the experimental observations. *Hint:* Consider two mechanisms, one stoichiometric and the other nonstoichiometric.

7.12. The electrical conductivity σ of a solid is predicted to vary as

$$\sigma = \frac{C}{T}\exp\left(-\frac{Q}{kT}\right)$$

where C is a constant and k is Boltzmann's constant. Measurements of σ, in arbitrary units, for ice as a function of T were as follows:

σ	31	135	230	630
T, K	200	220	230	250

Based on these data, what do you expect the conduction mechanism in ice to be if the band gap of ice is 0.1 eV, proton transport involving the breaking of a hydrogen bond is 0.25 eV, and a transfer of complex ions requiring simultaneous breaking of four hydrogen bonds is 1 eV?

Answer: Activation energy for conduction $= 0.3 \Rightarrow b$

7.13. The functional dependence of the electrical conductivity of an oxide on the oxygen partial pressure and temperature is shown in Fig. 7.22a. The temperature dependence of the conductivity is shown in Fig. 7.22b. Answer the following questions.

(a) Is this oxide stoichiometric or nonstoichiometric? Explain.

(b) Develop the defect reaction or reactions that would explain this behavior. Pay special attention to the slopes.

(c) What type of conductor (ionic, *p*-type, *n*-type, etc.) do you expect this oxide to be? Elaborate, using appropriate equations.

(d) To which energy does the slope of the line in Fig. 7.22b correspond? Explain, stating all assumptions.

(e) Label the curves in Fig. 7.22a in terms of increasing temperature. Explain.

(*f*) Do these figures assume equilibrium of any kind? Elaborate briefly.

(*g*) Describe what changes, if any, would occur to the defects in this crystal if the temperature were suddenly changed from, say, T_1 to T_2 (assuming $T_1 > T_2$). Elaborate on the atomic mechanisms that would be occurring to affect the changes, if any.

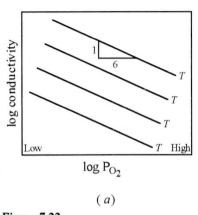

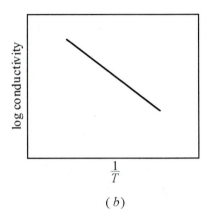

(*a*)   (*b*)

Figure 7.22
(*a*) Functional dependence of log σ on log P_{O_2}. (*b*) Effect of temperature on conductivity at a fixed P_{O_2}.

7.14. (*a*) The tracer diffusion of coefficient of oxygen in calcia-stabilized zirconia (CSZ) was measured to fit the relationship

$$D = 1.8 \times 10^{-6} \exp\left(-\frac{1.35 \text{ eV}}{kT}\right) \quad \text{m}^2/\text{s}$$

Assuming the transport number of oxygen is unity, estimate the electrical conductivity at 1000°C. Assume the unit cell side of 513 pm. State all assumptions.

Answer: $\approx 2.0 \text{S/m}$

(*b*) CSZ membranes are currently being used as solid electrolytes in fuel cells. To get maximum efficiency of the fuel cell, however, it is imperative to reduce the permeation of oxygen across the membrane. Assume that CSZ in part (*a*) has a hole transport number of 10^{-3} at 1000°C. If the thickness of the CSZ is 1 mm, estimate the molar flux of oxygen permeating through, if the fuel cell is operating between a P_{O_2} of 10^{-10} atm and air. State all assumptions.

Answer: 5.2×10^{-5} mol of $O_2/\text{m}^2\text{s}$

7.15. One side of an oxygen sensor is exposed to *air* and the other to an equilibrium mixture of Ni and NiO. The following results were obtained at the various temperatures indicated:

T, K	1200	1300	1400
emf, V	0.644	0.6	0.55

   (*a*) Calculate the partial pressure of oxygen in the Ni/NiO side at 1300 K.
   (*b*) Calculate the standard free energy of enthalpy and entropy of formation of NiO at 1300 K.
Answer: P_{O_2} at 1300 K $= 1 \times 10^{-10}$ atm, $\Delta G_{1300} = -124 \, \text{kJ/mol}$

7.16. (*a*) Given the results shown in Fig. 7.17, do they obey Eq. (7.62)?
    (*b*) If your answer is yes calculate K_w, and compare to those listed in Table 7.1.

7.17. (*a*) Show that in the case of oxidation, the rational rate constant is related to the parabolic rate constant by

$$K_r = \frac{K_x}{\Omega_{M_a O_b} \phi}$$

    where ϕ is the number of equivalents (see footnote 112).
   (*b*) Further show that K_r and K_w are related by

$$K_r = \frac{1}{2} \frac{|z|^2 \, \phi V_{MO}}{(M_O A)^2} \frac{\Delta w^2}{t} = \frac{1}{2} \frac{|z|^2 \, \phi V_{MO}}{(M_O A)^2} K_w$$

    where A is the area of the sample and M_O is the atomic weight of oxygen. In this case z is the valence on the anion.
   (*c*) The time dependence of the weight gain of 6×3 mm Gd metal foils at two different temperatures is shown in Fig. 7.23a. Calculate K_w and K_r for this oxide.
   (*d*) How long would it take to grow an oxide layer 25 μm thick at 1027°C.
   (*e*) In your own words explain the oxygen partial pressure dependence of the data shown in Fig. 7.23b. In other words, why does the sample gain more weight at higher partial pressures? Information you may find useful: molecular weight of $Gd_2O_3 = 362.5$ g/mol, density $= 7.41 \text{g/cm}^3$.

7.18. Based on their P-B ratio, predict which of the following metals will form a protective oxide layer and which will not: Be, Nb, Ni, Pd, Pb, Li, and Na.

7.19. Derive Eqs. (7.94) and (7.95).

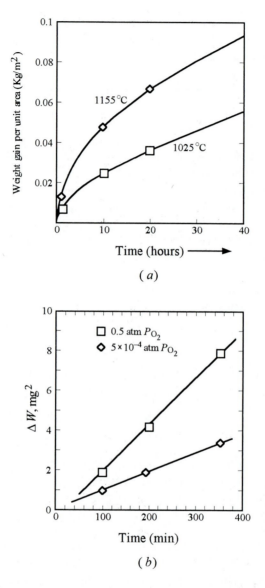

Figure 7.23
(*a*) Weight gain during the oxidation of Gd metal at two different temperatures.
(*b*) Weight gain during the oxidation of Gd metal at 1027°C at two different oxygen partial pressures.[120]

---

120 Adapted from D. B. Basler and M. F. Berard, *J. Amer. Cer. Soc.*, **57**:447 (1974).

ADDITIONAL READING

1. R. Bube, *Electrons in Solids*, Academic Press, New York, 1988.
2. L. Solymar and D. Walsh, *Lectures on the Electrical Properties of Materials*, 4th ed., Oxford University Press, New York, 1988.
3. F. A. Kröger, *The Chemistry of Imperfect Crystals*, 2d ed. rev., vols. 1 to 3, North-Holland, Amsterdam, 1973.
4. L. Heyne, in *Topics in Applied Physics*, S. Geller, ed., Springer-Verlag, Berlin, 1977.
5. P. Kofstad, *Nonstoichiometry, Diffusion and Electrical Conductivity in Binary Metal Oxides*, Wiley, New York, 1972.
6. O. Johannesen and P. Kofstad, "Electrical Conductivity in Binary Metal Oxides, Parts 1 and 2," *J. Mater. Ed.*, **7**: 910–1005 (1985).
7. C. Kittel, *Introduction to Solid State Physics*, 6th ed., Wiley, New York, 1986.
8. P. Kofstad, *High Temperarure Oxidation of Metals*, Wiley, New York, 1966.
9. H. Tuller in *Nonstoichiometric Oxides*, O. T. Sorensen, ed., Academic Press, New York, 1981, p. 271.
10. R. E. Hummel, *Electronic Properties of Materials*, Springer-Verlag, Berlin, 1985.
11. L. Azaroff and J. J. Brophy, *Electronic Processes in Materials*, McGraw-Hill, New York, 1963.
12. H. Rickert, *Electrochemistry of Solids*, Springer-Verlag, Heidelberg, 1982.
13. H. Schmalzreid, *Solid State Reactions*, 2d rev. ed., Verlag Chemie, Weinheim, 1981.
14. P. G. Shewmon, *Diffusion in Solids*, 2d ed., TMS, Warrendale, Pennsylvania, 1989.
15. A. R. Allnatt and A. B. Lidiard, *Atomic Transport in Solids*, Cambridge University Press, Cambridge, Massachusetts, 1993.
16. J. Philibert, *Atom Movements, Diffusion and Mass Transport in Solids*, Les Editions de Physique, in English, trans. S. J. Rothman, 1991.
17. R. J. Borg and G. J. Dienes, *An Introduction to Solid State Diffusion*, Academic Press, New York, 1988.
18. *Diffusion Data*, 1967–1973, vols. 1–7, Diffusion Information Center, Cleveland, Ohio.
19. F. H. Wohlbier and D. J. Fisher, eds., *Diffusion and Defect Data (DDD)*, 1974, vol. 8, Trans. Tech. Pub., Aerdermannsdorf, Switzerland.

CHAPTER 8

Phase Equilibria

Like harmony in music; there is a dark inscrutable
workmanship that reconciles discordant elements,
makes them cling together in one society.

William Wordsworth

8.1
INTRODUCTION

Phase diagrams are graphical representations of what equilibrium phases are present in a material system at various temperatures, compositions, and pressures. A **phase** is defined as a region in a system in which the properties and composition are spatially uniform. The condition for equilibrium is one where the electrochemical gradients of all the components of a system vanish. A system is said to be at equilibrium when there are no observable changes in either properties or microstructure with the passing of time, provided, of course, that no changes occur in the external conditions during that time.

The importance of knowing the phase diagram in a particular system cannot be overemphasized. It is the roadmap without which it is very difficult to interpret and predict microstructure distribution and evolution, which in turn have a profound effect on the ultimate properties of a material.

In principle, phase diagrams provide the following information:

1. The phases present at equilibrium
2. The composition of the phases present at any time during heating or cooling
3. The fraction of each phase present
4. The range of solid solubility of one element or compound in another

Like Chaps. 2 and 5, this chapter is not intended to be a comprehensive treatise on phase equilibria and phase diagrams. It is included more for the sake of completeness and is to be used as a reminder to the reader of some of the more important concepts invoked. For more information, the reader is referred to the references listed at the end of this chapter.

The subject matter is introduced by a short exposition of the Gibbs phase rule in Sec. 8.2. Unary component systems are discussed in Sec. 8.3. Binary and ternary systems are addressed in Secs. 8.4 and 8.5, respectively. Sec. 8.6 makes the connection between free energy, temperature, and composition, on one hand, and phase diagrams, on the other.

8.2
PHASE RULE

As noted above, phase diagrams are equilibrium diagrams. J. W. Gibbs showed that the condition for equilibrium places constraints on the degrees of freedom F that a system may possess. This constraint is embodied in the phase rule which relates F to the number of phases P present and the number of components C

$$F = C + 2 - P \tag{8.1}$$

where the 2 on the right-hand side denotes that two external variables are being considered, usually taken to be the temperature and pressure of the system.

The number of phases P is the number of physically distinct and, in principle, mechanically separable portions of the system. One of the easiest and least ambiguous methods to identify a phase is by analyzing its X-ray diffraction pattern — every phase has a unique pattern with peaks that occur at very well defined angles (see Chap. 4). For solid solutions and nonstoichiometric compounds, the situation is more complicated; the phases still have a unique X-ray diffraction pattern, but the angles at which the peaks appear depend on composition.

In the liquid state, the number of phases is much more limited than in the solid state, since for the most part liquid solutions are single-phase (alcohol and water are a common example). However, in some systems, most notably the silicates, liquid-liquid immiscibility results in the presence of two or more phases (e.g., oil and water). The gaseous state is always considered one phase because gases are miscible in all proportions.

The number of components C is the minimum number of constituents needed to fully describe the compositions of all the phases present. When one is dealing with binary systems, then perforce the number of components is identical to the number of elements present. Similarly, in ternary systems, one would expect C to be 3.

There are situations, however, when C is only 2. For example, for any binary join in a ternary phase diagram the number of components is 2, since one element is common.

The number of degrees of freedom F represents the number of variables, which include temperature, pressure, and composition, that have to be specified to completely define a system at equilibrium.

8.3
ONE-COMPONENT SYSTEMS

For a one-component system $C = 1$, and the degrees of freedom $F = 2$. In other words, to completely define the system, both temperature and pressure must be specified. If two phases are present, $F = 1$, and either pressure or temperature needs to be specified, but not both. For example, at 1 atm pressure, water and ice can coexist at only one temperature ($0°C$). At the triple point, three phases coexist, and there are no degrees of freedom left — the three phases must coexist at a unique temperature and pressure.

As single-phase substances are heated or cooled, they can undergo a number of polymorphic transformations. **Polymorphs** are different crystalline modifications of the same chemical substance. These transformations are quite common and include crystallization of glasses, melting, and many solid-solid phase transformations, some of which are described below. In general, there are two types of polymorphic transformations, displacive and reconstructive.

8.3.1   Reconstructive Transformations

As shown schematically in Fig. 8.1*a*, reconstructive transformations involve the breaking and rearrangement of bonds. Such transformations usually occur by nucleation and growth, which in turn usually depend on the rate at which atoms diffuse and consequently are relatively sluggish and easily suppressed (see Chap. 9). The reconstructive transformations that occur in quartz, specifically the α-β transformation (see below), are good examples.

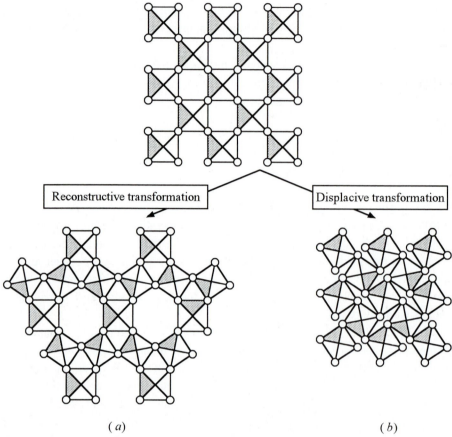

Figure 8.1

Schematic of (*a*) reconstructive and (*b*) displacive transformations.

8.3.2 Displacive Transformations

In contrast to reconstructive transformations, displacive transformations do not involve the breaking of bonds, but rather occur by the displacement of atomic planes relative to one another, as illustrated in Fig. 8.1*b*. These reactions occur quite rapidly, and the resulting microstructures are usually heavily twinned. In these transformations, the role of thermal entropy is very important since the enthalpies of the phases on either side of the transformation temperature are quite comparable. It follows that the transformation *usually* results in the formation of more open (less dense) structures at higher temperatures, for reasons that were touched upon in Chap. 5, namely that the more open structures have higher thermal entropies.[121]

Martensitic transformations in steel are probably the most studied of these transformations. Examples in ceramic systems of technological importance include the tetragonal-to-monoclinic transformation in ZrO_2, the cubic-to-tetragonal transformation in $BaTiO_3$, and numerous transformations in silica. In the remainder of this section, for the sake of illustration, each is discussed in greater detail.

Zirconia

Upon heating under 1 atm pressure, zirconia goes through the following transformations:

$$\text{Monoclinic} \xrightarrow{1170°C} \text{tetragonal} \xrightarrow{2370°C} \text{cubic} \xrightarrow{2680°C} \text{liquid}$$

It exhibits three well-defined polymorphs: a monoclinic phase, a tetragonal phase, and a cubic phase. The low-temperature phase is monoclinic, stable to 1170°C at which temperature it changes reversibly to the tetragonal phase, which in turn is stable to 2370°C. Above that temperature the cubic phase becomes stable up to the melting point of 2680°C. The tetragonal-to-monoclinic $(t \Rightarrow m)$ transformation is believed to occur by a diffusionless shear process that is similar to the formation of martensite in steels. This transformation is associated with a large volume change and undergoes extensive shear which is the basis for transformation toughening of zirconia, addressed in greater detail in Chap. 11.

Barium titanate

Barium titanate goes through the following phase transitions upon heating:

$$\text{Rhombohedral} \xrightarrow{-90°C} \text{orthorhombic} \xrightarrow{0°C} \text{tetragonal} \xrightarrow{130°C} \text{cubic}$$

Above 130°C, the unit cell is cubic, and the Ti ions are centered in the unit cell. Between 130 and 0°C, however, $BaTiO_3$ has a distorted perovskite structure with an eccentricity of the Ti ions. As discussed in greater detail in Chaps. 14 and 15, it is this eccentricity that is believed to be the origin of the main technical application of $BaTiO_3$ as a capacitor material with a high dielectric constant.

---

[121] Note that there are exceptions; for example, the tetragonal to monoclinic transformation of ZrO_2 is one where the more "open" structure is more stable at lower temperatures.

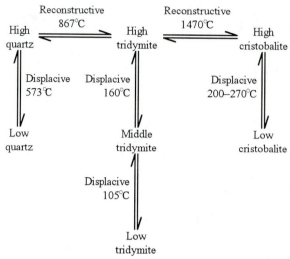

Figure 8.2
Polymorphic transformations in silica.

Silica

Silica has a multitude of polymorphs that undergo a number of both displacive and reconstructive transformations, the most important of which are summarized in Fig. 8.2. The displacive transformation from high to low quartz is associated with a large volume change (see Fig. 4.5) which, upon cooling, can create large residual stresses and result in a loss of strength (see Chap. 13). The best way to avoid the problem is to ensure that during processing all the quartz is converted to cristobalite, which because of its sluggish reconstructive transformation is metastable at room temperature. The volume change from high to low cristobalite is not as severe as that for quartz.

8.4
BINARY SYSTEMS

A binary system consists of two components and is influenced by three variables: temperature, pressure, and composition. When two components are mixed together and allowed to equilibrate, three outcomes are possible:

1. Mutual solubility and solid solution formation over the *entire* composition range, also known as *complete solid solubility*.
2. Partial solid solubility *without* the formation of an intermediate phase
3. Partial solid solubility with the formation of intermediate phases

One objective of this section is to qualitatively describe the relationship between these various outcomes and the resulting phase diagrams. First, however, it is important to appreciate what is meant by a solid solution in a ceramic system and the types of solid solutions that occur — a topic that was dealt with indirectly and briefly in Chap. 6. The two main types of solid solutions, described below, are substitutional and interstitial.

Substitutional solid solutions

In a substitutional solid solution, the solute ion directly substitutes for the host ion nearest to it in electronegativity, which implies, as noted in Chap. 6, that cations will substitute for cations and anions for anions. Needless to say, the rules for defect incorporation reactions (see Chap. 6) have to be satisfied at all times. For instance, the incorporation reaction of NiO in MgO would be written as

$$NiO \underset{MgO}{\rightarrow} O_O^x + Ni_{Mg}^x$$

where the Ni^{2+} ions substitute for Mg^{2+} ions. The resulting substitutional solid solution is denoted by $(Ni_{1-x}Mg_x)O$. The factors that determine the extent of solid solubility are discussed shortly.

Interstitial solid solutions

If the solute atoms are small, they may dissolve interstitially in the host crystal. The ease with which interstitial solid solutions form depends on the size of the interstitial sites in the host lattice relative to that of the solute ions. For example, in a close-packed structure such as rock salt, the only available interstitial sites are small tetrahedral sites, and interstitial solid solubility is not very likely. In contrast, in ThO_2 with its fluorite structure and TiO_2 where the interstitial sites are quite large, interstitial solid solutions form more easily. For example, it has been established that when YF_3 is dissolved in CaF_2, the appropriate incorporation reaction is

$$YF_3 \underset{CaF_2}{\rightarrow} Y_{Ca}^{\bullet} + F_i' + 2F_F^x$$

In other words, to maintain charge neutrality, the F ions reside on interstitial sites (see Fig. 8.3b). Another example involves the dissolution of ZrO_2 in Y_2O_3, where it has been established that the appropriate defect reaction is

$$2ZrO_2 \xrightarrow[Y_2O_3]{} 2Zr_Y^{\bullet} + 3O_O^X + O_i''$$

WORKED EXAMPLE 8.1. (*a*) Draw a representative unit cell for an $(Mg_{0.5}Ni_{0.5})O$ solid solution. (*b*) Repeat part (*a*) for a 0.25 mol fraction of YF_3 in CaF_2.

Answer

(*a*) A representative unit cell for this solid solution must contain $2Ni^{2+}$ and $2Mg^{2+}$ cations. It is left as an exercise to the reader to show that the unit cell in Fig. 8.3a fulfills that condition.

(*b*) A representative unit cell, the chemistry of which must reflect the composition of the solid solution, that is, $Y_1Ca_3F_9$, is shown in Fig. 8.3b. Note that the excess F ion occupies the large interstitial octahedral site that is normally vacant in the fluorite structure.

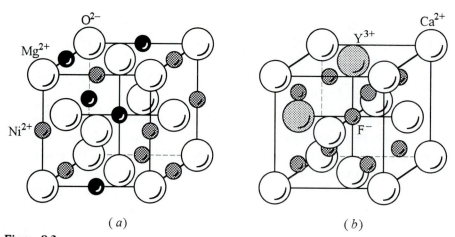

(*a*) (*b*)

Figure 8.3
(*a*) Representative unit cell of an $(Mg_{0.5}Ni_{0.5})O$ solid solution. (*b*) Representative unit cell for a 0.25 mol fraction of YF_3 in CaF_2 solid solution.

After this brief introduction to solid solutions, it is instructive to consider the type of phase diagrams expected for each of the three possible outcomes outlined above.

8.4.1   Complete Solid Solubility

For *complete* solid solubility to occur between two end members, the following conditions have to satisfied:

1. *Structure type.* The two end members must have the same structure type. For instance, SiO_2 and TiO_2 would not be expected to form complete solid solubility.
2. *Valency factor.* The two end members must have the same valence. If this condition is not satisfied, compensating defects form in the host crystal in order to maintain charge neutrality. Given that the entropy increase associated with defect formation is not likely to be compensated for by the energy required to form them over the entire composition range, complete solid solubility is unlikely.
3. *Size factor.* As a result of the mismatch in size of the solvent and solute ions, strain energy will develop as one is substituted for the other. For complete solid solubility to occur, that excess strain energy has to be low. Hence, in general, the size difference between the ions has to be less than 15 percent.
4. *Chemical affinity.* The two end members cannot have too high a chemical affinity for each other. Otherwise the free energy of the system will be lowered by the formation of an intermediate compound.

A typical phase diagram for two compounds that form a complete solid solubility over their entire composition range is shown in Fig. 8.4. Both NiO and MgO crystallize in the rock salt structure, and their cationic radii are very similar.

To illustrate the use and usefulness of phase diagrams, it is instructive to take a composition in Fig. 8.4, say, 60 mol % NiO, and examine what happens as it is cooled from the melt. At T_1, a solid solution of MgO and NiO (roughly 80 mol % Ni^{2+}) will start solidifying. At $T_2 \approx 2500°C$, two phases coexist: a solid solution of composition Z (see top of Fig. 8.4) and a liquid solution of composition X. The relative amounts of the each phase are given by the **lever rule**:

$$\text{Mole fraction liquid} = \frac{yz}{xz} \text{ and mole fraction solid} = \frac{xy}{xz}$$

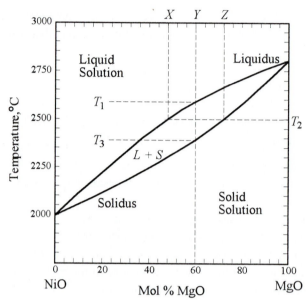

Figure 8.4

MgO-NiO phase diagram exhibiting solid solubility over entire composition range. Note liquidus and solidus lines.

Note that as the temperature is lowered, the composition of the solid solution moves along the **solidus** line toward NiO, while that of the liquid moves along the **liquidus** line.[122] At $T_3 \approx 2400°C$, the final liquid solidifies, and the composition of the solid solution is now the same as the initial composition.

Sometimes systems that exhibit complete solid solubility will also exhibit either a maximum (very rare in ceramic systems) or a minimum, as shown in Fig. 8.5.

8.4.2   Eutectic Diagrams with Partial Solid Solubility and No Intermediate Compounds

Given the numerous restrictions needed for the formation of complete solid solutions, they are the exception rather than the rule — most ceramic binary phase diagrams exhibit partial solubility instead. Furthermore, the addition of one component to another will lower the freezing point of the mixture relative to the melting point of the end members. The end result is lowered liquidus curves for

---

[122] The line separating the single-phase liquid region from the two-phase $(S+L)$ region is the liquidus line. Similarly, the line separating the single-phase solid region from the two-phase region is the solidus line.

both end members which intersect at a point. The point of intersection defines the lowest temperature at which a liquid can exist and is known as the **eutectic temperature** T_E. This type of diagram is well illustrated by the MgO–CaO system shown in Fig. 8.6, where MgO dissolves some CaO and vice versa. The limited solubility comes about mostly because the size difference between the Ca and Mg ions is too large for complete solid solubility to occur. Beyond a certain composition, the increase in strain energy associated with increasing solute content can no longer be compensated for by the increase in configuration entropy.

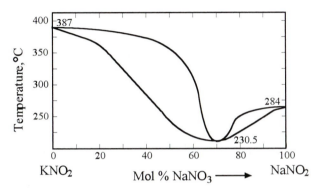

Figure 8.5

System with complete solid solubility with a minimum in temperature.

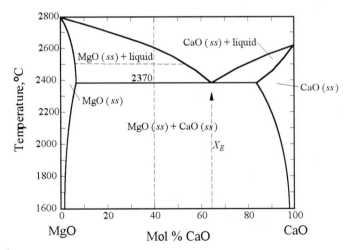

Figure 8.6

MgO-CaO phase diagram, which exhibits partial solid solubility of the end members for each other and a single eutectic.

To illustrate the changes that occur upon cooling consider what happens when a 40 mol % CaO composition, depicted by the dotted vertical line in Fig. 8.6, is cooled from the melt. Above 2600°C the liquid phase is stable. Just below 2600°C, a MgO solid solution ($\approx$ 95 mol % MgO saturated with CaO) will start to precipitate out. At 2500°C, two phases will coexist: an MgO–CaO solid solution and a liquid that is now richer in CaO ($\approx$ 55 mol % CaO) than the initial composition. Upon further cooling, the composition of the liquid follows the liquidus line toward the eutectic composition, whereas the composition of the precipitating solid follows the solidus line toward the point of maximum solubility. Just above T_E, that is, at $T_E + \delta$, a solid solution of CaO in MgO and a liquid with the eutectic composition $X_E \approx 65\%$ CaO coexist.

Just below T_E, however, the following reaction

$$L \Rightarrow S_1 + S_2 \tag{8.2}$$

known as a **eutectic reaction**, occurs, and the liquid disproportionates into two phases of very different compositions[123] — a calcia-rich and a magnesia-rich solid solution.

It is important to note here that the solution of one compound in another is unavoidable — a perfectly pure crystal is a thermodynamic impossibility for the same reason that a defect-free crystal is impossible.[124] The only legitimate question therefore is, How much solubility is there? In many binary systems, the regions of solid solution that are necessarily present do not appear on the phase diagrams. For example, according to Fig. 8.7a or 8.8, one could conclude incorrectly that neither Na_2O nor Al_2O_3 dissolves in SiO_2. This is simply a reflection of the scale over which the results are plotted — expanding the x axis will indicate the range of solubility that must be present. Note that in many applications and processes, this is far from being a purely academic question. For example, as noted in the previous chapter, the electrical and electronic properties of a compound can be dramatically altered by the addition of a few parts per million of impurities. Optical properties and sintering kinetics are also strongly influenced by small amounts of impurities. This is especially true when, as noted in Chap. 6, these impurities tend to segregate at grain boundaries.

---

[123] At the eutectic temperature, three phases coexist and there are no degrees of freedom left. In other words, the coexistence of three phases in a two-component system can occur only at a *unique* temperature, pressure, *and* composition.

[124] The decrease in free energy due to the increase in entropy associated with the mixing process is infinitely steep as the concentration of the solute goes to zero; that is, $\partial \Delta G / \partial n$ goes to $-\infty$ as $n \rightarrow 0$ [see Eq. (6.6)].

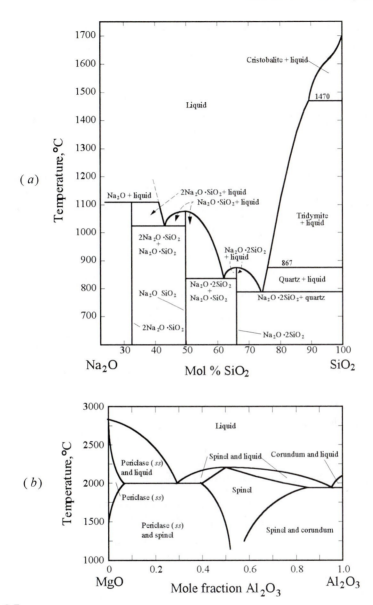

Figure 8.7
(a) $Na_2O - SiO_2$ phase diagram. (b) $MgO - Al_2O_3$ phase diagram.

8.4.3 Partial Solid Solubility with Formation of Intermediate Compounds

One of the conditions for the existence of a wide solid solution domain is the absence of a strong affinity of the end members for one another. That is not always the case — in many instances, the two end members react to form intermediate compounds. For instance, the compound $A_xB_yO_2$ can be formed by the reaction

$$xAO_{1/x} + yBO_{1/y} \Rightarrow A_xB_yO_2$$

where the free energy change for the reaction exceeds that for the simple mixing of the two end members to form a solid solution. Under these conditions, intermediate compounds appear in the phase diagram, which in analogy to the end members either can be line compounds (i.e., solubility of end members in the intermediate compound is small) or can have a wide range of stoichiometry. Furthermore, these intermediate phases can melt either congruently or incongruently.

Congruently melting intermediate phases

$Na_2O \cdot 2SiO_2$ and $Na_2O \cdot SiO_2$, shown in Fig. 8.7a, are examples of line compounds that melt **congruently**, i.e., without a change in composition. Note that in this case the resulting phase diagram is simply split into a series of smaller simple eutectic systems (e.g., Fig. 8.7a, for compositions greater than 50 mol % silica).

Spinel, $MgO \cdot Al_2O_3$, however, which also melts congruently and splits the phase diagram into two simple eutectic systems (Fig. 8.7b), is not a line compound but readily dissolves significant amounts of alumina, MgO, and Al_2O_3.

Incongruently melting intermediate phases

If the intermediate compound melts **incongruently**, i.e., the compound dissociates before melting into a liquid and another solid, then the phase diagram becomes slightly more complicated. A typical example of such a system is the $SiO_2 - Al_2O_3$ system (Fig. 8.8) where mullite, $2SiO_2 \cdot 3Al_2O_3$, melts at $\approx 1828°C$, by the formation of a liquid containing ≈ 40 mol % Al_2O_3 and "pure" alumina according to the reaction

$$S_1 \Rightarrow L + S_2 \tag{8.3}$$

This reaction is known as a **peritectic** reaction and is quite common in ceramic systems. Other examples of incongruently melting ternary compounds are $2Na_2O \cdot SiO_2$ (Fig. 8.7a) and $3Li_2O \cdot B_2O_3$ (Fig. 8.9).

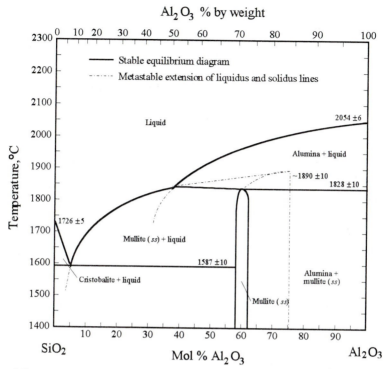

Figure 8.8

$SiO_2 - Al_2O_3$ phase diagram.[125]

This brings up the topic to be considered in this section which is nothing but a variation of the aforementioned case, and in which a ternary compound will dissociate into two other solid phases upon either cooling or heating. For example, according to Fig. 8.9, at about 700°C, $2Li_2O \cdot 5B_2O_3$ will dissociate into the 1 : 2 and 1 : 3 compounds.

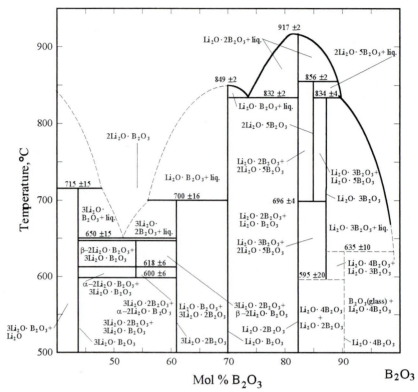

Figure 8.9

$Li_2O - B_2O_3$ phase diagram.

8.5
TERNARY SYSTEMS

Ternary phase diagrams relate the phases to temperature in a three-component system, and the four variables to be considered are temperature, pressure, and the concentration of two components (the composition of the third is fixed by the other two). A graphical representation is possible if the three components are represented by an equilateral triangle, where the apexes of the triangle represent the pure components, and temperature on a vertical axis as shown in Fig. 8.10a. The two-dimensional representation of the same diagram is shown in Fig. 8.10c, where the intersection of two surfaces is a line, the intersection of three surfaces is a point, and the temperature is represented by isotherms. The boundary curves represent equilibrium between two solids and the liquid, and the intersection of the boundary curves represents four phases in equilibrium (three solid phases and a liquid). This point is the lowest temperature at which a liquid can exist and, in complete analogy to the binary case, is called the **ternary eutectic.**

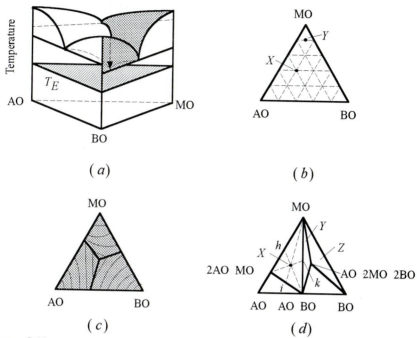

Figure 8.10

(*a*) Three-dimensional representation of a ternary phase diagram. (*b*) Triangular grid for representing compositions in a three-component system. (*c*) Two-dimensional representation of part (*a*) where boundary curves between two surfaces are drawn as heavy lines and temperature is represented by a series of lines corresponding to various isotherms.[126] (*d*) Isothermal section of a ternary system that includes a ternary $AO \cdot BO$ and a quaternary phase $AO \cdot 2MO \cdot BO$. The compatibility triangles are drawn with solid lines.

---

[126] To those familiar with topographical maps, height and temperature are analogous.

The composition at any point is found by drawing lines parallel to the three sides of the triangle. Thus, e.g., the composition of point X in Fig. 8.10b is 40 mol % AO, 20 mol % BO, and 40 mol % MO, while that at point Y is 80 mol % MO, 10 mol % AO, with the balance of BO. The temperature of the liquidus surface is depicted by isothermal contours, as shown in Fig. 8.10c.

Any ternary phase diagram that does not include a solid solubility region will consist of a number of compatibility triangles, shown as solid lines in Fig. 8.10d, with the apexes of the triangles representing the solid phases that would be present at equilibrium. For example, if the starting mixture is point X in Fig. 8.10d, then at equilibrium the phases present will be MO, 2AO·MO, and AO·BO. Similarly, for an initial composition at point Z, MO, BO, and the quaternary phase AO·2BO·2MO are the equilibrium phases, and so forth. This does not mean that the composition of any of the phases that appear or disappear during cooling has to remain inside the triangle, but simply that at the end, any phases that are not within the boundaries of the original triangle have to disappear.

Once the compatibility triangles are known, both the phases present at equilibrium and their relative amounts can be determined. Refer once more to Fig. 8.10d. At equilibrium, composition X would comprise the phases MO, 2AO·MO, and AO·BO in the following proportions:

$$\text{Mol fraction MO} = \frac{Xi}{i - \text{MO}}$$

$$\text{Mol fraction of } 2\text{AO} \cdot \text{MO} = \frac{Xk}{k - 2\text{AO} \cdot \text{MO}}$$

$$\text{Mol fraction of AO} \cdot \text{BO} = \frac{Xh}{h - \text{AO} \cdot \text{BO}}$$

Note that in going from a ternary to a binary representation, a dimension is lost; planes become lines and lines become points. Thus a quaternary phase is a point, and the edges of the triangles represent the corresponding binary phase diagrams (compare Fig. 10.8a and c).

8.6
FREE-ENERGY COMPOSITION AND TEMPERATURE DIAGRAMS

The previous sections dealt with various types of phase diagrams and their interpretations. What has been glossed over, however, is what determines their shape. In principle, the answer is simple: The phase or combination of phases for which the free energy of the system is lowest is by definition the equilibrium state.

However, to say that a phase transformation occurs because it lowers the free energy of the system is a tautology, since it would not be observed otherwise — thermodynamics forbids it. The more germane question, and one that is much more difficult to answer, asks, Why does any given phase have the lower free energy at any given temperature, composition, or pressure? The difficulty lies in the fact that to answer the questions, precise knowledge of all the subtle interactions between all the atoms that make up the solid and their vibrational characteristics, etc., is required. It is a many-body problem that is very sensitive to many variables, the least of which is the nature of the interatomic potentials one chooses to carry out the calculations.

The objective of this section is much less ambitious and can be formulated as follows: If the free-energy function for all phases in a given system were known as a function of temperature and composition, how could one construct the corresponding phase diagram? In other words, what is the relationship between free energies and phase diagrams? Two examples are considered below: polymorphic transformation in unary systems and complete solid solubility.

8.6.1 Polymorphic Transformations in Unary Systems

Congruent melting of a compound, or any of the polymorphic transformations discussed earlier, is a good example of this type of transformation. To illustrate, consider the melting of a compound. The temperature dependence of the free-energy functions for the liquid is

$$G_{T,\,\text{liq}} = H_{T,\,\text{liq}} - TS_{\text{liq}}$$

while that for the solid phase is

$$G_{T,s} = H_{T,s} - TS_s$$

where H and S are the enthalpy and entropy of the solid and liquid phases, respectively. The two functions are plotted in Fig. 9.1, assuming they are linear functions of temperature, which is only valid as long as (1) the heat capacities are not strong functions of temperature and (2) the temperature range considered is not too large.

$G_{T,\,\text{liq}}$ is steeper than $G_{T,s}$ because the entropy content of the liquid is larger (more disorder) than that of the solid. The salient point here is that at the temperature above which the lines intersect, the liquid has the lower energy and thus is the more stable phase, whereas below that temperature the solid is. Not surprisingly, the intersection temperature is the melting point of the solid.

8.6.2 Complete Solid Solutions

The free-energy versus composition diagram for a system that exhibits complete solid solubility is shown in Fig. 8.11. The components of the diagram are the two vertical axes that represent pure AO (left) and pure BO (right). The point labeled μ_{AO}° represents the molar free energy of formation ΔG_{form} of AO from its elements, and similarly, μ_{BO}° for BO. In this case AO has a lower free energy of formation than BO. If, for simplicity's sake, the solution is assumed to be ideal, that is, $\Delta H_{mix} = 0$, then the free energy of mixing of AO and BO is given by

$$\Delta G_{mix} = X_{AO}\mu_{AO}^{\circ} + X_{BO}\mu_{BO}^{\circ} - T\Delta S_{mix} \qquad (8.4)$$

where X_i represents the mole fraction of i. The entropy of mixing ΔS_{mix} [see Eq. (5.11)] is given by

$$\Delta S_{mix} = -RT(X_{AO}\ln X_{AO} + X_{BO}\ln X_{BO}) \qquad (8.5)$$

Combining these two equations and plotting ΔG_{mix} versus composition yields the curve $\mu_{AO}^{\circ} - M - \mu_{BO}^{\circ}$ shown in Fig. 8.11.

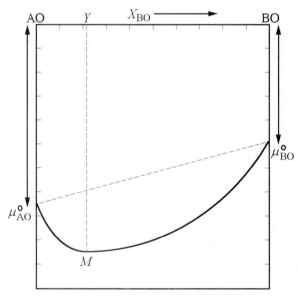

Figure 8.11
Free-energy versus composition diagram for an AO-BO mixture exhibiting complete solid solubility at a temperature that is lower than the solidus line.

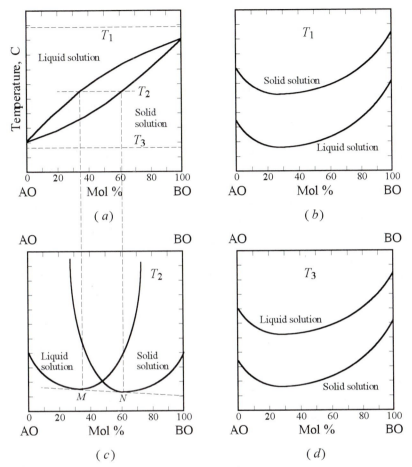

Figure 8.12
Temperature versus composition phase diagram and corresponding free-energy composition diagrams at various temperatures. In the two-phase region, a mixture of the solid and liquid solution is the lowest-energy configuration.

By using the same arguments, a free-energy versus composition function for the liquid solution can be determined. Superimposing the two functions as a function of temperature results in the curves depicted in Fig. 8.12b to d. It is, in principle, from these types of curves that the corresponding phase diagram shown in Fig. 8.12a can be plotted. At T_1 the free energy of the liquid solution is lowest at all compositions (Fig. 8.12b) and is the only phase that exists at that temperature. Conversely, at T_3, the solid solution is the most stable phase (Fig. 8.12d). At some intermediate temperature T_2, the free-energy versus composition curves have to intersect (Fig. 8.12c), from which it is obvious that, as depicted in Fig. 8.12a,

- Between pure AO and point M, the lowest energy of the system is that of the liquid solution.
- Between BO and N, the solid solution has the lowest energy.
- Between compositions M and N, the system's lowest energy state is given by the common tangent construction. In other words, the system's lowest free energy occurs when two phases (a solid phase and a liquid phase) coexist.

8.6.3   Stoichiometric and Nonstoichiometric Compounds Revisited

In Chap. 6, the notions of stoichiometry and nonstoichiometry were discussed at some length, and it was noted that a nonstoichiometric compound was one in which the composition range over which the compound was stable was not negligible. In the context of this chapter, the pertinent question is, How does one represent such a compound on a free-energy versus composition diagram? To answer the question, consider Fig. 8.13, where a nonstoichiometric compound $A_{1/2}B_{1/2}O$ is presumed to exist between two stoichiometric compounds, namely, $A_{3/4}B_{1/4}O$ and $A_{1/4}B_{3/4}O$. The latter are drawn as straight vertical lines, indicating that they exist only over a very narrow composition range; i.e., they are stoichiometric or **line compounds**. Note that the two tangents to the nonstoichiometric phase from the adjacent phases do not meet at a point, implying that there is a range of compositions over which the nonstoichiometric phase has the lowest free energy and thus exists.

In comparing Figs. 8.11 and 8.13, the similarities between the nonstoichiometric compound and solid solution free-energy versus composition curves should be obvious. It follows that an instructive way to look at the nonstoichiometric phase $A_{1/2}B_{1/2}O$ is to consider it to be for $X_{BO} < 1/2$ a solid solution between $A_{3/4}B_{1/4}O$ and $A_{1/2}B_{1/2}O$, and for $X_{BO} > 1/2$ a solid solution between $A_{1/4}B_{3/4}O$ and $A_{1/2}B_{1/2}O$. Note that for this to occur, the cations in the nonstoichiometric phase must exist in *more* than one oxidation state.

EXPERIMENTAL DETAILS: DETERMINING PHASE DIAGRAMS

Probably the simplest method to determine phase diagrams is to hold a carefully prepared mixture of known composition isothermally at elevated temperatures until equilibrium is achieved, quench the sample to room temperature rapidly enough to prevent phase changes during cooling, and then examine the specimen to determine the phases present. The latter is usually carried out by using a combination of X-ray diffraction and microscopy techniques.

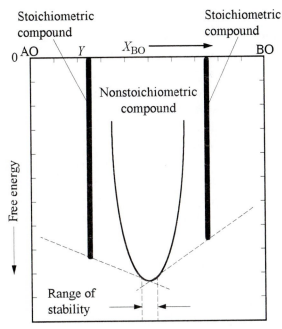

Figure 8.13
Free-energy versus composition curves of a nonstoichiometric compound $A_{1/2}B_{1/2}O$ that exists between two stoichiometric or line compounds, namely, $A_{3/4}B_{1/4}O$ and $A_{1/4}B_{3/4}O$.

And while in principle the procedure seems straightforward enough, it is a major problem encountered in all phase diagram determinations to ensure that equilibrium has actually been achieved. The most extensive and up-to-date set of phase diagrams for ceramists is published by the American Ceramic Society.

8.7
SUMMARY

Equilibrium between phases occurs at specific conditions of temperature composition and pressure. Gibbs' phase rule provides the relationship between the number of phases that exist at equilibrium, the degrees of freedom available to the system, and the number of components making up the system.

Phase diagrams are the roadmaps from which the number of phases, their compositions, and their fractions can determined as a function of temperature. In general, binary-phase diagrams can be characterized as exhibiting complete solid solubility between the end members, or partial solid solubility. In case of the latter, they will contain one or both of the following reactions depending on the species present. The first is the eutectic reaction is which a liquid becomes saturated with respect to the end members such that at the eutectic temperature two solids precipitate out of the liquid simultaneously. The second reaction is known as the peritectic reaction in which a solid dissociates into a liquid and a second solid of a different composition at the peritectic temperature. The eutectic and peritectic transformations also have their solid state analogues, which are called eutectoid and peritectoid reactions, respectively.

Ternary-phase diagrams are roadmaps for three component systems, where the major difference between them and binary-phase diagrams lies in how the results are presented. In ternary diagrams, the apexes of an equilateral triangle represent the compositions of the pure components, and the temperature appears as contour lines.

In principle, were one to know the dependence of the free energy of each phase as a function of temperature and composition, it would be possible to predict the phase diagram. The number of phases present at any temperature are simply the ones for which the total free energy of the system is at a minimum. Given that the free-energy versus composition information is, more often than not, lacking it follows that to date most phase diagrams are determined experimentally.

PROBLEMS

8.1. Which of the following transformations can be considered displacive and which can be considered reconstructive? Explain.
   (*a*) Melting
   (*b*) Crystallization
   (*c*) The tetragonal-to-monoclinic transformation in zirconia

8.2. (*a*) Explain why complete solid solubility can occur between two components of a substitutional solid solution but not an interstitial solid solution.
   (*b*) Can NaCl and CsCl form an extensive solid solution? Explain.
   (*c*) What type of solid solutions would you expect to be more likely in yttria? Magnesia? Explain.

8.3. The $CaO-ZrO_2$ phase diagram is shown in Fig. 8.14. What range of initial compositions can be used to manufacture toughened zirconia? Explain. See Chap. 11.

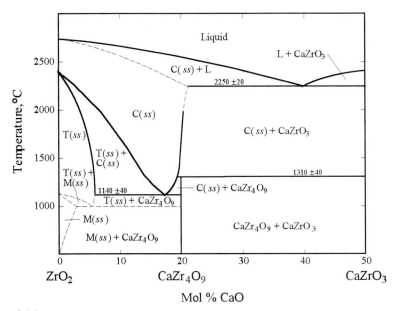

Figure 8.14
$ZrO_2 - CaZrO_3$ phase diagram.

8.4. Starting with stoichiometric spinel, write the incorporation reactions for both alumina and magnesia. Why do you think that spinel has such a wide range of solubility for its end members as compared to, say, $Na_2O \cdot SiO_2$?

8.5. (*a*) Show geometrically that the range of nonstoichiometry of a compound is related to the sharpness of the free-energy versus composition diagram; i.e., show that as $\partial G/\partial X$ approaches infinity, one obtains a line compound.
(*b*) From a structural point of view, what factors do you think are likely to determine $\partial G/\partial X$? Consider strain effects and defect chemistry.

ADDITIONAL READING

1. W. D. Kingery, H. K. Bowen, and D. R. Uhlmann, *Introduction to Ceramics*, 2d ed., Wiley, New York, 1976.
2. P. Gordon, *Principles of Phase Diagrams in Materials Systems*, McGraw-Hill, New York, 1968.
3. J. E. Ricci, *The Phase Rule and Heterogeneous Equilibrium*, Dover, New York, 1968.
4. A. Muan and E. F. Osborne, *Phase Equilibria among Oxides in Steelmaking*, Addison-Wesley, Reading, Massachusetts, 1965.

5. A. M. Alper, ed. *Phase Diagrams: Materials Science and Technology*, vols. 1 to 3, Academic Press, New York, 1970.

6. L. S. Darken and Gurry, *Physical Chemistry of Metals*, McGraw-Hill, New York, 1953.

7. C. G. Bergeron and S. H. Risbud, *Introduction to Phase Equilibria in Ceramics*, American Ceramic Society, Columbus, Ohio, 1984.

8. M. F. Berard and D. R. Wilder, *Fundamentals of Phase Equilibria in Ceramic Systems*, R. A. N. Publications, Marietta, Ohio, 1990.

9. F. A. Hummel, *Introduction to Phase Equilibria in Ceramic Systems*, Marcel Dekker, New York, 1984.

Phase diagram information

The most comprehensive compilation of ceramic phase diagrams is published by the American Ceramic Society, Columbus, Ohio. Additional volumes are anticipated.

10. E. M. Levin, C. R. Robbins, and H. F. McMurdie, *Phase Diagrams for Ceramists*, vol. 1, 1964; vol. 2, 1969.

11. E. Levin and H. McMurdie, eds., vol. 3, 1975.

12. R. Roth, T. Negas, and L. Cook, eds., vol. 4, 1981.

13. R. Roth, M. Clevinger, and D. McKenna, eds., vol. 5, 1983, and cumulative index, 1984.

14. R. Roth, J. Dennis, and H. McMurdie, eds., vol. 6, 1987.

CHAPTER 9

Formation, Structure, and Properties of Glasses

Prince Glass, Ceramic's son though crystal-clear
Is no wise crystalline. The fond Voyeur
And Narcissist alike devoutly peer
Into Disorder, the Disorderer
Being Covalent Bondings that prefer
Prolonged Viscosity and spread loose nets
Photons slip through. The average Polymer
Enjoys a Glassy state, but cools, forgets
To slump, and clouds in closely patterned Minuets.

John Updike, *Dance of the Solids*[†]

9.1
INTRODUCTION

From the time of its discovery, thousands of years ago (perhaps on a beach somewhere in ancient Egypt after the campfire was put out) to this day, glass has held a special fascination. Originally, the pleasure was purely aesthetic — glasses, unlike gems and precious stones for which the colors were predetermined by nature, could be fabricated in a multitude of shapes and vivid, extraordinary colors. Today that aesthetic appeal is further enhanced, scientifically speaking, by the challenge of trying to understand their structures and properties.

Numerous X-ray diffraction studies of glasses have shown that while glasses have short-range order, they clearly lack long-range order and can therefore be classified as solids in which the atomic arrangement is more characteristic of

---

[†] J. Updike, **Midpoint and other Poems**, A. Knopf, Inc., New York, New York, 1969. Reprinted with permission.

liquids. This observation suggests that if a liquid is cooled rapidly enough such that the atoms do not have enough time to rearrange themselves in a crystalline pattern before their motion is arrested, then a glass is formed. As a consequence of their structure, glasses exhibit many properties that crystalline solids do not; most notably, glasses do not have unique melting points but rather soften over a temperature range. Similarly, their viscosity increases gradually as the temperature is lowered.

This chapter will focus on why glasses form, their structure, and the properties that make them unique, such as their glass transition temperature and viscosity. In Sec. 9.2 the question of how rapidly a melt would have to be cooled to form a glass is addressed. Section 9.3 briefly describes glass structure. In Sec. 9.4, the focus is on trying to understand the origin of the glass transition temperature and the temperature and composition dependence of viscosity. Section 9.5 deals with another technologically important class of materials, namely glass-ceramics, their processing, advantages, and properties. Other properties such as mechanical, optical, and dielectric, that show similarities to those of crystalline solids are dealt with in the appropriate chapters.

9.2
GLASS FORMATION

Most liquids, when cooled from the melt, will, at a very well-defined temperature, namely, their melting point, abruptly solidify into crystalline solids. There are some liquids, however, for which this is not the case; when cooled, they form amorphous solids instead. Typically, the transformation of a liquid to a crystalline solid occurs by the formation of nuclei and their subsequent growth — two processes that require time. Consequently, if the rate of removal of the thermal energy is faster than the time needed for crystallization, the latter will not occur and a glass will form. It follows that it is only by understanding the nucleation and growth kinetics that the critical question concerning glass formation, namely, how fast a melt must be cooled to result in a glass, can be answered.

9.2.1 Nucleation

The two main mechanisms by which a liquid crystallizes are homogeneous and heterogeneous nucleation. *Homogeneous nucleation* refers to nucleation that occurs without the benefit of preexisting heterogeneities. It is considered first and in some detail because of its simplicity. *Heterogeneous nucleation* occurs at heterogeneities in the melt such as container walls, insoluble inclusions, and free surfaces. And even though the vast majority of nucleation occurs heterogeneously, it is not as well understood or amenable to analysis as homogeneous nucleation, a fact reflected in the following discussion.

Homogeneous nucleation

Consider the crystallization of a melt with a freezing point T_m. At T_m the free-energy change per mole ΔG_f associated with the solid-to-liquid transformation is zero, and

$$\Delta S_f = \frac{\Delta H_f}{T_m}$$

where ΔH_f and ΔS_f are the enthalpy and entropy of fusion per unit mole, respectively.

For temperatures $T < T_m$, the solid phase with its lower free energy will be more stable and will tend to form. The free-energy change ΔG_v for the transformation is the difference in energy between the undercooled liquid and the solid (Fig. 9.1). Assuming that for small undercooling ΔH_f and ΔS_f remain essentially unchanged, ΔG_v at T is given by

$$\Delta G_v = \Delta H_f - T\Delta S_f \approx \Delta H_f - \frac{\Delta H_f T}{T_m}$$
$$= \Delta H_f \left[\frac{\Delta T}{T_m} \right]$$

(9.1)

which implies that the driving force increases linearly with increasing undercooling, $\Delta T = T_m - T$.

The energy changes that have to be considered during homogeneous nucleation include

- The volume or bulk free energy released as a result of the liquid-to-solid transformation at $T < T_m$.
- The surface energy required to form the new solid surfaces
- The strain energy associated with any volume changes resulting from the transformation

If one assumes spherical nuclei with a solid-liquid interfacial energy between the growing nucleus and the melt γ_{sl} and radius r, and if one ignores strain effects, the energy changes that accompany their formation are

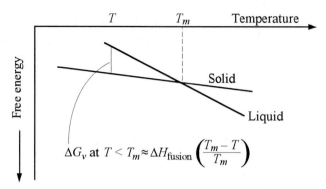

Figure 9.1
Schematic of free-energy changes as temperature is lowered below the equilibrium temperature T_m.

$$\text{Volume free energy} = -\frac{4}{3}\pi r^3 \frac{\Delta G_v}{V_m} = -\frac{4}{3}\pi r^3 \frac{\Delta H_f}{V_m}\left[\frac{\Delta T}{T_m}\right] \tag{9.2}$$

$$\text{Surface energy} = 4\pi r^2 \gamma_{sl} \tag{9.3}$$

where V_m is the molar volume of the crystal phase. The sum of Eqs. (9.2) and (9.3) represents the excess free-energy change (denoted by ΔG_{exc}) resulting from the formation of a nucleus, or

$$\Delta G_{exc} = 4\pi r^2 \gamma_{sl} - \frac{4}{3}\pi r^3 \frac{\Delta H_f}{V_m}\left[\frac{\Delta T}{T_m}\right] \tag{9.4}$$

The functional dependence of ΔG_{exc} on r is plotted in Fig. 9.2. Since the energy needed for the creation of new surfaces (top curve in Fig. 9.2) scales with r^2 whereas the volume energy term scales with r^3, this function clearly goes through a maximum at a critical radius r_c, which implies that the formation of small clusters with $r < r_c$ *locally* increases the free energy of the system.[127] Differentiating Eq. (9.4), equating to zero, and solving for r_c gives

$$r_c = \frac{2\gamma_{sl}V_m}{\Delta H_f(1-T/T_m)} \tag{9.5}$$

---

[127] Equation (9.4) represents the local increase in free energy due to the formation of a nucleus, and not the total free energy of the system. The latter must include the configurational entropy of mixing of n nuclei in the liquid. When that term is included, the total free energy of the system decreases as it must (see App. 9A).

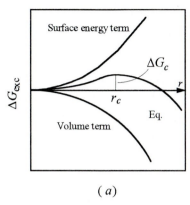
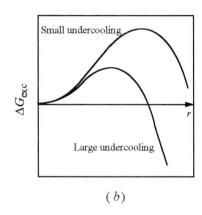

(a)   (b)

Figure 9.2
(a) Free energy versus embryo radius for a given $T < T_m$. Note that ΔG_{exc} goes through a maximum at $r = r_c$. (b) Effect of undercooling on ΔG_{exc}. With larger undercooling both ΔG_c and r_c decrease.

which, when substituted back into Eq. (9.4), yields the height of the energy barrier ΔG_c (Fig. 9.2a)

$$\Delta G_c = \frac{16\pi \gamma_{sl}^3 V_m^2}{3\Delta H_f^2 (1 - T/T_m)^2} \qquad (9.6)$$

Small clusters with $r < r_c$, are called *embryos* and are more likely to redissolve than grow. Occasionally, however, an embryo becomes large enough ($r \approx r_c$) and is then called a *nucleus* with an equal probability of growing or decaying. It is important to note that both ΔG_c and r_c are strong functions of undercooling (compare Fig. 9.2a and b).

By minimizing the free energy of a system containing N_v, total number of molecules or formula units of nucleating phase per unit volume, it can be shown (see App. 9A) that the metastable equilibrium concentration (per unit volume) of nuclei N_n^{eq} is related to ΔG_c by

$$N_n^{eq} = N_v \left[\exp\left(-\frac{\Delta G_c}{kT} \right) \right] \qquad (9.7)$$

The rate of nucleation per unit volume (number of nuclei per cubic meter per second) can be expressed by

$$I_v = \nu N_n^{eq} \qquad (9.8)$$

where ν is the frequency of successful atom jumps across the nucleus liquid interface or the rate at which atoms are added onto the critical nucleus, given by

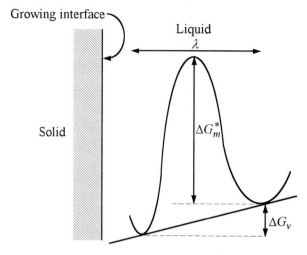

Figure 9.3
Schematic of growing nucleus or growing crystal. The atoms or molecules that result in growth of the crystal or nucleus jump a distance λ across an energy barrier ΔG_m^* down the chemical potential gradient $\Delta G_v/\lambda$ to the growing interface.

$$V = V_0 \exp\left(-\frac{\Delta G_m^*}{kT}\right) \tag{9.9}$$

where V_0 is the vibrational frequency of an atom and ΔG_m^* is the free energy of activation needed for an atom to jump across the nucleus-liquid interface[128] (Fig. 9.3). Combining Eqs. (9.7) – (9.9) yields the final expression for the rate of homogeneous nucleation

$$I_v = V_0 N_v \left[\exp\left(-\frac{\Delta G_m^*}{kT}\right)\right]\left[\exp\left(-\frac{\Delta G_c}{kT}\right)\right] \tag{9.10}$$

The first exponential term is sometimes referred to as the *kinetic barrier to nucleation*, whereas the second exponential term is known as the *thermodynamic barrier to nucleation*. And although it is not immediately obvious from Eq. (9.10), I_v goes through a maximum as a function of undercooling for the following reason: Increased undercooling reduces both r_c and ΔG_c (Fig. 9.2b), which in turn strongly *enhances* the nucleation rate but simultaneously severely *reduces* atomic mobility

---

[128] The implicit assumption here is that every atom that makes the jump sticks to the interface and contributes to the growth of the nucleus. In other words, a sticking coefficient of 1 was assumed.

and the rate of attachment of atoms to the growing embryo (see Prob. 9.2*a*). The net effect is that a maximum is expected and is well established experimentally (see Fig. 9.4*a*).

Experimentally, it is much easier to measure the viscosity η of an undercooled liquid than it is to measure ν. It is therefore useful to relate the nucleation rate to the viscosity, which is done as follows: Given the similarity between the elementary jump shown in Fig. 9.3 and a diffusional jump, the two can be assumed to be related by

$$D_{liq} = \text{const.}\, \nu\, \lambda^2 = \text{const.}\, \nu_0\, \lambda^2 \exp\left(-\frac{\Delta G_m^*}{kT}\right) \qquad (9.11)$$

where D_{liq} is the diffusion coefficient of the "formula units" in the liquid and λ is the distance advanced by the growing interface (Fig. 9.3) in a unit kinetic process usually taken as that of a molecular or formula unit diameter. Usually λ is taken to equal $(V_m/N_{Av})^{1/3}$, where V_m is the molar volume of the crystallizing phase. If it is further assumed that D_{liq} is related to the viscosity of the melt by the **Stokes-Einstein** relationship, namely,

$$D_{liq} = \frac{kT}{3\pi\lambda\eta} \qquad (9.12)$$

then by combining Eqs. (9.10) to (9.12), one obtains

$$I_v = (\text{const}) \frac{N_v kT}{3\pi\lambda^3\eta} \exp\left(-\frac{\Delta G_c}{kT}\right) \qquad (9.13)$$

With both the viscosity (see below) and the exponential term in Eq. (9.13) increasing at differing rates with decreasing temperature, it should once again be apparent why I_v goes through a maximum as a function of undercooling.

In deriving Eqs. (9.10) and (9.13), it is important to keep in mind that the following assumptions were made:

- Nucleation is homogeneous. This is very seldom the case. As discussed below, crystal nucleation occurs almost always heterogeneously on impurity particles or container walls at relatively low undercoolings.
- The rate given in Eq. (9.7) is a steady-state rate. In other words, density fluctuations develop and maintain an equilibrium distribution of subcritically sized embryos. As nucleation drains off critically sized embryos, new ones are produced at a rate sufficiently fast to maintain the equilibrium distribution.

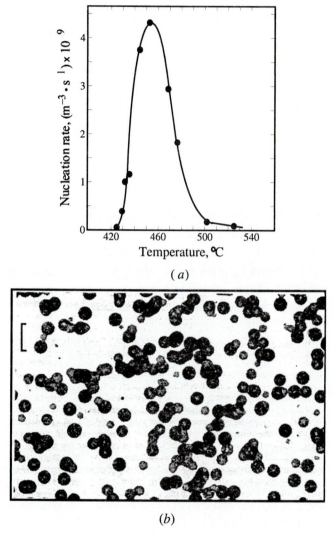

(a)

(b)

Figure 9.4

(a) Steady state nucleation rate as a function of temperature for a glass close to the $Li_2O \cdot 2SiO_2$ composition. (b) Reflection optical micrograph of a $BaO \cdot 2SiO_2$ glass after heat treatment.[129]

129 P. James, Chap. 3 in *Glasses and Glass-Ceramics*, M. H. Lewis, ed., Chapman & Hall, New York, 1989. Reprinted with permission.

- Nucleation occurs without a change in composition. If a change in composition accompanies the formation of a nucleus, the expression for the energy gained [i.e., Eq. (9.2)] changes and is no longer simply ΔH_f but must now include the free-energy change associated with the formation of the new phase(s) (see Chap. 8).
- Nucleation occurs without a change in volume. In other words, the strain energy was ignored. If there is a volume change and the associated strain energy is known, it is simply added to Eq. (9.3).

WORKED EXAMPLE 9.1. (*a*) If the enthalpy of fusion of a glass-forming liquid is 53 kJ/mol, its molar volume is 60 cm³/mol, the solid liquid interface energy is 150 mJ/m² , and its melting point is 1034°C, calculate the size of the critical radius, the height of the barrier to nucleation, and the steady-state metastable equilibrium concentration of nuclei at an undercooling of 500°C. (*b*) Repeat part (*a*) for a solid-liquid interface energy of 180 mJ/m² .

Answer

(*a*) Converting to SI units and applying Eq. (9.5), one obtains

$$r_c = \frac{2\gamma_{sl}V_m}{\Delta H_f (1 - T/T_m)} = \frac{2(150 \times 10^{-3})(60 \times 10^{-6})}{53,000(0.38)} = 8.93 \times 10^{-10} \text{ m}$$

Given that the Si-Si distance in silicates is on the order of 3×10^{-10} m, this result is not unreasonable — the critical radius would appear to be made of a few SiO_4 tetrahedra. The corresponding ΔG_c [(Eq. (9.6)] is

$$\Delta G_c = \frac{16\pi\gamma_{sl}^3 V_m^2}{3\Delta H_f^2 (1 - T/T_m)^2} = \frac{16(3.14)(150 \times 10^{-3})^3 (60 \times 10^{-6})^2}{3(53,000)^2 (0.38)^2} \approx 5 \times 10^{-19} \text{ J}$$

With 1 mol as a basis, N_v is simply N_{Av}/V_m, or 1×10^{22} molecular units per cubic centimeter; it follows [Eq. (9.7)] that

$$N_n^{eq} = N_v \left[\exp\left(-\frac{\Delta G_c}{kT}\right) \right] = 1 \times 10^{22} \exp\left(-\frac{5 \times 10^{-19}}{1.38 \times 10^{-23} \times 807} \right) = 317 \text{ nuclei/cm}^3$$

(*b*) Repeating the calculation while using the slightly lower interface energy of 130 mJ/m² yields $N_n^{eq} \approx 1.8 \times 10^9$, which is more than 5 orders of magnitude higher than that calculated in part (*a*)! This simple calculation makes amply clear the paramount importance of the surface energy term during nucleation. It is worth mentioning at this point that surface energies, in general, and solid-liquid interface energies, in particular, are fiendishly difficult to measure accurately and repeatedly.

Heterogeneous nucleation

Technologically, the vast majority of nucleation occurs heterogeneously at defects such as dislocations, interfaces, pores, grain boundaries, and especially free surfaces. These sites present preferred nucleation sites for three reasons. First, they are regions of higher free energy, and that excess energy becomes available to the system upon nucleation. Second, and more importantly, the heterogeneities tend to reduce γ, which allows nucleation to occur at relatively small undercoolings where homogeneous nucleation is unlikely. Third, the presence of pores or free surfaces will reduce any strain energy contribution that may suppress the nucleation or growth process.

It can be shown that the steady-state heterogeneous rate of nucleation of a supercooled liquid on a flat substrate is given by[130]

$$I_v = v_0 N_s \left[\exp\left(-\frac{\Delta G_m^*}{kT} \right) \right] \left[\exp\left(-\frac{\Delta G_{het}}{kT} \right) \right]$$

where N_s is the number of atoms or formula units of the liquid in contact with the substrate per unit area, and $\Delta G_{het} = \Delta G_c \left(2 - 3\cos\theta + \cos^3\theta \right)/4$, where θ is the contact angle between the crystalline nucleus and the substrate (see Chap. 10). Note that in the limit of a complete wetting, that is, $\theta = 0$, the thermodynamic barrier to nucleation vanishes.

EXPERIMENTAL DETAILS: MEASURING NUCLEATION RATES

As noted above, the overwhelming majority of glasses usually nucleate heterogeneously at surfaces; homogeneous or volume nucleation is rarely observed. There are a few glass systems, however, which nucleate homogeneously, and they have been studied in order to test the validity of Eq. (9.10) or (9.13). Of these, probably lithium disilicate, $Li_2O \cdot 2SiO_2$, has been one of the most intensively studied. In a typical nucleation experiment, the glass is heat-treated to a certain

---

[130] For more details, see J. W. Cahn, *Acta Met.*, **4**:449 (1956) and **5**:168 (1957). See also J. W. Christian, *The Theory of Transformations in Metals and Alloys*, 2d ed., Pergamon Press, London, 1975.

temperature for a given time, cooled, and sectioned. The number of nuclei is then counted by using optical or electron microscopy, and, assuming steady-state nucleation, the nucleation rate is calculated.[131] When the nucleation rate is plotted versus temperature, the typical bell-shaped curve (Fig. 9.4a) predicted from nucleation theory is obtained. A typical reflection optical micrograph of a glass after heat treatment to induce nucleation is shown in Fig. 9.4b.

It should be pointed out, however, that while Eq. (9.13) correctly represents the temperature dependence of the nucleation rate, the measured rates are ≈ 20 orders of magnitude larger than predicted! The reason for this huge discrepancy is not entirely clear, but it has been explained by allowing the surface energy term to be weakly temperature-dependent.[132]

9.2.2   CRYSTAL GROWTH

Once the nuclei are formed, they will tend to grow until they start to impinge upon each other. The growth of the crystals depends on the nature of the growing interface which has been related to the entropy of fusion.[133] It can be shown that for crystallization processes in which the entropy change is small, that is, $\Delta S_f < 2R$, the interface will be rough and the growth rate will be more or less isotropic. In contrast, for large entropy changes $\Delta S_f > 4R$ the most closely packed faces should be smooth and the less closely packed faces should be rough, resulting in large-growth-rate anisotropies. Based on these notions, various models of crystal growth have been developed, most notably:

Standard growth, $\Delta S_f < 2R$

In this model, the interface is assumed to be rough on the atomic scale, and a sizable fraction of the interface sites are available for growth to take place. It follows that under these circumstances, the rate of growth is solely determined by the rate of atoms jumping across the interface (that is, the assumption is that the process is controlled by the surface reaction rate and not diffusion). Using an analysis that is almost identical to the one carried out in Sec. 7.2.3, where the net

---

131 Sometimes if the size of the nuclei that form is too small to observe, a second heat treatment at a higher temperature is carried out to grow the nuclei to an observable size. Implicit in this latter approach is that the nuclei formed at the lower temperatures do not dissolve during the second heat treatment.

132 P. James, Chap. 3 in *Glasses and Glass-Ceramics*, M. H. Lewis, ed., Chapman and Hall, New York, 1989.

133 K. A. Jackson in *Progress in Solid State Chemistry*, vol. 3, Pergamon Press, New York, 1967.

rate of atom movement down a chemical potential gradient was shown to be [Eq. (7.26)]

$$v_{net} = v_0 \exp\left(-\frac{\Delta G_m^*}{kT}\right)\left\{1 - \exp\left(-\frac{\Xi}{kT}\right)\right\}$$

where Ξ was defined by Eq. (7.23), it is possible to derive an expression for the growth rate as follows. Comparing Figs. 7.5 and 9.3, the equivalence of Ξ and ΔG_v is obvious. Hence, the growth rate u of the interface is given by

$$u = \lambda v_{net} = \lambda v_0 \left\{\exp\left(-\frac{\Delta G_m^*}{kT}\right)\right\}\left\{1 - \exp\left(-\frac{\Delta G_v}{RT}\right)\right\} \qquad (9.14)$$

where ΔG_v is given by Eq. (9.1). It is left as an exercise to the reader to show that in terms of viscosity, this equation can be rewritten as

$$u \approx (\text{const}) \frac{kT}{3\pi\eta\lambda^2}\left[1 - \exp\left(-\Delta H_f \frac{\Delta T}{T_m TR}\right)\right] \qquad (9.15)$$

when the growth is occurring at temperature T, with an undercooling of ΔT.

This is an important result because it predicts that the growth rate, like the nucleation rate, should also go through a maximum as a function of undercooling. The reason, once more, is that with *increasing undercooling, the driving force for growth ΔG_v increases, while the atomic mobility, expressed by η, decreases exponentially with decreasing temperature.* It is important to note that the temperature at which the maximum growth rate occurs is usually different from that at which the nucleation rate peaks.

For small ΔT values, a linear relation exists between the growth rate and undercooling (see Prob. 9.2b). Conversely, for large undercooling the limiting growth rate

$$u = (\text{const}) \frac{kT}{3\pi\eta\lambda^2} = (\text{const}) \frac{D_{liq}}{\lambda} \qquad (9.16)$$

is predicted. Once a stable nucleus has formed, it will grow until it encounters other crystals or until the molecular mobility is sufficiently reduced that further growth is cut off.

Surface nucleation growth, $\Delta S_f > 4R$

In the normal growth model, the assumption is that all the atoms which arrive at the growing interface can be incorporated in the growing crystal. This will occur only when the interface is rough on an atomic scale. If, however, the interface is smooth, then growth will take place only at preferred sites such as ledges or steps. In other words, growth will occur by the spreading of a monolayer across the surface.

Screw dislocation growth

Here the interface is viewed as being smooth but imperfect on an atomic scale. Growth is assumed to occur at step sites provided by screw dislocations intersecting the interface. The growth rate is given by

$$u = f\lambda v\left[1 - \exp\left(-\Delta H_f \frac{\Delta T}{T_m RT}\right)\right] \tag{9.17}$$

where f is the fraction of preferred growth sites. It can shown that[134] the fraction of such sites is related to the undercooling by

$$f \approx \frac{\Delta T}{2\pi T_m}$$

Hence in this model, at small undercoolings, the growth rate is expected to be proportional to ΔT^2.

WORKED EXAMPLE 9.2. Empirically, it has been determined that for a given oxide glass, the constant in Eq. (9.15) is 10. Furthermore, the temperature dependence of the viscosity of that same glass is measured to be:

T, °C	1400	1300	1200	1000
η, Pa·s	10	250	1000	10^4

and its melting point is 1300C. The entropy of fusion was 8 J/(mol·K). Given a concentration of nucleation sites to be a constant equal to 10^6 cm^{-3}, for how long could a 1 cm^3 sample be held at 1000C without sensible bulk crystallization? Assume a molar volume of ≈ 10 cm^3/mol. State all assumptions.

---

134 W. B. Hillig and D. Turnbull, *J. Chem. Phys.*, **24**: 914 (1956).

Answer

Since the number of nuclei is fixed and constant, the growth rate u of the nuclei will determine the extent of crystallization. Once u is calculated, the size of the nuclei after a given time can easily be calculated. By rewriting Eq. (9.15) in terms of ΔS_f

$$u = \frac{10kT}{3\pi\eta\lambda^2}\left[1 - \exp\left(-\frac{\Delta S_f \Delta T}{RT}\right)\right]$$

the jump distance can be approximated by $\lambda = [10/(6.02 \times 10^{23})]^{1/3} = 2.55 \times 10^{-8}$ cm $= 2.55 \times 10^{-10}$ m. Inserting the appropriate values gives a linear growth rate of

$$u = \frac{10(1.38 \times 10^{-23})(1273)}{3(3.14)(10^4)(2.55 \times 10^{-10})^2}\left[1 - \exp\left(-\frac{8 \times 300}{8.314 \times 1273}\right)\right] = 5.7 \times 10^{-6} \text{ m/s}$$

Taking 1 cm^3 as a basis and assuming that the nuclei would be detected when they reached a volume fraction of, say, 10^{-4}, that is, when their volume $= 10^{-4} \times 1 = 10^{-4}$ cm^3. Further assume that the nuclei grow as spheres, with a volume $(4/3)\pi r^3$. Solving for r shows that the nuclei would be detected when they reached a radius of ≈ 3 μm. Based on their growth rate, the time to reach that size would be $(3 \times 10^{-6})/(5.7 \times 10^{-6}) \approx 0.5$ s.

9.2.3   Kinetic of Glass Formation

At this point, the fundamental question, posed at the outset of this section, namely, How fast must a melt be cooled to avoid the formation of a detectable volume fraction of the crystallized phase? can be addressed somewhat more quantitatively. The first step entails the construction of a **time-temperature-transformation** (TTT) curve for a given system. Such a curve defines the time required, at any temperature, for a given volume fraction to crystallize. Here the procedure, not unlike that used to solve Worked Example 9.2, is generalized.

If at any time t, in a total volume V, the nucleation rate is I_v, it follows that the number N_t of new particles formed in time interval $d\tau$ is

$$N_t = I_v V d\tau$$

For a time-independent constant growth rate u and assuming isotropic growth (i.e., spheres), the radius of the sphere after time t will be

$$r = \begin{cases} u(t - \tau) & \text{for } t > \tau \\ 0 & \text{for } t < \tau \end{cases}$$

and its volume will be

$$V_\tau = \frac{4}{3}\pi u^3 (t-\tau)^3$$

where τ is the time at which a given nucleus appears. Hence the total volume transformed after time t, denoted by V_t, is given by the number of nuclei at time t multiplied by their volume at that time, or

$$V_t = V_\tau N_t = \int V_\tau I_v V d\tau = \int_{t=0}^{t=t} V I_v \left(\frac{4}{3}\pi u^3\right)(t-\tau)^3 d\tau$$

Upon integration and rearranging, this gives

$$\frac{V_t}{V} = \frac{\pi}{3} I_v u^3 t^4 \qquad (9.18)$$

It is important to note that an implicit assumption made in deriving this expression is that the transformed regions do not interfere or impinge on one another. In other words, this expression is valid only for the initial stages of the transformation. A more exact and general analysis which takes impingement into account, but which will not be derived here, yields

$$\boxed{\frac{V_t}{V} = 1 - \exp\left(-\frac{\pi}{3} I_v u^3 t^4\right)} \qquad (9.19)$$

This is known as the *Johnson-Mehl-Avrami equation.*[135] This equation reduces to Eq. (9.18) at small values of time. These assumptions are made in deriving this equation:

1. Both the nucleation rate and the growth rate follow Boltzmann distributions.
2. The growth rate is isotropic and linear (i.e., surface reaction rate controlled) and three-dimensional with time. If the growth were diffusion-limited, the growth rate would be not linear with time but parabolic.
3. Nucleation rate is random and continuous.

Given the nucleation and growth rates at any given temperature, the fraction crystallized can be calculated as a function of time from Eq. (9.19). Repeating the process for other temperatures and joining the loci of points having the same volume fraction transformed yield the familiar TTT diagram, shown schematically in Fig. 9.5. Once constructed, an estimate of the **critical cooling rate** (CCR) is given by

135 For an excellent derivation, see K. Tu, J. Mayer, and L. Feldman, *Electronic Thin Film Science for Electrical and Materials Engineers*; Macmillan, New York, 1992, Chap. 10. For the original references, see W. L. Johnson and R. F. Mehl, *Trans. AIME*, **135**:416 (1936), and M. Avrami, *J. Chem. Phys.*, **7**:1103 (1937), **8**:221 (1940), **9**:177 (1941).

$$CCR \approx \frac{T_L - T_n}{t_n}$$

where T_L is the temperature of the melt and T_n and t_n are the temperature and time corresponding to the nose of the TTT curve, respectively (see Fig. 9.5). The critical cooling rates in degrees Celsius per second for a number of silicate glasses are shown in Fig. 9.6, where the salient feature is the strong (note the log scale on y axis) functionality of the CCR on glass composition.

9.2.4 Criteria for Glass Formation

The question of glass formation can now be restated: Why do some liquids form glasses while others do not? Based on the foregoing discussion, for a glass to form, the following conditions must exist:

1. A low nucleation rate. This can be accomplished by having either a small ΔS_f or a large crystal/liquid interfacial energy. The lower ΔS_f and/or the higher γ_{sl}, the higher ΔG_c and consequently the more difficult the nucleation.

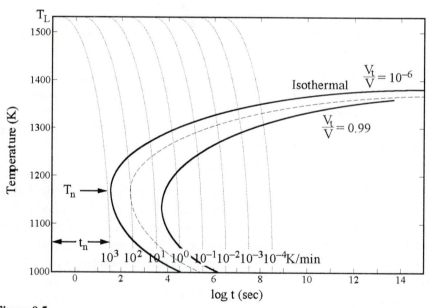

Figure 9.5
Isothermal TTT diagram.

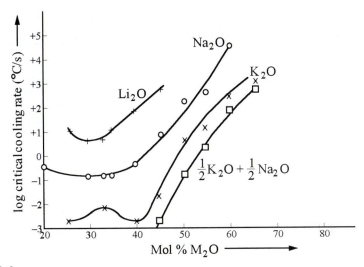

Figure 9.6

Critical cooling rates as a function of glass composition.[136]

2. High viscosity η_m at or near the melting point. This ensures that the growth rate will be small.
3. The absence of nucleating heterogeneities that can act as nucleating agents. The presence of nucleating agents can greatly reduce the size of the critical nucleus and greatly enhance the nucleation kinetics.

    Based on items 1 and 2 above, a useful criterion for the formation of a glass is the ratio

$$\Delta S_f \left[\frac{1}{\eta_m} \right]$$

and the smaller the product, the more likely a melt will form a glass, and vice versa. Table 9.1 shows that to be the case, indeed. Further inspection reveals that atom mobility as reflected in η_m at the melting point is by far the dominant factor. It follows that a *melt must have a high viscosity at its liquidus temperature or melting point if it is to form a glass.*

136 A. C. Havermans, H. N. Stein, and J. M. Stevels, *J. Non-Cryst. Solids*, **5**:66–69 (1970).

TABLE 9.1
Summary of glass-forming ability of various compounds

Compound	Melting point °C	ΔS_f J/(mol·K)	η_m, Pa·s	$\Delta S_f \times 1/\eta_m$	Comments
B_2O_3	450	33.2	5000	0.0066	Excellent glass former
SiO_2	1423	4.6	2.3×10^5	2.0×10^{-5}	Excellent glass former
$Na_2Si_2O_5$	874	31.0	200	0.155	Good glass former
Na_2SiO_3	1088	38.5	20	1.9	Poor glass former
GeO_2	1116	10.8	71,428	2×10^{-4}	Excellent glass former
P_2O_5	569				Glass former
$NaAlSi_3O_8$			3.2×10^5		Glass former
$CaSiO_3$	1544	31.0	1	31.0	Difficult to form glass
$NaCl$	800	25.9	2×10^{-3}	1.3×10^4	Not a glass former

9.3
GLASS STRUCTURE

In principle, if the requisite data were available, the TTT diagram for any material could be generated, and the CCR that would be required to keep it from crystallizing could be calculated. In other words, if cooled rapidly enough, any liquid will form a glass, and indeed glasses have been formed from ionic, organic, and metallic melts. What is of interest here, however, is the so-called inorganic glasses formed from covalently bonded, and for the most part silicate-based, oxide melts. These glass-forming oxides are characterized by having a continuous three-dimensional network of linked polyhedra and are known as **network formers**. They include silica, boron oxide (B_2O_3), phosphorous pentoxide (P_2O_5), and germania (GeO_2). Commercially, silicate-based glasses are by far the most important and the most studied and consequently are the only ones discussed here.[137]

Since glasses possess only short-range order, they cannot be as elegantly and succinctly described as crystalline solids — e.g., there are no unit cells. So the best way to describe a glass is to describe the building block that possesses the short-range order (i.e., the coordination number of each atom) and then how these blocks are put together. The simplest of the silicates is vitreous silica (SiO_2), and understanding its structure is fundamental to understanding the structure of other silicates.

---

[137] See Kingery et al. for descriptions of the structure of other glasses or most of the references listed at the end of this chapter.

Vitreous silica SiO$_2$

The basic building block for all crystalline silicates is the SiO$_4$ tetrahedron (see Chap. 3). In the case of quartz, every silica tetrahedron is attached to four other tetrahedra, and a three-dimensional *periodic* network results (see top of Table 3.4). The structure of vitreous silica is very similar to that of quartz, except that the network lacks symmetry or long-range periodicity. This so-called random network model, first proposed by Zachariasen,[138] is generally accepted as the best description of the structure of vitreous or fused silica and is shown schematically in two-dimensions in Fig. 1.1*b*. Quantitatively it has been shown that the Si–O–Si bond angle in vitreous silica while centered on 144°, which is the angle for quartz (see Prob. 9.5), has a distribution of roughly ±10 percent. In other words, most of the Si–O–Si bond angles fall between 130° and 160°, which implies once again that the structure of fused silica is quite uniform at a short range, but that the order does not persist beyond several layers of tetrahedra.

Multicomponent silicates

In Sec. 3.6, the formation of nonbridging oxygens upon the addition of alkali or alkaline earth oxides to silicate melts was discussed in some detail. Because, as discussed shortly, these oxides usually strongly modify the properties of a glass, they are referred to as **network modifiers**. The resulting structure is not unlike that of pure silica, except that now the continuous three-dimensional network is broken up due to the presence of nonbridging oxygens, as shown in Fig. 9.7.

Table 9.2 lists typical compositions of some of the more common commercial glasses and their softening points. Most of these glasses are predominantly composed of oxygen and silicon. Alumina is interesting in that it behaves sometimes as a glass network and sometimes as a glass modifier; either the aluminum ion can substitute for a Si ion and become part of the network, or it can form nonbridging oxygens and thus act as a modifier (see Sec. 3.6). Which role the alumina plays is usually a complex function of glass chemistry.

---

[138] W. H. Zachariasen, *J. Amer. Chem. Soc.*, **54**:3841 (1932).

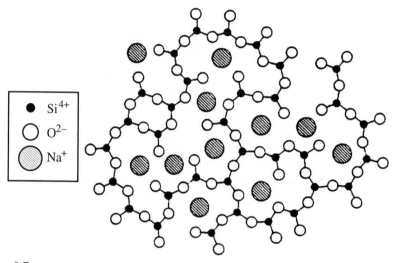

Figure 9.7

Two-dimensional schematic diagram of the silicate glass structure in the presence of modifier ions such as Na^+ and the formation of nonbridging oxygens.

TABLE 9.2

Approximate compositions (wt %) and softening temperatures of some common glasses

	η formers			η modifiers					Softening
	SiO_2	B_2O_3	Al_2O_3	Na_2O	K_2O	MgO	CaO	PbO	temp, °C
Fused silica	99.8					0.1	0.1		1600
Vycor	96.0	3	1						
Pyrex	81.0	13	2	3.5	0.5				830
Soda silica	72.0		1	20.0		3.0	4.0		
Lead silica	63.0		1	8.0	6.0		1.0	21	
Window	72.0	1	2	15.0	1.0	4.0	5.0		700
E glass	55.0	7	15	1.0	1.0		21.0		830

It should be pointed out that whereas this network model of silicate structures was very useful during the earlier stages of development of the theory of glasses, it does not fully explain several experimental facts. For example, it has been observed that significant structural changes occur at about 10 mol % alkali, and it is well documented that the molar volume of silicate melts remains fairly constant over a wide range of alkali concentrations — both observations are at odds with the simple network model described above. As a result, other models have been suggested such as the discrete polyanions and "iceberg" models which seem to better fit the experimental results. In the discrete polyanion model, it is assumed that between 0 and 0.1 mol fraction alkali oxide bonds are broken by the formation of nonbridging oxygens, whereas between 0.1 and 0.33 mol fraction M_2O, discrete six-member rings $(Si_6O_{15})^{6-}$ exist. Between 0.33 and 0.5, a mixture of $(Si_6O_{15})^{6-}$ and $(Si_3O_9)^{6-}$ or $(Si_4O_{12})^{8-}$ and $(Si_6O_{20})^{8-}$ rings are presumed to exist.

A further complication, which is beyond the scope of this book but is mentioned for the sake of completeness, is the fact that in the composition range between 12 and 33 percent, M_2O and SiO_2 are not completely miscible in the liquid state, further complicating the analysis.

9.4
GLASS PROPERTIES

The noncrystalline nature of glasses endows them with certain characteristics unique to them as compared to their crystalline counterparts. Once formed, the changes that occur in a glass upon further cooling are quite subtle and different from those that occur during other phase transitions such as solidification or crystallization. The change is not from disorder to order, but rather from disorder to disorder with less empty space. In this section, the implication of this statement on glass properties is discussed.

9.4.1   The Glass Transition Temperature

The temperature dependence of several properties of crystalline solids and glasses is compared schematically in Fig. 9.8. Typical crystalline solids will normally crystallize at their melting point, with an abrupt and significant decrease in the specific volume and configuration entropy (Fig. 9.8a and b). The changes in these properties for glasses, however, are more gradual, and there are not abrupt changes at the melting point, but rather the properties follow the liquid line up to a temperature where the slope of the specific volume or entropy versus temperature curve is markedly decreased. The point at which the break in slope occurs is known as the **glass transition temperature** and denotes the temperature at which a glass-forming liquid transforms from a rubbery, soft plastic state to a rigid, brittle, glassy state. In other words, the temperature at which a supercooled liquid becomes a glass, i.e., a rigid, amorphous body, is known as the glass transition temperature or T_g. In the range between the melting and glass transition temperatures, the material is usually referred to as a **supercooled liquid**.

Thermodynamic considerations

Given that (see Fig. 9.8) at the glass transition temperature, the specific volume V_s and entropy S are continuous, whereas the thermal expansivity α and heat capacity c_p are discontinuous, at first glance it is not unreasonable to characterize the transformation occurring at T_g as a second-order phase transformation. After all, recall that, by definition, second-order phase transitions require that the properties that depend on the first derivative of the free energy G such as

$$V_s = \left(\frac{\partial G}{\partial P} \right)_T \text{ and } S = -\left(\frac{\partial G}{\partial P} \right)_P$$

be continuous at the transformation temperature, but that the ones that depend on the second derivative of G, such as

$$\alpha = \frac{1}{V} \left(\frac{\partial V}{\partial T} \right)_P = \frac{1}{V} \left(\frac{\partial^2 G}{\partial P\, \partial T} \right) \text{ and } c_p = \left(\frac{\partial H}{\partial T} \right)_P = -T \left(\frac{\partial^2 G}{\partial T^2} \right)_P$$

be discontinuous.

What is occurring at T_g, however, is more complex, because it is experimentally well established that T_g is a function of the cooling rate, as shown in Fig. 9.8a; the transition temperature T_g shifts to lower temperatures with decreasing cooling rates. This implies that with more time for the atoms to rearrange, a denser glass will result and strongly suggests that T_g is not a thermodynamic quantity, but rather a kinetic one.

Further evidence for this conclusion includes the changes in α and c_p. The abrupt decrease in these properties at T_g has to be related to a sudden inability of some molecular degrees of freedom to contribute to these thermodynamic quantities. It is this "freezing out" of molecular degrees of freedom that is responsible for the observed behavior. As discussed below, the viscosity of a glass at T_g is quite large and on the order of 10^{15} Pa·s, which in turn implies that atomic mobility is quite low. It follows that if the time scale of the experiment is smaller than the average time for an atom to move, then that atom will not contribute to the property being measured and, for all practical purposes, the glass transition would appear as a relatively abrupt phenomenon, as observed.

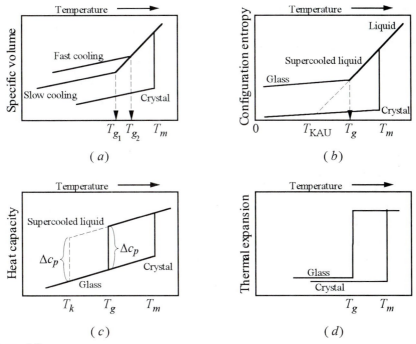

Figure 9.8

Schematic of property changes observed as a glass is cooled through T_g. (*a*) Specific volume; (*b*) configurational entropy; (*c*) heat capacity; and (*d*) thermal expansion coefficient.

Interestingly enough, if a glass-forming liquid were cooled slowly enough (at several times the age of the universe!) such that it follows the dotted line shown in Fig. 9.8*b* at a temperature T_{KAU} the entropy of the supercooled liquid would become lower than that of the crystal — a clearly untenable situation first pointed out by Kauzmann and referred to since as the Kauzmann paradox. This paradox is discussed in greater detail in Sec. 9.4.2.

Effect of composition on T_g

In a very real sense, T_g is a measure of the rigidity of the glass network; in general, the addition of network modifiers tends to reduce T_g, while the addition of network formers increases it. This observation is so universal that experimentally one of the techniques of determining whether an oxide goes into the network or forms nonbridging oxygen is to follow the effect of its addition on T_g.

EXPERIMENTAL DETAILS: MEASURING THE GLASS TRANSITION TEMPERATURE

Temperature T_g can be determined by measuring any of the properties shown in Fig. 9.8 as a function of the cooling rate. The temperature at which the property changes slope whether continuously or abruptly is defined as the glass transition temperature.

One of the more common techniques used to measure T_g employs differential thermal analysis (DTA) (see Experimental Details in Chap. 4). Upon heating of a glass, T_g, being a weak endothermic process, appears as a broad anomaly in the baseline of the DTA curves, as shown in Fig. 9.9a. If a glass does not crystallize upon further heating, then this anomaly is the only signature of the glass. If, however, the glass crystallizes at some temperature between T_g and the melting point, then two extra peaks emerge (Fig. 9.9b); the first is due to the crystallization or devitrification of the supercooled liquid and the second to the melting of these same crystals.

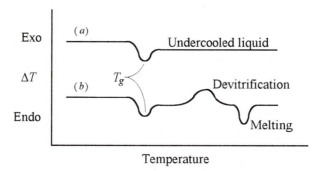

Figure 9.9
Typical DTA traces for the heating of a glass that (a) does not crystallize (note lack of endotherm at melting temperature) and (b) crystallizes somewhere between T_g and the melting point.

9.4.2 Viscosity

The viscosity of a glass and its temperature dependence are very important from a manufacturing point of view because they determine the melting conditions, the time and temperature required to homogenize a melt, the working and annealing temperatures, the rate of devitrification and thus the critical cooling rate, as well as the temperature of annealing of residual stresses.

Viscosity η is the ratio of the applied shear stress to the rate of flow, v, of a liquid. If a liquid contained between two parallel plates of area A and a distance d apart is subjected to a shear force F, then

$$\eta = \frac{Fd}{Av} = \frac{\tau_s}{\dot{\varepsilon}} \qquad (9.20)$$

where $\dot{\varepsilon}$ is the strain rate (s^{-1}) and τ_s the applied shear stress. The units of viscosity are pascal-seconds.

As noted above, upon solidification, the viscosity of a crystalline solid will vary abruptly, and over an extremely narrow temperature range, change by orders of magnitude. The viscosities of glass-forming liquids, however, change in a more gradual fashion. A schematic of the effect of temperature on the viscosity of a glass-forming liquid is shown in Fig. 9.10, where in addition to T_g, four other temperatures are of practical importance. The **strain point** is defined as the temperature at which $\eta = 10^{15.5}$ Pa·s. At this temperature, any internal strain is reduced to an acceptable level within 4 h.[139] The **annealing point** is the temperature at which the viscosity is 10^{14} Pa·s and any internal strains are reduced sufficiently within about 15 min. The **softening point** is the temperature at which the viscosity is $10^{8.6}$ Pa·s. At that temperature, a glass article elongates at roughly 3 percent per second. Finally the **working point** is the temperature at which the viscosity is 10^5 Pa·s, and glass can be readily shaped, formed, or sealed, etc.

Effect of temperature on viscosity

The functional dependence of viscosity on temperature has been measured in a large number of glass-forming liquids, and phenomenologically it has been determined that the most accurate three-parameter fit for the data over a wide temperature range is given by the Vogle-Fulcher-Tamman (V-F-T) equation[140]

$$\ln \eta = A + \frac{B}{T - T_0} \qquad (9.21)$$

where A, B, and T_0 are temperature-independent adjustable parameters and T is the temperature of interest.

---

[139] For the sake of comparison, water, motor oil no. 10, chocolate syrup, and caulking paste have viscosities of 0.001, 0.5, 50, and > 1000 Pa·s, respectively. In cgs units, viscosity is given in poise (abbreviated P). 1 centipoise (cP) = 0.01 P = 0.1 Pa·s.

[140] G. Fulcher, *J. Amer. Cer. Soc.*, **75**:1043–1059 (May 1992). Commemorative reprint.

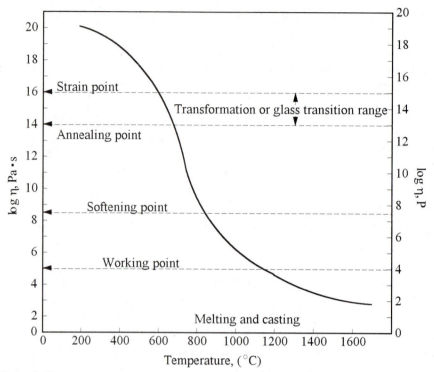

Figure 9.10

Functional dependence of viscosity on temperature. Note the log scale on the y axis.

A number of theories have been proposed to explain this behavior, most notable among them being the free volume theory[141] and the configuration entropy theory.[142] The former predicts that the transition occurring at T_g is a first-order phase transition and cannot account for the fact that in some systems T_g has been found to increase with increasing pressure, and thus will not be discussed further. The remainder of this section is devoted to the configuration entropy model, which while not perfect, succeeds rather well in explaining many experimental observations and at this time appears to be the most promising.

It was established in Chaps. 5 and 6 that the entropy of a crystal is the sum of the vibration and configuration entropies — the latter due to mixing of either defects and/or impurities. The entropy of a liquid or glass contains, in addition, a term reflecting its ability to change configurations. In the configuration entropy

141 M. Cohen and D. Turnbull, *J. Chem. Phys.*, **31**:1164–169 (1959), and D. Turnbull and M. Cohen, *J. Chem. Phys.*, **34**:120–125 (1961).

142 G. Adams and J. Gibbs, *J. Chem. Phys.*, **43**:139 (1965).

model, a simplified version of which is presented here, the liquid is divided into N_c blocks, each containing $n = N/N_c$ atoms, where N is the total number of atoms in the system. These blocks are termed *cooperatively rearranging regions* and are defined as the smallest region that can undergo a transition to a new configuration without a requisite simultaneous configuration change at its boundaries. It is further assumed, for the sake of simplicity, that for each block only two configurations exist. The total entropy of each block is thus $k \ln 2$, and the total configuration entropy of the supercooled liquid (SCL) is simply $\Delta S_{config} = N_c k \ln 2$. Replacing N_c by N/n and rearranging give

$$n(T) = \frac{Nk \ln 2}{\Delta S_{config}(T)} \tag{9.22}$$

If it is assumed that at $T = T_k$ the entropies of the SCL and the glass are identical and, further, that the heat capacity difference between the glass and the SCL, denoted by Δc_p, is a constant and independent of temperature (that is, Δc_p at T_g is equal to its value at T_k, see Fig. 9.8c), then it can be shown that for any temperature $T > T_k$ (see Prob. 9.7)

$$\Delta S = S_{config}^{SCL} - S_{config}^{glass} = \Delta c_p \ln \frac{T}{T_k} \tag{9.23}$$

Combining Eqs. (9.22) and (9.23) yields

$$n(T) = \frac{Nk \ln 2}{\Delta c_p \ln(T/T_k)} \tag{9.24}$$

This expression predicts that $n(T)$ increases with decreasing temperature. The situation is depicted schematically in Fig. 9.11, for three different temperatures $T > T_g$, $T \approx T_g$, and $T = T_k$. As the temperature is lowered and various configurations are frozen out, the cooperatively rearranging regions decrease in number and increase in volume until at T_k only one configuration remains, at which point there is a total loss of configurational entropy.

The major tenet of the model is that there is a direct relationship between the configuration entropy and the rate of molecular transport. In other words, it is postulated that as the blocks become larger, it takes more time for them to switch configurations. It follows that the relaxation time[143] τ can be assumed to be proportional to $n(T)$,

---

[143] The barrier to rearrangement increases in proportion to n because the potential energy increase of barrier ΔE scales as $n\Delta\mu$, where $\Delta\mu$ is the potential barrier per molecule hindering rearrangement.

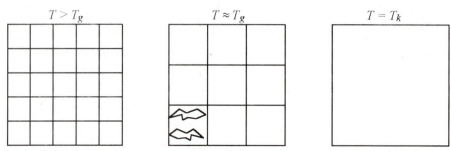

Figure 9.11
Schematic model of effect of temperature on number of blocks and their size. As temperature decreases from left to right, the number of blocks decreases but the number of atoms in each block increases. Shown in the corner of the middle diagram is what is meant by the two configurations that the atoms in a given cell block can have. That number was chosen to be 2 for simplicity; the original model does not make that simplifying assumption.

$$\tau = (\text{const.})\exp[\zeta n(T)] \tag{9.25}$$

where ζ is an undetermined factor. Since τ represents a characteristic time for structural relaxation, it is not unreasonable to assume that it is proportional to the viscosity. Putting all the pieces together, one sees that the viscosity can be related to temperature by

$$n(T) = K^{\wedge} \exp\frac{\zeta Nk\ln 2}{\Delta c_p \ln(T/T_k)} \tag{9.26}$$

where $K^{\wedge}$ is a constant. Taking the natural logarithm of both sides yields

$$\ln\eta = \ln K^{\wedge} + \frac{\zeta Nk\ln 2}{\Delta c_p \ln(T/T_k)} = A' + \frac{B'}{\ln(T/T_k)} \tag{9.27}$$

where A' and B' are constants. And while at first glance this equation does not appear to have the same temperature dependence as Eq. (9.21), it can be easily shown that when $T \approx T_k$, that dependence is indeed recovered (see Prob. 9.7).

Regardless of whether the two expressions are mathematically equivalent, both are equally good at describing the temperature dependence on viscosity. This is clearly shown in Fig. 9.12, where the temperature dependence of the viscosity is plotted for a number of sodium silicate melts and B_2O_3 glass. The data points of the silicates were generated by using Eq. (9.21), which in turn were best fits of the experimental results (see Fulcher). The lines are plotted by using Eq. (9.27). In the case of B_2O_3, the points are experimental points, and the line is again plotted by using Eq. (9.27). The fit in all cases is excellent, using the single adjustable parameter, namely, T_k.

Finally, it is worth noting that the values of T_0 or T_k needed to fit the viscosity data are close to the temperature at which the Kauzmann temperature, T_{KAU} is estimated from extrapolations of other properties such as those shown in Fig. 9.8, lending credence to the model. This model also provides a natural way out of the Kauzmann paradox, since not only do the relaxation times go to infinity as T approaches T_k, but also the configuration entropy vanishes since in glass at $T = T_k$ only one configuration is possible.

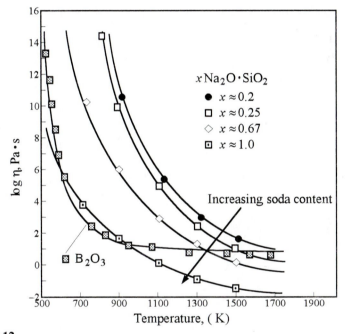

Figure 9.12

Temperature dependence of viscosity for a number of glasses. Data for the silicates are from Fulcher[144] and for B_2O_3 from Macedo and Napolitano.[145] The solid lines are fit according to Eq. (9.27). For all except the highest alkali concentration (for which T_k was chosen to be 136 K, compared to 0 K by Fulcher), T_k in Eq. (9.27) was identical to that used by Fulcher. For B_2O_3, the best fit was obtained for $T_k = 445$ K.

---

144 G. S. Fulcher, *J. Amer. Cer. Soc.*, **8** (6):339–355 (1925). Reprinted in *J. Amer. Cer. Soc.*, **75** (5):1043–1059 (1992).

145 P. B. Macedo and A. Napolitano, *J. Chem. Phys.*, **49**:1887–1895 (1968).

Effect of composition on viscosity

For pure SiO_2 to flow, the high-energy directional Si–O–Si bonds have to be broken. The activation energy for this process is quite large (565 kJ/mol), and consequently, the viscosity of pure liquid SiO_2 at 1940°C is 1.5×10^4 Pa·s, which is quite high, considering the temperature.

As discussed earlier, the addition of **basic oxides** to a melt (a basic oxide is an oxide which when dissolved in a melt contributes an oxygen ion to the melt, such as Na_2O and CaO) will break up the silicate network by the formation of nonbridging oxygens. Increasing additions of the base oxides thus result in breakdown of the original three-dimensional network into progressively smaller discrete ions. As a result, the number of Si–O bonds needed to be broken during viscous flow decreases, and the shear process becomes easier. The dramatic effects of basic oxide additions on viscosity are shown in Fig. 9.13. (Note the log scale of the y axis.)

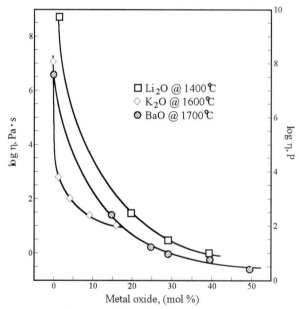

Figure 9.13

Dramatic effect of modifier metal oxide content on the viscosity of fused silica at various temperatures. The addition of about 10 mol % of modifier results in a 5 order-of-magnitude drop in the viscosity of pure silica.[146]

---

146 J. O'M. Bockris, J. D. Mackenzie, and J. A. Kitchener, *Faraday Soc.*, **51**:1734 (1955).

EXPERIMENTAL DETAILS: MEASURING VISCOSITY

The technique used to measure viscosity usually depends on the viscosity range of the glass. In the range up to $\approx 10^7$ Pa·s, one uses what is known as a *viscometer*, shown schematically in Fig. 9.14a. The fluid for which the viscosity is to be measured is placed between two concentric cylinders of length L that are rotated relative to each other. Usually, one of the cylinders, say, the inner one, is rotated at an angular velocity ω_a, while the outer cylinder is held stationary by a spring that measures the torque T acting upon it. It that case, it can be shown that the viscosity is given by[147]

$$\eta = \frac{(b^2 - a^2)T}{4\pi a^2 b^2 L \omega_a} \tag{9.28}$$

where a and b are defined in Fig. 9.14a.

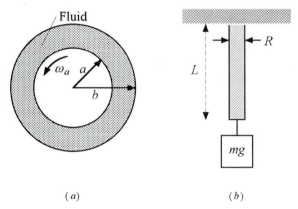

(a)   (b)

Figure 9.14
(a) Schematic cross section of viscometer. (b) Fiber elongation method.

As noted above, viscometers are usually good up to about 10^7 Pa·s. For higher values the fiber elongation method is usually used. In this method (Fig. 9.14b), a load is attached to the material for which the viscosity is to be measured and the material is heated to a given temperature. The strain rate at which the fiber elongates is then measured. It can be shown[148] that the rate of

---

147 R. Feynman, R. Leighton, and M. Sands, *The Feynman Lecture on Physics*, vol. 2, Addison-Wesley, Reading, Massachusetts, 1964, pp. 41–43.

148 J. Frenkel, *J. Phys. (Moscow)*, **9** (5):385–391 (1945).

energy dissipation $\dot{E}_v$ as a result of viscous flow of a cylinder of height L and radius R is given by

$$\dot{E}_v = \frac{3\pi\eta R^2}{L}\left(\frac{dL}{dt}\right)^2 \tag{9.29}$$

Integrating this equation with respect to time, assuming a constant strain rate, equating it to the decrease in potential energy of the system, and ignoring surface energy changes, one can show that

$$\eta = \frac{mg}{3\pi R^2 \dot{\varepsilon}} = \frac{mgL_0 t}{3\pi R^2 \Delta L} \tag{9.30}$$

where ΔL is the elongation in a given time t and L_0 is the original length of the fiber. Here it was also assumed that $\Delta L << L_0$.

9.4.3   Other Properties

These properties will not be discussed here except to point out a very important advantage of using glasses to tailor properties. As any phase diagram will show, nonstoichiometry notwithstanding, most crystalline phases exist over a very narrow range of compositions with limited solubility for other compounds. All this tends to limit the possibilities for property tailoring. But glasses are not subject to this constraint — they can be thought of a "garbage can" for other compounds or as a "universal solvent." Needless to say, it is this degree of freedom that has rendered glasses very useful as well as fascinating materials.

9.5
GLASS-CERAMICS

Glass-ceramics are an important class of materials that have been commercially quite successful. They are polycrystalline materials produced by the controlled crystallization of glass and are composed of randomly oriented crystals with some residual glass, typically between 2 and 5 percent, with no voids or porosity.

9.5.1   Processing

A typical temperature versus time cycle for the processing of a glass-ceramic is shown in Fig. 9.15, and it entails four steps.

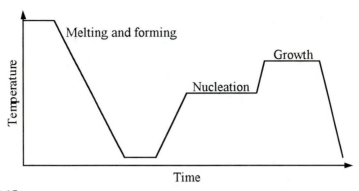

Figure 9.15
Temperature versus time cycle for controlled crystallization of a glass-ceramic body.

1. *Mixing and melting.* Raw materials such as quartz, feldspar, dolomite, and spodumene are mixed with the nucleating agents, usually TiO_2 or ZrO_2, and are melted.
2. *Forming.* As noted below, one of the major advantages of glass-ceramics lies in the fact that they can be formed by using conventional glass-forming techniques such as spinning, rolling, blowing, and casting; so complex-shaped, pore-free articles can be easily manufactured. The cooling rate during the formation process, however, has to be rapid enough to avoid crystallization or growth.
3. *Ceraming.* Once formed, the glass body is heated to a temperature high enough to obtain a very large nucleation rate. Efficient nucleation is the key to success of the process. The nucleation is heterogeneous, and the crystals grow on the particles of the nucleating agents, typically TiO_2 or ZrO_2, that are added to the melt. To obtain crystals on the order of 1 μm, the density of nucleating agents has to be on the order of 10^{12} to 10^{15} cm^{-3}.
4. *Growth.* Following nucleation, the temperature is raised to a point where growth of the crystallites occurs readily. Once the desired microstructure is achieved, the parts are cooled. During this stage the body usually shrinks slightly — by about 1 to 5 percent.

9.5.2   Properties and Advantages

Glass-ceramics offer several advantages over both the glassy and crystalline phases, including these:

1. The most important advantage of glass-ceramics over crystalline ceramics is the ease of processing. As discussed in Chap. 10, viscous sintering is much easier and faster than solid-state sintering. The motive for using glass-ceramics is to take advantage of the ease of processing inherent in the glass to shape and form complex shapes, followed by transforming the glass phase to a more refractory solid in which the properties can be tailored by judicious crystallization. Unlike ceramic bodies made by conventional pressing and sintering, glass-ceramics tend to be pore-free. This is because during crystallization the glass can flow and accommodate changes in volume.

2. Usually the presence of the crystalline phase results in much higher deformation temperatures than the corresponding glasses of the same composition. For example, many oxides have T_g values of 400 to 450°C and soften readily at temperatures above 600°C. A glass-ceramic of the same composition, however, can retain its mechanical integrity and rigidity to temperatures as high as 1000 to 1200°C.

3. The strength and toughness of glass-ceramics are usually higher than those of glasses. For example, the strength of a typical glass plate is on the order of 100 MPa, while that of glass-ceramics can be several times higher. The reason, as discussed in greater detail in Chap. 11, is that the crystals present in the glass-ceramics tend to limit the size of the flaws present in the material, increasing its strength. Furthermore, the presence of the crystalline phase enhances toughness (see Chap. 11).

4. As with glasses, the properties — most notably the thermal expansion coefficients — of glass-ceramics can be controlled by adjusting the composition. In many applications, such as glass-metal seals and the joining of materials, it is very important to match the thermal expansion coefficients to avoid the generation of thermal stresses.

The most important glass-ceramic compositions are probably based on lithium silicates. The phase diagram of the Li_2O–SiO_2 system is shown in Fig. 9.16. The commercial compositions usually contain more than about 30 percent lithia which upon crystallization yields $Li_2Si_2O_5$ as the major phase with some SiO_2 and Li_2SiO_3.

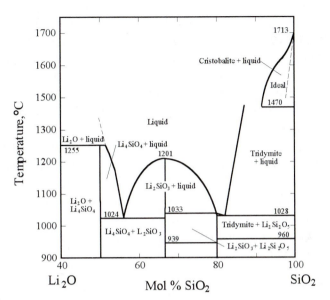

Figure 9.16

$LiO_2 - SiO_2$ phase diagram.[149]

9.6
SUMMARY

Glasses are supercooled liquids that solidify without crystallizing. They are characterized by having short-range but no long-range order. To form a glass, a melt has to be cooled rapidly enough that there is insufficient time for the nucleation and growth of the crystalline phases to occur.

Low atom mobility, i.e., high η, at or around the melting point, together with the absence of potent nucleating agents, is a necessary condition for glasses to form at moderate cooling rates. By far the most important predictor of whether a glass will form at a given cooling rate is the viscosity at or near the melting point.

Upon cooling of a glass melt, the driving force for nucleation increases, but the atom mobility decreases. These two counteracting forces result in maxima for both the nucleation and growth rates. The convolution of the two functions results in the familiar temperature-time-transformation diagrams, from which one can, in principle, quantify the critical cooling rate that would yield a glass.

---

149 F. C. Kracek, *J. Phys. Chem.*, **34**:2645, Part II (1930).

Glass structure is best described by the random network model.

At the glass transition temperature T_g the supercooled liquid transforms to a solid. The transformation that occurs at that point is kinetic and reflects the fact that, on the time scale of the observation, the translational and rotational motions of atoms or molecules which contribute to various properties "freeze out." In other words, they cease to contribute to the properties measured. Below T_g the glass behaves as an elastic solid.

Glass viscosity is a strong, but smoothly changing, function of temperature. The relationship between viscosity and temperature, for the most part, cannot be described by a simple Arrhenian equation. The gradual change of glass viscosity with temperature is very important from a processing point of view and allows glasses to be processed rapidly and relatively easily into pore-free complex shapes.

Both the glass transition temperature and the viscosity are reduced when nonbridging oxygens form.

Glass-ceramics are processed in the same way as glasses, but then they are given a further heat treatment to nucleate and grow a crystalline phase, such that the final microstructure is composed of crystals with a glass phase in between. The possibility of tailoring both the initial composition and size and the volume fraction of the crystalline phase allows for precise tailoring of properties.

APPENDIX 9A

Derivation of Eq. (9.7)

Consider a homogeneous phase containing N_v atoms per unit volume in which in a smaller volume, containing n atoms, the density fluctuates to form a new phase. As discussed in Sec. 9.2, the formation of these embryos results in a *local* increase in the free energy ΔG_c. So the reason the nuclei form must be related to an increase in the entropy of the system. This increase is configurational and comes about because once the nuclei have formed, it is possible to distribute N_n embryos on any of N_v possible sites.

The free energy of the system can be expressed as

$$\Delta G_{sys} = N_n \delta G_c - kT \ln \Omega \qquad (9A.1)$$

where Ω is the number of independent configurations of embryos and host atoms, with each configuration having the same energy, and k and T have their usual meaning. The number of configurations of distributing N_n embryos on N_v sites is

$$\Omega = \frac{N_v!}{N_n!(N_v - N_n)!} \tag{9A.2}$$

By combining Eqs. (9A.1) and (9A.2) it can be easily shown that at equilibrium (i.e., $\partial G_{sys}/\partial N_n = 0$) the number of nuclei is

$$N_n^{eq} \approx N_v \exp\left(-\frac{\Delta G_c}{kT}\right)$$

The similarity between this problem and that of determining the equilibrium number of defects should be obvious at this point.

PROBLEMS

9.1. (a) The water ice interface tension was measured to be 2.2×10^{-3} J/m^2. If the water is exceptionally clean, it is possible to undercool it by 40°C before it crystallizes. Estimate the size of the critical nucleus if the enthalpy of fusion of ice is 6 kJ/mol.

Answer: 0.9 Å

(b) Discuss what you think would happen if the water were not clean. Would the undercooling increase or decrease? Explain.

9.2. (a) Making use of the values used in Worked Example 9.1 and assuming $\Delta G_m^* = 50$ kJ/mol, plot I_v as a function of temperature.

(b) Starting with Eq. (9.14) or (9.15), show that for small undercoolings, a linear relationship should exist between the growth rate and the degree of undercooling.

9.3. (a) Take ΔS to be on the order of $2R$, and assume growth is occurring at 1000° C. For a liquid that melts at 1500°C, estimate what would represent a small undercooling. State all assumptions.

Answer: $\Delta T < 100°$ C

(b) Repeat part (a) for NaCl. Based on your answer, decide whether it would be easy or difficult to obtain amorphous NaCl. Explain. State all assumptions.

9.4. Based on your knowledge of silicate structure, would you expect the viscosity of $Na_2Si_2O_5$ to be greater or smaller than that of Na_2SiO_3 at their respective melting points? Which would you expect to be the better glass former? Explain.

9.5. (a) Show that for quartz the Si–O–Si bond angle is 144°.

(b) Classify the following elements as modifiers, intermediates, or network formers in connection to their role in oxide glasses: Si_____ ; Na _____ ; P _____ ; Ca _____ ; Al _____ .

9.6. The nucleation rate of an amorphous solid [i.e., Eq. (9.13)] can be expressed as $I_v = I_{v,0} \exp(-\Delta H_N / kT)$. For a given solid, $I_{v,0}(\mathrm{m}^{-3} \cdot \mathrm{s}^{-1})$ was measured to be

$$I_{v,0} = 0.8 \times 10^5$$

and the nucleation rate was measured to be $16.7\,\mathrm{m}^{-3} \cdot \mathrm{s}^{-1}$ at 140°C. The growth rates of the crystals were measured to be 7×10^{-7} and $3 \times 10^{-6}\,\mathrm{m/s}$ at 140 and 160°C, respectively. How long would it take to crystallize 95 percent of this solid at 165°C? Assume that growth rate is isotropic and linear with time and that nucleation is random and continuous.

Answer: 1.7 h

9.7. Show that the entropy change between the supercooled liquid and the glass can be expressed by Eq. (9.23). Also show that when $T \approx T_k$, Eq. (9.21) is recovered from Eq. (9.27). *Hint:* $\ln(T/T_k) \approx (T - T_k)/T_k$.

9.8. The activation energy for viscosity of pure silica drops from 565 to 163 kJ/mol upon the addition of 0.5 mol fraction MgO or CaO. The addition of alkali oxides has an even more dramatic effect, lowering the activation energy to 96 kJ/mol for 0.5 mol fraction additions. Explain, using sketches, why this is so.

9.9. Shown in Fig. (9.17) are the results of a two-dimensional simulation of nucleation and growth. The nucleation rate was assumed to be constant and equal to 0.0015 per square millimeter. The growth rate was assumed to be 1 mm/s. Repeat the experiment on your computer, and compare the surface fraction crystallized that you obtain from the simulation to one that you would derive analytically for this particular problem.

9.10 Material A has two allotropic forms: a high temperature β-phase and a lower temperature α-phase. The equilibrium transformation occurs at 1000 K. Consider two separate cases in each of which A is initially at some temperature above 1000 K and as such exists purely as β.

Case 1: The sample of A contains inclusions in the amount of $10^8\,\mathrm{cm}^{-3}$. These inclusions act as heterogeneous nuclei in the phase transformation. The sample is quenched to 980 K and held at that temperature while the transformation to the α-phase occurs.

Case 2: The sample of A is sufficiently clean that nucleation of α is homogeneous. The sample is quenched to 800 K and held at that temperature while the transformation to α occurs.

(a) In which case will the sample take longer to transform, i.e., go to 99% completion? You can assume that the thermal conductivity is high and that normal growth occurs.

(b) If the growth were not normal, would this change your ranking in part (a)? Explain.

Data you may find useful:

enthalpy of transformation	$\Delta H = 0.5RT_e$ where T_e is transformation temp.
molar volume of α	$V_m = 10 \text{ cm}^3/\text{mol}$
lattice parameter in α	$a = 3 \times 10^{-8}$ cm
surface energy along α-β	$\gamma_{\alpha-\beta} = 0.05 \text{ J/m}^2$
lattice diffusivities	$D = 10^{-8} \text{ cm}^2/\text{s}$ at 980 K
	$D = 10^{-10} \text{ cm}^2/\text{s}$ at 800 K
Diffusivity across α-β interface	$D = 10^{-7} \text{ cm}^2/\text{s}$ at 980 K
	$D = 10^{-9} \text{ cm}^2/\text{s}$ at 800 K

$t = 1$ s $t = 2$ s $t = 3$ s

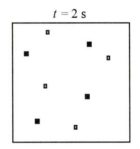

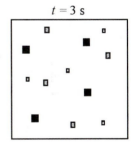

$t = 4$ s $t = 5$ s

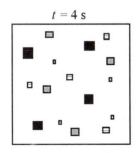

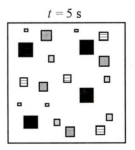

Figure 9.17

Two-dimensional simulation of nucleation and growth of a second phase as a function of time.

9.11. Loehman[150] reported on the formation of oxynitride glasses in the system Y–Si–Al–O–N. The silica-rich glasses contained up to 7% nitrogen. The changes in thermal expansion coefficient and the glass transition temperature are tabulated below.

(a) Discuss the incorporation of nitrogen in the oxide glass. Where is it located, and how does it affect the properties?

(b) Does the data have any technological implications?

Atomic % Nitrogen	0.0	1.5	7
Thermal expansion $\left(\times 10^{-6}\,°C^{-1} \right)$	7.5	6.8	4.5
Glass transition, °C	830	900	920
Microhardness (Vickers)	1000	1050	1100

ADDITIONAL READING

1. H. Rawson, *Properties and Applications of Glass*, vol. 3, *Glass Science and Technology*, Elsevier, Amsterdam, 1980.
2. R. Doremus, *Glass Science*, Wiley, New York, 1973.
3. G. Beall, "Synthesis and Design of Glass-Ceramics," *J. Mat. Ed.*, **14**:315 (1992).
4. G. Scherer, "Glass Formation and Relaxation," in *Materials Science and Technology, Glasses and Amorphous Materials*, vol. 9, R. Cahn, P. Haasen, and E. Kramer, eds., VCH, New York, 1991.
5. G. O. Jones, *Glass*, 2d ed., Chapman & Hall, London, 1971.
6. J. Jackle, "Models of the Glass Transition," *Rep. Prog. Phys.*, **49**:171 (1986).
7. G. W. Morey, *The Properties of Glass*, Reinhold, New York, 1954.
8. J. W. Christian, *The Theory of Transformations in Metals and Alloys*, 2d ed., Pergamon Press, London, 1975.
9. M. H. Lewis, ed., *Glasses and Glass-Ceramics*, Chapman & Hall, London, 1989.
10. J. Zarzycki, *Glassses and the Vitrous State*, Cambridge University Press, Cambridge, England, 1975.

---

[150] R. Loehman, *J. Amer. Cer. Soc.*, **62**: 491 (1979).

CHAPTER 10

Sintering and Grain Growth

All sintering data can be made to fit all sintering models.

Anonymous

10.1
INTRODUCTION

The effect of microstructure on properties has been neglected up to this point, mainly because the properties discussed so far, such as Young's modulus, thermal expansion, electrical conductivity, melting points, and density, are to a large extent microstructure-insensitive. In the remainder of this book, however, it will become apparent that microstructure can and does play a significant role in determining the properties. For example, as shown in Table 10.1, the optimization of various properties requires various microstructures.

And while metals and polymers are usually molten, cast, and, when necessary, machined or forged into the final desired shape, the processing of ceramics poses more of a challenge on account of their refractoriness and brittleness. With the notable exception of glasses, few ceramics are processed from the melt — the fusion temperatures are simply too high. Instead, the starting point is usually fine powders that are milled, mixed, and molded into the desired shape by a variety of processes and subsequently heat-treated or fired to convert them to dense solids. Despite the fact that the details of shaping and forming of the green (unfired) bodies can have a profound influence on the final microstructure, they are not directly addressed here, but will be touched upon later. The interested reader is referred to a number of excellent books and monographs that have been written on the subject, some of which are listed in Additional Reading at the end of this chapter.

TABLE 10.1
Desired microstructures for optimizing properties

Property	Desired microstructure
High strength	Small grain size, uniform microstructure, and flaw-free
High toughness	Duplex microstructure with high aspect ratios
High creep resistance	Large grains and absence of an amorphous grain boundary phase
Optical transparency	A pore-free microstructure with grains that are either much smaller or much larger than wavelength of light being transmitted
Low dielectric loss	Small, uniform grains
Good varistor behavior	Control of grain boundary chemistry
Catalyst	Very large surface area

As noted above, once shaped, the parts are fired or sintered. Sintering is the process by which a powder compact is transformed to a strong, dense ceramic body upon heating. In an alternate definition given by Herring,[151] **sintering** is ". . . understood to mean any changes in shape which a small particle or a cluster of particles of uniform composition undergoes when held at high temperature." As will become clear in this chapter, sintering is a complex phenomenon in which several processes are occurring simultaneously. There are many papers in the ceramic literature devoted to understanding and modeling of the sintering process; and if there were such a thing as the holy grail of ceramic processing science, it probably would be how consistently to obtain theoretical density at the lowest possible temperature. The main difficulty in achieving this goal, however, lies in the fact that the driving force for sintering is quite small, usually on the order of a few joules per mole, compared to a few kilojoules per mole in the case of chemical reactions (see Worked Example 10.1). Consequently, unless great care is taken during sintering, full density is difficult to achieve.

Sintering can occur in the presence or absence of a liquid phase. In the former case, it is called *liquid-phase sintering*, where the compositions and firing temperatures are chosen such that some liquid is formed during processing, as shown schematically in Fig. 10.1a. This process is of paramount importance and is technologically the process of choice. In the absence of a liquid phase, the process is referred to as *solid-state sintering* (Fig. 10.1b).

This chapter is mainly devoted to understanding the science behind the sintering process. In the next section, the driving forces and atomic mechanisms responsible for solid sintering are examined. Section 10.3 deals with sintering kinetics, and the factors affecting solid-state sintering are elucidated on the basis of

---

[151] C. Herring, *J. Appl. Phys.*, **21**:301–303 (1950).

the ideas presented in Sec. 10.2. In Sec. 10.4, liquid-phase sintering is discussed, and in Sec. 10.5, hot pressing and hot isostatic pressing are briefly touched upon.

10.2
SOLID-STATE SINTERING

The macroscopic driving force operative during sintering is the reduction of the excess energy associated with surfaces. This can happen by (1) reduction of the total surface area by an increase in the average size of the particles, which leads to coarsening (Fig. 10.2*b*), and/or (2) the elimination of solid/vapor interfaces and the creation of grain boundary area, followed by grain growth, which leads to densification (Fig. 10.2*a*). These two mechanisms are usually in competition. If the atomic processes that lead to densification dominate, the pores get smaller and disappear with time and the compact shrinks. But if the atomic processes that lead to coarsening are faster, both the pores and grains coarsen and get larger with time.

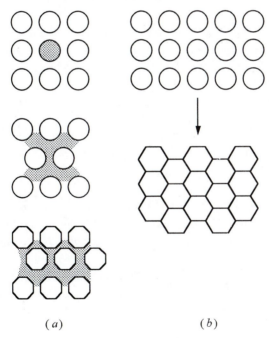

(*a*)　　　　　　　　(*b*)

Figure 10.1
(*a*) Liquid-phase sintering; (*b*) Solid-state sintering.

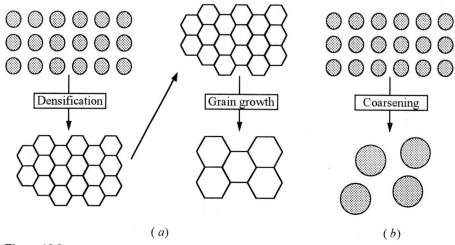

(a)   (b)

Figure 10.2
Schematic of two possible paths by which a collection of particles can lower its energy.
(a) Densification followed by grain growth. In this case, shrinkage of the compact has to occur. (b) Coarsening where the large grains grow at the expense of the smaller ones.

A necessary condition for sintering to occur is that the grain boundary energy γ_{gb} be less than twice the solid/vapor surface energy γ_{sv}, which implies that the equilibrium dihedral angle ϕ shown in Fig. 10.3a and defined

$$\gamma_{gb} = 2\gamma_{sv}\cos\frac{\phi}{2} \tag{10.1}$$

has to be less than 180° for densification to occur. For many oxide systems,[152] the dihedral angle is around 120°, implying that $\lambda_{gb}/\gamma_{sv} \approx 1.0$, in contrast to metallic systems where that ratio is closer to between 0.25 and 0.5.

WORKED EXAMPLE 10.1. (a) Calculate the enthalpy change for an oxide as the average particle diameter increases from 0.5 to 10 μm. Assume the molar volume of the oxide to be 10 cm^3/mol and a surface energy of 1 J/m^2. (b) Recalculate the enthalpy change if, instead of coarsening, the 0.5 μm spheres are sintered together as cubes, given that the dihedral angle for this system was measured to be 100.

---

[152] See, e.g., C. Handwerker, J. Dynys, R. Cannon, and R. Coble, *J. Amer. Cer. Soc.*, **73**:1371, (1990).

Figure 10.3

(a) Equilibrium dihedral angle between grain boundary and solid/vapor interfaces.

(b) Equilibrium dihedral angle between grain boundary and liquid phase.

Answer

(a) Taking 1 mol or 10 cm^3 to be the basis and assuming monosize spheres of 0.5 μm diameter, one can see that the number N of these spheres is

$$N = \frac{10}{4/3 \pi r^3} = 1.5 \times 10^{14}$$

Their corresponding surface area S is

$$S = (4\pi r^2)(1.5 \times 10^{14}) = 120 \text{ m}^2$$

The total surface energy is

Surface energy of 0.5 μm spheres $= S \times N = (120 \text{ m}^2)(1 \text{ J/m}^2) = 120 \text{ J}$.

Similarly, it can be shown that the total surface energy of the 10 μm spheres

$\approx$

6 J. The change in enthalpy associated with the coarsening is

$$6 - 120 = -114 \text{ J/mol}$$

In other words, the process is exothermic.

(b) Given that the dihedral angle is 100, applying Eq. (10.1), one obtains

$$\gamma_{gb} = 2\gamma_{sv} \cos\frac{\phi}{2} = 2 \times 1 \times \cos 50 = 1.28 \text{ J/m}^2$$

Mass conservation requires that $a^3 = 4/3 \pi r^3$, or $a \approx 0.4 \ \mu$m, where a is the length of the side of the cubes. The total grain boundary area (neglecting free surface) is

$$S_{gb} \cong \frac{6}{2}\left(0.4 \times 10^{-6}\right)^2\left(1.5 \times 10^{14}\right) = 72 \text{ m}^2$$

Thus the energy of the system after sintering is $1.28 \times 72 \approx 92.16$ J, which is less than the original of 120 J. The difference between the two is the driving force for densification.

EXPERIMENTAL DETAILS: SINTERING KINETICS

Based on the foregoing discussion, to understand what is occurring during sintering, one needs to measure the shrinkage, grain, and pore sizes as a function of the sintering variables, such as time, temperature, and initial particle size. If a powder compact shrinks, its density will increase with time. Hence, densification is best followed by measuring the density of the compact (almost always reported as a percentage of the theoretical density) as a function of sintering time. This is usually carried out dilatometrically (see Fig. 4.9), where the length of a powder compact is measured as a function of time at a given temperature. Typical shrinkage curves are shown in Fig. 10.4 for two different temperatures $T_2 > T_1$. For reasons that will become clear shortly, the densification rate is a strong function of temperature, as shown in the figure.

In contradistinction, if a powder compact coarsens, no shrinkage is expected in a dilatometric experiment. In that case, the coarsening kinetics are best followed by measuring the average particle size as a function of time via optical or scanning electron microscopy.

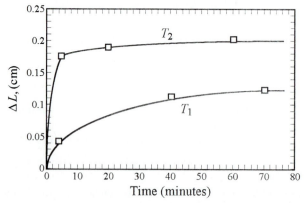

Figure 10.4

Typical axial shrinkage curves during sintering as a function of temperature, where $T_2 > T_1$.

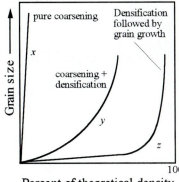
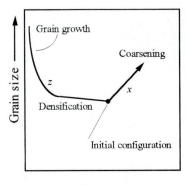

Figure 10.5

(*a*) Grain size versus density trajectories for densification (curve *z*) and coarsening (curve *x*). Curve *y* shows a powder for which both coarsening and densification are occurring simultaneously. (*b*) Alternate scheme to represent data in terms of grain and pore size trajectories.

It is useful to plot the resultant behavior in what is known as **grain size versus density trajectories**, such as those shown in Fig. 10.5*a* and *b*. Typically, a material will follow the path denoted by curve *y*, where both densification and coarsening occur simultaneously. However, to obtain near-theoretical densities, coarsening has to be suppressed until most of the shrinkage has occurred; i.e., the system should follow the trajectory denoted by curve *z*. A powder that follows trajectory *x*, however, is doomed to remain porous — the free energy has been expended, large grains have formed, but more importantly so have large pores. Once formed, these pores are kinetically very difficult to remove, and as discussed below, they may even be thermodynamically stable, in which case they would be impossible to remove.

An alternate method of presenting the sintering data is shown in Fig. 10.5*b*, where the time evolution of the grain and pore sizes is plotted; coarsening leads to an increase in both, whereas densification eliminates pores.

In Fig. 10.6*a-f*, the time dependence of the microstructural development of an MgO-doped Al_2O_3 compact sintered in air at 1600°C is shown. Note that as time progresses, the average grain size increases whereas the average pore size decreases. The corresponding grain size versus density trajectory is shown in Fig. 10.6*g*; it is typical for many ceramics that sinter to full density.

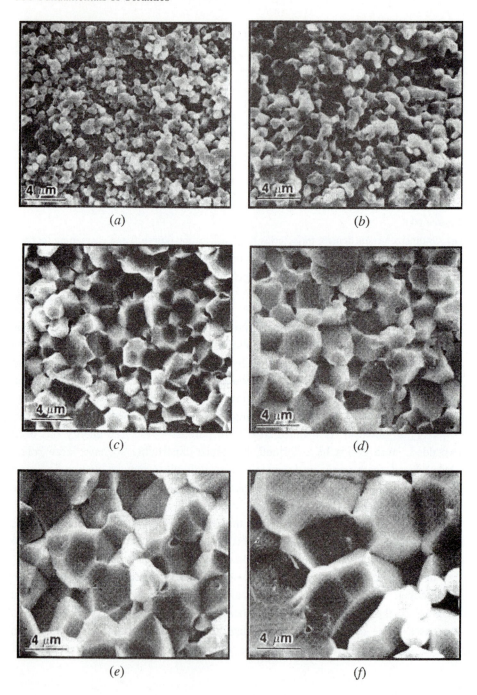

(a)

(b)

(c)

(d)

(e)

(f)

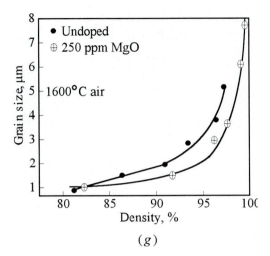

(g)

Figure 10.6

(a-f) Microstructure development for an MgO-doped alumina sintered in air at 1600C as a function of time. (g) Grain size trajectory for undoped and 250 ppm MgO-doped alumina sintered in air at 1600C.[153]

A good example of a powder that coarsens without densification is Fe_2O_3 sintered in HCl-containing atmospheres. The final microstructures after firing for 5 h at 1200°C in air and Ar-10% HCl are shown in Fig. 10.7a and b, respectively. The corresponding time dependence of the relative density as a function of HCl content in the gas phase is shown Fig. 10.7c. These results clearly indicate that whereas Fe_2O_3 readily sinters to high density in air, in an HCl atmosphere it coarsens instead.

Full density is thus obtained only when the atomic processes associated with coarsening are suppressed, while those associated with densification are enhanced. It follows that in order to understand and be able to control what occurs during sintering the various atomic processes responsible for each of the aforementioned outcomes are identified and described. Before that, however, it is imperative to understand the effect of curvature on the chemical potential of the ions or atoms in a solid.

153 K. A. Berry and M. P. Harmer, *J. Amer. Cer. Soc.*, **69**:143–149 (1986). Reprinted with permission.

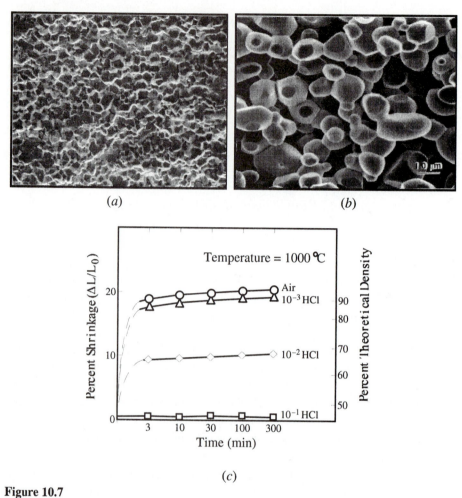

(a) (b)

(c)

Figure 10.7
(a) Microstructure of air-sintered Fe_2O_3. (b) Microstructure of Fe_2O_3 sintered in HCl-containing atmospheres. Note that in the latter case, significant coarsening of the microstructure occurred. (c) Effect of atmosphere on relative density versus time for Fe_2O_3 sintered at 1000°C.[154]

[154] D. Ready, *Sintering of Advanced Ceramics, Ceramic Trans.*, vol. 7, C. A. Handwerker, J. E. Blendell, and W. Kayser, eds., American Ceramic Society, Westerville, Ohio, p. 86, 1990. Reprinted with permission.

10.2.1 Local Driving Force for Sintering

As mentioned earlier, the global driving force operating during sintering is the reduction in surface energy, which manifests itself locally as curvature differences. From the Gibbs-Thompson equation (see App. 10A for derivation), it can be shown that the chemical potential difference *per formula unit* $\Delta\mu$ between atoms on a flat surface and under a surface of curvature κ is

$$\Delta\mu = \mu_{\text{curv}} - \mu_{\text{flat}} = \gamma_{\text{sv}}\Omega_{\text{MX}}\kappa \tag{10.2}$$

where Ω_{MX} is the volume of a formula unit. For simplicity in the following discussion, it will be assumed that one is dealing with an MX compound. Curvature κ depends on geometry; e.g., for a sphere of radius ρ, $\kappa = 2/\rho$ (see App. 10B for further details). Equation (10.2) has two very important ramifications that are critical to understanding the sintering process. The first is related to the partial pressure of a material above a curved surface, and the second involves the effect of curvature of vacancy concentration.

Effect of curvature on partial pressure

At equilibrium, this chemical potential difference translates to a difference in partial pressure above the curved surface, i.e.,

$$\Delta\mu = kT \ln \frac{P_{\text{curv}}}{P_{\text{flat}}} \tag{10.3}$$

Combining the two equations reveals that

$$\ln \frac{P_{\text{curv}}}{P_{\text{flat}}} = \kappa \frac{\Omega_{\text{MX}}\gamma_{\text{sv}}}{kT} \tag{10.4}$$

If $P_{\text{curv}} \approx P_{\text{flat}}$, then Eq. (10.4) simplifies to the more common expression, namely,

$$\frac{\Delta P}{P_{\text{flat}}} = \frac{P_{\text{curv}} - P_{\text{flat}}}{P_{\text{flat}}} = \kappa \frac{\Omega_{\text{MX}}\gamma_{\text{sv}}}{kT} \tag{10.5}$$

As noted above for a sphere of radius ρ, $\kappa = 2/\rho$, and Eq. (10.5) can be written as

$$\boxed{P_{\text{curv}} = P_{\text{flat}}\left(1 + \frac{2\Omega_{\text{MX}}\gamma_{\text{sv}}}{\rho T}\right)} \tag{10.6}$$

Given that the *radius of curvature is defined as negative for a concave surface and positive for a convex surface*, this expression is of fundamental importance because it predicts that the pressure of a material above a convex surface is *greater* than that over a flat surface, and vice versa for a concave surface. For example, the pressure inside a pore of radius ρ would be *less* than that over a flat surface; conversely, the pressure surrounding a collection of fine spherical particles will be *greater* than that over a flat surface.[155]

It is only by appreciating this fact that sintering can be understood. Given the importance of this conclusion, it is instructive to explore what occurs on the atomic level that allows this to happen. To do so, consider the following thought experiment: Place each of three different-shaped surfaces of the same solid in a sealed and evacuated chamber, as shown in Fig. 10.8, and heat until an equilibrium vapor pressure is established. Examining the figures shows that $P_1 < P_2 < P_3$, since on average the atoms on a convex surface are less tightly bound to their neighbors than atoms on a concave surface and will thus more likely escape into the gas phase, resulting in a higher partial pressure.

Effect of curvature on vacancy concentrations

The other important ramification of Eq. (10.2) is that the equilibrium vacancy concentration is also a function of curvature. In Chap. 6, the relationship between the equilibrium concentration of vacancies C_0, their enthalpy of formation Q, and temperature was given by

$$C_0 = K' \exp\left(-\frac{Q}{kT}\right) \tag{10.7}$$

Concave	Flat	Convex
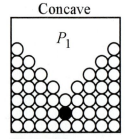	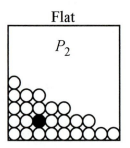	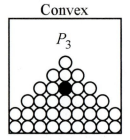
P_1	P_2	P_3

Figure 10.8
Effect of surface curvature on equilibrium pressure. At this scale it is easy to appreciate why $P_1 < P_2 < P_3$.

---

[155] Implicit in this result is that the MX compound is evaporating as MX molecules.

In this equation, the entropy of formation and all preexponential terms are included in the constant K'. An implicit assumption made to derive this expression was that the vacancies formed under a flat, stress-free surface.

Since the chemical potential of an atom under a curved surface is either greater or smaller than that over a flat surface by $\Delta\mu$, this energy has to be accounted for when one is considering the formation of a vacancy. Hence it follows that

$$C_{curv} = K' \exp\left(-\frac{Q+\Delta\mu}{kT}\right) = C_0 \exp\left(-\frac{\kappa\Omega_{MX}\gamma_{sv}}{kT}\right) \tag{10.8}$$

And since for the most part $\gamma_{sv}\Omega_{MX}\kappa \ll kT$, then with little loss in accuracy

$$\boxed{C_{curv} = C_0\left(1 - \frac{\kappa\Omega_{MX}\gamma_{sv}}{kT}\right)} \tag{10.9}$$

which is identical to

$$\boxed{\Delta C_{vac} = C_{curv} - C_0 = -C_0\frac{\kappa\Omega_{MX}\gamma_{sv}}{kT}} \tag{10.10}$$

It is left as an exercise for the reader to show that the *vacancy concentration under a concave surface is greater than that under a flat surface, which in turn is greater than that under a convex* surface.[156] The physics can be explained as follows: Given that a good measure of the enthalpy of formation of a vacancy is the difference in bonding between an atom in the bulk of the solid versus when it is on the surface, referring once again to Fig. 10.8 and focusing on the shaded atoms will make it obvious why it costs less energy to create a vacancy in the vicinity of a concave surface than a convex one.

To recap: Curvature causes local variations in partial pressures and vacancy concentrations. The partial pressure over a convex surface is higher than that over a concave surface. Conversely, the vacancy concentration under a concave surface is higher than that below a convex surface. In either case, a driving force is present that induces the atoms to migrate from the convex to the concave areas, i.e., from the mountaintops to the valleys. Given these conclusions, it is now possible to explore the various atomic mechanisms taking place during sintering.

---

[156] An important assumption made in deriving Eq. (10.10) is that under the curved surface, the defects form in their stoichiometric ratios; i.e., the defect concentrations are dictated by the Schottky equilibrium $[V_M] = [V_X]$.

10.2.2 Atomic Mechanisms Occurring during Sintering

There are basically five atomic mechanisms by which mass can be transferred in a powder compact:

1. Evaporation-condensation, depicted as path 1 in Fig. 10.9a.
2. Surface diffusion, or path 2 in Fig. 10.9a.
3. Volume diffusion. Here there are two paths. The mass can be transferred from the surface to the neck area — path 3 in Fig. 10.9a — or from the grain boundary area to the neck area — path 5 in Fig. 10.9b.
4. Grain boundary diffusion from the grain boundary area to the neck area — path 4 in Fig. 10.9b.
5. Viscous or creep flow. This mechanism entails either the plastic deformation or viscous flow of particles from areas of high stress to low stress and can lead to densification. However, since it is essentially the same thing that occurs during creep, it is dealt with not here but rather in Chap. 12.

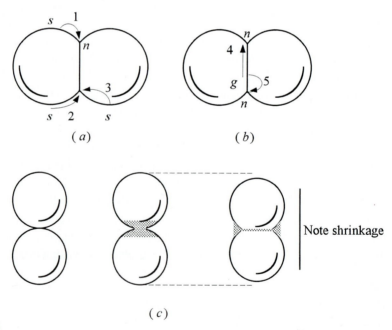

(a)   (b)

(c)

Figure 10.9

Basic atomic mechanisms that can lead to (a) Coarsening and change in pore shape and (b) densification. (c) Thought experiment illustrating how removal of material from the area between particles into the pore leads to shrinkage and densification.

Consider now which of these mechanisms leads to coarsening and which to densification.

Coarsening

At the outset, it is important to appreciate that any mechanism in which the source of material is the surface of the particles and the sink is the neck area *cannot* lead to densification, because such a mechanism does not allow the particle centers to move closer together. Consequently, evaporation-condensation, surface diffusion, and lattice diffusion from the surface to the neck area cannot lead to densification. They do, however, result in a change in the shape of the pores, a growth in the neck size, and a concomitant increase in compact strength. Moreover, the smaller grains, with their smaller radii of curvature, will tend to "evaporate" away and plate out on the larger particles, resulting in a coarsening of the microstructure.

The driving force in all cases is the partial pressure differential associated with the local variations in curvature. For instance, the partial pressure at point s in Fig. 10.9a is greater than that at point n, which in turn results in mass transfer from the convex to the concave surfaces. The actual path taken will depend on the kinetics of the various paths, a topic that will be dealt with shortly. At this point, it suffices to say that since the atomic processes are occurring in parallel, at any given temperature, it is the fastest mechanism that will dominate.

Densification

If mass transfer from the surface to the neck area or from the surface of smaller to larger grains does not lead to densification, other mechanisms have to be invoked to explain the latter. *For densification to occur, the source of material has to be the grain boundary or region between powder particles, and the sink has to be the neck or pore region.* To illustrate why this is the case, consider the thought experiment illustrated in Fig. 10.9c: Cut a volume (shaded area in Fig. 10.9c) from between two spheres, bring the two spheres closer together, and then place the extra volume removed in the pore area. Clearly such a process leads to shrinkage and the elimination of pores. Consequently, the only mechanisms, apart from viscous or plastic deformation, that can lead to densification are grain boundary diffusion and bulk diffusion from the grain boundary area to the neck area (Fig. 10.9b).

Atomistically, both these mechanisms entail the diffusion of ions from the grain boundary region toward the neck area, for which the driving force is the curvature-induced vacancy concentration. Because there are more vacancies in the neck area than in the region between the grains, a vacancy flux develops away from the pore surface into the grain boundary area, where the vacancies are eventually annihilated. Needless to say, an equal atomic flux will diffuse in the opposite direction, filling the pores.

10.3   Sintering Kinetics

Based on the foregoing discussion, a powder compact can reduce its energy by following various paths, some of which can lead to coarsening, others to densification. This brings up the central and critical question in sintering: What governs whether a collection of particles will densify or coarsen? To answer the question, models for each of the paths considered above must be developed and compared, with the fastest path determining the behavior of the compact. For instance, a compact in which surface diffusion is much faster than bulk diffusivity would tend to coarsen rather than densify.

In practice, the question is much more difficult to answer, however, because the kinetics of sintering are dependent on many variables, including particle size and packing, sintering atmosphere, degree of agglomeration, temperature, and presence of impurities. The difficulty of the problem is best illustrated by comparing the densification kinetics of an "as-received" yttria-stabilized zirconia powder compact to that of the same powder that was rid of agglomerates before sintering (Fig. 10.10). The marked reduction in the temperature, by about 300°C, needed to fully densify the compact which was prepared from an agglomerate-free compact is obvious.

In this section, some of the many sintering models proposed over the years to model this complex process are outlined. Because of the complex geometry of the problem, analytic solutions are only possible by making considerable geometric and diffusion flow field approximations, which are rarely realized in practice. Consequently, the models discussed below have limited validity and should be used with extreme care when one is trying to predict the sintering behavior of real powders. Much of the usefulness of these sintering models thus lies more in appreciating the general trends that are to be expected and identifying the critical parameters than in their predictive capabilities.

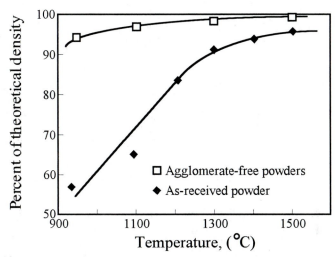

Figure 10.10
Temperature dependence of sintered density for an agglomerated or "as-received" and agglomerate-free yttria-stabilized zirconia powder (1 h). Eliminating the agglomerates in the green body resulted in a powder compact that densified much more readily.[157]

Sintering stages

Coble[158] described a sintering stage as an "interval of geometric change in which pore shape is totally defined (such as rounding of necks during the initial stage sintering) or an interval of time during which the pore remains constant in shape while decreasing in size." Based on that definition, three stages have been identified: an initial, an intermediate, and a final stage.

During the **initial stage**, the interparticle contact area increases by neck growth (Fig. 10.11b) from 0 to ≈ 0.2, and the relative density increases from about 60 to 65 percent.

The **intermediate stage** is characterized by continuous pore channels that are coincident with three-grain edges (Fig. 10.11c). During this stage, the relative density increases from 65 to about 90 percent by having matter diffuse toward, and vacancies away from the long cylindrical channels.

---

157 W. H. Rhodes, *J. Amer. Cer. Soc.*, **64**:19 (1981). Reprinted with permission.
158 R. L. Coble, *J. Appl. Phys.*, **32**:787–792 (1961); R. L. Coble, *J. Appl. Phys.*, **36**:2327 (1965).

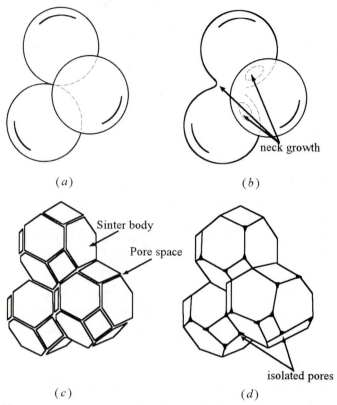

Figure 10.11

(*a*) Initial stage of sintering model represented by spheres in tangential contact. (*b*) Near end of initial stage; spheres have begun to coalesce. (*c*) Intermediate stage; grains adopted shape of dodecahedra, enclosing pore channels at grain edges. (*d*) Final stage; pores are tetrahedral inclusions at corners where four dodecahedra meet.

The **final stage** begins when the pore phase is eventually pinched off and is characterized by the absence of a continuous pore channel (Fig. 10.10*d*). Individual pores are either of lenticular shape, if they reside on the grain boundaries, or rounded, if they reside within a grain. An important characteristic of this stage is the increase in pore and grain boundary mobilities, which have to be controlled if the theoretical density is to be achieved.

Clearly, the sintering kinetics will be different during each of the aforementioned stages. To further complicate matters, in addition to having to treat each stage separately, the kinetics will depend on the specific atomic mechanisms operative. Despite these complications, most, if not all, sintering models share the following common philosophy:

1. A representative particle shape is assumed.
2. The surface curvature is calculated as a function of geometry.
3. A flux equation that depends on the rate-limiting step is adopted.
4. The flux equation is integrated to predict the rate of geometry change.

In the following subsections, this approach is used to predict the rates of various processes occurring during various stages. In particular, Sec. 10.3.1 deals with the initial stage, while Sec. 10.3.2 addresses the kinetics of densification. Coarsening and grain growth, because of their similarities, are discussed in Sec. 10.3.3.

10.3.1 Initial-Stage Sintering

Given the multiplicity of paths available to a powder compact during this stage, it is impossible to address them all in detail. Instead, the following approach has been adopted: The rate of neck growth by evaporation condensation (path 1 in Fig. 10.9a) is worked out in detail; the final results for the other mechanisms, namely, surface, grain boundary, lattice diffusion, and viscous sintering, are given without proof.[159]

Evaporation condensation (path 1 in Fig. 10.9a)

In this mechanism, the pressure differential between the surface of the particle and the neck area results in a net matter transport, via the gas phase from the surface to the neck. The evaporation rate (in molecules of MX per square meter per second), is given by the Langmuir expression

$$j = \frac{\alpha \Delta P}{\sqrt{2\pi m_{MX} kT}} = K_r \Delta P \qquad (10.11)$$

where α and m_{MX} are, respectively, the evaporation coefficient and the mass of the evaporating gas molecules; ΔP is the pressure differential between the surface and neck areas. By applying Eq. (10.5), it can be easily shown that the pressure differential between these two regions is

$$\Delta P = \frac{\Omega_{MX} P_{flat} \gamma_{sv}}{kT} \left[\frac{1}{\rho} - \frac{1}{r} \right] \approx \frac{\Omega_{MX} P_{flat} \gamma_{sv}}{\rho kT} \qquad (10.12)$$

159 For a summary and a critical analysis of the initial-stage sintering models, see, e.g., W. S. Coblenz, J. M. Dynys, R. M. Cannon, and R. L. Coble, in *Sintering Processes*, G. C. Kuczynski, ed., Plenum Press, New York, 1980.

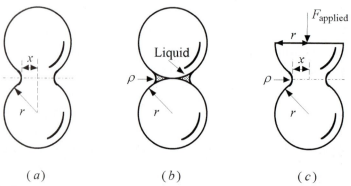

Figure 10.12

(*a*) Sphere tangency construction used for initial-stage sintering modeling. (*b*) Spherical particles held together by liquid capillary pressure. (*c*) Sphere tangency construction during hot pressing. The ratio of applied stress to boundary stress is proportional to $(x/r)^2$ (see Sec. 10.5).

where ρ and r, defined in Fig. 10.12*a*, are, respectively, the radius of curvature of the neck area and the sphere radius. Furthermore, according to this figure,

$$(r + \rho)^2 = (x + \rho)^2 + r^2$$

where x is the neck radius. For $x \ll r$, this equation simplifies to

$$\rho = \frac{x^2}{2(r - x)} \approx \frac{x^2}{2r} \tag{10.13}$$

Multiplying the flux of material arriving to the neck area by Ω_{MX} yields the rate at which the neck will grow, or

$$\frac{dx}{dt} = j\Omega_{MX} \tag{10.14}$$

Combining Eqs. (10.11) to (10.14) and integrating yield

$$\left(\frac{x}{r}\right)^3 = \left[\frac{6\alpha\gamma_{sv}\Omega_{MX}^2 P_{flat}}{\sqrt{2\pi m_{MX}kT}\,[kT]r^2}\right]t \tag{10.15}$$

This equation predicts that the rate of growth of the neck region (1) is initially quite rapid but then flattens out, (2) is a strong function of initial particle size, and (3) is a function of the partial pressure P_{flat} of the compound, which in turn depends exponentially on temperature.

To recap: Equation (10.15) was derived by assuming a representative shape (Fig. 10.12a), from which the surface curvature was calculated as a function of geometry [Eq. (10.13)]. A flux equation [Eq. (10.11)] was then assumed and integrated to yield the final result. By using essentially the same procedure, the following results for other models are obtained.

Lattice diffusion model (path 5, Fig. 10.9b)

If it is assumed that material diffuses away from the grain boundary area through the bulk and plates out in an area of width 2ρ, the following expression for the neck growth is obtained:[160]

$$\left(\frac{x}{r}\right)^4 = \left[\frac{64 D_{ambi} \gamma_{sv} \Omega_{MX}^2}{kTr^3}\right] t \tag{10.16}$$

where D_{ambi} is the ambipolar diffusivity, given by Eq. (7.87). It is important to note that the use of D_{ambi} in Eq. (10.16) implies that the compound is pure and stoichiometric, with the dominant defects being Schottky defects.[161] It is also implied that the two component fluxes are both diffused through the bulk. In general, however, it is important to appreciate that it is the *slowest species diffusing along its fastest path that is rate-limiting*, a conclusion that applies to all mechanisms discussed below as well.

Grain boundary diffusion model (path 4, Fig. 10.9b)

In this model, the mass is assumed to diffuse from the grain boundary area radially along the grain boundary of width δ_{gb} and plate out at the neck surface. The neck growth is given by[162]

$$\left(\frac{x}{r}\right)^6 = \left[\frac{192 \delta_{gb} D_{gb} \gamma_{sv} \Omega_{MX}^2}{kTr^4}\right] t \tag{10.17}$$

which leads to the following linear shrinkage:

$$\left(\frac{\Delta L}{L}\right)^3 = \left[\frac{3 \delta_{gb} D_{gb} \gamma_{sv} \Omega_{MX}}{kTr^4}\right] t$$

160 D. L. Johnson, *J. Appl. Phys.*, **40**:192 (1969).

161 The situation gets much more complicated very rapidly if the oxide is impure. See, e.g., D. W. Ready, *J. Amer. Cer. Soc.*, **49**:366 (1966).

162 W. S. Coblenz, J. M. Dynys, R. M. Cannon, and R. L. Coble, in *Sintering Processes*, G. C. Kuczynski, ed., Plenum Press, New York, 1980. See also: R. L. Coble, *J. Am. Cer. Soc.*, **41**:55 (1958).

where D_{gb} is the grain boundary diffusivity of the rate-limiting ion.

Surface diffusion model (path 2, Fig. 10.9b)

In this model, the atoms are assumed to diffuse along the surface from an area that is near the neck region toward the neck area. The appropriate expression for the growth of the neck with time is[163]

$$\left(\frac{x}{r}\right)^5 = \left[\frac{225\,\delta_s D_s\,\gamma_{sv}\Omega_{MX}}{kTr^4}\right]t \tag{10.18}$$

where D_s and δ_s are, respectively, the surface diffusivity and surface thickness.

Viscous sintering

The shrinkage of two glass spheres during the initial stage is given by the Frenkel equation[164]

$$\frac{\Delta L}{L} = \frac{3}{4}\frac{\gamma_{sv}}{\eta}t \tag{10.19}$$

where η is the viscosity of the glass. A micrograph of glass spheres that sinter by viscous flow upon heating is shown in Fig. 10.13.

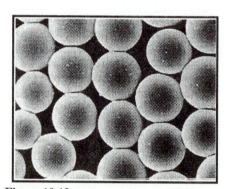 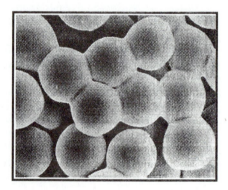

Figure 10.13
Micrograph of glass spheres that sinter by heating in air at a temperature range at which viscous flow can occur.

[163] W. S. Coblenz, J. M. Dynys, R. M. Cannon, and R. L. Coble, in *Sintering Processes*, G. C. Kuczynski, ed. Plenum Press, New York, 1980.

[164] J. Frenkel, *J. Phys.* (USSR), **9**:305 (1945).

General remarks

Usually the activation energies for surface, grain boundary, and lattice diffusivity increase in that order. Thus surface diffusion is favored at lower temperatures and lattice diffusion at higher temperatures. By comparing Eqs. (10.16), (10.17), and (10.19), it should be obvious why grain boundary and surface diffusion are preferred over lattice diffusion for smaller particles. Lattice diffusion, however, is favored at long sintering times, high sintering temperatures, and larger particles. However, by far the most forgiving mechanism with respect to particle size is viscous sintering (see Worked Example 10.2). It is important to note that these general trends also extend through the intermediate- and final-stage sintering stages.

WORKED EXAMPLE 10.2. It takes 0.2 h for the relative density of a 0.1 μm average diameter powder to increase from 60 to 65 percent. Estimate the time it would take for a powder of 10 μm average diameter to achieve the same degree of densification if the rate-controlling mechanism were (a) lattice diffusion and (b) viscous flow.

Answer

(a) If one assumes that the ratio x/r does not vary much as the density increases from 60 to 65 percent, which is not a bad assumption, it follows that Eq. (10.16) for lattice diffusion can be recast to read

$$\Delta t = \left(\frac{x}{r}\right)^4 \frac{kTr^3}{64 D_{ambi} \gamma_{sv} \Omega_{MX}^2} \approx K' \cdot r^3$$

If the 0.1 μm particles densify by lattice diffusion, the time needed to go from 60 to 65 percent will be given by $\Delta t = K' \cdot r^3$, from which

$$K' = 0.2 / \left(0.05 \times 10^{-6}\right)^3 = 1.6 \times 10^{21} \ h/m^3$$

The 10 μm particles will densify by the same amount after a time Δt given by

$$\Delta t = K' r^3 = 1.6 \times 10^{21} \left(5 \times 10^{-6}\right)^3 = 2 \times 10^5 \ \text{hours!}$$

(b) In the case of viscous flow a similar analysis shows that since t scales with r [Eq. (10.19)], Δt in this case is of about 20 h.

This calculation makes clear that viscous phase sintering is much more forgiving concerning the initial particle size. It also explains the absolute need to start with very fine crystalline powders if densification is to occur in reasonable times.

10.3.2 Densification Kinetics

Intermediate sintering model

Most of the densification of a powder compact occurs during the intermediate stage. Unfortunately, this stage is the most difficult to tackle because it depends strongly on the details of particle packing — a variable that is quite difficult to model. To render the problem tractable, Coble made the following assumptions:

1. The powder compact is composed of ideally packed tetrakaidecahedra of length a_p separated from each other by long porous channels of radii r_c (Fig. 10.14a).
2. Densification occurs by the bulk diffusion of vacancies away from the cylindrical pore channels toward the grain boundaries (solid arrows in Fig. 10.14b).
3. A linear, steady-state profile of the vacancy concentration is established between the source and the sink.
4. The vacancies are annihilated at the grain boundaries; i.e., the grain boundaries act as vacancy sinks. It is also assumed that where the vacancies are annihilated, their concentration is given by C_0 [Eq. (10.7)], i.e., under a stress-free planar interface.

Making these assumptions, one can show (App. 10C) that during the intermediate-stage sintering, the fractional porosity P_c should decrease linearly with time according to

$$P_c \approx (\text{const}) \frac{D_{\text{ambi}}}{d^3} \frac{\gamma_{\text{sv}} \Omega_{\text{MX}}}{kT} (t_f - t) \tag{10.20}$$

where t_f is the time at which the cylindrical channels vanish and d is the average diameter of sintering particles, which is assumed to scale with a_p.

Repeating the procedure used to arrive at Eq. (10.20) but assuming densification takes place by grain boundary diffusion (dashed arrows in Fig. 10.14b), one can easily show that (see Prob. 10.6)

$$P_c = (\text{const}) \left[\frac{D_{\text{gb}} \delta_{\text{gb}}}{d^4} \left(\frac{\gamma_{\text{sv}} \Omega_{\text{MX}}}{kT} \right) (t_f - t) \right]^{2/3} \tag{10.21}$$

where δ_{gb} is the width of the grain boundary shown in Fig. 10.14b.

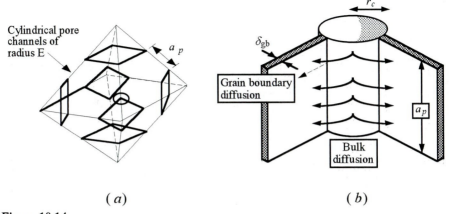

Figure 10.14

(*a*) Tetrakaidecahedron model of intermediate-stage sintering. (*b*) Expanded view of one of the cylindrical pore channels. The vacancies can diffuse down the grain boundary (dashed arrow) or through the bulk (solid arrows). Note that in both cases the vacancies are annihilated at the grain boundaries.

It is rather unfortunate that of all the sintering stages the most important is also the most difficult to model. For example, any intermediate-stage model that does not take into account the details of particle packing has very limited validity. A cursory examination of the results shown in Fig. 10.10 should make this point amply clear.

What is interesting about this process is that it is self-accelerating, since as the cylinder gets smaller in diameter, its curvature increases and the vacancy concentration gradient also increases. This process cannot and does not go on indefinitely; as the cylindrical pores get longer and thinner, at some point they become unstable and break up into smaller spherical pores along the grain boundary and/or at the triple points between grains (see Fig. 10.11*d*). It is at this point that the intermediate sintering stage gives way to the final-stage sintering, where both the annihilation of the last remnants of porosity and the simultaneous coarsening, i.e., grain growth, of the microstructure occur. The next subsection deals with the elimination of porosity; grain growth is dealt with later in Sec. 10.3.3.

Pore elimination

Pores can be eliminated by various mechanisms, which have in common the fact that atoms diffuse toward the pores and vacancies are transported away from the pores to a sink that is assumed to be the grain boundaries, dislocations, or external surface of the crystal. Consider the two following representative mechanisms.

Volume diffusion. In this model, the vacancy source is the pore surface of radius ρ_p and the sink is the spherical surface of radius R, where $R \gg \rho_p$ (Fig. 10.15a). By solving the flux equation, subject to the appropriate boundary conditions, it can be shown that (see App. 10D)

$$\rho_p^3 - \rho_{p,0}^3 = -\frac{6 D_{\text{ambi}} \gamma_{\text{sv}} \Omega_{\text{MX}}}{kT} t \tag{10.22}$$

where $\rho_{p,0}$ is the initial size of the pores at $t = 0$. To relate the radius of the pore to the porosity P_c, use is again made of the model shown in Fig. 10.14a. A pore is assumed to be present at the vertices of each tetrakaidecahedron; and since there are 24 vertices and each pore is shared by four polyhedra, it follows that the fraction porosity is given by

$$P_c = \frac{8 \pi \rho_p^3}{8\sqrt{2} a_p^3} = \frac{\pi \rho_p^3}{\sqrt{2} a_p^3} \tag{10.23}$$

where the denominator represents the volume of the tetrakaidecahedron. Combining this result with Eq. (10.22), and noting that the grain size d scales with a_p yields

$$P_c - P_0 = -(\text{const}) \frac{D_{\text{ambi}} \gamma_{\text{sv}} \Omega_{\text{MX}}}{d^3 kT} t \tag{10.24}$$

where P_0 is the porosity at the beginning of the final stage of sintering. This model predicts that the porosity during this stage will decrease linearly with time and inversely as d^3. In other words, smaller grains should result in much faster porosity elimination, which was expected since it was assumed that grains boundaries act as sinks.

Grain boundary diffusion. Equation (10.24) has limited validity, however, because as discussed in greater detail below, the elimination of the last remnants of vacancies usually occurs only if they remain attached to and are eliminated at the grain boundaries. The appropriate geometry is shown schematically in Fig. 10.15b, and by following a derivation similar to the one carried out in App. 10D, it can be shown (see Prob. 10.7) that

$$\rho_p^4 - \rho_{p,0}^4 = -\frac{8 \delta_{\text{gb}} D_{\text{gb}} \gamma_{\text{sv}} \Omega_{\text{MX}}}{kT} \frac{1}{\log(R_{\text{gb}}/\rho_p)} t \tag{10.25}$$

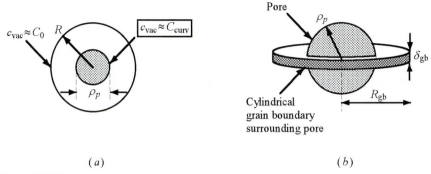

(a) (b)

Figure 10.15

Geometric constructions used to model porosity elimination, assuming (a) bulk diffusion, where the vacancies are assumed to be eliminated at the grain surface, and (b) grain boundary diffusion, where vacancies are restricted to diffusing along a grain boundary of width δ_{gb}.

where R_{gb} is defined in Fig. 10.15b. In other words, if grain boundary diffusion is the operative mechanism, the average pore size should shrink with $t^{1/4}$ rather than the $t^{1/3}$ dependence expected if bulk diffusion were important [i.e., Eq. (10.22)]. Unfortunately, no simple analytic expression exists relating the porosity to ρ_p, and numerical methods have to be used instead.

Effect of dihedral angle on pore elimination

It should be clear by now that pore shape and volume fraction continually evolve during sintering, and understanding that evolution is critical to understanding how high theoretical densities can be achieved. An implicit and fundamental assumption made in the foregoing analysis was the existence of a driving force to shrink the pores at all times — an assumption that is not always valid. As discussed below, at some conditions, the pores are thermodynamically stable.

To demonstrate the conditions for which this is the case, consider the four grains that intersect as shown in Fig. 10.16a. Referring to the figure, if the pore is allowed to shrink by an amount equal to the shaded area, the excess energy eliminated is proportional to $2\Gamma\gamma_{sv}$ while the excess energy gained will be proportional to $\lambda\gamma_{gb}$, where Γ and λ are defined in the figure. It follows that the ratio of the energy gained to that lost is

$$\frac{\text{Energy gained}}{\text{Energy lost}} = \frac{2\Gamma\gamma_{sv}}{\lambda\gamma_{gb}} \tag{10.26}$$

Combining this result with Eq. (10.1) and the fact that $\cos(\varphi/2) = \Gamma/\lambda$ (see Fig. 10.16a), one can easily show that the right-hand side of Eq. (10.26) equals 1.0, when $\varphi = \phi$. In other words, when the grains around a pore meet such that $\gamma_{gb} = 2\gamma_{sv}\cos(\phi/2)$, the *driving force for grain boundary migration and pore shrinkage goes to zero*.

This is an important conclusion since it implies that if pores are to be completely eliminated, their coordination number has be *less* than a critical value n_c. It can be further shown (see Prob. 10.8) that n_c is related to the dihedral angle by $n_c < 360/(180 - \phi)$. It is left as an exercise for the reader to determine which of the three pores in Fig. 10.16 is stable and which is not.

Based on these results, one may conclude that increasing the dihedral angle should, in principle, aid in the later stages of sintering. The situation is not so simple, however, since it can also be shown that the attachment of pores to boundaries is stronger for lower ϕ's. Given that in order to eliminate pores, they have to remain attached to the grain boundary, this latter property would tend to suggest that low dihedral angles would aid in the prevention of boundary-pore breakaway and thus be beneficial. Finally, note that channel breakup at the end of the intermediate-stage sintering occurs at smaller volume fractions of pores as the dihedral angle decreases, which again is beneficial.

10.3.3   Coarsening and Grain Growth Kinetics

Any collection of particles will coarsen with time, kinetics permitting, where coarsening implies an increase in the ensemble's average particle size with time. Comparing Fig. 10.2a and b shows the clear similarity between coarsening and grain growth. This section deals with the kinetics of the microstructural evolution of a collection of particles during sintering (Fig. 10.2b) and the grain growth kinetics associated with the final stages of sintering (Fig. 10.2a).

Coarsening

To model coarsening, consider a powder compact consisting of a distribution of particles, with an average particle radius r_{av}. Assuming all particles to be spheres, the *average* partial pressure over the ensemble is given by Eq. (10.6), or

$$P_{av} = P_{flat}\left(1 + \frac{2\Omega_{MX}\gamma_{sv}}{r_{av}kT}\right) \tag{10.27}$$

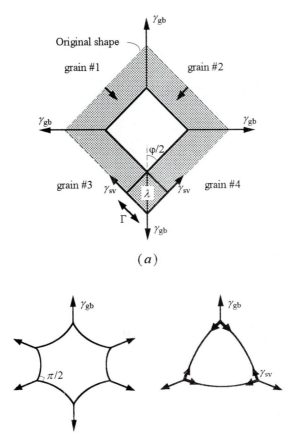

(a)

Actual curvature of boundary needed to maintain dihedral angle equilibrium assumed to be $\dfrac{\pi}{2}$.

(b)   (c)

Figure 10.16

Effect of pore coordination on pore shrinkage for a system for which the dihedral angle $\phi = \pi/2$. (a) Intersection of four grains around a pore. (b) Intersection of six grains; that is, $\varphi = 120°$. Note that to maintain the equilibrium dihedral angle, the surfaces of the grains surrounding the pore have to be convex. (c) Same system with $\varphi = 60°$. Here the pore surface has to be concave in order to maintain the equilibrium dihedral angle.

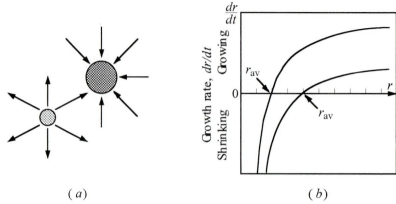

(a)   (b)

Figure 10.17
(a) Schematic of coarsening problem, showing how grains that are smaller than the average shrink at the expense of larger grains that grow. (b) Plot of Eq. (10.29). The lower curve has an average radius that is twice that of the top curve.

Similarly, the partial pressure P_r over any particle of radius $r \neq r_{av}$ is also given by Eq. (10.6), using the appropriate radius. Consequently, grains that are smaller than the average will "evaporate" away, while those that are larger will grow with time (Fig. 10.17a).

If the interface kinetics are rate-limiting, it follows that the velocity v of the solid/gas interface is linearly dependent on the driving force, and

$$v = \frac{dr}{dt} = K_r(P_{av} - P_r) \tag{10.28}$$

where K_r is a proportionality constant related to the mobility of the interface.[165] Combining Eqs. (10.6), (10.27) and (10.28), one obtains

$$\frac{dr}{dt} = K_r(P_{av} - P_r) = \frac{2\Omega_{MX}P_{flat}K_r\gamma_{sv}}{kT}\left(\frac{1}{r_{av}} - \frac{1}{r}\right) \tag{10.29}$$

A plot of this equation, shown in Fig. 10.17b for two different average particle sizes, in addition to demonstrating that small particles shrink and larger ones grow, yields another valuable insight, namely, that smaller grains $(r < r_{av})$ disappear much faster than larger grains grow. Furthermore, as time progresses and the average grain size increases, the growth rate for all particles is reduced and eventually goes to zero.

---

[165] The reader will note that for a process that is evaporation-controlled, $K_r = \alpha/\sqrt{2\pi m_{MX}kT}$ [see Eq. (10.11)].

The model can be taken further by making the simplifying assumption that the rate of increase of the average particle size is identical to that of particles that are twice the average size. In other words, by assuming

$$\frac{dr_{av}}{dt} = \frac{dr}{dt} \text{ at } r = 2r_{av}$$

Eq. (10.29) can be integrated to yield the final result

$$\boxed{r_{av}^2 - r_{0,\,av}^2 = \frac{2\gamma_{sv}\Omega_{MX}P_{flat}K_r}{kT}t}$$ (10.30)

where $r_{0,av}$ is the average particle size at $t = 0$. A more rigorous treatment[166] gives a value of 64/81 instead of 2. Equation (10.30) predicts a parabolic increase in the average grain size with time. It also predicts that the coarsening kinetics are enhanced for solids with high vapor/surface interface energies and high vapor pressures, both predictions in fair agreement with experimental observations. For example, it is now well established that covalently bonded solids such as Si_3N_4, SiC, and Si coarsen, rather than densify, because they have relatively high vapor pressures.

Grain growth

As noted above, during the final stages of sintering, in addition to the elimination of pores, a general coarsening of the microstructure by grain growth occurs. During this process the average grain size increases with time as the smaller grains are consumed by larger grains as shown in Fig. 10.18. Controlling and understanding the processes that lead to grain growth are important for two reasons. The first, discussed in greater detail in subsequent chapters, is related to the fact that grain size is a major factor determining many of the electrical, magnetic, optical, and mechanical properties of ceramics. The second is related to suppressing what is known as **abnormal grain growth**, which is the process whereby a small number of grains grow very rapidly to sizes that are more than an order of magnitude larger than average in the population (Fig. 10.21b). In addition to the detrimental effect that the large grains have on the mechanical properties (see Chap. 11), the walls of these large grains can pull away from porosities, leaving them trapped within them, which in turn limits the possibility of obtaining theoretical densities in reasonable times.

---

[166] C. Wagner, Z. Electrochem., 65:581 (1961).

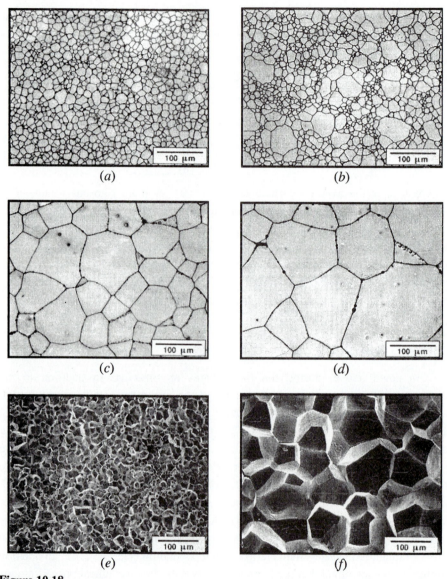

Figure 10.18

Time evolution of microstructure of CsI hot-pressed at 103 MPa at 100°C for (*a*) 5 min, (*b*) 20 min, (*c*) 1 h, and (*d*) 120 min. (*e*) Fractured surface of *a*, (*f*) fractured surface of *d*.[167]

---

[167] H-E. Kim and A. Moorhead, *J. Amer. Cer. Soc.*, **73**:496 (1990). Reprinted with permission.

Before one proceeds with the model, it is important to appreciate the origin of the driving force responsible for coarsening. Consider the schematic microstructure composed of cylindrical grains of varying curvatures, shown in Fig. 10.19a. Since in this structure the dihedral equilibrium angle has to be 120°, it follows that grains with more than six sides will tend to grow, while those with less than six sides will tend to shrink.[168] This occurs by migration of grain boundaries in the direction of the arrows (Fig. 10.19a). To appreciate the origin of the driving force, consider the atomic-scale schematic of such a boundary (Fig. 10.19b). At this level, it should be obvious why an atom on the convex side of the boundary would rather be on the concave side — it would, on average, be more tightly bound, i.e., have lower potential energy. Consequently, the atoms will jump from right to left, which means that the grain boundary will move from left to right, as shown in Fig. 10.19a. Looking at the problem at that level, one can easily see why straight (i.e., no curvature) grain boundaries would be stable and would not move.

More quantitatively, the driving forces per MX molecule $\Delta\mu_{gb}$ across the grain boundary is given by

$$\Delta\mu_{gb} = \gamma_{gb}\Omega_{MX}\kappa \tag{10.31}$$

where γ_{gb} is the grain boundary energy and κ its radius of curvature.

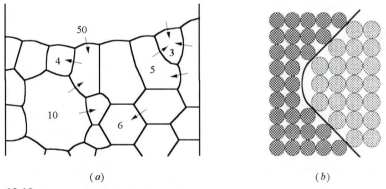

(a)   (b)

Figure 10.19
(a) Grain shape equilibrium and direction of motion of grain boundaries in a two-dimensional sheet (the grains are cylinders in this case). Note that grains with six sides are stable, while those with less than six sides shrink and those with more than six sides will grow. (b) Atomic view of a curved boundary. Atoms will jump from right to left, and the grain boundary will move in the opposite direction.

---

[168] The reason a grain that has less than six sides shrinks is the same as the reason the pore shown in Fig. 10.16c does.

To model the process, one needs to obtain a relationship between the grain boundary velocity u_{gb} and the driving force acting on the boundary. And since the situation is almost identical to that encountered during the growth of a solid/liquid interface (Fig. 9.3), Eq. (9.14) is directly applicable and

$$u_{gb} = \lambda v_{net} = \lambda v_0 \exp\left(-\frac{\Delta G_m^*}{kT}\right)\left\{1 - \exp\left(-\frac{\Delta \mu_{gb}}{kT}\right)\right\} \qquad (10.32)$$

where ΔG_v is replaced by $\Delta \mu_{gb}$. Expanding the term within braces in Eq. (10.32), for the usual case when $\Delta \mu_{gb} \ll kT$, one obtains

$$u_{gb} = \lambda v_{net} = \frac{\lambda v_0}{kT} \exp\left(-\frac{\Delta G_m^*}{kT}\right) \Delta \mu_{gb} \qquad (10.33)$$

This expression is usually abbreviated to

$$u_{gb} = M \Delta \mu_{gb} \qquad (10.34)$$

where

$$M = \frac{\lambda v_0}{kT} \exp\left(-\frac{\Delta G_m^*}{kT}\right) \qquad (10.35)$$

Comparing Eqs. (10.33) and (10.28) reveals the similarities between this problem and that of coarsening worked out in the previous section. So by modifying Eq. (10.30) to the problem at hand, the final result for grain growth is

$$\boxed{d_{av}^2 - d_{av,0}^2 = \frac{4 M \gamma_{gb} \Omega_{MX}}{\beta} t} \qquad (10.36)$$

where $d_{av,0}$ is the average grain size at $t = 0$, and β is a geometric factor that depends on the curvature of the boundary. For example, for a solid that is made up entirely of straight, noncurved grain boundaries, no grain growth would occur and β would be infinite. The process just described is sometimes referred to as **Ostwald ripening** and is characterized by a parabolic dependence of grain size on time.

Effect of microstructure and grain boundary chemistry on boundary mobility

In deriving Eq. (10.36), the implicit assumption that the grain boundaries were pore-, inclusion-, and essentially solute-free — a very rare occurrence, indeed — was made, and as such Eq. (10.36) predicts the so-called intrinsic grain growth kinetics. Needless to say, the presence of "second phases" or solutes at the boundaries can have a dramatic effect on their mobility, and from a practical point of view it is usually the mobility of these phases that is rate-limiting. To illustrate the complexity of the problem, consider just a few possible rate-limiting processes:

1. Intrinsic grain boundary mobility discussed above.
2. Extrinsic or solute drag. If the diffusion of the solute segregated at the grain boundaries is slower than the intrinsic grain boundary mobility, it becomes rate-limiting. In other words, if the moving grain boundary must drag the solute along, that tends to slow it down.
3. The presence of inclusions (basically second phases) at the grain boundaries. It can be shown that larger inclusions have lower mobilities than smaller ones, and that the higher the volume fraction of a given inclusion, the larger the resistance to boundary migration.
4. Material transfer across a continuous boundary phase. For instance, in Si_3N_4 boundary movement can occur only if both silicon and oxygen diffuse through the thin, glassy film that usually exists between grains.
5. In some cases, the redissolution of the boundary-anchoring second phase inclusions into the matrix can be rate limiting.

In addition to these, the following interactions, between pores and grain boundaries can occur

1. What is true of second phases is also true of pores. Pores *cannot* enhance boundary mobility; they only leave it unaffected or reduce it. During the final stages of sintering as the pores shrink, the mobility of the boundaries will increase (see below).
2. The pores do not always shrink — they can also coarsen as they move along or intersect a moving grain boundary.
3. The pores can grow by the Ostwald ripening mechanism.
4. Pores can grow by reactive gas evolution and sample bloating.

To discuss even a fraction of these possibilities in any detail is clearly not within the scope of this book. What is attempted instead is to consider in some detail one of the more important grain boundary interactions namely, that between the grain boundary and pore.

As the grains get larger and the pores fewer, the grain mobility increases accordingly. In some cases, at a combination of grain size and density, the mobility of the grain boundaries becomes large enough that the pores can no longer keep up with them; the boundaries simply move too fast for the pores to follow and consequently unpin themselves. This region is depicted on the grain size versus density trajectory in the upper right-hand corner of Fig. 10.20a.

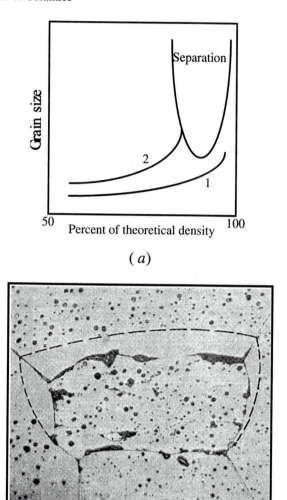

(a)

(b)

Figure 10.20

(a) Grain size versus densification trajectory including region where separation of boundaries and pores will occur. In that region, the grain boundaries will break away from the pores, entrapping them within the grains. To achieve full density, path 1 has to followed. Path 2 will result in entrapped pores. (b) Micrograph of the sweeping out of pores by the migration of grain boundaries. The original position of the boundary is depicted by the dotted line.[169]

[169] J. E. Burke, *J. Amer. Cer. Soc.*, **40**:80 (1957). Reprinted with permission.

If theoretical density is to be achieved, it is important that the grain boundary versus density trajectory not intersect this separation region. The importance of having the pores near grain boundaries is illustrated in Fig. 10.20*b*; the migration of the boundary to the right has swept and eliminated all the pores in its wake. Pores that are trapped within the grains will remain there because the diffusion distances between sources and sinks become too large.

There are essentially two strategies that can be employed to prevent pore breakaway, namely, reduce grain boundary mobility and/or enhance pore mobility. An example of the slowing grain boundary mobility enhancing the final density is shown in Fig. 10.6*b*, where the grain size versus density trajectories for two aluminas, one pure and the other doped with 250 ppm MgO, are compared. It is obvious from the results that the doped alumina achieves higher density — the reason is believed to be the result of impurity drag on the boundary by the MgO.

Abnormal grain growth

In some systems, it has been observed that a small number of grains in the population grow rapidly to very large sizes relative to the average size of the population (see Fig. 10.21*b*). This phenomenon is referred to as *abnormal grain growth* (AGG). AGG is to be avoided for the same reason as pore–grain boundary unpinning.

Although it is not entirely clear as to what results in AGG, there is mounting evidence that it is most likely associated with the formation of a liquid phase, or very thin liquid films at the grain boundaries. These can result from dopants intentionally added or simply from impurities in the starting powder. The effect of having small amounts of liquid during solid state sintering and its effect on the sintering and grain growth kinetics are discussed in the next section. There is little doubt, however, that small amounts of liquid can result in substantial coarsening of the microstructure, as shown in Fig. 10.24.

10.3.4 Factors Affecting Solid State Sintering

Typically a solid-state sintered ceramic is an opaque material containing some residual porosity and grains that are much larger than the starting particle sizes. On the basis of the discussion and models just presented, it is useful to summarize the more important factors that control sintering. Implicit in the following arguments is that theoretical density is desired.

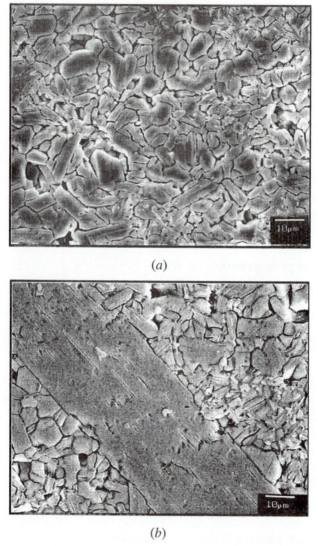

(a)

(b)

Figure 10.21

Microstructures showing (a) normal and (b) abnormal grain growth in Ti_3SiC_2

1. *Temperature*. Since diffusion is responsible for sintering, clearly increasing temperature will greatly enhance the sintering kinetics, because D is thermally activated. As noted earlier, the activation energies for bulk diffusion are usually higher than those for surface and grain boundary diffusion. Therefore, increasing the temperature usually enhances the bulk diffusion mechanisms which lead to densification.

2. *Green density*. Usually a correlation exists between the green (prior to sintering) density and the final density, since the higher the green density, the less pore volume that has to be eliminated.

3. *Uniformity of green microstructure*. More important than the green density is the uniformity of the green microstructure and the lack of agglomerates (see Fig. 10.10). The importance of eliminating agglomerates is discussed in greater detail below.

4. *Atmosphere*. The effect of atmosphere can be critical to the densification of a powder compact. In some cases, the atmosphere can enhance the diffusivity of a rate-controlling species, e.g., by influencing the defect structure. In other cases, the presence of a certain gas can promote coarsening by enhancing the vapor pressure and totally suppressing densification. An excellent example of the effect of atmosphere was shown in Fig. 10.7: Fe_2O_3 will readily densify in air but not in HCl-containing atmospheres.

    Another important consideration is the solubility of the gas in the solid. Because the gas pressure within the pores increases as they shrink, it is important to choose a sintering atmosphere gas that readily dissolves in the solid.

5. *Impurities*. The role of impurities cannot be overemphasized. The key to many successful commercial products has been the identification of the right pinch of magic dust. The role of impurities has been extensively studied, and to date their effect can be summarized as follows:

    a. *Sintering aids*. They are purposefully added to form a liquid phase (discussed in the next section). It is also important to note at this point that the role of impurities is not always appreciated. The presence of impurities can form low-temperature eutectics and result in enhanced sintering kinetics, even in very small concentrations.

    b. Suppress coarsening by reducing the evaporation rate and lowering surface diffusion. A classic example is boron additions to SiC, without which SiC will not densify.

    c. Suppress grain growth and lower grain boundary mobility (Fig. 10.6).

d. Enhance diffusion rate. Once the rate-limiting ion during sintering is identified, the addition of the proper dopant that will go into solution and create vacancies on that sublattice should, in principle, enhance the densification kinetics.

6. *Size distribution*. Narrow grain size distributions will decrease the propensity for abnormal grain growth.

7. *Particle size*. Since the driving force for densification is the reduction in surface area, the larger the initial surface area, the greater the driving force. Thus it would seem that one should use the finest initial particle size possible, and while in principle this is good advice, in practice very fine particles pose serious problems. As the surface/volume ratio of the particles increases, electrostatic and other surface forces become dominant, which leads to agglomeration. Upon heating, the agglomerates have a tendency to sinter together into larger particles, which not only dissipates the driving force for densification but also creates large pores between the partially sintered agglomerates which are subsequently difficult to eliminate. The dramatic effect of ridding a powder of agglomerates on the densification kinetics was illustrated in Fig. 10.10.

The solution lies in working with nature instead of against it. In other words, make use of the surface forces to colloidally deflocculate the powders and keep them from agglomerating.[170] However, once dispersed, the powders should *not* be dried, but piped directly into a mold or a device that gives them the desired shape. The reason is simple. In many cases, drying reintroduces the agglomerates and defeats the purpose of colloidal processing.

To avoid excessive shrinkage during fluid removal requires pourable slurries with a high volume fraction of particles. Once the slurry has been molded, its rheological properties must be dramatically altered to allow shape retention during unmolding. What is required in this stage is to change the viscous slurry to an elastic body without fluid-phase removal. The basic idea is to avoid, at all costs, passing through a stage where a liquid/vapor interface exists. The reason is simple and is discussed in the next section. The presence of liquid/vapor interfaces can result in strong capillary forces that can cause particle rearrangement and agglomeration. And whereas this is desirable during liquid-phase sintering, it is undesirable when a slurry is dried because it is uncontrollable and can result in shrinkage stresses, which in turn can result in the formation of either agglomerates or large cracks between areas that

---

170 See, e.g., F. Lange, *J. Amer. Cer. Soc.*, **72**:3 (1989).

shrink at different rates. The crazing of mud upon drying is a good example of this phenomenon.

Another possible source of flaws can be introduced during the cold pressing of agglomerated powders as a result of density differences between the agglomerates and the matrix. When the pressure is removed, the elastic dilation of the agglomerates and the matrix may be sufficiently different to cause cracks to form; i.e., the springback of the agglomerates will be different from the matrix as a result of differences in their density.

10.4
LIQUID-PHASE SINTERING

The term *liquid-phase sintering* is used to describe the sintering process when a proportion of the material being sintered is in the liquid state (Fig. 10.1*a*). Liquid-phase sintering of ceramics is of major commercial importance since a majority of ceramic products are fabricated via this route and include ferrite magnets, covalent ceramics such as silicon nitride, ferroelectric capacitors, and abrasives, among others. Even in products that are believed to be solid-state sintered, it has been demonstrated that in many cases the presence of liquid phases at grain boundaries probably plays a significant role.

Liquid-phase sintering offers two significant advantages over solid-state sintering. First, it is much more rapid; second, it results in uniform densification. As discussed below, the presence of a liquid reduces the friction between particles and introduces capillary forces that result in the dissolution of sharp edges and the rapid rearrangement of the solid particles.

During liquid-phase sintering, the composition of the starting solids is such as to result in the formation of a liquid phase upon heating. The liquid formed has to have an appreciable solubility of the solid phase and wet the solid. In the next sections, the reasons these two requirements should be met and the origin of the forces at play during liquid-phase sintering are elucidated.

10.4.1 Surface Energy Considerations

As discussed in detail in the preceding sections, the driving force during sintering is the overall reduction in surface energy of the system. In solid-state sintering, the lower-surface-energy grain boundaries replace the higher-energy solid/vapor surfaces. The presence of a liquid phase introduces a few more surface energies that have to be considered, namely, the liquid/vapor γ_{lv} and the liquid/solid γ_{ls} interface energies.

When a liquid is placed on a solid surface, either it will spread and wet that surface (Fig. 10.22a) or it will bead up (Fig. 10.22b). The degree of wetting and whether a system is wetting or nonwetting are quantified by the equilibrium contact angle θ that forms between the liquid and the solid and is defined in Fig. 10.22a and b. A simple balance of forces indicates that at equilibrium

$$\gamma_{sv} = \gamma_{ls} + \gamma_{lv} \cos \theta \qquad (10.37)$$

from which it is clear that high values of γ_{sv} and low values of γ_{ls} and/or γ_{lv} promote wetting. By using an argument not unlike the one used to derive Eq. (10.26), it can be shown that a necessary condition for liquid-phase sintering to occur is that the contact angle must lie between 0 and $\pi/2$, that is, the system must be wetting. For nonwetting systems, the liquid will simply bead up in the pores, and sintering can only occur by one of the solid-state mechanisms discussed previously.

The complete penetration of grain boundaries with a liquid is also important for microstructural development (e.g., it would lead to the breakup of agglomerates). It can be shown (Prob. 10.11b) that a necessary condition for the penetration and separation of the grains with a thick liquid film is $\gamma_{gb} > 2\gamma_{ls}$. This implies that the equilibrium dihedral angle Ψ, shown in Fig. 10.3b, and defined as

$$\gamma_{gb} = 2\gamma_{sl} \cos \frac{\Psi}{2} \qquad (10.38)$$

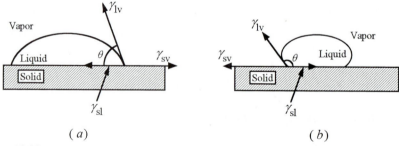

(a)   (b)

Figure 10.22
(a) Wetting system showing forces acting on the liquid drop. (b) Nonwetting system with $\theta > 90°$.

must be zero. In this context, it follows that high values of γ_{gb} and low values of γ_{ls} are desirable.

10.4.2 Capillary Forces

When solids and liquids are present together, capillary forces that result from surface tension are generated. These forces, as discussed below, can give rise to strong attractive forces between neighboring particles, which when combined with the lubricating potential of the liquid, can lead to very rapid and significant particle rearrangement and densification.

The origin of the attractive forces is twofold: the force exerted by the pressure differential across the meniscus that results from its curvature and the component of the liquid/vapor surface energy normal to the two surfaces. To better appreciate these two forces, consider the thought experiment illustrated in Fig. 10.23. Here a solid cylinder of radius X is placed between two plates, and the system is heated so as to melt the solid. Consider what happens in the following three cases.

1. Contact angle $= 90°$ (Fig. 10.23b). Here it is assumed that the cylindrical shape is retained upon melting (it is a thought experiment, after all). Upon melting, γ_{lv} will exert an attractive force given by

$$F_{att} = -2\pi X \gamma_{lv} \qquad (10.39)$$

Simultaneously, as a result of its radius of curvature, a *positive* pressure (see App. 10B)

$$\Delta P = \frac{\gamma_{lv}}{X}$$

will develop inside the cylinder, which will repel the plates by a force

$$F_{repuls} = \pi X^2 \frac{\gamma_{lv}}{X} = \pi X \gamma_{lv} \qquad (10.40)$$

Subtracting Eq. (10.40) from (10.39) yields a net attractive force $F_{net} = -\pi X \gamma_{lv}$ that will pull the plates together.

2. Contact angle $> 90°$ (Fig. 10.23c). In this case the liquid will bead up and push the plates apart. This simple experiment makes it clear why a wetting system is a necessary condition for liquid-phase sintering to occur.

3. Contact angle $< 90°$ (Fig. 10.23a). This is the most interesting case. Upon melting, the liquid will spread and pull the plates closer as a result of two forces. The first component is due to negative pressure that develops in the miniscus as a result of its negative curvature, or

$$F_{att} = \pi X^2 \Delta P$$

where ΔP is the pressure across the curved surface (App. 10B)

$$\Delta P = -\frac{\gamma_{lv}}{\rho_2} \qquad (10.41)$$

where ρ_2 is shown in Fig. 10.23a.[171] The second component, also attractive, is once again due to the component of γ_{lv} in the direction normal to the two plates, given by

$$F_2 = -2\pi X \gamma_{lv} \sin\theta \qquad (10.42)$$

In general, this component is always small compared to the first and can be ignored, except when θ approaches $90°$.

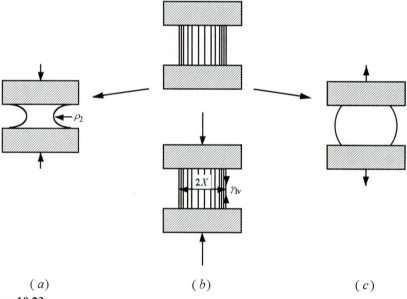

(a)   (b)   (c)

Figure 10.23

Thought experiment illustrating the origin of the two forces acting on a liquid drop for (a) wetting, $\theta < 90°$, (b) $\theta = 90°$, and (c) $\theta > 90°$. For the first two cases the net force is attractive.

---

[171] In general, ΔP is related to the two principal radii of curvature of the saddle, as shown in Fig. 10.26b in App. 10B. In most cases, however, $\rho_1 \gg \rho_2$ and Eq. (10. 41) is a very good approximation.

Another way to think of the problem is to assume that the negative pressure in the pores is pulling the liquid into them and hence pulling the particles together.[172] Typically, the pores during liquid-phase sintering are on the order of 0.1 to 1 μm. For a liquid with a γ_{lv} on the order of 1 J/m^2, this translates to compressive stresses [Eq. (10.45)] between particles on the order of 1 to 10 MPa. These stresses together with the greatly enhanced diffusion rates in the liquid (see below) are key to the process.

10.4.3 Liquid-Phase Sintering Mechanisms

Upon melting, a wetting liquid will penetrate between grains, as shown in Fig. 10.12b, and exert a attractive force, pulling them together. The combination of these forces and the lubricating effect of the liquid as it penetrates between grains leads to the following three mechanisms that operate in succession:

Particle rearrangement

Densification results from particle rearrangement under the influence of capillary forces and the filling of pores by the liquid phase. This process is very rapid, and if during the early stages of sintering, the liquid flows and completely fills the finer pores between the particles, 100 percent densification can result almost instantaneously.

Solution reprecipitation

At points where the particles touch, the capillary forces generated will increase the chemical potential of the atoms at the point of contact relative to areas that are not in contact. The chemical potential difference between the atoms at the two sites is given by (see Chap. 12)

$$\mu - \mu_0 = kT \ln \frac{a}{a_0} = \Delta P \, \Omega_{MX} \qquad (10.43)$$

where ΔP is given by Eq. (10.41), a is the activity of the solid in the liquid at the point of contact, and a_0 is the same activity under no stress, i.e., at the surface of the pore. This chemical potential gradient induces the dissolution of atoms at the contact points and their reprecipitation away from the area between the two particles, which naturally leads to shrinkage and densification. Furthermore, the kinetics of densification will be much faster than in the case of solid-state sintering,

---

[172] For example, it is this force that results in the strong cohesion of two glass slides when a thin layer of water is inserted between them.

because diffusion is now occurring in the liquid where the diffusivities are orders of magnitude higher than those in the solid state. Obviously, for this process to occur, there must be some solubility of the solid in the liquid — but not vice versa — and wetting. It is worth noting here that in addition to densification, coarsening or Ostwald ripening will occur simultaneously by the dissolution of the finer particles and their reprecipitation on the larger particles. The dramatic effect of having a small amount of liquid on the final microstructure is shown in Fig. 10.24.

Solid-state sintering

Once a rigid skeleton is formed, liquid-phase sintering stops and solid-state sintering takes over, and the overall shrinkage or densification rates are significantly reduced.

A schematic of a typical shrinkage curve for the three phases of sintering is shown in Fig. 10.25 (note the log scale on the x axis). Relative to the other two processes, particle rearrangement is the fastest, occurring in the time scale of minutes. The other two processes take longer since they depend on diffusion through the liquid or solid.

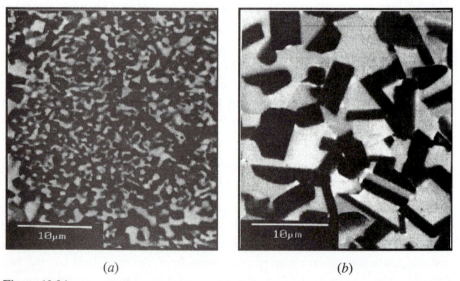

(a)   (b)

Figure 10.24
Effect of small amounts of liquid on coarsening of microstructure.[173]

---

[173] T. Lien, MSc. thesis, Drexel University, 1992.

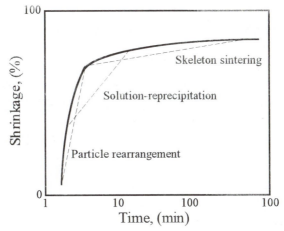

Figure 10.25

Time dependence of shrinkage evolution as a result of the mechanisms discussed in text.

Based on the discussion so far and numerous studies, it is now appreciated that for rapid densification to occur during liquid-phase sintering, the following conditions have to be met:

- There must be an appreciable solubility of the solid in the liquid in order for material transfer away from the contact areas to occur.
- Since the capillary pressure is proportional to $1/\rho_2$, which in turn scales with the particle diameter, it follows that the finer the solid phase, the higher the capillary pressures that can develop and the faster the densification rate.
- Wetting of the solid phase by the liquid is needed.
- Sufficient amount of liquid to wet the solid phase must be present.

Clearly liquid-phase sintering of ceramics is more forgiving in terms of powder packing, more rapid, and hence more economical than the solid-state version. Indeed, most commercial ceramics are liquid-phase sintered. And even some of the more advanced materials such as Si_3N_4 could not be densified without the presence of a liquid phase. If the properties required of the part are not adversely affected by the presence of a liquid, then it is the preferred route. However, for many applications, the presence of a glassy film at the grain boundaries can have a very detrimental effect on properties. For example, creep resistance can be significantly compromised by the presence of a glassy phase that softens upon heating. Another application in which the presence of an ionically conducting glassy phase is intolerable is in the area of ceramic insulators. Consequently, in the processing of today's electronic ceramics, liquid phases and residual porosities are avoided as much as possible.

10.5
HOT PRESSING AND HOT ISOSTATIC PRESSING

By now it should be clear that the driving force for densification is the chemical potential gradient between the atoms in the neck region and that at the pore. This can be accomplished by application of a compressive pressure to the compact during sintering (Fig. 10.12c); in other words, the simultaneous application of heat and pressure. If the applied pressure is uniaxial, the process is termed **hot pressing**, whereas if it is hydrostatic, the process is termed **hot isostatic pressing** or **HIP**.

The effect of applied pressure on chemical potential and vacancy concentrations is discussed in detail in Chap. 12. Here it will only be noted that the concentration of vacancies in an area subjected to a stress C_{stress} is related to C_0 by an equation similar to Eq. (10.9), that is,

$$C_{\text{stress}} = \left(1 + \frac{V_m \sigma_b}{RT}\right) C_0 \qquad (10.44)$$

where σ_b is the effective stress at the boundary due to the applied stress (see Worked Example 10.3). For a compressive applied stress, σ_b is negative[174] and the concentration of vacancies at the boundary (i.e., between the particles) is less than that at the edges, which results in a net flux of vacancies from the neck into the boundary areas and leads to densification.

The major advantage of hot pressing and HIP is the fact that the densification occurs quite readily and rapidly, minimizing the time for grain growth, which results in a finer and more uniform microstructure. The major disadvantages, however, are the costs associated with tooling and dies and the fact that the process does not lend itself to continuous production, since the pressing is usually carried out in a vacuum or an inert atmosphere.

WORKED EXAMPLE 10.3. Calculate the effect of a 10 MPa applied pressure on the vacancy concentration in a powder compact when $x/r = 0.2$, that is, during the initial stages of sintering at 1500C. State all assumptions.

Answer

Assume a surface energy of 1 J/m^2, a molar volume of 10 cm^3, and a particle radius of 2 μm. From Eq. (10.13), $\rho \approx x^2/(2r) = 4 \times 10^{-8}$ m. The vacancy concentration at

---

[174] Here and throughout the book, the chosen convention is that applied tensile forces are positive and compressive stresses are negative.

the neck area due to curvature is given by Eq. (10.9), or (note that for the neck region $\kappa = -1/\rho$).

$$C_{neck}^{curv} = \left(1 + \frac{V_m \gamma_{sv}}{\rho RT}\right) C_0 = \left(1 + \frac{10 \times 10^{-6} \times 1}{4 \times 10^{-8} \times 8.314 \times 1773}\right) C_0 \approx 1.017 C_0$$

Assuming the particle arrangement shown in Fig. 10.12c and a perfect cubic array, one can use a simple balance-of-forces argument, namely,

$$F_{app} = \sigma_a (2r)^2 = F_{boundary} = \sigma_b \pi x^2$$

from which it follows that $\sigma_b = -(4\sigma_a/\pi)(r/x)^2 \approx -318 \times 10^6$ Pa. Consequently, the vacancy concentration due to the applied stress at the boundary between grains is given by Eq. (10.44):

$$C_{boundary}^{stress} = \left(1 + \frac{V_m \sigma_b}{RT}\right) C_0 = \left(1 - \frac{10 \times 10^{-6} \times 318 \times 10^6}{8.314 \times 1773}\right) C_0 \approx 0.8 C_0$$

From this simple calculation it should be obvious why the application of a moderate pressure can result in a significant increase in the densification rate during all stages of sintering.

The vacancy concentration gradient that results from the curvature and stress is $\Delta C/x = (1.017 - 0.8)/x = 0.217/x$, which is greater than either alone. The effect of stress, however, is much more significant.

10.6
SUMMARY

1. Local variations in curvature result in mass transfer from areas of positive curvature (convex) to areas of negative curvature (concave). Quantitatively, this chemical potential differential is given by

$$\Delta \mu = \mu_{conv} - \mu_{conc} = \gamma_0 V_m \kappa$$

which has to be positive if sintering is to occur.

2. On the atomic scale, this chemical potential gradient results in *a local increase in the partial pressure of the solid and a local decrease in the vacancy concentration at the convex areas relative to the concave areas.* Looked at from another perspective, matter will always be displaced from the peaks into the valleys.

3. High vapor pressures and small particles will tend to favor gas transport mechanisms which lead to coarsening, whereas low vapor pressure and fast bulk or grain boundary diffusivities will tend to favor densification. If the atomic flux is from the surface of the particles to the neck region, or from the surface of smaller to larger particles, this leads to, respectively, neck growth and coarsening. However, if atoms diffuse from the grain boundary area to the neck region, densification results. Hence, all models that invoke shrinkage invariably assume that the grain boundary area or the free surface is the vacancy sink and that the neck surface is the vacancy source.

4. Sintering kinetics are dependent on the particle size and relative values of the transport coefficients, with smaller particles favoring grain boundary and surface diffusion and larger particles favoring bulk diffusion.

5. During the intermediate stage of sintering, the porosity is eliminated by the diffusion of vacancies from porous areas to grain boundaries, free surfaces, or dislocations. The uniformity of particle packing and lack of agglomerates are important for the achievement of rapid densification during this stage.

6. In the final stages of sintering, the goal is usually to eliminate the last remnants of porosity. This can only be accomplished, however, if the pores remain attached to the grain boundaries. One way to do this is to slow down grain boundary mobility by doping or by the addition of inclusions or second phases at the boundaries.

7. During liquid-phase sintering, the capillary forces that develop can be quite large. These result in the rearrangement of the particles as well as enhance the dissolution of matter between them, resulting in fast shrinkage and densification. Most commercial ceramics are manufactured via some form of liquid-phase sintering.

8. The application of an external force to a powder compact during sintering can greatly enhance the densification kinetics by increasing the chemical potential gradients of the atoms between the particles, inducing them to migrate away from these areas.

APPENDIX 10A

Derivation of the Gibbs-Thompson Equation

The work of expansion of a bubble must equal the increase in surface energy, or

$$\Delta P \, dV = \gamma \, dA$$

For a sphere of radius ρ, $dA/dV = 8\pi\rho/(4\pi\rho^2) = 2/\rho$. It follows that $\Delta P = 2\gamma/\rho$.
The Gibbs free energy change is given by

$$dG = VdP - SdT$$

For an isothermal process $dT = 0$ and $dG = VdP$. Integrating yields $\Delta G = V\Delta P$. Substituting for the value of ΔP given above and assuming 1 mol, one obtains

$$\Delta\mu = 2\gamma V_m/N_{Av}\rho = 2\gamma\Omega/\rho$$

where V_m is the molar volume, which is related to the atomic volume by $\Omega = V_m/N_{Av}$.

APPENDIX 10B

Radii of Curvature

In order to understand the various forces that arise as a result of curvature it is imperative to understand how curvature is defined and its implication. Any surface can be defined by its radius of curvature, κ, which for any surface can be defined as having two orthogonal radii of curvature, ρ_1 and ρ_2 (see Fig. 10.26) where

$$\kappa = \frac{1}{\rho_1} + \frac{1}{\rho_2} \tag{10B.1}$$

For a spherical particle, the two radii of curvature are defined as positive, equal to each other, and equal to the radius of the sphere ρ_{sphere} (see Fig. 10.26a). Thus for a sphere, $\kappa = 2/\rho_{sphere}$ and ΔP is positive. Conversely, for a spherical pore, the two radii are equal, but since in this case the surface of the pore is concave, it follows that $\kappa = -2/\rho_{pore}$ and ΔP is negative, which renders ΔC_{vac} [Eq. (10.10)] positive. That is, the vacancy concentration just below a concave surface is less than that under a flat surface.

In sintering, the geometry of the surface separating particles is modeled to be a saddle with two radii of curvature, as shown in Fig. 10.26b. It follows that

$$\kappa = \frac{1}{\rho_1} + \frac{1}{\rho_2} \approx \frac{2}{d} - \frac{1}{\rho_{neck}} \approx \frac{1}{\rho_{neck}} \tag{10B.2}$$

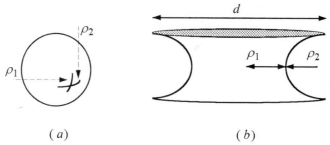

Figure 10.26
Definition of radii of curvature for (*a*) a sphere and (*b*) a saddle.

In most sintering problems, ρ_1 is on the same order as the particle diameter d, which is usually much larger than the radius of curvature of the neck ρ_{neck} and thus $\kappa \approx -1/\rho_{neck}$.

APPENDIX 10C

Derivation of Eq. (10.20)

Refer to Fig. 10.14*a*. The volume of each polyhedron is

$$V = 8\sqrt{2}a_p^3$$

and the volume of the porous channel per polyhedron is given by

$$V_p = \frac{1}{3}\left(36a_p\pi r_c^2\right)$$

where r_c is the radius of the channel. The factor 1/3 comes about because each cylinder is shared by three polyhedra. The fraction of pores is thus simply

$$P_c = \frac{V_p}{V} = \frac{12\pi a_p r_c^2}{8\sqrt{2}a_p^3} \approx 3.33\frac{r_c^2}{a_p^2} \tag{10C.1}$$

The total flux of vacancies diffusing away from the cylinder of surface area, S is given by[175]

---

[175] In cylindrical coordinates, Fick's second law is given by

$$\frac{\partial c}{\partial t} = \frac{D}{r}\frac{\partial}{\partial r}\left(r\frac{\partial c}{\partial r}\right)$$

$$J_X \cdot S = -\frac{D_{v,X}\Delta C_{v,X}}{a_p}(2\pi r_c a_p) = -2\pi D_{v,X}\Delta C_{v,X} r_c \qquad (10C.2)$$

where $D_{v,X}$ is the diffusivity of the anion vacancies assumed here to be rate limiting. Since for a cylindrical pore, $\kappa = -1/r_c$, $\Delta C_{v,X}$ in this case is given by Eq. (10.10):

$$\Delta C_{v,X} = C_0\left(\frac{\Omega_X \gamma_{sv}}{r_c kT}\right) \qquad (10C.3)$$

where Ω_X is the volume of the anion vacancy. Combining Eqs. (10C.2) and (10C.3) and noting that $D_{v,X}\Omega_X C_0 = D_X$ where D_X is the diffusivity of the anions, it follows that

$$J_X \cdot S = -2\pi D_{v,X}C_0\frac{\Omega_X \gamma_{sv}}{kT} = -\frac{2\pi D_{v,X}\gamma_{sv}}{kT} \qquad (10C.4)$$

Since it was assumed that the rate-limiting step was the diffusion of the anions, it follows that the ambipolar diffusion coefficient, $D_{ambi} \approx D_X$. Make that substitution in Eq. (10C.4) and note that the total volume of matter transported per unit time is $J \cdot S \cdot \Omega_{MX}$, or

$$\frac{dV}{dt} = -2\pi D_{ambi}\frac{\gamma_{sv}\Omega_{MX}}{kT}$$

Integrating this equation, while noting that for a cylinder of radius r and length a_p, $dV = 2\pi r a_p dr$ and rearranging terms yields

$$\frac{r_c^2}{a_p^2} = \frac{2D_{ambi}}{a_p^3}\left(\frac{\Omega_{MX}\gamma_{sv}}{kT}\right)(t_f - t) \qquad (10C.5)$$

where t_f is the time at which the pore vanishes. Experimentally, it is much easier to measure the average porosity than the actual radii of the pores. By combining Eqs. (10C.1) and (10C.5) and noting that the particle diameter d scales with a_p, this model predicts that

---

which at steady state, i.e., $\partial c/\partial t = 0$, has solution of the form $A + B\log r$, where A and B are constants that are determined from the boundary conditions. Thus strictly speaking, Eq. (10C.2), is incorrect since it assumes a planar geometry. It is used here for the sake of simplicity. Using the more exact expression, namely

$$J = D_v \frac{\Delta C}{\log(r/a_p)}\frac{1}{r}$$

does not greatly affect the final result.

$$P_c - P_0 = (\text{const.}) \frac{D_{\text{ambi}} \gamma_{\text{sv}} \Omega_{\text{MX}}}{d^3 kT} t \qquad (10C.6)$$

where P_0 is the porosity at the end of the intermediate stage of sintering.

APPENDIX 10D

█████████

Derivation of Eq. (10.22)

Given the spherical symmetry of the problem, it can be easily shown that

$$c_{\text{vac}} = B + \frac{A}{r} \qquad (10D.1)$$

is a solution to Fick's first law when the latter is expressed in spherical coordinates,[176] and c_{vac} is the vacancy concentration at any location. By assuming that the following boundary conditions (Fig. 10.15a) apply — at $r = \rho_p$, $c_{\text{vac}} = C_{\text{curv}}$ given by Eq. (10.9), and at $r = R$, $c_{\text{vac}} = C_0$, — it can be easily shown (see Prob. 10.7a) that

$$\left(\frac{dc_{\text{vac}}}{dr} \right)_{r=\rho_p} = -\frac{1}{\rho_p} \Delta C_{\text{vac}} \frac{R}{R - \rho_p} \qquad (10D.2)$$

satisfies the boundary and steady-state conditions, and ΔC_{vac} is given by Eq. (10.10).

Consequently the total flux of vacancies moving radially away from the pore is

$$J \cdot S = -4 \pi \rho_p^2 D_v \left(\frac{dc_{\text{vac}}}{dr} \right)_{r=\rho_p} = 4 \pi \rho_p^2 D_v \Delta C_{\text{vac}} \frac{R}{R - \rho_p} \qquad (10D.3)$$

where S is the surface area of the pore. The total volume eliminated per unit time is thus

---

[176] Fick's second law in spherical coordinates is:

$$\frac{\partial c}{\partial t} = \frac{D}{r^2} \frac{\partial}{\partial r} \left(r^2 \frac{\partial c}{\partial r} \right)$$

which at steady state becomes

$$\frac{d}{dr} \left(r^2 \frac{dc}{dr} \right) = 0$$

$$\frac{dV}{dt} = -J_a S\Omega_{MX} = -D_a \frac{8\pi\gamma_{sv}\Omega_{MX}}{kT} \frac{R}{R-\rho_p} \tag{10D.4}$$

where the product $D_v\Omega_aC_0$ was replaced by the diffusivity of the rate-limiting ion D_a in the bulk. For a spherical pore, $dV = 4\pi\rho_p^2 d\rho_p$; when this is combined with Eq. (10D.4), integrating the latter by neglecting ρ_p with respect to R, one obtains the final solution, namely

$$\rho_p^3 - \rho_{p,0}^3 = -\frac{6D_a\gamma_{sv}\Omega_{MX}}{kT}t \tag{10D.5}$$

PROBLEMS

10.1. (a) Explain, in your own words, why a necessary condition for sintering to occur is that $\gamma_{gb} < 2\gamma_{sv}$. Furthermore, show why this condition implies that $\phi < 180°$.

(b) The dihedral angles ϕ for three oxides were measured to be 150°, 120°, and 60°. If the three oxides have comparable surface energies, which of the three would you expect to densify most readily? Explain.

(c) At 1850°C, the surface energy of the interface between alumina and its vapor is approximately 0.9 J/m^2. The average dihedral angle for the grain boundaries intersecting the free surface was measured as 115°. In an attempt to toughen the alumina, it is dispersed with ZrO$_2$ particles that are left at the grain boundaries. Prolonged heating at elevated temperatures gives the particles their equilibrium shape. If the average dihedral angle between the particles and at the grain boundaries was measured to be 150°, estimate the interface energy of the alumina/zirconia interface. What conclusions, if any, could be reached concerning the interfacial energy if the particles had remained spherical?

Answer: 1.87 J/m^2

10.2. First 8.5 g of ZnO was pressed into a cylindrical pellet (diameter of 2 cm, height of 1 cm). Then the pellet was placed in a dilatometer and rapidly heated to temperature T_2. The isothermal axial shrinkage was monitored as a function of time, and the results are plotted in Fig. 10.4. Calculate the relative theoretical density of the pellet at the end of the run, i.e., after 80 min at T_2. State all assumptions. *Hint:* The radial shrinkage cannot be ignored.

Answer: 0.94

10.3. (*a*) Estimate the value of P/P_{flat} for a sphere of radius 1 nm at room temperature if the surface tension is 1.6 J/m^2 and the atomic volume is 20×10^{-30} m^3.

Answer: $P/P_{\text{flat}} = 4.88 \times 10^6$

(*b*) Calculate the relative change in average partial pressure at 1300 K as the average particle size increases from 0.5 to 10 μm.

Answer: $P_{0.5\ \mu m}/P_{10\ \mu m} = 1.00053$

(*c*) Would your answer in part (*b*) have changed much if the final diameter of the particle were that of a 10 cm^3 sphere?

Answer: $P_{0.5\ \mu m}/P_{10\ cm^3} = 1.0011$

(*d*) Calculate the vapor pressure of liquid silica over a flat surface at 2000 K, and compare it to the equilibrium vapor pressure inside a 0.5-μm-diameter bubble of silica vapor suspended in liquid silica at the same temperature.

10.4. Derive an expression for the pressure differential between points s and n in Fig. 10.9.

10.5. A 1 cm^2 surface is covered by a bimodal distribution of hemispherical clusters, one-half of which are 10 nm and the other half are 30 nm in diameter.

(*a*) Describe, in your own words, what you believe will happen as the system is held at an elevated temperature for a given time.

(*b*) Assuming that the large clusters consume all the small clusters, at equilibrium, what will be the reduction in energy of the system, given that the surface energy of the clusters is 2.89 J/m^2 and the initial total number of clusters is 4×10^{10}? Ignore the changes occurring to the substrate.

Answer: 9×10^{-5} J

10.6. (*a*) Using a similar approach to the one used to arrive at Eq. (10.20), derive Eq. (10.21).

(*b*) In the intermediate-phase sintering model, the grains were assumed to be tetrakaidecahedra (Fig. 10.14*a*). Repeat the analysis, using cubic particles. Does the final dependence of shrinkage on the dimensions of the cube that you obtain differ from Eq. (10.20) or (10.21)?

10.7. Derive Eq. (10.25). Plot this equation for various values of R_{gb} and ρ_p, and comment on the importance of the log term.

10.8. (*a*) Develop an expression relating the equilibrium coordination number of a pore n to the dihedral angle ϕ of the system.

Answer: $n = 360/(180 - \phi)$

(*b*) For a given packing of particles, will increasing the dihedral angle aid or retard pore elimination? Explain.

(c) Which pore(s) is thermodynamically stable in Fig. 10.16? Explain.

10.9. Given is an oxygen-deficient MO oxide that is difficult to densify, for which it is known that $D_{cat} < D_{an}$. Detail a strategy for enhancing the densification kinetics of this oxide.

10.10. The sintering of a compound MO is governed by diffusion of oxygen ions. If this compound is cation-deficient, propose a method by which the sintering rate may be enhanced.

10.11. (a) What is the range of contact angles for the following conditions?

    1. $\gamma_{lv} = \gamma_{sv} < \gamma_{ls}$
    2. $\gamma_{lv} > \gamma_{sv} > \gamma_{sl}$
    3. $\gamma_{lv} > \gamma_{sv} = \gamma_{ls}$

(b) Redraw Fig. 10.3b for the following cases: $\gamma_{gb} = \gamma_{sl}$, $\gamma_{gb} = 2\gamma_{sl}$, and $\gamma_{gb} = 0.1\gamma_{sl}$.

Make sure that you pay special attention to how far the liquid penetrates into the grain boundary as well as the dihedral angle. Which of the three cases would result in faster densification? Explain.

10.12. (a) Referring to Fig. 10.23b, qualitatively describe, using a series of sketches, what would happen to the shape of the cylinder as it melted such that the liquid could not spread on the solid surface (that is, X is fixed in Fig. 10.23b), but the plates were free to move vertically.

(b) Assuming that an equilibrium shape is reached in part (a) and that the radius of curvature of the resulting shape is $X/1.5$, calculate the wetting angle.

Answer: 132°

(c) Refer once again to Fig. 10.23b. If one assumes that upon melting the resulting shape is a cube of volume V, derive an expression for the resulting force. Is it attractive or repulsive? Explain.

Answer: $4\gamma_{lv}V^{1/3}$

(d) Repeat part (c), assuming the liquid between the plates takes the shape of a tetrakaidecahedron (Fig. 10.14a). Does the answer depend on which face of the tetrakaidecahedron is wetting? Explain.

10.13. (a) Show that the total surface energy of the "spherical cap" in Fig. 10.27a is given by

$$E_{tot} = \gamma_{sv}A + \gamma_{lv}\left[\frac{2V}{h} + \frac{2\pi h^2}{3}\right] + (\gamma_{sl} - \gamma_{sv})\left[\frac{6V - \pi h^3}{3h}\right]$$

where V is the volume of the droplet, given by $V = (\pi/6)(h^3 + 3h^2a)$, and A is the total area of the slab. *Hint*: The surface area of the spherical cap is $S = 2\pi Rh$, and $R = (a^2 + h^2)/2h$.

(b) Show that the surface energy is a minimum when

$$h^3 = \frac{3V}{\pi} \frac{\gamma_{sl} + \gamma_{lv} - \gamma_{sv}}{\gamma_{sv} + 2\gamma_{lv} - \gamma_{sl}}$$

(c) Show that this expression is consistent with Eq. (10.37); that is, show that for complete wetting, $\theta = 0°$, $h = 0$, and for $\theta = 90°$, $h = R$.

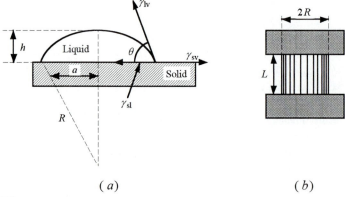

(a) (b)

Figure 10.27
(a) Schematic of sessile drop assuming it is a spherical cap. (b) Liquid cylinder between two plates.

10.14. (a) Long cylinders of any material and long cylindrical pores are inherently unstable and will tend to break up into one or more spheres. Explain.

(b) For the liquid cylinder shown in Fig. 10.27b, show that the length beyond which it will become mechanically unstable is given by $L = 2\pi R$. Is this the same value at which the cylinder becomes thermodynamically unstable? Explain. You may ignore solid-liquid interactions and gravity.

10.15. (a) If a long wire of radius r and length l is heated, it will tend to shorten. Explain.

(b) To keep it from shrinking, a force F must be applied to the wire. Derive an expression for F in terms of the surface tension σ of the wire and its dimensions.

(c) Write an expression for the change in energy associated with an incremental increase in length of the wire in terms of γ_{sv} and the dimensions of the wire. Relate these results to that obtained in part (b) and show that $\sigma = \gamma_{sv}$.

(d) Explain how this technique can be used to measure γ_{sv} of solids.

10.16. One criterion for the densification by sintering in the presence of a liquid is wetting of the solid by the liquid.

(a) Discuss the wetting phenomena which are generally relevant to the liquid-phase sintering of any system.

(b) If the surface energy of Al_2O_3 (s) is 0.9 J/m^2 and the surface tension of Cr (lq) is 2.3 J/m^2 at its melting point of 1875C, what is the value of the interfacial energy between the Cr liquid and alumina if the liquefied metal completely wets the oxide during the sintering process?

(c) Calculations for the interfacial energy between the Cr liquid and alumina yield a value of 0.323 J/m^2. Assuming this value is accurate, discuss its implications to the liquid-phase sintering of this cermet system.

ADDITIONAL READING

1. R. M. German, *Liquid Phase Sintering,* Plenum, New York, 1985.
2. V. N. Eremenko, Y. V. Naidich, and I. A. Lavrinenko, *Liquid Phase Sintering,* Consultants Bureau, New York, 1970.
3. W. D. Kingery, "Densification and Sintering in the Presence of Liquid Phase, I. Theory.," *J. of Appl. Phys.,* **30**:301–306 (1959).
4. J. Philibert, *Atom Movements, Diffusion and Mass Transport in Solids*, in English, S. J. Rothman, trans., Les Editions de Physique, Courtabeouf, France, 1991.
5. J. W. Martin, B. Cantor, and R. D. Doherty, *Stability of Microstructures in Metallic Systems*, Cambridge University Press, 2d ed., Cambridge, England, 1996.
6. I. J. McColm and N. J. Clark, *High Performance Ceramics*, Blackie, Glasgow, Scotland, 1988.
7. R. L. Coble, "Diffusion Models for Hot Pressing with Surface Energy and Pressure Effects as Driving Forces," *J. Appl. Phys.,* **41**:4798 (1970).
8. D. L. Johnson, "A General Method for the Intermediate Stage of Sintering," *J. Amer. Cer. Soc.,* **53**:574–577 (1970).
9. W. D. Kingery and B. Francois, "The Sintering of Crystalline Oxides. I. Interactions between Grain Boundaries and Pores," in G. C. Kuczynski, N. A. Hooten, and C. F. Gibbon, eds., *Sintering and Related Phenomena*, Gordon and Breach, New York, 1967, pp. 471–498.

10. S. Somiya and Y. Moriyoshi, eds., *Sintering Key Papers*, Elsevier, New York, 1990.[177]

11. J. Reed, *Principles of Ceramic Processing*, 2d ed., Wiley, New York, 1995.

---

[177] This compilation of the most important papers in the field of sintering is noteworthy. In addition to reproducing many of the classic papers in the field, there are several more recent papers that critically assess the validity of the various models and summarize the current sintering paradigms.

CHAPTER 11

Mechanical Properties: Fast Fracture

The careful text-books measure
(Let all who build beware!)
The load, the shock, the pressure
Material can bear.
So when the buckled girder
Lets down the grinding span.
The blame of loss, or murder,
Is laid upon the man.
Not on the stuff — the Man!

R. Kipling, *"Hymn of the Breaking Strain"*

11.1
INTRODUCTION

Sometime before the dawn of civilization, some hominid discovered that the edge of a broken stone was quite useful for killing prey and warding off predators. This seminal juncture in human history has been recognized by archeologists who refer to it as the stone age. C. Smith[178] goes further by stating, "Man probably owes his very existence to a basic property of inorganic matter, the brittleness of certain ionic compounds." In this context, Kipling's hymn and J. E. Gordon's statement[179] that "The worst sin in an engineering material is not lack of strength or lack of stiffness, desirable as these properties are, but lack of toughness, that is to say, lack of resistance to the propagation of cracks" stand in sharp contrast. But it is this contrast that in a very real sense summarizes the short history of technical ceramics; what was good enough for millennia now falls short. After all, the

---

178 C. S. Smith, *Science*, **148**:908 (1965).
179 J. E. Gordon, *The New Science of Engineering Materials*, 2d ed., Princeton University Press, Princeton, New Jersey, 1976.

consequences of a broken mirror are not as dire as those of, say, an exploding turbine blade. It could be argued, with some justification, that were it not for their brittleness, the use of ceramics for structural applications, especially at elevated temperatures, would be much more widespread since they possess other very attractive properties such as hardness, stiffness, and oxidation and creep resistance.

As should be familiar to most, the application of a stress to any solid will initially result in a reversible elastic strain that is followed by either fracture without much plastic deformation (Fig. 11.1a) or fracture that is preceded by plastic deformation (Fig. 11.1b). Ceramics and glasses fall in the former category and are thus considered brittle solids, whereas most metals and polymers above their glass transition temperature fall into the latter category.

The theoretical stress level at which a material is expected to fracture by bond rupture was discussed in Chap. 4 and estimated to be on the order of $Y/10$, where Y is Young's modulus. Given that Y for ceramics (see Table 11.1) ranges between 100 and 500 GPa, the expected "ideal" fracture stress is quite high — on the order of 10 to 50 GPa. For reasons that will become apparent shortly, the presence of flaws, such as shown in Fig. 11.2, in brittle solids will greatly reduce the stress at which they fail. Conversely, it is well established that extraordinary strengths can be achieved if they are flaw-free. For example, a defect-free silica glass rod can be elastically deformed to stresses that exceed 5 GPa! Thus it may be concluded, correctly one might add, that certain flaws within a material serve to promote fracture at stress levels that are well below the ideal fracture stress.

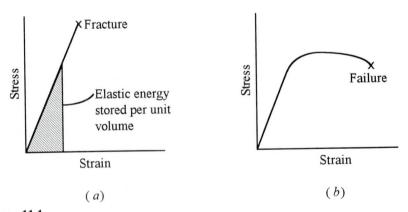

Figure 11.1
Typical stress-strain curves for (a) brittle solids and (b) ductile materials.

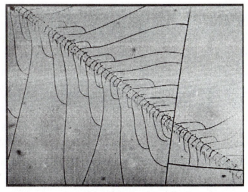

Figure 11.2
Surface cracks caused by the accidental contact of a glass surface with dust particles or another solid surface can result in significant reductions in strength.

The stochastic nature of the flaws present in brittle solids together with their flaw sensitivity has important design ramifications as well. Strength variations of ±25 percent from the mean are not uncommon and are quite large when compared to, say, the spread of flow stresses in metals, which are typically within just a few percent. Needless to say, such variability, together with the sudden nature of brittle failure, poses a veritable challenge for design engineers considering using ceramics for structural and other critical applications.

Flaws, their shape, and their propagation are the central theme of this chapter. The various aspects of brittle failure are discussed from several viewpoints. The concepts of fracture toughness and flaw sensitivity are discussed first. The factors influencing the strengths of ceramics are dealt with in Sec. 11.3.[180] Toughening mechanisms are dealt with in Sec. 11.4. Section 11.5 introduces the statistics of brittle failure and a methodology for design.

11.2
FRACTURE TOUGHNESS

11.2.1   Flaw Sensitivity

To illustrate what is meant by flaw or notch sensitivity, consider the schematic of what occurs at the base of an atomically sharp crack upon the application of a load

---

[180] The time-dependent mechanical properties such as creep and subcritical crack growth are dealt with separately in the next chapter.

F_{app}. For a crack-free sample (Fig. 11.3a), each chain of atoms will carry its share of the load F/n, where n is the number of chains, i.e., the applied stress σ_{app} is said to be uniformly distributed. The introduction of a surface crack results in a stress redistribution such that the load that was supported by the severed bonds is now being carried by only a few bonds at the crack tip (Fig. 11.3b). Said otherwise, the presence of a flaw will *locally amplify the applied stress at the crack tip* σ_{tip}. As σ_{app} is increased, σ_{tip} increases accordingly and moves up the stress versus interatomic distance curve, as shown in Fig. 11.3c. As long as $\sigma_{tip} < \sigma_{max}$, the situation is stable and the flaw will not propagate. However, if at any time σ_{tip} exceeds σ_{max}, the situation becomes catastrophically unstable (not unlike the bursting of a dam). Based on this simple picture, the reason why brittle fracture occurs rapidly and without warning, with cracks propagating at velocities approaching the speed of sound, is now obvious. Furthermore, it is also obvious why ceramics are much stronger in compression than in tension.

To be a little more quantitative in predicting the applied stress that would lead to failure, σ_{tip} would have to be calculated and equated to σ_{max} or $Y/10$. Calculating σ_{tip} is rather complicated (only the final result is given here) and is a function of the type of loading, sample, crack geometry, etc.[181] However, for a thin sheet, it can be shown that σ_{tip} is related to the applied stress by

$$\sigma_{tip} = 2\sigma_{app}\sqrt{\frac{c}{\rho}} \qquad (11.1)$$

where c and ρ are, respectively, the crack length and its radius of curvature[182] (Fig. 11.4).

Since, as noted above, fracture can be reasonably assumed to occur when $\sigma_{tip} = \sigma_{max} \approx Y/10$, it follows that

$$\sigma_f \approx \frac{Y}{20}\sqrt{\frac{\rho}{c}} \qquad (11.2)$$

where σ_f is the stress at fracture. This equation predicts that (1) σ_f is inversely proportional to the square root of the flaw size and (2) sharp cracks, i.e., those with a small ρ, are more deleterious than blunt cracks. Both predictions are in good agreement with numerous experimental observations.

---

181 C. E. Inglis, *Trans. Inst. Naval Archit.*, **55**:219 (1913).

182 This equation strictly applies to a surface crack of length c, or an interior crack of length $2c$ in a thin sheet. Since the surface of the material cannot support a stress normal to it, this condition corresponds to the plane stress condition (the stress is two-dimensional). In thick components, the situation is more complicated, but for brittle materials the two expressions vary slightly.

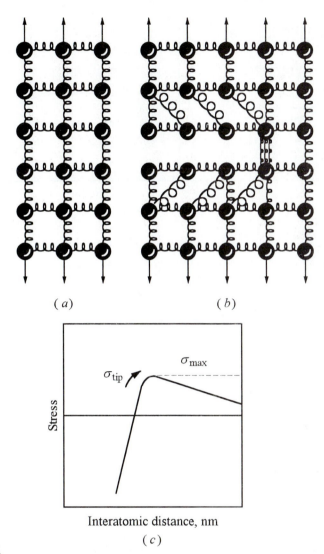

Figure 11.3

(*a*) Depiction of a uniform stress. (*b*) Stress redistribution as a result of the presence of a crack. (*c*) For a given applied load, as the crack grows and the bonds are sequentially ruptured, σ_{tip} moves up the stress versus displacement curve toward σ_{max}. When $\sigma_{tip} \approx \sigma_{max}$, catastrophic failure occurs. Note that this figure is identical to Fig. 4.6, except that here the *y* axis represents the stress on the bond rather than the applied force.

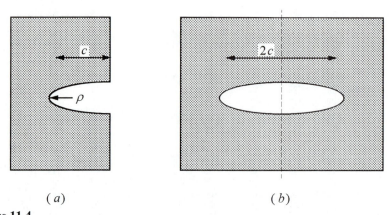

(a)   (b)

Figure 11.4
(a) Surface crack of length c and radius of curvature ρ. (b) Interior crack of length $2c$.
Note that from a fracture point of view, they are equivalent.

11.2.2 Energy Criteria for Fracture — The Griffith Criterion

An alternate and ultimately more versatile approach to the problem of fracture was
developed in the early 1920s by Griffith.[183] His basic idea was to balance the
energy consumed in forming new surface as a crack propagates against the elastic
energy released. The critical condition for fracture, then, occurs when the rate at
which energy is released is greater than the rate at which it is consumed. The
approach taken here is a simplified version of the original approach, and it entails
deriving an expression for the energy changes resulting from the introduction of a
flaw of length c in a material subjected to a uniform stress σ_{app}.

Strain energy

When a solid is uniformly elastically stressed, all bonds in the material elongate
and the work done by the applied stress is converted to elastic energy that is stored
in the stretched bonds. The magnitude of the elastic energy stored per unit volume
is given by the area under the stress-strain curve[184] (Fig. 11.1a), or

---

183 A. A. Griffith, *Phil. Trans. R. Acad.*, **A221**:163 (1920).
184 When a bond is stretched, energy is stored in that bond in the form of elastic energy. This
energy can be converted to other forms of energy as any schoolboy with a slingshot can attest;
the elastic energy stored in the rubber band is converted into kinetic energy of the projectile. If
by chance a pane of glass comes in the way of the projectile, that kinetic energy will in turn be
converted to other forms of energy such as thermal, acoustic, and surface energy. In other
words, the glass will shatter and some of the kinetic energy will have created new surfaces.

$$U_{elas} = \frac{1}{2}\varepsilon\sigma_{app} = \frac{1}{2}\frac{\sigma_{app}^2}{Y} \tag{11.3}$$

The total energy of the parallelopiped of volume V_0 subjected to a uniform stress σ_{app} (Fig. 11.5a) increases to

$$U = U_0 + V_0 U_{elas} = U_0 + \frac{V_0\sigma_{app}^2}{2Y} \tag{11.4}$$

where U_0 its free energy in the absence of stress.

In the presence of a surface crack of length c (Fig. 11.5b), it is fair to assume that some volume around that crack will relax (i.e., the bonds in that volume will relax and lose their strain energy). Assuming — it is not a bad assumption, as will become clear shortly — that the relaxed volume is given by the shaded area in Fig. 11.5b, it follows that the strain energy of the system in the presence of the crack is given by

$$U_{strain} = U_0 + \frac{V_0\sigma_{app}^2}{2Y} - \frac{\sigma_{app}^2}{2Y}\left[\frac{\pi c^2 t}{2}\right] \tag{11.5}$$

where t is the thickness of the plate. The third term represents the *strain energy released* in the relaxed volume.

Surface energy

To form a crack of length c, an energy expenditure of

$$U_{surf} = 2\gamma ct \tag{11.6}$$

is required, where γ is the intrinsic surface energy of the material. The factor 2 arises because two (bottom and top) new surfaces are created by the fracture event.

The total energy change of the system upon introduction of the crack is simply the sum of Eqs. (11.5) and (11.6), or

$$U_{tot} = U_0 + \frac{V_0\sigma_{app}^2}{2Y} - \frac{\sigma_{app}^2}{2Y}\left[\frac{\pi c^2 t}{2}\right] + 2\gamma ct \tag{11.7}$$

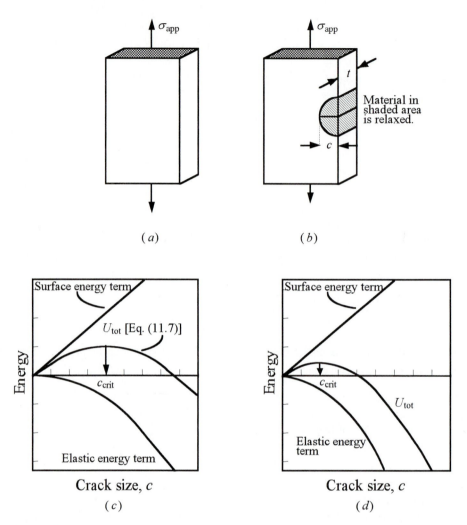

Figure 11.5

(a) Uniformly stressed solid. (b) Relaxed volume in vicinity of crack of length c. (c) Plot of Eq. (11.7) as a function of c. The top curve represents the surface energy term, and the lower curve represents the strain energy release term. Curve labeled U_{tot} is sum of the two curves. The critical crack length c_{crit} at which fast fracture will occur corresponds to the maximum. (d) Plot of Eq. (11.7) on the same scale as in part (c) but for $\sqrt{2}$ times the applied stress applied in (c). Increasing the applied stress by that factor reduces c_{crit} by a factor of 2.

Since the surface energy term scales with c and the strain energy term scales with c^2, U_{tot} has to go through a *maximum* at a certain critical crack size c_{crit} (Fig. 11.5c). This is an important result since it implies that extending a crack that is smaller than c_{crit} *consumes rather than liberates energy and is thus stable.* In contrast, *flaws that are longer than c_{crit} are unstable since extending them releases more energy than is consumed.* Note that increasing the applied stress (Fig. 11.5d) will result in failure at smaller critical flaw sizes. For instance, a solid for which the largest[185] flaw size lies somewhere between those shown in Fig. 11.5c and d will *not* fail at the stress shown in Fig. 11.5c, but will fail if that stress is increased (Fig. 11.5d).

The location of the maximum is determined by differentiating Eq. (11.7) and equating it to zero. Carrying out the differentiation, replacing σ_{app} by σ_f, and rearranging terms, one can show that the condition for failure is

$$\sigma_f \sqrt{\pi c_{crit}} = 2\sqrt{\gamma Y} \tag{11.8}$$

A more exact calculation yields

$$\boxed{\sigma_f \sqrt{\pi c_{crit}} \geq \sqrt{2\gamma Y}} \tag{11.9}$$

and is the expression used in subsequent discussions.[186] This equation predicts that a critical combination of *applied stress and flaw size is required to cause failure.* The combination $\sigma\sqrt{\pi c}$ occurs so often in discussing fast fracture that it is abbreviated to a single symbol K_I with units $\text{MPa} \cdot \text{m}^{1/2}$, and is referred to as the **stress intensity factor.** Similarly, the combination of terms on the right-hand side of Eq. (11.9), sometimes referred to as the **critical stress intensity factor,** or more commonly the **fracture toughness,** is abbreviated by the symbol K_{Ic}. Given these abbreviations, the condition for fracture can be succinctly rewritten as

$$\boxed{K_I \geq K_{Ic}} \tag{11.10}$$

Equations (11.9) and (11.10) were derived with the implicit assumption that the only factor keeping the crack from extending was the creation of new surface. This is only true, however, for extremely brittle systems such as inorganic glasses. In general, however, when other energy dissipating mechanisms, such a plastic deformation at the crack tip, are operative, K_{Ic} is defined as

---

[185] The largest flaw is typically the one that will cause failure, since it becomes critical before other smaller flaws (see Fig. 11.8a).

[186] Comparing Eqs. (11.8) and (11.9) shows that the estimate of the volume over which the stress is relieved in Fig. 11.5b was off by a factor of $\sqrt{2}$, which is not too bad.

$$K_{Ic} = \sqrt{YG_c}$$ (11.11)

where G_c is the **toughness** of the material in joules per square meter. For purely brittle solids,[187] the toughness approaches the limit $G_c = 2\gamma$. Table 11.1 lists Young's modulus, Poisson's ratio, and K_{Ic} values of a number of ceramic materials. It should be pointed out that since (see below) K_{Ic} is a material property that is also microstructure-dependent, the values listed in Table 11.1 are to be used with care.

Finally it is worth noting that the Griffith approach, Eq. (11.10), can be reconciled with Eq. (11.2) by assuming that ρ is on the order of $10r_0$, where r_0 is the equilibrium interionic distance (see Prob. 11.3). In other words, the Griffith approach implicitly assumes that the flaws are atomically sharp, a fact that must be borne in mind when one is experimentally determining K_{Ic} for a material.

To summarize: Fast fracture will occur in a material when the product of the applied stress and the square root of the flaw dimension are comparable to that material's fracture toughness.

WORKED EXAMPLE 11.1. (*a*) A sharp edge notch 120 μm deep is introduced in a thin magnesia plate. The plate is then loaded in tension normal to the plane of the notch. If the applied stress is 150 MPa, will the plate survive? (*b*) Would your answer change if the notch were the same length by was as internal notch (Fig. 11.4*b*) instead of an edge notch?

Answer

(*a*) To determine whether the plate will survive the applied stress, the stress intensity at the crack tip needs to be calculated and compared to the fracture toughness of MgO, which according to Table 11.1 is 2.5 MPa $\cdot$ m$^{1/2}$.

K_I in this case is given

$$K_I = \sigma\sqrt{\pi c} = 150\sqrt{3.14 \times 120 \times 10^{-6}} = 2.91 \text{ MPa} \cdot \text{m}^{1/2}$$

Since this value is greater than K_{Ic} for MgO, it follows that the plate will fail.

(*b*) In this case, because the notch is an internal one, it is not as detrimental as a surface or edge notch and

$$K_I = \sigma\sqrt{\pi\frac{c}{2}} = 150\sqrt{3.14 \times 60 \times 10^{-6}} = 2.06 \text{ MPa} \cdot \text{m}^{1/2}$$

---

[187] Under these conditions, one may calculate the surface energy of a solid from a measurement of K_{Ic} (see the section on measuring surface energies in Chap. 4).

TABLE 11.1

Data for Young's modulus Y, Poisson's ratio, and K_{Ic} values of selected ceramics at ambient temperatures[†]

	Y, (GPa)	Poisson's ratio	K_{Ic}, MPa·m$^{1/2}$	Vickers hardness, GPa
	Oxides			
Al_2O_3	390	0.200–0.250	2.0–6.0	19.0–26.0
Al_2O_3 (single crystal, $10\bar{1}2$)	340		2.2	
Al_2O_3 (single crystal, 0001)	460		> 6.0	
$BaTiO_3$	125			
BeO	386	0.340		0.8–1.2
HfO_2 (monoclinic)	240			
MgO	250–300	0.180	2.5	6.0–10.0
$MgTi_2O_5$	250			
$MgAl_2O_4$	248–270		1.9–2.4	14.0–18.0
Mullite [fully dense]	230	0.240	2.0–4.0	15.0
Nb_2O_5	180			
$PbTiO_3$	81			
SiO_2 (quartz)	94	0.170		12.0 (011)
SnO_2	263	0.290		
TiO_2	282–300			10.0±1.0
ThO_2	250		1.6	10.0
Y_2O_3	175		1.5	7.0–9.0
$Y_3Al_5O_{12}$				18.0±1.0
ZnO	124			2.3±1.0
$ZrSiO_4$ (zircon)	195	0.250		≈15.0
ZrO_2 (cubic)	220	0.310	3.0–3.6	12.0–15.0
ZrO_2 (partially stabilized)	190	0.300	3.0–15.0	13.0
	Carbides, Borides, and Nitrides and Silicides			
AlN	308	0.250		12.0
B_4C	417–450	0.170		30.0–38.0
BN	675			
Diamond	1000			
$MoSi_2$	400			
Si	107	0.270		10.0
SiC [hot pressed]	440±10	0.193	3.0–6.0	26.0–36.0
SiC (single crystal)	460		3.7	
Si_3N_4 Hot Pressed (dense)	300–330	0.220	3.0–10.0	17.0–30.0
TiB_2	500–570	0.110		18.0–34.0
TiC	456	0.180	3.0–5.0	16.0–28.0
WC	450–650		6.0–20.0	
ZrB_2	440	0.144		22.0

TABLE 11.1 (continued)

Data for Young's modulus Y, Poisson's ratio, and K_{Ic} values of selected ceramics at ambient temperatures†

	Y, (GPa)	Poisson's ratio	K_{Ic}, MPa·m$^{1/2}$	Vickers hardness, GPa
Halides and Sulphides				
CaF$_2$	110		0.80	1.800
KCl (forged single crystal)	24		≈0.35	0.120
MgF$_2$	138		1.00	6.000
SrF$_2$	88		1.00	1.400
Glasses and Glass Ceramics				
Aluminosilicate (Corning 1720)	89	0.24	0.96	6.6
Borosilicate (Corning 7740)	63	0.20	0.75	6.5
Borosilicate (Corning 7052)	57	0.22		
LAS (glass-ceramic)	100	0.30	2.00	
Silica (fused)	72	0.16	0.80	6.0–9.0
Silica (96%)	66		0.70	
Soda Lime Silica Glass	69	0.25	0.82	5.5

† The fracture toughness is a function of microstructure. The values given here are mostly for comparison's sake.

Since this value is < 2.5 MPa·m$^{1/2}$ it follows that the plate would survive the applied load.

Before one explores the various strategies to increase the fracture toughness of ceramics, it is important to appreciate how K_{Ic} is measured.

EXPERIMENTAL DETAILS: MEASURING K_{Ic}

There are several experimental techniques by which K_{Ic} can be measured. The two most common methods entail measuring the fracture stress for a given geometry and known initial crack length and measuring the lengths of the cracks emanating from hardness indentations.

Fracture Stress

Equation (11.9) can be recast in its most general form

$$\Psi \sigma_{frac} \sqrt{\pi c} \geq K_{Ic} \tag{11.12}$$

where Ψ is a dimensionless constant on the order of unity that depends on the sample shape, the crack geometry, and its relative size to the sample dimensions. This relationship suggests that to measure K_{Ic}, one would start with an *atomically sharp crack* [an implicit assumption made in deriving Eq. (11.10) — see Prob. 11.3] of length c and measure the stress at which fracture occurs. Given the sample and crack geometries, Ψ can be looked up in various fracture mechanics handbooks, and then K_{Ic} is calculated from Eq. (11.12). Thus, in principle, it would appear that measuring K_{Ic} is fairly straightforward; experimentally, however, the difficulty lies in introducing an atomically sharp crack.

Two of the more common test configurations are shown in Fig. 11.6. A third geometry not shown here is the **double torsion test**, which in addition to measuring K_{Ic} can be used to measure crack velocity versus K curves. This test is described in greater detail in the next chapter.

Single-edge notched beam (SENB) test

In this test, the specimen for which is depicted schematically in Fig. 11.6a, a notch of initial depth c is introduced, usually by using a diamond wheel, on the tensile side of a flexure specimen. The sample is loaded until failure, and then c is taken as the initial crack length. Fracture toughness K_{Ic} is calculated from

$$K_{Ic} = \frac{3\sqrt{c}\,(S_1 - S_2)\xi\,F_{\text{fail}}}{2\,BW^2}$$

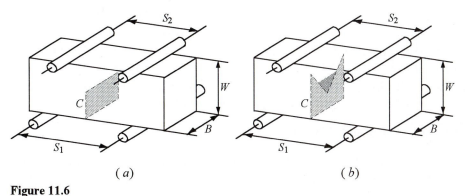

Figure 11.6

(a) Schematic of single-edge notched beam specimen; (b) Chevron notch specimen.

where F_{fail} is the load at which the specimen failed and ξ is a calibration factor. The other symbols are defined in Fig. 11.6a. The advantage of this test lies in its simplicity — its major drawback, however, is that the condition that the crack be atomically sharp is, more often than not, unfulfilled, which causes one to overestimate K_{Ic}.

Chevron notch (CN) specimen[188]

In this configuration, shown schematically in Fig. 11.6b, the chevron notch specimen looks quite similar to the SENB except for the vital difference that the shape of the initial crack is not flat but V- or chevron-shaped, as shown by the shaded area. The constant widening of the crack front as it advances causes crack growth to be stable *prior* to failure. Since an increased load is required to continue crack extension, it is possible to create an atomically sharp crack in the specimen *before* final failure, which eliminates the need to precrack the specimen. The fracture toughness[189] is then related to the maximum load at fracture F_{fail} and the minimum of a compliance function ξ^*.

$$K_{Ic} = \frac{(S_1 - S_2)\xi^* F_{fail}}{BW^{3/2}}$$

General remarks

It is worth noting that unless care is taken in carrying out the fracture toughness measurements, different tests will result in different values of K_{Ic}. There are three reasons for this: (1) The sample dimensions were too small, compared to the process zone (which is the zone ahead of the crack tip that is damaged). (2) The internal stresses generated during machining of the specimens were not sufficiently relaxed before the measurements were made. (3) The crack tip was not atomically sharp. As noted above, if the fracture initiating the flaw is not atomically sharp, apparently higher K_{Ic} values will be obtained. Thus although simple in principle, the measurement of K_{Ic} is fraught with pitfalls, and care must be taken if reliable and accurate data are to be obtained.

Hardness Indentation Method

Due to its simplicity, its nondestructive nature, and the fact that minimal machining is required to prepare the sample, the use of the Vickers hardness indentations to measure K_{Ic} has become quite popular. In this method, a diamond indenter is applied to the surface of the specimen to be tested. Upon removal, the sizes of the

---

188 A *chevron* is a figure or a pattern having the shape of a V.

189 For more information, see J. Sung and P. Nicholson, *J. Amer. Cer. Soc.*, **72 (6)**:1033–1036 (1989).

cracks that emanate (sometimes) from the edges of the indent are measured, and the Vickers hardness H in GPa of the material is calculated. A number of empirical and semiempirical relationships have been proposed relating K_{Ic}, c, Y, and H, and in general the expressions take the form

$$K_{Ic} = \Phi \sqrt{a} H \left(\frac{Y}{H} \right)^{0.4} f\left(\frac{c}{a} \right) \qquad (11.13)$$

where Φ is a geometric constraint factor and c and a are defined in Fig. 11.7. The exact form of the expression used depends on the type of crack that emanates from the indent.[190] A cross-sectional view and a top view of the two most common types of cracks of interest are shown in Fig. 11.7. At low loads, Palmqvist cracks are favored, while at high loads fully developed median cracks result. A simple way to differentiate between the two types is to polish the surface layers away; the median crack system will always remain connected to the inverted pyramid of the indent while the Palmqvist will become detached, as shown in Fig. 11.7b.

It should be emphasized the K_{Ic} values measured by using this technique are usually not as precise as those from other more macroscopic tests.

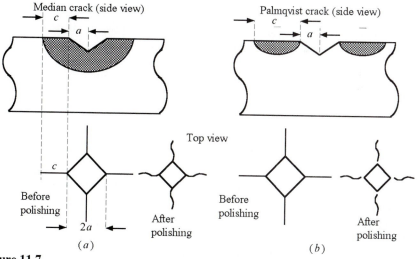

Figure 11.7
Crack systems developed from the Vickers indents. (a) Side and top views of a median crack. (b) Top and side views of a Palmqvist crack.

---

[190] For more information, see G. R. Anstis, P. Chantikul, B. R. Lawn, and D. B. Marshall, *J. Amer. Cer. Soc.*, **64**:533 (1981), and R. Matsumoto, *J. Amer. Cer. Soc.*, **70**(C):366 (1987). See also Problem 11.9.

11.2.3 Compressive and Other Failure Modes

Whereas it is now well established that tensile brittle failure usually propagates unstably when the stress intensity at the crack tip exceeds a critical value, the mechanics of compressive brittle fracture are more complex and not as well understood. Cracks in compression tend to propagate stably and twist out of their original orientation to propagate parallel to the compression axis, as shown in Fig. 11.8b. Fracture in this case is caused not by the unstable propagation of a single crack, as would be the case in tension (Fig. 11.8a), but by the slow extension and linking up of many cracks to form a crushed zone. Hence it is not the size of the largest crack that counts, but the size of the average crack c_{av}. The compressive stress to failure is still given by

$$\sigma_{fail} \approx Z \frac{K_{Ic}}{\sqrt{\pi c_{av}}} \tag{11.14}$$

but now Z is a constant on the order of 15.

Finally, in general there are three modes of failure, known as modes I, II, and III. Mode I (Fig. 11.9a) is the one that we have been dealing with so far. Modes II and III are shown in Fig. 11.9b and c, respectively. The same energy concepts that apply to mode I also apply to modes II and III. Mode I, however, is by far the more pertinent to crack propagation in brittle solids.

11.2.4 Atomistic Aspects of Fracture

Up to this point, the discussion has been mostly couched in macroscopic terms. Flaws were shown to concentrate the applied stress at their tip which ultimately led to failure. No distinction was made between brittle and ductile materials, and yet experience clearly indicates that the different classes of materials behave quite differently — after all, the consequences of scribing a glass plate are quite different from those of a metal one. Thus the question is, What renders brittle solids notch-sensitive, or more directly, why are ceramics brittle?

The answer is related to the crack tip plasticity. In the foregoing discussion, it was assumed that intrinsically brittle fracture was free of crack-tip plasticity, i.e., dislocation generation and motion. Given that dislocations are generated and move under the influence of *shear stresses*, two limiting cases can be considered:

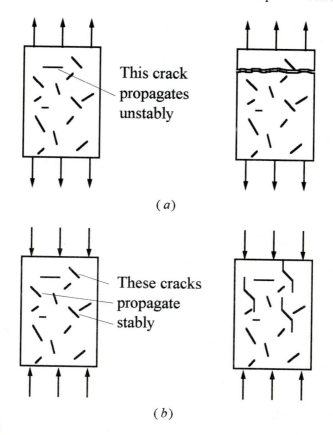

Figure 11.8

(*a*) Fracture in ceramics due to preexisting flaws tested in tension. Failure occurs by the unstable propagation of the worst crack that is also most favorably oriented. (*b*) During compressive loading, many cracks propagate stably, eventually linking up and creating a crush zone.[191]

1. The cohesive tensile stress ($\approx Y/10$) is *smaller* than the cohesive strength in shear, in which case the solid can sustain a sharp crack and the Griffith approach is valid.
2. The cohesive tensile stress is *greater* than the cohesive strength in shear, in which case shear breakdown will occur (i.e., dislocations will move away from the crack tip) and the crack will lose its atomic sharpness. In other words, the emission of dislocations from the crack tip, as shown in Fig. 11.10*a*, will move material away from the crack tip, absorbing energy and causing crack blunting, as shown in Fig. 11.10*b*.

---

[191] Adapted from M. F. Ashby and D. R. Jones, *Engineering Materials*, vol. 2, Pergamon Press, New York, 1986.

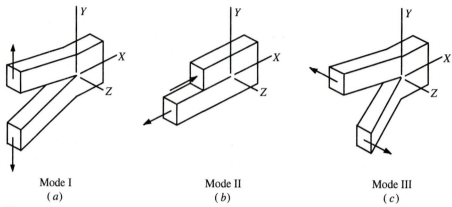

Mode I   Mode II   Mode III
(a)   (b)   (c)

Figure 11.9

The three modes of failure: (a) opening mode, or mode I, characterized by K_{Ic};
(b) sliding mode, or mode II, K_{IIc}; (c) tearing mode, or mode III, K_{IIIc}.

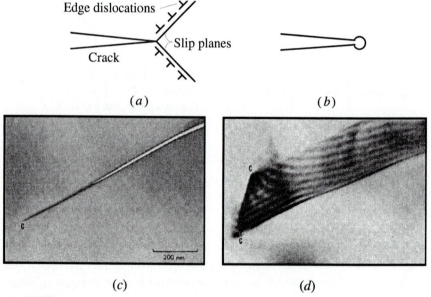

Figure 11.10

(a) Emission of dislocations from crack tip. (b) Blunting of crack tip due to dislocation
motion. (c) Transmission electron micrograph of cracks in Si at 25°C. (d) Another crack
in Si formed at 500°C, where dislocation activity in vicinity of crack tip is evident.[192]

[192] B. R. Lawn, B. J. Hockey, and S. M. Wiederhorn, *J. Mat. Sci.*, **15**:1207 (1980). Reprinted with permission.

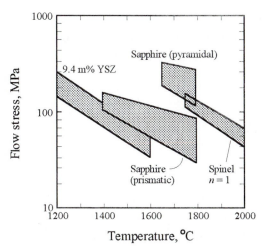

Figure 11.11

Temperature dependence of flow stress for yttria-stabilized zirconia (YSZ), sapphire, and equimolar spinel.[193]

Theoretical calculations have shown that the ratio of theoretical shear strength to tensile strength diminishes as one proceeds from covalent to ionic to metallic bonds. For metals, the intrinsic shear strength is so low that flow at ambient temperatures is almost inevitable. Conversely, for covalent materials such as diamond and SiC, the opposite is true: The exceptionally rigid tetrahedral bonds would rather extend in a mode I type of crack than shear.

Theoretically, the situation for ionic solids is less straightforward, but direct observations of crack tips in transmission electron microscopy tend to support the notion that most covalent and ionic solids are truly brittle at room temperature (see Fig. 11.11c). Note that the roughly order-of-magnitude difference between the fracture toughness of metals (20 to 100 MPa $\cdot$ m$^{1/2}$) and ceramics is directly related to the lack of crack-tip plasticity in the latter — moving dislocations consumes quite a bit of energy.

The situation is quite different at higher temperatures. Since dislocation mobility is thermally activated, increasing the temperature will tend to favor dislocation activity, as shown in Fig. 11.10d, which in turn increases the ductility of the material. Thus the condition for brittleness can be restated as follows: Solids are brittle when the energy barrier for dislocation motion is large relative to the thermal energy kT available to the system. Given the large flow stresses required to move dislocations at elevated temperatures in oxide single crystals (Fig. 11.11), it is

---

193 A. H. Heuer, cited in R. Raj, *J. Amer. Cer. Soc.*, **76**:2147–2174 (1993).

once again not surprising that ceramics are brittle at room temperatures. Finally, note that dislocation activity is not the only mechanism for crack blunting. At temperatures above the glass transition temperature viscous flow is also very effective in blunting cracks.

11.3
STRENGTH OF CERAMICS

Most forming methods that are commonly used in the metal and polymer industries are not applicable for ceramics. Their brittleness precludes deformation methods; and their high melting points, and in some cases (e.g., Si_3N_4, SiC) decomposition prior to melting, preclude casting. Consequently, as discussed in the previous chapter, most polycrystalline ceramics are fabricated by either solid- or liquid-phase sintering, which can lead to flaws. For example, how agglomeration and inhomogeneous packing during powder preparation often led to the development of flaws in the sintered body was discussed in Chap. 10. Inevitably, flaws are always present in ceramics. In this section, the various types of flaws that form during processing and their effect on strength are discussed. The subsequent section deals with the effect of grain size on strength, while Sec. 11.3.3 deals briefly with strengthening ceramics by the introduction of compressive surface layers. Before one proceeds much further, however, it is important to briefly review how the strength of a ceramic is measured.

EXPERIMENTAL DETAILS: MODULUS OF RUPTURE

Tensile testing of ceramics is time-consuming and expensive because of the difficulty in machining test specimens. Instead, the simpler transverse bending or flexure test is used, where the specimen is loaded to failure in either three- or four-point bending. The maximum stress or stress at fracture is commonly referred to as the *modulus of rupture* (MOR). For rectangular cross sections, the MOR in *four-point* bending is given by

$$\sigma_{MOR} = \frac{3(S_1 - S_2)F_{fail}}{2BW^2} \qquad (11.15)$$

where F_{fail} is the load at fracture and all the other symbols are defined in Fig. 11.6a. Note that the MOR specimen is unnotched and fails as a result of preexisting surface or interior flaws.

Once again a word of caution: Although the MOR test appears straightforward, it is also fraught with pitfalls.[194] For example, the edges of the samples have to be carefully beveled before testing since sharp corners can act as stress concentrators and in turn significantly reduce the measured strengths.

11.3.1 Processing and Surface Flaws

The flaws in ceramics can be either internal or surface flaws generated during processing or surface flaws introduced later, during machining or service.

Pores

Pores are usually quite deleterious to the strength of ceramics not only because they reduce the cross-sectional area over which the load is applied, but more importantly because they act as stress concentrators. Typically the strength and porosity have been related by the following empirical relationship:

$$\sigma_p = \sigma_0 e^{-BP} \tag{11.16}$$

where P, σ_p, and σ_0 are, respectively, the volume fraction porosity and the strength of the specimen with and without porosity; B is a constant that depends on the distribution and morphology of the pores. The exponential dependence of strength on porosity is clearly demonstrated in Fig. 11.12 for reaction-bonded Si_3N_4, which is formed by exposing a Si compact to a nitrogen atmosphere at elevated temperatures. The large scatter in the results mostly reflects the variability in the pore sizes, morphology, and distribution.

Usually, the stress intensities associated with the pores themselves are insufficient to cause failure, and as such the role of pores is indirect. Fracture from pores is typically dictated by the presence of other defects in their immediate vicinity. If the pore is much larger than the surrounding grains, atomically sharp cusps around the surface of the former can result. The critical flaw thus becomes comparable to the dimension of the pores. If the pores are spherical, as in glasses, they are less detrimental to the strength. Thus both the largest dimension of the pore and the smallest radius of curvature at the pore surface are what determine the effect on the strength. A typical micrograph of a pore that resulted in failure is shown in Fig. 11.13a.

---

[194] For a recent and comprehensive review of the MOR test, see G. Quinn and R. Morrell, *J. Am. Cer. Soc.*, **74**(9):2037–2066 (1991).

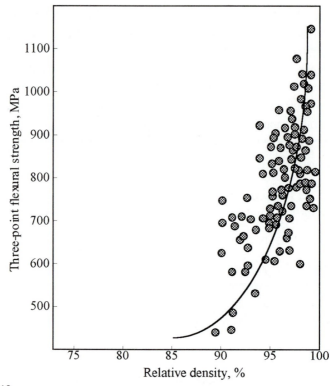

Figure 11.12

Functional dependence of strength on porosity for a reaction-bonded Si_3N_4.[195]

Inclusions

Impurities in the starting powders can react with the matrix and form inclusions that can have different mechanical and thermal properties from the original matrix. Consequently, as a result of the mismatch in the thermal expansion coefficients of the matrix α_m and the inclusions α_i, large residual stresses can develop as the part is cooled from the processing temperature. For example, a spherical inclusion of radius R in an infinite matrix will result in both radial (σ_{rad}) and tangential (σ_{tan}) residual stresses at a radial distance r away from the inclusion/matrix interface given by

$$\sigma_{rad} = -2\sigma_{tan} = \frac{(\alpha_m - \alpha_i)\Delta T}{[(1-2\nu_i)/Y_i + (1+\nu_m)/2Y_m]}\left(\frac{R}{r+R}\right)^3 \qquad (11.17)$$

---

[195] Data taken from O. Kamigaito, in *Fine Ceramics*, S. Saito, ed., Elsevier, New York, 1988.

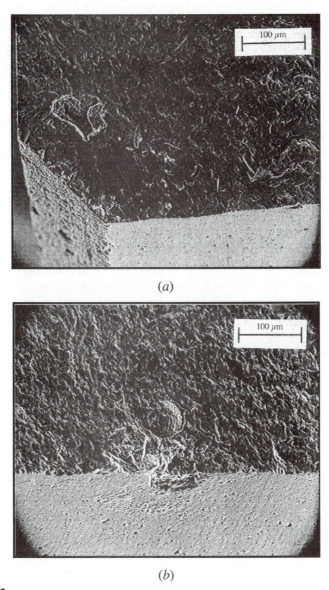

(a)

(b)

Figure 11.13

(a) Large pore associated with a large grain in sintered α-SiC. (b) An agglomerate with associated porosity in a sintered α-SiC.[196]

---

[196] G. Quinn and R. Morrell, *J. Am. Cer. Soc.*, **74**(9):2037–2066 (1991). Reprinted with permission.

where ν is Poisson's ratio; m and i refer to the matrix and inclusion, respectively; and ΔT is the difference between the initial and final temperatures (i.e., it is defined as positive during cooling and negative during heating). On cooling, the initial temperature is the maximum temperature below which the stresses are not relaxed. (See Chap. 13 for more details.)

It follows from Eq. (11.17) that upon cooling, if $\alpha_i < \alpha_m$, large tangential tensile stresses develop that, in turn, could result in the formation of radial matrix cracks. Conversely, if $\alpha_i > \alpha_m$, the inclusion will tend to detach itself from the matrix and produce a porelike flaw.

Agglomerates and large grains

The rapid densification of regions containing fine particles (agglomerates) can induce stresses within the surrounding compact. Voids and cracks usually tend to form around agglomerates, as shown in Fig. 11.13b. These voids form as a result of the rapid and large differential shrinkage of the agglomerates during the early stages of sintering. Since these agglomerates form during the fabrication of the green bodies, care must be taken at that stage to avoid them.

Grain boundary cracks   Cleavage cracks within grain

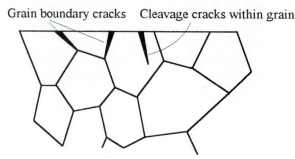

Figure 11.14
Schematic of cleavage and grain boundary cracks that can form on the surface of ceramics as a result of machining. The flaws are usually limited to one grain diameter, however, because they are deflected at the grain boundaries.

Similarly, large grains caused by exaggerated grain growth during sintering often result in a degradation in strength. Large grains, if noncubic, will be anisotropic with respect to such properties as thermal expansion and elastic modulus, and their presence in a fine-grained matrix essentially can act as inclusions in an otherwise homogeneous matrix. The degradation in strength is also believed to be partly due to residual stresses at grain boundaries that result from thermal expansion mismatches between the large grains and the surrounding matrix. The magnitude of the residual stresses will depend on the grain shape factor and the grain size, but can be approximated by Eq. (11.17). The effect of grain size on the residual stresses and spontaneous microcracking will be dealt with in greater detail in Chap. 13.

Surface flaws

Surface flaws can be introduced in a ceramic as a result of high-temperature grain boundary grooving, postfabrication machining operations, or accidental damage to the surface during use, among others. During grinding, polishing, or machining, the grinding particles act as indenters that introduce flaws into the surface. These cracks can propagate through a grain along cleavage planes or along the grain boundaries, as shown in Fig. 11.14. In either case, the cracks do not extend much farther than one grain diameter before they are usually arrested. The machining damage thus penetrates approximately one grain diameter from the surface. Consequently, according to the Griffith criterion, the fracture stress is expected to decrease with increasing grain size — an observation that is commonly observed. This brings up the next important topic, which relates the strength of ceramics to their grain size.

11.3.2 Effect of Grain Size on Strength

Typically, the strength of ceramics shows an inverse correlation to the average grain size G. A schematic of the dependence is shown in Fig. 11.15a, where the fracture strength is plotted versus $G^{-1/2}$. The simplest explanation for this behavior is that the intrinsic flaw size scales with the grain size, a situation not unlike the one shown in Fig. 11.14. The flaws form at the grain boundaries, which are weak areas to begin with, and propagate up to about one grain diameter. Thus once more invoking the Griffith criterion, one expects the strength to be proportional to $G^{-1/2}$, as is observed. It is worth noting that the strength does not keep on increasing with decreasing grain size. For very fine-grained ceramics, fracture usually occurs from preexistent process or surface flaws in the material, and thus the strength becomes relatively grain-size-insensitive. In other words, the line shown in Fig. 11.15 becomes much less steep for smaller grain sizes.

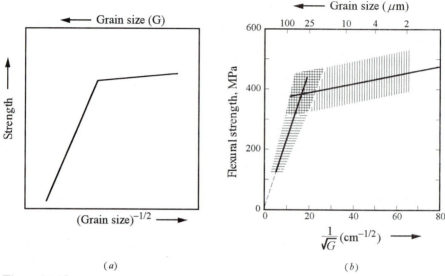

Figure 11.15
(*a*) Schematic relationship between grain size and strength for a number of ceramics.
(*b*) Actual data for $MgAl_2O_4$. Courtesy of R. W. Rice.

11.3.3 Effect of Compressive Surface Residual Stresses

The introduction of surface compressive layers can strengthen ceramics and is a well-established technique for glasses (see Sec. 13.5 for more details). The underlying principle is to introduce a state of compressive surface residual stress, the presence of which would inhibit failure from surface flaws since these compressive stresses would have to be overcome before a surface crack could propagate. These compressive stresses have also been shown to enhance thermal shock resistance and contact damage resistance.

There are several approaches to introducing a state of compressive residual stress, but in all cases the principle is to generate a surface layer with a higher volume than the original matrix. This can be accomplished in a variety of ways:

- Incorporation of an outer layer having a lower coefficient of thermal expansion, as in glazing or tempering of glass. These will be discussed in greater detail in Chap. 13.
- Using transformation stresses in certain zirconia ceramics (see next section).
- Physically stuffing the outer layer with atoms or ions such as by ion implantation.

- Ion-exchanging smaller ions for larger ions. The larger ions that go into the matrix place the latter in a state of compression. This is similar to physical stuffing and is most commonly used in glasses by placing a glass in a molten salt that contains the larger ions. The smaller ions are exchanged by the larger ions, which in turn place the surface in compression.

One aspect of this technique is that to balance the compressive surface stresses, a tensile stress develops in the center of the part. Thus if a flaw actually propagates through the compressive layer, the material is then weaker than in the absence of the compressive layer, and the release of the residual stresses can actually cause the glass to shatter. This is the principle at work in the manufacture of tempered glass for car windshields which upon impact shatter into a large number of small pieces that are much less dangerous than larger shards of glass, which can be lethal.

11.3.4 Effect of Temperature on Strength

The effect of temperature on the strength of ceramics depends on many factors, the most important of which is whether the atmosphere in which the testing is being carried out heals or exacerbates preexisting surface flaws in the material. In general, when a ceramic is exposed to a corrosive atmosphere at elevated temperatures, one of two scenarios is possible: (1) A protective, usually oxide, layer forms on the surface, which tends to blunt and partially heal preexisting flaws and can result in an increase in the strength. (2) The atmosphere attacks the surface, either forming pits on the surface or simply etching the surface away at selective areas; in either case, a drop in strength is observed. For ceramics containing glassy grain boundary phases, at high enough temperatures the drop in strength is usually related to the softening of these phases.

11.4
TOUGHENING MECHANISMS

Despite the fact that ceramics are inherently brittle, a variety of approaches have been used to enhance their fracture toughness and resistance to fracture. The essential idea behind all toughening mechanisms is to increase the energy needed to extend a crack, that is, G_c in Eq. (11.11). The basic approaches are crack deflection, crack bridging, and transformation toughening.

11.4.1 Crack Deflection

It is experimentally well established that the fracture toughness of a polycrystalline ceramic is appreciably higher than that of single crystals of the same composition. For example, K_{Ic} of single-crystal alumina is about 2.2 MPa·m$^{1/2}$, whereas that for polycrystalline alumina is closer to 4 MPa·m$^{1/2}$. Similarly, the fracture toughness of glass is ≈ 0.8 MPa·m$^{1/2}$, whereas the fracture toughness of a glass-ceramic of the same composition is closer to 2 MPa·m$^{1/2}$. One of the reasons invoked to explain this effect is crack deflection at the grain boundaries, a process illustrated in Fig. 11.16. In a polycrystalline material, as the crack is deflected along the weak grain boundaries, the average stress intensity at its tip K_{tip} is reduced, because the stress is no longer always normal to the crack plane [an implicit assumption made in deriving Eq. (11.9)]. In general, it can be shown that K_{tip} is related to the applied stress intensity K_{app} and the angle of deflection (defined in Fig. 11.16a) by

$$K_{tip} = \left(\cos^3 \frac{\theta}{2} \right) K_{app} \tag{11.18}$$

Based on this equation, and assuming an average θ value of, say, 45°, the increase in fracture toughness expected should be about 1.25 above the single-crystal value. By comparing this conclusion with the experimental results listed above, it is clear that crack deflection by itself accounts for some of, but not all, the enhanced toughening. In polycrystalline materials, crack bifurcation around grains can lead to a much more potent toughening mechanism, namely, crack bridging — the topic tackled next.

11.4.2 Crack Bridging:

In this mechanism, the toughening results from bridging of the crack surfaces behind the crack tip by a strong reinforcing phase. These bridging ligaments (Fig. 11.16b and c) generate closure forces on the crack face that reduce K_{tip}. In other words, by providing some partial support of the applied load, the bridging constituent reduces the crack-tip stress intensity. The nature of the ligaments varies but they can be whiskers, continuous fibers (Fig. 11.16c), or elongated grains (Fig.11.16b). A schematic of how these elastic ligaments result in a closure force is seen in Fig. 11.16c. A useful way to think of the problem is to imagine the unbroken ligaments in the crack wake as tiny springs that have to be stretched, and hence consume energy, as the crack front advances.

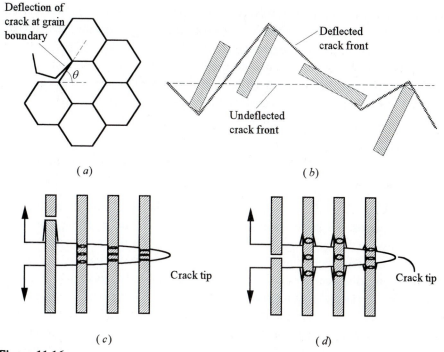

Figure 11.16

(a) Schematic of crack deflection mechanism at grain boundaries. (b) Schematic indicating deflection of crack front around rod-shaped particles.[197] (c) Schematic of ligament bridging mechanism (a) with no interfacial debonding and (d) with debonding. Note that in this case the strain on the ligaments is delocalized, and the toughening effect is enhanced.

It can be shown that the fracture toughness of a composite due to elastic stretching of a partially debonded reinforcing phase in the crack tip with no interfacial friction is given by[198]

$$K_{Ic} = \sqrt{Y_c G_m + \sigma_f^2 \left(\frac{r V_f Y_c \gamma_f}{12 Y_f \gamma_i} \right)} \qquad (11.19)$$

---

[197] Adapted from A. G. Evans and R. M. Cannon, *Acta. Met.*, **34**:761 (1986).

[198] See, e.g., P. Becher, *J. Amer. Cer. Soc.*, **74**:255–269 (1991).

where the subscripts c, m, and f represent the composite, matrix, and reinforcement, respectively; Y, V, and σ_f are the Young's modulus, volume fraction, and strength of the reinforcement phases, respectively; r is the radius of the bridging ligament, and G_m is the toughness of the unreinforced matrix; and γ_f/γ_i represents the ratio of the fracture energy of the bridging ligaments to that of the reinforcement/matrix interface. Equation (11.19) predicts that the fracture toughness increases with

- Increasing fiber volume fraction of reinforcing phase
- Increasing Y_c/Y_f ratio
- Increasing γ_f/γ_i ratio (i.e., the toughness is enhanced for weak fiber/matrix interfaces)

Comparing Fig. 11.16c and d reveals how the formation of a debonded interface spreads the strain displacement imposed on the bridging reinforcing ligament over a longer gauge length. As a result, the stress supported by the ligaments increases more slowly with distance behind the crack tip, and greater crack-opening displacements are achieved in the bridging zone, which in turn significantly enhances the fracture resistance of the composite. An essential ingredient of persistent bridge activity is that substantial pullout can occur well after whisker rupture. The fiber bridging mechanism is thus usually supplemented by a contribution of pullout of the reinforcement from fibers that fail away from the crack plane (Fig. 11.16c). As the ligaments pull out of the matrix, they consume energy that has to be supplied to the advancing crack, further enhancing the toughness of the composite.

That toughening contributions obtained by crack bridging and pullout can yield substantially increased fracture toughness is demonstrated in Fig. 11.17a for a number of whisker-reinforced ceramics. The solid lines are predicted curves and the data points are the experimental results; the agreement is quite good. A similar mechanism accounts for the high toughnesses achieved recently in Si_3N_4 with acicular grains, coarser grain-sized aluminas, and other ceramics.

11.4.3 Transformation Toughening

Transformation-toughened materials owe their very large toughness to the stress-induced transformation of a metastable phase in the vicinity of a propagating crack. Since the original discovery[199] that the tetragonal-to-monoclinic $(t \Rightarrow m)$ transformation of zirconia (see Chap. 8) has the potential for increasing both the

---

[199] R. Garvie, R. Hannick, and R. Pascoe, *Nature*, **258**:703 (1975).

fracture stress and the toughness of zirconia and zirconia-containing ceramics, a large effort has been dedicated to understanding the phenomenon.[200]

To understand the phenomenon, it is useful to refer to Fig. 11.18, where fine tetragonal zirconia grains are dispersed in a matrix. If these tetragonal particles are fine enough, then upon cooling from the processing temperatures, they can be constrained from transforming by the surrounding matrix and consequently can be retained in a *metastable tetragonal phase*. If, for any reason, that constraint is lost, the transformation which is accompanied by a relatively large volume expansion or dilatation ($\approx$ 4 percent) and shear strain ($\approx$ 7 percent) is induced. In transformation toughening, the approaching crack front, being a free surface, is the catalyst that triggers the transformation, which in turn places the zone ahead of the crack tip in compression. Given that the transformation occurs in the vicinity of the crack tip, extra energy is required to extend the crack through that compressive layer, which increases both the toughness and the strength of the ceramic.

The effect of the dilation strains is to reduce the stress intensity at the crack tip K_{tip} by a shielding factor K_s such that

$$K_{\text{tip}} = K_a - K_s \tag{11.20}$$

It can be shown that if the zone ahead of the crack tip contains a uniform volume fraction V_f of transformable phase that transforms in a zone of width w, shown in Fig. 11.18a, from the crack surface, then the shielding crack intensity factor is given by[201]

$$K_s = A' Y V_f \, \varepsilon^T \sqrt{w} \tag{11.21}$$

where A' is a dimensionless constant on the order of unity that depends on the shape of the zone ahead of the crack tip and ε^T is the transformation strain. A methodology to calculate ε^T is discussed in Chap. 13.

Fracture will still occur when $K_{\text{tip}} = K_{\text{Ic}}$ of the matrix in the absence of shielding; however, now the enhanced fracture toughness comes about by the shielding of K_{tip} by K_s. Careful microstructural characterization of crack-tip zones in various zirconias has revealed that the enhancement in fracture toughness does in fact scale with the product $V_f \sqrt{w}$, consistent with Eq. (11.21).

---

[200] See, e.g., A. G. Evans and R. M. Cannon, *Acta. Metall.*, **34**:761–800 (1986). For more recent work, see D. Marshall, M. Shaw, R. Dauskardt, R. Ritchie, M. Ready, and A. Heuer, *J. Amer. Cer. Soc.*, **73**:2659–2666 (1990).

[201] R. M. McMeeking and A. G. Evans, *J. Amer. Cer. Soc.*, **63**:242–246 (1982).

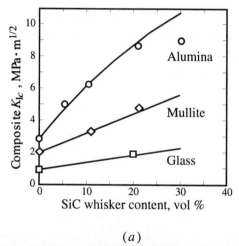

(a)

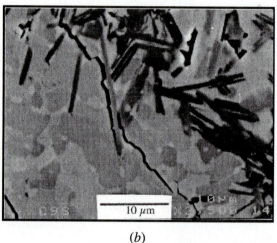

(b)

Figure 11.17

(a) The effect of SiC whisker content on toughness enhancement in different matrices.[202]
(b) Toughening is associated with crack bridging and grain pullout of elongated matrix grains.

It is unfortunate that the reason transformation toughening works so well at ambient temperatures — mainly the metastability of the tetragonal phase — is the same reason it is ineffective at elevated temperatures. Increasing the temperature reduces the driving force for transformation and consequently the extent of the transformed zone, leading to less tough materials.

[202] P. Becher, "Microstructural Design of Toughened Ceramics," *J. Amer. Cer. Soc.*, **74**:255–269 (1991).

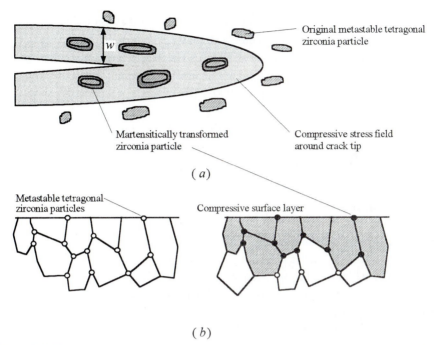

Figure 11.18

(*a*) Transformation zone ahead and around crack tip. (*b*) Surface grinding induces the martensitic transformation, which in turn creates compressive surface layers and a concomitant increase in strength.

It is worth noting that the transformation can be induced any time the hydrostatic constraint of the matrix on the metastable particles is relaxed. For example, it is now well established that compressive surface layers are developed as a result of the spontaneous transformation. The process is shown schematically in Fig. 11.18*b*. The fracture strength can be almost doubled by simply abrading the surface, since surface grinding has been shown to be an effective method for inducing the transformation. Practically this is of paramount importance, because we now have a ceramic that, in principle, becomes stronger as it is handled and small scratches are introduced on its surface.

At this stage, three classes of toughened zirconia-containing ceramics have been identified:

- *Partially stabilized zirconia* (PSZ). In this material the cubic phase is less than totally stabilized by the addition of MgO, CaO, or Y_2O_3. The cubic phase is then heat-treated to form coherent tetragonal precipitates. The heat treatment is such as to keep the precipitates small enough so they do not spontaneously transform within the cubic zirconia matrix but only as a result of stress.

- *Tetragonal zirconia polycrystals* (TZPs). These ceramics contain 100 percent tetragonal phase and small amounts of yttria and other rare-earth additives. With bend strength exceeding 2000 MPa, these ceramics are among the strongest known.
- *Zirconia-toughened ceramics* (ZTCs). These consist of tetragonal or monoclinic zirconia particles finely dispersed in other ceramic matrices such as alumina, mullite, and spinel.

11.4.4   *R* Curve Behavior

One of the important consequences of the toughening mechanisms described above is that they result in what is known as *R curve behavior*. In contrast to a typical Griffith solid where the fracture toughness is independent of crack size, *R* curve behavior refers to a fracture toughness which increases as the crack grows, as shown schematically in Fig. 11.19*a*. The main mechanisms responsible for this type of behavior are the same as are operative during crack bridging or transformation toughening, i.e., the closure forces imposed by either the transformed zone or the bridging ligaments. For example, referring once again to Fig. 11.16*c*, one sees that as the number of bridging ligaments increases in the crack wake, so will the energy required to extend the crack. It is worth noting that the increase in fracture toughness does not increase indefinitely but reaches a steady-state value, where the number of ligaments in the crack wake reach a steady-state with increasing crack extension since farther away from the crack tip, the ligaments tend to break and pull out completely.

There are four important implications for ceramics that exhibit *R* curve behavior:

1. The degradation in strength with increasing flaw size is less severe for ceramics without *R* curve behavior. This is shown schematically in Fig. 11.19*b*.
2. The reliability of the ceramic increases. This will be discussed in detail in Sec. 11.5.
3. On the down side, there is now an increasing body of evidence that seems to indicate that ceramics that exhibit *R* curve behavior are more susceptible to fatigue than ceramics that do not exhibit *R* curve behavior. This is discussed in greater detail in Chap. 12.

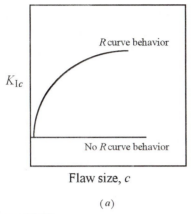
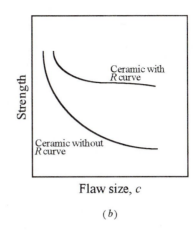

(a)   (b)

Figure 11.19
(a) Functional dependence of fracture toughness on flaw size for a ceramic exhibiting R curve behavior (top curve) and one that does not (lower curve). (b) Effect of R curve behavior on strength degradation as flaw size increases. Ceramics exhibiting R curve behavior are more flaw-tolerant than those that do not.

4. There is some recent evidence to suggest that R curve behavior enhances the thermal shock resistance of some ceramics. The evidence at this point is not conclusive, however, and more work is needed in this area.

To summarize, fracture toughness is related to the work required to extend a crack and is determined by the details of the crack propagation process. Only for the fracture of the most brittle solids is the fracture toughness simply related to surface energy. The fracture toughness can be enhanced by increasing the energy required to extend the crack. Crack bridging and martensitic transformations are two mechanisms that have been shown to enhance K_{Ic}.

11.5
DESIGNING WITH CERAMICS

In light of the preceding discussion, one expects that the failure stress, being as sensitive as it is to flaw sizes and their distributions, will exhibit considerable variability or scatter. This begs the question: Given this variability, is it still possible to design critical load-bearing parts with ceramics? In theory, if the flaws in a part were fully characterized (i.e., their size and orientation with respect to the applied stresses) and the stress concentration at each crack tip could be calculated, then given K_{Ic}, the exact stress at which a component would fail could be determined, and the answer to the question would be yes. Needless to say, such a procedure is quite impractical for several reasons, least among them the difficulty of characterizing all the flaws inside a material and the time and effort that would entail.

An alternative approach, described below, is to characterize the behavior of a large number of samples of the same material and to use a statistical approach to design. Having to treat the problem statistically has far-reaching implications since now the best that can be hoped for in designing with brittle solids is to state the *probability* of survival of a part at a given stress. The design engineer must then assess an acceptable risk factor and, using the distribution parameters described below, estimate the appropriate design stress.

Other approaches being taken to increase the reliability of ceramics are nondestructive testing and proof testing, both of which are briefly discussed in Secs. 11.5.2 and 11.5.3.

11.5.1 The Statistical Approach

Weibull distributions

One can describe the strength distribution of a ceramic in a variety of formalisms. The one most widely used today is the *Weibull distribution.*[203] This two-parameter semiempirical distribution is given by

$$f(x) = m(x)^{m-1} \exp\left(-x^m\right) \tag{11.22}$$

where $f(x)$ is the frequency distribution of the random variable x and m is a shape factor, usually referred to as the *Weibull modulus*. When Eq. (11.22) is plotted (see Fig. 11.20a), a bell-shaped curve results, the width of which depends on m; as m gets larger, the distribution narrows.

Since one is dealing with a strength distribution, the random variable x is defined as σ/σ_0, where σ is the failure stress and σ_0 is a normalizing parameter, whose physical significance will be discussed shortly, that is required to render x dimensionless.

Replacing x by σ/σ_0 in Eq. (11.22), one finds the survival probability, i.e., the fraction of samples that would survive a given stress level, is simply

$$S = \int_{\sigma/\sigma_0}^{\infty} f\left(\frac{\sigma}{\sigma_0}\right) d\left(\frac{\sigma}{\sigma_0}\right)$$

---

[203] W. Weibull, *J. Appl. Mech.*, **18**:293–297 (1951); *Mat. Res. & Stds.*, May 1962, pp. 405–411.

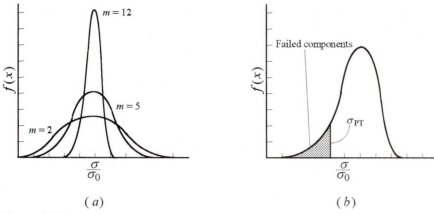

Figure 11.20
(a) The effect of m on the shape of the Weibull distribution. As m increases, the distribution narrows. (b) Truncation of Weibull distribution as a result of proof testing.

or

$$S = \exp\left[-\left(\frac{\sigma}{\sigma_0}\right)^m\right]$$ (11.23)

Rewriting Eq. (11.23) as $1/S = \exp(\sigma/\sigma_0)^m$ and taking the natural log of both sides twice yields

$$\ln\ln\frac{1}{S} = m\ln\frac{\sigma}{\sigma_0} = m\ln\sigma - m\ln\sigma_0$$ (11.24)

Multiplying both sides of Eq. (11.24) by -1 and plotting $-\ln\ln(1/S)$ versus $\ln\sigma$ yield a straight line with slope $-m$. The physical significance of σ_0 is now also obvious: It is the stress level at which the survival probability is equal to $1/e$, or 0.37. Once m and σ_0 are determined from the set of experimental results, the survival probability at any stress can be easily calculated from Eq. (11.23) (see Worked Example 11.2).

The use of Weibull plots for design purposes has to be handled with extreme care. As with all extrapolations, a small uncertainty in the slope can result in large uncertainties in the survival probabilities, and hence to increase the confidence level, the data sample has to be sufficiently large ($N > 100$). Furthermore, in the Weibull model, it is implicitly assumed that the material is homogeneous, with a single flaw population that does not change with time. It further assumes that only one failure mechanism is operative and that the defects are randomly distributed and are small relative to the specimen or component size. Needless to say, whenever any of these assumptions is invalid, Eq. (11.23) has to be modified. For instance, bimodal distributions that lead to strong deviations from a linear Weibull plot are not uncommon.

WORKED EXAMPLE 11.2. The strengths of 10 nominally identical ceramic bars were measured and found to be 387, 350, 300, 420, 400, 367, 410, 340, 345, and 310 MPa. (*a*) Determine m and σ_0 for this material. (*b*) Calculate the design stress that would ensure a survival probability higher than 0.999.

Answer

(*a*) To determine m and σ_0, the Weibull plot for this set of data has to be made. Do as follows:

- Rank the specimens in order of increasing strength, 1, 2, 3, ..., j, $j+1$, ..., N, where N is the total number of samples.
- Determine the survival probability for the jth specimen. As a first approximation, the probability of survival of the first specimen is $1-1/(N+1)$; for the second, $1-2/(N+1)$, for the jth specimen $1-j/(N+1)$, etc. This expression is adequate for most applications. However, an alternate and more accurate expression deduced from a more detailed statistical analysis yields

$$S_j = 1 - \frac{j - 0.3}{N + 0.4} \tag{11.25}$$

- Plot $-\ln\ln(1/S)$ versus $\ln\sigma$. The least-squares fit to the resulting line is the Weibull modulus.

The last two columns in Table 11.2 are plotted in Fig. 11.21. A least-squares fit of the data yields a slope of 10.5, which is typical of many conventional as-finished ceramics. From the table, $\sigma_0 \approx 387\,\text{MPa}$.

(*b*) To calculate the stress at which the survival probability is 0.999, use Eq. (11.23), or

$$0.999 = \exp\left\{-\left(\frac{\sigma}{385}\right)^{10.5}\right\}$$

from which $\sigma = 200\,\text{MPa}$. It is worth noting here that the error in using the average stress of 366 MPa instead of σ_0 changes the end result for the design stress only slightly. For most applications, it is sufficient to simply use the average stress.

TABLE 11.2
Summary of data needed to find *m* from a set of experimental results

Rank j	S_j	σ_j	$\ln\sigma_j$	$-\ln\ln(1/S_j)$
1	0.932	300	5.700	2.6532
2	0.837	310	5.734	1.7260
3	0.740	340	5.823	1.2000
4	0.644	345	5.840	0.8200
5	0.548	350	5.860	0.5080
6	0.452	367	5.905	0.2310
7	0.356	387	5.960	−0.0320
8	0.260	400	5.990	−0.2980
9	0.160	410	6.016	−0.6060
10	0.070	420	6.040	−0.9780

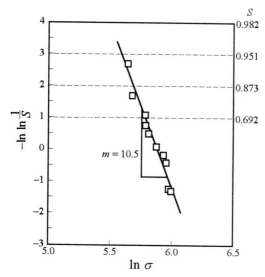

Figure 11.21
Weibull plot of data shown in Table 11.2. Slope of the line is the Weibull modulus *m*. The actual survival probability is shown on the right-hand side. At low stresses, S is large (left-hand corner of figure).[204]

---

[204] The reason that $-\ln\ln(1/S)$ is plotted rather than $\ln\ln(1/S)$ is simply aesthetic, such that the high survival probabilities appear on the upper left-hand sides of the plots.

Factors affecting the Weibull modulus

Clearly, from a design point of view, it is important to have high m's. Note that m should not be confused with strength, since it is possible to have a weak solid with a high m and vice versa. For instance, a solid with large defects that are all identical in size would be weak but, in principle, would exhibit large m. It is the *uniformity* of the microstructure, including flaws, grain size, and inclusions, that is critical for obtaining large m values.

Interestingly enough, increasing the fracture toughness for a truly brittle material will not increase m. This can be shown as follows: By recasting Eq. (11.24), m can be rewritten as

$$m = \frac{\ln\ln(1/S_{max}) - \ln\ln(1/S_{min})}{\ln(\sigma_{max}/\sigma_{min})} \qquad (11.26)$$

For any set of samples, the numerator will be a constant that depends only on the total number of samples tested [that is, N in Eq. (11.25)]. The denominator depends on the ratio $\sigma_{max}/\sigma_{min}$, which is proportional to the ratio c_{min}/c_{max}, which is clearly independent of K_{Ic}, absent R curve effects. Thus toughening of a solid per se will often not result in an increase in its Weibull modulus. However, it can be easily shown that if a solid exhibits R curve behavior, then an increase in m should, in principle, follow (see Prob. 11.12).

Effect of size and test geometry on strength

One of the important ramifications of brittle failure, or *weak-link statistics*, as it is sometimes referred to, is the fact that strength becomes a function of volume: *Larger specimens will have a higher probability of containing a larger defect, which in turn will cause lower strengths.* In other words, the larger the specimen, the weaker it is likely to be. Clearly, this is an important consideration when data obtained on test specimens, which are usually small, are to be used for the design of larger components.

Implicit in the analysis so far has been that the volumes of all the samples tested were the same size and shape. The probability of a sample of volume V_0 surviving a stress σ is given by

$$S(V_0) = \exp\left\{-\left[\frac{\sigma}{\sigma_0}\right]^m\right\} \qquad (11.27)$$

The probability that a batch of n such samples will all survive the same stress is lower and is given by[205]

$$S_{\text{batch}} = [S(V_0)]^n \tag{11.28}$$

Placing n batches together to create a larger body of volume V, where $V = nV_0$, one sees that the probability $S(V)$ of the larger volume surviving a stress σ is identical to Eq. (11.28), or

$$S(V) = S_{\text{batch}} = [S(V_0)]^n = [S(V_0)]^{V/V_0} \tag{11.29}$$

which is mathematically equivalent to

$$S = \exp\left\{ -\left(\frac{V}{V_0}\right)\left(\frac{\sigma}{\sigma_0}\right)^m \right\} \tag{11.30}$$

This is an important result since it indicates that the survival probability of a ceramic depends on both the volume subjected to the stress and the Weibull modulus. Equation (11.30) states that as the volume increases, the stress level needed to maintain a given survival probability has to be reduced. This can be seen more clearly by equating the survival probabilities of two types of specimens — test specimens with a volume V_{test} and component specimens with volume V_{comp}. Equating the survival probabilities of the two types of samples and rearranging Eq. (11.30), one can easily show that

$$\frac{\sigma_{\text{comp}}}{\sigma_{\text{test}}} = \left(\frac{V_{\text{test}}}{V_{\text{comp}}}\right)^{1/m} \tag{11.31}$$

A plot of this equation is shown in Fig. 11.22, where the relationship between strength and volume is plotted. The salient point here is that as either the volume *increases* or the Weibull modulus *decreases*, the more severe the downgrading of the design stress is required to maintain a given survival probability.

It is important to note that an implicit assumption made in deriving Eq. (11.31) is that only one flaw population (i.e., those due to processing rather than, say, machining) is controlling the strength. Different flaw populations will have different strength distributions and will scale in size differently. Also implicit in deriving Eq. (11.31) is that volume defects are responsible for failure. If, instead, surface flaws were suspected of causing failure, by using a derivation identical to the one used to get to Eq. (11.31), it can be shown that

---

[205] An analogy here is useful: The probability of rolling a given number with a six-sided die is 1/6. The probability that the same number will appear on n dice rolled simultaneously is $(1/6)^n$.

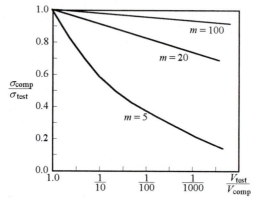

Figure 11.22
The effect of volume on strength degradation as a function of the Weibull modulus. The strength decreases as V increases and is more severe for low m.

$$\frac{\sigma_{comp}}{\sigma_{test}} = \left(\frac{A_{test}}{A_{comp}}\right)^{1/m} \tag{11.32}$$

in which case the strength will scale with area instead of volume.

Finally, another important ramification of the stochastic nature of brittle fracture is the effect of the stress distribution during testing on the results. When a batch of ceramics is tested in tension, the entire volume and surface are subjected to the stress. Thus a critical flaw *anywhere* in the sample will propagate with equal probability. In three- or four-point flexure tests, however, only one-half the sample is in tension, and the other one-half is in compression. In other words, the effective volume tested is, in essence, reduced. It can be shown that the ratio of the tensile to flexural strength for an equal probability of survival is

$$\frac{\sigma_{3\text{-point bend}}}{\sigma_{tension}} = \left[2(m+1)^2\right]^{1/m} \tag{11.33}$$

In other words, the samples subjected to flexure will appear to be stronger, by a factor that depends on m. For example, for $m = 5$, the ratio is about 2, whereas increasing m to 20 reduces the ratio to 1.4.

11.5.2 Proof Testing

In proof testing, the components are briefly subjected to a stress level σ_{PT} which is in excess of that anticipated in service. The weakest samples fail and are thereby eliminated. The resulting truncated distribution, shown in Fig. 11.20b, can be used with a high level of confidence at any stress that is slightly lower than σ_{PT}.

One danger associated with proof testing is subcritical crack growth, discussed in the next chapter. Since moisture is usually implicated in subcritical crack growth, effective proof testing demands inert, i.e., moisture-free, testing environments and rapid loading/unloading cycles that minimize the time at maximum stress.

11.6
SUMMARY

1. Ceramics are brittle because they lack a mechanism to relieve the stress buildup at the tips of notches and flaws. This makes them notch-sensitive, and consequently their strength will depend on the combination of applied stress and flaw size. The condition for failure is

$$K_I = \sigma_f \sqrt{\pi c} \geq K_{Ic}$$

where K_{Ic} is the fracture toughness of the material. The strength of ceramics can be increased by either increasing the fracture toughness or decreasing the flaw size.

2. Processing introduces flaws in the material that are to be avoided if high strengths are to be achieved. The flaws can be pores, large grains in an otherwise fine matrix, and inclusions, among others. Furthermore, since the strength of a ceramic component decreases with increasing grain size, it follows that to obtain a high-strength ceramic, a flaw-free, fine microstructure is desirable.

3. It is possible to toughen ceramics by a variety of techniques, which all make it energetically less favorable for a crack to propagate. This can be accomplished either by having a zone ahead of the crack martensitically transform, thus placing the crack tip in compression, or by adding whiskers or fibers or large grains (duplex microstructures) that bridge the crack faces as it propagates.

    Comparing the requirements for high strength (uniform, fine microstructure) to those needed to improve toughness (nonhomogeneous, duplex microstructure) reveals the problem in achieving both simultaneously.

4. The brittle nature of ceramics together with the stochastic nature of finding flaws of different sizes, shapes, and orientations relative to the applied stress will invariably result in some scatter to their strength. According to the Weibull distribution, the survival probability is given by

$$S = \exp\left\{ -\left(\frac{\sigma}{\sigma_0} \right)^m \right\}$$

where m, known as the *Weibull modulus*, is a measure of the scatter. Large scatter is associated with low m values, and vice versa.

5. If strength is controlled by defects randomly distributed within the volume, then strength becomes a function of volume, with the survival probability decreasing with increasing volume. However, if strength is controlled by surface defects, strength will scale with area instead.

6. Proof testing, in which a component is loaded to a stress level higher than the service stress, eliminates the weak samples, truncating the distribution and establishing a well-defined stress level for design.

PROBLEMS

11.1. (*a*) Following a similar analysis used to arrive at Eq. (11.7), show that an internal crack of length c is only $\sqrt{2}$ as detrimental to the strength of a ceramic as a surface crack of the same length.

(*b*) Why are ceramics usually much stronger in compression than in tension?

(*c*) Explain why the yield point of ceramics can approach the ideal strength σ_{theo}, whereas the yield point in metals is usually much less than σ_{theo}. How would you attempt to measure the yield strength of a ceramic, given that the fracture strength of ceramics in tension is usually much less than the yield strength?

11.2. (*a*) Estimate the size of the critical flaw for a glass that failed at 102 MPa if $\gamma = 1 \ \text{J/m}^2$ and $Y = 70 \ \text{GPa}$.

Answer: 4.3 μm

(*b*) What is the maximum stress this glass will withstand if the largest crack is on the order of 100 μm and the smallest on the order of 7 μm?

Answer: 21 MPa

11.3. Show that Eqs. (11.2) and (11.9) are equivalent, provided the radius of curvature of the crack $\rho \approx 14r_0$, where r_0 is the equilibrium interatomic distance; in other words, if it is assumed that the crack is atomically sharp. *Hint*: Find expressions for γ and Y in terms of n, m, and r_0 defined in Chap. 4. You can assume $n = 9$ and $m = 1$.

11.4. Al_2O_3 has a fracture toughness K_{Ic} of about 4 $MPa \cdot m^{1/2}$. A batch of Al_2O_3 samples were found to contain surface flaws about 30 μm deep. The average flaw size was more on the order of 10 μm. Estimate (a) the tensile strength and (b) the compressive strength.
Answer: 412 MPa, 10 GPa

11.5. To investigate the effect of pore size on the strength of reaction-bonded silicon nitride, Heinrich[206] introduced artificial pores (wax spheres that melt during processing) in his compacts prior to reaction bonding. The results he obtained are summarized below. Are these data consistent with the Griffith criterion? Explain clearly, stating all assumptions.

Wax grain size, μm	Average pore size, μm	Bend strength, MPa
0–36	48	140±12
63–90	66	119±12
125–180	100	101±14

11.6. The tensile fracture strengths of three different structural ceramics are listed below: hot-pressed silicon nitride (HPSN), reaction-bonded silicon nitride (RBSN), and chemical vapor-deposited silicon carbide (CVDSC), measured at room temperature.
  (a) Plot the cumulative *failure probability* of these materials as a function of fracture strength.
  (b) Calculate the mean strength and standard deviation of the strength distributions, and determine the Weibull modulus for each material.
  (c) Estimate the design stress for each material.
  (d) On the basis of your knowledge of these materials, why do you think they behave so differently?

HPSN (MPa)   521, 505, 500, 490, 478, 474,471,453, 452, 448, 444, 441, 439, 430, 428, 422, 409, 398, 394, 372, 360,341, 279

CVDSC   386, 351, 332, 327, 308, 296, 290, 279, 269, 260, 248, 231, 219, 199, 178, 139

RBSN   132, 120, 108, 106, 103, 99, 97, 95, 93, 90, 89, 84, 83, 82, 80, 80, 78, 76

206 J. Heinrich, *Ber. Dt. Keram. Ges.*, **55** (1978).

11.7. (a) When the ceramic shown in Fig. 11.23 was loaded in tension (along the length of the sample), it fractured at 20 MPa. The heavy lines denote cracks (two internal and one surface crack). Estimate K_{Ic} for this ceramic. State all assumptions.

Answer: 3.5 MPa·m$^{1/2}$

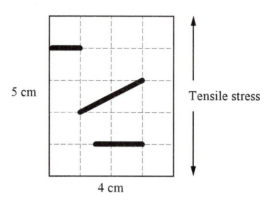

5 cm

4 cm

Tensile stress

Figure 11.23

Cross section of ceramic part loaded in tension as shown. The heavy lines denote flaws.

11.8. For silicon nitride, K_{Ic} is strongly dependent on microstructure, but can vary anywhere from 3 to 10 MPa·m$^{1/2}$. Which of the following silicon nitrides would you choose, one in which the largest flaw size is on the order of 50 μm and the fracture toughness is 8 MPa·m$^{1/2}$, or one for which the largest flaw size was 25 μm, but was only half as tough. Explain.

11.9. Evans and Charles[207] proposed the following equation for the determination of fracture toughness from indentation:

$$K_{Ic} \approx 0.15\left(H\sqrt{a}\right)\left(\frac{c}{a}\right)^{-1.5}$$

where H is the Vickers hardness in Pa and c and a were defined in Fig. 11.7. A photomicrograph of a Vickers indention in a glass slide and the cracks that emanate from it is shown in Fig. 11.24. Estimate the fracture toughness of this glass if its hardness is ≈ 5.5 GPa.

Answer: ≈ 1.2 MPa·m$^{1/2}$

---

207 A. G. Evans and E. A. Charles, *J. Amer. Cer. Soc.*, **59**:317 (1976).

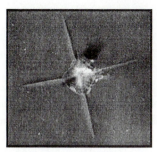

Figure 11.24
Optical photomicrograph of indentation in glass. 200X.

11.10. A manufacturer wishes to choose between two ceramics for a certain application. Data for the two ceramics tested under identical conditions were as follows:

Ceramic	Mean fracture stress	Weibull modulus
A	500 MPa	12
B	600 MPa	8

The service conditions are geometrically identical to the test conditions and impose a stress of 300 MPa. By constructing Weibull graphs with $S = 1/2$ for mean fracture stress or any other method, decide which ceramic will be more reliable and compare the probabilities of failure at 300 MPa. At what stress would the two ceramics give equal performance?
Answer: Stress for equal performance = 349 MPa

11.11. The MORs of a series of cylindrical samples ($l = 25$ mm and diameter of 5 mm) were tested and analyzed using Weibull statistics. The average strength was 100 MPa, with a Weibull modulus of 10. Estimate the stress required to obtain a survival probability of 95 percent for cylinders with diameters of 10 mm but the same length. State all assumptions.

11.12. Show why ceramics that exhibit R curve behavior should, in principle, also exhibit larger m values.

11.13. (*a*) In deriving Eq. (11.30) the flaw population was assumed to be identical in both volumes. However, sometimes in the manufacturing of ceramic bodies of different volumes and shapes, different flaw populations are introduced. What implications, if any, does this statement have on designing with ceramics? Be specific.

(*b*) In an attempt to address this problem, Kschinka et al.[208] measured the strength of different glass spheres in compression. Their results are summarized in Table 11.3, where D_0 is the diameter of the glass spheres, N is the number of samples tested, m is the Weibull modulus, σ_f is the average strength, and V is the volume of the spheres.

(i) Draw on *one graph* the Weibull plots for spheres of 0.051, 0.108, and 0.368 cm. Why are they different?

(ii) For the 0.051 cm spheres, what would be your design stress to ensure a 0.99 survival probability?

(iii) Estimate the average strength of glass spheres of 1-cm diameter.

(iv) If the effect of volume is taken into account, then it is possible to collapse all the data on a master curve. Show how that can be done. *Hint*: Normalize data to 0.156-cm spheres, for example.

TABLE 11.3

D_0, cm	N	m	σ_f (50%)	V, cm^3
0.368	47	6.19	143	2.61×10^{-2}
0.305	48	5.96	157	1.49×10^{-2}
0.241	53	5.34	195	7.33×10^{-3}
0.156	30	5.46	229	1.99×10^{-3}
0.127	45	5.37	252	1.07×10^{-3}
0.108	38	5.18	303	6.60×10^{-4}
0.091	47	3.72	407	3.95×10^{-4}
0.065	52	4.29	418	1.44×10^{-4}
0.051	44	6.82	435	6.95×10^{-5}

ADDITIONAL READING

1. R. W. Davidge, *Mechanical Behavior of Ceramics*, Cambridge University Press, New York, 1979.

2. R. Warren, ed., *Ceramic Matrix Composites*, Blackie, Glasgow, Scotland, 1992.

---

[208] B. A. Kschinka, S. Perrella, H. Nguyen, and R. C. Bradt, *J. Amer. Cer. Soc.*, **69**:467 (1986).

3. B. Lawn, *Fracture of Brittle Solids*, 2d ed., Cambridge University Press, New York, 1993.

4. A. Kelly and N. H. Macmillan, *Strong Solids*, 3d ed., Clarendon Press, Oxford, England, 1986.

5. G. Weaver, "Engineering with Ceramics, Parts 1 and 2," *J. Mat. Ed.*, **5**:767 (1983) and **6**:1027 (1984).

6. T. H. Courtney, *Mechanical Behavior of Materials*, McGraw-Hill, New York, 1990.

7. A. G. Evans, "Engineering Property Requirements for High Performance Ceramics," *Mat. Sci. & Eng.*, **71**:3 (1985).

8. S. M. Weiderhorn, "A Probabilistic Framework for Structural Design," in *Fracture Mechanics of Ceramics*, vol. 5, R. C. Bradt, A. G. Evans, D. P. Hasselman, and F. F. Lange, eds., Plenum, New York, 1978, p. 613.

9. M. F. Ashby and B. F. Dyson, in *Advances in Fracture Research*, S. R. Valluri, D. M. R. Taplin, P. Rama Rao, J. F. Knott, and R. Dubey, eds., Pergamon Press, New York, 1984, p. 3.

10. P. F. Becher, "Microstructural Design of Toughened Ceramics," *J. Amer. Cer. Soc.*, **74**:225 (1991).

CHAPTER 12

Creep, Subcritical Crack Growth, and Fatigue

The fault that leaves six thousand tons a log upon the sea.

R. Kipling, *McAndrew's Hymn*

12.1
INTRODUCTION

As discussed in the previous chapter, at low and intermediate temperatures, failure typically emanated from a preexisting flaw formed during processing or surface finishing. The condition for failure was straightforward: Fracture occurred rapidly and catastrophically when $K_I > K_{Ic}$. It was tacitly implied that for conditions where $K_I < K_{Ic}$, the crack was stable, i.e., did not grow with time, and consequently, the material would be able to sustain the load indefinitely. In reality, the situation is not that simple — preexisting cracks can and do grow slowly under steady and cyclic loadings, even when $K_I < K_{Ic}$. For example, it has long been appreciated that in metals, cyclic loadings, even at small loads, can result in crack growth, a phenomenon referred to as **fatigue**. In contrast, it has long been accepted that ceramics, because of their lack of crack-tip plasticity or work hardening, were not susceptible to fatigue. More recently, however, this has been shown to be false: Some ceramics, especially those that exhibit R curve behavior, are indeed susceptible to cyclic loading.

Another phenomenon that has been well appreciated for a long time is that the exposure of a ceramic to the combined effect of a steady stress and a corrosive environment results in slow crack growth. In this mode of failure, a preexisting subcritical crack, or one that nucleates during service, grows slowly by a stress-enhanced chemical reactivity at the crack tip and is referred to as **subcritical crack growth (SCG)**. Unfortunately, this phenomenon is also sometimes termed **static**

441

fatigue, seemingly to differentiate it from the dynamic fatigue situation just alluded to, but more to create confusion.

Last, **creep**, or the slow deformation of a solid subjected to a stress at high temperatures, also occurs in ceramics. Sooner or later, a part experiencing creep will either fail or undergo shape and dimensional changes that in close-tolerance applications would render a part useless.

Despite the fact that the atomic processes and micromechanisms that are occurring during each of these phenomena are quite different, there is a commonality among them. In each case, a nucleated or preexisting flaw grows with time, leading to the eventual failure of the part, usually with disastrous consequences. In other words, the ceramic now has a lifetime that one has to contend with.

In the sections that follow, each phenomenon is dealt with separately. In Sec. 12.5, the methodology for estimating lifetimes and the choice of appropriate design criteria are developed.

12.2
CREEP

Creep is the slow and continuous deformation of a solid with time that only occurs at higher temperatures, that is, $T > 0.5T_m$, where T_m is the melting point in Kelvins. In metals, it is now well established that grain boundary sliding and related cavity growth are the mechanisms most detrimental to creep resistance, which led to the development of single-crystal superalloy turbine blades that are very resistant to creep. In ceramics, the situation is more complex because several mechanisms, some of which are not sufficiently well understood, can lead to creep deformation. The problem is further complicated by the fact that different mechanisms may be operative over different temperature and stress regimes. In general, creep is a convoluted function of stress, time, temperature, grain size and shape, microstructure, volume fraction and viscosity of glassy phases at the grain boundaries, dislocation mobility, etc. Before one tackles the subject in greater detail, it is instructive to briefly review how creep is measured.

EXPERIMENTAL DETAILS: MEASURING CREEP

Typically, the creep response of a solid is found by measuring the strain rate as a function of applied load. This, most simply, can be done by attaching a load to a sample, heating it, and measuring its deformation as a function of time. The resulting strain is plotted versus time, as shown in Fig. 12.1a, where three regions are typically observed: (1) There is an initial, almost instantaneous response, followed by a decreasing rate of increase in strain with time. This region is known as the **primary creep** region. (2) There is a region where the strain increases linearly with time. This is known as the **steady-state** or **secondary creep** stage which, from a practical point of view, is the most important stage and is of major concern here. (3) There is a region known as the **tertiary creep** stage which occurs just before the specimen fails, where the strain rate increases rapidly with time.

Increasing the temperature and/or stress (Fig. 12.1b) results in an increase in both the instantaneous strain and the steady-state creep rates and a decrease in the time to failure.

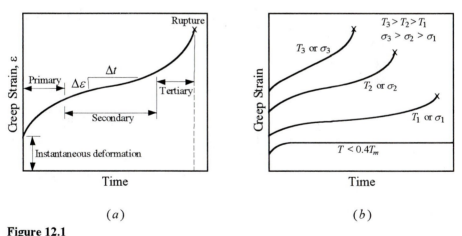

(a)   (b)

Figure 12.1
(a) Typical strain versus time creep curves. Experimentally not all three regions are always observed. (b) Effect of increasing stress and/or temperature on the creep response of the material.

Data such as shown in Fig. 12.1*b* can be further reduced by plotting the log of the *steady-state creep rate* $\dot{\varepsilon}_{ss}$ versus the log of the applied stress σ at a constant T. Such curves usually yield straight lines, which in turn implies that

$$\dot{\varepsilon}_{ss} = \frac{d\varepsilon}{dt} = \Gamma \sigma^{p} \tag{12.1}$$

where Γ is a temperature-dependent constant, and p is called the **creep law exponent** and usually lies between 1 and 8. For $p > 1$, this sort of creep is commonly referred to as *power law creep*.

Over a dozen mechanisms have been proposed to explain the functional dependence described by Eq. (12.1), but in general they fall into one of three categories: *diffusion, viscous,* or *dislocation* creep. To cover even a fraction of these models in any detail is clearly beyond the scope of this book. Instead, diffusion creep is dealt with in some detail below, followed by a brief mention of the other two important, but less well-understood and more difficult to model, mechanisms. For more comprehensive reviews, consult the references at the end of this chapter.

12.2.1 Diffusion Creep

For permanent deformation to occur, atoms have to move from one region to another, which requires a driving force of some kind. Thus, before one can even attempt to understand creep, it is imperative to appreciate the origin of the driving forces involved.

Driving force for creep

In general, the change in the Helmholtz[209] free energy A is given as

$$dA = -S \, dT - p \, dV \tag{12.2}$$

If the changes are occurring at constant temperature, as in a typical creep experiment, it follows that $dA = -p \, dV$. Upon rearrangement,

$$p = -\frac{\partial A}{\partial V}$$

---

[209] The Helmholtz free energy A represents the changes in free energy of a system when they are carried out under constant *volume*. In contrast, ΔG represents the free-energy changes occurring at constant *pressure*. However, since the volume changes in condensed phases, and the corresponding work against atmospheric pressure, are small, they can be neglected, and in general, $\Delta G \approx \Delta A$.

Multiplying both sides by the atomic volume Ω and noting that V/Ω is nothing but the number of atoms per unit volume, one finds that

$$p\Omega = -\frac{\partial A}{\partial V}\Omega = -\frac{\partial A}{\partial(V/\Omega)} = -\frac{\partial A}{\partial N} \tag{12.3}$$

Thus $\partial A/\partial N$ represents the excess (due to stress) chemical potential $\tilde{\mu} - \tilde{\mu}_0$ per atom, and μ_0 is the standard chemical potential of atoms in a stress-free solid (see Chap. 5). By equating p with an applied stress σ, it follows that the chemical potential of the atoms in a stressed solid is given by

$$\tilde{\mu} = \tilde{\mu}_0 - \sigma\Omega \tag{12.4}$$

By convention, σ is considered *positive when the applied stress is tensile and negative when it is compressive*.

To better understand the origin of Eq. (12.4), consider the situation depicted schematically in Fig. 12.2, where four pistons are attached to four sides of a cube of material such that the pressures in the pistons are unequal with, say, $P_A > P_B$. These pressures will result in normal compressive forces $-\sigma_{11}$ and $-\sigma_{22}$ on faces A and B, respectively. If an atom is now removed from surface A (e.g., by having it fill in a vacancy just below the A surface), piston A will move by a volume Ω, and the work done *on the system* is $\Omega P_A = \Omega \sigma_{11}$. By placing an atom on surface B (e.g., by having an atom from just below the surface diffuse to the surface), work is done by the system: $\Omega P_B = -\Omega \sigma_{22}$. The net work is thus

$$\Delta W_{A \Rightarrow B} = \Omega(P_B - P_A) = \Omega(\sigma_{11} - \sigma_{22}) \tag{12.5}$$

which is a fundamental result because it implies that energy can be recovered (that is, ΔW is negative) if atoms diffuse from higher to lower compressive stress areas (see Worked Example 12.1a).

Note that for the case where $\sigma_{11} = -\sigma_{22} = \sigma$, the energy recovered will be

$$\Delta W = -2\Omega\sigma \tag{12.6}$$

and it is a direct measure of the driving force available for an atom to diffuse from an area that is subjected to a compressive stress to an area subjected to the same tensile stress.

It is worth noting that the energy recovered when an atom moves from *just below* interface A to *just below* interface B is orders of magnitude lower than that given by Eq. (12.5). In other words, the strain energy contribution to the process is *not* the driving force — it is only when the atoms "*plate out*" onto the surface that the energy is recovered. The fundamental conclusion is that *atom movements that result in shape changes are much more energetically favorable than ones that do not result in such changes* (see Worked Example 12.1*b*).

WORKED EXAMPLE 12.1. (*a*) Refer to Fig. 12.2. If P_A is 20 MPa and P_B is 10 MPa, calculate the energy changes for an atom that diffuses from interface A to interface B. Assume the molar volume is 10 cm³/mol. (*b*) Show that for a typical ceramic $\pm\sigma\Omega$ is on the order of 1000 J/mol. Compare that value to the elastic strain energy term of an atom subjected to the same stress.

Answer

(*a*) If the molar volume is 10 cm³/mol, then $\Omega = 1.66 \times 10^{-29}$ m^{-3}. According to convention, $\sigma_{11} = -20$ MPa and $\sigma_{22} = -10$ MPa, and the net energy recovered is given by Eq. (12.5), or

$$\Delta W_{A \Rightarrow B} = 1.66 \times 10^{-29} \times 10^6 \{-20 - (-10)\} = -1.66 \times 10^{-22} \text{ J/atom} = -100 \text{ J/mol}$$

(*b*) Assuming the applied stress is 100 MPa, one obtains

$$\Omega\sigma = 1.66 \times 10^{-29} \times 100 \times 10^6 = -1.66 \times 10^{-21} \text{ J/atom} = 1000 \text{ J/mol}$$

For the second part, assume Young's modulus to be 150 GPa. The elastic energy associated with a volume Ω is given by [see Eq. (11.3)]

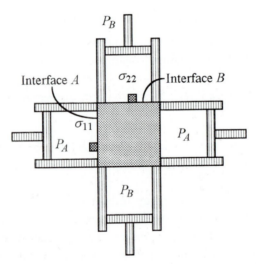

Figure 12.2
Schematic of thought experiment invoked to arrive at Eq. (12.5).

$$U_{elas} = \frac{\Omega}{2} \frac{\sigma^2}{Y} = \frac{1}{2} \frac{(1.66 \times 10^{-29})(100 \times 10^6)}{150 \times 10^9} = 5.53 \times 10^{-25} \text{ J/atom} = 0.33 \text{ J/mol}$$

which is roughly 3 orders of magnitude smaller than the $\sigma\Omega$ term.

Although Eqs. (12.4) to (12.6) elucidate the nature of the driving force operative during creep, they do not shed any light on *how* the process occurs at the atomic level. To do that, one has to go one step further and explore the effect of applied stresses on vacancy concentrations. For the sake of simplicity, the following discussion assumes creep is occurring in a pure elemental solid. The complications that arise from ambipolar diffusion in ionic compounds are discussed later. The equilibrium concentration of vacancies C_0 under a flat and stress-free surface is given by (Chap. 6)

$$C_0 = K' \exp\left(-\frac{Q}{kT}\right) \tag{12.7}$$

where Q is the enthalpy of formation of the vacancies and the entropy of formation and all preexponential terms are included in K'. Since the chemical potential of an atom under a surface subjected to a stress is either greater or smaller than that over a flat surface by $\Delta\mu$ [Eq. (12.4)], this energy has to be accounted for when one is considering the formation of a vacancy. It follows that

$$C_{11} = K' \exp\left(-\frac{Q + \Delta\mu}{kT}\right) = C_0 \exp\frac{\Omega\sigma_{11}}{kT} \tag{12.8}$$

and similarly,

$$C_{22} = C_0 \exp\frac{\Omega\sigma_{22}}{kT} \tag{12.9}$$

where C_{ii} is the concentration of vacancies just under a surface subjected to a normal stress σ_{ii}. Subtracting these two equations and noting that in most situations $\Omega\sigma << kT$, one obtains

$$\boxed{\Delta C = C_{22} - C_{11} = \frac{C_0\Omega(\sigma_{22} - \sigma_{11})}{kT}} \tag{12.10}$$

which is a completely general result. In the special case where $\sigma_{11} = -\sigma_{22} = \sigma$, it simplifies to

$$\boxed{\Delta C_{t-c} = C_{tens} - C_{comp} = \frac{2\Omega\sigma C_0}{kT}} \tag{12.11}$$

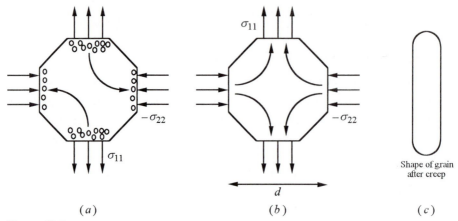

Figure 12.3
(*a*) Vacancy concentration gradients that develop as a result of stress gradients. The vacancy concentrations are higher below the tensile surface. Curved arrows denote direction of *vacancy* fluxes. (*b*) Schematic of a grain of diameter *d* subjected simultaneously to a tensile and a compressive stress. Curved arrows denote direction of *atomic* fluxes. (*c*) Shape of grain after creep has occurred.

Equations (12.10) and (12.11) are of fundamental importance since they predict that *the vacancy concentrations in tensile regions are higher than those in compressive regions* (Fig. 12.3*a*). In other words, stress or pressure gradients result in vacancy gradients, which in turn result in atomic fluxes carrying atoms or matter away in the opposite direction (Fig. 12.3*b*). It is only by appreciating this fact that sintering, creep, and hot pressing, among others, can be truly understood.

Diffusional fluxes

In Eq. (7.30), it was shown that the flux of atoms is related to the driving force by

$$J_i = \frac{c_i D_i}{kT} f \tag{12.12}$$

where c_i, D_i, and f are, respectively, the concentration, diffusivity, and driving force per atom. Once again by assuming that $\sigma_{11} = -\sigma_{22} = \sigma$, the chemical potential *difference* per atom between the top and side faces of the grain boundary shown in Fig. 12.3*b* is simply $\Delta\mu = -2\sigma\Omega$ [Eq. (12.6)]. This chemical potential difference acts over an average distance $d/2$, where d is the grain diameter; that is, $f = -d\tilde{\mu}/dx = 4\sigma\Omega/d$, which when combined with Eq. (12.12), results in

$$J_i = \frac{c_i D_i}{kT} \frac{4\sigma\Omega_i}{d} \tag{12.13}$$

The total number of atoms transported in a time t, crossing through an area A, is $N = J_i At$. Given that the volume associated with these atoms is $\Omega_i N$, the resulting strain from the displacement of the two opposite faces is given by

$$\varepsilon = \frac{\Delta d}{d} = \frac{2(\Omega_i N/A)}{d} = \frac{2\Omega_i J_i t}{d} \tag{12.14}$$

Combining Eq. 12.13 and 12.14 and noting that $c_i \Omega_i = 1$, one obtains the corresponding strain rate

$$\boxed{\dot{\varepsilon} = \frac{8\sigma\Omega_i D_i}{kTd^2}} \tag{12.15}$$

This expression is the well-known **Nabbaro-Herring** expression for creep, and it predicts that

1. The creep rate is inversely proportional to the square of the grain size d. Thus large-grain-size materials are more resistant to creep than fine-grained ceramics. This is well documented experimentally.
2. The creep rate is proportional to the applied stress, which is also experimentally observed, but as discussed in greater detail below, only at lower stresses. At higher stresses, the stress exponent is usually much greater than 1.
3. The slope of a plot of $\ln(T d\varepsilon/dt)$ versus $1/kT$ should yield the activation energy for creep. If creep occurs by lattice diffusion, that value should be the same as that measured in a diffusion experiment. This is often found to be the case.
4. Compressive stresses result in negative strains or shrinkage, while tensile strains result in elongation parallel to the direction of applied stress (Fig. 12.3c).

In deriving Eq. (12.15), the diffusion path was assumed to be through the bulk, which is usually true at higher temperatures where bulk diffusion is faster than grain boundary diffusion. However, at lower temperatures, or for very fine-grained solids, grain boundary diffusion may be the faster path, in which case the expression for the creep rate, known as **Coble creep**, becomes

$$\boxed{\dot{\varepsilon} = \psi \frac{\sigma\Omega_i \delta_{gb} D_{gb}}{kTd^3}} \tag{12.16}$$

where δ_{gb} is the grain boundary width and ψ is a numerical constant $\approx 14\pi$. Here D_i in Eq. (12.15) is replaced by $D_{gb} \delta_{gb}/d$. The term $1/d$ represents the density or number of grain boundaries per unit area; consequently, δ_{gb}/d can be considered to be a "grain boundary cross-sectional area."

It should be emphasized that Eqs. (12.15) and (12.16) are valid under the following conditions:

- The grain boundaries are the main sources and sinks for vacancies.
- Local equilibrium is established for the temperature and stress levels used; i.e., the sources and sinks are sufficiently efficient.
- Cavitation does not occur either at triple junctions or at grain boundaries.

Since both volume and grain boundary diffusion can contribute independently to creep, the overall creep rate can be represented by the sum of Eqs. (12.15) and (12.16). Note, however, that these expressions are strictly true only for pure metals or elemental crystals, since only one diffusion coefficient is involved. In general, for a binary or more complex compound, one should use a complex diffusivity $D_{complex}$ which takes into account the various diffusion paths possible for each of the charged species in the bulk and along the grain boundaries, as well as the effective widths of the latter.[210] Always remember that in ionic ceramics, the rate-limiting step is always *the slower-diffusing species moving along its fastest possible path*. For most practical applications, however, $D_{complex}$ simplifies to the diffusivity of the rate-limiting ion (see Prob. 12.1).

One final note: In Chap. 7 it was stated that it was of no consequence whether the flux of atoms or their defects were considered. To illustrate this important notion once again, it is worthwhile to derive an expression for the creep rate based on the flux of defects. Substituting Eq. (12.11) in the appropriate flux equation for the diffusion of vacancies, i.e.,

$$J_v = -D_v \frac{\Delta C_{t-c}}{\Delta x} = -\frac{C_0 D_v}{kT} \frac{4\sigma\Omega}{d} \tag{12.17}$$

results in an expression which, but for a negative sign (which is to be expected since the atoms and vacancies are diffusing in opposite directions), is identical to Eq. (12.13) (recall that $D_i c_i = C_0 D_v$). Thus once more it is apparent that it is

---

[210] See, e.g., R. S. Gordon, *J. Amer. Cer. Soc.*, **56**:174 (1973). For a compound M_aX_b

$$D_{complex} \approx \frac{\left(D^M d + \pi\delta_{gb}^M D_{gb}^M\right)\left(D^X d + \pi\delta_{gb}^X D_{gb}^X\right)}{\pi\left[a\left(D^M d + \pi\delta_{gb}^M D_{gb}^M\right) + b\left(D^X d + \pi\delta_{gb}^X D_{gb}^X\right)\right]}$$

where d is the grain size, δ_{gb} is the grain boundary width, and D^i and D_{gb}^i are the bulk and grain boundary diffusivities of the appropriate species, respectively.

equivalent whether one considers the flux of the atoms or the defects; the final result has to be and is the same, which is comforting.

12.2.2 Viscous Creep

Many structural ceramics often contain significant amounts of glassy phases at the grain boundaries, and it is now well established that for many of them the main creep mechanism is not diffusional, but rather results from the softening and viscous flow of these glassy phases. Several mechanisms have been proposed to explain the phenomenon, most notable among them being these three:

Solution reprecipitation

This mechanism is similar to the one occurring during liquid-phase sintering, where the dissolution of crystalline material into the glassy phase occurs at the interfaces loaded in compression and their reprecipitation on interfaces loaded in tension. The rate-limiting step in this case can be either the dissolution kinetics or transport through the boundary phase, whichever is slower. This topic was discussed in some detail in Chap. 10, and will not be repeated here.

Viscous flow of a glassy layer

As the temperature rises, the viscosity of the glass falls, and viscous flow of the glassy layer from between the grains can result in creep. The available models predict that the effective viscosity of the material is inversely proportional to the cube of the volume fraction f of the boundary phase, i.e.,

$$\eta_{\text{eff}} = (\text{const}) \frac{\eta_i}{f^3} \tag{12.18}$$

where η_i is the intrinsic or bulk viscosity of the grain boundary phase. Since the shear strain rate and shear stress are related by

$$\dot{\varepsilon} = \frac{\tau}{\eta_{\text{eff}}} \tag{12.19}$$

this model predicts a stress exponent of 1, which has sometimes been observed. One difficulty with this model is that as the grains slide past one another and the fluid is squeezed out from between them, one would expect the grains to eventually interlock. Thus, unless significant volume fractions of the glassy phase exist, this process must, at best, be a transient process. One interesting way to test whether this process is operative is to compare the creep rates of the same material in both tension and compression. If this process is operative, a large difference in the creep rates should be observed.

Viscous creep cavitation[211]

In some ceramic materials, notably those that contain glassy phases, failure commonly occurs intergranularly by a time-dependent accumulation of creep damage in the form of grain boundary cavities. The exact mechanism by which the damage accumulates depends on several factors such as microstructure, volume of glassy phase, temperature, and applied stress, but two limiting mechanisms have been identified: bulk and localized damage.

Low stresses and long exposure times tend to favor bulk damage, where cavities are presumed to nucleate and grow throughout the bulk, with failure occurring by the coalescence of these cavities to form a critical crack. A typical intergranular failure revealing the presence of numerous cavities along two-grain facets is shown in Fig. 12.4a. In ceramics that are essentially devoid of glassy phases, the cavities are believed to occur by vacancy diffusion. More commonly, however, cavitation is thought to occur by viscous hole growth within the intergranular glassy phase.

High stresses and short exposure times, however, tend to favor a more localized type of damage where the nucleation and growth of the cavities occur locally within the field of influence of a local stress concentrator, such as a preexisting flaw. Here two crack growth mechanisms have been identified: (1) direct extension of the creep crack along grain boundaries by diffusive or viscous flow and (2) cavitation damage ahead of a crack tip and its growth by the coalescence of these cavities, followed by the nucleation and growth of fresh cavities, and so forth (Fig. 12.4b).

The problem is further complicated by the fact that there are three time scales to worry about: sliding of grains with respect to each other, which in turn creates the negative pressure at the triple points that is responsible for the nucleation of the cavities, followed by the time needed to nucleate a cavity, and finally the growth and coalescence of these cavities — any one of which can be rate-limiting.

The understanding of this particular creep mechanism has been fueled by the emergence of Si_3N_4 as a serious candidate for structural applications at high temperatures. Usually Si_3N_4 is fabricated by liquid-phase sintering, and thus it almost always contains some glassy phase, which naturally renders cavitation creep important. How to best solve the problem is not entirely clear, but choosing glass compositions that can be easily crystallized in a postfabrication step and/or introducing second phases such as SiC, which could control grain boundary sliding, have been attempted, and the results to date appear promising.

---

211 For a recent review, see K. Chan and R. Page, *J. Amer. Cer. Soc.*, **76**:803 (1993).

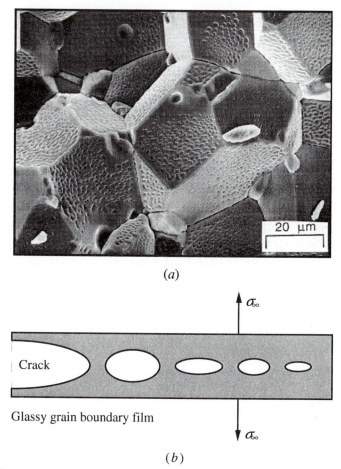

(a)

(b)

Figure 12.4

(a) Intergranular fracture of an alumina sample showing creep cavitation due to compressive creep at 1600°C.[212] Note closely spaced cavities along the two-grain facets. (b) Schematic of cavity formation in viscous grain boundary films as a result of applied tensile stress.

---

[212] Courtesy of R. Page, C. Blanchard, and R. Railsback, Southwest Research Institute, San Antonio, Texas.

12.2.3 Dislocation Creep

As noted above, the experimentally observed creep power law exponents, especially at higher temperatures and applied stresses, are in the range of 3 to 8, which none of the aforementioned models predict. Thus to explain the high-stress exponents, it has been proposed that the movement of atoms from regions of compression to tension occurs by the coordinated movement of "blocks" of material via dislocation glide or climb. In this mechanism, the creep rate can be formally expressed as

$$\dot{\varepsilon}_{ss} = \mathbf{b}\left\{\rho v(\sigma) + \frac{d\rho(\sigma)\lambda}{dt}\right\} \qquad (12.20)$$

in which $\mathbf{b}$ is Burger's vector, ρ the dislocation density, $v(\sigma)$ the average velocity of a dislocation at an applied stress σ, $d\rho/dt$ is the rate of nucleation of the dislocations at stress σ, and λ is the average distance they move before they are pinned. The main difficulty in developing successful creep models and checking their validity stems primarily from the fact that these various parameters are unknown and are quite nonlinear and interactive. Progress has been achieved recently for some materials, however.[213]

12.2.4 Generalized Creep Expression

In general, the steady-state creep of ceramics may be expressed in the form[214]

$$\dot{\varepsilon} = \frac{(\text{const})DG\mathbf{b}}{kT}\left(\frac{\mathbf{b}}{d}\right)^r\left(\frac{\sigma}{G}\right)^p \qquad (12.21)$$

where G is the shear modulus, $\mathbf{b}$ is Burger's vector, r is the grain size exponent, and p is the stress exponent defined in Eq. (12.1). It can be easily shown (see Prob. 12.3) that Eqs. (12.15) and (12.16) are of this form if $p = 1$.

Based on Eq. (12.21), the creep behavior of ceramics can be divided into two regimes:

1. A low-stress, small-grain-size regime where the creep rate is a function of grain size and the stress exponent is unity. Consequently, a plot of $\log\{\dot{\varepsilon}kT/(DG\mathbf{b})\}(d/\mathbf{b})^r$ versus $\log(\sigma/G)$ should yield a straight line with slope 1. The grain size exponent r will depend on the specific creep mechanism and is 2 for Nabbaro-Herring creep and 3 for Coble creep. Figure 12.5 compares the experimental normalized creep rate of alumina to σ/G collected from several sources (shaded area) to that predicted if the Nabarro-Herring creep was the operative mechanism. The

213 O. A. Ruano, J. Wolfenstine, J. Wadsworth, and O. Sherby, *J. Amer. Cer. Soc.*, **75**:1737 (1992).
214 W. R. Cannon and T. G. Langdon, *J. Mat. Sci.*, **18**:1–50 (1983).

agreement is rather good, considering the uncertainties in the diffusion coefficients, etc. Note that since it is d^2 that is plotted, and the slope is unity, this confirms that creep is a Nabbaro-Herring type of creep.

2. A high stress level where the creep rate becomes independent of grain size, that is, $r = 0$, and the stress exponent lies between 3 and 7. In this regime, a plot of $\log\{\dot{\varepsilon}kT/(DGb)\}(d/\mathbf{b})^r$ versus $\log(\sigma/G)$ should once again yield straight lines. In Fig. 12.6, the normalized creep rate for a number of ceramics is plotted as a function of $\log(\sigma/G)$, where it is obvious that all the data fall on straight lines with slopes between 3 and 7. Figure 12.7 schematically summarizes the creep behavior of ceramics over a wide range of applied stresses as a function of grain size. From the figure it is obvious that small grains are detrimental to the creep rates at low stresses, but that at higher stresses the intragranular movement of dislocations by climb or glide is the operative and more important mechanism. In Fig. 12.7, the role of intergranular films and the formation of multigrain junction cavitation are not addressed, but the stress exponents obtained experimentally from such mechanisms also fall in the range of 2 to 7, which, needless to say, can further cloud the interpretation of creep results.

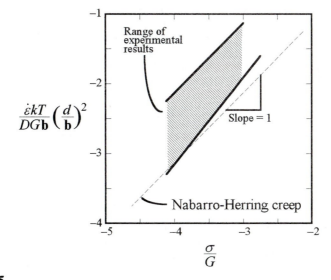

Figure 12.5

Summary of normalized creep rate versus normalized stress for alumina. The dotted line is what one would predict based on Eq. (12.15).[215]

---

215 Data taken from W. R. Cannon and T. G. Langdon, *J. Mat. Sci.*, **18**:1–50 (1983).

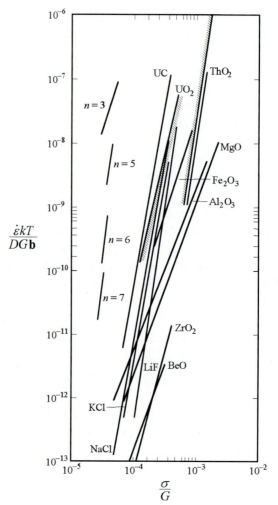

Figure 12.6
Summary of power law creep data for a number of ceramics.[216]

12.3
SUBCRITICAL CRACK GROWTH

Subcritical crack growth (SCG) refers to the slow growth of a subcritical flaw as a result of its exposure to the combined effect of stress and a corrosive

216 Data taken from W. R. Cannon and T. G. Langdon, *J. Mat. Sci.*, **18**:1–50 (1983).

environment.[217] As discussed in greater detail below, the combination of a reactive atmosphere and a stress concentration can greatly enhance the rate of crack propagation. For instance, silica will dissolve in water at a rate of 10^{-17} m/s, whereas the application of stress can cause cracks to grow at speeds greater than 10^{-3} m/s. The insidiousness of, and hence the importance of understanding, this phenomenon lies in the fact that as the crack tip advances, the material is effectively weakened and eventually can give way suddenly and catastrophically, after years of service.

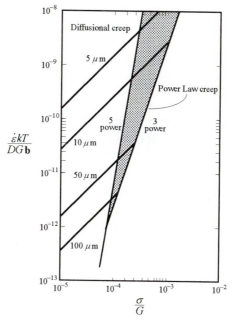

Figure 12.7
Effect of grain size on normalized creep rate versus normalized stress. As grains and consequently diffusion distances become smaller, diffusional creep becomes more important.[218]

The objective of this section is twofold: to describe the phenomenon of SCG and to relate it to what is occurring at the atomic level at the crack tip. Before one proceeds much further, however, it is important to briefly outline how this effect is quantified.

---

[217] A good example of this phenomenon that should be familiar to many is the slow crack growth over time in a car's windshield after it has been damaged.

[218] Data taken from W. R. Cannon and T. G. Langdon, *J. Mat. Sci.*, **18**:1–50 (1983).

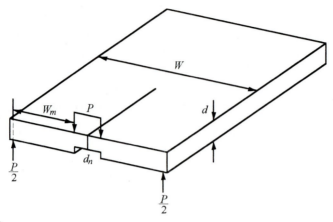

Figure 12.8
Schematic of double torsion specimen.

EXPERIMENTAL DETAILS: MEASURING SUBCRITICAL CRACK GROWTH

The techniques and test geometries that have been used to measure subcritical crack growth in ceramics are several, but they share a common principle, namely, the subjection of a well-defined crack to a well-defined stress intensity K_I, and measurement of its velocity v. The technique considered here, the advantages of which are elaborated upon below, is the double torsion geometry shown in Fig. 12.8. For the double torsion specimen, K_I is given by

$$K_I = PW_m \sqrt{\frac{3(1+\nu)}{Wd^3 d_n}} \tag{12.22}$$

where P is the applied load and ν is Poisson's ratio; all the other symbols and dimensions are defined in Fig. 12.8.
The measurements are carried out as follows:

1. A starter crack is introduced in a specimen, and a load is applied, as shown in Fig. 12.8. As a result, the starter crack will grow with time.
2. The rate of crack growth is measured, usually optically. For instance, two marks are placed on the specimen surface, the time required for the crack to propagate that distance is measured, and the crack velocity is simply $v = \Delta c / \Delta t$. The major advantage of using this test geometry should now be obvious: Since K_I is *not* a function of crack length, it follows that v is also not

a function of crack length; i.e., a constant crack velocity should be observed for any given load, which greatly simplifies the measurement and analysis. The major disadvantage of the technique, however, is that it requires reasonably large samples that require some machining.

3. The experiment is repeated under different loading conditions, either on the same specimen, if it is long enough, or on different specimens, if not.

If it is explored over a wide enough spectrum, a $\ln v$ versus K_I plot will exhibit four regions, as shown in Fig. 12.9a:

*   A threshold region below which no crack growth is observed
*   Region I, where the crack growth is extremely sensitive to K_I and is related to it by an exponential function of the form

$$v = A^* \exp \alpha K_I \qquad (12.23)$$

    where A^* and α are empirical fitting parameters
*   Region II, where the crack velocity appears to be independent of K_I
*   Region III, where the crack growth rate increases even more rapidly with K_I than in region I

Another measure of subcritical crack growth is the exponent n defined in Eq. (12.30). To measure n directly from velocity versus K_I curves is often difficult and time consuming. Fortunately, a simpler and faster technique, referred to as a **dynamic fatigue test** (not to be confused with the normal fatigue test discussed below), is available. In this method, the strain rate dependence of the average strength to failure is measured; i.e., the samples are loaded at varying rates and their strength at failure is recorded. It can be shown (see App. 12A) that the mean failure stress σ_1 at a constant strain rate $\dot{\varepsilon}_1$ is related to the mean failure stress σ_2 at a different strain rate $\dot{\varepsilon}_2$ by

$$\left(\frac{\sigma_1}{\sigma_2}\right)^{n+1} = \frac{\dot{\varepsilon}_1}{\dot{\varepsilon}_2} \qquad (12.24)$$

Hence by measuring the time to failure at different strain rates, n can be directly calculated from this relationship.

Yet another variation of this test, which is also used to obtain creep information, is to simply attach a load to a specimen and measure its time to failure. The results are then plotted in a format identical to the one shown in Fig. 12.11a for cyclic fatigue, and are referred to as **static fatigue** or **stress/life curves**.

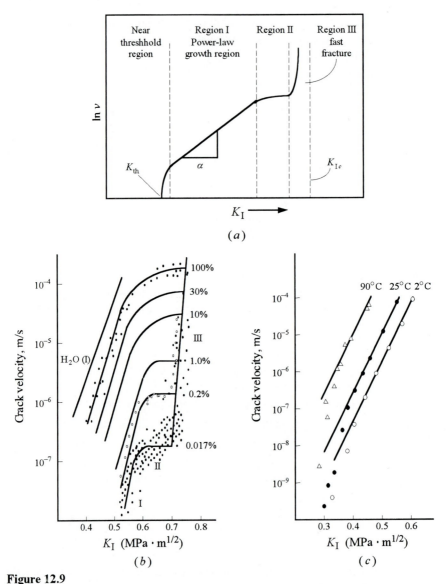

Figure 12.9

(*a*) Schematic of crack propagation rates as a function of K_I, where four stages are identified. (*b*) Actual data for soda-lime glass tested in N_2 gas of varying relative humidity shown on the right-hand side.[219] (*c*) Temperature dependence of crack propagation in same glass in water.

219 S. M. Wiederhorn, *J. Amer. Cer. Soc.*, **50**:407 (1967).

Typical v versus K data for soda-lime silicate glasses tested as a function of humidity are shown in Fig. 12.9b, where the following salient points are noteworthy:

- The rate of crack propagation is a strong function of the applied stress.
- Humidity has a dramatic effect on v; increasing the relative humidity in the ambient atmosphere from 0.02 to 100 percent, results in a greater than 3 orders-of-magnitude increase in v.
- Clear identification of the three regions just described is possible.
- The presence of a threshold K_I is not clearly defined because of the difficulty of measuring crack velocities that are much smaller than 10^{-7} m/s.
- The dramatic effect of increasing the temperature of water on the crack velocity is shown in Fig. 12.9c. The velocity increases by about 2 orders of magnitude over a temperature range of $\approx 100°C$, typical of thermally activated processes.

To understand this intriguing phenomenon, it is imperative to appreciate what is occurring at the crack tip on the atomic scale. Needless to say, the details will depend on several factors, including the chemistry of the solid in which SCG is occurring, the nature of the corrosive environment, temperature and stress levels applied. However, given the ubiquity of moisture in the atmosphere and the commercial importance of silicate glasses, the following discussion is limited to SCG in silicate-based glasses, although the ideas presented are believed to have general validity. Furthermore, since, as discussed in greater detail in Sec. 12.5.1, it is region I that determines the lifetime of a part, it is dealt with in some detail below. The other regions will be briefly touched upon at the end of this section.

In general, the models that have been suggested to explain region I for glasses in the presence of moisture can be divided into three categories: diffusional (i.e., a desintering of the material along the fracture plane), plastic flow, and chemical-reaction theories. Currently the chemical-reaction approach appears to be the most consistent with experimental results and is the one developed here.

It is now generally accepted that the stressed Si–O–Si bonds at the crack tip react with water, and other polar molecules, to form two Si–OH bonds according to the following chemical reaction:

$$Si - O - Si + H_2O = 2Si - OH \tag{I}$$

In this process, illustrated in Fig. 12.10 and referred to as **dissociative chemisorption**, a water molecule is assumed to diffuse and chemisorb to the crack tip (Fig. 12.10a) and rotate so as to align the lone pairs of its oxygen molecules with the unoccupied electron orbitals of the Si atoms (Fig. 12.10b). Simultaneously, the hydrogen of the water molecule is attracted to a bridging oxygen. Eventually, to relieve the strain on the Si–O–Si bond, the latter ruptures and is replaced by two Si–OH bonds (Fig. 12.10c).

Experimentally, it is now fairly well established that water is not the only agent that causes SCG in glasses; other polar molecules such as methanol and ammonia have also been found to cause SCG, provided that the molecules are small enough to fit into the extending crack (i.e., smaller than about 0.3 nm).

The basic premise of the chemical reaction theory is that the rate of reaction (I) is a strong function of K_I at the crack tip. According to the absolute reaction rate theory,[220] the *zero-stress* reaction rate of a chemical reaction is given by

$$K_{r,0} = \kappa \exp\left(-\frac{\Delta G^*_{P=0}}{RT}\right) = \kappa \exp\left(\frac{-\Delta H^* + T\Delta S^*}{RT}\right) \tag{12.25}$$

where κ is a constant and ΔH^* and ΔS^* are, respectively, the differences in enthalpy and entropy between the reactants in their ground and activated states.[221] In the presence of a hydrostatic pressure P, this expression has to be modified to read

$$K_r = \kappa \exp\left(-\frac{\Delta G^*}{RT}\right) \tag{12.26}$$

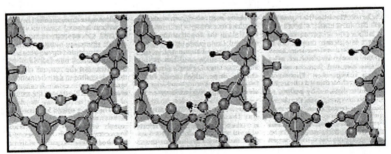

Figure 12.10
Steps in the dissociative chemisorption of water at the tip of a silica glass. (*a*) Tip of crack with approaching water molecule; (*b*) chemisorption of water and its alignment; (*c*) the breaking of an Si–O–Si bond and the formation of two Si–OH bonds.[222]

[220] S. Glasstone, K. Laidler, and H. Eyring, *The Theory of Rate Processes*, McGraw-Hill, New York, 1941.

[221] The implicit assumption here is that the forward rate of the reaction is much greater than the backward rate.

[222] Adapted from T. Michalske and B. Bunker, *Sci. Amer.*, **257**:122 (Dec. 1987).

where ΔG^* is now given by

$$\Delta G^* = \Delta H^* - T\,\Delta S^* + P\,\Delta V^*$$

and ΔV^* is the difference in volume between the reactants in their ground and activated states. Combining Eqs. (12.25) and (12.26), one sees that the effect of stress on chemical reactivity is simply

$$K_r = K_{r,0}\exp\left(-\frac{P\Delta V^*}{RT}\right) \tag{12.27}$$

This result is important because it predicts that applying a hydrostatic $(P > 0)$ pressure to a reaction should slow it down, and vice versa. Physically, this can be more easily appreciated by considering a diatomic molecular bond: That a tensile stress along the axis of the bond must enhance the rate at which rupture occurs, and consequently its chemical reactivity, is obvious.

If the crack velocity is assumed to be directly related to the reaction rate K_r and P is proportional to K_I at the crack tip, then Eq. (12.27) can be recast as [223] (see Prob. 12.5)

$$v = v_0\exp\left(\frac{-\Delta H^* + \beta K_I}{RT}\right) \tag{12.28}$$

where v_0, ΔH^*, and β are empirical constants determined from $\log v$ versus K_I types of plots. This formalism accounts, at least qualitatively, for the principal external variables since

1. It predicts a strong dependence of v on K_I consistent with the behavior in region I.
2. It also predicts an exponential temperature effect, as observed (Fig. 12.10c). Note that for constant temperatures this result and Eq. (12.23) are of the same form.
3. Although not explicitly included in Eq. (12.28), the effect of moisture is embedded in the preexponential factor. This comes about by recognizing that the rate of reaction [Eq. (12.26)] is proportional to the concentration of water (see Prob. 12.5b).

---

[223] R. J. Charles and W. B. Hillig, pp. 511–527 in *Symposium on Mechanical Strength of Glass and Ways of Improving It*, Florence, Italy, September 1961, *Union Scientifique Continentale du Verre*, Charleroi, Belgium, 1962. See also S. W. Weiderhorn, *J. Amer. Cer. Soc.*, **55**:81–85 (1972).

It is important to note that in much of the literature, an empirical power-law expression of the form

$$v = A\left(\frac{K_I}{K_{I_c}}\right)^n \tag{12.29}$$

which is usually further abbreviated to[224]

$$v = A'' K_I^n \tag{12.30}$$

is sometimes used to describe SCG, instead of Eq. (12.28). Note that in this formalism the temperature dependence is embedded in the A term [see Eq. (12.46)]. The sole advantage of using Eq. (12.29) or (12.30) over Eq. (12.28) is the ease with which the former can be integrated (see Sec. 12.5.1). Equation (12.28), however, must be considered more fundamental, first, because it has some scientific underpinnings and, second, because it explicitly predicts the exponential temperature dependence of the phenomenon. Finally, note that the experimental results can usually be fitted equally well by either of these equations; extrapolations, however, can lead to considerable divergences.[225]

Before one moves on to the next topic, consider briefly the other regions observed in the v versus K plots:

1. *Threshold region.* Although it is difficult to unequivocally establish experimentally that there is a threshold stress intensity K_{th} below which no crack growth occurs from v versus K_I types of curves, probably the most compelling results indicating that it indeed exists come from crack healing studies. At very low values of K_I, the driving force for crack growth is low, and it is thus not inconceivable that its rate of growth at some point would equal the rate at which it heals, the driving force for which would be the reduction in surface energy. In other words, a sort of dynamic equilibrium is established, and a threshold results.

2. Region II. The hypothesis explaining the weak dependence of crack velocity on K in region II is that the crack velocity is limited by the rate of arrival of the corroding species at the crack tip.

3. Region III. This stage is not very well understood, but once again a combination of stress and chemical reaction is believed to accelerate the crack.

---

224 Note that while A has the dimensions of meters per second, A'' has the unwieldy dimensions of $ms^{-1}\left(Pa \cdot m^{1/2}\right)^{-n}$.

225 T. Michalske and B. Bunker, *J. Amer. Cer. Soc.*, **76**:2613 (1993).

12.4
FATIGUE OF CERAMICS

It has long been assumed that because dislocation motion in ceramics is limited, strain hardening and consequent crack extension during cyclic loading would not occur, and hence ceramics were not susceptible to fatigue damage. And indeed, ceramics with homogeneous microstructures such as glass or very fine-grained single-phase ceramics do not appear to be susceptible to cyclic loadings.

However, more recently, with the development of tougher ceramics that exhibit R curve behavior (see Chap. 11), such as transformation-toughened zirconia and whisker- and fiber-reinforced ceramics, it is becoming clear that the situation is not as simple as first thought. Recent data seem to suggest that R curve behavior can be detrimental to fatigue life. Before one tackles the micromechanisms of fatigue, however, a brief description is warranted of what is meant by fatigue and what the relevant parameters are.

EXPERIMENTAL DETAILS: MEASURING FATIGUE

In a typical fatigue test, a sample is subjected to an alternating stress of a given amplitude and frequency. The *cyclic stress amplitude* is defined as

$$\sigma_{amp} = \frac{\sigma_{max} - \sigma_{min}}{2} \tag{12.31}$$

whereas the load ratio R is defined as

$$R = \frac{\sigma_{min}}{\sigma_{max}} \tag{12.32}$$

where σ_{min} and σ_{max} are, respectively, the minimum and maximum stress to which the sample is subjected (Fig. 12.11a). The experiments can be carried out either in tension-tension, compression-compression, or tension-compression, in which case R would be negative.

Two types of specimens are typically used, smooth "crack-free" specimens or specimens containing long cracks, i.e., cracks of dimensions that are large with respect to the structural features of the material.

For the smooth or crack-free specimens, the experiments are run until the sample fails. The results are then used to generate **S/N curves** where the applied stress amplitude is plotted versus the cycles to failure (which are equivalent to the time to failure if the frequency is kept constant), as shown in Fig. 12.11a.

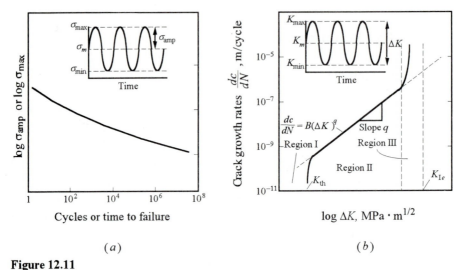

Figure 12.11
(a) The stress amplitude versus cycles to failure curve (S/N curve). Inset shows definition of stress amplitude σ_{amp}. (b) Curves of $\log(dc/dN)$ versus $\log \Delta K_I$. Slope of curve in region II is q, and ΔK is defined in inset.

For the specimens with long cracks, the situation is not unlike that dealt with in the previous section, except that instead of measuring v versus K, the crack growth rate per cycle dc/dN is measured as a function of ΔK_I, defined as

$$\Delta K_I = \xi(\sigma_{max} - \sigma_{min})\sqrt{\pi c} \qquad (12.33)$$

where ξ is a geometric factor of the order of unity.

Typical crack growth behavior of ceramics is represented schematically in Fig. 12.11b as $\log \Delta K_I$ versus $\log(dc/dN)$. The resulting curve is sigmoidal and can be divided into three regions, labeled I, II, and III. Below K_{th}, that is, region I, the cracks will not grow with cyclic loading. But just prior to rapid failure, the crack growth is accelerated once more (region III).

In the midrange, or region II, the growth rates are well described by

$$\boxed{\frac{dc}{dN} = B(\Delta K)^q} \qquad (12.34)$$

where B and q are empirically determined constants.

Given the similarity in behavior between fatigue and SCG [compare Figs. 12.9a and 12.11b or Eqs. (12.34) and (12.30)], one of the major experimental difficulties in carrying out fatigue experiments lies in ascertaining that the degradation in strength observed is truly due to the cyclic nature of the loading and not due to SCG. Even more care must be exercised when the tests are carried out at higher temperatures, since as noted above, SCG is a thermally activated process and hence becomes more important at elevated temperatures.

Typical long-crack data for a number of ceramics are shown in Fig. 12.12, where the $\log(dc/dN)$ versus $\log \Delta K_I$ curves are linear and very steep, implying high values of q. These studies also indicate that under cyclic loading, the thresholds for crack growth can be as low as 50 percent of the fracture toughness measured under monotonically increasing loads.

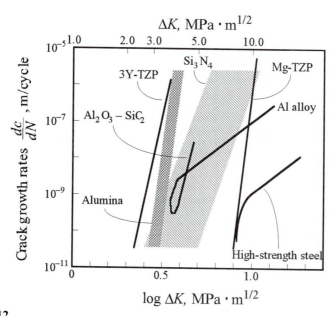

Figure 12.12
Cyclic fatigue long-crack propagation data for several ceramics compared to those for some typical metals.[226] TZPs are tetragonal zirconia polycrystals (see Chap. 11).

226 Data taken from R. O. Ritchie and R. H. Dauskardt, "Cyclic Fatigue of Ceramics," *J. Cer. Soc. Japan.*, **99**:1047–1062 (1991).

Micromechanisms of fatigue

At this point, the micromechanics of what is occurring at the crack tip in ceramic materials are not fully understood; the models put forth are still tentative, and more work needs to be carried out in this area to clearly establish the various mechanisms and their applicability to various systems. The recent results, however, have established that (1) no one micromechanical model can successfully explain all fatigue data in ceramics; (2) fatigue in ceramics appears to be fundamentally different from that of metals, where crack propagation results from dislocation activity at the crack tip; and (3) ceramics that exhibit R curve behavior appear to be the most susceptible to fatigue, indicating that the cyclic nature of the loading somehow diminishes the effect of the crack-tip shielding mechanisms discussed in Chap. 11. For instance, in the case of fiber- or whisker-reinforced ceramics, it is believed that unloading induces fracture or buckling of the whiskers in the crack wake, which in turn reduces their shielding effect. If the toughening, on the other hand, is achieved primarily by grain bridging or interlocking, then the unloading cycle is believed to cause cracking and/or crushing of asperities between crack faces and the progressive frictional sliding at bridging interfaces.

Finally, it is interesting to note that the few studies on cyclic fatigue of ceramics at elevated temperatures seem to indicate that at high homologous temperatures (i.e., in the creep regime) cyclic fatigue does not appear to be as damaging as SCG or static fatigue. In the cases where it has been observed, the improved cyclic fatigue behavior has been attributed to bridging of the crack surfaces by grain boundary glassy phases.

12.5
LIFETIME PREDICTIONS

Creep, fatigue, and SCG are dangerous in that if they are not taken into account, they can result in sudden and catastrophic failure with *time*. Thus, in addition to the probabilistic aspects of failure discussed in Chap. 11, from a design point of view, the central question is, How long can a part serve its purpose reliably? The conservative approach, of course, would be to design with stresses that are below the thresholds discussed above. An alternative approach is to design a part to last for a certain lifetime, after which it would be replaced or at least examined for damage. In the following sections, the methodology is described that is used to calculate lifetime for each of these three phenomena.

12.5.1 Lifetime Predictions during SCG

Replacing v in Eq. (12.29) by dc/dt, rearranging, and integrating, one obtains

$$\int_0^{t_f} dt = \int_{c_i}^{c_f} \frac{dc}{v} = \frac{K_{Ic}^n}{A} \int_{c_i}^{c_f} \frac{1}{K_I^n} dc \qquad (12.35)$$

where c_i and c_f are the initial and final (just before failure) crack lengths, respectively, and t_f is the time to failure. Recalling that $K_I = \Psi \sigma_a \sqrt{\pi c}$ [Eq. (11.12)], one sees that dc in Eq. (12.35) can be eliminated and recast in terms of K_I as

$$t_f = \frac{2 K_{Ic}^n}{A \Psi^2 \sigma_a^2 \pi} \int_{K_i}^{K_{Ic}} \frac{dK_I}{K_I^{n-1}} \qquad (12.36)$$

which upon integration yields

$$t_f = \frac{2 K_{Ic}^n}{A \Psi^2 \pi \sigma_a^2 (n-2)} \left[\frac{1}{K_i^{n-2}} - \frac{1}{K_{Ic}^{n-2}} \right] \qquad (12.37)$$

Given that $n \approx 10$, even if K_I is as high as $0.5 K_{Ic}$, the second term is less than 0.5 percent of the first, and thus with great accuracy[227]

$$\boxed{t_f = \frac{2 K_{Ic}^n}{A \Psi^2 \pi \sigma_a^2 (n-2) K_i^{n-2}}} \qquad (12.38)$$

When a similar integration is carried out on Eq. (12.28), the lifetime is given by (see Prob. 12.5)

$$t_f = \frac{2}{v_0'} \left[\frac{RT}{\beta \sigma_a \Psi} \sqrt{\pi c_i} + \left(\frac{RT}{\beta \sigma_a \Psi} \right)^2 \right] e^{\left(-\frac{\beta \sigma_a \Psi \sqrt{\pi c_i}}{RT} \right)} \qquad (12.39)$$

where $v_0' = v_0 \exp[-\Delta H^*/(RT)]$. Both Eqs. (12.38) and (12.39) predict that the lifetime of a component that is susceptible to SCG is a strong function of K_I (see Worked Example 12.2).

---

[227] Ignoring the second term in Eq. (12.37) says that the fraction of the lifetime the crack spends as its size approaches c_f is insignificant compared to the time taken by the crack to increase from c_i to $c_i + \delta c$, when its velocity is quite low.

WORKED EXAMPLE 12.2. For silica glass tested in ambient temperature water, $v_0 = 3 \times 10^{-22}$ m/s and $\beta = 0.182$ m$^{5/2}$. Estimate the effect of increasing K_I from 0.4 to 0.5 MPa·m$^{1/2}$ on the lifetime.

Answer

By noting that $K_I = \Psi \sigma_a \sqrt{\pi c}$, Eq. (12.39) can be rewritten as

$$t_f = (\text{const}) e^{-\beta K_I/(RT)}$$

where the dependence of the preexponential term on K_I was ignored, which is a very good approximation relative to the exponential dependence. Substituting the appropriate values for K_I and β in this expression, one obtains

$$\frac{t_{f@0.4}}{t_{f@0.5}} = \frac{\exp\left(-\dfrac{0.182 \times 0.4 \times 10^6}{8.314 \times 300}\right)}{\exp\left(-\dfrac{0.182 \times 0.5 \times 10^6}{8.314 \times 300}\right)} = 1476$$

In other words, a decrease in K_I from 0.5 to 0.4 MPa·m$^{1/2}$ increases the lifetime by a factor of about 1500!

The methodology just described, while useful in predicting the lifetime of a sample with a well-defined starter crack (t_f cannot be calculated without a knowledge of K_I or c_i), is not amenable to a probabilistic analysis such as the one discussed in Sec. 11.5. However, as discussed below, the relationships derived for the lifetimes, particularly Eq. (12.38), permit the straightforward construction of strength/probability/time diagrams that can be used for design purposes.

Strength/probability/time (SPT) diagrams

For a set of identical specimens, i.e., given that A, K_I or equivalently c_i, and Ψ are constant, it follows directly from Eq. (12.38) that

$$\frac{t_2}{t_1} = \left(\frac{\sigma_1}{\sigma_2}\right)^n \tag{12.40}$$

where t_1 and t_2 are the lifetimes at σ_1 and σ_2, respectively. If one assumes that a sample was tested such that it failed in 1 s at a stress of σ_1 [clearly not an easy task, but how that stress can be estimated from constant strain rate experiments is outlined in App. 12A — see Eq. (12A.8)], it follows from Eq. (12.40) that the stress σ_2 at which these same samples would survive a lifetime of $t_2 = 10^\theta$ is

$$\frac{10^{\theta}}{1} = \left(\frac{\sigma_1}{\sigma_2}\right)^n$$

where $t_1 = 1$ s. Taking the log of both sides and rearranging, one obtains

$$\log \sigma_2 = \log \sigma_1 - \frac{\theta}{n} \tag{12.41}$$

This is an important result in that it allows for the straightforward construction of an SPT diagram from a Weibull plot (Fig. 11.21). The procedure is as follows (see Worked Example 12.3):

- Convert the stress at any failure probability to the equivalent stress σ_{1s} that would have resulted in failure in 1 s, using Eq. (12A.8).
- Plot $\log \sigma_{1s}$ versus $\log\log(1/S)$, as shown in Fig. 12.13 (line labeled 1 s). Since the Weibull modulus m is not assumed to change, the slope of the line will remain the same as the one shown in Fig. 11.21.
- Draw a series of lines parallel to the original line, with a spacing between the lines equal to $1/n$, as shown in Fig. 12.13. Each line represents a decade increase in lifetime.

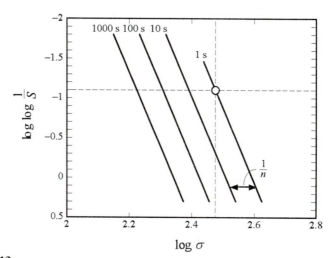

Figure 12.13

Effect of SCG rate exponent n on Weibull plots. The Weibull plots are shifted by $1/n$ for every decade of life required of the part. Data are the same as plotted in Fig. 11.21, but are converted to a log plot.

WORKED EXAMPLE 12.3. The data collected in Worked Example 11.2 were measured at a strain rate of 1×10^{-3} s^{-1} and n for this material was measured to be 10. (a) Construct the SPT diagram for the data listed in Table 11.2. (b) Calculate the design stress that would result in a lifetime of 10^4 s and still maintain a probability of survival of 0.999. Assume $Y = 350$ GPa.

Answer

(a)   Referring to Table 11.2, the failure stress at, say, $S = 0.837$ [i.e., $\log \log(1/S) = -1.11$] was 310 MPa. Converting this stress to the stress that would have caused the sample to fail in 1 s [Eq. (12A.8)] yields

$$\sigma_1 = 310 \left(\frac{310 \times 10^6}{350 \times 10^9 \times 1 \times 10^{-3}} \right)^{1/11} = 306 \text{ MPa}$$

for which $\log \sigma = 2.48$. The intersection of the two dotted lines establishes a point on the Weibull plot (Fig. 12.13). A line is then drawn through this point, with the same slope m as the original one. The other lines are plotted parallel to this line but are shifted to the right by $1/n$ or 0.1 for every decade of lifetime, as shown in Fig. 12.13.

(b) In Worked Example 11.2, the stress for which the survival probability was 0.999 was calculated to be 200 MPa. The corresponding 1 s failure stress is

$$\sigma_1 = 200 \left(\frac{200 \times 10^6}{350 \times 10^9 \times 1 \times 10^{-3}} \right)^{1/11} = 190 \text{ MPa}$$

To calculate the design stress that would result in a lifetime of 10^4 s, use is made of Eq. (12.40), or

$$\frac{10^4}{1} = \left(\frac{190}{\sigma_2} \right)^{10}$$

Solving for σ_2 yields 75 MPa. In other words, because of SCG, the applied stress would have to be downgraded by a factor of ≈ 3 in order to maintain the same survival probability of 0.999. It is left as an exercise to readers to convince themselves that for *higher* values of n, the reduction in design stress would have been *less* severe.

12.5.2 Lifetime Predictions during Fatigue

Given the similarities between the shape of the curves in the intermediate region for SCG and fatigue, one can design for a given fatigue lifetime by using the aforementioned methodology (see Prob. 12.13). However, given the large values of q, there is little gain in doing so; design based on the threshold fracture toughness

ΔK_{th} alone suffices. This can be easily seen in Fig. 12.12: To avoid fatigue failure, the stress intensity during service should simply lie to the left of the lines shown.

The actual situation is more complicated, however, because the results shown in Fig. 12.12 are only applicable to long cracks. Short cracks have been shown to behave quite differently from long ones; furthermore, the very high values of q imply that marginal differences in either the assumed initial crack size or component in-service stresses can lead to significant variations in projected lifetimes (see Prob. 12.13).

The more promising approach at this time appears to be to use S/N curves such as shown in Fig. 12.11a and simply to design at stresses below which no fatigue damage is expected, i.e., use a fatigue limit approach. The major danger of this approach, however, lies in extrapolating data that were evaluated for simple and usually small parts to large, complex structures where the defect population may be quite different.

12.5.3 Lifetime Predictions during Creep

The starting point for predicting lifetimes during creep is the **Monkman-Grant equation**, which states that the product of the time to failure t_f and the strain rate $\dot{\varepsilon}$ is a constant, or

$$\dot{\varepsilon} t_f = K_{MG} \qquad (12.42)$$

What this relationship, in effect, says is that every material will fail during creep when the strain in that material reaches a certain value K_{MG}, independent of how slow or how fast that strain was reached. That the Monkman-Grant expression is valid for Si_3N_4 is shown in Fig. 12.14, where the range of data obtained for the vast majority of tensile stress rupture tests lies in the hatched area. On such a curve, Eq. (12.42) would appear as a straight line with a slope of 1, which appears to be the case.

If one assumes the creep rate is either grain-size-independent or for a set of samples with comparable grain sizes, Eq. (12.21) can be recast as

$$\dot{\varepsilon}_{ss} = A_0 \left(\frac{\sigma_a}{\sigma_0} \right)^p \exp\left(-\frac{Q_c}{kT} \right) \qquad (12.43)$$

where Q_c is activation energy for creep, A_0 is a constant, σ_a is the applied stress that causes failure, and σ_0 is a normalizing parameter that defines the units of σ_a. Since the stresses one deals with are normally in the megapascal range, σ_0 is usually taken to be 1 MPa. Combining these two equations, one obtains

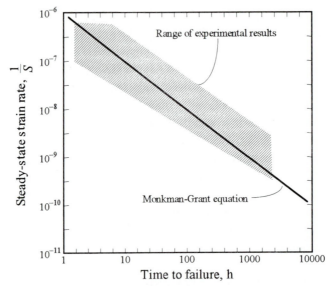

Figure 12.14

Summary of a large number of tensile rupture data collected from various sources for Si_3N_4 plotted to verify the Monkman-Grant equation.[228]

$$t_f = \frac{K_{MG}}{A_0}\left(\frac{\sigma_0}{\sigma_a}\right)^p \exp\frac{Q_c}{kT} \tag{12.44}$$

In other words, the lifetime of a part should decrease exponentially with increasing temperature as well as with increasing applied stress; both predictions are borne out by experiments. It is worth noting that Eqs. (12.42) to (12.44) are only valid if the rate of damage generation was controlled by the bulk creep response of the material and steady-state conditions are established during the experiment.

12.5.4 Fracture Mechanism Maps

High-temperature failure of ceramics typically occurs by either subcritical crack growth or creep. In an attempt to summarize the data available so as to be able to quickly and easily identify the mechanisms responsible for failure and their relative importance, Ashby suggested plotting the data on what is now known as *fracture*

228 Adapted from S. M. Wiederhorn, *Elevated Temperature Mechanical Behavior of Ceramic Matrix Composites*, S. V. Nair and K. Jakus, eds., Butterworth, Stoneham, Massachusetts, 1993.

mechanism maps.[229] The starting point for constructing such a map is to recast Eqs. (12.44) and (12.38), respectively, as

$$\sigma_a = \left(\frac{K_{MG}}{A_0}\right)^{1/p} \frac{\sigma_0}{t_f^{1/p}} \exp\frac{Q_c}{pkT} \tag{12.45}$$

$$\sigma_a = B_0\sigma_0 \left(\frac{t_0}{t_f}\right)^{1/n} \exp\frac{Q_{SCG}}{nkT} \tag{12.46}$$

at which point their similarities become quite obvious.[230] Variables t_0 and σ_0 are introduced in Eq. (12.46) to keep B_0 dimensionless and define the scales — most commonly, t_0 is chosen to be 1 h and σ_0 is again 1 MPa.

To construct such a map (see Worked Example 12.4), SCG and creep rupture data must be known for various temperatures. The temperature dependence of the stress levels required to result in a given lifetime are then calculated from Eqs. (12.45) and (12.46). The mechanism that results in the lowest failure stress at a given temperature thus defines the threshold stress or highest applicable stress for the survival of a part for a given time. In other words, the lifetime of the part is determined by the fastest possible path. Such maps are best understood by actually plotting them.

WORKED EXAMPLE 12.4. Using the following information, construct a fracture deformation map for Si_3N_4: $K_{MG} \approx 5.4 \times 10^{-3}$ (from Fig. 12.13), $p = 4$, $Q_c = 800$ kJ/mol, $A_0 = 1.44 \times 10^{19}$ h^{-1}, $\sigma_0 = 1$ MPa (these are typical values for Si_3N_4, except p which is usually closer to 8 or 9; see Prob. 12.11 for another set of data). For subcritical crack growth in Si_3N_4, assume $B_0 = 80$, $n = 55$, $Q_{SCG} = 760$ kJ/mol, $\sigma_0 = 1$ MPa, and $t_0 = 1$ h.

Answer

Plugging the appropriate numbers in Eq. (12.45), one obtains

$$\sigma_a(\text{MPa}) = \sigma_0\left(\frac{K_{MG}}{A_0t_f}\right)^{1/p} \exp\frac{Q_c}{pkT} = \frac{4.4 \times 10^{-6}}{t_f^{1/4}} \exp\frac{24,055}{T}$$

---

[229] M. F. Ashby, C. Gandhi, and D. M. R. Taplin, *Acta Metall.*, **27**:1565 (1979).

[230] Note that the temperature dependence which was buried in A in Eq. (12.38) is now spelled out. Also note that the exponent of σ_a in Eq. (12.38) once K_I is replaced by $\Psi\sigma_a\sqrt{\pi c}$ is n (see Prob. 12.5).

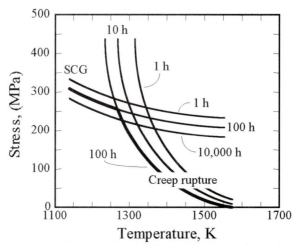

Figure 12.15

Fracture mechanism map for Si_3N_4 using data given in Worked Example 12.4.

When this equation is plotted as a function of temperature, a series of steep lines emerge that are shifted toward the left with increasing lifetimes, as shown in Fig. 12.15.

Similarly, a plot of Eq. (12.46), that is,

$$\sigma_a(\text{MPa}) = \sigma_0 B_0 \left(\frac{t_0}{t_f}\right)^{1/n} \exp\frac{Q_{SCG}}{nkT} = \frac{80}{t_f^{1/55}} \exp\frac{1662}{T}$$

gives another series of almost parallel lines labeled SCG in Fig. 12.15. Note that the large value of n makes Si_3N_4 much less susceptible to SCG at higher temperatures.

The advantage of such maps is that once they have been constructed, the stress-temperature regime for which a part would survive a given lifetime is easily delineated. Referring to Fig. 12.15, one sees that to design a part to withstand 100 h of service, the design should be confined to the domain encompassed by the 100-h lines. In other words, a part subjected to a combination of stress and temperature that lies within the heavy lines (lower left-hand corner of Fig. 12.15) will survive for at least 100 h — any other combination would result in a shorter lifetime.

12.6
SUMMARY

1. The removal of atoms from regions that are in compression and placing them in regions that are in tension reduces the free energy of the atoms by an amount $\approx 2\Omega\sigma$, where σ is the applied stress and Ω is the atomic volume. This reduction in energy is the driving force for creep.

2. Diffusional creep is a thermally activated process that depends on the diffusivity of the slower species along its fastest path. When the material is subjected to relatively low stresses and/or temperatures, the creep rate typically increases linearly with stress; i.e., the stress exponent is unity. Because the diffusion path scales with grain size d, finer-grained material is usually less resistant to diffusional creep than large-grain material. If the ions diffuse through the bulk (Nabbarro-Herring creep), the creep rate is proportional to d^{-2}. But if they diffuse along the grain boundaries (Coble creep), then the creep rate scales as d^{-3}.

3. At high stresses, the creep rate is much more sensitive to the applied stress, with stress exponents anywhere between 3 and 8, and is independent of grain size.

4. The presence of glassy phases along grain boundaries can lead to cavitation and stress rupture.

5. Subcritical crack growth can also occur by the combined effect of stresses and corrosive environment and/or the accumulation of damage at crack tips. The basic premise at ambient and near-ambient temperatures is that SCG results from a stress-enhanced reactivity of the chemical bonds at the crack tip. This phenomenon is thus a strong function of the stress intensity at the crack tip and typically is thermally activated. At high temperatures, the phenomenon of SCG is believed to occur by the formation of cavities ahead of a crack tip. The crack then grows by the coalescence of these cavities.

6. In some ceramic materials, most notably those that exhibit R curve behavior, data have been reported that confirm fatigue effects. These effects are believed to result from a weakening of the shielding elements, such as whiskers or large grains. The results also seem to indicate that short cracks behave differently from long ones, indicating that perhaps, from a the design point of view, it is more promising to use S/N curves (Fig. 12.11a) than crack growth rate curves (Fig. 12.11b).

APPENDIX 12A

Derivation of Eq. (12.24)

Equation (12.40), while appearing to be a useful equation to estimate the lifetime of a part, has to be used with caution since it assumes that c_i is identical in all samples. And while that is possible in a laboratory setting where well-defined cracks can be introduced in a sample, its usefulness in practice is limited. What is thus needed is a methodology to transform data (i.e., one with a distribution of failure times) to an equivalent stress that would have caused failure in, say, 1 s; in other words, a renormalization of the time-to-failure data.

Figure 12.16 illustrates the problem. The sloping line represents the stress on a sample as a result of some increasing strain rate $\dot{\varepsilon}$. The specimen fails at stress σ_f after time t_ε. Had the stress been applied instantaneously, the sample would have failed after a shorter time t_σ because the average stress is initially higher.

Equation (12.40) can be recast to read

$$t\sigma^n = \Gamma' \tag{12A.1}$$

where Γ' is a constant. Invoking the notion that the sum of the fractional times the material spends at any stress has to be unity, i.e.,

$$\sum \frac{dt}{t} = 1 \tag{12A.2}$$

implies that for a constant-strain-rate test

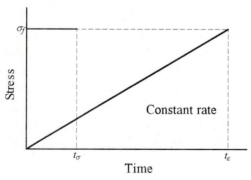

Figure 12.16
Comparison of stress versus time response for constant-stress and constant-rate tests.

$$\int_0^{t_\varepsilon} \frac{dt}{t} = 1 \tag{12A.3}$$

where t_ε is the time for the sample to fail when the strain rate is $\dot{\varepsilon}$ (Fig. 12.16). It follows directly from Fig. 12.16 that

$$t_\varepsilon = \sigma_f \frac{dt}{d\sigma} \tag{12A.4}$$

Combining Eqs. (12A.4) and (12A.3), eliminating t by using Eq. (12A.1), and integrating Eq. (12A.3), one obtains

$$1 = \int_0^{\sigma_f} \frac{t_\varepsilon \sigma^n d\sigma}{\Gamma' \sigma_f} = \frac{t_\varepsilon \sigma_f^n}{\Gamma'(n+1)} \tag{12A.5}$$

where σ_f is the fracture stress. Making use of Eq. (12A.1) once again, one sees that

$$\frac{t_\varepsilon}{t_\sigma} = n + 1 \tag{12A.6}$$

Rewriting Hooke's law as $\sigma_f = Y \dot{\varepsilon} t_\varepsilon$, one obtains

$$\sigma_f = Y \dot{\varepsilon}_1 (n+1) t_\sigma \tag{12A.7}$$

Combining this equation with Eqs. (12.40), (12A.6), and (12A.7) yields the sought-after result

$$\sigma_1 = \sigma_f \left(\frac{\sigma_f}{Y \dot{\varepsilon} t_\varepsilon} \right)^{1/n} = \sigma_f \left(\frac{\sigma_f}{Y \dot{\varepsilon}(n+1)} \right)^{1/n} \tag{12A.8}$$

This result thus allows for the calculation of the constant stress σ_{1s} which would have caused failure in 1 s from the stress σ_f at which the sample failed at a constant strain rate $\dot{\varepsilon}$, both of which are experimentally accessible.

By rewriting Eq. (12A.8) for two different strain rates, it is easy to show that

$$\frac{\overline{\sigma}_{\varepsilon_1}}{\overline{\sigma}_{\varepsilon_2}} = \left(\frac{\dot{\varepsilon}_1}{\dot{\varepsilon}_2} \right)^{1/n} \tag{12A.9}$$

where $\overline{\sigma}_{\varepsilon_i}$ is the mean value of the failure stress measured at strain $\dot{\varepsilon}_i$.

PROBLEMS

12.1. (a) Consider an oxide for which $dD^M \gg \delta_{gb}D^M_{gb} \gg dD^X \gg \delta_{gb}D^X_{gb}$. Which ion do you think will be rate-limiting and which path will it follow? See footnote for definition of terms.

(b) Repeat part (a) for $dD^X \gg \delta_{gb}D^M_{gb} \gg \delta_{gb}D^X_{gb} \gg dD^M$.

12.2. If the grain boundary diffusivity is given by

$$D_{gb} = 100\exp(-40 \text{ kJ}/RT)$$

and the bulk diffusion coefficient is

$$D_{latt} = 300\exp(-50 \text{ kJ}/RT)$$

at 900 K determine whether grain boundary or lattice diffusion will dominate. How about at 1300 K? At which temperature will they be equally important?

Answer: $T = 1095$ K

12.3. For what values of r and p does Eq. (12.21) become similar to Eqs. (12.15) and (12.16)?

12.4. (a) Derive an expression for the vacancy concentration difference between the side and bottom surfaces of a cylindrical wire of radius ρ subjected to a normal tensile stress σ_{nn}.

(b) Calculate the vacancy concentration at two-thirds of the absolute melting point of Al and Cu subjected to a tensile strain of 0.2 percent. The enthalpy of vacancy formation for Cu and Al is, respectively, 1.28 and 0.67 eV. The Young's moduli at two-thirds of their respective melting points are 110 and 70 GPa, respectively. State all assumptions.

Answer: $C_{Cu} \approx 3.72 \times 10^{21}$ m^{-3}, $C_{Al} \approx 2.84 \times 10^{23}$ m^{-3}

(c) At what radius of curvature of the wire will the surface energy contribution be comparable to the applied stress contribution? Assume the applied stress is 10 MPa. State all other assumptions. Comment on the implications of your solution to the relative importance of externally applied stresses versus those that result from curvature.

Answer: $r_c \approx 0.01$ μm

12.5. (a) Derive Eq. (12.28).

(b) Rederive Eq. (12.28), but include a term that takes into account the concentration of moisture.

(c) Derive Eqs. (12.38) and (12.39).

(d) For $n=10$, estimate the error in neglecting the second term within brackets in Eq. (12.37). Why can this term be safely neglected? And does that imply that the lifetime is independent of K_{Ic}? Explain.

(e) Show that the lifetime of a part subjected to SCG [Eq. (12.38)] can be equally well expressed in terms of the initial crack length c_i by

$$t_f = \frac{2K_{Ic}^n}{A\left(\Psi\sqrt{\pi\sigma_a}\right)^n (n-2)c_i^{n/2-1}}$$

12.6. Typical crack growth data for a glass placed in a humid environment are listed below. Calculate the values of A'', A, and n, given that $K_{Ic} = 0.7 \text{ MPa} \cdot \text{m}^{1/2}$. What are the units of A''?

Stress intensity, MPa·m$^{1/2}$	0.4	0.5	0.55	0.6
Crack velocity, m/s	1×10^{-6}	1×10^{-4}	1×10^{-3}	1×10^{-2}

Answer: $n=22.5$, $A''=772$, $A \approx 0.25 \text{ m/s}$

12.7. If the specimen in Prob. 11.7 was loaded in tension under a stress of 10 MPa and all cracks but the surface crack shown on the left-hand side were ignored, calculate the lifetime for the part. Assume $n=15$ and $A = 0.34 \text{ m/s}$. State all other assumptions.

Answer: 123 s

12.8. (a) The Weibull plots shown in Fig. 12.17 were generated at two different strain rates $\dot{\varepsilon}_1 = 2\times10^{-6} \text{ s}^{-1}$ and $\dot{\varepsilon}_2 = 2\times10^{-5} \text{ s}^{-1}$. Which strain is associated with which curve? Briefly explain your choice.

(b) Calculate the stress needed to obtain a 0.9 survival probability and a lifetime of 10^7 s. Information you may find useful: $K_{Ic} = 3 \text{ MPa} \cdot \text{m}^{1/2}$, $Y = 100 \text{ GPa}$.

Answer: 42 MPa

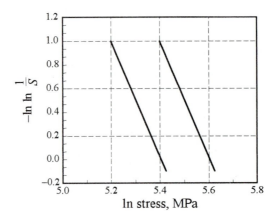

Figure 12.17
Effect of strain rate on Weibull plots.

12.9. *(a)* A design engineer wants to design an engine component such that the probability of failure will be at most 0.01, and the choice is between the following two materials:

Ceramic A	Mean strength 600 MPa	$m = 25$	$n = 11$
Ceramic B	Mean strength 500 MPa	$m = 17$	$n = 19$

Which would you recommend and why? What maximum design stress would be allowable to ensure the survival probability required? State all assumptions.

Answer: 499 MPa

(b) The values shown above were measured for samples that had the following dimensions: 1 cm by 1 cm by 8 cm. If the component part is 10 times that volume, would your recommendation change concerning which material to use? Would the design stress change?

Answer: 455 MPa

(c) If the part were to be used in a moist environment, do you get concerned? If the lifetime of the part is to be 4 years or $\approx 10^9$ s, what changes, if any, would you recommend for the design stress? Does your recommendation as to which material to use change? Explain. Assuming, for the sake of simplicity, that the data reported in part *(a)* above were obtained in 1 s, determine which material to use by calculating the design stress needed for the lifetime noted for both.

Answer: Material A, 76 MPa; Material B, 295 MPa

12.10 *(a)* To study the degradation in strength of a ceramic component, the average flexural strength of the material was measured *following* exposure under stress in a corrosive environment. The Weibull modulus, measured to be

10, did not change with time, but the average strength was found to decrease from 350 MPa after 1 day to 330 MPa after 3 days. Qualitatively explain what is happening.

(b) Calculate the average strength one would expect after a 10 weeks exposure.
Answer: 279 MPa

(c) Post-test examination of the samples that failed after 1 day showed that the average size of the cracks that caused failure was of the order of 120 μm. Calculate the average crack size that was responsible for failure after 3 days. Information you may find useful: $K_{Ic} = 3$ MPa·m$^{1/2}$.
Answer: 135 μm

12.11. You are currently using the Si$_3$N$_4$ whose properties are listed in Worked Example 12.4, and a new Si$_3$N$_4$ appears on the market with the following properties: $K_{MG} = 5.4 \times 10^{-3}$, $p = 9$, $Q_c = 1350$ kJ/mol, $A_0 = 4 \times 10^{19}$ h^{-1}, and $\sigma_0 = 1$ MPa. For subcritical crack growth you can assume $B_0 = 100$, $n = 50$, $Q_{SCG} = 900$ kJ/mol, $\sigma_0 = 1$ MPa, and $t_0 = 1$ h. By constructing a fracture mechanism map and comparing it to Fig. 12.15, which material would you use and why?
Answer: See Fig. 12.18

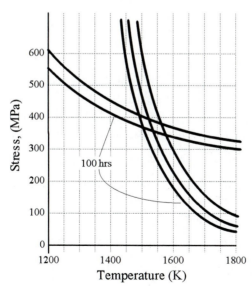

Figure 12.18
Fracture mechanism map for Si$_3$N$_4$, for which the properties are listed in Prob. 12.11.

12.12. The surface energy of a glass can be measured by carrying out a zero creep experiment, where a glass fiber is suspended in a furnace and is heated. As a result of gravity, the wire will extend to a final equilibrium length l_{eq}, beyond which no further increase in length is measured.

(a) What do you suppose keeps the fiber from extending indefinitely?

(b) Show that at equilibrium $l_{eq} = 2\gamma_{sv}/\rho r_{eq} g$ where g and r are, respectively, the gravitational constant and the density of the glass; r_{eq} is the equilibrium radius of the wire. *Hint*: Write an expression for the total energy of the system that includes gravitational and surface energy terms, and minimize that function with respect to l.

12.13. (a) Show that the number of cycles to failure during fatigue is given by

$$N_f = \frac{2}{B\left(\xi\sqrt{\pi\Delta\sigma_a}\right)^q (q-2)c_i^{q/2-1}}$$

where c_i is the initial crack size. All other terms are defined in the text. Note the similarity of this expression to the one derived in Prob. 12.5e.

(b) Estimate the values of B and q for Mg-TZP shown in Fig. 12.11. What are the units of B?

Answer: $B = 1.7 \times 10^{-48}$; $q \approx 40$

(c) Estimate the effect of increasing the applied stress by a factor of 2 on N_f.

Answer: 4.3×10^{12} cycles!

(d) Estimate the effect of doubling the assumed initial crack size on N_f.

Answer: 5×10^5 cycles!

12.14. The maximum design stress for an alumina shaft was calculated to be 100 MPa, assuming the applied load was static. If the shaft were to be subjected to a cyclic load such that K_{max} corresponds to 100 MPa, should the design stress be altered? If so, by how much? State all assumptions.

Answer: 60 MPa

12.15. Rectangular glass slides were tested in 3-point bending as a function of surface finish. The results obtained are listed below. Qualitatively explain the trends.

Treatment	As-received	HF etched	Abraded normal to applied stress	Abraded parallel to applied stress
Strength, MPa	87	106	42	71

ADDITIONAL READING

1. W. Cannon and T. Langdon, "Creep of Ceramics, Parts I and II," *J. Mat. Sci.*, **18**:1–50 (1983).
2. R. Ritchie and R. Daukardt, "Cyclic Fatigue of Ceramics: A Fracture Mechanism Approach to Subcritical Crack Growth and Life Prediction," *J. Cer. Soc. Japan.*, **99**:1047–1062 (1991).
3. R. Raj, "Fundamental Research in Structural Ceramics for Service Near 2000°C," *J. Am. Cer. Soc.*, **76**:2147–2174 (1993).
4. N. A. Fleck, K. J. Kang, and M. F. Ashby, "The Cyclic Properties of Engineering Materials," *Acta. Met.*, **42**:365–381 (1994).
5. S. Suresh, *Fatigue of Materials*, Cambridge University Press, Cambridge, England, 1991.
6. H. Reidel, *Fracture at High Temperatures*, Springer-Verlag, Heidelberg, Germany, 1987.
7. K. Chan and R. Page, "Creep Damage in Structural Ceramics," *J. Amer. Cer. Soc.*, **76**:803 (1993).
8. R. L. Tsai and R. Raj, "Overview 18: Creep Fracture in Ceramics Containing Small Amounts of Liquid Phase," *Acta Met.*, **30**:1043–1058 (1982).
9. M. F. Ashby, C. Gandhi, and D. M. R. Taplin, "Fracture Mechanism Maps for Materials Which Cleave: FCC, BCC and HCP Metals and Ceramics," *Acta Met.*, **27**:1565 (1979).
10. R. W. Davidge, *Mechanical Behavior of Ceramics*, Cambridge University Press, Cambridge, England, 1979.
11. B. Lawn, *Fracture of Brittle Solids*, 2d ed., Cambridge University Press, Cambridge, England, 1993.
12. R. Warren, ed., *Ceramic Matrix Composites*, Blackie, Glasgow, Scotland, 1992.
13. A. Kelly and N. H. Macmillan, *Strong Solids*, 3d ed., Clarendon Press, New York, 1986.
14. G. Weaver, "Engineering with Ceramics, Parts 1 and 2," *J. Mat. Ed.*, **5**:767 (1983), and **6**:1027 (1984).
15. A. G. Evans, "Engineering Property Requirements for High Performance Ceramics," *Mat. Sci & Eng.*, **71**:3 (1985).
16. S. M. Weiderhorn, "A Probabilistic Framework for Structural Design," in *Fracture Mechanics of Ceramics*, vol. 5, R. C. Bradt, A. G. Evans, D. P. Hasselman, and F. F. Lange, eds., Plenum, New York, 1978, p. 613.
17. M. F. Ashby and B. F. Dyson, *Advances in Fracture Research*, S. R. Valluri, D. M. R. Taplin, P. Rama Rao, J. F. Knott, and R. Dubey, eds., Pergamon Press, New York, 1984, p. 3.
18. T. H. Courtney, *Mechanical Behavior of Materials*, McGraw-Hill, New York, 1990.

CHAPTER 13

Thermal Properties

> *What happens in these Lattices when* Heat
> *Transports Vibrations through a solid mass?*
> $T = 3Nk$ *is much too neat;*
> *A rigid Crystal's not a fluid Gas.*
> Debye *in 1912 proposed Elas-*
> *Tic Waves called* phonons *that obey Max Planck's*
> $E = h\nu$. *Though amorphous Glass,*
> Umklapp *Switchbacks, and Isotopes play pranks*
> *Upon his Formulae,* Debye *deserves warm Thanks.*

> John Updike, *The Dance of the Solids*[†]

13.1
INTRODUCTION

As a consequence of their brittleness and their low thermal conductivities, ceramics are prone to thermal shock; i.e., they will crack when subjected to large thermal gradients. This is why it is usually not advisable to pour a very hot liquid into a cold glass container, or cold water on a hot ceramic furnace tube — the rapidly cooled surface will want to contract, but will be restrained from doing so by the bulk of the body, so stresses will develop. If these stresses are large enough, the ceramic will crack.

Thermal stresses will also develop because of thermal contraction mismatches in multiphase materials or anisotropy in a single phase. It thus follows that thermal stresses exist in all polycrystalline ceramics with noncubic structures that undergo phase transformations or include second phases with differing thermal expansion characteristics. These stresses can result in the formation of stable microcracks and

---

[†] J. Updike, **Midpoint and other Poems**, A. Knopf, Inc., New York, New York, 1969. Reprinted with permission.

can strongly influence the strength and fracture toughness of ceramics. In a worst-case scenario, these stresses can cause the total disintegration of a ceramic body. Used properly, however, they can enhance the strength of glasses. The purpose of this chapter is to explore the problem of thermal residual stresses, why they develop and how to quantify them.

Another important thermal property dealt with in Sec. 13.6 is thermal conductivity. It is the low thermal conductivity of ceramics, together with their chemical inertness and oxidation resistance, that renders them as a class of materials uniquely qualified to play an extremely demanding and critical role during metal smelting and refining. Many ceramics such as diaspore, alumina, fosterite, and periclase are used for the fabrication of high-temperature insulative firebrick without which the refining of metals would be impossible.

13.2
THERMAL STRESSES

The Origin of Thermal Residual Stresses

As noted above, thermal stresses can be induced by differential thermal expansion in multiphase materials or anisotropy in the thermal expansion coefficients of single-phase solids. The latter is treated in Sec. 13.4. To best illustrate the idea of how differential thermal expansion in multiphase materials leads to thermal stresses, consider the simple case shown schematically in Fig. 13.1a, where a solid disk is placed inside of a ring of a different material. To emphasize the similarity of this problem to that of an inclusion in a matrix, which was discussed in Chap. 11 and is one of practical significance, the disk will henceforth be referred to as the *inclusion*, and the outside ring as the *matrix*, with thermal expansion coefficients α_i and α_m, respectively.

Before one attempts to find a quantitative answer, it is important to qualitatively understand what happens to such a system as the temperature is varied. Needless to say, the answer will depend on the relative values of α_i and α_m, and whether the system is being heated or cooled. To illustrate, consider the case where $\alpha_i > \alpha_m$ and the system is heated. Both the inclusion and the matrix will expand[231] (Fig. 13.1b); however, given that $\alpha_i > \alpha_m$, the inclusion will try to

---

[231] Note that the expansion of the matrix implies that the internal diameter of the ring increases with increasing temperature.

expand at a faster rate, but will be radially restricted from doing so by the outside ring. It follows that upon heating, both the inclusion and the matrix will be in radial compression. It is left as an exercise to the reader to show that if the assembly were cooled, the inclusion would develop radial tensile stresses. It should be noted here, and is discussed in greater detail below, that stresses other than radial also develop.

The quantification of the problem is nontrivial and is usually carried out today by using finite-element and other numerical techniques. However, for simple geometries, a powerful method developed by Eshellby[232] exists, which in principle is quite simple, elegant, and ingenious. The problem is solved by carrying out the following series of imaginary cuts, strains, and welding operations illustrated in Fig. 13.1:

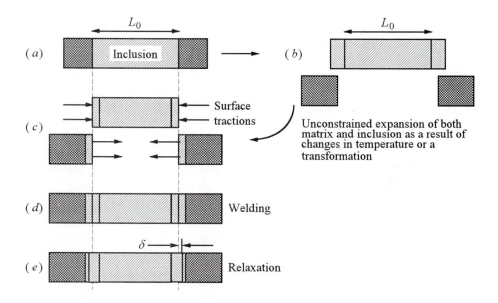

Figure 13.1

Steps involved in Eshellby's method. (*a*) Initial configuration. (*b*) Cutting and allowing for free expansion of both inclusion and matrix as a result of heating. Note that the radius of the outside ring increases upon heating. (*c*) Application of surface forces needed to restore elements to original shape. (*d*) Weld pieces together. (*e*) Allow the system to relax. Note displacement of original interface as a result of relaxation.

---

232 J. D. Eshellby, *Proc. Roy. Soc.*, **A241**:376–396 (1957).

1. Cut the inclusion out of the matrix.
2. Allow both the inclusion and the matrix to expand or contract as a result of either heating or cooling (or as a result of a phase transformation) (Fig. 13.1b).
3. Apply sufficient surface traction to restore the elements to their original shape (Fig. 13.1c).
4. Weld the pieces together (Fig. 13.1d).
5. Allow the system to relax (Fig. 13.1e).

To apply this technique to the problem at hand, do the following:

1. Cut the inclusion, and allow both it and the matrix to freely expand (Fig. 13.1b). The thermal strain in the inclusion is given by [Eq. (4.2)]:

$$\frac{\Delta L}{L_0} = \varepsilon_i = \alpha_i \Delta T = -\alpha_i (T_{final} - T_{init})$$

$$\boxed{\varepsilon_i = \alpha_i (T_{final} - T_{init})} \tag{13.1}$$

Similarly, for the matrix

$$\varepsilon_m = \alpha_m \Delta T \tag{13.2}$$

Note that as defined here, ΔT is positive during heating and negative during cooling. On cooling, T_{final} is usually taken to be room temperature; T_{init}, however, is more difficult to determine unambiguously, but it is the highest temperature below which the residual stresses are not relieved, which, depending on the material in question, may or may not be identical to the processing or annealing temperature. At high enough temperatures, stress relaxation by diffusive or viscous flow will usually relieve some, if not most, of the residual stresses; it is only below a certain temperature that these stress relaxation mechanisms become inoperative and local elastic residual stresses start to develop from the contraction mismatch.

2. Apply a stress to each element to restore it to its original shape[233] (Fig. 13.1c). For the inclusion,

$$\sigma_i = -Y_i \varepsilon_i = -Y_i \alpha_i \Delta T \tag{13.3}$$

where Y is Young's modulus. For the matrix:

$$\sigma_m = Y_m \varepsilon_m = Y_m \alpha_m \Delta T \tag{13.4}$$

---

[233] Equations (13.2) and (13.3) are strictly true only for a one-dimensional problem. Including the other dimensions does not generally greatly affect the final result [see Eq. (13.8)].

Note that the applied stress needed to restore the inclusion to its original shape is compressive (see Fig. 13.1c), which accounts for the minus sign in Eq. (13.3).

3. Weld the two parts back together (Fig. 13.1d), and allow the stresses to relax. Since the stresses are unequal, one material will "push" into the other, and the location of the *original interface will shift by a strain* δ in the direction of the larger stress until the two stresses are equal (Fig. 13.1e). At equilibrium the two radial stresses are equal and are given by

$$\sigma_{i,\mathrm{eq}} = Y_i[\varepsilon_i + \delta] = \sigma_{m,\mathrm{eq}} = Y_m[\varepsilon_m - \delta] \tag{13.5}$$

Solving for δ, plugging that back into Eq. (13.5), and making use of Eqs. (13.1) to (13.4), one can easily show (see Prob. 13.2) that

$$\sigma_{i,\mathrm{eq}} = \sigma_{m,\mathrm{eq}} = \frac{\Delta\alpha\,\Delta T}{1/Y_i + 1/Y_m} = \frac{(\alpha_m - \alpha_i)\Delta T}{1/Y_i + 1/Y_m} \tag{13.6}$$

This is an important result, and it predicts that

- If $\Delta\alpha$ is zero, no stress develops, which makes sense since the matrix and the inclusion would be expanding at the same rate.
- For $\alpha_i > \alpha_m$, upon heating (positive ΔT), the stresses generated in the inclusion and matrix should be compressive or negative, as anticipated.
- If the inclusion is totally *constrained from moving* (that is, $\alpha_m = 0$ and Y_m is infinite), then Eq. (13.6) simplifies to the more familiar equation

$$\sigma_{i,\mathrm{eq}} = -Y_i\alpha_i\,\Delta T \tag{13.7}$$

which predicts that upon heating, the stress generated will be compressive, and vice versa upon cooling.

In treating the system shown in Fig. 13.1, for simplicity's sake, only the radial stresses were considered. The situation in three dimensions is more complicated, and it is important at this stage to be able to at least qualitatively predict the nature of these stresses. Since the problem is no longer one-dimensional, in addition to the radial stresses, the **axial** and **tangential** or **hoop stresses** have to be considered.

To qualitatively predict the nature of these various stresses, a useful trick is to assume the lower thermal expansion coefficient of the two components to be zero and to carry out the Eshellby technique. To illustrate, consider the nature of the thermal residual stresses that would be generated if a fiber with expansion coefficient α_f were embedded in a matrix (same problem as the one shown in Fig. 13.1, except that now the three-dimensional state of stress is of interest), densified, and *cooled* from the processing temperature for the case when $\alpha_m > \alpha_f$.

Given that $\alpha_m > \alpha_f$ and by making use of the aforementioned trick, i.e., by assuming $\alpha_f = 0$ (which implies its dimension does not change with temperature changes), it follows that upon cooling, the matrix will shrink both axially and radially (the hole will get smaller). Consequently, the stress required to fit the matrix to the fiber will have to be axially tensile; when the matrix is welded to the fiber and allowed to relax, this will place the fiber in a state of axial residual compressive stress, which, in turn, is balanced by an axial tensile stress in the matrix. Radially, the matrix will clamp down on the fiber, resulting in radial compressive stresses in both the fiber and the matrix, in agreement with the conclusions drawn above. In addition, the system will develop tensile tangential stresses, as shown in Fig. 13.2a.[234] These stresses, if sufficiently high, can cause the matrix to crack radially as shown in Fig. 13.2c. It is left as an exercise to readers to determine the state of stress when $\alpha_m < \alpha_f$, and to compare their results with those summarized in Fig. 13.2b.

Finally, in this section the problem of a spherical inclusion in an infinite matrix is considered. It can be shown that the radial (σ_{rad}) and tangential (σ_{tan}) stresses generated for a spherical inclusion of radius R at a distance r away from the interface are given by:

$$\sigma_{rad} = -2\sigma_{tan} = \frac{(\alpha_m - \alpha_i)\Delta T}{(1 - 2\nu_i)/Y_i + (1 + \nu_m)/(2Y_m)}\left(\frac{R}{r+R}\right)^3 \qquad (13.8)$$

where ν_i and ν_m are, respectively, Poisson's ratio for the inclusion and matrix. The stress is a maximum at the interface, i.e., at $r = 0$, and drops rapidly with distance. Note that the final form of this expression is similar to Eq. (13.6). It is worth noting here that the Eshellby technique is not restricted to calculating thermal stresses; also, it can be used to calculate transformation stresses.

---

[234] To appreciate the nature of tangential stresses, it helps to go back to the Eshellby technique and ask, What would be required to make the hole in the matrix, which is now smaller than the fiber it surrounds, larger? The answer is, One would have to stretch the matrix in a manner similar to fitting a smaller-diameter hose around a larger-diameter pipe. This naturally results in a tangential stress in the hose. Experience tells us that if the hose is too small, it will develop radial cracks similar to the one shown in Fig. 13.2c.

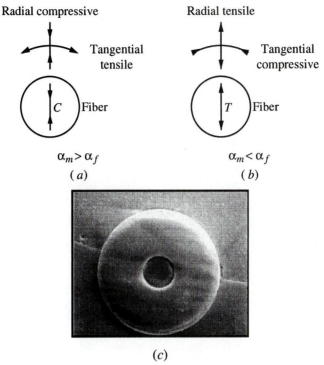

Figure 13.2

Radial and tangential stresses developed upon cooling of a fiber embedded in a matrix for (a) $\alpha_m < \alpha_f$ and (b) $\alpha_m > \alpha_f$. (c) Micrograph of radial cracks generated around fibers upon cooling when $\alpha_m > \alpha_f$.

13.3
THERMAL SHOCK

Generally speaking, thermal stresses are to be avoided since they can significantly weaken a component. In extreme cases, a part can spontaneously crumble during cooling. As noted earlier, *rapid* heating or cooling of a ceramic will often result in its failure. This kind of failure is known as *thermal shock* and occurs when thermal gradients and corresponding thermal stresses exceed the strength of the part. For instance, as a component is rapidly cooled from a temperature T to T_0, the surface will tend to contract but will be prevented from doing so by the bulk of the component that is still at temperature T. By using arguments similar to the ones made above, it is easy to appreciate that in such a situation surface tensile stresses would be generated that have to be counterbalanced by compressive ones in the bulk.

EXPERIMENTAL DETAILS: MEASURING THERMAL SHOCK RESISTANCE

Thermal shock resistance is usually evaluated by heating samples to various temperatures T_{max}. The samples are rapidly cooled by quenching them from T_{max} into a medium, most commonly ambient temperature water. The postquench retained strengths are measured and plotted versus the severity of the quench, or $\Delta T = T_{max} - T_{ambi}$. Typical results of such experiments are shown in Fig. 13.3a, where the salient feature is the occurrence of a rapid decrease in retained strength around a critical temperature difference ΔT_c below which the original strength is retained. As the quench temperature is further increased, the strength decreases but more gradually. Actual data for single-crystal and polycrystalline alumina are shown in Fig. 13.3b.

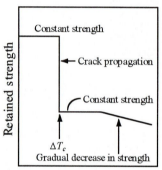

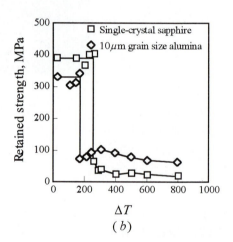

(a)   (b)

Figure 13.3

(a) Schematic of strength behavior as a function of severity of quench ΔT. (b) Actual data for single-crystal and polycrystalline alumina[235] (error bars were omitted for the sake of clarity).

From a practical point of view, it is important to be able to predict ΔT_c. Furthermore, it is only by understanding the various parameters that affect thermal shock that successful design of solids which are resistant to thermal shock can be carried out. In the remainder of this section, a methodology is outlined for doing

235 T. K. Gupta, *J. Amer. Cer. Soc.*, **55**:249 (1972).

just that, an exercise that will by necessity highlight the important parameters that render a ceramic resistant to thermal shock.

To estimate ΔT_c, the following assumptions are made[236]

1. The material contains N identical, uniformly distributed, Griffith flaws per unit volume.
2. The flaws are circular with radii c_i.
3. The body is uniformly cooled with the external surfaces rigidly constrained to give a well-defined triaxial tensile state of stress given by[237]

$$\sigma_{\text{ther}} = -\frac{\alpha Y \Delta T}{(1-2\nu)} \tag{13.9}$$

4. Crack propagation occurs by the simultaneous propagation of these N cracks, with negligible interactions between the stress fields of neighboring cracks.

The derivation is straightforward and follows the one carried out in deriving Eq. (11.9). The total energy of the system can be expressed as

$$U_{\text{tot}} = U_0 - U_{\text{strain}} + U_{\text{surf}}$$

where U_0 is the energy of the stress- and crack-free crystal of volume V_0; U_{surf} and U_{strain} are, respectively, the surface and strain energies of the system. Since it was assumed that the stress fields were noninteracting, in the presence of N cracks U_{tot} is modified to read

$$U_{\text{tot}} = U_0 + \frac{V_0 \sigma_{\text{ther}}^2}{2Y} - \frac{N \sigma_{\text{ther}}^2}{2Y} \frac{4\pi c_i^3}{3} + NG_c \pi c_i^2 \tag{13.10}$$

where the third term on the right-hand side represents the strain energy released by the existence of the cracks and the last term is the energy needed to extend them.

Differentiating this expression with respect to c_i, equating the resulting expression to zero, and rearranging terms, one can easily show (see Prob. 13.6a) that for $\Delta T > \Delta T_c$, where ΔT_c is given by

$$\Delta T_c \geq \sqrt{\frac{G_c(1-2\nu)^2}{\alpha^2 Y c_i}} \tag{13.11}$$

the cracks will grow and consume the strain energy of the system. Conversely, for $\Delta T \leq \Delta T_c$, the strain energy that develops in the system is insufficient to extend the

---

[236] The derivation shown here is a simplified version of one originally outlined by D. P. H. Hasselman, *J. Amer. Cer. Soc.*, **46**:453 (1963) and **52**:600 (1969).

[237] Note similarity of this equation to Eq. (13.7).

cracks, which in turn implies that the strength should remain unchanged, as experimentally observed.

In contrast to the situation of a flaw propagating as a result of a constant applied stress, in which the flaw will extend indefinitely until fracture, the driving force for crack propagation during thermal shock is finite. In the latter case, the cracks will only extend up to a certain length c_f that is commensurate with the strain energy available to them and then stop. To estimate c_f, one simply equates the strain energy available to the system to the increase in surface energy, or

$$\pi N G_c \left(c_f^2 - c_i^2 \right) = \frac{(\alpha \Delta T_c)^2 Y}{2(1 - 2\nu)^2} \tag{13.12}$$

For short initial cracks, that is, $c_f \gg c_i$, substituting for the value of ΔT_c from Eq. (13.11), one obtains

$$c_f \cong \sqrt{\frac{1}{\pi N c_i}} \tag{13.13}$$

which interestingly enough does not depend on any material parameters.

For the sake of clarity, the model used to derive Eqs. (13.11) and (13.13) was somewhat simplified. Using a slightly more sophisticated approach, Hasselman obtained the following relationships:

$$\Delta T_c = \sqrt{\frac{\pi G_c (1 - 2\nu)^2}{Y \alpha^2 (1 - \nu^2) c_i}} \left[1 + \frac{16 N c_i^3 (1 - \nu)^2}{9(1 - 2\nu)} \right] \tag{13.14}$$

$$c_f = \sqrt{\frac{3(1 - 2\nu)}{8(1 - \nu^2) N c_i}} \tag{13.15}$$

And while at first glance these expressions may appear different from those derived above, on closer examination, their similarity becomes obvious. For example, for small cracks of low density, the second term in brackets in Eq. (13.14) can be neglected with respect to unity, in which case, but for a few terms including Poisson's ratio and π, Eq. (13.14) is similar to Eq. (13.10). The same is true for Eqs. (13.13) and (13.15).

Before one proceeds further, it is worthwhile to summarize the physics of events occurring during thermal shock. Subjecting a solid to a rapid change in temperature results in differential dimensional changes in various parts of the body and a buildup of stresses within it. Consequently, the strain energy of the system will increase. If that strain energy increase is not too large, i.e., for small ΔT

values, the preexisting cracks will not grow and the solid will not be affected by the thermal shock. However, if the thermal shock is large, the many cracks present in the solid will extend and absorb that excess strain energy. Since the available strain energy is finite, the cracks will extend only until most of the strain energy is converted to surface energy, at which point they will be arrested. The final length to which the cracks will grow will depend on their initial size and density. If only a few, small cracks are present, then their final length will be quite large and the degradation in strength will be high. Conversely, if there are numerous small cracks, then each will extend by a small amount and the corresponding degradation in strength will not be that severe. In the latter case, the solid is considered to be **thermal-shock-tolerant**.

It is this latter approach that is used in fabricating insulating firebricks for furnaces and kilns. The bricks are fabricated such that the final product is porous and contains many flaws which, perforce, makes them weak. However, because of the very large number of flaws and pores within them, these bricks can withstand severe thermal cycles without structural failure.

Inspecting Eq. (13.11) or (13.14), it is not difficult to conclude that a good figure of merit for thermal shock resistance is

$$R_H = (\text{const})(\Delta T_c) = (\text{const})\sqrt{\frac{G_c}{\alpha^2 Y}} = \frac{K_{Ic}}{\alpha Y} \qquad (13.16)$$

from which it is clear that ceramics with low thermal expansion coefficients, low elastic moduli, but high fracture toughnesses should be resistant to thermal shock.

Kingery's[238] approach to the problem was slightly different. He postulated that failure would occur when the thermal stress, given by Eq. (13.7), was equal to the tensile strength σ_t of the specimen (see Prob. 13.4). By equating the two, it can be shown that the figure of merit in this case is

$$R_{TS} = (\text{const})(\Delta T_c) = (\text{const})\frac{(1-2\nu)\sigma_t}{\alpha Y} \qquad (13.17)$$

However, given that σ_t is proportional to $(G_c Y/c_{max})^{1/2}$, it is a trivial exercise to show that R_{TS} is proportional to $R_H/c_{max}^{1/2}$, implying that the two criteria are related.[239]

One parameter which is not included in either model, and which clearly must have an important effect on thermal shock resistance, is the thermal conductivity of

---

238 W. D. Kingery, *J. Amer. Cer. Soc.*, **38**:3–15 (1955).

239 It is interesting to note that the Hasselman solid is a highly idealized one where all the flaws are the same size.

the ceramic k_{th} (see Sec. 13.6). Given that thermal gradients are ultimately responsible for the buildup of stress, it stands to reason that a highly thermally conductive material would not develop large thermal gradients and would thus be thermal shock resistant. For the same reason, the heat capacity and the heat-transfer coefficient between the solid and the environment must also play a role. Thus an even better indicator of thermal shock resistance is to multiply Eq. (13.16) or (13.17) by k_{th}. These values are calculated for a number of ceramics and listed in Table 13.1 in columns 7 and 8. Also listed in Table 13.1 are the experimentally determined values. A correlation between the two sets of values is apparent, giving validity to these aforementioned models.

Note that in general the nitrides and carbides of Si, with their lower thermal expansion coefficients, are more resistant to thermal shock than oxides. In theory, a material with zero thermal expansion would not be susceptible to thermal shock. In practice, a number of such materials do actually exist commercially, including some glass-ceramics that have been developed which, as a result of thermal expansion anisotropy, have extremely low α's. Another good example is fused silica which also has an extremely low α and thus is not prone to thermal shock.

TABLE 13.1

Comparison of thermal shock parameters for a number of ceramics. Poisson's ratio was taken to be 0.25 for all materials

Material	MOR, MPa	Y, GPa	α 10^6 K^{-1}	k_{th}, W/(m·K)	K_{Ic}, MPa·m$^{1/2}$	$k_{th}R_{TS}$, W/m	$R_H k_{th}$, W/m^2	ΔT_c, exper.
SiAlON	945	300	3.0	21	7.7	16,500	180.00	900
HP†–Si$_3$N$_4$	890	310	3.2	15–25	5.0	16,800	126.00	500–700
RB‡–Si$_3$N$_4$	240	220	3.2	8–12	$\approx$2.0	2,557	28.00	$\approx$500
SiC (sintered)	483	410	4.3	84	3.0	17,300	143.00	300–400
HP†–Al$_2$O$_3$	380	400	9.0	6–8	3.9	633	8.00	200
HP†–BeO	200	400	8.5	63		2,800		
PSZ§	610	200	10.6	2	$\approx$10.0	435	9.43	$\approx$500
Ti$_3$SiC$_2$	300	320	10.0	43	$\approx$7.0			>1400

† Hot-pressed

‡ Reaction-bonded

§ Partially stabilized zirconia

13.4
SPONTANEOUS MICROCRACKING OF CERAMICS

In the previous section, the emphasis was on thermal shock, where failure of a ceramic part was initiated by a *rapid and/or severe temperature change*. This is not always the case; both single- and multiphase ceramics have been known to spontaneously microcrack upon cooling. Whereas thermal shock can be avoided by slow cooling, the latter phenomenon is *unavoidable* regardless of the rate at which the temperature is changed.

Spontaneous microcracking results from the buildup of residual stresses which can be caused by one or more of the following three reasons:

- Thermal expansion anisotropy in single-phase materials
- Thermal expansion mismatches in multiphase materials
- Phase transformations and accompanying volume changes in single- or multiphase materials

In the remainder of this section each of these cases is explored in some detail.

13.4.1   Spontaneous Microcracking due to Thermal Expansion Anisotropy

Noncubic ceramics with high thermal expansion anisotropy have been known to spontaneously microcrack upon cooling.[240] The cracking, which occurs along the grain boundaries, becomes progressively less severe with decreasing grain size, and below a certain "critical" grain size, it is no longer observed. The phenomenon has been reported for various solids such as Al_2O_3, graphite, Nb_2O_5, and many titania-containing ceramics such as TiO_2, Al_2TiO_5, Mg_2TiO_5, and Fe_2TiO_5. Data for some anisotropic crystals are given in Table 13.2.

Before one attempts to quantify the problem, it is important once again to understand the underlying physics. Consider the situation shown in Fig. 13.4a, where the grains, assumed to be cubes, are arranged in such a way that adjacent *grains* have different thermal expansion coefficients along their x and y axes as shown, with $\alpha_1 < \alpha_2$. To further elucidate the problem, use the aforementioned trick of equating the lower thermal expansion to zero, or $\alpha_1 = 0$. If during cooling the grains were unconstrained, the shape of the assemblage would be that shown in Fig. 13.4b. But the cooling is *not* unconstrained, which implies that a buildup of stresses at the boundaries will occur. It is this stress that is ultimately responsible for the failure of these ceramics.

---

240   The thermal expansion coefficients of cubic materials are isotropic and hence do not exhibit this phenomenon.

TABLE 13.2
Thermal expansion coefficients for some ceramic crystals with anisotropic thermal expansion behavior

Material	Normal to c axis	Parallel to c axis
Al_2O_3	8.3	9.0
Al_2TiO_5	−2.6	11.5
$3Al_2O_3 \cdot 2SiO_2$ (mullite)	4.5	5.7
$CaCO_3$	−6.0	25.0
$LiAlSi_2O_6$ (β-spodumene)	6.5	−2.0
$LiAlSiO_4$ (β-eucryptite)	8.2	−17.6
$NaAlSi_3O_8$ (albite)	4.0	13.0
SiO_2 (quartz)	14.0	9.0
TiO_2	6.8	8.3
$ZrSiO_4$	3.7	6.2

To estimate the critical grain size above which spontaneous microcracking would occur, once again the various energy terms have to be considered. For the sake of simplicity, the grains are assumed to be cubes with grain size d in which case the total energy of the system is[241]

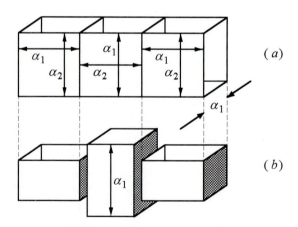

Figure 13.4
Schematic of how thermal expansion anisotropy can lead to the development of thermal stresses upon cooling of a polycrystalline solid. (*a*) Arrangement of grains prior to cooling shows relationship between thermal expansion coefficients and grain axis.
(*b*) Unconstrained contraction of grains. Here it was assumed that $\alpha_1 = 0$.

---

[241] The treatment here is a slightly simplified version of that carried out by J. J. Cleveland and R. C. Bradt, *J. Amer. Cer.*, **61**:478 (1978).

$$U_{\text{tot}} = U_s - NU_g d^3 + 6Nd^2 G_{c,\text{gb}} \tag{13.18}$$

where N is the number of grains relieving their stress and $G_{c,\text{gb}}$ is the grain boundary fracture toughness; U_s is the energy of the unmicrocracked body, and U_g is the strain energy per unit volume stored in the grains. Differentiating Eq. (13.18) with respect to d and equating to zero yields the critical grain size

$$d_{\text{crit}} = \frac{4G_{c,\text{gb}}}{U_g} \tag{13.19}$$

U_g is estimated as follows: For a totally constrained grain, the stress developed is given by Eq. (13.7). Extending the argument to two adjacent grains, one can see that the residual stress can be approximated by

$$\sigma_{\text{th}} = \frac{1}{2} Y \Delta\alpha_{\max} \Delta T \tag{13.20}$$

where $\Delta\alpha_{\max}$ is the maximum anisotropy in thermal expansion between two crystallographic directions. Substituting Eq. (13.20) in the expression for the strain energy per unit volume, that is, $U_g = \sigma^2/(2Y)$, and combining with Eq. (13.19), one obtains

$$d_{\text{crit}} = \frac{32G_{c,\text{gb}}}{Y \Delta\alpha_{\max}^2 \Delta T^2} \tag{13.21}$$

In general, however,

$$\boxed{d_{\text{crit}} = (\text{const}) \frac{G_{c,\text{gb}}}{Y \Delta\alpha_{\max}^2 \Delta T^2}} \tag{13.22}$$

where the value of the numerical constant one obtains depends on the details of the models. This model predicts that the critical grain size below which spontaneous microcracking will *not* occur is a function of the thermal expansion anisotropy, the grain boundary fracture toughness, and Young's modulus. Experimentally, the functional relationship among d_{crit}, ΔT, and $\Delta\alpha_{\max}$ is reasonably well established (see Prob. 13.8).

EXPERIMENTAL DETAILS: DETERMINATION OF MICROCRACKING

Unless a ceramic component totally falls apart in the furnace as the sample is cooled from the sintering or processing temperature, it is experimentally difficult to observe directly grain boundary microcracks. There are, however, a number of indirect techniques to study the phenomenon. One is to fabricate ceramics of varying grain sizes and measure their flexural strengths after cooling. A dramatic decrease in strength over a narrow grain size variation is usually a good indication that spontaneous microcracking has occurred.

13.4.2   Spontaneous Microcracking due to Thermal Expansion Mismatches in Multiphase Materials

Conceptually there is little difference between this situation and the preceding one; the similarity of the two cases is easily appreciated by simply replacing one of the grains in Fig. 13.4 by a second phase with a different thermal expansion coefficient from its surroundings.

13.4.3   Spontaneous Microcracking due to Phase-Transformation-Induced Residual Stresses

Here the residual stresses do not develop as a result of thermal expansion mismatches or rapid variations in temperature, but as a result of phase transformations. Given that these transformations entail atomic rearrangements, they are always associated with a volume change (e.g., Fig. 4.5). Conceptually, the reason why such a volume change should give rise to residual stresses should at this point be obvious. Instead of using $\Delta\alpha$, however, the resultant stresses usually scale with $\Delta V/V_0$, where ΔV is the volume change associated with the transformation. The stress can be approximated by

$$\sigma \approx \frac{Y}{3(1-2\nu)} \frac{\Delta V}{V_0} \tag{13.23}$$

which can be quite large. For example, a 3 percent volumetric change in a material having a Y of 200 GPa and Poisson's ratio of 0.25 would provide a stress of about 4 GPa!

Residual stresses are generally deleterious to the mechanical properties and should be avoided. This is especially true if a part is to be subjected to thermal cycling. In some situations, however, residual stresses can be used to advantage.

A case in point is the transformation toughening of zirconia discussed in Chap. 11, and another excellent example is the tempering of glass discussed in the next section.

13.5
THERMAL TEMPERING OF GLASS

Because of the transparency and chemical inertness of inorganic glasses, their uses in everyday life are ubiquitous. However, for many applications, especially where safety is concerned, as it is manufactured, glass is deemed to be too weak and brittle. Fortunately, glasses can be significantly strengthened by a process referred to as *thermal tempering*, which introduces a state of compressive residual stresses on the surface that can significantly strengthen the glass (see Sec. 11.3.3).

The appropriate thermal process, illustrated in Fig. 13.5, involves heating the glass body to a temperature above its glass transition temperature, followed by a two-step quenching process. During the first quenching stage, initially the surface layer contracts more rapidly than the interior and becomes rigid while the interior is still in a viscous state. This results in a tensile state of stress at the surface, shown in Fig. 13.5c. However, since the interior is viscous these stresses will relax, as shown in Fig. 13.5d.

During the second quenching step, the entire glass sample is cooled to room temperature. Given that on average the glass interior will have cooled at a *slower* rate than its exterior, its final specific volume will be *smaller* than that of the exterior.[242] The situation is shown in Fig. 13.5e and leads directly to the desired final state of stress (Fig. 13.5f) in which the external surfaces are in compression and the interior is in tension.

By using this technique, the mean strength of soda-lime silicate glass can be raised to the range of 150 MPa, which is sufficient to permit its use in large doors and windows as well as safety lenses. Tempered glass is also used for the side and rear windows of automobiles. In addition to being stronger, tempered glass is preferred to untempered glass for another reason: The release of large amounts of stored elastic energy upon fracture tends to shatter the glass into a great many fragments which are less dangerous than larger shards. Windshields, however, are made of two sheets of tempered glass in between which a polymer layer in

---

[242] This effect was discussed briefly in Sec. 9.4.1 and illustrated in Fig. 9.8a. Simply put, the more time the atoms have to arrange themselves during the cooling process (slow cooling rate), the denser the glass that results.

embedded. The function of that layer is to hold the fragments of glass together in case of fracture and to prevent them from becoming lethal projectiles.

13.6
THERMAL CONDUCTIVITY

The conduction of heat through solids occurs as a result of temperature gradients. In analogy to Fick's first law, the relationship between the heat flux and temperature gradients $\partial T / \partial x$ is given by

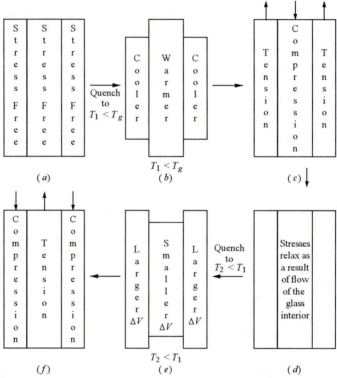

Figure 13.5

Thermal process that results in tempered glass. (a) Initial configuration. (b) The glass is quenched to a temperature that is below T_g, which results in the rapid contraction of the exterior. (c) Resulting transient state of stress. (d) The relaxation of these stresses occurs by the flow and deformation of the interior. (e) Second quenching step results in a more rapid cooling rate for the exterior than for the interior. This results in a glass with a smaller specific volume in the center than on the outside. (f) Final state of stress at room temperature.

TABLE 13.3
Approximate values for thermal conductivities of selected ceramic materials

Material	k_{th} W/(m·K)	Material	k_{th} W/(m·K)
Al_2O_3	30.0–35.0	Spinel ($MgAl_2O_4$)	12.0
AlN	200.0–280.0	Soda-lime silicate glass	1.7
BeO	63.0–216.0	TiB_2	40.0
MgO	37.0	Ti_3SiC_2	43.0
PSZ	2.0		
SiC	84.0–93.0	Cordierite (Mg-aluminosilicate)	4.0
SiAlON	21.0	Glasses	0.6–1.5
SiO_2	1.4	Forsterite	3.0
Si_3N_4	25.0		

$$\frac{\partial Q}{\partial t} = k_{th} A \frac{\partial T}{\partial x} \qquad (13.24)$$

where $\partial Q/\partial t$ is the heat transferred per unit time across a plane of area A normal to the flow of the thermal energy; and k_{th} is a material property (analogous to diffusivity) that describes the ability of a material to transport heat. Its units are $J/(s \cdot m \cdot K)$ or equivalently $W/(m \cdot K)$. Approximate values for k_{th} for a number of ceramics are listed in Table 13.3.

Thermal conduction mechanisms

Describing the mechanisms of conduction in solids is not easy. Here only a brief qualitative sketch of some of the physical phenomena is given. In general, thermal energy in solids is transported by lattice vibrations, i.e., phonons, free electrons, and radiation. Given that the concentration of free electrons in ceramics is low and that most ceramics are not transparent, phonon mechanisms dominate and are the only ones discussed below.

Imagine a small region of a solid being heated. Atoms in that region will have large amplitudes of vibration and will vibrate violently around their average positions. Given that these atoms are bonded to other atoms, it follows that their motion must also set their neighbors into oscillation. As a result the disturbance caused by the application of heat propagates outward in a wavelike disturbance.[243] These waves, in complete analogy to electromagnetic waves, can be scattered by imperfections, grain boundaries, and pores or even reflected at other internal surfaces. In other words, every so often the disturbance will have the direction of its propagation altered. The average distance that the disturbance travels before

---

[243] A situation not unlike the propagation of light or sound through a solid.

being scattered is analogous to the average distance traveled by a gas molecule and is referred to as the *mean free path* λ_{th}.

By assuming the number of these thermal energy carriers to be N_{th}, their average velocity v_{av}, and the average distance λ_{th} traveled by a carrier before being scattered, it is reasonable to assume that, in analogy to the electrical conductivity equation of $\sigma = n\mu q$, k_{th} is given by

$$k_{th} = (const)(N_{th}\lambda_{th}v_{th})$$

In general, open, highly ordered structures made of atoms or ions of similar size and mass tend to minimize phonon scattering and hence lead to increased values of k_{th}. An excellent example of such a solid is diamond, which has one of the highest thermal conductivity values of any known material. Other good examples are SiC, BeO, and AlN. More complex structures, such as spinels, and ones where there is a large difference in mass between ions, such as UO_2 and ZrO_2, tend to have lower values of k_{th}. Similar arguments suggest that the thermal conductivity of a solid will be decreased by the addition of a second component in solid solution. This effect is well known, as shown, e.g., by the addition of NiO to MgO or Cr_2O_3 to Al_2O_3.

Furthermore, the lack of long-range order in amorphous ceramics results in more phonon scattering than in crystalline solids and consequently leads to lower values of k_{th}.

Finally, it is important to mention the effect of porosity. Since the thermal conductivity of air is negligible compared to the solid phases, the addition of large (> 25 percent) volume fractions of pores can significantly reduce k_{th}. This approach is used in the fabrication of firebrick. As noted above, the addition of large-volume fractions of porosity has the added advantage of rendering the firebricks thermal-shock-tolerant. Note that heat transfer by radiation, which scales as T^3, across the pores has to be minimized. Hence for optimal thermal resistance, the pores should be small and the pore phase should be continuous.

EXPERIMENTAL DETAILS: MEASURING THERMAL CONDUCTIVITY

Several techniques are used to measure k_{th}. One method that has gained popularity recently is the laser flash technique. In principle the technique attempts to measure the time evolution of the temperature on one side of the sample as the other side is very rapidly heated by a laser pulse. As it passes through the solid, the signal will be altered in two ways: There will be a time lag between the time at which the solid was pulsed and the maximum in the response. This time lag is directly proportional to the thermal diffusivity, D_{th}, of the material. The second effect will be a reduction in the temperature spike, which is directly related to the heat capacity, c_p, of the solid. The heat capacity, thermal diffusivity, and thermal conductivity and density, ρ, are related by:

$$k_{th} = \rho c_p D_{th}$$

Hence k_{th} can be calculated if the density of the solid is known and D_{th} and c_p are measured.

13.7
SUMMARY

Temperature changes result in dimensional changes which in turn result in thermal strains. Isotropic, unconstrained solids subjected to uniform temperatures can accommodate these strains without the generation of thermal stresses. The latter will develop, however, if one or more of the following situations is encountered:

- Constrained heating and cooling.
- Rapid heating or cooling. This situation can be considered a variation of that above. By rapidly changing the temperature of a solid, its surface will usually be constrained by the bulk and will develop stresses. The magnitude of these stresses depends on the severity of thermal shock or rate of temperature change. In general, the higher the temperature from which a ceramic is quenched the more likely it is to fail or thermal shock. Thermal shock can be avoided by slow heating or cooling. Solids with high thermal conductivities, fracture toughnesses and/or low thermal expansion coefficients are less prone to thermal shock.
- Heating or cooling of multiphase ceramics in which the various constituents have differing thermal expansion coefficients. The stresses generated in this case will depend on the mismatch in thermal expansion coefficients of the various phases. These stresses cannot be avoided by slow heating or cooling.
- Heating or cooling of ceramics for which the thermal expansion is anisotropic. The magnitude of the stresses will depend on the thermal expansion anisotropy, and can cause polycrystalline bodies to spontaneously microcrack. This damage cannot be avoided by slow cooling, but can be avoided if the grain size is kept small.

- Phase transformations in which there is a volume change upon transformation. In this case, the stresses will depend on the volume change during the transformation. They can only be avoided by suppressing the transformation.

If properly introduced, thermal residual stresses can be beneficial, as in the case of tempered glass.

Finally, in the same way that solids conduct sound, they also conduct heat, i.e., by lattice vibrations. Heat conduction occurs by the excitation and interaction of neighboring atoms.

PROBLEMS

13.1. Give an example for each of (*a*) thermal strain but no stress, (*b*) thermal stress but no strain, and (*c*) a situation where both exist.

13.2. (*a*) Derive Eq. (13.6).

(*b*) A metallic rod ($\alpha = 14 \times 10^{-6}\,°C^{-1}$, $Y = 50$ GPa at 800°C) is machined such that it perfectly fits inside an alumina tube. The assembly is then slowly heated; at 800°C the alumina tube cracked. Assume Poisson's ratio to be 0.25 for both materials.

(i) Describe the state of stress that develops in the system as it is heated.

(ii) Estimate the strength of the alumina tube.

Answer: 170 MPa

(iii) In order to increase the temperature at which this system can go, several strategies have been proposed (some of which are wrong): Use an alumina with a larger grain size; use another ceramic with a higher thermal expansion coefficient; use a ceramic that does not bond well with the metal; and use a metal with a higher stiffness at 800°C. Explain in some detail (using calculations when possible) which of these strategies you think would work and which would not. Why?

(iv) If the situation were reversed (i.e., the alumina rod were placed inside a metal tube), describe in detail the three-dimensional state of stress that would develop in that system upon heating.

(v) It has been suggested that one way to bond a ceramic rotor to a metal shaft is to use the assembly described in part (iv). If you were the engineer in charge, describe how you would do it. This is not a hypothetical problem but is used commercially and works quite well.

13.3. Consider a two-phase ceramic in which there are spherical inclusions B. If upon cooling, the inclusions go through a phase transformation that causes them to expand, which of the following states of stress would you expect, and why?

(a) Hydrostatic pressure in B; radial, compressive, and tangential tensile hoop stresses.

(b) Debonding of the interface and zero stresses everywhere.

(c) Hydrostatic pressure in B; radial, tensile, and tangential compressive hoop stresses.

(d) Hydrostatic pressure in B; radial, compressive, and tangential compressive hoop stresses.

(e) Hydrostatic pressure in B; radial, tensile, and tangential tensile hoop stresses.

13.4. (a) Plot the radial stress as a function of r for an inclusion in an infinite matrix, given that $\Delta \alpha = 5 \times 10^{-6}$, $\Delta T = 500^{\circ} C$, $Y_i = 300$ GPa, $Y_m = 100$ GPa, and $v_i = v_m = 0.25$.

(b) If the size of the inclusions were 10 μm, for what volume fraction would the "infinite" matrix solution be a good one? What do you think would happen if the volume fraction were higher? State all assumptions.

Answer: ≈ 5 to 10 vol. % depending on assumptions

13.5. (a) Is thermal shock more likely to occur as a result of rapid heating or rapid cooling? Explain.

(b) A ceramic component with Young's modulus of 300 GPa and a K_{Ic} of 4 MPa$\cdot$m$^{1/2}$ is to survive a water quench from 500°C. If the largest flaw in that material is on the order of 10 μm, what is the maximum value of α for this ceramic for it to survive the quench? State all assumptions.

Answer: $5 \times 10^{-6}\,^{\circ}C^{-1}$

13.6. (a) Derive Eq. (13.11).

(b) Which of the materials listed below would be best suited for an application in which a part experienced sudden and severe thermal fluctuations while in service?

Material	MOR, MPa	k_{th}, W/(m$\cdot$K)	Modulus, GPa	K_{Ic}, MPa$\cdot$m$^{1/2}$	α, K^{-1}
1	700	290	200	8	9×10^{-6}
2	1000	50	150	4	4×10^{-6}
3	750	100	150	4	3×10^{-6}

13.7. (a) Explain how a glaze with a different thermal expansion can influence the effective strength of a ceramic component. To increase the strength of a component, would you use a glaze with a higher or lower thermal expansion coefficient than the substrate? Explain.

   (b) Fully dense, 1-cm-thick alumina plates are to be glazed with a porcelain glaze ($Y = 70$ GPa, $\nu = 0.25$) of 1-mm thickness with a thermal expansion coefficient of 4×10^{-6}°C. Assuming the "stress-freezing" temperature of the glaze to be 800°C, calculate the stress in the glaze at room temperature.

13.8. Using acoustic emission and thermal contraction data, Ohya et al.[244] measured the functional dependence of the microcracking temperature of aluminum titanate ceramics on grain size as the samples were cooled from 1500°C. The following results were obtained:

Grain size, μm	3	5	9
Microcracking temperature upon cooling, °C	500	720	900

   (a) Qualitatively explain the trend observed.

   (b) Are these data consistent with the model presented in Sec. 13.4.1? If so, calculate the value of the constant that appears in Eq. (13.22), given that $G_{c,gb} = 0.5$ J/m^2, $Y = 250$ GPa, and $\Delta\alpha_{max} = 15 \times 10^{-6}$°C.
   Answer: ≈ 184 μm·(°C)^2

   (c) Based on these results, estimate the grain size needed to obtain a crack-free aluminum titanate body at room temperature. State all necessary assumptions.
   Answer: ≈ 0.7 μm

13.9. Explain why volume changes as low as 0.5 percent can cause grain fractures during phase transformations of ceramics. State all assumptions.

13.10. (a) If a glass fiber is carefully etched to remove "all" Griffith flaws from its surface, estimate the maximum temperature from which it can be quenched in a bath of ice water without failure. State all assumptions. Information you may find useful: $Y = 70$ GPa, $\nu = 0.25$, $\gamma = 0.3$ J/m^2, and $\alpha = 10 \times 10^{-6}$°C.
   Answer: 5000°C

   (b) Repeat part (a) assuming 1-μm flaws are present on the surface.
   Answer: 82°C

---

244 Y. Ohya, Z. Nakagawa, and K. Hamano, *J. Amer. Cer. Soc.*, **70**:C184–C186 (1987).

(c) Repeat part (b) for Pyrex, a borosilicate glass for which $\alpha \approx 3 \times 10^{-6} \, ^\circ C$. Based on your results, explain why Pyrex is routinely used as labware.

13.11. Qualitatively explain how the following parameters would affect the final value of the residual stresses in a tempered glass pane: (a) thickness of glass, (b) thermal conductivity of glass, (c) quench temperature, (d) quench rate.

13.12. Rank the following three solids in terms of their thermal conductivity: MgO, $MgO \cdot Al_2O_3$, and window glass. Explain.

13.13. (a) Estimate the heat loss through a 0.5-cm-thick, 1000 cm^2 window if the inside temperature is 25°C and the outside temperature is 0°C. Information you may find useful: k_{th} conductivity of soda lime is 1.7 W/(m·K).

(b) Repeat part (a) for a porous firebrick that is used to line a furnace running at 1200°C. Typical values of k_{th} for firebricks are 1.3 W/(m·K). State all assumptions.

ADDITIONAL READING

1. W. D. Kingery, H. K. Bowen, and D. R. Uhlmann, *Introduction to Ceramics*, 2d ed., Wiley, New York, 1976.
2. C. Kittel, *Introduction to Solid State Physics,* 6th ed., Wiley, New York, 1986.
3. W. D. Kingery, "Thermal Conductivity of Ceramic Dielectrics," *Progress in Ceramic Science*, vol. 2, J. E. Burke, ed., Pergamon Press, New York, 1961.
4. D. P. H. Hasselman and R. A. Heller, eds., *Thermal Stresses in Severe Environments*, Plenum, New York, 1980.
5. H. W. Chandler, "Thermal Stresses in Ceramics," *Trans. J. Brit. Cer. Soc.*, **80**:191 (1981).
6. Y. S. Touloukian, R. W. Powell, C. Y. Ho, and P. G. Klemens, eds., *Thermophysical Properties of Matter*, vol. 2, *Thermal Conductivity — Nonmetallic Solids*, IFI/Plenum Press, New York, 1970.
7. D. W. Richerson, *Modern Ceramic Engineering*, 2d ed., Marcel Dekker, New York, 1992.

CHAPTER 14

Dielectric Properties

*It serves to bring out the actual mechanical connexions
between the known electro-magnetic phenomena; so I
venture to say that any one who understands the
provisional and temporary character of this hypothesis
will find himself rather helped than hindered by it in
his search after the true interpretation of the phenomena.*

James Maxwell, *Phil. Mag.*, **21**:281 (1861).

14.1
INTRODUCTION

Dielectric materials will not conduct electricity and as such are of critical importance as capacitive elements in electronic applications and as insulators. It could be argued, with some justification, that without the discovery of new compositions with very high charge-storing capabilities, i.e., relative dielectric constants $k' > 1000$, the impressive miniaturization of semiconductor-based devices and circuits would not have been as readily implemented. In addition, the traditional use of ceramics as insulators in high-power applications is still a substantial economic activity.

In contrast to electrical conductivity, which involves long-range motion of charge carriers, the dielectric response results from the *short-range* motion of these carriers under the influence of an externally applied electric field. Inasmuch as all solids are comprised of positive and negative entities, the application of an electric field to any solid will result in a separation of its charges. This separation of charge is called **polarization**, defined as the *finite displacement of bound charges* of a dielectric in response to an applied electric field, and the orientation of their molecular dipoles if the latter exist.

The dielectric properties can vary widely between solids and are a function of temperature, frequency of applied field, humidity, crystal structure, and other external factors. Furthermore, the response can be either linear or nonlinear. This chapter examines linear dielectrics from a microscopic point of view as well as the effects of temperature and frequency on the dielectric response. Materials for which the response is nonlinear are discussed in the next chapter.

14.2
BASIC THEORY

Before polarization is discussed, it is imperative to understand how one measures polarization and to gain a qualitative understanding of how readily or not so readily polarizable a solid is. Consider two metal parallel plates of area A separated by a distance d in vacuum (Fig. 14.1a). Attaching these plates to the simple electric circuit, shown in Fig. 14.1a, and closing the circuit will result in a transient surge of current that rapidly decays to zero, as shown in Fig. 14.1b. Given that

$$Q = \int I\,dt \qquad (14.1)$$

the area under the I versus t curve is the total charge that has passed through the circuit and is now stored on the capacitor plates.

Repeating the experiment at different voltages V and plotting Q versus V should yield a straight line, as shown in Fig. 14.2. In other words, the well-known relationship

$$Q = CV \qquad (14.2)$$

is recovered. The slope of the Q versus V curve is the **capacitance C_{vac}** of the parallel plates in vacuum, given by

$$C_{vac} = \frac{\varepsilon_0 A}{d} \qquad (14.3)$$

where ε_0 is the permittivity of free space, which is a constant[245] equal to 8.85×10^{-12} $C^2/(J \cdot m)$. The units of capacitance are farads (F), where $1\,F = 1\,C/V = 1\,C^2/J$.

---

[245] If cgs electrostatic units are used, then $\varepsilon_0 = 1$ and $\varepsilon = k'$ and a factor $1/4\pi$ appears in all the equations. In SI where $\varepsilon_0 = 8.85 \times 10^{-12}$ $C^2/(J \cdot m)$, the factor $1/4\pi$ is included in ε_0 and omitted from the equations. Here only SI units are used.

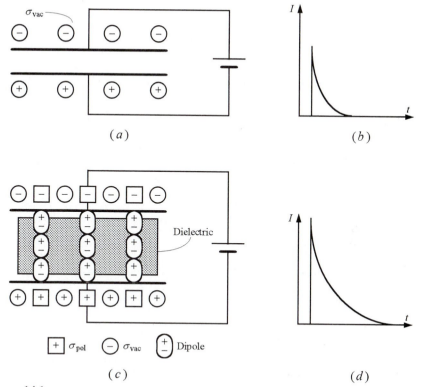

Figure 14.1
(a) Parallel-plate capacitor of area A and separation d in vacuum attached to a voltage source. (b) Closing of the circuit causes a transient surge of current to flow through the circuit. Charge stored on the capacitor is equal to the area under the curve. (c) Same as (a) except that now a dielectric is placed between the plates. (d) Closing of the circuit results in a charge stored on the parallel plates that has to be greater than that stored in (b).

If a dielectric (which can be a gas, solid, or liquid) is introduced between the plates of the capacitor (Fig. 14.1c) and the aforementioned experiment is repeated, the current that flows through the external circuit and is stored on the capacitor plates will increase (Fig. 14.1d). Repeating the experiment at different voltages and plotting the total charge stored on the capacitor versus the voltage applied will again result in a straight line but with a larger slope than that for vacuum, as shown in Fig. 14.2. In other words, Eq. (14.3) is now modified to read

$$C = \frac{\varepsilon A}{d} \tag{14.4}$$

where ε is the *dielectric constant* of the material between the plates.

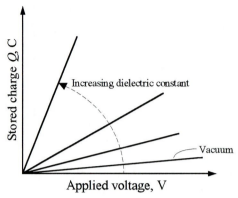

Figure 14.2
Functional dependence of Q on applied voltage. Slope of curve is related to the dielectric constant of the material.

The **relative dielectric constant** of a material k' is defined as

$$k' = \frac{\varepsilon}{\varepsilon_0} \tag{14.5}$$

Since ε is always greater than ε_0, the minimum value for k' is 1. By combining Eqs. (14.4) and (14.5), the capacitance of the metal plates separated by the dielectric is

$$C = \frac{k'\varepsilon_0 A}{d} = k' C_{\text{vac}} \tag{14.6}$$

Thus k' is a dimensionless parameter that compares the charge-storing capacity of a material to that of vacuum.

The foregoing discussion can be summarized as follows: When a voltage is applied to a parallel-plate capacitor in vacuum, the capacitor will store charge. In the presence of a dielectric, an additional "something" happens within that dielectric which allows the capacitor to store more charge. The purpose of this chapter is to explore the nature of this "something." First, however, a few more concepts need to be clarified.

Polarization charges

By combining Eqs. (14.2) and (14.3), the surface charge in vacuum σ_{vac} is

$$\sigma_{\text{vac}} = \left[\frac{Q}{A}\right]_{\text{vac}} = \frac{\varepsilon_0 V}{d} = \varepsilon_0 E \tag{14.7}$$

where E is the applied electric field. Similarly, by combining Eqs. (14.2) and (14.4), in the presence of a dielectric, the surface charge on the metal plates increases to

$$\left[\frac{Q}{A}\right]_{\text{die}} = \frac{\varepsilon_0 k'V}{d} = \sigma_{\text{vac}} + \sigma_{\text{pol}} \tag{14.8}$$

where σ_{pol} is the excess charge per unit surface area present on the *dielectric surface* (Fig.14.1c). Also σ_{pol} is numerically equal to and has the same dimensions $\left(C/m^2\right)$ as the **polarization P** of the dielectric, i.e.,

$$P = \sigma_{\text{pol}} \tag{14.9}$$

Electromagnetic theory defines the **dielectric displacement D** as the surface charge on the metal plates, that is, $D = [Q/A]$. Making use of this definition and combining Eqs. (14.7) to (14.9), one finds that

$$D = \varepsilon_0 E + P \tag{14.10}$$

In other words, the total charge stored on the plates of a parallel-plate capacitor D is the sum of the charge that would have been present in vacuum $\varepsilon_0 E$ and an extra charge that results from the polarization of the dielectric material P. The situation is depicted schematically in Fig. 14.1c. Note that if $P = 0$, D is simply given by Eq. (14.7).

Further combining Eqs. (14.7) to (14.10), and noting that the electric field is the same in both cases, one sees that:

$$P = (k' - 1)\varepsilon_0 E = \chi_{\text{die}} \varepsilon_0 E \tag{14.11}$$

where

$$\chi_{\text{die}} = \frac{\sigma_{\text{pol}}}{\sigma_{\text{vac}}}$$

and χ_{die} (chi) is known as the *dielectric susceptibility* of the material. The next task is to relate P to what occurs at the atomic scale.

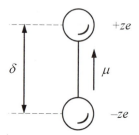

$+ze$

δ

μ

$-ze$

Figure 14.3

Definition of an electric dipole moment.

Microscopic approach. A **dipole moment** μ, shown in Fig. 14.3, is defined as[246]

$$\mu = q\delta$$

where δ is the distance separating the centers of the $+ve$ and $-ve$ charges $q = \pm ze$. μ is a vector with its positive sense directed from the negative to the positive charge.

If there are N such dipoles per unit volume, it can be shown that P is simply

$$P = N\mu = Nq\delta \tag{14.12}$$

Combining Eqs. (14.11) and (14.12), one sees that

$$\boxed{k' - 1 = \frac{P}{\varepsilon_0 E} = \frac{Nq\delta}{\varepsilon_0 E}} \tag{14.13}$$

This result is fundamental to understanding the dielectric response of a solid. It basically says that the greater the separation of the charges of a dipole δ for a given applied field, the greater the relative dielectric constant. In other words, the more polarizable a medium, the greater its dielectric constant.

One can further define the **polarizability** of an atom or ion as

---

[246] The dipole moment of a charge $q = ze$ relative to a fixed point is defined as the vector $ze\zeta_i$, where ζ_i is the radius vector from the fixed point to the position of the charge. The total dipole moment of a system is the vector sum of all the individual dipoles

$$\mu = \sum z_i e \zeta_i$$

This quantity is *independent* of the position of the fixed point. In the absence of a field, $\sum z_i e \zeta_0 = 0$. Upon the application of an electric field, that results in the displacement of the charges by an amount δ_i from their equilibrium position, that is, $\zeta = \zeta_0 + \delta_i$. It follows that $\mu = \sum z_i e \zeta_i = \sum z_i e \delta_i$. Practically, it follows that to calculate the dipole moment of any ion, we need only know its position relative to its equilibrium position.

$$\alpha = \frac{P}{NE_{\text{loc}}} \tag{14.14}$$

where E_{loc} is the *local electric field* to which the atom is subjected. The SI units of polarizability are $C \cdot m^2 V^{-1}$ or Fm^2.

For dilute gases, where the molecules are far apart, the locally applied field can be assumed to be identical to the externally applied field E and by combining Eqs. (14.13) and (14.14), it follows that

$$k' - 1 = \frac{N\alpha}{\varepsilon_0} \tag{14.15}$$

However, in a solid, polarization of the surrounding medium can, and will, substantially affect the magnitude of the local field. It can be shown that (see App. 14A), for cubic symmetry, the local field is related to the applied field by

$$E_{\text{loc}} = \frac{E}{3}(k' + 2)$$

which when combined with Eqs. (14.13) and (14.14) gives

$$\boxed{k' - 1 = \frac{N\alpha/\varepsilon_0}{1 - N\alpha/(3\varepsilon_0)}} \tag{14.16}$$

which can be rearranged as

$$\frac{k' - 1}{k' + 2} = \frac{\alpha N}{3\varepsilon_0}$$

preferred by some. This expression is known as the **Clausius-Mossotti relation**,[247] and it provides a valuable link between the macroscopic k' and the microscopic α. It follows from this relationship that a measurement of k' can, in principle, yield information about the relative displacement of the positive and negative charges making up that solid. It should be emphasized here that this expression is only valid for linear dielectrics and is not applicable to ferroelectrics, discussed in the next chapter. It is also worth noting that whenever $N\alpha/3\varepsilon_0 \ll 1$ Eq. (14.16) simplifies to Eq. (14.15), as expected.

Up to this point, the discussion was restricted to static electric fields. In most electrical applications, however, the applied electric field is far from static — with

---

[247] Written in terms of the refractive index $n = \sqrt{k_e}$, this relation is known as the **Lorentz-Lorenz relation**. Since electromagnetic radiation, if one ignores the magnetic component, is nothing but a time-varying electric field, it should come as no surprise later, in Chap. 16, when it is discovered that the dielectric and optical responses of insulators are intimately related.

frequencies that range from 60 Hz for standard ac power to gigahertz and higher for communication networks. It is thus important to introduce now a formalism by which one can describe not only the static response of a dielectric which is represented by k' or α, but also the effect of frequency on both k' and any losses that occur in the dielectric as a result of the application of a time-varying electric field. This is typically done by representing the dielectric constant as a complex quantity that depends on frequency, as described in the following section.

14.3
EQUIVALENT CIRCUIT DESCRIPTION OF LINEAR DIELECTRICS

Ideal dielectric

The application of a sinusoidal voltage[248] $V = V_0 \exp i\omega t$ to a dielectric, i.e., one without losses, will result in a charging current (see Prob. 14.1) given by

$$I_{chg} = \frac{dQ}{dt} = C\frac{dV}{dt} = i\omega CV = \omega CV_0 \exp i\left(\omega t + \frac{\pi}{2}\right)$$

or

$$I_{chg} = -\omega \kappa' C_{vac} V_0 \sin \omega t \qquad (14.17)$$

In other words, the resulting *current* will be $\pi/2$ rad or 90° out of phase from the applied field, which implies that the oscillating *charges* are in phase with the applied voltage.[249]

Nonideal dielectric

As noted above, Eq. (14.17) is only valid for an ideal dielectric. In reality, the charges are never totally in phase for two reasons: (1) the dissipation of energy due to the inertia of the moving species and (2) the long-range hopping of the charged species, i.e., ohmic conduction. The total current is thus the vectorial sum of I_{chg} and I_{loss}, or

---

[248] Remember $e^{i\omega} = \cos \omega + i \sin \omega$, where ω is the angular frequency in units of radians per second. To convert to hertz, divide by 2π, since $\omega = 2\pi\nu$, where ν is the frequency in hertz or s^{-1}.

[249] When the charges are *in phase* with the applied field, this automatically implies that the current is $\pi/2$ rad ahead of the applied voltage. This comes about since $I = dQ/dt$. Interestingly enough, the loss current is that in which the charges are oscillating $\pi/2$ out of phase with the applied voltage.

$$I_{\text{tot}} = I_{\text{chg}} + I_{\text{loss}} = \{i\omega C + G_L(\omega)\}V + G_{\text{dc}}V \qquad (14.18)$$

where G is the conductance of the material (see below). I_{loss} is defined as

$$I_{\text{loss}} = \{G_L(\omega) + G_{\text{dc}}\}V \qquad (14.19)$$

and is written in this form to emphasize that G_L is a function of frequency, whereas G_{dc} is not. In the limit of zero frequency, $G_L \Rightarrow 0$ and one recovers Ohm's law or

$$I_{\text{tot}} = I_{\text{loss}} = G_{\text{dc}}V$$

since $G_{\text{dc}} = 1/R$, where R is the direct-current (dc) resistance of the material.

The total current in the dielectric is thus made up of two components that are $90°$ out of phase with each other and have to be added vectorially, as shown in Fig. 14.4. The total current in a nonideal dielectric will thus lead the applied voltage by an angle of $90° - \phi$, where ϕ is known as the **loss angle** or the **loss tangent** or **dissipation factor**.

It is important to note that the dielectric response of a solid can be succinctly described also by expressing the relative dielectric constant as a complex quantity, made up of a real k' component and an imaginary k'' component, i.e.,

$$k^* = k' - ik'' \qquad (14.20)$$

Replacing k' in Eq. (14.6) by k^*, making use of Eq. (14.2), and noting that the $I_{\text{chg}} = I_{\text{tot}} - I(\omega = 0) = dQ/dt$, one obtains

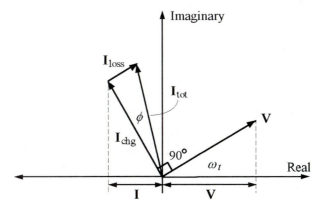

Figure 14.4

Vectorial representation of applied voltage, charging, loss, and total currents. Note that when $\phi = 0$, $I_{\text{tot}} = I_{\text{chg}}$, whereas when $\phi = \pi/2$, $I_{\text{tot}} = I_{\text{loss}}$.

$$I_{tot} - I(\omega = 0) = \frac{dQ}{dt} = k^* C_{vac} i\omega V = (k' - ik'') C_{vac} i\omega V \qquad (14.21)$$

which, when rearranged, gives

$$I_{tot} = (i\omega C_{vac} k' + \omega k'' C_{vac})V + G_{dc}V \qquad (14.22)$$

Comparing this expression to Eq. (14.18) reveals immediately that

$$G_{ac} = G_{dc} + G_L = G_{dc} + \omega k'' C_{vac}$$

where G_{ac} is the **ac conductance** of the material, with units of siemens, denoted by S, or Ω^{-1}. The corresponding **ac conductivity** (S/m) is given by

$$\boxed{\sigma_{ac} = \sigma_{dc} + \omega k'' \varepsilon_0} \qquad (14.23)$$

Furthermore, from Fig. 14.4 one can define $\tan \phi$ as

$$\boxed{\tan \phi = \frac{I_{loss}}{I_{chg}} = \frac{G_{dc} + \omega k'' C_{vac}}{\omega k' C_{vac}}} \qquad (14.24)$$

Note that for a dielectric for which $G_{dc} \ll \omega k'' C_{vac}$, $\tan \phi \approx k''/k'$.

Power dissipation in a dielectric

In general, loss currents are a nuisance since they tend to heat up the dielectric and retard electromagnetic signals. The average power dissipated in a dielectric is

$$P_{av} = \frac{1}{T} \int_0^T I_{tot} V \, dt$$

where $T = 2\pi/\omega$ is the time period. For an *ideal* dielectric, $I_{tot} = I_{chg}$ and

$$P_{av} = \frac{1}{T} \int_0^T -\omega k' C_{vac} V_0^2 \sin \omega t \cos \omega t \, dt = 0$$

During one-half of the cycle, the capacitor is being charged and the power source does work on the capacitor; in the second half of the cycle, the capacitor is discharging and does work on the source. Consequently, the average power drawn from the power source is *zero*, which is an important result because it shows that an ideal dielectric is loss-free.

In a nonideal dielectric, however, the *loss current* and voltage are in phase, and

$$P_{av} = \frac{1}{T} \int_0^T I_{loss} V \, dt = \frac{1}{T} \int_0^T (\omega k'' C_{vac} + G_{dc}) V_0^2 \cos \omega t \cos \omega t \, dt = \frac{1}{2} G_{ac} V_0^2$$

Note that for dc conditions, or $\omega = 0$, this expression is idential to the well-known expression for the power loss under dc conditions, or $I^2 R$.

The corresponding power loss *per unit volume* is given by

$$P_V = \frac{1}{2} \sigma_{\mathrm{ac}}(\omega) E_0^2 \qquad (14.25)$$

where $E_0 = V_0/d$ is the amplitude of the electric field. It follows that the power loss per unit volume $(\mathrm{W/m}^3)$ in a dielectric is directly related to σ_{ac} or k'' and frequency.

Although the mathematical representation, at first glance, may not appear to be simple, the physics of the situation is more so: When a time-varying electric field is applied to a dielectric, the charges in the material will respond. Some of the bound charges will oscillate in phase with the applied field and will result in charge storage and contribute to k'. Another set of charges, both bound and those contributing to the dc conductivity, will oscillate $90°$ out of phase with the applied voltage and result in energy dissipation in the dielectric.[250] This energy dissipation ends up as heat (the temperature of the dielectric will increase). In an ideal dielectric, the loss angle ϕ is zero.

The remainder of this chapter is concerned with the various polarization mechanisms operative in ceramics and their temperature and frequency dependencies.

EXPERIMENTAL DETAILS: MEASURING DIELECTRIC PROPERTIES

There are many techniques used to measure the dielectric properties of solids. One of the more popular ones is known as *ac impedance spectroscopy*, described below. Another technique compares the response of the dielectric to that of a calibrated variable capacitor. In this method, the capacitance of a parallel-plate capacitor in vacuum is compared with one in the presence of the material for which the dielectric properties are to be measured. Then k' is simply calculated from Eq. (14.6). A typical circuit for carrying out such an experiment is shown in

---

[250] This is a rather simplistic interpretation but one that is easily visualized. More realistically, the charges will oscillate slightly out of phase, with an angle ϕ out of phase to be exact, with respect to the applied field. It is worth emphasizing once more that k' describes the behavior of the *bound* charges and that σ_{ac} has two contributions to it: the bound charges that are out of phase with the applied field for which the conductance is $k'' \omega C_{\mathrm{vac}}$ and the "free" charges whose conductance is simply G_{dc}. Whether a charge will jump back when the field reverses sign, and would thus be considered a bound charge, or whether it will continue to drift when the field reverses sign can only be distinguished in responses to dc fields.

Fig. 14.5a. Varying the capacitance of the calibrated capacitor to keep the resonance frequency $\omega_0 = \{L(C_s + C)\}^{-1/2}$ constant when vacuum is between the plates, versus when the substance is inserted, allows C_{vac} and C_{solid} to be determined and, in turn, k'.

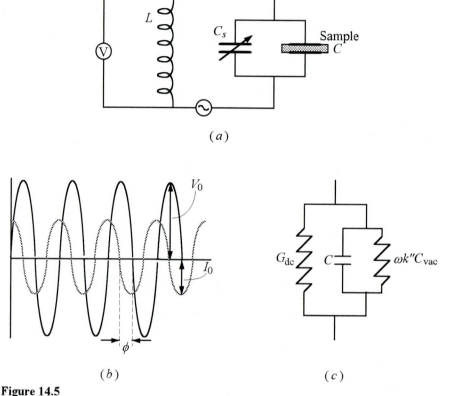

(a)

(b)   (c)

Figure 14.5

(a) Apparatus for measuring the dielectric constant of a material; L is the inductance of the coil. (b) The actual response of a nonideal dielectric to an applied voltage is such that the angle between the current and voltage is not $\pi/2$, but $\pi/2 - \phi$. (c) Equivalent circuit used to model the dielectric response of a solid. Here G_{dc} represents the dc response of the material, whereas $\omega k''C_{vac}$ is the conductance of the bound charges, which vanishes as ω goes to zero.

AC impedance

Here a sinusoidal voltage is applied to the sample, and the magnitude and phase shift of the resulting current are measured by using sophisticated electronics. From the ratio of the magnitude of the resulting current I_0 to the imposed voltage V_0, and the magnitude of the phase difference ϕ between the two, all defined in Fig. 14.5b, k' and k'' can be obtained. It can be shown (see Prob. 14.1b) that if one assumes the equivalent circuit shown in Fig. 14.5c, k' and k'' are given by

$$k' = \frac{I_0(\omega)d}{V_0 A\, \omega \varepsilon_0} \sin\left\{\frac{\pi}{2} - \phi(\omega)\right\}$$

and

$$k'' = \frac{\sigma_{\text{ac}} - \sigma_{\text{dc}}}{\omega \varepsilon_0}$$

where

$$\sigma_{\text{ac}} = \frac{I_0(\omega)d}{V_0 A} \cos\left\{\frac{\pi}{2} - \phi(\omega)\right\}$$

and d and A are, respectively, the thickness and cross-sectional area of the sample. It is important to remember, as is emphasized in these equations, that both I_0 and ϕ depend on the frequency of the applied field ω. It is interesting here to look at the limits; for dc conditions the current that passes through the capacitor will be determined by its dc conductivity. As the frequency increases, more and more of the bound charges will start to oscillate out of phase with the applied voltage and will contribute to σ_{ac}.

In a typical experiment, the frequency of the applied voltage is varied over the range between a few hertz and 100 MHz. Measurements in the frequency range between 10^9 and 10^{12} Hz are more complex and beyond the scope of this book. However, in the IR and UV frequencies, the dielectric constant and loss can once again be measured from measurements of the reflectivity of the samples and the refractive index (see Chap. 16).

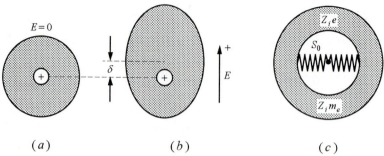

(a)   (b)   (c)

Figure 14.6

Electronic polarization of the atomic cloud surrounding a nucleus. (*a*) At equilibrium, i.e., in the absence of an external electric field (*b*) In the presence of an external electric field. (*c*) Schematic of model assumed in text. S_0 represents the stiffness of bond between the electrons and the nucleus.

14.4
POLARIZATION MECHANISMS

Up to this point, the discussion was couched in terms of polarization, or the displacement of charges with respect to each other. In this section, the specifics of particle separation are considered. In solids, especially in ceramics, various charged entities are capable of polarization, such as electrons, protons, cations, anions, and charged defects. The following mechanisms, most of which are described in detail in the coming sections, represent the most important polarization mechanisms in ceramics.

Electronic polarization

This mechanism entails the displacement of the electron cloud relative to its nucleus (Fig. 14.6*a*).

Ionic displacement polarization

In this case, there are two types of ionic displacements:

1. Displacements where the displaced charge is bound elastically to an equilibrium position (Fig. 14.9) and referred to below as *ionic polarization*.
2. Displacements that occur between several equilibrium sites for which the probability of occupancy of each site depends on the strength of the external field. This mechanism is also known as *dipolar* or *ion jump polarization* and

is depicted schematically in Fig. 14.10. Another definition of ion jump polarization is the preferential occupation of equivalent or near-equivalent lattice sites as a result of the applied field biasing one site with respect to the other. If the alignment occurs *spontaneously and cooperatively, nonlinear* polarization results and the material is termed *ferroelectric*. Because of the relatively large displacements, relative dielectric constants on the order of 5000 can be attained in these materials. Nonlinear dielectrics are dealt with separately in Chap. 15. But if the polarization is simply due to the motion of ions from one site to another adjacent site, then the polarization behavior is linear with voltage. These solids are discussed below.

Space charge polarization

In Chap. 5, the notion of a Debye length was briefly alluded to, and it was shown that whenever two dissimilar phases come into contact with each other, an electrified interface will result. This so-called double layer acts as a capacitor with properties and responses different from those of the bulk material. The behavior and interpretation of interfacial phenomena are quite complex and not within the scope of this book; they fall more in the realm of solid-state electrochemistry and will not be discussed further.

The total polarizability is the sum of the contributions from the various mechanisms, or

$$\frac{k_{\text{tot}}^* - 1}{k_{\text{tot}}^* + 2} = \frac{1}{3\varepsilon_0}\left[N_e\alpha_e + N_{\text{ion}}\alpha_{\text{ion}} + N_{\text{dip}}\alpha_{\text{dip}} + N_{\text{space chg}}\alpha_{\text{space chg}}\right] \quad (14.26)$$

where N_i represents the number of polarizing species per unit volume. In the remainder of this section, electronic, ionic, and ion jump polarization are discussed in some detail.

14.4.1 Electronic Polarization

Electronic polarization, shown schematically in Fig. 14.6*a* and *b*, occurs when the electron cloud is displaced relative to the nucleus it is surrounding. It is operative at most frequencies and drops off only at very high frequencies ($\approx 10^{15}$ Hz). Since all solids consist of a nucleus surrounded by electrons, electronic polarization occurs in all solids, liquids, and gases; and since it does not involve hopping of ions or atoms between lattice sites, it is temperature-insensitive.

The simplest classical theory for electronic polarization treats the atom or ion, of atomic number Z_i, as an electrical shell of charge $Z_i e$ and mass $Z_i m_e$, attached to an undeformable ion nucleus[251] (Fig. 14.4c). If the natural frequency of vibration of the system is ω_0, it follows that the corresponding restoring force is

$$F_{restor} = M_r \omega_0^2 \delta \qquad (14.27)$$

where M_r is the **reduced mass** of the oscillating system, defined as

$$M_r = \frac{Z_i m_e m_n}{Z_i m_e + m_n} \qquad (14.28)$$

where m_e and m_n are, respectively, the mass of the electrons and of the nucleus. Since in this case $m_e << m_n$, it follows that for electronic polarization $M_r \approx Z_i m_e$.

The application of the electric field, as discussed above, will result in the separation of charges and the creation of an electric dipole moment, since now the center of negative charge and the center of positive charge will no longer coincide, as shown schematically in Fig. 14.6b.

An oscillator displaced by an amount δ by a driving force $F = Z_i e E = Z_i e E_0 \exp(i\omega t)$ with a restoring force and a **damping constant** or **friction factor**, f, must obey the equation of motion

$$M_r \left(\frac{d^2\delta}{dt^2} + f \frac{d\delta}{dt} + \omega_0^2 \delta \right) = Z_i e E_0 \exp i\omega t \qquad (14.29)$$

which is nothing but Newton's law with a restoring force and a friction factor f (rad/s). If f is small, there is little friction, while for large f the frictional forces are also large.[252] It can easily be shown (see Prob. 14.3) that

$$\delta = \frac{e E_0}{m_e \sqrt{\left(\omega_0^2 - \omega^2\right)^2 + f^2 \omega^2}} \exp i(\omega t - \phi) \qquad (14.30)$$

and

$$\delta = \frac{e E_0}{m_e \left\{ \left(\omega_0^2 - \omega^2\right) + i\omega f \right\}} \exp i\omega t \qquad (14.31)$$

---

[251] Clearly, this is a gross oversimplification. The restoring force, and consequently the resonance frequency of each electron, has to be different. See Eq. (14.40) for a more accurate expression.

[252] Damping constant f is related to the anharmonicity of the vibrations — as they become more anharmonic, f increases.

are identical solutions to Eq. (14.29), provided ϕ, which represents the phase difference between the forced vibration and the resulting polarization, takes the value

$$\tan \phi = \frac{f\omega}{\omega_0^2 - \omega^2} \tag{14.32}$$

Note that δ is a measure of the displacement of the electron cloud as a whole relative to its equilibrium position in the absence of a field, as shown schematically in Fig. 14.6a and b.

By replacing k_e' in Eq. (14.13) by k_e^*, and substituting for δ, it can be shown that, *assuming the applied field is identical to the local field* (see Prob. 14.5), the real and imaginary parts of k_e^* are, respectively,

$$k_e'(\omega) = 1 + \frac{Z_i e^2 N \left(\omega_0^2 - \omega^2\right)}{\varepsilon_0 m_e \left\{\left(\omega_0^2 - \omega^2\right)^2 + f^2 \omega^2\right\}} \tag{14.33}$$

and

$$k_e''(\omega) = \frac{Z_i e^2 N \omega f}{\varepsilon_0 m_e \left\{\left(\omega_0^2 - \omega^2\right)^2 + f^2 \omega^2\right\}} \tag{14.34}$$

The frequency dependencies of k_e' and k_e'' are plotted in Fig. 14.7 and are characteristic of typical dispersion curves experimentally observed for dielectrics. It is important to note that Eqs. (14.33) and (14.34) are only valid for a dilute gas, since it was implicitly assumed that the local field was identical to the applied field. To solve the problem more accurately for solids, the local rather than the applied field would have to considered in Eq. (14.29). Fortunately, doing this does not change the general forms of the solutions; it only modifies the value of the resonance frequency ω_0 (see App. 14A).

Based on Eqs. (14.33) and (14.34), the frequency response can be divided into three domains:

1. $\omega_0 \gg \omega$. Here the charges are oscillating in phase with the applied electric field and contribute to k_e'. Under dc conditions,

$$k_e' - 1 = \frac{Z_i e^2 N}{\varepsilon_0 m_e \omega_0^2} \tag{14.35}$$

and k_e'' is zero.

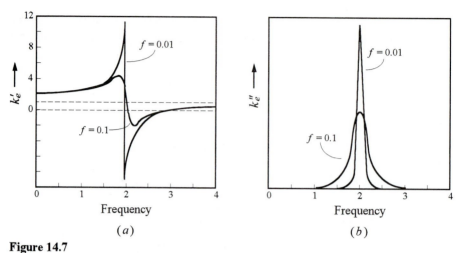

Figure 14.7

General frequency dependence of (a) k'_e, that is, Eq. (14.33) (note that as f increases, the width of the resonance peak also increases); (b) k''_e, that is, Eq. (14.34). ω_0 was assumed to be 2.

2. $\omega_0 \approx \omega$. When the frequency of the applied field approaches the natural frequency of vibration of the system, the system is said to be at *resonance*. The displacements, were it not for the frictional forces, would go to infinity; and thus just before resonance, k'_e goes through a maximum. Exactly at resonance the charges are 90° out of phase with the applied field and thus are not contributing to the dielectric constant. Furthermore, at resonance k''_e and the energy losses are at a maximum.

3. $\omega_0 \ll \omega$. In this region, the electric field is changing direction too fast for the electric charges to respond; no polarization results, and k'_e goes to 1.

To summarize: When a varying electric field is applied to a solid, the charges will start dancing to the tune of that externally applied field — they will oscillate with the same frequency of vibration as the applied field. The amplitude of the vibrations, however, will vary and will depend on the relative values of ω_0 and ω. The charges that are in phase with the applied field will not absorb energy, but will contribute to k'_e. Another set of charges will oscillate out of phase with the applied field, will absorb energy, and will contribute to dielectric loss. When the frequency of the applied field is much smaller than the natural frequency of vibration or the relaxation time of the polarization process involved, then the charges can follow the field nicely, k'_e, which is directly related to the amplitude of vibration, is a weak function of frequency, and the loss is negligible. As the frequency of the applied field becomes comparable to the natural frequency of vibration of the system, the system goes into resonance, the amplitudes of vibration tend to be very large, and consequently so is k'_e. Further, this large increase in the amplitude of vibration results in the large losses observed at resonance.[253] At very high frequencies, the charges cannot keep up with the applied field, their amplitude of vibration will be uncorrelated, and k'_e will approach 1. In other words, if the frequency of the applied field is so large that the polarization mechanism cannot follow, then no polarization ensues.

Micrsocopic factors affecting k'_e

Comparing Eqs. (14.15) and (14.35) gives

$$\alpha_e = \frac{Z_i e^2}{m_e \omega_0^2} \tag{14.36}$$

immediately. This leads to the final, but not surprising, result that electronic polarizability is related to the restoring force or the strength of the bond holding the electrons in place to their nucleus, reflected in ω_0. It was shown in Sec. 4.4 [see Eq. (4.6)] that for small displacements, the restoring force can be assumed to be proportional to the displacement, i.e.,

$$F_{\text{restor}} = S_0(r - r_0) = S_0 \delta \tag{14.37}$$

where S_0 is the *stiffness* of the bond and δ is the displacement from its equilibrium position. Recall that S_0 was defined as

$$S_0 = \left(\frac{dF}{dr}\right)_{r=r_0} \approx \frac{Z_i^2 e^2}{4\pi\varepsilon_0 r_0^3} \tag{14.38}$$

where F is the net force between the unlike charges. In other words, the assumption is made that the electron cloud is attached to its nucleus by a spring of stiffness S_0, as shown in Fig. 14.6c. Combining Eqs. (14.36) to (14.38) together with Eq. (14.27) yields

$$\alpha_e \approx 4\pi\varepsilon_0 r_0^3 \tag{14.39}$$

It follows that, according to this result, α_e should scale with the volume of the atom or ion. Simply put, *the larger the atom or ion, the less bound the electrons are to their nucleus and the more amenable they would be to polarization.* According to Eq.

---

[253] It is worth noting here that this description of resonance is applicable for any resonance phenomenon, be it mechanical, electrical, or magnetic. The nature of the resonating species and the driving forces may vary, but the physics and the interpretation really do not.

(14.39), a plot of α_e versus the cube of the atomic or ionic radii should yield a straight line — a fact that, to a first approximation, is borne out by experimental results, as shown in Fig. 14.8. For example, the noble gases do indeed fall on a straight line; anions, however, tend to lie above that line, and cations below it. Thus, in addition to the size effect, there are two other factors (see also Sec. 4.2.2, where polarizability was first encountered) that determine the polarizability of an ion or atom:

1. *Charge.* The polarizability of an ion is a strong function of its net charge, as shown in Fig. 14.8. Anions are usually more polarizable than cations, and the effect is greater than a simple volume argument. For example, Kr, Cl^-, and S^{2-} are similar in size, yet S^{2-} with its double negative charge is almost 3 times more polarizable than Cl^-, which in turn is more polarizable than Kr. This is understandable, since the outer electrons are less tightly bound in a negative ion and thus are expected to contribute the most to the polarizability.

2. *Nucleus shielding* and the configuration of outer electrons. In general, *d* electrons do not shield the nucleus as well as *s* or *p* electrons; hence the polarizability of atoms with *d* electrons is less than that of similarly sized atoms with *s* or *p* electrons.

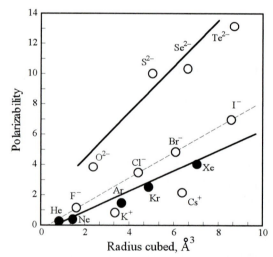

Figure 14.8
Relationship between ionic or atomic radius and polarizability.[254]

---

254 For atomic radii, see R. D. Shannon, *Acta Crystallogr.*, **B32**:925 (1976). (See also App. 3A). For polarizability data, see C. Kittel, *Introduction to Solid State Physics*, 4th ed., Wiley, New York, 1971. The polarizabilities were determined empirically from refractive-index data.

WORKED EXAMPLE 14.1. The relative dielectric constant of 1 mole of Ar gas was measured at 0°C and 1 atm pressure to be 1.00056. (*a*) Calculate the polarizability of Ar. (*b*) Calculate the relative dielectric constant if the pressure is increased to 2 atm. (*c*) Estimate the radius of an Ar atom and compare to that of tabulated data.

Answer

(*a*) At 0°C, the number of atoms per cubic meter (assuming ideal gas behavior, i.e., $N = N_{Av}P/RT$)

$$N = \frac{(6.02 \times 10^{23})(1.013 \times 10^5)}{8.31 \times 273} = 2.69 \times 10^{25} \text{ m}^{-3}$$

Substituting this value in Eq. (14.15) and solving for α_e yields

$$\alpha_e = \frac{(1-1.00056)(8.85 \times 10^{-12})}{2.69 \times 10^{25}} = 1.84 \times 10^{-40} \text{ F} \cdot \text{m}^2$$

(*b*) It is important to note here that *polarizability is an atomic property*. The dielectric constant, on the other hand, depends on the density of the atoms or the manner in which these atoms are assembled to form a crystal. Doubling the pressure will double N, and once again applying Eq. (14.15), it follows that

$$k'_e - 1 = \frac{2(2.69 \times 10^{25})(1.84 \times 10^{-40})}{8.85 \times 10^{-12}} = 0.0011$$

Note that doubling the pressure almost doubles $k'_e - 1$, but does not affect the polarizability.

(*c*) To obtain the radius use is made of Eq. (14.39), from which it follows that

$$r_0 = \left(\frac{1.84 \times 10^{-40}}{4\pi \times 8.85 \times 10^{-12}} \right)^{1/3} \approx 1.2 \times 10^{-10} \text{ m}$$

The experimental value for the diameter of Ar is somewhere between 3×10^{-10} m and 4×10^{-10} m, depending on how it is measured. Thus the estimate of r_0 is off by a factor of ≈ 3, which is not too surprising given the simple model used to derive Eq. (14.39) in which the discreteness of the electron cloud was not accounted for (see below).

Note that in the cgs system, the unit of polarizability is the cubic centimeter. To convert from SI to cgs:

$$\alpha(\text{F} \cdot \text{m}^2) = 4\pi\varepsilon_0 \times 10^{-6} \times \alpha(\text{cm}^3)$$

In the discussion so far and for the sake of simplicity, the electron cloud was treated as one unit — an obvious oversimplification. In reality, each atom has j ($j = Z_i$) oscillators associated with it, each having an oscillator strength γ_j. The jth oscillator vibrates with its own natural frequency and damping constant f_j. The total electronic polarizability of such an atom or ion is given by the sum of all the oscillators

$$k_e^* = 1 + \frac{e^2}{\varepsilon_0 m_e} \sum_j \frac{\gamma_j \left(\omega_{0,j}^2 - \omega^2\right)}{\left(\omega_{0,j}^2 - \omega^2\right)^2 + i\omega^2 f_j^2} \qquad (14.40)$$

The oscillator strength γ_j is related to (from quantum mechanics) the probability of transition of an electron from one band to the next.

For covalent solids, by far the major contribution to the dielectric constant results from electronic polarization. In ionic solids, the situation is more complicated, as discussed below. It should also be pointed out that the electronic polarizability of a compound can, to a very good approximation, be taken as the sum of the polarizabilities of the atoms or ions making up that compound.

14.4.2 Ionic Polarization

Electron clouds are not the only species that can respond to an applied electric field. Ionic charges in a solid can respond equally well and can significantly contribute to the dielectric constant. *Ionic polarization* is defined as the displacement of positive and negative ions toward the negative and positive electrodes, respectively, as shown schematically in Fig. 14.9. Ionic resonance occurs in the infrared frequency range (10^{12} to 10^{13} Hz), and consequently this phenomenon will be encountered once again in Chap. 16. Since it does not entail migration, ionic polarization is quite temperature-insensitive.

The equation of motion to solve is similar to Eq. (14.29) except that

- The ions are assumed to be attached to *one another* by a spring having a natural frequency of vibration ω_{ion} which is directly related to the coulombic attraction holding the ions together.
- The reduced mass of the system is now given by $M_r = m_c m_a / (m_c + m_a)$ where m_c and m_a are the cation and anion masses, respectively.
- The friction factor f_{ion} will also be different in this case.

It is not surprising that the final result for ionic polarizability is very similar to that for electronic polarization and is given by

$$k'_{ion}(\omega) = 1 + \frac{(ze)^2 N_{ion}\left(\omega_{ion}^2 - \omega^2\right)}{\varepsilon_0 M_r\left\{\left(\omega_{ion}^2 - \omega^2\right)^2 + f_{ion}^2\omega^2\right\}}$$ (14.41)

$$k''_{ion}(\omega) = \frac{(ze)^2 N_{ion}\omega f_{ion}}{\varepsilon_0 M_r\left\{\left(\omega_{ion}^2 - \omega^2\right)^2 + f_{ion}^2\omega^2\right\}}$$ (14.42)

where N_{ion} is the number of ion pairs per cubic meter. Note the similarity between Eqs. (14.33) and (14.41); as noted above, the only differences are the appearance of ω_{ion} rather than ω_0, of f_{ion} rather than f, and the use of M_r instead of m_e. Moreover, for $\omega \gg \omega_{ion}$, the ions can no longer follow the applied field and drop out, that is, $k'_{ion} \Rightarrow 1$, as expected.

Under dc conditions, Eq. (14.41) reduces to

$$k'_{ion} = 1 + \frac{(ze)^2 N_{ion}}{\varepsilon_0 M_r\omega_{ion}^2}$$ (14.43)

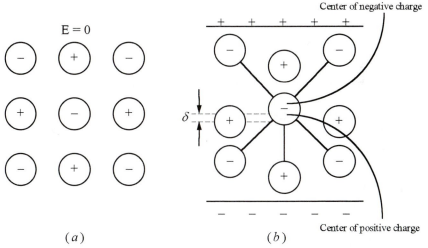

(a)   (b)

Figure 14.9
Ionic polarizability. (a) Ion positions at equilibrium; (b) Upon the application of the electric field, the center of negative charge is no longer coincident with the center of positive charge, i.e., polarization occurs.

In other words, the ionic polarization is inversely proportional to the ionic masses and the square of natural frequency of vibration of the ions ω_{ion}, which depends on the strength of the ionic bond. *Stronger bonds between ions with higher ω_{ion} values will be less readily polarized.* The factors that influence the strength of an ionic bond were discussed in detail in Chap. 4 and will not be repeated here. Finally, it is important to note that ceramics in which the bond is predominantly covalent (i.e., the atoms are not charged) exhibit little or no ionic polarization.

WORKED EXAMPLE 14.2. (*a*) In a similar approach to that used to derive Eq. (14.39), derive an expression for k_{ion}. (*b*) Based on the expression derived in part (*a*) calculate k_{ion} of NaCl and MgO, given that the Born exponent for NaCl is ≈ 8 and that for MgO is ≈ 7. State all assumptions, and compare to experimental results listed in Table 14.1.

Answer

(*a*) Replacing ω_{ion}^2 in Eq. (14.43) by S_0/M_r and noting that for the ionic bond (see Prob. 4.2)

$$S_0 \approx \frac{(ze)^2}{4\pi\varepsilon_0 r_0^3}(n-1)$$

where r_0 is the equilibrium interionic spacing and n is the Born exponent, it is trivial to show that

$$\alpha_{ion} \approx \frac{4\pi\varepsilon_0 r_0^3}{n-1}$$

which when combined with the Clausius-Mossotti equation results in

$$\frac{k_{ion}-1}{k_{ion}+2} \approx \frac{\alpha_{ion}N_{ion}}{3\varepsilon_0} \approx \frac{4\pi N_{ion} r_0^3}{3(n-1)}$$

which is the sought-after result. It is interesting to note that the charge on the ions does not appear in the final expression, because both the electric force and the restoring force scale with $z_1 z_2$.

(*b*) The number of ion pairs in NaCl $(\rho = 2.165 \text{ g/cm}^3)$ and in MgO $(\rho = 3.6 \text{ g/cm}^3)$ is, respectively,

$$N_{ion}(\text{NaCl}) = \frac{2.165 \times 6.02 \times 10^{23} \times 10^6}{23 + 35.45} = 2.23 \times 10^{28} \text{ ion pairs/m}^3$$

$$N_{ion}(\text{MgO}) = \frac{3.6 \times 6.02 \times 10^{23} \times 10^6}{24.31 + 16} = 5.38 \times 10^{28} \text{ ion pairs/m}^3$$

From App. 3A, $r_{ion}(NaCl) = 116 + 167 = 283$ pm, and $r_{ion}(MgO) = 86 + 126 = 212$ pm. Substituting these values in the expression given above and solving for k_{ion}, one obtains

$$k_{ion}(NaCl) = 2.3 \text{ and } k_{ion}(MgO) = 2.67$$

Experimentally (see Table 14.1), $k_{ion}(NaCl) = 5.89 - 2.41 = 3.48$, and $k_{ion}(MgO) = 9.83 - 3 = 6.83$. And although the agreement between theory and experiment is not excellent, given the simplicity of the model used (i.e., assuming the ions to be hard spheres, etc.), it is quite satisfactory. This is especially true when it is appreciated that, as discussed in the following section, the values of of the static dielectic constants listed in Table 14.1 include other contributions in addition to ionic polarization.

14.4.3 Dipolar Polarization

In contrast to electronic polarization and ionic polarization, which occur at high frequencies $(\omega > 10^{10} \text{ Hz})$, dipolar polarization occurs at lower frequencies and is thus important because it can greatly affect the capacitive and insulative properties of glasses and ceramics in low-frequency applications.

As noted above, the **ion jump polarization** is the preferential occupation of equivalent or near-equivalent lattice sites as a result of the applied fields biasing one site over the other.[255] The situation is depicted schematically in Fig. 14.10, where an ion is localized in a deep energy well, but within that well two equivalent sites, labeled A and B in Fig. 14.10b, exist. The sites are separated from each other by a distance λ_s and an energy barrier ΔH_m. In the absence of an electric field (Fig. 14.10a), each site has an equal probability of being occupied and no net polarization will result. In the presence of a field (Fig. 14.10b) the two sites are no longer equivalent — the electric field will bias the B sites, resulting in a net polarization.[256]

---

[255] The discussion in this section is applicable to any polar solid (i.e., one that has a permanent dipole) in which relaxation of the permanent dipoles occurs. In principle this approach could be applicable to piezoelectric and ferroelectric solids at temperatures below their transition temperature as well (see next chapter). However, the same response, as shown in Fig. 14.11, can also occur as a result of heavily damped resonance. This is easily seen in Fig. 14.7a; as f, which is a measure of the damping, increases, the resultant resonance curves become flatter. Experimentally it is not easy to distinguish between the two phenomena.

[256] The following analogy should help: Think of the crystal as a ship in which the ions are passengers, confined to cabins with two beds each, where the beds are arranged parallel to the

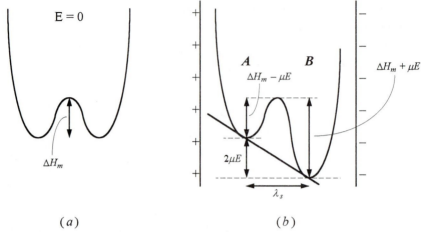

Figure 14.10

Dipolar polarization; (*a*) Energy versus distance diagram in the absence of applied field; the two sites are equally populated. (*b*) The application of an electric field will bias one site relative to the other.

As discussed in Chap. 7, the probability of an ion's making a jump in the absence of a bias is given by the Boltzmann factor,

$$\Theta = K \exp\left(-\frac{\Delta H_m}{kT}\right) \tag{14.44}$$

The potential energy of a dipole depends on the orientation of the dipole moment μ_{dip} with respect to the applied field E or

$$U = -\mu_{\text{dip}} \bullet E = -\mu_{\text{dip}} E \cos\theta \tag{14.45}$$

where θ is the angle between μ and E. For $\theta = 0°$ or $180°$, the potential energy is simply $\pm\mu_{\text{dip}}E$ depending on whether the moment aligns with or against the field. It follows that the energy difference between the two locations is $2\mu_{\text{dip}}E$. Assuming N_{dip} bistable dipoles per unit volume, in the presence of a field, the jump probability $A \Rightarrow B$ is

$$\Theta_{A \to B} = K \exp\left(-\frac{\Delta H_m - \mu_{\text{dip}} E}{kT}\right)$$

---

ship axis. In the absence of a bias, it is fair to assume the passengers will occupy either bed equally such that the center of gravity of the ship remains at its center. If now the ship tilts to either side, the passengers will tend to favor that side and so the center of gravity of the ship will no longer be at its center.

For the most part, the applied fields are small enough that $\mu_{dip}E/(kT) << 1$ and this equation simplifies to

$$\Theta_{A \to B} = \left(1 + \frac{\mu_{dip}E}{kT}\right)\Theta \qquad (14.46)$$

Similarly, the jump probability $B \to A$ is given by

$$\Theta_{B \to A} = \left(1 - \frac{\mu_{dip}E}{kT}\right)\Theta \qquad (14.47)$$

At steady state,

$$N_A\Theta_{A \to B} = N_B\Theta_{B \to A} \qquad (14.48)$$

where N_A and N_B are the number of ions in each well. Combining Eqs. (14.46) to (14.48) and rearranging yield

$$N_B - N_A = (N_B + N_A)\frac{\mu_{dip}E}{kT} = N_{dip}\frac{\mu_{dip}E}{kT} \qquad (14.49)$$

The **static polarization per unit volume P_s** is defined as

$$P_s = (N_B - N_A)\mu_{dip} = N_{dip}\frac{\mu_{dip}^2 E}{kT} \qquad (14.50)$$

from which it is obvious that if $N_B = N_A$, there would be no polarization — a result that is not too surprising. Combining Eqs. (14.13) and (14.50), one obtains

$$\boxed{k'_{dip} - 1 = \frac{N_{dip}\mu_{dip}^2}{kT\varepsilon_0}} \qquad (14.51)$$

It is important to note the following:

- Now k'_{dip} is a function of the total number of dipoles per unit volume, the charge on the ions that are jumping, and the distance of the jump. Note here that neither ΔH_m nor the frequency of the applied field plays a role because Eq. (14.51) represents the equilibrium situation under *static* (i.e., dc) conditions. It gives the equilibrium value, but says nothing about how rapidly or slowly equilibrium is reached (see below).
- Increasing the temperature will reduce k'_{dip} as a *result of thermal randomization*. This functionality on temperature is known as *Curie's law* and will be encountered again in Chap. 15. In analogy to paramagnetism (see Chap. 15), any solid for which the susceptibility is proportional to $1/T$ can be termed a **paraelectric** solid.

- It is worth noting here, once again, that in deriving Eq. (14.51), the assumption was made [Eq. (14.13)] that the local field is the same as the applied field — not necessarily a good assumption in this case.

Dynamic response and the Debye equations

To understand and model the dynamic response of dipolar polarization is quite a complicated affair. Debye, however, rendered the problem tractable by making the following assumptions:

- At high frequencies, that is, $\omega \gg 1/\tau$, the relative dielectric constant is given by κ_{∞}', where $\kappa_{\infty}' = k_{ion}' + k_e'$ (i.e., the sum of the ionic and electronic contributions).
- As $\omega \Rightarrow 0$, the relative dielectric constant is given by k_{static}' where $k_{static}' = k_{dip}' + \kappa_{\infty}'$.
- The polarization decays exponentially as

$$P(t) = P_0 \exp\left(-\frac{t}{\tau}\right) \tag{14.52}$$

where τ is the **relaxation time** of the system or the *average residence time of an atom or ion at any given site*. From these assumptions it can be shown that[257]

$$k_{dip}'(\omega) = k_{\infty}' + \frac{k_{static}' - k_{\infty}'}{1 + \omega^2 \tau^2} \tag{14.53}$$

$$k_{dip}''(\omega) = \frac{\omega\tau}{1 + \omega^2 \tau^2}(k_{static}' - k_{\infty}') \tag{14.54}$$

$$\tan\phi = \frac{k_{dip}''}{k_{dip}'} = \frac{(k_{static}' - k_{\infty}')\omega\tau}{k_{static}' + k_{\infty}'\omega^2 \tau^2} \tag{14.55}$$

These are known collectively as the **Debye equations**, and are plotted in Fig. 14.11. At low frequencies, all polarization mechanisms can follow the applied field, and the total dielectric constant is $\approx k_{static}'$ which includes the dipolar, ionic, and electronic contributions. When $\omega\tau = 1$, then k_{dip}' goes through an inflection point and k_{dip}'' is at a maximum. At higher applied frequencies, the dipolar

---

257 See, e.g., L. L. Hench and J. K. West, *Principles of Electronic Ceramics*, Wiley-Interscience, New York, 1990, p. 198.

polarization component to k_{static} drops out and only the ionic and electronic contributions remain.[258]

It is important to note that an implicit assumption made in deriving the Debye equations is that of a single relaxation time. In other words, the heights of the barriers are identical for all sites. And while this may be true for some crystalline solids, it is less likely to be so for an amorphous solid such as a glass, where the random nature of the structure will likely lead to a distribution of relaxation times.

Temperature dependence of dipolar polarization

As noted above, τ is a measure of the average time an ion spends at any one site. In other words, $\tau \propto 1/\Theta$, which according to Eq. (14.44) renders τ exponentially dependent on temperature, or

$$\tau = \tau_0 \exp \frac{\Delta H_m}{kT} \tag{14.56}$$

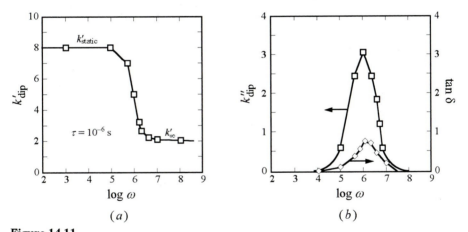

Figure 14.11

Frequency dependence of dielectric parameters for dipolar polarization on (a) k', assuming a time constant τ of 10^{-6} s, and (b) k'' and $\tan\delta$. Note that the maximum for k'' is coincident with the inflection point in k', whereas the maximum for $\tan\delta$ is shifted to higher frequencies.

---

[258] If the electric field switches polarity in a time that is much *shorter* than an ion's residence time on a site, the average energies of the two sites become equivalent (i.e., a bias no longer exists — the sites become energetically degenerate).

and immediately implies that the resonance frequency should also be an exponential function of temperature. As the temperature increases, the atoms vibrate faster and are capable of following the applied field to higher frequencies. This is indeed found to be the case, as shown in Fig. 14.12a, where the dielectric loss peaks are plotted as a function of temperature. As the temperature increases, the maximum in the loss angle shifts to higher temperatures, as expected. Furthermore, when the frequency at which the peaks occur is plotted as a function of reciprocal temperature, the expected Arrhenian relationship is observed (Fig. 14.12b).

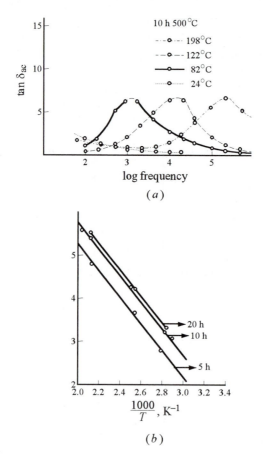

Figure 14.12

(a) Effect of temperature on dielectric loss peak in an $Li_2O - SiO_2$ glass. As the temperature increases, the frequency at which the maximum in loss occurs also increases, since the mobility of the ions increases. (b) Temperature dependence of frequency at which maximum in dielectric-loss peak occurs.

14.4.4 Dielectric Spectrum

From the foregoing discussion, it is clear that the dielectric response of solids is a complex function of frequency, temperature, and type of solid. Under dc conditions, all mechanisms are operative, and the dielectric constant is at its maximum and is given by the sum of all the mechanisms. As the frequency of the applied electric field increases, various mechanisms will be unable to follow the field and will drop off, as shown in Fig. 14.13. At very high frequencies, none of the mechanisms is capable of following the field, and the relative dielectric constant approaches 1.0.

Temperature, however, will influence only the polarization mechanisms that depend on long-range ionic displacement such as dipolar polarization. Ionic polarization is not strongly affected by temperature since long-range mobility of the ions is not required for it to be operative.[259]

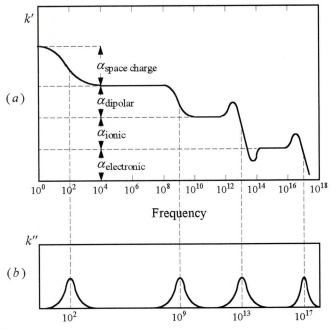

Figure 14.13
Variation of (*a*) relative dielectric constant and (*b*) dielectric loss with frequency.

---

259 This should not be confused with the effect of temperature on the dielectric *loss* (see the next section).

14.5
DIELECTRIC LOSS

The **dielectric loss** is a measure of the energy dissipated in the dielectric in unit time when an electric field acts on it. Combining Eqs. (14.23) and (14.25), it can be shown that the power loss per unit volume dissipated in a dielectric is related to k'', the frequency of the applied field, and its dc conductivity, by

$$P_V = \frac{1}{2} \{ \sigma_{dc} + \omega \varepsilon_0 k''(\omega) \} E_0^2 \qquad (14.57)$$

This power loss represents a wastage of energy as well as attendant heating of the dielectric. If the rate of heat generation is faster than it can be dissipated, the dielectric will heat up, which, as discussed below, could lead to dielectric breakdown and other problems. Furthermore, as the temperature increases, the dielectric constant is liable to change as well, which for finely tuned circuits can create severe problems. Another reason for minimizing k'' is related to the sharpness of the tuning circuit that would result from using the capacitor — lower losses give rise to much sharper resonance frequencies.

From Eq. (14.57) it is immediately apparent that in order to reduce power losses, it is imperative to

- Use solids that are highly insulating, $\sigma_{dc} \Rightarrow 0$. In other words, use very pure materials with large band gaps such that the number of free carriers (whether they are impurity ions, free electrons, or holes) is as low as possible.
- Reduce k''. Thus, almost by definition, a good dielectric must have very low dc conductivity and a low k'' — hence the need to understand what contributes to k''.

Since temperature usually tends to increase the conductivity of a ceramic exponentially (Chap. 7), its effect on dielectric loss can be substantial. This is demonstrated in Fig. 14.14a, where the loss tangent is plotted as a function of temperature for different glasses that have varying resistivities. In all cases, the increased mobility of the cations results in an increase in the dielectric loss tangent. The effect of impurities, inasmuch as they increase the conductivity of a ceramic, can also result in large increases in the dielectric loss. This is shown in Fig. 14.14b, for very pure NaCl (lower curve) and one that contains lattice impurities (top curve).

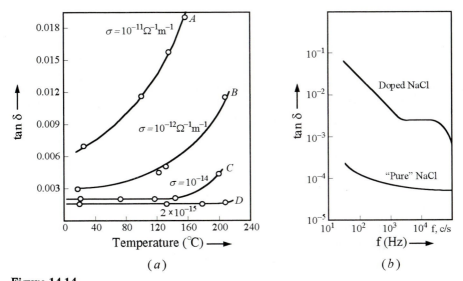

Figure 14.14

(a) Variation of loss angle with temperature for alkali glasses as a function of their resistivity. The measurements were carried out at 1 Mc/s. (b) Effect of impurities and frequency on $\tan\phi$ of NaCl. Curve labeled 1 is for a nominally pure sample; curve 2 is for one that was doped so as to introduce lattice defects.

Needless to say, the frequency at which a dielectric is to be used must be as far removed from a resonance frequency as possible, since near resonance, k'' can increase substantially.

Interestingly enough, the crystal structure can have an effect on k'' and the loss tangent. In general, for close-packed ionic solids, the dielectric loss is quite small, whereas loosely packed structures tend to have higher dielectric loss. This is nicely demonstrated when values of $\tan\phi$ for α– and γ–aluminas are compared. For α–alumina, $\tan\phi < 0.0003$ at 100 °C and 10^{-6} s^{-1}, whereas the loss tangent for the less dense γ modification is greater than 0.1.

14.6
DIELECTRIC BREAKDOWN

When a dielectric is subjected to an ever-increasing electric field, at some point a short circuit develops across it. **Dielectric breakdown** is defined as the voltage gradient or electric field sufficient to cause the short circuit. This phenomenon depends on many factors, such as sample thickness, temperature, electrode composition and shape, and porosity.

In ceramics, there are two basic types of breakdown; intrinsic and thermal.

- *Intrinsic*. In this mechanism, electrons in the conduction band are accelerated to such a point that they start to ionize lattice ions. As more ions are ionized and the number of free electrons increases, an avalanche effect is created. Clearly, the higher the electric field applied, the faster the electrons will be accelerated and the more likely this breakdown mechanism will be.
- *Thermal breakdown*. The criterion for thermal breakdown is that the rate of heat generation in the dielectric, as a result of losses, must be greater than the rate of heat removal from the sample. Whenever this condition occurs the dielectric will heat up, which in turn will increase its conductivity, which causes further heating, etc. This is termed *thermal breakdown* or *thermal runaway*.

14.7
CAPACITORS AND INSULATORS

Ceramic dielectric materials are typically used in electric circuits as either capacitors or insulators. Capacitors act as electrical buffers, diverting spurious electric signals and storing surges of charge that could damage circuits and disrupt their operation. By blocking dc signals and allowing only ac signals, capacitors can separate ac and dc signals and couple alternating currents from one part of a circuit to another. They can discriminate between different frequencies as well as store charge.

Whether a dielectric solid is to be used as a capacitor or as an insulator will depend on its characteristics. For capacitive functions, high relative dielectric constants are required together with low losses. The perfect dielectric would have a very large k' and no losses. But if the dielectric is used for its insulative properties, whether in high-power applications or as a substrate for integrated circuits, then it is desirable to have as low a dielectric constant as possible and once again minimal losses. It is worth noting that the need for low-loss insulators has grown significantly recently with the advent of high-frequency telecommunications networks. Since the power losses [Eq. (14.57)] are proportional to frequency, the need for lower and lower loss insulators is more crucial than ever.

Table 14.1 lists the values of k'_{static} and k'_e (which, as discussed in Chap. 16, is nothing but the square of the refractive index) together with $\tan\delta$ of a number of ceramics.

In general, dielectrics are grouped into three classes:

Class 1 dielectrics include ceramics with relatively low and medium dielectric constants and dissipation factors of less than 0.003. The low range covers $k'_{static} = 5$ to 15, and the medium k'_{static} range is 15 to 500.

Class II dielectrics are high-permittivity ceramics based on ferroelectrics (see Chap. 15) and have values of k' between 2000 and 20,000.

Class III dielectrics (not discussed here) contain a conductive phase that effectively reduces the thickness of the dielectric and results in very high capacitances. Their breakdown voltages are quite low, however.

Low-permittivity ceramics are widely used for their insulative properties. The major requirements are good mechanical, thermal, and chemical stability; good thermal shock resistance; low-cost raw materials; and low fabrication costs. These include the clay- and talc-based ceramics also known as *electrical porcelains*. A large-volume use of these materials is as insulators to support high tension cables that distribute electric power. Other applications include lead-feedthroughs and substrates for some types of circuits, terminal connecting blocks, supports for high-power fuse holders, and wire-wound resistors.

Another very important low-permittivity, low-loss ceramic is alumina. Alumina has such an excellent combination of good mechanical properties, high thermal conductivity, and ease of metallization that it is widely used today for thick-film circuit substrates and integrated electronic packaging.

Other low-permittivity ceramics that are emerging as likely candidates to replace alumina are BeO and AlN. Broadly speaking, these compounds have properties that are quite comparable to those of alumina, except that their thermal conductivities are roughly 5 to 10 times that of alumina. AlN has a further advantage that its thermal expansion coefficient of 4.5 is a better match with that of Si $\left(2.6\times10^{-6\circ}C^{-1}\right)$ than that of alumina. With these properties, AlN, despite its higher cost, may replace alumina as the size, number, and density of chips increase and more heat needs to be dissipated.

Medium-permittivity ceramics are widely used as class I dielectrics, but only if they have low dissipation factors. This precludes the use of most ferroelectric compounds that tend to have higher loss tangents. The three principal areas in which these low-loss class I materials are used are high-power transmission capacitors in the megahertz frequency range, stable capacitors for general electronic use, and microwave-resonant cavities that operate in the gigahertz range.

TABLE 14.1

Dielectric properties of some ceramic materials[†]

Compound	k'_{static}	$k'_e = n^2$	$\tan\delta$ $(\times 10^4)$	Compound	k'_{static}	$k'_e = n^2$	$\tan\delta$ $(\times 10^4)$
			Halides				
AgCl	12.30	4.00		LiF	8.90	1.92	2
AgBr	13.10	4.60		LiI	11.00	3.80	
CsBr	6.67	2.42		NaBr	6.40	2.64	
CsCl	7.20	2.62		NaCl	5.89	2.40	2
CsI	5.65	2.62		NaF	5.07	1.74	
KBr	4.90	2.34	2	NaI	7.28	2.93	
KCl	4.84	2.19	10	RbBr	4.86	2.34	
KF	5.46	1.85		RbCl	4.92	2.19	
KI	5.10	2.62		RbF	6.48	1.960	
LiBr	9.0–13.00	3.22		RbI	4.91	2.60	
LiCl	11.86	2.79		TlBr	30.00	5.40	
			Binary oxides				
Al_2O_3	9.40	3.13	0.4–2	MnO	18.10		
BaO		3.90		Sc_2O_3		3.96	
BeO	6.80	2.95	2	SiO_2	3.80	2.30	4
CaO	12.00	3.40		SrO	13.00	3.31	
Cr_2O_3	11.80	6.50		TiO_2 rutile	114.00	6.40–7.40	2–4
Eu_2O_3		4.45		TiO_2 (∥ c)[‡]	170.00	8.40	16
Ga_2O_3		3.72		TiO_2 (∥ a)[§]	86.00	6.80	2
Gd_2O_3		4.41		Y_2O_3		3.72	
MgO	9.83	3.00	3	ZnO	9.00	4.00	
			Ternary oxides				
$BaTiO_3$	3000.00	5.76	1–200	$MgTiO_3$	16.00		2
$CaTiO_3$	180.00	6.00		$SrTiO_3$	285.00	6.20	
$MgAl_2O_4$	8.20	2.96	5–8				
			Glasses				
Pb-silica glass	19.00		57	Soda-lime glass	7.60	2.30	100
Pyrex	4.00–6.00			Vycor			8
			Others				
AlN	8.80		5–10	α–SiC	9.70	6.70	
C (diamond)	5.68	5.66		Si	11.70	11.70	
				ZnS	8.32	5.13	

[†] The values quoted in the literature are quite variable especially for k'_s which depends strongly on sample purity and quality.

[‡] Parallel to the c axis in a single crystal rutile.

[§] Parallel to the a axis.

14.8
SUMMARY

1. The application of an electric field E across a linear dielectric material results in polarization P or the separation of positive and negative charges. The relative dielectric constant k' is a measure of the capacity of a solid to store charge relative to vacuum and is related to the extent to which the charges in a solid polarize. Atomically there are four main polarization mechanisms: electronic, ionic, dipolar, and space charge.

   For linear dielectrics, it is assumed that P is linearly related to E, with the proportionality constant related to k'. When a sinusoidal electric field of frequency ω is applied to a dielectric, some of the bound charges will move in phase with the applied field and will contribute to k'. Another set of bound charges, however, will oscillate out of phase with the applied field, will result in energy dissipation, and will contribute to the dielectric loss factor k''. In addition to these bound charges, there will always be a dc component to the total current which contributes to the total conductivity of the sample and is a loss current.

2. Electronic polarization involves the displacement of the electrons relative to their nucleus and exhibits a resonance when the frequency of the applied field is comparable to the natural frequency of vibration of the electronic cloud. The latter is determined by several factors, the most important being the volume of the ion involved. Electronic polarization is quite insensitive to temperature and is exhibited by all matter.

3. Ionic polarization involves the displacement of cations relative to anions. Resonance occurs when the frequency of the applied field is close to the natural frequency of vibration of the ions. The latter is determined by the strength of the bond holding the ions, which in turn is related to, among other things, the net charges on the ions and their equilibrium interatomic distance. Since it does not involve ions jumping from site to site, ionic polarization is also quite insensitive to temperature.

4. For dipolar polarization to occur, two or more adjacent sites separated by an energy barrier must exist. The preferential occupancy of one site relative to the other as a result of the application of an electric field results in solids that can have quite large k' values. Increasing the temperature increases the randomness of the system and tends to decrease k'_{dip}. Such solids possess a relaxation time τ that is a measure of the average time an ion spends at any given site. When $\omega\tau = 1$, the loss is maximized and k'_{dip} has an inflection point.

5. Power dissipation in a dielectric depends on both its dc conductivity and k''. In general, a dielectric should be used at temperatures and frequencies that are as far removed as possible from a resonance or relaxation frequency. The composition should also be such as to minimize the dc conductivity.
6. For a capacitor k' should be maximized, whereas for an insulator it should be minimized. In both applications, however, the losses should be minimized.

APPENDIX 14A

LOCAL ELECTRIC FIELD

To estimate the local field, refer to Fig. 14.15 where a reference atom is surrounded by an imaginary sphere of such an extent that beyond it the material can be treated as a continuum. If the reference atom is removed while the surroundings remain frozen, the total field at point A will stem from three sources:

- E_1, the free charges due to the applied electric field E
- E_2, the field that arises from the free ends of the dipole chains that line the cavity
- E_3, the field due to atoms or molecules in the near vicinity of the reference molecule.

In highly symmetric crystals such as cubic crystals, it can be assumed that the additional individual effects of the surrounding atoms mutually cancel, or $E_3 = 0$.

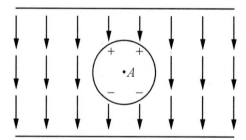

Figure 14.15
Model for calculation of internal field.

By applying Coulomb's law to the surface of the sphere, it can be shown that[260]

$$E_2 = \frac{E}{3}(k' - 1) \qquad (14A.1)$$

The local field is thus given by

$$E_{\text{loc}} = E_1 + E_2 = E + \frac{E}{3}(k' - 1) = \frac{E}{3}(k' + 2) \qquad (14A.2)$$

When combined with Eq. (14.13), one obtains

$$E_{\text{loc}} = E_1 + E_2 = E + \frac{P}{3\varepsilon_0} \qquad (14A.3)$$

Substituting the local field instead of the applied field in Eq. (14.29) yields

$$M_r \left\{ \frac{d^2\delta}{dt^2} + f\frac{d\delta}{dt} + \left(\omega_0^2 - \frac{Ne^2}{3M_r\varepsilon_0} \right)\delta \right\} = Z_i e E_0 \exp i\omega t \qquad (14A.4)$$

It follows that the effect of polarization on the surroundings is to lower the resonance frequency of the individual oscillator from ω_0 to

$$\omega_0' = \sqrt{\omega_0 - \frac{Ne^2}{3m_e\varepsilon_0}} \qquad (14A.5)$$

PROBLEMS

14.1. (a) Show that Eq. (14.17) can be written as

$$i\omega CV = \omega CV_0 \exp i\left(\omega t + \frac{\pi}{2}\right)$$

and that consequently $I_{\text{chg}} = -\omega\kappa' C_{\text{vac}} V_0 \sin \omega t$.

(b) The **admittance** of a circuit is defined as

$$Y^* = \frac{I_0}{V_0}\cos\left(\frac{\pi}{2} - \phi\right) - i\frac{I_0}{V_0}\sin\left(\frac{\pi}{2} - \phi\right)$$

---

[260] See, e.g., N. Ashcroft and N. Mermin, *Solid State Physics*, Holt-Saunders, International Ed., 1976, p. 534, or C. Kittel, *Introduction to Solid State Physics*, 6th ed, Wiley, New York, 1988.

where I_0, V_0, and ϕ are defined in Fig. 14.3b.   Here the first term represents the loss and the second term the charging current. By equating this equation with the charging and loss currents derived in text, show that

$$\sigma_{ac} = \frac{I_0(\omega)d}{V_0 A}\cos\left[\frac{\pi}{2} - \phi(\omega)\right]$$

14.2. A parallel-plate capacitor with plates separated by 0.5 cm and with a surface area of 100 cm^3 is subjected to a potential difference of 1000 V across the plates.

(a) Calculate its capacitance.

Answer: $C = 18$ pF

(b) A glass plate with a relative dielectric constant of 5.6, which just fills the space between the plates, is inserted between them. Calculate the surface charge density on the glass plate.

Answer: 8.14 $\mu C/m^2$

(c) What voltage is required to store a charge of 5×10^{-10} C on a capacitor with plates 20 mm $\times$ 20 mm and separated by 0.01 mm of (i) vacuum and (ii) BaTiO$_3$?

(d) A material is placed in an electric field of 2000 V/m that causes a polarization of 5×10^{-8} C/m^2. What is the dielectric constant of this material?

14.3. Show that either Eq. (14.30) or Eq. (14.31) is a solution to Eq. (14.29).

14.4. (a) When an external field is applied to an NaCl crystal, a 5 percent expansion of the lattice occurs. Calculate the dipole moment for each Na$^+$ – Cl$^-$ pair. The ionic radii of Na and Cl are 0.116 and 0.167 m, respectively.

Answer: 2.30×10^{-30} C·m

(b) Calculate the dipole moment of an NaCl molecule in a vapor if the separation between the ions is 2.5 Å. State all assumptions.

14.5 (a) Starting with Eq. (14.30) or (14.31), derive Eqs. (14.33) and (14.34).

(b) Plot Eqs. (14.33) and (14.34) for various values of f.

14.6. (a) Discuss the possible polarization mechanisms for (a) Ar gas, (b) LiF, (c) water, and (d) Si.

(b) The dielectric constant of a soda-lime glass at very high frequencies $(> 10^{14}$ Hz$)$ was measured to be 2.3. At low frequency $(\approx 1$ MHz$)$ constant k' was 6.9. Explain.

(c) The static dielectric constants of the following solids are given:

NaCl	5.9	MgO	9.6
SiO_2	3.8	$BaTiO_3$	1600.0
Soda-lime glass	7.0		

Give a brief explanation for the different values. Discuss the various contributions to k'_{static}. Would you expect the ranking of these materials to change at a frequency of 10^{14} s^{-1}? Explain.

14.7. The relative dielectric constant of O_2 gas at 0°C was measured to be 1.000523. Using this result, predict the dielectric constant of liquid O_2 if its density is 1.19 g/cm^3. Compare your answer with the experimentally determined value of 1.507. State all assumptions and discuss implications of your results vis à vis the assumptions made.
Answer: 1.509

14.8. Using the ion positions for the tetragonal $BaTiO_3$ unit cell shown in Fig. 14.16, calculate the electric dipole moment per unit cell and the saturation polarization for $BaTiO_3$. Compare your answer with the observed saturation polarization (that is, $P_s = 0.26$ C/m^2). Hint: Use the Ba ions as a reference.
Answer: $P_s = 0.16$ C/m^2

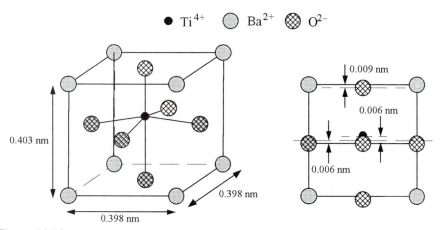

Figure 14.16
(a) In the tetragonal unit cell of $BaTiO_3$ the Ba ions occupy the unit cell corners, while Ti is near the center of the cell. The oxygens are near the centers of the faces. (b) Projection of the (100) face. Because the ions are displaced with respect to the symmetric position, the center of negative charge does not coincide with the center of positive charge, resulting in a net dipole moment per unit cell.

14.9. The temperature variation of the static dielectric constant for some gases is shown in Fig. 14.17.

    (*a*) Why do the dielectric constants for CCl_4 and CH_4 not vary with T?

    (*b*) What type of polarization occurs in CCl_4 and CH_4?

    (*c*) Why is k' greater for CCl_4 than for CH_4?

    (*d*) What type of polarization leads to the inverse temperature dependence of k' shown in CH_3Cl, CH_2Cl_2, and $CHCl_3$?

    (*e*) Why is the temperature variation of k' greater for CH_3Cl than for $CHCl_3$?

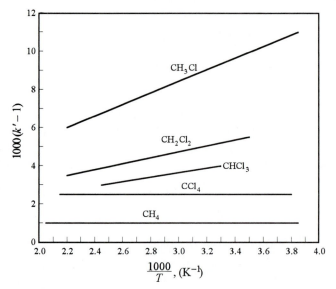

Figure 14.17

Temperature dependence of relative dielectric constants of some gases.

14.10. (*a*) Derive the following expression for dipolar polarization.

$$k'_{dip} - 1 = \frac{(ze)^2 N_{dip} \lambda_s^2}{4kT\varepsilon_0}$$

where λ_s is defined in Fig. 14.10, and z is the charge on the dipole. Hint: μ_{dip} in this case is $ze/2\lambda_s$.[261]

    (*b*) Where do you expect to have more ions at equilibrium in Fig. 14.10? Explain in your own words why the unequal distribution of ions in the two sites would give rise to polarization.

---

[261] This comes about because in the absence of a field the ions would be situated on average at $\lambda_s/2$, and the dipole moment would then be 0. See footnote 246.

(c) The static relative dielectric constant of a ceramic was measured at 200°C and found to be 140. Estimate the jump distance. State all necessary assumptions. Information you may find useful: Molar volume $=10 \text{ cm}^3/\text{mol}$; charge on cations is +4.

Answer: 0.0362 nm

14.11. (a) The electric dipole moment of water is 6.13×10^{-30} C·m. Calculate the dipole moment of each O–H bond.

Answer: 5.15×10^{-30} C·m

(b) The dielectric constant of water was measured as a function of temperature in such a way that the number of water molecules was kept constant. The results are tabulated below. Do the results behave according to Curie's law, i.e., as $1/T$? Explain.

(c) From the results estimate the density of the water molecules in the experiment.

Temp. (K)	119	148	171	200
k'	1.004	1.0037	1.0035	1.0032

Answer: $N = 1.735 \times 10^{25}$ m^{-3}

14.12. (a) Show that the frequency at which $\tan \phi$ for dipolar polarization is maximized is given by

$$\omega_{\max} = \frac{(k_s/k_\infty)^{1/2}}{\tau}$$

(b) The dielectric loss for thoria was measured as a function of temperature and frequency, and the results are tabulated below.[262] The static and high-frequency permittivities have been found from other measurements to be

$$k_s' = 19.2 \text{ and } k_\infty' = 16.2$$

Assuming the ion jump polarization is responsible for the variation in $\tan \phi$, estimate both τ_0 and ΔH_m [defined in Eq. (14.56)].

---

262 Data taken from PhD. Thesis of J. Wachtman, U. of Maryland, 1962. Quoted in *Lectures on the Electrical Properties of Materials*, 4th ed., Solymar and Walsh, Oxford Science Publications, 1989.

$\omega = 695$ Hz		$\omega = 6950$ Hz	
T, K	tan ϕ	*T*, K	tan ϕ
475	0.029	498	0.010
485	0.042	518	0.025
494	0.063	543	0.055
503	0.081	568	0.086
509	0.086	581	0.086
516	0.092	590	0.073
524	0.086	604	0.055
532	0.070	612	0.043
543	0.042	621	0.036
555	0.023	631	0.026

14.13. (*a*) Derive Eq. (14.24).

    (*b*) The following data[263] were determined for a technical ceramic as a function of temperature. The ceramic was in the form of a parallel plate capacitor with thickness 1/2 cm and diameter 2.54 cm.

    (i) Plot k' versus $\log \omega$ for all three temperatures on same plot.

    (ii) Plot k'' versus $\log \omega$ for all three temperatures on same plot.

    (iii) Plot $\tan \delta$ versus $\log \omega$ for all three temperatures on same plot.

    (iv) From your results, calculate the activation energy for the relaxation process, and compare it to the activation energy for dc conductivity. How do these activation energies compare? What conclusions can you reach regarding the basic atomic mechanisms responsible for conductivity and those for polarization?

    (v) What is the most probable polarization mechanism operative in this ceramic?

---

[263] Problem adapted from L. L. Hench and J. K. West, *Principles of Electronic Ceramics*, Wiley-Interscience, New York, 1990. The frequencies as reported in Hench and West are incorrect at the low end, and have been corrected here.

Frequency, Hz	72°C		90°C		112°C	
	G_{ac}, $\mu\Omega^{-1}$	C, pF	G_{ac}, $\mu\Omega^{-1}$	C, pF	G_{ac}, $\mu\Omega^{-1}$	C, pF
20 K	0.885	4.76	3.764	5.72	6.827	6.36
15 K					6.781	6.83
10 K	0.792	5.44	3.672	6.47	6.732	8.04
5 K	0.748	6.12	3.636	7.90	6.678	12.85
3 K					6.628	21.80
2 K	0.732	7.11	3.605	15.21	6.557	39.40
1.5 K					6.534	62.20
1.2 K					6.480	91.10
1 K	0.706	8.91	3.572	36.12	6.405	118.70
500	0.676	14.50	3.514	104.70	6.058	353.40
340	0.672	22.53				
260	0.603	32.40				
200	0.598	47.08	3.292	455.40		
100	0.574	139.00				

14.14. Clearly stating all assumptions, calculate the power loss for a parallel-plate capacitor ($d = 0.02$ cm, $A = 1$ cm^2) made of $MgAl_2O_4$ subjected to

(a) A dc voltage of 120 V

(b) An ac signal of 120 V and a frequency of 60 Hz

Answer: 0.05 μW

(c) An ac signal of 120 V and a frequency of 60 MHz

Answer: 0.05 W

14.15. (a) Explain why microwaves are very effective at rapidly and efficiently heating water or water-containing substances.

(b) In the microwave, the food is heated from the inside out, whereas in a regular oven the food is heated from the outside in. Explain.

ADDITIONAL READING

1. H. Frohlich, *Theory of Dielectrics*, 2d ed., Oxford Science Publications, 1958.

2. L. L. Hench and J. K. West, *Principles of Electronic Ceramics*, Wiley-Interscience, New York, 1990.

3. A. J. Moulson and J. H. Herbert, *Electroceramics*, Chapman & Hall, London, 1990.

4. N. Ashcroft and N. Mermin, *Solid State Physics*, Holt-Saunders International Ed., 1976.

5. C. Kittel, *Introduction to Solid State Physics*, 6th ed, Wiley, New York, 1988.

6. J. C. Anderson, *Dielectrics*, Chapman & Hall, London, 1964.
7. L. Azaroff and J. J. Brophy, *Electronic Processes in Materials,* McGraw-Hill, New York, 1963.
8. A. Von Hippel, ed., *Dielectric Materials and Applications*, Wiley, New York, 1954.
9. R. C. Buchanan, ed., *Ceramic Materials for Electronics*, Marcel Dekker, New York, 1986.
10. "Capacitors," *Scientific American*, July 1988, p. 86.

CHAPTER 15

Magnetic and Nonlinear Dielectric Properties

Magnetic *Atoms, such as Iron, keep*
Unpaired Electrons in their middle shell,
Each one a spinning Magnet that would leap
The Bloch *Walls whereat antiparallel*
Domains converge. Diffuse Material
Becomes Magnetic *when another Field*
Aligns domains like Seaweed in a swell.
How nicely microscopic forces yield
In Units growing visible, the World we wield!

John Updike, *Dance of the Solids*[†]

15.1
INTRODUCTION

The first magnetic material exploited by man as a navigational tool was a ceramic — the natural mineral magnetite (Fe_3O_4) also known as *lodestone*. It is thus somewhat paradoxical that it was as late as the mid-fifties of this century that magnetic ceramics started making significant commercial inroads. Since then magnetic ceramics have acquired an ever-increasing share of world production and have largely surpassed metallic magnets in tonnage. With the advent of the information communication revolution, their use is expected to become even more ubiquitous.

In addition to dealing with magnetic ceramics, this chapter also deals with dielectric ceramics, such as ferroelectrics, for which the dielectric response is

---

[†]   J. Updike, **Midpoint and other Poems**, A. Knopf, Inc., New York, New York, 1969. Reprinted with permission.

nonlinear. Ferroelectricity was first discovered in 1921 during the investigation of anomalous behavior of Rochelle salt. A second ferroelectric material was not found until 1935. The third major ferroelectric material, $BaTiO_3$, was reported in 1944. Ferroelectric ceramics possess some fascinating properties that have been exploited in a number of applications such as high permittivity capacitors, displacement transducers and actuators, infrared detectors, gas igniters, accelerometers, wave filters, color filters, and optical switches, and the generation of sonic energy, among others.

And while at first glance magnetic and nonlinear dielectric properties may not appear to have much in common, they actually do. The term *ferroelectric*[264] was first used in analogy to ferromagnetism. It is thus instructive at the outset to point out some of the similarities between the two phenomena. In both cases, the properties that result can be traced to the presence of permanent dipoles, magnetic in one and electric in the other, that respond to externally applied fields. An exchange energy exists between the dipoles that allows them to interact with one another in such a way as to cause their spontaneous alignment, which in turn gives rise to nonlinear responses. The orientation of all the dipoles with the applied field results in saturation of the polarization, and the removal of the field results in a permanent or residual polarization. The concept of domains is valid in both, and they both respond similarly to changes in temperature.

This chapter is structured as follows: Section 15.2 introduces the basic principles and relationships between various magnetic parameters. Section 15.3 deals with magnetism at the atomic level. In Sec. 15.4, the differences and similarities between para-, ferro-, antiferro-, and ferrimagnetism are discussed. Magnetic domains and hysteresis curves are dealt with in Sec. 15.5, while Sec. 15.6 deals with magnetic ceramics. The remainder of the chapter deals with the nonlinear dielectric response of ceramics in light of both the discussion in Chap. 14 and magnetic properties. Hopefully, the similarities will become clear, resulting in a deeper understanding of both phenomena.

15.2
BASIC THEORY

A **magnetic field intensity** H is generated whenever electric charges are in motion. The latter can be simply electrons flowing in a conductor or the orbital

---

264 As will become clear shortly, there is nothing "ferro" about ferroelectricity. It was so named because of the similarities between it and ferromagnetism, which was discovered first.

motion and spins of electrons around nuclei and/or around themselves. For example, it can be shown that the magnetic field intensity at the center of a circular loop of radius r through which a current i is flowing is

$$H = \frac{i}{2r} \qquad (15.1)$$

and it is normal to the plane of the loop as shown in Fig. 15.1a. The units of magnetic field are amperes per meter: 1 A/m is the field strength produced by a 1-A current flowing through a loop of 1 m.

In vacuum, H will result in a **magnetic field**[265] **B** given by

$$B = \mu_0 H \qquad (15.2)$$

where the constant μ_0 is the permeability of free space which is $4\pi \times 10^{-7}$ Wb/(A·m). B can be expressed in a number of equivalent units, such as $V \cdot s/m^2 = Wb/m^2 = T(\text{tesla}) = 10^4$ G(gauss) (see Table 15.1). A magnetic induction of 1 T will generate a force of 1 N on a conductor carrying a current of 1 A perpendicular to the direction of induction.

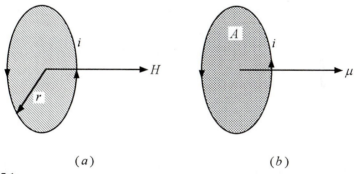

(a)   (b)

Figure 15.1
(a) Magnetic field intensity resulting from passage of a current through a loop of radius r.
(b) Magnetic moment of current loop of area A.

---

[265] The magnetic field B has been known by a variety of names: *magnetic induction, magnetic field strength*, and *magnetic flux density*. Here B is referred to strictly as the *magnetic field*. In many cases, H is also called the magnetic field, which, needless to say, can be quite confusing. This came about historically, because in the cgs system μ_0 was 1 and so B and H were numerically identical. To avoid confusion, H is strictly referred to here as the *magnetic field intensity* or *magnetizing field*.

TABLE 15.1
Definitions, dimension of units, and symbols used in magnetism[†]

Symbol	Quantity	Value	Units
H	Magnetic field intensity or magnetizing field		A/m[‡]
M	Magnetization		A/m
B	Magnetic field		$Wb/m^2 = T = V \cdot s/m^2 = 10^4$ G
μ_0	Permeability of free space	$4\pi \times 10^{-7}$	$Wb/(A \cdot m) = V \cdot s/(A \cdot m)$
μ	Permeability of a solid		$Wb/(A \cdot m) = V \cdot s/(A \cdot m)$
μ_r	Relative permeability		Dimensionless
χ_{mag}	Relative susceptibility		Dimensionless
μ_{ion}	Net magnetic moment of an atom or ion		$A \cdot m^2 = C \cdot m^2/s$
μ_s	Spin magnetic moment		$A \cdot m^2$
μ_{orb}	Orbital magnetic moment		$A \cdot m^2$
μ_B	Bohr magneton	9.274×10^{-24}	$A \cdot m^2$

[†] It is unfortunate that both μ_0 and μ_{ion} have the same symbol, but they will be clearly marked at all times to avoid confusion.

[‡] 1 A/m = 0.01257 Oersted(Oe).

In the presence of a solid, B will be composed of two parts — that which would be observed in the absence of the solid plus that due to the solid, or

$$B = \mu_0(H + M) \tag{15.3}$$

where M is the **magnetization** of the solid, defined as the net magnetic moment μ_{ion} per unit volume,

$$M = \frac{\mu_{ion}}{V} \tag{15.4}$$

The origin of μ_{ion} is discussed in detail in the next section. The units of μ_{ion} are ampere-square meters. A magnetic moment of $1 \ A \cdot m^2$ experiences a maximum torque of $1 \ N \cdot m$ when oriented perpendicular to a magnetic induction of 1 T.

In paramagnetic and diamagnetic solids (see below), B is a linear function of H such that

$$B = \mu H \tag{15.5}$$

where μ is the permeability of the solid (not to be confused with μ_{ion}). For ferro- and ferrimagnets, however, B and H are no longer linearly related, but as discussed below, μ can vary rapidly with H.

The **magnetic susceptibility** is defined as

$$\chi_{\text{mag}} = \frac{M}{H} \tag{15.6}$$

The **relative permeability** μ_r is given as

$$\mu_r = \frac{\mu}{\mu_0} \tag{15.7}$$

and it compares the permeability of a medium to that of vacuum. This quantity is analogous to the relative dielectric constant k'. And μ_r and χ_{mag} are related by

$$\mu_r = \chi_{\text{mag}} + 1 \tag{15.8}$$

Note that M and H have the same units, but in contradistinction to H, which is generated by electric currents or permanent magnets *outside* the material (which is why it is sometimes referred to as the magnetizing field), M is generated from the uncompensated spins and angular momenta of electrons *within* the solid. In isotropic media, B, H, and M are vectors that point in the same direction, whereas χ and μ are scalars.

EXPERIMENTAL DETAILS: MEASURING MAGNETIC PROPERTIES

Consider the experimental setup[266] shown in Fig. 15.2a, which is composed of four elements:

- A sample for which the magnetic properties are to be measured
- A sensitive balance from which the sample is suspended
- A permanent bar magnet with its north pole pointing upward
- A solenoid of n turns per meter, through which a current i flows so as to produce a magnetic field that is in the *same* direction as that of the permanent magnet (i.e., with the north pole pointing up)

---

[266] In reality, a shaped pole piece which creates a uniform magnetic field *gradient* dB/dx is used instead of the solenoid. The setup shown in Fig. 15.2a is simpler and is used to clarify the various concepts.

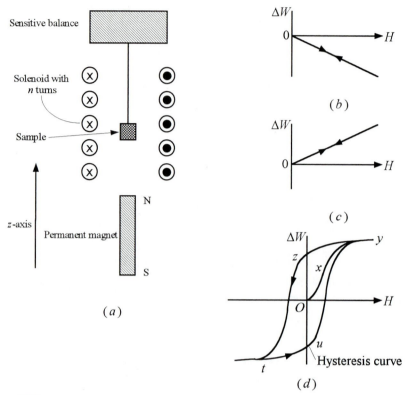

Figure 15.2
(*a*) Schematic of arrangement by which the magnetic properties of a material can be measured. The solenoid results in a uniform magnetic field and hence by itself will not create a force on the sample. It is only the nonuniformity of the magnetic field (that is, $dB/dz \neq 0$) of the permanent magnet that will create the force. (*b*) Typical response of a diamagnet. (*c*) Typical response of a paramagnet. (*d*) Typical response of a ferromagnet. Note that upon removal of the field a remnant magnetization remains.

To measure the magnetic properties of a given solid, a small cylinder of that solid is suspended from a sensitive balance into the center of the solenoid (Fig. 15.2*a*). According to Ampere's law, passing a current i through the solenoid will create an axial uniform magnetic field intensity H of strength

$$H = ni \tag{15.9}$$

This magnetizing field will in turn induce the magnetic moments (what that means is discussed later) in the material to align themselves either with or against the applied field. It can be shown that the magnetic force on a material with magnetization M and volume V is given by[267]

$$F_z = \mu_{\text{ion}} \frac{dB}{dz} = VM \frac{dB}{dz} \tag{15.10}$$

where dB/dz is the gradient in magnetic field along the z axis due to the *permanent magnet*. Combining Eq. (15.10) with Eq. (15.6), shows that

$$F_z = V \chi_{\text{mag}} H \frac{dB}{dz} \tag{15.11}$$

In other words, the force on the sample is directly proportional to its susceptibility and the applied field. Since in this case dB/dz is negative, if the sample is attracted to the permanent magnet, that implies χ_{mag} is positive and vice versa (see Worked Example 15.1).

There are four possible outcomes of such an experiment:

1. The sample is very weakly repelled by the permanent magnet, and the weight of the sample will appear to diminish (Fig. 15.2*b*), implying that χ_{mag} is *negative.* Increasing H by increasing the current in the solenoid will linearly increase the repulsive force. Such a material is termed a **diamagnetic** material, and most ceramics fall in this category.

2. The sample is weakly attracted to the permanent magnet, with a force of attraction that is proportional to H. The sample will appear to have gained weight (Fig. 15.2*c*), implying a positive χ_{mag}. Such solids are known as **paramagnets**. Repeating the experiment at different temperatures will quickly establish that the force of attraction, or χ_{mag}, decreases with increasing T. When the field intensity is removed, the sample will return to its original weight; in other words, all the changes that occurred in the presence of the field are completely reversible. The same is true for diamagnetic materials.

3. The sample is strongly attracted to the permanent magnet (Fig. 15.2*d*). The shape of the curve obtained, however, will depend on the sample's history. If one starts with a virgin sample, i.e., one that was never exposed to a field before, the weight gain will follow the line *Oxy*, shown in Fig. 15.2*d*. At low applied field intensities, the sample is initially weakly attracted, but as H is further increased, the rate of weight increase will be quite rapid, until finally at high magnetizing fields, the force will appear to be saturated and further increases in H will have a minimal effect (Fig. 15.2*d*). The plateau is known as the **saturation magnetization.**

---

[267] If the permanent magnet were not present, no force would be exerted on the sample, since to experience a force, the magnetic field has to be have a gradient (i.e., it cannot be uniform).

Furthermore, as H is reduced, the sample's response is nonreversible in that it follows the line yz. When the applied field intensity is zero, the sample will appear to have permanently gained weight! In other words, a permanent magnet with a **remnant magnetization** M_r has been created. Upon further cycling of the applied field, the sample's response will follow the loop $yztu$. Such so-called hysteresis loops represent energy losses and are typical of all ferromagnetic materials. The reason for their behavior is described more thoroughly below. Such behavior is termed **ferromagnetic** or **ferrimagnetic**.

Repeating the same experiment at increasing temperatures would result in essentially the same behavior except that the response of the magnet would weaken and M_r would also diminish. At a critical temperature, the material will totally lose its ferromagnetism and will behave as a paramagnet instead.

4. An **antiferromagnetic** material behaves similarly to a paramagnetic one, i.e., it is weakly attracted. However, to differentiate between the two, the experiment would have to be carried out as a function of temperature: The susceptibility of an antiferromagnetic material will appear to go through a maximum as the temperature is lowered (Fig. 15.5b); in contrast to that of a paramagnet which will continually increase with decreasing temperature (Fig. 15.3b).

Before this plethora of phenomena can be satisfactorily explained, it is imperative to understand what occurs at the atomic level that gives rise to magnetism, which is the topic of the following section.

WORKED EXAMPLE 15.1. A chunk of a magnetic ceramic weighing 10 g is attached to the sensitive balance shown in Fig. 15.1a and is suspended in the center of a toroidal solenoid with 10 turns per centimeter. A current of 9 A is passed through the coils. The magnetic field gradient due to the permanent magnet was measured to be 100 G/cm. When the current was turned on, such that the field was in the same direction as the permanent magnet, the weight of the sample was found to increase to 10.00005 g. The density of the solid is 5 g/cm^3. (*a*) Calculate the susceptibility of this material. (*b*) Calculate the magnetization M of the solid. (*c*) What conclusions can be inferred concerning the class of magnet to which this ceramic belongs?

Answer

(*a*) The force in the z direction, H, dB/dz, and V in SI units are, respectively,

$$F_z = \Delta W \cdot g = 0.00005 \times 10^{-3} \times 9.8 = -4.9 \times 10^{-7} \text{ N}$$

$$H = ni = \frac{9 \times 10}{10^{-2}} = 9000 \text{ A/m}$$

$$\frac{dB}{dz} = \frac{-100 \times 10^{-4}}{10^{-2}} = -1.0 \text{ T/m}$$

$$V = \frac{10}{5} = 2 \text{ cm}^3 = 2 \times 10^{-6} \text{ m}^3$$

Substituting these values in Eq. (15.11) and solving for χ_{mag}, one obtains

$$\chi_{mag} = \frac{4.9 \times 10^{-7}}{2 \times 10^{-6} \times 9000 \times 1} = 2.7 \times 10^{-5}$$

(b) $M = \chi_{mag} H = 2.7 \times 10^{-5} \times 9000 = 0.245$ A/m. Note that in this case, because χ_{mag} is small with very little loss in accuracy,

$$B = \mu_0(H + M) \approx \mu_0 H \tag{15.12}$$

(c) Given that the sample was attracted to the magnet, it must be paramagnetic, ferromagnetic, or antiferromagnetic. However, given the small value of χ_{mag}, ferromagnetism can be safely eliminated, and the material must either be paramagnetic or antiferromagnetic. To narow the possibilities further, the measurement would have to be repeated as a function of temperature.

15.3
MICROSCOPIC THEORY

For a solid to interact with a magnetic field, it must possess a net magnetic moment which, as discussed momentarily, is related to the angular momentum of the electrons, as a result of either their revolution around the nucleus and/or their revolution around themselves. The former gives rise to an **orbital angular moment** μ_{orb}, whereas the latter is the **spin angular moment** μ_s. The sum of these two contributions is the **total angular moment** of an atom or ion, μ_{ion}.

15.3.1 Orbital Magnetic Moment

It can be easily shown from elementary magnetism that a current i going around in a loop of area A' will produce an orbital magnetic moment μ_{orb} given by

$$\mu_{orb} = iA' \tag{15.13}$$

that points normal to the plane of the loop (Fig. 15.1b). A single electron rotating with an angular frequency ω_0 gives rise to a current

$$i = \frac{e\omega_0}{2\pi} \tag{15.14}$$

Assuming the electron moves in a circle of radius r, then combining Eqs. (15.13) and (15.14), one obtains

$$\mu_{\text{orb}} = \frac{e\omega_0 r^2}{2} \tag{15.15}$$

But since $m_e\omega_0 r^2$ is nothing but the **orbital angular momentum** Π_0 of the electron, it follows that

$$\mu_{\text{orb}} = \frac{e\Pi_0}{2m_e} \tag{15.16}$$

where m_e is the rest mass of the electron. This relationship clearly indicates that it is the angular momentum that gives rise to magnetic moments.

Equation (15.16) can be slightly recast in units of $h/(2\pi)$ to read

$$\mu_{\text{orb}} = \frac{eh}{4\pi m_e} \frac{2\pi\Pi_0}{h} = \frac{eh}{4\pi m_e} l \tag{15.17}$$

where the integer $l(= 2\pi\Pi_0/h)$ is the orbital angular momentum quantum number (see Chap. 2). Note that this result is consistent with quantum theory predictions that the angular momentum has to be an integral multiple of $h/(2\pi)$.

The ratio $eh/4\pi m_e$ occurs quite frequently in magnetism and has a numerical value of 9.27×10^{-24} A·m^2. This value is known as the **Bohr magneton** μ_B, and, as discussed below, it is the value of the orbital angular momentum of a single electron spinning around the Bohr atom. In terms of μ_B, Eq. (15.17) can be succinctly recast as:

$$\mu_{\text{orb}} = \mu_B l \tag{15.18}$$

In deriving Eq. (15.18), it was assumed that the angular momentum was an integral multiple of l; quantum mechanically, however, it can be shown[268] (not very easily, one should add) that the relationship is more complicated. The more accurate expression given here without proof is

$$\boxed{\mu_{\text{orb}} = \mu_B \sqrt{l(l+1)}} \tag{15.19}$$

---

268 See, e.g., R. P. Feynman, R. B. Leighton, and M. Sands, *The Feynman Lectures on Physics*, vol. 2, Chap. 34, Addison-Wesley, Reading, Massachusetts, 1964.

15.3.2   Spin Magnetic Moment

This moment arises from the spin of the electrons around themselves. Quantitatively, the **spin magnetic moment** μ_s is given by

$$\mu_s = \frac{e\Pi_s}{m_e} \qquad (15.20)$$

where Π_s is the **spin angular momentum**. Given that Π_s has to be an integer multiple of $h/2\pi$, it can be shown that

$$\mu_s = 2\mu_B s \qquad (15.21)$$

where s is the *spin quantum number*. Since $s = \pm 1/2$, it follows that the μ_s of an electron is one Bohr magneton. It is important to note, however, that Eq. (15.21) is not quite accurate. Quantum mechanically, it can be shown that the correct relationship — and the one that should be used — is

$$\boxed{\mu_s = 2\mu_B \sqrt{s(s+1)}} \qquad (15.22)$$

15.3.3   Total Magnetic Moment of a Polyelectronic Atom or Ion

The total angular moment of an ion, with *one* unpaired electron, $\mu_{\text{ion}, 1}$, is simply

$$\mu_{\text{ion}, 1} = \mu_s + \mu_{\text{orb}} = \frac{e\Pi_s}{m_e} + \frac{e\Pi_0}{2m_e} \qquad (15.23)$$

The suffix 1 was added to emphasize that this expression is only valid for one electron. Combining terms and introducing a factor g, known as the **Lande splitting factor,** one arrives at

$$\mu_{\text{ion}, 1} = \mu_s + \mu_{\text{orb}} = g\left(\frac{e}{2m_e}\right)\Pi_{\text{tot}} \qquad (15.24)$$

where Π_{tot} is the total angular momentum. If only the spin is contributing to $\mu_{\text{ion}, 1}$, then $g = 2$. Conversely, if only the orbital momentum is contributing to the total, then $g = 1$. Thus in general, g lies between 1 and 2 depending on the relative contribution of μ_s and μ_{orb} to $\mu_{\text{ion}, 1}$.

Equations (15.23) and (15.24) are only valid for ions that possess only one electron. If an atom or ion has more than one electron the situation is more complicated. For one, it can be shown (App. 15A) that the orbital magnetic momenta of the electrons add vectorially such that

$$L = \sum m_l \tag{15.25}$$

where m_l is the orbital magnetic quantum number (see Chap. 2). Similarly, the spins add such that the total spin angular momentum of the ion is given by

$$S = \sum s \tag{15.26}$$

where $s = \pm 1/2$.

The total angular momentum J of the atom is then simply the *vector* sum of the two noninteracting momenta L and S such that

$$J = L + S$$

This is known as **Russell-Saunders coupling**, and it will not be discussed any further because, for the most part, the orbital angular momentum of the transition-metal cations of the $3d$ series that are responsible for most of the magnetic properties exhibited by ceramic materials is *totally quenched*, that is, $L = 0$. Needless to say, this greatly simplifies the problem because now $\mathbf{J} = \mathbf{S}$ and μ_{ion} is given by

$$\boxed{\mu_{\text{ion}} = 2\mu_B \sqrt{S(S+1)}} \tag{15.27}$$

Thus the total magnetic moment of an ion (provided that the angular momentum is quenched) is related to the sum of the individual contributions of the unpaired electrons. Needless to say, to predict the magnetic moment of an atom or ion, one needs to know how many electrons are unpaired, or which quantum states are occupied.[269] The following examples make that point clear.

WORKED EXAMPLE 15.2. Calculate the spin and total magnetic moment of an isolated Mn^{2+} cation, assuming that the orbital angular momentum is quenched, that is, $L = 0$.

---

[269] A set of empirical rules known as *Hund's rules* determines the occupancy of the available electronic state within an atom. Very briefly stated, the electrons will occupy states with all spins parallel within a shell as far as possible. In other words, they will only pair up if absolutely necessary. They will also start by occupying the state with the largest orbital momentum, followed by the next largest, etc. The splitting of the d orbitals due to the presence of the ligands (see Chap. 16) further complicates the problem. Suffice it to say here that if the splitting is large, the electrons will violate Hund's rule and pair up, rather than populate the higher d orbitals.

Answer

Mn^{2+} has five d electrons (see Table 15.2) that occupy the following orbitals:

m_l	2	1	0	−1	−2	2	1	0	−1	−2
m_s	1/2	1/2	1/2	1/2	1/2	−1/2	−1/2	−1/2	−1/2	−1/2
s	↓	↓	↓	↓	↓					

It follows that $S = \sum m_s = 5 \times 1/2 = 2.5$. Since the angular momentum is quenched, $J = S$, and according to Eq. (15.27), the total magnetic moment for the ion is

$$\mu_{\text{ion}} = 2\mu_B \sqrt{S(S+1)} = 5.92\mu_B$$

which is in excellent agreement with the measured value of $5.9\mu_B$ (see Table 5.2).

WORKED EXAMPLE 15.3. Show that the angular momentum of any atom or ion with a closed shell configuration is zero.

Answer

A good example is Cu^+. It has 10 d electrons arranged as follows:

m_l	2	1	0	−1	−2	2	1	0	−1	−2
m_s	1/2	1/2	1/2	1/2	1/2	−1/2	−1/2	−1/2	−1/2	−1/2
s	↓	↓	↓	↓	↓	↑	↑	↑	↑	↑

Thus $L = \sum m_l = 0$ and $S = \sum m_s = 0$, and consequently, $J = 0$.

Two important conclusions can be drawn from these worked examples:

1. When an electronic shell is completely filled, all the electrons are paired, their magnetic moments cancel, and consequently, their net magnetic moment vanishes. Hence in dealing with magnetism, only partially filled orbitals need to be considered. Said otherwise, the existence of unpaired electrons is a necessary condition for magnetism to exist.

2. The fact that the calculated magnetic moment assuming *only* spin orbital momentum for the isolated cations of the $3d$ transition series compares favorably with the experimentally determined values (see Table 15.2) implies that the orbital angular momentum for these ions is indeed quenched.

TABLE 15.2
Magnetic moments of isolated cations of 3d transition series

Cations	Electronic configuration	Calculated moments $2\mu_B\sqrt{S(S+1)}$	Measured moments in μ_B
Sc^{3+}, Ti^{4+}	$3d^0$ [ ][ ][ ][ ][ ]	0.00	0.0
V^{4+}, Ti^{3+}	$3d^1$ [↑][ ][ ][ ][ ]	1.73	1.8
V^{3+}	$3d^2$ [↑][↑][ ][ ][ ]	2.83	2.8
V^{2+}, Cr^{3+}	$3d^3$ [↑][↑][↑][ ][ ]	3.87	3.8
Mn^{3+}, Cr^{2+}	$3d^4$ [↑][↑][↑][↑][ ]	4.90	4.9
Mn^{2+}, Fe^{3+}	$3d^5$ [↑][↑][↑][↑][↑]	5.92	5.9
Fe^{2+}	$3d^6$ [↓↑][↑][↑][↑][↑]	4.90	5.4
Co^{2+}	$3d^7$ [↓↑][↓↑][↑][↑][↑]	3.87	4.8
Ni^{2+}	$3d^8$ [↓↑][↓↑][↓↑][↑][↑]	2.83	3.2
Cu^{2+}	$3d^9$ [↓↑][↓↑][↓↑][↓↑][↑]	1.73	1.9
Cu^+, Zn^{2+}	$3d^{10}$ [↓↑][↓↑][↓↑][↓↑][↓↑]	0.00	0.0

15.4
PARA-, FERRO-, ANTIFERRO-, AND FERRIMAGNETISM

As noted above, the vast majority of ceramics are diamagnetic with negative susceptibilities $\left(\approx 10^{-6}\right)$. The effect is thus small and of very little practical significance, and will not be discussed further. Instead, in the following sections, the emphasis will be on by far the more useful and technologically important classes of magnetic ceramics, namely, the ferro- and ferrimagnetic ones. Before one considers these, however, it is important to understand the behavior of another class of materials, namely, paramagnetic materials, not because of their practical importance (very little) but because they represent an excellent model system for understanding the other, more complex phenomena.

15.4.1 Paramagnetism

Paramagnetic solids are those in which the atoms have a permanent magnetic dipole (i.e., unpaired electrons). In the absence of a magnetizing field, the magnetic moments of the electrons are randomly distributed, and the net magnetic moment per unit volume is zero (Fig. 15.3a). An applied field tends to orient these moments in the direction of the field such that a net magnetic moment in the same sense as the applied field develops (Fig. 15.3b). The susceptibilities are thus positive but small, usually in the range of 10^{-3} to 10^{-6}. This tendency for order is naturally counteracted, as always, by thermal motion. It follows that the susceptibility decreases with increasing temperature, and the relationship between the two is given by

$$\chi_{\text{mag}} = \frac{C}{T} \tag{15.28}$$

where C is a constant known as the **Curie constant.** This $1/T$ dependence, shown schematically in Fig. 15.3c, is known as **Curie's law**; the remainder of this section is devoted to understanding the origin of this dependence.

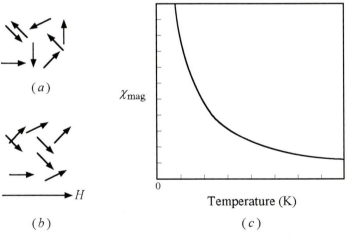

(a)

χ_{mag}

Temperature (K)

(b)

(c)

Figure 15.3

(a) In the absence of an applied magnetic field intensity, the magnetic moments of the individual atoms are pointing in random directions, resulting is a net magnetic moment of zero for the solid. (b) The application of a magnetizing field tends to align the moments in the direction of the field, resulting in a net moment. (c) Variation of χ_{mag} with temperature for a paramagnetic solid.

The situation is almost identical to that worked out in Chap. 14 for dipolar polarization, and the problem can be tackled by following a derivation nearly identical to the derivation of Eq. (14.51). To simplify the problem, it is assumed here that the ions in the solid possess a total magnetic moment μ_{ion} given by Eq. (15.27) (i.e., it is assumed that the orbital angular momentum is quenched). Furthermore, it is assumed that the interaction of these moments can align themselves either parallel or antiparallel to the applied field B. The magnetic energy is thus $\pm\mu_{ion}B$, with the plus sign corresponding to the case where the electron is aligned against the field and the minus sign when it is aligned with the field. For $\mu_{ion}B/(kT) \ll 1$, it can be shown that (Prob. 15.2a) the net magnetization is given by[270]

$$M = (N_1 - N_2)\mu_{ion} = \frac{N\mu_{ion}^2 B}{kT} \tag{15.29}$$

where N is total number of magnetic atoms or ions per unit volume, that is, $N = N_1 + N_2$. Here N_1 and N_2 represent, respectively, the number of electrons per unit volume aligned with and against the applied field. In other words, the net magnetization is proportional to the *net* number of electrons aligned with the field, that is, $N_1 - N_2$. For paramagnetic solids, one can further assume that $B \approx \mu_0 H$ [Eq. (15.12)] which, when combined with Eqs. (15.29) and (15.6), results in

$$\boxed{\chi_{mag} = \frac{N\mu_0\mu_{ion}^2}{kT} = \frac{C}{T}} \tag{15.30}$$

which is the sought-after result, since it predicts that χ_{mag} should vary as $1/T$.

It should be pointed out that Eq. (15.30) is slightly incorrect because the electrons were assumed to be aligned either with or opposite to the applied field. In reality, the electron momenta may have any direction in between, and an angular dependence should be accounted for in deriving Eq. (15.30). This problem was tackled by Langevin, with the final result being

$$\boxed{\chi_{mag} = \frac{N\mu_0\mu_{ion}^2}{3kT} = \frac{C}{T}} \tag{15.31}$$

which, but for a factor of 3 in the denominator (which comes about from averaging the moments over all angles), is identical to Eq. (15.30).

---

[270] The assumption that $\mu_{ion}B/(kT) \ll 1$ is for the most part an excellent one. See Prob. 15.3.

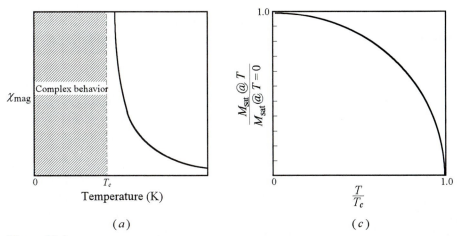

(a)   (c)

Figure 15.4
(a) Temperature dependence of χ_{mag} for a ferromagnetic solid that undergoes a transition at the Curie temperature T_C. (b) Spontaneous magnetization ($H = 0$) of a ferromagnetic crystal as a function of temperature.

15.4.2 Ferromagnetism

In a certain class of magnetic materials, namely, ferro- and ferrimagnets, the temperature dependence of χ_{mag} obeys not Curie's law, but rather the modified version

$$\chi_{mag} = \frac{C}{T - T_C} \tag{15.32}$$

known as the **Curie-Weiss law**, a plot of which is shown in Fig. 15.4a. Above a critical temperature T_C, known as the **Curie temperature**, the material behaves paramagnetically, whereas below T_C *spontaneous* magnetization sets in. Furthermore, the extent of that spontaneous magnetization is a function of temperature and reaches a maximum at absolute zero, as shown in Fig. 15.4b.

Qualitatively, this is explained as follows: At high temperatures, thermal disorder rules and the solid is paramagnetic. As the temperature is lowered, however, a **magnetic interaction energy** comes into play that tends to align the magnetic moments parallel to one another and produce a macroscopic magnetic moment with maximum ordering occurring at 0 K. The temperature at which this ordering appears is called the *Curie temperature*, with high Curie temperatures corresponding to strong interactions and vice versa.[271]

---

[271] The situation is quite analogous to melting. The stronger the bond between the atoms the higher the melting points. The Curie temperature can be considered to be the temperature at which the magnetic ordering "melts."

Given that ferromagnetism exists up to a finite temperature above absolute zero and then disappears, one is forced to postulate that in these materials:

- Some of the spins on the atoms must be unpaired.
- There is some *interaction* between neighboring electronic spins that tends to align them and keep them aligned even in the *absence* of a field.
- This ordering energy, at a sufficiently high temperature, is no longer capable of counteracting the thermal disordering effect, at which point the material loses its ability to spontaneously magnetize.

Before one proceeds much further, it should be emphasized that the nature of this interaction energy, also known as the **exchange energy**, is nonmagnetic and originates from quantum mechanical electrostatic interactions between neighboring atoms. Suffice it to say here that ferromagnetism is caused by a strong internal or local magnetic field aligning the magnetic moments on individual ions.

To understand the temperature dependence of ferromagnetic materials, one needs to find an expression for the local field B_{loc} which an electron inside a ferromagnetic material placed in a magnetizing field H experiences. This is a nontrivial problem and only the end result, namely,

$$B_{loc} = \mu_0(H + \lambda M) \qquad (15.33)$$

is given here; λ is the known as the **mean field constant** or **coupling coefficient** and is a measure of the strength of the interaction between neighboring moments — as noted above, the larger λ, the stronger the interaction. Replacing B in Eq. (15.29) by B_{loc} one obtains

$$M = N\mu_{ion}^2 \frac{B_{loc}}{kT} = N\mu_{ion}\frac{\mu_{ion}\mu_0 H + \lambda\mu_{ion}\mu_0 M}{kT} \qquad (15.34)$$

By further noting that $M_{sat} = N\mu_{ion}$ (see below) and defining

$$T_C = \frac{\lambda\mu_{ion}\mu_0 M_{sat}}{k} \qquad (15.35)$$

Equation (15.34) can be recast as

$$\frac{M}{M_{sat}} = \frac{\mu_{ion}\mu_0 H}{kT} + \frac{M}{M_{sat}}\frac{T_C}{T} \qquad (15.36)$$

which can be further simplified to

$$\frac{M}{M_{sat}} = \frac{\mu_{ion}\mu_0 H}{k(T - T_C)} \qquad (15.37)$$

Finally, since by definition $M/H = \chi_{mag}$, one obtains the final sought-after temperature dependence of χ_{mag}, namely, the Curie-Weiss law:

$$\chi_{mag} = \frac{\mu_{ion}\mu_0 M_{sat}}{k(T - T_C)} = \frac{C}{T - T_C} \tag{15.38}$$

Note that if one neglects the interaction between neighbors, i.e., if λ, and consequently T_C, is assumed to be zero, then Eq. (15.28) is recovered, as one would expect. It is thus easy to appreciate at this point that it is the *interaction of neighboring electrons that gives rise to* T_C and ferromagnetism.

Also, note that Eq. (15.38) only applies above T_C. Below T_C the material behaves quite differently in that it will spontaneously (i.e., even when $H = 0$) magnetize. The behavior below T_C is shown schematically in Fig. 15.4b, where M/M_{sat} is plotted versus T/T_C. As T approaches absolute zero, M approaches M_{sat}. The physics of the situation can be summarized as follows: When the thermal motions are small enough, the coupling between the atomic magnets causes them to all line up parallel to one another, even in the absence of an externally applied field, which gives rise to a permanently magnetized material.

To recap: By invoking that neighboring electrons interact in such a way as to keep their spins pointing all in the same direction and in the same direction as the applied magnetizing field, it is possible to, at least qualitatively, explain the general response of ferromagnetic materials to temperature and magnetizing fields.

15.4.3 Antiferromagnetism and Ferrimagnetism

In some materials, the coupling coefficient is *negative*, which implies that the magnetic moments on adjacent ions are antiparallel, as shown schematically in Fig. 15.5a. If these moments are equal, they cancel and the net moment is zero — such solids are known as **antiferromagnets**. According to Eq. (15.35), a negative λ gives rise to a negative T_C, and the resulting susceptibility versus temperature curve is shown in Fig. 15.5b, where maximum susceptibility is observed at temperature T_N, known as the **Neel temperature.** Above T_N, the Curie-Weiss law holds again, except that it is modified to read

$$\chi_{mag} = \frac{C}{T + T_C} \tag{15.39}$$

which takes into account the fact that T_C is negative.

Antiferromagnetism has been established in a number of compounds, primarily the fluorides and oxides of Mn, Fe, and Co such as MnF_2, MnO, FeF_2, CoF_2, NiO, CoO, FeO, MnS, MnSe, and Cr_2O_3. The type of ordering that occurs depends on the crystal structure of the compound in question. For instance, it is now well established that the spins in NiO and MnO are arranged as shown in Fig. 15.5c; where the spins in a single (111) plane are parallel but adjacent (111) planes are antiparallel.

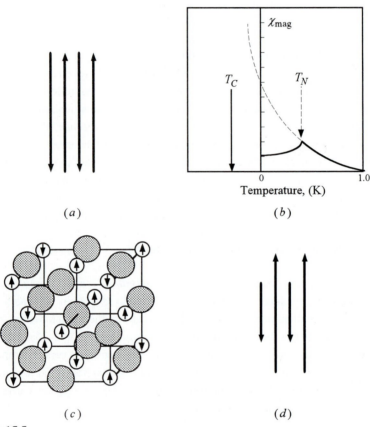

(a) (b)

(c) (d)

Figure 15.5
(a) In antiferromagnetic materials, nearest-neighbor moments are aligned antiparallel to one another, and the net moment is zero. (b) Temperature dependence of the susceptibility of an antiferromagnetic material. Maximum in χ_{mag} occurs at the Neel temperature T_N. (c) Antiparallel magnetic spins in MnO. (d) Unequal magnetic moments on adjacent sites give rise to a net magnetic moment. This is termed *ferrimagnetism*.

A variation of antiferromagnetism is seen in the situation depicted in Fig. 15.5*d*, where the coupling is negative but the *adjacent moments are unequal*, which implies that they do not cancel, and a net moment equal to the difference between the two submoments results. Such materials are termed **ferrimagnets** and, for reasons discussed later, include all magnetic ceramics. Interestingly enough, the temperature dependence of their properties is the same as that for ferromagnets.

15.5
MAGNETIC DOMAINS AND THE HYSTERESIS CURVE

15.5.1 Magnetic Domains

In the foregoing discussion of ferromagnetism, it was concluded that the material behaves parmagnetically above some temperature T_C, with M being proportional to H (or B), but that below a critical temperature spontaneous magnetization occurs. However, in the experiment described at the beginning of this chapter, it was explicitly stated that a virgin slab of magnetic material had a zero net magnetization. At face value, these two statements appear to contradict each other. The way out of this apparent dilemma is to appreciate that spontaneous magnetization occurs only within small regions $(\approx 10^{-5} \text{ m})$ within a solid. These are called **magnetic domains**, defined as regions where all the spins are pointing in the same direction. As discussed in greater detail below, these domains form in order to reduce the overall energy of the system and are separated from one another by **domain** or **Bloch walls**, which are high-energy areas,[272] defined as a transition layer that separates adjacent regions magnetized in different directions (Fig. 15.6*d*). The presence of these domain walls and their mobility, both reversibly and irreversibly, are directly responsible for the *B-H* hysteresis loop discussed below.

The reason magnetic domains form is best understood by referring to Fig. 15.6*a* to *c*. The single domain configuration (Fig. 15.6*a*) is a high-energy configuration because the magnetic field has to exit the crystal and close back on itself. By forming domains that close on themselves, as shown in Fig. 15.5*b* and *c*, the net macroscopic field is zero and the system has a lower energy. However, this

---

[272] The situation is not unlike grain boundaries in a polycrystalline material, with the important distinction that whereas a polycrystalline solid will always attempt to eliminate these areas of excess energy, in a magnetic material an equilibrium is established.

reduction in energy is partially offset by the creation of domain walls. For instance, the structure of a 180° domain wall is shown schematically in Fig. 15.6d. Some of the energy is also offset by the **anisotropy energy**, which is connected with the energy difference that arises when the crystal is magnetized in different directions. As noted below, the energy to magnetize a solid is a function of crystallographic direction — there are "easy" and "difficult" directions.

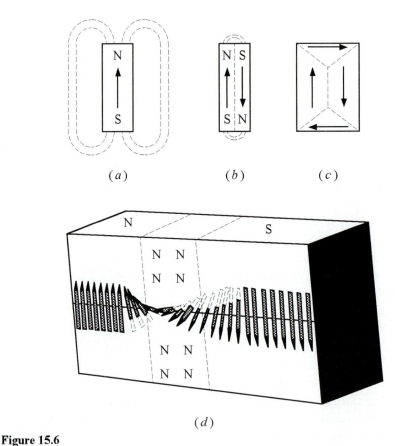

(a)   (b)   (c)

(d)

Figure 15.6
Schematic showing how formation of domains lowers the energy of the system. (a) No domains. (b) Two domains separated by a 180° wall. (c) The 90° domains are called *closure domains* because they result in the flux lines being completely enclosed within the solid. Closure domains are much more common in cubic crystals than in hexagonal ones because of the isotropy of the former. (d) Alignment of individual magnetic dipoles within a 180° wall.

15.5.2 Hysteresis Loops

The relationship between the existence of domains and hysteresis is shown in Fig. 15.7. The dependence of M on H for a virgin sample is shown in Fig. 15.7a. What occurs at the microscopic level is depicted in Fig. 15.7b to e. Initially the virgin sample is not magnetized, because the moments of the various domains cancel (Fig. 15.7b). The change in M very near the origin represents magnetization by reversible Bloch wall displacements, and the tangent OI to this initial permeability is called the **initial relative permeability** μ_i. As the magnetizing field H is increased, the domains in which the moments are favorably oriented with the applied field grow at the expense of those that are not. (Compare shaded areas in Fig. 15.7b and c.) This occurs by the *irreversible* movement of the domain walls,[273] up to a certain point (point X in Fig. 15.7a), beyond which wall movement more or less ceases. Further magnetization occurs by the rotation of the moments within the domains that are not aligned with the field, as shown in Fig. 15.7d. At very high H, all the domains will have rotated, and magnetization is said to have *saturated* at M_s. This **saturation magnetization** is simply the product of the magnetic moment on each ion μ_{ion} and the total number of atoms N per unit volume, or

$$M_{sat} = N\mu_{ion} \tag{15.40}$$

Upon removal of H, M does not follow the original curve, but instead follows the line $M_s - M_r$ and intersects the Y axis at the point labeled M_r, which is known as the **remnant magnetization.** Note that upon removal of the magnetizing field, the size of the domains and their orientation do not change. The difference between M_s and M_r simply reflects the recovery of the rotational component of the domains.

In order to completely rid the material of its remnant magnetization, the polarity of the magnetizing field has to be reversed. The value of H at which M goes to zero is called the **coercive magnetic field intensity** H_c (Fig. 15.8a).

Based on the shape of their hysteresis loops, magnetic materials have been classified as either soft or hard. The B-H hysteresis loops for each are compared in Fig. 15.8. Broadly speaking, **soft magnetic** materials have coercive fields below about 1 kA/m, whereas **hard magnetic** materials have H_c above about 10 kA/m. In addition to this classification, the shape of the hysteresis curve is also used to

---

273 At low fields, the movement of the domain walls is very much like an elastic band (or a pinned dislocation line) that stretches reversibly. If the field is removed at that point, the "unloading" curve is coincident with the loading curve and the process is completely reversible.

estimate the magnetic energy stored per unit volume in a permanent magnet. This value is given by the product $(BH)_{max}$ (Fig. 15.8a).

This difference in behavior between the two types of materials is related once again to the presence of domains and the ease or difficulty with which they can be induced to migrate and/or demagnetize. In the discussion up to this point, M was treated as if it were a unique function of H, but the actual situation is more complicated; M depends on the relative orientation of the various crystallographic planes to the direction of the applied field intensity. In other words, it exhibits **orientation anisotropy.** Also M depends on the shape of the crystal being magnetized; i.e., it exhibits **shape anisotropy**. This shape factor is quite important; e.g., it is much easier to magnetize a thin, long needle if its long axis is aligned parallel to the magnetizing field than if it is perpendicular to it.

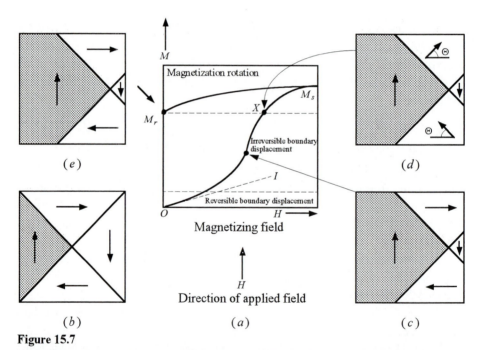

Figure 15.7
Relationship between domains and hysteresis. (*a*) Typical hysteresis loop for a ferromagnet. (*b*) For a virgin sample, $H = 0$ and $M = 0$ due to closure domains. (*c*) With increasing H, the shaded domain which was favorably oriented to H grows by the irreversible movement of domain walls up to point X. (*d*) Beyond point X, magnetization occurs only by the rotation of the moments. (*e*) Upon removal of the field, the irreversibility of the domain wall movement results in a remnant magnetization; i.e., the solid is now a permanent magnet.

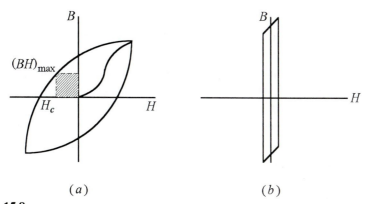

Figure 15.8

Comparison of hysteresis loops for (*a*) hard and (*b*) soft magnets. Note high H_c values for hard magnets.

Orientational anisotropy is related to **magnetostriction.** When a material is magnetized, it changes shape, which in turn introduces elastic strains in the material. And since elastic properties are tensors, it follows that the penalty for magnetizing crystals in different crystallographic directions is not equivalent. Some crystallographic directions are easier to magnetize than others. One measure of this effect is the magnetostriction constant λ_m, defined as the strain induced by a saturating field which is positive if the field causes an increase in dimension in the field direction. Table 15.3 lists some values for λ_m for a number of polycrystalline ferrites. The values listed represent the average of the single-crystal values.

The details of these intriguing phenomena are beyond the scope of this book, but the interested reader is referred to Additional Reading for further details. What is important here is to appreciate how these phenomena can and have been exploited to increase magnetic energy density by an order of magnitude for every decade since the turn of this century!

TABLE 15.3
Saturation magnetostriction constants of some polycrystalline ferrites

Composition	$\lambda_m(\times 10^6)$
Fe_3O_4	+40
$MnFe_2O_4$	−5
$CoFe_2O_4$	−110
$NiFe_2O_4$	−26
$MgFe_2O_4$	−6

15.6
MAGNETIC CERAMICS AND THEIR APPLICATIONS

As noted above, soft magnetic materials are characterized by large saturation magnetizations at low H values and low coercive fields and are typically used in applications where rapid reversal of the magnetization is required, such as electromagnets, transformer cores and relays. The major advantage of soft magnetic ceramics, compared to their metal counterparts, is the fact that they are electrical insulators. This property is fundamental in keeping eddy current losses low and is one of the main reasons why the major applications of magnetic ceramics have been in areas where such losses have to be minimized. Hard magnets, however, are characterized by high saturation magnetization as well as high coercive forces; i.e., they are not easily demagnetized. Hard magnetic solids are thus used to make permanent magnets and recording media.

Magnetic ceramics are further classified according to their crystal structures into spinels, garnets, and hexagonal ferrites. Typical compositions and some of their magnetic properties are listed in Table 15.4.

15.6.1 Spinels or Cubic Ferrites

Spinels were first encountered in Chap. 3 (Fig. 3.10), and their structure was described as having an oxygen ion sublattice arranged in a cubic close-packed arrangement with the cations occupying various combinations of the octahedral and tetrahedral sites. The cubic unit cell is large, comprising 8 formula units and containing 32 O and 64 T sites (only one-eighth of that unit cell is shown in Fig. 15.9). In normal spinels, for which the general chemical formula is $A^{2+}B^{3+}O_4$ (or equivalently $AO \cdot B_2O_3$), the divalent cations A are located on the tetrahedral (or T) sites and the trivalent cations B on the octahedral (or O) sites. In inverse spinels, the A cations and one-half the B cations occupy the O sites, with the remaining B cations occupying the T sites.[274]

---

[274] It is the crystal field energy or ligand field splitting (see Chap. 16) that stabilizes the inverse spinel.

TABLE 15.4

Magnetic properties of a number of magnetic ceramics. Magnetic moments are given in Bohr magnetons per formula unit at 0 K

Material	Curie temp., K	B_{sat} (T) @ RT	Calculated moments[†]			Experimental
			T site	O site	Net	
Fe[‡]	1043	2.14			2.14	2.22
Spinel ferrites [$AO \cdot B_2O_3$]						
$Zn^{2+}\vert Fe^{3+}Fe^{3+}\vert O_4$			0	5.–5	0	(Antiferro.)
$Fe^{3+}\vert Cu^{2+}Fe^{3+}\vert O_4$	728	0.20	-5§	1.73+5	1	1.30
$Fe^{3+}\vert Ni^{2+}Fe^{3+}\vert O_4$	858	0.34	-5§	2+5	2	2.40
$Fe^{3+}\vert Co^{2+}Fe^{3+}\vert O_4$	1020	0.50	-5§	3+5	3	3.70–3.90
$Fe^{3+}\vert Fe^{2+}Fe^{3+}\vert O_4$	858	0.60	-5§	4+5	4	4.10
$Fe^{3+}\vert Mn^{2+}Fe^{3+}\vert O_4$	573	0.51	-5§	5+5	5	4.60–5.0
$Fe^{3+}[Li_{0.5}Fe_{1.5}]O_4$	943		-5§	0+0.75		2.60
$Mg_{0.1}Fe_{0.9}[Mg_{0.9}Fe_{1.1}]O_4$	713	0.14	0–4.5	0+5.5	1	1.10
Hexagonal ferrites						
$BaO:6Fe_2O_3$	723	0.48				1.10
$SrO:6Fe_2O_3$	723	0.48				1.10
$Y_2O_3:5Fe_2O_3$	560	0.16				5.00
$BaO:9Fe_2O_3$	718	0.65				
Garnets						
YIG $\{Y_3\}[Fe_2]Fe_3O_{12}$	560	0.16			5	4.96
$\{Gd_3\}[Fe_2]Fe_3O_{12}$	560				16	15.20
Binary oxides						
EuO	69					6.80
CrO_2	386	0.49				2.00

† For the sake of simplicity, the moments were calculated by using the classical expression [Eq. (15.21)] rather than the more accurate quantum mechanical result given in Eq. (15.27).

‡ Fe is included for comparison purposes.

§ The minus sign denotes an antiferromagnetic coupling.

As noted in Chap. 6, the spinel structure is very amenable to large substitutional possibilities which has led to considerable technological exploitation of ferrites. The simplest magnetic oxide, magnetite,[275] or Fe_3O_4, is a naturally occurring ferrite that has been used for hundreds of years as a lodestone for navigational purposes. There are quite a number of other possible compositions with the general formula $MeO \cdot Fe_2O_3$, some of which are listed in Table 15.4. The Me ion represents divalent ions such as Mn^{2+}, Co^{2+}, Ni^{2+}, and Cu^{2+} or a combination of ions with an average valence of $+2$. In general, the divalent ions

275 Its structural relationship to spinel becomes apparent when its formula is rewritten as $FeO \cdot Fe_2O_3$.

prefer the octahedral sites, and thus most ferrites form inverse spinels. However, Zn and Cd ions prefer the tetrahedral sites, forming normal spinels.

In spinels the interaction between the A and B sublattices is *almost* always antiferromagnetic[276] (i.e., they have opposite spins), as shown in Fig. 15.9. The spin ordering is of the "superexchange" type, so called because it occurs via the agency of the intervening oxygen ions. The net magnetic moment will depend on the electronic configuration of the cations that populate each type of site. The following examples clarify this concept.

> **WORKED EXAMPLE 15.4.** (*a*) Calculate the net magnetic moment[277] of the inverse spinel Fe_3O_4. Also calculate its saturation magnetization and magnetic fields, given that the lattice parameter of the unit cell is 837 pm.
>
> (*b*) The addition of nonmagnetic ZnO to a spinel ferrite such as Ni ferrite leads to an *increase* in the saturation magnetization. Explain.
>
> (*c*) Repeat part (*a*) for the normal spinel $ZnO \cdot Fe_2O_3$. It is worth noting that in this spinel the octahedral or O-sites are coupled antiferromagnetically.

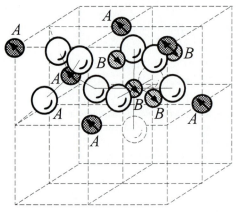

Figure 15.9

Spinel unit cell showing antiferromagnetic coupling between A and B sublattices.

---

276 There are exceptions, however. (See Worked Example 15.4c.)

277 In a typical spinel, solid solutions, quenching, and redox equilibria can very rapidly complicate the simple analysis presented here. Additional Reading contains a number of references that the interested reader can consult.

Answer

(a) Fe_3O_4 can be written as $FeO \cdot Fe_2O_3$ or $Fe^{3+}[Fe^{2+}Fe^{3+}]O_4$. Because it is an inverse spinel, one-half the Fe^{3+} cations occupy the T sites, and the other half the O sites. These cations interact *antiferromagnetically* which implies that their net moment is zero. The Fe^{2+} cations occupy the remaining O sites, and their net magnetic moment is (see Table 15.2) $4.9\mu_B$. The calculated net moment is thus $4.9\mu_B$, which is in reasonably good agreement with the measured value of 4.1 (Table 15.4). Note that this agreement implies that for the most part the orbital angular momentum of the ions in these solids is indeed quenched.

Since each unit cell contains eight Fe^{2+} ions (see Fig. 15.9), the saturation magnetization is given by Eq. (15.4), or

$$M_s = \frac{8 \times 4.9 \times 9.274 \times 10^{-24}}{\left(8.37 \times 10^{-10}\right)^3} = 6.2 \times 10^5 \text{ A/m}$$

It follows that the saturation magnetic field is given by

$$B_{sat} = \mu_0 M_s = 6.2 \times 10^5 \times 4 \times \pi \times 10^{-7} = 0.78 \text{ T}$$

which compares favorably with the measured value of 0.6 (see Table 15.4).

It is interesting to note that, for reasons that are not entirely clear, even better agreement between the measured and theoretical values is obtained if the classical expression for μ_{ion}, that is, $\mu_{ion} = 2\mu_B S$, is used instead of the more exact expression [Eq. (15.27)].

(b) According to Table 5.4, the saturation magnetization of $NiO \cdot Fe_2O_3$ is $2\mu_B$. The substitution of Ni by Zn, which prefers the tetrahedral sites, results in an occupancy given by

$$\left(Fe_{1-\delta}^{3+}Zn_{\delta}^{2+}\right)\left(Fe_{1+\delta}^{3+}Ni_{1-\delta}^{2+}\right)O_4$$

which results in diminishing the number of magnetic moments on the tetrahedral sites (first set of parentheses) and increasing the number on octahedral sites, resulting in a higher magnetization. Furthermore, as the occupancy of the A sites by magnetic ions decreases, the antiparallel coupling between the A and B sites is reduced, which lowers the Curie temperature.

(c) Being a normal spinel implies that the Zn^{2+} ions occupy the tetrahedral A sites and the Fe^{3+} ions occupy the octahedral sites. The Zn ions are diamagnetic and do not contribute to the magnetic moment (Table 15.2). Given that the Fe^{3+} cations on the O sites couple antiferromagnetically, their moments cancel and the net magnetization is zero, as observed.

In commercial and polycrystalline ferrites processing, variables and resultant microstructures have important consequences on measured properties. Only a few will be mentioned here. For example, the addition of a few percent of cobalt to Ni-ferrite can increase its resistivity by several orders of magnitude by ensuring that the iron is maintained in the Fe^{3+} state. Similarly, it is important to sinter MnZn ferrites under reducing atmospheres to ensure that the manganese is maintained in the Mn^{2+} state but not too reducing so as to convert the Fe^{3+} to Fe^{2+}.

Changes in the microstructure in the form of additional inclusions such as second-phase particles or pores introduce pinning sites that impede domain wall motion and thus lead to increased coercivity and hysteresis loss. Conversely, for high-permeability ceramics, very mobile domain wall motion is required. It is now well established that one of the most significant microstructural factors that influence domain motion and hence the shape of the hysteresis curves in magnetic ceramics is the grain boundaries. For example, increasing the average grain size of some Mn-Zn ferrites from 10 to 30 μm increases μ_i from 10^4 to 2.5×10^4.

Applications of spinel ferrites can be divided into three groups: low-frequency, high-permeability applications; high-frequency, low-loss applications; and microwave applications. It is important to note that the properties of magnetic materials are as much a function of frequency as dielectric materials. The ideas of resonance and loss and their frequency dependencies, which were discussed in detail in Chap. 14, also apply to magnetic materials. The coupling in this case is between the applied magnetic field and the response of the magnetic moment vectors. And while these topics are beyond the scope of this book, it is important to be cognizant of them when choosing magnetic materials for various applications.[278]

15.6.2 Garnets

The general formula of magnetic garnets,[279] is $P_3Q_2R_3O_{12}$ or $3Me_2O_3 \cdot 5Fe_2O_3$, where Me is typically yttrium but can also be other rare earth ions. The basic crystal structure is cubic with an octahedron, a tetrahedron, and a two dodecahedra (a distorted or skewed cube) as building blocks arranged as shown in Fig. 15.10a (here only one dodecahedron is shown for clarity's sake). The Q and R cations occupy the octahedral a and tetrahedral d sites, respectively. The c sites are

---

[278] For more information see, e.g., A. J. Moulson and J. M. Herbert, *Electroceramics*, Chapman & Hall, London, 1990.

[279] Magnetic garnets are isostructural with the semiprecious mineral $Ca_3Al_2(SiO_4)_3$. In natural garnets, which are nonmagnetic, the R ions are always Si^{4+}, the divalent cations such as Ca or Mn are the P cations, whereas the trivalent cations such as Al^{3+} or Fe^{3+} are the Q cations.

occupied by the P cations. Each oxygen lies at the vertex that is common to four polyhedra, i.e., one tetrahedron, one octahedron, and two dodecahedra.

The most investigated and probably the most important magnetic garnet is the yttrium-iron garnet, $\{Y_3\}[Fe_2]Fe_3O_{12}$ or $3Y_2O_3 \cdot 5Fe_2O_3$, commonly referred to as YIG. In YIG, the Y^{3+} cations occupy the c sites and because of their closed-shell configuration are diamagnetic. The Fe^{3+} cations are distributed on the a and d sites, and the net magnetization is due to the difference between their respective moments. Given that there are 3 Fe^{3+} ions on the d sites for every 2 Fe^{3+} ions on the a sites, the net magnetic moment per formula unit (at $T = 0$ K) is $3 \times 5.92 - 2 \times 5.92 = 5.9 \mu_B$, which is in reasonable agreement with the measured value of 4.96. It is worth noting once again that if the classical value for μ_B, that is, $5\mu_B$, is used, even better agreement is obtained between theory and experiment.

The situation becomes more complicated when magnetic rare-earth ions are substituted for the Y, as shown in Fig. 15.10b. For these so-called rare-earth garnets, the M^{3+} ions are paramagnetic trivalent ions that occupy the c sites. The magnetization of these ions is opposite to the *net* magnetization of the ferric ions on the $a+d$ sites (see inset in Fig. 15.10c). At low temperatures, the net moment of the rare-earth ions can dominate the moment of the Fe^{3+} ions (Fig. 15.10b). But because of the c-a coupling and c-d coupling, the rare-earth lattice loses its magnetization rapidly with increasing temperature (Fig. 15.10b). The total moment can thus pass through zero, switch polarity, and increase once again as the Fe^{3+} moment starts to dominate. Typical magnetization versus temperature curves for various iron garnets are shown in Fig. 15.10c. The point at which the magnetization goes to zero is known as a **compensation point**. One consequence of having this compensation point is that the magnetization is quite stable with temperature, an important consideration in microwave devices. It is this property that renders Fe garnets unique and quite useful.

15.6.3 Hexagonal Ferrites

Here the material is not ferromagnetic in the sense that all adjacent spins are parallel; rather, all the spins in one layer are parallel and lie in the plane of the layer.[280] In the adjacent layer, all the spins are once again parallel within the layer but pointing in a different direction from the first layer, etc. Commercially the most important hexagonal ferrite is $BaO \cdot 6Fe_2O_3$, which is isostructural with a mineral known as magneto-plumbite and is the reason hexagonal ferrites are sometimes called magnetoferrites.

---

[280] For more details, see L. L Hench and J. K. West, *Principles of Electronic Ceramics*, Wiley, New York, 1990, p. 321.

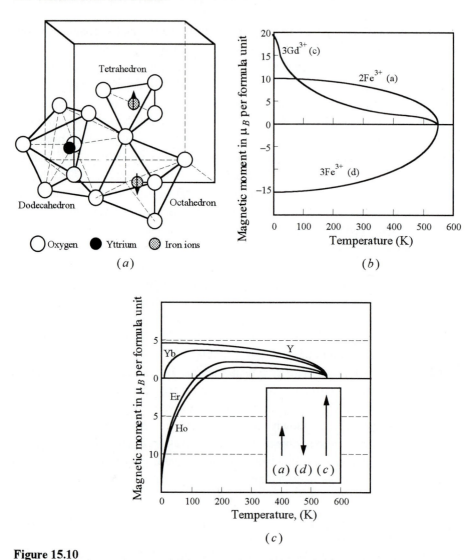

Figure 15.10

(*a*) Structural units and positions of cations in *a*, *c*, and *d* sites. The oxygen ions are shared by four polyhedra — an octahedron, a tetrahedron, and two (one not shown) dodecahedra. (*b*) Magnetization of the sublattices in Gd-iron garnet or GdIG as a function of temperature. Because of weak coupling between the Gd and the Fe, its magnetization drops more rapidly with temperature. Note compensation point at ≈ 290 K. (*c*) Resulting magnetization versus temperature curves for some Fe garnets.

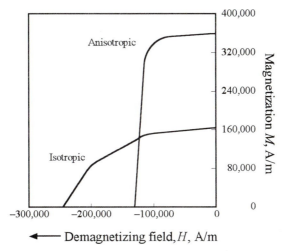

Figure 15.11

Demagnetization curves for oriented (top) and isotropic or random (bottom) hexagonal ferrites.[281]

The crystallographic structure of hard ferrites is such that the magnetically preferred (i.e., easy) orientation is the c axis, i.e., perpendicular to the basal plane. Consequently, hexagonal ferrites have been further classified as being either isotropic or anisotropic depending on whether the grains are arranged randomly or aligned. The latter is attained by compacting the powder in a magnetic field. The effect of aligning the grains on the B-H loop is shown in Fig. 15.11, from which it is obvious that the energy product is significantly improved by the particle orienting during the fabrication process. Other important microstructural factors include particle size and shape and the volume fraction of the ferrite phase. Typical values for the latter are 0.9 for sintered materials and 0.6 for plastic-bonded materials.

One of the major attributes of hexagonal ferrites is their very high crystal anisotropy constants, and consequently, they are used to fabricate hard magnets with high coercive fields. Typical hexagonal ferrites have $(BH)_{max}$ values in the range of 8 to 27 kJ/m^3, quite a bit lower than those of good metallic permanent magnets $(\approx 80 \; kJ/m^3)$. Despite this disadvantage, their low conductivity $(10^{-18} \; S/m)$, together with high coercive forces (0.2 to 0.4 T), low density, ease of manufacturing, availability of raw materials, and especially low cost per unit of available magnetic, energy renders them one of the most important permanent magnetic materials available. They are mainly used in applications where large demagnetizing fields are present such as flat loudspeakers and compact dc motors.

---

281 Adapted from F. Esper, in *High Tech Ceramics*, G. Kostorz, ed., Academic Press, London, 1989.

They are also used, for example, to produce "plastic" magnets, in which the magnetic particles are embedded in a polymer matrix.

15.7
PIEZO- AND FERROELECTRIC CERAMICS

The solids discussed in the remainder of this chapter have one thing in common: They exhibit various polar effects, such as piezoelectricity, pyroelectricity, and ferroelectricity. **Piezoelectric** crystals are those that become electrically polarized or undergo a change in polarization when subjected to a stress, as shown in Fig. 15.12c to f. The application of a compressive stress results in the flow of charge in one direction in the measuring circuit and in the opposite direction for tensile stresses. Conversely, the application of an electric field will stretch or compress the crystal depending on the orientation of the applied field to the polarization in the crystal.

Pyroelectric crystals are ones that are spontaneously polarizable (see below) and in which a change in temperature produces a change in that spontaneous polarization. A limited number of pyroelectric crystals have the additional property that the direction of spontaneous polarization can be reversed by application of an electric field, in which case they are known as ferroelectrics. Thus a **ferroelectric** is a spontaneously polarized material with reversible polarization. Before proceeding much further it is important to appreciate that not all crystal classes can exhibit polar effect.

15.7.1 Crystallographic Considerations

Of the 32 crystal classes or point groups (see Chap. 1), 11 are centrosymmetric and thus nonpolar (e.g., Fig. 15.12a and b). Of the remaining 21 noncentrosymmetric, 20 have one or more polar axes and are piezoelectric, and of these 10 are polar.[282] Crystals in the latter group are called **polar** crystals because they are spontaneously polarizable, with the magnitude of the polarization dependent on temperature. In the polar state, the center of positive charge does **not** coincide with the center of negative charge (Fig. 15.12c); i.e., the crystal possesses a permanent dipole. Each of these 10 classes is pyroelectric with a limited number of them being ferroelectric. It thus follows that all ferroelectric crystals are pyroelectric and all pyroelectric

---

[282] One of the noncentrosymmetric point groups (cubic 432) has symmetry elements which prevent polar characteristics.

crystals are piezoelectric, but not vice versa. To appreciate the difference between a piezoelectric and a ferroelectric crystal it is instructive to compare Fig. 15.12c and e. An unstressed piezoelectric (Fig. 15.12c) crystal only develops a dipole when stressed (Fig. 15.12d) as a consequence of its symmetry. A ferroelectric crystal, on the other hand, possess a dipole even in the unstressed state (Fig. 15.12e). The application of a stress only changes the value of the polarization (Fig. 15.12f). For example, quartz is piezoelectric but not ferroelectric, whereas $BaTiO_3$ is both. In the remainder of this chapter, ferro- and piezoelectricity are described in some detail.

15.7.2 Ferroelectric Ceramics

Given the definition of ferroelectricity as the *spontaneous* and reversible polarization of a solid, it is not surprising that ferroelectricity and ferromagnetism have a lot in common (Table 15.5). Ferroelectricity usually disappears above a certain critical temperature T_C; above that temperature the crystal is said to be in a paraelectric (in analogy with paramagnetism) state and obeys a Curie-Weiss law. Below T_C, spontaneous polarization occurs in domains. A typical plot of polarization versus electric field will exhibit a hysteresis loop (see below).

TABLE 15.5
Comparison of dielectric and magnetic parameters

	Magnetic	Dielectric
GENERAL		
Applied field	$H\,(A/m)$	$E\,(V/m)$
Material response	$M\,(A/m)$	$P\,(C/m^2)$
Field equations	$B = \mu_0(H + M)$	$D = \varepsilon_0 E + P$
	$B_{loc} = \mu_0(H + \lambda M)$	$E_{loc} = E + \dfrac{\beta P}{\varepsilon_0}$
Susceptibility	$\mu_r - 1 = \chi_{mag} = \dfrac{M}{H}$	$k' - 1 = \chi_{die} = \dfrac{P}{\varepsilon_0 E}$
Energy of moment	$U = -\mu_{ion} \bullet B$	$U = -\mu \bullet E$
	Paramagnetism	*Dipolar polarization*
	$M = \dfrac{N\mu_{ion}^2 B}{kT}$	$P = \dfrac{N_{dip}\mu_{dip}^2 E}{kT}$
Curie constant (K)	$C = \dfrac{\mu_0 \mu_{ion}^2 N}{3k}$ (Eq. (15.30))	$C = \dfrac{N_{dip}\mu_{dip}^2}{k\varepsilon_0}$ (Eq. (14.51))
	Ferromagnetic	*Ferroelectric*
Curie-Weiss law $(T > T_C)$	$\chi_{mag} = \dfrac{C}{T - T_C}$	$\chi_{die} = \dfrac{C}{T - T_C}$
Saturation	$M_s = N\mu_{ion}$	$P_s = N_{dip}\mu_{dip}$

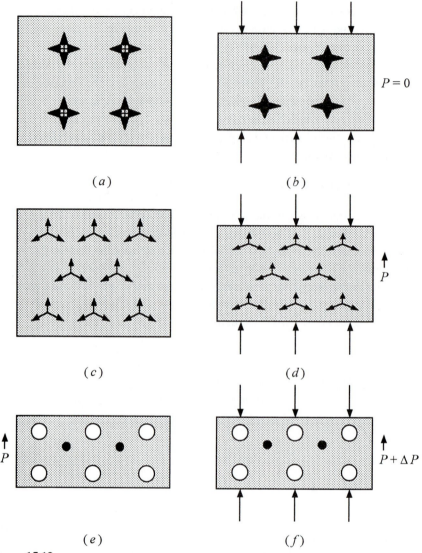

Figure 15.12

(*a*) Unstressed centrosymmetric crystal. The arrows represent dipole moments.
(*b*) Applying a stress to such a crystal *cannot* result in polarization. (*c*) Unstressed
noncentrosymmetric crystal, i.e., piezoelectric. Note that this structure is *not* ferroelectric
because it does not possess a permanent dipole. (*d*) Stressed crystal develops a
polarization as shown. (*e*) Unstressed *polar* crystal, i.e., a ferroelectric, possesses a
permanent dipole even in the unstressed state. (*f*) Stressed ferroelectric crystal. The
applied stress changes the polarization by ΔP.

Structural origin of the ferroelectric state

Commercially, the most important ferroelectric materials are the titania-based ceramics with the perovskite structure such as $BaTiO_3$ and $PbTiO_3$. In contradistinction to magnetism, ferroelectric materials go through a phase transition from a centrosymmetric nonpolar lattice to a noncentrosymmetric polar lattice at T_C. Typically these perovskites are cubic at elevated temperatures and become tetragonal as the temperature is lowered. The crystallographic changes that occur in $BaTiO_3$ as a function of temperature and the resulting polarization are shown in Fig. 15.13a. In the cubic structure, the TiO_6 octahedron has a center of symmetry, and the six Ti–O dipole moments cancel in antiparallel pairs. Below T_C, the position of the Ti ions moves off center, which in turn results in a permanent dipole for the unit cell (see Worked Example 15.5). The resulting changes in the dielectric constant are shown in Fig. 15.13b, the most salient feature of which is the sharp increase in k' around the same temperature that the phase transition from the cubic to the tetragonal phase occurs.

In order to understand the origin of the paraelectric to ferroelectric transition and the accompanying structural phase transitions it is important to understand how the local field is affected by the polarization of the lattice. Equation (14A.2), which relates the local field E_{loc} to the applied field E, and the polarization P can be generalized to read

$$E_{loc} = E + \beta \frac{P}{\varepsilon_0} \qquad (15.41)$$

where β is a measure of the enhancement of the local field.[283] By postulating that the polarizability α varies inversely with temperature, i.e., $N\alpha = C'/T$, and combining Eqs. (15.41), (14.13), and (14.14), one can show that:

$$k' - 1 = \frac{N\alpha}{\varepsilon_0 - \beta N\alpha} = \frac{(T_C/\beta)}{T - T_C} \qquad (15.42)$$

where $T_C = \beta C'/\varepsilon_0$. Comparing this expression to the Curie-Weiss law [Eq. (15.32)], it follows that

$$T_C = \beta C \qquad (15.43)$$

---

[283] In App. 14A, β was found to be 1/3. This value is only valid, however, if the material is a linear dielectric and has cubic symmetry.

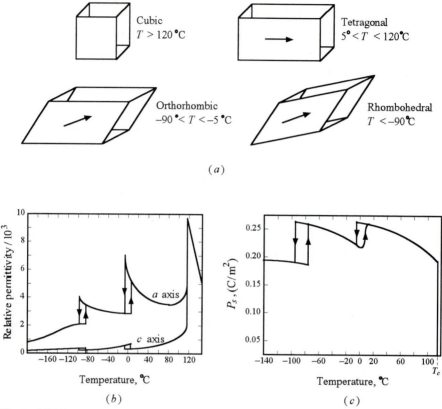

Figure 15.13
(a) Crystallographic changes in BaTiO$_3$ as a function of temperature. (b) Temperature dependence of relative dielectric constant of BaTiO$_3$ single crystal, for the c and a axis. (c) Temperature dependence of the saturation polarization for polycrystalline BaTiO$_3$. Note similarity between this figure and Fig. 15.4b.

where C is the Curie constant for ferroelectric ceramics. A perusal of Table 15.6 shows that for the titanate and niobate-based ceramics the Curie constant, C, is of the order of $10^5\,°C$.[284]

---

[284] It is instructive once again to compare ferromagnetic and ferroelectric solids. In both cases, an exchange energy or interaction energy is responsible for the spontaneous polarization. For ferromagnetic solids typical values of the exchange coupling coefficient λ are of the order of 350 (see Prob. 15.5). For ferroelectric solids, on the other hand, the interaction, as measured by β, is of the order of 2×10^{-3} (see Prob. 15.13). It follows that the interaction factor is about 5 orders of magnitude larger in ferromagnets than in ferroelectrics.

TABLE 15.6
Summary of dielectric data for a number of ferroelectric ceramics

Material	T_C, °C	Curie constant, °C	k' at T_C	P_{sat}, C/m^2
Rochelle salt	24	2.2×10^2	5000	0.25 (RT)
BaTiO$_3$	120	1.7×10^5	1600	0.26 (RT)
				0.18 @ T_C
SrTiO$_3$ [is paraelectric down to about 1 K]		7.8×10^5		
PbTiO$_3$	490	1.1×10^5		0.50 (RT)
PbZrO$_3$ (antiferroelectric)	230		3500	
LiNbO$_3$	1210			0.71–3.00 (RT)
NaNbO$_3$	−200			0.12 @ T_C
KNbO$_3$	434	2.4×10^5	4200	0.26 @ T_C
LiTaO$_3$	630			0.50 @ T_C
				0.23 @ 450
PbTa$_2$O$_5$	260			0.10 @ T_C
PbGeO$_{11}$	178			0.05 @ T_C
SrTeO$_3$	485			0.40 @ T_C

Equation (15.42) is important because it predicts that in the absence of a phase transition, the crystal would fly apart as T approached T_C, or equivalently when $\varepsilon_0 = \beta N \alpha$. As discussed in Chap. 2, for every bond there is an attractive component and a repulsive component to the total bond energy. If as the temperature is lowered, the repulsive component becomes weaker or softer, it follows that the anharmonicity of the bond will increase, which, as seen in Chap. 4, will increase the magnitude of the displacements of the ions, which in turn increases the dielectric constant as observed. The anharmonicity of the bond cannot increase indefinitely, however, and at some critical temperature the energy well for the Ti^{4+} ions in the center of the unit cell bifurcates into two sites, as shown in Fig. 14.10a. As the ions populate one site or the other, the interaction between them ensures that all other ions occupy the same site, giving rise to spontaneous polarization.

WORKED EXAMPLE 15.5. Using the ion positions for the tetragonal BaTiO$_3$ unit cell shown in Fig. 14.16, calculate the electric dipole moment per unit cell and the saturation polarization for BaTiO$_3$. Compare your answer with the observed saturation polarization (that is, $P_s = 0.26$ C/m^2).

Answer

The first step is to calculate the moment of each ion in the unit cell. Taking the corner Ba ions are the reference ions:

Ion	Q, C	d, m	$\mu = Qd$
Ba^{2+} (reference)	$+2(1.6 \times 10^{-19})$	0	0.0
Ti^{4+}	$+4(1.6 \times 10^{-19})$	$+0.006 \times 10^{-9}$	3.84×10^{-30}
$2O^{2-}$	$-4(1.6 \times 10^{-19})$	-0.006×10^{-9}	3.84×10^{-30}
O^{2-}	$-2(1.6 \times 10^{-19})$	-0.009×10^{-9}	2.88×10^{-30}
		Sum	10.56×10^{-30}

Thus the dipole moment per unit cell is 10.56×10^{-30} C·m.

Since $P_s = \mu/V$, where V is the volume of the unit cell, it follows that the saturation polarization is given by:

$$P_s = 10.56 \times 10^{-30} \times (0.403 \times 0.389 \times 0.389) \times 10^{-27} = 0.16 \text{ C/m}^2$$

This value is less than the observed value because it does not take the contribution of electronic polarization of the ions into account.

Hysteresis

In addition to resulting in very large k' at T_C, spontaneous polarization will result in hysteresis loops, as shown in Fig. 15.14. At low applied fields, the polarization is reversible and almost linear with the applied field. At higher field strengths, the polarization increases considerably due to switching of the ferroelectric domains. Further increases in the electric field continue to increase the polarization as a result of further distortions of the TiO_6 octahedra.[285]

Upon removal of the applied field, P does not go to zero but remains at a finite value called the **remnant polarization** P_r. As in the ferromagnetic case, this remnant is due to the fact that the oriented domains do not return to their random state upon removal of the applied field.[286] In order to do that, the field should be reversed to a **coercive field** E_c.

---

[285] In contrast, in ferromagnetic materials the application of a magnetic field greater that that required for M_s does not increase the net magnetization; i.e., a true saturation is observed.

[286] It seems natural to assume that polar crystals would be a source of electric fields around them just as magnets are a source of magnetic fields. In practice, however, the net dipole moment is not detectable in ferroelectrics because the surface charges are usually rapidly neutralized by ambient charged particles.

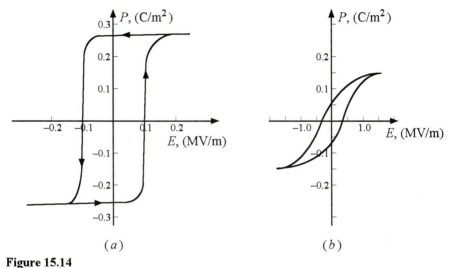

Figure 15.14
(*a*) Ferroelectric hysteresis loop for a single-crystal. (*b*) Polycrystalline sample.

Ferroelectric domains

As defined above, a domain is a microscopic region in a crystal in which the polarization is homogeneous. However, in contrast to domain walls in ferromagnetic materials that can be relatively thick (Fig. 15.6*d*), the ferroelectric domain walls are exceedingly thin (Fig. 15.15). Consequently, the wall energy is highly localized and the walls do not move easily.

Practically, it is important to reduce the sharp dependence of k' on temperature. In other words, it is important to broaden the permittivity versus temperature peaks as much as possible. One significant advantage of ceramic ferroelectrics is the ease with which their properties can be modified by adjusting composition and/or microstructure. For example, the substitution of Ti by other cations results in a shift in T_C, as shown in Fig. 15.16. Replacing Ti^{4+} by Sr^{2+} ions reduces T_C while the substitution of Pb^{2+} increases it. This is very beneficial because it allows for the tailoring of the peak permittivity in the temperature range for which the ferroelectric capacitor is to be used. Furthermore, certain additions (for example, $CaZrO_3$) to $BaTiO_3$ would result in regions of variable composition that contribute a range of Curie temperatures so that the high permittivity is spread over a wider temperature range.

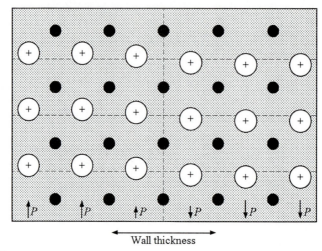

Figure 15.15
Ferroelectric domain wall thickness.

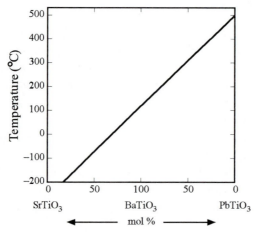

Figure 15.16
Effect of cationic substitutions in BaTiO$_3$ on T_C.

HIGH PERFORMANCE CONCRETES AND APPLICATIONS

Edited by

S P Shah
*Walter P Murphy Professor of Civil Engineering, and
Director of NSF Center for Science and Technology
of Advanced Cement Based Materials
North Western University, Evanston, IL, USA*

S H Ahmad
*Professor of Civil Engineering
North Carolina State University
Raleigh, NC, USA*

Edward Arnold
A member of the Hodder Headline Group
LONDON   MELBOURNE   AUCKLAND

© 1994 P Shah and S H Ahmad

First published in Great Britain 1994

British Library Cataloguing in Publication Data

High Performance Concretes and Applications
  I. Shah, Surendra P. II. Ahmad, S. H.
  691

  ISBN 0–340–58922–1

Typeset in 10/12 Times by Wearset, Boldon, Tyne and Wear.
Printed in Great Britain for Edward Arnold, a division of Hodder Headline PLC, Mill Road, Dunton Green, Sevenoaks, Kent TN13 2YA by St Edmundsbury Press Ltd, Bury St Edmunds, Suffolk, and bound by Hartnolls Ltd, Bodmin, Cornwall.

Contents

Preface

High performance concretes (HPC) represents a rather recent development in concrete materials technology. HPC is not a commodity but a range of products, each specifically designed to satisfy in the most effective way the performance requirements for the intended application.

Concrete has large number of properties or attributes. These attributes can be grouped into three general categories: (1) attributes which benefit the construction process; (2) enhanced mechanical properties; (3) enhanced non-mechanical properties such as durability etc. For hardened concrete, strength and durability are the two most important attributes.

In the last three or four years, several national-scale research programs have been established to study various aspects of high performance concretes. These include the two in the US: Center for Science and Technology for Advanced Cement-Based Materials (ACBM), Strategic Highway Research Program (SHRP); The Canadian Network of Centers of Excellence (NCE) Program on High Performance Concrete; the Royal Norwegian Council for Scientific and Industrial Research Program; the Swedish National Program on HPC; the French National Program called 'New Ways for Concrete' and the Japanese New Concrete Program. As the results from these programs start to be disseminated and digested in the concrete industry, the concrete technology will experience a significant advancement.

Historically, more attention has been given to the strength attribute and concrete performance has been specified and evaluated in terms of compressive strength — the higher the compressive strength, the better the expected performance. However, experience has shown that durability considerations become more important for structures exposed to hostile environments (e.g. marine structures and sanitary structures) and for structures such as bridges and pavements which are designed for longer service life. The SHRP program on High Performance Concrete has defined HPC for highway applications in terms of strength and durability attributes and water-cementious materials ratio. HPC is defined as concrete that meets the following criteria:

- It shall have one of the following strength characteristics:
  - 4-hour strength $\geqslant$2500 psi (17.5 MPa)
  - 24-hour strength $\geqslant$5000 psi (35 MPa)
  - 28-day strength $\geqslant$10,000 psi (70 MPa)

- It shall have a durability factor >80% after 300 cycles of freezing and thawing
- It shall have a water-cementitious materials ratio ≤0.35

During the last decade, developments in mineral and chemical admixtures have made it possible to produce concretes with relatively much higher strengths than was thought possible. Presently concretes with strengths of 14,000 to 16,000 psi (98 to 112 MPa) are being commercially produced and used in the construction industry in USA. Other countries such as England, Canada, Norway, Sweden, France, Italy, Japan, Hong Kong and South Korea are aggressively employing the high strength concrete technology in their construction practice.

The aim of this book is to summarize the developments of the last decade in the area of materials development for producing higher strength concrete, production methods, mechanical properties evaluation, non mechanical properties such as durability, and the implication of material properties on the structural design and performance. The use of higher strength concretes in the construction industry has steadily increased during the last decade, and therefore two chapters have been devoted to summarize the applications of higher strength concretes. Experts from USA, Canada, France, Norway, Spain and Japan have contributed individual chapters so as to give the book a broad perspective of the prevailing state-of-the-art in different parts of the world. The book is intended for the academics, engineers, consultants, contractors and researchers.

The eleven chapters in the book are arranged so that the reader can be selective. Chapter 1 provides the background for the selection of materials and proportions. This chapter also provides information on quality control aspects for concretes with higher strengths.

Chapter 2 addresses the short term mechanical properties such as compressive strength, modulus of elasticity, tensile strength etc. The long term mechanical properties such as creep, shrinkage and temperature effects are discussed in Chapter 3. Chapter 4 provides the information on the bond and fatigue characteristics. The important aspect of the durability and its implication for the performance of concrete are discussed in Chapter 5.

The fracture mechanics approach to the understanding of the structural response is outlined in Chapter 6. The behavior of the structural members such as beam, columns and slabs is detailed in Chapter 7. The ductility issues of the structural members and the structural ductility is presented in Chapter 8.

Chapter 9 addresses structural design considerations and the structural applications with special emphasis on high-rise buildings and bridges. This chapter also summarizes the special construction considerations needed for these concretes. Chapter 10 is dedicated to high strength lightweight aggregate concrete and its applications. The last chapter is devoted to the applications of HPC in Japan and South East Asia.

List of Contributors

P Acker
Head, Division: 'Betons et Ciments pour Ouvrages d'Art', Laboratoire
Central des Ponts et Chaussees, Paris, France

S H Ahmad
Professor, Department of Civil Engineering, North Carolina State
University, Raleigh, NC, USA

P N Balaguru
Professor, Civil Engineering Department, Rutgers University, Piscataway,
NJ, USA

T W Bremner
Professor of Civil Engineering, University of New Brunswick, Fredericton,
Canada

F de Larrard
Senior Scientist, Division: 'Betons et Ciments pour Ouvrages d'Art',
Laboratoire Central des Ponts et Chaussees, Paris, France

A S Ezeldin
Assistant Professor, Department of Civil, Environmental and Coastal
Engineering, Stevens Institute of Technology, Hoboken, NJ, USA

R Gettu
Senior Researcher, Technical University of Catalunya, Barcelona, Spain

S K Ghosh
Director, Engineered Structures and Codes, Portland Cement
Association, Skokie, IL, USA

O E Gjørv
Professor, Division of Building Materials, Norwegian Institute of
Technology – NTH, Trondheim – NTH, Norway

T A Holm
Vice President of Engineering, Solite Corporation, PO Box 27211, Richmond, VA, USA

R Le Roy
Research Engineer, Division: 'Betons et Ciments pour Ouvrages d'Art', Laboratoire Central des Ponts et Chaussees, Paris, France

S Mindess
Professor, Department of Civil Engineering, University of British Columbia, Vancouver, Canada

S Nagataki
Professor of Civil Engineering, Tokyo Institute of Technology, O-okayama, Meguru-ku, Tokyo 152, Japan

A H Nilson (Professor)
162 Round Pound Road, HC-60, Box 162, Medomak, Maine, USA (formerly of Cornell University)

H G Russell
Vice President, Construction Technology Laboratories Inc., 5420 Old Orchard Road, Skokie, IL, USA

M Saatcioglu
Associate Professor of Civil Engineering, University of Ottawa, Canada

E Sakai
Manager, Special Cement Additives Division, Denki Kagaku Kogyo Co. Ltd, Yuraku-cho, Chiyoda-ku, Tokyo 100, Japan

S P Shah
Walter P Murphy Professor of Civil Engineering; Director, NSF Center for Science and Technology of Advanced Cement-Based Materials; and Director, Center for Concrete and Geomaterials, Northwestern University, Evanston, IL, USA

1 Materials selection, proportioning and quality control

S Mindess

1.1 Introduction

High performance concretes (HPC) are concretes with properties or attributes which satisfy the performance criteria. Generally concretes with higher strengths and attributes superior to conventional concretes are desirable in the construction industry. For the purposes of this book, HPC is defined in terms of strength and durability. The researchers of Strategic Highway Research Program SHRP-C-205 on High Performance Concrete[1] defined the high performance concretes for pavement applications in terms of strength, durability attributes and water-cementitious materials ratio as follows:

- It shall have one of the following strength characteristics:
  4-hour compressive strength $\geq$2500 psi (17.5 MPa) termed as very early strength concrete (VES), or
  24-hour compressive strength $\geq$5000 psi (35 MPa) termed as high early strength concrete (HES), or
  28-day compressive strength $\geq$10,000 psi (70 MPa) termed as very high strength concrete (VHS).
- It shall have a durability factor greater than 80% after 300 cycles of freezing and thawing.
- It shall have a water-cementitious materials ratio $\leq$0.35.

High strength concrete (HSC) could be considered as high performance if other attributes are satisfactory in terms of its intended application. Generally concretes with higher strengths exhibit superiority of other attributes. In North American practice, high strength concrete is usually considered to be a concrete with a 28-day compressive strength of at least 6000 psi (42 MPa). In a recent CEB-FIP State-of-the-Art Report on High

Strength Concrete[2] it is defined as concrete having a minimum 28-day compressive strength of 8700 psi (60 MPa). Clearly then, the definition of 'high strength concrete' is relative; it depends upon both the period of time in question, and the location.

The proportioning (or mix design) of normal strength concretes is based primarily on the w/c ratio 'law' first proposed by Abrams in 1918. At least for concretes with strengths up to 6000 psi (42 MPa), it is implicitly assumed that almost any normal-weight aggregates will be stronger than the hardened cement paste. There is thus no explicit consideration of aggregate strength (or elastic modulus) in the commonly used mix design procedures, such as those proposed by the American Concrete Institute.[3] Similarly, the interfacial regions (or the cement-aggregate bond) are also not explicitly addressed. Rather, it is assumed that the strength of the hardened cement paste will be the limiting factor controlling the concrete strength.

For high strength concretes, however, *all* of the components of the concrete mixture are pushed to their critical limits. High strength concretes may be modelled as three-phase composite materials, the three phases being (i) the hardened cement paste (hcp); (ii) the aggregate; and (iii) the interfacial zone between the hardened cement paste and the aggregate. These three phases must all be optimized, which means that each must be considered explicitly in the design process. In addition, as has been pointed out by Mindess and Young,[4]

> 'it is necessary to pay careful attention to all aspects of concrete production (i.e. selection of materials, mix design, handling and placing). It cannot be emphasized too strongly that quality control is an essential part of the production of high-strength concrete and requires full cooperation among the materials or ready-mixed supplier, the engineer, and the contractor'.

In essence then, the proportioning of high strength concrete mixtures consists of three interrelated steps: (1) selection of suitable ingredients – cement, supplementary cementing materials, aggregates, water and chemical admixtures, (2) determination of the relative quantities of these materials in order to produce, as economically as possible, a concrete that has the desired rheological properties, strength and durability, (3) careful quality control of every phase of the concrete-making process.

1.2   Selection of materials

As indicated above, it is necessary to get the maximum performance out of all of the materials involved in producing high strength concrete. For convenience, the various materials are discussed separately below. However, it must be remembered that prediction with any certainty as to how they will behave when combined in a concrete mixture is not feasible. Particularly when attempting to make high strength concrete, any material incompatibilities will be highly detrimental to the finished product. Thus, the culmination of any mix design process must be the extensive testing of trial mixes.

High strength concrete will normally contain not only portland cement, aggregate and water, but also superplasticizers and supplementary cementing materials. It is possible to achieve compressive strengths of up to 14,000 psi (98 MPa) using fly ash or ground granulated blast furnace slag as the supplementary cementing material. However, to achieve strengths in excess of 14,000 psi (100 MPa), the use of silica fume has been found to be essential, and it is frequently used for concretes in the strength range of 9000–14,000 psi (63–98 MPa) as well.

Portland cement

There are two different requirements that any cement must meet: (i) it must develop the appropriate strength; and (ii) it must exhibit the appropriate rheological behaviour.

High strength concretes have been produced successfully using cements meeting the ASTM Standard Specification C150 for Types I, II and III portland cements. Unfortunately, ASTM C150 is very imprecise in its chemical and physical requirements, and so cements which meet these rather loose specifications can vary quite widely in their fineness and chemical composition. Consequently, cements of nominally the same type will have quite different rheological and strength characteristics, particularly when used in combination with chemical admixtures and supplementary cementing materials. Therefore, when choosing portland cements for use in high strength concrete, it is necessary to look carefully at the cement fineness and chemistry.

Fineness

Increasing the fineness of the portland cement will, on the one hand, increase the early strength of the concrete, since the higher surface area in contact with water will lead to a more rapid hydration. On the other hand, too high a fineness may lead to rheological problems, as the greater amount of reaction at early ages, in particular the formation of ettringite, will lead to a higher rate of slump loss. Early work by Perenchio[5] indicated that fine cements produced higher early age concrete strengths, though at later ages differences in fineness were not significant. Most cements now used to produce high strength concrete have Blaine finenesses that are in the range of 1467 to 1957 ft^2/lb (300 to 400 m^2/kg), though when Type III (high early strength) cements are used, the finenesses are in the range of 2201 ft^2/lb (450 m^2/kg).

Chemical composition of the cement

The previously cited work of Perenchio[5] indicates that cements with higher C_3A contents leads to higher strengths. However, subsequent work[6] has shown that high C_3A contents generally leads to rapid loss of flow in the fresh concrete, and as a result high C_3A contents should be avoided in cements used for high strength concrete. Aitcin[7] has shown that the C_3A should be primarily in its cubic, rather than its orthorhombic, form. Further, Aitcin[7] suggests that attention must be paid not only to the total amount of SO_3 in the cement, but also to the amount of soluble sulfates. Thus, the degree of sulfurization of the clinker is an important parameter.

In addition to commercially available cements conforming to ASTM Types I, II and III, a number of cements have been formulated specifically for high strength concrete. For instance, in Norway, Norcem Cement has developed two special cements for high strength concrete, in addition to their ordinary portland cement. The characteristics of these cements are given in Table 1.1.[8] Note that for the two special cements (SP30-4A and SP30-4A MOD), the C_3A contents were held to 5.5%.

Table 1.1   Composition of special cements for high strength concrete (developed by Norcem Cement[8])

	SP30*	SP30-4A	SP30-4A MOD
C_2S (%)	18	28	28
C_3S (%)	55	50	50
C_3A (%)	8	5.5	5.5
C_4AF (%)	9	9	9
MgO (%)	3	1.5–2.0	1.5–2.0
SO_3 (%)	3.3	2–3	2–3
Na_2O equivalent (%)	1.1	0.6	0.6
Blaine fineness (m^2/kg)	300	310	400
heat of hydration (kcal/kg)	71	56	70
setting time (min): initial	120	140	120
final	180	200	170

* Ordinary portland cement, for comparison
$1 \, m^2/kg = 4.89 \, ft^2/lb$

Supplementary cementing materials

As indicated above, most modern high strength concretes contain at least one supplementary cementing material: fly ash, blast-furnace slag, or silica fume. Very often, the fly ash or slag is used in conjunction with silica fume. These materials are all specified in the Canadian CSA Standard A23.5.[9] In the United States, fly ash is specified in ASTM C618,[10] and blast furnace slag in ASTM C989[11]; there is, as yet, no U.S. standard for silica fume. These materials are described in detail in *Supplementary Cementing Materials for Concrete*.[12]

Using a somewhat different approach, a high silica modulus portland

Table 1.2   Bogue composition and other properties
of HTS cement (after Aitcin *et al.*[13])

C_2S (%)	22
C_3S (%)	62
C_3A (%)	3.6
C_4AF (%)	6.9
Na_2O equivalent (%)	0.38
lime saturation factor	92.7
silica modulus	4.8
Blaine fineness, m^2/kg	320

$1 \, m^2/kg = 4.89 \, ft^2/lb$

cement (referred to as HTS, or Haute Teneur en Silica, or high silica
content) was developed,[13] with the composition shown in Table 1.2. Note
that, compared to more conventional cements (such as the SP-30 of
Table 1.1), there is a very high total silicate content (84%), and C_3A
content of only 3.6%. The cement is rather coarsely ground (Blaine
fineness of $1565 \, ft^2/lb$ ($320 \, m^2/kg$)). It is made from a clinker composed of
small alite and belite crystals, and minute C_3A crystals. It is capable of
producing concretes with excellent 28-day compressive strengths, as indicated in Table 1.3, when used in conjunction with 10% silica fume.

Table 1.3   28 day compressive strengths of concrete
made with HTS cement and 10% silica fume[13]

w/c	f_c' (MPa)
0.31	74
0.23	106
0.20	115
0.17	124

$1 \, ksi = 6.89 \, MPa$
$1 \, MPa = 0.145 \, ksi$

Silica fume

It is possible to make high strength concrete without silica fume, at
compressive strengths of up to about 14,000 psi (98 MPa). Beyond that
strength level, however, silica fume becomes essential, and even at lower
strengths 9000–14,000 psi (63–98 MPa), it is easier to make HSC with silica
fume than without it. Thus, when it is available at a reasonable price, it
should generally be a component of the HSC mix.

Silica fume[14] is a waste by-product of the production of silicon and
silicon alloys, and is thus not a very well-defined material. Consequently, it
is important to characterize any new source of silica fume, by determining
the specific surface area by nitrogen adsorption, and the silica, alkali and
carbon contents. In addition, it is desirable to minimize the content of

Table 1.4   Some Canadian specifications for silica fume (taken from CSA Standard A23.5[9])

Chemical requirements	
SiO_2, min (%)	85
SO_3, max (%)	1.0
Loss in ignition, max (%)	6.0
Physical requirements	
Accelerated pozzolanic activity index, min, (%) of control	85
Fineness, max, (%) retained on 45 μm sieve	10
Soundness – autoclave expansion or contraction (%)	0.2
Relative density, max variation from average (%)	5
Fineness, max variation from average (%)	5
Optional physical requirements	
Increase of drying shrinkage, max (%) of control	0.03
Reactivity with cement alkalis: min reduction (%)	80

crystalline material. The acceptance limits for silica fume, taken from CSA-A23.5[9] are given in Table 1.4.

Silica fume is available in several forms. In its bulk form, its unit weight is in the range of 118 to 147.5 pcf (200–250 kg/m^3), which makes handling difficult. More commonly now, silica fume is available in a densified form, in which the bulk densities are about twice as great as those of the bulk form (i.e. 400–500 kg/m^3). In general, this makes it easier to handle. In addition, silica fume is available in slurry form (often in conjunction with superplasticizers in the liquid phase), with a solids content of about 50%. This form of silica fume requires special equipment for its use. Finally, silica fume is available already blended with portland cement (at percentages of the total mass of cementitious material in the range of 6.7 to 9.3%) in Canada, France and Iceland. In spite of this apparently wide selection, however, in any one location the choice of silica fumes will be very limited, and one must use what is locally available.

Fly ash

Fly ash has, of course, been used very extensively in concrete for many years. Fly ashes are, unfortunately, much more variable than silica fumes in both their physical and chemical characteristics. Any fly ash which works well in ordinary concrete mixes is likely to work well in high strength concrete as well. However, most fly ashes will result in strengths of not much more than 10,000 psi (70 MPa), though there have been a few reports of high strength concretes with strengths of up to 14,000 psi (98 MPa) in which fly ash has been used. For higher strengths, silica fume must be used in conjunction with the fly ash, though this practice has not been common in the past.

In general, for high strength concrete applications, fly ash is used at dosage rates of about 15% of the cement content. Because of the variability of the fly ash produced even from a single plant, however, quality control is particularly important. This involves determinations of the Blaine specific surface area, as well as the chemical composition (in particular the contents of SiO_2, Al_2O_3, Fe_2O_3, CaO, alkali, carbon and sulfates). And, as with silica fume, it is important to check the degree of crystallinity; the more glassy the fly ash, the better.

Blast furnace slag

In North America, slag is not as widely available as in Europe, and hence there is not much information available as to its performance in high strength concrete. However, the indications are that, as with fly ash, slags that perform well in ordinary concrete are suitable for use in high strength concrete, at dosage rates between 15% and 30%. The lower dosage rates should be used in the winter, so that the concrete develops strength rapidly enough for efficient form removal. For very high strengths, in excess of 14,000 psi (98 MPa), it will likely be necessary to use the slag in conjunction with silica fume.

The chemical composition of slags does not generally vary very much. Therefore, routine quality control is generally confined to Blaine specific surface area tests, and X-ray diffraction studies to check on the degree of crystallinity (which should be low).

Limitation on the use of silica fume, fly ash or slag

There appear to be no particular deleterious effects when silica fume is used in concrete. However, the use of fly ash and slag may lead to some problems:

(i) The *early strength development* of mixes in which some of the portland cement has been replaced by slag or fly ash is less rapid than that when only portland cement is used, and this may adversely affect the time at which the forms can be stripped, particularly at low temperatures. One way of dealing with this problem is by further reductions in the w/c ratio, through the use of even more superplasticizer. Clearly, this is not economically very attractive; if high *early* strength is needed, it may well be necessary to reduce the fly ash or slag content.

(ii) The existing test data are rather ambiguous with regard to the *free-thaw durability* of high strength concrete made with supplementary cementitious materials. This is true both for air-entrained and non-air-entrained mixes. Therefore, until more data are available, designers should be cautious when using high strength concrete in an environment in which the concrete will be subjected to many freeze-thaw cycles in a saturated state.

(iii)   At the substitution levels used (15–30%), fly ash or slag will have very little effect on the *maximum temperature development* in mass concrete pours.

Superplasticizers

In modern concrete practice, it is essentially impossible to make high strength concrete at adequate workability in the field without the use of superplasticizers. Unfortunately, different superplasticizers will behave quite differently with different cements (even cements of nominally the same type). This is due in part to the variability in the minor components of the cement (which are not generally specified), and in part to the fact that the acceptance standards for superplasticizers themselves are not very tightly written. Thus, some cements will simply be found to be incompatible with certain superplasticizers.

There are, basically, three principal types of superplasticizer: (i) *lignosulfonate-based*; (ii) polycondensate of formaldehyde and melamine sulfonate (often referred to simply as *melamine sulfonate*; and (iii) polycondensate of formaldehyde and naphthalene sulfonate, (often referred to as *naphthalene sulfonate*).

In addition, a variety of other molecules might be mixed in with these basic formulations. It may thus be very difficult to determine the precise chemical composition of most superplasticizers; certainly manufacturers try to keep their formulations as closely guarded secrets.

It should be noted that much of what we know about superplasticizers comes from tests carried out on normal strength concretes, at relatively low superplasticizer contents. This does not necessarily reflect their performance at very low w/c ratios and very high superplasticizer addition rates.

Lignosulfonate-based superplasticizers

In high strength concrete, lignosulfonate superplasticizers are generally used in conjunction with either melamine or naphthalene superplasticizers. They tend not to be efficient enough for the economic production of very high strength concretes on their own. Sometimes, lignosulfonates are used for initial slump control, with the melamines or naphthalenes used subsequently for slump control in the field.

Melamine sulfonate superplasticizers

Until recently, only one melamine superplastizer was available (tradename Melment), but now other melamine-based superplasticizers are likely to become commercially available.

Melamine superplasticizers are clear liquids, containing about 22% solid particles; they are generally in the form of their sodium salt. These superplasticizers have been used for many years now with good results, and so they remain popular with high strength concrete producers.

Naphthalene sulfonate superplasticizers

Naphthelene superplasticizers have been in use longer than any of the others, and are available under a greater number of brand names. They are available as both a powder and a brown liquid; in the liquid form they typically have a solids content of about 40%. They are generally available as either calcium salts, or more commonly, sodium salts. (Calcium salts should be used in case where a potentially alkali-reactive aggregate is to be used.)

The particular advantages of naphthalene superplasticizers, apart from their being slightly less expensive than the other types, appears to be that they make it easier to control the rheological properties of high strength concrete, because of their slight retarding action.

Superplasticizer dosage

There is no *a priori* way of determining the required superplasticizer dosage; it must be determined, in the end, by some sort of trial and error procedure. Basically, if strength is the primary criterion, then one should work with the lowest w/c ratio possible, and thus the highest superplasticizer dosage rate. However, if the rheological properties of the high strength concrete are very important, then the highest w/c ratio consistent with the required strength should be used, with the superplasticizer dosage then adjusted to get the desired workability. In general, of course, some intermediate position must be found, so that the combination of strength and rheological properties can be optimized. Typical superplasticizer dosages for a number of high strength concrete mixes are given below, in Tables 1.5 to 1.10.

Table 1.5 Mix proportions for Interfirst Plaza, Dallas (adapted from Cook[15])

	1 cm max size aggregate	*25 cm max size aggregate*
water (kg/m^3)	166	148
cement, Type I (kg/m^3)	360	357
fly ash, Class C (kg/m^3)	150	149
coarse aggregate (kg/m^3)	1052	1183
fine aggregate (kg/m^3)	683	604
water reducer L/m^3	1.01	1.01
superplasticizer L/m^3	2.54	2.52
w/cementitious ratio	0.33	0.29
f_c' 28-day (MPa)- moist cured	79.5	85.8
f_c' 91-day (MPa)- moist cured	89.0	92.4

1 lb/yd^3 = 0.59 kg/m^3 or	1 kg/m^3 = 1.69 pcf
1 in. = 25.4 mm or	1 in. = 0.0393 mm

Table 1.6   High strength concrete mix design guidelines (after Peterman and Carrasquillo[16])

	H-H-00	*H-H-01*	*H-H-10*	*H-H-11*
water (kg/m^3)	195	143	173	134
cement (kg/m^3)	558	474	391	335
fly ash (kg/m^3)	–	–	167	144
coarse agg./fine agg. ratio	2.0	2.0	2.0	2.0
superplasticizer	–	yes*	–	yes*
w/cementitious ratio	0.34	0.30	0.31	0.27
f_c' 56-day (MPa)	66	72	69	76

* Use highest dosage of superplasticizer which will not lead to segregation or excessive retardation.

1 lb/yd^3 = 0.59 kg/m^3 or   1 kg/m^3 = 1.69 pcf
1 in.   = 25.4 mm   or   1 in.   = 0.0393 mm

Table 1.7   Mix proportions for high strength concrete at Pacific First Center, Seattle (adapted from Randall and Foot[17])

water (kg/m^3)	131
cement – Type II (kg/m^3)	534
fly ash – Type F (kg/m^3)	59
silica fume (kg/m^3)	40
coarse aggregate – 1 cm max. size (kg/m^3)	1069
fine aggregate – F.M. = 3.2 (kg/m^3)	623
water reducer I (L/m^3)	1.77
water reducer II (L/m^3)	7.39
w/cementitious ratio	0.21
f_c' 56-day (MPa)	124

1 lb/yd^3 = 0.59 kg/m^3 or   1 kg/m^3 = 1.69 pcf
1 in.   = 25.4 mm   or   1 in.   = 0.0393 mm

Table 1.8   Five examples of commercially produced high strength concrete mix designs (after Aitcin, Shirlaw and Fines[18])

water (kg/m^3)	195	165	135	145	130
cement (kg/m^3)	505	451	500	315	513
fly ash (kg/m^3)	60	–	–	–	–
slag (kg/m^3)	–	–	–	137	–
silica fume (kg/m^3)	–	–	30	36	43
coarse aggregate (kg/m^3)	1030	1030	1110	1130	1080
fine aggregate (kg/m^3)	630	745	700	745	685
water reducer (L/m^3)	0.975	–	–	0.9	–
retarder (L/m^3)	–	–	4.5	1.8	–
superplasticizer (L/m^3)	–	11.25	14	5.9	15.7
w/cementitious ratio	0.35	0.37	0.27	0.31	0.25
f_c' 28-day (MPa)	64.8	79.8	42.5	83.4	119
f_c' 91-day (MPa)	78.6	87.0	106.5	93.4	145

1 lb/yd^3 = 0.59 kg/m^3 or   1 kg/m^3 = 1.69 pcf
1 in.   = 25.4 mm   or   1 in.   = 0.0393 mm

Table 1.9 High strength mixtures in the Chicago area (adapted from Burg and Ost[19])

	1	*2*	*3*	*4*	*5*
			Mix number		
water (kg/m^3)	158	160	155	144	151
cement (kg/m^3)	564	475	487	564	475
fly ash (kg/m^3)	–	59	–	–	104
silica fume (kg/m^3)	–	24	47	89	74
coarse aggregate, SSD 12 mm max size	1068	1068	1068	1068	1068
fine aggregate, SSD (kg/m^3)	647	659	676	593	593
superplasticizer – Type F (L/m^3)	11.61	11.61	11.22	20.12	16.45
retarder – Type D (L/m^3)	1.12	1.04	0.97	1.47	1.51
w/cementitious ratio	0.281	0.287	0.291	0.220	0.231
f_c' 28-day (MPa) – moist cured	78.6	88.5	91.9	118.9	107.0
f_c' 56-day (MPa) – moist cured	81.4	97.3	94.2	121.2	112.0
f_c' 91-day (MPa) – moist cured	86.5	100.4	96.0	131.8	119.3

Table 1.10 Mix design for a high strength concrete designed for a low heat of hydration (adapted from Burg and Ost[19])

water (kg/m^3)	141
cement – Type I (kg/m^3)	327
fly ash – Type F (kg/m^3)	87
silica fume (kg/m^3)	27
coarse aggregate – 25 mm max. size (kg/m^3)	121
fine aggregate (kg/m^3)	742
superplasticizer, ASTM Type F (L/m^3)	6.31
superplasticizer, ASTM Type G (L/m^3)	3.25
water/cementitious ratio	0.32
f_c' 28-day (MPa) – moist cured	3.1
f_c' 91-day (MPa) – moist cured	88.6

Retarders

At one time retarders were recommended for some high strength concrete applications, to minimize the problem of over rapid slump loss. However, it is difficult to maintain a compatibility between the retarder and the superplasticizer, i.e. to minimize slump loss without excessively reducing early strength gain. In modern practice, retarders are recommended only as a last resort; the rheology is better controlled by the use of the appropriate supplementary cementing materials described above.

Aggregates

The aggregate properties that are most important with regard to high strength concrete are: particle shape, particle size distribution, mechanical properties of the aggregate particles, and possible chemical reactions between the aggregate and the paste which may affect the bond. Unlike their use in ordinary concrete, where we rarely consider the strength of the aggregates, in high strength concrete the aggregates may well become the strength limiting factor. Also, since it is necessary to maintain a low w/c ratio to achieve high strength, the aggregate grading must be very tightly controlled.

Coarse aggregate

It goes without saying that, for high strength concrete, the coarse aggregate particles themselves must be strong. A number of different rock types have been used to make high strength concrete; these include limestone, dolomite, granite, andesite, diabase, and so on. It has been suggested[1] that in most cases the aggregate strength itself is not usually the limiting factor in high strength concrete; rather, it is the strength of the cement–aggregate bond. As with ordinary concretes, however, aggregates that may be susceptible to alkali–aggregate reaction, or to D-cracking, should be avoided if at all possible, even though the low w/c ratios used will tend to reduce the severity of these types of reaction.

From both strength and rheological considerations, the coarse aggregate particles should be roughly equi-dimensional; either crushed rock or natural gravels, particularly if they are of glacial origin, are suitable. Flat or elongated particles must be avoided at all costs. They are inherently weak, and lead to harsh mixes. In addition, it is important to ensure that the aggregate is clean, since a layer of silt or clay will reduce the cement–aggregate bond strength, in addition to increasing the water demand. Finally, the aggregates should not be highly polished (as is sometimes the case with river-run gravels), because this too will reduce the cement–aggregate bond.

Not much work has been carried out on the effects of aggregate mineralogy on the properties of high strength concrete. However, a detailed study by Aitcin and Mehta,[20] involving four apparently hard strong aggregates (diabase, limestone, granite, natural siliceous gravel) revealed that the granite and the gravel yielded much lower strengths and E-values than the other two aggregates. These effects appeared to be related both to aggregate strength and to the strength of the cement–aggregate transition zone. Cook[15] has also pointed out the effect of the modulus of elasticity of the aggregate on that of the concrete. However, much work remains to be done to relate the mechanical and mineralogical properties of the aggregate to those of the resulting high strength concrete.

It is commonly assumed that a smaller maximum size of coarse aggregate will lead to higher strengths,[1,2,5,6,21] largely because smaller sizes will improve the workability of the concrete. However, this is not necessarily the case. While Mehta and Aitcin[6] recommend a maximum size of 10–12 mm, they report that 20–25 mm maximum size may be used for high strength concrete. On the other hand, using South African materials, Addis[22] found that the strength of his high strength concrete increased as the maximum size of aggregate increased from 13.2 to 26.5 mm. This, then, is another area which requires further study.

Fine aggregate

The fine aggregate should consist of smooth rounded particles,[2] to reduce the water demand. Normally, the fine aggregate grading should conform to the limits established by the American Concrete Institute[3] for normal strength concrete. However, it is recommended that the gradings should lie on the coarser side of these limits; a fineness modulus of 3.0 or greater is recommended,[1,6] both to decrease the water requirements and to improve the workability of these paste-rich mixes. Of course, the sand too must be free of silt or clay particles.

1.3 Mix proportions for high strength concrete

Only a few formal mix design methods for high strength concrete have been developed to date.[7,22,23] Most commonly, purely empirical procedures based on trial mixtures are used. For instance, according to the Canadian Portland Cement Association, 'the trial mix approach is best for selecting proportions for high-strength concrete'.[24] In other cases, mix design 'recipes' are provided for different classes of high strength concrete; an example of this approach is given by Peterman and Carrasquillo.[16]

In this section, it is not the intention to provide a canonical mix proportioning method. Much work remains to be done before any mix proportioning method for high strength concrete becomes as universally accepted, at least in North America, as has the *ACI Standard 211.1*[3] for normal strength concretes. Rather, the principles on which such a mix design method should be based will be discussed, and some general guidelines (and a number of empirically derived mixes drawn from the literature) will be presented.

Proportions of materials

Water/cementitious ratio

For normal strength concretes, mix proportioning is based to a large extent on the w/c ratio 'law'. For these concretes, in which the aggregate strength is generally much greater than the paste strength, the w/c ratio does indeed determine the strength of the concrete for any given set of raw materials. For high strength concretes, however, in which the aggregate strength, or the strength of the cement–aggregate bond, are often the strength-controlling factors, the role of the w/c ratio is less clear. To be sure, it is necessary to use very low w/c ratios to manufacture high strength concrete. However, the relationship between w/c ratio and concrete strength is not as straightforward as it is for normal strength concretes.

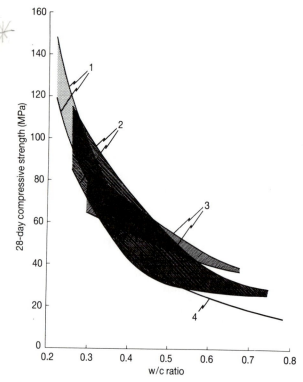

Fig. 1.1   Compressive strength versus w/c material ratio: (1) after Aitcin[7]; (2) after Fiorato[25]; (3) after Cook[15]; (4) normal strength concrete from CPCA[24]

Figure 1.1 shows a series of w/cementitious material vs compressive strength curves for high strength concrete. The sets of curves numbered 1, 2 and 3 show the strength range that might be expected for a given w/cementitious ratio. (Curve 1 is from Aitcin[7]; curve 2 is from Fiorato[25]; curve 3 is from Cook.[15]) For comparison, the w/c ratio vs strength curve for normal strength concrete is shown as curve 4.[24] Figure 1.2 shows a similar series of w/cementitious vs strength curves obtained by other investigators. Curve 1 is from Addis and Alexander,[23] who used high early strength cement. Curve 2 is from Hattori.[25] Curves 3 and 4 are from Suzuki[27]; curve 3 is for ordinary portland cement, and curve 4 for high early strength cement.

Several conclusions may be drawn from Figs. 1.1 and 1.2. First, while strength clearly increases as the w/cementitious ratio decreases, there is a considerable scatter of the results, which must be due to variations in the materials, used in the different investigations. Second, and more important, the range of strengths for a given w/cementitious ratio increases as the w/cementitious ratio decrease. If one looks at all of the curves in Figs. 1.1 and 1.2, at a w/cementitious ratio of 0.45, the range in strength is from 5400 psi (37 MPa) to 9500 psi (66 MPa); at a ratio of 0.26, the range is from 11,300 psi (78 MPa) to 17,400 psi (120 MPa). Therefore, the

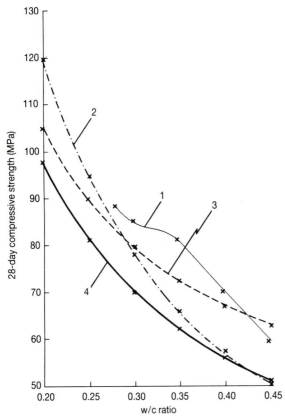

Fig. 1.2 Compressive strength versus w/c material ratio: (1) high early strength cement, after Addis and Alexander[23]; (2) after Hattori[26]; (3) ordinary Portland cement, after Suzuki[27]; (4) high early strength cement after Suzuki[24]

w/cementitious ratio by itself is not a very good predictor of compressive strength. The w/cementitious vs strength relationship must thus be determined for any given set of raw materials.

Cementitious materials content

For normal strength concretes, cement contents are typically in the range of 590 to 930 pcf (350 to 550 kg/m^3). For high strength concretes, however, the content of cementitious materials (cement, fly ash, slag, silica fume) is higher, ranging from about 845 to 1090 pcf (500 to 650 kg/m^3). The quantity of supplementary cementing materials may vary considerably, depending upon workability, economy and heat of hydration considerations.

Supplementary cementing materials

As indicated earlier, it is possible to make high strength concrete without using fly ash, slag or silica fume. For higher strengths, however, supplementary cementing materials are generally necessary. In particular, the use of silica fume is required for strengths much in excess of 14,000 psi (98 MPa). In any event, the use of silica fume (which is now readily available in most areas) makes the production of high strength concrete much easier; it is generally added at rates of 5% to 10% of the total cementitious materials.

Superplasticizers

With very careful mix design and aggregate grading, it is possible to achieve strengths of about 14,000 psi (98 MPa) without superplasticizers. However, as they are readily available they are now almost universally used in high strength concrete, since they make it much easier to achieve adequate workability at very low w/cementitious ratios.

Ratio of coarse to fine aggregate

For normal strength concretes, the ratio of coarse to fine aggregate (for a 0.55 in., 14 mm max size of aggregate) is in the range of 0.9 to 1.4.[24] However, for high strength concrete, the coarse/fine ratio is much higher. For instance, Peterman and Carrasquillo[16] recommend a coarse/fine ratio of 2.0. And, as seen in Tables 1.5 to 1.10, coarse/fine ratios used in practice vary in the range of 1.5 to 1.8.

Examples of high strength concrete mixes

As stated earlier, there is yet no generally agreed upon method of mix proportioning. Mix designs for high strength concrete have, hitherto, been developed empirically, depending on the raw material available in any location. In this section, a number of typical mix designs, drawn from the recent literature, will be presented.

Table 1.5 shows the mix proportions for Interfirst Plaza, Dallas,[15] in which the concrete achieved compressive strength of about 11,500 psi (80 MPa). Table 1.6 gives high strength concrete mix design guidelines originally developed for the Texas State Department of Highways and Public Transportation.[16] The expected 56-day strengths for these four mixes range from 9500 to 11,000 psi (66 to 76 MPa). It should be noted that the mix designs in Tables 1.5 and 1.6 do not involve the use of silica fume. Table 1.7 shows the mix proportions of Pacific First Center, Seattle[17] in which the concrete reached a 56-day compressive strength of 18,000 psi (126 MPa).

Table 1.8 gives a series of mix designs for a number of high strength concrete projects,[18] while Table 1.9 describes high strength concrete mixes commercially available in Chicago.[19] In Tables 1.7, 1.8 and 1.9, it should be noted that the higher strength mixes all contained silica fume. Finally,

Table 1.10 presents a mix design for a high strength, low heat of hydration concrete.[19]

From Tables 1.5 to 1.10, it may be seen that the mix designs, even for concretes of approximately the same strength, vary considerably. This reflects the differences in the quality of *all* of the raw materials available for each specific mix. So, while these examples may serve as a general guideline for the production of high strength concrete, copying a mix design used in one location is unlikely to produce the same concrete properties in another area.

In the end, as with conventional concrete, mix design will require the production of a number of trial mixes, though the examples given above may provide reasonable guidance for the first trial batch. In particular, it is essential first to ensure that the available raw materials are *capable* of producing the desired strengths, and that there are no incompatibilities between the cements, the admixture(s) and the supplementary cementing materials. With materials for which there is not much field experience, it may be necessary to try different brands of cement, different brands of superplasticizers, and different sources of fly ash, slag, or silica fume, in order to optimize both the materials and the concrete mixture. This sounds like a lot of work, and in general it is. At present, there is simply no straightforward procedure for proportioning a high strength concrete mixture with unfamiliar materials.

1.4   Quality control and testing

Conventional normal strength concrete is a relatively forgiving material; it can tolerate small changes in materials, mix proportions or curing conditions without large changes in its mechanical properties. However, high strength concrete, in which all of the components of the mix are working at their limits, is not at all a forgiving material. Thus, to ensure the quality of high strength concrete, every aspect of the concrete production must be monitored, from the uniformity of the raw materials to proper batching and mixing procedures, to proper transportation, placement, vibration and curing, through to proper testing of the hardened concrete.

The quality control procedures, such as the types of test on both the fresh and hardened concretes, the frequency of testing, and interpretation of test results are essentially the same as those for ordinary concrete. However, Cook[15] has presented data which indicate that for his high strength concrete, the compressive strength results were *not* normally distributed, and the standard deviation for a given mix was not independent of test age and strength level. This led him to conclude that the 'quality control techniques used for low to moderate strength concretes may not necessarily be appropriate for very high strength concretes.' To this date, however, separate quality control/quality assurance procedures for high strength concrete have not been developed.

The remainder of this section deals primarily with the determination of the compressive strength, f_c', since this is the basis on which high strength concrete is designed and specified.

Age at test

Traditionally, the acceptance standards for concrete involve strength determinations at an age of 28 days. Although there is, of course, nothing magical about this particular test age, it has been used universally as the reference time at which concrete strengths are reported. However, for high strength concretes, it has become common to determine compressive strengths at 56 days, or even 90 days. The justification for this is that concrete in structures will rarely, if ever, be loaded to anything approaching its design strength in less than 3 months, given the pace of construction. The increase in strength between 28 and 56 or 90 days can be considerable (10% to 20%), and this can lead to economies in construction. It is thus perfectly reasonable to measure strengths at later ages, and to specify the concrete strength in terms of these longer curing times.

There are, however, two drawbacks to this approach. First, it can be misleading to compare the compressive strengths of normal and high strength concretes, if these are measured at different times. Of more importance, there is a certain margin of safety when concrete strengths are measured at 28 days, since the concrete will generally be substantially stronger when it finally has to carry its design loads, perhaps at the age of one year for a typical high-rise concrete building. If strengths are specified at later ages, this margin is reduced (by an unknown amount), and hence there is an implicit reduction in the factor of safety. And, of course, finding higher strengths at later test ages does not in any way imply that the concrete has somehow become 'better' than a concrete whose strength was measured in the conventional way at 28 days.

Curing conditions

In general, the highest concrete strengths will be obtained with specimens continuously moist cured (at 100% relative humidity) until the time of testing. Unfortunately, the available data on this point are ambiguous. Carrasquillo, Nilson and Slate[28] found that high strength concrete, moist-cured for 7 days and then allowed to dry at 50% relative humidity till 28 days showed a strength loss of about 10% when compared to continuously moist-cured specimens. However, in subsequent work, Carrasquillo and Carrasquillo[29] found that up to an age of 15 days, specimens treated with a curing compound and allowed to cure in the field under ambient conditions yielded slightly higher strengths than moist-cured specimens. At 28 days, moist-cured specimens and field-cured specimens (with or without curing compounds) yielded approximately the same results. Only at later ages (56 and 91 days) did the strengths of the moist-cured specimens surpass those of the field-cured specimens treated with a curing compound. Similarly, for the mixes shown in Table 1.9, Burg and Ost[19] found that, when specimens that had been moist cured for 28 days were then subjected to air curing, their strengths at 91 days *exceeded* those of continuously moist-cured specimens; however, by 426 days, the continuously moist-cured specimens were from about 3% to 10% higher in strength than the air-cured ones.

On the other hand, several investigators have reported that, as long as a week or so of moist curing is provided, subsequent curing under ambient conditions is not particularly detrimental to strength development. Peterman and Carrasquillo[16] have stated that 'the 28-day compressive strength of high strength concrete which has been cured under ideal conditions for 7 days after casting is not seriously affected by curing in hot or dry conditions from 7 to 28 days after casting.'

Finally, contrary results were reported by Moreno[30] who indicated that air-cured specimens were about 10% stronger than moist-cured specimens at all ages up to 91 days.

Type of mold for casting cylindrical specimens

ASTM C470: *Molds for Forming Concrete Test Cylinders Vertically*, describes the requirements for both reusable and single-use molds, and ASTM C31: *Making and Curing Concrete Test Specimens on the Field* permits both types of mold to be used. However, it has long been known that different molds conforming to ASTM C470 will result in specimens with different measured strengths. This is true for both normal strength and high strength concretes. In general, more flexible molds will yield lower strengths than very rigid molds, because the deformation of the flexible molds during rodding or vibration leads to less efficient compaction than when using rigid molds. The experimental data largely bear this out. It should be noted that, whatever the mold materials, the molds must be properly sealed to prevent leakage of the mix water. If any significant leakage does occurs, the apparent strength will generally increase, because of the lower effective w/c ratio, and increased densification of the specimens.

For the standard 6×12 in. $(150 \times 300$ mm) molds, Carrasquillo and Carrasquillo[29] found that steel molds gave strengths about 5% higher than plastic molds, while Hester[31] found about a 10% difference. Similar results were reported by Howard and Leatham.[32] Peterman and Carrasquillo[16] reported that steel molds gave strengths about 10% higher than those obtained with cardboard molds, and Hester[31] showed that steel molds gave strengths about 6% higher than tin molds.

On the other hand, Cook[15] reported that 'good success was experienced on the use of single-use rigid plastic molds', while Aitcin[33] reports increasing use of rigid, reusable plastic molds. In addition, Carrasquillo and Carrasquillo[29] have reported that for the smaller 4×8 in. $(100 \times 200$ mm) molds, there were no strength differences between steel, plastic or cardboard molds.

In view of the above results, it would be prudent to use rigid steel molds whenever practicable, particularly for concrete strengths in excess of about 14,000 psi (98 MPa), at least until more test data become available for the smaller molds.

Specimen size

For most materials, including concrete, it has generally been observed that the smaller the test specimen, the higher the strength. For high strength concrete, however, though this effect is often observed, there are contradictory results reported in the literature. The results of a number of studies are compared in Table 1.11. It may be seen that the observed strength ratios of 4×8 in. (100×200 mm) cylinders to 6×12 in. (150× 300 mm)cylinders range from about 1.1 to 0.93. These contradictory results may be due to differences in testing procedures amongst the various investigators.

It must be noted that while for a given set of materials and test procedures, it may be possible to increase the apparent concrete strength by decreasing the specimen size, this does not in any way change the strength of the concrete in the structure. One particular specimen size does not give 'truer' results than any other. Thus, one should be careful to specify a particular specimen size for a given project, rather than leaving it as a matter of choice.

Specimen end conditions

According to ASTM C39: *Compressive Strength of Cylindrical Concrete Specimens*, the ends of the test specimens must be plane within 0.002 in. (0.05 mm). This may be achieved either by capping the ends (usually with a sulfur mortar) or by sawing or grinding. Unfortunately, different end

Table 1.11   Effect of specimen size on the compressive strength of high strength concrete

Investigator	$\dfrac{f_c' \ (100 \times 200 \text{ mm cylinder})}{f_c' \ (150 \times 300 \text{ mm cylinder})}$
Peterman and Carrasquillo[16]	~1.1
Carrasquillo, Slate and Nilson[34]	~1.1
Howard and Leatham[32]	~1.08
Cook[15]	~1.05
Burg and Ost[19]	~1.01
Aitcin[33]	ambiguous results
Moreno[30]	
83 MPa concrete	~1.0
119 MPa concrete	~0.93
Carrasquillo and Carrasquillo[29]	~0.93

conditions can lead to different measured strengths, and so the end preparation for testing high strength concrete specimens should be specified explicitly for any given project.

The most common method for preparing the ends of normal strength concrete is to use sulfur caps; for high strength concrete, high strength sulfur mortars are commercially available. However, if the strength of the cap is less than the strength of the concrete, the compressive load will not be transmitted uniformly to the specimen ends, leading to invalid results. Thus, for high strength concrete, in addition to high strength capping compounds, a number of other end preparation techniques are being investigated. These include grinding the specimen ends, or using unbonded systems, consisting of a pad constrained in a confining ring which fits over the specimen ends.

Most compressive strength tests on high strength concrete are still carried out using a high strength capping compound. The materials available in North America will achieve compressive strengths of 12,000 psi to 13,000 psi (84 MPa to 91 MPa) when tested as 2 in. (50 mm) cubes.[33] Peterman and Carrasquillo[21] recommend the use of such capping compounds, since they give higher concrete strengths than ordinary capping compounds. Cook[16] has used such compounds for concrete strengths up to 10,000 psi (70 MPa), while Moreno[30] considers them to be satisfactory at strengths up to 17,000 psi (119 MPa).

Burg and Ost[19] report that a high strength capping material may be used with concrete strengths of up to 15,000 psi (105 MPa); beyond that, the mode of failure of the cylinders changed from the normal cone failure of a columnar one. They recommend grinding of the cylinder ends for strengths beyond 15,000 psi (105 MPa). Similarly, Aitcin[33] has reported that above about 17,000 psi (119 MPa), the high strength capping material is pulverized as the specimens fail, which might well affect the measured strength. He too recommends grinding of the specimen ends for very high strength concretes. (It might be noted that end grinders for concrete cylinders are now commercially available. In 1992, the cost of such a machine was approximately US$12,000.)

Because of the uncertainty with high strength capping compounds, and the costs and time involved in end grinding, a considerable amount of research has been carried out on unbonded capping systems. These consist of metal restraining caps into which elastomeric inserts are placed; the assemblies then fit over the ends of the cylinder. As the elastomeric inserts deteriorate with repeated use, they are replaced from time to time.

Richardson[35] used a system of neoprene inserts in aluminium caps for testing normal strength concretes in the range of 3000 psi to 6000 psi (21 MPa to 42 MPa). He found that below 4000 psi (28 MPa), the neoprene pads gave somewhat lower strengths than conventional sulfur caps, while above 4000 psi (28 MPa) they gave somewhat higher strengths. Overall, however, the mean compressive strengths were not significantly different between the two systems.

Carrasquillo and Carrasquillo[29] compared a high strength sulfur capping compound to an unbonded system consisting of a polyurethane pad in an aluminium restraining ring. They found that up to about 10,000 psi (70 MPa), the unbonded system gave strengths that were 97% of those obtained with the capping compound. Beyond 10,000 psi (70 MPa), however, the unbonded system gave much higher strengths; they hypothesized that this might be due to greater end restraint of the cylinders with such a system. In subsequent work,[36] they found that up to 10,000 psi (70 MPa), polyurethane pads in an aluminium cap gave results within 5% of those achieved with high strength sulfur caps, while up to 11,000 psi (77 MPa), neoprene pads in steel caps gave results within 3% of those obtained with the sulfur end caps. However, they concluded that the use of either unbonded system was questionable; substantial differences in test results were obtained when two sets of restraining caps (from the same manufacturer) were used.

To improve the results obtained with unbonded systems, Boulay[37] developed a system in which, instead of elastomeric inserts, a mixture of dry sand and wax is used. It was found[38] that the sand mixture gave results which were intermediate between those obtained with ground ends or with sulfur mortar caps.

In summary, then, below about 14,000 psi (98 MPa), a thin, high strength sulfur mortar cap may be used successfully. Beyond that strength level, it would appear that grinding specimen ends is currently the only way to ensure valid test results.

Testing machine characteristics

In general, for normal strength concrete, the characteristics of the testing machine itself are assumed to have little or no effect on the peak load. However, for very high strength concretes the machine may well have some effect on the response of the specimen to load. From a review of the literature, Hester[31] concluded that the longitudinal stiffness of the testing machine will not affect the maximum load, and this view is shared also by Aitcin.[33] However, if the machine is not stiff enough, the specimens may fail explosively. and, of course, a very stiff machine (with servo-controls) is required if one wishes to determine the post-peak response of the concrete. On the other hand, Hester[31] also reports that if the machine is not stiff enough *laterally*, compressive strengths may be adversely affected.

One must also be concerned about the capacity of the testing machine when testing very high strength concretes. Aitcin[33] calculated the required machine capacities for different strength levels and specimen sizes, using the common assumption that the failure load should not exceed ~2/3 of the machine capacity. Some of his results are reproduced in Table 1.12. Relatively few commercial laboratories are equipped to test high strength concrete, since a common capacity of commercial testing machine is 292,500 lbs (1.3 MN). To test a 6 × 12 in. (150 × 300 mm) cylinder of

Table 1.12 Machine capacity required for high strength concrete[33]

	Failure load		*Machine capacity*	
Specimen size	$f_c' = 100$ MPa	$f_c' = 150$ MPa	$f_c' = 100$ MPa	$f_c' = 150$ MPa
100×200 mm	0.785 MN	1.18 MN	1.2 MN	1.75 MN
150×300 mm	1.76 MN	2.65 MN	2.65 MN	4.0 MN

Note: 1 MN = 225,000 lbf

21,400 psi (150 MPa) concrete requires a 900,000 lb (4.0 MN) testing machine, and relatively few machines of this size are available in commercial laboratories. This then, is probably the driving force behind the move to the smaller 4×8 in. (100×200 mm) cylinders.

Effect of loading platens

Again, for ordinary concrete, the effects of the spherically seated bearing blocks (platens) are not explicitly considered, as long as they meet the requirements of ASTM C39: *Compressive Strength of Cylindrical Concrete Specimens.* However, recent work at the Construction Technology Laboratories in Skokie, Illinois[39] has shown that, for high strength concrete, even this cannot be ignored. Spherical bearing blocks which deform in such a way that the stresses are higher around the periphery of the specimen than at the centre, yield higher compressive strengths than blocks which deform so that the highest stresses are at the centre of the specimen, and fall off towards the edges (i.e. a 'concave' rather than a 'convex' stress distribution). Measured differences can be as high as 15% for concretes with compressive strengths greater than 16,000 psi (112 MPa).

1.5 Conclusions

In conclusion, then, it has been shown that the production of high strength concrete requires careful attention to details. It also requires close cooperation between the owner, the engineer, the suppliers and producers of the raw materials, the contractor, and the testing laboratory.[32] Perhaps most important, we must remember that the well-known 'laws' and 'rules-of-thumb' that apply to normal strength concrete may well not apply to high strength concrete, which is a distinctly different material. Nonetheless, we now know enough about high strength concrete to be able to produce it consistently, not only in the laboratory, but also in the field. It is to be hoped that codes of practice and testing standards catch up with the high strength concrete technology, so that the use of this exciting new material can continue to increase.

Acknowledgements

This work was supported by the Canadian Network of Centres of Excellence on High-Performance Concrete.

References

1 SHRP-C/FR-91-103 (1991) *High performance concretes, a state of the art report*. Strategic Highway Research Program, National Research Council, Washington, DC.
2 FIP/CEB (1990) *High strength concrete, state of the art report*. Bulletin d'Information No. 197.
3 ACI Standard 211.1 (1989) *Recommended practice for selecting proportions for normal weight concrete*. American Concrete Institute, Detroit.
4 Mindess, S. and Young, J.F. (1981) *Concrete*. Prentice Hall Inc., Englewood Cliffs.
5 Perenchio, W.F. (1973) An evaluation of some of the factors involved in producing very high-strength concrete. *Research and Development Bulletin*, No. RD014-01T, Portland Cement Association, Skokie.
6 Mehta, P.K. and Aitcin, P.-C. (1990) Microstructural basis of selection of materials and mix proportions for high-strength concrete, in *Second International Symposium on High-Strength Concrete, SP-121*. American Concrete Institute, Detroit, 265–86.
7 Aitcin, P.-C. (1992) private communication
8 Ronneburg, H. and Sandvik, M. (1990) High Strength Concrete for North Sea Platforms, *Concrete International*, **12**, 1, 29–34
9 CSA Standard A23.5-M86 (1986) *Supplementary cementing materials*. Canadian Standards Association, Rexdale, Ontario.
10 ASTM C618 *Standard specification for fly ash and raw or calcined natural pozzolan for use as a mineral admixture in portland cement concrete*. American Society for Testing and Materials, Philadelphia, PA.
11 ASTM C989 *Standard specification for ground iron blast-furnace slag for use in concrete and mortars*. American Society for Testing and Materials, Philadelphia, PA.
12 Malhotra, V.M. (ed) (1987) *Supplementary cementing materials for concrete*. Minister of Supply and Services, Canada.
13 Aitcin, P.-C., Sarkar, S.L., Ranc, R. and Levy, C. (1991) A High Silica Modulus Cement for High-Performance Concrete, in S. Mindess (ed.), *Advances in cementitious materials*. Ceramic Transactions **16**, The American Ceramic Society Inc., 102–21.
14 Malhotra, V.M., Ramachandran, V.S., Feldman, R.F. and Aitcin, P.-C. (1987) *Condensed silica fume in concrete*. CRC Press Inc., Boca Ratan, Florida.
15 Cook, J.E. (1989) 10,000 psi Concrete. *Concrete International*, **11**, 10, 67–75.
16 Peterman, M.B. and Carrasquillo, R.L. (1986) *Production of high strength concrete*. Noyes Publications, Park Ridge.
17 Randall, V.R. and Foot, K.B. (1989) High strength concrete for Pacific First Center. *Concrete International: Design and Construction*, **11**, 4, 14–16.
18 Aitcin, P.-C., Shirlaw, M. and Fines, E. (1992) High performance concrete: removing the myths, in *Concrescere*, Newsletter of the High-Performance Concrete Network of Centres of Excellence (Canada), 6, March.
19 Burg, R.G. and Ost, B.W. (1992) *Engineering properties of commercially available high-strength concretes*. Research and Development Bulletin RD104T, Portland Cement Association, Skokie.

20 Aitcin, P.-C. and Mehta, P.K. (1990) Effect of coarse aggregate type or mechanical properties of high strength concrete. *ACI Materials Journal*, American Concrete Institute, Detroit, **87**, 2, 103–107.

21 ACI Committee 363 (1984) *State-of-the-art report on high strength concrete* (ACI 363R-84). American Concrete Institute, Detroit.

22 Addis, B.H. (1992) *Properties of High Strength Concrete Made with South African Materials*, Ph.D. Thesis, University of the Witwatersrand, Johannesburg, South Africa.

23 Addis, B.J. and Alexander, M.G. (1990) A method of proportioning trial mixes for high-strength concrete, in ACI Sp-121, *High strength concrete, Second International Symposium*, American Concrete Institute, Detroit, 287–308.

24 Canadian Portland Cement Association (1991) *Design and control of concrete*. Edition CPCA, Ottawa.

25 Fiorato, A.E. (1989) PCA research on high-strength concrete. *Concrete International*, **11**, 4, 44–50.

26 Hattori, K. (1979) Experiences with mighty superplasticizer in Japan, in ACI SP-62, *Superplasticizers in concrete*, American Concrete Institute, Detroit, 37–66.

27 Suzuki, T. (1987) Experimental studies on high-strength superplasticized concrete, in *Utilization of high strength concrete, Symposium proceedings*. Stavanger, Norway: Tapis Publishers, Trondheim, 53–4.

28 Carrasquillo, R.C., Nilson, A.H. and Slate, F.O. (1981) Properties of high strength concrete subject to short-term loads. *Journal of American Concrete Institute*, **78**, 3, 171–8.

29 Carrasquillo, P.M. and Carrasquillo, R.L. (1988). Evaluation of the use of current concrete practice in the production of high-strength concrete. *ACI Materials Journal*, **85**, 1, 49–54.

30 Moreno, J. (1990) 225 W. Wacker Drive. *Concrete International*, **12**, 1, 35–9.

31 Hester, W.T. (1980) Field testing high-strength concretes: a critical review of the state-of-the-art. *Concrete International*, **2**, 12, 27–38.

32 Howard, N.L. and Leatham, D.M. (1989) The production and delivery of high-strength concrete. *Concrete International*, **11**, 4, 26–30.

33 Aitcin, P.-C. (1989) Les essais sue les betons a tres hautes performances, in *Annales de L'Institut Technique du Batiment et des Travaux Publics*, No. 473. Mars-Avril. Serie: Beton 263, 167–9.

34 Carrasquillo, R.L., Slate, F.O. and Nilson, A.H. (1981) Microcracking and behaviour of high strength concrete subjected to short term loading. *American Concrete Institute Journal*, **78**, 3, 179–86.

35 Richardson, D.N. (1990) Effects of testing variables on the comparison of neoprene pad and sulfur mortar-capped concrete test cylinders. *ACI Material Journal*, **87**, 5, 489–95.

36 Carrasquillo, P.M. and Carrasquillo, R.L. (1988) Effect of using unbonded capping systems on the compressive strength of concrete cylinders. *ACI Materials Journal*, **85**, 3, 141–7.

37 Boulay, C. (1989) La boite a sable, pour bien ecraser les betons a hautes performances. *Bulletin de Liaison des Laboratoires des Ponts et Chausses*, Nov/Dec.

38 Boulay, C., Belloc, A., Torrenti, J.M. and De Larrard, F. (1989) *Mise au point d'un nouveau mode operatoire d'essai de compression pour les betons a haute performances*. Internal report, Laboratoire Central des Ponts et Chaussees, Paris, December.

39 *CTL Review* (1992) Construction Technology Laboratories, Inc., Skokie, Illinois, **15**, 2.

2 Short term mechanical properties

S H Ahmad

2.1   Introduction

Chapter 1 discussed the production of concrete and the effects of a large number of constituent materials – cement, water, fine aggregate, coarse aggregate (crushed stone or gravel), air and other admixtures on the production process. Some quality control issues were also addressed. In the present chapter, the mechanical properties of hardened concrete under short term conditions or loadings are discussed.

Concrete must be proportioned and produced to carry imposed loads, resist deterioration and be dimensionally stable. The quality of concrete is characterized by its mechanical properties and ability to resist deterioration. The mechanical properties of concrete can be broadly classified as short-term (essentially instantaneous) and long-term properties. Short-term properties include strength in compression, tension, modulus of elasticity and bond characteristics. The long-term properties include creep, shrinkage, behavior under fatigue, and durability characteristics such as porosity, permeability, freeze-thaw resistance and abrasion resistance. The creep and shrinkage characteristics are discussed in Chapter 3, the behavior under fatigue and the bond characteristics is addressed in Chapter 4. The important aspect of durability is presented in Chapter 5.

While information on high performance concretes (HPC) as defined in Chapter 1 is scarce, there is a substantial body of information on the mechanical properties of high strength concrete and additional information is being developed rapidly. One class of high performance concretes are the early strength concretes. The mechanical properties of these types of high performance concretes are being investigated under the Strategic Highway Research Program SHRP C-205 which is in progress at North Carolina State University. Since high performance concretes typically have low water/cementitious materials (w/c) ratios and high paste contents, characteristics will in many cases be similar to those of high strength concrete. Much of the discussion in this chapter will therefore concentrate on high strength concretes.

A significant difference in behavior between the early strength and the high strength concretes is in the relationship of compressive strength to mechanical properties. Strength gain in compression is typically much faster than strength gain in aggregate–paste bond, for instance. This will lead to relative differences in elastic modulus and tensile strength of early strength concretes and high strength concretes, expressed as a function of compressive strength. The relationships of mechanical properties to 28-day compressive strength developed in other studies cannot necessarily be expected to apply to early strength concretes. The information developed under the SHRP program will be useful to fill this knowledge gap.

2.2   Strength

The strength of concrete is perhaps the most important overall measure of quality, although other characteristics may also be critical. Strength is an important indicator of quality because strength is directly related to the structure of hardened cement paste. Although strength is not a direct measure of concrete durability or dimensional stability, it has a strong relationship to the w/c ratio of the concrete. The w/c ratio, in turn, influences durability, dimensional stability and other properties of the concrete by controlling porosity. Concrete compressive strength, in particular, is widely used in specifying, controlling and evaluating concrete quality.

The strength of concrete depends on a number of factors including the properties and proportions of the constituent materials, degree of hydration, rate of loading, method of testing and specimen geometry.

The properties of the constituent materials which affect the strength are the quality of fine and coarse aggregate, the cement paste and the paste–aggregate bond characteristics (properties of the interfacial, or transition, zone). These, in turn, depend on the macro- and microscopic structural features including total porosity, pore size and shape, pore distribution and morphology of the hydration products, plus the bond between individual solid components. A simplified view of the factors affecting the strength of concrete is shown in Fig. 2.1.

Testing conditions including age, rate of loading, method of testing, and specimen geometry significantly influence the measured strength. The strength of saturated specimens can be 15% to 20% lower than that of dry specimens. Under impact loading, strength may be as much as 25% to 35% higher than under a normal rate of loading (10 to 20 microstrains per second). Cube specimens generally exhibit 20% to 25% higher strengths than cylindrical specimens. Larger specimens exhibit lower average strengths.

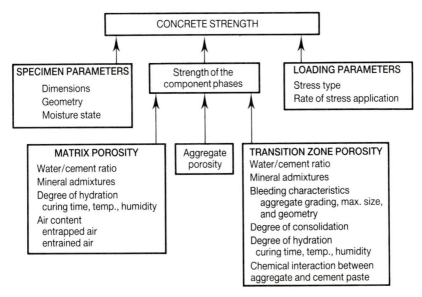

Fig. 2.1 An oversimplified view of the factors influencing strength of plain concrete[53]

Constituent materials and mix proportions

Concrete composition limits the ultimate strength which can be obtained and significantly affects the levels of strength attained at early ages. A more complete discussion of the effects of constituent materials and mix proportions is given in Chapter 1. However, a review of the two dominant constituent materials on strength is useful at this point. Coarse aggregate and paste characteristics are typically considered to control maximum concrete strength.

Coarse aggregate

The important parameters of coarse aggregate are its shape, texture and the maximum size. Since the aggregate is generally stronger than the paste, its strength is not a major factor for normal strength concrete, or in early strength concrete. However, the aggregate strength becomes important in the case of higher-strength concrete or lightweight aggregate concrete. Surface texture and mineralogy affect the bond between the aggregates and the paste and the stress level at which microcracking begins. The surface texture, therefore, may also affect the modulus of elasticity, the shape of the stress-strain curve and, to a lesser degree, the compressive strength of concrete. Since bond strength increases at a slower rate than compressive strength, these effects will be more pronounced in early strength concretes. Tensile strengths may be very sensitive to differences in aggregate surface texture and surface area per unit volume.

The effect of different types of coarse aggregate on concrete strength has been reported in numerous articles. A recent paper[12] reports results of four different types of coarse aggregates in a very high strength concrete mixture (w/c = 0.27). The results showed that the compressive strength was significantly influenced by the mineralogical characteristics of the aggregates. Crushed aggregates from fine-grained diabase and limestone gave the best results. Concretes made from a smooth river gravel and from crushed granite that contained inclusions of a soft mineral were found to be relatively weaker in strength.

The use of larger maximum size of aggregate affects the strength in several ways. Since larger aggregates have less specific surface area, the bond strength between aggregates and paste is lower, thus reducing the compressive strength. Larger aggregate results in a smaller volume of paste thereby providing more restraint to volume changes of the paste. This may induce additional stresses in the paste, creating microcracks prior to application of load, which may be a critical factor in very high strength concretes.

The effect of the coarse aggregate size on concrete strength was discussed by Cook *et al.*[22] Two sizes of aggregates were investigated: a 3/8 in. (10 mm) and a 1 in. (25 mm) limestone. A superplasticizer was used in all the mixes. In general, the smallest size of the coarse aggregate produces the highest strength for a given w/c ratio, see Figs 2.2–2.6. It may be noted that compressive strengths in excess of 10,000 psi (70 MPa) can be produced using a 1 in. (25 mm) maximum size aggregate when the mixture is properly proportioned.

Although these studies[12,22] provide useful data and insight, much more research is needed on the effects of aggregate mineral properties and

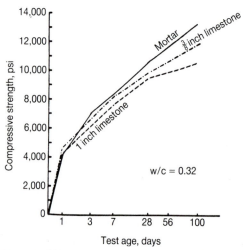

Fig. 2.2   Effect of aggregate type on strength at different ages for a constant w/c materials ratio without superplasticizer[22]

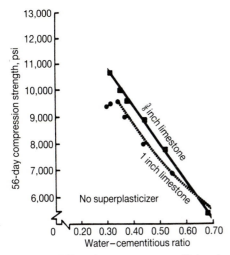

Fig. 2.3   Effect of aggregate type on 56 day strength for concrete for different w/c materials ratio[22]

particle shape on the strength and durability of higher strength concrete. This was recognized as one of the research needs by the ACI 363 Committee.[3]

Paste characteristics

The most important parameter affecting concrete strength is the w/c ratio,

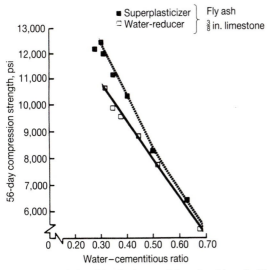

Fig. 2.4   Relationship of w/c materials ratio with and without a high range water-reducing admixture for coarse aggregate size not exceeding $\frac{3}{8}$ in. (10 mm)[22]

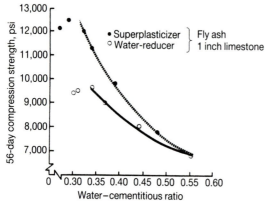

Fig. 2.5   Relationship of w/c materials ratio with and without a high range water-reducing admixture for coarse aggregate size not exceeding 1 in. (25.4 mm)[22]

sometimes referred to as the w/b (binder) ratio. Even though the strength of concrete is dependent largely on the capillary porosity or gel/space ratio, these are not easy quantities to measure or predict. The capillary porosity of a properly compacted concrete is determined by the w/c ratio and degree of hydration. The effect of w/c ratio on the compressive strength is shown in Fig. 2.7. The practical use of very low w/c ratio concretes has been made possible by use of both conventional and high range water reducers, which permit production of workable concrete with very low water contents.

Supplementary cementitious materials (fly ash, slag and silica fume) have been effective additions in the production of high strength concrete. Although fly ash is probably the most common mineral admixture, on a volume basis, silica fume (ultra-fine amorphous silica, derived from the production of silicon or ferrosilica alloys) in particular, used in combina-

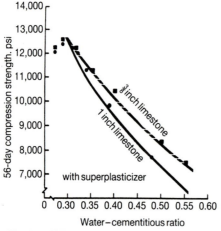

Fig. 2.6   Effect of aggregate type on strength at different ages for a constant w/c materials ratio, with superplasticizer[22]

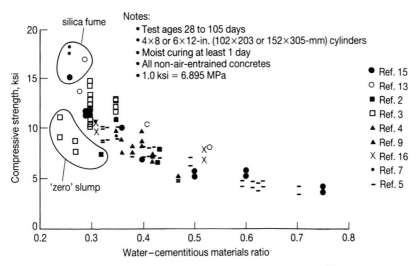

Fig. 2.7   Summary of strength data as a function of w/c materials ratio[29]

tion with high-range water reducers, has increased achievable strength levels dramatically (Fig. 2.7).[10,51,52]

The effect of condensed silica fume on the strength of concrete was reported in a very comprehensive study.[28] The beneficial effect of using up to 16% (by weight of cement) condensed silica on the compressive strength is shown in Fig. 2.8. The data indicate that to achieve 10,000 psi (70 MPa) 28 day $4 \times 4 \times 4$ in. $(100 \times 100 \times 100$ mm) cube strength, the w/c ratio

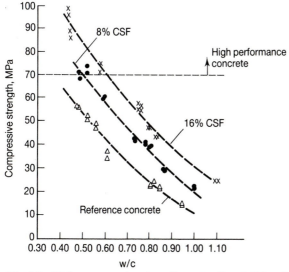

Fig. 2.8   28-day compressive strength versus w/c materials ratio for concrete with different condensed silica fume contents[28]

required is about 0.35 if no silica fume is used; however, with 8% silica fume, the w/c needed is about 0.50, and with 16% silica fume content the w/c ratio requirement increases to about 0.65. This indicates that higher compressive strength can be achieved very easily with high silica fume content at relatively higher w/c ratios.

The efficiency of silica fume in producing concrete of higher strength depends on water/cement + silica fume ratio, dosage of silica fume, age and curing conditions. Yogenendram *et al.*[85] investigated the efficiency of silica fume at lower w/c ratio. Their results indicated that the efficiency is much lower at w/c ratio of 0.28 as compared to the efficiency at w/c ratio of 0.48.

The performance of chemical admixtures is influenced by the particular cement and other cementitious materials. Combinations which have been shown to be effective in many cases may not work in all situations, due to adverse cement and admixture interaction (see Fig. 2.9). Substantial testing should be conducted with any new combination of cements, and mineral or chemical admixtures prior to large scale use.

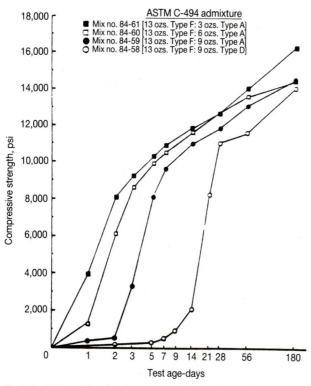

Fig. 2.9   Effect of varying dosage rates of normal retarding water-reducing admixtures on the strength development of concrete[22]

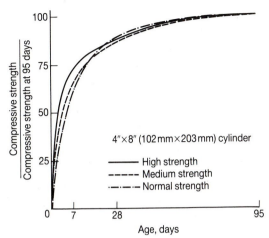

Fig. 2.10   Normalized strength gain with age for limestone concretes moist-cured until testing[16]

Strength development and curing temperature

The strength development with time is a function of the constituent materials and curing techniques. An adequate amount of moisture is necessary to ensure that hydration is sufficient to reduce the porosity to a level necessary to attain the desired strength. Although cement paste will never completely hydrate in practice, the aim of curing is to ensure sufficient hydration. In pastes with lower w/c ratios, self-desiccation can occur during hydration and thus prevent further hydration unless water is supplied externally.

The strength development with time up to 95 days for normal, medium and high strength concretes utilizing limestone aggregate sand moist cured until testing are shown in Fig. 2.10. The results indicate a higher rate of strength gain for higher strength concrete at early ages. At later ages the difference is not significant. The compressive strength development of 9000 psi, 11,000 psi, and 14,000 psi (62 MPa, 76 MPa, 97 MPa) concretes up to a period of 400 days is shown in Fig. 2.11. The results shown in the figure are for mixes containing cement only or cement and fly ash, with some mixes using high range water-reducing agents. The data indicate that for moist-cured specimens, strengths at 56 days are about 10% greater than 28 day strengths. Strengths at 90 days are about 15% greater than 28 day strengths. While it is inappropriate to generalize from such results, they do indicate the potential for strength gain at later ages.

In a recent study[45] at North Carolina State University (NCSU), concretes utilizing a number of different aggregates and mineral admixtures, with strengths from 7000 psi to 12,000 psi (48 MPa to 83 MPa) at 28 days and from 10,000 psi to almost 18,000 psi (69 MPa to 124 MPa) at one year were tested. On examining the absolute strength gain against the percentage strength gain with time, it was concluded that there appears to be no

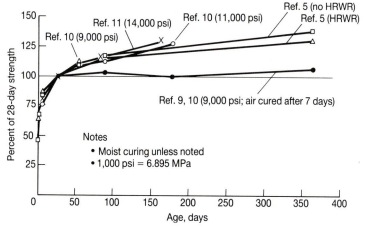

Fig. 2.11   Compressive strength development for concretes with and without high range water reducers[29]

single, constant factor which can be used to predict later strengths accurately from early strengths except in a very general sense. This is no doubt due to the contributions of not only the ultimate strength of the aggregate and the mortar, but to the strength of the transition zone. The transition zone strength, or interfacial bond strength of the mortar to the aggregate, of concretes of higher strengths, is typically affected by the binder composition as well as the ultimate strength of the mortar. Results for splitting tensile strength and modulus of rupture were similar.

The effect of condensed silica fume (CSF) on concrete strength development at 20 °C generally takes place from about 3 to 28 days after mixing. Johansen[40] measured strength up to 3 years and concluded that there was little effect of CSF on either the strength gain between 28 days and 1 year or between 1 and 3 years for water-stored specimens.

The effect of cement types on the strength development is presented in Table 2.1. At ordinary temperatures, for different types of portland and

Table 2.1   Approximate relative strength of concrete as affected by cement type

Type of portland cement		Compressive strength (percent of strength of Type I or normal portland cement concrete)			
ASTM	Description	1 day	7 days	28 days	90 days
I	Normal or general purpose	100	100	100	100
II	Moderate heat of hydration and moderate sulfate resisting	75	85	90	100
III	High early strength	190	120	110	100
IV	Low heat of hydration	55	65	75	100
V	Sulfate resisting	65	75	85	100

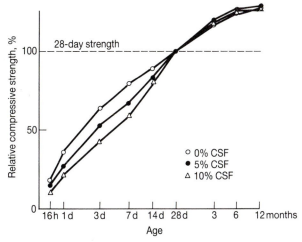

Fig. 2.12 Compressive strength development of concrete cured at 20 °C with different dosages of condensed silica fume[48]

blended cements, the degree of hydration at 90 days and above is usually similar; therefore, the influence of cement composition on the porosity of the matrix and strength is primarily a concern at early ages. The effect of condensed silica fume on the strength development of concretes with four different types of cement was investigated by Maage and Hammer.[48] The four cement types were ordinary portland cement, 10% and 25% pulverized fuel ash (fly ash) blends, and a 15% slag blend. Concrete mixes without CSF and with 0%, 5%, and 10% CSF were made at 5 °C, 20 °C and 35 °C and maintained at these temperatures in water for up to one year. The compressive strengths were measured from 16 hours up to a period of one year. Mixes in three strength classes were made: 2000 psi, 3500 psi and 6500 psi (15 MPa, 25 MPa and 45 MPa). Figure 2.12 shows the compressive strength development of concrete water-cured at 20 °C, with various CSF dosages and utilizing different cement types. In the figure each curve represents a mean value for four cement types, and relative compressive strength of 100% represents 28 day strength for each mix type. From the figure, it can be seen that at 20 °C curing, regardless of the cement type, the CSF had the same influence on the strength–age relationship. Figures 2.13 and 2.14 show relative strength development at 5 °C with and without 10% CSF for the four cement types, and similar data for 35 °C curing are shown in Figs. 2.15 and 2.16. At 5 °C curing, the blended cement lags behind ordinary portland cement concrete (OPC) up to 28 days; with 10% CSF the lag increases which indicates that the pozzolanic reactions have not contributed much to the strength in the 28 day period. At 35 °C the CSF mix is more strongly accelerated (in comparison with 20 °C curing) than the reference mixes, particularly between the first and the seventh day.

Curing at elevated temperatures has a greater accelerating effect on condensed silica fume (CSF) concrete than on control concrete.

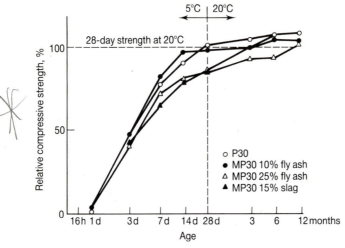

Fig. 2.13 Development of compressive strength in reference concrete cured in water at 5 °C for 28 days then at 20 °C. 100% represents 28-day strength at 20 °C for each cement type[48]

Evidence[28] indicates that a curing temperature of roughly 50 °C is necessary for CSF concrete to equal one day strength of an equivalent control mix. Curing at temperatures below 20 °C retards strength development more for CSF concrete than for control concrete. CSF makes it possible to design low-heat concrete over a wide range of strength levels.[28] Therefore the condensed silica fume concrete is more sensitive to curing temperature than ordinary portland cement concrete. The effect of curing on the condensed silica fume and fly ash concrete was studied in a recent investigation,[68] in which concrete was exposed to six different curing

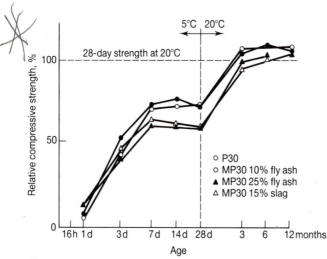

Fig. 2.14 Development of compressive strength in concrete containing 10% condensed silica fume and cured in water at 5 °C for 28 days then at 20 °C. 100% represents 28-day strength at 20 °C for each cement type[48]

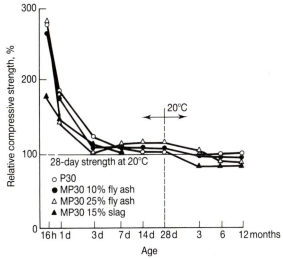

Fig. 2.15 Development of compressive strength in reference concrete cured in water at 35 °C for 28 days then at 20 °C. 100% represents 28-day strength at 20 °C for each cement type[48]

conditions. It was concluded that concrete cured at 20 °C continuously in water(reference) exhibited increasing strengths at all ages; concrete cured in water for 3 days before exposure to 50% RH showed higher initial strength, but the strength decreased after 2–4 months with respect to the reference; and concrete exposed to 50% RH showed lower strength after 28 days of curing than that cured in water.

Curing techniques have significant effects on the strength. The key concerns in curing, especially for concrete of higher strength, are maintaining adequate moisture and temperatures to permit continued cement hydration. Water curing of higher strength concrete is highly recommended[2] due to its low w/c ratio. At w/c ratio below 0.40, the

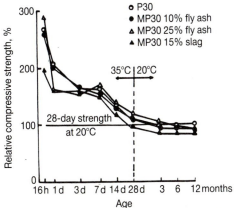

Fig. 2.16 Development of compressive strength in concrete containing 10% condensed silica fume and water-cured at 35 °C for 28 days then at 20 °C. 100% represents 28-day strength at 20 °C for each cement type[48]

Table 2.2   Effect of drying on compressive strength[16]

Moist cured, days	Drying period, days	Test age, days	Strength attained after drying* / Strength attained when moist cured until test age			
			Compressive strength f_c'		Modulus of rupture f_r'	
			Normal strength	High strength	Normal strength	High strength
0–7	8–28	28	0.98	0.91	0.83	0.74
0–7	8–28	28	0.94	0.89	0.86	0.74
0–7	8–28	28	0.95	0.88	0.88	0.74
0–28	29–95	95	0.99	0.95	0.97	0.91
0–28	29–95	95	1.01	0.96	0.96	0.93
0–28	29–95	95	0.99	0.96	0.99	0.91

Normal strength: $f_c' = 3330$ psi at 28 days; $f_c' = 3750$ psi at 95 days
High strength: $f_c' = 10,210$ psi at 28 days; $f_c' = 11,560$ psi at 95 days
* Average of three tests

ultimate degree of hydration is significantly reduced if free water is not provided. The effects of two different curing conditions on concrete strength were investigated.[16] The two conditions were moist curing for seven days followed by drying at 50% relative humidity until testing at 28 days, and moist curing the 28 days followed by drying at 50% relative humidity until testing at 95 days. Higher strength concrete showed a larger reduction in compressive strength when allowed to dry before completion of curing. The results are shown in Table 2.2. It has been reported that the strength is higher with moist curing as compared to field curing.[19]

Compressive strength

Conventionally, in the USA, concrete properties such as elastic modulus, tensile or flexural strength, shear strength, stress-strain relationships and bond strength are usually expressed in terms of uniaxial compressive strength of 6×12 in. (150×300 mm) cylinders, moist cured to 28 days. Compressive strength is the common basis for design for most structures, other than pavements, and even then is the common method of routine quality testing. The terms 'strength' and 'compressive strength' are used virtually interchangeably. The discussion above generally applies equally well to all measures of strength, although most results and conclusions were based either primarily or exclusively on compressive strength results.

Maximum, practically achievable, compressive strengths have increased steadily in the last decade. Presently, 28 day strengths of up to 12,000 psi (84 MPa) are routinely obtainable. The trend for the future has been examined in a recent ACI Committee 363 article[3] which identified development of concrete with compressive strength in excess of 20,000 psi (138 MPa) as one of the research needs.

Testing variables have a considerable influence on the measured compressive strength. The major testing variables are: mold type, specimen size, end conditions and rate of loading. The sensitivity of measured compressive strength to testing variables varies with level of compressive strength.

Since the compressive strength of early strength concretes are at conventional levels, conventional testing procedures can be used for the most part, although curing during the first several hours can affect test results dramatically. Testing of very high strength concretes is much more demanding. However, in all concretes, not just high performance concrete, competent testing is critical.

The effect of mold type on strength was reported in a recent paper by Carrasquillo and Carrasquillo.[18] Their results indicated that use of 6 × 12 in. (150 × 300 mm) plastic molds gave strengths lower than steel molds, and use of 4 × 8 in. (102 × 203 mm) plastic molds gave negligible difference with steel molds. They concluded that steel molds should be used for concrete with compressive strengths up to 15,000 psi (103 MPa). It seems appropriate that steel molds should also be used for concrete of higher strengths. The specimen size effect on the strength is shown in Fig. 2.17, which shows the relationship between the compressive strength of 4 × 8 in. (102 × 203 mm) cylinders and 6 × 12 in. (150 × 300 mm) cylinders. The figure indicates that 4 × 8 in. (102 × 203 mm) cylinders exhibit approximately 5% higher strengths than 6 × 12 in. (150 × 300 mm) cylinders. Similar results were also obtained in a recent study at North Carolina

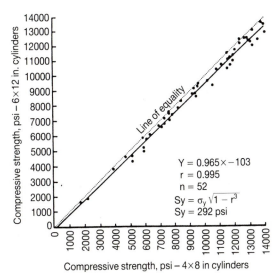

Fig. 2.17 Relationship between the compressive strength of 4 × 8 in. (102 × 203 mm) cylinders and 6 × 12 in (152 × 304 mm) cylinders[22]

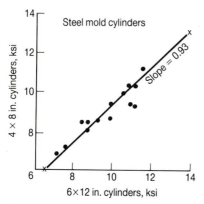

Fig. 2.18   Compressive strength of concrete cylinders cast in 4 × 8 in. (102 × 203 mm) steel molds versus 6 × 12 in. (152 × 304 mm) steel molds[18]

State University.[45] A contradictory result[19] is reported, however, which indicates that the compressive strength of 4 × 8 in. (102 × 203 mm) cylinders is slightly lower than 6 × 12 in. (150 × 300 mm) cylinders, see Fig. 2.18. The strength gain for 17,000 psi (117 MPa) concrete as shown by 6 × 12 in. (150 × 300 mm) and 4 × 8 in. (102 × 203 mm) cylinders has been reported by Moreno[54] and the results are shown in Fig. 2.19. His study also showed that the specimen size effect on the compressive strength is negligible on the basis of 29 tests, see Table 2.3. Another study[16] concluded that the ratio of 6 × 12 in. (150 × 300 mm) cylinder to 4 × 8 in. (102 × 203 mm) cylinder was close to 0.90 regardless of the strength of concrete for the ranges tested between 3000 and 11,000 psi (21 and 76 MPa).

The relationship between the compressive strength of 6 × 12 in. (150 × 300 mm) and cores from a column was studied for concrete with a strength of 10,000 psi (69 MPa), see Table 2.4. It was concluded that the 85% criterion specified in the ACI Building Code (ACI 318–89)[1] would be

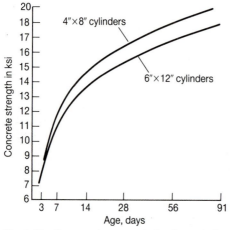

Fig. 2.19   Compressive strength development for 17,000 psi (117 MPa) concrete[54]

Table 2.3 Comparison of compressive strength test results of 12,000 psi (83 MPa) concrete at 56 days as obtained by 6 × 12 in. (152 × 304 mm) and 4 × 8 in. (102 × 203 mm) cylinders[54]

Cylinder size	6 × 12 in.	4 × 8 in.
Mean	13,444	13,546
Standard deviation	463	515
Coefficient of variation, percent	3.4	3.8
Number of tests	29	29

applicable to high strength concrete. This study also confirms the belief that job cured specimens do not give accurate measurements of the in-place strength. The reason for lower strength in the middle portion of the columns is probably due to temperature rise, i.e. 100 °F (38 °C) for high strength mixtures. In a recent study,[11] it was shown that the strength of 4 in. (100 mm) cores taken from a mock column at two and a half years after casting was nearly identical to that of specimens cured for 28 days in lime-saturated water at room temperature. The strength of the concrete tested was 12,300 psi (85 MPa).

The effect of end conditions on the compressive strength of concrete is summarized in a recent paper.[18] More than five hundred 6 × 12 in. (150 × 300 mm) cylinders from concretes having compressive strengths from 2500 to 16,500 psi (17 MPa to 114 MPa) were tested with either unbonded caps (two types) or sulfur mortar caps. It was concluded that use of unbonded caps (with a restraining ring and elastomeric insert) could provide a cleaner, safer and more cost-effective alternative to sulfur mortar for capping concrete cylinders. For concretes between 4000 and 10,000 psi (28 and 69 MPa), the use of polyurethane inserts with aluminium restraining rings in testing concrete cylinders yielded average test results within 5% of those obtained using sulfur mortar. For concrete strengths below 11,000 psi (76 MPa), the use of neoprene inserts with steel restraining rings in testing concrete cylinders yielded average test results within 3% of those obtained using sulfur mortar. For higher strength concrete, the use of either unbonded capping system is questionable. Substantial differences in compressive strength test results were obtained when two sets of restraining rings obtained from the same manufacturer were used. It was recommended that prior to acceptance, each set should be tested for correlation to results obtained from cylinders capped according to ASTM C617 for all strength levels of concrete for which the unbonded caps are to be used. (Equipment now exists for parallel grinding the ends of concrete cylinders prior to compression testing, thereby eliminating the need for any type of end cap.)

Measured compressive strength increases with higher rates of loading. This trend has been reported in a number of studies[14,15,26,38,49,83] for concrete with strengths in the range of 2000 to 5500 psi (14 to 49 MPa). However, only one study[8] has reported the effect of strain rate on concretes with compressive strengths in excess of 6000 psi (41 MPa). Based

Table 2.4 Column core strengths versus 6 × 12 in. (152 × 304 mm) cylinders*[22]

Test age Maximum size stone	7-Days		28-Days		56-Days		180-Days		1 Year	
	⅛ in	1 in	⅛ in	1 in	⅛ in	1 in	⅛ in	1 in	⅛ in	1 in
Compressive strength of 6 × 12 in cylinders, psi										
Field-cured	8,596	8,139	10,177	10,204	10,775	10,542	11,514	11,546	12,444	11,772
Moist-cured	9,228	9,277	11,522	11,236	12,376	12,448	13,852	13,776	14,060	13,951
Compressive strength of cores, psi										
West face	9,407	9,312	11,118	10,743	11,598	10,964	12,970	12,400	15,080	13,775
Middle	8,660	8,959	9,674	9,724	9,833	9,656	11,635	10,720	13,404	13,003
East face	9,180	9,706	10,584	10,575	10,756	10,603	11,626	11,589	14,088	13,695
All cores	9,083	9,326	10,459	10,347	10,729	10,408	12,077	11,570	14,190	13,490
Cores/6 × 12 in moist-cured cylinders, percent										
West face	102	101	97	96	94	88	94	90	107	99
Middle	94	97	84	87	79	78	84	78	95	93
East face	99	105	92	94	87	85	87	84	100	98
All cores	98	101	91	92	87	84	87	84	101	97

*Reported strengths are average of two specimens.

on their research and other reported data.[14,15,26,38,49,83] Ahmad and Shah[8] proposed an equation to estimate the strength under very fast loading conditions. The recommended equation is

$$(f_c')_{\dot{\varepsilon}} = f_c' \left[0.95 + 0.27 \log \frac{\dot{\varepsilon}}{f_c'} \right] \alpha \qquad (2.1)$$

where $\dot{\varepsilon}$ is the strain rate in microstrains per sec ($\mu\varepsilon$/sec).

The shape factor α accounts for the different sizes of the specimens tested by different researchers and is given by

$$\alpha = 0.85 + 0.95 (d) - 0.02 (h) \quad \text{for } \frac{h}{d} \leqslant 5 \qquad (2.2)$$

where d = diameter or least lateral dimension (in.), h = height (in.)

No information is available on the effect of rate of loading on the strength for concrete with strengths in excess of 10,000 psi (70 MPa).

Tensile strength

The tensile strength governs the cracking behavior and affects other properties such as stiffness, damping action, bond to embedded steel and durability of concrete. It is also of importance with regard to the behavior of concrete under shear loads. The tensile strength is determined either by direct tensile tests or by indirect tensile tests such as flexural or split cylinder tests.

Direct tensile strength

The direct tensile strength is difficult to obtain. Due to the difficulty in testing, only limited and often conflicting data is available. It is often assumed that direct tensile strength of concrete is about 10% of its compressive strength.

Two recent studies[23,31] have reported the direct tensile strength of concrete. The study at Delft University[23] utilized 4.7 in. (120 mm) diameter cylinders having a length of 11.8 in. (300 mm). The study at Northwestern[31] employed $3 \times 0.75 \times 12$ in. ($76 \times 19 \times 304$ mm) and $3 \times 1.5 \times 12$ in. ($76 \times 38 \times 304$ mm) thin plates having a notch in the central region for creating a weak section for crack initiation and propagation, and used special wedge like frictional grips. The study at Delft tested concrete of one strength which had either been sealed for four weeks or moist-cured for two weeks and air-dried for two weeks. The results indicated 18% higher tensile strength for the sealed concrete compared to the air-dried concrete. The investigation at Northwestern included different concrete strengths up to 7000 psi (48 MPa) strength, and it was concluded that the uniaxial tensile strength can be estimated by the expression $6.5 \sqrt{f_c'}$.

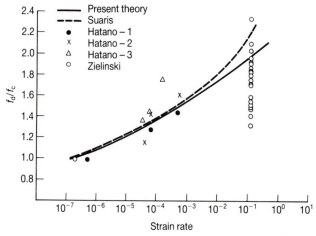

Fig. 2.20   Effect of strain rate on tensile strength of concrete[59]

Direct tensile strength data is not available for concrete with strengths in excess of 8000 psi (55 MPa).

The effect of rate of loading on the tensile strength has been the focus of some studies by Hatano,[34] Suaris and Shah,[80] and Zielinski et al.[87] The effect of fast strain rate on the tensile strength of concrete as observed by these studies is shown in Fig. 2.20. Also shown in the figure is a comparison of the predictions per a constitutive theory for concrete subjected to static uniaxial tension[59] and the experimental results.

The effect of sustained and cyclic loading on the tensile properties of concrete was investigated by Cook and Chindaprasirt.[21] Their results indicate that prior loading of any form reduces the strength of concrete on reloading. Strain at peak stress and the modulus on reloading follows the same trend as strength. This behavior can be attributed to the cumulative damage induced by repetitive loadings. Saito[69] investigated the microcracking phenomenon of concrete understatic and repeated tensile loads, and concluded that cumulative damage occurs in concrete due to reloading beyond the stage at which interfacial cracks are formed.

The effect of uniaxial impact in tension was investigated by Zielinski et al.[87] Their results indicated an increase in the tensile strength similar to the phenomenon generally observed under uniaxial impact in compression.

Indirect tensile strength

The most commonly used tests for estimating the indirect tensile strength of concrete are the splitting tension test (ASTM C496) and the third-point flexural loading test (ASTM C78).

(a)   Splitting tensile strength   As recommended by ACI Committee 363,[2]

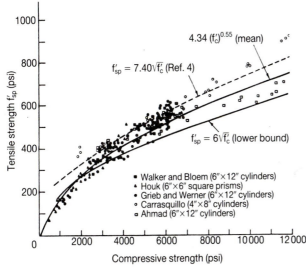

Fig. 2.21 Variation of splitting tensile strength of normal weight concrete with the compressive strength[4]

the splitting tensile strength (f_{ct}) for normal weight concrete can be estimated by

$$f_{ct} = 7.4 \sqrt{f_c'} \text{ psi} \quad 3000 \leqslant f_c' \leqslant 12{,}000 \text{ psi} \\ (21 \leqslant f_c' \leqslant 83 \text{ MPa})$$ (2.3)

In 1985, based on the available experimental data of split cylinder tests on concretes of low-, medium-,[32,37,81] and high strengths,[7,16,25] an empirical relationship was proposed by Ahmad and Shah[4] as

$$f_{ct} = 4.34 (f_c') \text{ psi} \quad f_c' \leqslant 12{,}000 \text{ psi} \\ (f_c' \leqslant 83 \text{ MPa})$$ (2.4)

Figure 2.21 shows the experimental data, with the predictions using the above equation and the recommendations of the ACI Committee 363. The latter appears to overestimate values of tensile strength. Recommendations of ACI Committee 363 were based on work performed at Cornell University.[16]

Figure 2.22 shows the aging effect on splitting tensile strength, which is similar to that under compressive loading. In an investigation by Ojdrovic on cracking modes,[60] it was concluded that at early ages, tensile strength of concrete is the property of the matrix which governs the cracking mode.

The effect of prior compressive loading on the split tensile strength was investigated by Liniers[46] and the results are shown in Fig. 2.23. From this figure, he concluded that limiting the compressive stresses to 60% of the strength is essential if only tolerable damage is to be accepted.

The tensile strength of condensed silica fume (CSF) concrete is related to the compressive strength in a manner similar to that of normal concrete.

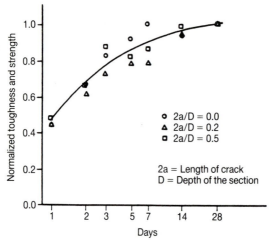

Fig. 2.22   Normalized splitting tensile strength as a function of the age at testing[60]

However if CSF concrete is exposed to drying after one day of curing in the mold, the tensile strength is reduced more than the control concrete.[28]

(b)   Flexural strength or modulus of rupture   Flexural strength or modulus of rupture is measured by a beam flexural test and is generally taken to be a more reliable indicator of the tensile strength of concrete. The modulus of rupture is also used as the flexural strength of concrete in pavement design. It is often assumed that flexural strength of concrete is about 15% of the compressive strength.

In the absence of actual test data, the modulus of rupture may be estimated by

$$f_r = k\sqrt{f_c'}$$
(2.5)

typically in the range of 7.5 to 12. For high strength concrete, the ACI Committee 363 *State-of-the-Art Report*[2] recommends a value of $k = 11.7$ as

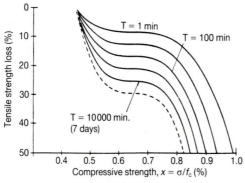

Fig. 2.23   Tensile strength loss as a function of compressive stress fraction for different duration of loading[46]

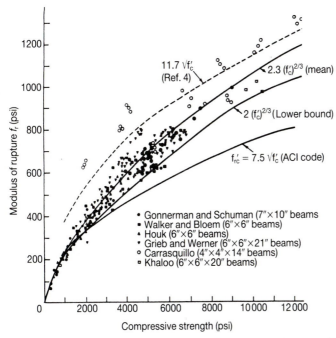

Fig. 2.24 Variation of modulus of rupture with the compressive strength[4]

appropriate for concrete with compressive strength in the range of 3000 psi to 12,000 psi (21 MPa to 83 MPa).

Based on the available data of beam flexural tests on concretes of low, medium[32,37,81] and high strengths,[7,16,25] an empirical equation to predict the flexural strength (modulus of rupture) was proposed[4] as

$$f_r = 2.30(f_c')^{2/3} \tag{2.6}$$

where f_c' is the compressive strength in psi.

The above equation is of the same form as proposed by Jerome,[39] which was developed on the basis of data for concretes of strengths up to 8000 psi (56 MPa). Figure 2.24 shows the plot of the experimental data and the proposed equation[4] for predicting the modulus of rupture of concretes with strengths up to 12,000 psi (83 MPa). Also shown in the figure is the expression recommended by Carrasquillo and Nilson.[16]

The results of uniaxial and biaxial flexural tests[86] indicated that the tensile strength was 38% higher in the uniaxial stress state than in the biaxial stress state.

Flexural strength is higher for moist-cured as compared to field cured specimens.[19] However, wet-cured specimens containing condensed silica fume (CSF) exhibit a lower ratio of tensile to compressive strength than dry-stored concrete specimens with silica fume.[47] For all concretes, allowing a moist cured beam to dry during testing will result in lowered measured strength, due to the addition of applied load and drying shrinkage stresses on the tensile face. The flexural strength of condensed silica fume (CSF) concrete is related to the compressive strength in a manner similar to that of concrete without silica fume; however, if CSF concrete is exposed to drying after only one day of curing in the mold, the flexural strength reduces more than the control concrete.[28]

2.3   Deformation

The deformation of concrete depends on short-term properties such as the static and dynamic modulus, as well as strain capacity. It is also affected by time dependent properties such as shrinkage and creep.

Static and dynamic elastic modulus

The modulus of elasticity is generally related to the compressive strength of concrete. This relationship depends on the aggregate type, the mix proportions, curing conditions, rate of loading and method of measurement. More information is available on the static modulus than on the dynamic modulus since the measurement of elastic modulus can be routinely performed whereas the measurement of dynamic modulus is relatively more complex.

Static modulus

The static modulus of elasticity can be expressed as secant, chord or tangent modulus. According to the ACI Building Code (ACI-318-89),[1] E_c, the static, secant modulus of elasticity, is defined as the ratio of the stress at 45% of the strength to the corresponding strain. Static, chord modulus of elasticity, as determined by ASTM C469, is defined as the ratio of the difference of the stress at 40% of the ultimate strength and the stress at 50 millionths strain to the difference in strain corresponding to the stress at 40% of ultimate strength and 50 millionths strain.

At present there are two empirical relationships that can be used for design when the static modulus of elasticity has not been determined by tests. They are the ACI Code formula[1]

$$E_c = 33\omega^{1.5}\sqrt{f_c'}\ \text{psi} \qquad (2.7)$$

where ω = unit weight in pounds per cubic foot (pcf) and the formula recommended by the ACI Committee 363 on High Strength Concrete[2] for concrete with unit weight of 145 pcf.

$$E_c = 1.0 \times 10^6 + 40,000\sqrt{f_c'}\ \text{psi} \qquad (2.8)$$

This formula is based on work performed at Cornell University.[16]

Figure 2.25 shows the range of scatter of data with the predictions of the ACI equation and the ACI Committee 363 equation. A third equation was

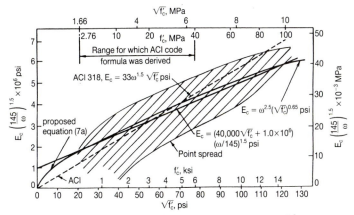

Fig. 2.25 Secant modulus of elasticity versus concrete strength[4]

recommended by Ahmad and Shah[4] which seems to be more representative of the trend of the data. The equation is

$$E_c = \omega^{2.5}(\sqrt{f_c'})^{0.65} \text{ psi} \qquad (2.9)$$

where ω = unit weight of concrete in pcf.

Figure 2.26 gives a comparison of experimental values of elastic modulus collected by Cook[22] with the predictions by the ACI 318-89 Code and the ACI Committee 363 equations. The concrete contained aggregates from South Carolina, Tennessee, Texas and Arizona. Aggregate sizes varied from 3/8 in. to 1 in. (10 to 25 mm) and consisted primarily of crushed limestones, granites and native gravels. Cook recommended the following equation which gives a better fit for the particular set of experimental data.

$$E_c = \omega^{2.5}(\sqrt{f_c'})^{0.315} \text{ psi} \qquad (2.10)$$

where ω = 151 pcf.

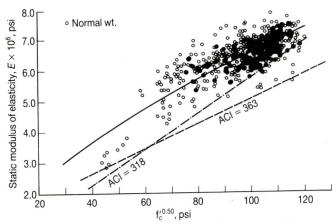

Fig. 2.26 Secant modulus of elasticity versus concrete strength for normal weight concrete[22]

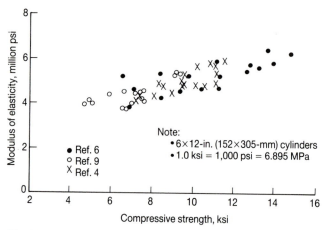

Fig. 2.27   Secant modulus of elasticity as a function of strength[29]

Figure 2.27 summarizes test results of modulus of elasticity as a function of compressive strength. These results confirm the increased stiffness at higher strengths. Modulus of elasticity of very high strength concrete up to 17,000 psi (117 MPa) is shown in Fig. 2.28. According to Moreno,[54] the results are generally closer to the predictions of the ACI Code (ACI 318-89) equation. However, at strength higher than 15,000 psi (105 MPa), the ACI Code equation overestimates the test results. Moreno also contends that ACI Committee 363 equation[2] always predicts results lower than the test data even for 17,000 psi (117 MPa) concrete, and hence it was concluded that the equation recommended by the ACI Committee 363 is more appropriate for higher-strength concrete.

In a recent study at NCSU[45] based on the results of 16 specimens with strengths varying between 8000 psi (55 MPa) at 28 days and 18,000 psi (124 MPa) at one year, it was concluded that ACI Committee 363[2] formula gave closer predictions of experimental results obtained from 6 × 12 in. (150 × 300 mm) cylinders.

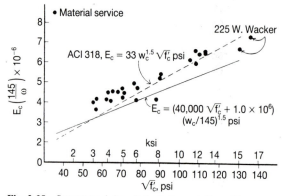

Fig. 2.28   Secant modulus of elasticity variation with square root of the compressive strength[54]

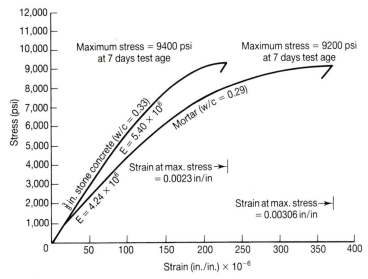

Fig. 2.29 Effect of coarse aggregate and mix proportions on the modulus of elasticity[22]

The modulus of elasticity of concrete is affected by the properties of the coarse aggregate. The higher the modulus of elasticity of the aggregate the higher the modulus of the resulting concrete. The shape of the coarse aggregate particles and their surface characteristics may also influence the value of the modulus of elasticity of concrete. Figure 2.29 shows the effect of the coarse aggregate type and the mix proportions on the modulus of elasticity. From this figure it can be concluded that, in general, the larger the amount of coarse aggregate with a high elastic modulus, the higher would be the modulus of elasticity of concrete. The use of four different types of coarse aggregates in a very high strength concrete mixture (w/c = 0.27) showed that elastic modulus was significantly influenced by the mineralogical characteristics of the aggregates.[12] Limestone and crushed aggregates from fine-grained diabase gave higher modulus than a smooth river gravel and crushed granite that contained inclusions of a soft mineral.

It is generally accepted that regardless of the mix proportions or curing age, concrete specimens tested in wet conditions show about 15% higher elastic modulus than the corresponding specimens tested in dry conditions.[53] This is attributed to the effect of drying on the transition zone. Because of drying, there is microcracking in the transition zone due to shrinkage, which reduces the modulus of elasticity.

As strain rate is increased, the measured modulus of elasticity increases. Based on the available experimental data for concrete with strength up to 7000 psi (48 MPa),[8,14,15,26,38,49,83] the following empirical equation was proposed by Ahmad and Shah[4] for estimating the modulus of elasticity under very high strain rates.

$$(E_c)_{\dot{\varepsilon}} = E_c \left[0.96 + 0.038 \frac{\log \varepsilon}{\log \varepsilon_s} \right] \qquad (2.11)$$

where $E_c = 27.5\omega^{1.5}\sqrt{f_c'}$, $\dot{\varepsilon}$ is the strain rate in microstrains per second ($\mu\varepsilon$/sec), $\varepsilon_s = 32\ \mu\varepsilon$/sec.

A recent paper[43] has suggested that if internal strains are measured by means of embeddable strain gauges, the measured modulus is 50% higher than that from strain measurements made on the surface. The author concluded that the reason for this observation is the non-uniform strain field across the section of the cylinders.

Dynamic modulus

The measurement of dynamic modulus corresponds to a very small instantaneous strain. Therefore the dynamic modulus is approximately equal to the initial tangent modulus. Dynamic modulus is appreciably higher than the static (secant) modulus. The difference between the two moduli is due in part to the fact that heterogeneity of concrete affects the two moduli in different ways. For low, medium and high strength concretes, the dynamic modulus is generally 40%, 30% and 20% respectively higher than the static modulus of elasticity.[53]

Popovics[66] has suggested that for both lightweight and normal weight concretes, the relation between the static and dynamic moduli is a function of density of concrete, just as is the case with relation between the static modulus and strength.[66] Popovics expressed E_c as a linear function of $E_d^{1.4}/\rho$ where ρ is the density of concrete, and E_d is the dynamic modulus.

The ratio of static to dynamic modulus is also affected by the age at testing as shown by Philleo[65] in Fig. 2.30. The figure indicates that at early ages (up to 6 months) the ratio of the two moduli increases from 0.4 to about 0.8 and becomes essentially constant thereafter.

A typical relationship between the dynamic modulus determined by the vibration of the cylinders and their compressive strength is shown in Fig. 2.31. It has been reported by Sharma and Gupta[77] that the relationship between the strength and the dynamic modulus is unaffected by air entrainment, method of curing, condition at test, or type of cement.

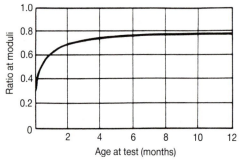

Fig. 2.30   Ratio of static and dynamic modulus of elasticity of concrete at different ages[65]

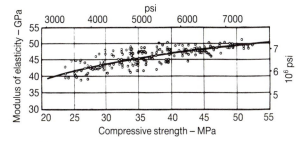

Fig. 2.31   Relation between the dynamic modulus of elasticity, determined by transverse vibration of cylinders, and their compressive strength[77]

It should also be noted that no information is available regarding the relationship between the static and dynamic modulus of elasticity for concrete with strength in excess of 8000 psi (55 MPa).

2.4   Strain capacity

The usable strain capacity of concrete can be measured either in compression or in tension. In the compression mode, it can be measured by either concentric or eccentric compression testing. In the tensile mode, the strain capacity can be either for direct tension or indirect tension. The behavior under multiaxial stress states if outside the scope of this chapter, and only the behavior under uniaxial stress condition will be discussed.

Stress-strain behavior in compression

The stress-strain behavior is dependent on a number of parameters which include material variables such as aggregate type and testing variables such as age at testing, loading rate, strain gradient and others noted above.

The effect of the aggregate type of the stress-strain curve is shown in Fig. 2.32 which indicates that higher strength and corresponding strain are achieved for crushed aggregate from fine-grained diabase and limestone, as compared to concretes made from smooth river gravel and from crushed granite that contained inclusions of a soft mineral.

A number of investigations[5,35,41,58,75,76,82,84] have been undertaken to obtain the complete stress-strain curves in compression. Axial stress-strain curves for concretes with compressive strengths up to 14,000 psi (98 MPa) concrete as obtained by different researchers are shown in Fig. 2.33.

It is generally recognized that for concrete of higher strength, the shape of the ascending part of the curve becomes more linear and steeper, the strain at maximum stress is slightly higher, and the slope of the descending part becomes steeper. The existence of the postpeak descending part of the stress-strain curve has been the focus of a recent paper.[79] It was concluded that the postpeak behavior can be quantified for inclusion in finite element

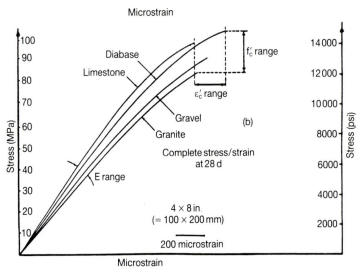

Fig. 2.32 Effect of the aggregate type on the ascending portion of the stress-strain curves of concrete at 28-days[12]

analysis and that it can have considerable influence on the predicted structural behavior and strength.[67]

To obtain the descending part of the stress-strain curve, it is necessary to avoid specimen-testing machine interaction. One approach is to use a closed-loop system with a constant rate of axial strain as a feedback signal for closed-loop operation. The difficulties of obtaining the postpeak behavior experimentally and methods of overcoming these difficulties are

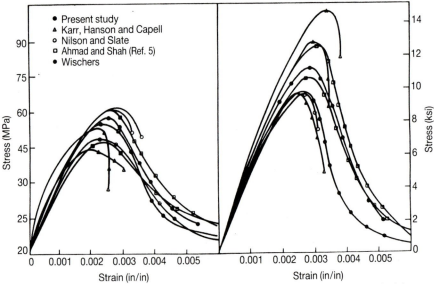

Fig. 2.33 Different stress strain curves reported for high strength concrete under uniaxial compression[4]

described in a study by Ahmad and Shah.[9] For very high strength concretes it may be necessary to use the lateral strains as a feedback signal rather than the axial strains.[74] In a paper by Kotsovos,[44] it is argued that a more realistic description of the postpeak specimen behavior may be a complete and immediate loss of load-carrying capacity as soon as the peak load is exceeded. A different point of view is reflected in another recent paper[79] which suggests that there is usable strength for concrete after peak stress. Based on the above mentioned experimental investigations, different analytical representations for the stress-strain curve have been proposed. They include use of a fractional equation,[6,70,82] or a combined power and exponential equation[75] and serpentine curve. The fractional equation is a comprehensive, yet simple way of characterizing the stress-strain response of concrete in compression.[4] The fractional equation can be written as

$$f_\varepsilon = (f_c')\frac{A(\varepsilon/\varepsilon_c') + (B-1)(\varepsilon/\varepsilon_c')^2}{1 + (A-2)(\varepsilon/\varepsilon_c') + B(\varepsilon/\varepsilon_c')} \tag{2.12}$$

(for $f > 0.1\varepsilon, f_c'$, when $\varepsilon > \varepsilon_c'$)

where f_ε is the compressive stress at strain ε, f_c' and ε_c' the maximum stress and corresponding strain,
A and B are parameters which determine the shape of the curve.

The values of the parameters A and B, which control the shape of the ascending and the descending parts, respectively, may be estimated by

$$A = E_c\frac{\varepsilon_c'}{f_c'} \tag{2.13}$$

$$B = 0.88087 - 0.57 \times 10^{-4}(f_c') \tag{2.14}$$

$$\varepsilon_c' = 0.001648 + 1.14 \times 10^{-7}(f_c') \tag{2.15}$$

$$E_c = 27.55\omega^{1.5}\sqrt{f_c'} \tag{2.16}$$

where f_c' is the compressive strength in psi and ω is the unit weight in pcf.

The parameters A, B, ε_c' and E_c are as recommended by Ahmad and Shah[4] and were determined from the statistical analysis of the experimental results on 3×6 in. $(75 \times 152$ mm) concrete cylinders.[5,6] These cylinders were tested under strain controlled conditions in a closed-loop testing machine and had compressive strengths ranging from 3000 to 11,000 psi (20 to 75 MPa).

Stress-strain behavior in tension

The direct tensile stress-strain curve is difficult to obtain. Due to difficulties in testing concrete in direct tension, only limited and often conflicting data are available.

Direct tensile tests were carried out on tapered cylindrical specimens of 4.7 in. diameter and 11.8 in. length (120 mm diameter × 300 mm).[24] For

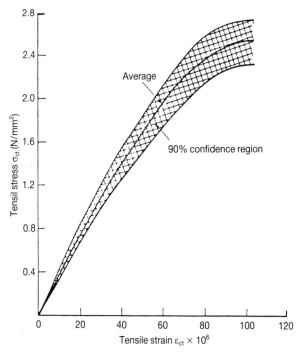

Fig. 2.34 A typical stress strain curve for concrete under uniaxial tension determined under static load controlled tests[24]

the application of the load, steel platens were glued to the top and bottom of the specimens. In order to provide plane-parallel and axial connection of these platens, a special gluing press was designed. Some 500 direct tensile tests, 300 compressive and 300 splitting tests were performed. A typical stress-strain curve with a 95% confidence region for concrete subjected to direct tension is shown in Fig. 2.34. The stress-strain curve shown in the figure is for dry specimens. The results may vary slightly for specimens tested in moist conditions.

A study at Northwestern by Gopalaratham and Shah[31] points out that due to the localized nature of the post-cracking deformations intension, no unique tensile stress-strain relationship exists. According to this study, the uniaxial tensile strength can be estimated by $\sqrt{f_c'}$, and the tangent modulus of elasticity is identical in tension and compression. The stress-strain relationship in tension before peak is less nonlinear than in compression.

Laser speckle interferometry was employed in a recent study,[13] to investigate the behavior of concrete subjected to uniaxial tension. Unique post-peak stress-strain and stress-deformation behavior were not observed. The stress-strain response of concrete was found to be sensitive to gauge-length. Strains measured within a gauge length inside the microcracking zone were two orders of magnitude higher than values previously reported.[27]

In a recent study,[20] it was shown that while the use of strain gauges would lead to non-objective constitutive stress-strain relations, interferometric measurements on notched specimens allow an indirect determination of the local stress-strain and stress-separation (deformations) relations. Guo and Zhang[33] tested 29 specimens in direct tension and obtained complete stress-deformation curves. Based on the experimental results an equation was also derived for the stress-displacement curves.

Flexural tension

While the information on the stress-strain behavior in tension is severely limited, virtually no data are available regarding the strain capacity in flexural tension. This is an area for which research is sorely needed to provide a basis for design where flexural cracking is an important consideration.

2.5 Poisson's ratio

Poisson's ratio under uniaxial loading conditions is defined as the ratio of lateral strain to strain in the direction of loading. In the inelastic range, due to volume dilation resulting from internal microcracking, the apparent Poisson's ratio is not constant but is an increasing function of the axial strain.

Experimental data on the values of Poisson's ratio for high strength concrete is very limited.[16,64] Based on the available experimental information, Poisson's ratio of higher strength concrete in the elastic range appears comparable to the expected range of values for lower-strength concrete. In the inelastic range, the relative increase in lateral strains is less for higher-strength concrete compared to concrete of lower strength.[6] That is, higher-strength concrete exhibits less volume dilation than lower-strength concrete (see Fig. 2.35). This implies less internal microcracking for concrete of higher strength.[17] The lower relative expansion during the

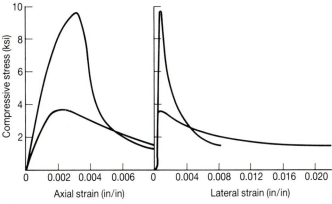

Fig. 2.35   Axial stress versus axial strain and lateral strain for normal and high-strength concrete[4]

inelastic range may mean that the effects of triaxial stresses will be proportionately different for higher-strength concrete. For example, the effectiveness of hoop confinement is reported to be less for higher-strength concrete.[6]

Information on Poisson's ratio of concrete with strength greater than 12,000 psi (83 MPa) is not available in the literature.

Acknowledgements

This work was supported by the Strategic Highway Research Program (SHRP) at North Carolina State University, Raleigh, NC.

References

1   Committee 318 (1989) *Building code requirements for reinforced concrete.* American Concrete Institute.
2   Committee 363 (1984) State-of-the-art report on high strength concrete. *ACI Journal*, **81**, 4, July–Aug, 364–411.
3   Committee 363 (1987) Research needs for high-strength concrete. *ACI Materials Journal*, November–December, 559–61.
4   Ahmad, S.H. and Shah, S.P. (1985) Structural properties of high strength concrete and its implications for precast prestressed concrete. *PCI Journal*, Nov–Dec, 91–119.
5   Ahmad, S.H. (1981) *Properties of confined concrete subjected to static and dynamic loading.* Ph.D. Thesis, University of Illinois, Chicago, March.
6   Ahmad, S.H. and Shah, S.P. (1982) Stress-strain curves of concrete confined by spiral reinforcement. *ACI Journal*, **79**, 6, Nov–Dec, 484–90.
7   Ahmad, S.H. (1982) *Optimization of mix design for high strength concrete.* Report No. CE 001-82, Department of Civil Engineering, North Carolina State University, Raleigh, NC.
8   Ahmad, S.H. and Shah, S.P. (1985) Behavior of hoop confined concrete under high strain rates. *ACI Journal*, **82**, 5, Sept–Oct, 634–47.
9   Ahmad, S.H. and Shah, S.P. (1979) Complete stress-strain curves of concrete and nonlinear design. Progress Report PFR 79-22878 to the National Science Foundation, University of Illinois, Chicago, August and also *Non-linear design of concrete structures.* University of Waterloo Press, 1980, 222–30.
10 Aitcin, P.-C. (1983) *Condensed silica fume.* University of Sherbrooke, Sherbrooke, Quebec.
11 Aitcin, P.-C., Sarkar, S.L. and Laplante, P. (1990) Long-term characteristics of a very high strength concrete. *Concrete International*, Jan, 40–4.
12 Aitcin, P.-C. and Metha, P.K. (1990) Effect of coarse-aggregate characteristics on mechanical properties of high-strength concrete. *ACI Materials Journal*, March–April, 103–107.
13 Ansari, F. (1987) Stress-strain response to microcracked concrete in direct tension. *ACI Materials Journal*, **84**, 6, Nov–Dec, 481–90.
14 Atchley, B.L. and Furr, H.L. (1967) Strength and energy absorption capabilities of plain concrete under dynamic and static loadings. *ACI Journal*, **64**, 11, Nov, 745–56.
15 Bresler, B. and Bertero, V.V. (1975) Influences of high strain rate and cyclic loading on behavior of unconfined and confined concrete in compression. *Proceedings, Second Canadian Conference on Earthquake Engineering*, MacMaster University, Hamilton, Ontario, Canada, June.

16 Carrasquillo, R.L., Nilson, A.H. and Slate, F.O. (1981) Properities of high strength concrete subject to short-term loads. *ACI Journal*, May–June, 171–8.

17 Carrasquillo, R.L., Slate, F.O. and Nilson, A.H. (1981) Microcracking and behaviour of high strength concrete subjected to short term loading. *ACI Journal*, **78**, 3, May–June, 179–86.

18 Carrasquillo, P.M. and Carrasquillo, R.L. (1988) Effect of using unbonded capping systems on the compressive strength of concrete cylinders. *ACI Materials Journal*, May–June, 141–7.

19 Carrasquillo, P.M. and Carrasquillo, R.L. (1988) Evaluation of the use of current concrete practice in the production of high-strength concrete. *ACI Materials Journal*, **85**, 1, Jan–Feb, 49–54.

20 Cedolin, L., Poli, S.D. and Iori, I. (1987) Tensile behavior of concrete. *Journal of Engineering Mechanics*, ASCE, **113**, 3, March, 431–49.

21 Cook, D.J. and Chindaprasirt, P. (1981) Influence of loading history upon the tensile properties of concrete. *Magazine of Concrete Research*, **33**, 116, Sept, 154–60.

22 Cook, J.E. (1989) Research and application of high-strength concrete: 10,000 PSI concrete. *Concrete International*, Oct, 67–75.

23 Cornelissen, H.A.W. (1984) Fatigue failure of concrete in tension. *Heron*, **29**, 4, 1–68.

24 Cornelissen, H.A.W. and Reinhardt, H.W. (1984) Uniaxial tensile fatigue failure of concrete under constant-amplitude and programme loading. *Magazine of Concrete Research*, **36**, 129, Dec, 216–26.

25 Dewar, J.D. (1964) The indirect tensile strength of concrete of high compressive strength. *Technical Report*, No. 42.377. Cement and Concrete Association, Wexham Springs, England, March.

26 Dilger, W.H., Koch, R. and Andowalczyk, R. (1984) Ductility of plain and confined concrete under different strain rates. *ACI Journal*, **81**, 1, Jan–Feb, 73–81.

27 Evans, R.H. and Marathe, M.S. (1968) Microcracking and stress-strain curves for concrete in tension. *Materials and Structures, Research and Testing*, (RILEM, Paris), **1**, 1, Jan–Feb, 61–4.

28 FIP Commission on Concrete (1988) *Condensed Silica Fume in Concrete*. State of Art Report, Federation Internationale de la Precontrainte, London.

29 Fiorato, A.E. (1989) PCA research on high-strength concrete. *Concrete International*, April, 44–50.

30 Freedman, S. (1970/71) High strength concrete. *Mocern Concrete*, **34**, 6, Oct 1970, 29–36; 7, Nov 1970, 28–32; 8, Dec 1970, 21–24; 9, Jan 1971, 15–22; and 10, Feb 1971, 16–23.

31 Gopalaratham, V.S. and Shah, S.P. (1985) Softening response of concrete in direct tension. *Research Report*, Technological Institute, Northwestern University, Chicago, June 1984 also *ACI Journal*, **82**, 3, May–June 1985, 310–23.

32 Grieb, W.E. and Werner, G. (1962) Comparison of splitting tensile strength of concrete with flexural and compressive strengths. *Public roads*, **32**, 5, Dec.

33 Guo, Z-H. and Zhang, X-Q. (1987) Investigation of complete stress-deformation curves for concrete in tension. *ACI Materials Journal*, **84**, 4, July–Aug, 278–85.

34 Hatano, T. (1960) Dynamic behavior of concrete under impulsive tensile load. *Technical Report*, No. C-6002, Central Research Institute of Electric Power Industry, Tokyo, 1–15.

35 Helland, S. *et al.* (1983) Hoyfast betong. Presented at Norsk Betongdag, Trondheim, Oct. (In Norwegian).

36 (1977) *High-strength concrete in Chicago, high-rise buildings*. Task Force Report No. 5, Chicago Committee on High-Rise Buildings, Feb.

37 Houk, H. (1965) Concrete aggregates and concrete properties investigations. *Design Memorandum*, No. 16, Dworshak Dam and Reservoir, U.S. Army Engineer District, Walla, WA.

38 Hughes, B.P. and Gregory, R. (1972) Concrete subjected to high rates of loading and compression. *Magazine of concrete Research*, **24**, 78, London, March.

39 Jerome, M.R. (1984) Tensile strength of concrete. *ACI Journal*, **81**, 2, March–April, 158–65.

40 Johansen, R. (1979) Silicastov Iabrikksbetong. Langtidseffekter. *Report*, STF65 F79019, FCB/SINTEF, Norwegian Institute of Technology, Trondheim. (In Norwegian), and Johansen R, (1981) Report 6: Long-term effects. *Report*, STF65 A81031, FCB/SINTEF, Norwegian Institute of Technology, Trondheim, 1981.

41 Kaar, P.H., Hanson, N.W. and Capell, H.T. (1977) Stress-strain characteristics of high strength concrete. *Research and Development Bulletin*, RD051-01D, Portland Cement Association, Skokie, Illinois, 11 pp. also *Douglas McHenry International Symposium on Concrete and Concrete Structures*, ACI special publication, SP-55, Detroit 1978, 161–85.

42 Klieger, P. (1958) Effect of mixing and curing temperatures on concrete strength. *ACI Journal*, **54**, 12, June, 1063–81.

43 Klink, S.A. (1985) Actual elastic modulus of concrete. *ACI Journal*, Sept–Oct, 630–3.

44 Kotsovos, M.D. (1983) Effect of testing techniques on the post-ultimate behaviour of concrete in compression. *Materiaux et Constructions* (RILEM, Paris), Jan–Feb, **16**, 91, 3–12.

45 Leming, M.L. (1988) Properties of high strength concrete: an investigation of high strength concrete characteristics using materials in North Carolina. Research Report FHWA/NC/88-006, Department of Civil Engineering, North Carolina State University, Raleigh, N.C., July.

46 Liniers, A.D. (1987) Microcracking of concrete under compression and its influence on tensile strength. *Materiaux et Constructions* (RILEM, Paris), **20**, 116, Mar, 111–16.

47 Loland, K.E. and Gjørv, O.E. (1981) Silikabetong. *Nordisk Betong*, 6, 1–6 (In Norwegian).

48 Maage, M. and Hammer, T.A. (1985) Modifisert Portlandsement. Detrapport 3. Fasthetsutvikling Og E-Modul. *Report*, STF65 A85041, FCB/SINTEF, Norwegian Institute of Technology, Trondheim (In Norwegian).

49 Mainstone, R.J. (1975) Properties of materials at high rates of straining or loading. *Materieaux et Constructions* (Rilem, Paris), **8**, 44, March–April.

50 Malhotra, H.L. (1956) The effect of temperature on the compressive strength of concrete. *Magazine of Concrete Research*, V. 8, No. 23, Aug, 85–94.

51 Malhotra, V.M. (ed.) (1983) *Fly ash, silica fume, slag and other mineral by-products in concrete*, SP 79, Vol. II, American Concrete Institute.

52 Malhotra, V.M. (ed.) (1986) *Fly ash, silica fume, slag and natural pozzolans in concrete*, SP 91, Vol. II, American Concrete Institute.

53 Metha, P.K. (1986) *Concrete structures, properties and materials*. Prentice Hall, Inc, Englewood Cliffs, New Jersey.

54 Moreno, J. (1990) The state of the art of high-strength concrete in Chicago: 225 W. Wacker Drive. *Concrete International*, Jan, 35–9.

55 Nawy, E.G. *Reinforced concrete: a fundamental approach*, Second edition. Prentice Hall, New Jersey.

56 Neville, A.M. (1981) *Properties of concrete*, Third edition. Pitman Publishing Ltd., London.

57 Ngab, A.S., Nilson, A.H. and Slate, F.O. (1981) Shrinkage and creep of high strength concrete. *ACI Journal*, **78**, 4, July–Aug, 255–61.

58 Nilson, A.H. and Slate, F.O. (1979) Structural design properties of very high strength concrete. *Second Progress Report*, NSF Grant ENG 7805124, School of Civil and Environmental Engineering, Cornell University, Ithaca, New York.

59 Oh, B.H. (1987) Behavior of concrete under dynamic tensile loads. *ACI Materials Journal*, Jan–Feb, 8–13.

60 Ojorv, O.E. (1988) High strength concrete. *Nordic Betong*, **32**, 1, Stockholm, 5–9.

61 Parrot, L.J. (1969) The properties of high-strength concrete. Technical Report No. 42.417, *Cement and Concrete Association*, Wexham Springs, 12 pp.

62 Pentalla, V. (1987) Mechanical properties of high strength concretes based on different binder compositions. *Proceedings Symposium on Utilization of High Strength Concrete*, Norway, June 15–18, 123–34.

63 Perenchio, W.F. (1973) An evaluation of some of the factors involved in producing very high-strength concrete. *Research and Development Bulletin*, No. RD014.01T, Portland Cement Association, Skokie.

64 Perenchio, W.F. and Klieger, P. (1978) Some physical properties of high strength concrete. *Research and Development Bulletin*, No. RD056.01T, Portland Cement Association, Skokie, IL.

65 Philleo, R.E. (1955) Comparison of results of three methods for determining Young's Modulus of elasticity of concrete. *ACI Journal*, **51**, Jan, 461–9.

66 Popovics, S. (1975) Verification of relationships between mechanical properties of concrete-like materials. *Materials and Structures* (RILEM, Paris), **8**, 45, May–June, 183–91.

67 Pramono, E. (1988) Numerical simulation of distributed and localized failure in concrete. *Structural Research Series*, No. 88-07, C.E.A.E. Department, University of Colorado, Boulder.

68 Ronne, M. (1987) Effect of condensed silica fume and fly ash on compressive strength development of concrete. *American Concrete Institute, SP-114-8*, 175–89.

69 Saito, M. (1987) Characteristics of microcracking in concrete under static and repeated tensile loading. *Cement and Concrete Research*, **17**, 2, March, 211–18.

70 Sargin, M. (1971) *Stress-strain curves relationships for concrete and analysis of structural concrete sections*. Study No. 4, Solid Mechanics Division, University of Waterloo, Ontario, Canada.

71 Saucier, K.L. (1984) High strength concrete for peacekeeper facilities. *Final Report*, Structures Laboratory, U.S. Army Engineer Waterways Experimental Station, March.

72 Saucier, K.L., Tynes, W.O. and Smith, E.F. (1965) High compressive strength concrete-report 3, Summary Report. *Miscellaneous Paper No. 6-520, U.S. Army Engineer Waterways Experiment Station*, Vicksberg, Sept.

73 Sellevold, E.J. and Radjy, F.F. (1983) *Condensed silica fume (microsilica) in concrete: water demand and strength development*. Publication SP-79, American Concrete Institute, Vol. II, 677–94.

74 Shah, S.P., Gokos, U.N. and Ansari, F. (1981) An experimental technique for obtaining complete stress-strain curves for high strength concrete. *Cement, Concrete and Aggregates*, CCAGDP, **3**, Summer.

75 Shah, S.P., Fafitis, A. and Arnold, R. (1983) Cyclic loading of spirally reinforced concrete. *Journal of Structural Engineering*, ASCE, **109**, ST7, July, 1695–710.

76 Shah, S.P. and Sankan, R. (1987) Internal cracking and strain suffering response of concrete under uniaxial compression. *ACI Materials Journal*, **84**, May–June, 200–12.

77 Sharma, M.R. and Gupta, B.L. Sonic modulus as related to strength and static

modulus of high strength concrete. *Indian Concrete Journal*, **34**, 4, 139–41.

78   Skalny, J. and Roberts, L.R. (1987) High-strength concrete. *Ann. Rev. Mater. Sci.*, **17**, 35–56.

79   Smith, S.S., William, K.J., Gerstle, K.H. and Sture, S. (1989) Concrete over the top, or: is there life after peak? *ACI Materials Journal*, **86**, 5, Sept–Oct, 491–7.

80   Suaris, W. and Shah, S.P. (1983) Properties of concrete subjected to impact. *Journal of Structural Engineering*, ASCE, **109**, 7, July, 1727–41.

81   Walker, S. and Bloem, D.L. (1960) Effects of aggregate size on properties of concrete. *ACI Journal*, **32**, 3, September, 283–98.

82   Wang, P.T., Shah, S.P. and Naaman, A.E. (1978) Stress-strain of normal and lightweight concrete in compression. *ACI Journal*, **75**, 11, November, 603–11.

83   Watstein, D. Effect of straining rate on the compressive strength and elastic properties of concrete. *ACI Journal*, **49**, 8, April, 729–44.

84   Wischers, G. (1979) Application and effects on compressive loads on concrete. Betontechnische Berichte, 1978, Betonverlag Gmbh, Dusseldorf, 31–56.

85   Yogenendran, V., Langan, B.W. and Ward, M.A. (1987) Utilization of silica fume in high strength concrete. *ACI Materials Journal*, **84**, 2, March–April, 85–97, also *Proceedings Symposium on Utilization of High Strength Concrete*, Norway, June 15–18, 85–97.

86   Zielinski, Z.A. and Spiropoulos, I. (1983) An experimental study on the uniaxial and biaxial flexural tensile strength of concrete. *Canadian Journal of Civil Engineering*, **10**, 104–15.

87   Zielinski, A.J., Reinhardt, H.W. and Kormeling, H.A. (1981) Experiments on concrete under uniaxial impact tensile loading. *Materiaux et Constructions* (RILEM, Paris), **14**, 80, Mar–Apr, 103–12.

88   Zielinski, A.J., Reinhardt, H.W. and Kormeling, H.A. (1981) Experiments on concrete under repeated uniaxial impact tensile loading. *Materiaux et Constructions* (RILEM, Paris), **14**, 81, May–June, 163–9.

3 Shrinkage creep and thermal properties

F de Larrard, P Acker and R Le Roy

3.1 Introduction

Shrinkage and creep are time-dependent deformations that, along with cracking, provide the greatest concerns for the designers because of uncertainty associated with their prediction. Concrete exhibits elastic deformations only under loads of short duration and, due to additional deformation with time, the effective behavior is that of an inelastic and time-dependent material. A quantitative knowledge of mechanical behavior, including delayed deformations and thermal effects, is necessary for a number of structures: bridges, buildings etc. In other cases, control of short- and long-term cracking requires an accurate modelling of strains and stresses at all ages of the structure.

High performance concretes (HPC) are concretes with attributes (including creep, shrinkage and thermal effects) that are superior to conventional concretes. It is generally recognized that the higher the compressive strength of concrete, the better the other attributes. For the purpose of this chapter, high strength concrete (HSC) is defined as concrete with 28 day compressive strength in excess of 6000 psi (42 MPa). With the increasing use of concretes of higher strengths, the knowledge of time dependent behavior and the thermal properties is becoming important to ascertain the long term behavior of structures utilizing high strength concrete.

In this chapter, the mechanisms of shrinkage and creep are presented, along with experimental data which enunciate the effects of different parameters on the time dependent properties of concrete. The thermal properties are summarized. Also some field case histories are presented in which concretes to meet the higher performance requirements with special attention to shrinkage, creep and strength characteristics were designed and used. In these field cases, thermal effects and long-term strains have been monitored, and compared with predictions based on laboratory data.

3.2   Shrinkage

Shrinkage is the decrease of concrete volume with time. This decrease is due to changes in the moisture content of the concrete and physio chemical changes, which occur without stress attributable to actions external to the concrete. Swelling is the increase of concrete volume with time. Shrinkage and swelling are usually expressed as a dimensionless strain (in./in. or mm/mm) under given conditions of relative humidity and temperature. Shrinkage is primarily a function of the paste, but is significantly influenced by the stiffness of the coarse aggregate. The interdependence of many factors creates difficulty in isolating causes and effectively predicting shrinkage without extensive testing. The principal variables that affect shrinkage are summarized in Table 3.1.[1] The key factors affecting the magnitude of shrinkage are:

Aggregate

The aggregate acts to restrain the shrinkage of cement paste; hence concrete with higher aggregate content exhibits smaller shrinkage. In addition, concrete with aggregates of higher modulus of elasticity or of rougher surfaces is more resistant to the shrinkage process.

Water-cementitious material ratio

The higher the w/c ratio is, the higher the shrinkage. This occurs due to two interrelated effects. As w/c increases, paste strength and stiffness decrease; and as water content increases, shrinkage potential increases.

Member size

Both the rate and the total magnitude of shrinkage decrease with an increase in the volume of the concrete member. However, the duration of shrinkage is longer for large members since more time is needed for shrinkage effects to reach the interior regions.

Medium ambient conditions

The relative humidity greatly affects the magnitude of shrinkage; the rate of shrinkage is lower at higher values of relative humidity. Shrinkage becomes stabilized at low temperatures.

Table 3.1 Factors affecting concrete creep and shrinkage and variables considered in the recommended prediction method[1]

	Factors	Variables Considered	Standard Conditions
Concrete (Creep & Shrinkage) — Concrete Composition	Cement Paste Content	Type of cement	Type I and III
	Water-Cement Ratio	Slump	2.7 in. (70 mm)
	Mix Proportions	Air Content	≤ 6 percent
	Aggregate Characteristics	Fine Aggregate Percentage	50 percent
	Degree of Compaction	Cement Content	470 to 752 lb/cu.yd (279 to 446 kg/m³)
Initial Curing	Length of Initial Curing	Moist Cured	7 days
		Steam Cured	1–3 days
	Curing Temperature	Moist Cured	73.4 ± 4°F (23 ± 2°C)
		Steam Cured	≤ 212°F, (≤ 100°C)
	Curing Humidity	Relative Humidity	≥ 95 percent
Member Geometry & Environment (Creep & Shrinkage) — Environment	Concrete Temperature	Concrete Temperature	73.4 ÷ 4 °F, (23 ÷ 2 °C)
	Concrete Water Content	Ambient Relative Humidity	40%
Geometry Size and Shape	Volume-Surface Ratio, (V/s) or Minimum Thickness		v/s = 1.5 in (v/s = 38 mm) 6in. (150 mm)
Loading (Only Creep) — Loading History	Concrete Age at Load Application	Moist Cured	7 days
		Steam Cured	1–3 days
	Duration of Loading Period	Sustained Load	Sustained Load
	Duration of Unloading Period Number of Load Cycles	—	—
Stress Conditions	Type of Stress and Distribution Across the Section	Compressive Stress	Axial Compression
	Stress/Strength Ratio	Stress/Strength Ratio	≥0.50

Admixtures

Admixture effect varies from admixture to admixture. Any material which substantially changes the pore structure of the paste will affect the shrinkage characteristics of the concrete. In general, as pore refinement is enhanced shrinkage is increased.

Pozzolans typically increase the drying shrinkage, due to several factors. With adequate curing, pozzolans generally increase pore refinement. Use of a pozzolan results in an increase in the relative paste volume due to two mechanisms; pozzolans have a lower specific gravity than portland cement and, in practice, more slowly reacting pozzolans (such as Class F fly ash) are frequently added at better than one-to-one replacement factor, in order to attain specified strength at 28 days. Additionally, since pozzolans such as fly ash and slag do not contribute significantly to early strength, pastes containing pozzolans generally have a lower stiffness at earlier ages as well, making them more susceptible to increased shrinkage under standard testing conditions. Silica fume will contribute to strength at an earlier age than other pozzolans but may still increase shrinkage due to pore refinement.

Chemical admixtures will tend to increase shrinkage unless they are used in such a way as to reduce the evaporate water content of the mix, in which case the shrinkage will be reduced. Calcium chloride, used to accelerate the hardening and setting of concrete, increases the shrinkage. Air-entraining agents, however, seem to have little effect.

Cement type

The effects of cement type are generally negligible except as rate-of-strength-gain changes. Even here the interdependence of several factors make it difficult to isolate causes. Rapid hardening cement gains strength more rapidly than ordinary cement but shrinks somewhat more than other types, primarily due to an increase in the water demand with increasing fineness. Shrinkage compensating cements can be used to minimize or eliminate shrinkage cracking if they are used with restraining reinforcement.

Carbonation

Carbonation shrinkage is caused by the reaction between carbon dioxide (CO_2) present in the atmosphere and calcium hydroxide ($CaOH_2$) present in the cement paste. The amount of combined shrinkage varies according to the sequence of occurrence of carbonation and drying process. If both phenomena take place simultaneously, less shrinkage develops. The process of carbonation, however, is dramatically reduced at relative humidities below 50%.

The effect of the aggregate content and the w/c ratio on the shrinkage

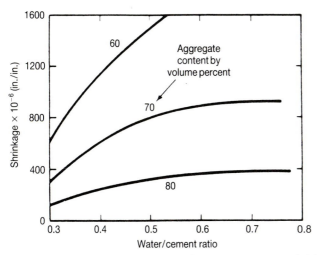

Fig. 3.1 Effect of w/c materials ratio and aggregate cement on shrinkage[23]

deformations is shown in Fig. 3.1. The figure reinforces the generally recognized fact that shrinkage deformations decrease with a higher aggregate content and a lower w/c ratio.

The shrinkage properties of concretes with higher compressive strengths are summarized in an ACI State-of-the-Art Report.[2] The basic conclusions were: (1) Shrinkage is unaffected by the w/c ratio[3] but is approximately proportional to the percentage of water by volume in concrete, (2) Laboratory[4] and field studies[5,6] have shown that shrinkage of higher strength concrete is similar to that of lower-strength concrete, (3) Shrinkage of high strength concrete containing high range water reducers is less than for lower strength concrete, (4) Higher strength concrete exhibits relatively higher initial rate of shrinkage,[7,8] but after drying for 180 days, there is little difference between the shrinkage of higher strength concrete and lower strength concrete made with dolomite or limestone. Shrinkage of high performance concrete may be expected to differ from conventional concrete in three broad areas: plastic shrinkage, autogenous shrinkage, and drying shrinkage.

Plastic shrinkage occurs during the first few days after fresh concrete is placed. During this period, moisture may evaporate faster from the concrete surface than it is replaced by bleed water from lower layers of the concrete mass. Paste-rich mixes, such as high performance concretes, will be more susceptible to plastic shrinkage than conventional concretes. Drying shrinkage occurs after the concrete has already attained its final set and a good portion of the chemical hydration process in the cement gel has been accomplished. Drying shrinkage of high strength concretes, although perhaps potentially larger due to higher paste volumes, do not, in fact, appear to be appreciably larger than conventional concretes. This is probably due to the increase in stiffness of the stronger mixes. Data for early strength high performance concretes is limited. Autogenous shrinkage due to self-desiccation is perhaps more likely with very low w/c ratio concretes, although there is little data outside indirect evidence with certain high strength concrete research.[9]

Mechanisms of setting and hardening

The main physical mechanisms that occur during the setting and the hardening process of concrete include:

Sedimentation

Before setting, concrete constituents are in suspension and, in certain cases nonoptimized packing of granular skeleton. A vertical displacement of the constituents occurs by gravity:[10] downward for the larger grains, upward for the water which entraps the elements having a lower sedimentation velocity; a film of very clean water appears at the top surface (bleeding).

Le Chatelier's contraction

The volume of hyrdrates formed in the hydration reaction is substantially smaller than the sum of the volumes of the two components (anhydrous cement and water) entering into the reaction. The range is generally between 8 to 12%, depending on the properties of the cement. The potential lineic shrinkage of the cement paste is assumed to be about 3 to 4% (these values cannot be produced experimentally, due to the mechanical stiffness of the hardened paste itself).

Heat of hydration

The hydration reaction is highly exothermic (150 to 350 joules per gram of cement) which elevates the temperature under adiabatic conditions to values between 25 and 55 K in the concrete.

Self-desiccation

Only a fraction of the hydration is completed during the setting process. The hydration continues inside a rigid porous skeleton, resulting in a reduction in the water content in the pore space, a reduction that has the same mechanical effect as drying.[11]

Desiccation

This occurs in ordinary concrete, when about twice as much water as is strictly necessary for the hydration of the cement is used for reasons of workability, whereas this is not the case for high performance concretes

(HPC). After demolding, a drying process begins from the surfaces in contact with the ambient atmosphere.[12]

Setting and early hardening kinetics

At the start of setting process, there are isolated solid grains in a connected liquid phase. Hydration starts from the surface of the cement grains; they are covered with a layer of hydrates, a crust that grows thicker and increasingly retards the very hydration reaction that causes it to form.

The formation of the hydrates around the grains then leads to the establishment of contacts, and the crystals coalesce; in less than one hour, the concrete changes from a suspension to a continuous solid.[13]

In ordinary or conventional concretes hydration never ends; the layer of hydrates that forms around the cement grains becomes thicker and thicker and more and more impermeable, slowing down the reaction, but never actually cuasing it to stop. Anhydrous cores which are residues of cement grains can be found in very old concrete, explaining that physical and chemical properties continue to evolve over a long time.[14] From the structural standpoint, this post-setting hydration process has two consequences. First an internal growth of the skeleton (hardening) and second a simultaneous reduction of the evaporate water content in the pore space (self-desiccation). This occurs due to the negative volumetric balance of the chemical reaction. From a mechanical point of view the reduction of the evaporated water has the same effect as the drying, which is essentially the shrinkage of the skeleton of the hydrating materials.

For low water/cement ratios, the values of this autogenous shrinkage are of the same order as those of a drying shrinkage resulting from a reduction in water content equal to that produced by the hydration itself. Therefore, the observation of a macroscopic shrinkage of the hardened cement paste indicates that the mineral matrix as a whole is under compression.[15] This compression can be attributed to the force exerted by the pore water, through the surface tensions developed at the liquid–vapor interface.

Just after setting, the porous network is completely filled with water, and the self-desiccation process begins. During a long time, *the liquid water is connected* (there exists a continuous path from each point to each other in this phase). This does not mean that the gaseous phase is not also connected. Several connected networks can coexist in three dimensions, and are therefore relatively free to move under a pressure gradient (Darcy's process). The mobility of the liquid phase however decreases as fast as the water content decreases.[16]

As self-desiccation and drying proceed, the mobility of the liquid phase decreases rapidly and the dominant mechanism of transfer is no longer in the liquid phase, but in the gaseous phase. Then the phenomenon becomes completely different. It is no longer the total pressure that sets the water molecules in motion, as in the liquid phase, but their concentration. This is because of the trapped air, and any disturbance of total pressure equilibrium in the gaseous phase sets the air in motion with no distinction among molecules. If water moves in the gaseous phase, it is because there is a disturbance of equilibrium of the concentrations (Fick's process). This results in much slower movements. The tendency toward the equilibrium of concentrations results only from the fact that the probability of the presence of a molecule in the space accessible to it tends toward a uniform function, and its kinetics mainly depend on thermal random agitation.

High strength concrete (HSC) in general appears to be more sensitive to early drying. During the recent construction of a HSC nuclear containment, cracks were seen to form before setting.[17] These cracks completely disappeared when water curing was applied, showing that here, unlike the previous case, desiccation alone was involved (in addition to the effect of curing, it should be pointed out that the cracks were not very deep). The effect was very similar to the desiccation cracking of a soil, with the presence of ultrafines. Before setting, this gives the concrete material a structure close to that of the soils sensitive to this type of cracking. It should be emphasized that, for each type of HSC, an effective curing is essential.

High strength concrete exhibits several specific aspects such as high sensitivity to early drying (and the absolute necessity of an efficient curing) and faster autogenous shrinkage.

Heat development

The setting of cement is a highly exothermic chemical process which generates about 150 to 350 Joules per gram of cement. This generates to a temperature rise of between 25 and 55 K in the concrete under adiabatic conditions. Setting may then occur (in the core of massive parts, for instance) at a higher temperature and the differential contractions that occur during the subsequent cooling may lead to very significant cracking.[18] Hydration heat in concrete is roughly proportional to the amount, fineness and chemical composition (C_3A content) of the cement used. It increases with the amount, the fineness of the cement and the C_3A content in the cement. For the same cement content, less water leads to a reduction in the degree of hydration and an increase in the tensile strength of the hardened concrete.

Heat of hydration is affected not only by the cement content, but also by the water/cementitious ratio $w/(c+s)$ and the silica fume content.[19] Smeplass and Maage investigated the heat of hydration for a number of high strength concrete mixes by the means of a so-called semi-adiabatic calorimeter test. The results of Smeplass and Maage[19] indicate that the heat of hydration can be affected within a relatively wide range by the utilization of traditional mix-design parameters (Figs. 3.2 and 3.3). The heat of hydration per cement unit decreases approximately by 9% when the $w/(c+s)$ ratio is reduced from 0.36 to 0.27. At a constant paste/aggregate ratio, the resulting temperature rise increases by 6% (3 °C).

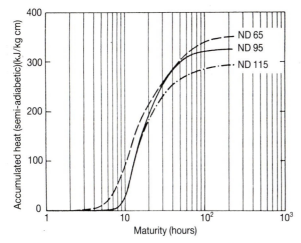

Fig. 3.2 Semi-isothermal heat development of three high strength concretes with different values of w/(c + s) ratio: 0.50 (ND65), 0.36 (ND95) and 0.27 (ND115)

Consequently, the effect of an increase in cement content normally is stronger than the effect of a reduced degree of hydration.

Smeplass and Maage[19] also found that at w/(c + s) ratio of 0.50, the replacement of cement by silica fume on a 1:1 basis induces a signifcant increase in the heat evolved per cement unit. The use of silica fume reduces the retarding effect of the dispersing agent, but also the magnitude of the secondary accelerating effect. The effect is observed for w/(c + s) ratios ranging from 0.50 to 0.27. At w/(c + s) ratios about 0.50 the effect of the silica fume on the total heat evolvement is just as strong as the effect of the cement. The effect weakens with decreasing w/(c + s) ratio. At w/(c + s) ratio 0.27, silica fume does not affect the heat evolvement significantly. Below w/(c + s) ratio 0.40, cement replacements by silica fume result in a reduced 'final' semi-adiabatic temperature.

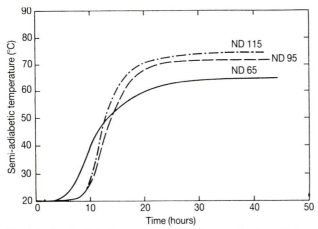

Fig. 3.3 Semi-adiabatic temperature development for three high strength concretes with different values of w/(c + s) ratio: 0.50 (ND65), 0.36 (ND95) and 0.27 (ND115)

From these results, it appears that HSC is not necessarily more sensitive to thermal cracking under all conditions. Furthermore, it should be noticed that thermal effects are not significant for concrete thickness less than 1.6 ft (40 cm). This is because of a drastic size effect due to the fact that the rate of heat diffusion increases like the square of the thickness (and this effect is accentuated by thermo-activation).[20] For specific industrial applications more sensitive to thermal cracking, like reservoirs, appropriate HSC mix-designs may be utilized with relatively low heat of hydration.[17]

Figure 3.4 shows that for a given temperature rise, the setting occurs sooner in high strength concretes than in normal strength concretes (NSC).

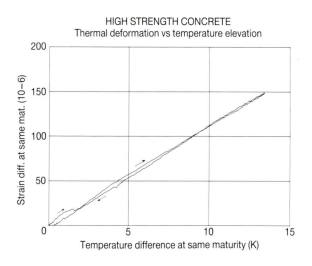

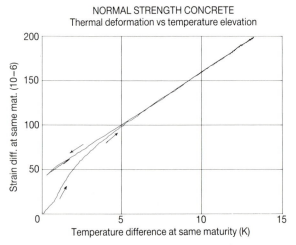

Fig. 3.4 Strain difference between two samples (at the same maturity but with different curing contractions) versus temperature difference, for a normal strength concrete (4a) and a high strength concrete (4b)

Figure 3.4a and 3.4b shows the thermal strain difference versus the temperature difference for normal as well as high strength concretes. The slope of curves shown in these figures is the coefficient of thermal expansion. The time over which the coefficient of thermal expansion becomes constant is essentially the time required to complete the transition from the suspension state to the solid state. In an experiment where a concrete specimen is hardening while being completely restrained, the final stress will be controlled by the difference between the temperature at the time of setting and the temperature at equilibrium with the ambient medium. Therefore, if the setting is lower, the cracking hazard during the setting and the early age is reduced.

Hardening kinetics

In order to obtain a parametric description of the mechanical behavior such as strength, modulus of elasticity, autogenous shrinkage and basic creep of high strength concrete, a total of ten mixes were examined by de Larrard and Le Roy.[21] The variables were the water/cement ratio (w/c = 0.28, 0.33, 0.38, 0.42), the paste volume ratio (V_p = 0.269, 0.286, 0.313, 0.325), and the silica fume ratio (s/c = 0.00, 0.05, 0.10, 0.15). The mix proportions and the results are given in Table 3.2.[21]

The hardening kinetics during early age and up to 28 days were investigated by studying the effect of aging (between 1 and 28 days) on the compressive strength and the modulus of elasticity for the ten HSC mixes and control reference mix [Figs. 3.5 and 3.6]. From these figures, it clearly appears that the hardening kinetics of high strength concretes significantly differs from that of ordinary concrete during the early age (up to 1 day), because of the reduction of water/cement ratio, and after one day, due to the hydration process of silica.

Autogenous shrinkage

The high autogenous (or self-desiccation) shrinkage of silica-fume HSC was first pointed out by Paillere, Buil and Serrano.[22] A very early shrinkage, even for non-drying specimens, was monitored in free specimens, leading to early cracking in restrained ones. More recently, de Larrard and Le Roy[21] have carried out measurements of autogenous shrinkage on ten HSC mixes (Table 3.2), and they proposed a relationship between mix-design and autogenous shrinkage, for HSC in the range of 60 to 100 MPa.[23] The final form of the equation can be written as

$$\varepsilon = \frac{(1 + E_d/E_g)(1 - g/g^*) + [4(1 - g^*)g/g^*]/[(1 - 2g^*) + g^*E_g/E_d]}{1 + g/g^* + (1 - g/g^*)E_d/E_g}$$

$$\frac{(1 - 0.65 \exp[-11s/c])(1 - 1.43w/c)K_c}{E_c} \tag{3.1}$$

Table 3.2 Compositions and results of tests on 10 HSC mixes. Concrete C_0 is a control concrete, without admixture nor silica fume

Formulae lb/yd³ (kg/m³)	C_0	C_1	C_2	C_3	C_4	C_5	C_6	C_7	C_8	C_9	C_{10}
Coarse aggregate	2022 (1200)	2049 (1216)	2088 (1239)	1975 (1172)	1914 (1136)	2029 (1204)	2022 (1200)	2025 (1202)	2022 (1200)	2022 (1200)	2022 (1200)
Sand	1129 (670)	1127 (669)	1144 (679)	1083 (643)	1050 (623)	1112 (660)	1109 (658)	1110 (659)	1109 (658)	1109 (658)	1109 (658)
Cement	576 (342)	671 (398)	617 (366)	711 (422)	770 (457)	598 (355)	723 (429)	629 (373)	718 (426)	694 (412)	650 (386)
Silica Fume	0 (0)	67.0 (39,8)	61.7 (36,6)	71.1 (42,2)	77.0 (45,7)	59.8 (35,5)	72.3 (42,9)	62.9 (37,3)	0 (0)	34.7 (20,6)	97.6 (57,9)
Superplast.*	0 (0)	32.5 (19,3)	30.0 (17,8)	34.5 (20,5)	37.4 (22,2)	29.0 (17,2)	33.9 (20,1)	30.5 (18,1)	34.9 (20,7)	33.7 (20,0)	31.7 (18,8)
Added water	288 (171)	199 (118)	182 (108)	211 (125)	229 (136)	231 (137)	179 (106)	217 (129)	212 (126)	206 (122)	192 (114)
Density	2,36	2,43	2,45	2,42	2,42	2,41	2,43	2,41	2,43	2,43	2,43
Entrap. air %	1,9	0,6	1,2	0,7	0,4	0,7	0,9	0,5	1,2	0,8	0,6
Slump in (mm)	2.4 (60)	7.9 (200)	7.1 (180)	8.7 (220)	9.8 (250)	8.7 (220)		8.7 (220)	7.9 (200)		
Aggregate proportions	0.705	0.714	0.731	0.687	0.675	0.712	0.711	0.715	0.708	0.712	0.714
w/c**	0.50	0.33	0.33	0.33	0.33	0.42	0.28	0.38	0.33	0.33	0.33
s/c	0.0	0.1	0.1	0.1	0.1	0.1	0.1	0.1	0.0	0.05	0.15
Compressive strengths in ksi (MPa)											
fc_1	1.65 (11,4)	3.68 (25,4)	3.88 (26,8)	3.31 (22,8)	4.54 (31,3)	2.99 (20,6)	4.96 (34,2)	2.07 (14,3)	3.68 (25,4)	3.70 (25,5)	4.18 (28,8)
fc_3	3.68 (25,4)	7.47 (51,5)	7.21 (49,7)	6.97 (48,1)	7.12 (49,1)	5.16 (35,6)	7.79 (53,7)	5.48 (37,8)	5.54 (38,2)	6.21 (42,8)	6.18 (42,6)
fc_7	4.6 (32,0)	10.3 (70,7)	10.0 (69,1)	10.1 (69,5)	10.2 (70,3)	8.21 (56,6)	10.9 (75,6)	8.38 (57,8)	8.29 (57,2)	9.35 (64,5)	9.71 (67,0)
fc_{28}	6.31 (43,5)	13.4 (92,1)	13.7 (94,3)	13.5 (93,3)	14.4 (99,4)	10.8 (74,6)	14.1 (97,3)	11.5 (79,5)	9.74 (67,2)	10.8 (74,6)	13.7 (94,3)

Young's modulus in ksi × 10³ (GPa)

Ei_1	3.93 (27,1)	4.47 (30,8)	4.60 (31,7)	4.16 (28,7)	5.15 (35,5)	4.23 (29,2)	5.60 (38,6)	3.35 (23,1)	4.38 (30,2)	4.52 (31,2)	4.92 (33,9)
Ei_3	4.95 (34,1)	6.25 (43,1)	6.26 (43,2)	5.97 (41,2)	6.13 (42,3)	5.96 (41,1)	6.82 (47,0)	5.90 (40,7)	6.05 (41,7)	6.15 (42,4)	6.38 (44,0)
Ei_7	5.28 (36,4)	6.73 (46,4)	7.09 (48,9)	6.77 (46,7)	6.77 (46,7)	6.47 (44,6)	7.40 (51,0)	6.34 (43,7)	6.67 (46,0)	6.93 (47,8)	6.93 (47,8)
Ei_{28}	5.99 (41,3)	7.31 (50,4)	7.51 (51,8)	7.70 (53,1)	7.05 (48,6)	6.80 (46,9)	7.74 (53,4)	6.57 (45,3)	6.67 (46,0)	7.22 (49,8)	7.50 (51,7)
Autogenous shrinkage between 72 and 5000 hours (10^{-6})	41	89	76	111	108	67	82	91	44	64	94

* Melamine resin at 30.9% dry extract
** w/c = total water/cement alone

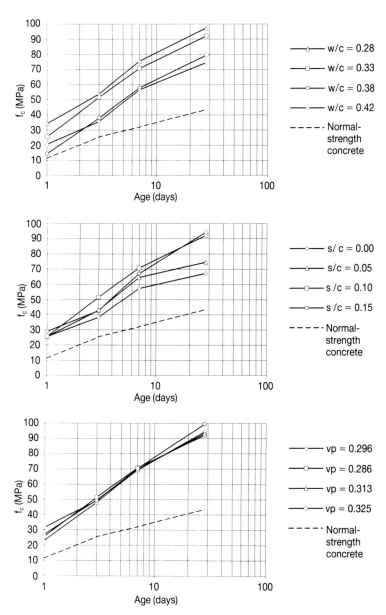

Fig. 3.5   Compressive strength development for the ten concrete mixes given in Table 3.2.
Dash lines represent average values of the control mix, a 5000 psi (35 MPa) normal strength
concrete made with the same components

where E_d = delayed modulus of the paste = $E_p/6$, where E_p is the paste
   E_g = modulus
    g = modulus of the aggregate
   g^* = actual aggregate volume

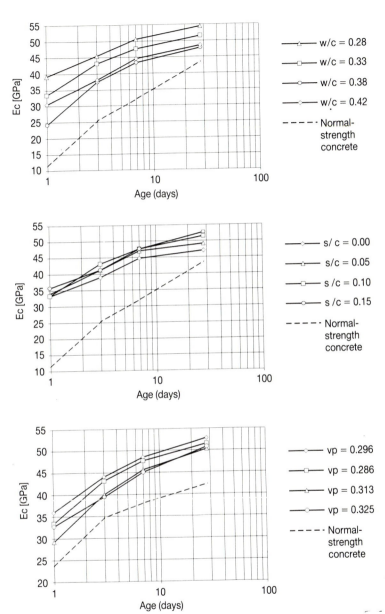

Fig. 3.6 Young's modulus development for the ten concrete mixes given in Table 3.2. Dash lines represent average values of the control mix, a 5000 psi (35 MPa) normal strength concrete made with the same components

s/c = maximum aggregate volume
w/c = silica-cement ratio
K_c = water-cement ratio
    Calibrating coefficient.

The value of K_c depends on the cement and the time over which the shrinkage is evaluated. The value of K_c recommended was 1.23 ksi (8.5 MPa) which was computed for the shrinkage occurring between 72 and 5000 hours for the test data of de Larrard and Le Roy.[21]

It can be concluded that a high autogenous shrinkage leading to early age cracking occurs in concretes with both low w/c and presence of silica fume. Therefore, in the cases where this cracking is likely and harmful, it is better not to use high content of using silica fume, or concretes with very low water/cement ratios.[17]

3.3 Creep

When a viscoelastic material is subjected to a stress, from a time t_0, its strain, measured parallel to the axis of the stress, changes over time. This time dependent increase in strain of hardened concrete subjected to sustained stress is termed creep. It is usually determined by subtracting, from the total measured strain in a loaded specimen, the sum of the initial instantaneous strain (usually considered elastic) due to sustained stress, the shrinkage, and any thermal strain in an identical load-free specimen, subjected to the same history of relative humidity and temperature conditions. The principal variables that affect creep are summarized in Table 3.1.

Creep is closely related to shrinkage and both phenomenon are related to the hydrated cement paste. As a rule, a concrete that is resistant to shrinkage also has a low creep potential. The principal parameter influencing creep is the load intensity as a function of time; however, creep is also influenced by the composition of the concrete, the environmental conditions and the size of the specimen.

A distinction must first be made between *basic* creep, which occurs in the absence of hygrometric exchanges with the ambient medium, and *drying* creep, that appears when the material dries. The former corresponds to the creep of a very thick actual structural element, whereas the latter, generally measured in the laboratory on small specimens – transverse dimensions less than 0.5 ft (15 cm) – reflects the maximum creep exhibited by thin structures. The basic creep and the total creep are the limits, between which lies the creep of a part having any shape, drying at between 50 and 100% relative humidity.

There is an abundant literature on creep of normal strength concretes and Neville[23] has given an excellent summary. Information on creep of concretes of higher strengths is limited.[2,24]

Relation between the basic creep and microstructure

Concrete consists of aggregate and cement paste, which itself consists of hydrates (CSH), anhydrous grains, free water and air bubbles. The basic

Table 3.3 Degrees of hydration and remaining anhydrous minerals in various cement pastes after 150 days (data taken from Sellevold and Justnes[25]). The compressive strengths have been evaluated thanks to Feret's modified formula[17,45,71,84] with Kg = 4.91 and R_c = 55 MPa

Pastes	W/C+S	W/C	S/C	W	C	SF	α_c	α_{sf}	Anhyd.	f_c ksi (MPa)
		Initial prop.					D° of hydr.			
A0	0.2	0.200	0.00	0.387	0.613	0.000	0.46	–	0.331	14.8 (102)
A8	0.2	0.216	0.8	0.379	0.557	0.064	0.46	0.87	0.308	16.2 (112)
A16	0.2	0.232	0.16	0.373	0.510	0.117	0.42	0.80	0.319	16.2 (112)
B0	0.3	0.300	0.00	0.486	0.514	0.000	0.61	–	0.200	10.3 (71)
B8	0.3	0.324	0.8	0.478	0.468	0.054	0.57	0.95	0.204	11.7 (81)
B16	0.3	0.348	0.16	0.472	0.430	0.099	0.53	0.90	0.212	11.7 (81)
C0	0.4	0.400	0.00	0.558	0.442	0.000	0.82	–	0.080	7.7 (53)
C8	0.4	0.432	0.08	0.550	0.404	0.046	0.72	1.00	0.113	8.8 (61)
C16	0.4	0.464	0.16	0.544	0.372	0.085	0.70	1.00	0.111	8.8 (61)

creep is intrinsic to the hydrates (the creep of the aggregates being zero or negligible with respect to that of the matrix; the same for the anhydrous cement). Its physical origin is poorly understood, and in any case no consensus has been reached concerning it. But it must be controlled to a great extent by the total volume of hydrates in the concrete microstructure. The amplitude of the basic creep is also directly influenced by the presence of free water in the microstructure of the material: a dry concrete exhibits no creep.[25]

Therefore, two features of HSC microstructure are expected to modify the creep behavior (compared with the one of NSC): the volume of hydrates and the free water content. In Table 3.3, one can see the volume of hydrates for cement pastes with water/binder ratio ranging between 0.2 to 0.4, and silica/cement ratio ranging from 0 to 16% (data taken from Sellevold and Justnes[26]). Obtaining a higher strength concrete entails imposing a low water/binder ratio; this gives a reduced volume of the hydrate and the free water content. Moreover, self-desiccation appears. These are factors which reduce creep.

Basic creep versus mix design and age at loading

In a recent study,[21] a series of ten concretes were tested in order to quantify the influence of mix-design on basic creep of HSC. The mix proportions are presented in Table 3.2. Data on the materials are given elsewhere.[21] For each concrete, 6.3 in. × 39.5 in. (160 × 1000 mm) cylindrical specimens were loaded at 1, 3, 7 and 28 days, and the creep strains were measured from the time of loading up to 18 months. Based on the results the following equation was proposed:

$$\varepsilon_{cr}(t, t_0) = \frac{K_{cr} \cdot \sigma \cdot [(t - t_0)^\alpha / \beta + (t - t_0)^\alpha]}{E_{i28}} \qquad (3.2)$$

where t is the time (in hours), t_0 is the age of concrete at loading, σ is the applied stress, E_{i28} is the elastic modulus, ε_{cr} is the creep strain, K_{cr} is the creep coefficient, α and β are material parameters depending on t_0.

The results of the creep tests, with specific creep values at the end of the tests are given in Table 3.4. Creep curves for concrete C1 are shown in Fig. 3.7. For each mix, it can be noted that the amplitude of specific creep decreases with the age of the material. The ratio of specific creep of concrete loaded at 1 day to the one of concrete loaded at 28 days is approximately equal to 2. Another general feature is the lowering of the kinetics: the concrete creeps faster when it is loaded at early age. As a consequence, for some mixes, the specimens loaded at 28 days continued to creep quite quickly at the end of the tests, so that the extrapolation (K_{cr} value) is not very reliable. The results also indicate that creep is sensitive to the paste content of concrete (Fig. 3.8) and it decreases when the water/cement ratio decreases (Fig. 3.9). A low dosage of silica fume (5% by weight of cement) leads to a decrease of the specific creep, but higher dosage (up to 15%) increases the deformation (Fig. 3.10).

In summary, any change of the mix-design parameters involving an increase of strength also leads to a decrease of creep, except when the silica fume amount is more than 10%. On the other hand, the kinetics of deformations do not show any obvious tendency when the values of mix-design parameters are changed.

Creep under high stress

The creep of concrete exhibits a quasi-linear domain in which the delayed strain (not counting shrinkage) is proportional to the applied stress. The limit of this domain, of the order of 40% of the strength at failure for conventional concretes, seems to be a little higher for HSC,[27,28] as, moreover, is its instantaneous behavior. Above this threshold, creep increases faster than applied stress. Above 75%, a delayed failure may occur.[29]

Models for basic creep

A number of analytical models are available for estimating the creep behavior and are summarized in the ACI committee report.[1] In addition,

Table 3.4 Results of creep tests for high performance concretes (the mix-compositions for which appear in Table 3.2)

Concrete	Age of loading (days)	Béta (ħalpha)	Alpha	Kcr	Specific creep at 24 months [10–6/mPa]
C1	1	7.3	0.569	1.09	20.49
	3	37.9	0.698	0.49	9.14
	7	30.7	0.609	0.70	12.53
	28	26.2	0.474	0.64	9.96
C2	1	1.7	0.425	0.93	17.08
	3	51.1	0.822	0.60	11.26
	7	42.1	0.686	0.55	9.95
	28	21.7	0.345	0.85	9.21
C3	1	3.7	0.677	1.14	22.54
	3	24	0.662	1.16	22.15
	7	49.6	0.781	0.60	11.65
	28	18.2	0.385	0.65	9.09
C_4	1	5.1	0.539	1.18	22.56
	3	–	–	–	–
	7	16.4	0.567	0.81	14.92
	28	7.1	0.342	0.33	5.15
C5	1	6	0.625	1.45	29.41
	3	5.1	0.521	0.67	13.30
	7	10.3	0.474	0.93	17.36
	28	21.9	0.306	1.47	14.42
C6	1	5.2	0.579	1.28	23.17
	3	97.1	1.052	0.70	12.85
	7	10.1	0.494	0.73	12.42
	28	32.5	0.39	0.67	7.14
C7	1	15.7	0.788	0.88	18.24
	3	46	0.754	0.64	13.01
	7	44.8	0.837	0.56	11.54
	28	35.7	0.371	1.04	11.16
C8	1	17.1	0.846	1.04	22.02
	3	23.7	0.534	0.58	10.88
	7	18	0.459	0.76	13.44
	28	22.8	0.306	1.46	14.43
C9	1	2.6	0.596	1.06	21.38
	3	14.8	0.541	1.16	22.02
	7	20.5	0.585	0.64	12.21
	28	42	0.433	0.74	9.39
C10	1	3.9	0.616	1.09	20.51
	3	55.1	0.79	0.74	13.76
	7	10.5	0.453	0.91	15.30
	28	31.1	0.328	1.25	10.52

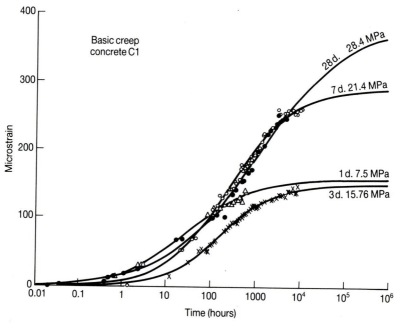

Fig. 3.7   Basic creep of concrete C1 (mix composition, see Table 3.2)

there are other available models.[29,30,31,32] The model of Bazant and Chern based on log double-power law appears to be the quite effective analytically.

At LCPC (Paris), Auperin, de Larrard *et al.*[33] have conducted in the past a preliminary program of experiments on a silica fume concrete having a strength of 11.6 ksi (80 MPa) at 28 days, with a view to characterizing its creep vs. the age of loading. Based on the trend of the experimental data. The following expression was proposed:

$$\varepsilon_{cr}(t, t_0) = K_{cr}[\varepsilon/E_i(28)]f(t - t_0) \tag{3.3}$$

where K_{cr} is the creep coefficient, E_i is the modulus and $f(t - t_0)$ a kinetic

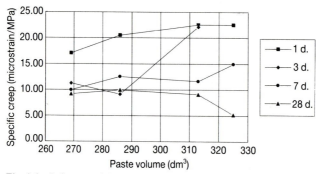

Fig. 3.8   Influence of the paste volume on the basic creep of high strength concrete (specific creep after 2 years)

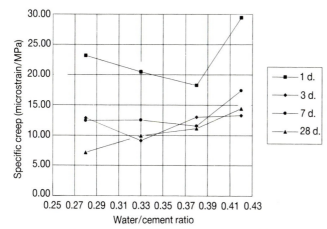

Fig. 3.9   Influence of the w/c ratio on the basic creep of high strength concrete (specific creep after two years)

function of time t, which is expressed in days. For the modulus, the creep coefficient and the kinetic function, the following expressions were proposed.

$$E_i(t_0) = 1.132\exp\left[-0.42/(t_0-0.4)^{1/2}\right]E_{i28} \qquad (3.4)$$

$$K_{cr}(t_0) = 0.363\exp\left[9.3/(t_0-0.34)\right]^{1/2} \qquad (3.5)$$

$$f(t-t_0) = \frac{\exp\left[-1.7/(t-t_0+0.027)\right]^{1/3} - \exp(-4)}{1-\exp(-4)} \qquad (3.6)$$

Investigation of different loading paths (for HSC concrete 760 lbs/cyd, 484.4 kg/m³, of cement, 8% silica fume and w/c of 0.38) showed that superposition was also valid for higher strength concretes (Fig. 3.11). Comparison of data of this high strength concrete with other data from the literature revealed that its creep kinetics was exceptionally fast. The creep amplitude for early loading (at one day) was also unusually high.

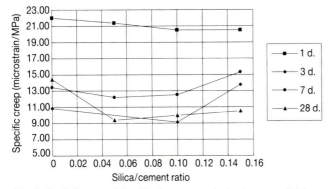

Fig. 3.10   Influence of the silica/cement on the basic creep of high strength concrete (specific creep after two years)

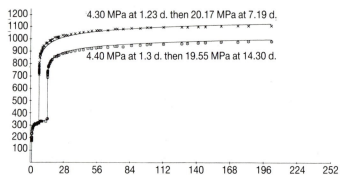

Fig. 3.11   Creep of high strength concrete subjected to two stress steps, and application of the superposition method. fC 28 = 12 ksi (83áMPa), cement: 760 lb/yd (450 kg/m), silica fume: 8%, w/c = 0.38; deformation in microstrains, time in days (linear scale)

A more comprehensive program has been recently carried out at LCPC.[21] As already pointed that the compressive strength appears to be a convenient parameter in order to evaluate the basic creep.[29] After a smoothing of the test results of 40 specimens,[21] the following equations were proposed in the LCPC model:

$$\varepsilon_{cr}(t, t_0) = (\sigma/E_{i28})\,K_{cr}(t - t_0)\,\alpha/[(t - t_0)^\alpha + \beta] \qquad (3.7)$$

$$K_{cr} = 1.2 - 3.93 \times 10^{-2} f_c(t_0) \qquad (3.8)$$

$$\beta = 1.5\exp 3 f_c(t_0)/f_{c28}] \qquad (3.9)$$

where $\alpha = 0.5$ and $f_c(t_0)$ is the compressive strength (mean cylinder value in ksi) at the age at loading t_0. To utilize the model, one must evaluate the E-modulus at 28 days. A sophisticated formula, based upon the homogenization theory, has been recently proposed by the authors[21] and can be used if no experimental data is available. More simple models are available in the building codes. The basic creep of high strength concretes seems to strongly decrease for loading after 28 days, while the evolution of strength is poor. This trend is similar to normal strength concretes.

3.4   Drying shrinkage and drying creep

Drying process

There have so far been few investigations on the drying of high strength concretes (HSCs). As pointed out earlier, HSC is subjected to self-desiccation. Thus, the water consumed by hydration cannot go out of the material, and the water loss of hardened HSC is generally smaller than that of NSC, but not zero. For continuation of the drying process, external humidity must be lower than the internal one. In the case of very-high strength silica fume concrete, the internal humidity falls below 80%.[34] Therefore, such a concrete will not show any drying shrinkage when exposed to an external humidity above 80% relative humidity.

For most high strength concretes, the pore structure is very fine, due to the low porosity of the paste, and to the presence of cementitious admixtures. This leads to a very low gas-permeability.[35] Thus, the drying kinetics is expected to be very slow. In order to check these assumptions, the drying process of two concretes – a NSC and a silica fume very high strength concrete (VHSC) – has been monitored by means of gammadensimetry.[36] Two different curing regimes were employed on 6.3 in. (160 mm) diameter-cylinders: drying at 50% relative humidity and 68 °F (20 °C) after demolding (at 24 hours), and sealed during 28 days, then dried in the same conditions as the previous regime. The distributions of water losses are shown in Figs. 3.12 and 3.13.

With the latter curing regime, the hydration of the covercrete is more complete, so that the drying kinetics is slower than in the former case. After four years, only three centimeters of the very high strength concrete (VHSC) had begun to dry, while the NSC had reached its hygral equilibrium with the ambient air. For long-term extrapolation, say 50 or 100 years, it is not easy to evaluate the future water field in the VHSC. A conservative hypothesis is that the whole specimen will dry at the same level than the one of the covercrete (in this example, the uniform water loss would be about 2%, the value of 3.5% corresponding to a region modified by the wall effect). However, as the material continues to hydrate, and becomes tighter after 28 days, this hypothesis could be pessimistic.

Drying shrinkage

Very little information is available concerning drying shrinkage of HSC, as most tests in the literature are performed on drying specimens, without sealed companions (which should allow to separate the part due to self-desiccation). Moreover, it is difficult to propose mathematical models giving sound extrapolations, because of the great slowness of the drying process. At the moment, it is only possible to indicate some tendencies, related to strains measured during a limited time on small specimens.

Field measurements of surface shrinkage strains on a mock column, fabricated with high strength concrete, after two and four years and the comparison with measurements on specimens under laboratory conditions[9] showed that the surface shrinkage strains under field conditions are considerably lower than those measured under laboratory conditions(Fig. 3.14).

There is conflicting information on the drying shrinkage for concrete with high range water reducers.[37,38,39] Flowing concrete, for a given strength, is likely to require a slightly higher cement content and therefore will exhibit somewhat higher shrinkage.[40] Use of a HRWR to reduce water content can be expected to reduce shrinkage in most cases. Tests over a period of one year showed that the effect of naphthalene-based HRWR, with 840 to 1000 pcy (500 to 600 kg/m^3-cement) was to reduce shrinkage.[37]

However, there was an increase in swelling after a year of storage in water. The swelling of concrete with superplasticizer was approximately 50% greater than that of control concrete. Since swelling increases with an increase in cement content, it can be postulated that the higher swelling of the concrete with superplasticizer is due to a higher hydrated-cement paste content because of a rapid early-age development of strength. An alternative explanation is that the admixture modifies the paste structure so that its swelling capacity is increased.

The drying shrinkage in high strength concrete with silica fume is either equal to or somewhat less than that of concrete without silica fume but containing fly ash. This was concluded by Luther and Hansen,[41] based on their results on five high strength concrete mixtures which were monitored for 400 days. The use of high fly ash content (50% cement replacement by weight with low calcium, Class F fly ash) for 5800 psi and 8700 psi (40 MPa and 60 MPa) concretes,[42] resulted in ultimate shrinkage values from 400 microstrains to 500 microstrains, with swelling amounting to 40% to 55% of the shrinkage value. The study also indicated the importance of continued water curing for full pozzolanic reaction of fly ash. The data[42] shows that shrinkage of concrete properly proportioned with high fly ash content compares favorably with that of portland cement concrete.

The influence of drying and of sustained compressive stresses, at and in excess of normal working stress levels, on shrinkage properties of high, medium and low strength concretes were investigated by Smadi, Slate and Nilson.[27] The 28 day compressive strength ranged from 3000 psi to 10,000 psi (21 to 69 MPa). The long-term shrinkage was found to be greater for low strength concrete and smaller for medium and high strength concretes. The study also indicated that the effect of aggregate content on the shrinkage of low strength and medium strength concrete is less significant than the effect of w/c ratio.

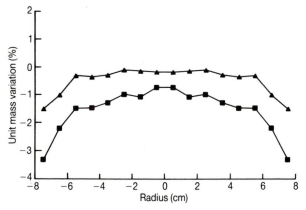

Water content distribution in control concrete
demoulded at 1 day. ▲ 28 days, ■ 80 days

(a)

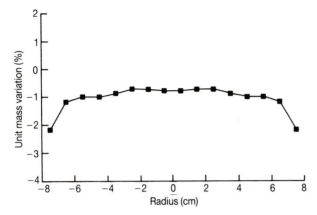

Water content distribution in control concrete
at 90 days (demoulded at 28 days)

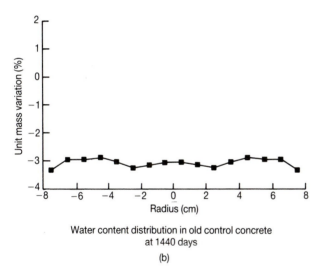

Water content distribution in old control concrete
at 1440 days

(b)

Fig. 3.12 Water content distribution of NSC with two curing regimes. (a) Drying from 1 day at 20 ÁC and 50% RH; (b) sealed during 28 days then drying

In a recent study,[43] it was shown that for five mixes with 28-day design strengths ranging from 8700 psi to 9300 psi (60 to 64 MPa), the shrinkage deformation was inversely proportional to the moist-curing time (the longer the curing time, the lower the shrinkage). It was also concluded that shrinkage was somewhat less for concrete mixtures with lower cement paste and larger (1.5 in. or 38 mm) aggregate size. In addition, the use of a high range water reducing admixture did not have a significant effect on the shrinkage deformation.

Observed shrinkage deformations of higher strength concrete (12,000 psi to 19,700 psi or 83 MPa to 136 MPa) were compared by Pentalla[44] to the predictions based on CEB Code recommendation.[30] It was concluded that shrinkage deformation of higher strength concrete took place much faster than predicted by the CEB formula.

The drying shrinkage characteristics for seven HSC mixes were recently investigated.[45] The data of seven HSC mixes are presented in Table 3.5. Mix No. 1 is a control NSC. Mix No. 2 is a HSC without silica fume (same composition as the one used for Joigny bridge). Mixes No. 3 to 5 are silica fume HSCs with constant binder dosages, and variable water/content ratios. Mix No. 6 is a VHSC, and mix No. 7 is a particular silica fume HSC, where a part of the cement has been replaced by a limestone filler (in order to minimize the thermal cracking). Shrinkage of mixes No. 1 and 6 are plotted in Fig. 3.15. The drying shrinkage obtained for mix No. 2 is quite low, but was still continuing at the end of the test. With mixes No. 3 to 5, the balance between the two kinds of shrinkage is emphasized. When the water/cement ratio increases, there is less autogenous shrinkage and more drying shrinkage, the sum remaining roughly constant. The optimal mix-design of mix No. 6 leads to a moderate autogenous shrinkage (in spite of the very low water/cement ratio), and a very low drying shrinkage. As for the last mix (No. 7), the negligible autogenous shrinkage entails a quite large drying one (the highest of the seven concretes). However, gammadensimetry measurements have shown, in this particular case, that the drying process was very rapid, with practically no hygral gradients.[36] Therefore, this high value of drying shrinkage is propbably near the asymptotic one, unlike the other mixes.

Alfes[47] recently proposed a quantitative model for total shrinkage of HSC, taking the aggregate restraining effect into account. This model is similar to the model of de Larrard and Le Roy,[21] except that the latter predicts the autogenous shrinkage from the whole mix composition, with

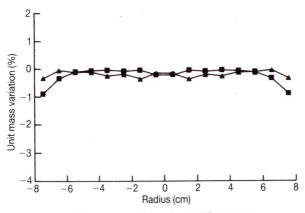

Water content distribution in VHS concrete

demoulded at 1 day. ▲28 days, ■90 days

(a)

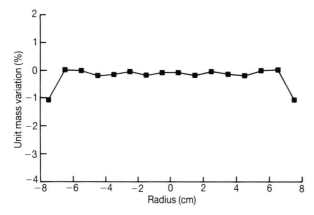

Water content distribution in VHS concrete

at 90 days (demoulded at 28 days)

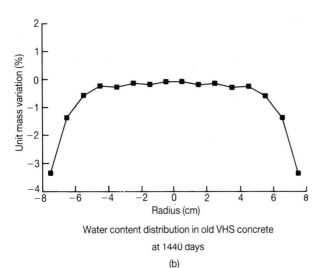

Water content distribution in old VHS concrete

at 1440 days

(b)

Fig. 3.13   Water content distribution of VHSC with two curing regimes. (a) Drying from 1 day at 20 ÁC and 50% RH; (b) sealed during 28 days then drying

only one parameter to fit, while Alfes' model bases the computations on the paste shrinkage. The equation for Alfes' model is:

$$\varepsilon_s = \varepsilon^{sm}.V_m{}^n \qquad (3.10)$$

$$n = \frac{1}{[0.23 + 1.1/[1 + V_m(E_g/E_m - +)]} \qquad (3.11)$$

where ε_s is the concrete shrinkage, ε_{sm} the paste shrinkage, V_m is the paste volume, E_g and E_m are the moduli of aggregate and paste respectively.

In recent papers[48,49] a large amount of data on carefully controlled

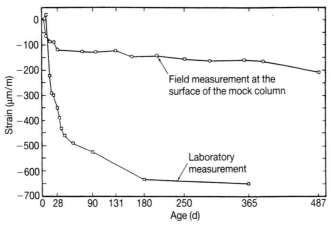

Fig. 3.14   Comparison of field and laboratory drying shrinkage[9]

shrinkage tests of concrete involving a large number of identical specimens were used to compare the predictive equations of ACI, CEB-FIP, and Bazant and Panula.[31,50–52] Bazant and Panula claimed that their equation (which has a large number of empirical constants) gave the best agreement with the experimental data. The validity of the Bazant and Panula equation for high strength concrete was explored by Bazant *et al.*,[52] and it was reported that the equation could be made applicable to higher strength concrete with minor adjustments. Similar conclusions by Almudaiheem and Hansen[53] also showed good correlations with the experimental data.

Drying creep

There is no consensus on the physico-chemical origin of the phenomenon of drying creep. According to Bazant and Chern,[32] the increase in the delayed *intrinsic* strain of concrete caused by drying is greater under load than in the absence of loading. A reduction of drying creep should, therefore, be expected. According to other researchers,[15,20,54,55] drying creep is related primarily to a *structural effect*: differential drying induces a state of skin cracking that relaxes part of the self-stresses and so decreases the deformation caused by self-desiccation) (i.e. drying shrinkage).

The composition of concrete can essentially be defined by the w/c ratio, aggregate and cement types and quantities. Therefore, as with shrinkage, an increase in w/c ratio and in cement content generally results in an increase in creep. Also, as with shrinkage, the aggregate induces a restraining effect so that an increase in aggregate content reduces creep. Numerous tests have indicated that creep deformations are proportional to the applied stress at low stress levels. The valid upper limit of the relationship can vary between 0.2 and 0.5 of the compressive strength. This range of the proportionality is expected due to the large extent of microcracks in concrete at about 40 to 45% of the strength.

Table 3.5   Shrinkage of various HSC mixes[45]

Mixes no.	1 NSC	2 HSC	3 HSC	4 HSC	5 HSC	6 VHSC	7 HSC
Cement type	BPC*	OPC	OPC	OPC	OPC	OPC	BPC*
Strength in ksi (MPa)	7.97 (55)	9.43 (65)	9.43 (65)	9.43 (65)	9.43 (65)	7.97 (55)	7.97 (55)
Dosage lb/yd³ (kg/m³)	590 (350)	758 (450)	768 (456)	763 (453)	763 (453)	709 (421)	448 (266)
Limestone filler lb/yd³ (kg/m³)	– –	– –	– –	– –	– –	– –	111 (66)
Silica fume lb/yd³ (kg/m³)	– –	– –	60.7 (36)	60.7 (36)	60.7 (36)	70.8 (42)	67.4 (40)
Superplasticizer** lb/yd³ (kg/m³)	– –	7.58 (4,5)	11.8 (7,0)	11.1 (6,6)	6.06 (3,6)	13.3 (7,9)	6.07 (3,6)
Retarder** lb/yd³ (kg/m³)	– –	1.52 (0,9)	0.84 (0,5)	0.84 (0,5)	0.84 (0,5)	– –	– –
Water lb/yd³ (kg/m³)	329 (195)	283 (168)	254 (151)	295 (175)	316 (188)	189 (112)	280 (166)
W/C	0,56	0,37	0,33	0,39	0,42	0,27	0,62
Slump, in (mm)	2.7 (70)	>7.9 (>200)	>7.1 (>180)	>7.1 (>180)	>7.1 (>180)	>7.9 (>200)	>7.1 (180)
Mean cylinder strength at 28 days in ksi (MPa)	5.8 (40)	11.3 (78)	13.6 (94)	12.0 (83)	10.7 (74)	14.6 (101)	9.7 (67)
Autogenous shrinkage at 1 year (10^{-6})	90	90	290	200	140	150	30
Drying shrinkage at 1 year (10^{-6})	290	90	120	190	260	110	310
Total shrinkage for drying specimens (10^{-6})	380	180	410	390	400	260	340

*Blended Portland cement containing 9% of limestone
**Equivalent dry extract

Very few data are reported on creep of concrete containing condensed silica fume. In one study,[56] creep tests were performed on a concrete in which 25% of the cement was replaced by silica fume and a naphthalene-based high-range water reducer was added. The results showed that total deformation was reduced under drying conditions, with no significant reduction in basic creep.

Results of studies on creep of concrete containing fly ash[57–59] indicate that creep of sealed specimens can be reduced in the same proportions as the ratio of replacement of portland cement by fly ash, ranging between 0 and 30%, if water content is reduced substantially. However, fly ash mixes

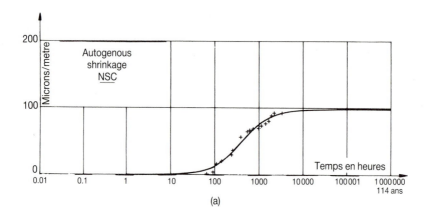

(a)

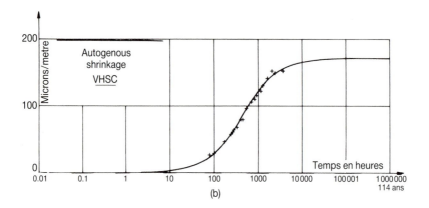

(b)

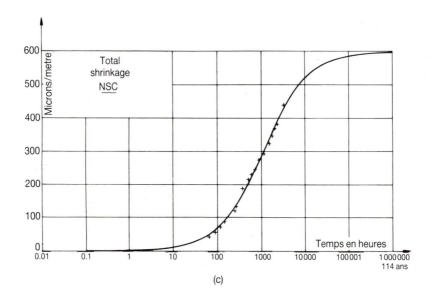

(c)

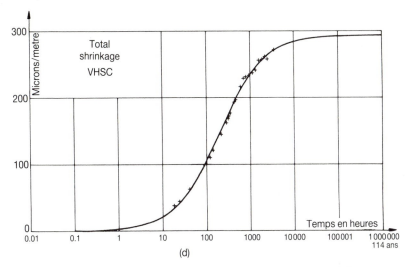

Fig. 3.15 (a) Autogenous shrinkage of NSC after demolding at 24 hours[46]; (b) autogenous shrinkage of VHSC after demolding at 24 hours; (c) total shrinkage of drying specimens of NSC at 50%, 68 ÁF (20 ÁC); (d) total shrinkage of drying specimens of VHSC at 50% RH, 68 ÁF (20 ÁC)

frequently show slightly higher creep under drying conditions than control mixes at the same 28-day strength, due to somewhat slower initial strength gain combined with early drying associated with standard test procedures.

Information on creep of concrete containing a high-range water reducer[39,60] is restricted in scope. The creep of concrete with melamine-based high-range water reducer was reported to be 10% higher than control concrete.[61] Tests over a period of one year[37] showed that mixes with 840 lbs/cu yd (500 kg/m^3) cement containing a naphthalene-based high-range water reducer exhibited creep characteristics similar to control concrete; mixes with 1000 lbs/cu yd (600 kg/m^3) cement and the same high-range water reducer exhibited greater creep than control mixes. Flowing concrete for a given strength is likely to require a slightly higher cement content and it exhibits lower creep, while its creep recovery and post creep recovery elastic deformation are generally comparable to control concrete.[40]

In a recent study by Luther and Hansen,[41] the specific creep of five high strength concrete mixtures ($7350 < f_c' < 69$ MPa) was monitored for 400 days. It was concluded that creep of the silica fume (SF) concrete was not significantly different from that of fly ash concrete. The relationship between the specific creep and compressive strength is shown in Fig. 3.16, which also includes other data.[4,62–65] The data in the figure show a hyperbolic relationship between specific creep and compressive strength. Also, these data[41] nestle between the Neville data (dashed curve on the left) and the Perenchio and Klieger data (dashed curve on the right), and they agree with the applicable Ngab, Nilson and Slate results. Furth-

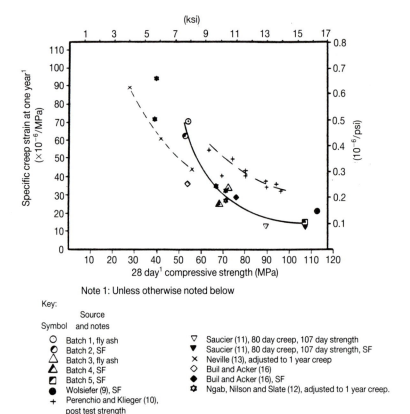

Fig. 3.16   Relationship between specific creep and compressive strength[41]

ermore, these data near high-strength levels are very close to Wolsiefer's[62] and Saucier's[64] results. Thus all of the concretes show similar specific creep to compressive strength relationships. Therefore, there is no apparent difference between the specific creep of silica fume (SF) concrete, portland cement concrete, or fly ash concrete.

It has been reported by Pentalla[44,66] that the creep deformation of higher-strength concrete with strengths from 12,000 psi to 19,700 psi (83 to 136 MPa) takes place much faster than the prediction by the CEB Code recommendations.[30]

The influence of drying and of sustained compressive stresses, at and in excess of normal working stress levels, on creep properties of high, medium and low strength concretes was investigated by Smadi, Slate and Nilson.[27] The 28 day compressive strength ranged from 3000 psi to 10,000 psi (21 MPa to 69 MPa). Creep strains, creep coefficient and specific creep were all smaller for high strength concrete than for concretes of medium and low strengths at different stress levels, and at any time after loading. The creep-to-stress proportionality limit was higher for the high strength concrete than for the others by about 20%.[27]

The effect of mix proportions on creep characteristics was investigated in a study,[43] in which five mixes with 28-day strengths ranging from 8700 psi to 9300 psi (60 MPa to 64 MPa) were used. The results indicated that creep is somewhat less for concrete mixtures with lower cement paste and large aggregate size. It was also shown that the use of high-range water-reducing admixture did not show a significant effect on the creep deformations.

A very limited amount of work is reported on tensile creep tests.[67] Tensile creep tests at 35% of the ultimate short-term strength show that specific creep increases with an increasing w/c ratio and decreasing aggregate-cement ratio. These trends are similar to those of compressive creep and the levels of specific creep are similar. Sealed concrete creeps less than immersed concrete.[67] Also, tensile creep generally increases with concurrent shrinkage and swelling. These effects are similar to those that occur in compression.[67] In another study relations between the relative stress level and the time of failure were derived, from uniaxial tensile creep tests.[68] These relations were not found to be affected by temperature, cement type or concrete quality.

For high strength concrete with silica fume, drying creep is poor in cylindrical specimens with 4.33 in. (110 mm) diameter[29,44] and practically non-existent for 6.3 in. (160 mm) diameter cylindrical specimens.[33,69,70] In HSC without silica fume, the drying effect remains, but is smaller than in NSC. In the same way as for drying shrinkage, the data are too scarce and the phenomenon too slow for allowing the development of reliable and comprehensive mathematical models, which could be applicable for different mixture proportions, different sizes of specimens and different abient conditions.

Figure 3.17 shows the basic and the total creep of several high strength concrete mixes reported in the literature along with the prediction of the LCPC model.[21] As expected, it can be seen that the contribution to overall shrinkage due to drying decreases when the strength of concrete increases (the drying process becoming slower, as the concrete is tighter). It can be noted that the compressive strength is far from being the only variable controlling the total creep of HSC. It appears that information regarding this relationship between the mix-design and creep phenomenon for HSC is lacking and there is a need for further investigations.

3.5 Thermal properties

The thermal properties of concrete are of special concern in structures where thermal differentials may occur from environmental effects, including solar heating of pavements and bridge decks. The thermal properties of concrete are more complex than for most other materials, because not only is concrete a composite material the components of which have different thermal properties, but its properties also depend on moisture content and porosity. Data on thermal properties of high performance concrete are limited, although the thermal properties of high strength concrete fall

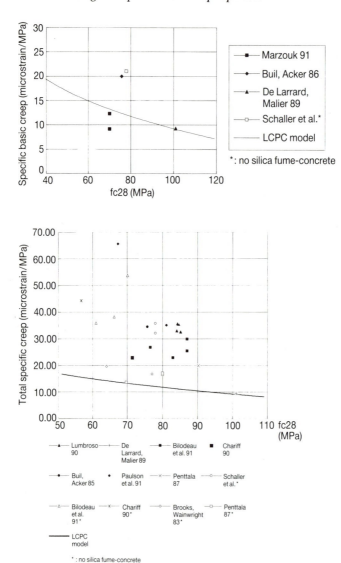

Fig. 3.17 Comparison of creep strains at 1 year reported in the literature with the prediction of LCPC model; (a) for specimens loaded near 28 days, and sealed; (b) specimens loaded at 28 days and kept at 50% RH

approximately within the same range as those of lower strength concrete,[7,71] for characteristics such as specific heat, diffusivity, thermal conductivity and coefficient of thermal expansion.

Three types of test are commonly used to study the effect of transient high temperature on the stress-strain properties of concrete under axial compression: (1) unstressed tests where specimens are heated under no initial stress and loaded to failure at the desired elevated temperature; (2)

stressed tests, where a fraction of the compressive strength capacity at room temperature is applied and sustained during heating and, when the target temperature is reached, the specimens are loaded to failure; and (3) residual unstressed tests, where the specimens are heated without any load, cooled down to room temperature, and then loaded to failure.

For normal strength concretes, exposed to temperatures above 450 °C, the residual unstressed strength has been observed to drop sharply due to loss of bond between the aggregate and cement paste. If concrete is stressed while being heated, the presence of compressive stresses retards the growth of cracks, resulting in a smaller loss of strength.[72]

The moisture content at the time of testing has a significant effect on the strength of concrete at elevated temperatures. Tests on sealed and unsealed specimens have shown that higher strength is obtained if the moisture is allowed to escape.[73,74]

For a given temperature, as the preload level is increased, the ultimate strength and the stiffness of normal strength concrete has been observed to decrease while the ultimate strain also decreases.[75]

In a recent study, Castillo and Durrani[76] tested concrete with strengths from 4500 psi to 12,900 psi (31 to 89 MPa) under temperatures ranging from 23 °C to 800 °C. The presence of loads in real structures were simulated by preloading the specimens before exposure to elevated temperatures. The strength of stressed and unstressed specimens of both normal and high strength concretes at different temperatures is shown in Fig. 3.18. Each point represents an average of at least three specimens normalized with respect to maximum compressive strength at room temperature. From this figure it can be seen that exposure to temperatures in the range of 100 °C to 300 °C decreased the compressive strength of high strength concrete by 15% to 20%. As the strength increased, the loss of strength from exposure to high temperature also increased. At tempera-

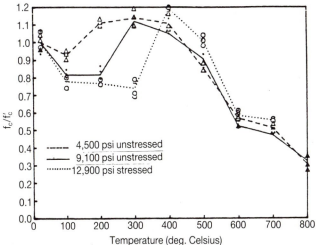

Fig. 3.18   Variation of compressive strength with increase in temperature[76]

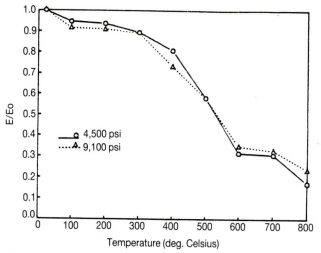

Fig. 3.19   Variation in modulus of elasticity of normal and high-strength concrete with increase in temperature[76]

tures above 400 °C, the high strength concrete progressively lost its compressive strength which at 800 °C dropped to about 30% of the room-temperature strength. The study also observed that none of the preloaded specimens were able to sustain the load beyond 700 °C. About one third of these specimens failed in explosive manner in the temperature range of 320 °C to 360 °C while being heated under a constant preload. The variation of the modulus of elasticity of normal and high strength concretes with increasing temperature is shown in Fig. 3.19. The modulus decreases between 5% to 15% when exposed to temperatures in the range of 100 °C to 300 °C. This trend is similar for normal and high strength concretes. At 800 °C, for both the normal and high strength concretes, the modulus of elasticity was only 20% to 25% of the value at room temperature.

In a recent study at Helsinki University,[77] high temperature behavior of three high strength concretes made with different binder combinations was investigated. Ordinary cements, silica fume and class F fly ash with superplasticizers were used as binders. The aggregates were granite-based sand and crushed diabase. Compressive strengths of 4 in. (100 mm) cubes at 28 days were 12,300 psi to 16,000 psi (85 MPa to 111 MPa). The study consisted of investigations of mechanical properties at elevated temperatures, and attention was also given to the chemical and physical background of their alteration due to heating. The thermal stability and alterations were also investigated. The mechanical properties at high temperatures were studied by determining the stress-strain relationship. The risk of spalling was also studied primarily with small specimens. The three high strength concretes showed very similar temperature behavior. They all show, in the temperature region from 100 °C to 350 °C, more loss of strength than normal strength concrete. This is caused by temperature-dependent destruction of cement paste. Its influence on the strength of high strength concrete is more decisive than on the normal strength concrete, because the cement paste matrix of high strength concrete must carry higher loads than in normal strength concrete (more homogeneous stress distribution between the aggregate and cement paste). The denser cement paste and overall microstructure of the higher strength concrete result in slower drying. So the higher risk of destructive spalling must be taken into account in structures exposed to fire.

It is generally recognized that concrete cast and cured at low temperatures develops strengths at a significantly slower rate than similar concrete placed at room temperature. Lee[78] conducted a study on mechanical properties of high strength concrete in the temperature range between +20 °C and −70 °C (68 °F and −94 °F) without considering the effect of freezing cycles. Test results showed that the values of compressive and tensile strength, modulus of elasticity and Poisson's ratio increased as the temperature decreased. The ratio of bond strength at low temperature is generally larger under reversed cyclic loading than under monotonic or repeated cyclic loads. The rate of increase in compressive, splitting tensile strength, Young's modulus, local bond strength and Poisson's ratio for high strength concrete at corresponding low temperature (−10 °C, −30 °C, −50 °C, −70 °C) is generally lower than for normal strength concrete. The effect of decreasing temperature on the compressive strength and the tensile strength is shown in Figs. 3.20 and 3.21, respectively. From these figures it can be seen that higher strength concrete shows less susceptibility to decreasing temperatures compared to normal strength concrete.

Price,[79] and Klieger[80] determined that concrete mixed and placed at 4 °C had a 28-day compressive strength which was 22% lower than concrete cast and continuously cured at 21 °C. However, recent work[81,82] indicated that expected slow strength development at low temperatures was not realized for cold cast and cured concretes.

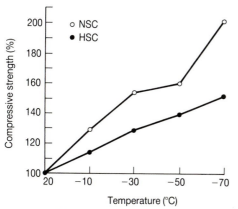

Fig. 3.20 Effect of low temperature variation on the percentage of compressive strength increase for normal and high-strength concrete[78]

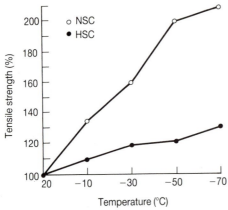

Fig. 3.21   Effect of low temperature on the percentage of tensile stress increase for normal and high-strength concrete[78]

3.6   Structural effects: case studies

In this section, two examples of field applications where the concrete had to be designed to meet the performance specifications are presented. The examples are of a nuclear containment structure in Civaux (France) and a prestressed concrete bridge at Joigny (France). For the nuclear containment structure, there was a great need to improve the air-tightness and reduction of thermal cracking. For the prestressed concrete bridge, there was a need for higher strength concrete with reduced heat of hydration, shrinkage and creep properties.

Thermal cracking – Civaux nuclear containment structure (France)

The reactors of the nuclear power plants currently being built in France are enclosed by double concrete containments. The purpose of these prestressed concrete structures are to protect the nuclear reactor from possible outside dangers and most important of all, to protect the environment (and its population) from any radioactive release should an event or accident occur during the operation of the plant.

   In practice, the critical requirement is of limiting the leaks to 1.5% of the total mass of the gas contained in the vessel, for 24 hours under an internal absolute pressure of 72.5 psi (0.5 MPa) and a temperature close to 300 °F (150 °C) (design accident conditions). The axisymmetric structure is therefore massively prestressed (along the meridians and the parallels), so much so that the concrete remains in compression in all design conditions and particularly when the vessel is pressurized to the critical pressure. In the absence of internal pressure, the concrete must be able to withstand this prestress and this results in wall thickness of 4 ft to 5 ft (120 to 15 cm). As for the raft, earthquake considerations result in thickness of about 10 ft (3 m).

Nevertheless, this massive amount of concrete does not guarantee air-tightness, primarily because of systematic thermal cracking. The numerical analysis of computed thermal stresses clearly showed that this damage is not a skin cracking phenomenon (due to the local gradient) but crossing cracks phenomenon, which is due to the restraint of the mean thermal contraction of the last layer of fresh concrete by the previous one. For this reason, cracks are vertical, widely spaced and visible. After complete cooling of the structure, the opening of these cracks is reduced, but only to a small extent, for several reasons. These include the setting (and initialization of strains) which does not occur at the ambient temperature, the changing value of the modulus of concrete (it being quite a lot higher during cooling than during temperature rise, the duration of cooling of concrete during which thermal stresses are generated due to visco-elastic behavior of concrete.

The specifications required for a concrete to meet the performance criteria for this application were: (a) high stability and workability, to avoid formation of porous zones in concrete; (b) limited shrinkage, including autogenous shrinkage and thermal contraction; (c) improved resistance to tensile stresses resulting from restrained deformations; (d) lower material permeability to air, which would cut losses through the plain concrete (i.e. outside of the cracked zones; (e) improved anchorage on passive reinforcements, which should reduce crack opening; and (f) higher compressive strength, making it possible to alter the design of the structure, starting with the elimination of such extra thickness as the wall/raft and the wall/dome gussets. Cost was also a consideration and the aim was to keep the extra cost per cubic meter to a minimum, given the large quantities involved in the structure, about 16,000 yd^3 (12,000 m^3).

A comprehensive approach (described by de Larrard[17,45,71,83]) led to a high performance concrete (HPC) for the Civaux containment project with silica fume and a large reduction in the cement content, compensated, for its 'grading' role, by a calcareous filler (HSC1). Finally, two HPC mixes and two conventional concrete mixes (OC1 and OC2) were designed for this project (Table 3.6). The contractor preferred a concrete without any retarder, i.e. mix HSC2. The heat of hydration of OC2 and HSC2 were investigated, by the means of a semi-adiabatic calorimeter, on the basis of the same spproach similar to Smeplass.[19] The results (Fig. 3.22) show a significant reduction (20 to 25% on the final heat values per cubic meter).

Comparative testing of concretes

The principal mechanical and physical properties of these four concretes are shown in Table 3.7. The compressive strength values of the HPC mixes for this project increased rather slowly with time because of the high proportion of the binder constituents having a slow rate of hydration and

Table 3.6   Composition of the four concretes investigated: masses given in lb/yd³ (kg/m³)

	OC1	OC2	HSC1	HSC2
Aggregates 12.5/25[1]	1499	1296	1591	1499
	(890)	(769)	(944)	(890)
Aggregates 4/12.5[1]	350	511	372	352
	(208)	(303)	(221)	(209)
Sand 0/5[2]	1222	1178	1296	1353
	(725)	(699)	(769)	(803)
Cement[3]	632	590	448	448
	(375)	(350)	(266)	(266)
Filler[4]		102	112	118
		(60.8)	(66.4)	(69.8)
Silica fume			67.9	67.9
			(40.3)	(40.3)
Superplasticizer[5]	6.32[6]	2.06[5]	58.3[7]	15.3
	(3.75)	(1.22)	(34.6)	(9.08)
Retarder[8]			3.59	
			(2.13)	
Mixing water	320	329	214	271
	(190)	(195)	(127)	(161)
Slump in in (mm)	3.1	2.8	3.5	7.1
	(80)	(70)	(90)	(180)

[1] 'Arlaut' crushed calcareous aggregates
[2] Crushed calcareous sand containing 6% filler
[3] 'Airvault CPJ 55' Portland cement with 9% calcareous filler
[4] 'Cical' calcareous filler
[5] Melamine type 'Melment' 20% dry content in weight
[6] Lignosulfonate type plasticizer
[7] 'Rhéobuild 1000' naphtalene type superplasticizer
[8] 'Melretard'

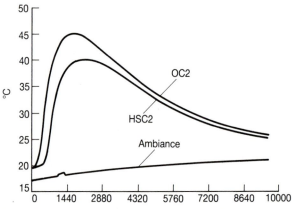

Fig. 3.22   Semi-adiabatic temperature of normal strength concrete, OC2, and high strength concrete, HSC2. (The mix designs are shown in Table 3.6)

Table 3.7 Comparative tests on the four concretes given in Table 3.6

	OC1	OC2	HSC1	HSC2
Compressive strength in ksi (MPa)				
f_{c1}	–	–	1.74	–
	–	–	(12)	–
f_{c3}	–	–	5.22	–
	–	–	(36)	–
f_{c7}	–	–	7.25	–
	–	–	(50)	–
f_{c28}	5.8	6.67	10.2	9.72
	(40)	(46)	(70)	(67)
Splitting strength in ksi (MPa)				
f_{t1}	–	–	0.17	–
	–	–	(1.2)	–
f_{t3}	–	–	0.45	–
	–	–	(3.1)	–
f_{t7}	–	–	0.57	–
	–	–	(3.9)	–
f_{t28}	0.54	0.58	0.65	0.59
	(3.7)	(4.0)	(4.5)	(4.1)
Modulus in ksi $\times 10^3$ (GPa)				
E_{i3}	–	–	4.93	–
	–	–	(34)	–
E_{i8}	–	–	5.37	–
	–	–	(37)	–
E_{i28}	4.93	4.93	5.51	5.22
	(34)	(34)	(38)	(36)
Flexural strength in ksi (MPa)				
F_{f18h}	0.36	0.32	0.09	0.32
	(2.5)	(2.2)	(0.6)	(2.2)
f_{f1}	0.44	0.45	0.39	0.41
	(3.0)	(3.1)	(2.7)	(2.8)
f_{f3}	0.61	0.58	0.61	0.59
	(4.2)	(4.0)	(4.2)	(4.1)
f_{f8}	0.65	0.59	0.78	0.68
	(4.5)	(4.1)	(5.4)	(4.7)
f_{f28}	–	0.67	–	0.80
	–	(4.6)	–	(5.5)
Tensile strain at failure (10^{-6})				
ϵ_{18h}	120	120	105	100
ϵ_1	130	125	130	120
ϵ_3	155	160	155	150
ϵ_8	150	140	165	160
ϵ_{28}	–	155	–	160
Shrinkage between 28 and 90 days (10^{-6})				
– of protected specimen		20		10
– of a drying specimen		150		80

Table 3.7   *cont.*

	OC1	OC2	HSC1	HSC2
Creep + shrinkage (10^{-6})				
– of protected specimen		250		280
– of a drying specimen		700		360
under a stress of (MPa):		12		20
Basic creep (10^{-6})				
– per ksi (MPa):		230		270
		132		93.1
		(19.2)		(13.5)
Additional drying dreep (10^{-6})				
– per ksi (MPa):		320		10
		184		3.45
		(26.7)		(0.5)
Creep/elastic strain ratio K_C				
– without drying		0.65		0.49
– with exchange of water		1.56		0.50
Air permeability ($10^{-18} ft^2$ (m^2)				
– under 4.3 psi	69.9		8.82	
(0.03 MPa)	(6.5)		(0.82)	
– under 29 psi	32.3		1.61	
(0.20 MPa)	(3.0)		(0.15)	
– under 58 psi	153		1.29	
(0.40 MPa)	(14.3)		(0.12)	

because of the relatively low cement content. The increase in the splitting tensile strength values with time was proportional to the compressive strength gain with time. The modulus of elasticity for the high strength mixes was relatively smaller than expected, in spite of the small volume of binder paste in these concretes.

The amplitude of the autogenous shrinkage is governed by the ability of the binder to continue to hydrate and consume water after setting. In the HPC mixes designed for this project, part of the cement was replaced by a chemically inert filler, and by silica fume, which reacts more slowly, and probably without consuming water. Therefore for these concrete mixes, the self-desiccation was low, unlike most silica-fume high strength concrete mixes. This was confirmed experimentally by using embedded gauges to measure early age shrinkage on 6.3×39.5 in. (160×1000 mm) cylinders, at $68 \pm 1.8\,°C$ ($20 \pm 1\,°F$) and $50 \pm 10\%$ relative humidity. For each concrete, one of two specimens was protected against drying by two coats of resin separated by aluminium foil; the other specimen was dried freely. The autogenous shrinkage for HPC mixes for this project was lower than that for ordinary concrete and practically zero (Table 3.7). In the first month,

HPC mixes for this project exhibited more drying shrinkage, but the opposite is observed after 28 days.

In order to obtain basic and drying creep of OC2 and HSC2 concretes, the companion specimens were loaded at 28 days under 30% of their instantaneous failure load. Creep results (where autogenous and drying shrinkage were respectively deduced) are shown in Table 3.7. Like all silica-fume concrete results reported in the literature, the HSC exhibits negligible drying creep. It should be pointed out that the reduction of total creep resulting from the use of HSC is therefore between 30 and 70%, depending on the thickness of the structure.

Temperature and restrained deformations fields were calculated using the CESAR-LCPC finite-element program[83]. Temperatures were calculated with constant diffusivity coefficient and a second term representing the heat of hydration, the rate of which depends on the temperature according to Arrhenius's law. An on-site model representing two 66 ft (20 m)-sections of a containment shell 4 ft (1.20 m)-thick with its reinforcements and prestressing sheets were built, one with the conventional concrete, the other with the HPC. The maximum temperature rise in the first core was 72 °F (40 °C), and the temperature rise in the high strength concrete core was 54 °F (30 °C). The conventional concrete wall showed vertical cracks, at half of the vertical prestressing sheets, three of these cracks exceeding 0.008 in. (200 μm). By contrast, only one microcrack, with an opening of 0.004 in. (100 μm), occurred in the shell with HPC. These results confirm the hypothesis regarding thermal cracking, i.e. if young concrete is able to support the tensile stresses which pass through a maximal value at a very early age during its maturing, thermal cracking is avoided.

HPC without silica fume – Joigny bridge (France)

In order to obtain a better knowledge of the serviceability of HSC and its interest for prestressed structures, the construction of the Joigny bridge (1988–1989) was accompanied by laboratory tests and on site monitoring, particularly regarding the effects of hydration heat, shrinkage and creep.

The composition and strength values of the concrete are given in Table 3.8. The high cement content was used to obtain the desired high strength

Table 3.8 Composition of the Joigny Bridge concrete: in lb/yd³ (kg/m³)

Yonne coarse aggregates 5/20	1730	(1027)
Yonne sand 0/4	1092	(648)
Fine sand 0/1	177	(105)
'CPA HP' Portland cement from Cormeilles	758	(450)
'Melment' superplasticizer (40% dry content)	18.96	(11.25)
'Melretard' retarder	7.58	(4.5)
Added water	266	(158)
Slump	9 in	(230 mm)

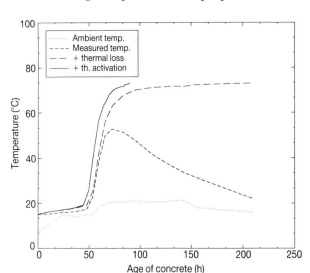

Fig. 3.23   Semi-adiabatic temperature of the high strength concrete used in the construction of Joigny Bridge, France

values, which also made it necessary to use a large proportion of superplasticizer. A retarding agent was also used to prevent the poured concrete from setting before the casting of the deck was completed (about 24 h). The aim was in fact to allow the concrete to adapt to the deformations of the formwork during the casting of the three spans of the structure. Four 6.3×39.5 in. (150×1000 mm) concrete cylinders were cast on site and raw materials were stored for additional laboratory tests.

The heat of hydration was measured by the means of a semi-adiabatic calorimeter. The final values are very high, due to the high cement content (Fig. 3.23). Autogenous and drying shrinkage, measured on protected (with self-adhesive aluminium sheets) and non-protected 6.3×39.5 in. (160×1000 cm) concrete cylinders at $68\,°F$ ($20\,°C$) and 50% relative humidity are shown in Fig. 3.24. The final values are close to current values of normal strength concrete. As pointed out previously, high autogenous shrinkage of certain HSC is mainly related to the presence of silica fume, together with low water-cement ratio.

Experimental results of creep and shrinkage results allow us to predict the delayed behavior of the structure, and can be compared to the values measured on site. Figure 3.25 shows the mean deformation of the mid-span section versus time along with the predicted results. The difference between the calculated and the measured deformation can be attributed in part to the climatic effect (annual variations).

3.7   Summary and conclusions

The physical phenomenon controlling the shrinkage and creep deformations in concrete properties can be broadly categorized in the following

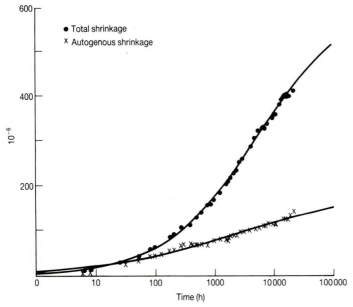

Fig. 3.24 Autogenous and drying shrinkage of the high strength concrete used in the construction of Joigny Bridge, France

stages. These phenomena allow the main features of delayed strains in higher strength concretes to be understood.

Stage 1 – Just after casting, the concrete remains in a suspension state, where little or no segregation occurs due to the very compact state of the cement paste. This dormant period is lengthened by the retarding effect of superplasticizers.

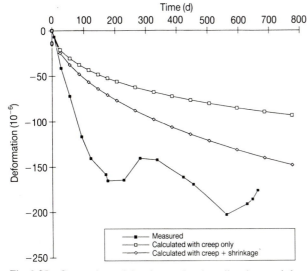

Fig. 3.25 Comparison of the observed and predicted mean deformation of the mid-span section of Joigny Bridge, France

Stage 2 – When the cement hydration appears, the setting starts sooner due to the closeness of cement grains; self-desiccation of the internal porosity, and a rapid heat development occurs at the same time.

Stage 3 – Hydraton continues until the internal humidity falls below a certain level (say, 70–80%). During the hardening phase, concrete becomes rapidly watertight. If the concrete is submitted to drying, the effect will be harmful on fresh or plastic concrete (because of the lack of bleeding), but the effect is very limited after the hydration has begun. The endogenous phenomena of higher strength concretes indicate that autogenous shrinkage is higher in HSC than normal strength concretes, especially when silica fume is incorporated. The heat development is higher in HSC because of the higher cement dosage, but limited by incomplete hydration. The basic creep is reduced due to the decreased volume of hydrates, and to the self-desiccation process.

The deformations for higher strength concretes related to drying are generally reduced, particularly for short and medium term. Even, the drying creep of silica-fume HSC could be practically insignificant. However, the very long-term drying shrinkage of HSC needs more investigations.

The laboratory experimental observations have been validated by monitoring of some full-scale structures (during several years). Parametric models have been proposed for autogenous shrinkage and basic creep. More accurate models, and models for drying-related deformations, are still to be developed. These will provide tools for engineers to accurately predict long term time dependent performance for designing and maintaining reinforced and prestressed concrete structures.

References

1   ACI Committee 209 (1990) Prediction of creep, shrinkage and temperature effects in concrete structures, *Manual of concrete practice*, Part 1. American Concrete Institute, 209R 1–92.
2   ACI Committee 363 (1984) State-of-the-art report on high strength concrete. *ACI Journal*, **81**, 4, 364–411.
3   Freedman, S. (1970/71) High strength concrete. *Modern Concrete*, **34**, 6, 1970, 29–36; 7, 1970, 28–32; 8 1970, 21–4; 9, 1971, 15–22; 10, 1971, 16–23.
4   Ngab, A.S., Nilson, A.H. and Slate, F.O. (1981) Shrinkage and creep of high strength concrete. *ACI Journal*, **78**, 4, 255–61.
5   Chicago Committee on High-Rise Buildings (1977) *High strength concrete in Chicago, high-rise buildings*.
6   Pfeifer, D.W., Nagura, D.D., Russell, H.G. and Corley, W.G. (1971) Time dependent deformations in a 70 story structure, in *Designing for effects of creep, shrinkage, temperature in concrete structures, SP-37*. American Concrete Institute, Detroit, 159–85.
7   Parrot, L.J. (1969) *The properties of high-strength concrete*, Technical Report No. 42.417. Cement and Concrete Association, Wexham Springs.
8   Swamy, R.N. and Anand, K.L. (1973) Shrinkage and creep in high strength concrete. *Civil Engineering and Public Works Review* (London), **68**, 807, 859–65; 867–8.
9   Aitcin, P.-C., Sarkar, S.L. and Laplante, P. (1990) Long-term characteristics of a very high strength concrete. *Concrete International*, 40–44.

10 Powers, T.C. (1968) *The properties of fresh concrete*. Wiley & Sons, New York.

11 Buil, M. (1979) *Contribution à l'étude du retrait de la pâte de ciment durcissante*. Rapport de Recherche No. 92, LCPC, Paris.

12 L'Hermite, R.G., Mamillan, M. (1973) Répartition de la teneur en eau dans le béton durci. *Annales de l'ITBTP*, 309–10, 30–34.

13 Scrivener, K.L. (1989) The microstructure of concrete, in *Materials science of concrete, I*, Skalny, J.P. (ed.), The American Ceramic Society Inc.

14 Byfors, J. (1980) *Plain concrete at early ages*, Report of the Swedish Cement and Concrete Research Institute, Stockholm.

15 Acker, P. (1992) Physicochemical mechanisms of concrete cracking, in *Materials science of concrete, II*, Skalny, J.P. (ed.), The American Ceramics Society Inc.

16 Daian, J.F. and Saliba, J. (1990) Using a pore-network model to simulate the experimental drying of cement mortar. *Proceedings of the 7th International Drying Symposium*, IDS'90/CHISA'90, Prague.

17 de Larrard, F., Ithurralde, G., Acker, P. and Chauvel, D. (1990) High performance concrete for a nuclear containment. *Second International Symposium on Utilization of High-Strength Concrete*, Berkeley, ACI SP 121–27.

18 Bamforth, P.B. (1982) *Early age thermal cracking in concrete*. Slough Institute of Concrete Technology, Tech. No. TN/2.

19 Smeplass, S. and Maage, M. (1990) Heat of hydration of high-strength concretes, in *Proceedings IABSE Colloquium on High-Strength Concrete*, Berkeley, 433–56.

20 Acker, P., Foucrier, C. and Malier, Y. (1986) Temperature-related mechanical effects in concrete elements and optimization of the manufacturing process, in *Properties of concrete at early ages*, Young, J.F. (ed.), ACI Publ, SP-95, Detroit, 33–47.

21 de Larrard, F. and Le Roy, R. (1992) *The influence of mix-composition on the mechanical properties of silica-fume high-performance concrete*. Fourth International ACI-CANMET Conference on Fly Ash, Silica Fume, Slag and Natural Pozzolans in Concrete, Istanbul.

22 Paillere, A.M., Buil, M. and Serrano, J.J. (1987) Durabilité du béton à très hautes performances: incidence du retrait d'hydratation sur la fissuration au jeune âge [Durability of VHSC: effect of hydration shrinkage on early-age cracking], First International RILEM Conference, *From Materials Science to Construction Materials Engineering*, Versailles, Chapman and Hall, Vol. 3, 990–97 [in French].

23 Neville, A.M., Dilger, W.H. and Books, J.J. (1983) *Creep of plain and structural concrete*. Construction Press.

24 ACI Committee 363 (1987) Research needs for high-strength concrete. *ACI Materials Journal*, 559–61.

25 Bazant, Z.P., Asghari, A.A. and Schmidt, J. (1976) Experimental study of creep of Portland cement paste at variable water content. *Materials and Structures, RILEM*, **9**, 52, 279–90.

26 Sellevold, E. and Justnes, H. (1992) *High-strength concrete binders: nonevaporable water, self-desiccation and porosity of cement pastes with and without condensed silica fume*. Fourth International ACI-CANMET Conference on Fly Ash, Silica Fume, Slag and Natural Pozzolans in Concrete, Istanbul.

27 Smadi, M.M., Slate, F.O. and Nilson, A.H. (1987) Shrinkage and creep of high-, medium- and low-strength concretes, including overloads. *ACI Materials Journal*, 224–34.

28 Bjerkeli, L., Tomaszewicz, A. and Jensen, J.J. (1990) Deformation properties and ductility of high-strength concrete. *Second International Conference on Utilization of High Strength Concrete*, ACI Special Publication, Berkeley.

29   Lumbroso, V. (1990) 'Réponse différée du béton à hautes performances soumis à un chargement stationnaire – influence des conditions d'environnement et de la composition' [Delayed behavior of high-strength concrete submitted to stationary loading]. Ph.D. doctoral thesis, Institut National des Sciences Appliquées, Toulouse.

30   Comite European du Béton Federation Internationale de la Precontrainte (1978) *CEB-FIP Model Code for Concrete Structures*, 3rd edn, Paris.

31   Bazant, Z.P. and Panula, L. (1980) Creep and shrinkage characterization for analyzing prestressed concrete structures. *PCI Journal*, **25**, 3, 86–122.

32   Bazant, Z.P. and Chern, J.C. (1985) Log double power law for concrete creep. *ACI Journal*, **82**, 5, 665–75.

33   Auperin, M., De Larrard, F., Richard, P. and Acker, P. (1989) Retrait et fluage de bétons à hautes performances – influence de l'âge au chargement [Shrinkage and creep of high-strength concretes – influence of the age at loading]. *Annales de l'Institut Technique du Bâtiment et des Travaux Publics*, 474, 50–75.

34   Buil, M. (1990) Le comportement physicochimique du système ciment-fumée de silice [Physico-chemical behavior of the cement/silica fume system]. *Annales de l'Institut Technique du Bâtiment et des Travaux Publics*, 483 [in French].

35   Perraton, D. and Aitcin, P.-C. (1990) La perméabilité au gaz des BHP [Gas-permeability of HSC] in *Les Bétons à hautes Performances*, Malier, Y. (ed.), Presses de l'ENPC, Paris [in French].

36   De Larrard, F. and Bostvironnois, J.L. (1991) On the long-term strength losses of silica-fume high-strength concretes. *Magazine of Concrete Research*, **43**, 155, 109–19.

37   Books, J.J. and Wainwright, P.J. (1983) Properties of ulta-high-strength concrete containing a superplasticizer. *Magazine of Concrete Research*, **35**, 125.

38   Johansson, A. and Petersons, A. (1979) *Flowing concrete, advances in concrete slab technology*. Pergamon Press, Oxford, 58–65.

39   Omojola, A. (1974) *The effect of cormix SPI super workability aid on creep and shrinkage of concrete*. Research and Development Report No. 014J/74/1683, Taylor Woodrow, London.

40   Dhir, R.K. and Yap, A.W.F. (1984) Superplasticized flowing concrete: strength and deformation properties. *Magazine of Concrete Research*, **36**, 129, 203–15.

41   Luther, M.D. and Hansen, W. (1989) Comparison of creep and shrinkage of high-strength silica fume concretes with fly ash concretes of similar strengths, in *Fly ash, silica fume, slag and natural pozzolans in concrete, SP 114, Vol. 1*. American Concrete Institute, 573–91.

42   Swamy, R.N. and Mahmud, H.B. (1989) Shrinkage and creep behaviour of high fly ash content concrete, in *Fly ash, silica fume, slag and natural pozzolans in concrete, SP 114, Vol. 1*. American Concrete Institute, 453–75.

43   Collins, T.M. (1989) Proportioning high-strength concrete to control creep and shrinkage. *ACI Materials Journal*.

44   Pentalla, V. (1987) Mechanical properties of high-strength concretes based on different binder combinatons. *Stavanger Conference Utilization of High-Strength Concrete*, Tapir Ed., Trondheim, Norway.

45   De Larrard, F. and Acker, P. (1990) *Déformations Libres des Bétons à Hautes Performances* [Free deformations of HSC], Séminaire sur la Durabilité des BHP, Cachan [in French].

46   De Larrard, F. and Malier, Y. (1989) *Propriétés constructives des bétons à très hautes performances – de la micro- à la macrostructure* [Engineering properties of very-high-strength concretes, from the micro- to the macrostructure],

Annales de l'Institut Techniques des Travaux Publics, No. 479.

47 Alfes, C. (1989) High-strength, silica-fume concretes of low deformability. *Betonwerk + Fertigteil-Technik*, No. 11.

48 Bazant, Z.P., Kim, J.K., Wittmann, F.H. and Alou, F. (1987) Statistical extrapolation of shrinkage data – Part II: Bayesian updating. *ACI Materials Journal*, **84**, 2, 83–91.

49 Bazant, Z.P., Wittmann, F.H., Kim, J.K. and Alou, F. (1987) Statistical extrapolation of shrinkage data – Part I: Regression. *ACI Materials Journal*, 20–34.

50 Bazant, Z.P. and Panula, L. (1978/1979) Practical prediction of time-dependent deformations of concrete. *Materiaux et Constructions*, (RILEM) Paris, **11**, 65, 307–28 and 66, 415–34, 1978; **12**, 69, 169–83, 1979.

51 Bazant, Z.P. and Panula, L. (1982) New model for practical 'Prediction of creep and shrinkage', in *Designing for creep and shrinkage in concrete structures*. SP 76, American Concrete Institute, 7–23.

52 Bazant, Z.P. and Panula, L. (1984) Practical preediction of creep and shrinkage of high strength concrete. *Materiaux et Constructions*, (RILEM) Paris, **17**, 101, 375–78.

53 Almudaiheem, J.A. and Hansen, W. (1989) Prediction of concrete drying shrinkage from short-term measurements. *ACI Materials Journal*, 401–8.

54 Acker, P. (1980) The drying of concrete – consequences on creep tests interpretation, *Re-evaluation of the time-dependent behavior of concrete*. CEB Task Group, Lausanne.

55 Wittmann, F.H. and Roelfstra, P.E. (1980) Total deformation of loaded drying concrete. *Cement and Concrete Research*, **10**, 5.

56 Buil, M. and Acker, P. (1985) Creep of a silica fume concrete. *Cement and Concrete Research*, **15**, 463–6.

57 Bamforth, P.B. (1980) In situ measurements of the effect of partial Portland cement replacement using either fly ash or granulated blast-furnace slag on the performance of mass concrete. *Proceedings of Institution of Civil Engineers, Part 2*, **69**, 777–800.

58 FIP Commission on Concrete (1988) *Condensed silica fume in concrete*, State of art report. Federation Internationale de la Precontrainte, London.

59 Yamato, T. and Sugita, H. (1983) Fly ash, silica fume, slag and other mineral by-products in concrete, in *Fly ash, silica fume, slag and other mineral by-products in concrete, Vol. 1*, SP 79. American Concrete Institute, 87–102.

60 Brooks, J.J., Wainwright, P.J. and Neville, A.M. (1979) Time-dependent properties of concrete containing a superplasticizing admixture, *Superplasticizers in concrete*, SP 62. American Concrete Institute, 293–314.

61 Alexander, R.M., Bruere, G.M. and Ivanusec, I. (1980) The creep and related properties of very high-strength syper-plasticized concrete. *Cement and Concrete Research*, **10**, 2, 131–7.

62 Wolsiefer, J. (1984) Ultra high-strength field placeable concrete with silica fume admixture. *Concrete International*, **6**, 4, 25–31.

63 Perenchio, W.F. and Klieger, P. (1978) *Some physical properties of high strength concrete*. Research and Development Bulletin No. RD 056.01T, Portland Cement Association, Skokie, IL.

64 Saucier, K.L. (1984) *High strength concrete for peacekeeper facilities*, Final report. Structures Laboratory, US Army Engineer Waterways Experimental Station.

65 Neville, A.M. (1981) *Properties of concrete*, 3rd edn, Pitman Publishing Ltd, London.

66 Pentalla, V. and Rautanen, T. (1990) Microporosity, creep and shrinkage of high-strength concretes. *Second International Conference on Utilization of High-Strength Concrete*, Berkeley, ACI Special Publication.

67   Domone, P.L. (1974) Uniaxial tensile creep and failure of concrete. *Magazine of Concrete Research*, **20**, 88, 144–52.

68   Cornelissen, H.A.W. and Siemes, A.J.M. (1985) Plain concrete under sustained tensile or tensile and compressive fatigue loadings, in *Behaviour of offshore structures*. Elsevier Science Publishers, Amsterdam, 487–98.

69   De Larrard, F., Acker, P., Malier, Y. and Attolou, A. (1988) *Creep of very-high-strength concretes*, 13th IABSE Conference, Helsinki.

70   De Larrard, F. (1990) *Creep and shrinkage of high-strength field concretes*, Second International Conference on Utilization of High-Strength Concrete, Berkeley, ACI SP 121–28.

71   Saucier, K.L., Tynes, W.O. and Smith, E.F. (1965) *High compressive strength concrete – report 3, summary report*, Miscellaneous Paper No. 6-520, US Army Engineer Waterways Experiment Station, Vicksburg.

72   Abrams, M.S. (1971) Compressive strength of concrete at temperatures to 1600 °F, in *Temperature and concrete*, SP 25. American Concrete Institute, 33–58.

73   Hannant, D.J. Effects of heat on concrete strength. *Engineering*, London, **203**, 21, 302.

74   Lankard, D.R., Birkimer, D.L., Fondfriest, F.F. and Snyder, M.J. (1971) Effects of moisture content on the structural properties of Portland cement concrete exposed to temperatures up to 500 °F, in *Temperature and concrete*, S 25. American Concrete Institute, 59–102.

75   Schneider, U. (1976) Behavior of concrete under thermal steady state and non-steady state conditions. *Fire and Materials*, **1**, 103–15.

76   Castillo, C. and Duranni, A.J. (1990) Effect of transient high temperature on high-strength concrete. *ACI Materials Journal*, **87**, 1, 47–53.

77   Diederichs, U., Jumppanen, U.-M. and Pentalla, V. (1989) *Behaviour of high strength concrete at high temperatures*, Report 92. Department of Structural Engineering, Helsinki University, Espoo.

78   Lee, G.C., Shih, T.S. and Chang, K.C. (1988) Mechanical properties of high-strength concrete at low temperature. *Journal of Cold Regions Engineering*, **2**, 4, 169–79.

79   Price, W.H. (1951) Factors influencing concrete strength. *ACI Journal*, **47**, 6, 417–31.

80   Klieger, P. (1958) Effect of mixing and curing temperatures on concrete strength. *ACI Journal*, **54**, 12, 1063–81.

81   Aitcin, P.-C., Cheung, M.S. and Sha, V.S. (1985) Strength development of concrete cured under Arctic sea conditions, in *Temperature effects on buildings*, STP 858, ASTM, Philadelphia, 3–20.

82   Gardner, N.J., Sau, P.L. and Cheung, M.S. (1988) Strength development and durability of concretes cast and cured at 0 °C. *ACI Materials Journal*, **85**, 6, 529–36.

83   De Larrard, F. (1990) Creep and shrinkage of high-strength field concretes. *Second International Conference in Utilization of High-Strength Concrete*, Berkeley, ACI SP 121–28.

4 Fatigue and bond properties

A S Ezeldin and P N Balaguru

4.1 Introduction

Many concrete structural members are subjected to repeated fluctuating loads the magnitude of which is well below the maximum load under monotonic loading. This type of loading is typically known as fatigue loading. Contrary to static loading where sustained loads remain constant with time, fatigue loading varies with time in an arbitrary manner. Fatigue is a special case of dynamic loading in which inertia forces do not influence the stresses. Examples of structures that are subjected to fatigue loading include bridges, offshore structures and machine foundations.

Fatigue is one property of concrete that is not well understood, specially in terms of the mechanism of failure, because of the difficult and tedious experiments required for conducting research investigations. Detailed presentations of research findings concerning the fatigue behavior of concrete are included in the references.[2,3,21,38,40] Use of high performance concrete in structures subjected to fatigue loading requires knowledge about its behavior under such loading. As mentioned previously (Chapter 2) the performance characteristics of concrete generally improve with the strength attribute. Unfortunately, the available data on the fatigue behavior of high-strength concrete is very limited.[4,5] This chapter presents an overview of the results available in the published literature on concrete fatigue in general, with emphasis on the behavior of high-strength concrete (compressive strength >6000 psi (42 MPa)).

Fatigue is the process of cumulative damage that is caused by repeated fluctuating loads. Fatigue loading types are generally distinctly divided between high-cycle low amplitude and low-cycle high amplitude. Hsu[20] has summarized the range of cyclic loading into a spectrum of cycles. He classified the fatigue loads into three ranges, Table 4.1. The low-cycle fatigue loading occurs with less than 1000 cycles. The high-cycle fatigue loading is defined in the range of 10^3 to 10^7 cycles. This range of fatigue

Table 4.1   Fatigue load spectrum[20]

Low-cycle fatigue	High-cycle fatigue			Super-high-cycle fatigue
Structures subjected to earthquake	Airport pavements and bridges	Highway and railway bridges, highway pavements, concrete railroad ties	Mass rapid transmit structures	Sea structures

Number of cycles

loading occurs in bridges, highways, airport runways and machine founda-tions. The super-high-cycle fatigue loading is characterized by even higher cycles of fatigue loads. This category was established in recent years for the newly developed sophisticated modern structures such as elevated sections on expressways and offshore structures.

Fatigue damage occurs at non-linear deformation regions under the applied fluctuating load. However, fatigue damage for members that are subjected to elastic fluctuating stresses can occur at regions of stress concentrations where localized stresses exceed the linear limit of the material. After a certain number of load fluctuation, the accumulated damage causes the initiation and/or propagation of cracks in the concrete matrix. This results in an increase in deflection and crack-width and in many cases can cause the fracture of the structural member.

The total fatigue life N, is the number of cycles required to cause failure of a structural member or a certain structure. Many parameters affect the fatigue strength of concrete members. These parameters can be related to stress (state of stress, stress range, stress ratio, frequency, and maximum strength), geometry of the element (stress concentration location), con-crete properties (linear and non-linear behavior), and external environ-ment (temperature and aggressive elements).

Structural members are usually subjected to a variety of stress histories. The simplest form of these stress histories is the constant-amplitude cyclic-stress fluctuation shown in Fig. 4.1. This type of loading, which usually occurs in heavy machinery foundations, can be represented by a constant stress range, Δf; a mean stress f_{mean}; a stress amplitude, f_{amp}; and a stress ratio R. These values can be obtained using the following equations, Fig. 4.1.

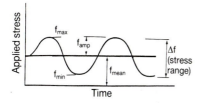

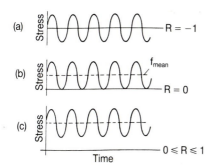

Fig. 4.1   Constant amplitude fatigue loading. (a) $R = -1$; (b) $R = 0$; (c) $0 < R < 1$

$$\Delta f = f_{max} - f_{min} \tag{4.1}$$

$$f_{mean} = \frac{f_{max} - f_{min}}{2} \tag{4.2}$$

$$f_{amp} = \frac{f_{max} - f_{min}}{2} \tag{4.3}$$

$$R = \frac{f_{min}}{f_{max}} \tag{4.4}$$

Thus, a complete reversal of load from a minimum stress to an equal maximum stress corresponds to an $R = -1$ and a mean stress of zero, as shown in Fig. 4.1(a). A cyclic stress from zero to a peak value corresponds to an $R = 0$ and a mean stress equal to half the peak stress value, Fig. 4.1(b). $R = 1$ represents a case of constant applied stress with no intensity fluctuation. Generally, range of fatigue load for concrete structures is between $R = 0$ and $R = 1$, Fig. 4.1(c). However, special cases such as lateral wave loading on offshore structures can produce a reversed loading condition with a mean stress close to zero ($R = -1$).

Variable-amplitude random-sequence stress histories are very complex, (Fig. 4.2). This type of stress history is experienced by offshore structures. During this stress history, the probability of the same sequence and magnitude of stress ranges to occur during a particular time interval is very small. To predict the fatigue behavior under such loading a method known as the Palmgren–Miner hypothesis has been proposed. This hypothesis, first suggested by Palmgren[35] and then used by Miner[32] to test notched

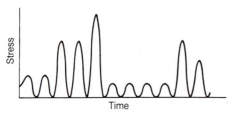

Fig. 4.2   Variable-amplitude random sequence fatigue loading

aluminium specimens, is quite simple because it assumes that damage accumulates linearly with the number of cycles applied at a particular load level. The failure equation is represented as:

$$\sum_{i=1}^{k} \frac{n_i}{N_i} = 1.0 \tag{4.5}$$

where n_i is the number of constant amplitude cycles at stress level i; N_i is the number of cycles that will cause failure at that stress level i, and k is the number of stress levels. Several researchers have checked the validity of the P–M hypothesis for concrete.[18,22,28] Their results indicated that the equality to the unity is not always true. Holmen[19] proposed a modified P–M hypothesis where an interaction factor w that depends on the loading parameters was introduced

$$\sum_{i=1}^{k} \frac{n_i}{N_i} = w \tag{4.6}$$

The factor w has been expressed as a function of the ratio (S_{min}/S_c), Fig. 4.3.

Concern with fatigue damage of concrete was recognized early in this century. Van Ornum[50,51] observed that cementitious composites possessed the properties of progressive failure, which become total under the repetition of load well below the ultimate strength of the material. He also noticed that the stress-strain curve of concrete varies with the number of repetitions, changing from concave towards the strain axis (with a hysteresis loop on unloading) to a straight line, which shifts at a decreasing rate (plastic permanent deformation) and finally to concave toward the stress axis, Fig. 4.4. The degree of this latter concavity is an indication of how near the concrete is to failure.

The fatigue strength can be represented by means of *S-N* curves (known also as the *f–N* curve or the Wohler's curve of the fatigue curve). In these curves (Fig. 4.5.) *S* is a characteristic stress of the loading cycle, usually indicating a stress range or a function of the maximum and minimum stress and *N* is the number of cycles to failure. *S* is expressed with a linear scale while *N* is presented using log scale. Using this format, usually the data can be approximated to a straight regression line. At any point on the curve, the stress value is the 'fatigue strength' (i.e., the value of stress range that

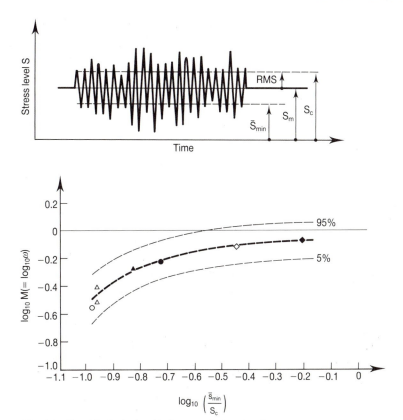

Fig. 4.3 Empirical relationship between fatigue loading parameters and the factor w (= Miner sum at failure)[19]

will cause failure at a given number of stress cycles and at a given stress ratio) and the number of cycles is the 'fatigue life' (i.e., the number of stress cycles that will cause failure at a given stress range and stress ratio). To include the effect of the minimum stress f_{min}, and the stress range $f_{max} - f_{min}$, the fatigue strength can also be represented by means of a

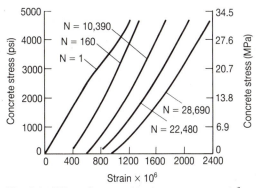

Fig. 4.4 Effect of repeated load on concrete strain[3]

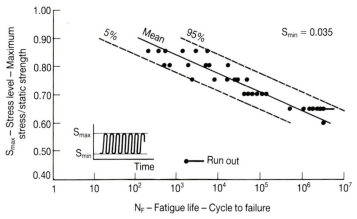

Fig. 4.5   Typical *S-N* relationship for concrete in compression[28]

$f_{max} - f_{min}$ modified Goodman diagram, shown in Fig. 4.6. The maximum and minimum stress levels in this diagram are expressed in terms of the percentage of the static strength. The diagram is for a given number of cycles to failure (e.g., 2 million cycles). From such a diagram, for a specified f_{min}/f_c' value, the allowable ratio f_{max}/f_c' can be determined.

4.2   Mechanism of fatigue

Considerable research has been done to study the nature of fatigue failure. However, the mechanism of fatigue failure is still not clearly understood. Researchers have measured surface strains, change in pulse velocity, internal and surface cracks in an attempt to understand the phenomenon of fatigue fracture. Large increase in the longitudinal and transverse strains and decrease in pulse velocity have been reported prior to fatigue failure. No satisfactory theory of the mechanics of fatigue failure has yet been

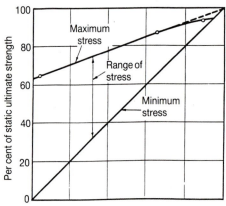

Fig. 4.6   Modified Goodman diagram for concrete subjected to repeated axial loading[16]

proposed for either normal or high strength concrete. However, with the present stage of knowledge the following two observations can be stated.[42]

1   The fatigue of concrete is associated with initiation and propagation of internal microcracks at the cement paste-aggregate interface and/or within the cement paste itself.
2   Cracks due to fatigue failure are more extensive than cracks initiated by static compressive failure.

4.3   Cyclic compression

Graf and Brenner[16] studied the effect of the minimum stress and stress range on the fatigue strength of concrete. Based on a fatigue criteria of 2 million cycles of loading, they developed a modified Goodman diagram for the repeated compressive loading, Fig. 4.6. Both maximum stress and minimum stress level were expressed in terms of the percentage of the static compressive strength.

Aas-Jakobson[1] studied the effect of the minimum stress. He observed that the relationship between f_{max}/f_c' and f_{min}/f_c' was linear for fatigue failure at 2 million cycles of loads. Based on statistical analysis of data, he proposed a general model between $\log N$ and the cycles of stresses using a factor β. He gave a value of $\beta = 0.064$.

$$\log N = \frac{1}{\beta} \left[\frac{1 - f_{max}/f_c'}{1 - f_{min}/f_{max}} \right] \qquad (4.7)$$

Tepfers and Kutti[48] compared their experimental fatigue strength data of plain normal and lightweight concrete with the equation proposed by Aas-Jakobson.[1] Their results indicated the use of $\beta = 0.0679$ for normal weight concrete and $\beta = 0.0694$ for lightweight concrete when $R < 0.80$. They recommended to use a mean value of $\beta = 0.0685$ for both normal and lightweight concrete.

Kakuta *et al.*,[23] based mainly on the Japanese tests, proposed the following expression for the fatigue of concrete

$$\log N = 17 \left[1 - \frac{(f_{max} - f_{min})/f_c'}{1 - f_{min}/f_c'} \right] \qquad (4.8)$$

Gray *et al.*[17] conducted an experimental investigation on the fatigue properties of high strength lightweight aggregate concrete under cyclic compression. They tested 150 3×6 in. (75×150 mm) cylinders using maximum stress levels of 40, 50, 60, 70, and 80% of the static ultimate strength and minimum stress levels of 70 and 170 psi (0.5 to 1.2 MPa). Their test results indicated that there was no difference in the fatigue properties between normal and high strength lighweight concrete. The variation of rate of loading between 500–1000 cycles/minute had no effect on the fatigue properties. They also observed no fatigue limit up to 10 million cycles of loading.

Bennett and Muir[7] studied the fatigue strength in axial compression of high strength concrete using 4-in. (100 mm) cubes. The compressive strength was as high as 11,155 psi (78 MPa). They found that after one million cycles, the strength of specimens subjected to repeated load varied between 66% and 71% of the static strength for a minimum stress level of 1250 psi (8.75 MPa). The lower values were found for the higher-strength concretes and for concrete made with smaller-size coarse aggregate. However, the actual magnitude of the difference was small.

Equations (4.7) and (4.8) represent a significant contribution towards summarizing the available test data on normal strength concrete. However, as pointed out by Hsu,[20] the equation has two main limitation, namely: the static strength f_c' which is used to normalize the maximum stress f_{max} is time-dependent, and the rate of loading is not included. Hsu[20] introduced the effect of time (T) into the (stress-number of cycles) relationship where T is the period of the repetitive loads expressed in sec/cycle. With this approach, a three-dimensional space is created consisting of non-dimensional stress ratio, f, as the vertical axis with $\log N$ and $\log T$ as the two orthogonal horizontal axes. A graphical representation of such a space is shown in Fig. 4.7. Based on the available experimental data in literature, Hsu proposed two equations, one for high-cycle fatigue and the other for low-cycle fatigue.

For high-cycle fatigue

$$\frac{f_{max}}{f_c'} = 1 - 0.0662\,(1 - 0.566R)\log N - 0.0294\log T \qquad (4.9a)$$

For low-cycle fatigue

$$\frac{f_{max}}{f_c'} = 1.2 - 0.2R - 0.133(1 - 0.779R)\log N - 0.05301\,(1 - 0.445R)\log T$$
$$(4.9b)$$

These equations were found to be applicable for the following conditions:

(a)   normal weight concrete with f_c' up to 8000 psi (56 MPa)
(b)   for stress range $0 < R < 1$
(c)   for load frequency range between 0 to 150 cycles/sec
(d)   for number of cycles from 1 to 20 million cycles
(e)   for compression and flexure fatigue.

Chimamphant[10] conducted series of experiments on uniaxial cyclic compression of high-strength concrete. Concrete strength varied from 7500 psi (52 MPa) to 12,000 psi (84 MPa). Maximum stress level varied from $0.4 f_c'$ to $0.9 f_c'$ while the minimum stress level was kept constant at $0.1 f_c'$. Two different rates of loading were used, namely: 6 cycles/sec and 12 cycles/sec. He observed no significant difference in the fatigue behavior of high strength concrete when compared to normal strength concrete. He reported that up to 1 million cycles the S-N curve of high-strength concrete was linear. He also observed no measurable effect on the fatigue strength

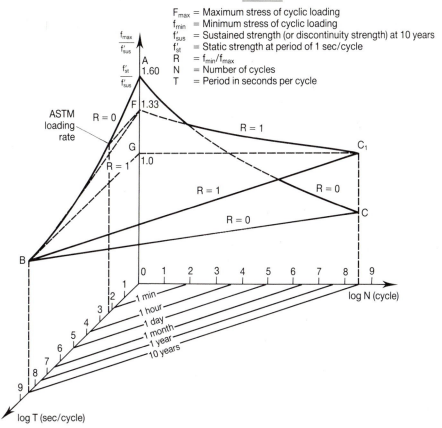

Notations

F_{max} = Maximum stress of cyclic loading
f_{min} = Minimum stress of cyclic loading
f'_{sus} = Sustained strength (or discontinuity strength) at 10 years
f'_{st} = Static strength at period of 1 sec/cycle
R = f_{min}/f_{max}
N = Number of cycles
T = Period in seconds per cycle

Fig. 4.7 Graphical representation of f-N-T-R relationship[20]

of high-strength concrete when the loading rate was changed from 6 to 12 cycles/sec.

Petkovic *et al.*[36] performed studies on the fatigue properties of high-strength concrete in compression. Two types of normal-weight concrete with compressive strengths of 8000 psi (56 MPa) and 10,900 psi (76 MPa) and one type of lightweight aggregate concrete with a compressive strength of 11,600 psi (81 MPa) were tested. Their experimental results gave an indication of the existence of a fatigue limit, which cannot be defined as one level of loading, since its correlation to different loading parameters must be taken into account. They presented design rules for fatigue in compression, which could be applied to the three types of concrete tested (Fig. 4.8). The following three ranges can be distinguished based on the number of cycles. Retion 1: from the beginning of loading to $\log N = 6$, the *S-N* curve lines follow the expression

$$\log N = (1 - S_{max}) \times (12 + 16 S_{min} + 8 S_{min}^2) \qquad (4.10a)$$

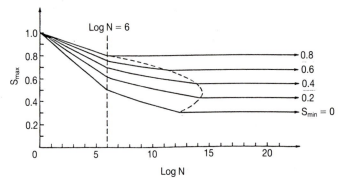

Fig. 4.8   The *S-N* diagram for failure of high strength concrete in compression[36]

Region 2: beyond the value of $\log N = 6$, the inclination of these lines is changed and the $\log N$ value is found by multiplying the basic expression in Equation (4.10a) by the coefficient from

$$C = 1 + 0.2(\log N - 6) \qquad (4.10b)$$

Region 3: the fatigue limit for different values of S_{min}. The position of S_{min} the transfer to the fatigue limit is given by the stippled curve. Its shape was only a consequence of the chosen expressions and hence, has no physical significance.

4.4   Cyclic tension

Because of difficulties encountered in applying direct tensile cyclic loads on concrete specimens and avoiding eccentricity of loading, most tension fatigue studies were conducted using indirect tension tests such as splitting tests or beams in flexure. Tepfers[44] performed splitting fatigue tests on 6 in. (150 mm) concrete cubes. Two types of concrete with strengths of 5900 psi (41 MPa) and 8200 psi (57 MPa) were used. The selected stress ratios were 0.20, 0.30 and 0.40. He observed that the fatigue strength equation similar to Equations (4.7) and (4.8) could be used for tension fatigue. He also found that the concrete strength had no effect on fatigue strength when the non-dimensional form (f_{max}/f_c') was used.

Clemmer[12] studied the flexure fatigue of plain concrete. He reported that the fatigue limit for concrete was 55% of the static ultimate flexural strength. Kesler[25] found no fatigue limit but established fatigue strength at 10 million cycles of stress ranging from a small tension value to a maximum value. He reported a fatigue strength of 62% of the static ultimate flexural strength. Williams[52] found no fatigue limit for lightweight aggregate beams. A comprehensive experimental study was conducted by Murdock and Kesler[33] on 175 concrete $6 \times 6 \times 60$ in. ($152 \times 152 \times 1520$ mm) prisms having a compressive strength of 4500 psi (31 MPa). Specimens were loaded at the third points in order to avoid shear stresses at the middle span. Three stress ratios were used, namely: 0.25, 0.50 and 0.75. They observed no fatigue limit for plain normal weight concrete subjected to repeated flexure loading of at least 10 million cycles. They found also that stress range has a significant influence on fatigue strength. They proposed the following fatigue strength equation in terms of stress range at 10 million cycles

$$F_{10} = 0.56 + 0.44M \quad \text{or} \quad F_{10} = \frac{1.3}{(2.3 - R)} \qquad (4.11)$$

where $M = \dfrac{f_{r_{min}}}{f_r}$ and $R = \dfrac{f_{r_{max}}}{f_{r_{min}}}$

for values of M and R between 0 and 1.
for stress reversal, they proposed

$$F_{10} = 0.56$$

where $-0.56 < M < 0$ and $-1 < R < 0$

In order to evaluate the probability of fatigue failure, McCall[29] performed studies on air-entrained $3 \times 3 \times 14.5$ in. $(75 \times 75 \times 368$ mm$)$ concrete prisms. The mean modulus of rupture was 0.68 ksi. Using a loading rate of 1800 cycles/minute, the beams were tested to failure or to 20 million cycles whichever occurred first. The maximum flexure stress varied from 45% to 70% of the concrete modulus of rupture. He observed no fatigue limit in a range up to 20 million cycles. Based on his results, he proposed the following mathematical model for flexure fatigue strength

$$L = 10^{-0.0957} R^{3.32} (\log n)^{3.17} \qquad (4.12)$$

where L = probability of survival = $1 - P$
P = probability of failure
$R = S/S_{rp}$
S = stress range used in the test
S_{rp} = mean static strength

They found the probability of failure at 20 million cycles to be slightly less than 0.5 for concrete tested at a stress level of 50% of modulus of rupture.

Direct tensile fatigue tests were first conducted by Kolias and Williams.[27] Using constant-amplitude stress range and a constant minimum stress level, they conducted tests on concrete mixes with different coarse aggregates. Their results showed that finely graded concrete has a shorter fatigue life. They suggested that this behavior is due to the more brittle nature of concrete with small aggregate size. Cornelissen[13] conducted an extensive experimental study on 250 necked cylindrical specimens $(4.8 \times 12$ in., 122×305 mm$)$. Using a constant amplitude loading at a frequency of 6 cycles/sec., he studied the effect of maximum and minimum stress levels. The direct uniform tensile stress was applied by bonding (using epoxy) the loading plates on to the top and bottom ends of the cylinders. He derived two stress-number of cycles equations; one for dried specimens (4.13a) and one for sealed specimens (4.13b)

$$\log N = 14.81 - 14.42 \left| \frac{\sigma_{max}}{f_c'} \right| + 2.79 \left| \frac{\sigma_{min}}{f_c'} \right| \tag{4.13a}$$

$$\log N = 13.92 - 14.42 \left| \frac{\sigma_{max}}{f_c'} \right| + 2.79 \left| \frac{\sigma_{min}}{f_c'} \right| \tag{4.13b}$$

Using these equations, the tensile fatigue strength would be 60% and 54%, respectively, of the static strength for 2 million cycles of loading.

Saito and Imai[39] used friction grips to conduct direct tension fatigue tests on $2.8 \times 2.8 \times 29$ in. ($71 \times 71 \times 736$ mm) concrete prisms with enlarged ends having a compressive strength of 5600 psi (39 MPa). Sinusoidal pulsating loads were applied at a constant rate of 240 cycles/minute. Maximum stress levels varied from 75% to 87.5% of the static strength while minimum stress level was maintained at 8%. The ratio of minimum to maximum stress, R, was in the range of 0.09 to 0.11. The surfaces of all specimens were coated with paraffin wax to prevent drying during fatigue test. Based on their results, they proposed the following S-N relationship for a 50% probability of failure

$$S = 98.73 - 4.12 \log N \tag{4.14}$$

where S = maximum applied stress range (as percentage of f_c')
         N = number of cycles to failure.

Using their equation, they estimated the fatigue strength for 2 million cycles under direct tensile loading to be 72.8% of the static strength. They observed that this fatigue strength was considerably higher than fatigue strength under indirect tension tests.

4.5   Reversed loading

Tepfers[45] studied the fatigue of plain concrete with static compressive strength in the range of 3000 to 10,000 psi (21 to 70 MPa) subjected to cyclic compression-tension stresses. He used two different testing configurations. One was transversely compressed concrete cubes subjected a pulsating splitting load, and the other was concrete prisms with axial pulsating compressive loads and central splitting line loads. He observed a slight reduction in the fatigue strength of concrete subjected to reversed cyclic loading when compared to the fatigue strength of concrete in compression. He suggested that this reduction could be due to the difficulties in loading the specimens precisely on the tensile side of the pulse. He concluded that the fatigue strength equation proposed by other investigators[1,23,47] could be used to predict the fatigue strength due to reversed stresses.

4.6 Effect of loading rate

Kesler[25] evaluated the effect of loading rate on concrete fatigue strength. Using three different rate of loading, namely: 70, 230 and 440 cycles/ minute, and two different concrete strength ($f_c' = 3600$ and 4600 psi (25 to 32 MPa)), he concluded that the rate of loading within the range used in the investigation had little or no effect on the fatigue strength.

Sparks and Menzies[41] used a triangular wave form with constant loading and unloading to study the effect of loading rate on the fatigue compression strength of plain concrete made of three different types of coarse aggregate. The rate of loading were 70 and 7000 psi (0.5 to 49 MPa)/s. The minimum stress was constant ($0.33f_c'$) while the maximum stress varied between $0.70f_c'$ and $0.90f_c'$. They observed an increase in the fatigue strength with an increase in rate of loading of the fatigue load.

4.7 Effect of stress gradient

Ople and Hulsbos[34] studied the effect of stress gradient on the fatigue strength of plain concrete. They tested $4 \times 6 \times 12$ in. ($100 \times 152 \times 304$ mm) prisms under repeated compression at arate of 500 cycles/minute with three different eccentricities (0, 1/3 and 1 in. (0, 8.4, 25.4 mm)). The tests were performed until failure or up to 2 million cycles. Keeping the minimum stress constant at 10% of the static strength, they varied the maximum stress from 65% to 95% the static strength. They found that the mean *S-N* curves of both concentrically and eccentrically loaded samples were parallel. They concluded that the fatigue strength of eccentrically stresses specimens was higher than that of concentrically stressed speimens by about 17% of the static strength. They also reported that the fatigue life of both type of specimens was highly sensitive to small variations in maximum stress levels.

4.8 Effect of rest periods

Hilsdorf and Kesler[18] investigated the fatigue strength of 185 $6 \times 6 \times 60$ in. ($150 \times 150 \times 1500$ mm) plain concrete prisms subjected to varying flexure stresses. Using a rate of 450 cycle/minute and a ratio of minimum/ maximum stress of 0.17, they loaded the specimens until failure or 1 million cycles with five different rest periods of 1, 5, 10 and 27 minutes. Their results indicated that the increase in the rest period increased the fatigue strength for a specified fatigue life. This was clear when the length of rest period increased from 1 to 5 minutes. From 5 to 27 minutes of rest periods, the fatigue strength did not show any variation.

4.9   Effect of loading waveform

Tepfers *et al.*[46] studied the effect of loading waveforms on the fatigue strength. They used three different waveforms, sinusoidal, triangular and rectangular. The results indicated that the triangular waveform was less damaging than the sinusoidal, while the rectangular waveform was the most damaging.

4.10   Effect of minimum stress: comparison of normal and high strength concrete

Petkovic *et al.*[36] conducted tests on high strength concrete (8000 psi (64 MPa) to 11,600 psi (87.2 MPa)) subjected to constant amplitude but different levels of stress. The minimum stress level varied from 0.05 to 0.6 and the maximum stress level varied from 0.6 to 0.95 of the compressive strength. The results of the tests showed no reason to distinguish between the fatigue properties of normal and high strength concrete when the stress levels are expressed relatively to the static strength of the concrete.

4.11   Effect of concrete mixture properties and curing

Raithby and Galloway[37] studied the effects of moisture condition, and age on fatigue of plain concrete with static strength in the range of 3000 psi (21 MPa) to 6400 psi (44.8 MPa). They observed that the moisture condition significantly affected the fatigue life. Oven-dried concrete showed the longest fatigue life while partially dried concrete gave the shortest. The fully saturated concrete exhibited intermediate fatigue life. They suggested that the difference in strains generated by moisture gradient with the concrete could be the cause of such performance. They also found that the mean fatigue life increased with the age of concrete. The mean fatigue life of 2 years old concrete was 2000 times the fatigue life of 4 weeks old concrete.

Klaiber and Lee[26] studied the effect of air content, water-cement ratio, coarse-aggregate type and fine aggregate type. The concrete static strength ranged from 2000 to 7400 psi (14 to 51 MPa). After testing 350 $6 \times 6 \times 36$ in. ($152 \times 152 \times 912$ mm) beams under flexural fatigue, they found that of the variables investigated, air content and coarse aggregate type had the greatest effect on flexural fatigue strength. The fatigue strength decreased with the increase of the air content, and concrete made of gravel yielded higher fatigue strength than concrete made with limestone. Water/cement ratio also affected fatigue strength but to a lesser degree. The fatigue strength decreased for low water/cement ratio (less than 0.32) but seemed not affected for higher water/cement ratios.

In their study on the fatigue behavior of high strength concrete, Petkovic *et al.*[36] investigated the influence of different moisture conditions and size of test specimens on fatigue. Cylinder sizes of 2×6 in. (50×150 mm) and 4×12 in. (100×300 mm) for three different concrete types were studied under three moisture conditions: in air, sealed and in water. The cyclic loading was sinusoidal with the maximum and minimum load levels equal to 70% and 5%, respectively, of the static strength. They found the moisture effects on fatigue to be scale dependent. Dried specimens of small dimensions gave generally longer fatigue lives. The sealed conditions was found to give results closer to the immersed specimens than to the specimens exposed to air. They concluded that sealed cylinders are preferable for fatigue tests carried out using relatively small specimens.

4.12 Biaxial state

Takhar *et al.*[43] studied the fatigue behavior of 96 concrete cyclinders with a compressive strength of 5000 psi (35 MPa) subjected to three different confining pressures (0, 1000 and 2000 psi (14 MPa)). They used sinusoidal load at a rate of 60 cycle/minute. Keeping the minimum stress level at $0.2f_c'$ they varied the maximum stress level (0.8, 0.85 and $0.9f_c'$). They found that the increase in confining pressure prolonged the fatigue life of concrete. The effect of the lateral confining pressure was dependent on the maximum stress level of the fatigue load. For a maximum stress level of $0.90f_c'$, the difference in fatigue behavior with or without the lateral confining pressure was not significant, while for a maximum stress level of $0.80f_c'$ the difference was significant.

Traina and Jeragh[48] performed an experimental investigation to study the behavior of plain concrete with compressive strength of 4000 psi (28 MPa) subjected to slow cyclic loading in compressive biaxial states. Three inch concrete cubes were subjected to two types of biaxial stress states. The first was a proportional loading type in which two loading paths are used, namely $\sigma_2/\sigma_1 = 1.0$ and $\sigma_2/\sigma_1 = 0.5$. The second loading type consisted of a constant stress in the direction with a cyclic stress in the σ_1 direction. The cyclic stress varied from zero up to 1.2 of the unconfined ultimate compressive strength. They found that concrete tested under all biaxial states of stress exhibited higher fatigue strength than uniaxial states of stress for any given number of cycles. They also observed that the stress-strain response of concrete is dependent on the stress level and number of load repetitions for both uniaxial and biaxial states of stress. They noticed a limiting value of volume change per unit volume at which concrete may be considered either failed or near failure. This limiting value was found to be higher for all biaxial states of stress and independent of the stress level at which concrete was subjected to fatigue loading.

4.13   Bond properties

In addition to portland cement, water, aggregates, reinforcing bars and/or prestressing reinforcerment, fresh high-strength concrete usually contain chemical admixtures and pozzolanic materials. Hence, in cured high-strength concrete several types of interface exist, namely: (a) interfaces between the various chemical components that make up the hydrated cement paste (hcp), (b) interfaces between (hcp) and other unhydrated cement particles and added pozzolanic materials, (c) interfaces between (hcp) and coarse aggregates, and (d) interfaces between the concrete matrix and the steel reinforcement.

The bond at the interface at any of the preceding levels is the outcome of a combination of mechanical interlock, physical bonding involving van der Waals' forces and chemical ionic reactions between the different phases of the hydrated paste. Mindess[30,31] discussed the importance of these types of bonds with respect to the behavior of concrete. The following sections cover the first three types of interfacial bonds.[30,31]

Hydrated cement paste interfaces

Hydrated cement paste consists of individual chemical components of hydration products. The paste derives its strength from: (a) intraparticle bonds, represented by the inter-atomic forces within the individual chemical components resulting from hydration (C-S-H and $Ca(OH)_2$); chemical ionic-covalent bonds are considered to be the major 2 source of intraparticle bonds; and (b) interparticle bonds, originated due to the atoms forces which attract the individual paste particles to each other. Physical bonds of the van der Waals' type act primarily between particles.

Micro-structure studies of cement paste indicate that the C-S-H is very well bonded to the various hydration phases. In addition, a strong adhesion seems to exist between the C-S-H and the $Ca(OH)_2$. These studies suggest that the strength of the cement mortar is controlled by the total porosity, the pore size distribution and any existing macroscopic flaws, rather than by the destruction of the bond between its components.

Hydrated cement paste–pozzolanic materials interfaces

Most of the high-strength concrete currently being produced contains silica fume and/or fly ash. The micro-structure that is obtained is distinctly different from that of traditional cement pastes. The structure becomes much denser and more amorphous. Boundaries between C-S-H particles are not clearly identified. However, the strength of the resulting hydrated cement paste is not significantly improved by this interparticle bond enhancement. It was demonstrated that the cement paste with and without silica fume yields the same strength at equal water/binder ratios. This would confirm that enhancing interparticle bonds would have only a minor effect on the strength of hydrated cement paste.

Hydrated cement paste–aggregate interfaces

Many investigations have dealt with the cement–aggregate interfaces. In general, there is an agreement that for normal strength concrete this interfacial region is the weak link in the concrete matrix. This is because bleed water accumulates at the lower surface of the coarse aggregate particles creating a porous paste and planes of weakness. This interface zone is generally composed of a duplex film and a transition zone. The duplex film which is usually no more than 1 μm thick is formed of about 0.5 μm thin layer of $Ca(OH)_2$ in contact with the aggregate followed by another thin layer of C-S-H. The composition of this 'transition zone' is deeply different from the bulk cement paste.

Addition of pozzolanic materials increase the strength of concrete. This is achieved primary because they are capable of producing a great reduction in the relative thickness of the transition zone. This would yield a better overall homogeneity producing higher strength for the paste–aggregate link. The extent to which this enhancement is achieved varies according to the characteristics of the individual pozzolanic material, its addition percentage and the age of curing. The silica fume was found to be the most effective pozzolanic additive in reducing the thickness of the transition zone and in achieving better overall homogeneity of the matrix. Hence, pozzolanic admixtures in general improves the bond at the hcp–aggregate interfaces. By strengthening this weak link, higher strength concrete is obtained. This bond at the interface is, sometimes, greatly improved, resulting in strength of the coarse aggregate becoming the limiting factor of the high strength that can be achieved. There is one major disadvantage which arises from an increase in the cement–aggregate bond strength. That is the ductility of the concrete decreases. This could be attributed to the decrease in extensive microcracking at the interface before failure.

While some knowledge about the micro-structure at the cement–aggregate interfaces has been accumulated, considerable research is still required to provide a more general and global understanding of interfacial micro-structure behavior of normal and high strength concrete.

Reinforcing steel–concrete bond

Reinfoced concrete is a structural material whose effectiveness depends on the interaction between the concrete matrix and reinforcement rebars. Three mechanisms can be identified that contribute to the bond between concrete and steel reinforcement:[6]

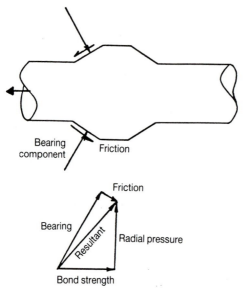

Fig. 4.9   Bond strength components for bar embedded in concrete[50]

1   Adhesive bond between steel and concrete matrix.
2   Frictional bond between the steel and the surrounding matrix.
3   Mechanical anchoring of the steel to the concrete through the bearing
    stresses that develop between the concrete and the deformations of the
    steel bars.

When a bar is pulled, the rib bears against the surrounding concrete. Friction and adhesion between the concrete and steel along the face of the rib act to prevent the rebar from sliding. This force adds vectorially to the bearing stress acting perpendicular to the rib to yield the bond strength (Fig. 4.9). The bearing stress is controlled by the radial pressure that the concrete cover and lateral reinforcement can resist before splitting and the effective shear strength capable of shearing the concrete surrounding the rebar. Pullout failure occurs when the steel bar is well confined by concrete cover or transverse reinforcement prevents a splitting failure. The pullout failure is primarily due to bearing of the ribs against the concrete causing the key between ribs to shear from the surrounding concrete.

Clark's[11] study had firmly established the effectiveness of using deformed reinforcing bars in concrete. Chapman and Shah[9] conducted an investigation to determine the bond strength between reinforcing steel and concrete at early age. They found that smooth bars did not exhibit any age effect, while the bond behavior of deformed bars was highly age dependent. They concluded that adhesion and friction contribution to the bond strength is relatively small compared to the bond strength that derives from the bearing stresses that develop between the deformations on the steel and the surrounding concrete.

Brettmann *et al.*[8] studied the effect of superplasticizers, extensively used when producing high strength concrete, on concrete–steel bond strength. They included the effect of degree of consolidation, concrete slump, concrete temperature and bar position on the bond strength of #8 deformed reinforcing bars embedded in concrete with and without super-plasticizer. The bond tests were conducted at concrete strength between 4000 and 4800 psi (28 to 33 MPa) using a modified cantilever beam.

The experimental results indicated that high-slump superplasticized concrete provided a lower bond strength than low-slump concrete of the same strength. They also observed that vibration of high-slump concrete increased the bond strength compared to high-slump concrete without vibration.

Treece and Jirsa[49] conducted experimental investigation to study the bond strength and epoxy coated reinforcing bars embedded in normal and high strength concrete, and compared it to that of uncoated bars. Twenty-one beams with splices in a constant moment region were tested to evaluate the effect of bar size, concrete strength, casting position and coating thickness. Concrete strength varied from 3860 to 10,510 psi (27 to 74 MPa). The results showed that epoxy coating significantly reduced the bond strength of reinforcing bars; for splitting failure the bond strength was about 65% of the bond strength of uncoated bars while for a pullout failure, the bond strength was about 85% of that for uncoated bars. The results also indicated that the reduction in bond strength was independent of bar size and concrete strength, and that the bond strength was not affected by the variations in the coating thickness when the average coating thickness was between 5 and 14 mil (1 mil = 0.025 mm).

Kemp[24] conducted a comprehensive experimental research plan that included a study of the influence of reinforcing bar, embedment length and spacing, stirrups, concrete cover and associatd concrete strength (up to 6000 psi (42 MPa)), and the interaction of shear and flexural bond behavior. He proposed the following design equation which he found suitable for ultimate load design.

$$(F_b)_{\text{ult}} = 232.2 + 2.716\left(\frac{C_{bs}}{D_{ia}}\sqrt{f_c'}\left[+0.201\right)\frac{A_{sst}f_{yst}}{S_pD_{ia}}\right] \qquad (4.15)$$

$$+ 195.0I_{\text{aux}} + 21.06(F_dN)^{0.66}$$

where
A_{sst} = area of transverse reinforcement, in.^{2}
C_{bs} = the smallest concrete cover, in.
D_{ia} = diameter of reinforcing bar, in.
f_c' = concrete compressive strength, psi
F_d = dowel force psi bar, kip/bar.
f_{yst} = yield strength of transverse reinforcement, psi
I_{aux} = parameter for auxiliary reinforcement (=1 when the member has auxiliary reinforcement and 0

when the specimen is without auxiliary reinforcement)

S_p = center-to-center spacing between two adjacent transverse reinforcement, in.

N = number of load cycles.

Gjørv et al.[15] studied the effect of silica fume on the mechanical behavior of the steel reinforcement-concrete bond, the concrete compressive strength from 3000 psi (42 MPa) to 12,000 psi (84 MPa) with and without silica fume. Using a #6 bar, they found that for the same compressive strength increased addition of silica fume up to 16% by weight of cement showed an improving effect on the bond strength, especially in the high compressive stress range. They justified this effect by several mechanisms: reduced accumulation of free water at the interface during casting of the specimens, reduced preferential orientation of CH crystals at the transition zone, and densification of the transition zone due to pozzolanic reaction between CH and silica fume.

Chimamphant[10] conducted pull-out tests on three different reinforcing bar diameters (#3, #4 and #5) embedded in concrete having a compressive strength ranging from 7500 psi (52 MPa) to 12,000 psi (84 MPa). He observed that the larger the bar diameter, the lower the average bond strength. In this study, the average bond strength factor (average bond strength/compressive strength) was about 0.212. Most reported values for that factor when using nominal strength concrete range from 0.15 to 0.26. These values get higher if lateral confinement is provided. The results of this study indicated that normalized bond in both high strength concrete and normal strength concrete are essentially the same.

Using 20% (by weight) of silica fume to obtain high strength concrete, Ezeldin and Balaguru[14] conducted experimental studis on the bond behavior of bars embedded in high tensile concrete with and without steel fibers (compressive strength up to 11,800 psi (82 MPa)). Four bar diameters were used, namely, #3, #5, #6 and #8. They found that the addition of silica fume increased the bond strength of concrete. However, in presence of fibers, the proportionality constant between bond strength and square root of the compressive strength seemed to be constant. They concluded that the bond strength equations used for normal strength concrete (without silica fume) could be used for high strength fiber reinforced concrete (with silica fume).

4.14   Summary

The use of high strength concrete in modern structures and the widespread adoption of ultimate strength design procedures renewed the interest in studying the fatigue and bond properties of high strength concrete.

Most of the fatigue investigations have been performed on concretes of normal strength. These investigations have established the effect of the range of stress, load history, rate of loading, stress gradient, curing and material properties on the fatigue behavior. The limited fatigue studies conducted on high strength concrete covered the effects of these variables on the fatigue properties. Their results indicated no signifcant difference between the fatigue behavior of normal and high strength concretes when the stress levels are expressed relatively to the static strength of concrete.

The bond at the hydrated cement paste–aggregate interfaces is greatly improved in high strength concrete because of the addition of pozzolanic materials. This could result in the coarse aggregate characteristics becoming the limiting factor of the high strength that can be achieved. The average reinforcing steel–concrete matrix bond is increased for high strength concrete when compared to normal strength concrete. More research is needed on the morphology and micro-structure of the steel–cement paste transition zone in order to characterize the effect of including pozzolanic materials in high strength concrete on the bond properties.

References

1 Aas-Jakobson, K. (1970) *Fatigue of concrete beams and columns*, Bulletin No. 70-1. NTH Institute of Betonkonstruksjoner, Trondheim.
2 ACI Committee 215 (1974) *Fatigue of concrete*. American Concrete Institute, SP-41.
3 ACI Committee 215 (1974) Consideration for design of concrete structures subject to fatigue loading. *Journal of the American Concrete Institute, Proceedings*, **71**, 3, 97–121. Revised in 1986.
4 ACI Committee 363 (1984) State-of-the-art report on high-strength concrete. *ACI Journal*, Title No. 81-34, 364–411.
5 ACI Committee 363 (1987) Research needs for high-strength concrete. *ACI Journal*, Title No. 84-M49, 559–61.
6 Bartos, P. (ed.) (1982) *Bond in concrete*. Applied Science Publishers.
7 Bennett, E.W. and Muir, S.E.st.J. (1967) Some fatigue tests on high-strength concrete in axial compression. *Magazine of Concrete Research*, London, **19**, No. 59, 113–17.
8 Brettmann, B., Darwin, D. and Donahey, R. (1986) Bond of reinforcement to superplasticized concrete, Proceedings. *ACI Journal*, 98–107.
9 Chapman, R.A. and Shah, S.P. (1987) Early age bond strength in reinforced concrete. *ACI Journal*, **84**, No. 6, 501–10.
10 Chimamphant, S. (1989) Bond and Fatigue Characteristics of High-Strength Cement-Based Composites, Ph.D. Dissertation, New Jersey Institute of Technology, Newark, New Jersey.
11 Clark, P. (1949) Bond of concrete reinforcing bars. *ACI Journal, Proceedings*, **46**, 3, 161–84.
12 Clemmer, H.E. (1922) Fatigue of concrete. *Proceedings, ASTM*, **22**, Part II, 408–19.
13 Cornelissen, H.A.W. and Timmers, G. (1981) *Fatigue of plain concrete in uniaxial tension and alternating tension- compression*, Report No. 5-81-7, Stevin Laboratory, University of Technology, Delft.
14 Ezeldin, A.S. and Balaguru, P.N. (1989) Bond behavior of normal and high-strength fiber reinforced concrete. *ACI Materials Journal, Proceedings*, **86**, 5, September–October, 515–24.
15 Gjørv, O.E., Monteiro, P.J. and Mehta, P.K. (1990) Effect of condensed silica

fume on the steel–concrete bond. *ACI Materials Journal, Proceedings*, **87**, 6, Nov–Dec, 573–80.

16 Graf, O. and Brenner, E. (1934/1936) *Experiments for investigating the resistance of concrete under often repeated loads* (Versuche Zur Ermittlung der Widerstandsfahigkeit Von Beton gegen oftmals Wiederholte Druck-belastung), Bulletins No. 76 and No. 83, Deutscher Ausschuss fur Eisen-beton.

17 Gray, W.H., McLaughlin, J.F. and Antrim, J.D. (1961) Fatigue properties of lightweight aggregate concrete. *ACI Journal*, Title No. 58-6, August, 149–61.

18 Hilsdorf, H.K. and Kesler, C.E. (1966) Fatigue strength of concrete under varying flexural stresses. *Journal of American Concrete Institute, Proceedings*, **63**, 10, Oct, 1069–76.

19 Holmen, J.O. (1979) *Fatigue of concrete by constant and variable amplitude loading*, Bulletin No. 79-1, Division of Concrete Structures, Norwegian Institute of Technology, University of Trondheim.

20 Hsu, T.C. (1981) Fatigue of plain concrete. *ACI Journal, Proceedings*, Title No. 78-27, July–August, 292–305.

21 (1982) *International association for bridge and structural engineering*, Proceedings of Colloquium, Lausanne, IABSE Reports, **37**.

22 Jinawath, P. (1974) Cumulative Fatigue Damage of Plain Concrete in Compression, Ph.D. Thesis, University of Leeds.

23 Kakuta, Y. *et al.* (1982) New concepts for concrete fatigue design procedures in Japan. *Proceedings, IABSE*, Lausanne, **37**, 51–8.

24 Kemp, E. (1986) Bond on reinforced concrete: behavior and design criteria. *ACI Journal, Proceedings*, **82**, 1, Jan–Feb, 49–57.

25 Kesler, C.E. (1953) Effects of speed of testing on flexural fatigue strength of plain concrete. *Proceedings, Highway Research Board*, **32**, 251–8.

26 Klaiber, F.W. and Lee, D.Y. (1982) *The effects of air content, water-cement ratio, and aggregate type on the flexural fatigue strength of plain concrete.* American Concrete Institute, Special Publications, SP-75, 111–32.

27 Kolias, S. and Williams, R.I.T. (1978) *Cement-bound road materials: strength and elastic properties measured in the laboratory*, TRRL Report No. 344. Transport and Research Laboratory, Crowthorne, Berkshire.

28 Leeuwen, J.V. and Siemes, J.M. (1979) Miner's rule with respect to plain concrete. *Heron*, **24**, 1, 34 pp.

29 McCall, J.T. (1958) Probability of fatigue failure of plain concrete. *ACI Journal*, Aug, 233–44.

30 Mindess, S. (1989) Interfaces in concrete, in *Materials science of concrete I*, edited by Skalny, J.P. The American Ceramic Society, 163–180.

31 Mindess, S. (1988) Bonding in cementitious composites – how important is it, in *Bonding in cementitious composites*, edited by Mindess, S. and Shah, S.P. Materials Research Society, **114**, 3–10.

32 Miner, M.A. (1945) Cumulative damage in fatigue. *Transactions, American Society of Mechanical Engineers*, **67**, A159–A164.

33 Murdock, J.W. and Kesler, C.E. (1958) Effect of range of stress on fatigue strength of plain concrete beams. *ACI Journal*, August, 221–31.

34 Ople, Jr, F.S. and Hulsbos, C.L. (1966) Probable fatigue life of plain concrete with stress gradient, Research Report. *ACI Journal*, Title No. 63-2, January, 59–81.

35 Palmgren, A. (1924) 'Die Lebensdauer von Kugellagern, VDI.' *Zeitschrift Verein Deutscher Ingenieur*, **68**, 339–41.

36 Petkovic, G., Lenschow, R., Stemland, H. and Rosseland, S. (1991) *Fatigue of high-strength concrete*. American Concrete Institute, Special Publication, SP 121-25, 505–25.

37 Raithby, K.D. and Galloway, J.W. (1974) *Effects on moisture condition, age,*

and rate of loading on fatigue of plain concrete. ACI Publications, SP-41, 15–34.

38 RILEM Committee 36-RDL (1984) Long term random dynamic loading of concrete structures. *RILEMs Materials and Structures*, **17**, 97, Jan, 1–27.

39 Saito, M. and Imai, S. (1983) Direct tensile fatigue of concrete by the use of friction grips. *ACI Journal*, Title No. 80-42, Sept–Oct, 431–8.

40 Shah, S.P. (1982) *Fatigue of concrete structures.* American Concrete Institute, SP-75.

41 Sparks, P.R. and Menzies, J.B. (1973) The effect of rate of loading upon the static and fatigue strengths of plain concrete in compression. *Magazine of Concrete Research*, **25**, 83, June, 73–80.

42 Su, E.C.M. and Hsu, T.T.C. (1986) *Biaxial compression fatigue of concrete*, Research Report UHCE 86-17. University of Houston, December.

43 Takhar, S.S., Jordaan, I.J. and Gamble, B.R. (1974) *Fatigue of concrete under lateral confining pressure.* ACI Publications, SP-41, 59–69.

44 Tepfers, R. (1979) Tensile fatigue strength of plain concrete. *ACI Journal*, Title No. 76-39, August, 919–33.

45 Tepfers, R. (1986) *Fatigue of plain concrete subjected to stress reversals.* ACI Publications, SP-75, 195–215.

46 Tepfers, R., Gorlin, J. and Samuelsson, T. (1973) Concrete subjected to pulsating load and pulsating deformation of different pulse wave-form. *Nordisk Betong*, No. 4, 27–36.

47 Tepfers, R. and Kutti, T. (1979) Fatigue strength of plain, ordinary, and lightweight concrete. *ACI Journal*, Title No. 76-29, May, 635–53.

48 Traina, L.A. and Jeragh, A.A. (1982) *Fatigue of plain concrete subjected to biaxial-cyclical loading.* American Concrete Institute, Special Publications, SP-75, 217–234.

49 Treece, R.A. and Jirsa, J.O. (1989) Bond strength of epoxy-coated reinforcing bars. *Proceedings, ACI Materials Journal*, **86**, 2, March–April, 167–74.

50 Van Ornum, J.L. (1903) Fatigue of cement products. *Transactions, ASCE*, **51**, 443.

51 Van Ornum, J.L. (1907) Fatigue of concrete. *Transactions, ASCE*, **58**, 294–320.

52 Williams, H.A. (1943) Fatigue tests of lightweight aggregate concrete beams. *ACI Journal, Proceedings*, **39**, April, 441–8.

5 Durability

O E Gjørv

5.1 Introduction

Recent developments in concrete technology have made it possible to produce concrete mixtures with strength properties that are beyond the strengths that are currently used by the structural design practice. For high-quality natural mineral aggregates, compressive strengths of up to 33,000 psi (230 MPa) can now be produced under laboratory conditions.[1] If the mineral aggregate is replaced by high-quality ceramic aggregate, compressive strengths of up to 65,000 psi (460 MPa) can be achieved under controlled laboratory conditions.[2] Even with lightweight aggregate, compressive strengths of more than 14,000 psi (100 MPa), with a density of less than 3200 lb/yd^3 (1900 kg/m^3), can be obtained.[3] However, very often it is not the improved strength which is the primary objective but rather the improved durability and overall performance. Therefore, the term 'high-performance concrete', is inclusive of the term 'high-strength concrete'. In this chapter, a brief summary of the most recent developments on durability of high-strength concrete is presented.

5.2 Permeability

One of the main characteristics of high-strength concrete compared to that of the normal-strength concrete is the more uniform and homogeneous microstructure. When the portland cement is combined with the ultrafine particles of silica fume in low w/c ratios, the microstructure of such systems consists mainly of poorly crystalline hydrates forming a more dense matrix of low porosity. With increasing silica fume content, larger content of the calcium hydroxide is transformed into calcium silcicate hydrates (Fig. 5.1), while the remaining calcium hydroxide tends to form smaller crystals compared to that in pure portland cement pastes. From Table 5.1 it can be seen that with the addition of higher percentage of silica fume, the

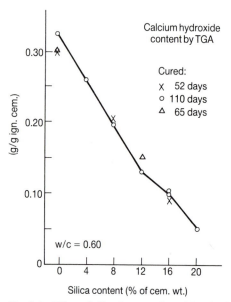

Fig. 5.1   Effect of silica fume on the calcium hydroxide content of ordinary Portland cement[4]

calcium/silicate ratio of the hydrates decreases, which allows the hydrates to incorporate ions, such as alkalis and aluminium. As a consequence, it appears that an increased resistance to aggressive ions and alkali-aggregate reaction is obtained with an increase in the silica fume content. Also, the electrical resistivity is increased.[5]

For concrete with very low w/c ratios, microcracking due to self-desiccation may affect the concrete permeability. As can be seen from Table 5.2 it is primarily the presence of silica fume in high-strength concrete, which appears to increase the autogenous shrinkage. While high-strength concrete without silica fume has an autogenous shrinkage of the same order as that of normal-strength concrete, the presence of silica fume may increase the autogenous shrinkage by up to twice that of normal-strength concrete.

One of the main effects of silica fume in high-strength concrete, however, is to improve the microstructure of the transition zone between the aggregate and the cement paste. For normal-strength concrete on a pure portland cement basis, the transition zone around the aggregate, which is 20 to 100 μm wide, has a very different microstructure compared to that of the bulk matrix.[6,7] This transition zone which is inferior in quality

Table 5.1   Effect of silica fume on the calcium/silicate-ratio of the hydrates[5]

Cement	Ca/Si-ratio
Portland cement (OPC)	1.6
OPC + 13% silica fume	1.3
OPC + 28% silica fume	0.9

Table 5.2 Effect of silica fume on the autogenous shrinkage

Type of concrete	Autogenous shrinkage μ m/m	References
NSC	90–120	6, 7
HSC	80–90	7, 8
HSC with silica fume		6, 8

and thus leads to a poorer bond between the aggregate and the cement paste, is typically characterized by the following key elements:

1 The transition zone is richer in calcium hydroxide and ettringite than the bulk phase, and the calcium hydroxide is oriented. A rim of massive calcium hydroxide can often be observed around the aggregates.
2 The porosity of the transition zone is greater than that of the bulk phase, and a gradient in porosity can be observed with a declining trend as the distance from the aggregate surface increases.

For normal-strength concrete, the special microstructure of the transition zone is apparently related to the formation of water-filled spaces around the aggregates in the fresh concrete, due to internal bleeding and to a 'wall effect', which interferes with inefficient packing of the cement particles around the aggregates. Calcium hydroxide and ettringite are known to preferably grow in large pores, which accounts for the greater contents of these phases in the transition zone.[8,9] Also, this zone has larger w/c ratio relative to the bulk[10] and is therefore characterized by higher porosity.

When silica fume is introduced into the system, and in particular in high-strength concrete, considerable changes in the microstructure of the transition zone take place. Regourd *et al.*[11,12] and Aitcin *et al.*[13,14] observed that high-strength concrete with silica fume was not as crystallized and porous as normal-strength concrete, and all of the space in the vicinity of the aggregate was occupied with amorphous and dense calcium silicate hydrates. Also, direct contact was formed between the aggregate and the calcium silicate hydrates rather than with calcium hydroxide as in normal-strength concrete. Scrivener *et al.*[15] quantified the interfacial microstructure and demonstrated that in high-strength concrete with silica fume, the porosity of the transition zone was practically eliminated (Fig. 5.2), and practically no gradient in porosity was observed, in contrast to normal-strength concrete.

As demonstrated in Table 5.3 the addition of silica fume may have a substantial effect on the permeability of concrete. By adding 20% of silica fume to 169 lb/yd^3 (100 kg/m^3) of cement (OPC), the same permeability as that of 421 lb/yd^3 (250 kg/m^3) cement is obtained. Addition of 10% silica fume to 421 lb/yd^3 (250 kg/m^3) cement gives a permeability as low as $1.8 . 10^{-14}$ m/s. For meeting the higher durability performance requirements for offshore concrete platforms in the North Sea, the permeability is limited to 10^{-2} m/s.[17]

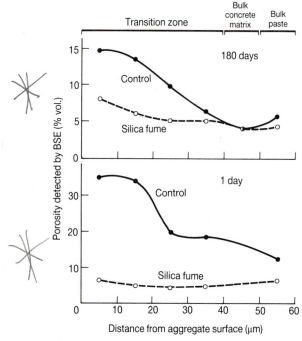

Fig. 5.2   Effect of silica fume on the porosit of the transition zone between aggregate and cement paste[15]

For cement pastes with very low w/c ratio in the range of 0.20 to 0.30, experiments[18] have shown that a 10% replacement of the cement with silica fume did only reduce the total porosity to a small extent. However, a refinement of the pore size distribution took place in such a way that the content of larger pores was reduced for decreasing w/c ratio. The effect of w/c ratio on chloride diffusivity was substantially higher at high than at low w/c ratios, but a 10% replacement with silica fume reduced the diffusivity so much that the effect of w/c ratio became less significant.

The above qualitative and quantitative observations indicate that in high-strength concrete the bulk matrix becomes very dense and typically, this dense matrix extends up to the aggregate surface, in such a way that the inhomogeneity of the transition zone is largely eliminated. It is now

Table 5.3   Effect of silica fume on permeability of concrete[16]

Cement (OPC) lb/yd³ (kg/m³)	Silica lb/yd³ (kg/m³)	Permeability m/s
168.6 (100)	0 (0)	$1,6 \cdot 10^{-7}$
168.6 (100)	16,9 (10)	$4,0 \cdot 10^{-10}$
168.6 (100)	33,7 (20)	$5,7 \cdot 10^{-11}$
168.6 (100)	0 (0)	$4,8 \cdot 10^{-11}$
421.5 (250)	42,1 (25)	$1,8 \cdot 10^{-14}$

well documented that this improved microstructure is closely related to the reduced permeability and improved performance of high-strength concrete.

5.3   Corrosion resistance

Since high-strength concrete is characterized by a low porosity and a more uniform microstructure compared to that of normal-strength concrete, this indicates a high resistance to penetration of carbon dioxide and chloride ions into the concrete. However, during production of high-strength concrete, both macrocracking due to plastic shrinkage and microcracking due to self-desiccation may represent potential problems from a corrosion protection point of view. Also, somewhat depending on the amount of mineral admixture used, both the reserve basicity and the ability of the cement paste to bind chlorides may be affected.

By replacing part of the portland cement with a mineral admixture such as fly ash, only modest effects on the pore solution chemistry have been observed,[19] but this may vary with the type of fly ash. Unlike fly ash, however, the presence of silica fume has an extensive and profound effect on the pore solution chemistry.[19,20]

For an increasing replacement of the cement by silica fume, the concentrations of both K^+ and OH^- ions are substantially lowered. However, by replacing up to 20% it appears from Fig. 5.3 that the pH does not drop below that of a saturated calcium hydroxide solution which is approximately 12.5. Even at 30% cement replacement the pH does not drop below 11.5 which is considered to be a threshold value for maintaining a good passivity of embedded steel. According to Diamond[19] it appears that the removal of alkalis from the pore solution with consequent lowering

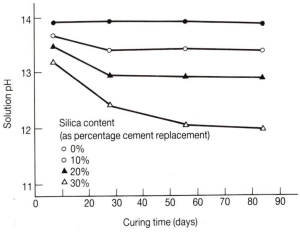

Fig. 5.3   Effect of silica on the alkalinity of cement paste with w/c = 0.50[19]

Table 5.4   Effect of slag on chloride penetration[21]

Type of mix	Mix proportion			Chloride concentration (ppm)		
Material %	W/C	C/S	HRWR %	Surface layer 0–0.27 in (0–7 mm)	Medium layer 0.27–0.5 in (7–14 mm)	Deep layer 0.55–0.79 in (14–20 mm)
Control 0	0.4	1/1.5	1.0	3970	521	42
OR slag 40	0.4	1/1.5	1.0	4570	259	62
VF slag 40	0.4	1/1.5	1.0	2910	111	34
VFG slag 40	0.4	1/1.5	1.0	3350	52	28
Silica fume 10	0.4	1/1.5	1.0	5110	312	43

of pH is less complete at low w/c ratios. Thus, it appears that it is carbonation of the concrete and penetration of chlorides which should be the controlling factors for the passivity of embedded steel. There are no data reported in the literature, however, that carbonation of high-strength concrete either with or without mineral admixtures represents any problem.

Nakamura et al.[21] have investigated chloride binding of slag mortars with w/c = 0.40 using alternative 24 hr. periods of immersing in seawater and oven drying at 60 °C. The specimens were then analysed for chlorides at three levels, the test results of which are given in Table 5.4. The high surface chloride content of the ordinary slag specimens is attributed to the greater binding capacity for chlorides in this blend than in the ordinary portland cement. The penetration to the middle layer suggests that there is some beneficial effect of the slag, the greatest being for the very finely ground slag with additions of anhydrite. The chloride contents of the inner layers were of the same order as of the reference specimens without exposure to chlorides. It should be noted, however, that these were immature specimens; they were only 7 days old at the beginning of the testing which lasted for only a further 14 days. A greater effect of the slag would be expected for older specimens.

When silica fume is used as a partial replacement for the cement, there is not yet a general consensus of the influence of chloride-binding. Page and Vennesland[20] determined the chloride binding capacity to be substantially reduced as shown in Fig. 5.4, whereas Byfors et al.[22] observed an increase in the amount of bond Cl⁻ in ordinary portland cement with 10% silica fume additions relative to the ordinary portland cement. If the concrete becomes carbonated, it is well established that the capacity for chloride binding is distinctly reduced.[23] It appears that, since carbonation is not a problem for high-strength concrete, the effect of carbonation on a possible chloride penetration should not represent any problem.

Recent investigations on offshore concrete platforms in the North Sea indicate that the rate of chloride penetration into concrete made with portland cement alone is much higher than expected earlier, but not much information on chloride penetration through high-strength concrete is

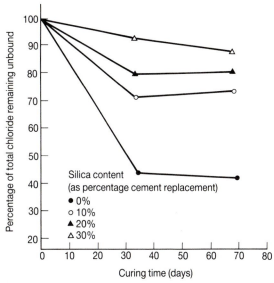

Fig. 5.4 Effect of silica fume on chloride binding capacity of hydrated cement paste of w/c = 0.50 with 0.40% chloride by weight of cement and silica fume[20]

available. Since the rate of chloride penetration is generally reduced with reduced permeability, it appears that high-strength concrete should provide a better protection against chloride penetration than normal strength concrete. Addition of mineral admixtures such as silica fume, slag or fly ash also increases the resistance to chloride penetration.[24,25] If sufficient amounts of chlorides reach the embedded steel, however, electrical resistivity and availability of oxygen are the additional factors which control the corrosion rate.

Since an electrical current passes through the concrete in the form of charged ions, it is reasonable to assume that there is close relationship between electrical resistivity, ion concentration and porosity. The moisture content of concrete is also an important factor. If the concrete is dry enough, the resistivity may be too high to allow any significant transport of ions, and significant corrosion rate with not occur.

Figure 5.5 illustrates the effect of moisture content on the electrical resistivity of normal strength concrete. By successively reducing the moisture content from 100 to 20%, it can be seen that the resistivity increases by approximately three orders of magnitude. In high-strength concrete, the w/c ratio is less than what is theoretically necessary for complete hydration, where self-desiccation of the concrete is a real possibility. Consequently, the mixing water in high-strength concrete will be used up relatively rapidly and long before the clinker hydration is complete. Excluding any interaction with the atmosphere, the relative humidity inside high-strength concrete will be lower than that in normal-strength concrete. If silica fume is also used in the concrete, both the reduced permeability and the changed pore solution chemistry will affect

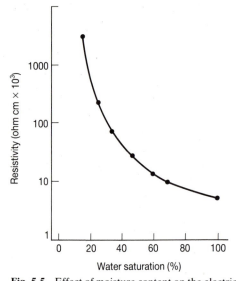

Fig. 5.5   Effect of moisture content on the electrical resistivity of concrete with w/c = 0.40[26]

the concrete in such a way that a dramatic increase in electrical resistivity may be observed, in particular at high cement contents (Fig. 5.6).

For high-strength concrete resistivities of up to 1000 ohm m even for water-saturated conditions have been observed.[27] Observations from existing concrete structures indicate that corrosion of embedded steel hardly represents any practical problem if the electrical resistivity exceeds a threshold level of 5000 to 700 ohm m.[28] Other information indicates that resistivities as low as 200 ohm m can reduce the rate of corrosion to a

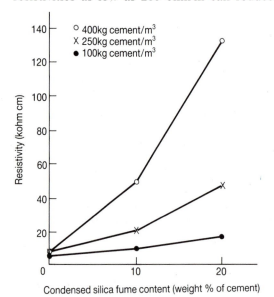

Fig. 5.6   Effect of silica fume on the electrical resistivity of concrete[27]

negligible level.[29] Therefore, it appears that high-strength concrete typically will have an electrical resistivity which is above the level where corrosion of embedded steel will represent any practical problem. If the steel should become depassivated, the rate of corrosion will also be controlled by the oxygen availability.

Even for normal strength concrete very few investigations on oxygen availability have been reported,[30] and for high-strength concrete, no particular information is available. Since the permeability of high-strength concrete is very low, the rate of oxygen transport through high-strength concrete must also be very low. For a cathodic reaction to take place, enough moisture must also be available in order to dissolve the oxygen. For a homogeneous high-strength concrete without any cracks, it is reasonable to assume, therefore, that oxygen availability will generally be very low.

Although the risk of corrosion is generally higher in cracked concrete than in uncracked concrete, the only information available on effect of cracks is based on normal concrete. Even with the comprehensive and realistic experiments carried out by Houston and Furguson[31] on normal concrete, however, it was difficult to establish a simple relationship between crack-width and risk of corrosion. When the concrete elements were investigated at an early stage of exposure, there appeared to be a distinct effect of crack width on corrosion. After longer periods of exposure, however, the effect of crack width was very small or almost non-existent. This is also demonstrated in Fig. 5.7, which shows results obtained for beams with 25 mm concrete cover after 10 years of exposure in different types of atmospheric environments. Also for submerged concrete, recent information has shown that the risk of cracks on corrosion is less than previously expected.[32,33]

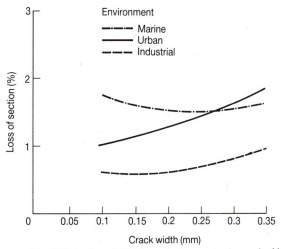

Fig. 5.7 Effect of crack width on steel corrosion in cracked beams after 10 years of exposure[32]

As far as the ability of high-strength concrete to protect embedded steel from corrosion is concerned, not much systematic research has been carried out so far. Based on the general information available, however, it appears that high-strength concrete has a high ability to protect embedded steel. This general conclusion is supported by the good performance observed for offshore concrete platforms in the North Sea, where concrete with compressive strengths of 45 to 70 MPa has been used. For these structures, no problems due to corrosion of embedded steel has been reported so far, even after 15 to 20 years of the combined exposure to heavy mechanical loading and a severe marine environment.

5.4 Frost resistance

Even for normal-strength concrete, production of concrete with a good and stable air-void system is normally a problem,[34,35] but in the presence of a high dosage of superplasticizer, the establishment of a good and stable air-void system may be an even bigger problem.[36] For production of high-strength concrete, it may also be a conflicting requirement to entrain air which will decrease the strength. Therefore, much attention has been given to finding out whether a frost resistant high-strength concrete can be produced without any air entrainment.

The general problem of assessing the frost resistance of concrete is the lack of correlation between existing laboratory test methods and field performance. Also, different test methods and different test conditions appear to give different and conflicting test results.

In 1981, Okada *et al.*[37] reported very good frost resistance (ASTM C666A-A) of non-air-entrained high-strength concrete with w/c ratios in the range of 0.25 to 0.35. Similar observations were later on reported by Foy *et al.*[38] and Gagne *et al.*[39] who observed good frost resistance of non-air-entrained concrete with w/c ratios of 0.25 and 0.30, respectively.

Using ASTM C666-A, Malhotra *et al.*[40] also tested a number of concretes with different types of cement and w/c ratios of 0.30 and 0.35. They concluded, however, that air entrainment was necessary for these concretes to be frost resistant. Hammer and Sellevold[41] tested non-air-entrained concrete with 0 and 10% silica fume and w/c ratios varying from 0.25 to 0.40. Even for the lowest w/c ratios some of the specimens were damaged during testing. These observations were in conflict with low-temperature calorimeter data, which clearly demonstrated a very low freezable water content. Based upon this, Hammer and Sellevold[41] suggested, therefore, that the observed damage could be due to thermal fatigue caused by too large differences between the thermal expansion coefficients of aggregate and binder rather than ice formation.

As far as salt scaling is concerned, there are also some conflicting results reported in the literature. In 1984 Petersson[42] reported that deterioration of high-strength concrete due to salt scaling was small for the first 50 to 100 cycles, but increased very rapidly to total destruction in the following 10 to

20 cycles. Foy *et al.*,[38] however, observed that the resistance to salt scaling of concrete with a w/c ratio of 0.25 was very good even after 150 cycles. Both Hammer and Sellevold[41] and Gagne *et al.*[39] have demonstrated that it is possible to produce high-strength concrete which is resistant to salt scaling without any air entrainment. These test results included w/c ratios of up to 0.37.

5.5 Chemical resistance

According to Biczók[43] chemical deterioration of concrete can be classified into three types of process depending on the predominant chemical reaction taking place. Leaching corrosion of concrete is the process where parts or all of the hardened cement paste are removed from the concrete. Normally, this is caused by the action of water of low carbonate hardness or carbonic acid content. The next process is corrosion by exchange reactions and by removal of readily soluble compounds from the hardened cement paste. This process occurs as a result of a base exchange reaction between the readily soluble compounds of hardened cement paste and the aggressive solution. The third process is the swelling corrosion, largely due to the formation of new and stable compounds in the hardened cement paste. This process is primarily the result of attack by certain salts. Also alkali-aggregate reactions cause expansion, where the concrete eventually is destroyed by a swelling pressure.

For all of the above deteriorating processes, the permeability of the concrete is the key factor governing the rate of deterioration (Fig. 5.8). In addition, calcium hydroxide is also an easily soluble constituent which is very vulnerable to chemical attack.

In sulfate-containing solutions the calcium hydroxide reacts with the sulfates to produce gypsum which may further react with the aluminates to form ettringite. Both gypsum and ettringite can cause disruptive expansion. Therefore, the pozzolans normally used in high-strength concrete are very effective in reducing the calcium hydroxide (Fig. 5.1), and hence also greatly enhance the resistance against sulfate attack. In addition to

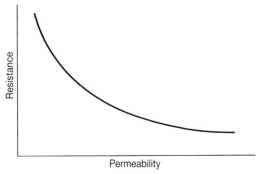

Fig. 5.8 Relationship between resistance to chemical deterioration and permeability[16]

reducing the calcium hydroxide, pozzolans, such as silica fume, also form calcium silicate hydrates which are able to incorporate aluminum ions, thus reducing the amount of aluminum available for ettringite formation.

The beneficial effect of silica fume in high sulfate-containing environments has been reported in several investigations.[44,45] In these investigations, the performance achieved has been equal or better than that obtained by use of sulfate resistant cements. Mehta[46] exposed a number of high-strength concretes to solutions of both 5% sodium sulfate and 1% concentrations of sulfuric and hydrochloric acid for a period of up to 182 days. Although the portland cement contained 7% C_3A, the results showed that w/c ratios of 0.33 to 0.35 gave too low a permeability to cause any deterioration. In more aggressive environments pure portland cements have shown some deterioration, whereas addition of silica fume has given practically unaffected performance.[45] Even for normal-strength concrete the presence of silica fume will improve the long-term performance of concrete in very aggressive environments. In 1952 a large number of concrete specimens were exposed in a field station in the underground of Oslo city, which consists of very aggressive alum shale. In spite of a sulfate content of up to 5 g/l of SO_3 and a pH of 2.8, a 15% replacement of ordinary portland cement with silica fume gave the same good performance after 26 years of exposure as that by sulfate resisting cements (ASTM Type V).[47]

The presence of pozzolans such as silica fume can also be used to control the expansion caused by alkali-aggregate reaction.[48,49] Porewater analyses of cement paste with silica fume have demonstrated the ability of silica fume to reduce the alkali concentration in the pore water rapidly, thus making it unavailable for the slower reaction with reactive silica in the aggregate.[19,20] Also, for high-strength concrete the effect of self-desiccation may reduce the moisture content to a level where no alkali-aggregate reaction can take place.

Asgeirsson and Gudmundsson[50] used silica fume with Icelandic high-alkali cement and reactive sand in mortar bars to demonstrate the ability of silica fume to reduce the expansion. Later on, field experience with Icelandic cement blended with silica fume in 200 houses constructed during the period of 1979–86 has confirmed these observations.[51]

5.6   Fire resistance

Diederichs *et al.*[52] have shown that high-strength concrete is more vulnerable to elevated temperatures compared to that of normal-strength concrete (Fig. 5.9). For high-strength concrete a distinct loss of strength (30%) was observed already at 150 °C, while normal strength concrete retained its strength up to 250 °C. The effects on compressive strength and modulus of elasticity are shown in Figs. 5.10 and 5.11, respectively.

Investigations carried out on the fire resistance of high-strength concrete

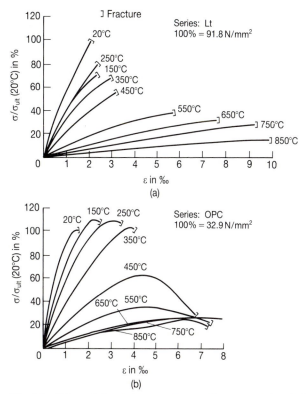

Fig. 5.9   Effect of elevated temperature on the stress-strain relationship for (a) high-strength concrete with ordinary Portlane cement and fly ash, and (b) ordinary Portlane cement[52]

are rather limited. In 1981 Pedersen[53] reported the results of an investigation where 4 in. (100 mm) cylinders of high-strength concrete with a w/c ratio of 0.16 and 20% silica fume were tested. During heating up at a rate of 33.8 °F/min (1 °C/min), several of the specimens suddenly disintegrated

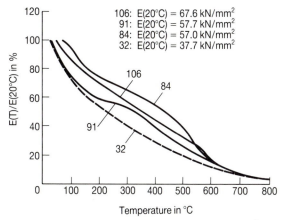

Fig. 5.10   Effect of elevated temperature on compressive strength[52]

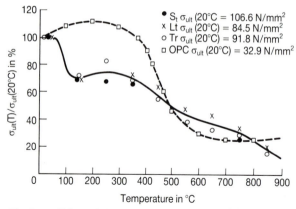

Fig. 5.11   Effect of elevated temperature on modulus of elasticity[52]

at a temperature of approximately 570 °F (300 °C). Shirley *et al.*,[54] however, tested concrete slabs according to ASTM E119,[55] where five different types of concrete with strengths varying from 7145 to 71,145 psi (50 to 120 MPa) revealed no significant difference in behavior. There was no explosive behavior, and none of the concretes showed even minor spalling on the exposed surface. Jensen *et al.*[56] reported a number of test series where different types of concrete elements were exposed to a hydrocarbon fire, where 2012 °F (1100 °C) is reached in 30 min. The strength of the elements varied from 5715 to 11,430 psi (40 to 80 MPa). The test results showed that spalling and damage occurred earlier than expected and that increasing moisture content increased the severity of the effects. Concrete with silica fume was also more sensitive to spalling. Williamson[57] tested concrete slabs according to ASTM 119 with two reference mixes (4286 and 11,429 psi) (30 and 80 MPa) and two mixes with silica fume (7143 and 14,286 psi) (50 and 100 MPa). The relative humidity in the concrete was monitored during the storage period of about six months, where a level of 71–74% RH was reached at the time of testing. No external spalling or internal damage to any of the specimens was observed.

Although the above test results appear to be contradictory, it seems that increasing moisture content leads to increased damage. Also, concrete with silica fume is more dense and dries out more slowly in such a way that a higher vapour pressure can build up internally and cause spalling.[58]

For offshore structures for oil and gas explorations where high-strength concrete may be exposed to the combination of a moist environment and fire, a passive fire-protection material may provide a safety precaution against concrete spalling.[56,59]

Intentionally induced relief for building up vapour pressure may be another approach for increasing the resistance to concrete spalling. Thus, some promising experiments have been carried out by incorporating polymer fibers[60,61] or polymer particles,[62] which melts already at a low temperature and thus provide relief channels for the vapour pressure.

5.7 Abrasion-erosion resistance

Although some building codes have recently increased their upper strength level for structural utilization of concrete to about 14,290 psi (100 MPa),[63] very few structures have so far been built with concrete strengths of more than 10,000 psi (70 MPa). At the same time, mineral aggregate-based concrete with very high strengths of up to approximately 33,000 psi (230 MPa) can now be produced.[1] There appears, therefore, to be a great potential for utilization of high-strength concrete to applications where high abrasion resistance is needed.

Mechanical abrasion is the dominant abrasion for pavements and bridge decks exposed to studded tires. This effect has been studied under controlled laboratory conditions; and accelerated load facility for full-scale testing of abrasion resistance of highway pavements exposed to heavy traffic by studded tires was built in Norway in 1985 (Fig. 5.12). Figure 5.13 presents some data from the test facility. By increasing the concrete strength from 7145 (50%) up to 14,290 psi (100 MPa) the abrasion of the concrete was reduced by roughly 50%. At 21,430 psi (150 MPa) the abrasion of the concrete was reduced to the same low level as that of high quality massive granite. Compared to an Ab 16t type of asphalt for a typical Norwegian highway traffic, this represents an increased service life of the highway pavement by a factor of approximately ten.

In the Scandinavian countries where studded tires are extensively used, even high quality asphalt pavements may have a service life of only two to

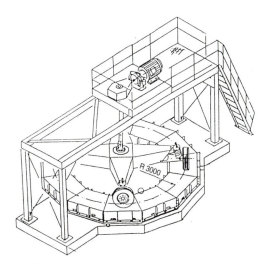

Fig. 5.12 Accelerated load facility for testing the abrasion resistance of highway pavements exposed to studded tires[64]

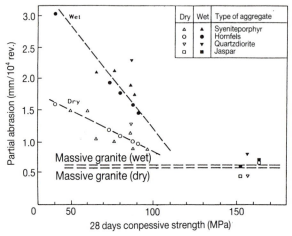

Fig. 5.13   Relationship between abrasion resistance and compressive strength[64]

three years or less. Encouraged by the above test results several high-strength concrete pavements have been completed over recent years.[65]

The wear of non-pavement surfaces such as concrete floors, sidewalks, stairs, etc. is caused primarily by foot traffic, light vehicular traffic, and the skidding, scraping, or sliding of objects on the surface. In some types of industrial operation, the use of steel wheeled vehicles, forks, buckets of lift trucks and loaders inflicts very severe damage to the concrete surfaces they operate on. The movement of abrasive granular material in and out of concrete storage facilities such as silos and bins also creates significant abrasion problems. In the Scandinavian countries the utilization of high-strength concrete for such problems has already started.

Hydraulic abrasion, or abrasion erosion is recognizable by the smooth, worn concrete surface in addition to cavitation erosion, where the surface is full of small holes and pits. Spillway aprons, stilling basins, sluiceways and tunnel linings are particularly susceptible to abrasion erosion.

Most concretes used in hydraulic structures in the past could not be classified as high-strength concrete. The concrete provided the mass, and strength was a secondary consideration only. Concerns about thermal cracking at early ages in the large concrete sections used in these structures have also led to low cement contents to minimize heat of hydration, with subsequent low concrete strengths. Where higher quality concrete is used, the resistance to high water velocities has been satisfactory for many years, but even these concretes do not fully resist the abrasive action, grinding or repeating impacts of the debris. Observations of abrasion-erosion of concrete surfaces in the stilling basins of several major USA dams have varied from 2 in. (50 mm) to 118 in. (3000 mm). At Dworshak Dam, 2000 yd³ (1530 m³) of concrete and bedrock were eroded from the stilling basin. In many of these instances, the abrasion-erosion is accelerated because of the impact forces of large rocks and boulders caught in turbulent flows. These forces weaken the concrete surface and make it more susceptible to removal.

In 1983 a major repair on the stilling basin of Kinzua Dam in Pennsylvania, USA, was carried out by using high-strength concrete.[66] The structure which has been in operation since 1967, had already had an extensive repair carried out in 1973–74. In 1983 approximately 1960 yd^3 (1500 m^3) of 10 in. (250 mm) slump concrete was placed, the 28 day compressive strength of which was 12,857 psi (90 MPa). Diver inspection of the concrete after one year of service including a period with very large volume of debris in the stilling basin, showed that the concrete was performing as intended.

For marine or offshore concrete structures exposed to 'ice abrasion', the actual mechanism that results in loss of surface is more complex than the simple act of ice rubbing on the concrete surface. Research work and field observations suggest that the concrete deterioration at or near the waterline is due to a combination of environmental causes plus the impact of loading of the concrete surface by repeated impacts from ice floes.

Pieces of ice, driven by wind and current, can possess significant kinetic energy, much of which is dissipated into the concrete when the ice collides with the concrete structure. Some of the energy is lost in the crushing of the ice. The frequency of the loading is dependent on the circumstances which occur at any given time at a particular structure and can vary from an occasional impact to repeated impacts every few seconds. A large ice floe in open waters will, upon initial contact with a structure, both load the structure and begin to crush itself at the point of contact with the structure. As the driving forces of the floe continue to move it forward against the structure, the resistance of the structure continues to increase to a point where the floe experiences a local failure in the ice, usually in flexure, some distance from the initial point of contact with the structure. This momentarily releases the load on the structure. The original ice, now damaged by crushing and cracking, is shunted away by the moving flow and new, undamaged ice in the floe then collides with the structure. The characteristics of the ice and the floe, combined with the dynamic response of the structure, will establish a 'ratcheting' effect on the concrete surface, repeatedly loading and unloading it. With time, this repetitive loading behavior can destroy the effectiveness of the aggregate bond near the surface of the concrete and both cause and propagate existing microcracks in the concrete.

Regardless of type of abrasion, both laboratory and field experience indicate that compressive strength is the single most important factor in determining the abrasion resistance of the concrete. Also, the abrasion resistance can be significantly improved by the use of hard and dense aggregate both in the upper and lower part of the grading curve. By replacing the 0.079 to 0.157 in. (2 to 4 mm) fraction of the natural sand with crushed high quality material, the compressive strength in the Norwegian highway pavement investigations[65] decreased from 23,471 to

21,900 psi (164.3 to 153.3 MPa), while the service life of the pavement increased by 50%. Generally, the abrasion of concrete is higher in wet than in dry condition, but the experience indicates that also this effect is reduced by increased strength level. From Fig. 5.13 it can be seen that at 7143 psi (50 MPa), the wet abrasion loss was approximately 100% higher than the dry abrasion loss, while at 14,286 psi (100 MPa) the wet abrasion loss was only 50% higher. At 214,429 psi (150 MPa), only a small difference between the wet and dry abrasion loss was observed. For such a dense concrete, it appears that the effect of moisture becomes more negligible.

5.8   Concluding remarks

In recent years more rapid developments in the field of concrete technology have taken place. Increasing construction challenges in combination with new innovations in materials and construction techniques have strengthened the stature of concrete as a major construction material.

There are several reasons for the above development. A rapid development in the general field of materials science has taken place. New cementitious materials and admixtures have been developed, and advancement in processing of aggregates has also taken place. Reinforced and prestressed concrete is being utilized in new areas, such as offshore structures for oil and gas explorations. From the first concrete structure produced for the North Sea operations in 1973 (the Ekofisk tank) to the most recent offshore structures currently under design, the strength of concrete has increased from 5715 to 11,430 psi (40 to 80 MPa). In North America high-strength concrete is being used in tall buildings, where concrete with strengths up to 18,570 psi (130 MPa) has been used in the columns.[67]

For highway pavements and bridge decks subjected to studded tires, experimental work has shown that utilization of high-strength concrete may increase the service life by a factor of up to ten, compared to that of a high-quality asphalt pavement. Industrial floors, hydraulic structures and structures exposed to ice abrasion also represent a great potential for utilization of high-strength concrete. In many areas, large quantities of resources are being spent on maintenance and rehabilitation of concrete structures due to lack of durability. There is a great challenge, therefore, for the engineering profession to utilize and further develop the technology of high-performance concrete for the benefit of society.

References

1   Division of Building Materials, The Norwegian Institute of Technology, NTH, Trondheim, Norway (unpublished data).
2   Elkem A/S Materials, Kristiansand, Norway (unpublished data).
3   Zhang, M.H. and Gjørv, O.E. (1990) Development of high-strength lightweight concrete. *High-Strength Concrete*, ACI SP-121, 667–81.

4 Sellevold, E.J., Bager, D.H., Klitgaard, J. and Knudsen, T. (1982) *Silica fume-cement pastes: hydration and pore structure*, Report BML 82.610. Division of Building Materials, the Norwegian Institute of Technology, NTH, 19–50.

5 Regourd, M., Mortureux, B. and Gautir, E. (1981) Hydraulic reactivity of various pozzolans. *Fifth International Symposium on Concrete Technology*, Monterey, Mexico.

6 Diamond, S. (1986) The microstructure of cement paste in concrete. *Proceedings, 8th Int. Congress on Chemistry of Cement, Vol. I*, Rio de Janeiro, 122–47.

7 Maso, J.C. (1980) The bond between aggregates and hydrated cement paste. *Proceedings, 7th Int. Congress on Chemistry of Cement, Vol. 1*, Paris, VII 1/3 to VII 1/15.

8 Monteiro, P.J.M. and Mehta, P.K. (1985) Ettringite formation on the aggregate-cement paste interface. *Cement and Concrete Research*, **15**, 2, 378–80.

9 Olliver, J.P. and Grandet, J. (1982) Sequence of formation of the aureole of transition. *Proceedings, RILEM Coloq. Liaisons Pates de Ciment Materiaux Associes*, Tolouse, A14 to A22.

10 Hoshino, M. (1988) Difference of the w/c ratio, porosity and microscopical aspect between the upper boundary paste and the lower boundary paste of the aggregate in concrete. *Materials and Structures*, RILEM, **21**, 175, 336–40.

11 Regourd, M. 'Microstructure of high strength cement paste systems,' Very High-Strength Cement-Based Materials. *Materials Research Society, Proceedings*, **42**, 3–17.

12 Regourd, M., Mortureux, B., Aitcin, P.C. and Pinsonneault, P. (1983) Microstructure of field concretes containing silica fume. *Proceedings, 4th Int. Symp. on Cement Microscopy*, Nevada, USA, 249–60.

13 Sarkar, S.L. and Aitcin, P.C. (1987) Comparative study of the microstructure of very high strength concretes. *Cement, Concrete and Aggregates*, **9**, 2, 57–64.

14 Aitcin, P.C. (1989) From gigapascals to nanometers. *Proceedings, Engr. Foundation Conf. on Advances in Cement Manufacture and Use*. The Engineering Foundation, 105–30.

15 Scrivener, K.L., Bentur, A. and Pratt, P.L. (1988) Quantitative characterization of the transition zone in high strength concretes. *Advances in Cement Research*, **1**, 2, 230–7.

16 Gjørv, O.E. (1983) Chemical processes related to concrete. *Proceedings, CEB-RILEM International Workshop on Durability of Concrete Structures*, Copenhagen, 341–4.

17 (1976) *Regulations for the structural design of fixed structures on the Norwegian continental shelf*. Norwegian Petroleum Directorate, Stavanger.

18 Zhang, M.H. and Gjørv, O.E. (1991) Effect of silica fume on pore structure and diffusivity of low porosity cement pastes. *Cement and Concrete Research*, Vol. 21, 800–8.

19 Diamond, S. (1983) Effects of microsilica (silica fume) on pore-solution chemistry of cement pastes. *Journal of the American Ceramic Society*, **66**, 5, C82–C84.

20 Page, C.L. and Vennesland, Ø. (1983) Pore solution composition and chloride binding capacity of silica-fume cement pastes. *Materials and Structures*, RILEM, **16**, 91, 19–25.

21 Nakamura, N., Sakai, M., Koibuchi, K. and Iijima, Y. (1986) Properties of high-strength concrete incorporating very finely ground granulated blast furnace slag. *ACI SP-91*, 1361–80.

22 Byfors, J., Hansson, C.M. and Tritthart, J. (1986) Pore solution expression as a method to determine influence of mineral additives on chloride binding. *Cement and Concrete Research*, **16**, 760–70.

23   Tuutti, K. (1982) *Corrosion of steel in concrete*, Report fo. 4-82. Swedish Cement and Concrete Research Institute, Stockholm.

24   (1990) *High-strength concrete, state-of-the-art-report*, FIP/CEB.

25   CEB Bulletin d'Information No. 182 (1989) *Durable concrete structures – CEB Design Guide*, Second edition.

26   Gjørv, O.E., Vennesland, Ø. and El-Busaidy, A.H.S. (1977) Electrical resistivity of concrete in the oceans. *Offshore Technology Conference, Proceedings*, OTC 2803, Houston, Texas, 581–8.

27   Vennesland, Ø. and Gjørv, O.E. (1983) Silica concrete – protection against corrosion of embedded steel. *ACI SP-79*, Vol. II, 719–29.

28   Danish Great Belt Link (unpublished data).

29   Gewertz, M.W., Tremper, B., Beaton, J.L. and Stratfull, R.F. (1958) Causes and repair of deterioration to a California bridge due to corrosion of reinforcing steel in a marine environment. *Highway Research Board, Bulletin 182*.

30   Browne, R.D. (1980) Mechanisms of corrosion of steel in concrete in relation to design, inspection, and repair of offshore and coastal structures. *ACI SP-65*, 169–204.

31   Houston, A. and Furguson, P.M. (1972) *Corrosion of reinforcing steel embedded in structural concrete*. Centre for Highway Research, The University of Texas at Austin, Research Report No. 112-1-F, 148 pp.

32   Schiessl, P. (1975) 'Admissible crack width in reinforced concrete structures', Contribution II 3-17, Inter-Association Colloquium on the Behaviour in Service of Concrete Structures. *Preliminary Reports, Vol. II*, Liege.

33   Vennesland, Ø. and Gjørv, O.E. (1981) Effect of cracks in submerged concrete sea structures on steel corrosion. *Materials Performance*, **20**, 49–51.

34   Gjørv, O.E. and Bathen, E. (1987) Quality control of the air-void system in hardened concrete. *Nordic Concrete Research*, 95–110.

35   Gjørv, O.E., Okkenhaug, K., Bathen, E. and Husevåg, R. (1988) Frost resistance and air-void characteristics in hardened concrete. *Nordic Concrete Research*, 89–104.

36   Siebel, E. (1989) Air-void characteristics and freezing and thawing resistance of superplasticized air-entrained concrete with high workability. *ACI SP 119*, 297–319.

37   Okada, E., Hisaka, M., Kazama, Y. and Hattori, K. (1981) Freeze-thaw resistance of superplasticized concretes. *Developments in the Use of Superplasticizers, ACI SP-68*, 269–82.

38   Foy, C., Pigeon, M. and Bauthia, N. (1988) Freeze-thaw durability and deicer salt scaling resistance of a 0.25 water-cement ratio concrete. *Cement and Concrete Research*, **18**, 604–14.

39   Gagne, R., Pigeon, M. and Aitcin, P.C. (1990) Durabilité au gel bétons de hautes performances mécaniques. *Materials and Structures, RILEM*, **23**, 103–9.

40   Malhotra, V.M., Painter, K. and Bilodeau, A. (1987) Mechanical properties and freezing and thawing resistance of high-strength concrete incorporating silica fume. *Proceedings, CANMET – ACI International Workshop on Condensed Silica Fume in Concrete*, Montréal, Canada, 25 p.

41   Hammer, T.A. and Sellevold, E.J. (1990) Frost resistance of high strength concrete. *ACI SP-121*, 457–87.

42   Petersson, P.-E. (1984) *Inverkan av salthaltiga miljøer på betongens frostbestandighet*, Technical Report SP-RAPP 1984: 34 ISSN 0280-2503. National Testing Institute, Borås, Sweden.

43   Biczók, I. (1972) *Concrete corrosion, concrete protection*. Akademiai Kiado, Budapest.

44 Mather, K. (1982) Current research in sulfate resistance at the Waterways Experiment Station. *George Verbeck Symposium on Sulfate Resistance of Concrete, ACI SP-77*, 63–74.

45 Cohen, M.D. and Bentur, A. (1988) Durability of portland cement – silica fume pastes in magnesium and sodium sulfate solutions. *ACI Materials Journal*, **85**, 148–57.

46 Mehta, P.K. (1985) Studies on chemical resistance of low water-cement ratio concretes. *Cement and Concrete Research*, **15**, 6, 969–78.

47 Gjørv, O.E. (1983) Durability of concrete containing condensed silica fume. *ACI SP-79, Vol. II*, 695–708.

48 Davis, G. and Oberholster, R.E. (1987) Use of the NBRI accelerated test to evaluate the effectiveness of mineral admixtures in preventing the alkali-silica reaction. *Cement and Concrete Research*, **16**, 2, 97–107.

49 Hooton, R. (1987) Some aspects of durability with condensed silica fume in concretes. *Proceedings, CANMET – ACI International Workshop on Silica Fume in Concrete*, Montréal, Canada.

50 Asgeirsson, H. and Gudmundsson, G. (1979) Pozzolanic activity of silica dust. *Cement and Concrete Research*, **9**, 249–52.

51 Sveinbjørnsson, S. (1987) *Alkali-aggregate demages in concrete with microsilica – field study*. Icelandic Building Research Institute, Reykjavik.

52 Diederichs, U., Jumppanen, U.M., Penttala, V. (1988) Material properties of high strength concrete at elevated temperatures. *IABSE 13th Congress*, Helsinki, June.

53 Pedersen, S. (1981) *Beregningsmetoder for varmepåvirkede betonkonstruktioner*. Institute of Building Design, Technical University of Denmark, Lyngby.

54 Shirley, S.T., Burg, R.G. and Fiorato, A.E. (1988) Fire endurance of high strength concrete slabs. *ACI Materials Journal*, **85**, 2, 102–8.

55 American Society for Testing and Materials, ASTM, Philadelphia, E119.

56 Jensen, J.J., Danielsen, U., Hansen, E.Aa. and Seglem, S. (1987) Offshore structures exposed to hydrocarbon fire. *Proceedings, First International Conference on Concrete for Hazard Protection*, Edinburgh, Sept, 113–25.

57 Williamson, R.B. and Rashed, A.I. (1987) *A comparison of ASTM E119 fire endurance exposure of two EMSAC concretes with similar conventional concretes*. Fire Research Laboratory, University of California, Berkeley, July.

58 Hertz, K. (1984) *Heat-induced explosion of dense concretes*. Institute of Building Design, Technical University of Denmark, Lyngby, Report no. 166.

59 Danielsen, U. (1989) Marine concrete structures exposed to hydrocarbon fires, Nordic Seminar on Fire Resistance of Concrete. *SINTEF Report STF65 A89036*, Trondheim, 56–76.

60 (1984) *Investigation of surface burning characteristics of fibermesh I fiber reinforcement in regular and lightweight concrete*. Southwest Research Laboratory, San Antonio, USA.

61 (1985) *Small scale fire tests of fiber-enhanced concrete*. Under-Writers Laboratories Inc., USA.

62 Chandra, S., Berntsen, L. and Anderberg, Y. (1980) Some effects of polymer addition on the fire resistance of concrete. *Cement and Concrete Research*, **10**, 367–75.

63 (1989) *Norwegian Standard NS 3473, Concrete structures, Design rules*. Oslo.

64 Gjørv, O.E., Bærland, T. and Rønning, H.R. (1990) Abrasion resistance of high-strength concrete pavements. *Concrete International*, **12**, 1, 45–8.

65 Helland, S. (1990) High-strength concrete used in highway pavements. *ACI SP-121*, 757–66.

66 Holland, T.C., Krysa, A., Luther, M.D. and Lin, T.C. (1986) Use of

silica-fume concrete to repair abrasion-erosion damage in the Kinzua dam stilling basin. *ACI SP-91*, 841–63.

67   Godfrey, K. (1987) Concrete strength record jumps 36%. *Civil Engineering*, 84–6 and 88.

6 Fracture mechanics

R Gettu and S P Shah

6.1 Introduction

Fracture mechanics is the study of crack propagation and the consequent structural response. Tremendous research interest in the 1980s led to fracture models that have been tailored to represent the quasi-brittle behavior of concrete. The validation of these approaches has opened two important avenues of application – materials engineering and structural analysis.

The mechanical behavior of concrete that is designed to have high strength is different in many aspects from that of normal concrete. These differences have yielded some characteristics that are not beneficial, such as brittleness. Fracture models can be used to understand the microstructural mechanics that control brittleness and crack resistance (toughness), and to provide reliable means of quantifying them. New high-performance concretes can then be engineered to possess higher toughness and lower brittleness. Increased resistance to cracking may also lead to better durability, long-term reliability and seismic resistance.

The application of fracture mechanics to structural analysis and design is motivated by the fact that the failure of concrete structures is primarily due to cracking, and several types of failures could occur catastrophically, especially in high-strength concrete structures. Certain aspects of such failures cannot be predicted satisfactorily by empirical relations obtained from tests, but can be explained rationally through fracture mechanics. In general, structural analysis based on fracture principles can lead to better estimates of crack widths and deformations under service loads, the safety factors under ultimate loads, and post-failure response during collapse.

The present chapter reviews the classical theory of fracture, its extensions for the modeling of concrete behavior, and the microstructural features that necessitate such models. Material characterization based on fracture mechanics and its implications are also discussed. Some examples of the applications are presented to demonstrate the scope of the fracture approach. More pertinent details can be found in recent publications such as the two RILEM reports,[1,2] the ACI report,[3] two detailed bibliographies[4,5] and several conference and workshop proceedings.[6–13]

6.2   Linear elastic fracture mechanics

Consider a panel (Fig. 6.1a) made of an ideal-elastic material. When it is uniformly stressed in tension, the stress-flow lines (imaginary lines showing the transfer of load from one loading-point to another are straight and parallel to the direction of loading. If the panel is cut as in Fig. 6.1b, the stress-flow lines are now forced to bend around the notch producing a stress-concentration (i.e., a large local stress) at the tip. The magnitude of the local stress depends mainly on the shape of the notch, and is larger when the notch is narrower. Also, since the stress-flow lines change direction a biaxial stress field is created at the notch-tip.

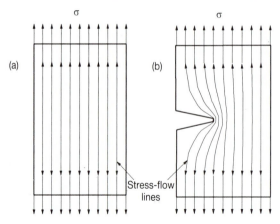

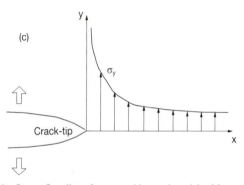

Fig. 6.1   Stress-flow lines for a panel in tension: (a) without crack; (b) with crack; and (c) stress concentration ahead of the crack

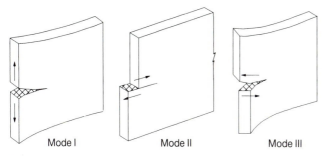

Fig. 6.2   Basic fracture modes

For a sharp notch or a crack, the stress (σ_y) in the loading direction and along the crack plane ($y = 0$) is of the form:

$$\sigma_y = \frac{K_I}{\sqrt{2\pi x}} + C_1 + C_2\sqrt{x} + C_3 x + \ldots \tag{6.1}$$

where C_1, C_2, C_3, ... depend on geometry and loading. Note that the near-tip ($x \to 0$) stresses depend only on the first term, and at the tip ($x = 0$) this term and the theoretical stress become infinite (see Fig. 6.1c). The parameter K_I in Equation (6.1) characterizes the 'intensity' of the stress-field in the vicinity of the crack, and is called the stress intensity factor. It depends linearly on the applied stress (or load), and is a function of the geometry of the structure (or specimen) and the crack length:

$$K_I = \sigma F \sqrt{\pi a} \tag{6.2}$$

where σ = applied (nominal) stress as in Fig. 6.1b, a = crack length, and F = function of geometry and crack length. The subscript of K represents the mode of crack-tip deformation; the three basic modes (see Fig. 6.2) are Mode I – tensile or opening, Mode II – in-plane shear or sliding, and Mode III – anti-plane shear or tearing. When two or all three modes occur simultaneously, the cracking is known as mixed-mode fracture. Only tensile cracks are discussed in this chapter (unless mentioned otherwise) since crack propagation in concrete is dominated by Mode I fracture. Moreover, analytical treatments of the three modes are quite similar.

A more useful relation for K_I, in terms of the load and structural dimensions, can be obtained by writing Equation (6.2) as:

$$K_I = \frac{P}{b\sqrt{d}} f(\alpha) \tag{6.3}$$

where P = applied load, b = panel thickness, d = panel width (see Fig. 6.3a), $f(\alpha)$ = function of geometry, and $\alpha = a/d$ = relative crack length. When panel length $W = 2d$, the geometry function is (Chapter 3, Broek[14]):

$$f(\alpha) = \sqrt{\pi\alpha}(1.12 - 0.23\alpha + 10.56\alpha^2 - 21.74\alpha^3 + 30.42\alpha^4) \tag{6.4}$$

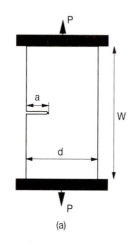

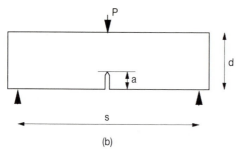

Fig. 6.3 Fracture specimens: (a) single-edge-notched panel in tension; and (b) single-edge-notched three points bend specimen

The advantage in Equation (6.3) is that it can also be used for several other geometries, such as the beam in Fig. 6.3b with span/depth = $s/d = 4$. Function f, however, is different for this beam:[15]

$$f(\alpha) = 6\sqrt{\alpha}\left\{\frac{1.99 - \alpha(1-\alpha)(2.15 - 3.93\alpha + 2.7\alpha^2)}{(1+2\alpha)(1-\alpha)^{3/2}}\right\} \qquad (6.5)$$

From Equation (6.1) it is obvious that a stress-based criterion for crack propagation is meaningless since the stress at the crack-tip is always infinite, irrespective of the applied load. Griffith,[16] in his landmark paper, suggested a criterion based on energy: Crack propagation occurs only if the potential energy of the structure is thereby minimized. This thermodynamic criterion forms the basis of fracture mechanics theory. Denoting the potential energy per unit thickness as U, the energy release rate is dU/da, which is usually designated as G (after Griffith). If energy (per unit length of crack-front) is consumed during fracture at the rate dW/da, denoted as R (for resistance of the material against fracture), the fracture criterion is:

$$\frac{dU}{da} = \frac{dW}{da} \quad \text{or} \quad G = R \qquad (6.6)$$

There is no fracture when $G<R$, and the fracture is unstable (i.e., sudden or catastrophic) when $G>R$. This criterion does not, however, imply that crack propagation is independent of the stress-field at the crack-tip. Irwin[17] showed that the energy release rate and the crack-tip fields are directly related by the equations suggested by Knott[18]:

$$G = K_I^2/E' \tag{6.7}$$

where $E' = E$ for plane stress, $E' = E/(1-v^2)$ for plane strain, $E =$ modulus of elasticity, and $v =$ Poisson's ratio. Using Equation (6.7), the fracture criterion of Equation (6.6) can be written as:

$$K_I = K_{IR}, \quad K_{IR} = \sqrt{E'R} \tag{6.8}$$

In the classical theory of linear elastic fracture mechanisms (LEFM), R is a constant (i.e., independent of the structural geometry and crack length), usually denoted as G (for critical strain energy release rate). Then, G and the associated critical stress intensity factor $K_{Ic}(= \sqrt{E'G_c}$; also known as fracture toughness) are material properties. When LEFM governs, the behavior is termed as Griffith or ideal-brittle fracture. More details of LEFM are given in textbooks such as Knott[18] and Broek.[14]

The LEFM parameter K_{Ic} (or G_c) can be determined from tests of specimens such as the single edge-notched tension specimen (Fig. 6.3a) and the three-point bend (3PB) specimen (Fig. 6.3b). Substitution of the experimentally determined maximum load (fracture load) for P in Equation (6.3) yields $K_I = K_{Ic}$. Geometry function $f(\alpha)$ can be determined using elastic analysis techniques including the finite element method.[13,14] For several common geometries, $f(\alpha)$ can be found in fracture handbooks such as Tada *et al.*[15] and Murakami.[19] Note that in the above-mentioned specimens, failure occurs immediately, i.e., the crack length is the notch length (a_0) and $\alpha = \alpha_0$ $(=$ relative notch length $= a/d)$, at failure.

The main features of LEFM can be summarized as:

1 The fracture criterion involves only one material property, which is related to the energy in the structure and the near-tip stress field.
2 The stresses near the crack-tip are proportional to the inverse of the square root of distance from the tip and become infinite at the tip.
3 During fracture, the entire body is elastic and energy is dissipated only at the crack-tip; i.e., fracture occurs at a point.

In reality, stresses do not become infinite, and some inelasticity always exists at the crack-tip. However, LEFM can still be applied as long as the inelastic region, called the fracture process zone, is of negligible size. Materials where such conditions exist include glass, layered silicate, diamond, and some high strength metals and ceramics (Chapter 4, Knott[18]). For other materials, the applicability of LEFM depends on the size of the cracked body relative to the process zone. In general, LEFM solutions can be used when the structural dimensions and the crack length are much larger than the process zone size. The process zone in concrete

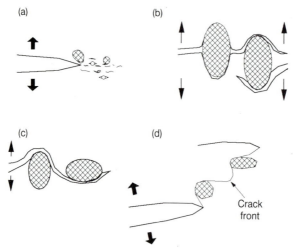

Fig. 6.4   Fracture process zone mechanisms: (a) microcracking; (b) crack bridging; (c) crack deflection; and (d) crack bowing

and the applicability of LEFM to concrete structures is discussed in the following sections.

6.3   The fracture process zone

Flaws, such as water-filled pores, air voids, lenses of bleed water under coarse aggregates and shrinkage cracks, exist in normal concrete even before loading. Due to the stress concentration ahead of a crack, microcracks could initiate at nearby flaws and propagate (Fig. 6.4a) forming a fracture process zone as suggested first by Glucklich.[20] Discrete microcracking at the tip of a propagating crack in hardened cement paste (hcp) has been observed by Struble *et al.*[21] with a scanning electron microscope. Also, acoustic waves generated during microcracking have been monitored using nondestructive techniques,[22,23,24] and used to determine the location, size and orientation of individual microcracks. Depending on the size and orientation of the microcracks, their interaction with the main crack could significantly decrease the K_I at its crack-tip.[25,26] This beneficial effect is known as crack-shielding or toughening. However, an amplification of K_I is also possible, especially if the main crack can propagate by coalescing with the microcracks. When the matrix is more compact and the flaws are less in number, as in high-strength concrete, the extent of microcracking is reduced.[27,28]

The most effective shielding mechanism in cement-based materials is crack-bridging (see Fig. 4b). A propagating crack is arrested when it encounters a relatively strong particle, such as an unhydrated cement grain, sand grain, gravel piece or steel fiber. Upon increase of load, the crack may be forced around and ahead of the particle which then bridges the crack-faces.[21,29] At this stage, crack-branching could also occur.[30,31]

Until the bridging particle is debonded (or broken), it ties the crack-faces together and thereby, decreases the K_I at the crack-tip.[32,33] In addition, energy is dissipated through friction during the detachment of the bridges and the separation of the crack-faces, which increases the fracture resistance R.

Microcracking, crack-bridging and frictional separation cause the process zone in concrete and other cementitious materials. Several investigations, using a wide range of techniques, have focused on whether and where these phenomena exist. However, results have been influenced significantly by methodology, and have been contradictory at times.[34] The various techniques used to detect and study the process zone have been reviewed by Mindess.[35] The most sensitive of the non-destructive methods are based on optical interferometry using laser light. Researchers using moiré interferometry[36,37] or holographic interferometry[38,39] observed well-defined crack-tips surrounded by extensive process zones with large strains. They concluded that strains behind the crack-tip (i.e., in the wake) decrease gradually to zero, and that energy dissipation near the crack-faces is most significant. Castro-Montero *et al.*[39] defined the process zone as the region where the experimentally determined strain-field deviates considerably from LEFM. Their results are shown in Fig. 6.5 for three stages in the fracture of a center-notched mortar plate. The zones labeled A have strains less than the LEFM solution, and appear to be of constant size. The zones B (in the wake) have strains greater than LEFM, and seem to increase with crack length and load. It was concluded, therefore, that most toughening occurs in the wake zone. Note that the dimensions of the crack are not constant through the thickness of the specimen,[40,41] and therefore the process zone has a complex three-dimensional character. Measurements of surface deformations should be interpreted with this feature in mind.

In some analytical fracture models, such as the micromechanics

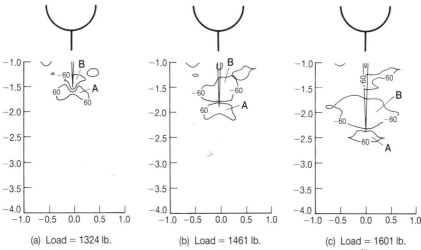

(a) Load = 1324 lb.   (b) Load = 1461 lb.   (c) Load = 1601 lb.

Fig. 6.5   Fracture process zone from holographic interferometry (1 lb = 4.45 N)[39]

approach of Horii,[42] each process zone mechanism is described individual-ly. In general, however, the treatment of the process zone need not distinguish between the different mechanisms; i.e., the toughening effect of the bridging (wake) zone and the microcracked zone are equivalent.[43] Therefore, all the shielding mechanisms can be lumped into a conceptual fracture process zone that lies ahead of the traction-free zone of the crack or in the wake of the 'actual' crack-front. This zone can also include the toughening effects of other inelastic mechanisms such as crack-deflection (tortuosity of the crack path; Fig. 6.4c) and crack-bowing (unevenness of the crack front; Fig. 6.4d). Cracks in concrete follow paths of least resistance, and subsequently, are considerably tortuous with rough crack-faces. In the hardened cement paste, the crack usually passes around unhydrated cement grains and along calcium hydroxide cleavage planes.[21] The crack-face roughness appears to be less when the cement paste contains silica fume.[44] In mortar, the crack follows the interfaces between the sand grains and the hardened cement paste.[30] Since the weakest phase in normal concrete is the aggregate-mortar interface, cracks tend to avoid the aggregates and propagate through the interfaces.[45] In silica-fume concrete the interfaces and the mortar are much stronger, and therefore, cracks are less tortuous and sometimes pass through the gravel.[27,29,46] Also, cracks propagate through coarse aggregates that are weaker than the mortar, as in lightweight concrete.[46,47] The tortuosity of the crack gives rise to a higher R due to a larger surface area. Also, further shielding occurs since the non-planar crack experiences a lower K_I than the corresponding planar crack.[47] Similarly, shielding exists when the crack front is trapped by bridging particles and bows between the bridges until they break or slip. This crack-bowing effect (Fig. 6.4d) can be an important toughening mechanism for very-high strength concrete with strong inter-faces where cracks propagate through the aggregates. The bowed crack-front has a lesser K_I than a straight front.[32,33]

In conclusion, a sizable fracture process zone occurs ahead of a propagating crack in cementitious materials. Energy is dissipated through-out this zone causing crack-shielding or toughening. Its size and effective-ness depend on the microstructure and inherent material heterogeneity; therefore, toughening in mortar is greater than in hardened cement paste and less than in concrete. The stress distribution (Fig. 6.6) in the process

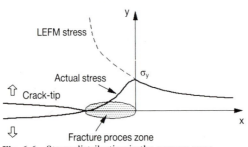

Fig. 6.6   Stress distribution in the process zone

zone differs considerably from the LEFM distribution (Fig. 6.1c). Within the process zone, the stress and strain undergo 'softening' with a gradual decrease to zero instead of an abrupt (brittle) drop. Models of fracture in cementitious materials should account for these factors in order to predict their behavior satisfactorily.

6.4 Notch sensitivity and size effects

In the 1960s and 1970s, several researchers examined the use of LEFM for concrete with tests of notched panels and beams.[4,48] They generally obtained K_{Ic} (and/or G_c) from LEFM relations by taking the initial notch length as the critical crack length at failure (i.e., at the observed maximum load). Early works[20,49,50] concluded that the critical parameters thus determined varied with notch length, and specimen size and shape. From these and later investigations[51–55] three significant trends can be identified: (1) the fracture behavior of materials with a finer microstructure is closer to LEFM (e.g., hcp vs. concrete); (2) for a given specimen shape and size, K_{Ic} depends on notch length; and (3) for a given specimen shape and relative notch length (α_0), K_{Ic} increases with specimen size. Also, the effects of specimen thickness and notch width on K_{Ic} are negligible within a certain range.[54,56–58]

One aspect of fracture is called notch-sensitivity, and is defined as the loss in net tensile strength due to the stress-concentration at a notch. Considering the 3PB specimen (Fig. 6.3b) with notch length a_0, the net tensile strength (σ_u) is calculated from bending theory using the failure load (P_u) and net depth ($d - a_0$), as $\sigma_u = 1.5 P_u s / b (d - a_0)^2$. The σ_u values corresponding to the 3PB ($s/d = 6.25$, $d = 64$ mm) tests of concrete by Nallathambi[59] are plotted in Fig 6.7a, along with the scatter. For determining the LEFM behavior, the above equation for σ_u is substituted in Equation (6.3) and K_I at $P = P_u$ is taken as K_{Ic}. The resulting curve is shown in Fig. 6.7a (for an assumed value of $K_{Ic} = 15.8$ MPa$\sqrt{\text{mm}}$). The observed response seems to be satisfactorily described by LEFM for $\alpha_0 = 0.1 \sim 0.5$. However, other tests do not yield such a correlation with LEFM. Data from tests of larger specimens ($s/d = 6$) by Nallathambi[59] are shown in Fig. 6.7b. It appears that for small notch lengths, σ_u is almost constant (notch insensitive) as in failure governed by limit stress criteria. Another important feature is the strong dependence of σ_u on specimen size. Notch sensitivity, therefore, is not a material property but depends on the structure and the flaw size. Also, decrease in σ_u with increase in crack length is seen only for short notches (Fig. 6.7a). Therefore, notch-sensitivity cannot always be used to judge the proximity of fracture behavior to LEFM.

The influence of notch (or flaw) length is illustrated better by studying the experimentally determined values of K_{Ic}. Data from four-point bend tests ($40 \times 40 \times 160$ mm beams) by Ohgishi et al.[57] are plotted in Fig. 6.8a.

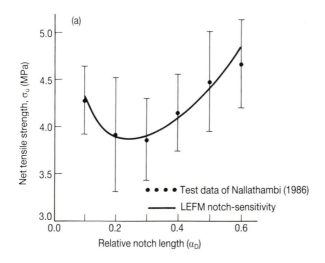

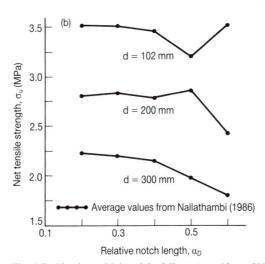

Fig. 6.7   Notch sensitivity of the failure stress (data of Nallathambi)

For hcp, practically constant values of K_{Ic} are obtained (as in other works such as Gjørv *et al.*[55]), but the values for mortar (with sand of maximum size 0.3 mm) increase with notch length. It appears that the behavior of hcp is much closer to LEFM than mortar. Unless very large specimens are tested, the observed values increase with notch length until a certain point and then decrease.[60] This trend can be seen in the tests of 3PB specimens ($50 \times 100 \times 40$ mm) of high strength concrete (compressive strength $f_c = 124$ MPa) by Biolzi *et al.*[61] Their data, in Fig. 6.8b, shows the variation in LEFM-based K_{Ic} with increase in notch depth. It is obvious that concrete, even of high strength, does not behave according to LEFM in usual test specimens. It should be mentioned that the effect of notch length on K_{Ic} varies considerably with specimen geometry and size. From tests

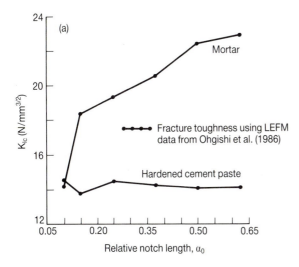

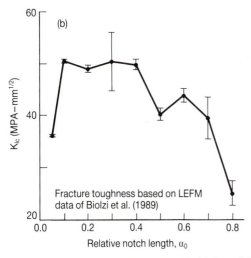

Fig. 6.8   Influence of notch length on K: (a) data of Ohgishi *et al.*[57]; and (b) data of Biolzi *et al.*[61]

conducted by Kesler *et al.*[62] on large wedge-loaded center-cracked plates of hcp, mortar and concrete, Saouma *et al.*[63] obtained K_{Ic}-values that were independent of crack length.

In LEFM, K_{Ic} does not depend on the size of the specimen or structure. However, the failure stress is size-dependent. This can easily be demonstrated by substituting $P = P_u$ and $K_I = K_{Ic}$ in Equation (6.3), which yields $K_{Ic} = P_u f(\alpha)/b\sqrt{d}$. Assuming that failure occurs with negligible crack extension ($\alpha = \alpha_0$), and defining the nominal failure stress as $\sigma_N = P_u/bd$,

$$\sigma_N = K_{Ic} f(\alpha_0) \frac{1}{\sqrt{d}} \qquad (6.9)$$

Since K_{Ic} is constant in LEFM and $f(\alpha_0)$ is a shape-dependent constant, Equation (6.9) implies that $\sigma_N = (\text{constant}) x 1/\sqrt{d}$ in geometrically similar structures. This is the size effect according to LEFM where the failure stress decreases with increase in certain structural dimensions. Tests of concrete specimens exhibit a more complex behavior, yielding K_{Ic}-values that may be size-dependent. Therefore, the size effect on failure stress is not always that of LEFM. From tests of notched beams, Walsh[53] recognized a significant trend in this size effect. For small specimens, the failure stress is constant, as in failure criteria based on limit stress. For very large specimens, the size effect is the strongest; $\sigma_N \propto 1\sqrt{d}$ as in LEFM. Therefore, the structure should be greater than a certain size for LEFM to apply. Such a transition in failure mode has also been observed in hcp and mortar.[57] The data of Tian *et al.*[58] are shown in Figs. 6.9a and 6.9b for compact tension specimens of concrete (maximum aggregate size = 20 mm). It is clear that the failure stress and K_{Ic} are significantly size-dependent.

In summary, the fracture of cementitious materials that have a finer microstructure (e.g., hcp) is closer to LEFM than others (e.g., concrete, fiber-reinforced mortar). The mode of structural failure generally lies between two limits: the strength limit, and LEFM behavior. When the size of the initial crack or the size of the uncracked ligament is of the same order as the inherent material heterogeneity (i.e., much smaller than the structure), failure is governed by limit criteria. This failure mode also occurs when the characteristic structural dimensions are small (i.e., of the same order as the size of the heterogeneities). When the structure is very large, failure is governed by LEFM.

In order to model concrete failure, several methods have been developed for determining size-independent fracture properties using practical-size specimens. Also, it is essential that analysis techniques are able to simulate the observed geometry-effects on cracking and failure before they can be applied to predict actual structural behavior.

6.5   Fracture energy from work-of-fracture

The resistance of the material to fracture, R, was defined earlier (Equation (6.6)) as the energy needed to create a crack of unit area. When a notched specimen, such as the beam in Fig. 6.3b, is tested until failure, the total work done gives the total energy dissipated. Assuming that all the energy has been consumed in extending the crack, determination of the work done and the crack area would yield R. This value is usually called fracture energy, and is denoted here as G. Nakayama[64] tested beams with triangular notches and obtained the load versus load-point displacement curves. (The triangular or V-shaped notches prevented catastrophic failure.) The area under a curve provided the total energy. Taking the unnotched part of the cross-section (i.e., the ligament area) as the area of

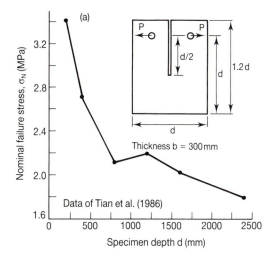

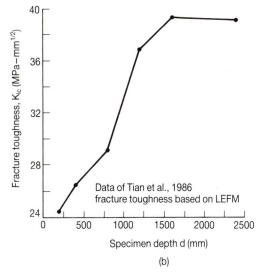

Fig. 6.9   Size effect on failure stress (data of Tian *et al.*[58])

the crack, the fracture energies for several brittle materials were determined.

For LEFM-type behavior, the fracture energy is a constant equal to the critical strain energy release rate, i.e. $G_f = G_c$. In practice however, G_f is not always constant. Tattersall and Tappin[65] found that the fracture energy increased with specimen size and decreased with a significant decrease in stiffness of the testing machine. In the first application of the work-of-fracture method to concrete, Moavenzadeh and Kuguel[66] showed that the determination of G_f was not straightforward. Since cracking in concrete is tortuous, they argued that the crack area should be determined exactly instead of simply using the ligament area. They obtained the crack area from microscopy techniques, and thereby computed fracture energies that were almost the same for hcp, mortar and concrete. Since the measurement of 'true' crack area is difficult, and to a certain extent subjective, later researchers have usually circumvented the problem by taking the ligament (or projected) area as the nominal area of the crack. Then, the fracture energy is also a nominal value. Moavenzadeh and Kuguel[66] used straight notches for their concrete beams, as have later researcher who have been able to obtain the entire load-displacement curve using stiff servo-hydraulic testing machines.

Petersson[67] proposed that the work-of-fracture from a notched beam test would provide a material constant for G_f if the following conditions are satisfied: (1) energy consumption outside the fracture process zone is negligible, (2) energy consumption is independent of specimen geometry, and (3) the fracture is always stable. Due to the effect of toughening or the increase in R with crack extension (see Section 6.6.2), condition (2) can never be satisfied exactly. However, as the sizes of ligament and specimen increase, G_f would tend asymptotically towards a constant value. Therefore, G_f approaches a material property when the fracture zone is negligible compared to the specimen size.[68] Accordingly, the work-of-fracture method was recommended by RILEM[69] with a lower limit on the specimen size. For concrete with a maximum aggregate size of 16–32 mm, the required beam depth is 200 mm and the length is 1.2 m. The notch length should be half the depth of the beam. If the total energy consumed (including the work done by the weight of the beam) is W_f, the fracture energy is:[69,70]

$$G_f = W_f/A_{\text{lig}} \tag{6.10}$$

where A_{lig} = area of the ligament. In the case of a beam with a straight notch, $A_{\text{lig}} = b(d - a_0)$. The corresponding value of fracture toughness can be calculated from Equation (6.7). Condition (3) is satisfied by using a testing machine with high stiffness and/or servo-control.

It has been shown that the specimen sizes recommended by RILEM do not always provide size-independent values of G_f. For concrete, Hillerborg (1983) concluded on the basis of the fictitious crack model (see Section 6.6.1) that the work-of-fracture method would provide constant values only at the LEFM limit when the beam was 2–6 m deep and has a notch longer than 200–400 mm. This would be the point at which condition (2) is practically satisfied. However, with increasing loads there is some crushing at the loading points and possible energy dissipation outside the process zone. Due to this, Petersson's condition (1) may not be satisfied in certain cases. Nallathabi *et al.*[71] conducted a detailed experimental study of concrete with a maximum aggregate size of 20 mm. The work-of-fracture of beams with depths ranging from 150 mm to 300 mm was determined. It was shown that G_f increased with an increase in beam depth, and decreased with an increase in notch depth. All the values obtained were much greater than G_c-values calculated from LEFM (Equation (6.3)). By comparing several studies, Hillerborg[72] concluded that in some cases the work-of-fracture method yielded size-independent values, but in others doubling the specimen size resulted in a 20% increase in G_f. A 20% increase in G was also observed for high strength concrete beams by Gettu et al.[73] by increasing the beam depth from 200 mm to 400 mm. Such a size effect on G_f has been observed in compact tension specimens by Wittmann et al.[74] and was attributed to the increase in the width of the fracture zone with increase in the ligament length. Other possible sources of size effects and errors in the work-of-fracture method have been described in detail by Planas and Elices,[75] and Brameshuber and Hilsdorf.[76]

Though the method yields size-dependent values, it has widely been used due to its simplicity. Availability of extensive data has led to empirical 'code-type' relations linking fracture energy to conventional design properties. One such equation is provided by the CEB-FIP Model Code:[77,78]

$$G_f = x_F f_{cm}{}^{0.7} \tag{6.11}$$

where x_F is a tabulated coefficient that depends on the aggregate size (e.g., for a maximum aggregate size $= 16$ mm, $x_F = 6$), $f_{cm} =$ mean compressive strength of the concrete in MPa, and G_f is obtained in N/m. The commentary to the code cautions that Equation (6.11) gives values that may be size dependent with deviations up to $\pm 20\%$. Nevertheless, the relation is useful when experimental data is lacking.

It should be emphasized that G_f, by itself, is not a reliable measure of toughness or ductility, and that using G_f as the sole fracture parameter could lead to erroneous conclusions. If one were to conclude from the observed increase in G_f with compressive strength (as suggested by Equation (6.11)) that ductility increases with the strength, this would be wrong. With the use of additional parameters (see Section 6.7), the higher brittleness in high-strength concrete can be adequately characterized.

6.6   Nonlinear fracture mechanics of concrete

Since LEFM (or any theory with only one fracture parameter) cannot adequately characterize the cracking and failure of concrete, several nonlinear techniques have been proposed. These approaches are, in general, modifications of LEFM, with fracture criteria based on two or more material parameters that account for the process zone. Based on whether the fracture zone is modeled explicitly or implicitly, the approaches may be categorized as cohesive crack models and effective crack models, respectively. All the models represent the actual three-dimensional fracture zone through an 'equivalent' line crack which simulates only a certain part of the actual process. Since the relation between the model and reality is different in each approach, the various models are not equivalent in all aspects.

Cohesive crack models

In the cohesive crack models, the fracture zone is simulated by a set of line forces imposed on the crack faces that tend to close the crack. The magnitude and distribution of these forces, or cohesive stresses, are related to material parameters in a prescribed manner. The cohesive stresses cause toughening by decreasing the G at the crack tip. The stresses have also been taken to represent the bridging action of aggregates and fibers. The fictitious crack model, the crack band model and several other closing pressure models that are discussed here can be classified as cohesive crack models. All cohesive crack models have to be used along with an appropriate constitutive model for the uncracked concrete. Normally, it is reasonable to assume that the uncracked concrete is elastic.

Fictitious crack model (FCM)

Hillerborg *et al.*[79] proposed a discrete-crack approach for the finite element analysis of fracture in concrete, which became known as the fictitious crack model. It is similar in principle to the Barenblatt model where there is a zone at the crack-tip with varying stresses. According to the FCM, a crack is initiated when the tensile stress reaches the material strength f_t, and it propagates in a direction normal to the stress. As the crack-opening (w) increases, the stresses (f) in the fracture zone behind the crack-tip gradually decrease from f_t to zero. The stress-opening relation is assumed to be a property of the material. The fracture criteria of the FCM are similar to the final stretch model of Wnuk[80] for elastic-plastic fracture.

The fracture criteria of the FCM (see Fig. 6.10) can be summarized as:

$$w = 0, \quad f = f_t \tag{6.12a}$$

$$w = w_c, \quad f = 0 \tag{6.12b}$$

$$\int_0^{w_c} f \, dw = G_f \tag{6.12c}$$

where w_c is the critical crack opening. The conditions in Equations (6.12a) and (6.12b) occur at the tips of the fictitious crack (a_f) and the traction-free crack (a_0), respectively. Note that Equation (6.12c) gives the energy absorbed for producing unit area of a traction-free crack. If the shape of $f(w)$ is known, there are only two independent material parameters which usually are f_t and G_f. Several different shapes have been assumed for the $f(w)$ relation including linear, bilinear and smoothly varying functions. The experimental determination and the significance of the $f(w)$ curve are discussed in on Constitutive Relations below.

In a parametric study of the FCM, Carpinteri *et al.*[81] used a linear $f(w)$ function in the finite element analysis of a notched beam. They showed that with an increase in G_f, the peak load and the deformation at the peak increase, and the steepness of the post-peak load-deformation curve

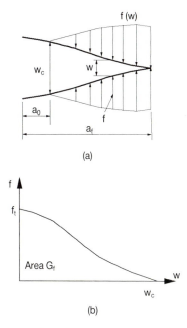

(a)

(b)

Fig. 6.10   The fictitious crack model

decreases. Increasing f_t caused an increase in the peak load but no change in the final part of the load-deformation curve. Increasing the specimen size produced an effect identical to that obtained by decreasing G_f.

The FCM has been extended to include non-planar fracture[82–84] where the direction of crack extension is defined at each step by remeshing. Other discrete crack models with mixed-mode considerations have also been proposed.[85–87] The effects of suatained and cyclic loading have been predicted with a discrete crack model by Akutagawa *et al.*[88]

Crack band model (CBM)

A convenient approach for the finite element analysis of fracture in concrete structures is to assume that cracking is distributed throughout the element. Crack propagation is then represented by a stress-strain relation that exhibits softening. The crack band model of Bazant differs from earlier smeared crack models due to the use of fracture mechanics criteria.[89,90] This eliminates spurious mesh dependence that is obtained when the criterion for crack propagation is a stress or strain limit.

In the analysis with the CBM, finite elements of width equal to that of the fracture zone have to be used in the region where cracking is expected. Excluding this aspect, analyses with the CBM and the FCM are practically equivalent. The strain due to cracking, ε_c, in the constitutive relation of the CBM can be related to the crack opening of the FCM (see Equation (6.12)) as:

$$\varepsilon_c = w/w_h \qquad (6.13)$$

where w_h is the element width. Note that w_h is not the actual process zone width but an effective width of the softening region needed for the numerical analysis. To allow the use of finite elements that are wider than w_h, which is necessary in the analysis of large structures, certain modifications have to be made as described by Bazant.[91]

When cracking is non-planar (i.e., not parallel to the usually square mesh lines), the crack pattern can be roughly predicted with the CBM.[91] For obtaining accurate results step-wise remeshing is needed as in the FCM. Recently, smeared-cracking models that are more general have been proposed[92,93] to analyze cases where mixed-mode cracking occurs with the transfer of tensile and shear stresses across the process zone. However, in a comparative study of smeared and discrete crack models, Rots and Blaauwendraad[85] suggested that, in non-planar crack problems, the smeared approach should be used as a qualitative predictor, followed by a corrector analysis with the discrete approach.

Singular closing pressure models

The process zone of concrete is treated in several approaches like the plastic zone in metals. Accordingly, the stresses at the crack-tip are taken to be finite (i.e., non-singular), and therefore, $K_I = 0$. Another class of cohesive crack models include a singular crack-tip (i.e., $K_I \neq 0$) as in LEFM, with the criterion for crack extension as $K_I = K_{Ic}$. Here K_{Ic} represents the fracture toughness of the matrix phase (hcp or mortar).

The use of a singular closing pressure models is justified by the fact that, in concrete, the crack profile at the tip is parabolic[39,94] as in the case of LEFM. On the other hand, a non-singular model with $K_I = 0$ produces a crack profile having zero slope at the tip which does not match the experimentally observed profile (see Fig. 6.11a). Singular models have been proposed by Jenq and Shah,[95] Foote *et al.*,[96] and Yon *et al.*[97] A closing pressure model has also been proposed for mixed-mode fracture by Tasdemir *et al.*[98] In these models, toughening is simulated by closing stresses or tractions imposed behind the singular crack-tip (see Fig. 6.11b). Consequently, the strain energy release rate has the following form until the critical state:

$$G = \frac{K_i^2}{E'} + \int_0^{\text{CTOD}} f(w)\,dw \qquad (6.14)$$

where CTOD = opening at the intial crack-tip. The first term in Equation (6.14) is due to the singularity and the second term gives the energy dissipated by the cohesive stresses. The use of a singular model will lead to realistic predictions of crack profiles and near-tip deformations. However, in the analysis of global structural behavior, singular and non-singular models give practically the same results.

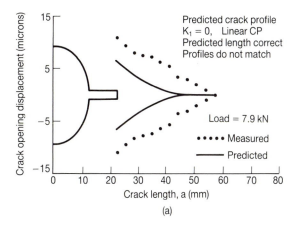

(a)

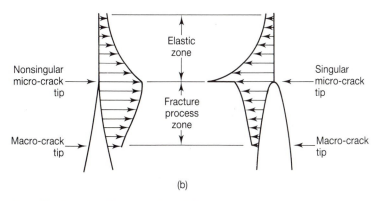

(b)

Fig. 6.11 Singular cohesive crack models

Constitutive relations for the cohesive crack

For the experimental determination of the tensile stress-strain relation or stress-opening relation for a crack, a uniaxial tension test seems to be the ideal method. However, in a concrete specimen under direct tension, the stress field and the cracking, especially after the peak, are not always uniform. Also, the transverse compression applied for gripping the specimen, and the dimensions of the specimen can influence the results considerably. In spite of these problems, reliable $f(w)$ relations have been obtained from tension tests by some investigators.[99] Under specific test conditions, Gopalaratnam and Shah,[100] Cornelissen *et al.*,[101] Guo and Zhang,[102] and Hordijk[99] could determine the $f(w)$ relation for monotonic and cyclic loading. Their results show that the envelope of the cyclic behavior is not affected by the cycling of the load, and is the same as that obtained for monotonic loading.

A generalizing stress-operating relation, that is applicable to several cement-based materials including normal and high-strength concrete, has been proposed by Wecharatana.[103] From a detailed study of the tensile softening response of notched dog-bone and end-tapered specimens, he obtained an equation of the form:

$$\left(\frac{f}{f_t}\right)^m + \left(\frac{w}{w_c}\right)^{2m} = 1 \tag{6.15}$$

where m is a material parameter; for 24 MPa concrete, $m = 0.27$ and for 83 MPa concrete, $m = 0.2$.

Several other experimental techniques have been proposed for determining the stress-opening relation. The method of Li,[104] based on the J-integral (Chapter 4 of Broek[14]), requires the load, displacement and crack-tip opening of two fracture specimens to be monitored until failure. Using this data, the $f(w)$ curve including its parameters, f_t, w_c and G_f, can be obtained.[104]

The ability to measure deformations continuously over a wide field with sufficient accuracy by means of optical interferometry has led to experimental-numerical approaches for determining the constitutive relations of cracking. In these investigations, deformations near the propagating crack in a test specimen were obtained using moiré interferometry[37,105] or sandwich-hologram interferometry.[38,39] From these data, the parameters of the $f(w)$ curve having a prescribed shape were obtained. In the works of Castro-Montero et al.[39] and Miller et al.,[106] holograms of an area near the notches of center-cracked plates were made after each loading step, with a laser light source. The interference fringes observed in a sandwich of two consecutive holograms, with the same illumination direction, represent one component of the relative displacement undergone during the load step. To obtain the correct displacement vectors three holograms have to be made at each step with different illumination directions. An image analysis system was used to analyze the fringes objectively. The crack-opening and the strain field were thereby computed at each load step. Using finite element analysis with the FCM, the parameters of the $f(w)$ relation were determined such that the calculated crack-openings matched those computed from the holograms.

Other experimental-numerical approaches (e.g. Roelfstra and Wittmann[107]) use the global load-deformation response of fracture specimens to obtain the parameters of the $f(w)$ relation. Wittmann et al.[108] assumed a bilinear form for $f(w)$ characterized by four constants f_t, f_c, f_1 and w_1, with a change of slope at (f_1, w_1). By fitting the load-displacement curve from tests of compact tension specimens, they found that several $f(w)$ relations could produce the same global behavior. However, the computed fracture energy G_f (as in Equation (6.12c)) was almost the same. They proposed that, in order to get unique results, the tensile strength f_t should be determined independently and the ratio f_t/f_1 should be set at a certain value (in the range of 3–5). The resulting stress-opening relations seem to be independent of specimen size and loading rate.

The global structural behavior that is predicted by fracture analysis can be strongly influenced by the shape of the $f(w)$ relation.[107,109,110] From available data, several code-type formulations have been proposed for the stress-opening curve. The CEB-FIP Model Code[77,78] recommends a bilinear relation whose parameters are f_t, G_f, w_c and x_f, where w_c and x_F (see Equation (6.11)) have tabulated values that depend on the aggregate size. For a maximum aggregate size of 16 mm, $w_c = 0.15$ mm. Values for G_f can be obtained from Equation (6.11). The change in slope is at $f = 0.15f_t$, and the mean tensile strength can be estimated from:

$$f_t = 0.30f_c^{2/3} \qquad (6.16)$$

where f_c is the characteristic compressive strength in MPa. A similar relation has also been formulated by Liaw et al.[111]

Some recent works have focused on establishing rational relations between the microstructure of the concrete and the stress-opening behavior of the crack. The relation of Li and Huang[112] for $f(w)$ is formulated in terms of the aggregate content, maximum aggregate size, fracture toughness of the hcp and the characteristics of the aggregate-hcp interface. Their model assumes that microcracking, frictional debonding of the aggregates and crack-deflection are the main toughening mechanisms. Another model proposed by Duda[113] incorporates also the aggregate size distribution through a probabilistic formulation. Note that these models are only valid for concrete with weak interfaces, and with aggregates that are stronger than the hcp.

Effective crack models

Several models have been proposed where the fracture analysis is performed on an elastic structure that is geometrically identical to the actual structure. The length of the effective (LEFM) crack is initially the same as the crack or flaw in the actual structure. Further equivalence between the actual and effective cracks is prescribed by the model. The nonlinearity of the fracture process is usually represented by fracture criteria that are extensions of LEFM.

R-curve models

Consider a concrete specimen with an initial crack or flaw. As the load increases, a process zone at the tip will form and increase in size until the traction-free crack propagates. An effective crack can be defined such that its compliance is equal to that of the actual crack including its fracture zone. Accordingly, this effective crack will extend as the process zone grows. Since toughening increases with process zone sizes, its effect can now be modeled very simply by assuming that LEFM relations can be applied to the effective crack, and that the crack resistance varies with the effective crack extension. Then, the fracture criterion is $G = R$ as in Equation (6.6) (or $K_I = K_{IR}$ as in Equation (6.8), where R is not constant

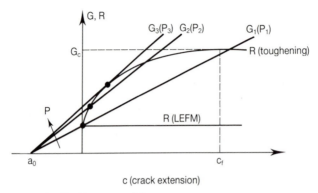

Fig. 6.12   The *R*-curve model

(as in LEFM) but is an increasing function of the effective crack growth c, which is denoted as $R(c)$ or the *R*-curve.

R-curve models can also be motivated by other concepts. Clarke and Faber[114] suggested that increase in crack resistance is due to statistical variability of the material microstructure. They argued that as the crack lengthens, the probability of encountering features of largest resistance increases. Therefore, the overall fracture resistance increases until the crack attains a length beyond which the probability is unity. Subsequently, the *R*-curve rises gradually only when there is a wide distribution of microstructural resistance. When the mortar is as strong as the aggregates, as in several high strength concretes, the *R*-curve may be expected to rise very sharply. This also occurs when the aggregates are of constant size and equi-spaced, or when the material is practically homogeneous.

The implications of the *R*-curve can be seen from Fig. 6.12, where the variation of *G* is shown for a specimen such as a beam (Fig. 6.3b) with a notch of length a_0. (The discussion would be identical if *R* and *G* were replaced by K_{IR} and K_I, respectively.) For a certain load *P*, *G* can be obtained as a function of crack length (with $a = a_0 + c$) from LEFM. Two *R*-curves are donsidered in Fig. 6.12: the 'LEFM' *R*-curve for an ideal-brittle material where *R* is practically constant; and the 'Toughening' *R*-curve which rises monotonically to a horizontal asymptote, for a material such as concrete. For constant *R*, fracture occurs when $G = R$ at $c = 0$, and subsequent crack propagation (i.e., $c > 0$) is unstable or catastrophic since $G > R$. For the rising $R(c)$, fracture is again initiated when $G = R$ at $c =$ but subsequntly fracture cannot occur at the same load since $G < R$. With increasing *P*, stable fracture occurs until the slope of $R(c)$ is less than that of $G(a)$. Therefore, with a rising $R(c)$ catastrophic failure occurs only after some stable fracture, and the load capacity increases even after fracture is initiated, i.e., the structure is flaw tolerant. Note that increasing $G(a)$ functions (as in Fig. 6.12) are exhibited by several geometries, called positive geometries. For negative geometries, the slopes maybe negative over a certain range of a. Discussion in this work is limited to positive geometries, and the reader is referred to Jenq and Shah,[115] Planas and Elices,[116] Ouyang *et al.*[117] and Bazant *et al.*[118] for details on negative geometries.

Usually R is taken to be a constant after a certain amount of crack extension. In Fig. 6.12, R asymptotically reaches a constant value G_c at $c = c_f$, where G_c is the critical strain energy release rate in LEFM. This reiterates that when R is constant, the R-curve model is identical to LEFM. It is, however, possible that after a horizontal plateau, R may decrease, especially due to the interaction of the crack with the specimen boundary.[119] Also, the plateau value may depend on structural dimensions.

The R-curve behavior of concrete has been studied by several investigators. From fracture tests, Brown[51] found that for mortar the crack resistance increased with crack length, while R was constant for hcp. Wecharatana and Shah[120] obtained R-curves from load-displacement curves of test specimens and the corresponding crack lengths measured optically. They concluded that mortar and concrete exhibited rising R-curves that were geometry-dependent. Generally, it is not possible to accurately measure the crack extension in a concrete specimen. Therefore, indirect procedures such as the compliance methods have been used. In the multi-cutting method of Hu and Wittmann,[121] the notch of a fractured specimen is extended several times and the compliance is determined at each step. The total crack length is that beyond which the compliance is the same as that of a virgin specimen. The traction-free crack length is that where notch extension does not change the compliance. Based on these crack lengths, Hu[122] obtained R-curves that decrease after reaching a maximum.

R-curves can also be derived from stress-opening relations of the cohesive crack models. For a given $f(w)$ relation, Foote *et al.*[96] obtained geometry-dependent K_{IR}-curves using the Green's function approach. Their curves were independent of the initial crack length for a given specimen but the values of c_f (see Fig. 6.12) varied with specimen geometry and size. Also, in their approach, c_f is approximately equal to the length of the fully developed cohesive zone. It was shown later that these R-curves reach a plateau only for large specimens, and that for smaller specimens the slope may become constant and even increase.

Planas *et al.*[124] proposed an $R(\mathrm{CTOD})$ curve that is directly related to $f(w)$. CTOD (for crack tip opening displacement) is the opening of the effective crack at the location of its initial tip. In their model,

$$R(\mathrm{CTOD}) = \sum\nolimits_0^{\mathrm{CTOD}} f(w)\,dw \qquad (6.17)$$

The $R(\mathrm{CTOD})$ curve is geometry-independent unlike the $R(c)$ curve. The CTOD can be related to c through LEFM for obtaining a conventional $R(c)$ relation.

Bazant and Kazemi[125] considered c_f and G_c (see Fig. 6.12) to be material properties, and derived an R-curve given in parametric form as:

$$\text{for } c < c_f, \quad R(c = G_c \frac{g'(\gamma)}{g'(\alpha_0)} \frac{c}{c_f}, \quad \frac{c}{c_f} = \frac{g'(\alpha_0)}{g(\alpha_0)} \left(\frac{g(\gamma)}{g'(\gamma)} - \gamma + \alpha_0 \right) \tag{6.18a}$$

$$\text{for } c \geq c_f, \quad R(c) = G_c \tag{6.18b}$$

where $g(\alpha) = \{f(\alpha)\}^2$ (see Equation (6.3)), $g'(a)$ = derivative of $g(\alpha)$ with respect to α, G_c and c_f are the fracture energy and process zone size defined by the size effect model (see below), and γ is a dummy parameter. Equation (6.18) was derived for an infinitely large specimen where the process zone develops without restrictions. For finite structures, the maximum load occurs at $c < c_f$ and $R < G_c$. Bazant et al.[118] suggested later that for finite size, Equation (6.18) is valid until the maximum load and subsequently the R-curve is constant. With this modification, Gettu et al.[73] satisfactorily predicted load-deflection curves of high strength concrete, using Equation (6.18).

Ouyang et al.[117] proposed an R-curve that is the envelope of several specimens that are geometrically similar and different in size, but with the same initial notch length. For infinite size, their R-curve is given as:

$$R = \xi \left\{ 1 - \left(\frac{d_2 \kappa - \kappa + 1}{d_1 \kappa - \kappa + 1} \right) \left(\frac{\kappa a_0 - a_0}{a - a_0} \right)^{d_2 - d_1} \right\} (a - a_0)^{d_2} \tag{6.19a}$$

$$d_{1,2} = \frac{1}{2} + \frac{\kappa - 1}{\kappa} \pm \left\{ \frac{1}{4} + \frac{\kappa - 1}{\kappa} - \left(\frac{\kappa - 1}{\kappa} \right)^2 \right\}^{1/2} \tag{6.19b}$$

where κ and ξ are functions of a_0, E, K_{Ic}, $COTD_c$ and specimen geometry.[126] K_{Ic} and $COTD_c$ are the critical stress intensity factor and critical CTOD, respectively, defined according to the two parameter fracture model (see the next Section). For finite specimens, the plateau of the R-curve begins at the maximum load. Using this model, the load-deformation response of several fracture specimens have been predicted.[117]

The R-curve can be easily used in nonlinear fracture analysis of structures when the $R(c)$ and $G(a)$ functions are available. However, the R-curve approach has some limitations: (1) $R(c)$ is not a true material property but also depends on structual geometry, and (2) $G(a)$ functions cannot be determined without knowing the crack pattern.

Two parameter fracture model (TPFM)

The TPFM of Jenq and Shah[115] proposes an effective crack that is equivalent in compliance to the elastic component of the actual crack. As shown in Fig. 6.13a, two independent material parameters K_{Ic}^s and $COTD_c$ are defined in terms of the critical state of the effective crack. The critical state is usually at the maximum load and corresponds to the

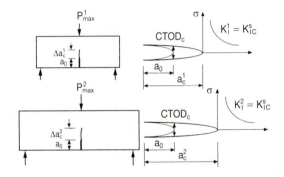

(a) Fracture criteria: $K_I = K_{IC}^s$ and CTOD = $CTOD_c$
(Superscripts 1 and 2 correspond to small and large specimens, respectively)

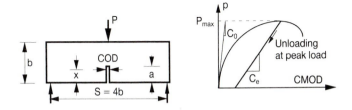

(b) Determination of K_{Ic} and $CTOD_c$ from C_0 and C_e obtained in a notched-beam test (C_0 and C_e are the initial compliance and the unloading compliance at the peak load, respectively)

Fig. 6.13   The two parameter fracture model

effective crack length a_c. The associated critical stress intensity factor is K_{Ic}^s, and the critical CTOD is $CTOD_c$. The fracture criteria of the model are:

$$K_I = K_{Ic} \text{ and } CTOD = CTOD_c \qquad (6.20)$$

These criteria have to be satisfied simultaneously for fracture to occur (see Fig. 6.13a). The TPFM has also been extended to mixed mode fracture.[127]

A RILEM[128] recommendation describes the procedure for obtaining the parameters of the TPFM from the load-CMOD response of a 3PB specimen, where CMOD (for crack-mouth opening displacement) is the opening of the notch mouth. As shown in Fig. 6.13b, the unloading compliance C_e is determined just after the peak load P_{max} (within 95% of P_{max}), and used along with the initial compliance C_i in LEFM relations to get a_c. The parameters K_{Ic}^s and $CTOD_c$ are then computed for the effective crack at load P_{max}. It has been demonstrated that this method yields parameters that are practically size-independent.[110] Though several specimen geometries can be used, the RILEM recommendation proposes the 3PB specimen since the method has been extensively verified with this geometry. Typical values of the TPFM parameters are given in Section 6.7.

An effective crack model has been proposed by Karihaloo and Nallathambi[129,130] that is similar in principle to the TPFM. In their model, the critical length of the effective crack is determined such that the deflection of the effective specimen is the same as that of the actual specimen, under peak load P_{max}. The K_{Ic}-values determined from this method are comparable to those of the TPFM; typical values are about 31 MPa$\sqrt{mm}$ for normal concrete with $f_c = 27$ MPa, and about 58 MPa$\sqrt{mm}$ for 78 MPa high strength concrete.[131] See Section 6.7 for more details.

Size effect method (SEM)

In Section 6.4, the size effect observed in fracture tests of concrete was discussed. As a result of this phenomenon, tests on geometrically similar specimens (with same shape and relative notch length) of different sizes yield failure stresses that vary with specimen size. This trend has been modeled by Bazant[132] with the relation:

$$\sigma_N = \frac{B}{\sqrt{1+\beta}}, \quad \beta = \frac{d}{d_0} \tag{6.21}$$

where σ_N = nominal failure stress (see Equation (6.9)), d = characteristic structural dimension (for the 3PB specimen, d = depth), and B and d_0 are empirical parameters. Equation (6.21) has the form shown graphically in Fig. 6.14, which implies that for small sizes, the failure is governed by limit stress criteria (no size effect), and that for large sizes, the failure is governed by LEFM ($\sigma_N \propto 1/\sqrt{d}$).

Since Equation (6.21) relates the failure stress of small specimens to LEFM behavior, Bazant proposed that the LEFM asymptote be used to define fracture parameters unambiguously for an infinitely large size. Test data could then be extrapolated to the limit of $d \to \infty$, where effects of the specimen geometry are theoretically absent. This approach was used to determine fracture parameters for concrete and mortar by Bazant and

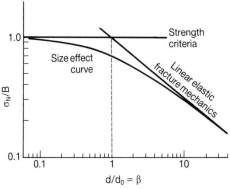

Fig. 6.14   The size effect model

Pfeiffer.[133] This also lead to a reformulation of Equation (6.21) in terms of two fracture parameters:[125]

$$\sigma_N = \frac{K_{Ic}}{\sqrt{g'(\alpha_0)c_f + g(\alpha_0)d}} \qquad (6.22)$$

where $K_{Ic}(=\sqrt{E'G_c})$ and c_f are the fracture toughness and the maximum process zone length, for the LEFM limit of finite size. (See Fig. 6.12 and Equation (6.18).) The functions g and g' are the same as in Equation (6.18). In this model, the effective crack is similar to that of the TPFM. The SEM has been used to model the fracture of several types of specimens and structures.[73,125]

The procedure for obtaining the fracture parameters of the SEM has been proposed as a RILEM[134] recommendation. The peak loads of at least three sizes of 3PB specimens are needed to calibrate Equations (6.21) and (6.22), and consequently determine K_{Ic}, G_c and c_f. Typical values for these parameters are given in Section 6.7.

Comments on the effective crack models

Planas and Elices[135,136] conducted comparative studies of several effective crack models, and have concluded that their fracture criteria and their performance within the practical size range are indistinguishable. The models, however, may exhibit differences in their asymptotic behavior when they are used in the analysis of structures far larger than those used for determining their parameters (see also Karihaloo and Nallathambi[130]).

It should be emphasized that consistency in the choice of a model is of utmost importance; i.e., the same model should be used for both the material characterization and the analysis. Also, results from the same model should be used for comparing the characteristics of different materials or structures.

It should also be noted that G_c used in the effective crack models is not equal to G_f obtained from the work-of-fracture method (see Section 6.5) unless the fracture is governed by LEFM.

6.7 Material characterization

For the characterization of concrete, fracture mechanics provides parameters that are complementary to traditional measures such as strength and limit strain. In some applications where cracking governs the response, fracture parameters may become the only data that are needed for analysis. In general, two aspects of fracture behavior can be identified: the resistance against cracking, and the brittleness of crack propagation. This section deals with the quantification of both these aspects and the review of certain factors that influence them.

Crack resistance

In most of the nonlinear fracture models, one of the parameters quantifies resistance of the material against cracking, and the other denotes the brittleness or ductility of the material (see below). In cohesive crack models such as the FCM and CBM, the crack resistance is measured by the tensile strength f_t. It should be noted that when f_t is determined from a tensile test it could vary with the size and shape of the specimen. In general, f_t increases with an increase in f (see Equation (6.16)). The other parameter of the FCM, the fracture energy G_f determined from the work-of-fracture method, depends on both the crack resistance and the process zone deformation (see Equation (6.11)).

The parameters of the effective crack models that quantify crack resistance are K_{Ic} and G_c. The $K_{Ic}{}^s$ of the TPFM varies between 30–50 MPa$\sqrt{}$mm for normal concrete and between 25–30 MPa$\sqrt{}$mm for mortar.[130] The variation of $K_{Ic}{}^s$ with strength, as determined by Nallathambi and Karihaloo,[131] is shown in Table 6.1. The values of K_{Ic}

Table 6.1 Variation of K_{Ic}^s with compressive strength

f_c (MPa)	27	39	49	68	68
K_{Ic}^s (MPa$\sqrt{}$mm)	31	40	44	48	58

obtained from their effective crack model are almost the same. The increase in $K_{Ic}{}^s$ with strength has also been demonstrated by John and Shah.[137] Lange[138] determined $K_{Ic}{}^s$ for several compositions of hcp and mortar, with and without silica fume and the results are given in Table 6.2.

K_{Ic} from the SEM usually varies between 25–60 MPa$\sqrt{}$mm for normal concrete, and its value compares well with $K_{Ic}{}^s$ from the TPFM.[130] From an analysis of the tests of Bazant and Pfeiffer,[133] values of 36 MPa$\sqrt{}$mm for concrete $f_c = 34$ MPa and 27 MPa$\sqrt{}$mm for mortar ($f_c = 48$ MPa) were obtained by Bazant and Kazemi[125] using Equation (6.22). For 86 MPa high-strength concrete, Gettu *et al.*[73] obtained K_{Ic} of 30 MPa$\sqrt{}$mm. Chern and Tarng[139] conducted size effect tests on fiber-reinforced concrete to determine G_c. They used 20 mm long steel fibers and crushed limestone aggregates. Their results are summarized in Table 6.3.

Table 6.2 K_{Ic}^s for hcp and mortar, with and without silica fume

Material	f_c (MPa)	K_{Ic}^s (MPa$\sqrt{}$mm)
hcp	62	14
hcp + 5% sf *	65	14
hcp + 10% sf *	65	15
coarse mortar 1:1	59	20
coarse mortar 1:2	47	22
fine mortar 1:1	63	23

*silica fume

Table 6.3 G_c of fiber-reinforced concrete

Max. aggregate size (mm)	Fiber content (%)	f_c (MPa)	G_c (N/m)
25	0	33	39
	1	37	59
	2	40	112
13	0	31	29
	1	40	67
	2	44	118
5	0	40	20
	1	45	75
	2	48	124

K_{Ic} of high-strength concretes has also been determined by other means. Using a compliance method, de Larrard *et al.*[140] found that by increasing the concrete strength from 54 to 105 MPa, the K_{Ic} increased by about 30%. A general conclusion is that the crack resistance increases with increase in the conventional compressive strength, but at a lesser rate. However, when fibers are added to concrete, the crack resistance increases much more than the strength.

Brittleness

The failure of plain concrete is generally brittle, but not as brittle as that of glass. This ductility, or rather the 'pseudo-ductility', that concrete possesses can be quantified through fracture mechanics. In this work (as in others; cf. Hucka and Das[141]), ductility is taken to be the inverse of brittleness.

In almost all of the nonlinear fracture models, the brittleness of the material can be related to parameters that depend on the dimensions or the deformations of the (effective) fracture process zone. Some brittleness quantifiers have been defined by combining G_f with other properties. Hillerborg[68] defined a characteristic length l_{ch} that is proportional to the process zone length:

$$l_{ch} = \frac{EG_f}{f_t^2} \qquad (6.23)$$

The relation is similar to that used by Irwin for the size of the plastic zone, which is of significant importance with respect to the ductile-brittle transition in the fracture of metals. In the context of concrete, a smaller l_{ch} implies that the material is more brittle. Typical values for l_{ch} are given in Table 6.4.

A code-type relation has also been proposed for l_{ch} by Hilsdorf and Brameshuber[78] that is valid for concrete with f_c in the range of 10–100 MPa:

$$l_{ch} = 600 x_F f_{cm}^{-0.3} \qquad (6.24)$$

Table 6.4   Typical values for l_{ch}

Material	l_{ch}	Reference
glass	1 micron	Bache[142]
dense silica cement paste	1 mm	Bache[142]
hcp	5–15 mm	Hillerborg[68]
mortar	100–200 mm	Hillerborg[68]
high-strength concrete (50–100 MPa)	150–300 mm	Hillsdorf and Brameshuber[78]
normal concrete	200–55 mm	Hillerborg[68]
dam concrete max. aggregate size = 38 mm	0.7 m	Brühwiler *et al.*[143]
glass fiber reinforced mortar	0.5–3 m	Hillerborg[68]
steel fiber reinforced concrete	2–20 m	Hillerborg[68]

where x_F and f_{cm} are defined below Equation (6.11).

The brittleness of the material can also be judged from the shape of the $f(w)$ curve (Fig. 6.10). A material with a steep incline in f is more brittle than one that exhibits a more gradual decline. This aspect is quantified by the model of Wecharatana[103] in Equation (6.15), where parameter m is a brittleness index. The critical separation w_c of the FCM also reflects the ductility of a material, but its value is affected significantly by the shape chosen for the $f(w)$ curve (see Section 6.6).

The R-curve has traditionally represented the brittleness of a toughening material. A gradually rising R-curve (see Section 6.6) implied that the material was less brittle than one where the R-curve rises rapidly. Accordingly, when two materials have almost the same G_c, that with a smaller c_f would be more brittle (see Fig. 6.12).

Three quantities derived from the TPFM have been used as measures of ductility. The higher the pre-peak crack extension, the lower is the brittleness of a quasi-brittle material. Therefore, the critical effective crack length a_c is an indicator of the material ductility. It was used by John and Shah[137] to compare the brittleness of high-strength mortar to that of normal concrete. It can be seen in Fig. 6.15 that a_c decreases with an

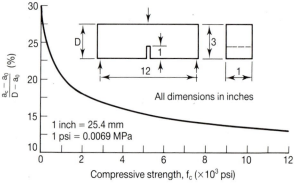

Fig. 6.15   Decrease in ductility with the increase of strength (results of John and Shah[137])

Table 6.5   Typical values for $CTOD_c$ and Q

Material	Q (mm)	$CTOD_c$ (mm)	Reference
hcp	25	0.0035	Lange[138]
hcp + 5% sf *	10	0.0025	Lange[138]
hcp + 10% sf *	5	0.0017	Lange[138]
coarse mortar 1:2	57	0.0046	Lange[138]
concrete	300	0.02	Jenq and Shah[115]
high strength mortar (110 MPa)	90	0.011	John and Shah[137]

* silica fume

increase in compressive strength demonstrating that brittleness increases with strength. Since a_c depends on the size and shape of the specimen tested, two other size-independent quantifiers of ductility have also been defined (Jenq and Shah[115]) – one is $CTOD_c$ and the other is a length parameter Q:

$$Q = \left(\frac{E\,CTOD_c}{K_{Ic}{}^s} \right)^2 \tag{6.25}$$

Both $CTOD_c$ and Q are smaller for a more brittle material. Typical values of these parameters are given in Table 6.5 (see Table 6.2 also).

In the SEM of Bazant, the process zone size, c_f, defined in Equation (6.22) is a measure of the material ductility. Bazant and Kazemi[125] obtained c_f-values of 10–25 mm for concrete ($f_c = 34$ MPa) and 6–15 mm for mortar ($f_c = 48$ MPa). For a high strength concrete ($f_c = 86$ MPa), Gettu *et al.*[73] found that c_f was about 3–6 mm at 14 days.

From a review of the test data, it may be concluded that brittleness increases with increase in the strength of concrete. For normal concrete, brittleness decreases with increase in aggregate size and with the addition of fibers. The presence of silica fume seems to increase the brittleness significantly.

Previous discussion in this section has dealt with the brittleness of the material. This can be described simply as the lack of flaw-tolerance or the inability of the material to prevent unstable and catastrophic crack propagation. The brittleness of structural failure, however depends on both the material and the structural geometry. Hillerborg[68] and Bache[142] have taken the quantity d/l_{ch} to indicate structural brittleness. Carpinteri[48] considered the square-root of this quantity as a brittleness number.

Carpinteri *et al.*[81] proposed a dimensionless brittleness number that increases with decreasing structural brittleness. This is based on the parameters of the FCM (see Section 6.6), as given as:

$$s_E = \frac{G_f}{f_t d} \tag{6.26}$$

For a linear $f(w)$ relation, $s_E = w_c/2d$.

In the brittleness numbers s_E and d/l_{ch}, the structural brittleness increases with increase in the material brittleness characterized by w_c and l_{ch}, respectively. They also imply that the structural brittleness increases with the dimensions of the structure. The second effect causes the characteristic increase in steepness of the post-peak part of load-deflection curve with increase in size (see Bosco *et al.*[144]). Bazant has argued that the effects of specimen geometry and relative length of the initial flaw should also be included in the structural brittleness number. Accordingly, the brittleness number β of the SEM (see Equations 6.21 and 6.22) is defined as:[125],[133]

$$\beta = \frac{g(\alpha_0)}{g'(\alpha_0)} \frac{d}{c_f} \qquad (6.27)$$

where the first term accounts for the shape of the structure, and the second accounts for the structural dimensions and the material brittleness. β can be directly determined from size effect tests by fitting Equation (6.21). The size effect curve in Fig. 6.14 shows the implications of β. When the size or material brittleness is large (i.e., β is large), the behavior is close to LEFM, and when β is small the behavior is governed by limit criteria. The position of the data on the size effect curve also gives an idea of the brittleness. Chern and Tarng[139] found that with the addition of fibers, the behavior shifted away from LEFM implying a decrease in structural brittleness.

Factors influencing the fracture parameters

Discussion in previous sections assumed that the concrete was subjected to normal conditions of temperature, loading rate, humidity, etc., and that its composition did not vary significantly. In this section, the effects of these factors on the fracture parameters will be briefly reviewed.

Concrete composition

The presence of aggregates influences the fracture properties of a cement-based material considerably (as already mentioned above). This was first studied by Moavenzadeh and Kuguel,[66] and Naus and Lott[145] using LEFM. It was demonstrated that K_{Ic} for concrete was larger than that of mortar, and K_{Ic} for mortar was larger than that of hcp. In normal concrete, where cracks are arrested and deflected by the aggregates, the crack resistance generally increases with increase in the size of the aggregates.[71,146] This trend is, at times, opposite to that observed for the compressive strength. From studies based on the FCM, Roelfstra and Wittmann,[107] and Wittmann *et al.*[108] found that increase in the aggregate size caused the increase of G_f and w_c. These trends are reflected in the relations proposed by Hilsdorf and Brameshuber[78] for G_f and l_{ch} (see Equations (6.11) and (6.24)). The increase in crack resistance and the decrease in brittleness, with an increase in aggregate size, is due to the larger fracture process which arises when there is a wider range of heterogeneities, as explained in Section 6.3. Mihashi *et al.*[23] observed this effect in a study of the acoustic emissions recorded during fracture. It should also be mentioned that in a study of dam concrete, Saouma *et al.*[147] concluded that the crack resistance of concrete was higher for angular aggregates than for round aggregates, but was practically independent of the aggregate size.

In some high-strength concretes where the aggregates rupture during fracture, aggregate size is not expected to influence the fracture parameters. However, the presence of fibers will increase both the resistance and the ductility.

It has also been found that K_{Ic} decreases with an increase in water-cement ratio[71,145,148] and air content.[71,149] This implies that a denser microstructure would exhibit higher crack resistance. This also explains the increase in K_{Ic} with longer curing periods,[66,71] and with aging in the first few weeks.[54,148–150] Consequently, high strength concrete has a higher crack resistance than normal concrete, but also a higher brittleness due to fewer inhomogeneities.

Loading rate

Under dynamic and impact loading, the failure stress of concrete increases considerably with an increase in loading rate.[151] Several investigators have studied rate effects on fracture parameters of concrete in order to explain this characteristic increase in strength. By applying the TPFM, John and Shah[152,153] showed that there is a small increase in K_{Ic}^s with increase in loading rate, but a_c and $CTOD_c$ decrease considerably. Consequently, it was concluded that the brittleness of concrete increases with loading rate in the dynamic regime. Using work-of-fracture methods, increase in G_f with rate has been observed by Wittmann *et al.*[108] and Oh.[154]

Rate effects in the static range involve a strong interaction of creep and fracture.[155,156] At slow rates, Wittmann *et al.*[148] found that while f_t increases with loading rate, G_f and w_c seem to decrease. From size effect tests, Bazant and Gettu[157,158] obtained a shift in failure behavior towards LEFM, implying an increase in brittleness, when the deformation rate was decreased. Their results show a decrease in K_{Ic} and c_f with a decrease in rate. They also showed from relaxation tests that the effect of creep is much stronger in the presence of cracks.

The effect of loading rate on the fracture of high-strength concrete has not yet been investigated thoroughly. However, it appears that time-dependent effects are lesser in high strength concrete than in normal concrete (e.g., Banthia *et al.*[159]).

Temperature and humidity

The crack resistance of concrete generally increases with the decrease of temperature.[151] Using the SEM, Bazant and Prat[160] showed that at high temperatures, G_c decreases more when the concrete is wet than when it is dry. This effect was observed on G_f, even at sub-zero temperatures by Planas *et al.*[161] and Ohlsson *et al.*[162] At very low temperatures, Maturana *et al.*[163] also found that the freezing of the water in the concrete gave rise to a considerable decrease in the brittleness in terms of l_{ch}.

The detrimental effect of free water on the crack resistance of concrete was attributed to stress-corrosion by Shah and Chandra.[164] Michalske and Bunker[165] suggested that this phenomenon arises due to the weakening of strained silicate bonds at the crack-tip by water. Rossi[166] has proposed that free water is also the cause of rate effects in concrete, and therefore, there were lesser effects of loading rate in dry concrete and high-strength concrete due to the absence of moisture.

The effect of water in the curing stage is, however, different. As in the case of compressive strength, proper curing with water results in K_{Ic} that is significantly higher than for concrete cured in air.[149]

Loading history

The influence of fatigue loading on fracture parameters is of importance since cracks can propagate under repeated loads to cause failure.[167] Pons *et al.*[168] found that under low-frequency cyclic loading, G_f and LEFM-based K_{Ic} decreased considerably with increase in the number of cycles and the amplitude of the loading. They concluded that characterization of concrete through its fracture parameters is not independent of the loading history. Using the SEM, Schell[169] found that fatigue fracture of high-strength concrete was similar to that of normal concrete, and that it could be modeled adequately through nonlinear fracture mechanics coupled with the Paris' law.

Another important aspect of loading history is the effect that prior compressive loads may have on subsequent fracture. Tinic and Brühwiler[170] showed that when concrete is subjected to repeated compressive loading, the tensile strength could decrease by about 50%. Such losses of crack resistance seem to be lesser in high-strength concrete, but still significant.[171]

6.8   Other aspects of fracture in concrete

Other aspects of fracture characterization in concrete include the mixed mode fracture, interfacial fracture, field and core-based fracture tests, and stochastic models for fracture. These are summarized in this section.

Mixed-mode fracture

Since the crack resistance of concrete in Mode I is much lower than in other modes, several investigators have doubted whether mixed- mode fracture can exist in concrete.[172] In this work, a distinction is made between non-planar Mode I cracking, and mixed-mode cracking where the crack undergoes both opening and sliding at its tip. It appears that mixed-mode fracture occurs only when normal and shear loads are simultaneously applied on the crack.[173,174] Otherwise, the crack chooses a non-planar path where K_I dominates. Its behavior is, however, influenced by crack-face friction.[127,175,176] In either case, the fracture is less brittle than planar Mode I, especially for concrete where the crack path is tortuous.

Interfacial fracture

The bond between aggregates and hcp dictates several aspects of the behavior of concrete (cf. Mindess and Shah[177]). In normal concrete, the crack resistance of the interface is lower than that of the hcp and the aggregates,[178] and therefore, fracture occurs through these interfaces. This also causes crack tortuosity and a wider fracture process zone. In some high-strength concretes, the interfaces are stronger, and consequently, the brittleness is higher. Due to its importance in materials engineering and micromechanical modeling, researchers[179,180] are focusing more attention on the fracture mechanics and characterization of interfaces.

Field and core-based fracture tests

Fracture tests may sometimes have to be performed on existing structures, such as dams, for characterizing the actual material. For this purpose, in situ methods are being validated (e.g., by Saouma *et al.*[181,182]). Core-based fracture tests for concrete are also being explored[183,184] since concrete is usually extracted from structures in the form of cores. It should be noted that core-based tests have been used extensively for the characterization of rocks.[185]

Stochastic models for fracture

As concrete is a hectrogeneous material, the randomness in its microstructure plays an important role in its behavior. However, most of the prevalent fracture models for concrete are deterministic. In future these models may be extended to also include stochastic effects. Nevertheless, there have been several other studies that account for the randomness in concrete; for example, the random crack model of Zaitsev and Wittmann[186] which is applicable for normal and high-strength concrete, and the stochastic theory of Mihashi.[187]

6.9   Applications

Fracture mechanics can be used as a tool in the materials engineering of concrete and in structural analysis. The motivations for such applications and some illustrative examples are presented in this section.

Materials engineering

The engineering of cement-based materials based on fracture mechanics has been emphasized by several researchers. The introduction of weaker inclusions, which function as flaws causing irregular crack surfaces and high dissipation of energy, has been suggested as a means of improving crack resistance and brittleness. For example, Mai *et al.*[188] showed that the resistance of cement mortar could be increased more than 10 times by embedding pieces of paper. The resulting loss in bending strength was less than 50%. Beaudoin[189] found that when a small volume fraction of mica flakes were added to high alumina cement paste, the fracture toughness and the bending strength increased considerably. The compressive strength, however, decreased with increase in mica content. Not much work has been done on the engineering of structural concrete using fracture principles, but the ideas discussed in the section establish the basis for such approaches.

Some of the flaws that are usually present in concrete are pores. They influence fracture behavior significantly, especially in the hcp. Alford *et al.*[190] conducted fracture tests on macro-defect free (MDF) hcp plates that were prepared by pressing a mixture of cement paste and a water soluble polymer. Their results show that while the compressive strength and crack resistance increase appreciably, the brittleness also increases due to the reduced porosity.[191] Several other methods for modifying the microstructure of high-strength hcp to enhance the fracture performance have been suggested by Beaudoin and Feldman.[192]

Usually, high-strength concretes are materials with higher strength and good workability. Unfortunately, they are also highly brittle. This increase in brittleness with strength has been discussed in Section 6.7 and is illustrated in Fig. 6.15.[193] The results from a study[73] where concretes with f_c of 86 MPa and 33 MPa were tested, are shown in Fig. 6.16. It can be seen for a 160% increase in f_c, K_{Ic} increases only 25% and the ductility decreases by about 60% (see Section 6.6). Generally, the problem of brittleness is handled by confining the high-strength concrete with steel; for example, to avoid catastrophic failures of columns under seismic loading.

The best available method for increasing the ductility of concrete is through the use of fibers. Naaman[194] showed that an efficient high-performance material with superior strength and ductility is obtained when a high-strength matrix is reinforced with fibers. It appears that when a large volume (8–12%) of fibers is used the tensile strength of the matrix also

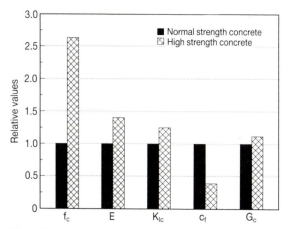

Fig. 6.16   Fracture parameters for normal and high-strength concrete (data of Gettu *et al.*[73])

increases significantly.[195] The fracture behavior of fiber-reinforced concrete involves several concepts discussed in this work, and has been treated more thoroughly elsewhere.[196–198]

The use of fracture mechanics principles in the design of materials with higher toughness and ductility has produced encouraging results in the ceramics industry,[199] and a similar approach is needed for high-performance concretes. Other aspects of concrete behavior will also benefit from an increase in crack resistance and a decrease in brittleness. These include the bond between steel and concrete, which is more brittle in silica fume concrete than normal concrete.[200] The durability, long-term reliability and thermal performance, which are greatly affected by cracking, would also be enhanced.

Structural analysis and design

The application of fracture models is most straightforward to concrete structures with negligible reinforcement, such as dams,[201–203] plinths of bridge piers,[204] and thin-walled pipes and shells.[205] Fracture analysis has been used successfully to study the collapse mechanisms of such structures and to make their design safer. In usual structures, fracture analysis is more difficult due to the influence of the steel reinforcement and distributed cracking. Nevertheless, several aspects of structural behavior have been identified where fracture mechanics would lead to rational and conservative design.[206,207] Some of these are discussed in this section.

In the calculation of beam and slab deflections, the modulus of rupture is usually taken to be a material property. However, it has been shown that the flexural strength decreases considerably with an increase in the depth, more so for high-strength concrete.[208] A similar trend has also been observed for the shear strength (see below). In the presence of shrinkage stresses this effect is even greater, and could result in a large underestimation of the deflections. Fracture analysis accounts for the dependence of failure stress on brittleness, and can be used to determine the value of the strength that should be used in the design. Hillerborg[209,210] has also argued that the compressive stress-strain diagram used in flexural design should depend on the brittleness of the concrete and on the beam depth. This could lead to better estimates of the rotational capacity, the ultimate moment of over-reinforced beams, and the balanced reinforcement ratio.

Fracture principles have also been used to study failure mechanisms of reinforced concrete. For example, the bond-slip behavior of reinforcing steel was simulated with a nonlinear mixed-mode model by Ingraffea *et al.*[211] They concluded that secondary radial cracks that emanate from the ribs of the bar allow stable bond-slip before the occurrence of primary debonding cracks. Also, failure mechanisms in the pullout of anchor bolts have been modeled using fracture mechanics.[212,213]

Bosco *et al.*[144,214] have proposed that the minimum flexure reinforcement should be determined based on fracture mechanics principles. Their approach is based on a structural brittleness number N_p derived by Carpinteri:[215]

$$N_p = \frac{f_y \sqrt{d}}{K_{Ic}} \frac{A_s}{A}$$

(6.28)

where A_s/A = steel reinforcement percentage, and f_y = yield strength of the steel. The moment-rotation response is similar for beams with the same N_p. The condition for design is that $N_p = N_{p_c}$, where N_{p_c} corresponds to the critical case when the steel yielding moment is equal to the first cracking moment. From tests on various sizes of reinforced concrete beams with different steel ratios, they obtained $N_{p_c} = 0.14$ for concrete with $f_c = 30$ MPa, and $N_{p_c} = 0.26$ for high-strength concrete with $f_c = 76$ MPa. The minimum steel percentage increases with an increase in concrete strength and a decrease in beam depth. Since design codes specify constant values, they may be unconservative for smaller beams, especially at higher concrete strengths.

Shear failure

One type of failure that deserves more discussion is shear failure. Diagonal tension and torsional failure of beams, and punching shear failure of slabs occur in a brittle manner. To avoid such catastrophic collapses, reinforcement is provided across potential crack locations. In the design against such failures, the resistance is taken to be the sum of the contributions of steel and concrete. However, since failure is due to fracture, there is a size effect on the shear strength of concrete.[205,206,216] The effect of beam depth on shear failure has been studied extensively by Bazant using the SEM.[217,218] He has proposed that nonlinear fracture mechanics be used to model the shear failure, and that the design shear strength of concrete should decrease with increase in structural brittleness. Similar conclusions have been made by Gustafsson and Hillerborg,[219] Shah,[193] and Thorenfeldt and Drangsholt.[220] Bazant and Kazemi[218] also found that the brittleness decreases when there is bond-slip. Walraven[221] showed from tests on normal and lightweight concrete that the size effect is important in both slender and short beams.

Shear failure has been analyzed numerically using several fracture models. Saouma and Ingraffea[222] employed a discrete crack approach based on LEFM, with aggregate interlock and a nonlinear constitutive model for the uncracked concrete. A crack band model was used by de Borst and Nauta[223] who obtained load-displacement behavior and crack patterns that matched the experimentally obtained results. The same model was also used to analyze the punching shear failure of reinforced concrete slabs. The approach of Wang and Blaauwendraad[224] involves two stages. First, a predictor analysis is performed using a crack band model with shear transfer across the crack. The resulting crack pattern is utilized in the choice of possible crack paths for discrete analyses from which the load-displacement response is determined. The path that gives the lowest load corresponds to the actual failure pattern. They concluded that shear failure can be modeled with just Mode I fracture, and that the size effect was caused only by the fracture of concrete.

Jenq and Shah[225] proposed an approximate method, based on mixed-mode fracture criteria and the TPFM, to calculate the shear failure loads of beams. They assume that Mode I cracks initiate due to flexure at the tensile face and propagate up to the longitudinal rebars. Beyond the rebars, one of the cracks propagates towards the compression face in mixed-mode. The fracture criteria are the same as in Equation (6.20), except that K_I and $CTOD_c$ are replaced by vector sums of the stress intensity factors and crack-tip displacements for the decoupled Modes I and II, respectively. An experimentally calibrated relation is used for modeling the bond-slip behavior of the rebars. The analyses is carried out for different crack paths, and that which gives the lowest failure load is taken to be the critical case. The predicted dependence of the failure load on beam depth and steel ratio (ρ) is given in Fig. 6.17 for normal and high-strength concretes.

The decrease of shear strength of concrete with increase in brittleness is of significant importance to high-strength concrete. Several works have shown that empirical relations based on tests of normal concrete can be unconservative for concretes of higher strengths.[226–228] The increased brittleness could also pose problems in other types of shear failure, such as the punching of slabs[229] and the failure of moment-resistant joints.[230] Such failures are potential applications where fracture mechanics can lead to rational design that accounts for the brittleness of the structure.

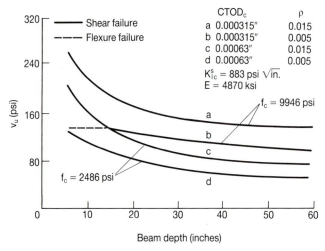

Fig. 6.17   Influence of beam size and compressive strength on shear failure loads. v = shear strength; r = reinforcement ration (1 in. = 25.4 mm), 1 psi = 0.0069 MPa, 1 ksi = 6.9 MPa) (results of Jenq and Shah[225])

References

1   Elfgren, L. (ed.) (1989) *Fracture mechanics of concrete structures: from theory to application*, Report of RILEM Technical Committee 90-FMA. Chapman and Hall, London.

2   Shah, S.P. and Carpinteri, A. (eds) (1991) *Fracture mechanics test methods for concrete*, Report of RILEM Technical Committee 89-FMT Fracture mechanics of concrete: test methods. Chapman and Hall, London.

3   ACI Committee 446 (1990) *Fracture mechanics of concrete: concepts, models and determination of material properties*, Report 446.1R, ACI, Detroit; also in *Fracture mechanics of concrete structures*, Bazant, Z.P. (ed.); Abstract in *Concrete International* ACI, **12**, 67–70.

4   Mindess, S. (1983) The cracking and fracture of concrete: an annotated bibliography, 1928–1981, in Wittmann, F.H. (ed.) *Fracture mechanics of concrete*, Elsevier Science, Amsterdam, 539–661.

5   Mindess, S. (1986) The cracking and fracture of concrete: an annotated bibliogrphy, 1982–1985, in Wittmann, F.H. (ed.) *Fracture toughness and fracture energy of concrete* (International Conference, Lausanne). Elsevier Science, Amsterdam, 539–661.

6   Shah, S.P. and Swartz, S.E. (eds) (1989) *Fracture of concrete and rock*, SEM-RILEM International Conference, Houston, 1987. Springer Verlag, New York.

7   Mihashi, H., Takahashi, H. and Wittmann, F.H. (eds) (1989) *Fracture toughness and fracture energy: test methods for concrete and work*, International Workshop, Sendai, Japan, 1988. A.A. Balkema, Rotterdam.

8   Mazars, J. and Bazant, Z.P. (eds) (1989) *Cracking and damage: strain localization and size effect*, France-USA Workshop, Cachan, France, 1988. Elsevier Applied Science, London.

9   Rossmanith, H.P. (ed.) (1990) *Fracture and damage of concrete and rock*, International Conference, Vienna, 1988. Pergamon Press, Oxford.

10 Elfgren, L. and Shah, S.P. (eds) (1991) *Analysis of concrete structures by fracture mechanics*, International RILEM Workshop, Abisko, Swden, 1989. Chapman and Hall, London.

11 Shah, P. (ed.) (1991) *Toughening mechanisms in quasi-brittle materials*, NATO Workshop, Evanston, USA, 1990. Kluwer Academic, Dordrecht, Netherlands.

12 van Mier, J.G.M., Rots, J.G. and Bakker, A. (eds) (1991) *Fracture processes in concrete, rock and ceramics*, International RILEM/ESIS Conference, Noordwijk, Netherlands. Spon, London.

13 Bazant, Z.P. (ed.) (1992) *Fracture mechanics of concrete structures*, International Conference, Breckenridge, USA. Elsevier Applied Science, London.

14 Broek, D. (1989) *The practical use of fracture mechanics*. Kluwer Academic, Dordrecht.

15 Tada, H., Paris, P.C. and Irwin, G.R. (1985) *The stress analysis of cracks handbook*. Paris Productions, St Louis, Missouri, USA.

16 Griffith, A.A. (1920) The phenomena of rupture and flow in solids. *Phil. Trans. Roy. Soc.*, **221A**, 163.

17 Irwin, G.R. (1957) Analysis of stresses and strains near the end of a crack traversing a plate. *J. Applied Mechanics*, **24**, 361–4.

18 Knott, J.F. (1979) *Fundamentals of fracture mechanics*. Butterworth, London.

19 Murakami, Y. (editor-in-chief) (1990) *Stress intensity factors handbook*. Pergamon Press, Oxford.

20 Glucklich, J. (1963) Fracture of plain concrete. *J. Engineering Mech. Div. (ASCE)*, **89**, EM6, 127–38.

21 Struble, L.J., Stutzman, P.E. and Fuller, E.J., Jr. (1989) Microstructural aspects of the fracture of hardened cement paste. *J. American Ceramic Society*, **72**, 12, 2295–9.

22 Maji, A.K., Ouyang, C. and Shah, S.P. (1990) Fracture mechanisms of brittle materials based on acoustic emission. *J. Materials Research*, **5**, 207–17.

23 Mihashi, H., Nomura, N. and Niiseki, S. (1991) Influence of aggregate size on fracture process zone of concrete detected with three dimensional acoustic emission technique. *Cement and Concrete Research*, **21**, 737–44.

24 Ouyang, C., Landis, E. and Shah, S.P. (1991) Damage assessment in concrete using quantitative acoustic emission. *J. Engineering Mechanics*, **117**, 11, 2681–98.

25 Kachanov, M. (1986) Interaction of a crack with some microcrack systems, in Wittmann, F.H. (ed.) *Fracture toughness and fracture energy of concrete*, International Conference, Lausanne, 1985. Elsevier Science, Amsterdam.

26 Hutchinson, J.W. (1987) Crack tip shielding by micro-cracking in brittle solids. *Acta. Metall.*, **35**, 1605–19.

27 Carrasquillo, R.L., Slate, F.O. and Nilson, A.H. (1981) Microcracking and behavior of high strength concrete subject to short-term loading. *ACI Journal*, **78**, 3, 179–86.

28 Mihashi, H., Nomura, N., Izumi, M. and Wittmann, F.H. (1991) Size dependence of fracture energy of concrete, in van Mier, J.G.M., Rots, J.G. and Bakker, A. (eds) *Fracture processes in concrete, rock and ceramics*. Spon, London.

29 van Mier, J.G.M. (1991) Crack face birdging in normal, high strength and Lytag concrete, in van Mier, J.G.M., Rots, J.G. and Bakker, A. (eds) *Fracture processes in concrete, rock and ceramics*. Spon, London.

30 Mindess, S. and Diamond, S. (1982) The cracking and fracture of mortar. *Materials Structure*, **15**, 86, 107–13.

31 Diamond, S. and Bentur, A. (1985) On the cracking in concrete and fiber reinforced cements, in Shah, S.P. (ed.) *Application of fracture mechanics to*

cementitious composites, NATO Workshop, Evanston, USA. 1984. Martinus Nijhoff, Dordrecht, Netherlands, 87–140.

32   Li, V.C. and Huang, J. (1990) Crack trapping and bridging as toughening mechanisms in high strength concrete, in Shah, S.P., Swartz, S.E. and Wang, M.L. (eds) *Micromechanisms of failure of quasi-brittle material*, International Conference, Albuquerque, USA. Elsevier Applied Science, London, 579–88.

33   Bower, A.F. and Ortiz, M. (1991) Three-dimensional analysis of crack trapping and bridging, in van Mier, J.G.M., Rots, J.G. and Bakker, A. (eds) *Fracture processes in concrete, rock and ceramics*. Spon, London, 110–28.

34   Mindess, S. (1991) The fracture process zone in concrete, Shah, S.P. (ed.) *Toughening mechanisms in quasi-brittle materials*. Kluwer Academic, Dordrecht, Netherlands, 271–86.

35   Mindess, S. (1991) Fracture process zone detection, in Shah S.P. and Carpinteri, A. (eds) *Fracture mechanics test methods for concrete*. Chapman and Hall, London, 231–61.

36   Cedolin, L., Dei Poli, S. and Iori, I. (1983) Experimental determination of the fracture process zone in concrete. *Cement and Concrete Research*, **13**, 557–67.

37   Cedolin, L., Dei Poli, S. and Iori, I. (1987) Tensile behavior of concrete. *J. Engineering Mechanics*, **113**, 3, 431–49.

38   Miller, R.A., Shah, S.P. and Bjelkhagenm, H.I. (1988) Crack profiles in mortar measured by holographic interferometry. *Experimental Mechanics*, **28**, 4, 388–94.

39   Castro-Montero, A., Shah, S.P. and Miller, R.A. (1990) Strain field measurement in fracture process zone. *J. Engineering Mechanics*, **116**, 11, 2463–84.

40   Swartz, S.E. and Go, C.G. (1984) Validity of compliance calibration to cracked concrete beams in bending. *Experimental Mechanics*, **24**, 2, 129–34.

41   Bascoul, A., Kharchi, F. and Maso, J.C. (1989) Concerning the measurement of the fracture energy of microconcrete according to the crack growth in a three-points bending test on notched beans, in Shah, S.P. and Swartz, S.E. (eds) *Fracture of concrete and rock*. Springer-Verlag, New York, 396–408.

42   Horii, H. (1991) Mechanisms of fracture in brittle disordered materials, in van Mier, J.G.M., Rots, J.G. and Bakker, A. (eds) *Fracture processes in concrete, rock and ceramics*. Spon, London, 95–110.

43   Thouless, M.D. (1988) Bridging and damage zones in crack growth. *J. American Ceramic Society*, **71**, 6, 408–13.

44   Diamond, S. and Mindess, D. (1992) SEM investigations of fracture surfaces using stereo pairs: I Fracture surfaces of rocks and of cement paste. *Cement and Concrete Research*, **22**, 67–78.

45   Hsu, T.T.C., Slate, F.O., Sturman, G.M. and Winter, G. (1963) Microcracking of plain concrete and the shape of the stress-strain curve. *ACI Journal*, **60**, 209–24.

46   Bentur, A. and Mindess, S. (1986) The effect of concrete strength on crack patterns. *Cement and Concrete Research*, **16**, 59–66.

47   Faber, K.T. and Evans, A.G. (1983) Crack deflection processes – I: Theory. *Acta Metall.*, **31**, 4, 565–76.

48   Carpinteri, A. (1982) Application of fracture mechanics to concrete structures. *J. Struct. Div. (ASCE)*, **108**, No. ST4, 833–48.

49   Kaplan, M.F. (1961) Crack propagation and the fracture of concrete. *J. ACI*, **58**, 5, 591–610.

50   Romualdi, J.P. and Batson, G.B. (1963) Mechanics of crack arrest in concrete. *J. Engineering Mech. Div. (ASCE)*, **89**, EM3, 147–68.

51 Brown, J.H. (1972) Measuring the fracture toughness of cement paste and mortar. *Magazine of Concrete Research*, **24**, 81, 185–96.

52 Shah, S.P. and McGarry, F.J. (1971) Griffith fracture criterion and concrete. *J. Engineering Mech. Div. (ASCE)*, **97**, EM6, 1663–76.

53 Walsh, P.F. (1972) Fracture of plain concrete. *Indian Concrete Journal* (Bombay), **46**, 11, 469–70; 476.

54 Higgins, D.D. and Bailey, J.E. (1976) Fracture measurements on cement paste. *J. Materials Science*, **11**, 11, 1995–2003.

55 Gjørv, O.E., Sorensen, S.I. and Arnesen, A. (1977) Notch sensitivity and fracture toughness of concrete. *Cement and Concrete Research*, **7**, 3, 333–44.

56 Mindess, S. and Nadeau, J.S. (1976) Effect of notch width on *K* for mortar and concrete. *Cement and Concrete Research*, **6**, 4, 529–34.

57 Ohgishi, S., Ono, H., Takatsu, M. and Tanahashi, I. (1986) Influence of test conditions on fracture toughness of cement paste and mortar, in Wittmann, F.H. (ed.) *Fracture toughness and fracture energy of concrete*, International Conference, Lausanne, 1985. Elsevier Science, Amsterdam, 281–90.

58 Tian, M., Huang, S., Liu, E., Wu, L., Long, K. and Yang, Z. (1986) Fracture toughness of concrete, in Wittmann, F.H. (ed.) *Fracture toughness and fracture energy of concrete*, International Conference, Lausanne, 1985. Elsevier Science, Amsterdam, 281–90.

59 Nallathambi, P. (1986) 'A Study of Fracture of Plain Concrete', Doctoral thesis, University of Newcastle, NSW, Australia.

60 Carpinteri, A. (1982) Sensitivity and stability of progressive cracking in plain and reinforced cement composites. *International Journal of Cement Composites and Lightweight Concrete*, **4**, 1, 47–56.

61 Biolzi, L., Cangiano, S., Tognon, G. and Carpinteri, A. (1989) Snap-back softening instability in high-strength concrete beams. *Materials Structure*, 22, 429–36.

62 Kesler, C., Naus, D. and Lott, J. (1972) Fracture mechanics – its applicability to concrete. *Proceedings of the 1971 International Conference on Mechanical Behavior of Materials* (Japan), IV, 113–24.

63 Saouma, V.E., Ingraffea, A.R. and Catalona, D.M. (1982) Fracture toughness of concrete: K revisited. *J. Engineering Mech. Div. (ASCE)*, **108**, EM6, 1152–66.

64 Nakayama, J. (1965) Direct measurement of fracture energies of brittle heterogeneous materials. *J. American Ceramic Society*, **48**, 11, 583–7.

65 Tattersall, H.G. and Tappin, G. (1966) The work of fracture and its measurement in metals, ceramics and other materials. *J. Materials Science*, **1**, 299–301.

66 Moavenzadeh, F. and Kuguel, R. (1969) Fracture of concrete. *J. Materials (ASTM)*, **4**, 3, 497–519.

67 Petersson, P.E. (1980) Fracture energy of concrete: method of determination. *Cement and Concrete Research*, **10**, 78–89.

68 Hillerborg, A. (1983) Analysis of one single crack, in Wittmann, F.H. (ed.) *Fracture mechanics of concrete*. Elsevier Science, Amsterdam, 223–49.

69 RILEM Committee of Fracture Mechanics of Concrete (1985) Determination of the fracture energy of mortar and concrete by means of three-point bend tests on notched beams, RILEM Draft Recommendation. *Materials Structures*, **18**, 106, 285–90.

70 Hillerborg, A. (1985) The theoretical basis of a method to determine the fracture energy *G* of concrete. *Materials Structures*, **118**, 106, 291–6.

71 Nallathambi, P., Karihaloo, B.L. and Heaton, B.S. (1984) Effect of specimen and crack sizes, water/cement ratio and coarse aggregate texture upon fracture toughness of concrete. *Magazine of Concrete Research*, **36**, 129, 227–36.

72   Hillerborg, A. (1989) Existing methods to determine and evaluate fracture toughness of aggregative material – RILEM recommendation on concrete, in Mihashi, H., Takahashi, H. and Wittmann, F.H. (eds) *Fracture toughness and fracture energy, test methods for concrete and rock*. A.A. Balkema, Rotterdam, 145–52.

73   Gettu, R., Bazant, Z.P. and Karr, M.E. (1990) Fracture properties and brittleness of high-strength concrete. *ACI Materials Journal*, **87**, 6, 608–18.

74   Wittmann, F.H., Mihashi, H. and Nomura, N. (1990) Size effect on fracture energy of concrete. *Engineering Fracture Mechanics*, **35**, 1, 2, 3, 107–15.

75   Planas, J. and Elices, M. (1989) Conceptual and experimental problems in the determination of the fracture energy of concrete, in Mihashi, H., Takahashi, H. and Wittmann, F.H. (eds) *Fracture toughness and fracture energy, test methods for concrete and rock*. A.A. Balkema, Rotterdam, 165–82.

76   Brameshuber, W. and Hilsdorf, H.K. (1990) Influence of ligament length and stress state on fracture energy of concrete. *Engineering Fracture Mechanics*, **35**, 1, 2, 3, 95–106.

77   CEB (1990) *CEB-FIP model code*, Chapter 2 – Material properties, Bulletin No. 195. Ciomite Euro-International du Béton, Lausanne.

78   Hilsdorf, H.K. and Bramshuber, W. (1991) Code-type formulation of fracture mechanics concepts for concrete. *International Journal of Fraction*, **51**, 61–72.

79   Hillerborg, A., Modwer, M. and Petersson, P.-E. (1976) Analysis of crack formation and crack growth in concrete by means of fracture mechanics and finite elements. *Cement and Concrete Research*, **6**, 773–82.

80   Wnuk, M.P. (1974) Quasi-static extension of a tensile crack contained in a viscoelastic-plastic solid. *Journal of Applied Mechanics*, **41**, 1, 234–42.

81   Carpinteri, A., Di Tommaso, A. and Fanelli, M. (1986) Influence of material parameters and geometry on cohesive crack propagation, in Wittmann, F.H. (ed.) *Fracture toughness and fracture energy of concrete*, International Conference, Lausanne, 1985. Elsevier Science, Amsterdam, 117–35.

82   Bocca, P., Carpinteri, A. and Valente, S. (1991) Mixed mode fracture of concrete. *International Journal of Solids Structure*, **27**, 9, 1139–53.

83   Gerstle, W.H. and Xie, M. (1992) FEM modeling of fictitious crack propagation in concrete. *J. Engineering Mechanics*, **118**, 2, 416–34.

84   Bittencourt, T.N., Ingraffea, A.R. and Llorca, J. (1992) Simulation of arbitrary, cohesive crack propagation, in Bazant, Z.P. (ed.) *Fracture mechanics of concrete structures*. Elsevier Applied Science, London, 339–50.

85   Rots, J.G. and Blaauwendraad, J. (1989) Crack models for concrete: discrete or smeared? Fixed, multi-directional or rotating? *HERON*, **34**, 1.

86   Schellekens, J.C.J. (1990) *Interface elements in finite element analysis*, TU-Delft report 25-2-90-5-17. Stevin Laboratory, Delft University of Technology, Delft, Netherlands.

87   Feenstra, P.H., de Borst, R. and Rots, J.G. (1990) Stability analysis and numerical evaluation of crack-dilatancy models, in Bicanic, N. and Mang, H. (eds) *Computer aided analysis and design of concrete structures*, Second International Conference, Zell am See, Austria. Pineridge Press, Swansea, UK, 987–99.

88   Akutagawa, S., Jeang, F.L., Hawkins, N.M., Liaw, B.M., Du, J. and Kobayashi, A.S. (eds) (1991) Effects of loading history on fracture properties of concrete. *ACI Materials Journal*, **88**, 2, 170–80.

89   Bazant, Z.P. and Cedolin, L. (1979) Blunt crack band propagation in finite element analysis. *J. Engineering Mech. Div. (ASCE)*, **105**, EM2, 297–315.

90   Bazant, Z.P. and Oh, B.H. (1983) Crack band theory for fracture of concrete. *Materials Structure*, **16**, 93, 155–77.

91 Bazant, Z.P. (1985) Mechanics of fracture and progressive cracking in concrete structures, in Sih, G.C. and Di Tommaso, A. (ed.) *Fracture mechanics of concrete: structural application and numerical calculation.* Martinus Nijhoff, Dordrecht, Netherlands, 1–93.

92 Rots, J.G. and de Borst, R. (1987) Analysis of mixed-mode fracture in concrete. *J. Engineering Mechanics*, **113**, 11, 1739–58.

93 Carol, I. and Prat, P.C. (1990) A statically constrained microplane model for the smeared analysis of concrete cracking, in Bicanic, N. and Mang, H. (eds) *Computer aided analysis and design of concrete structures*, Second International Conference, Zell am See, Austria. Pineridge Press, Swansea, UK, Vol. 2, 919–30.

94 Du, J., Yon, J.H., Hawkins, N.M. and Kobayashi, A.S. (1990) Analysis of the fracture process zone of a propagating crack using moire interferometry, in Shah, S.P., Swartz, S.E. and Wang, M.L. (eds) *Micromechanics of a failure of quasi-brittle materials*, International Conference, Albuquerque, USA. Elsevier Science, London, 146–55.

95 Jenq, Y.S. and Shah, S.P. (1985) A fracture toughness criterion for concrete. *Engineering Fracture Mechanics*, **21**, 5, 1055–69.

96 Foote, R.M.L., Mai, Y.-W. and Cotterell, B. (1986) Crack growth resistance curves in strain-softening materials. *J. Mech. Phys. Solids*, **34**, 6, 593–607.

97 Yon, J.-H., Hawkins, N.M. and Kobayashi, A.S. (1991) Numerical simulation of Mode I dynamic fracture of concrete. *J. Engineering Mechanics*, **117**, 7, 1595–610.

98 Tademir, M.A., Maji, A.K. and Shah, S.P. (1990) Crack propagation in concrete under compression. *J. Engineering Mechanics*, **116**, 5, 1058–76.

99 Hordijk, D.A. (1991) 'Local Approach to Fatigue of Concrete', Doctoral Thesis, Delft University of Technology, Delft, Netherlands.

100 Gopalaratnam, V.S. and Shah, S.P. (1985) Softening response of plain concrete in direct tension. *ACI Journal*, **82**, 3, 310–23.

101 Cornelissen, H.A.W., Hordijk, D.A. and Reinhardt, H.W. (1986) Experimental determination of crack softening characteristics of normalweight and lightweight concrete. *HERON*, **31**, 2, 45–56.

102 Guo, Z. and Zhang, X. (1987) Investigation of complete, stress-deformation curves for concrete in tension. *ACI Materials Journal, 4*, **84**, 278–85.

103 Wecharatana, M. (1990) Britteleness index of cementitious composites, in Suprenant, B.A. (ed.) *Serviceability and durability of construction material*, First Materials Engineering Congress, Denver, USA. ASCE, New York, 966–75.

104 Li, V.C., Chan, C.-M. and Leung, C.K.Y. (1987) Experimental determination of the tension-softening relations for cementitious composites. *Cement and Concrete Research*, **17**, 441–52.

105 Du, J., Yon, J.H., Hawkins, N.M. and Kobayashi, A.S. (1990) Analysis of the fracture process zone of a propagating crack using moire interferometry, in Shah, S.P., Swartz, S.E. and Wang, M.L. (eds) *Micromechanics of a failure of quasi-brittle materials*, International Conference, Albuquerque, USA. Elsevier Science, London, 146–55.

106 Miller, R.A., Castro-Montero, A. and Shah, S.P. (1991) Cohesive crack models for cement mortar examined using finite element analysis and laser holographic measurement. *J. American Ceramics Society*, **74**, 130–8.

107 Roelfstra, P.E. and Wittmann, F.H. (1986) Numerical method to link strain softening with failure of concrete, in Wittmann, F.H. (ed.) *Fracture toughness and fracture energy of concrete*, International Conference, Lausanne, 1985. Elsevier Science, Amsterdam, 163–75.

108 Wittmann, F.H., Rokugo, K., Brühwiler, E., Mihashi, H. and Simonin, P. (1988) Fracture energy and strain softening of concrete as determined by

means of compact tension specimens. *Materials Structure*, **21**, 21–32.

109   Alvaredo, A.M. and Torrent, R.J. (1987) The effects of the strain-softening diagram on the bearing capacity of concrete beams. *Materials Structure*, **20**, 448–54.

110   Ratanalert, S. and Wecharatana, M. (1989) Evaluation of the fictitious crack and two-parameter fracture models, in Mihashi, H., Takahashi, H. and Wittmann, F.H. (eds) *Fracture toughness and fracture energy, test methods for concrete and rock*. A.A. Balkema, Rotterdam, 345–66.

111   Liaw, B.M., Jeang, F.L., Du, J.J., Hawkins, N.M. and Kobayashi, A.S. (1990) Improved nonlinear model for concrete fracture. *J. Engineering Mechanics*, **116**, 2, 429–45.

112   Li, V.C. and Huang, J. (1990) Relation of concrete fracture toughness to its internal structure. *Engineering Fracture Mechanics*, **35**, 1, 2, 3, 39–46.

113   Duda, H. (1991) Grain-model for the determination of the stress-crack-width relation, in Elfgren, L. and Shah, S.P. (eds) *Analysis of concrete structures by fracture mechanics*. Chapman and Hall, London, 88–96.

114   Clarke, D.R. and Faber, K.T. (1987) Fracture of ceramics and glasses. *J. Physical Chemistry of Solids*, **48**, 11, 1115–57.

115   Jenq, Y.S. and Shah, S.P. (1985) A two parameter fracture model for concrete. *J. Engineering Mechanics*, **111**, 4, 1227–41.

116   Planas, J. and Elices, M. (1990) Anomalous size effect in cohesive materials like concrete, in Suprenant, B.A. (ed.) *Serviceability and durability of construction materials*, First Materials Engineering Congress, Denver, USA. ASCE, New York, Vol. 2, 1345–56.

117   Ouyang, C. and Shah, S.P. (1991) Geometry-dependent *R*-curve for quasi-brittle materials. *J. American Ceramics Society*, **74**, 11, 2831–36.

118   Bazant, Z.P., Gettu, R. and Kazemi, M.T. (1991) Identification of nonlinear fracture properties from size effect tests and structural analysis based on geometry-dependent *R*-curves. *International Journal of Rock Mechanics and Mineral Science*, **28**, 1, 43–51; corrigenda: **28**, 2, 3, 233.

119   Sakai, M. and Bradt, R.C. (1986) Graphical methods for determining the nonlinear fracture parameters of silica and graphite refractory composites, in Bradt, R.C., Evans, A.G., Hasselman, D.P.H. and Lange, F.F. (eds) *Fracture mechanics of ceramics*. Plenum, New York, Vol. 7, 127–42.

120   Wecharatana, M. and Shah, S.P. (1983) Predictions of nonlinear fracture process zone in concrete. *J. Engineering Mechanics*, **109**, 5, 1231–46.

121   Hu, X. and Wittmann, F.H. (1990) Experimental method to determine extension of fracture-process zone. *J. Materials in Civil Engineering*, **2**, 1, 15–23.

122   Hu, X. (1990) *Fracture process zone and strain-softening in cementitious materials*, Research report 1, Institute of Building Materials. Swiss Federal Institute of Technology, Zurich, 1990.

123   Cottrell, B. and Mai, Y.-W. (1987) Crack growth resistance curve and size effect in the fracture of cement paste. *J. Materials Science*, **22**, 2734–8.

124   Planas, J., Elices, M. and Toribio, J. (1989) Approximation of cohesive crack models by R-CTOD curves, in Shah, S.P., Swartz, S.E. and Barr, B. (eds) *Fracture of concrete and rock: recent development*, International Conference, Cardiff. Elsevier Applied Science, London, 203–12.

125   Bazant, Z.P. and Kazemi, M.T. (1990) Determination of fracture energy, process zone length and brittleness number from size effect, with application to rock and concrete. *International Journal of Fracture*, **44**, 111–31.

126   Ouyang, C and Shah, S.P. (1991) Geometry-dependent *R*-curve for quasi-brittle materials. *J. American Ceramics Society*, **74**, 11, 2931–6.

127   Jenq, Y.S. and Shah, S.P. (1988) Mixed-mode fracture of concrete. *International Journal of Fracture*, **38**, 123–42.

128 RILEM Committee on Fracture Mechanics of Concrete – Test methods (1990) Determination of fracture parameters (K and CTOD) of plain concrete using three-point bend tests, RILEM Draft Recommendation. *Mater. Struct.*, **23**, 457–60.

129 Karihaloo, B.L. and Nallathambi, P. (1989) An improved effective crack model for the determination of fracture toughness of concrete. *Cement and Concrete Research*, **19**, 603–10.

130 Karihaloo, B.L. and Nallathambi, P. (1991) Notched beam test: Mode I fracture toughness, in Shah, S.P. and Carpinteri, A. (eds) *Fracture mechanics test methods for concrete*. Chapman and Hall, London, 1–86.

131 Nallathambi, P. and Karihaloo, B.L. (1990) Fracture of concrete: application of effective crack model, *Proceedings of Ninth International Conference on Experimental Mechanics*, Lyngby, Denmark, Vol. 4, 1413–22.

132 Bazant, Z.P. (1984) Size effect in blunt fracture: concrete, rock, metal. *J. Engineering Mechanics*, **110**, 518–35.

133 Bazant, Z.P. and Pfeiffer, P.A. (1987) Determination of fracture energy from size effect and brittleness number. *ACI Materials Journal*, **84**, 463–79.

134 RILEM Committee on Fracture Mechanics of Concrete – Test methods (1990) Size-effect method for determining fracture energy and process zone size of concrete, RILEM Draft Recommendation. *Materials Structure*, **23**, 461–65.

135 Planas, J. and Elices, M. (1990) Fracture criteria for concrete: mathematical approximations and experimental validation. *Engineering Fracture Mechanics*, **35**, 1, 2, 3, 87–94.

136 Planas, J. and Elices, M. (1990) The approximation of a cohesive crach by effective elastic cracks, in Firrao, D. (ed.) *Fracture behaviour and design of materials and structures*, Eighth European Conference on Fracture, Turin, Italy. Engineering Materials Advisory Services, Warley, UK, Vol. 2, 605–11.

137 John, R. and Shah, S.P. (1989) Fracture mechanics analysis of high-strength concrete. *J. Materials in Civil Engineering*, **1**, 4, 185–98.

138 Lange D.A. (1991) 'Relationship Between Microstructure, Fracture Surfaces and Material Properties of Portland Cement', Ph.D. Thesis, Northwestern University, Evanston, USA.

139 Chern, J.-C. and Tarng, K.-M. (1990) Size effect in fracture of steel fiber reinforced concrete, in Shah, S.P., Swartz, S.E. and Wang, M.L. (eds) *Micromechanics of failure of quasi-brittle materials*, International Conference, Albuquerque, USA. Elsevier Applied Science, London, 244–53.

140 de Larrard, F., Boulay, C. and Rossi, P. (1987) Fracture toughness of high-strength concretes, in Holand, I., Helland, S., Jakobsen, B. and Lenschow, R. (eds) *Utilization of high strength concrete*, Symposium, Stavanger, Norway. Tapir, Trondheim, Norway, 215–23.

141 Hucka, V. and Das, B. (1974) Brittleness determination of rocks by different methods. *Int. J. Rock Mech. Min. Sci.*, **11**, 3289–392.

142 Bache, H.H. (1986) Fracture mechanics in design of concrete and concrete structures, in Wittman, F.H. (ed.) *Fracture toughness and fracture energy of concrete*, International Conference, Lausanne, 1985. Elsevier Science, Amsterdam, 577–86.

143 Brühwiler, E., Broz, J.J. and Saouma, V.E. (1991) Fracture model evaluation of dam concrete. *J. Materials in Civil Engineering*, **3**, 4 235–51.

144 Bosco, C., Carpinteri, A. and Debernardi, P.G. (1990) Minimum reinforcement in high-strength concrete. *J. Structural Engineering*, **116**, 2, 427–37.

145 Naus, D.J. and Lott, J.L. (1969) Fracture toughness of Portland cement concretes. *J. ACI*, **66**, 6, 481–9.

146 Strang, P.C. and Bryant, A.H. (1979) The role of aggregate in the fracture of concrete. *J. Materials Science*, **14**, 1863–8.

147   Saouma, V.E., Broz, J.J., Brühwiler, E. and Boggs, H.L. (1991) Effect of aggregate and specimen size on fracture properties of dam concrete. *J. Materials in Civil Engineering*, **3**, 3, 204–18.

148   Wittmann, F.H., Roelfstra, P.E., Mihashi, H., Huang, Y.-Y., Zhang, X.-H. and Nomura, N. (1987) Influence of age of loading, water-cement ratio and rate of loading on fracture energy of concrete. *Materials Structure*, **20**, 103–10.

149   Bascoul, A., Detriche, C.H. and Ramoda, S. (1991) Influence of the characteristics of the binding phase and the curing conditions on the resistance to crack propagation of mortar [in French]. *Materials Structure*, **24**, 129–36.

150   Ojdrovic, R.P., Stojimirovic, A.L. and Petroski, H.J. Effect of age on splitting tensile strength and fracture resistance of concrete. *Cement and Concrete Research*, **17**, 70–76, 150.

151   Reinhardt, H.W. (1991) Loading rate, temperature and humidity effects, in Shah, S.P. and Carpinteri, A. (eds) *Fracture mechanics test methods for concrete*. Chapman and Hall, London, 199–230.

152   John, R. and Shah, S.P. (1986) Fracture of concrete subjected to impact loading. *Cement and Concrete Aggregates* (ASTM), **8**, 1, 24–32.

153   John, R., Shah, S.P. and Jenq, Y.-S. (1987) A fracture mechanics model to predict the rate sensitivity of Mode I fracture of concrete. *Cement and Concrete Research*, **17**, 249–62.

154   Oh, B.H. (1990) Fracture behavior of concrete under high rates of loading. *Engineering Fracture Mechanics*, **35**, 1, 2, 3, 327–32.

155   Liu, Z.-G., Swartz, S.E., Hu, K.K. and Kan, Y.-C. (1989) Time-dependent response and fracture of plain concrete beams, in Shah, S.P., Swartz, S.E. and Barr, B. (eds) *Fracture of concrete and rock: recent developments*, International Conference, Cardiff. Elsevier Applied Science, London, 577–86.

156   Zhou, F. and Hillerborg, A. (1992) Time-dependent fracture of concrete: testing and modelling, in Bazant, Z.P. (ed.) *Fracture mechanics of concrete structures*. Elsevier Applied Science, London, 906–11.

157   Bazant, Z.P. and Gettu, R. (1990) Size effect in concrete structures and influence of loading rate, in Suprenant, B.A. (ed.) *Serviceability and durability of construction materials*, First Materials Engineering Congress, Denver, USA. ASCE, New York, Vol. 2, 1113–23.

158   Bazant, Z.P. and Gettu, R. (1992) Rate effects and load relaxation in static fracture of concrete. *ACI Materials Journal* (in press).

159   Banthia, N.P., Mindess, S. and Bentur, A. (1987) Impact behaviour of concrete beams. *Materials Structure*, **20**, 293–302.

160   Bazant, Z.P. and Prat, P.C. (1988) Effect of temperature and humidity on fracture energy of concrete. *ACI Materials Journal*, **85**, 262–71.

161   Planas, J., Maturana, P., Guinea, G. and Elices, M. (1989) Fracture energy of water saturated and partially dry concrete at room and at cryogenic temperatures, in Salama, K., Ravi-Chandar, K., Taplin, D.M.R. and Rama Rao, P. *Advances in fracture research*, Seventh International Conference on Fracture, Houston, USA. Pergamon Press, Oxford, Vol. 2, 1809–17.

162   Ohlsson, U., Daerga, P.A. and Elfgren, L. (1990) Fracture energy and fatigue strength of unreinforced concrete beams at normal and low temperatures. *Engineering Fracture Mechanics*, **35**, 1, 2, 3, 195–203.

163   Maturana, P., Planas, J. and Elices, M. (1990) Evolution of fracture behaviour of saturated concrete in the low temperature range. *Engineering Fracture Mechanics*, **35**, 4, 5, 827–34.

164   Shah, S.P. and Chandra, S. (1970) Fracture of concrete subjected to cyclic and sustained loading. *ACI Journal*, **67**, October, 816–25.

165 Michalske, T.A. and Bunker, B.C. (1984) Slow fracture model based on strained silicate structures. *Journal of Applied Physics*, **56**, 10, 2686–93.

166 Rossi, P. (1991) Influence of cracking in the presence of free water on the mechanical behaviour of concrete. *Magazine of Concrete Research*, **43**, 154, 53–7.

167 Swartz, S.E., Hu, K.-K. and Jones, G.L. (1978) Compliance monitoring of crack growth in concrete. *J. Engineering Mech. Div. (ASCE)*, **104**, EM4, 789–800.

168 Pons, G., Ramoda, S.A. and Maso, J.C. (1988) Influence of the loading history on fracture mechanics parameters of microconcrete: effects of low-frequency cyclic loading. *ACI Materials Journal*, **85**, 341–6.

169 Schell, W.F. (1992) 'Fatigue Fracture of High Strength Concrete under High Frequency Loading', M.S. thesis, Northwestern University, Evanston, IL, USA; also Bazant, Z.P. and Schell, W.F. Fatigue fracture of high strength concrete and size effect. *ACI Materials Journal* (submitted).

170 Tinic, C. and Brühwiler, E. (1985) Effect of compressive loads on the tensile strength of concrete at high strain rates. *Int. J. Cem. Comp. Lightweight Conc.*, **7**, 103–8.

171 Gettu, R., Oliveira, M.O.F., Carol, I. and Aguado, A. (1992) Influence of transverse compression on Mode I fracture of concrete, in Bazant, Z.P. (ed.) *Fracture mechanics of concrete structures*. Elsevier Science, London, 193–7.

172 Carpinteri, A. and Swartz, S. (1991) Mixed-mode crack propagation in concrete, in Shah, S.P. and Carpinteri, A. (eds) *Fracture mechanics test methods for concrete*. Chapman and Hall, London, 129–90.

173 Reinhardt, H.W., Cornelissen, H.A.W. and Horijk, D.A. (1989) Mixed mode fracture tests on concrete, in Shah, S.P. and Swartz, S.E. (eds) *Fracture of concrete and rock*. Springer-Verlag, New York, 117–30.

174 van Mier, J.G.M., Nooru-Mohamed, M.B. and Timmers, G. An experimental study of shear fracture and aggregate interlock in cement-based composites. *HERON*, **36**, 4.

175 Swartz, S.E. and Taha, N.M. (1990) Mixed mode crack propagation and fracture in concrete. *Engineering Fracture Mechanics*, **35**, 1, 2, 3, 137–44.

176 Ballatore, E., Carpinteri, A., Ferrara, G. and Melchiorri, G. (1990) Mixed mode fracture energy of concrete. *Engineering Fracture Mechanics*, **35**, 1, 2, 3, 145–57.

177 Mindess, S. and Shah, S.P. (eds) (1988) *Bonding in cementitious composites*, Symposium, Boston, 1987. Materials Research Society, Pittsburgh, USA.

178 Hollemeier, B. and Hilsdorf, H.K. (1977) Fracture mechanics studies on concrete compounds. *Cement and Concrete Research*, **7**, 5, 523–36.

179 Buyukozturk, O. and Lee, K.M. (1992) Interface fracture mechanics of concrete composites, in Bazant, Z.P. (ed.) *Fracture mechanics of concrete structures*. Elsevier Applied Science, London, 163–8.

180 Maji, A.K., Wang, J. and Cardiel, C.V. (1992) Fracture mechanics of concrete, rock and interface, in Bazant, Z.P. (ed.) *Fracture mechanics of concrete structures*. Elsevier Applied Science, London, 413–18.

181 Saouma, V.E., Broz, J.J. and Boggs, H.L. (1991) In situ testing for fracture properties of dam concrete. *J. Materials in Civil Engineering*, **3**, 3, 219–34.

182 Saouma, V.E., Brühwiler, E., Keating, S., Ryan, J. and Shulz, J. (1991) Innovative fracture testing techniques for dam engineering, in Saouma, V.E., Dungar, R. and Morris, D. (eds) *Dam fracture*, Proceedings, International Conference, Boulder, USA. Electric Power Research Institute, Palo Alto, USA, 459–75.

183 Brühwiler, E. (1991) Determination of fracture properties of dam concrete from core samples, in Saouma, V.E., Dungar, R. and Morris, D. (eds) *Dam*

fracture, Proceedings, International Conference, Boulder, USA. Electric Power Research Institute, Palo Alto, USA, 427–43.

184   Linsbauer, H.N. (1991) Fracture mechanics material parameters of mass concrete based on drilling core tests – review and discussion, in van Mier, J.G.M., Rots, J.G. and Bakker, A. (eds) *Fracture processes in concrete, rock and ceramics*. Spon, London, Vol. 2, 661–3.

185   Ouchterlony, F. (1990) Fracture toughness testing of rock with core based specimens. *Engineering Fracture Mechanics*, **35**, 1, 2, 3, 351–66.

186   Zaitsev, Y.B. and Wittmann, F.H. (1981) Simulation of crack propagation and failure of concrete. *Materials Structure*, **14**, 83, 357–65.

187   Mihashi, H. (1985) Stochastic approach to study the fracture and fatigue of concrete, in Eggwertz, S. and Lind, N.C. (eds) *Probabilistic methods in the mechanics of solids and structures*, IUTAM Symposium, Stockholm, 1984. Springer-Verlag, Berlin, 307–17.

188   Mai, Y.-W., Hakeem, M. and Cotterell, B. (1982) Imparting fracture resistance to cement mortar by intermittent interlaminar bonding. *Cement and Concrete Research*, **12**, 661–3.

189   Beaudoin, J.J. (1982) Properties of high alumina cement paste reinforced with mica flakes. *Cement and Concrete Research*, **12**, 157–66.

190   Alford, N.M.N., Groves, G.W. and Double, D.D. (1982) Physical properties of high strength cement pastes. *Cement and Concrete Research*, **12**, 349–58.

191   Mai, Y.-W. and Cotterell, B. (1985) Porosity and mechanical properties of cement mortar. *Cement and Concrete Research*, **15**, 995–1002.

192   Beaudoin, J.J. and Feldman, R.F. (1985) High strength cement pastes – a critical appraisal. *Cement and Concrete Research*, **15**, 105–16.

193   Shah, S.P. (1990) Fracture toughness for high-strength concrete. *ACI Materials J.*, **87**, 3, 260–65.

194   Naaman, A.E. (1985) High strength fiber reinforced cement composites, in Young, J.F. (ed.) *Very high strength cement-based materials*, Materials Research Society Symposium Proceedings, Vol. 42. Materials Research Society, Pittsburgh, USA, 219–29.

195   Shah, S.P. (1991) Do fibers increase the tensile strength of cement-based matrices? *ACI Materials Journal*, **88**, 6, 595–602.

196   Gopalaratnam, V. and Shah, S.P. (1987) Failure mechanisms and fracture of fiber reinforced concrete, in Shah, S.P. and Batson, G.B. (eds) *Fiber reinforced concrete properties and applications*, SP-105. ACI, Detroit, USA, 1–25.

197   Shah, S.P. and Ouyang, C. (1991) Mechanical behavior of fiber-reinforced cement-based composites. *J. American Ceramics Society*, **74**, 11, 2727–38 and 2947–53.

198   Balaguru, P.N. and Shah, S.P. (1992) *Fiber-reinforced cement composites*. McGraw Hill, New York.

199   Becher, P.F. (1991) Microstructural design of toughened ceramics. *J. American Ceramics Society*, **74**, 2, 255–69.

200   Ezeldin, A.S. and Balaguru, P.N. (1989) Bond behavior of normal and high-strength fiber reinforced concrete. *ACI Materials Journal*, **86**, 6, 515–24.

201   Linsbauer, H.N. (1991) Design and construction of concrete dams under consideration of fracture mechanics aspects, in Elfgren, L. and Shah, S.P. (eds) *Analysis of concrete structures by fracture mechanics*. Chapman and Hall, London, 160–8.

202   Saouma, V.E., Brühwiler, E. and Boggs, H.L. (1982) A review of fracture mechanics applied to concrete dams. *Dam Engineering*, **1**, 1, 41–57.

203   Martha, L.F., Llorca, J., Ingraffea, A.R. and Elices, M. (1991) Numerical simulation of crack initiation and propagation in an arch dam. *Dam Engineering*, **2**, 3, 193–213.

204 Swenson, D.V. and Ingraffea, A.R. (1990) The collapse of the Schoharie Creek Bridge: a case study in the fracture mechanics of concrete, in Bicanic, N. and Mang, H. (eds) *Computer aided analysis and design of concrete structures*, Second International Conference, Zell am See, Austria. Pineridge Press, Swansea, Vol. 1, 403–24.

205 Gustafsson, P.J. and Hillerborg, A. (1985) Improvements in concrete design achieved through the application of fracture mechanics, in Shah, S.P. (ed.) *Application of fracture mechanics to cementitious composites*, NATO Workshop, Evanston, USA, 1984. Nijhoff, Dordrecht, Netherlands, 639–80.

206 Hawkins, N.M. (1985) The role for fracture mechanics in conventional reinforced concrete design, in Shah, S.P. (ed.) *Application of fracture mechanics to cementitious composites*, NATO Workshop, Evanston, USA, 1984. Nijhoff, Dordrecht, Netherlands, 639–80.

207 Bazant, Z.P., Ozbolt, J. and Eligehausen, R. (1992) Fracture size effect: I. Review of evidence for concrete structures. *J. Structural Engineering* (submitted).

208 Hillerborg, A. (1989) Fracture mechanics and the concrete codes, in Li, V.C. and Bazant, Z.P. (eds) *Fracture mechanics: application to concrete*, SP-118. ACI, Detroit, 157–69.

209 Hillerborg, A. (1990) Fracture mechanics concepts applied to moment capacity and rotational capacity of reinforced concrete beams. *Engineering Fracture Mechanics*, **35**, 1, 2, 3, 233–40.

210 Hillerborg, A. (1991) Size dependency of the stress-strain curve in compression, in Elfgren, L. and Shah, S.P. (eds) *Analysis of concrete structures by fracture mechanics*. Chapman and Hall, London, 171–8.

211 Engaffea, A.R. Gerstle, W., Gergely, P. and Saouma, V. (1984) Fracture mechanics of bond in reinforced concrete. *J. Structural Engineering*, **110**, 4, 871–90.

212 Ballarini, R. and Shah, S.P. (1991) Fracture mechanics based analyses of pull-out tests and anchor bolts, in Elfgren, L. and Shah, S.P. (eds) *Analysis of concrete structures by fracture mechanics*. Chapman and Hall, London, 245–80.

213 Ohlsson, U. and Elfgren, L. (1991) Anchor bolts analyzed with fracture mechanics, in van Mier, J.G.M., Rots, J.G. and Bakker, A. (eds) *Fracture processes in concrete, rock and ceramics*. Spon, London, 887–97.

214 Bosco, C., Carpinteri, A. and Debernardi, P.G. (1991) Use of the brittleness number as a rational approach to minimum reinforcement design, in Elfgren, L. and Shah, S.P. (eds) *Analysis of concrete structures by fracture mechanics*. Chapman and Hall, London, 133–51.

215 Carpinteri, A, A fracture mechanics model for reinforced concrete collapse, in *Advanced mechanics of reinforced concrete* (IABSE Colloquium, Delft, Netherlands), Reports of the Working Commissions, Vol. 34, IABSE, Zurich, 17–30, 1981

216 Shioya, T., Iguro, M., Nojiri, Y., Akiyama, H. and Okada, T. (1989) Shear strength of large reinforced concrete beams, in Li, V.C. and Bazant, Z.P. (eds) *Fracture mechanics: applications to concrete*, SP-118. ACI, Detroit, USA, 259–79.

217 Bazant, Z.P. and Kim, J.-K. (1984) Size effect in shear failure of longitudinally reinforced beams. *ACI Journal*, **81**, 456–68.

218 Bazant, Z.P. and Kazemi, M.T. (1991) Size effect on diagonal shear failure of beams without stirrups. *ACI Structural Journal*, **88**, 3, 268–76.

219 Gustafsson, P.J. and Hillerborg, A. (1988) Sensitivity in shear strength of longitudinally reinforced concrete beams to fracture energy of concrete. *ACI Structural Journal*, **85**, 3, 286–94.

220 Thorenfeldt, E. and Drangsholt, G. (1990) Shear capacity of reinforced high

strength concrete beams, in Hester, W.T. (ed.) *High-strength concrete*, Second International Symposium, SP-121. ACI, Detroit, USA, 129–54.

221   Walraven, J. (1990) Scale effects in beams with unreinforced webs, loaded in shear, in *Progress in concrete research*, Annual Report. Delft University of Technology, Netherlands, Vol. 1, 103–12.

222   Saouma, V.E. and Ingraffea, A.R. Fracture mechanics analysis of discrete cracking, in *Advanced mechanics of reinforced concrete* (IABSE Colloquium, Delft, Netherlands), Reports of the Working Commissions, Vol. 34, IABSE, Zurich, 413–36, 1981

223   de Borst, R. and Nauta, P. (1984) Smeared crack analysis of reinforced concrete beams and slabs failing in shear, in Damjanic, F., Hinton, E., Owen, D.R.J., Bicanic, N. and Simonvic, V. (eds) *Computer-aided analysis and design of concrete structures*, International Conference, Split, Yugoslavia. Pineridge Press, Swansea, Vol. 1, 261–73.

224   Wang, Q.B. and Baauwendraad, J. (1991) Consequence of concrete fracture: brittle failure and size effect of RC beams under shear, in van Mier, J.G.M., Rots, J.G. and Bakker, A. (eds) *Fracture processes in concrete, rock and ceramics*. Spon, London, 909–18.

225   Jenq, Y.-S. and Shah, S.P. (1989) Shear resistance of reinforced concrete beams – a fracture mechanics approach, in Li, V.C. and Bazant, Z.P. (eds) *Fracture mechanics: application to concrete*, SP-118. ACI, Detroit, USA, 237–58.

226   Mphonde, A.G. and Frantz, G.C. (1984) Shear tests of high- and low-strength concrete beams. *ACI Journal*, **81**, 350–7.

227   Elzanaty, A.H., Nilson, A.H. and Slate, F.O. (1986) Shear capacity of reinforced concrete beams using high-strength concrete. *ACI Journal*, **83**, 2, 290–6.

228   Johnson, M.K. and Ramirez, J.A. (1989) Minimum shear reinforcement in beams with higher strength concrete. *ACI Structural Journal*, **86**, 4, 376–82.

229   Marzouk, H. and Hussein, A. (1991) Experimental investigation on the behavior of high-strength concrete slabs. *ACI Structural Journal*, **88**, 6, 701–13.

230   Ehsani, M.R. and Alameddine, F. (1991) Design recommendations for Type 2 high-strength reinforced concrete connections. *ACI Structural Journal*, **8**, 3, 277–91.

7 Structural members

Arthur H Nilson

7.1 Introduction

High performance concretes with higher strengths, having mix proportions quite different from ordinary concretes and modified by several additives, have mechanical properties differing in important ways from ordinary concretes.

It has been well established, for example, that for high-strength concrete the internal microcracking that occurs in concrete as load is applied is delayed until a higher fraction of ultimate load is reached. One result is that the range of linear elastic response to compression is extended. On the other hand, the extensive ductility exhibited by lower strength concretes after maximum stress, as microcracks spread to form an interconnected network, is not characteristic of higher strength concretes, and material ductility is less.[1–3] The response to sustained loading is also different. The much lower creep coefficient of high-strength concretes has also been correlated with differences in internal microcracking.[4–9]

Given the empirical nature of many of the equations in current use for the design of reinforced concrete members and structures, it is clear that a re-examination of those equations is imperative, in order to insure the safe and economical use of high-strength concrete. It is remarkable that the use of the material has preceded a full knowledge of its properties and behavior.

The emphasis in this chapter will be placed on the practical design of members for short-term and sustained load. While much work remains to be done, great progress has been made in the past decade in developing the needed information on material and member performance.[10–17] Design procedures have been checked and, in the USA, provisions of the American Concrete Institute (ACI) Code reviewed. Recommendations for change or for additional research have been presented where appropriate.

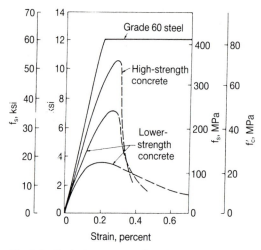

Fig. 7.1   Typical concrete and steel stress-strain curves

7.2   Axially loaded columns

Columns subject to purely axial loads are rare in practice. Generally moments exist concurrently, due to rigid frame action or load eccentricity. Most design codes recognize this, requiring that column moments be accounted for in design. However, in examining the differences in performance of high-strength concrete members, it is helpful to treat axially loaded columns first.

Contribution of steel and concrete

In most present design practice, the nominal strength of a column is calculated using the direct addition law, i.e., summing the individual strength contributions of the concrete and steel. The justification for this is evident in Fig. 7.1, which superimposes typical stress-strain curves in compression for three concretes with that for reinforcing steel having 60,000 psi (414 MPa) yield strength. The curve for reinforcing steel is drawn to a different vertical scale for convenience.

For low-strength concrete, when the concrete reaches the range of significant nonlinearity (about 0.001 strain), the steel is still in the elastic range and consequently starts to pick up a larger share of the load. When the strain is close to 0.002, the slope of the concrete curve is nearly zero, and there is little change in stress. The steel reaches its yield point at about this strain. Both materials deform plastically, and the strength of the column is accurately predicted by

$$P_n = 0.85f_c'A_c + f_yA_s \tag{7.1}$$

where f_c' = cylinder compressive strength of concrete
f_c' = yield strength of steel
A_c = area of concrete section
A_s = total area of steel

The factor 0.85 is used to account for the observed difference in the strength of concrete in a column compared with concrete of the same mix in a standard test cylinder.

A similar analysis holds for high-strength concrete, except that the steel will yield somewhat before the concrete reaches its peak stress. However, the steel will yield at constant stress until the concrete reaches its maximum stress. Prediction of strength may therefore still be based on Equation (7.1). Test evidence confirms that use of the factor 0.85 is still appropriate.

Lateral confinement

Lateral steel, preferably in the form of continuous spirals, has two beneficial effects on column behaviour: (a) it greatly increases the strength of the core concrete through confinement against lateral expansion, and (b) it increases the axial strain capacity of the concrete, permitting a more gradual and ductile failure. Equating the increase in strength due to the spiral confinement and the loss of capacity in spalling and incorporating a suffery factor of 1.2, the following minimum spiral reinforcement ratio ρ is specified in the ACI 318–89 Code:

$$\rho = 0.45[A_g/A_c - 1]f_c'/f_y \tag{7.2}$$

where ρ = ratio of volume of spiral reinforcement to volume of con-
    $A_g$crete core
    A_c = gross area of concrete cross section
    f_c' = area of concrete core (measured to outside of spiral)
    f_y = cylinder compressive strength of concrete
       = yield strength of spiral steel
The increase in compressive strength of columns provided with spiral steel is based on an experimentally derived relationship for strength gain:

$$f_c - f_c'' = 4.0f_2' \tag{7.3a}$$

where f_c = compressive strength of spirally reinforced concrete column
    f_c'' = compressive strength of unconfined concrete column
    f_2' = concrete confinement stress produced by spiral
Equation (7.3a) can be shown to lead directly to Equation (7.2). The concrete confinement stress produced by the spiral is calculated on the basis that the spiral steel has yielded, using the familiar hoop tension equation, from which

$$f_2' = [2A_{sp}f_y]/d_c s \tag{7.4}$$

where A_{sp} = area of spiral steel
    d_c = diameter of concrete core
    s = pitch of spiral
and other terms are as already defined.

Research by Ahmad and Shah[18] has shown that spiral reinforcement is less effective for high-strength concrete columns, and that the stress in the spiral steel at maximum column load is often significantly less than yield stress, as assumed above. These results are consistent with research by Martinez.[19,20] Martinez *et al.*, used an 'effective' confinement stress $f_2(1 - s/d_c)$ in evaluating their experimental results, where f_2 is the confinement stress in the concrete, calculated using the *actual* stress (not necessarily the yield stress) in the spiral steel. The term $(1 - s/d_c)$ reflects the reduction in effectiveness of spirals associated with increased spacing of the wires. Thus an improved version of Equation (7.3a) is:

$$f_c - f_c'' = 4.0f_2(1 - s/d_c) \tag{7.3b}$$

Figure 2, from Martinez *et al.*,[20] shows results ot tests of columns having different concrete strengths. Clearly the strength gain predicted by Equation (7.3b) is valid for normal-weight concrete of all strengths for confinement stress up to about 300 psi (21 MPa). A similar plot based on Equation (7.3a) would show a somewhat unconservative prediction for higher confinement stresses, but it can be shown that confinement stresses for practical column spirals are seldom more than about 1000 psi (7 MPa). For this range, the ACI Code basis derived using Equation (7.3a) gives satisfactory results. It follows that Equation (7.2) can be used without change for high-strength normal-weight concrete columns as well as for columns using lower strength concrete.

It is important to note, however, that a spiral has much less confining effect in lightweight concrete columns, probably due to the crushing of the lightweight concrete under the spirals at heavy loads. This results in a greatly reduced strength gain in the core, compared with that for normal density concrete (see Fig. 7.2). Lightweight columns would require about

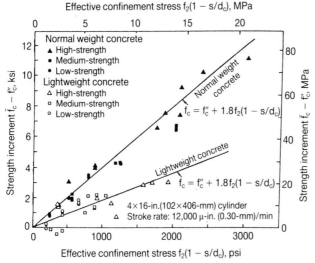

Fig. 7.2 Strength increment provided by spiral reinforcement action

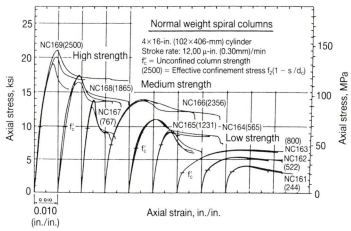

Fig. 7.3 Experimental stress-strain curves for spirally reinforced columns

2.5 times more spiral steel than corresponding normal-weight concrete columns. Whether or not such heavy spirals are practical is questionable.

There is not yet general agreement on the effectiveness of spirals for improving the *ductility* of high-strength concrete columns, that is, for increasing the strain limit and flattening the negative slope of the stress-strain curve past peak stress. Ahmad and Shah[18] conclude that confining spirals are about as effective in flattening the negative slope for high-strength as for low-strength concrete columns. This conclusion was reached on the basis of the results from their analytical model. The analytical model employs the basic constitutive relationship of concrete and the confining spiral steel and assume perfect compatibility in the lateral direction. However, the experimental results of Martinez *et al.*[20] showed significant differences, summarized in Fig. 7.3. Three groups of curves are identical by the three concrete strengths. Each of these groups consists of three sets of curves corresponding to three different amounts of lateral reinforcement. (Indicated in each set of curves with a short horizontal line is the average unconfined column strength corresponding to that particular set of confined columns.) Different behavior for different confinement stress is evident. Not only is the strain at peak stress much less for higher strength concrete, but the stress falls off sharply after peak value. This is true even for high-strength columns with very large confinement stress.

It is concluded that normal weight concrete columns with spiral steel show strength gain resulting from spirals that is well-predicted by present equations, but past peak stress they are likely to show much less ductility.[18–24] Furthermore, spirals in lightweight concrete columns are much less effective than predicted by present equations, and reserve strength past spalling of the concrete shell may be lacking.[20,25]

A separate issue is the relative merit of column ties versus spirals.[26,27] Ties are relatively ineffective in developing confinement pressure on the core concrete in columns of all strengths, and tied columns for all concrete strengths typically show less ductility than spirally-reinforced columns. In applications where severe overloads are likely, and where column toughness is critical, as in seismic zones, spirally-reinforced columns are much preferred.

Long-term loads

In concrete structures, particularly tall buildings, the larger part of the column load may be dead load, sustained continuously over the life of the structure. Significant creep occurs, which not only results in column shortening but may also modify all moments, thrusts and shears in the building if differential shortening occurs. Creep is conveniently described in terms of the creep coefficient:

$$C_c = \text{creep strain/initial elastic strain} \qquad (7.5)$$

There is general agreement that the coefficient of creep for high-strength concrete is significantly less than that of low-strength concrete.[4-9] The exact values of the creep coefficients for concretes of varying strengths depend on many factors other than compressive strength, including mix components and proportions, water/cementitious ratio, additives, temperature and humidity, age at first loading, and stress level with respect to strength. For common mixes, for sustained stress levels not more than about one-half the ultimate strength, and for normal ambient conditions, the values of Table 7.1 are representative.

It would seem, therefore, that the use of high-strength concrete in tall building columns, for example, would provide a means for reducing long-term column shortening. On the other hand, the sustained load stress for the high strength concrete columns would normally be higher, hence the initial elastic strain higher, tending toward creep strains that may be as high as before, even though the creep coefficient is less. Russell and Corley[28] present results of a study of time-dependent column behavior for a tall cast-in-place concrete structure.

Cyclic loading and fatigue

High-strength concrete may be relatively free of internal microcracking

Table 7.1   Variation of creep coefficient with compressive strength

Compressive strength psi (MPa)	Creep coefficient
3000 (21)	3.1
4000 (28)	2.9
6000 (41)	2.4
8000 (55)	2.0
10,000 (69)	1.6
12,000 (83)	1.4

even up to about 75% of ultimate load when loaded monotonically. On the other hand, it is known to be more brittle than low-strength concrete, lacking much of the ductility that accompanies progressive crack growth. Some experimental research indicates that fatigue strength is essentially independent of compressive strength.[29] Other research indicates that failure of concrete subject to repeated loading can be predicted approximately by the concept of the envelope curve, directly related to the short-term monotonic stress-strain curve.[24,30,31] For high-strength concrete, each load application causes relatively less incremental damage. However, the number of cycles to failure may not necessarily be larger because of the greater slope of the post-peak envelope curve.

7.3 Flexure in beams

The special mechanical properties of high-strength concrete affect the behavior of beams. In most ways, high-strength concrete beams behave according to the same rules that provide the basis for design of beams using lower strength concrete, but some differences have been found, mostly relating to deflection calculations, to be discussed in the following section. Topics pertaining to flexural calculations will be treated here.

Compressive stress distribution

The shape of the flexural compressive stress distribution in beams is directly related to (although not necessarily identical to) the shape of the compressive stress-strain curve in uniaxial compression. Considering the differences in the uniaxial test curves shown in Fig. 7.1, it is reasonable to expect corresponding differences in the flexure stress block shape.

Figure 7.4a shows the generally parabolic stress block typical of lower strength concrete. Three stress block parameters, k_1, k_2, and k_3 are sometimes used to define the characteristics of the block, where k_1 = ratio of average to maximum stress, k_2 = ratio of depths of compressive resultant to neutral axis, and k_3 = ratio of maximum stress in beam to maximum stress in axial compression test. In ordinary design in the USA and many countries, it is recognized that the exact shape of the stress block is unimportant provided that one knows (a) the magnitude of the compression result, and (b) the level at which it acts. These may be defined using an 'equivalent rectangular distribution' shown in Fig. 7.4b, with only two parameters: the first, β_1, defining the ratio of the rectangular block depth to the actual neutral axis depth, and the second (shown as a constant 0.85 in Fig. 7.4b) defining an equivalent constant stress that produces the proper value of the compressive resultant. The two parameters of Fig. 7.4b are easily related to the three parameters of Fig. 7.4a.[32]

For high strength concrete, which responds to compression in a more linear way, and to a higher fraction of ultimate stress, the actual compress-

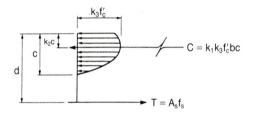

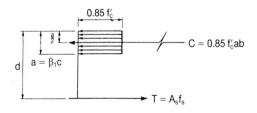

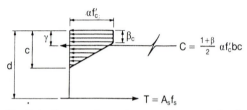

Fig. 7.4   Concrete flexural compressive stress distributions: (a) parabolic; (b) equivalent rectangular; (c) trapezoidal

ive stress distribution in the beam might be more closely represented by the trapezoidal distribution of Fig. 7.4c. However, for practical beams, which are invariably design as underreinforced with failure triggered by yielding of the tensile steel, comparative studies[33–42] indicate that the usual rectangular stress block is perfectly satisfactory. This holds even for compressive strength of 12,000 psi (83 MPa) and above, with differences of only a few percentage points between strengths predicted by mathematically continuous representation of stress, the trapezoidal block and the rectangle. It follows that the nominal flexural strength of an underreinforced high-strength concrete beam can be calculated by the usual equations:

$$M_n = A_s f_y (d - a/2) \tag{7.6}$$

$$a = A_s f_y / 0.85 f_c'' \tag{7.7}$$

where A_s = total tension steel area, f_y = steel yield stress, f_c'' = concrete compressive strength, d = effective depth to tensile steel, b = width of compression zone.

Limiting compressive strain and balanced steel ratio

While high-strength concrete reaches its maximum stress at a strain that is somewhat higher than that for lower-strength concrete, its ultimate strain is lower. However, for concretes in the present-day strength range of interest, say from 4000 psi to 12,000 psi (28 MPa to 83 MPa) the differences are not great, and tests indicate that the assumption of limiting strain of 0.003, as found in the 1989 ACI Code, is satisfactory. This strain limit, coupled with the yield strain of the tensile steel, provides the basis for calculating a balanced steel ratio, for which steel yielding and concrete crushing would theoretically occur simultaneously. In USA design practice, the maximum steel ratio is set at 0.75 times the balanced value. Tests have confirmed that this provides adequate ductility for high-strength concrete beams as well as beams of normal concrete. Thus the maximum steel ratio is:

$$\rho_{max} = 0.75 \times 0.85 \beta_1 \left(f_c'/f_y \right) \times \left(\varepsilon_{cu}/\varepsilon_{cu} + \varepsilon_y \right) \tag{7.8}$$

in which $\beta_1 = a/c$, $c =$ depth to the actual neutral axis, $\varepsilon_{cu} =$ ultimate concrete strain $= 0.003$, and $\varepsilon_y =$ steel yield strain.

Minimum tensile steel ratio

In lightly reinforced beams, if the flexural strength of the newly cracked section is less than the moment that produced cracking, the beam may fail immediately and without warning of distress upon formation of the first flexural crack. To avoid this, a lower limit steel ratio is established. This is done by equating the cracking moment, computed from the concrete modulus of rupture, to the flexural strength of the cracked section. The minimum steel ratio clearly depends upon the modulus of rupture, which in USA practice is related to the square root of the compressive strength. It can be shown that, for rectangular cross section beams, the minimum steel ratio should be:

$$\rho_{min} = 1.8\sqrt{f_c'}/f_y \qquad \text{in psi units} \tag{7.9a}$$
$$\rho_{min} = 0.149\sqrt{f_c'}/f_y \qquad \text{in MPa units}$$

and for T beams of normal proportions with flanges in compression

$$\rho_{min} = 2.7\sqrt{f_c'}/f_y \qquad \text{in psi units} \tag{7.9b}$$
$$\rho_{min} = 0.224\sqrt{f_c'}/f_y \qquad \text{in MPa units}$$

In the 1989 ACI code, concrete strength is not included as a variable, and the minimum steel ratio continues to be set at

$$\rho_{min} = 200/f_y \qquad \text{in psi units} \tag{7.9c}$$
$$\rho_{min} = 1.4/f_y \qquad \text{in MPa units}$$

It is easily confirmed that the ACI minimum steel ratio is conservative for rectangular beams for all but very high concrete strengths, and conservative for T beams for strengths to about 5000 psi (35 MPa). For other cases, use of Equation (7.9a) or (7.9b) is preferable.

7.4   Beam deflections

Immediate deflections

The main uncertainties in predicting elastic deflections of reinforced concrete beams are (a) elastic modulus E_c, (b) modulus of rupture f_r, and (c) effective moment of inertia I_e, which depends on the extent of cracking of the beam.

For the elastic modulus, research studies indicate the following expression may be used for concrete strengths to 12,000 psi (83 MPa):

$$E_c = 40,000\sqrt{f_c'} + 1,000,000 \text{ psi} \qquad (7.10)$$
$$E_c = 3320\sqrt{f_c'} + 6900 \text{ MPa}$$

Equation (7.10) should be multiplied by $(\omega_c/145)^{1.5}$ in psi units and by $(\omega_c/2300)^{1.5}$ in MPa units for concrete densities other than 145 lb/ft^3 or 2320 kg/m^3. The ACI Code value for elastic modulus, $E_c = 57,000\sqrt{f_c'}$ psi or $E_c = 4700\sqrt{f_c'}$ MPa, gives values that may be as much as 15% too high for high-strength concrete.[43]

For deflection prediction, a value for modulus of rupture of $7.5\sqrt{f_c'}$ psi or $0.7\sqrt{f_c'}$ MPa may be used. Test evidence confirms that the ACI Code equation for effective moment of inertia gives good results for high-strength beams as well as normal-strength beams:

$$I_e = [M_{cr}/M_a]^3 I_g + [1 - (M_{cr}/M_a)^3]I_{cr} \qquad (7.11)$$

in which M_{cr} = cracking moment, M_a = maximum moment in the span, I_g = gross moment of inertia of section, and I_{cr} = moment of inertia of cracked transformed section.

Deflection under sustained loads

It is the general practice to calculate time-dependent deflections of beams due to creep and shrinkage by applying multipliers to computed elastic deflections. This procedure is valid for high-strength concrete beams, but because high-strength concrete has a creep coefficient that is significantly below that of ordinary concrete, it follows that creep deflections of high-strength concrete beams should be less, relative to initial elastic deflections, than for otherwise identical low-strength beams. It can further be expected that compression steel, which reduces time-dependent beam deflections by limiting creep displacement of the concrete on the compression side of a beam, would be less important in high-strength than in

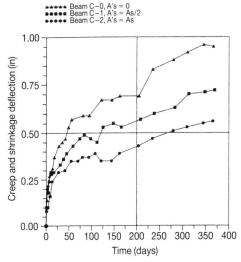

Fig. 7.5 Creep and shrinkage deflection of beams using 5400 psi (37 MPa) concrete

low-strength concrete beams. Experimental research confirms both observations.[44–46]

One-year data for tensile-reinforced beams of concrete compressive strength 5400 psi and 13,000 psi, are summarized in Figs 7.5 and 7.6, respectively.[45,46] First comparing beams with no compressive steel, Beam C-0 versus Beam A-0, it is seen that use of high strength concrete reduced creep and shrinkage deflection by about 32%. Comparing beams with compressive steel area half the tensile area, Beam C-1 versus Beam A-1, using high-strength concrete produced a deflection reduction of about 17%. For Beam C-2 versus Beam A-2, both with compressive steel area

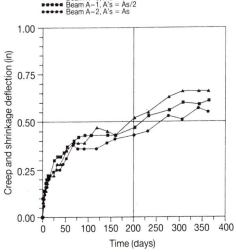

Fig. 7.6 Creep and shrinkage deflection of beams using 13,100 psi (90 MPa) concrete

equal to the tensile steel area, use of high-strength concrete produced no reduction in creep and shrinkage deflection.

From these data it is concluded that use of high-strength concrete is an effective means to reduce time-dependent deflections of beams with no compression steel. On the other hand, if compressive steel is present, there is little additional advantage gained, with regard to deflection reduction, through the use of high-strength concrete.

According to the 1989 ACI Code, additional long-term deflection resulting from creep and shrinkage is found by multiplying the immediate deflection caused by the sustained load by the factor:

$$\lambda = \zeta/(1 + 50\rho') \tag{7.12}$$

where ρ' is the compression steel ratio A_s'/bd at midspan for simple and continuous spans and ζ is a time-dependent factor to be taken as follows: time 5 years or more $= 2.0$; 12 months $\zeta = 1.4$; 6 months $\zeta = 1.2$; and 3 months $\zeta = 1.0$. These factors were based on tests where concrete strength was mostly in the range from 3000 psi to 4000 psi (21 to 28 MPa) and consequently do not recognize concrete compressive strength as a variable. With high-strength concrete beams now having strength of 12,000 psi (83 MPa) or more, a revised equation is needed.

Review of available test data for low-, medium- and high-strength concrete beams suggests the following equation,[46] recommended as a replacement for Equation (7.12):

$$\lambda = \mu\zeta/(1 + 50\mu\rho') \tag{7.13}$$

in which:

$$\mu = 1.4 - f_c'/10,000 \tag{7.14}$$

with the limits that: $0.4 < \mu < 1.0$. It is evident that including the additional factor μ in the numerator of Equation (7.13) accounts for the lower creep coefficient of high-strength concrete, and including it in the second term of the denominator accounts for the reduced influence on deflections of compression steel in high-strength beams. Analysis of test data indicates that while two separate factors might be used, the differences are small, and this resulted in recommendation of the single factor of Equation (7.14). Note that for concrete strength of 4000 psi (28 MPa) or lower, Equation (7.13) gives a long-term deflection multiplier identical with that resulting from use of Equation (7.12) from the 1989 ACI Code.

7.5   Shear in beams

In current practice in the USA, total shear resistance is made up of two parts: V_s provided by the stirrups and V_c, nominally the 'concrete contribution'. The concrete contribution includes, in an undefined way, the contributions of the still uncracked concrete at the head of a hypothetical diagonal crack, the resistance provided by aggregate interlock along the diagonal crack face, and the dowel resistance provided by the main reinforcement. The 1989 (and earlier) ACI Code includes two equations for, V_c, either one of which may be used at the designer's option. The simple Code Eq. (11-3):

$$V_c = 2\sqrt{f_c'}\,b_w d \tag{7.15}$$

is widely used in practice. For greater refinement, Code Eq. (11-6) is available:

$$V_c = [1.9\sqrt{f_c'} + 2500\,\rho_w V_u/M_u]\,b_w d \tag{7.16}$$

In Equations (7.15) or (7.16), V_u and M_u are, respectively, the factored shear and moment at the section, b_w is the web width, d the effective depth to the flexural steel, and ρ the longitudinal flexural steel ratio $A_s/b_w d$.

High-strength concrete loaded in uniaxial compression fractures suddenly and, in doing so, may form a failure surface that is smooth and nearly a plane.[1-3] This is in contrast to the rugged failure surface characteristic of low-strength concrete. In beams controlled by shear strength, the state of stress is biaxial, combining diagonal compression in the direction from the load point to the support with diagonal tension in the perpendicular direction. Diagonal tension cracks formed in high-strength concrete beam tests have been found to have relatively smooth surfaces.[47-52] Tests confirm that aggregate interlock decreases as concrete strength increases. Thus a shear strength deficiency may exist. Data from tests at the University of Connecticut[47,48] indicated that the value of V_c calculated using the 1983 ACI Code equation should be reduced for high-strength concrete. Data from Cornell University tests[49,50] confirmed that those equations do not properly account for the influence of the several parameters. Further research at North Carolina State University[51,52] confirmed that 1983 ACI Code provisions for shear overestimated the benefits of increasing concrete strength, and may be unconservative, particularly for beams in the normal range of a/d ratios, and normal (not heavy) amounts of flexural tension steel.

A comparison of some test results with predicted concrete shear contribution from the ACI Code Eqs. (11-3) and (11-6), for beams without web steel, is shown in Figs. 7.7, 7.8 and 7.9.[50] Figure 7.7 clearly shows that for beams with a moderate amount of flexural tensile steel, Code Eq. (11-6) is unconservative above about 6000 psi (41 MPa). Increasing the flexural steel ratio improves the value of V_c, as seen in Fig. 7.8, but the predictions remain unconservative for normally-proportioned beams with shear span ratio a/d of 4 to 6, except for very heavy flexural steel ratios. The influence of increasing shear span is further seen in Fig. 7.9.

On the other hand, for beams with web reinforcement, ACI Code Eq. (11-6) gives conservative results, at least for strengths tested, up to 9500 psi (66 MPa), as shown in Fig. 7.10. From this it is concluded that the provision of extra vertical stirrups may make up, in some way, for

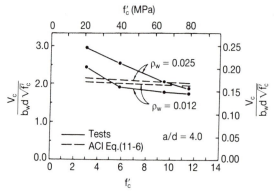

Fig. 7.7   Effect of f_c' on shear strength of beams without stirrups

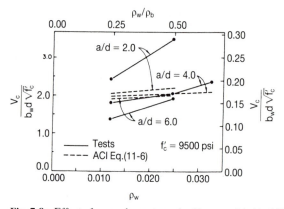

Fig. 7.8   Effect of ρ_ω on shear strength of beams with f_c' of 9500 psi (66 MPa)

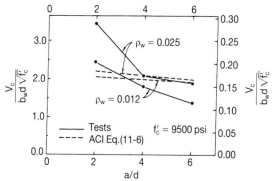

Fig. 7.9   Effect of shear span ratio a/d on shear strength of beams with $\sqrt{}$ *f9* of 9500 psi (66 MPa)

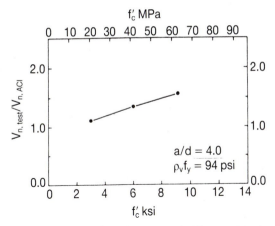

Fig. 7.10   Effect of f_c' on shear strength of beams with stirrups.

deficiencies in the contribution that is called V_c in beams using high-strength concrete.

While important progress has been made in recent years toward developing a rational method for design of web reinforcement that would take account of the actual behavior of reinforced concrete beams,[53] changes were made in the 1989 ACI Code to account for some of the deficiencies mentioned above on an entirely ad hoc basis.[54] The earlier shear design methodology was modified as follows:

1   The values of $\sqrt{f_c'}$ used in the shear design equations shall not exceed 100 psi (0.69 MPa), except that,
2   Values of $\sqrt{f_c'}$ greater than 100 psi (0.69 MPa) are permitted in calculating V_c for reinforced concrete beams in which the minimum web reinforcement equals $f_c'/5000$ times ($f_c'/35$ times), but not more than three times, the amount normally required.

Clearly these provisions affect only members with concrete strengths greater than 10,000 psi (69 MPa). While comparison with the test data of Roller and Russell[54] indicates that these changes are appropriate, it should be noted that many of the beams tested in that investigation contained extraordinarily large amounts of flexural tension steel (a feature that was established in other investigations to greatly enhance shear strength), and could not be considered practical beams.

7.6   Bond and anchorage

Present methods of design for development length and anchorage of tensile steel are based on tests, generally using concrete with compressive strength not greater than about 4000 psi (28 MPa). Although some information has recently become available for high strength concrete,[55] not enough data have been obtained to permit a definitive statement. Pending further research, a new limitation was introduced with the 1989 ACI Code to the effect that values of $\sqrt{f_c'}$ used with the normal bond and anchorage equations must not be taken greater than 100 psi (0.69 MPa). Clearly this limit affects only members with concrete strength greater than 10,000 psi (69 MPa). It is to be hoped that current research will provide the information needed, on the one hand, to verify the safety of the present ad hoc approach, and on the other hand, to permit optimum use of the undoubtedly greater bond strength of the new high-performance concretes.

7.7   Flexural and shear cracking

The modulus of rupture, which is an appropriate measure of concrete tensile strength for use in predicting flexural cracking load, has been reported to be in the range from 8 to $12\sqrt{f_c'}$ psi (0.66 to $0.99\sqrt{f_c'}$ MPa) for normal density concrete, and from 6 to $8\sqrt{f_c'}$ psi (0.50 to $0.66\sqrt{f_c'}$ MPa) for lightweight concrete.[25] Thus it appears that the values stated or implied in the 1989 ACI Code, $7.5\sqrt{f_c'}$ psi ($0.62\sqrt{f_c'}$ MPa) for normal density and $5.6\sqrt{f_c'}$ psi ($0.46\sqrt{f_c'}$ MPa) for lightweight concretes are quite conservative for higher concrete compressive strengths. However, curing conditions in the field are seldom as ideal as in the laboratory, and the ACI values may be closer to actual tensile strength of the concrete in beams on the job site.

It may be observed that the assumption of modulus of rupture lower than the actual value is neither conservative nor unconservative, but simply will result in an inaccurate prediction of cracking load. This results, in turn, in an inaccurate prediction of both initial and time-dependent deflections.

The direct tensile strength is seldom measured, but is of interest in studying web-shear cracking, mainly in prestressed concrete beams. This form of shear cracking initiates in the web, spreading both upward and downward to form the complete diagonal tension crack, and contrasts with flexure-shear cracking, initiating with a flexure crack and diagonally upward through the web. The second type of crack is typical in reinforced concrete beams, and often occurs in prestressed beams as well.

In USA practice, web-shear strength is predicted by empirical equations involving a number of terms, only one of which relates to the direct tensile strength. As noted above, the prediction of shear shrength of high-strength concrete reinforced beams is not firmly based on theory; shear strength for prestressed beams is even less so. While significant experimental research has been done,[47–54] the development and adoption of rational code provisions, perhaps based on the concepts of Schlaich *et al.*,[53] is still ahead.

7.8 Prestressed concrete beams

Characteristics of high-strength concrete, discussed previously in this chapter in the context of axially loaded members and reinforced concrete beams, affect the behavior of prestressed concrete beams in corresponding ways. There appears to be no reason why the normal procedures for flexural design cannot be applied to prestressed concrete beams using high-strength concrete, as they were to reinforced concrete beams. In fact, it is interesting to note that use of relatively high-strength concrete, with compressive strength in the range from 8000 psi to 10,000 psi (55 MPa to 69 MPa) was almost routine in the precast prestressed concrete industry long before such concretes were proposed for cast-in-place construction. No special difficulties have been encountered, although with strengths trending rapidly upward, questions do arise, mainly in connection with shear strength prediction.

The tests of high-strength concrete beams upon which shear provisions of the 1989 ACI Code are based were of reinforced, not prestressed, members. However, the code provision permitting values of $\sqrt{f_c'}$ greater than 100 psi (0.69 MPa) if web reinforcement minimums are increased is stated to apply to both reinforced beams *and* prestressed beams. While this may be valid, there is no direct experimental justification.

An extensive program of shear research on prestressed concrete beams is described in el-Zanaty *et al.*[49] Results confirmed that the basis of the 1989 ACI Code approach for calculating the 'concrete contribution' V_c, by which the web shear-cracking load and the flexure-shear cracking load are calculated separately, and V_c taken as the lower of the two values, is sound in the sense that both types of shear cracking were observed in 34 prestressed beam tests, and the basic behavior was different in each mode. It was found that the ACI Code equations conservatively predicted the shear cracking load for both flexure-shear and web-shear type cracks for beams without stirrups. It was further disclosed from test results for beam with stirrups that the shear strength predicted by ACI Code equations was on the safe side, although the degree of safety was variable. An overall conclusion was that several important parameters, namely concrete compressive strength, shear-span ratio, shear reinforcement index, and degree of prestress are not properly considered in the 1989 ACI Code equations. The basic flaw in the '$V_c + V_s$' approach is once again evident.

7.9 Slabs

Very little research has been done to date on structural slabs made using high-strength concrete, and in fact concrete for slabs in practice has seldom been over about 6000 psi (41 MPa). Probably this reflects the difficulty that would likely be experienced with deflections for the very thin structural slabs that could be designed using higher strength concrete, although prestressing offers the possibility of controlling deflections using the load-balancing method.

Meanwhile, there seems to be no reason that flexural behavior determined from beam testing could not be applied with safety to slab design. Similarly, shear design methods for beams could probably be applied safely to one-way and two-way edge supported slabs. While stirrups normally would not be used, shear stress in such elements would usually be far below that value permitted on the concrete.

Shear in beamless column-supported slabs, or flat plates, has long been a concern however. Present ACI Code design equations, again largely empirical and based on testing with concrete strengths in the range below 6000 psi (41 MPa), require re-evaluation before they are applied above that level. An experimental investigation[56] studying punching shear in concrete flat plates having concrete with compressive strength of about 10,000 psi (69 MPa) concluded that ACI Code equations, which relate slab shear strength to $\sqrt{f_c'}$, overestimate the benefit of increasing concrete strength on shear capacity. It was suggested that the British Codes CP 110 and BS 8110, which base strength on $\sqrt[3]{f_c'}$, may give a better representation. Clearly, more research is needed over the full range of parameters.

7.10   Eccentrically loaded columns

Compressive stress distribution

As pointed out in Section 7.2, columns in practice must be designed to resist not only axial loads but bending moments as well, either as a result of continuous frame action or from eccentric application of the load. In discussing beams in Section 7.3, it was pointed out that the shape of the compressive stress distribution in high-strength concrete beams is different from that in beams using low-strength concrete, because of the difference in shape of the stress-strain curve in compression. For reinforced concrete beams, which typically are under-reinforced, with strength controlled by yielding of the tensile reinforcement, the actual shape of the concrete compressive stress block used to calculate flexural strength is of little importance as long as the internal lever arm between compressive and tensile resultant is nearly correct. Over-reinforced beams are not permitted in present practice. It follows that present procedures for predicting flexural strength of beams are satisfactory, regardless of whether low- or high-strength concrete is used.

However, in the case of bending combined with axial load, i.e. eccentrically loaded columns, members for which failure would be in flexural compression cannot be avoided. For members with low eccentricity of loading, failure will be initiated when the concrete reaches its compressive strain limit, while the steel on the far side of the column may be well below its tensile yield, may be nearly unstressed, or loaded in compression. It is apparent that the shape of the compressive stress-strain curve used in the column analysis for such cases could assume much greater importance.

Short-column strength

An analytical study was made[57] to predict the failure load of eccentrically loaded columns using concrete strengths to 12,000 psi (83 MPa), with steel ratios from 1 to 4%, and with square or rectangular cross sections, the latter studied for both strong-axis and weak-axis bending. Three different concrete stress distributions were studied: (a) a continuous function based on actual stress-strain curves from uniaxial compression tests, (b) the ACI Code equivalent rectangular stress block, and (c) a trapezoidal stress distribution proposed in Pastor *et al.*[36] to provide a closer approximation to the actual stress distribution in bending than does the rectangular block.

For each of the 54 columns studied, the complete column nominal strength and design strength curves were constructed, as the results from using the three alternative compressive stress distributions were compared.

The continuous stress block can be assumed to be the most accurate predictor of column strength. As expected, the strength predicted by the ACI equivalent rectangular stress block showed the greater difference when compared with the continuous function for all cases, and on the unconservative side, while the trapezoidal stress block gave values between the continuous representation and the rectangular block. The difference between the rectangular and continuous function results were as large as 12% when 12,000 psi (83 MPa) concrete was used, but only up to 6% with 4000 psi (28 MPa) concrete. The difference between the trapezoidal and continuous function results were 3% or less for the high-strength concrete and 5% or less for low-strength concrete. Maximum differences were generally found when the eccentricity ratio e/h was in the range from 0.1 to 0.5.

The overall conclusion from the study was that either the continuous stress block or the trapezoidal approximation might be used in practice for columns with eccentricities that would produce failure in the compression failure range with significant bending, but with the rectangular stress block as much as 12% in error on the unconservative side. For low eccentricities, there was little difference in the predictions of the three representations, and for large eccentricities that would produce failure in the tension range there was almost no difference.

Slenderness effects

Most columns in present practice are short columns, in the sense that their strength is governed solely by material properties and the geometry of the cross section. However, with the increasing use of higher strength steel as well as concrete, and with improved methods for dimensioning of compression members, it is now possible to design much smaller cross sections than in the past. This clearly leads to columns with higher slenderness, and rational and reliable design procedures have become increasingly important.

The slender column design procedures of the 1989 ACI Code, while considerably more complicated than earlier methods, have been generally accepted by the profession. They appear to be suitable for use with high-strength concrete columns as well, although certain adjustments to the materials parameters may be appropriate. The modified value for elastic modulus given by Equation (7.10) is recommended in substitution for the 1989 ACI Code value, which may be as much as 15% too high for high-strength concrete. In addition, some modification of the factor β_d used in the slender column analysis is appropriate, to reflect the much lower creep coefficient typical of high-strength concrete. No more specific recommendation is possible at this time.

Beam-column connections

Current USA practice regarding joint design follows the recommendations of ACI Committee 352.[58] While not a part of the 1989 ACI Code, these recommendations provide a basis for the safe design, using ordinary concretes, of beam-column joints for both common construction and for buildings subject to seismic forces. According to the committee recommendations, joints are classified as Type 1 joints, connecting members in an ordinary structure designed on the basis of strength, according to the main body of the ACI Code, and Type 2 joints connecting members designed to have sustained strength under deformation reversals into the inelastic range, such as members in a structure designed for earthquake motions.

There appear to be no special considerations introduced with the use of high-strength concrete in the design of Type 1 joints, other than a requirement for careful detailing for clearance of bars in the smaller members that are typical. However, for Type 2 joints, in which ductile behavior must be maintained through a number of load reversals, some modifications to certain of the present Committee 352 recommendations have been suggested.[59]

Present recommendations limit the joint shear stress to $\gamma \sqrt{f_c'}$, where the coefficient γ is a function of the type of joint and loading condition, and the value of f_c' used is not to exceed 6000 psi (41 MPa). New test results show that with high-strength concrete, this equation becomes seriously unconservative, and a modified equation is proposed by Ehsani and Alameddine.[59] The γ factor remains unchanged, differentiating between corner, exterior and interior joints.

Further attention has been directed to the joint confinement requirements of the Committee 352 report, because present design recommendations result in excessive congestion of reinforcement for higher-strength concrete frames. Test results reported in Ehsani and Alemeddine[59] indicate that much lower amounts of steel may be adequate for high-strength concrete.

7.11 Summary and conclusions

A brief review has been presented of the special characteristics of high-strength concrete, and how they influence the behavior of structural concrete members of various types. Because of the generally empirical nature of many of the important design equations incorporated in present-day codes and used in practice, it is imperative that these equations be reviewed, and predicted performance compared against the results obtained in comprehensive test programs.

Important experimental research on high-strength concrete has been done over the past 15 years, both on the material itself and on structural members making use of it. This research has in some cases confirmed that present codified procedures are suitable. In other cases serious problems have been disclosed, indicating that design equations should be modified or even discarded in favor of new approaches. Code changes have been slow, and not all the necessary changes have yet been incorporated. For the design engineer proposing to exploit the many advantages of the new material, it is essential to seek out the best current experimental information. To this end, the extensive reference list that follows may be of particular value.

References

1 Carrasquillo, R.L., Nilson, A.H. and Slate, F.O. (1980) Microcracking and engineering properties of high-strength concrete. *Research Report* No. 80-1, Department of Structural Engineering, Cornell University, Ithaca, Feb. 1980, 254 pp.
2 Carrasquillo, R.L., Nilson, A.H. and Slate, F.O. (1981) Properties of high-strength concrete subject to short-term loads. *ACI Journal*, Proceedings **78**, No. 3, May–June, 171–8.
3 Carrasquillo, R.L., Slate, F.O. and Nilson, A.H. (1981) Microcracking and behavior of high-strength concrete subject to short-term loading. *ACI Journal*, Proceedings **78**, No. 3, May–June, 179–86.
4 Ngab, A.S., Slate, F.O. and Nilson, A.H. (1980) Behavior of high-strength concrete under sustained compressive stress. *Research Report* No. 80-2, Department of Structural Engineering, Cornell University, Ithaca, Feb, 201 pp.
5 Ngab, A.S., Nilson, A.H. and Slate, F.O. (1981) Shrinkage and creep of high-strength concrete. *ACI Journal*, Proceedings **78**, No. 4, July–Aug, 255–61.
6 Ngab, A.S., Slate, F.O. and Nilson, A.H. (1981) Microcracking and time-dependent strains in high-strength concrete. *ACI Journal*, Proceedings **78**, No. 4, July–Aug, 262–8.
7 Smadi, M.M., Slate, F.O. and Nilson, A.H. (1982) Time-dependent behavior

of high-strength concrete under high sustained compressive stresses. *Research Report* No. 82-16, Department of Structural Engineering, Cornell University, Ithaca, Nov.

8   Smadi, M.M., Slate, F.O. and Nilson, A.H. (1987) Shrinkage and creep of high-, medium- and low-strength concretes, including overloads. *ACI Materials Journal*, Proceedings **84**, No. 3, May–June, 224–34.

9   Smadi, M.M., Slate, F.O. and Nilson, A.H. (1985) High-, medium- and low-strength concrete subject to sustained overloads – strains, strengths and failure mechanisms. *ACI Journal*, Proceedings **82**, No. 5, Sept–Oct, 657–64.

10   (1977) High-strength concrete in Chicago high-rise buildings. *Task Force Report* No. 5, Chicago Committee on High-Rise Buildings, 63 pp.

11   Kaar, P.H., Hanson, N.W. and Capell, H.T. (1978) Stress-strain characteristics of high-strength concrete, Douglas McHenry International Symposium on Concrete and Concrete Structures, *Special Publication SP-55*, American Concrete Institute, Detroit, 161–185. Also, Research and Development Bulletin No. RD051.01D, Portland Cement Association.

12   Perenchio, W.F. and Klieger, P. (1978) Some physical properties of high-strength concrete. *Research and Development Bulletin* No. RD056.01T, Portland Cement Association, Skokie, 7 pp.

13   Shah, S.P. (1979) *Proceedings national science foundation workshop on high-strength concrete*. University of Illinois at Chicago Circle, Dec, 226 pp.

14   Wang, P.T., Shah, S.P. and Naaman, A.E. (1978) Stress-strain curves of normal and lightweight concrete in compression. *ACI Journal*, Proceedings **75**, No. 11, Nov, 603–11.

15   (1987) Research needs for high-strength concrete, Report by ACI Committee 363, *ACI Materials Journal*, Proceedings **84**, No. 6, Nov–Dec, 559–61.

16   (1987) *Proceedings symposium on utilization of high-strength concrete*, Stavanger, Norway, June 15–18. Tapir Publishers, N-7034 Trondheim-NTH, Norway, 688 pp.

17   Nilson, A.H. (1985) Design implications of current research on High-Strength Concrete, high-strength concrete. *Special Publication SP-87*. American Concrete Institute, Detroit, 85–118.

18   Ahmad, S.H. and Shah, S.P. (1982) Stress-strain curves of concrete confined by spiral reinforcement. *ACI Journal*, Proceedings **79**, No. 6, Nov–Dec, 484–90.

19   Martinez, S., Nilson, A.H. and Slate, F.O. (1982) Spirally-reinforced high-strength concrete columns. *Research Report* No. 82-10, Department of Structural Engineering, Cornell University, Ithaca, Aug.

20   Martinez, S., Nilson, A.H. and Slate, F.O. (1984) Spirally-reinforced high-strength concrete columns. *ACI Journal*, Proceedings **81**, No. 5, Sept–Oct, 431–42.

21   Fafitis, A. and Shah, S.P. (1985) Lateral reinforcement for high-strength concrete columns, High-Strength Concrete, *Special Publication SP-87*. American Concrete Institute, Detroit, 213–32.

22   Yong, Y.K., Nour, M.G. and Nawy, E.G. (1988) Behavior of laterally confined high-strength concrete under axial loads. *Journal of Structural Engineering*, ASCE, **114**, No. 2, Feb, 332–51.

23   Iyenger, K.T., Sundara, R., Desayi, P. and Reddy, K.N. (1970) Stress-strain characteristics of concrete confined in steel binders. *Magazine of Concrete Research*, London, **22**, No. 72, Sept, 173–84.

24   Ahmad, S.H. (1981) 'Properties of Confined Concrete Subject to Static and Dynamic Loads', *Ph.D. Thesis,* University of Illinois at Chicago Circle, Mar.

25   Slate, F.O., Nilson, A.H. and Martinez, S. (1986) Mechanical properties of high-strength lightweight concrete. *ACI Journal*, Proceedings **83**, No. 4, July–Aug, 606–13.

26 Vallenas, J., Bertero, V.V. and Popov, E.P. (1977) Concrete confined by rectangular hoops and subjected to axial loads. *Research Report* No. UCB/EERC-77/13. Earthquake Engineering Research Center, University of California, Berkeley.

27 Sheikh, S.A. and Uzumeri, S.M. (1980) Strength and ductility of tied concrete columns. *Journal of Structural Division*, ASCE **106**, No. ST5, May, 1079–1102.

28 Russell, H.G. and Corley, W.G. (1978) Time-dependent behavior of columns in water tower place, Douglas McHenry International Symposium on Concrete and Concrete Structures, *Special Publication SP-55*. American Concrete Institute, Detroit, 347–73. Also, Research and Development Bulletin No. RD052.01B, Portland Cement Association.

29 Bennett, E.W. and Muir, S.E.S. (1967) Some fatigue tests on high-strength concrete in uniaxial compression. *Magazine of Concrete Research*, London **19**, No. 59, June, 113–17.

30 Bresler, B. and Bertero, V.V. (1975) Influence of high strain rate on cyclic loading behavior of unconfined and confined concrete in compression, *Proceedings* Second Canadian Conference on Earthquake Engineering. Department of Civil Engineering, McMaster University, Hamilton, June, 15-1–15-38.

31 Bertero, V.V., Bresler, B. and Liao, H. (1969) Stiffness degradation of reinforced concrete members subject to cyclic flexural moments. *Report No.* EERC-69/12, University of California, Berkeley, Dec.

32 Nilson, A.H. and Winter, G. (1991) *Design of concrete structures*, 11th ed. McGraw-Hill Inc., New York, 904 pp.

33 Leslie, K.E., Rajagopalan, K.S. and Everhard, N.J. (1976) Flexural behavior of high-strength concrete beams. *ACI Journal*, Proceedings **73**, No. 9, Sept, 517–21.

34 Nedderman, H. (1973) 'Flexural Stress Distribution in Very High Strength Concrete', *M.Sc. Thesis*, University of Texas at Arlington, Dec. 182 pp.

35 Zia, P. (1977) Structural design with high-strength concrete, *Research Report* No. PZIA-77-01. Civil Engineering Department, North Carolina State University, Raleigh, Mar, 65 pp.

36 Pastor, J.A., Nilson, A.H. and Slate, F.P. (1984) Behavior of high-strength concrete beams, *Research Report* No. 84-3. Department of Structural Engineering, Cornell University, Ithaca, Feb.

37 Shin, S., Ghosh, S.K. and Moreno, J. Flexural ductility of ultra-high-strength concrete members, accepted for publication in *ACI Structural Journal*.

38 Ahmad, S.H. and Barker, R. (1991) Flexural behavior of reinforced high-strength lightweight concrete beams. *ACI Structural Journal*, Proceedings **88**, No. 1, Jan–Feb, 69–77.

39 Wang, P.T., Shah, S.P. and Naaman, A.E. (1978) High-strength concrete in ultimate strength design, *Proceedings* ASCE **104**, No. ST11, Nov, 1761–73.

40 (1977) Discussion of 'Flexural behavior of high-strength concrete beams', by Leslie, K.E., Rajagopalan, K.S. and Everhard, N.J. *ACI Journal*, Proceedings **74**, No. 3, Mar, 104–45.

41 Ahmad, S.H. and Batts, J. (1991) Flexural behavior of doubly-reinforced high-strength lightweight concrete beams with web reinforcement. *ACI Structural Journal*, Proceedings **88**, No. 3, May–June, 351–8.

42 *Private communications* with ACI Committee 318, Subcommittee D on Flexure and Axial Loads.

43 (1984) State-of-the-art report on high strength concrete, Reported by ACI Committee 363. *ACI Journal*, **81**, No. 4, July–Aug, 364–411.

44 Leubkeman, C.H., Nilson, A.H. and Slate, F.O. (1985) Sustained load deflection of high-strength concrete beams, *Research Report* No. 85-2. Department of Structural Engineering, Cornell University, Ithaca, Feb. 164 pp.

45   Paulson, K.A., Nilson, A.H. and Hover, K.C. (1989) Immediate and long-term deflection of high-strength concrete beams, *Research Report* No. 89-3. Department of Structural Engineering, Cornell University, Ithaca, 230 pp.

46   Paulson, K.A., Nilson, A.H. and Hover, K.C. (1991) Long-term deflection of high-strength concrete beams. *ACI Materials Journal*, Proceedings **88**, No. 2, Mar–Apr, 197–206.

47   Mphonde, A.G. and Frantz, G.C. (1984) Shear tests of high and low strength concrete beams without stirrups. *ACI Journal*, Proceedings **81**, No. 4, July–Aug, 350–7.

48   Mphonde, A.G. and Frantz, G.C. (1985) Shear tests of high and low strength concrete beams with stirrups, High Strength Concrete, *Special Publication SP-87*. American Concrete Institute, Detroit, 179–96.

49   El-Zanaty, A.H., Nilson, A.H. and Slate, F.O. (1986) Shear capacity of prestressed concrete beams using high-strength concrete. *ACI Journal*, Proceedings **83**, No. 3, May–June, 359–68.

50   El-Zanaty, A.H., Nilson, A.H. and Slate, F.O. (1986) Shear capacity of reinforced concrete beams using high-strength concrete. *ACI Journal*, Proceedings **83**, No. 2, Mar–Apr, 290–6.

51   Ahmad, S.H., Khaloo, A.R. and Poveda, A. (1986) Shear capacity of reinforced high-strength concrete beams. *ACI Journal*, Proceedings **83**, No. 2, Mar–Apr, 297–305.

52   Ahmad, S.H. and Lue, D.M. (1987) Flexure-shear interaction of reinforced high-strength concrete beams. *ACI Structural Journal*, Proceedings **84**, No. 4, July–Aug, 330–41.

53   Schlaich, J., Shafer, K. and Jennewein, M. (1987) Toward a consistent design of structural concrete. *J. Prestressed Concrete Institute*, **32**, No. 3, May–June, 74–150.

54   Roller, J.J. and Russell, H.G. (1990) Shear strength of high-strength concrete beams. *ACI Structural Journal*, Proceedings **87**, No. 2, Mar–Apr, 191–8.

55   Treece, R.A. and Jirsa, J.O. (1989) Bond strength of epoxy-coated reinforcing bars. *ACI Materials Journal*, **86** (2), March–April, 167–74.

56   Marzouk, H. and Hussein, A. (1991) Experimental investigations on the behavior of high-strength concrete slabs. *ACI Structural Journal*, **88**, No. 6, Nov–Dec, 701–13.

57   Garcia, D.T. and Nilson, A.H. (1990) A comparative study of eccentrically-loaded high-strength concrete columns, *Research Report* No. 90-2. Department of Structural Engineering, Cornell University, Ithaca, Jan.

58   (1985) 'Recommendations for Design of Beam-Column Joints in Monolithic Reinforced Concrete Structures', Reported by ACI Committee 352. *ACI Structural Journal*, **82**, No. 3, May–June, 266–83.

59   Ehsani, M.R. and Alameddine, F. (1991) Design recommendations for Type 2 high-strength reinforced concrete connections. *ACI Structural Journal*, Proceedings **88**, No. 3, May–June, 277–91.

8 Ductility and seismic behavior

S K Ghosh and Murat Saatcioglu

8.1 Introduction

It has been shown for quite some time that concrete becomes less deformable or more brittle as its compressive strength increases. Figure 8.1 shows a high-strength concrete cylinder being tested in compression. The failure is obviously explosive, indicating that the material is brittle. The same fact is depicted in a different way by Fig. 8.2[1] which shows the

Fig. 8.1 Testing of a high-strength concrete cylinder

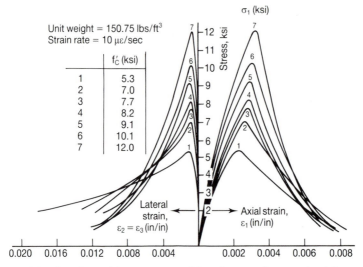

Fig. 8.2   Complete axial stress versus axial strain curves for normal weight concretes of different strengths[1]

axial stress-strain curves and axial-lateral strains curves in compression of normal weight concretes having different strength levels. Low-strength concrete obviously can develop only a modest stress level, but it can sustain that stress over a significant range of strains. High-strength concrete attains a much higher stress level, but then cannot sustain it over any meaningful range of strains. The load-carrying capacity drops precipitously beyond the peak of the stress-strain relationship.

Figure 8.3[1] shows the stress-strain curves of lightweight concretes having different compressive strengths. These curves were obtained by Ahmad in an investigation on mechanical properties of high strength lightweight

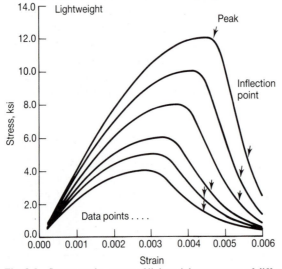

Fig. 8.3   Stress-strain curves of lightweight concretes of different strengths[1]

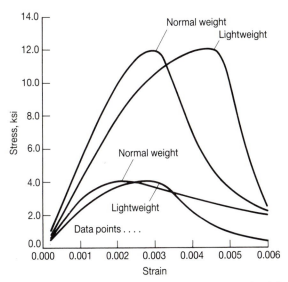

Fig. 8.4   Comparison of stress-strain curves of normal weight and lightweight concretes[1]

aggregate concrete which was conducted at North Carolina State University. In Fig. 8.4,[1] a selected comparison is made between the stress-strain curves of normal weight and lightweight concretes having essentially the same compressive strengths of about 4000 and 12,000 psi (27.6 and 82.8 MPa). It can be seen that for similar strengths, lightweight concrete exhibits a steeper drop of the descending part of the stress-strain curve than normal weight concrete. In other words, lightweight concrete is a more brittle material than normal weight concrete of the same strength.

The lack of deformability of high-strength concrete does not necessarily result in a lack of deformability of high-strength concrete members that combine this relatively brittle material with reinforcing steel. This interesting and important aspect is discussed in detail in this chapter. The application of high-strength concrete in regions of high seismicity is discussed at the end of the chapter. Such application depends, of course, on adequate inelastic deformability of high-strength concrete structural members under reversed cyclic loading of the type induced by earthquake excitation.

8.2   Deformability of high-strength concrete beams

Normal weight concrete beams under monotonic loading

Perhaps the earliest investigation on the deformability of high-strength concrete beams was carried out by Leslie, Rajagopalan and Everard.[2] They tested 12 under-reinforced rectangular beams with f_c' ranging between 9300 and 11,800 psi (64 and 81 MPa). The specimen details are shown in Fig. 8.5 and Table 8.1. It was observed that as the reinforcement ratio ρ increased, the 'maximum ultimate deflection' decreased, and the

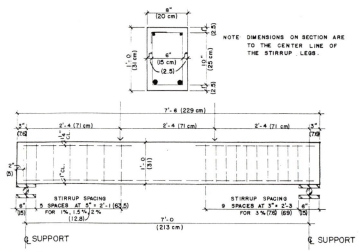

Fig. 8.5   Details of specimens tested by Leslie *et al.*[2]

ductility index μ (the ratio of maximum ultimate deflection to the deflection at the end of the initial linear portion of the load-deflection curve) decreased drastically. It is evident from Table 8.1 that the ductility index drops to quite low values for $\rho/\rho_b = 0.69$, whereas a ratio of up to 0.75 is allowed by the ACI Code.[3] However, this observation probably requires some qualification. Table 8.2 from Leslie *et al.*[2] lists the average ductility indices for increasing values of ρ. Also listed in the table are average values of ultimate deflection and yield deflection. First, neither of these terms was clearly defined in Leslie *et al.*[2] Secondly, as can be seen, the ductility index decreased with increasing ρ not so much because the ultimate deflection decreased as because the yield deflection increased. Thus, without knowing precisely how the yield deflection was determined, it is difficult to tell if this investigation was indicative of any lack of inelastic deformability on the part of high-strength concrete beams containing moderately high amounts of tension reinforcement (above 1.5%). It should be noted from Table 8.1 that μ values varied widely for the beams having the same tension reinforcement ratio, and that there was no correlation between this variation and the variation in concrete strength.

A comprehensive investigation on the deformability of high-strength concrete beams was carried out by Pastor *et al.*[4] Two series of tests, A and B, were conducted.

Series A consisted of four beams of high-strength concrete (HSC), one of medium strength concrete (MSC) and one of low-strength concrete (LSC). The scope was limited to singly reinforced, unconfined rectangular members subject to short-term 1/3 point loading. Beam Series A details are given in Fig. 8.6 and Table 8.3. Concrete compressive strength, f_c' (at the time of testing), and the tensile steel ratio, ρ, were the experimental variables.

The scope of Series B was limited to high-strength concrete rectangular

Table 8.1 Specimen details and ductility indices[2]

Specimen[a]	f_c', psi (MPa)	b, in. (mm)	d, in. (mm)	A_s[b]	ρ	ρ_b	ρ/ρ_b	μ
7.5–1	9310 (64.1)	8.25 (210)	10.63 (270)	2#6	0.01	0.0045	0.30	5.9
8.0–1	10,660 (73.5)	8.25 (210)	10.63 (270)	2#6	0.01	0.046	0.26	8.0
9.0–1	10,620 (73.2)	8.25 (210)	10.63 (270)	2#6	0.01	0.045	0.27	4.3
7.5–1.5	9720 (67.0)	8.00 (203)	10.56 (268)	2#7	0.014	0.051	0.44	4.5
8.0–1.5	11,400 (78.6)	8.13 (207)	10.56 (268)	2#7	0.014	0.051	0.31	2.5
9.0–1.5	11,630 (80.2)	8.50 (216)	10.56 (268)	2#7	0.013	0.051	0.30	4.2
7.5–2	10,850 (748)	8.50 (216)	10.50 (267)	2#8	0.018	0.040	0.69	3.2
8.0–2	10,610 (73.1)	7.88 (200)	10.50 (267)	2#8	0.019	0.040	0.63	2.4
9.0–2	11,780 (81.2)	8.13 (207)	10.50 (267)	2#8	0.019	0.039	0.56	2.7
7.5–3	11,650 (80.3)	8.38 (213)	10.50 (267)	3#8	0.027	0.039	0.82	1.9
8.0–3	11,730 (80.9)	8.25 (210)	10.50 (267)	3#8	0.027	0.039	0.71	2.1
9.0–3	11,210 (77.3)	8.25 (210)	10.50 (267)	3#8	0.027	0.039	0.77	1.5

[a] The first number indicates cement content in sacks/cu.yd. The second number indicates the nominal percentage of longitudinal reinforcement.
[b] The yield strengths for No. 6, No. 7 and No. 8 bars were 60.22 ksi (415 MPa), 55.83 ksi (385 MPa) and 66.88 ksi (461 MPa), respectively.

members, confined and doubly reinforced, under short-term 1/3 point loading. No low-strength concrete or medium strength concrete beams were tested. Beams in this series were modeled after beam A-4 of Series A to establish a basis for comparison between the behavior of unconfined singly reinforced, and confined doubly reinforced beams. Beam Series B details are given in Fig. 8.7 and Table 8.4. Compressive steel ratio, ρ', and lateral reinforcement ratio, ρ_s, were the controlled variables. Concrete strength was kept around 8500 psi (58.7 MPa), except for B-2a which had

Table 8.2 Comparison of ductility indices[2]

Reinforcement ratio ρ	Deflection at the end of initial strength portion, in. (mm)	Ultimate deflection, in. (mm)	Ductility index
0.01	0.28 (71)	1.70 (432)	6.0
0.014	0.32 (81)	1.18 (300)	3.7
0.019	0.36 (91)	1.05 (267)	2.9
0.027	0.55 (140)	1.00 (254)	1.8

Table 8.3   Beam Series A details[4]

Beam	Compressive strength at test, psi	Age at test, days	Test region dimensions			Tensile steel			ρ	ρ/ρ_b
			h, in.	b, in.	d, in.	Rebars	A_s, in.²	f_y, ksi		
A-1	3700	95	12.07	7.34	10.69	2#6	0.88	69.0	0.011	0.52
A-2	6500	113	12.01	7.22	10.63	3#6	1.32	69.0	0.017	0.53
A-3a	9284	7	12.03	7.13	10.65	2#6	0.88	66.0	0.012	0.26
A-4	8535	122	12.00	7.38	10.56	3#7	1.80	71.0	0.023	0.87
A-5	9264	122	12.00	6.56	10.50	3#8	2.37	71.0	0.034	0.87
A-6a	8755	137	12.00	6.94	9.75	6#7	3.60	59.5	0.053	1.10

1 in. = 25.4 mm
1000 psi = 6.895 MPa

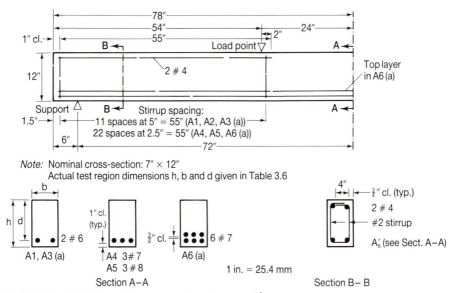

Fig. 8.6   Details of Series A beams tested by Pastor *et al.*[4]

$f_c' = 9284$ psi (64.1 MPa), while the amount of tensile steel was the same for all beams and consisted of three No. 7 (22 mm diameter) deformed bars.

To facilitate the evaluation of the moment-curvature and load-deflection data, experimental curves are represented as shown in Fig. 8.8 (Series A) and Fig. 8.9 (Series B). Key load-deformation values, identified in each

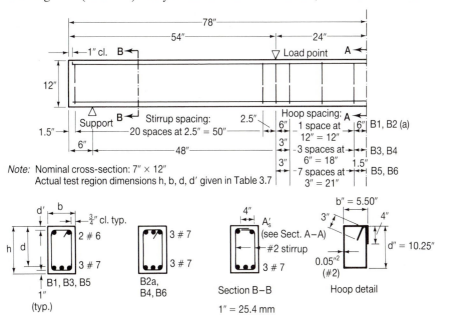

Fig. 8.7   Details of Series B beams tested by Pastor *et al.*[4]

Table 8.4 Beam Series B details[4]

Beam	Compressive strength at test, psi	Age at test, days	Test region dimensions				Tensile steel				Compressive steel			Tie steel[a]	
			h, in.	b, in.	d, in.	d', in.	A_s, in.^{2}	f_y, ksi	ρ	ρ/ρ_b	A_s', in.^{2}	f_y, psi	ρ'	A_s'', in.^{2}	ρ_s
B-1	8534	186	12.13	6.94	10.69	1.76		62	0.024	0.43	2#6 (0.88)	69	0.012	No. 2 at 12 in. (0.05)	0.0023
B-2a	9284	7	12.25	7.19	10.81	1.70		60	0.023	0.31	3#7 (1.80)	60	0.023	No. 2 at 12 in. (0.05)	0.0023
B-3	8578	189	12.25	6.88	10.81	1.88	3#7 (1.80)	62	0.024	0.46	2#6 (0.88)	69	0.012	No. 2 at 6 in. (0.05)	0.0047
B-4	8478	186	12.19	6.69	10.75	1.88		62	0.025	0.36	3#7 (1.80)	62	0.025	No. 2 at 6 in. (0.05)	0.0047
B-5	8516	194	12.06	7.00	10.63	1.69		62	0.024	0.43	2#6 (0.88)	69	0.012	No. 2 at 3 in. (0.05)	0.0093
B-6	8468	190	12.06	6.88	10.63	1.75		62	0.025	0.36	3#7 (1.80)	62	0.025	No. 2 at 3 in. (0.05)	0.0093

[a] Yield strength = 51 ksi
1 in. = 25.4 mm
1000 psi = 6.895 MPa

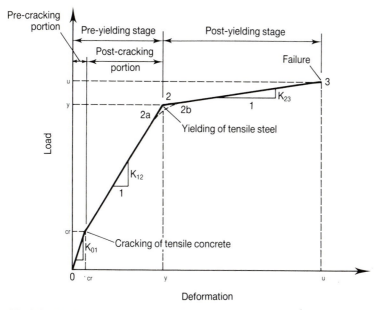

Fig. 8.8   Idealization of beam Series A load-deformation curves[4]

experimental figure, are listed in Table 8.5 (moment-curvature data) and
Table 8.6 (load-deflection data). Included in these tables are the corresponding values of the curvature (μ_c) and displacement (μ_d) ductility
indices.

Curvature and displacement ductility indices of Series A beams, as listed

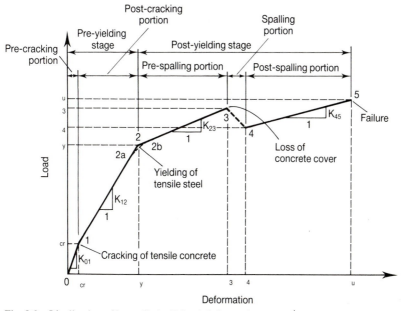

Fig. 8.9   Idealization of beam Series B load-deformation curves[4]

Table 8.5 Moment-curvature data and curvature ductility indices for Series A and B beams[4(a)]

Beam	Moments M (ft – kips)					Curvatures $\phi \times 10^{-6}$ (in.$^{-1}$)					Slopes $K = 10^{-3}$ (kip-in.^{2})				Ductility index
	M_{cr}	M_y	M_3	M_4	M_u	ϕ_{cr}	ϕ_y	ϕ_3	ϕ_4	ϕ_u	K_{01}	K_{12}	K_{23}	K_{45}	μ_c
A-1	8.0	43.0	$M_3 = M_u$	Not Applicable	50.0	50	350	$\phi_3 = \phi_4$	Not Applicable	1409	1920	1400	79	Not Applicable	4.03
A-2	10.2	68.5			75.0	20	388			1293[b]	6120	1901	86[c]		3.33[c]
A-3a	7.2	45.6			56.0	17	373			3394	5082	1294	41		9.10
A-4	12.2	95.0			100.0	30	390			1060	4880	2760	90		2.72
A-5	12.6	112.2			119.0	25	424			673	6048	2995	328		1.59
A-6a	15.0	124.0			142.0	35	427			630	5143	3337	1064		1.48
B-1	15.0	85.0	96.0	84.0	84.0	45	396	1181	1446	1446	4000	2393	168	–	3.65
B-2a	13.0	88.1	106.0	108.0	116.0	25	513	2558	4396	7148	6240	1847	105	35	13.93
B-3	15.0	85.0	100.0	94.0	97.2	30	259	1182	1225	2732	6000	3668	197	25	10.55
B-4[d]	7.6	85.0	103.0	97.0	119.0	17	357	1450	1700	4448	5365	2732	198	96	12.46
B-5[d]	10.0	85.0	97.0	94.0	101.0	30	410	1179	1981	2871	4000	2368	187	94	7.00
B-6[d]	14.4	85.0	97.0	97.0	110.0	50	439	1431	2130	3463	3456	2178	145	117	7.84

(a) See Figs. 8.8 and 8.9 for definition of symbols
(b) Extrapolated
(c) From extrapolated values
(d) Failed prematurely

1 in. = 25.4 mm

1 kip = 4.44822 kN

Table 8.6 Load-deflection data and displacement ductility indices for Series A and B Beams[4(a)]

Beams	Loads P (kips)					Deflections Δ (in.)					Slopes K (kips/in.)				Ductility index
	P_{cr}	P_y	P_3	P_4	P_u	Δ_{cr}	Δ_y	Δ_3	Δ_4	Δ_u	K_{01}	K_{12}	K_{23}	K_{45}	μ_d
A-1	4.0	23.5	$P_3 = P_u$	Not applicable	25.0	0.12	0.95	$\Delta_3 = \Delta_u$	Not applicable	3.50	33.3	23.5	0.6	Not applicable	3.68
A-2	5.1	34.2			37.5	0.06	0.94			2.50[b]	85.0	33.1	2.1[c]		2.66[c]
A-3a	3.6	24.0			28.0	0.05	0.71			3.63	72.0	31.0	1.4		5.11
A-4	6.1	47.5			50.0	0.08	1.05			1.88	76.3	42.7	3.0		1.79
A-5	6.3	56.1			59.5	0.08	1.05			1.21	78.8	51.3	21.3		1.15
A-6a	7.5	64.0			71.0	0.08	0.95			1.20	93.8	64.9	28.0		1.26
B-1	7.5	42.5	48.0	42.0	42.0	0.10	0.83	1.75	1.95	1.95	75.0	48.0	6.0	–	2.35
B-2a	6.5	43.8	53.0	54.0	58.0	0.08	0.79	3.00	4.55	5.63	81.3	52.5	4.2	3.7	7.13
B-3	7.5	43.5	50.0	47.0	48.6	0.11	0.89	2.48	2.60	4.04	68.2	46.2	4.1	1.1	4.54
B-4[d]	3.8	42.5	51.5	48.5	59.5	0.04	0.84	2.63	3.15	7.10	95.0	48.4	5.0	2.8	8.45
B-5[d]	5.0	44.2	48.5	47.0	50.5	0.07	0.79	1.95	2.95	4.36	71.4	54.4	3.7	2.5	5.52
B-6[d]	7.2	42.5	48.5	48.5	55.0	0.09	0.86	2.16	3.13	5.38	80.0	45.9	4.6	2.9	6.26

(a) See Figs. 8.8 and 8.9 for definition of symbols
(b) Extrapolated
(c) From extrapolated values
(d) Failed prematurely

1 in. = 25.4 mm
1 kip = 4.448822 kN

in Tables 8.8 and 8.6, are plotted as functions of the tensile steel ratio ρ in Fig. 8.10. Results show an overall reduction in both μ_c and μ_d with increasing ρ. The loss of ductility with increasing ρ is associated mainly with a decrease in the ultimate deformation of the member. In turn, ultimate deformations are inversely proportional to the neutral axis depth at failure.

Figure 8.11 plots the inverse of the Series A neutral axis depths at failure as a function of ρ. Since *c* values reflect the influence of fundamental material properties such as f_c' and f_y, the behavior shown in Fig. 8.11 provides a basic explanation to that shown in Fig. 8.10. For high-strength concrete beams, therefore, loss of ductility with increasing ρ can be traced back to the fact that *c* increases with increasing values of ρ.

Consider the upper half of Fig. 8.10. Points corresponding to the low-strength and medium strength concrete beams lie below the curve defined by the high-strength concrete data. When all other variables are held constant, μ_c clearly increases with f_c'.

Tognon et al.[5] arrived at the same conclusion from tests on singly reinforced model beams of every high-strength concrete. Reinforced concrete beams (4 in. × 8 in. × 6.5 ft or 100 mm × 200 mm × 2 m) were manufactured using very high-strength concrete (VHSC, $f_c' = 23{,}484$ psi or 162 MPa) and an ordinary concrete (LSC, $f_c' = 5797$ psi or 40 MPa) as a reference. The reinforcement consisted of three or six longitudinal steel

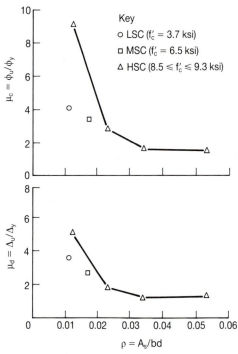

Fig. 8.10   Beam Series A curvature and displacement ductility indices versus tensile steel ratio[4]

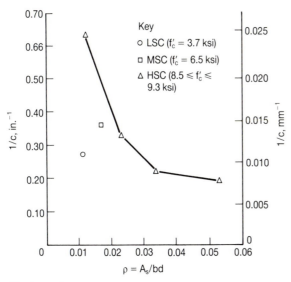

Fig. 8.11   Inverse values of beam Series A neutral axis depths at failure versus tensile steel ratio[4]

bars in the tensile zone with a concrete cover of approximately 0.4 in. or 10 mm. The amount of transverse reinforcement provided in the part of the beam subjected to variable moment was sufficient to prevent shear cracking.

To emphasize the different behavior of the two concretes with low, medium and high percentages of reinforcement, reinforced concrete beams were prepared with 0.87, 1.97 and 4.61% tension steel. 8.88% steel was also used for the very high-strength concrete. Balanced steel percentages were 3.85 and 13.49 for the ordinary and very high-strength concretes, respectively.

Ductility was represented by the ratio of ultimate curvature to yield curvature. The yield curvature was defined as corresponding to a tensile steel strain of 0.2%, and the ultimate curvature to a tensile steel strain of 0.1% or a compressive concrete strain of 0.35%.

Figure 8.12 shows that the beams made with VHSC are more ductile than those made with LSC when the reinforcement is about or over 1%. Moreover, the ductility of the beams made with LSC quickly decreases as the reinforcement ratio increases, whereas that of the beams made with VHSC declines gently up to a reinforcement ratio of about 5%.

It is worth noting that the curvature ductility of all heavily reinforced sections $(\rho > \rho_b)$ theoretically approaches unity regardless of concrete strength, f_c'. This suggests that the increase in μ_c with f_c' tends to decrease with ρ. There seems to be, therefore, a limiting tensile steel ratio beyond which μ_c values are practically the same regardless of f_c'.

Results shown in the lower half of Fig. 8.10 pertaining to μ_d exhibit the same general characteristics previously described for μ_c. Differences

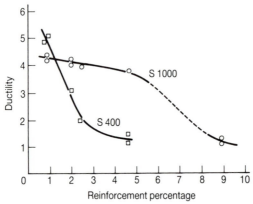

Fig. 8.12   Curvature ductility versus reinforcement percentage for LSC and VHSC beams[5]

between μ_d values due to differences in f_c' are, however, much smaller, mainly because the expected reduction from μ_c to μ_d is greater for the lightly to moderately reinforced HSC beams than for the LSC and MSC members. This suggests that the hypothetical ρ limit beyond which μ_d is no longer influenced by f_c' is comparatively smaller than that for μ_c. In any case, it seems reasonable to assume that μ_d increases with f_c', although much less significantly than μ_c.

The above conclusion emphasizes the difference between material, sectional and member ductility. The ductility of concrete as a material depends on the post-peak deformation of its stress-strain response, a property influenced primarily by f_c'. The ductility of a singly reinforced concrete section, on the other hand, depends on the post-yield deformation of its momentum-curvature response. Similarly, the ductility of a singly reinforced beam as a structural unit depends on the post-yield deformation of its load-deflection response. The two latter characteristics are associated with the position of the neutral axis at failure, a property influenced not only by f_c', but also by ρ and f_y.

As shown in Fig. 8.11, neutral axis depths at failure for the LSC and MSC beams are greater than the corresponding HSC values. It follows, therefore, that the decrease in c with f_c' more than compensates for the loss of material ductility in the HSC range, and helps explain the observed increase in sectional and member ductility with f_c' for constant ρ and f_y.

Consider Fig. 8.13[4] where μ_c and μ_d are plotted versus ρ/ρ_b. It is interesting to note that differences between LSC, MSC and HSC μ_c values shown in Fig. 8.13 are greatly reduced compared to those observed in the upper half of Fig. 8.10. This can be explained in terms of the inverse proportionality that exists between ρ/ρ_b and f_c'. In effect, everything else being equal, ρ_b will be higher for a beam with higher f_c'. Therefore, a beam of low f_c' with the same ρ/ρ_b ratio as that of a higher strength member necessarily has less area of steel. Less tensile ateel area implies a shallower neutral axis depth at failure. Consequently, the difference in ultimate

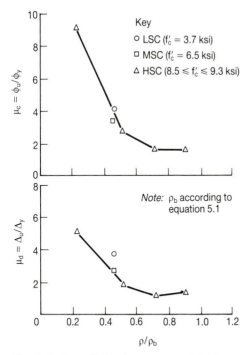

Fig. 8.13   Beam Series A curvature and displacement ductility indices versus tensile steel ratio expressed as a fraction of balanced steel ratio[4]

curvatures and therefore in curvature ductility indices due to differences in f_c' decreases, which is precisely what is illustrated in the upper half of Fig. 8.13.

Results shown in the lower half of Fig. 8.13 for $\rho/\rho_b = 0.45$ indicate a general tendency for μ_d to decrease with increasing f_c', although the difference between the values for MSC and HSC is insignificant. Insofar as the behavior suggested by the LSC and the corresponding HSC values, the observed decrease in μ_d with f_c' agrees with results reported by Leslie *et al.*[2] For reasons previously mentioned while discussing the relationship between μ_d and f_c', differences between LSC and HSC μ_d values should decrease with increasing values of ρ/ρ_b.

Inaccuracies in linear voltage differential transducer (LVDT) measurements introduced by the relatively large rotations of doubly reinforced beams at loads approaching ultimate resulted in erratic ϕ_u values which in turn resulted in unreliable μ_c values that exhibited large scatter. Consequently, curvature ductility indices for Series B beams, listed in Table 8.5, were not included in further analyses. On the other hand, data from related Cornell tests by Fajardo[4] were included to compensate for the loss of Series B data due to the premature failures of B-4, B-5, and B-6. Each of these beams had failed due to rupture of the tensil steel at the first stirrup weld outside the central test region (stirrups were spot welded to the longitudinal rebars). Although in all cases the tensile steel had already yielded when failure occurred and in some cases measured strains were well in the strain hardening portion of the steel stress-strain curve, this type of failure was considered premature. Full details of the Fajardo beams are given in Table 8.7.

Consider Fig. 8.14 where displacement ductility indices (μ_d) of nine doubly reinforced confined HSC beams and one singly reinforced unconfined HSC member (beam A-4) are plotted versus the quantity $\rho''f_y''/f_c''$. The symbol f_c'' is the yield strength of the lateral steel, f_c' is the compressive strength of the concrete, and ρ'' is the combined volumetric ratio of compressive and lateral reinforcement, defined as follows:

$$\rho'' = \rho_s + A_s'/b''d'' \tag{8.1}$$

where ρ_s = volumetric ratio of lateral reinforcement
  $= 2(b'' + d'')A_s''/b''d''s$
  A_s' = area of longitudinal compressive steel
  A_s'' = area of lateral tie steel
  b'' = width (outside-to-outside of tie steel vertical legs) of the d''confined core
  s = depth (outside-to-outside of tie steel horizontal legs) of the confined core
  = spacing (center-to-center) of tie steel

The best-fit linear regression curves shown interpolating the plotted data in Fig. 8.14 intersect at a $\rho''f_y''/f_c'$ value of about 0.11. They define two distinctly different modes of behavior that are briefly discussed below.

The relatively flat slope of the line to the left indicates that beams with

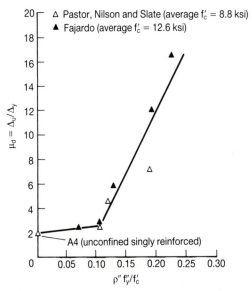

Fig. 8.14   Influence of compression and transverse reinforcement on the displacement ductility of doubly reinforced confined HSC beams[4]

Table 8.7 Properties of beams tested by Fajardo[4]

Beam	Age, days	f'_c, psi	f_y, ksi	$b \times h$, in.	L, ft	d, in.	d', in.	A_s, in.²	ρ, %	A'_s, in.²	ρ', %	ρ_b %	ρ/ρ_b	Confining hoops
B-7	121	12,800	65.0	7×12	12.0	10.50	1.44	2.35 (3#8)	3.2	1.20 (2#7)	1.6	7.7	0.42	#3 at 12 in.
B-8	134	12,650	65.0	7×12	12.0	10.50	1.50	2.35 (3#8)	3.2	2.35 (3#8)	3.2	9.2	0.35	#3 at 12 in.
B-9	128	12,800	65.0	7×12	12.0	10.50	1.44	2.35 (3#8)	3.2	1.20 (2#7)	1.6	7.7	0.42	#3 at 6 in.
B-10	133	12,000	65.0	7×12	12.0	10.50	1.50	2.35 (3#8)	3.2	2.35 (3#8)	3.2	8.9	0.36	#3 at 6 in.
B-11	132	12,650	65.0	7×12	12.0	10.50	1.25	2.35 (3#8)	3.2	0.40 (2#4)	0.50	6.6	0.48	#3 at 6 in.

1 in. = 25.4 mm
1000 psi = 6.895 MPa

$\rho''f_y''/f_c' < 0.11$ exhibited a load-deformation behavior with virtually no post-spalling response. These beams failed almost immediately after losing their cover concrete. Since significant inelastic dilatancy of the compression zone concrete occurs only after loss of the cover concrete, the presence of longitudinal compression and lateral tie steel has little or no influence on the ductility of these members. Consequently, the post-yielding load-deformation behavior and the displacement ductility indices were essentially the same as those of the corresponding singly reinforced unconfined member (A-4).

On the other hand, the relatively steep slope of the line to the right suggests that the post-spalling lateral expansion of the compressed concrete was restrained by the compression zone reinforcement. For beams with $\rho''f_y''/f_c' > 0.11$, therefore, ductility was influenced significantly by the presence of longitudinal compressive and lateral tie steel. It should be mentioned that testing of the two beams that gave the highest μ_d values (B-8 and B-10) was halted when excessive deflections could not be accommodated by the test setup.

The relative efficiency of the longitudinal compressive and the lateral tie steel in increasing μ_d is examined in Fig. 8.15. Figure 8.15(a) plots μ_d versus $\rho_s f_y''/f_c'$ for two values of ρ'/ρ, while Fig. 8.15(b) plots μ_d versus $\rho'f_y''/f_c'$ for three values of ρ_s. In both cases straight lines are used to represent the variation between data points.

Consider first Fig. 8.15(a). For a relatively low $\rho_s f_y''/f_c'$ value of 0.0175, increasing A_s' by a factor of two augments μ_d by a factor of almost four.

Consider now Fig. 8.15(b). For a significant range of $\rho'f_y''/f_c'$ values, increasing ρ_s 2.5 times results in a relatively small increase in μ_d. When ρ_s is increased by a factor of five, the effect of μ_d is, as expected, more pronounced. Nevertheless, the increase in μ_d is still smaller than the increase obtained by doubling the area of longitudinal compressive steel.

Results in Fig. 8.15 suggest that lateral ties are not as efficient in

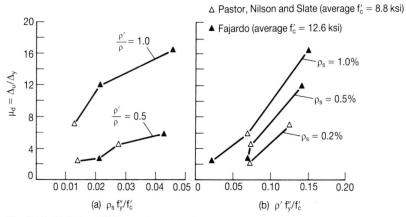

Fig. 8.15   Relative efficiency of compression and transverse reinforcement in doubly reinforced confined HSC beams[4]

increasing the post-yield deformations of beams as they are, for example, in increasing the post-peak deformations of concentrically loaded members. This is mainly because lateral deformations in beams tend to be large at the extreme compression fiber, and practically non-existent at the neutral axis location. Consequently, lateral confining stresses are unevenly and therefore inefficiently distributed across the depth of the compression zone.

On the other hand, the presence of properly restrained cxompression steel allows the member to behave (after loss of cover) much like a two-flanged steel beam (particularly when $A_s' = A_s$), the core concrete acting in this case as the connecting web. Consequently, significant inelastic deformations and large μ_d can be obtained. For such behavior to occur, however, the compression steel must remain stable in the strain hardening range. The stability of the compression bars is influenced strongly by the lateral restraint provided by the transverse ties. Hence, it may be reasonable to think that the primary role of the transverse steel in increasing beam ductility is not as a provider of concrete confinement, but rather as a lateral support mechanism for the longitudinal steel. Reducing the spacing of the ties did not, therefore, increase the confinement of the concrete core as much as it reduced the unsupported length of the compression bars.

Summarizing, the addition of longitudinal compression steel and lateral tie steel increases the displacement ductility of singly reinforced HSC beams (the former more efficiently than the latter), provided that the reinforcement index $\rho''f_y''/f_c'$ is greater than 0.11. Beams with $\rho''f_y''/f_c' < 0.11$ exhibit μ_d values very similar to that of the corresponding singly reinforced member. On the other hand, for indices > 0.11, a practically linear $\mu_d - \rho''f_y''/f_c'$ relationship develops in which μ_d increases noticeably for relatively small values of $\rho''f_y''/f_c'$.

Swartz *et al.*[6] tested four beams ($f_c' = 11,500$ psi or 79 MPa for the first two, and 12,300 psi or 85 MPa for the other two) with longitudinal reinforcing varying from $0.5\rho_b$ to $1.5\rho_b$ based on an assumed triangular stress block (ρ/ρ_b ratios would be somewhat lower, based on the ACI rectangular stress block). The amount of shear reinforcement also varied from zero to 100% based on ACI 318-83.[3] These designs are shown in Fig. 8.16. Load versus midspan displacement is plotted for each beam in Fig. 8.17. The beams with little or no shear reinforcement and also the over-reinforced beam exhibited very little ductility. No detail other than that shown on Fig. 8.17 was available.

The flexural ductility of ultra-high-strength concrete members (concrete strength ranging up to 15 ksi or 103.4 MPa) under monotonic loading was investigated by Shin.[7-9] All specimens were 6 in. (150 mm) × 12 in. (300 mm) in cross-section, and 10 ft (3 m) long (Fig. 8.18). Three sets of twelve specimens each were manufactured, using concrete compressive strengths of 4 ksi (27.6 MPa), 15 ksi (103.4 MPa) and 12 ksi (82.7 MPa) for sets A, B and C, respectively. There were six groups of two identical

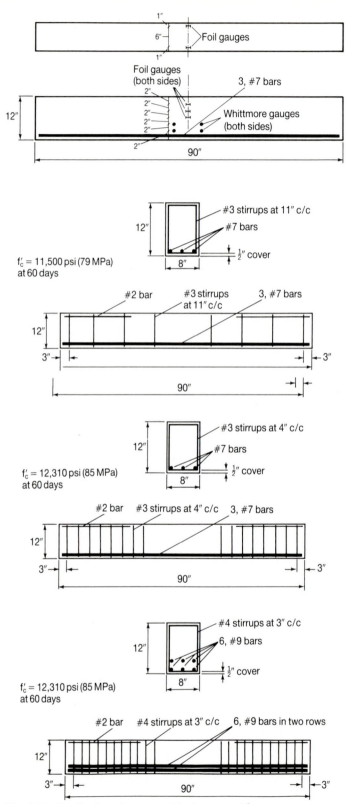

Fig. 8.16 Details of specimens tested by Swartz *et al.*[6]

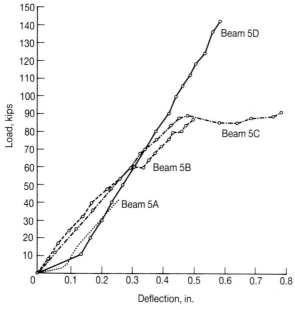

Fig. 8.17 Load versus midspan displacement of beams tested by Swartz *et al.*[6]

specimens in each set, two groups containing four No. 3 (10 mm diameter) bars, two more groups having four No. 5 (16 mm diameter) bars, and the last two groups being reinforced with four No. 9 (29 mm diameter) bars at the four corners. The difference between two groups of specimens with the same concrete strength and longitudinal reinforcement was in the spacing

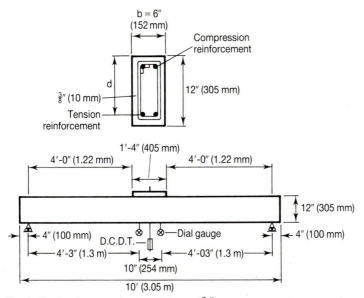

Fig. 8.18 Specimens and setup – Shin *et al.*[7–9]

of No. 3 (10 mm diameter) ties which was either 3 in. (75 mm) or 6 in. (150 mm).

The specimens were reinforced as if they were columns. They were cast horizontally under field conditions, and were also cured under field conditions. Each specimen was tested under two-point loading, which subjected a considerable portion of the specimen to pure flexure. The longitudinal reinforcement was divided equally into tension and compression reinforcement areas. While the tension reinforcement always yielded at advanced load stages, the compression reinforcement quite often developed only small stresses all the way up to beam failure.

Member ductility was first defined as:

$$\mu_0 = \Delta_0/\Delta_y \tag{8.2}$$

where Δ_0 is member deflection corresponding to the maximum load on the member, and Δ_y is member deflection at first yielding of the tension reinforcement.

In view of the fact that the beams continued to sustain substantial loads well beyond the peak of the load-deflection diagram (the load-carrying capacity of heavily reinforced beams temporarily dropped off immediately following the peak, but then picking up again), a second definition of ductility was also considered:

$$\mu_f = \Delta_f/\Delta_y \tag{8.3}$$

where Δ_f is the 'final' deflection corresponding to 80% of the maximum load along the descending branch of the load-deflection curve. Since the concept of ductility is related to the ability to sustain inelastic deformations without a substantial decrease in the load-carrying capacity, the definition of ductility as given by Equation (8.2) was felt to be logical and practical.

The ratio ρ/ρ_b turned out to be the most dominant factor influencing the magnitudes of the ductility indices. For a doubly reinforced section in which the compression reinforcement does not yield at the ultimate stage (defined by the extreme compression fiber concrete strain attaining a value of 0.3%), ρ_b is given by the following equation:

$$\rho_b = \rho_b + \rho'(f_b'/f_y) \tag{8.4}$$

where $$\rho_b = k_1 k_3 (f_c'/f_y)[E_s \varepsilon_u/(E_s \varepsilon_u + f_y)] \tag{8.5}$$

is the balanced reinforcement ratio for the corresponding singly reinforced section, and

where $$f_b' = E_s \varepsilon_u - (d'/d)(E_s \varepsilon_u + f_y) \leqslant f_y \tag{8.6}$$

is the stress in the compression reinforcement at balanced strain conditions.

Plots of μ_0 versus ρ/ρ_b and μ_f versus ρ/ρ_b are presented in Figs. 8.19 and 8.20, respectively. The figures clearly show that for the same concrete strength, the ductility indices decrease rather drastically as the ratio ρ/ρ_b increases. However, even at the large ρ/ρ_b ratio of 0.8, μ_f, which probably

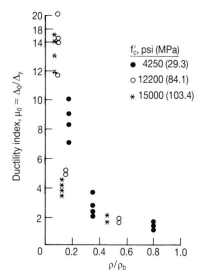

Fig. 8.19   Flexural ductility, as defined by Equation (8.2), under monotonic loading[7–9]

is a more practical measure of ductility for beams tested than μ_0, has rather substantial values.

Figures 8.19 and 8.20 generally show that the same amounts of longitudinal and confinement reinforcement, the ductility indices rise sharply as the concrete strength increases from 4 ksi (27.6 MPa) to 12 ksi (82.7 MPa) (nominal values), but then decrease somewhat as f_c' increases further from 12 ksi (82.7 MPa) to 15 ksi (103.4 MPa).

The confinement reinforcement spacing, within the range studied, did

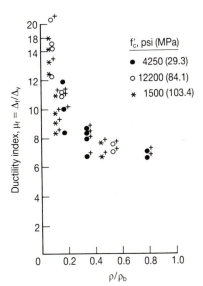

Fig. 8.20   Flexural ductility, as defined by Equation (8.3), under monotonic loading[7–9]

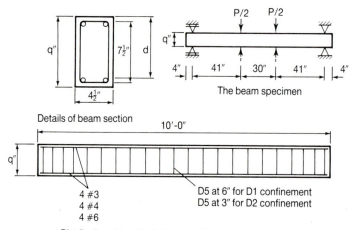

Fig. 8.21   Specimens and setup – Kamara *et al.*[9,10]

not have an appreciable effect on member ductility, for reasons discussed earlier and in Shin *et al.*[8]

Normal weight concrete beams under reversed cyclic loading

Shin[8] and Kamara[9,10] investigated the flexural ductility of ultra-high-strength concrete members (concrete strength ranging up to 15 ksi or 103.4 MPa) under fully reversed cyclic loading.

All specimens (Fig. 8.21) were 4.5 in. (112.5 mm) × 9 in. (225 mm) in cross-section, and 10 ft (3 m) long. Four sets of six specimens each were manufactured, using concrete compressive strengths of 5 ksi (34.5 MPa), 11 ksi (75.9 MPa), and 15 ksi (103.4 MPa) for sets A, B, and C, respectively. Set D was a duplicate of set C. In each set there were two beams reinforced with four No. 3 (10 mm diameter) bars, two beams reinforced with four No. 4 (13 mm diameter) bars and two beams reinforced with four No. 6 (19 mm diameter) bars. The difference between two specimens with the same concrete strength and longitudinal reinforcement was in the spacing of No. 2 (6 mm diameter) ties which was either 3 in. (75 mm) or 6 in. (150 mm).

The specimens, reinforced like columns, were cast horizontally under field conditions, and were also cured under field conditions. The specimens were tested under two-point loading which subjected a considerable portion of the specimens to pure flexure. The applied cyclic loading followed the displacement controlled schedule shown in Fig. 8.22.

The measured hysteretic load-deflection curves for the twenty-four beams tested have been presented in Kamara.[10] A sample is illustrated in Fig. 8.23. The envelopes of the hysteretic load-deflection curves for the beams were also included in Kamara.[10] It was observed that the envelope of the load-deflection curves for a cyclically loaded beam showed the same

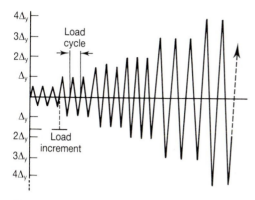

Fig. 8.22   Deformation sequence for reversed cyclic loading tests by Kamara *et al.*[9,10]

features as the load-deflection curve for a comparable monotonically loaded beam.

Member ductility, as defined by Equation (8.3), was investigated. The final deflection Δ_f was considered to be 2.4 in. (61 mm) which was the deflection at the end of cycling. The value of the applied load at that deflection was found to be equal to or slightly larger than the conservative value of the calculated ultimate load. The ductility ratios for downward and upward load cycles are plotted in Figs. 8.24 and 8.25, respectively.

In view of previous research work and building codes, a deflection ductility of 4 appears to represent a reasonably conservative minimum requirement for members subjected to gravity plus wind or moderate seismic loads. This requirement was more than met by all the specimens tested in Kamara's program. Thus it was concluded that, in the absence of axial loads acting simultaneously with flexure, high-strength reinforced concrete members subjected to reversed cyclic loading possess as much

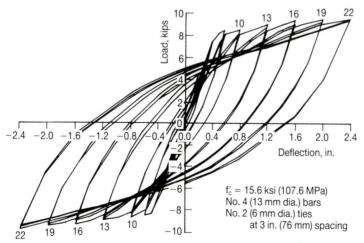

Fig. 8.23   Hysteretic load-deflection curve of specimen tested under reversed cyclic loading[9,10]

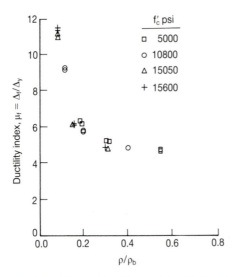

Fig. 8.24   Flexural ductility, as defined in Equation (8.3), under reversed cyclic loading – upward deflection[9,10]

ductility as is likely to be required of them in practical situations. It should be remembered, however, that the specimens tested in the course of Kamara's investigation were under zero axial load.

For the same amounts of longitudinal reinforcement and confinement reinforcement, the ductility ratios were found to increase with increasing concrete strength. For the same concrete strength, the ductility ratio decreased with increasing amounts of longitudinal reinforcement. Within the range studied in Kamara's work, the spacing between ties appeared to have virtually no effect on the ductility of the tested specimens.

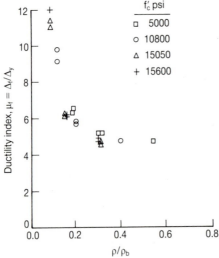

Fig. 8.25   Flexural ductility, as defined in Equation (8.3), under reversed cyclic loading – downward deflection[9,10]

Lightweight concrete beams under monotonic loading

Ahmad and Barker[11] reported limited experimental data on the flexural behavior of high-strength lightweight concrete beams. Flexural tests were conducted on six singly reinforced beams. Experimental variables were the compressive strength of concrete ($5200 < f_c' < 11,000$ psi or $35.9 < f_c' < 75.9$ MPa) and the reinforcement ratio, ρ/ρ_b ($0.18 < \rho/\rho_b < 0.54$). No compression or lateral reinforcement was used in the beams. A summary of the experimental program is presented in Table 8.8.

Deflection ductility was defined as the ratio of the deflection at ultimate to the deflection at yielding of the tensile steel. Ultimate was defined as the stage beyond which it was felt during testing that a beam would not be able to sustain additional deformation at the same load intensity.

The deflection ductility index μ_d decreased with an increase in tensile steel content ρ (Fig. 8.26 from Ahmad and Barker[11]). The results of Figs. 8.26 and 8.27[11] show that for an approximately equal ρ/ρ_b ratio, μ_d decreases with an increase in f_c'. As discussed earlier in connection with Pastor's investigation,[4] this is because the value of ρ_b increases with greater concrete strengths. Therefore, for a constant value of ρ/ρ_b, a beam with a higher strength concrete contains more steel than one with a lower strength concrete, which in turn decreases μ_d. For the beams with compressive strengths of 5000 and 8000 psi (34.5 and 55.9 MPa), a sharp reduction in μ_d occurred with an increase in ρ/ρ_b. However, for beams of 11,000 psi (75.9 MPa) concrete, the value of μ_d appeared to be less sensitive to changes in ρ/ρ_b. The trend indicated that beyond a certain range of ρ/ρ_b,

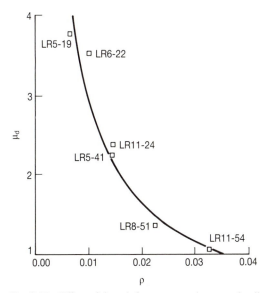

Fig. 8.26 Effect of the reinforcement ratio, ρ, on the displacement ductility of singly reinforced lightweight concrete beams tested by Ahmad and Barker[11]

Table 8.8 Test program[11]

Beam[a]	b, in. (mm)	h, in. (mm)	d, in. (mm)	Age at testing, days	$f_c'^{(b)}$ psi (MPa)	$f_c'^{(c)}$ psi (MPa)	E_c ksi (MPa)	Reinforcing bar detail	A_s, in.² (mm²)	ρ	ρ_b	ρ/ρ_b
LR5-19	6.0 (152.4)	12.0 (304.8)	10.25 (260)	2	5470 (37.7)	5200 (35.9)	3500 (2415)	2#4	0.4 (258)	0.0065	0.0344	0.189
LR5-41	6.0 (152.4)	12.0 (304.8)	9.25 (235)	2	5690 (39.3)	5410 (37.3)	3520 (2429)	4#4	0.8 (5160)	0.0144	0.0354	0.407
LR8-22	6.0 (152.4)	12.0 (304.8)	10.19 (259)	5	8770 (60.5)	8330 (57.5)	3760 (2594)	2#5	0.62 (400)	0.0101	0.0454	0.222
LR8-51	6.0 (152.4)	12.0 (304.8)	9.13 (231)	5	8550 (59.0)	8120 (56.0)	3750 (2588)	4#5	1.24 (800)	0.0226	0.0443	0.511
LR11-24	6.0 (152.4)	12.0 (304.8)	10.13 (257)	49	11560 (79.8)	10980 (75.8)	4410 (3043)	2#6	0.88 (568)	0.0145	0.0599	0.242
LR11-54	6.0 (152.4)	12.0 (304.8)	9.00 (229)	49	11590 (80.0)	11010 (76.0)	4770 (3291)	4#6	1.76 (1135)	0.0326	0.0600	0.543

[a] Beam nomenclature: for Beam LR5-19, '5' indicates the approximate concrete compressive strength in ksi and '19' indicates reinforcement ratio ρ/ρ_b

[b] Based on 4×8-in. cylinder strength

[c] Using equivalent 6×12-in. cylinder strength, assumed to be 95 per cent of 4×8-in. cylinder strength

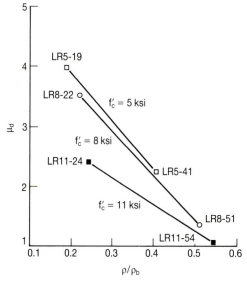

Fig. 8.27 Effect of the reinforcement ratio, ρ/ρ_b, on the displacement ductility of singly reinforced lightweight concrete beams tested by Ahmad and Barker[11]

deflection ductility μ_d becomes essentially independent of concrete strength, as suggested earlier by Pastor and Nilson.[4] Figure 8.28 shows the effect of concrete strength on μ_d for beams with different values of ρ/ρ_b. The comparison of deflection ductilities for lightweight and normal weight high-strength concrete beams (Fig. 8.29 from Ahmad and Barker[11])

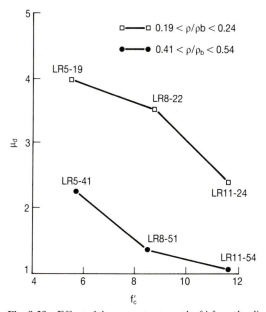

Fig. 8.28 Effect of the concrete strength, f_c' f_g on the displacement ductility of singly reinforced lightweight concrete beams tested by Ahmad and Barker

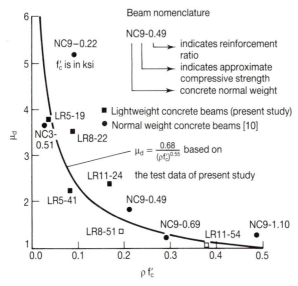

Fig. 8.29 Comparison of ductilities of lightweight and normal weight high-strength singly reinforced lightweight concrete beams tested by Ahmad and Barker[11]

indicates that ρ has a similar influence on μ_d for beams made of both types of concrete. The μ_d values obtained in Ahmad's study[11] were lower than those reported by Pastor and Nilson[4] for reinforced normal weight high-strength concrete beams.

Ahmad and Batts[12] developed limited experimental data on the flexural behavior of doubly reinforced high-strength lightweight concrete beams with web reinforcement. Flexural tests were conducted on six doubly reinforced beams. Experimental variables were the compressive strength of concrete ($6700 < f_c' < 11,060$ psi or $46.2 < f_c' < 76.3$ MPa) and the reinforcement ratio, ρ/ρ_b ($0.16 < \rho/\rho_b < 0.47$). All the beams had compression and web reinforcement. The compression reinforcement was kept to approximately half of the tension reinforcement, and web reinforcement was provided by No. 2 (6 mm diameter) smooth bars placed as stirrups at a spacing equal to half the depth of the section. A summary of the experimental program is presented in Table 8.9.

Deflection ductility was defined as in Ahmad and Barker.[11] The index μ_d decreased with an increase in the tensile steel content ρ (Fig. 8.30 from Ahmad and Batts[12]). The results in Figs. 8.31 and 8.32[12] show that for an approximately equal ρ/ρ_b ratio, μ_d decreases with an increase in f_c', for reasons discussed earlier. For beams with concrete strengths of 8000 and 11,000 psi (56 and 77 MPa), the ductility decreased relatively less with an increase in ρ/ρ_b than for beams with 5000 psi (35 MPa) concrete. Test results (Fig. 8.33 from Ahmad and Batts[12]) showed that the ductility decreases with increasing values of the product $\rho f_c'$. However, for larger values of the product $\rho f_c'$, the ductility essentially becomes constant.

A comparison of the results presented in Figs. 8.26–8.29 and those given

Table 8.9 Test program[12]

Beam	Compressive strength at test f_c' [a], psi	f_c' [b], psi	Test region dimensions				Tensile steel			Web steel			
			h, in.	b, in.	d, in.	d', in.	A_s, in.²	ρ	ρ/ρ_b, in.²	A_s', in.²	ρ'	A_s'', in.²	ρ_s
LJ-6-16	6700	6380	12	6	10.00	1.90	0.40	0.0067	0.16	0.22	0.0037	0.05	0.012
LJ-7-31	7720	7330	12	6	9.00	2.00	0.80	0.0148	0.31	0.40	0.0074	0.05	0.017
LJ-8-21	8080	7680	12	6	9.94	2.00	0.62	0.0104	0.21	0.40	0.0067	0.05	0.017
LJ-8-44	8360	7940	12	6	8.88	2.10	1.24	0.0233	0.44	0.62	0.0116	0.05	0.023
LJ-11-22	11740	11150	12	6	9.88	2.00	0.88	0.0148	0.22	0.40	0.0067	0.05	0.017
LJ-11-47	11060	10510	12	6	8.75	2.10	1.76	0.0335	0.47	0.88	0.0168	0.05	0.030

(a) Based on 4 × 8-in. cylinder strength
(b) Using equivalent 6 × 12-in. cylinder strength, assumed to be 95 per cent of 4 × 8-in. cylinder strength
The tension and compression reinforcement consisted of ASTM A 615 Grade 60 deformed bar
The stirrups used were No. 2 smooth bars with yield strength of 60 ksi

1000 psi = 6.895 MPa
1 in. = 25.4 mm

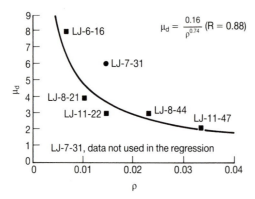

Fig. 8.30   Effect of reinforcement ratio, ρ, on the displacement ductility of doubly reinforced ($\rho' \approx 0.5\rho$) lightweight concrete beams tested by Ahmad and Batts[12]

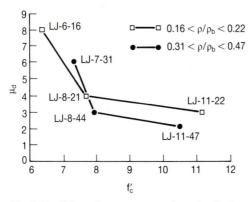

Fig. 8.31   Effect of concrete strength on the displacement ductility of doubly reinforced ($\rho' \approx 0.5\rho$) lightweight concrete beams tested by Ahmad and Batts[12]

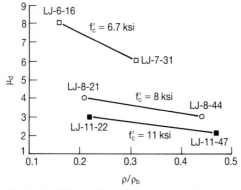

Fig. 8.32   Effect of the reinforcement ratio, ρ/ρ_b, on the displacement ductility of doubly reinforced ($\rho' \approx 0.5\rho$) lightweight concrete beams tested by Ahmad and Batts[12]

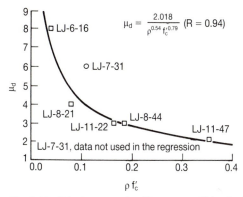

Fig. 8.33 Displacement ductility versus $\rho f_c'$ for doubly reinforced ($\rho' \approx 0.5\rho$) lightweight concrete beams tested by Ahmad and Batts[12]

in Figs. 8.30–8.33 clearly shows that the addition of compression reinforcement equal to approximately half the amount of tension reinforcement had a distinctly beneficial effect on deflection ductility.

Lightweight concrete beams under reversed cyclic loading

Ghosh *et al.*[13] conducted an experimental investigation aimed at gathering information on the flexural properties, including ductility, of high-strength lightweight concrete members (concrete with a dry unit weight of approximately 120 pcf and with compressive strength approaching 9 ksi at 56 days) under reversed cyclic loading.

Two sets of six specimens each were manufactured using lightweight aggregate concrete having compressive strengths of 5 ksi (34.5 MPa) at 28 days and 9 ksi (62 MPa) at 56 days. The test variables were the concrete strength, the amount of longitudinal reinforcement, and the spacing of ties. The test results, including hysteretic load-deflection curves, for the specimens representing columns under zero axial load are reported in Ghosh *et al.*[13]

The specimen dimensions, test procedure and loading history were identical to those used earlier by Kamara.[9,10] All specimens of each series were cast at the same time.

Except for one specimen that failed in shear, the moderate – as well as high-strength lightweight – concrete specimens exhibited stable hysteretic behavior all the way up to the limiting stroke of the testing machine. Flexure-dominated behavior could be ensured by supplying design shear strength in excess of the shear corresponding to the probable flexural strength.

The maximum deflection that could be imposed on the lightweight concrete beams of this investigation and on the normal weight concrete beams tested under reversed cyclic loading in the previous investigation by Kamara[9,10] was limited by the maximum stroke of the testing machine, so that full potential values of the ductility index μ_f (Equation (8.3)) could not be measured. The ductility indices of the lightweight concrete beams were lower than those of the corresponding normal weight concrete beams. The reason is that, because of the lower modulus of elasticity of lightweight concrete, the neutral axis depth at yield in a lightweight concrete beam was larger than that in a corresponding normal weight concrete beam. This made the yield curvature, and consequently the yield deflection, Δ_y, larger in the lightweight beam. In terms of the drop in maximum load carrying capacity at the maximum deflection that could be imposed, there was no significant difference between lightweight and normal weight concrete beams having the same longitudinal reinforcement and comparable concrete strengths, if the shear-dominated lightweight beam were excluded from consideration.

The ratio ρ/ρ_b once again turned out to be the most dominant factor influencing the magnitudes of the ductility indices. Plots of μ_0 versus ρ/ρ_b and μ_f versus ρ/ρ_b are presented in Figs. 8.34 and 8.35, respectively. The figures clearly show that for the same concrete strength, the ductility indices decrease rather drastically as the ratio ρ/ρ_b increases. However, all specimens, with the exception of the one that failed in shear, developed rather substantial values of μ_f, which is probably a more practical measure of ductility for the specimens tested than μ_0. It needs to be pointed out again that the values of μ_f in Fig. 8.35 are the largest values that could be measured, and not the largest values that could be attained.

Figures 8.34 and 8.35 generally show that the same amount of longitudinal reinforcement, ductility increases with increasing concrete strength.

The confinement reinforcement spacing, within the range studied, did not have an appreciable effect on member ductility, as is to be expected in view of earlier discussions.

Conclusions

The following conclusions can be drawn with respect to the deformability of high-strength concrete beams:

1 Although high-strength concrete is a less deformable material than lower strength concrete, the curvature ductility, μ_c, of a singly reinforced concrete section increases with f_c' for the same value of the reinforcement ratio, $\rho = A_s/bd$. This is because the neutral axis depth c decreases with increasing concrete strength, and the decrease in c with f_c' more than compensates for the loss of material ductility in the HSC range. For the same f_c', μ_c decreases with increasing ρ, because the neutral axis depth, c increases with increasing values of ρ.

2 The curvature ductility of all heavily reinforced sections $(\rho > \rho_b)$ theoretically approaches unity regardless of concrete strength, f_c'. Thus, the increase in μ_c with f_c' tends to decrease with ρ. There

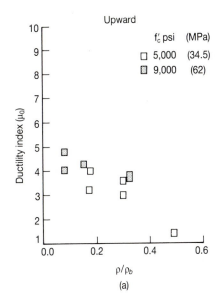

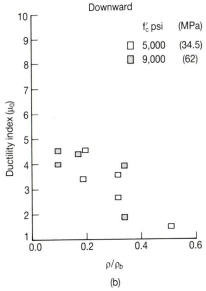

Fig. 8.34   Ductility index μ_0 versus ρ/ρ_b for doubly reinforced ($\rho = \rho'$) lightweight concrete beams[13]

appears to be a limiting tensile steel ratio beyond which μ_c values are practically the same regardless of f_c'.

3   The curvature ductility, μ_c, decreases with increasing values of the reinforcement ratio, ρ/ρ_b. At the same ρ/ρ_b, the differences between LSC, MSC and HSC curvature ductility (μ_c) values are greatly reduced compared to those observed for a constant value of ρ. This is

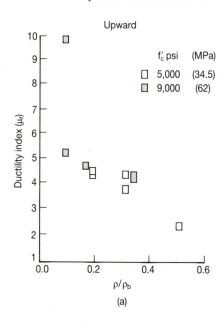

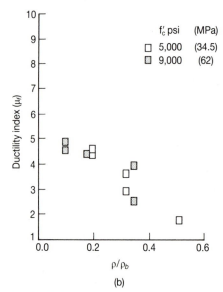

Fig. 8.35   Ductility index μ_f versus ρ/ρ_b for doubly reinforced ($\rho = \rho'$) lightweight concrete beams[13]

because the value of ρ_b increases with greater concrete strengths. For a constant value of ρ/ρ_b, a beam with a higher strength concrete contains more steel than one with lower strength concrete.

4   The member deflection ductility, μ_d, decreases with increasing values of ρ. For the same value of ρ, μ_d increases with increasing

concrete strength, f_c'. However, the magnitude of the increase is less than in the case of μ_c.

5   The deflection ductility, μ_d, also decreases with increasing values of ρ/ρ_b. For the same ρ/ρ_b, depending on that value of ρ/ρ_b, μ_d may decrease with increasing concrete strength. The differences between LSC and HSC μ_d values decrease with increasing values of ρ/ρ_b.

6   The ρ limit beyond which μ_d is no longer influenced by f_c' is comparatively smaller than that for μ_c.

7   The addition of longitudinal compression steel and lateral tie steel increases the displacement ductility of singly reinforced HSC beams (the former more efficiently than the latter), provided that the reinforcement index $\rho''f_y''/f_c'$ is greater than a certain critical value.

8   For low values of the above reinforcement index, beams fail amost immediately after losing their concrete cover. Since significant inelastic dilatancy of the compression zone concrete occurs only after loss of the cover concrete, the presence of longitudinal compression and lateral tie steel has little influence on the ductility of these members.

9   For moderate to larger values of the index $\rho''f_y''/f_c'$, displacement ductility is influenced significantly by the presence of longitudinal compressive and lateral tie steel, because the post-spalling lateral expenasion of the compressed concrete is restrained by the compression zone reinforcement.

10   Lateral ties are not as efficient in increasing the post-yield deformations of beams as they are in increasing the post-peak deformations of concentrically loaded members. This is because lateral confining stresses are unevenly and inefficiently distributed across the beam compression zone. The primary role of the transverse steel in increasing beam ductility is not as a provider of concrete confinement, but rather as a lateral support mechanism for the longitudinal steel.

11   The above conclusions drawn from tests on normal weight concrete beams under monotonic loading also apply, from all indications, to lightweight concrete beams under monotonic loading, except that, all other variables being the same, ductility is likely to be lower for the lightweight beam than for the corresponding normal weight member.

12   In the absence of axial loads acting simultaneously with flexure, high-strength reinforced concrete members subjected to reversed cyclic loading possess as much ductility as is likely to be required of them in practical situations. Under the same loading, the ductility indices of lightweight concrete beams are lower than those of the corresponding normal weight concrete beams. The reason is that, because of the lower modulus of elasticity of lightweight concrete, the neutral axis depth at yield is a lightweight concrete beam is larger than that in a corresponding normal weight concrete beam. This makes the yield curvature, and consequently the yield deflection larger in the lightweight beam.

8.3   Deformability of high-strength concrete columns

Deformability of high-strength concrete columns plays a major role in providing overall strength and stability to earthquake resistant structures. High deformability requirements in the first-story columns of multistory buildings can only be achieved through confinement of the core concrete. Inelastic deformability of high-strength concrete columns and associated confinement requirements are assessed in this section by reviewing test data from the available literature.[14] The data included the results of column tests conducted either under monotonically increasing concentric compression[15–19] or under lateral load reversals.[20–24] Concrete strength in the range of 4800 to 16,800 psi (33 to 116 MPa) has been considered. The tests include both normal weight and lightweight concrete columns with circular, square and rectangular cross-sections.

Deformability of concrete is evaluated in terms of a strain ductility ratio. Stress-strain relationships of confined high-strength concrete, obtained from column tests under concentric compression, are used for this purpose. The stress-strain relationship of the confined core is obtained by subtracting the contributions of longitudinal reinforcement and cover concrete from the recorded column capacity. The strain ductility ratio is defined as the ratio of concrete strain at 85% of the peak stress on the descending branch, to the strain corresponding to the peak stress. This is illustrated in Fig. 8.36.

Column deformability is evaluated in terms of a displacement ductility ratio. The displacement ductility ratio is defined as the ratio of maximum displacement recorded prior to exceeding 20% strength decay under cyclic loading, to the yield displacement. Cyclic loading consists of at least two cycles at each of the incrementally increasing deformation levels, where each increment is less than twice the yield displacement. Although approximately, these restrictions on imposed deformations reflect effects of loading history, and eliminate variations in column ductility resulting from distinctly different loading histories. Figure 8.37 depicts the definition of the displacement ductility ratio. The following notation is used in this section:

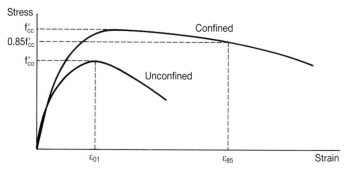

Fig. 8.36   Strain ductility ratio

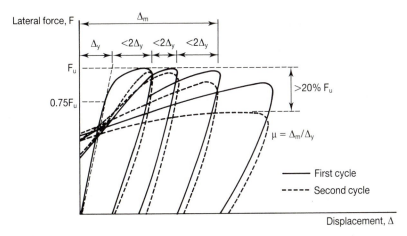

Fig. 8.37   Displacement ductility ratio

A_g : Gross area of concrete column section
b : Column cross-sectional dimension
D : Diameter of circular column section
f_c' : Concrete cylinder strength
f_{co}' : Unconfined concrete strength in column, determined by testing a specimen of the same size and shape as the column
f_{yt}' : Yield strength of confinement reinforcement
h : Column cross-sectional dimension
LWC   : Lightweight concrete
M : Maximum bending moment sustained by test specimen
NWC   : Normal weight concrete
P : Axial compression load
s : Tie spacing
s_1 : Spacing of legs of column ties in the horizontal plane
v : Maximum shear stress sustained by test specimen
V : Maximum shear force sustained by test specimen
ε_0 : Strain corresponding to unconfined strength of concrete
ε_{85} : Strain corresponding to 85% of confined strength on the descending branch of concrete stress-strain relationship
μ : Displacement ductility ratio
ρ_s : Volumetric ratio of confinement steel

Tables 8.10 and 8.11 provide a summary of the test data considered, for columns tested under concentric and lateral cyclic loadings, respectively. The tables also include strain and displacement ductility ratios. Figure 8.38 illustrates cross-sectional shapes and reinforcement arrangements considered in the experimental programs. The results indicate that deformability of confined high-strength concrete, and columns made with such concrete, are affected by *three* groups of parameters: (1) those related to applied loading, (2) those related to concrete confinement, and (3) those

Table 8.10   Columns tested under concentric compression

Column no.	Test label	Section type	Section dimensions	Ref.	f_{co}' (psi)	ρ_s (%)	f_{yt} (ksi)	s (in.)	$\varepsilon_{85}/\varepsilon_0$
1	III-3	1	D = 3 in.	[8.15]	5500	1.6	60	1.00	2.8
2	III-4				5500	3.2	60	0.50	7.2
3	V-2				9500	1.6	60	1.00	2.3
4	V-3				9500	3.2	60	0.50	4.5
5	PH5-5B	2	b = 5.8 in.	[8.16]	6200	2.1	198	2.00	4.4
6	PH5-9A		h = 5.8 in.		9000	2.1	198	2.00	2.5
7	PS5-5C				5900	2.1	198	2.00	5.1
8	PS5-7A				8100	2.1	198	2.00	3.5
9	PS5-9A				10,000	2.1	198	2.00	1.8
10	PD5-9A				11,000	4.2	198	2.00	7.6
11	NC164-1	1	D = 4 in.	[8.17]	7280	2.2	55	0.20	3.4
12	NC164-2				7280	2.2	55	0.20	3.2
13	NC164-3				7280	2.2	55	0.20	3.2
14	NC166-1				7281	7.5	60	0.25	12.6
15	NC166-2				7281	7.5	60	0.25	15.2
16	NC166-3				7281	7.5	60	0.25	11.9
17	NC168-1				9958	7.7	60	0.25	4.1
18	NC168-2				9958	7.7	60	0.25	4.3
19	NC168-3				9958	7.7	60	0.25	4.0
20	NC82-1				6191	2.3	60	0.36	4.9
21	NC82-2				6191	2.3	60	0.36	4.4
22	NC83-1				8447	3.4	60	0.25	3.3
23	NC83-2				8447	3.4	60	0.25	2.7
24	NC242-1	1	D = 5 in.	[8.17]	6108	2.2	60	0.65	3.3
25	NC242-2				6108	2.2	60	0.65	3.0
26	NC243-1				8255	3.1	60	0.65	2.0
27	NC243-2				8255	3.1	60	0.65	1.8
28	LC167-1	1	D = 4 in.	[8.17]	7990	2.7	55	0.16	2.1
29	LC167-2				7990	2.7	55	0.16	2.0
30	LC167-3				7990	2.7	55	0.16	1.9

					12,660	3.2	90	1.00	5.2
31	ND95-3		1	D = 6 in.	[8.18]	12,660	3.2	90	1.00	5.2
32	ND115-3					14,445	3.2	90	1.00	3.5
33	LWA75-3					10,210	3.2	90	1.00	2.9
34	ND95-10-1-3	3	3	b = 6 in.	[8.18]	13,230	3.2	90	1.00	3.0
35	LWA75-10-1-3			h = 6 in.		11,880	3.2	90	1.00	1.4
36	3		1	D = 3 in.	[8.19]	8200	2.4	85	1.25	5.6
37	9					8200	2.4	85	1.25	3.6

Notes: Section type and reinforcement arrangement are shown in Fig. 8.38. Columns No. 28, 29, 30, 33, and 35 have lightweight concrete. Column No. 37 was subjected to $5\sqrt{f_c'}$ psi shear stress.
1 in. = 25.4 mm; 1000 psi = 6.895 MPa

Table 8.11 Columns tested under lateral load reversals

Column no.	Test label	Section type	Section dimensions	Ref.	f_c' (psi)	ρ_s (%)	f_{yt} (ksi)	s (in.)	$P/A_g F_c'$ (%)	M/Vh	μ
38	600-02 77-1.2-1	4	b = 8 in. h = 6 in.	[8.20]	8500	3.00	50	20	2.75	2.5	10.8
39	600-02 79-1.2-10	4			8660	3.00	52	20	2.32	2.5	7.2
40	AL-2	5	b = 8 in. h = 8 in.	[8.21]	12,430	4.37	48	1.34	63	2.5	2.0
41	AH-2	5			12,430	4.37	115	1.34	63	2.5	7.3
42	BL-2	5			16,800	4.37	48	1.34	42	2.5	3.3
43	BH-2	5			16,800	4.37	115	1.34	42	2.5	8.0
44	1	5	b = 10 in. h = 10 in.	[8.22]	9670	2.00	121	1.77	31	2.0	6.7
45	2	5			12,260	2.57	121	1.38	28	2.0	6.3
46	3	5			5020	1.64	121	2.17	60	2.0	5.0
47	4	5			9670	2.60	46	1.97	57	2.0	2.5
48	5	5			9670	2.25	121	1.57	57	2.0	3.5
49	6	6			9670	2.08	198	1.77	57	2.0	5.0
50	7	5			12,260	2.57	121	1.38	51	2.0	5.0
51	N-1	7	b = 10 in. h = 10 in.	[8.23]	13,200	1.41	193	1.57	35	2.0	5.0
52	N-3	7			13,200	1.41	193	1.57	52	2.0	2.6
53	B1	5	b = 10 in. h = 10 in.	[8.24]	14,500	1.28	112	2.36	35	2.0	2.0
54	B2	5			14,500	1.92	112	1.57	35	2.0	3.3
55	B3	5			14,500	1.55	50	2.36	35	2.0	1.7
56	B4	5			14,500	1.28	163	2.36	35	2.0	3.3
57	B5	8			14,500	1.28	112	1.18	35	2.0	2.0
58	B6	8			14,500	1.26	124	2.36	35	2.0	2.0

Note: Section type and reinforcement arrangement are shown in Fig. 8.38.
1 in. = 25.4 mm, 1000 psi = 6.895 MPa

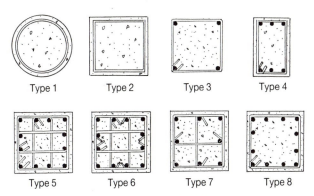

Fig. 8.38   Column cross-sections and reinforcement arrangements considered in tests

related to concrete type. These groups of parameters are discussed in the following sections.

Parameters related to applied loading

Effect of axial load

Axial compression reduces column deformability. Table 8.12 and Fig. 8.39 provide comparisons of columns tested under different levels of constant axial compression and incrementally increasing lateral load reversals. The results indicate that column deformability, as indicated by displacement ductility ratio, decreases with increasing axial compression.

The effect of axial tension on column deformability was investigated by Watanabe *et al.*[25] A specimen subjected to a constant level of axial tension showed reduced capacity but increased deformability, under lateral load reversals.

Table 8.12   Effect of axial compression on column deformability

Column no.	f_c' (psi)	ρ_s (%)	f_{yt} (ksi)	s/h	s_1/h	$v/\sqrt{f_c'}$ (psi)	$P/(A_g f_c')$ (%)	μ
45	12,260	2.57	121	0.14	0.25	8.4	28	6.3
50	12,260	2.57	121	0.14	0.25	8.4	51	5.0
44	9670	2.00	121	0.18	0.25	9.0	31	6.7
48	9670	2.25	121	0.16	0.25	9.6	57	3.5
40	12,430	4.37	48	0.17	0.26	8.9	63	2.0
42	16,800	4.37	48	0.17	0.26	8.9	42	3.3
41	12,430	4.37	115	0.17	0.26	9.0	63	7.3
43	16,800	4.37	115	0.17	0.26	8.9	42	8.0
51	13,200	1.41	193	0.16	0.38	9.4	35	5.0
52	13,200	1.41	193	0.16	0.38	10.0	52	2.6

1 in. = 25.4 mm; 1000 psi = 6.895 MPa

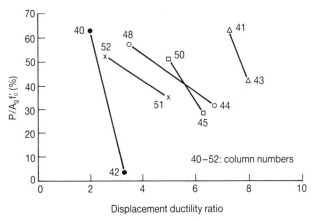

Fig. 8.39   Effect of axial compression on column deformability

Effect of shear stress

Data on the effect of shear on deformability of high-strength concrete is scarce. Limited tests reported by Abdel-Fattah and Ahmad,[19] conducted on concrete cylinders, indicate that the shear effect on high-strength concrete is less pronounced than that on normal-strength concrete. Table 8.13 shows that the strain ductility ratio decreases with imposed shear.

Table 8.13   Effect of shear on concrete deformability

Column no.	f_{co}' (psi)	ρ_s (%)	f_{yt} (ksi)	s/h	s_1/h	$v/\sqrt{f_c'}$ (psi)	$\varepsilon_{85}/\varepsilon_0$
36	8200	2.4	85	0.42	N/A	0	5.6
37	8200	2.4	85	0.42	N/A	5.1	3.6

1 in. = 25.4 mm; 1000 psi = 6.895 MPa

Effect of loading history

Two identical columns were tested by Chung *et al.*[20] under successively increasing lateral deformation cycles to investigate the effect of loading history. One of the columns was subjected to one cycle at each deformation level, while the other was subjected to ten cycles at each level. Table 8.14 indicates that the deformability of high-strength concrete columns decreases with increasing numbers of inelastic deformation cycles.

Table 8.14   Effect of load history on column deformability

Column no.	f_c' (psi)	ρ_s (%)	f_{yt} (ksi)	s/h	s_1/h	$v/\sqrt{f_c'}$ (psi)	$P/(A_g f_c')$ (%)	No. of cycles	μ
38	8500	3.00	50	0.35	0.66	5.9	20	1	10.8
39	8600	3.00	50	0.29	0.66	5.8	20	10	7.2

1 in. = 25.4 mm; 1000 psi = 6.895 MPa

Effect of rate of loading

Experimental data on high-strength concrete columns subjected to high strain rates is limited. Small-scale columns tested by Bjerkeli *et al.*[18] under strain rates of 3.33×10^{-5} per sec and 0.33×10^{-6} per sec indicate no significant effect of loading rate on the deformability of confined high-strength concrete.

Parameters related to concrete confinement

Volumetric ratio of confinement reinforcement

Volumetric ratio of confinement steel is one of the main parameters that affects concrete confinement. An increase in the volumetric ratio directly translated into a corresponding increase in confinement pressure, and resulting improvements in the deformability of concrete. This is shown in Table 8.15 and Fig. 8.40. While 7280 psi (50 MPa) concrete with 2.2% confinement steel shows a strain ductility ratio of approximately 3, the same concrete confined with 7.5% of a similar grade steel shows an increased ductility ratio of about 12. A comparison made with the deformability of normal-strength concrete[26] indicates that higher volumetric ratio of confinement steel is required for high-strength concrete to attain deformabilities usually expected of normal-strength concrete.

Strength of confinement reinforcement

Confinement pressure is a passive pressure that is activated by lateral expansion of concrete under axial compression. This pressure is dependent on the ability of concrete to expand laterally prior to failure. It is also limited by the strength of the lateral steel. If the transverse strain of high-strength concrete is not high enough to strain the confinement steel to its capacity, then the full capacity of steel is not utilized. It is speculated that high-strength concrete, being a brittle material, may not develop

Table 8.15   Effect of volumetric ratio of confinement steel on concrete deformability

Column no.	f_c' (psi)	ρ_s (%)	f_{yt} (ksi)	s/h	s_1/h	$\varepsilon_{85}/\varepsilon_0$
9	10000	2.1	198	0.34	N/A	1.8
10	10000	4.2	198	0.34	N/A	7.6
11	7280	2.2	55	0.05	N/A	3.4
14	7281	7.5	60	0.06	N/A	12.6
12	7280	2.2	55	0.05	N/A	3.2
15	7281	7.5	60	0.06	N/A	15.2
13	7280	2.2	55	0.05	N/A	3.2
16	7281	7.5	60	0.06	N/A	11.9
23	8447	3.4	60	0.06	N/A	2.7
16	7281	7.5	60	0.06	N/A	11.9

1 in. = 25.4 mm; 1000 psi = 6.895 MPa

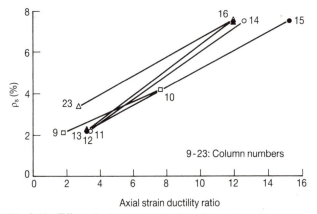

Fig. 8.40 Effect of volumetric ratio of confinement steel on concrete deformability

transverse strains high enough to strain the steel to its yield level. Research by Ahmad and Shah[15] has shown that spiral reinforcement is less effective for high-strength concrete columns, and that the stress in the spiral steel at maximum column load is often significantly less than yield stress, as assumed commonly. These results are consistent with Martinez *et al.*[17] who did not observe any yielding in the confinement reinforcement of high-strength concrete. Therefore, the same researchers recommended an upper limit for strength of confinement reinforcement. The use of high-strength steel for confinement of high-strength concrete is often questioned by researchers.

Experimental research by Muguruma *et al.*,[21] however, indicates that very high ductilities can be achieved in high-strength concrete when it is confined by high-strength steel. Table 8.16 and Fig. 8.41 show the improvements achieved in column deformability with the use of high-strength steel as confinement reinforcement. Columns with 12,500 to 16,800 psi (86 to 116 MPa) concretes, confined with 4.4% volumetric ratio of steel, show approximately 250% increase in displacement ductility ratios when the steel yield strength is increased by 140% from 48 to 115 ksi (328 to

Table 8.16 Effect on transverse steel strength on column deformability

Column no.	f_c' (psi)	ρ_s (%)	s/h	s_1/h	$v/\sqrt{f_c'}$ (psi)	$P/(A_g f_c')$ (%)	f_{yt} (ksi)	μ
40	12430	4.37	0.17	0.26	8.9	63	48	2.0
41	12430	4.37	0.17	0.26	9.0	63	115	7.3
42	16800	4.37	0.17	0.26	8.9	42	48	3.3
43	16800	4.37	0.17	0.26	8.9	42	115	8.0
47	9670	2.60	0.20	0.25	9.9	57	46	2.5
48	9670	2.25	0.16	0.25	9.6	57	121	3.5
49	9670	2.08	0.18	0.25	9.5	57	198	5.0
55	14500	1.55	0.24	0.25	9.5	35	50	1.7
53	14500	1.28	0.24	0.25	8.7	35	112	2.0
56	14500	1.28	0.24	0.25	8.7	35	163	3.3

1 in. = 25.4 mm; 1000 psi = 6.895 MPa

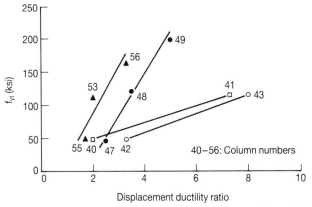

Fig. 8.41   Effect of transverse steel strength on column deformability

792 MPa). These columns develop displacement ductility ratios in excess of 7 when confined with high-strength steel. The results clearly show the beneficial effect of high-strength steel in confining high-strength concrete.

Further examination of the test data indicates that the improvement associated with the use of high-strength steel as confinement reinforcement is approximately the same as that obtained by increasing the volumetric ratio of confinement steel (ρ_s). This is to be expected if the confinement steel develops its strength prior to concrete strength decay, since in this case the tensile forcee in the confinement steel is directly related to the amount as well as the yield strength of steel (f_{yt}). It is also observed that higher cofinement pressure (i.e. higher $\rho_s f_{yt}$) is required for higher strength concretes to maintain the same level of deformability. Therefore the test data are re-evaluated based on the non-dimensional ratio $\rho_s f_{yt}/f_c'$, as shown in Table 8.17. Deformability of columns with normal-strength

Table 8.17   Effect of $\rho_s f_{yt}/f_c'$ on column deformability

Column no.	f_c' (psi)	ρ_s (%)	f_{yt} (ksi)	s/h	s_1/h	$v/\sqrt{f_c'}$ (psi)	$P/(A_g f_c')$ (%)	$\rho_s f_{yt}/f_c'$	μ
C9	4800	1.49	57	0.25	0.66	2.1	26	0.18	1.9
58	14500	1.26	124	0.24	0.73	9.4	35	0.11	2.0
C14	3300	0.95	34	0.17	0.72	3.8	0	0.10	2.0
57	14500	1.28	112	0.12	0.75	9.0	35	0.10	2.0
C45	5100	1.60	41	0.17	0.26	8.9	20	0.13	3.8
56	14500	1.28	163	0.24	0.25	8.7	35	0.14	3.3
C44	5260	1.60	41	0.17	0.26	8.4	32	0.13	2.8
54	14500	1.92	112	0.16	0.25	9.1	35	0.15	3.3
44	9670	2.00	121	0.18	0.25	9.0	31	0.25	6.7
45	12260	2.57	121	0.14	0.25	8.4	28	0.25	6.3

Note: Columns C9, C14, C44 and C45 are normal strength concrete columns, previously investigated by Saatcioglu[25]
1 in. = 25.4 mm; 1000 psi = 6.895 MPa

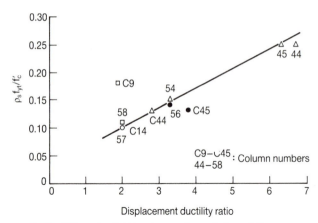

Fig. 8.42   Effect of $\rho_s f_{yt}/f_c'$ ratio on column deformability

concrete, previously evaluated by Saatcioglu,[26] is also included in Table 8.17 to extend the lower end of the strength range, and to relate deformability of high-strength concrete to that of normal-strength concrete for which extensive research data and knowledge already exist. Table 8.17 and Fig. 8.42 show the variation of column displacement ductility ratio with the $\rho_s f_{yt} f_c'$ ratio. Examination of the test data also indicates that higher strength concretes require proprtionately higher values of the product $\rho_s f_{yt}$. Higher values of this product can be achieved either by the use of higher volumetric ratio and/or higher strength of confinement steel. Figure 8.43 shows column pairs with constant $\rho_s f_{yt}/f_c'$ ratios, where the columns in a pair have distinctly different concrete strengths, and yet develop similar displacement ductility ratios. This implies that within the range of parameters considered in the test programs evaluated here, deformabilities usually expected of normal-strength concrete columns can be attained in high-strength concrete columns by maintaining the $\rho_s f_{yt}/f_c'$ ratio approximately constant.

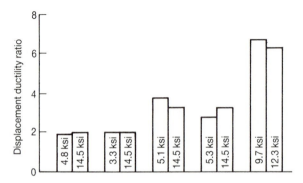

Fig. 8.43   Comparisons of displacement ductility ratios of columns with different concrete strengths but constant $\rho_s f_{yt}/fc'$ ratios

Table 8.18 Effect of tie spacing on concrete deformability

Column no.	f_{co}' (psi)	ρ_s (%)	f_{yt} (ksi)	s_1/h	s/h	$\varepsilon_{85}/\varepsilon_0$
26	8255	3.1	60	N/A	0.17	2.0
22	8447	3.4	60	N/A	0.06	3.3
27	8255	3.1	60	N/A	0.17	1.8
23	8447	3.4	60	N/A	0.06	2.7
24	6108	2.2	60	N/A	0.13	3.3
20	6191	2.3	60	N/A	0.09	4.9
25	6108	2.2	60	N/A	0.13	3.0
21	6191	2.3	60	N/A	0.09	4.4

1 in. = 25.4 mm; 1000 psi = 6.895 MPa

Spacing of the transverse reinforcement

The spacing of transverse reinforcement is one of the important parameters that affects the distribution of confinement pressure. Closer spacing
of transverse reinforcement increases uniformly of lateral pressure along
the column height and improves the effectiveness of confinement reinforcement. However, this improvement may not result in significantly
different column deformabilities unless the other confinement parameters
are favorable. For example, the volumetric ratio and/or yield strength steel
must be sufficiently high for the spacing to make a difference in column
deformability. Table 8.18. and Fig. 8.44 show the effect of tie spacing on
the strain ductility ratio of high-strength concrete.

Arrangement of transverse reinforcement

The arrangement of transverse reinforcement is another parameter that
affects the distribution of confinement pressure. The tension force that
develops in a transverse leg of a hoop or a crosstie is distributed on the side

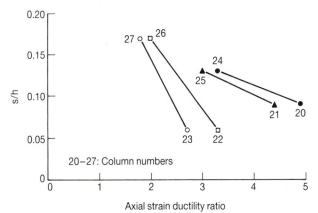

Fig. 8.44 Effect of tie spacing on concrete deformability

of the core concrete. If the transverse legs are closely spaced in the horizontal plane, the resulting confinement pressure becomes more uniform in this plane, improving effectiveness of the confinement reinforcement. This has been clearly shown for normal-strength concrete columns.[27] Tests on high-strength concrete columns with different reinforcement arrangements are scarce in the literature. Although the effect of transverse reinforcement arrangement is expected to be the same for high-strength concrete columns, the available data are not sufficient to illustrate this point clearly.

Section geometry and size

Circular spirals are more effective in confining concrete than rectilinear ties. The superiority of circular spirals comes from their geometric shape which produces uniform pressure around the circumference of the core. Rectilinear ties produce non-uniform pressure which peaks at loacations of transverse legs of tie steel. It was shown earlier for normal-strength concrete that full effectiveness of circular geometry can be achieved by rectilinear reinforcement if the tie arrangement includes close spacing of transverse legs in both longitudinal and transverse directions.[27] Table 8.19 provides a comparison of strain ductility ratios for high-strength concretes, confined by either circular spirals or square perimeter hoops. The results indicate that circular spirals produce higher deformabilities than square perimeter hoops.

Experimental data on the effect of size on the deformability of high-strength concrete are not sufficient. Most high-strength concrete columns tested in the past were small-scale specimens. However, tests conducted by Martinez[17] indicate that small specimens produce lower strength than specimens of larger sizes.

Parameters related to concrete type

Concrete strength

Strength and deformability of concrete are known to be inversely prop-

Table 8.19   Effect of section geometry

Column no.	f_{co}' (psi)	ρ_s (%)	f_{yt} (ksi)	s/h	s_1/h	Cross section	$\varepsilon_{85}/\varepsilon_0$
31	12660	3.2	90	0.17	N/A	Circular	5.2
34	13230	3.2	90	0.17	0.89	Square	3.0
33	10210	3.2	90	0.17	N/A	Circular	2.9
35	11880	3.2	90	0.17	0.89	Square	1.4

Note: Columns 33 and 35 are lightweight concrete columns
1 in. = 25.4 mm; 1000 psi = 6.895 MPa

Table 8.20 Effect of strength on concrete deformability

Column no.	f_{c0}' (psi)	ρ_s (%)	f_{yt} (ksi)	s/h	s_1/h	$\varepsilon_{85}/\varepsilon_0$
31	12660	3.2	90	0.17	N/A	5.2
32	14445	3.2	90	0.17	N/A	3.5
1	5500	1.6	60	0.33	N/A	2.8
3	9500	1.6	60	0.33	N/A	2.3
2	5500	3.2	60	0.17	N/A	7.2
4	9500	3.2	60	0.17	N/A	4.5
7	5900	2.1	198	0.34	N/A	5.1
8	8100	2.1	198	0.34	N/A	3.1
9	10000	2.1	198	0.34	N/A	1.8
5	6200	2.1	198	0.34	N/A	4.4
6	9000	2.1	198	0.34	N/A	2.5
14	7281	7.5	60	0.06	N/A	12.6
17	9958	7.7	60	0.06	N/A	4.1
15	7281	7.5	60	0.06	N/A	15.2
18	9958	7.7	60	0.06	N/A	4.3
16	7281	7.5	60	0.06	N/A	11.9
19	9958	7.7	60	0.06	N/A	4.0

1 in. = 25.4 mm; 1000 psi = 6.895 MPa

ortional. It is therefore reasonable to expect the deformability of unconfined high-strength concrete to decrease with increasing strength.
Table 8.20 includes test data obtained from concentrically tested confined
high-strength concrete columns. The data are arranged such that the strain
ductility of confined concrete with different unconfined concrete strengths
can be compared. The results show the expected trend, i.e. concrete
ductility decreases with increasing strength. This is also shown in Fig. 8.45.
A comparison between 5700 and 10,800 psi (40 MPa and 74 MPa) concretes indicates a 40% reduction in strain ductility ratio in the higher

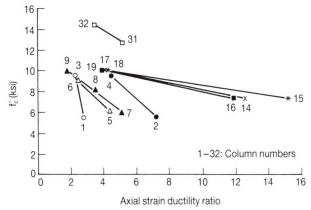

Fig. 8.45 Effect of strength on concrete deformability

Table 8.21   Effect of concrete strength on column deformability

Column no.	f_c' (psi)	ρ_s (%)	f_{yt} (ksi)	s/h	s_1/h	$v/\sqrt{f_c'}$ (psi)	$P/(A_g f_c')$ (%)	μ
46	5020	1.64	121	0.22	0.25	8.2	60	5.0
48	9670	2.25	121	0.16	0.25	9.6	57	3.5

1 in. = 25.4 mm; 1000 psi = 6.895 MPa

strength concrete. Similar observations can be made with regard to high-strength concrete columns tested under reversed cyclic loading. Table 8.21 and Fig. 8.46 illustrate the effect of unconfined concrete strength on the displacement ductility ratio. In all cases, ductility decreases with increasing concrete strength. This shows that the confinement steel requirements for high-strength columns must be more stringent than those for normal-strength concrete columns.

Unit weight of aggregate

The deformability of high-strength concrete decreases with the unit weight of the aggregate. Lightweight aggregate concrete shows a significantly lower ductility than normal weight concrete of the same strength. Martinez[17] reported that the confinement efficiency of circular spirals is approximately 60%, less for lightweight concrete than for normal weight concrete. Table 8.22 shows comparisons of strain ductility ratios for normal weight and lightweight confined high-strength concretes, indicating a significant drop in deformability with the use of lightweight aggregate in concrete.

Conclusions

The following conclusions can be made with respect to the deformability of high-strength concrete columns:

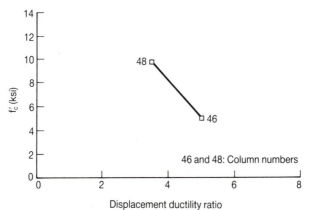

Fig. 8.46   Effect of concrete strength on column deformability

Table 8.22 Effect of unit weight of aggregate on concrete deformability

Column no.	f_{co}' (psi)	ρ_s (%)	f_{yt} (ksi)	s/h	s_1/h	Concrete type	$\varepsilon_{85}/\varepsilon_0$
31	12660	3.2	90	0.17	N/A	NWC	5.2
33	10210	3.2	90	0.17	N/A	LWC	2.9
34	13230	3.2	90	0.17	0.89	NWC	3.0
35	11880	3.2	90	0.17	0.89	LWC	1.4
11	7280	2.2	55	0.05	N/A	NWC	3.4
28	7990	2.7	55	0.04	N/A	LWC	2.1
12	7280	2.2	55	0.05	N/A	NWC	3.2
29	7990	2.7	55	0.04	N/A	LWC	2.0
13	7280	2.2	55	0.05	N/A	NWC	3.2
30	7990	2.7	55	0.04	N/A	LWC	1.9

1 in. = 25.4 mm; 1000 psi = 6.895 MPa

1 High-strength concrete columns are brittle members unless confined by proper reinforcement. For the same reinforcement arrangement, concrete strength and column deformability are inversely proportional.

2 High-strength concrete columns can be confined to develop deformabilities usually expected from seismic resistant elements. The parameters of confinement are essentially the same as those for normal strength concrete. However, the confinement requirements for high-strength concrete columns are more stringent than those for normal strength concrete columns.

3 The major parameters of confinement for high-strength concrete are the volumetric ratio, yield strength and spacing of confinement reinforcement. Experimental evidence points out that, with volumetric ratios and/or steel yield strengths higher than those used for normal strength concrete, it is possible to attain deformabilities usually achieved in normal strength concrete.

4 The use of high strength steel as confinement reinforcement appears to be an effective way of increasing confinement pressure to the level needed for improved deformability of high-strength concrete.

5 There is a strong indication that the extra confinement pressure required in high-strength concrete, relative to that required for normal strength concrete, is proportional to the difference in concrete strengths. Within the range of column tests investigated, columns with distinctly different concrete strengths show approximately the same displacement ductility ratio, if the ratio $\rho_s f_{yt}/f_c'$ is kept constant.

6 Axial compression reduces column deformability under reversed cyclic loading.

7 Lightweight high-strength concrete columns show significantly less deformability than those observed in corresponding normal weight concrete columns.

8   More research is needed to establish ductility and confinement requirements for seismic resistant high-strength concrete columns.

8.4   Deformability of high-strength concrete beam-column joints

ACI-ASCE 352 (1985)[28] recommends that for joints which are part of the
primary system for resisting seismic lateral loads, the sum of the nominal
moment-strengths of the column sections above and below the joint
(ΣM_c), calculated using the axial load which gives the minimum column-
moment strength, should not be less than 1.4 times the sum of the nominal
moment strengths of the beam sections at the joint (ΣM_g). It may be noted
that in ACI 318-89,[3] the minimum required flexural strength ratio $\Sigma M_c/$
ΣM_g is 1.2, instead of 1.4.

ACI-ASCE 351 (1985)[28] further recommends that where rectangular
hoop and crosstie reinforcement is used to confine the concrete within a
joint, the center-to-center spacing between layers of transverse reinforcement should not exceed the least of one-quarter of the minimum column
dimension (h_c), six times the diameter of the longitudinal bars to be
restrained, or 6 in. (150 mm). It may again be noted that in ACI 318-89,[3]
the maximum spacing of transverse reinforcement is restricted to the
smaller of $h_c/4$ or 4 in. (100 mm).

The 352 recommendations[28] specify maximum allowable joint shear
stresses in the form of $\gamma\sqrt{f_c'}$, where the joint shear stress factor γ is a
function of the joint type (i.e., interior, exterior or corner). The primary
function of the joint transverse reinforcement is to provide confinement for
the concrete in the joint region. Thus, if the joint shear stresses are in
excess of the maximum allowable limits, the problem cannot be remedied
by providing additional transverse reinforcement.

The above recommendations are based on tests of normal-strength
concrete connections with compressive strength, f_c', ranging between 3500
and 5500 psi (24 and 38 MPa). Several experimental investigations have
recently been conducted to check the validity of the recommendations for
high-strength concrete beam-column joints, and are discussed below.

Ehsani *et al.*[29] conducted a study to investigate the effect of various shear
stress levels on beam-column connections constructed with high-strength
concrete, and to compare the results with those of subassemblies constructed with ordinary-strength concrete.

Four reinforced concrete beam-column subassemblies were constructed
with high-strength concrete. The results were compared with those from a
similar specimen constructed with ordinary-strength concrete and reported
by Ehsani and Wight[30] (Specimen 5 of Ehsani *et al.*[29] is identical to
Specimen 4b of Ehsani and Wight[30]). The configuration of the specimens
qualifies them as corner connections.[28] As shown in Fig. 8.47, a corner
connection is one in which the beam frames into only one side of the

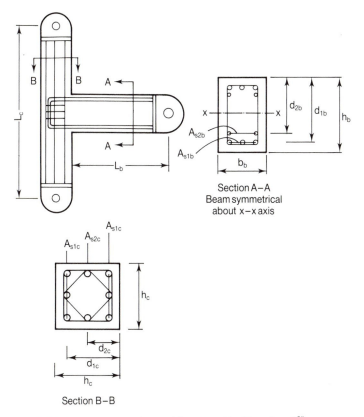

Section A–A
Beam symmetrical
about x–x axis

Section B–B

Fig. 8.47 Beam-column subassemblage tested by Ehsani *et al.*[29]

column and no spandrel beams are present. Reinforcing details for the specimens are presented in Table 8.23. Note that the beam and column of each specimen were properly reinforced to eliminate the possibility of shear failure of these elements. All specinents were cast flat rather than vertically as in actual construction.

The primary variable for the high-strength concrete specimens was the joint shear stress. According to ACI 352,[28] the maximum allowable joint shear stress for a corner connection constructed with ordinary-strength concrete is $12\sqrt{f_c'}$ psi ($1.0\sqrt{f_c'}$ MPa). As shown in Table 8.24, the joint shear stress for the high-strength specimens varied between $7.52\sqrt{f_c'}$ and $12.84\sqrt{f_c'}$ psi ($0.63\sqrt{f_c'}$ and $1.07\sqrt{f_c'}$ MPa). The joint shear stress in Specimen 5 was $12.55\sqrt{f_c'}$ psi ($1.04\sqrt{f_c'}$ MPa).

For the high-strength concrete specimens, the column-to-beam flexural strength ratio was held fairly constant between 1.7 and 1.9. The concrete compressive strength for these specimens was either 9380 or 9760 psi (64.7 or 67.3 MPa). This minor difference in concrete compressive strength was assumed to have no significant influence on the test results.

The joints of all specimens were reinforced with three layers of No. 4,

Table 8.23   Physical dimensions and properties of specimens[29]

	Specimen number				
	1	2	3	4	5
L_c, in.	136.0	136.0	136.0	136.0	84.0
L_b, in.	62.0	62.0	62.0	62.0	60.0
h_c, in.	13.4	13.4	11.8	11.8	11.8
d_{1c}, in.	11.4	11.4	9.8	9.8	9.6
d_{2c}, in.	6.7	6.7	5.9	5.9	5.9
A_{s1c}	2 No. 7 + 1 No. 6	2 No. 7 + 1 No. 6	2 No. 7 + 1 no. 6	2 No. 8 + 1 No. 7	4 No. 6
A_{s2c}	2 No. 6	2 No. 6	2 No. 6	2 No. 7	2 No. 6
h_b, in.	18.9	18.9	17.3	17.3	17.3
b_b, in.	11.8	11.8	10.2	10.2	10.2
d_{1b}, in.	16.9	16.9	15.4	15.4	15.4
d_{2b}, in.	15.0	15.0	13.4	13.4	13.4
A_{s1b}	2 No. 6 + 1 No. 5	3 No. 6	3 No. 6	3 No. 7	3 No. 7
A_{s2b}	2 No. 5	2 No. 6	2 No. 5	2 No. 5	3 No. 6

1 in. = 25.4 mm

Table 8.24   Design parameters of test specimens[29]

	Specimen number				
	1	2	3	4	5
Column axial load P, kips	30	76	86	73	50
Column balanced axial load, kips	498	513	362	319	267
$M_{n,col}$ at P, in.-kips	1638	1865	1582	1917	1478
$M_{n,beam}$ in.-kips	1729	2041	1663	2290	2101
M_R	1.89	1.83	1.90	1.67	1.41
Joint shear force V_j, kips	123	150	133	165	131
Joint shear stress v_j, psi	729	888	1027	1269	1008
f_c', psi	9380	9760	9380	9760	6470
Joint shear stress/$\sqrt{f_c'}$	7.52	8.99	10.61	12.84	12.55
h_b/column bar diameter	21.6	21.6	19.8	17.3	23.0
Required l_{dh}, in.	7.7	7.6	7.7	8.9	7.3
Provided l_{dh}, in.	10.8	10.8	9.2	9.3	8.6

1 in. = 25.4 mm; 1000 psi = 6.895 MPa; $1.0\sqrt{f_c'}$ psi = $0.083\sqrt{f_c'}$ MPa

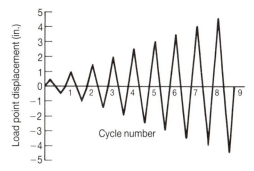

Fig. 8.48   Deformation sequence for reversed cyclic loading tests by Ehsani *et al.*[29]

Grade 60 hoops. As shown in Fig. 8.47, each layer of hoop consisted of a large square tie, enclosing all column longitudinal corner reinforcement, and a smaller square tie supporting the four intermediate longitudinal bars of the column. The area and the spacing of the hoops satisfied all recommended values.[28]

The selection of Specimen 5 for comparison with the high-strength concrete specimens was based on the fact that this specimen very closely satisfies all requirements of ACI 352.[28] The flexural strength ratio for Specimen 5 is 1.4. The joint shear stress for this specimen exceeds the maximum allowable value by only 4%. Other design parameters for this specimen, such as the spacing of the hoops and the development of longitudinal bars within the joint, satisfy or are very close to those recommended in ACI 352.[28] As a result, the performance of Specimen 5 can be interpreted as minimally acceptable by the joint ACI-ASCE Committee 352 criteria. This specimen can thus serve as a benchmark.

An axial load, less than 40% of the balanced axial load, was applied to each column and kept constant throughout the test. Displacement-controlled loads were applied to the free end of the beams, as shown in Fig. 8.48. Specimen 5 was loaded to 1.5 times the observed beam yield displacement during the first cycle of loading. The maximum displacement in each subsequent cycle of loading was increased by one-half of the observed yield displacement.

Plots of the applied shear versus the displacement at the free end of the beam for all specimens were presented in Ehsani *et al.*[29] The hysteresis diagrams indicated that Specimens 1, 2, 3 and 5, with large areas enclosed within each cycle of load-displacement curves, had the capacity to dissipate large amounts of energy. This was not true of Specimen 4, which had relatively smaller areas for each cycle.

Comparison of the loss of strength for the specimens is shown with the aid of Fig. 8.49. This figure shows the plot of percentage yield strength versus the displacement ductility. The yield load and displacement for each specimen were determined from strain gage data based on the yielding of the beam longitudinal reinforcement at the face of the column. The

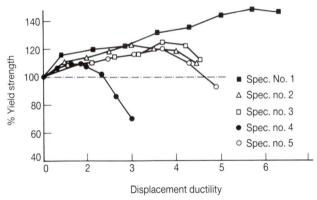

Fig. 8.49   Maximum load carried by each specimen at various displacement levels[29]

average of the maximum positive and negative loads during each cycle was divided by the yield strength to obtain percentage yield strength. The displacement ductility, plotted as the abscissa in Fig. 8.49, is defined as the displacement at the end of each cycle divided by the yield displacement of the specimen. It is clear that Specimens 1, 2, 3 and 5 performed very well by maintaining their yield strength for displacement ductilities of 4 or higher. For Specimen 4, however, the load-carrying capacity was sharply reduced to values below its yield strength after a displacement ductility of 2.5.

Within the limitations of the study by Ehsani *et al.*[29] the following conclusions could be drawn:

1   Properly detailed connections constructed with high-strength concrete exhibit ductile hysteretic response similar to those for ordinary-strength concrete connections.
2   The maximum permissible joint shear stress factor γ should probably be a function of the concrete compressive strength.
3   The mode of failure is determined by a combination of flexural strength ratio and joint shear stresses rather than by the flexural strength ratio alone.
4   High joint shear stresses significantly reduce the energy-absorption capability of subassemblies even in the presence of high flexural strength ratios.

Ehrani and Alameddine[31] investigated the effects of key variables on the behavior of high-strength reinforced concrete beam-column corner connections subjected to inelastic cyclic loading.

Twelve specimens were constructed and tested (Fig. 8.50 and Table 8.26). The configuration of the specimens (Fig. 8.50) qualified them as corner connections.[28] All specimens were cast flat rather than vertically as in actual construction. The variables studied were concrete compressive strength (8.1, 10.7, and 13.6 ksi or 55.8, 73.8, and 93.8 MPa), joint-shear stress (1100 and 1400 psi or 7.6 and 9.7 MPa), and the degree of joint

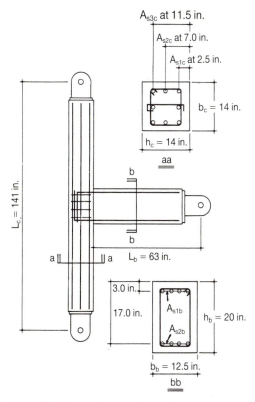

Fig. 8.50 Beam-column subassemblage tested by Ehsani and Alameddine[31]

confinement provided in the form of stirrups and crossties. For all specimens, the flexural strength ratio M_R, defined as the sum of the flexural capacities of the columns to that of the beam, was equal to 1.4, which is the minimum value allowed by the 253 recommendations.[28]

The 352 recommendations[28] limit the joint-shear stress in corner connections to $12\sqrt{f_c'}$ psi $(1.0\sqrt{f_c'}$ MPa). The low joint-shear stress investigated in this study is considered equivalent to $12\sqrt{8000} \approx 1100$ psi (7.6 MPa), while the high joint-shear stress is equivalent to $12\sqrt{8000} \approx 1400$ psi (9.7 MPa).

Table 8.25 Required and provided joint confinement[31]

f_c', psi	s, in., Eq. (21-3) Ref. [8.3]	s, in., Eq. (21-4) Ref. [8.3]	No. of sets of hoops, Eq. (21-3) Ref. [8.3]	No. of sets of hoops, Eq. (21-4) Ref. [8.3]	No. of sets of hoops used
8000	2.2	4.5	6	2	4 & 6
11,000	1.6	3.3	8	4	4 & 6
14,000	1.3	2.6	11	5	4 & 6

1 in. = 2.54 mm; 1000 psi = 6.895 MPa

Table 8.26 Physical dimensions and properties of specimens[31]

Specimen	LLs	LHs	HLs	HHs
L_c, in.	141	141	141	141
L_b, in.	63	63	63	63
h_c, in.	14	14	14	14
d_{1c}, in.	2.5	2.5	2.5	2.5
d_{2c}, in.	7.0	7.0	7.0	7.0
d_{3c}, in.	11.5	11.5	11.5	11.5
A_{s1c}	2 No. 8 & 1 No. 7	2 No. 8 & 1 No. 7	3 No. 8	3 No. 8
A_{s2c}	2 No. 7	2 No. 7	2 No. 8	2 No.8
A_{s3c}	2 No. 8 & 1 No. 7	2 No. 8 & 1 No. 7	3 No. 8	3 No. 8
h_b, in.	20	20	20	20
b_b, in.	12.5	12.5	12.5	12.5
d_{1b}, in.	3.00	3.00	3.00	3.00
d_{2b}, in.	17.00	17.00	17.00	17.00
A_{s1b}	4 No. 8	4 No. 8	4 No. 9	4 No. 9
A_{s2b}	4 No. 8	4 No. 8	4 No. 9	4 No. 9
No. of hoops	4	6	4	6
ρ_s	1.20	1.80	1.22	1.84
$h_b/d_{b,col}$	20	20	20	20
Required l_{dh}, in. for f_c' = 8000 psi	8.9	8.9	10.0	10.0
Provided l_{dh}, in.	10.5	10.5	10.5	10.5

1 in. = 25.4 mm; 1000 psi = 6.895 MPa

To ensure adequate confinement of the joint, the 352 recommendations[28] as well as ACI 318-89[3] requires that the total cross-sectional area of hoops and crossties A_{sh} be at least equal to that given by Eqs. (21-3) and (21-4) of ACI 318.[3]

The area of transverse reinforcement A_{sh} in each set included a Grade 60, No. 4, closed rectangular hoop with a 135-deg standard hook, in addition to a Grade 60, No. 4 crosstie, as shown in Fig. 8.50. The required and provided joint confinement for all specimens are given in Table 8.25. Satisfying these requirements would have resulted in joints congested with reinforcement and impractical to construct. The amount of joint transverse reinforcement was either low, corresponding to four layers of reinforce-

Table 8.27 Design parameters of test specimens[31]

Specimen	P_{col} psi	$P_{b,col}$ kips	$M_{n,col}$ in.-kips	$M_{pr,beam}$ in.-kips	$M_{n,beam}$ in.-kips	V_j, kips	v_j, psi
LL8	66	531	2119	3719	3027	210.7	1136
LH8	66	531	2119	3719	3027	210.7	1136
HL8	114	530	2546	4448	3637	268.4	1447
HH8	114	530	2546	4448	3637	268.4	1447
LL11	64	652	2183	3823	3118	209.8	1131
LH11	62	639	2157	3772	3081	210.2	1133
HL11	132	637	2692	4690	3845	266.3	1435
HH11	136	637	2710	4754	3872	266.3	1435
LL14	53	736	2178	3826	3112	209.8	1131
LH14	50	764	2178	3826	3112	209.8	1131
HL14	110	762	2689	4686	3842	266.7	1437
HH14	107	762	2681	4652	3830	267.0	1439

1 kip = 4.45 kN; 1 in.-kip = 0.133 kN-m

ment ($\rho_s \approx 1.2\%$), or high, corresponding to six layers of reinforcement ($\rho_s \approx 1.8\%$). The beam and the columns of the specimens were reinforced to eliminate any possibility of shear failure of these elements.

For all 12 specimens tested, the terminating beam longitudinal bars were hooked within the transverse reinforcement of the joint using 90-deg standard hooks. The provided development length from the critical secton defined by the 352 recommendations[28] was higher than that required by the recommendations.

An axial load, smaller than the balanced column load, was applied to the column (Table 8.27) and kept constant throughout the test. The magnitude of this load was determined from the column interaction diagram to satisfy a flexural strength ratio M_R of 1.4.

Reversed cyclic displacements were applied to the free end of the beam. Each test consisted of nine cycles of loading. The maximum displacement for the first cycle strarted as $\frac{1}{2}$ in. (13 mm) and was incremented by $\frac{1}{2}$ in. (13 mm) during each subsequent cycle. The final cycles of loading, corresponding to story drifts of 7%, are very unlikely to be experienced by any real structure.

With the flexural strength ratio M_R held constant at a minimum value of 1.4, the joint shear stress as well as the joint confinement level were both found to be key factors in achieving adequate strength and ductility of the joint. Specimens with low shear level and high joint confinement were able to develop the ultimate capacities in the beams. These same specimens had the least stiffness degradation and loss of load-carrying capacity at dis-

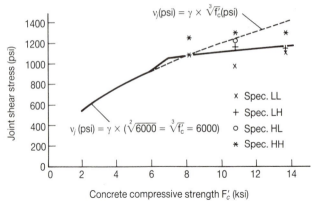

Fig. 8.51 Permissible shear stress in earthquake-resisting high-strength concrete beam-column joint[31]

placements beyond the yield displacement. Other specimens with high joint shear stresses and/or low joint confinement levels suffered greater strength loss and lower ductility.

Based on the test results, the maximum permissible joint shear stress was proposed to be modified as indicated in Fig. 8.51. In addition, modified joint confinement requirements for high-strength concrete were proposed.

Shin *et al.*[32] reported on the testing of one normal strength and ten high-strength concrete half-scale beam-column joint specimens. One of the high-strength specimens was tested under monotonic loading, all others were tested under reversed cyclic loading. According to the classification in the ACI-ASCE recommendations,[28] the configuration of the specimens qualified them as corner connections. The variables investigated were:

1   The concrete compressive strength ($f_c' = 4380$ and $11,380\,\text{psi}$ or 30.2 and 78.5 MPa).
2   Number of confining hoops within the joint core (3, 2, 1 and 0; corresponding spacing $= h_c/4$, $h_c/3$, $h_c/2$ and infinity, respectively).
3   Type of loading (cyclic, monotonic).
4   Column-to-beam flexural strength ratio ($M_R = \Sigma M_c/\Sigma M_g = 1.4$, 1.6, 1.8 and 2.0) and
5   Number of bent-up bars within the joint core (1 and 2).

Details of the specimens are given in Table 8.28 and Fig. 8.52. Each specimen was cast horizontally with one batch of concrete.

Figure 8.53 schematically shows the test setup. The column of each specimen was kept under a constant axial load equal to 40% of the balanced axial load. The vertical load on the beam was applied at 36 in. (980 mm) from the column face. All reversed cyclic loading tests were run under displacement control. The displacement sequence applied at the end of each beam is illustrated in Fig. 8.22. Δ_y is the vertical beam displacement at the point of loading, corresponding to yielding of the longitudinal beam

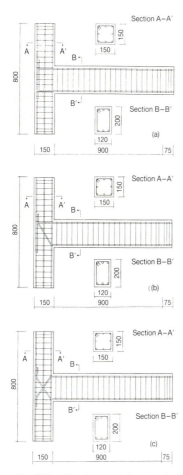

Fig. 8.52 Specimen reinforcing details [32] (2) all specimens except HJC3-R0-B1, HJCS-R0-B2, (b) HJC3-R0-B2, (c) HJCS-R0-B1

reinforcement at the column face. In the case of the specimen resisting monotonic loads, the test was conducted under load control until one-third of the expected maximum load was attained; a switch was made to displacement control thereafter.

In the normal strength concrete specimen NJC3-R0, many hairline flexural cracks formed uniformly along the beam at low displacement stages; failure occurred in the beam region between the column face and a section one-quarter of the beam depth away from the column face. In the corresponding high-strength concrete specimen HJC3-R0, wide cracks were concentrated at the beam-column joint face. Visually and otherwise, damage to the high-strength specimen was more severe than damage to the normal-strength specimen. Figure 8.54 graphically shows the effects of the other test variables on the failure mode.

A comparison between the hysteresis loops for NJC3-R0 ($f_c' = $

Table 8.28 Description of test specimens and selected test results[32]

Specimens	f'_c psi (MPa)	Loading type[a]	Number of hoops in core	M_R ($\Sigma M_c/\Sigma M_g$)	Number of bent-up bars in core	Column Section in. (mm)	Column Reinforcement	Beams Section in. (mm)	Beams Reinforcement	P_{max} kips (kN)	Shear stress constant[c] Theoretical	Shear stress constant[c] Experimental
NJC3-R0	4380 (30.2)	C	3	1.4	0	6 × 6 (150 × 150)	8 No. 3 (10 mm dia.)	4⅘ × 8 (120 × 120)	4 No. 4 (13 mm dia.) 2 No. 3 (10 mm dia.)	8.82 (39.23)	11.83	10.72
HJM3-R0	11,380 (78.5)	M	3	1.4	0	–	–	–	6 No. 4 (13 mm dia.)	13.78 (61.29)	8.21	6.35
HJC3-R0	–	C	3	1.4	0	–	–	–	–	10.03 (44.62)	8.21	7.86
HJC2-R0	–	C	2	1.4	0	–	–	–	–	12.68 (56.39)	8.21	6.80
HJC1-R0	–	C	1	1.4	0	–	–	–	–	10.36 (46.09)	8.21	7.73
HJC0-R0	–	C	0	1.4	0	–	–	–	–	9.56 (42.66)	8.21	8.05
HJC3-R1	–	C	3	1.6	0	–	–	–	4 No. 4 (13 mm dia.) 2 No. 3 (13 mm dia.)	10.77 (48.05)	7.24	5.78
HJC3-R2[b]	–	C	3	1.8	0	–	–	–	4 No. 3 (10 mm dia.) 2 No. 4 (13 mm dia.)	–		
HJC3-R3	–	C	3	2.0	0	–	–	–	6 No. 3 (10 mm dia.)	6.81 (30.40)	4.53	3.80

| HJC3-R0-B1 | – | C | 3 | 1.4 | 2 | – | – | – | – | 6 No. 4 (13 mm dia.) | 7.72 (34.32) | 8.21 | 8.79 |
| HJC3-R0-B2 | – | C | 3 | 1.4 | 1 | – | – | – | – | – | 10.03 (44.62) | 8.21 | 7.86 |

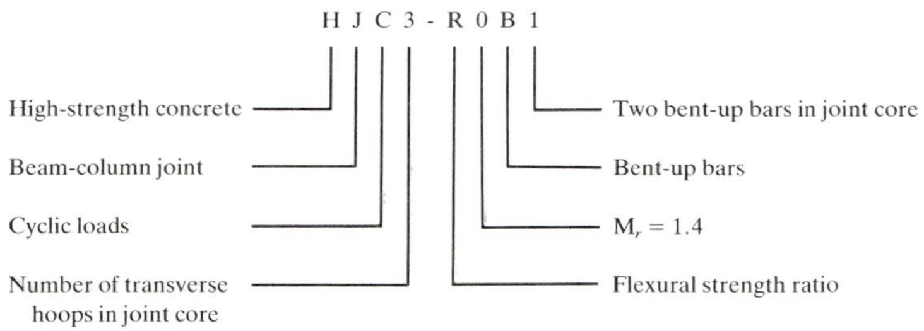

H J C 3 - R 0 B 1

- High-strength concrete
- Beam-column joint
- Cyclic loads
- Number of transverse hoops in joint core
- Two bent-up bars in joint core
- Bent-up bars
- $M_r = 1.4$
- Flexural strength ratio

(a) C—Cyclic load, M—Monotonic load
(b) HJC3-R2 failed prematurely
(c) $V_j/A_j\sqrt{f'_c}$

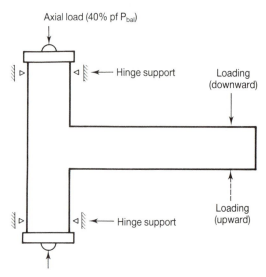

Fig. 8.53   Beam-column joint test setup[32]

430.2 MPa) and HJC3-R0 ($f_c' = 78.5$ MPa) showed distinctly pinched hysteresis loops for HJC3-R0, probably because failure of HCJ3-R0 was largely concentrated at the beam-column joint face. The amount of transverse confinement reinforcement within the joint core had little effect on hysteretic behavior of the specimens. However, at advanced deformation stages, the load resisting capacity decreased less for specimens with more confinement reinforcement within the joint. Specimen HJC3-R0-B1, with two bent-up bars within the joint core, exhibited distinctly pinched hysteresis loops.

The hysteresis loops were less pinched for HJC2-R0-B2, with one bent-up bar within the joint core. However, the hysteresis loops were more pinched for HJC3-R0-B2 than for HJC3-R0 with no bent-up bars. In HJC3-R0, damage was spread over the beam and the beam-column joint core; in HJC3-R0-B1, damage was concentrated between the beam-column joint face and the nearest layer of longitudinal column reinforcement. Energy dissipation capacity, as defined in Fig. 8.55, is plotted in Fig. 8.56 for specimens with different confinement reinforcements within the joint core, different bent-up bars within the joint core, different beam-colum flexural strength ratios, respectively. In all these figures, the energy dissipation capacities of high-strength concrete specimens should be compared to that of the normal-strength concrete specimen, NJC3-R0. Figure 8.56(a) shows that only HJC2-R0 exhibited higher energy dissipation capacity than NJC3-R0. It should be noted, however, that the column of HJC2-R0 was severely damaged through load cycles 10–15. Figure 8.56(b) shows that with the addition of bent-up bars within the joint core, the energy dissipation capacities of high-strength concrete specimens approach that of the comparable normal-strength concrete specimen.

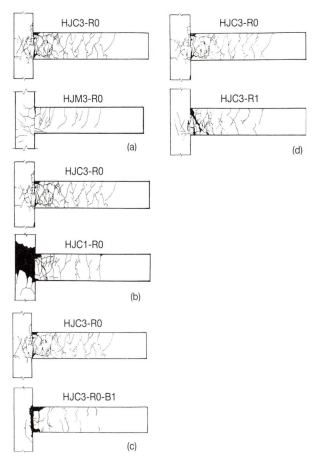

Fig. 8.54 Failure modes of specimens with (a) different loading patterns, (b) different confinement reinforcement, (c) different bent-up bars, and (d) different flexural strength ratios[32]

Figure 8.56(c) shows a definite correlation between energy dissipation capacity and column-beam flexural strength ratio, and indicates that a ratio of at least 1.6 was needed for the energy dissipation capacity of a high-strength concrete beam-column joint specimen to match that of a comparable normal-strength concrete specimen.

It should be noted that the column-to-beam flexural strength ratio, M_R, was varied in this investigation by varying the amount of flexural reinforcement in the beams. As can be seen from Table 8.28, this meant that as M_R increased, the intensity of shear stress within the joint decreased. Thus, what appears to be the effect of increasing M_R on energy dissipation capacity may very well have been the effect of a decreasing shear stress level within the joint.

The following observations could be made from the investigation by Shin et al.[32]

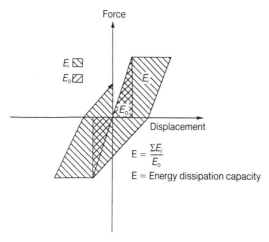

Fig. 8.55   Definition of energy dissipation capacity, as plotted in Fig. 8.56[32]

1   In high-strength concrete beam-column joints subjected to reversed cyclic loading, combined bending and shear was the final failure mode. Under monotonic loading, the final failure was dominated by bending.
2   In high-strength concrete beam-column joints designed with wider transverse reinforcement spacing than recommended by ACI-ASCE 352,[28] failure occurred in the beam-column joint core, and extended throughout the panel zone, into the upper and lower columns.
3   The behavior of high-strength concrete beam-column joint specimens with bent-up bars within the joint core was shear dominated; the hysteretic load-displacement loops were severely pinched due to stress concentration at the beam-column joint face. Thus it appeared desirable to avoid the use of bent-up bars.
4   Energy dissipation capacity increased with increasing values of M_R, beam-to-column flexural strength ratios of joints. This may have been the effect of simultaneous decreases in joint shear stress levels.

Conclusions

The following general conclusions can be drawn with respect to the deformability of high-strength concrete beam-column joints.

1   Properly detailed connections constructed with high-strength concrete exhibit ductile hysteretic response similar to those for ordinary-strength connections.
2   The maximum permissible joint shear stress factor γ should probably be a function of the concrete compressive strength.
3   For a given amount of transverse reinforcement within the joint, the mode of failure is determined by a combination of flexural strength ratio and joint shear stress rather than by the flexural strength ratio alone. High joint shear stresses significantly reduce the energy-

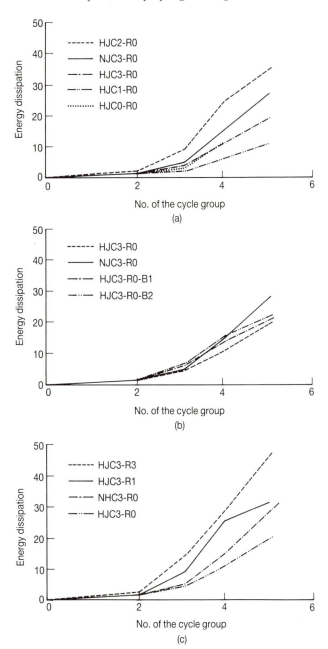

Fig. 8.56   Energy dissipation capacity, as affected by (a) confinement reinforcement within joints, (b) different bent-up bars within joints, and (c) different column-beam flexural strength ratios[32]

absorption capability of subassemblies even in the presence of high flexural strength ratios.

4   For a given column-to-beam flexural strength ratio, the joint shear

stress as well as the joint confinement level are key factors in achieving adequate strength and ductility of the joint. Specimens with low joint shear stresses and high joint confinement are able to develop the flexural strengths of the beams. These same specimens have the least stiffness degradation and loss of strength at post-yield displacements. Specimens with high joint shear stresses and/or low joint confinement levels suffer greater strength loss and lower ductility.

8.5 Application of high-strength concrete in regions of high seismicity

The application of high-strength concrete in highly seismic regions has lagged behind its application in regions of low seismicity. One of the primary reasons has been a concern with the inelastic deformability of high-strength concrete structural members under reversed cyclic loading of the type induced by earthquake excitation. This section discusses the current state of application of high-strength concrete (with specified compression strength in excess of 6000 psi or 40 MPa) in buildings across the USA, including major west coast cities.

As late as the early 1950s, the tallest concrete buildings were in the 20-story height range. By 1975, the 74-story high Water Tower Place, till recently the tallest concrete building in the world, had already been constructed. This virtual revolution within a very short time span was made possible by a number of factors, the most important amongst which were the availability of: new, improved construction methods; bigger cranes; high-strength materials; innovative structural systems; and high-storage, high-speed computer hardware plus the corresponding software that gave the structural engineer unprecedented analytical capabilities. It is futile to speculate which of the factors was more or less important than the others; all of them contributed to the dramatic growth in height of concrete buildings.

Figure 8.57 shows a series of nine concrete buildings, each of which, with the exception of Two Prudential Plaza, was the tallest concrete building in the world at the time of its completion. It is clear that the growth in the height of concrete buildings has gone hand-in-hand with the availability of higher and higher strength concretes.

Almost incredibly, seven of the nine record-setting buildings are located in Chicago, a city that in many ways has pioneered the evolution of high-strength concrete technology. However, very recently there has been an impressive spread in the availability of ultra-high-strength concrete (with specified compression strength in excess of 10,000 psi or 70 MPa). Figure 8.58 shows that 12,000 psi (80 MPa) of higher-strength concrete has been used in the last three or four years in Atlanta, Cleveland, Minneapolis, New York, and most significantly, Seattle which is in Uniform Building Code[33] Seismic Zone 3.

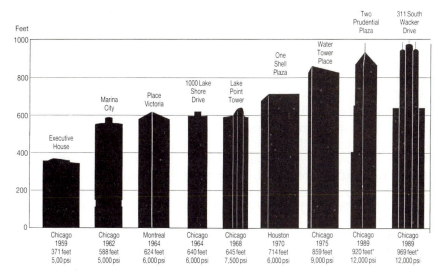

Fig. 8.57   High-strength concrete in high-rise construction

In fact, the highest concrete strength ever used in a building has been 19,000 psi (130 MPa) in the composite columns of Seattle's 62-story, 759-ft high Two Union Square (Skilling Ward Magnusson Barkshire Inc., Structural Engineers). The strength was obtained by use of: what may be a

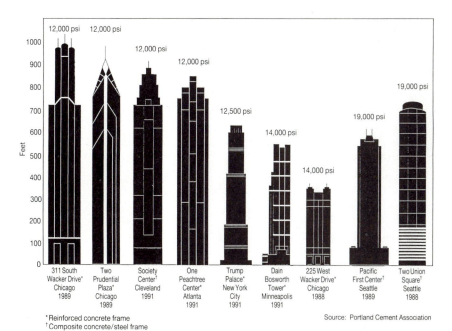

Fig. 8.58   Ultra-high-strength concrete shapes new skylines

record low water cementitious ratio of 0.22 (this is the single most important factor in increasing strength and reducing shrinkage and creep); the strongest of available cements; a superplasticizer which reduces the need for water, and provides the necessary workability; a very high cement content; a very strong, small (3/8 in. or 10 mm), round glacial aggregate available locally; silica fume (increasing strength by about 25%); a design strength obtained at 56 rather than the usual 28 days; and an extraordinarily thorough quality assurance program. The 19,000 psi (130 MPa) strength was the byproduct of the design requirement for an extremely high modulus of elasticity of 7.2 million psi (49,650 MPa). The stiffness was desired in order to meet the occupant-comfort criterion for the completed building. The same concrete strength was later used by Skilling Ward in the composite columns of the shorter 44-story Pacific First Center (Fig. 8.59).

Concrete with strengths of 95,000 psi (65 MPa) concrete has been used by Skilling Ward at 600 California in San Francisco, and at 1300 Clay in Oakland, both composite buildings. The Watry Design Group has used 8000 psi (55 MPa) concrete in several all-concrete Bay Area buildings, including the 19-story Fillmore Building (Fig. 8.60).

The spread in the use of high-strength concrete in Southern California has been hampered by the City of Los Angeles Code provision restricting the strength of concrete to a maximum of 6000 psi (40 MPa). Even then,

Fig. 8.59 44-story Pacific First Center in Seattle using 19,000 psi (130 MPa) concrete [photo courtesy of Skilling Ward Magnusson Barkshire Inc., structural engineers]

Fig. 8.60 19-story Fillmore Building in San Francisco using 8000 psi (55 MPa) concrete [photo courtesy of Watry Design Group, structural engineers]

concrete strength in excess of 6000 psi has been used in the Great American Plaza office-hotel-garage complex in San Diego, a 14-story residential building at 5th and Ash in San Diego, and in the 22-story Pacific Regent (senior citizen housing) at LaJolla.

The three biggest advantages of high-strength concrete that make its use attractive in high-rise buildings are that it provides more

- strength/unit cost
- strength/unit weight
- stiffness/unit cost

than most other building materials including normal weight concrete.

Commercially available 14,000 psi (96 MPa) concrete costs significantly less in dollars per cu. yd. (m^3) than $3\frac{1}{2}$ times the price of 4000 psi (27 MPa) concrete. In fact, the unit price goes up relatively little as concrete strength increases from 4000 to 10,000 psi (27 to 70 MPa). Thus, high-strength concrete gives the user more strength per dollar.

The unit weight of concrete goes up only insignificantly as concrete strength increases from moderate to very high levels. Thus, more strength per unit weight is obtained, which can be a significant advantage for construction in high seismic regions, since earthquake induced forces are directly proportional to mass.

The modulus of elasticity of concrete remains proportional to the square root of the compressive strength of concrete at the age of loading even for high-strength concrete.[34] The user thus obtains a higher stiffness per unit weight and unit cost. Indeed, it is quite common for a structural engineer to consider and specify high-strength concrete for its stiffness, rather than for its strength.

It is important to mention that the specific creep (ultimate creep strain per unit of sustained stress) of concrete decreases significantly as the concrete strength increases. The most recent verification of this is available in Bjerkeli *et al.*[18] This is indeed a fortunate coincidence without which the application of high strength concrete in highrise buildings would have been seriously hampered. Because of the lower specific creep, high-strength concrete columns with their high stress levels suffer no more total shortening than normal-strength concrete columns with their lower strength levels. Otherwise, the problem of differential shortening of vertical elements within highrise buildings would have been aggravated by the use of high-strength concrete in columns.

The use of high-strength concrete in taller buildings (and, of course in other applications not discussed here) is bound to increase across the world in the ensuing years, because of the very significant advantages discussed here.

8.6 Summary

Information available in the literature on the deformability of high-strength concrete beams, columns and beam-joints is reviewed in three separate sections of this chapter. At the end of each of those sections, relevant practical conclusions are drawn. The last section is devoted to the application of high-strength concrete in regions of high seismicity which has been on the increase in recent years. The latter trend should continue into the future.

References

1 Khaloo, A.R. and Ahmad, S.H. (1988) Behaviour of normal and high strength concrete under combined compression-shear loading. *ACI Materials Journal* **85**, No. 6, November–December, 551–9.
2 Leslie, K.E., Rajagopalan, K.S. and Everard, N.J. (1976) Flexural behavior of high-strength concrete beams. *ACI Journal*, Proceedings **73**, No. 9, September, 517–21.
3 ACI Committee 318 (1989) *Building code requirements for reinforced concrete and commentary*, ACI 318-83, ACI 318-89 and ACI 318 R-89. American Concrete Institute, Detroit, MI, 1983.
4 Pastor, J.A., Nilson, A.H. and Slate, F.O. (1984) Behavior of high-strength concrete beams, *Research Report 84-3*. Department of Structural Engineering, Cornell University, Ithaca, New York, February, 311 pp.
5 Tognon, G., Ursella, P. and Coppetti, G. (1980) Design and properties of

concretes with strength over 1500 kgf/cm². *ACI Journal*, Proceedings **77**, No. 3, May–June, 171–8.

6 Swartz, S.E., Nikaeen, A., Narayan Babu, H.D., Periyakaruppan, N. and Refai, T.M.E. (1985) Structural bending properties of higher strength concrete, *Special Publication 87*. American Concrete Institute, Detroit, MI, 147–78.

7 Shin, S.-W. (1986) 'Flexural Behavior including Ductility of Ultra-High-Strength Concrete Members', *Ph.D. Thesis*, University of Illinois at Chicago, Chicago, IL, 232 pp.

8 Shin, S.-W., Ghosh, S.K. and Moreno, J. (1989) Flexural ductility of ultra-high-strength concrete members. *ACI Journal*, Proceedings **86**, No. 4, July–August, 394–400.

9 Shin, S.-W., Kamara, M. and Ghosh, S.K. (1990) Flexural ductility, strength prediction and hysteretic behavior of ultra-high-strength concrete members, *Special Publication 121*. American Concrete Institute, Detroit, MI, 239–64.

10 Kamara, M.E. (1988) 'Flexural Behavior including Ductility of Ultra-High-Strength Concrete Members Subjected to Reversed Cyclic Loading', *Ph.D. Thesis*, University of Illinois at Chicago, Chicago, IL, 258 pp.

11 Ahmad, S.H. and Barker, R. (1991) Flexural behavior of reinforced high-strength lightweight concrete beams. *ACI Structural Journal*, **88**, No. 1, January–February, 69–77.

12 Ahmad, S.H. and Batts, J. (1991) Flexural behavior of doubly reinforced high-strength lightweight concrete beams with web reinforcement. *ACI Structural Journal*, **88**, No. 3, May–June, 351–8.

13 Ghosh, S.K., Narielwala, D.P., Shin, S.W. and Moreno, J. (1991) Flexural behavior including ductility of high-strength lightweight concrete members under reversed cyclic loading, Presented at the International Symposium on Performance of Structural Lightweight Concrete, ACI Fall Convention, Dallas, TX, November (to be published in an *ACI Special Publication*).

14 Razvi, S.R. and Saatcioglu, M. (1992) Confinement and deformability of high-strength concrete columns, Submitted for publication, *ACI Structural Journal*.

15 Ahmad, S.H. and Shah, S.P. (1982) Stress-strain curves of concrete confined by spiral reinforcement. *ACI Journal*, **79**, No. 6, November–December, 484–90.

16 Muguruma, H., Watanabe, F., Iwashimizu, T. and Mitsueda, R. (1983) Ductility improvement of high strength concrete by lateral confinement. *Transactions of the Japan Concrete Institute*, **5**, 403–10.

17 Martinez, S., Nilson, A.H. and Slate, F.O. (1984) Spirally reinforced high strength concrete columns. *ACI Journal*, **81**, No. 5, September–October, 431–42.

18 Bjerkeli, L., Tomaszewicz, A. and Jensen, J.J. (1990) Deformation properties and ductility of high strength concrete, *Special Publication 121*. American Concrete Institute, Detroit, MI, 215–38.

19 Abdel-Fattah, H. and Ahmad, S.H. (1989) Behavior of hoop confined high strength concrete under axial and shear loads. *ACI Structural Journal*, **86**, November–December, 652–9.

20 Chung, H., Hayashi, S. and Kokusho, S. (1990) Reinforced high strength columns subjected to axial forces, bending moments, and shear forces. *Transactions of the Japan Concrete Institute*, **2**, 335–42.

21 Muguruma, H., Watanabe, F. and Komuro, T. (1990) Ductility improvement of high strength concrete columns with lateral confinement, *Special Publication 121*. American Concrete Institute, Detroit, MI, 47–60.

22 Sugano, S., Nagashima, T., Kimura, H., Tamura, A. and Ichikawa, A. (1990) Experimental studies on seismic behavior of reinforced concrete members of

high strength concrete, *Special Publication 121*. American Concrete Institute, Detroit, MI, 61–87.

23   Kabeyasawa, T., Li, K.N. and Huang, K. (1990) Experimental study on strength and deformability of ultrahigh strength concrete columns. *Transactions of the Japan Concrete Institute*, **12**, 315–22.

24   Sakai, Y., Hibi, J., Otani, S. and Aoyama, H. (1990) Experimental study on flexural behavior of reinforced concrete columns using high-strength concrete. *Transactions of the Japan Concrete Institute*, **12**, 323–30.

25   Watanabe, F., Muguruma, H., Matsutani, T. and Sanda, D. (1987) Utilization of high strength concrete for reinforced concrete high-rise buildings in seismic area, *Proceedings of the Symposium* on Utilization of High Strength Concrete, Tapir, Publisher, Stavangar, Norway, 655–66.

26   Saatcioglu, M. (1991) Deformability of reinforced concrete columns, *Special Publication 127*. American Concrete Institute, Detroit, MI, 421–52.

27   Saatcioglu, M. and Razvi, S.R. (1992) Strength and ductility of confined concrete. *ASCE Journal of Structural Engineering*, **118**, No. 6, 1590–607.

28   (1985) Monolithic reinforced concrete structures, *ACI 352 R-85*. American Concrete Institute, Detroit, 19 pp.

29   Ehsani, M.R., Moussa, A.E. and Vallenilla, C.R. (1987) Comparison of inelastic behavior of reinforced ordinary- and high-strength concrete frames. *ACI Structural Journal*, **84**, No. 2, March–April, 161–9.

30   Ehsani, M.R. and Wight, J.K. (1985) Exterior reinforced concrete beam-to-column connections subjected to earthquake-type loading. *ACI Journal*, Proceedings **82**, No. 4, July–August, 492–9.

31   Ehsani, M.R. and Alameddine, F. (1991) Design recommendations for Type 2 high-strength reinforced concrete connections. *ACI Structural Journal*, **88**, No. 3, May–June, 277–91.

32   Shin, S.-W., Lee, L.-S. and Ghosh, S.K. (1992) High-strength concrete beam-column joints, Presented at the *10th World Conference on Earthquake Engineering*, Madrid, Spain, July.

33   (1991) *Uniform building code*. International Conference of Building Officials, Whittier, CA.

34   Russell, H.G. (1990) Shortening of high-strength concrete members, *Special Publication 121*. American Concrete Institute, Detroit, MI, 1–20.

9 Structural design considerations and applications

Henry G Russell

9.1 Introduction

The successful application of high-strength concrete requires the complete cooperation of the owner, architect, structural engineer, contractor, concrete supplier and testing laboratory. In locations where applications of high-strength concrete have been successfully accomplished, the construction team has worked together for their mutual benefit and for the benefit of the engineering community. In the case of buildings, the owner must be willing to allow the building to become a state-of-the-art structure. The architect must be willing to design a structure that will utilize the benefits of high-strength concrete. The structural engineer must have the knowledge and ability to adapt structural design concepts based on lower strength concretes for use with higher strength materials. At the same time, the designer must work within the boundaries of the codes and specifications. The contractor must be willing to work with different materials and must accept the need for a higher degree of quality control. The concrete supplier must be able to supply concrete of the specified strength. Finally, the quality control testing laboratory must have the capability to prepare the test specimens and to test them appropriately. This requires that all members of the team work together from the inception of the project until its completion. Where this approach has been adopted, the development and applications of high-strength concrete have been successful; higher strength materials have been utilized and taller buildings have been built.

As the development of high-strength concrete has continued, the definition of high-strength concrete in North America has changed. In the 1950s, a compressive strength of 5000 psi (34 MPa) was considered high strength. In the 1960s, commercial usage of 6000 and 7500 psi (41 and

52 MPa) concrete was achieved. In the early 1970s, 9000 psi (62 MPa) concrete was being used. In the 1980s, design strengths of 14,000 psi (97 MPa) were used for commercial applications in buildings. Strengths as high as 19,000 psi (131 MPa) have been used, although their commercial application has been limited to one geographic location. The primary applications of these higher strength concretes have been in the columns of high-rise buildings. However, there is an increasing interest in the use of higher strength concretes in long-span bridges and offshore structures. The following sections of this chapter describe structural design considerations, construction considerations and quality control aspects and some specific applications of high-strength concrete. Although the specific applications are predominantly in North America, there have been applications in other countries.[1-4]

9.2   Structural design considerations

Many of the design provisions that exist in the codes and standards of North America and Europe are based on experimental results obtained with conventional-strength concretes. As higher-strength concretes have become available, the applicability of the design provisions for higher-strength concrete has been questioned. This section highlights four areas where designers should give special consideration.[5]

Flexural strength

The ACI Building Code[6] allows the use of rectangular, trapezoidal, parabolic or other stress distribution in design provided that the predicted strength is in substantial agreement with the results of comprehensive tests. However, in most situations, it is convenient to utilize an equivalent rectangular compressive stress distribution. Based on published data,[7] it appears that, for under-reinforced beams, the present ACI methods can be used for concretes with compressive strengths up to 15,000 psi (103 MPa.) Additional information is required for concretes with compressive strengths in excess of 15,000 psi (103 MPa).

Shear strength

Recent tests[8] have indicated the need to modify the shear strength design provisions of the ACI Building Code.[6] Test results have indicated that, with the utilization of higher-strength concretes in reinforced concrete beams, the specified minimum amount of web reinforcement must be increased as the concrete compressive strength increases. This has been found necessary for concretes with compressive strengths in excess of

10,000 psi (69 MPa). This increase in the minimum amount of web reinforcement is needed to control the extent of shear cracking in the beams and to provide ductile behavior. The 1989 version of the ACI Building Code[6] contains provisions to achieve this.

Development length

Design provisions in the United States permit the use of shorter development lengths or anchorage lengths for reinforcing bars as the concrete compressive strength increases. However, due to a lack of experimental data, the ACI Building Code[6] was modified to limit the design provisions for development length to concrete with a compressive strength of less than 10,000 psi (69 MPa). Consequently, although a concrete with a compressive strength in excess of 10,000 psi (69 MPa) may be used, the design must be based on a concrete compressive strength of 10,000 psi (69 MPa). This limitation removes one of the advantages of utilizing higher-strength concretes. Currently, there does not seem to be any reason why the limitation should apply. However, data are needed to substantiate removal of the limitation.

Long-term deformations

The use of high-strength concretes in high-rise buildings requires that special attention be paid to the long-term length changes that occur in high-strength concrete members.[9] Long-term deformations result from creep and shinkage. In addition, instantaneous deformations occur whenever load is added to the building.

As with lower-strength concretes, creep deformations in a member depend on the creep properties of the concrete at age of loading, stress level in the concrete, size of member and amount of reinforcement. The creep per unit stress of high-strength concretes is less than the creep per unit stress of lower-strength concretes. This means that, at the same stress level, creep deformations will be less for higher-strength concretes. Alternatively, for concretes loaded to the same ratio of stress to strength, the creep deformations will be about the same irrespective of concrete strength.

Shrinkage of most high-strength concrete is about the same as that of lower-strength concretes and is more dependent on factors other than the strength level. Some admixtures are said to reduce shrinkage. However, the reduced shrinkage may be the result of lower water content in the mix rather than the use of a specific admixture.

Instantaneous deformations also constitute a major source of shortening in the lower story columns of high-rise buildings. Instantaneous deformations are primarily a function of the modulus of elasticity at the age of loading and can be calculated from the following equation:

$$\varepsilon = \frac{P}{A_c E_c + A_s E_s} \tag{9.1}$$

where A_c = area of concrete
  A_s = area of steel reinforcement
  E_c = modulus of elasticity of concrete at age of loading
  E_s = modulus of elasticity of steel reinforcement
  P = applied axial load.

The ACI Building Code[6] contains the following equation for calculation of the modulus of elasticity:

$$E_c = w_c^{1.5} 33 (f_c')^{1/2} \tag{9.2}$$

where w_c = unit weight of concrete in lb/cu ft
  f_c' = compressive strength of concrete as measured on 6×12-in. (152×305-mm) cylinders.

Equation (9.2) was developed by Pauw[10] on the basis of concretes with compressive strengths up to about 5500 psi (38 MPa). As additional data have become available on higher-strength concretes, various investigators have made comparisons between the equation and the data for higher-strength concretes. Based on published data, Martinez *et al.*[11] recommended a modified equation for the calculation of modulus of elasticity. This revised equation predicted a lower modulus of elasticity for higher-strength concretes when compared with Equation (9.2). However, more recent data published by Cook[12] indicate that Equation (9.2) underestimates the modulus of elasticity for the higher-strength concretes. It should be noted that the higher-strength concretes reported by Martinez *et al.* were obtained using a smaller aggregate size. However, Cook was able to produce higher-strength concretes with larger-size aggregates.

It should be recognized that Equation (9.2) was based on a statistical analysis of the available data. As such, there is considerable scatter in the relationship between modulus of elasticity and concrete compressive strength. Consequently, it is recommended that, when accurate calculations are required for high-strength concretes, the modulus of elasticity should be measured as part of the concrete mix design preparation. Alternatively, the engineer must assume that there is going to be some deviation from the values predicted by Equation (9.2) or any other selected equation.

The value of modulus of elasticity used in Equation (9.1) should be the modulus of elasticity of the concrete in the column. However, the modulus of elasticity is measured on smaller specimens. The question, therefore, arises whether the modulus of elasticity as measured on 6×12-in. (152×305-mm) concrete cylinders or other plain concrete specimens is applicable to large structural members. Currently, there are no published data on this topic related to high-strength concrete. However, Hester[4] and Cook[12] have extracted cores from large concrete members and measured their modulus of elasticity. For the strength levels used in his program,

Cook showed that the modulus of elasticity as measured on cores varied as the measured compressive strength varied. If the core strength was low, the modulus of elasticity was also low. Hester[4] also observed the same phenomena. However, in Hester's work, both strength and modulus of elasticity of cores were considerably lower than corresponding values measured on cylinders.

Despite the variations that exist in high-strength concrete and the lack of information about some of the properties, good correlation has been obtained between calculated and measured deformations on real structures.[13,14] These correlations have been conducted for structural members with concrete compressive strengths up to 10,000 psi (69 MPa). There is currently a need to extend these types of correlations to members with higher concrete compressive strengths.

In the design of 311 South Wacker Drive, special consideration was given to vertical shortening of different structural elements. In the design process, column loads were first calculated using a conventional analysis. The column loads were then used in the calculation of vertical shortening of column stacks. The calculated differential shortening was then utilized in subsequent analysis of the forces in the structural frame. If the differential movements were too large, concrete compressive strengths and percentages of reinforcement for the columns were revised to minimize the differential. This iterative analysis process was repeated several times. The calculated differential movements were then utilized to specify formwork chamber for the floor slabs. During construction differential movements were monitored. Very close agreement with design values was obtained.

9.3 Construction considerations

Most of the construction procedures used for high-strength concrete are similar to those for conventional concrete. However, high-strength concretes are less tolerant of errors than lower-strength concretes. Some special considerations are noted below.[5]

Vertical load transmission

The use of high-strength concrete in the columns of high-rise buildings, together with the use of lower-strength concrete in beams and slabs, gives rise to construction problems at the slab-column and beam-column intersections. The ACI Building Code[6] addressed this specific situation. The Code requires that when the specific compressive strength of the concrete in the column is more than 40% greater than the concrete strength specified for the floor system, transmission of column load through the floor system shall be provided by one of three alternatives.

In the first alternative, concrete of the strength specified for the column must be placed in the floor at the column location. The top surface of the concrete column must extend 2 ft (600 mm) into the slab from the face of the column. The second alternative requires that the design strength of the column through the floor system be based on the lower concrete strength. If necessary, additional vertical reinforcement through the intersection shall be provided. In the third alternative, when the column is supported laterally on all four sides by beams, a combination of strengths based on the concrete column strength and the slab concrete strength may be utilized.

In most instances, the designer and contractor will select the first alternative known as 'puddling'. Application of this procedure requires the placing of two different concrete mixes in the floor system. The lower-strength concrete must be placed while the higher-strength concrete is still plastic. Both concretes must be adequately vibrated to ensure that they are well integrated. This requires careful coordination of the concrete deliveries and the possible use of retarders. In some cases, additional inspection services will be required when this procedure is used. It is important that the higher-strength concrete in the floor of the region of the column be placed before the lower-strength concrete in the remainder of the floor. This procedure prevents accidental placing of the lower-strength concrete in the column area. The designer has responsibility to indicate on the drawings where the high- and low-strength concretes are to be placed. In some instances, designers have elected not to use this approach because of the difficulty of maintaining the required coordination. However, inspection of this procedure after the concrete has been placed and hardened is relatively easy. High-strength concrete contains higher cement contents and mineral admixtures. Consequently, the high-strength concretes tend to have a darker color than the lower-strength concretes. Therefore, location of the high-strength concrete can be observed visually. As higher and higher concrete strengths are being used in columns with no corresponding increase in the strength of concrete used in floor slabs, this design and construction provision is becoming more important.

At 311 South Wacker Drive, the contractor elected to use 9000 psi (62 MPa) concrete in some floor slabs to avoid puddling 12,000 psi (83 MPa) concrete. At the same time, the contractor was able to reduce the floor thickness by post-tensioning. This, in turn, reduced the dead load on the structure.

Heat of hydration

Heat development in cement-rich high-strength concretes can result in high internal temperatures. Field measurements on 30-in. (760-mm) thick columns have shown internal temperatures of 150 to 180 °F (66 to 82 °C). Thus, consideration must be given to minimizing thermal gradients as these can result in cracking. The most common solutions to this problem are to keep mixing temperatures as low as possible, use the lowest amount of cement needed to obtain the specified strength level, use mineral admixtures or low-heat-generation cement, and insulate forms as necessary to maintain a more uniform temperature distribution until concrete strengths are sufficient to resist thermally induced tensile stresses. Thermal blankets and expanded polystyrene insulation have been used to insulate the concrete, even in hot climates.

Consolidation and finishing

High-strength concretes are produced with low ratios of water-to-cementitious materials. They also have higher cement contents than conventional concretes and will contain mineral admixtures in the form of fly ash and/or silica fume. These combined factors would normally make high-strength concrete difficult to place. However, the use of high-range water reducers (HRWR) commonly known as superplasticizers has made high-strength concretes extremely workable and easy to consolidate. It is not unusual to have high-strength concretes with slumps as high as 10 in. (250 mm). Care is needed, however, to ensure that the HRWR is added at the optimum time and at the optimum rate.

With the use of silica fume in concrete, the mixtures have become stickier and difficulties have been encountered in obtaining an adequate surface finish. This is not important for the top surfaces of columns. However, it is very critical when high-strength concrete is used in beams or floor slabs.

Plastic shrinkage cracks

When high-strength concrete is used in floor slabs, plastic shrinkage cracks may result. The primary cause for plastic shrinkage cracks for freshly placed slabs is very rapid loss of moisture from the concrete caused by low humidity, high wind and/or high temperature. When moisture evaporates from the freshly placed concrete surface faster than it is replaced by bleedwater, the surface shrinks. With restraint, tensile stresses develop in the weak concrete, and the concrete cracks. With high-strength concrete and low water-cementitious ratios, the mixing water content is very low. Less water will generally mean less bleeding. If a concrete has the tendency not to bleed, then it will probably exhibit plastic shrinkage cracks if conditions of low humidity, high wind and/or high temperatures exist.

To eliminate plastic shrinkage cracks, several jobs have used fog nozzles to saturate the air above the slab surface and plastic sheeting to cover the surface between final finishing operations. While it may be desirable to increase bleeding for crack reduction, this may not be feasible due to other strength or durability requirements that the concrete must possess.

Plastic settlement cracks

Generally, plastic settlement cracking occurs after initial placement, vibration and finishing, as the concrete has a tendency to continue to consolidate or settle. During this time, the unhardened concrete may be locally restrained by reinforcing steel or formwork. The local restraint results in voids or cracks adjacent to the restraining element.

With the use of high-slump, high-strength concretes, there is a greater tendency for plastic settlement cracking. In many cases, this has been associated with reinforcing steel in columns. When this situation arises, high slump and low concrete increase the probability of cracking. Furthermore, plastic settlement cracking will increase with a marginal amount of vibration and with flexible forms.

These problems can be avoided by using the lowest practicable slump, preferably less than 4 in. (100 mm). The concrete should be adequately vibrated, the greatest concrete cover used (with the engineer's approval), and proper formwork design shuld be followed. In addition, it has been found that revibration (post-vibration or back-vibration) will reduce plastic settlement cracking and close cracks once they have formed. The concrete should be revibrated as late as possible before initial set and while the internal vibrator will sink under its own weight into the concrete and liquefy it momentarily. External or form vibrators can also be used for revibration to reduce plastic settlement cracking.

9.4   Quality control

Quality control measures for high-strength concrete are eessentially the same as for conventional strength concrete. However, as strengths increase, concrete becomes much less forgiving of improper sampling and testing procedures. In fact, inadequacies in sampling and testing procedures can generally only result in lower apparent concrete strengths. Marginal testing practices that have insignificant effects on results for conventional-strength concretes can give erroneous or erratic results for higher-strength concretes.

Testing procedures

As concrete strengths increase, the use of standard 6×12-in. (152×305-mm) cylinders becomes problematic for many laboratories in North America because test machine capabilities may be insufficient. However, since building code provisions are referenced to 6×12-in. (152×305-mm) cylinders, designers are naturally reluctant to accept test results from smaller specimens. A potential solution to this dilemma is to develop (by testing) correlation curves between larger and smaller specimens for the job concrete mixes. The smaller cylinders can then be used for quality control with direct correlation to equivalent larger specimen strengths. This approach was successfully used on the construction of Two Union Square in Seattle. It should also be noted that, in areas where the use of high-strength concrete is becoming more prevalent, testing laboratories are upgrading to higher-capacity machines.

Tests have shown that concrete specimens made in either cardboard, plastic or tin molds attain lower strengths than specimens made in steel molds. The Canadian Standards Association (CSA) requires that molds other than steel are acceptable, if documentation is available and the cylinders produced from non-steel molds have compressive strengths equivalent to those obtained using steel molds. The most common practice is to use plastic molds for lower-strength concretes. It is reasonable to continue this practice for high-strength concretes. If measured strengths come into question, the CSA correlation approach can be used to verify that the mold material is not causing the problem. Re-use of plastic molds should not be permitted for high-strength concretes.

For high-strength concrete, test strengths are particularly sensitive to specimen end conditions. The American Society for Testing and Materials (ASTM) C 39,[15] *Standard Test Method for Compressive Strengths of Cylindrical Concrete Specimens*, provides guidance on perpendicularity and planeness, and ASTM C 617,[16] *Standard Practice for Capping Cylindrical Concrete Specimens*, covers capping procedures. High-strength capping compounds should be used with a uniform thickness of $\frac{1}{8}$ in. (3 mm). As an alternative to capping, cylinders can be lapped to meet ASTM end-condition requirements, but this procedure is generally more costly. If lapping is used, it should be done a day or two after casting because lapping becomes more difficult as the concrete gains strength.

Compression testing machines should meet the requirements of ASTM C 39,[15] or other applicable standards. Careful attention should be paid to platen smoothness as small deviations can cause increased errors in measured strengths. It is recommended that machines be calibrated every six months rather than annually when used for high-strength concrete testing.

In-place strengths

Several national codes address the use of concrete core data when strengths of concrete quality control specimens do not exceed the specified strengths. For example, the ACI Building Code[6] allows concrete to be accepted if the average strength measured on the cores exceeds 85% of the specified strength and no single strength is less than 75% of the specified strength. The factors of 75 and 85% are based on lower-strength concretes and were determined from comparisons of core strengths with cylinder strengths. Data on concretes with compressive strengths up to 17,000 psi (83 MPa), have indicated that the factors are still valid. However, in at least two projects using concrete with specified strengths in excess of 12,000 psi (83 MPa), the concrete core strengths have been lower than 85% of the cylinder strengths. This has raised the question of the validity of the factors for high-strength concrete. Designers and contractors are cautioned to address this issue before construction begins.

9.5   High rise buildings

The development and availability of higher strength concretes in specific geographic locations and the desire to build taller concrete structures have been synonymous. As a result, growth in the usage of higher strength concretes and increases in the height of the tallest concrete building, have proceeded together. The availability of the higher strength concretes makes it economically feasible to achieve the structures. The increased costs of higher strength concrete materials and increased quality control are more than offset by the ability of the high-strength concrete columns to carry a higher load, to require less reinforcing steel, and to need less formwork. Schmidt and Hoffmann[17] were the first to publish data indicating that the most economical way to design a column was with the highest available strength concrete and the least amount of reinforcing steel. Similar data are shown in Fig. 9.1. This figure takes into account the higher costs of the higher strength concretes, the costs of the reinforcing steel and the cost of the formwork. As the figure illustrates, high-strength concrete with minimum reinforcing steel represents the most economical solution. Although today's unit costs are different from those used by Schmidt and Hoffman, the trend remains the same.

The use of high-strength concretes also permits the use of smaller size

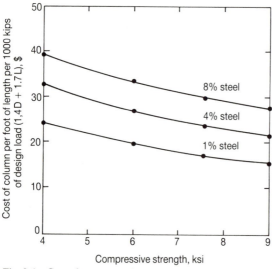

Fig. 9.1   Cost of concrete columns

columns. Since the columns in high-rise buildings are likely to be located in prime retail rentable floor space, minimization of the size of these columns is advantageous for the architect's layout as well as for the owner who wants to maximize rentable floor space. The desire to have smaller and smaller column sizes is facilitated by the availability of the high-strength concretes. In some structures, smaller size columns have also been required in order to minimize their interruption of the packing spaces in the lower stories of high-rise buildings. In the Richmond-Adelaide Centre in Toronto, Canada, the use of high-strength concrete columns enabled the architect to increase the use of the underground parking garage by approximately 30%. In a 15-story parking garage at 900 N. Michigan Avenue, Chicago, high-strength concrete was specified for the columns to reduce the lateral bending stiffness, yet provide sufficient capacity to carry the vertical load. Consequently, when the floor system was post-tensioned, the amount of post-tensioning force being resisted by the columns was minimized.

Although it was originally thought that high-strength concrete would only be used in the columns of very high-rise buildings, more recent applications have indicated that the same benefits can be achieved with shorter buildings.[18] As new applications are realized, it is anticipated that further economical benefits will be achieved.

Table 9.1 lists some of the more significant achievements in the applications of high-strength in high-rise buildings in North America. The table does not include every building that used concrete with a design compressive strength in excess of 6000 psi (41 MPa). However, it serves to illustrate the increase in height and concrete strength with time as well as the range of locations. Some of the structures are described in more detail in the following sections.

Lake Point Tower, Chicago

Lake Point Tower was constructed in Chicago in 1966–1967. Construction of this building represented the first major use of 7500 psi (52 MPa) concrete in a major commercial application. This required a carefully planned research and development program to optimize on all available materials in order to achieve the desired strength. As shown in Fig. 9.2, the building has three wings. Each wing is 65 ft (20 m) wide and extends 117 ft (36 m) from the center of the building. Total height of the structure above ground is 645 ft (197 m). The floors of the building are 8 in. (205 mm) thick flat plates reinforced with Grade 60 deformed bars. They contain light-weight aggregate concrete having a partial replacement of fines with sand to produce a density of 100 lb/ft^3 (1.60 Mg/m^3). The interior columns and core walls have diameters and thicknesses that vary over the height of the structure as listed in Table 9.2. The columns and walls were designed for normal weight concretes having compressive strengths of either 7500, 6000, 5000 or 3500 psi (52, 41, 34 or 24 MPa). Despite the availability of 7500 psi

Table 9.1   Buildings in North America with high-strength concrete

Building	Location	Year[1]	No. of stories	Maximum design strength	
				psi	MPa
One Shell Plaza	Houston	1968	52	6000L[2]	41L[2]
Pacific Park Plaza	Emeryville,	1983	30	6500	45
SE Financial Center	CA	1982	53	7000	48
Petrocanada Building	Miami	1982	34	7250	50
Lake Point Tower	Calgary	1966	70	7500	52
1130 S. Michigan Ave.	Chicago	–	–	7500	52
Texas Commerce Tower	Chicago	1981	75	7500	52
Helmsley Place Hotel	Houston	1978	53	8000	55
Trump Tower	New York	–	68	8000	55
City Center Project	New York	1981	52	8000	55
Larimar Place Condominiums	Minneapolis	1980	31	8000	55
NCNB Corporate Center	Denver	1990	68	8000	55
499 Park Avenue	Charlotte, NC	–	27	8500	59
Royal Bank Plaza	New York	1975	43	8800	61
Richmond-Adelaide Centre	Toronto	1978	33	8800	61
Midcontinental Plaza	Toronto	1972	50	9000	62
Frontier Towers	Chicago	1973	55	9000	62
Water Tower Place	Chicago	1975	76	9000	62
River Plaza	Chicago	1976	56	9000[3]	62[3]
Chicago Mercantile Exchange	Chicago	1982	40	9000[4]	62[4]
Columbia Center	Chicago	1983	76	9500	66
Interfirst Plaza	Seattle	1983	72	10,000	69
Scotia Plaza	Dallas	1987	68	10,150	70
311 S. Wacker Drive	Toronto	1988	70	12,000	83
One Peachtree Center	Chicago	1990	62	12,000	83
Bay Adelaide Center	Atlanta	1991	57	12,300	85
Society Tower	Toronto	1990	63	12,000	83
900 N. Michigan Annex	Cleveland	1986	15	14,000	97
Two Union Square	Chicago	1988	58	14,000[5]	97
255 W. Wacker Drive	Seattle	1988	31	14,000[6]	97
Dain Bosworth/ Niemann Marcus Plaza	Chicago Minneapolis	–	39	14,000	97

1.   Approximate year in which high-strength concrete was cast
2.   L = lightweight concrete
3.   Two experimental columns of 11,000 psi (76 MPa) strength were included
4.   Two experimental columns of 14,000 psi (97 MPa) strength were included
5.   19,000 psi (131 MPa) achieved because of high modulus of elasticity required by the designer
6.   One experimental column of 17,000 psi (117 MPa) strength was included

(52 MPa) concrete at the time Lake Point Tower was built, the use of reinforcement with a yield stress of 75 ksi (520 MPa) was also required.

Construction of the first story of the building began in June 1966 and the 70th story was completed in December 1967. Weather permitting, the building was built at the rate of one floor every three working days. The

Fig. 9.2   Lake Point Tower

building was extensively instrumented for purposes of verifying the time-dependent deformations in the columns.[13] In addition, an extensive program was used to develop basic information about the elastic properties, creep and shrinkage of the various concretes used in the building. Shrinkage of the high-strength concrete was found to be very similar to that of lower strength concretes. However, the specific creep in millionths/psi was found to be significantly lower than measured on the lower strength concretes. Although the maximum specified compressive strength concrete for the lower columns was 7500 psi (52 MPa), concrete strengths at an age of one year exceeded 10,000 psi (69 MPa).

Table 9.2   Properties of interior columns and core walls at Lake Point Tower

Story	Concrete design strength psi	Grade of reinforcement ksi	Column diameter in.	Core wall thickness in.
68–59	3,500	60	30	0, 12
58–44	5,000	60	30	12
43–35	5,000	60	36	12, 14
34–30	6,000	60	36	16
29–17	6,000	60	40	16, 18, 20
16–12	7,500	60	40	22
11–1	7,500	75	40	24, 30

Metric equivalents:
   1000 psi = 1 ksi = 6.895 MPa
   1 in. = 25.4 mm

Fig. 9.3   Water Tower Place

Water Tower Place, Chicago

Water Tower Place shown in Fig. 9.3 is a 76-story reinforced concrete building situated in Chicago, Illinois. When construction was completed in 1976, Water Tower Place was the tallest reinforced concrete building in the world and represented the most significant application of high-strength concrete. The building consists of a 13-story lower portion containing commercial and office space, and a 63-story tower that rises from one quadrant of the base structure. The lower portion of the building is 214×531 ft $(65 \times 162$ m). The tower structure measures 94×221 ft $(29 \times 67$ m) in plan.

In the lower portion of the structure, floor loads are carried by reinforced concrete flat slabs with bay sizes of 30×31 ft $(9 \times 9$ m). Light-weight concrete is used in all floor slabs above ground level. Normal weight concrete is used in the floor slabs of four basements below ground level. In the lower structure, floor loads are carried by flat slabs constructed with lightweight concrete. Horizontal forces on the building are resisted by the tubular design of the tower which consists of two equal tubes formed by the perimeter columns and a transverse wall across the shorter dimension of the rectangular tower.

Since the column framing system in the tower is different from the framing system in the lower portion, transfer between the two framing systems occurs at the 14th floor level. The transfer is achieved through the use of 15-ft deep (4.5-m) by 4-ft (1.2-m) wide reinforced concrete transfer

Table 9.3   Properties of columns and walls at Water Tower Place

Story	Concrete design strength psi	Grade of reinforcement ksi	Column size Interior* in.	Exterior in.	Wall thickness in.
75–73	4000**	60	18 × 24	25 × 44	16
73–72	4000**	60	18 × 24	10 × 48	16
72–63	4000	60	18 × 24	10 × 48	12
63–60	4000	60	18 × 30	10 × 48	12
60–55	5000	60	18 × 30	10 × 48	12
55–53	6000	60	18 × 30	10 × 48	12
53–50	6000	60	18 × 40	10 × 48	12
50–45	6000	60	18 × 40	14 × 48	12
45–40	6000	75	18 × 40	14 × 48	12
40–34	7500	75	18 × 40	14 × 48	12
34–33	7500	75	18 × 40	14 × 48	16
33–32	7500	75	18 × 54	16 × 44	16
32–31	7500	75	18 × 54	16 × 48	16
31–25	7500	75	18 × 54	16 × 48	14
25–15	9000	75	18 × 54	16 × 48	16
15–14	9000		Transfer girder		
14–B2	9000	75	48 × 48	48 × 48	12 and 18
B2–B4	9000	75	48 × 48	48 dia.	12 and 18

* Size of column supporting a full interior bay. Some individual columns change size at levels other than those shown
** 5000 psi concrete used in wall
Metric equivalents:
1 ksi = 6.89 MPa
1 in. = 25.4 mm

girders spanning 30 or 31 ft (9 m). These transfer girders contain concrete with a design compressive strength of 9000 psi (62 MPa).

Column sizes, wall thicknesses, and reinforcement details change continuously throughout the height of the structure. The highest specified concrete compressive strength which is used in the lower columns and walls is 9000 psi (62 MPa). The lowest strength concrete used at the top of the building is 4000 psi (28 MPa). Normal weight concrete is used in all columns and walls. Details of the floor levels where major changes in column and wall properties occur are given in Table 9.3. Concrete with a compressive strength of 9000 psi (62 MPa) is used from the fourth basement level through to story 25. Thereafter, 7500 psi (52 MPa) concrete is used through story 40. At the time of its construction, Water Tower Place represented the greatest use of concrete with a compressive strength in excess of 6000 psi (41 MPa). The complete building contains approximately 160,000 cu yd (122,000 cu m) of concrete, of which 90,000 cu yd (69,000 cu m) are lightweight. Water Tower Place, with a height of 859 ft (262 m), was the world's tallest reinforced concrete building until Two Prudential Plaza and 311 South Wacker Drive were completed. More importantly, however, the building was originally conceived as a structural steel frame building. Without the availability of high-strength concrete, Water Tower

Place would not have beeen built as a reinforced concrete structure.

An extensive program of field measurements of column shortening and laboratory programs to determine concrete properties was performed in connection with the construction of the building.[14] Field measurements for a total of 13 years have been reported.[19] Based on the observed creep and shrinkage deformations in laboratory specimens, the time-dependent shortening of the columns was calculated. The effects of loading history, column size, amount of reinforcement and concrete properties were taken into account. The calculated values were compared with those measured in the building and satisfactory agreement was obtained. In constructing the building, the designer had allowed for a shortening of $\frac{3}{8}$ in. (10 mm) per story height in determining joint sizes between the exterior marble panels. Measured shortening was approximately one half of this amount.

Texas Commerce Tower, Houston

The Texas Commerce Tower in United Energy Plaza, Houston, Texas is a 75-story composite steel and concrete building.[20,21] Although a composite structure, the building contains approximately 95,000 cu yd (73,000 cu m) of cast-in-place concrete and 8500 tons of reinforcing steel. Prior to the construction of the Texas Commerce Tower, the maximum concrete strength utilized in building construction in Houston was 6000 psi (42 MPa) lightweight concrete used for One Shell Plaza. However, most of the building construction in Houston utilized normal weight concrete with 5000 psi (35 MPa) as a maximum strength. From the owner's perspective, the use of these lower strength concretes would have resulted in extremely large and unacceptable column sizes. Consequently, extensive work was done to develop a higher strength concrete using limestone aggregate which had to be imported into the Houston area. This enabled the column sizes to be reduced to an acceptable level.

The structural frame of the building uses exterior columns consisting of steel erection columns and cast-in-place concrete columns. At the time of its construction, the Texas Commerce Tower was unique in that all the concrete placed on the project was pumped; the highest concrete place-ment being approximately 1000 ft (304 m) above street level. The 7500 psi (52 MPa) concrete had an average 28-day concrete compressive strength of 8146 psi (56 MPa) with a 56-day strength of 9005 psi (62 MPa).

In addition to providing the vertical load resisting system for the structure, high-strength concrete also provided increased stiffness for interstory drift. Since the modulus of elasticity of the high-strength concretes is greater than that for lower strength concretes, a structure with high-strength concrete and the same member sizes will have a greater stiffness compared to a similar structure with lower strength concretes. As a result, the maximum deflection of the building under design wind loads will be less with the high-strength concretes. Differential movements between the exterior composite frame and the interior columns of structural steel were also a major design consideration in this structure. The use of high-strength concrete, with a higher modulus of elasticity, reduced the actual shortening of the concrete columns. Consequently, there was less need for compensation between the interior steel frame and the exterior concrete columns. High-strength concrete was also advantageous from the standpoint of speed of construction. Construction of the concrete portion of the building proceeded at the rate of two floors per week, which at least equaled that for the structural steel construction. Concrete compressive strengths of 5200 psi (36 MPa) at three days for the 7500 psi (52 MPa) concrete allowed the columns to be stripped at an early age.

Interfirst Plaza, Dallas

Interfirst Plaza is a 72-story structure which utilizes composite steel and concrete columns.[22,23] The concrete design strength was 10,000 psi (69 MPa) for all columns in the building. High-strength concrete was specified to provide a high stiffness or high modulus of elasticity for the full height of the structure. This was needed because the height to width ratio of the building was 7.24. High-strength concrete was specified to provide maximum stiffness per unit dollar. It has been estimated that high-strength concrete in this structure provided six times as much stiffness per dollar compared to a structural steel framing system. The building has a height of 921 ft (281 m) and, at the time of its construction, was the tallest in Dallas. Vertical load in the building is carried by 16 columns which range in size from 8×8 ft (2.4×2.4 m) to 10×6 ft (3×1.8 m) and contain as much as 96 No. 18 bars. Even with 10,000 psi (69 MPa) concrete available, a large amount of reinforcing steel was still needed.

Two Prudential Plaza, Chicago

Two Prudential Plaza, shown in Fig. 9.4, is a 920-ft (280-m) tall building and, prior to the completion of 311 South Wacker Drive, had a short period in which it was the world's tallest reinforced concrete building.[24] The foundation system for the building consists of rock caissons containing 8000 psi (55 MPa) compressive strength concrete. The lower floor shear walls and columns contain 12,000 psi (83 MPa) concrete while the mid-range floors utilize 10,000 and 8000 psi (69 and 55 MPa) compressive strength concrete. The top floor columns contain 9000 psi concrete. Floor slabs and beams utilize 4000 psi (28 MPa) concrete. The cone-shaped portion of the top of the building uses a structural steel framing system.

Initially, the building was designed with a slip formed concrete core and structural steel perimeter columns. A value engineering study by the project team revealed that concrete was a more economical material in terms of both time and money. The concrete building was also more rigid against lateral loading than the structural steel equivalent.

Fig. 9.4   Two Prudential Plaza

311 South Wacker Drive, Chicago

311 South Wacker Drive is currently the tallest reinforced concrete building in the world.[24,25] The 70-story structure rises 969 ft (295 m) above street level. In plan, the tower design is trapezoidal up to the 51st floor where it becomes octagonal to offer additional corner officer space. During design, several alternatives were considered for the structural framing system.

These included reinforced concrete, composite steel frame with concrete shear wall, and steel structure with concrete shear wall core. The selected reinforced concrete framing system uses a continuous shear wall tied to the columns with beams at each floor level to form a diaphragm. Concrete with strengths up to 12,000 psi (83 MPa) are used in the lower columns of the building with strengths of 10,000 and 9000 psi (69 and 62 MPa) being used at the higher levels.

As illustrated in Fig. 9.5, the lower portions of the building consist of a six sided structure 225×135 ft (69×41 m). The building is then stepped back at different levels resulting in an octagonal tower that rises to the 70th level. The building is topped by a 65 ft (20 m) diameter, 75 ft (23 m) high ornamental drum. The vertical load carrying elements of the building consist of reinforced concrete columns and shear walls. The shear walls are located in the central area of the building. Lateral load is resisted by a combination of exterior columns and interior concrete shear walls.

Fig. 9.5   311 South Wacker Drive

311 South Wacker Drive contains approximately 110,000 cu yd (84,000 cu m) of reinforced concrete. A special feature of the construction was the decision to pump virtually all of the concrete.[19,26] Pumping rates as high as 100 cu yd per hour were achieved for the first 36 stories of the building. Another unique feature of the building was the decision to utilize 9000 psi (62 MPa) concrete in some of the floor slabs. This decision was made in order to avoid the more conventional option of puddling high-strength concrete into the floor slabs around the columns. In addition to the utilization of the 9000 psi (62 MPa) concrete, the floors were also post-tensioned to reduce their thickness and self weight of the structure. The mean strengths of the 9000, 10,000 and 12,000 psi (62, 69, 83 MPa) concrete at 90 days were 12,570, 11,240, and 13,900 psi (85.7, 77.5, 95.8 MPa). The high-strength concrete columns were generally stripped at an age of 16 to 18 hours and were wrapped in blankets for a period of five days to reduce temperature gradients. In general, a five-day cycle for construction was achieved for the building.

225 West Wacker Drive, Chicago

225 West Wacker Drive in Chicago is a 31-story building which utilizes 14,000 psi (97 MPa) compressive strength concrete in the columns of the basement levels and the first five stories above ground. Concrete strengths change at appropriate floor levels. Concrete strengths used in the building include 14,000, 12,000, 10,000 and 8000 psi (97, 83, 69 and 55 MPa).

Although 225 West Wacker is a relatively short building, high-strength concrete was used in the columns simply because it was still an economical means of satisfying the structural requirements.

Average compressive strengths measured at 56 days for the 14,000 psi(97 MPa) concrete were 16,140 psi (111 MPa) with a coefficient of variation of 5.8%. Maximum strength measured was 18,040 psi (124 MPa). In addition to the use of 14,000 psi (97 MPa) design strength, one experimental column was cast with 17,000 psi (117 MPa) compressive strength concrete. The casting of this column was part of ongoing research to address the practical problems of producing and placing higher strengths concretes in the Chicago area.[27] Measured compressive strengths at 56 days were in excess of 18,000 psi (124 MPa).

Two Union Square, Seattle

The highest strength concrete used in any large scale commercial application is the 19,000 psi (131 MPa) in the 58-story 720-ft (220-m) tall Two Union Square in Seattle.[28] The concrete compressive strength originally specified for the structure was 14,000 psi (97 MPa). However, the designer also wished to achieve a modulus of elasticity of 7.2×10^6 psi (50 GPa). To achieve the modulus of elasticity, testing showed that it was necessary to go to a concrete compressive strength of 19,000 psi (131 MPa). Some strengths in excess of 20,000 psi (138 MPa) were achieved for the concrete used in the structure.

The building frame consists of four 10-ft (3-m) diameter core columns and 14 perimeter columns that range in diameter from 3 to 4 ft (0.9 to 1.2 m). All concrete in all 18 columns including the top stories of the building has a 19,000 psi (131 MPa) compressive strength. The columns consists of reinforcing in the form of a permanent $\frac{5}{8}$ in. (16 mm) thick steel shells surrounding the perimeter. The steel skin is tied to the concrete using shear studs at 1 ft (300 mm) centers. The technique of using steel shells with an interior core of concrete was made to reduce the cost of the structure. Concrete was pumped into the steel shells of the columns from the bottom of each tier. No vibration of the concrete was found to be needed. Extensive quality control was also introduced on this project to ensure that all of the concrete would achieve the specified strengths.

Miglin-Beitler Building, Chicago

If construction is completed, the Miglin-Beitler Tower will establish a new record as the world's tallest building.[29,30] The tip of the tower of the structure will be 1,999 ft 6 in. (609.5 m) above street level. The structural frame consists of a cruciform tube structure to achieve structural efficiency, required dynamic performance, simplicity of construction and unobstructed integration of the structure into the leased office floor space. The structural solution consists of a 62 ft 6 in. (19 m) square concrete core with walls varying in thickness from 36 to 18 in. (0.9 to 0.5 m). On the exterior of the building, eight large columns extend outside the footprint of the building. These columns vary in dimension from 6.5 × 33 ft (1.9 × 10.0 m) at the base to 5.5 × 15 ft (1.7 × 4.6 m) at the middle to 4.5 × 30 ft (1.4 × 9.2 m) at the top. Link beams connect the four corners of the core to the eight fin columns at each core level. Link beams are made of reinforced concrete. In addition, 3 two-story deep outrigger walls located at three different levels in the building further connect the exterior columns to the concrete core. The floor system consists of a conventional structural steel composite system utilizing rolled steel sections, corrugated deck and concrete topping. During erection, the floor system is supported on light steel erection columns. The last structural component of the cruciform tube is the exterior virendeel truss which consists of horizontal spandrels and two vertical colummns at each of the 60 ft (18 m) faces on the four sides of the building. The foundation system of the project consists of caissons varying in diameter from 8 to 10 ft (2.4 to 3.0 m).

The availability of concrete with strengths up to 15,000 psi (103 MPa) was a critical factor in selecting concrete for the primary structural system. The columns of the building will contain concrete with compressive strengths of 14,000, 12,000 and 10,000 psi (97, 83 and 69 MPa). As illustrated in Fig. 9.6, construction of the Miglin-Beitler Tower will add a unique feature to the Chicago skyline. It is also likely to stand as the world's tallest building for a number of years.

Fig. 9.6 Miglin-Beitler Building

Fig. 9.7   Bridges utilizing high-strength concrete

9.6   Bridges

Without doubt, the largest use of high-strength concrete has been in the columns of buildings. However, high-strength concrete is receiving more and more attention for use in bridge structures. Three examples of the usage in North America are illustrated in Fig. 9.7. The tensile strength of high-strength concrete increases with compressive strength. This is beneficial in the design of prestressed concrete members such as bridge girders where the tensile strength may control the design. The reduced creep of high-strength concrete is also beneficial in reducing prestress losses in bridge girders. Consequently, utilization of high-strength concrete results in economies in prestressed concrete girders. Table 9.4 lists a selection of bridges that have utilized concrete strengths of 6000 psi (42 MPa) or greater.[7]

Prestressed concrete girders

Research studies have indicated the potential applications of high-strength concrete in solid section prestressed concrete girders.[31–33] In general, the studies have generally reached the same conclusions. For a given girder size, it is possible to increase the span capability by the utilization of the

Table 9.4 Bridges with high-strength concrete

Bridge	Location	Year	Maximum span ft	m	Maximum design strength psi	MPa
Willows Bridge	Toronto	1967	158	48	6000	41
Houston Ship Canal	Texas	1981	750	229	6000	41
San Diego to Coronado	California	1969	140	43	6000[1]	41[1]
Linn Cove Viaduct	North Carolina	1979	180	55	6000	41
Pasco-Kennewick	Washington	1978	981	299	6000	41
Coweman River Bridges	Washington	–	146	45	7000	48
East Huntington	W. VA to Ohio	1984	900	274	8000	55
Annacis Bridge	British Columbia	1986	1526	465	8000	55
Nitta Highway Bridge	Japan	1968	98	30	8500	59
Kaminoshima Highway Bridge	Japan	1970	282	86	8500	59
Joigny	France	1989	150	46	8700	60
Tower Road	Washington	–	161	49	9000	62
Esker Overhead	British Columbia	1990	164	50	9000	62
Fukamitso Highway Bridge	Japan	1974	85	26	10,000	69
Ootanabe Railway Bridge	Japan	1973	79	24	11,400	79
Akkagawa Railway Bridge	Japan	1976	150	46	11,400	79

1. Lightweight concrete

higher strength concretes. Tower Road Bridge shown in Fig. 9.7 is an example. For fixed girder dimensions and span lengths, it is possible to use fewer girders in a bridge when high-strength concretes are utilized. This also results in a lower unit cost for a given length structure. The utilization of longer span lengths for a multispan structure results in the need for less piers and foundations. This, in turn, reduces the cost of the substructure.

In general, prestressed concrete girders have been produced with compressive strengths in excess of 6000 psi (41 MPa) for many years. Although the design may be based on 6000 psi (41 MPa) at 28 days, strengths required for release of the prestressing result in 28-day strengths well in excess of the minimum specified. However, the higher strengths have not been utilized in design. It should be noted that the State of Washington in the United States has been utilizing high-strength concretes in prestressed concrete girders for many years.

Research has indicated that the advantages of utilizing higher and higher strength concretes do not continue forever. A point is reached at which the higher compressive strengths cannot be utilized.[33] It becomes impossible to induce sufficient prestressing force into the girders using existing prestressing strand dimensions and strengths to take advantage of the higher concrete strengths. This maximum occurs somewhere in the range of 8000 to 12,000 psi (55 to 83 MPa) depending upon the particular situation being analyzed.

East Huntington Cable stayed bridge

The East Huntington Bridge shown in Fig. 9.7 is a segmental prestressed concrete cable stayed bridge over the Ohio River between the states of West Virginia and Ohio.[34] Span lengths are 158, 300, 900 and 608 ft (48, 91, 274, 185 m). In a cable stayed bridge, the superstructure is a long compression member designed to resist the horizontal components of the forces in the cable stays. Also, in long span bridges, the superstructure weight is the predominant load for which the structure must be designed. Consequently, the selection of high-strength concrete results in a lighter structure to resist the longitudinal compressive forces. In the case of the East Huntington Bridge, high-strength concrete was also selected because of its improved durability and higher tensile strengths. Specified compressive strength of the superstructure elements was 8000 psi (55 MPa) at 28 days. Actual strengths averaged 9900 psi (68 MPa) at 28 days and 10,500 psi (72 MPa) at 90 days. In competitive bidding against a steel box girder alternate design, the concrete option was bid at 10 million US dollars less than the steel alternate.

Annacis cable stayed bridge

The Annacis Bridge also named the Alex Fraser Bridge crosses the Fraser River near Vancouver, British Columbia, Canada.[35,36] The cable stayed portion of the structure consists of five continuous spans with a center span of 1,526 ft (465 m). The composite superstructure consists of twin 83-in. (2.1-m) deep I-beams of constant depth, transverse floor beams that taper in depth, and a composite precast concrete deck with a cast-in-place overlay. The precast deck panels are typically 44 ft × 13 ft × 8.5 in. (13.5 × 4.0 × 0.215 m) with a specified compressive strength of 8000 psi (55 MPa) at 56 days. Composite action between the precast concrete elements and the steel framework is achieved through shear studs welded to the floor beam top flanges. Deck panels are integrated with the steel superstructure by cast-in-place strips of concrete. Required strength for the cast-in-place concrete was also specified at 8000 psi (55 MPa) at 56 days. A cost comparison between a composite concrete deck and an orthotropic steel deck indicated that the total cost of the structure with a concrete deck was substantially less.

Joigny Bridge

The Joigny Bridge is an experimental bridge designed to demonstrate the possibility of producing high-strength concrete bridges using existing batching plants and local aggregates.[37] The bridge is a three-span structure with span lengths of 111, 151 and 111 ft (34, 46 and 34 m). In cross section, the bridge is a double-T with a deck width of 49 ft (15 m). The bridge was designed based on a concrete with a characteristic compressive strength of

Fig. 9.8   Glomar Beaufort Sea 1

8700 psi (60 MPa) after 28 days and utilizes external longitudinal prestressing. This bridge was part of the French program to introduce high performance concretes into bridge construction.

9.7   Special applications

Offshore structures

Concretes with compressive strengths in excess of 6000 psi (41 MPa) have been used in offshore structures since the 1970s. High-strength concrete is important in offshore structures as a means to reduce self weight while providing strength and durability. Various applications have been described by Ronneberg[38] and CEB/FIP.[1]

In 1984, the Glomar Beaufort Sea 1, shown in Fig. 9.8, was placed in the Arctic.[39] This exploratory drilling structure contains about 12,000 cu yd (9200 cu m) of high-strength lightweight concrete with unit weights of about 112 lb/ft^3 (1.79 Mg/m^3) and 56-day compressive strengths of 9000 psi (62 MPa). The structure also contains about 6500 cu yd (5000 cu m) of high-strength normal weight concrete with unit weights of about 145 lb/ft^3 (2.32 Mg/m^3) and 56-day compressive strengths of about 10,000 psi (69 MPa).

Miscellaneous

Miscellaneous applications of high-strength concrete have been summarized by ACI Committee 363[7] and CEB/FIP.[1] For example:

Precast panels for dam	– 9000 psi (62 MPa)
Prestressed concrete poles	– 10,000 psi (69 MPa)
Grandstand roofs	– 7500 and 8850 psi (52 and 61 MPa)
Marine foundations	– 8000 psi (55 MPa)
Underwater bridge	– 9400 psi (65 MPa)

Grandstand elements	– 8700 psi (60 MPa)
Avalanche shelters	– 10,900 psi (75 MPa)
Piles	– 10,900 psi (75 MPa)

Acknowledgements

The author acknowledges information provided by the designers and contractors for the various structures described in this chapter. Portions of this chapter were written by S.H. Gebler and D.A. Whiting of Construction Technology Laboratories. Photographs were provided by the Portland Cement Association.

References

1 CEB/FIB Working Group on HSC (1990) *High strength concrete – state of the art*. The Institution of Structural Engineers, London.
2 Burnett, I.D. (1989) High-strength concrete in Melbourne, Australia. *Concrete International Design and Construction*. American Concrete Institute, **11**, No. 4, April, 17–25.
3 Holand, I., Helland, S., Jakobsen, B. and Lenschow, R. (1987) *Utilization of high strength concrete*, Proceedings, Symposium in Stavangar, Norway, June 15–18. Tapir, Trondheim, 688 pp.
4 Hester, W.T. (1990) *High-strength concrete, second international symposium*, Publication SP-121. American Concrete Institute, Detroit, 786 pp.
5 Russell, H.G. (1990) Use of high-strength concretes. *Building Research and Practice*, **18**, No. 3, May/June, 146–152.
6 ACI Committee 318 (1989) Building Code Requirements for Reinforced Concrete. American Concrete Institute, Detroit, 353 pp.
7 ACI Committee 363 (1984) State-of-the-art report on high strength concrete. *ACI Journal*, **81**, No. 4, July/August, 364–411.
8 Roller, J.J. and Russell, H.G. (1990) Shear strength of high-strength concrete beams with web reinforcement. *ACI Structural Journal*, **87**, No. 2, March–April, 191–8.
9 Russell, H.G. (1985) High-rise concrete buildings: shrinkage, creep and temperature effects, *Analysis and design of high-rise buildings*, Publication SP-97. American Concrete Institute, Detroit, 125–37.
10 Pauw, A. (1960) Static modulus of elasticity of concrete as affected by density. *Journal of the American Institute*, Proceedings **32**, No. 6, December, 679–787.
11 Martinez, S., Nelson, A.H. and Slate, F.O. (1982) *Spirally-reinforced high strength concrete columns*, Research Report No. 82-10. Department of Structural Engineering, Cornell University.
12 Cook, J.E. (1989) Research and application of 10,000psi (f_c') high-strength concrete, *Concrete International Design & Construction*. American Concrete Institute, **11**, No. 10, October, 67–75.
13 Pfeifer, D.W., Magura, D.D., Russell, H.G. and Corley, W.G. (1971) Time-dependent deformations in a 70-story structure, *Designing for effects of creep, shrinkage and temperature in concrete structures*, Publication SP27-7. American Concrete Institute, Detroit, 159–85.
14 Russell, H.G. and Corley, W.G. (1977) Time-dependent behavior of columns in Water Tower Place, *Douglas McHenry international symposium on concrete and concrete structures*, Publication SP-55-14. American Concrete Institute,

Detroit. Also printed as PCA Research and Development Bulletin RD052.01B, Portland Cement Association, Skokie, Illinois, 10 pp.

15 *Standard test method for compressive strengths of cylindrical concrete specimens*, ASTM C 39. American Society for Testing and Materials, Philadelphia.

16 *Standard practice for capping cylindrical concrete specimens*, ASTM C 617. American Society for Testing and Materials, Philadelphia.

17 Schmidt, W. and Hoffman, E.S. (1975) 9000-psi concrete – why?, why not? *Civil Engineering*, ASCE, **45**, No. 5, May, 52–5.

18 Giraldi, A. (1989) High-strength concrete in Washington, D.C. *Concrete International Design and Construction*. American Concrete Institute, **11**, No. 3, March, 52–5.

19 Russell, H.G. and Larson, S.C. (1989) Thirteen years of deformations in Water Tower Place. *ACI Structural Journal*, **86**, No. 2, March–April, 182–91.

20 Colaco, J.P. (1985) 75-Story Texas Commerce Plaza, Houston – the use of high-strength concrete. *High-Strength Concrete*, Publication SP-87. American Concrete Institute, Detroit, 1–8.

21 Pickard, S.S. (1981) Ruptured composite tube design for Houston's Texas Commerce Tower. *Concrete International Design & Construction*. American Concrete Institute, **3**, No. 7, July, 13–19.

22 (1983) Tower touches few bases. *Engineering News Record*, June 16, 24–25.

23 LeMessurier, W.J. (1982) Toward the ultimate in composite frames, *Building Design and Construction*. Cahners Publication Company, **23**, No. 11, November, 14–21.

24 Case history report – the world's tallest concrete skyscrapers, *Bulletin No. 40*. Concrete Reinforcing Steel Institute, Schaumburg, IL, 8 pp.

25 (1989) Tall concrete buildings come of age, *Engineering News Record*, November 30, 25–27.

26 Page, K.M. (1990) Pumping high-strength on world's tallest concrete building, *Concrete International Design & Construction*. American Concrete Institute, **12**, No. 7, January, 26–8.

27 Moreno, J. (1990) 225 W. Wacker Drive, *Concrete International Design & Construction*. American Concrete Institute, **12**, No. 1, January, 35–9.

28 (1989) Put that in your pipe and cure it. *ENR*, February 16, 44–53.

29 Thornton, C.H. (1990) The world's tallest building – Chicago's Miglin-Beitler Tower. *Engineered Concrete Structures*, **3**, No. 3, Portland Cement Association, December, 1–2.

30 Thornton, C.A., Hungspruke, O. and DeScena, R.P. (1991) Looking down at the Sears Tower. *Modern Steel Construction*. American Institute of Steel Construction, August, 27–30.

31 Jobse, H.J. (1981) Applications of high-strength concrete for highway bridges, Executive Summary, U.S. Department of Transportation, Federal Highway Administration, Washington, DC, Report No. FHWA/RD 81/096, October, 27 pp.

32 Jobse, H.J. and Moustafa, S.E. (1984) Applications of high strength concrete for highway bridges. *Journal of the Prestressed Concrete Institute*, **29**, No. 3, May–June, 44–73.

33 Zia, P., Schemmel, J.J. and Tallman, T.E. (1989) *Structural applications of high strength concrete*, Report No. FHWA/NC/89-006, Center for Transportation Engineering Studies, North Carolina State University, Raleigh, June, 330 pp.

34 (1984) Hybrid girder in cable-stay debut. *Engineering News Record*, November 15, 32–6.

35 Taylor, P.R. and Torrejon, J.E. (1987) Annacis Bridge – design and construction of the cable-stayed span. *Quarterly Journal of the Federation Internationale de la Precontrainte*, 4, 18–23.

36 (1986) Stayed girder reaches a record with simplicity. *Engineering News Record*, May 22.
37 Pliskin, L. and Malier, Y. (1990) The French R&D Project, 'New developments for concrete, the high strength concrete Bridge of Joigny', Preprint No. 89-0586, Transportation Research Board, 69th Annual Meeting, Washington, D.C., January.
38 Ronneberg, H. and Sandvik, M. (1990) High strength concrete for North Sea platforms, *Concrete International Design & Construction*. American Concrete Institute, **12**, No. 1, January, 29–34.
39 Fiorato, A.E., Person, A. and Pfeifer, D.W. (1984) The first large scale use of high-strength lightweight concrete in the Arctic environment, *Second Symposium on Arctic Offshore Drilling Platforms*, Houston, Texas, April.

10 High strength lightweight aggregate concrete

T A Holm and T W Bremner

10.1 Introduction

It may be argued that the first practical use of high strength concrete took place in World War I when the American Emergency Fleet Corporation built lightweight concrete ships with specified compressive strengths of 5000 psi when commercial normal weight concrete strengths of that time were 2000 psi. It was fully recognized by these forward looking engineers that high self-weight was the major impediment in the use of structural concrete.

Concrete density can be reduced in several ways, i.e., lightweight aggregates, cellular foams, high air contents, no fines mixes, etc., but only high quality structural grade lightweight aggregates can develop high strength lightweight aggregate concretes. As such, the letter 'A' in the abbreviation LAC for lightweight aggregate concrete will be dropped and similarly HSLC indicates high strength lightweight aggregate concrete. In a similar fashion, LWA, NWA, NWC and HSNWC represent lightweight aggregate, normal weight aggregate, normal weight concrete and high strength normal weight concrete respectively.

Structural efficiency

The entire hull structure of the USS *Selma* was constructed with HSLC in a shipyard in Mobile, Alabama and launched in 1919. The strength/density (S/D) ratio (structural efficiency) of 50 used in the USS *Selma* (+5000 psi/100 pcf) was extraordinary for that time.[1] Improvements in structural efficiency of concrete since that time are shown schematically in Fig. 10.1, revealing upward trends in the 1950s with introduction of prestressed

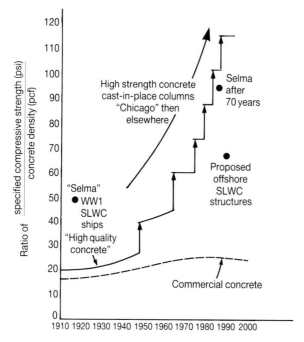

Fig. 10.1   The structural efficiency of concrete. The ratio of specified compressive strength density (psi/pcf) through the recent history of construction (from Holm and Bremner[1])

concrete, followed by production of HSNWC for columns of very tall cast-in-place concrete frame commercial buildings. It would appear that the strength/density ratio for the LC produced in the World War I ship construction program was only exceeded by HSNWC 40 years later.

Compression strength of the LC cores taken at the water line from the USS *Selma* and tested in 1980 were found to be twice the 28 day specified strengths, and from a structural efficiency standpoint are not appreciably different from the HSNWC of today. Analysis of the physical and engineering properties of the HSLC in the ships of World War I, the 104 HSLC World War II ships, as well as numerous recent bridges built, can be found in other reports that amply prove the long-term successful perform-ance of HSLC.[2–4]

Maximum strength 'ceiling'

The strength 'ceiling' of HSLC is reached when further additions of binder materials do not significantly increase strength. Figure 10.2 demonstrates that the compressive strength ceiling for the particular $\frac{3}{4}$ in. (20 mm) top size LWA tested was somewhat more than 8000 psi (55 MPa) at an age of 75 days. When the top size of this aggregate was reduced to $\frac{3}{8}$ in. (10 mm), the strength ceiling significantly increased to more than 10,000 psi

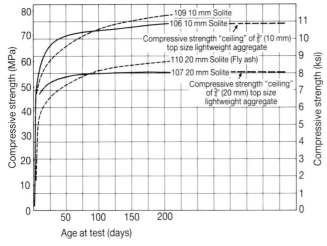

Fig. 10.2 Compressive strength versus age of lightweight concrete (1975–1979 Series)[5]

(69 MPa). Mixes incorporating fly ash demonstrated higher strength ceilings at later ages than non-fly ash control concretes. Results on concretes containing $\frac{1}{2}$ in. top (13 mm) size were intermediate between the $\frac{3}{4}$ in. (20 mm) and $\frac{3}{8}$ in. (10 mm) curves are reported in *Criteria for Designing Lightweight Concrete Bridges*[5] and are not shown for clarity.

Analyzing strength as a function of the quantity of cementitious binder as shown in Fig. 10.3, however, reveals that mixes incorporating binder quantities exceeding an optimum volume are not cost effective.

Pore system of structural lightweight aggregate

Strength ceilings of LWA produced from differing quarries and plants will

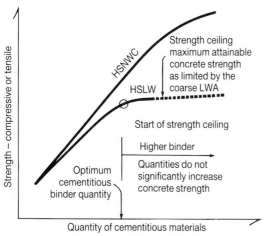

Fig. 10.3 Strength versus concrete binder content

Fig. 10.4   Contact zone – structural lightweight concrete, W P Lane Memorial Bridge over Chesapeake Bay, Annapolis, MD – constructed in 1952[6]

vary considerably. This variation is due to structural characteristics of the pore system developed during the production process. The producer's goal is to manufacture a high quality structural grade LWA which has non-interconnected, essentially spherical, well distributed pores surrounded by a strong, crack free vitrious ceramic matrix. The scanning electron micrograph in Fig. 10.4 demonstrates a well developed pore distribution system of a concrete sample cored from a highly exposed 30 year old bridge deck.[6]

Internal integrity at the contact zone

Micrographs of concretes obtained from mature structural LC ships, marine structures and bridge have consistently revealed minimal microcracking and a limited volume of unhydrated cement grains. The boundary between the cementitious matrix and coarse aggregates is essentially indistinguishable at the 'contact zone' transition between the two phases in all mature HSLCs. The contact zone in LC is enhanced by several factors including: pozzolanic reactivity of the surface of the lightweight aggregate developed during high temperature 2000 °F (1100 °C) production, surface roughness and, most importantly, the opportunity for th two-phase porous system to reach moisture equilibrium without developing the water gain lenses frequently observed under and on the sides of NWA.[7]

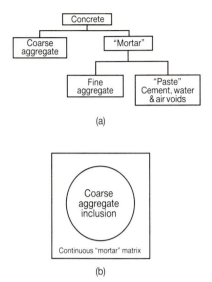

(a)

(b)

Fig. 10.5   Cement, water, air voide and fine aggregate combine in (a) to form the continuous mortar matrix that surrounds the coarse aggregate inclusion in (b) to produce concrete

Principles of elastic compatibility of a particulate composite

A particular composite is by its very definition heterogeneous, and concrete is perhaps the most heterogeneous of composites with size of inclusions varying from large aggregate down to unhydrated cement grains, and containing voids the size of entrapped air bubbles down to the gel pores in the cement paste. The general understanding of concrete as a particulate composite previously used in the analysis of regular strength LC may be extended to HSLC.[8]

Concrete can be considered as a two-phase composite composed of coarse aggregate particles enveloped in a continuous mortar matrix. This latter phase includes all the other concrete constituents including fine aggregate, mineral admixtures, cement, water and voids from all sources. This division, schematically shown in Fig. 10.5, is visible to the naked eye and may be used to explain important aspects of the strength and durability of concrete.

The elastic mismatch between coarse aggregate particles and the surrounding mortar matrix gives rise to stress concentrations when the composite is subjected to an applied stress. These stress concentrations are superimposed on a system already subjected to internal stresses arising from dissimilar coefficients of thermal expansion of the constituents and from aggregate restraint of matrix volume changes. The latter can be caused by drying shrinkage, thermal shrinkage from hydration temperatures, or changes that result from continued hydration of the cement paste. These inherent stresses are essentially self-induced and may be of a

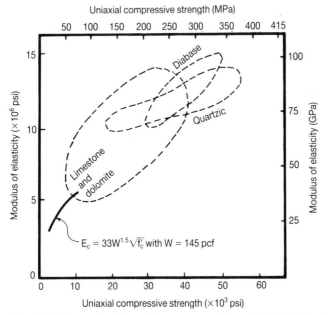

Fig. 10.6　Range of stuffness of concrete caused by variability in the stiffness of the aggregate (adapted from Stagg[9])

magnitude that extensive microcracking may take place before any super-imposed stress is applied.

Natural aggregates have an extremely wide range of elastic moduli resulting from large differences of mineralogy, porosity, flaws, lamina-tions, grain size and bonding. It is not uncommon for a fine-grained diabase rock to have an elastic modulus greater than 13×10^6 psi (90 GPa) while poorly bonded, highly porous natural aggregates have been known to have values lower than 3×10^6 psi (20 GPa). Aggregate description by name of rock is insufficiently precise, as demonstrated in one rock mechanics text which reported a range of elastic modulus of 3 to 10×10^6 psi (20 to 69 GPa) for one rock type.[9]

Figure 10.6 illustrates compressive strength and stiffness characteristics reported for several rock types and compares these wide ranges with the modulus of elasticity of concrete as suggested by equation $E_c = 33\omega^{1.5}\sqrt{f_c'}$ (in psi) of ACI 318–89 Code. The ratio of the coarse aggregate modulus to that of the concrete composite can be shown to be as much as 3, signaling a further difference between the two interacting phases (mortar and coarse aggregate) of as much as 5 to 1. That the strength-making potential of the stone or gravel is normally not fully developed is evident from visual examination of fracture surfaces of concrete cylinders after compression testing. The nature of the fracture surface of concretes is strongly influenced by the degree of heterogeneity between the two phases and the extent to which they are securely bonded together. Shah[10] reported on the profound influence exerted by the contact zone in compressive strength tests on concretes in which aggregate surface area was modified by coatings. The degree of heterogeneity and the behavior of the contact zone between the two phases are the principal reasons for the departure of some concretes from estimates of strength based upon the water-to-cement (w/c) ratio. As has been suggested,[8] undue preoccupation with the matrix w/c ratio may lead to faulty estimates of compressive strength and even greater misunderstanding of concrete's behavior from durability, permeability and tensile type loading conditions.

Obviously the characteristics of the NWA will have a major effect on elastic compatibility. The interaction between the absolute volume percentage of coarse aggregate ($\pm 35\%$) and the mortar phase ($\pm 65\%$) will result in a concrete with a modulus intermediate between the two fractions. At usual commercial strength levels the elastic mismatch within structural LC is considerably reduced due to the limited range of elastic properties of usual LWA particles.

Elastic matching of components of commercial strength lightweight concrete

Muller-Rochholz measured the elastic modulus of individual particles of LWA and NWA using ultrasonic pulse velocity techniques.[11] This report concluded that the modulus of elasticity of structural grade LWA exceeded values of the cementitious paste fraction, and suggested that instances when LC strength exceeded that of companion NWC at equal binder content, were understandable in light of the relative stress homogeneity.

The *FIP Manual of Lightweight Aggregate Concrete*[12] prepared by the Federation Internationale de la Precontrainte (FIP) reports that the modulus of elasticity of an individual particle of LWA may be estimated by the formula: $E_c = 0.008p^2$ (MPa), where p is the dry particle density. Usual North American structural grade LWA having dry particle densities of 1.2 to 1.5 (1200 to 1500 kg/m^3) would result in a particle modulus of elasticity from 1.7 to 2.6×10^6 psi (11.5 to 18 GPa). At these densities the modulus of elasticity of individual particles of LWA approaches that measured on the mortar fraction of air-entrained commercial strength LC.[8]

The elastic modulus of air-entrained and nonair-entrained mortars is shown as a function of compressive strength in Fig. 10.7. The modulus of a typical individual particle of coarse LWA, as well as a range of values of modulus for stone aggregates, is also shown. These results were obtained by testing concretes and equivalent mortars with the same composition found in concrete, with the exception that the coarse aggregate had been fractioned out.

Mortar matrix mixes were produced to cover usual ranges of cement contents at the same time as companion structural LCs were cast with all other mix constituents kept the same. Data and analysis of these tests are beyond the scope of this chapter.

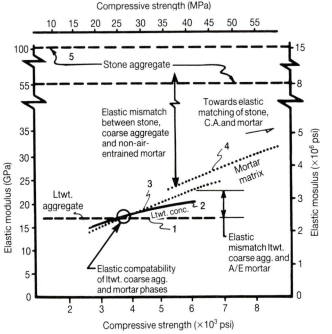

Fig. 10.7 Elastic mismatch in lightweight and normal weight concrete

Sand LC with compressive strength of approximately 4000 psi (28 MPa) made with typical North American structural grade LWAs has values of E_a/E_m approaching unity. From a stress concentration point of view, this combination of constituents would act as a homogeneous material resulting in concrete with minimum stress-induced microcracking. Thus, at ordinary commercial strengths, the elastic match of the two components will be close for air-entrained concrete made with high quality LWA. Matching of the elastic properties of ordinary concrete using a high modulus NWA such as a diabase will be possible only with the ultra-high quality matrix fractions recently developed incorporating superplasticizers and supplementary cementitious materials.

Air entrainment in concrete significantly reduces the stiffness of the mortar fraction and, as shown in Fig. 10.7, results in a convergence of elastic properties of the two phases of sanded structural LC while increasing the degree of elastic mismatch in ordinary concrete. This fact, combined with the slight reduction in mixing water caused by air entrainment, explains why the strength penalty caused by air entrainment is less significant for structural LC than for concretes using highly rigid NWA.

Elastic mismatching of components of high strength lightweight concrete

Combining ultra-high strength, low air content mortar matrix fractions with coarse LWA will produce an elastic mismatch resulting in fracture that starts with transverse splitting of the structural LWA particles. Splitting action stemming from lateral strains is indirectly responsible for the strength ceiling of structural LC observed when improvements in mortar matrix quality result in little or no increase in compressive strength.

In general, for concretes using high quality NWA, elastic compatibility between the two fractions will occur only at extremely high compressive strengths. Ultra-high strength mortar fractions developed by superplasticizers and mineral admixtures will increase the possibility of achieving elastic compatibility at higher compressive strengths when ordinary aggregates are used.

While elastic mismatching plays an important yet incompletely understood role in the compressive strength capabilities of the composite, the influence on other properties (tensile and shrinkage cracking, and particularly limitations on in-service permeability and durability due to microcracking) are far more significant.

10.2 Materials for high strength lightweight aggregate concrete

For the purpose of this chapter, HSLC should have a maximum equilibrium unit weight of 125 pcf (2000 kg/m^3), as defined in ASTM C567, and a specified compressive strength of at least 5000 psi (34.5 MPa). This level on unit weight may be necessary when developing compressive strengths approaching 10,000 psi (70 MPa) while still maintaining benefits of weight reduction. HSLC with compressive strengths ranging from 5000 psi (41 MPa) to 7000 psi (48 MPa) are commercially available in some areas and testing programs on HSLC with ultimate strengths approaching 10,000 psi (70 MPa) are ongoing.

With the exception of LWA, the materials (cement, admixtures, NW fine aggregate) and methods used to produce HSLC are essentially similar to those used in their NWC counterparts. As materials and proportioning methods common to HSNWCs are extensively reported on in other chapters, they will not be addressed except where differences require explanation.

Cementitious materials

Cement

Portland cements used for HSC should conform to the requirements of ASTM C150. Granulated iron blast furnace slags used as a replacement for portland cement should conform to ASTM C989. Compressive data on cements, supplementary cementitious materials, admixtures, etc. are reported on in ACI 363.[13]

Supplementary cementitious materials

Production of ultra-HSLC generally requires the use of supplementary cementitious materials. High quality fly ash meeting the requirements of ASTM C618 will reduce permeability, improve placing qualities, lower heat of hydration, and improve long term strength characteristics. Microsilica will improve compressive strength at all ages and also provide significantly improved resistance to chloride penetration.

Supplementary cementitious materials function very effectively in HSLC, because pozzolanic activity requires the combination of the calcium hydroxide liberated during cement hydration with finely divided silica in the presence of moisture. As shown in Fig. 10.2, mixes incorporating mineral admixtures achieved higher strength ceilings than the control concretes. Favorable hydrating environments will be provided for a longer time due to the internal curing provided by the LWA absorbed moisture, thus promoting increased activity of the pozzolanic materials.

Admixtures

When used in HSLC, admixtures offer reduced water demand, enhance durability, and improved workability in a manner comparable to that of HSNWC. Water reducers, retarders and high range water reducers should conform to ASTM C494 and be dosed according to manufacturers' recommendations.

Air entrainment

LC mixtures, normally contain entrained air. Entrained air serves to increase the cohesiveness of the fresh concrete mix, and to make concrete resistant to the effects of freezing and thawing when in a wet environment. When freezing and thawing is not a consideration, then small amounts of entrained air (3 to 5%) are adequate. Entrained air volumes should meet the requirements of ACI 201 according to the severity of the exposure conditions. While air entrainment will diminish strength making characteristics of the cementitious matrix, it will also lower water and sand volumes necessary to achieve satisfactory workability, with the net effect being only a modest reduction in the strength of HSLC.

Coarse aggregate

HSLCs normally require only coarse LWA. As reported earlier, most, but not all, HSLC mixes require a reduction of the LWA top size, particularly in the 7000 to 10,000 psi (48 to 70 MPa) range. Certain LWAs, however, because of the strength of the vitreous material enveloping the pores, have routinely used the 3/4 to #4 (20 to 5 mm) gradation in production of high strength precast concrete for more than four decades. Most LWA manufacturing plants will limit coarse aggregate to two sizes to minimize production and stockpiling problems, but these plants will entertain other gradations if project volumes warrant.

Gradations of 3/4 to #4 (20 to 5 mm) or 1/2 to #4 (13 to 5 mm) will normally be appropriate for usual size HSLC members while 3/8 to #8 (10 to 5 mm) gradations may be necessary in highly reinforced members to allow adequate placement conditions.

Fine aggregate

HSLC normally incorporates normal weight sand as the fine aggregate fraction. Quality criteria developed for sands used in HSNWC (e.g., FM of about 3.0 for optimum workability and strength, etc) are identical to those used in manufacturing HSLC.

10.3 High strength lightweight concrete laboratory testing programs

Systematic laboratory investigations into physical and engineering properties of HSLC are too numerous for all to be elaborated on here. Most early programs extending strength/density relationships were conducted by LWA manufacturers and innovative precast concrete producers striving for high early release strengths, longer span flexural members, or taller one-piece precast columns.[5] These in-house programs developed functional data directly focused on members supplied to real projects. In general, project lead times were short, practical considerations of shipping and erection immediate, and mixes targeted towards satisfying specific job requirements. This type of research brought about immediate incremental progress but, in general, was not sufficiently comprehensive.

Unfortunately some researchers did not use advanced admixture formulations or supplementary cementitious materials (i.e., high range water reducer, microsilica, fly ash, slag cement) that significantly improve matrix quality and as such provide data of no commercial value. These investigations, as well as others incorporating unrealistic mixtures, inappropriate LWA, or impractical density combinations, are not reported.

Special requirements of offshore concrete structures have now brought about an explosion of practical research into the physical and engineering properties of HSLC. Several large confidential joint industry projects are now becoming publicly available as the sponsors release data according to an agreed upon timetable.[14–16] These monumental studies, comprehensively summarized by Hoff and presented at the November 1991 American Concrete Dallas Symposium on the 'Performance of Structural Lightweight Concrete',[17] are widely preferred throughout this chapter. In addition to providing comprehensive physical property data on HSLC, these programs developed innovative testing methods: revolving disc, tumbler and sliding contact ice abrasion wearing tests, freeze/thaw reistsance to spectral cycles, and freeze bond testing techniques, etc., that measured properties unique to offshore applications in the Arctic.

In addition, Hoff summarizes and reports on four other major investigations that included data on HSLC.[18] These programs included HSLC data in investigations into structural qualities of high strength concrete and include SINTEF/FCB Trondheim, Norway,[19] Hovik, Norway,[20] Gerwick,[21] and National Institute of Standards and Technology, Washington, DC.[22]

Major North American laboratory studies into properties of HSLC include those conducted at or sponsored by Expanded Shale Clay & Slate Institute,[23–25] CANMET,[26–28] Ramakrishnan,[29] Berner,[30] and Luther.[31] Because of their special structure needs, much pioneering into this issue has been conducted by Norwegian sources[32–34] with additional important contributions from other Russian, German, and UK sources.[35–39]

It has been estimated that the cost for these commercially supported research programs into the physical and structural properties of HSLC has exceeded one million dollars.[40] While much research has been already effectively transferred into actual practice on current projects, there remains a formidable task of analyzing, digesting and codifying an immense body of data into design recommendations and code standards.

10.4　Physical properties of high strength lightweight aggregate concrete

Compressive strength

There is a substantial body of information, developed over a long period of time, demonstrating that high strength can be achieved with LC. As early as 1923, Duff Abrams[41] commented, 'The high strength secured from lightweight aggregates consisting of burnt shale, has shown the fallacy of the older views that the strength of the concrete is dependent upon the strength of the aggregate.' Moderately HSLC can be achieved without significantly higher binder contents than used with NWC when the designer selects the appropriate LWA. For every commercially available LWA there is, however, a strength ceiling which is reached when compression strength is limited by crushing the aggregate. It is uneconomical to endeavor to produce HSLC exceeding the strength ceiling by using greater amounts of cement. Aggregate and readymix concrete suppliers are generally aware of the strength making potential of both NWA and LWA and should be consulted early in the design process.

Tensile strength

Tensile strength of HSLC is limited by the fact that approximately 50% of the aggregate volume is pore space. The ACI code (ACI 318–89) requires

LWA producers to supply tensile splitting test data on concrete incorporating their aggregate, allowing structural engineers to modify code equations for shear, torsion and cracking.

High tensile splitting developed on mature specimens of HSLC have shown clearly visible high moisture contents on the split surface, demonstrating that well compacted mixes with high binder content, and particularly those incorporating mineral admixtures (microsilica, fly ash), are essentially impermeable and will release moisture very slowly. High strength specimens drying in laboratory air for over several months were still visibly moist over 90% of the split diameter.[5] The reductions in splitting strength observed in tests on air dried commerical strength LC that are caused by differential dry moisture gradients in the concrete prior to reaching hygro-equilibrium are significantly delayed and diminished in high binder content HSLC.

Elastic properties

Modulus of elasticity

Concrete is a composite material composed of a continuous matrix enveloping particulate inclusions. Stiffness of the composite is related to the stiffness of its constituents in a rather complex way, and it is surprising that the recommended ACI formula has been so effective. One factor affecting stiffness is the variation of aggregate modulus of elasticity within a particular density range. At the same specific gravity, LaRue found the modulus of elasticity of natural aggregates could vary by a factor of as much as three.[42] Concrete strengths also tended towards a maximum where the aggregate modules matched the modulus of the concrete made from them.

Although the ACI 318 formula, $E_c = 33\omega^{1.5}\sqrt{f_c'}$ ($E_c = 0.043\omega^{1.5}\sqrt{f_c'}$) has provided satisfactory results in estimating the elastic modulus of NWC and LC in the usual commercial strength range from 3000 to 5000 psi (20 to 35 MPa), it has not been adequately calibrated to predict the modulus of high strength concretes. Practical modification of the formula was first provided by ACI 213-77[43] to more reasonably estimate the modulus of HSLC:

$$E_c = C\omega^{1.5}\sqrt{f_c'}$$

$C = 31$ for 5000 psi   ($C = 0.40$ for 35 MPa)
$C = 29$ for 6000 psi   ($C = 0.38$ for 41 MPa)

When designs are controlled by elastic properties (e.g. deflections, buckling, etc), the specific value of E_c should be measured on the proposed concrete mixture in accordance with the procedure of ASTM C 469, *Standard Test Method for Static Modulus of Elasticity and Poisson's Ratio of Concrete in Compression*.

In general, structural grade rotary kiln produced LWAs have a comparable chemical composition and are manufactured under a similar temperature regime. They achieve low density by formation of a vesicular structure in which the vesicles are essentially spherical non-interconnected pores enveloped in a vitreous matrix. It would be expected that, with such similarities, the variability in stiffness of the aggregate would be principally due to the density as determined by the pore volume system. Aggregate density, in turn, is reflected in a reduced concrete unit weight which is accounted for in the first term in the above equation.

As with NWC, increasing matrix stiffness is directly related to matrix strength which, in turn, affects concrete strength. When large percentages of cementitious materials are used, the LC strength ceiling may be reached causing the above equation to overestimate the stiffness of the concrete.

Poisson's ratio

Testing programs investigating the elastic properties of HSLC have reported an average Poisson's ratio of 0.20, with only slight variations due to age, strength level, or aggregates used.[44,45]

Maximum strain capacity

Several methods of determining the 'complete' stress-strain curve of LC have been attempted. At Lehigh University,[15] the concrete cylinders were loaded by a beam in flexure, while at Illinois University, the approach was to load a concrete cylinder completely enclosed within a steel tube of suitable elastic properties.[46]

Despite formidable testing difficulties, both methods secured meaningful data; one of the more complete stress-strain curves obtained by loading the concrete through a properly proportioned beam in flexure is demonstrated in Fig. 10.8.

At failure, HSLC will cause the release of a greater amount of energy stored in the loading frame than an equal strength concrete composed of stiffer NWA. As energy stored in the test frame is proportional to the applied load moving through a deformation inversely proportional to the modulus of elasticity, it is not unusual for a HSLC to release almost 50% greater frame energy. To avoid shock damage to the testing equipment, it is recommended that a lower percentage of maximum usable machine capacity be used when testing HSLC and that suitable precautions are taken for testing technicians as well.[5]

Dimensional stability

Shrinkage

Figure 10.9 demonstrates the shape and ultimate shrinkage strains from

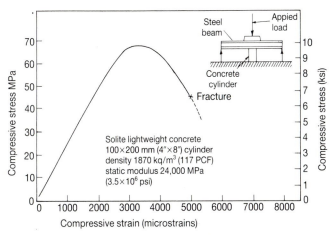

Fig. 10.8 Stress versus strain under uniaxial compression

one extensive testing program that incorporated both HSLC and HSNWC.[5] These results are consistent with data reported in other investigations.[17] Of interest is the relative equality of the maximum shrinkage of the HSNWC mix and the 4500 psi (30 MPa) LC mix introduced for comparative purposes, indicating an apparent trade-off between the rigid skeletal structure of the NWA and the contractive forces developed by the higher binder content. Shrinkage of the 10 mm top size HSLC mix lagged behind early values of the other mixes, equalled them at 90 to 130 days, and reached an ultimate value at one year approximately 14% higher than the reference HSNWC. Shrinkage values of mixes incorporating cement containing interground fly ash averaged somewhat greater than their high strength non-fly ash counterparts.

Shrinkage and density data were measured on $4 \times 4 \times 12$ in. ($100 \times 100 \times 300$ mm) concrete bars fabricated at the same time and from the same mix as the compressive strength cylinders. Curing was provided by damp cloth for 7 days, after which the specimens were stripped from the

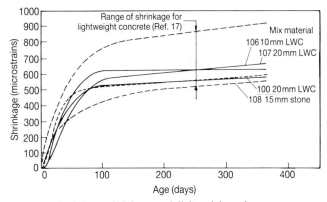

Fig. 10.9 Shrinkage of high strength lightweight and stone concretes

molds. Brass wafers were epoxied at one day to the bar surface at a 10 in. (250 mm) gage distance with mechanical measurement by a Whitemore gage. Reference readings were established 7 days after fabrication after which specimens were allowed to dry in laboratory air, 70 °F (21 °C), (50% ± 5 RH), with no further curing. Shrinkage and weight readings were taken weekly for three months then monthly with results shown to one year. Ten year shrinkage strains were only slightly higher than one year results and will be reported in another publication.[47]

Creep

Rogers reported that the one year creep strains measured on several high strength North Carolina and Virginia LCs were similar to those measured on companion NWC.[48] Greater creep strains measured on HSLC containing both fine and coarse LWA, when compared to reference HSNWC in reports by Reichard[44] and Shideler,[45] could be anticipated because of the larger matrix volume required because of the angular particle shape of the LWA fines.

While the Prestressed Concrete Institute provides recommendations for increasing stress losses due to creep when using HSLC, it may be advisable to obtain accurate design coefficients for long span HSLC structures by conducting pre-bid laboratory tests in accordance with the procedures of ASTM C512, *Standard Test Method for Creep of Concrete in Compression*.

As reported in ACI 213,[43] and shown in Fig. 10.10, specific creep values decrease significantly with increasing strength of LC. Additionally, at higher strength levels, the creep strain envelope developed from a wide range of LCs tested converged towards the performance of the reference NWC.

Thermal properties

Accurate physical property input data is essential when considering the thermal response of restrained members in exposed structures. Obvious

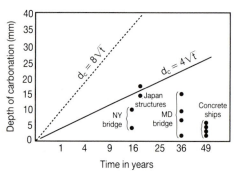

Fig. 10.10　Measured depth of carbonation (mm) of exposed lightweight concrete structures

cases in point include exposed exterior columns of multi-story cast-in-place concrete frames as well as massive offshore concrete structures constructed in temperate zones and then towed to harsh Arctic marine environments. As thermal behavior of concrete is a function of the thermal contributions of the separate components, the low expansions, conductivities and diffusivities of HSLC are predictable in the light of thermal measurement data reported on regular strength LC.[43] A summary of data relating to the coefficient of linear thermal expansion, thermal conductivity, and diffusivity data extracted from the results obtained in the joint industry investigation digested by Hoff[17] is shown below.

- *Coefficient of linear thermal expansion*
  At high moisture contents typical of marine exposed concrete, the coefficient of linear thermal expansion for HSLC ranged from 4 to 7 microstrain/°F (7 to 13 microstrain/°C).[17]
- *Thermal conductivity*
  Over a temperature range of −50 to 70 °F (−46 to 21 °C), thermal conductivity values of the HSLC varied from 5.6 to 7.6 Btu.in/hr.sf. F (0.8 to 0.9 W/m. K) According to density and mix composition factors.[17]
- *Specific heat*
  The specific heat of HSLC tested averaged 0.23 Btu/lb. F (1.0 J/kg K).[17]
- *Thermal diffusivity*
  The thermal diffusivity of the HSLC tested ranged from 0.020 to 0.022 sq ft/hr (18.3 to 20.8 sq cm/hr) for the concrete densities tested.[17]

Durability

Numerous laboratory durability testing programs evaluating freeze/thaw resistance have confirmed field observations of the long-term, proven performance of highly exposed HSLC. The United States Federal Highway Administration closely examined 12 study cases of more than 400 North American LC bridge decks and concluded:

'Some have questioned the durability, wear resistance, and long-term freeze/thaw qualities of lightweight concrete. No evidence was found that these properties differ from those of normal weight concrete. In fact, there is evidence that these properties could be better for lightweight concrete, especially if the normal weight concrete is of poor quality. This leads to the suggestion that the designer might consider specifying lightweight concrete if natural aggregates are not of high quality. Although lightweight aggregates vary depending on the raw material source, they are usually of a more consistent quality than some natural aggregates. Specified material tests will provide the necessary quality characteristics.'[4]

These findings supported similar conclusions reported earlier in the LC Bridge Deck Survey of the Expanded Shale Clay and Slate Institute.[24] Reports of the inspections of the durability of the HSLC used in World War I and World War II concrete ships also attested to excellent long-term performance.[2,3,6,7,49]

Freeze/thaw resistance

Most accelerated freeze/thaw testing programs conducted on structural LC have incorporated HSLC on the high end of the compressive strength range of the specimens tested. Investigations in North America[25,50] and in Europe[39,51] researching the influence of entrained air volume, cement content, aggregate moisture content, specimen drying times, and testing environments have arrived at essentially the same conclusion: air entrained LCs properly proportioned with high quality binders provide satisfactory results when tested under usual laboratory freeze/thaw testing programs.

Core samples taken from the hulls of 70 year old LC ships as well as the 30 to 40 year old LC bridge decks have demonstrated concretes with high internal integrity and low levels of microcracking. This proven record of high resistance to weathering and corrosion is due to physical and chemical mechanisms; they include superior resistance to microcracking developed by significantly higher aggregate/matrix contact zone adhesion as well as internal stress reduction due to the elastic matching of coarse aggregate and matrix phases. High ultimate strain capacity is also provided by concrete with a high strength/modulus ratio. In addition, because of elastic compatibility, the stress/strength ratio at which the disruptive disintegration of concrete begins is higher for LC than for equal strength NWC. A well dispersed void system provided by lightweight fine aggregates will assist the entrained air pore system, and may also serve an absorption function by reducing disruptive mechanisms in the matrix phase. Additionally, long-term pozzolanic action is provided by the silica rich expanded aggregate combining with calcium hydroxide liberated during cement hydration. This will reduce permeability and minimize leaching of soluble compounds.

It is widely recognized that while ASTM C666, *Standard Test Method for Resistance of Concrete to Freezing and Thawing*, provides a useful comparative testing procedure, there remains inadequate correlation between accelerated laboratory test results and the observed behavior of mature concretes exposed to natural freezing and thawing. Inadequate laboratory/field correlation observed when testing NWC is compounded when interpreting results from laboratory tests on structural LC prepared with high aggregate moisture contents. A proposed modification to ASTM C666 recommends a 14 day air drying period prior to the first freezing cycle, to improve correlation between laboratory test data and observed field performance.[2]

Durability characteristics of any concrete, both NWC and LC, are decisively influenced by the protective qualities of the paste fraction. It is imperative that the concrete matrix provide high quality, low permeability characteristics in order to protect steel reinforcing from corrosion, which is clearly the dominant form of structural deterioration observed in current construction. The protective quality of the matrix in concretes proportioned primarily for thermal resistance that incorporate high air contents and low cement quantities will be significantly reduced. Very low density, non-structural LC will not provide resistance to the intrusion of chlorides, carbonation, etc comparable to the long-term satisfactory performance demonstrated with high quality, structural grade LC.[2]

For a number of years field exposure testing programs have been conducted by the Canadian Department of Minerals, Energy and Technology (CANMET) on various types of concrete exposed to a cold marine environment at the Treat Island Severe Weather Exposure Station maintained by the U.S. Army Corps of Engineers at Eastport, Maine.[52,53] Concrete specimens placed on a mid-tide wharf experience alternating conditions of sea water immersion followed by cold air exposure at low tide. In typical winters the specimens experience about 100 cycles of freezing and thawing. In 1978, a series of prisms were cast using commercial NWAs with various cement types and including supplementary cementitious materials. W/c ratios of 0.40, 0.50, and 0.60 were used to produce 28 day compressive strengths of 4350, 3770, and 3480 psi (30, 26 and 24 MPa) respectively. In 1980, these mixes were essentially repeated with the exception being that the $1\frac{1}{2}$ in. (40 mm) gravel aggregate was replaced with a 1 in. (25 mm) expanded shale LWA. Fine aggregates used in both 1978 and 1980 were commercially available natural sands. Cement contents for the semi-LC mixtures were 800, 600, and 400 pcy (480, 360, and 240 kg/m^3), which produced compressive strengths of 5220, 4350 and 2755 psi (36, 30 and 19 MPa) respectively. All specimens continue to be evaluated annually for ultrasonic pulse velocity, and resonant frequence as well as visually ratings. Ultrasonic pulse velocities are measured centrally along the long axis of the prisms. Negligible differences exist between the structural LC (8 years) and NWC (10 years) after exposure to twice daily sea water submission and approximately 1000 cycles of freezing and thawing.[2]

Permeability

Conventional strength concrete employs a matrix with a w/c ratio significantly higher than that required for the chemical reaction associated with hydration of portland cement. This excess of uncombined water is free to either move due to an applied hydraulic gradient or, if allowed to evaporate, provide conduits through which gas can either diffuse into the concrete or flow in response to a pressure differential. Concentration gradients of chemicals in liquid form can also diffuse into saturated concrete or can be absorbed by an initially dry concrete.

The above defines a material which is porous, in that it contains both pores and conduits that communicate with a free surface and, as a result, is permeable to liquids and gases. The normal definition of the adjective 'porous' usually contains any, or all, of the following: possessing pores, or containing vessels and conduits, or being permeable to liquids and gases. In the case of concrete of usual strength, all of the above are applicable.

Both HSNWC and HSLC are usually made with water contents that only slightly exceed that required to hydrate the cement. This means that while the hydrated cement paste still contains pores and conduits, these channels are not fully continuous nor do all of them communicate with the surface. High strength concrete normally contains both silica fume and a superplasticizer, which leads to a densification of the cement paste matrix after a relatively short period of moist curing. Slag cement, fly ash, or both, are frequently incorporated into the mix which further densifies the cement paste matrix.

Aggregates inserted into the cementitious matrix are surrounded by the paste and isolated from one another by this essentially impermeable matrix. Microcracks, however, may form in the concrete as a result of the volume changes associated with hydration of the cement paste matrix. Microcracking can also result from stresses that arise in concrete as a result of differing aggregate/matrix thermal expansion coefficients when the concrete is heated or cooled. Aggregates and matrix fractions expand and contract at different rates as the concrete gains and loses water, further increasing prospects for microcracking. Microcracking will also result from the lack of elastic compatibility that exists between the aggregate inclusion and the cement paste matrix when the composite concrete is subjected to an applied stress. It is these crack networks that, more than any other factor, render concrete permeable to gases and liquids. The high stiffness of the matrix fraction of high strength concrete results in a closer elastic match and a lower propensity to form stress-induced microcracks.

For all concrete types, the ratio of aggregate to matrix stiffness starts out at infinity as hydration begins, therefore any volume change due to the hydration process, moisture change, or thermal changes may lead to microcracking in the concrete prior to the superposition of any design loads on the concrete. In most instances where liquid permeability is of concern, or where presence of moisture is associated with deterioration due to corrosion, autogenous healing also is operative. Laboratory testing is, however, not usually conducted on mature concrete and this beneficial effect is not observed. Field and early age laboratory tests on high strength concrete do indicate permeability coefficients substantially lower by several orders of magnitude, however, when compared with commercial concrete. In fact, satisfactory techniques to satisfactorily measure the permeability of high strength concrete in most cases have not been developed.

The role the aggregate plays in the liquid and gas permeability of concrete is minor for both LC and NWC. With both types of aggregate, small cracks formed during the final crushing to size of the aggregate particles may have some small effect. However, few of the cracks would be oriented in the direction of flow, and the matrix essentially seals the cracks at the surface of the aggregates. Scanning electron microscopy studies have indicated that structural grade LWAs have a vesicular structure with essentially no interconnection of pores so that flow of liquid or gas is insignificant. Porous aggregate particles will not be permeable because of the lack of continuous channels through the particles. Permeability investigations conducted on LC and NWCs exposed to the same testing criteria have been reported by Khokrin,[51] Nishi,[54] Keeton,[55] Bamforth,[56] and Bremner.[57] It is of interest that in every case, despite wide variations in concrete strengths, testing media (water, gas and oil) and testing techniques (specimen size, medium pressure and equipment) structural LC had equal or lower permeability than its heavier counterparts. This result has been attributed to the reduction of microcracks in elasticly compatible LC and the enhanced bond and superior contact zone present in structural LC.

Corrosion resistance

Corrosion resistance of HSLC is at least comparable to the performance of HSNWC. Investigations of mature bridges and marine structures report that internal integrity and minimal microcracking have effectively limited rapid intrusion of aggressive forces into the concrete.[2,3,6] Internal integrity effectively limits disruptive effects to diffusion mechanisms which are orders of magnitude slower in their deteriorating actions.

Carbonate resistance

Penetration of the carbonation front into concrete is primarily determined by the vapor diffusion characteristics of the mortar matrix. With high quality matrixes typical for all HSCs, this issue will not be a concern. Field measurements of carbonation depths in LC marine and bridge structures have demonstrated that the rate of carbonation is extremely low, with results shown in Figs. 10.10 and 10.11. Adequate protection against

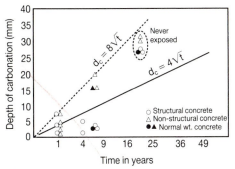

Fig. 10.11 Measured depth of carbonation (mm) of laboratory specimens of lightweight concrete

carbonation depassivating reinforcing steel in the service life of high quality structures is provided by covers recommended by ACI 318.

Abrasion resistance

Investigations into the abrasion resistance of LC using tests conducted in accordance with ASTM C779, *Standard Test Method for Abrasion Resistance of Horizontal Concrete Surfaces*, reported that the

> 'wear trends were similar for both normal weight and lightweight concretes and, because of the small differences in the wear amounts between both types of concretes, it appears that lightweight concrete could be used satisfactorily in abrasive situations where normal weight concrete might normally be specified.'[17]

Bare LC bridge decks exposed to more than 100 million vehicles crossing over a more than 20 year period have demonstrated wear patterns comparable to NWC.

Unit weight

The fresh unit weight of HSLC is a function of mix proportioning, air contents, water demand, and the specific gravity and moisture content of the LWA. Decrease in density of exposed concrete is due to moisture loss which, in turn, is a function of ambient conditions and surface area/volume ratio of the member. Design professionals should specify a maximum fresh unit weight for LC, as limits of acceptability should be controlled at time of placement.

Dead loads used for design should be based upon equilibrium density which, because of the very low permeability of HSLC, may be assumed to be reached after 180 days for moderate sized members. Extensive tests conducted during North American durability studies demonstrated that despite wide initial variations of aggregate moisture content, equilibrium unit weight for commercial strength LC was found to be 3.1 pcf ($50 \, \text{kg/m}^3$) above the oven dry unit weight. European recommendations for in-service density are similar.[12] For HSLC however, reduction in density from fresh to equilibrium will be significantly less due to the low permeability and high cementitious content which, because of internal curing periods, will serve to hydrate a larger fraction of the mix water. Because of the high cementitious content of HSLC, equilibrium density of non-submerged concrete will be close to the fresh density. Hoff reported that HSLC mixtures containing supplementary cementitious materials exposed to hydrostatic pressures equal to 200 ft. (61 m) of water demonstrated an increase of density of less than 4 pcf ($64 \, \text{kg/m}^3$).[17]

Ductility

The ductility of concrete structural frames should be analyzed as a composite system – that is, as reinforced concrete. Ahmad's studies indicate that the ACI rectangular stress block is adequate for strength predictions of HSLC beams and that the recommendation of 0.003 as the maximum usable concrete strain is an acceptable lower bound for HSLC members with strengths not exceeding 11,000 psi (76.5 MPa) and p/p_b values less than 0.54.[58] Moreno found that while LC exhibited a distinctive descending portion of the stress-strain curve, it was possible to obtain a flat descending curve with reinforced LC members that were provided with a sufficient amount of confining reinforcement slightly greater than that with NWC.[59] This report also included studies that showed that it was economically feasible to obtain desired ductility when increasing the amounts of steel confinement.

Rabbat *et al.*, came to similar conclusions when analyzing the seismic behavior of LC and NWC columns.[60] This report focused on how properly detailed reinforced concrete columns could provide ductility and maintain strength when subjected to inelastic deformations from moment reversals. These investigations concluded that properly detailed columns made with LC performed as well under moment reversals as NW columns.

10.5   Constructability of high strength lightweight concretes

Proportioning

Proportioning rules used for ordinary concrete mixes also apply to HSLC, with increased attention given to concrete density and the absorption characteristics of the LWA. Most structural grade LC is proportioned by absolute volume methods where the volume of fresh concrete produced is considered equal to the sum of the absolute volumes of cement, aggregates, net water and entrained air. Proportioning by this method requires determination of absorbed and adsorbed moisture contents and the as-batched specific gravity of separate aggregate sizes.

As with HSNWC, air entrainment is required for HSLC for durability and resistance to scaling. With moderate air contents, bleeding and segregation are reduced and mixing water requirements lowered while maintaining equivalent workability. In recognition of HSLC low permeability and the impact of air on strength properties, recommended air contents may be adequate when slightly lower than that required for usual exposed concretes. Because of the elastic matching of the LWA and the cementitious binder paste, strength reduction penalties due to high air contents are somewhat lower for LC than for NWC.[8] Air content of LC is determined in accordance with the procedures of ASTM C173, *Test for Air*

Content of Freshly Mixed Concrete by the Volumetric Method. Volumetric measurements assure reliable results while pressure meters may develop erratic data due to the influence of aggregate porosity.

When absorbed moisture levels are greater than that developed after a one day immersion, the rate of further aggregate absorption will be very low and for all practical purposes HSLC is batched, placed and finished with the same facility as their NWC counterparts. Under these conditions net w/c ratios may be established with a precision comparable to concretes containing NWA. Water absorbed within the LWA particle prior to mixing is not available for establishing the cement paste volume at the time of setting. This absorbed water is available, however, for continued hydration of the cement after external curing has ended.

Mixing, placing, finishing and curing

Properly proportioned HSLC can be delivered, placed and finished with the same facility and with less physical effort than that required for NWC. The most important consideration in handling any type of concrete is to avoid separation of coarse aggregate from the matrix fraction. Avoid excessive vibration as this practice serves to drive the heavier mortar fraction down from the surface of flat work where it is required for finishing. On completion of final finishing, curing operations similar to ordinary concrete should begin as soon as possible. LCs batched with aggregates having high absorptions carry their own internal water supply for curing within the aggregate and, as a result, are more forgiving to poor curing practices or unfavorable ambient conditions. This 'internal curing' water is transferred from the LWA to the matrix phase as evaporation takes place on the surface of the concrete, thus continuously maintaining moisture balance by replacing moisture essential for an extended continuous hydration period determined by ambient conditions and the as-batched moisture content of the LWA.

Pumping

LWA may absorb part of the mixing water when exposed to high pumping pressures. To avoid loss of line workability, it is essential to raise the level of absorption of the LWA prior to pumping. Presoaking is best accomplished at the aggregate production plant where a uniform moisture content is achieved by applying water by spray bars directly to the aggregate moving on belts. This moisture content can be maintained and supplemented at the concrete plant by stockpile hose and sprinkler systems for at least one, but preferably, three days.

Presoaking will significantly reduce LWA rate of absorption, minimizing water transfer from the matrix fraction which can cause slump loss during pumping. Higher moisture contents developed during presoaking will result in an increased specific gravity which, in turn, develops higher fresh concrete density. High water content due to presoaking will eventually diffuse out of the concrete, developing a longer period of internal curing and a larger fresh to equilibrium density differential than that usually associated with LC using aggregates of a lower moisture content. Aggregate suppliers should be consulted for mix design recommendations necessary for consistent pumpability. Mix designs and the physical properties measured on samples of HSLC pumped 830 ft. (250 m) to the 60th floor of the NationsBank project in Charlotte, NC (Fig. 10.12) are shown in Table 10.1.

Laboratory and field control

Changes in LWA moisture content, gradation or specific gravity as well as the ususal job site variation in entrained air suggest frequent fresh concrete

Fig. 10.12

Table 10.1   Mix design and physical properties for concretes pumped 830 ft. (268 m) on Nationsbank Building, Charlotte, NC, 1991

Mix #	1	2*	3
Mix proportions:			
Cement type III (lbs.)	550	650	750
Fly ash (lbs.)	140	140	140
Solite 3/4 to #4 (lbs.)	900	900	900
Sand (lbs.)	1370	1287	1203
Water (gals.)	35.5	36.5	37.2
WRA (oz.)	27.6	31.6	35.6
Superplasticizer	55.2	81.4	80.1
Fresh concrete properties:			
Initial slump (inches)	$2\frac{1}{2}$	2	$2\frac{1}{4}$
Slump after superplasticizer	$5\frac{1}{2}$	$7\frac{1}{2}$	$6\frac{3}{4}$
% air	2.5	2.5	2.3
Unit weight (pcf.)	117.8	118.0	118.0
Compressive strength (psi):			
4 days	4290	5110	5710
7 days	4870	5790	6440
28 days (avg.)	6270	6810	7450
Splitting tensile strength (psi):	520	540	565

* Mix selected and used on project

checks to facilitate adjustments necessary for consistent concrete characteristics. Standardized field tests for consistency, fresh unit weight and entrained air content should be employed to verify field concrete conformance with trial mixes and the project specification. Sampling should be conducted in accordance with ASTM C172, *Standard Practice for Sampling Freshly Mixed Concrete*, and ASTM C173, *Standard Test Method for Air Content*. The *Standard Test Method for Unit Weight of Structural Lightweight Concrete*, ASTM C567, describes methods for the determination of the in-service, equilibrium unit weight of structural LC. In general, when variations in fresh density exceed ±2%, adjustments in batch weights are required to restore specified concrete properties. To avoid adverse effects on durability, strength and workability, air content of HSLC should not vary more than ±1% from specified values.[4]

10.6   Applications of high strength lightweight aggregate concrete

Precast and prestressed structures

HSLC with compressive strength targets ranging from 5000 to 8000 psi (35 to 55 MPa) has been successfully used for almost four decades by North American precast and prestressed concrete producers. Presently there are

Fig. 10.13

ongoing investigations into somewhat longer span lightweight precast concrete bridges that may be feasible from a trcuking/lifting/logistical point of view.

Garage members with 50–63 ft. (15–20 m) spans are generally constructed with double tees composed of sanded LC with air dry density of approximately 115 pcf (1850 kg/m^3) (Fig. 10.13).

Weight reduction is primarily for lifting efficiencies and lower transportation costs. One prestressed garage project is of interest from the perspective of the precast producer's quality control adjustments of the mix components, when statistical studies of the plant's first use of HSLC indicated unduly high strengths. The first series of statistical test results with a mix that included 755 pcy (450 kg/m^3) of cement yielded seven day strengths of 7450 psi (51 MPa) and 28 day strengths in excess of 9000 psi (62 MPa). Cutting cement back to 705 pcy (429 kg/m^3) developed a 28 day strength of 7910 psi (54.5 MPa) and after a final cement reduction to 660 pcy (390 kg/m^3), the HSLC developed 7500 psi (52 MPa) compressive strength at 28 days.[5]

Buildings

The 450 ft. (140 m) multi-purpose Federal Post Office and Office Building constructed in 1967 with five post office floors and 27 office tower floors was the first major New York City building application of post-tensioned floor slabs. Concrete tensioning strengths of 3500 psi (24 MPa) were routinely achieved at 3 days for the 30 × 30 ft (9 × 9 m) floor slabs with a design target strength of 6000 psi (41 MPa) at 28 days. Approximately 30,000 cubic yd (23,000 m^3) of structural LC were incorporated into the floors, and the cast-in-place architectural envelope serves a structural as

Fig. 10.14

well as aesthetic function. Despite the highly polluted urban atmosphere, the buff colored concrete has maintained its handsome appearance (Fig. 10.14).

The recently completed (1991) North Pier Tower (Chicago) utilized HSLC in the floor slabs as an innovative structural solution to avoid construction problems associated with the load transfer from HSNWC columns through the floor slab system. ACI 318 requires differences in compressive strength between column concrete and the intervening floor slab concrete to be less than a ratio of 1.4. By using HSLC in the slabs with a strength greater than (9000/1.4) = 6430 psi (44 MPa) the floor slabs could be placed using routine placing and techniques thus avoiding scheduling and placing problems associated with the 'mushroom' technique (Fig. 10.15). In this approach, high strength column concrete is overflowed

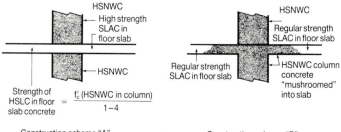

Fig. 10.15 Alternate construction schemes for transfer of HSNWC column loads through floor slabs

from the column and intermingled with the regular strength floor slab concrete. The technique used in the North Pier project avoids delicate timing considerations that are necessary to avoid cold joints.

Bridges

Of the more than 400 LC bridge decks constructed throughout North America, most have been produced with concretes at higher than usual commercial level. The Sebastian Inlet Bridge, which utilizes extra long HSLC drip-in spans during its construction in 1965, is included in one LWA supplier's listing of almost 100 completed bridges. Transportation engineers generally specify higher concrete strengths on bridge decks, primarily to insure high quality mortar fractions (high strength combined with high air content) that will minimize maintenance costs. One state authority has completed more than 20 bridges utilizing HSLC using a target strength of 5200 psi (36 MPa), 6–9% air content, and an air dry density of 115 pcf (1850 kg/m^3). Recent studies have identified tens of thousands of bridges in the United States which are functionally obsolete with low load capacity, unsound concrete, or insufficient number of traffic lanes. To remedy limited lane capacity, Washington, DC engineers have replaced a 4-lane bridge with 5 new lanes providing a 50% increase in one-way, rush hour traffic, without replacing the existing structure, piers, or foundations.

Marine structures

Because many offshore concrete structures will be constructed in shipyards located in lower latitudes and then floated and towed to the project site, there is a special need to reduce weight and improve the structural efficiency of the cast-in-place structure. Because shallow water conditions mandate lower draft structures the submerged density ratio of

$$\frac{\text{HSNWC}}{\text{HSLC}} \frac{2.50 - 1.00}{2.00 - 1.00} = 1.50$$

which is greater than the air density ratio

$$\frac{2.50}{2.00} = 1.25$$

becomes increasingly important.

These requirements have already been satisfied by several projects placed in the Arctic, e.g., HSLC was used in 1981 in the TARSUIT CAISSON Retained Island project constructed in Vancouver, Canada and transported to the Beaufort Sea on the Alaskan North Slope.[61] This project was followed in 1984 with the HSLC used in the construction of the Concrete Island Drilling System built in Japan and also towed to the

Table 10.2   Physical properties, strength and chloride ion permeability of structural lightweight concrete microsilica[1]

Lightweight concrete mix		K	O	F	C	N+ (With Microsilica)	N−[1] (Without Microsilica)
Fresh concrete properties	Wet density psi (kg/m³)	116.5 (1870)	120.5 (1930)	117.9 (1870)	116.5 (1870)	118.2 (1890)	116.8 (1870)
	Slump in (mm)	7.5 (190)	8.5 (215)	8.5 (215)	8.8 (225)	8 (205)	6 (150)
	Air content %	5.5	5.2	5.8	5.2	5.8	5.2
Compressive strength at age (days)	2	6360 (43.9)	4850 (33.4)	4180 (28.8)	5920 (40.8)	5800 (40.0)	2890 (19.9)
	7	8270 (57.0)	5740 (39.6)	5310 (36.6)	7470 (51.5)	6990 (48.2)	3960 (27.3)
	28	9648 (66.5)	6840 (47.2)	6050 (41.7)	8600 (59.3)	7460 (51.0)	5050 (34.8)
	90	9855 (68.0)	6550 (45.2)	6225 (42.6)	8990 (62.0)	7525 (51.9)	5270 (36.3)
W/B ratio[2]		.36	.33	.35	.31	.32	.43
Chare passed (coulombs)[3]		260	450	450	220	370	4800

(1) Mix N− identical to N+ with exception of no microsilica/superplasticizer.
(2) Includes water from slurry.
(3) Rapid determination of the chloride permeability of concrete. AASHTO T277. Microsilica Elkem, Pittsburgh, PA.

Beaufort Sea.[62] This project called for compressive strengths of 6500 psi (45 MPa) with density of 115 pcf (1840 kg/m³). In addition to reducing draft during construction and towing, use of HSLC in offshore gravity based structures can be justified by the improved floating stability as well as the opportunity to carry more topside loads.

Rehabilitation of bridges and parking decks

Numerous opportunities exist for the efficient rehabilitation of existing deteriorated bridges and parking decks. For example, replacing 3 in. of deteriorated NWC with 4 in. of low permeability HSLC will also provide opportunities to improve deficient surface geometry; for example, increasing slopes to drain and improved super elevation on curves.

Cooperative research testing programs with a microsilica producer[1] have demonstrated significant reductions in the permeability of concrete to chloride ions when measured by AASHTO T277, *Rapid Determination of the Chloride Permeability of Concrete*. Table 10.2 reports the physical properties, compressive strength, and shows the dramatic reduction of current passed by LC 'N−' (without microsilica) from 4800 to 370 coulombs when microsilica is added to the concrete.

References

1   Holm, T.A. and Bremner, T.W. (1990) *70 year performance record for high strength structural lightweight concrete*, Proceedings of First Materials Engineering Congress. Materials Engineering Division, ASCE, Denver, Colorado, Aug.

2   Holm, T.A. and Bremner, T.W. (1991) The durability of structural lightweight concrete, ACI SP-126, *Durability of Concrete*. Second International Conference, Montreal, Canada, August.

3   Holm, T.A. (1980) Performance of structural lightweight concrete in a marine environment, ACI Publication SP-65, *Performance of Concrete in a Marine Environment*. International Symposium, St. Andrews-By-The-Sea, Canada, August.

4   (1985) Criteria for designing lightweight concrete bridges, by U.S. Dept. of Transportation, Federal Highway Administration, Report No. FHWA/RD-85/045, August.

5   Holm, T.A. (1980) Physical properties of high strength lightweight aggregate concretes, Second International Congress of Lightweight Concrete, London, UK, April.

6   Holm, T.A., Bremner, T.W. and Newman, J.B. (1984) Lightweight aggregate concrete subject to severe weathering. *Concrete International*, June.

7   Bremner, T.W., Holm, T.A. and deSousa, H. (1984) Aggregate-matrix interaction in concrete subjected to severe exposure, FIP-CPCI International Symposium on Concrete Sea Structures in Arctic Regions. Calgary, Canada, August.

8   Bremner, T.A. and Holm, T.A. (1986) Elastic compatibility and the behavior of concrete. *Journal American Concrete Institute*, March/April.

9   Stagg, K.G. and Zienkiewicz, O.C. (1968) *Rock mechanics in engineering practice*. J. Wiley and Sons, New York.

10 Shah, S.P. and Chandra, S. (1968) Critical stress, volume change and microcracking of concrete. *Journal American Concrete Institute*, September.

11 Muller-Rochholz, J. (1979) Determination of the elastic properties of lightweight aggregate by ultrasonic pulse velocity measurements. *International Journal of Lightweight Concrete*, Lancaster, UK, Vol. 1, No. 2.

12 (1983) *FIP manual of lightweight aggregate concrete*, 2nd Edition. Federation Internationale de la Precontrainte/Surrey University Press, Glasgow.

13 ACI 363 (1984) State of the art report on high-strength concrete. *ACI Journal*, July/August.

14 ABAM Engineers, Inc. (1983) Developmental design and testing of high-strength lightweight concretes for marine Arctic structures, Program Phase I, Joint Industry Project Report, AOGA Project No. 198, Federal Way, Washington, May.

15 ABAM Engineers, Inc. (1984) Developmental design and testing of high-strength lightweight concretes for marine Arctic structures, Program Phase II, Joint Industry Project Report, AOGA Project No. 203, Federal Way, Washington, DC, September.

16 ABAM Engineers, Inc. (1986) Developmental design and testing of high-strength lightweight concretes for marine Arctic structures, Program Phase III, Joint Industry Project Report, AOGA Project No. 230, Federal Way, Washington, August.

17 Hoff, G.C. (1991) High strength lightweight concrete for Arctic applications, ACI Symposium on *Performance of lightweight concrete*. Dallas, Texas, November (To be published by ACI).

18 Hoff, G.C. High strength lightweight aggregate concrete – current status and future needs.

19 (1987–1989) High strength concrete, Joint Industry Project, SINTEF, Trondheim, Norway.

20 (1985–1988) Ductility performance of offshore concrete structures, Joint Industry Project, Veritec, Hovik, Norway.

21 (1984–1987) Design of peripheral concrete walls subjected to concentrated ice loading, Joint Industry Study, Ben C. Gerwick, Inc., San Francisco, California.

22 (1987–1988) Punching shear resistance of lightweight concrete offshore structures for the Arctic, Joint Industry Project, National Institute of Standards and Technology, Washington, DC.

23 (1960) Story of the *Selma* – expanded shale concrete endures the ravages of time. Expanded Shale Clay and Slate Institute, Bethesda, Maryland, June, Second Edition.

24 (1960) *Bridge deck survey*. Expanded Shale Clay and Slate Institute, Washington, DC.

25 (1970) *Freeze-thaw durability of structural lightweight concrete.* Lightweight Concrete Information Sheet No. 13, Expanded Shale Clay & Slate Institute, Salt Lake City, Utah.

26 Malhotra, V.M. (1987) CANMET investigations in the development of high-strength lightweight concrete, *Proceedings*, Symposium on the Utilization of High Strength Concrete, Stavanger, Norway, June 15–18, TAPIR, Trondheim, Norway.

27 Malhotra, V.M. (1981) *Mechanical properties and durability of superplasticized semi-lightweight concrete*, SP 68-16. American Concrete Institute.

28 Seabrook, P.I. and Wilson, H.S. (1988) High strength lightweight concrete for use in offshore structures – utilization of fly ash and silica fume. *International Journal of Cement Composites and Lightweight Concrete*, August, Lancaster, UK.

29 Ramakrishnan, V., Bremner, T.W. and Malhotra, V.M. (1991) Fatigue strength and endurance limit of lightweight concrete, ACI Symposium, *Performance of structural lightweight concrete*, Dallas, Texas, November. (To be published by ACI.)

30 Berner, D.E. (1991) High ductility, high strength lightweight aggregate concrete, ACI Symposium on *Performance of structural lightweight concrete*, Dallas, Texas, November. (To be published by ACI.)

31 Luther, M.D. (1991) Lightweight microsilica concrete, ACI Symposium, *Performance of structural lightweight concrete*, Dallas, Texas, November. (To be published by ACI.)

32 Zhang, M. and Gjørv, O.E. (1991) Characteristics of lightweight aggregates for high strength concrete. *ACI Materials Journal*, March/April, Detroit, Michigan.

33 Zhang, M. and Gjørv, O.E. (1991) Mechanical properties of high strength lightweight concrete. *ACI Materials Journal*, May/June, Detroit, Michigan.

34 Zhang, M. and Gjørv, O.E. (1990) Development of high strength lightweight concrete, *High strength concrete* – Second International Symposium, SP-121. American Concrete Institute, Detroit, Michigan.

35 Kudriatsev, A.A. (1973) The modulus of elasticity and the modulus of deformations of structural keramzit concrete, *Structure, strength and deformation of lightweight concrete*, Moscow (in Russian).

36 Dovzhik, V.G. and Dorf, V.A. (1973) The relationship between the compressive strength and modulus of elasticity of keramzit concrete and the strength and deformation properties of its components, *Structure, strength and deformation of lightweight concrete*, Moscow (in Russian).

37 Buzhevick, G.A. and Zhitkevich, R.K. (1973) Investigation of the distribution of average deformations and stresses in the components of high strength

keramzit concrete for various combinations of their moduli of elasticity, *The strength, structure and deformation of lightweight concrete*, Moscow (in Russian).

38 Weigler, H. and Karl, S. (1972) *Stahlleichtbeton*. Bauverlag, Wiesbaden, Germany.

39 Swamy, R.N. and Jiang, E.D. (1991) Pore structure and carbonation of lightweight concrete after ten year exposure, ACI Symposium, *Performance of structural lightweight concrete*, Dallas, Texas, November. (To be published by ACI.)

40 Hoff, G.C. (1991) Private communication, November.

41 Abrams, D. (1923) Influence of aggregates on the durability of concrete, *ASTM Proceedings*, **23**, Part II Technical Papers.

42 LaRue, H.A. (1946) Modulus of elasticity of aggregates and its effect on concrete, *Proceedings*, ASTM **46**.

43 ACI 213 (1989) *Guide for structural lightweight aggregate concrete*, Reported by ACI Committee 213. American Concrete Institute, Detroit, Michigan.

44 Reichard, T.W. (1964) *Creep and drying shrinkage of lightweight and normal-weight concretes*, Monograph No. 74, National Bureau of Standards, Washington, DC, March. (Available from Superintendent of Documents, US Government Printing Office, Washington, DC.)

45 Shideler J.J. (1957) Lightweight-aggregate concrete for structural use. *ACI Journal, Proceedings*, **54**, No. 4, October. Also, *Development Department Bulletin*, No. D17, Portland Cement Association, Skokie, Illinois.

46 Shah, S.P., Naaman, A.E. and Moreno, J. (1983) Effect of confinement on the durability of lightweight concrete. *International Journal of Cement Composites and Lightweight Concrete*, Lancaster, UK, February.

47 Holm, T.A. and Bremner, T.W. Long term shrinkage of structural lightweight concrete, unpublished.

48 Rogers, G.L. (1957) On the creep and shrinkage characteristics of solite concretes, *Proceedings, World Conference on Prestressed Concrete*, July, San Francisco, California.

49 Holm, T.A., Bremner, T.W. and Vaysburd, A. (1988) Carbonation of marine structural lightweight concrete, *Proceedings Second International Conference on Concrete in a Marine Environment*, St. Andrews-By-The-Sea, Canada, SP-109. American Concrete Institute, Detroit, Michigan.

50 Klieger, P. and Hansen, J.A. (1961) Freezing and thawing tests of lightweight aggregate concrete. *Journal American Concrete Institute*, Detroit, Michigan, January.

51 Khokrin, N.K. (1973) The durability of lightweight concrete structural members, SAMARA, USSR (in Russian).

52 Malhotra, V.M., Carette, G.G. and Bremner, T.W. (1987) Durability of concrete containing supplementary cementing materials in marine environment, Katherine and Bryant Mather International Conference, SP-100. American Concrete Institute, Detroit, Michigan.

53 Malhotra, V.M., Carette, G.G. and Bremner, T.W. (1988) Current status of CANMET's studies on the durability of concrete containing supplementary cementing materials in marine environment, *Second International Conference on Concrete in a Marine Environment*, St. Andrews-By-The-Sea, Canada, SP-109, ACI, Detroit, Michigan.

54 Nishi, S., Oshio, A., Sone, T. and Shirokuni, S. (1980) Water tightness of concrete against sea water. Onoda Cement Co., Ltd.

55 Keeton, J.R. (1970) Permeability studies of reinforced thin-shell concrete, Naval Civil Engineering Laboratory, Port Hueneme, California, Technical Report R692 YF 51.42.001, 01.001, August.

56 Bamforth, P.B. (1987) The relationship between permeability coefficient for

concrete obtained using liquid and gas. *Magazine of Concrete Research*, **39**, No. 138, March.

57 Bremner, T.W., Holm, T.A. and McInerney, J.M. (1991) Influence of compressive stress on the permeability of concrete, ACI *Symposium on Performance of Structural Lightweight Concrete*, Dallas, Texas, November. (To be published by ACI.)

58 Ahmad, S.H. and Batts, J. (1991) Flexural behavior of doubly reinforced high strength lightweight concrete beam with web reinforcement. *ACI Structural Journal*, May/June, Detroit, Michigan.

59 Moreno, J. (1986) Lightweight concrete ductility. *Concrete International*, American Concrete Institute, November, Detroit, Michigan.

60 Rabbat, B.G., Daniel, J.I., Weinmann, T.L. and Hanson, N.W. (1986) Seismic behavior of lightweight and normal weight concrete columns. *ACI Journal*, January/February, Detroit, Michigan.

61 (1982) Concrete island towed to Arctic, American Concrete Institute, *Concrete International*, March, Detroit, Michigan.

62 McNarey, J.F., Ono, Y., Okada, T., Imai, M. and Kuroki, (1984) Freeze-thaw durability of high strength lightweight concrete, American Concrete Institute Fall Convention, November, Detroit, Michigan.

11 Applications in Japan and South East Asia

Shigeyoshi Nagataki and Etsuo Sakai

11.1 Introduction

In Japan high strength concrete was first achieved as early as the 1930s. For example, Yoshida reported in 1930 that high strength concrete with 28-day compressive strength of 102 MPa was obtained.[1] This result was obtained by a combination of compression and vibration processes without chemical or mineral admixtures. This isolated development was not followed by systematic development in the production and use of high strength concrete till the mid 1960s.

In 1968, high-strength reinforced mortar piles were developed by a process of autoclave curing with the use of silica powder in cement. The compressive strength of concrete used was 11,300 psi (78 MPa).[2] In 1970, high strength concrete with a compressive strength of 12,700 psi (88 MPa) was developed and used for prestressed concrete piles.[3] This high strength concrete was produced by using superplasticizer and autoclave curing. In the 1970s high strength concrete was utilized in the construction of railway bridges,[4] and since then its use in the construction industry has continued to increase. The growth of the use of high strength concrete is primarily due to the development of naphthalene sulfonate condensed superplasticizer, which was developed in Japan in 1964.[5]

The definition of high strength concrete as defined by the Japan Society of Civil Engineering (JSCE) is different from that of Architectural Institute of Japan (AIJ). In 1980 JSCE published *Proposed Recommendations for the Design and Construction of High Strength Concrete*,[6] which defined high strength concrete as a concrete with a design compressive strength 8500 to 11,500 psi (59 to 79 MPa). On the other hand, AIJ defines high strength concrete as a concrete with strength of 3900 to 5100 psi (27 to 35 MPa) for normal weight concrete, and with strength of 3500 to 3900 psi (24 to 27 MPa) for a lightweight concrete.[7]

For civil engineering structures in Japan, high strengtrh strength concrete is used mainly for bridges, high-rise buildings, and piles.[4] Studies in the use of high strength concrete in super high-rise reinforced concrete buildings are being promoted under the leadership of the Japanese National Project by Ministry of Construction (MOC). Studies related to the development and production of high strength concrete of strengths up to 17,100 psi (118 MPa) are under way and techniques for making these concretes practical for high-rise construction are being investigated.[8]

In South East Asia, studies into the utilization of high strength concrete have just started. However, in some of the countries, especially in Singapore, practical applications of high strength concrete have already commenced in high-rise buildings.

11.2   Methods of strength development

In this section, different methods employed for developing concretes with higher strengths are summarized, with special emphasis on the methods being used or studied in Japan. In recent years, many studies aimed at developing practical methods for producing higher strength concretes have been undertaken, especially in connection with the Japanese National Project related to high-rise buildings. There are a number of techniques for attaining higher strength concretes and they are summarized in Fig. 11.1.[9]

The strength development of concrete primarily depends upon the characteristics of hardened cement paste, aggregates, and the interface boundary between the hardened cement paste and the aggregates. It is well known that the strength of hardened cement paste depends on its degree of porosity, and laws have been proposed to explain the relationship between porosity and strength, e.g., Knudsen's formula.[10] According to these laws, the smaller the pore ratio the greater the strength. Therefore, to achieve higher strengths, it is necessary to reduce capillary pores. This can be done by reducing the water cement ratio which results in an increase in the amount of hydrates generated during the hydration process. Another approach for reducing the capillary pores is to impregnate the concrete with polymers. Polymer impregnated concrete (PIC) is one type of high performance concrete which also has a high strength.

Reduction in water cement (w/c) ratio increases the strength. With the use of superplasticizers, the water cement ratio can be reduced while maintaining the workability of the concrete. For concretes with very low water cement ratios, special techniques such as harmonic vibration and centrifugal compacting process have been developed in Japan and are being used by the construction industry.

A combination of a high dosage of dispersing agents and ultra-fine particles make it possible to maintain the workability of concrete with a low water cement ratio, e.g., Densified Systems of homogeneously

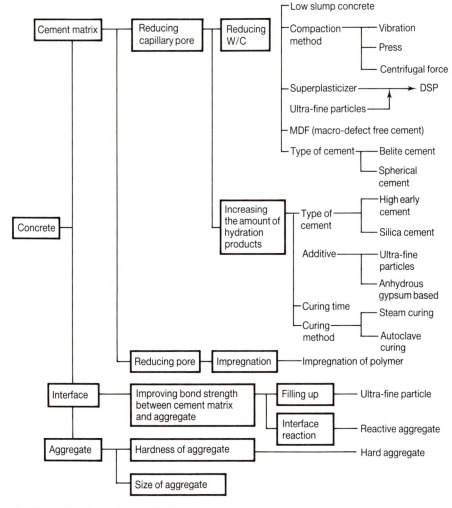

Fig. 11.1   Techniques for attaining high strength

arranged ultra-fine Particles (DSP). In these types of concretes, strengths of over 14,500 psi (100 MPa) are achievable.[11] The microstructure of the hydrate of this concrete is quite different from that of conventional concrete because the ultra-fine particles play a more important role in the reaction than the cement particles. It is reported that most of the cement particles which remain unaffected by the hydration reaction become inner-fillers which makes the hydrate of this concrete different from that of conventional concrete.[12] Another approach for reducing the water cement ratio without losing the workability is by using different kinds of cement such as belite cement[13] or spherical cement.[14] This approach is being investigated in Japan and the results are encouraging.

Increasing the amount of hydrates also results in a reduction of the total pore volume in the concrete. The use of high-early strength cement or silica cement is one method which has been used for some time to increase the proportions of the hydrates. In addition to this, like pozzolanic reactions or latent hydraulicity, the calcium hydroxide generated by hydration of cement or alkaline stimulation generates hydrates in the pores. Mineral additives with ultra-fine particles as the main ingredient are usually used for high-strength concrete. Silica fume, classified and ground blast furnace slag or fly ash,[15] and metakaolin[16] are the most popular mineral additives employed in the production of higher strength concretes. In addition, anhydrous gypsum based additives have also been used to produce ettringite actively,[17] which in turn can also reduce the pore volume.

The amount of hydrates in concrete largely depends on the curing method and the curing time. Steam curing or autoclave curing is effective in promoting the hydration within a certain period, which in turn helps in developing the strength.

The strength of the interface between the aggregate and the paste plays an important role in the achievement of higher strength. The interface strength is affected by the interface structure. Use of calcium hydroxide makes the spaces around the aggregates larger than those in other parts of the concrete continuum, and the greater the amount of calcium hydroxide, the lower the aggregate-paste interfacial bond strength, which in turn reduces the overall strength of the composite.

The replacement of cement by silica fume can increase the strength remarkably. This is attributable to the calcium silicate hydrates generated by both ultra-fine particles such as silica fume and the calcium hydroxide around the aggregates which improve the bond between the aggregates and the cement matrix.[18]

Higher strengths can also be achieved by improving the interfacial bond strength by using cement clinkers that are as reactive as the aggregates.[19] Also use of stronger aggregates with superior surface characteristics improves the strength. Figure 11.2 shows the relationship between water to binder ratio and the compressive strength of DSP mortars with various kinds of aggregates.[12] The results indicate that at a certain water binder ratio, the greater strength of the aggregates results in a greater compressive strength in the mortar, which is remarkable, especially in the higher strength zone. This is also the case with coarse aggregates, and one test result shows that at a constant strength of mortar, the maximum difference between the strengths of the concretes achieved by employing different kinds of coarse aggregates is 5800 psi (40 MPa).[20]

Various methods for examining the quality of aggregates have been proposed. These include, for example, a method for obtaining material factors of aggregates by examining the experimental relation between the strength of mortar and that of concrete for each absolute volume of coarse aggregate.[21] Another method is to determine the strength of coarse

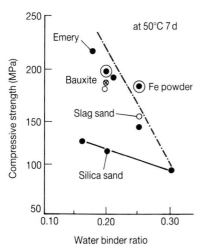

Fig. 11.2   Relation between the water to binder ratio and the compressive strength of DSP mortars with various types of aggregate

aggregates by examining the relation between the strength ratio of concrete to mortar and the reciprocal of the strength of mortar.[22] It is also reported that from the view point of mineralogy or petrology, strength of mineral, fineness of particles, micro cracks within particles, and cleavage should be examined[23] to maximize the intrinsic strength of the concrete composite. Since the quality of aggregates has recently been getting poorer, especially in Japan, it is thought that for producing higher strengths concretes in the future, careful studies of aggregates as well as of cement matrix, will become very important.

11.3   Applications

The applications of high strength concrete in Japan are summarized in Table 11.1. The details of these examples including the methods of strength development employed are described below.

Bridges

High strength concrete has been used for producing prestressed concrete (PC) girders for the purpose of reducing the dead load and for achieving longer span.[8] Japanese Industrial Standard (JIS) requires the compressive strength of concrete used for PC girders to be above 7100 psi (49 MPa) as shown in Table 11.1. Most of the high strength concrete used for railway bridges is usually a superplasticized concrete with a low water cement ratio. Examples include the Dai-ni-Ayaragigawa Bridge, Ootanabe Bridge, Iwahana Bridge, Kazuki Bridge and Akkagawa Bridge. The strengths of concretes used in the highway bridges are in the range of

Table 11.1   Application of high strength concrete in Japan

Concrete constructions and concrete products	Design strength (MPa)	Techniques for high strength	Notes
Railway bridges	59~79	Superplasticizer (+high early cement)	Lightened bridge weight prestressed concrete
Highway bridges	59~69	Superplasticizer (+high early cement)	
Prestressed concrete beams		Superplasticizer (+ high early cement)	
for slab bridges	>49		JISA5313
for beam bridges	>49		JISA5316
for light load slab bridges	>49		JISA5319
Diaphram walls	~49	Superplasticizer + low heat cement	Massive concrete
Oil drilling rigs	58~65	Superplasticizer + silica fume	Light weight concrete
Abrasion resistance concrete	39~79	Superplasticizer + silica fume or anhydrous gypsum based additive	Repair of dam (floor)
High rise RC buildings	35~47	Superplasticizer or new type superplasticizer (+ultra-fine particles)	National project 60~112 MPa
Reinforced spun concrete piles	>39	(Superplasticizer)	JISA5310 centrifugal force
Pretensioned spun concrete piles	>49	Superplasticizer	JISA5335 centrifugal force
Posttentioned spun concrete piles	>49	Superplasticizer	JISA5336 centrifugal force
Pretensioned spun high strength concrete piles	>79	Superplasticizer + autoclave curing or anhydrous gypsum based additive	JISA5337 centrifugal force
Steel concrete composite piles	>79	Superplasticizer + autoclave curing or anhydrous gypsum based additive + expansive additives	centrifugal force
Prestressed spun concrete poles	>49 (or 79)	Superplasticizer (+autoclave curing)	JISA5309 centrifugal force
Centrifugal reinforced concrete pipes	49~69	Superplasticizer + anhydrous gypsum based additive or low slump concrete	Centrifugal force Jacked pipe
Railway sleepers	39~49	Superplasticizer (+high early cement)	
Concrete segments	49	Superplasticizer	
Machine beds	79~108	Superplasticizer + silica fume (DSP)	Damping capability

Fig. 11.3 Iwahana Railway Bridge (Sanyo shinkan-Sen)

8400 psi to 9800 psi (59 to 69 MPa). Examples include the Nitto, Kami-noshima, Jodoji and Seto Highway Bridges. In addition to the above examples, concretes with design strengths in the range of 5600 to 6400 psi (39 to 44 MPa) has been utilized in many bridges in the form of prestressed or cast-in-place concrete construction.

The mix proportions of the concrete used for these bridges are listed in Table 11.2. In the Dai-ni-Ayaragigawa Railway Bridge, high strength concrete was employed in order to reduce the dead load. The weight of the main girder with 5600 psi (39 MPa) normal strength concrete would have been 170 tons. This was reduced to 150 tons by the use of 8500 psi (59 MPa) strength concrete. The strength of the concrete with job-site curing resulted in 9000 to 10,400 psi (62 to 72 MPa), with an average of 9400 psi (65 MPa). The production of the PC girders was done in a yard near the construction site, and the concrete used was produced in ready-mix concrete plant about 2 km from the yard.[24]

In the construction of Iwahana Bridge (Fig. 11.3), precast concrete members were primarily used but for the joints cast-in-place concrete was used. Concrete mix information such as cement type, design strength, w/c ratio, dosage of superplasticizer etc. is given in Table 11.2. The w/c ratio for cast-in-place concrete was about 0.29. The average strength attained was 10,700 psi (74 MPa).[25]

In the construction of the Akkagawa Bridge, concrete with a design strength of 11,400 psi (79 MPa) was employed, and it was produced with autoclave curing. The average concrete strength and the standard deviation of the strength after curing, were 14,100 psi (97 MPa) and 580 psi (4 MPa) respectively.[26]

The merit of using high strength concrete for piers was examined.[27] It was reported that by increasing the design strength of concrete from 3900 to 8600 psi (27 to 59 MPa), the cross section of a pier for monorail could be reduced by 36%. It was also estimated that by increasing the design strength from 3900 psi (27 MPa) to 5700 and 11,400 psi (39 and 79 MPa), the cross section of a highway bridge pier could be reduced by 13% and 56% respectively.[27]

Table 11.2 Mix proportions of high strength concrete for railway bridges

Name of bridges	Type of structure	Design strength (MPa)	Type of cement	W/C (%)	s/a (%)	Dosage of SP (CX%)	Average of strength (MPa)	Construction
Da-ni-Ayaragigawa	1 Girder bridge span: 49 m	59	Normal cement	30	40	0.75	65	1973
Iwahana	Span: 45 m	79	High early cement	23	38.5	1.5	83	1974
Akkagawa	Span: 45.9 m	79	Normal cement	30	39.5	1.5	93*	1976

(Slump: 12 + 2.5 cm, G_{max}: 20 mm, * autoclave curing)

Table 11.3 Mix proportions of the main tower for Aomori Oohashi Road Bridge (Type of structure: PC cable-stay bridge, design strength: 59 MPa)

G_{max} (mm)	W/C (%)	S/a (%)	Unit weight (kg/m³)		Target of slump flow	Dosage of new SP (%)	Average of compressive strength (MPa)
			Cement	Water			
25	31.4	39	430	135	40~45 cm	2.2 ~2.5	80
25	35.0	40	386	135	40~45 cm	1.85~2.8	74

Fig. 11.4   Aomori Bay Bridge

The main towers of Aomori Bay Bridge were constructed in 1989 and high strength concrete with a design strength of 8600 psi (59 MPa) was used.[28] Figure 11.4 shows the bridge and Table 11.3 lists the mix proportions of the concrete. In this prestressed concrete cable-stayed highway bridge, high strength concrete was employed to achieve a slender structure with a reduced dead load and attractive aesthetic appearance. The concrete incorporated a new type of superplasticizer. Due to the presence of carboxylate polymer, this new superplasticizer reduced the slump loss which is caused by conventional sulfonated naphthalene formaldehyde condensate or sulfonated melamine condensate superplasticizer. Among these new types of superplasticizer, the one with air-entraining agent is currently being used.

Oil drilling rigs

An oil drilling rig to be used in the Arctic Ocean was made in Japan, using high strength concrete for purposes of greater durability and a lighter dead load.[29] In Table 11.4, some of the mix proportions, related to lightweight high strength concrete with silica fume and superplasticizer are given. Some special methods were introduced into the production of the concrete for this oil drilling rig in order to enhance its freeze-thaw resistance. It should be noted that up to now, only one drilling rig of this type has been made in Japan.

Diaphragm walls

Usually, for diaphragm walls concrete of 3000 to 3500 psi (21 to 24 MPa) is used. However, it is desirable to use higher strength concretes for deep underground large scale walls, which are subjected to higher water and earth pressures. This is very important for Japan, since there is a strong need to develop underground space because of the limited land mass. The

Table 11.4 Mix proportions of high strength concrete using light weight aggregate for oil drilling rigs

W/C (%)	Unit weight (kg/m^3)					Slump (mm)	Air (%)	Compressive strength (MPa, for 28d)
	W	C	Silica fume	Fine aggregate	Coarse aggregate			
30.8	160	520	52	603	493	185	7.7	58
28.5	143	500	75	604	456	170	5.2	65

(with superplasticizer and air entraining agent)

concrete for a diaphragm wall does not need consolidation, because it is cast in bentonite slurry through a tremie pipe. Therefore, the fluidity and the resistance to segregation of concrete are important considerations for designing the concrete mixes.

Table 11.5 shows one of the mix proportions of the concrete used for the diaphragm wall of LNG tanks at the Tokyo Gas Company's Sodegaura Plant.[30] Usually, a diaphragm wall is for temporary use, however in this case the upper part of the diaphragm wall was combined with an inner lining concrete of the side wall to achieve a permanent structure. The inner lining concrete was cast first using the top down lining method in which the excavation was carried out without shoring. The foundation slab was then cast, followed by the construction of the side wall. The thickness of the side wall was 120 cm, and it utilized high strength concrete. Low-heat cement with three binding components was employed in order to solve the problem related to mass concreting. Figure 11.5 shows the uncovered diaphragm wall after excavation.

Abrasion resistant applications

Due to its superior abrasion resistance qualities, concretes with higher strength have been used in the industrial flooring and in the repair of dams.

Fig. 11.5 Diaphragm wall – LNG tank

Table 11.5 Mix proportions of high strength concrete for diaphragm walls

Design strength (MPa)	W/C (%)	s/a (%)	Unit weight (kg/m³)		Dosage of SP (CX%)		Average of compressive strength (MPa)		
			W	C	1	2*	Standard curing (91d)	Strength of core	Standard deviation
49	28.4	38.8	128	450	1.5	1.0	69	69	7

(*: Relayed Addition, Cement: Low heat type; Cement-BFS-FA)

It is known that concrete of over 5700 psi (39 MPa) is very effective as regards to abrasion-resistance.

Table 11.6 lists some of the mix proportions of the concrete used for the repair work on an intake of a dam. The concrete strength for this application varied between 8300 to 8600 psi (57 to 59 MPa). A mineral additive (anhydrous gypsum based additive) was added to the concrete.[31] This additive can reduce the unit cement content, and can also solve the serious problem of thermal cracking in massive concrete. The unit cement content of this concrete was about 508 lbs/cyd (300 kg/m^3).[32] Concrete of about 11,300 psi (78 MPa) with silica fume has seen limited use for abrasion or chemical-resistant floors.

High-rise reinforced concrete buildings

For the high rise construction in Japan, the strengths of the concretes utilized generally range between 5100 psi (35 MPa) to 9000 psi (63 MPa) and chemical additives such as superplasticizer is usually used. Concrete with 17,100 psi (118 MPa) strength is currently being used in a MOC sponsored national project into high-rise construction. All the high-rise buildings in Japan are designed for earthquake excitations. For these special design conditions, the arrangements for the reinforcement in the joints is rather congested, consequently the flow ability and the filling capacity of fresh concrete is very desirable. Slump loss in fresh concrete is a serious problem in relation to flow ability and filling capacity. Recently, a new type of superplasticizer whose main ingredient is carboxylate polymer with a low rate of slump loss, high water reduction and little retardation has been developed. An investigation to establish a standard for new types of superplasticizer for high strength concrete is in progress.

High strength concrete is used in the columns of high rise reinforced concrete buildings to carry the loads economically and to provide relatively more floor space by having smaller sections. Generally, the design strength of the concrete is 5100 to 5900 psi (35 to 41 MPa) for 25-story buildings, and 5900 to 6800 psi (41 to 47 MPa) for 30 to 40-story buildings.[33] A National Project sponsored by the MOC into high-rise buildings using high strength concrete of up to about 17,100 psi (118 MPa) is now in progress. In this project, the employment of DSP materials with ultra-fine particles like silica fume or blast furnace slag and additives like anhydrous gypsum based additive for high strength concrete are being examined. Only one full-scale experiment of construction has been carried out to date.[34]

Table 11.7 shows the mix proportions and the strength test results of high strength concrete with design strength of 5900 psi (41 MPa). This concrete was used for the recently completed 41-storied high rise building. The actual strength achieved in the field was 8100 psi (56 MPa).[36] Table 11.8 shows the mix proportions of the concrete used for a 8-story building.[36] The design strength for this concrete was of about 8600 psi (59 MPa). In this case, precast concrete was used for the columns and

11.6 Mix proportions of abrasion resistance concrete for dams

Execution place	Slump (cm)	W/C (%)	Unit weight (kg/m³)					Compressive strength (MPa, for 28d)
			W	C	Fine aggregate	Coarse aggregate	Additive*	
Slope of dam body	15	36.5	146	400	776	1049	40	59
Flat place of dam body	18	38.0	152	400	769	1040	40	57

(* Anhydrous gypsum based)

Table 11.7 Mix proportions of concrete for high rise RC buildings

Design strength (MPa)	W/C (%)	Unit weight (kg/m³)				Average of comprehensive strength (MPa, for 28d)	Standard deviation (MPa)
		W	C	Fine aggregate	Coarse aggregate		
41	35.5	175	493	673	1009	56	4
35	40.0	175	438	719	1009	50	3
29	44.5	175	393	755	1009	46	3

beams, and cast-in-place high strength concrete for the slabs and the joints of the columns.

The strengths of all these concretes mentioned above are over the upper limit of 5100 psi (35 MPa) which was established by AIJ as a reference strength to distinguish between the normal strength and high strength concrete. Presently there is no standard for selecting materials and mix proportions for the production of high strength concrete. Hence the high-rise reinforced concrete projects are authorized by the Ministry of Construction (MOC) on the basis of design, construction procedure and methodology.

Concrete piles

Concrete strengths for this application range between 5600 psi (39 MPa) to 11,500 psi (79 MPa). The types of concrete pile used in Japan include reinforced spun concrete piles, pretensioned spun concrete piles, post-tensioned spun concrete piles and pretensioned spun high strength concrete piles (Table 11.1). The production and use of Pretensioned High-strength Concrete (PHSC) piles is increasing very fast and presently exceeds 50% of the total market of piles. This is because: (1) A pretensioned high strength concrete pile can withstand higher vertical and horizontal loads compared with other kinds of pile which reduces the number of piles required, resulting in savings, (2) It has higher impact resistance, (3) The production period is shorter. Figure 11.6 shows a prestressed high strength concrete pile being driven. The typical mix proportions for high strength concrete used for concrete piles is given in Table 11.9. The high strength concrete piles are generally made using the autoclave curing method, and sometimes incorporate silica fine powders.[6]

An alternate method for producing high strength concrete piles in Japan, is by the use of ordinary steam curing with an anhydrous gypsum based

Fig. 11.6   PHC piles

Table 11.8 Mix proportions of high strength concrete for high rise RC buildings

| Design strength (MPa) | Target strength (MPa) | W/C (%) | s/a (%) | Slump (cm) | Air (%) | Unit weight (kg/m³) | | Compressive strength (MPa) | | | | |
						C	W	Standard curing (28d) Job	Site 56d	Core 56d	Core 28d
59	71	29.1	40.1	25	1.5	550	160	81	70	72	64
								86	87	87	62

Table 11.9 Mix proportions of high strength concrete for concrete piles

| Products | Design strength (MPa) | Target strength (MPa) | Type of cement | G_max | Slump (mm) | W/C (cm) | s/a (%) | Unit weight (kg/m³) | | | Dosage of SP (CX%) |
								(%)	C	W	
PHC piles*	78	88	Normal	Normal	20	7***	32	40	450	142	1.25
PC piles**	49	57	Normal	Normal	20	4	36	42	420	152	0.85

(*:Steam curing + autoclave curing, **:Steam curing, ***:By using of pump – 12 cm)

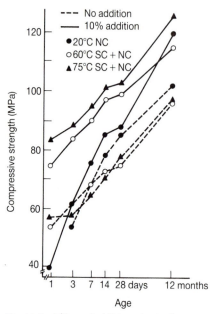

Fig. 11.7　Effect of addition of anhydrous gypsum based additive on compressive strength of concrete

mineral additive as an additional ingredient in concrete. This approach allows the production of larger diameter piles which are becoming popular. The effect of addition of anhydrous gypsum based mineral additive on the compressive strength of concrete is shown in Fig. 11.7. These results are for a mix with the water to cement ratio, the unit cement content and the sand aggregate ratio of 0.29, 470 kg/m³ and 0.42 respectively. This is one of the mix proportions that is being currently used in Japan for the production of high strength concrete piles. For this mix, the strength is about 11,400 psi (79 MPa) just after steam curing, and the 28-day strength is over 14,500 psi (100 MPa). Figure 11.8 shows the hydrated and unhydrated products of cement with anhydrous gypsum based additive in the concrete. During the curing, anhydrous gypsum is reduced and ettringite is produced actively, which causes a decrease in the total pore volume of hardened paste. The ettringite absorbs a greater amount of water than other hydrates resulting in a low water cement ratio.[18]

One enterprising application of high strength concrete in piles is a composite pile of a steel pipe and concrete. This steel concrete composite pile is made by lining a steel pipe with concrete using centrifugal force. The flexural ductility of this steel concrete composite pile is higher than prestressed high strength concrete piles. An example of mix proportion for the high strength concrete used in the production of composite pile is given in Table 11.10, which shows that the concrete is produced with an expansive additive.[37] The main ingredient of the additive is magnesium oxide or calcium oxide when autoclave curing is employed. On the other

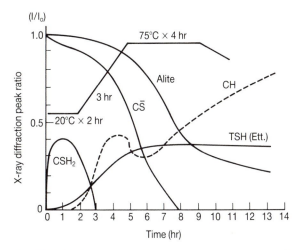

Fig. 11.8 Hydration and unhydration products of cement with anhydrous gypsum based additive

hand, when the composite piles are produced with ordinary steam curing, a combination of anhydrous gypsum based additive and expansive additive is used.[37]

In addition to the piles, high strength concrete is being used for the production of concrete poles. The mix proportions and the manufacturing method are basically similar to those used for the production of the piles.

Other concrete products

The other concrete products in which high strength concrete is being used include hume pipes, sleepers and concrete segments and the mix proportions for the concrete for these applications is given in Table 11.11.

Hume pipes are widely used for sewerage or agricultural canals. Usually, prestressed concrete with an expansive additive added is employed for the production of hume pipes. Because of the jacking method used for driving the concrete pipe, higher strengths of concrete are needed to withstand the static and inertial forces. Considering that longer distance jacking with be desirable in the future, the use of higher strength concretes will increase for this application. In Japan, hume pipes with concrete of up to about

Table 11.10 Mix proportions of high strength concrete for steel concrete composite piles

				Unit weight (kg/m³)			
G_{max} (mm)	Slump (cm)	W/Binder (%)	s/a (%)	W	C	Expansive additive	Dosage of SP (Binder X %)
20	13	30.7	42	132	404	26	1.5

(Autoclave curing, main component of expansive additive – CaO)

10,000 psi (69 MPa) strengths have been produced with centrifugal compaction and have been used in a number of applications. For these
concretes, sometimes only superplasticizer, sometimes a combination of
superplasticizer and anhydrous gypsum based additive is used. For the
placement of concretes with very low water cement ratio, vibratory
centrifugal compaction method is employed.

Prestressed concrete sleepers are used for railway tracks. In addition to
the prestressed concrete mentioned in Table 11.1, JIS specifications for
various kinds of prestressed concrete products have been established. The
specifications require that the design strength of the concrete, and the
strength at the time of prestressing must exceed 7100 psi (49 MPa) and
5700 psi (39 MPa) respectively. Because of this requirement, a combination of high-early-strength cement and superplasticizer is usually employed
in the manufacturing process.

Concrete shield segments are being used as lining in shield tunneling.
High-strength concrete is being used for these segments to make them
lighter hence reducing the transportation and erection costs and to make
them stronger to better withstand the high earth pressures. Figure 11.9
shows concrete segments ready for use in tunnel construction by the shield
tunneling method.

One of the unique applications of high strength concrete is the use of
15,600 psi (108 MPa) concrete with DSP material for the beds of machine
tools. Because of the high degree of rigidity and vibration damping
properties of higher strength concretes, it is very advantageous for this
application.[39] This application is an example of high performance concrete
use in which attributes other than strength govern its choice. Figure 11.10
(by Nippei Toyama Corporation) shows a machine tool, the bed for which
is made by using concrete with a strength of 15,600 psi (108 MPa).

Fig. 11.9 Concrete segment

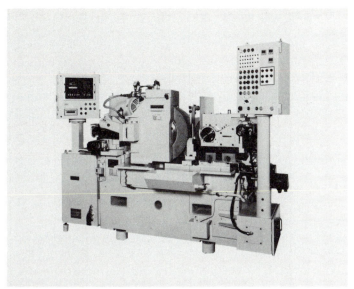

Fig. 11.10   Machine bed (Nippei Toyama Corporation)

Applications in South East Asia

The examples of applications of high strength concrete in the countries of South East Asia are summarized in Table 11.12. The applications of high strength concrete are basically limited high-rise reinforced concrete buildings. The concrete strength usually range from 7200 to 8700 psi (50 to 60 MPa). Figure 11.11 shows a high-rise building in Singapore, which was built by using high strength concrete in the columns from the 3rd underground story to the 7th floor. In the core walls up to the 45th floor, composite structural segments of steel and concrete with strengths of 8700 psi (60 MPa) and 7200 psi (50 MPa) respectively were employed. In addition to Singapore, Korea[39] and Hong Kong, investigations into possible applications of high strength concrete have already been initiated in Thailand.

11.4   Summary and conclusions

There are widespread applications of high performance concretes in Japan and South East Asia. High Performance concrete with very high compressive strength generally called high strength concrete has been used in bridges, oil drilling rigs, diaphragm walls in deep underground tanks etc. For special applications such as industrial flooring and repair of dams, high performance concrete has been used due to its superior abrasion resistance property. For high rise construction, the higher compressive strength of high performance concrete has been utilized in columns to maximize the

Table 11.11 Mix proportions of various concrete products

Products	Design strength (MPa)	Target strength (MPa)	Type of cement	G_{max} (mm)	Slump (cm)	W/C (%)	S/a (%)	Unit weight (kg/m³) W	C	Dosage of SP (C X %)
Centrifugal reinforced concrete pipes	–	54	Normal	20	7	37	43	168	460	0.95
PC railway sleepers	49	64	High early	20	4	32	40	140	440	0.85
PC beams	49	64	High early	20	3	33	35	147	450	0.85
Segments	49	61	Normal	20	5	33	40	146	450	1.05

(Pre-curing for 2~3 h, steam curing for 3~5 h at 65~70°C)

Table 11.12 Application of high strength concrete in South East Asia

Country	Constructions	Design strength (MPa)	Techniques for high strength	Notes
Singapore	Road link	50	Superplasticizer (+ silica fume)	Caisson and bridge
	High rise building	50~60	Superplasticizer or + silica fume	Hoechang building, UOB Plaza, Savu Tower
	Concrete spun piles for wharf	60	Superplasticizer	Expansion of Jurous Port
Korea	High rise building	>49 (for 28d, target strength)	Superplasticizer	Bridges – 40 MPa
Hong Kong	High rise RC building	60 (for 28d)	Superplasticizer + 25% PFA cement	78 floors the expected date of completion: 1992, Aug.

Fig. 11.11   High rise building: Singapore, UOB Plaza

resistance capacity without sacrificing the available floor space. Higher strength concretes have also been used for prestressed concrete piles. Other applications of high performance concrete in Japan and South East Asia include hume pipes for sewerage or agricultural canals, prestressed concrete sleepers, concrete shield segments as lining in shield tunneling, beds for machine tools.

In the last two decades or so, the construction industry in Japan has gained tremendous amount of experience in use of high performance concretes especially high strength concretes. As further development takes place in the high performance concretes, it appears that the use of high performance concretes will continue to increase in Japan and South East Asia.

References

1   Yoshida, T. (1930) *Proceedings Japan Society of Civil Engineering*, **26**, 997.
2   Kokubu, K. and Fukuzawa, K. (1987) *New Concrete Technology*, **8**, 90 pp.
3   Nishi, H., Ooshio, A. and Fukuzawa, K. (1972) *Cement Concrete*, No. 299, 22.
4   Nagataki, S. (1989) *Proceedings MRS International Meeting on Advanced Materials*. Tokyo, **13**, 3.
5   Hattori, K., Yamakawa, K., Tuji, A. and Akashi, T. (1964) *Semento Gijutu Nennpo*, **18**, 200.
6   (1980) Concrete Library, 47 *Japan Society of Civil Engineering*.
7   (1991) JASS-5, *Architectural Institute of Japan*.
8   Masuda, Y. (1990) Text of Concrete Lecture. *Japan Cement Association*, No. 249, 15.
9   Sakai, E. (1991) *Cement Concrete*, No. 535, 25.
10 Knudsen, F.P. (1959) *Journal, American Ceramic Society*, No. 42, 376.
11 Bache, H.H. (1980) WO 80/00959, 15 May.
12 Mino, I. and Sakai, E. (1989) *Proceedings, MRS International Meeting on Advanced Materials*, Tokyo, **3**, 247.
13 Hanehara, T. (1990) *Journal Ready Mixed Concrete*, **9**, 92.
14 Kitamura, M., Hitotsuya, K., Tanaka, I., Take, T. and Suzuki, N. (1991) *Proceedings of 45th General Meeting Cement Association of Japan*, 172.
15 Muguruma, H., Mino, I., Ashida, M. and Sakai, E. (1987) *Proceedings, Symposium on Utilization of High Strength Concrete*, Norway, 63.
16 Bredy, P., Chabamet, M. and Pera, J. (1989) *Proceedings, MRS Symposium on Pore structure and Permeability of Cementitious Materials*, **137**, 431.
17 Kageyama, H., Nakagawa, K. and Nagahuchi, T. (1978) *Journal of the Society of Materials Science*, Japan, **29**, 220.
18 Rosenberg, A.M. and Gaidis, J.M. (1986) Presented to *Second Conference on the Use of Fly Ash, Silica Fume, Slag and Natural Pozzolans in Concrete*, Madrid.
19 Ooshio, A. (1976) *Concrete Journal*, **14**, 34.
20 Kokubu, K. and Hisaka, M. (1990) *Concrete Journal*, **28**(2), 14.
21 Hanehara, T. (1990) *Journal of Ready Mixed Concrete*, **9**, 108.
22 Tanigawa, Y., Nakamura, M., Shibata, T. and Odaka, S. (1991) *Proceedings Annual Meeting, Japan Concrete Association*, No. 209.
23 Sarkar, S.L. and Aitcin, P.C. (1990) *ASTM Special Technical Publication*, No. 1061, 129.
24 Machida, F., Hirose, T., Miyasaka, Y. and Kitta, T. (1974) *Prestressed Concrete*, **16**, 36.
25 Machida, F., Suetugu, H., Yamamoto, T. and Fukumoto, Z. (1975) *Prestressed Concrete*, **17**, 4.
26 Matsumoto, Y., Saito, T., Miura, I. and Mine, Y. (1977) *Proceedings, Japan Society of Civil Engineering*, No. 264, 97.
27 Guide-line of Design and Execution for Constructions with High Strength Concrete, *Japan Cement Association*.
28 Ishibashi, T., Yoshida, H., Ooba, M. and Takeuchi, K. (1990) *Concrete International*, **28**(5), 59.
29 Ono, Y., Suzuki, T., Niwa, M. and Iguro, M. (1984) *Cement Concrete*, No. 450, 8.
30 Okada, T., Imai, M. and Kimura, K. (1987) *Foundation Work*, **111**, 11.
31 Nakajima, K., Matsunami, H., Shimizu, H. and Fukushima, I. (1991) *46th Annual Meeting of Japan Society of Civil Engineers*, V.500.
32 Sugita, H., Nagamatsu, T. and Fufimoto, H. (1989) *Journal Civil Engineering for Electric Power*, No. 223, 63.
33 Masuda, K. (1990) *Concrete International*, **28**(12), 14.

34 Kawai, T., Yamazaki, Y., Imai, M., Tachibana, D. and Inada, Y. (1989) *Cement Concrete*, No. 508, 31.

35 Yagi, S., Tabuchi, H., Hamano, Y., Senda, T. and Itinose, K. (1991) *Cement Concrete*, No. 536, 10.

36 Tomatsuri, K., Kuroha, K. and Iizima, M. (1991) *Cement Concrete*, No. 531, 22.

37 Matsumoto, Y. and Fukuzawa, K. (1980) *Proceedings of Japan Society of Civil Engineering*, No. 301, 125.

38 Nakayama, N. (1982) State of Art of Recent Admixtures. *Japan Concrete Association*, No. 54.

39 Shin, S.W. (1980) *Proceedings of Korea Society of Concrete*, **2**, 165.

Index

among townsmen and, for a male-dominated society, even among women. Merchants in the towns had to be literate, since bookkeeping demanded some education. Among the peasants, at least one head of each five-household group had to be able to read and to interpret regulations. Some estimates have pegged the level of literacy at 40 to 50 percent among males and considerably less for females. In his study of education in Tokugawa times, Ronald Dore concluded that the level of literacy among all classes in Japan was higher at that time than that for Europeans. It is doubtful that any developing nation in modern times has equaled that level.

Many Japanese scholars have emphasized the fact that, among the samurai leadership, training increasingly centered on Neo-Confucian strains of rationalism and pragmatism ("real learning"). One effect was the declaration of the unity of scholarship and politics and the resultant recruitment and training of "men of ability," to be active in politics. Religion reentered the Japanese scene with the way of the samurai (bushidō), a strange mix of Confucian loyalty-obedience and Zen Buddhist contemplation. Loyalty was strictly to the local lord and through him to the Edo shogunate, regardless of whether or not the samurai agreed with orders. Later, this feudal sense of loyalty at the local level was transformed into national loyalty by those who were pledged to lords who in turn supported a new national regime.

In a feudal system dedicated to an economy in kind, the authorities in Edo, the lord-vassals in the domains, and their samurai followers found themselves increasingly beset with problems of money. Horie Yasuzō of Kyōto University has traced a significant evolution: Whereas Confucian doctrine continued to lay primary emphasis on agriculture as the "foundation for the economy," agriculture became increasingly commercialized. The local domain economy was independent, so to speak, but its independence was eroded by increasing connections with the Japan-wide economy. Much to the mystification of samurai, merchants perfected a system of payment of rice stipends in advance, in other words, futures. Thus developed the commercial or commodity society.

When foreigners en masse returned to Japan in the nineteenth century, one of their first critical observations was that Tokugawa Japan had remained "backward" because of a policy known as seclusion. The Edo policy, enforced first in the seventeenth century, was never one of isolation, though. Japan always remained in the realm of East Asia. Contact with China continued. Exchange with the West was just strictly limited—the Japanese maintaining only a Dutch door to the outside world at Nagasaki. What came to be known as Dutch studies began as a trickle and, by the end of the Edo era, constituted a small flood that helped to undercut the Tokugawa bastion.

Despite seclusion, the slow, steady (one might even say subversive) permeation of the samurai society by Western ideas, through Dutch studies, was impressive. By the end of the eighteenth century, such learning had been added to the curricula of most domain schools. Dutch studies, which originally focused on medicine, astronomy, and the natural sciences, later led Japanese to mastery of military affairs. Watanabe Masao has concluded that the tradition of Dutch learning paved the way for the Japanese to make rapid strides in assimilating Western science.

There is a danger, of course, in such an analysis of cultural change—the possibility that one will conclude that nothing much of importance happened in the late nineteenth century. It is true that the impact of the West has been exaggerated by outsiders. Nonetheless, the Tokugawa system did have its own grave weaknesses. Godai Tomoatsu, one of the first samurai students to study overseas, likened Japanese to "frogs in a well" because of their ignorance of the outside world. In fact, it was the combination of internal cultural change with external cultural contact that shaped modern Japan.

3

Shibui: A Taste of Japanese Culture

Some art historians and critics have argued that the Japanese artistic tradition dates only from the sixth century, that is, after the introduction of Buddhism to Japan. Thus it has been stated that in painting, sculpture, and architecture, the Japanese found inspiration in traditions already refined in India, China, and Korea. Such was certainly the case in Japan's Buddhist art as well as classical art. The latter imitated Chinese style and was part of the Great Tradition expressive of the tastes of aristocratic classes.

Japan has, however, also enjoyed an older tradition of folk arts and crafts, predominantly secular and expressive of the tastes of ordinary Japanese. These works sprang from the Little Tradition, a lifestyle that has prompted some critics to find the Japanese culture one of the most aesthetically sensitive. Once again in the realm of art, as in history, it is a matter of where the emphasis should be placed: on the admittedly overwhelming influences from abroad, which did affect Japan's artistic tradition, or on the persistent indigenous style, which has apparently survived the effects of alien overlays.

Japanese art is certainly the major, possibly the only, form of the nation's culture that has directly influenced the West. Meissen and other eighteenth-century porcelain makers had Imari pottery as their model; impressionist and postimpressionist painters were profoundly influenced by *ukiyo-e* woodblock prints; architects beginning with Frank Lloyd Wright have expressed a debt to Japanese design. Japanese literature has had an indirect effect, through translation, on Western prose, poetry, and, particularly, theater. Such Japanese film masterpieces as *Ran*, *Gate of Hell*, and *Rashōmon* have won worldwide acclaim. Less well known abroad have been films depicting contemporary life, the best of which match in excellence award-winning cinema from Italy and the United States.

51

Of course, it is not possible in a short review to investigate all the rich veins of Japanese culture. One can, however, appreciate the contribution made by Japanese art, which has always had a great capacity to speak clearly about Japan's cultural values. Indeed, the direct impression to be gained from observation of an art object and the indirect impression to be derived from the written (if translated) word are often far more reliable indexes of Japanese character than the haze of description provided by the historian or the social scientist.

ORIGINS

As indicated in Chapter 2, many of the artifacts representative of Japan's prehistoric and semihistoric periods were found in or near burial sites. They were ritual objects, at first of interest mainly to archaeologists.

Archaeology and Art

Increasingly, archaeological finds were subjected to critical examination by art experts. Pottery vessels from the Jōmon period, for example, revealed a crude technology, but also an extraordinary variety of quite elaborate designs. In fact, Jōmon has been called one of the most interesting and creative of all prehistoric art traditions. So strange—one is tempted to say so modern—were some of the designs of the ceramics that contemporary critics have referred to them as surrealist. Apparently, the Japanese have always had a strong aesthetic instinct.

When they excavated artifacts from the Yayoi period (300 B.C.–A.D. 300), archaeologists uncovered an entirely different ceramic tradition. The pottery was wheel-made for the first time, and utensils obviously emanated from a settled rice culture that enjoyed irregular ties with the continent. The clearest mark of change was the mastery of metals: bronze bells, their surface decorations depicting contemporary life; weapons, including the sword; and metal mirrors, similar to those used in China's Han dynasty.

In similar fashion, archaeological finds from the Tomb (or Yamato) period were indicative of a changing culture from about A.D. 300. The tombs revealed the presence of distinct social units acting on the command of powerful ruling families.

The most remarkable finds from the huge burial mounds were quite small, the hand-fashioned clay cylinder-based sculptures called *haniwa*. These had been placed in a circle around the base and at the top of the mounds. At first simple, these clay figurines of entirely indigenous design later became more detailed. Among the statues were not only the various people of Yamato: noblemen, footsoldiers, priests, and peasants; but also their animals: horses, chickens, deer, monkeys,

The Grand Shrine of Ise: The Inner Shrine (*Naiku*), in classic Shintō style

and birds. Rich in humor and bewildering in variety, the *haniwa* illustrated people of various stations and their dwellings. Miniature farmhouses revealed a style remarkably similar to dwellings of later years. Sometimes the figures were of crude workmanship; but all, like caricatures, were deft and direct in expression. Later artists strove to return to this simple uncluttered style, symbolic of Japan's earliest aesthetic taste.

Besides the tombs, little of prehistoric Japan has remained in architecture. It has been possible to see original structures in reproduction, though, since Shintō shrines have been faithfully built according to the primitive designs and even with old tools. Two examples include the hallowed imperial sanctuary at Ise and the rival Shintō shrine at Izumo in Shimane Prefecture. One immediately notes the extensive use of wood in its natural state; the simple, almost severe, lines of the walls and roofs, especially the gates (torii) and fences; and the way that all of the structures harmonize with the natural forest environment.

Culture Contact

Rich as the indigenous tradition was, Japanese do admit that Japan was immensely enriched by the impact of continental (chiefly Chinese) influence filtered through Korea. At the core of the foreign impact was Buddhism. A converted imperial court and its Buddhist temples became the patrons of art. Korean and Chinese immigrants taught the Japanese carvers, carpenters, builders, and painters the secrets of their crafts as they all worked together on Buddhist structures and religious sculpture. Art historians have called this first period of Buddhist influence Asuka, after the district in which the capital was located at the time. The Asuka era conventionally is dated from 552—when the first image of Buddha is supposed to have arrived—to 710, when the capital was moved to Nara. Then and later, Buddhism played much the same role in maturing Japanese art as early Christianity did in the European tradition.

Fortunately much more than archaeological finds remain from the Asuka age, particularly in the realm of architecture. Its most spectacular monument is reputed to be the oldest wooden building in the world, the Hōryūji Seminary, which was constructed near Nara in 607 (damaged by fire, it was reconstructed in 1949). The main units include the gate (chūmon), the golden hall (kondō), the lecture hall (kōdō), and the pagoda (gojūnotō). This last five-story structure combines the tradition of the Han tower and that of the even older Indian stupa. The whole project reveals an obvious break from the more simple native style, with the employment of stone foundations, heavy pillars, vermilion paint, white plaster, and broad tiled roofs. The three-storied Yakushiji pagoda, completed in the early Nara period, illustrates the growing confidence and construction skills of Japanese.

ANCIENT JAPAN

As has been noted, under continuing Chinese influence the Japanese in 710 established their first permanent capital at Nara. Since then, Nara has remained a museum-city representing a burst of artistic activity, particularly in the field of eighth-century Buddhist art. Nara symbolized the life of an aristocracy that lived within a narrow radius of the capital. As John Hall described it, "Nara, in its grandeur, was an oasis of

The Hōryūji Temple complex, founded in 607, near Nara

cosmopolitan art in an otherwise traditional and rather folkish Japanese plain."[1]

It must be remembered, however, that in the eighth century Nara was not a museum-city, not a settlement created to celebrate the new art. It was a working capital, modeled after the great Chinese metropolis, Ch'ang-an, which has long since passed from the scene.

Continuing Culture Contact

Continental influence on Nara can be seen primarily in the art inspired by Buddhism. To many critics the most remarkable art created then was sculpture, perhaps the finest ever produced by the Japanese. The earliest works were rendered in wood or in bronze; later clay was used. At first the figures were stylized and symbolic rather than representational.

Certainly by the early Nara period Japanese craftsmen had risen above imitation and become skillful artisans in their own right. Witness the polished wood figure of the Miroku (Maitreya), Buddha of the Future. This religious icon, located in the Chūgūji nunnery near Hōryūji, symbolizes religious quietude but also reveals an unmistakable Japanese quality, hauntingly recalling the abstract forms of the *haniwa*. At the other extreme are the guardian kings, fashioned of clay, painted with lacquer, and placed at the temple entrances to ward off evil spirits. These are fierce warriors with muscular frames and contorted facial expressions, in marked contrast with the serene visages of the buddhas and bodhisattvas.

Perhaps the best symbol of eighth-century Nara was the Great Buddha (Daibutsu), a 53-foot-high (16 m) bronze figure, the biggest statue ever cast in Japan. Dedicated in 752, this sculpture represented a major technical achievement. (Since damaged by fire, it has been replaced by an even more cumbersome figure.) Ten years before, construction of the Great East Temple (Tōdaiji) to house the Buddha had begun. Both temple and statue revealed the growing power of the imperial family, which sponsored this central headquarters among a network of state-sponsored temples in the provinces.

Of even greater value to the cultural historian is the nearby Shōsōin, an imperial treasure house dedicated in 760 by the widow of the emperor who built Tōdaiji. Regularly "aired" and inventoried, the collection allows art experts to reconstruct the material culture and wealth of Nara and bears witness to the high development and sophistication of this culture. Nowhere else has such a superb collection of eighth-century objects been preserved. No other collection in Japan so clearly reveals the variety of early cultural influences, drawn from Korea, China, India, and even Persia. Included are textiles, gold and silver vessels, mirrors, lacquer boxes, musical instruments, screens, banners, hangings, writing tools, books, maps, saddles, stirrups, and swords, used by both priests and the imperial court.

After the capital was moved to Kyōto in 794, and until the beginning of the tenth century, the Heian era also showed the strong strains of continental influence. During this period the glories of the T'ang dynasty were imported directly to the new capital. Buddhism remained the major religion of the aristocracy; monasteries, the patrons of fine arts.

Early Heian architecture, however, began to show a distinct penchant for adapting monumental T'ang style to simpler Japanese needs. The best examples were the Fujiwara-sponsored Kasuga Shrine, actually in Nara; and the Heian Jingū (extant in a modern reconstruction) in Kyōto. Once again the woodwork painted vermilion, the complicated and bracketed eaves, and the slightly curved tile roof revealed Chinese influences, while construction methods showed the Japanese carpenters' skill in assimilating alien design. As for Buddhist temples, the Japanese began to build amidst the beauties of nature. For example, in 851 the Tendai sect began construction of Enryakuji in the mountain heights and forested valleys of Mt. Hiei, just north of Kyōto.

After the fall of the great T'ang (906), direct contact with China ceased, and Japanese designers were thrown on their own. Large public buildings like the Byōdōin, built at Uji in 1053 under Fujiwara patronage, continued to show Chinese influence. Yet the placement of the main Phoenix Hall (Hōōdō) was marked by Japanese insight. The building symbolized a huge bird settling on the irregularly shaped pond in the foreground. An even better example of reversion to native instincts was the shrine at Itsukushima, built out over the waters of the Inland Sea at Miyajima. (This setting provided the opening scenes for Japan's first great color movie, *Gate of Hell.*)

The structure that had a lasting effect was not the splendid palace or the towering temple, however, but the typical Japanese dwelling. Style was governed partly by materials available. Stone was rare. In its place, wood, paper, plaster, and thatch were used, following the native tradition of early Shintō structures. Most obvious were the designer's exploitation of contrasts and the carpenter's use of natural materials. The structures had raised wooden floors, removable walls, lightweight partitions, and deep thatched roofs.

Culture Change

After the early twelfth century, Japanese art was on its own. Japanese style began to reassert itself, and nowhere was this better revealed than in language. Our knowledge of archaic Japanese language derives from the first written forms dating from about the eighth century. The *Kojiki* (Records of ancient matters), for example, appeared to be pure Chinese until it was recognized that some of the ideographs made no sense in Chinese. Obviously, they were being used to indicate Japanese personal and place names. When first faced with the problem of writing, the Japanese had two choices: the semantic (Chinese) or the phonetic (Japanese). As one might guess, they began to use both. In other words, the basic Chinese characters suggested the skeleton and body of an idea, which was then clothed in an elaborate Japanese grammatical robe woven of ideographs adapted for phonetic purposes. The latter eventually evolved into a regular system of Japanese phonograms called "borrowed names" (*kana*).

It is not the purpose here to dwell in detail on what Edwin O. Reischauer has called one of the most intricate languages in the world. Suffice it to say that eventually order was created out of chaos: the Chinese ideographs were increasingly used for substantive units (for example, nouns); the *kana* for phonetics and inflections in a regular pattern. A good example of this system is found in the *Manyōshū* (Collection of myriad leaves), which dates from the late Nara period and consists of some 20 books and over 4,500 poems. The style, particularly of the poetry, differed substantially from that of the early (Chinese-style) dynastic histories.

In other words, linguistic confusion was not able to obscure certain deep native themes. Because of the regular meter and rhythm of this early poetry, the *Manyōshū* became a valuable index to archaic Japanese and a symbol of the persistence of Japanese tradition. Moreover, the collection, unlike officially inspired genealogies, was rich in the sentiment of ordinary people.

Among the various techniques employed in the *Manyōshū*, the following may be mentioned: alliteration, to be used in all forms of Japanese poetry thereafter; the pillow-word, used to modify the following word either through sound (in essence, a pun) or by association; and above all, the vivid word-picture. Thus, "grass-for-pillow" was an imaginative word signal for journey. The lyrical poetry displayed an intimacy with nature. Many of the short poems (*tanka*, 31 syllables in lines of 5-7-5-7-7) included a snapshot of a natural scene twisted into a hint of human emotion. Two examples may be cited:

> Oh how steadily I love you—
> > You who awe me
> Like the thunderous waves
> > That lash the sea-coast of Ise!

> To what shall I liken this life?
> > It is like a boat,
> Which, unmoored at dawn,
> > Drops out of sight
> And leaves no trace behind.[2]

In these thoughts, the material was not entirely rhetorical. As viewed also by Shintō, each natural phenomenon retained its individuality, an entity permeated by an awesome and spiritual quality.

The poetry was also imbued with a characteristic Japanese feeling of the vanity and evanescence of life: "life frail as foam," "all is in vain," and "nothing endures"—such phrases doubtless indicate an overlay of Buddhism in the mind of the poet.

The *Manyōshū*, along with the dynastic histories, was a mirror, reflecting society up to the late eighth century. The poems revealed a staid, settled, agrarian way of life in the countryside, but also political consolidation under the court and a measure of progress. Much has been made of the artificiality of the later Heian court and of the oppressive hierarchy of feudal systems even later on. From the *Manyōshū* collection emanated an optimistic spirit, a frankness, a genuineness of feeling, and sheer happiness. The martial spirit was only a minor theme.

Out of this early poetic tradition came Japanese writing in a kind of shorthand—a simplified cursive style called "grasshand" (*sōsho*). Tradition holds that a priest, Kōbō Daishi (or Kūkai, 774–835), chose forty-seven of the adapted ideographs to create a Japanese syllabary. This in turn led to a fine art, brush calligraphy. In the writing of the ideographs, the Chinese artist was regarded as paramount; in the rendering of script, or *kana*, the Japanese excelled in calligraphy. Basic skills in calligraphy profoundly affected Japanese painting. Thus were blended the various components of an artistic tradition: the abstract calligraphy, the word-pictures of the poems, and, eventually, black-and-white painting. The last was to share the basic canons of calligraphy: balanced tension, use of open space, quality of line, texture, and variation of a basic theme.

It was only one more step to what was called "woman's language" (or "easy *kana*"). In fact, there was nothing easy about the precursors of the modern novel, written by women in the tenth century. Somewhat neglected has been the *Pillow Book*, a shrewd, sometimes sarcastic set of observations on Heian court life prepared by Sei Shōnagon. Far more widely known was the masterpiece of a contemporary and rival in court, a lady since known by her pseudonym, Murasaki Shikibu. Ostensibly the biography of a brilliant sensualist (the shining prince), *The Tale of Genji*[3] is the first Japanese novel. Moreover, it has been called the world's first great work of prose fiction. We know that the author was born about 978. She finished the tale about 1020.

The novel concerns a hero, Prince Genji, who is the son of an emperor by a concubine. He is at once handsome, scholarly, witty, engaging, and sensitive, and yet daring, courageous, and skillful in the military arts. The author spins out the story in a long, episodic, often repetitious style, but also sets for herself the highest aesthetic standards (described in the novel itself). Although written as a romance, the work is full of ideas, complicated human motivations, and psychological insights into personality. In what was to become a characteristic of Japanese style, the episodes suggest pathos in high places and a sense that life is fleeting and all material things impermanent. The use of language is innovative, quite different from previous writing, which slavishly followed Chinese style.

Nor is *The Tale of Genji* without historical significance. Read with hindsight, the novel depicts late Heian court life as luxurious but obviously shallow. Like the *Manyōshū*, *Genji* was a mirror reflecting contemporary life. This time a foreground of Kyōto refinement is balanced by a background of uncourtliness and provincial vigor.

Never has there been a more fortunate coincidence in combining the written word and illustration than that found in the Genji scroll prepared by an eleventh-century artist. His work marked a reversion in painting, parallel to that in language, to a distinctly native style, since called *Yamato-e* (Japanese-style painting). Even the form of presentation was unique, a horizontally prepared and unrolled scroll (*emakimono*). This art displayed a growing secular concern with everyday life, a reliance on inspiration drawn from indigenous scenes, a stylized quality of representation, brilliant use of color, and a formalized perspective, which looked down and from a slight angle on the scenes as they unfolded.

THE ART OF FEUDAL JAPAN

Just as economic wealth and power in the late twelfth century shifted away from stiflingly confined Kyōto to the provincial vigor of Kamakura, so too Japanese art reflected changing social patterns, local influences, and new tastes and values. The transition was faithfully recorded in the increasingly popular *Yamato-e* style of painting and particularly in the horizontal scrolls. From this period came the *Heiji monogatari* scroll (about 1250), which depicted, in place of languid palace life, the exploits of warrior-samurai—choking smoke, a burning palace, warriors on horseback, the severing of heads, and scurrying aristocrats— all are portrayed with a wealth of narrative detail and vivid realism.

Continuing Culture Contact

During the Kamakura period there was renewed contact with China. The effects of new religious sects were felt in Japanese art, not only at the ancient capital of Kyōto and in the new headquarters at Kamakura, but also throughout the provinces. By this time artists were identified by schools, as the group-centered consciousness of Japanese life surfaced in the arts and crafts.

In architecture, evidence of a sort of classical revival was shown in the rebuilding of Kōfukuji, with its five-storied pagoda, in Nara. Another remarkable example was the Sanjūsangendō (hall of thirty-three bays), in Kyōto. The immense gate of Tōdaiji in Nara built in this period preserved the contemporary multi-story architectural style of Sung China.

Sculpture at Kamakura also demonstrated a renewal of religious spirit, a revival of Amidism (a certain Buddhist tradition), and, at the same time, a new realism. A 49-foot-high (15 m) bronze statue of Buddha was erected in Kamakura (in 1252) in emulation of the older Daibutsu of Nara.

Nowhere was the new emphasis on realism more clearly evident than in the sculpture of Unkei (1148–1223). He was aided by his six sons and a guild of carvers and carpenters. The team caught the tragedy of human life in marvelously lifelike wooden figures. The Unkei school injected an almost ferocious muscular vigor even into its Buddhist art like the two king-guardians (*niō*) at the south gate of Tōdaiji, reconstructed after the civil wars.

Typical of Chinese influence adapted to Japanese surroundings was the Relic Hall of the Engakuji, near Kamakura. The use of unfinished materials set amid natural surroundings reflected the profound influence of a new teaching, Zen Buddhism, the full impact of which will be described below.

Continuing Culture Change

The great strain of the Mongol invasions in the late thirteenth century eventually brought about the collapse of the Kamakura system. The epic struggle against the invaders was graphically depicted in narrative scrolls prepared in 1293 by Tosa Nagataka, who thus continued the *Yamato-e* tradition. An age of political uncertainty and, finally, endemic civil war followed. Turmoil marked the triumph of provincial forces— particularly the samurai—over the aristocracy and Kamakura nobility. Literature of the times faithfully captures the ethos of the samurai, which essentially consisted of two ideals: bravery in battle and loyalty to the lord. Much more accurate in its details, though, is the semiofficial history of the Kamakura shogunate, *Azuma kagami* (Mirror of the East). But closer to the semifictional war tales of an earlier era is the *Taiheiki* (Chronicle of the great peace), prepared shortly after the civil conflicts of the fourteenth century.[4]

Despite, or possibly because of, this time of troubles, an art not so much of ostentation but of quiet contemplation emerged. Possibly because of the turmoil from the early thirteenth through the fifteenth centuries, Japanese have looked back on the succeeding Tokugawa period to find a culture of peace and on the preceding Heian era to recapture "the classical age" in Japanese art. Nevertheless, the Muromachi period (1392–1567), even with uneasy Ashikaga leadership, provided what were later regarded as characteristically Japanese modes of artistic expression. These included: in painting, a transition from colorful narrative scrolls to quiet, contemplative black-and-white sketches (*sumi-e*); in literature, a change from precious court diaries to somber, classic Nō drama; in domestic arts, a shift of emphasis from aristocratic treasures to ceramics, utensils for the tea ceremony, and the skills of landscape gardening.

Zen Culture

Although the Muromachi era was a period of disorder and significant cultural change on the domestic front, it was also a time of strong continental influence. Nowhere was this more obvious than in the subtle but pervasive influence of a Buddhist sect called Zen in Japan (Ch'an in China). Zen monks became not only the leading scholars and writers, but also designers, artists, collectors, and critics—in short, those who set the aesthetic standards.

In architecture, this was the period when the great Zen temples of Kyōto—Daitokuji, Tōfukuji, and Nanzenji—were established. It was also the era of two masterpieces built by Ashikaga shōgun. The Golden Pavilion (Kinkakuji) was erected in 1397 as a villa at Kitayama, then outside Kyōto. (Completely destroyed by fire, it was reconstructed in 1950.[5]) The Silver Pavilion (Ginkakuji) was built in 1489 as a country retreat at Higashiyama, then north of Kyōto. Both recalled the aristocratic tastes of a classical age and, in their placement in natural surroundings, the deep introspective quality of Zen Buddhism.

Painting was doubtless the greatest art of the Muromachi period, and here, too, the aesthetic principles emanated from the Zen monasteries. Inspiration was derived from the landscape artists of Sung China but steadily evolved into a unique Japanese technique under the guidance of Josetsu and Shūbun (early fifteenth century), culminating with the master, Sesshū (1420–1506). The artists (many were Zen priests) perfected a new monochrome style known as water-and-ink (*suiboku*). Vivid colors were replaced by bold strokes (often perfected in calligraphy), flat washes, shades, line, and balance. Open spaces were used ingeniously. In the quality of stroke and space a direct approach, almost a religious perception, was portrayed.

Such principles were carried over into another great Japanese contribution to the world of art, landscape gardening. Gardens varied from the large and complex like Tenryūji to the cool and quiet like the moss garden at Saihōji (both in Kyōto). The latter cleverly evoked a waterfall and stream, entirely without water. Even the uninitiated have agreed, however, that the abstract rock garden of Ryōanji (constructed about 1500) in Kyōto best carried out the aesthetic spirit of Zen. Here was indeed the *suiboku* style rendered in sand and stone. In a severely limited space (which looks larger to the eye) an abstract sea of carefully waved sand, broken here and there by isolated rock islands, was conceived. The "program" of the garden was minimal, yet its pattern suggested meaning to all.

The impact of the Zen approach, an inner world of elegant but transient beauty, on now-famous Japanese folk art (*mingei*) was enormous.

Ryōanji, Kyōto: Zen-style rock garden built about 1500

An aesthetic vocabulary was built up: the solitary (*wabi*), the antique mellowed by use (*sabi*), and the astringent (*shibui*). Used to describe the taste of bitter green tea, *shibui* came to be heard frequently in Kyōto and perhaps less so throughout Japan. Astringent best described Zen style: subtle rather than direct effect; economy of means rather than opulence; the power of suggestion and meditation rather than blunt statement. Edwin O. Reischauer called it "the cultivation of the little"; Hugo Munsterberg, "the cult of the subdued"; and art expert Sherman Lee, "tea taste in Japanese art." Indeed, at the very center of aesthetic connoisseurship was the famous tea ceremony (*cha-no-yu*).

A Zen monk, Eisai, had first brought tea from China late in the twelfth century. The serving and consumption of tea in an ordered, slow ceremony were carried to their final formal style by teamaster Sen no Rikyū (1521–91). The greatest impact was, of course, on ceramics. At first, Chinese utensils were used. Later, rice bowls from Korea under the Yi dynasty were popular. Finally, the preference for indigenous and local wares, such as those made at Bizen, Iga, Tamba, Shigaraki, and Seto, won out. During these years, pieces made at Seto approached Chinese quality so closely that the kiln lent its name for the Japanese word for pottery in general (*setomono*). Other utensils used in the ceremony became objets d'art: cast-iron water kettles, the Bizen-ware tea caddy,

A Japanese meal: Arrangement for the eye as well as for the palate is essential.

the Oribe tile platform, the bamboo tea whisks, lacquer trays, and the delicate wooden scoop for the bitter green powdered tea.

The tea ceremony interacted with the other arts. The architecture of the tea pavilion, with its attendant garden, was designed to be quiet and meditative. In the background in an alcove would be a flower arrangement (*ikebana*) and a landscape modeled after, but not entirely imitative of, the Sung style. The cultivation of tiny trees (*bonsai*) or the arrangement just outside the pavilion of a small garden symbolized the whole universe of nature.

Although Nō drama belongs to the long tradition of Japanese literature, music, and dance, the visual effects of Nō picked up various artistic strains. From the indigenous cult came Shintō legends; from the classical era, the gorgeously embroidered costumes; and from the Zen contemplative tradition, the imaginatively carved masks. These masks represent various characters and in themselves are works of art.

Originally, mysterious and symbolic dances set to music adapted from the continent had been performed for the Nara court. At first Nō plays were concerned with men who were really gods and Shintō divinities. Later secular themes entered the repertory and involved ghosts of famous warriors, doomed to reenact their last battles in afterlife, elegant ladies in women-plays, and devils in drama concerned with the supernatural. *Hagoromo* (The robe of feathers), for example, contained many of these elements and used as its backdrop the beautiful Miho pine-clad seashore (just below Shimizu). "Crazy words" (*kyōgen*) interludes provided the comic relief. Nō was molded into what it remained essentially for 600 years by two great masters, Kanze Kanami (1333–84) and his son, Seami (or Zeami, 1363–1443), who founded the Kanze school of performers. Nō worked a powerful influence later on the grand Kabuki theater, which in turn represents Japan's greatest contribution to the world stage.

Japan's art did not remain for all time simple and understated, as expressed in Zen. During the era of the great unifiers, Oda Nobunaga, Toyotomi Hideyoshi, and Tokugawa Ieyasu, style returned to the grand, the opulent, and the monumental. The great castles built by Nobunaga at Azuchi and by Hideyoshi at Momoyama became not only the centers of political gravity, but also the focal points for patronage of the arts, lending their names to a cultural era, the Azuchi-Momoyama (1568–99). The castles had unusual architecture, combining knowledge of defense imported from the West with Japanese temple style. The one built at Ōsaka (restored in November 1931) was a symbol of the wealth and power of the unifiers. Some 30,000 workers were employed to construct the outworks and tower. Most of the castles, except two prominent ones—the famous White Heron at Himeji, west of Kōbe, and Nijō in Kyōto—were victims of modern bombardment. Nijō was built in the early years of the seventeenth century for Tokugawa Ieyasu as a residence for his stays in the old capital.

The interiors of the palaces and residences were richly endowed by the Momoyama decorators. The Kanō school, founded by Masanobu and Motonobu during the latter part of the Muromachi period, set the standards in multiple-panel screens (*byōbu*) and wall paintings. Similarly, weaving, dyeing, and embroidery were used to produce bold and colorful patterns. Momoyama textiles are probably the finest the Japanese have ever produced.

Despite this burst of opulence during the late sixteenth century, the Japanese never forgot the Zen heritage. Especially in the ensuing Tokugawa period, aside from flashes of brilliance, Japanese art withdrew to subdued expression and to an astringent quality described as *shibui*, the soul of Zen.

EDO STYLE

There is nostalgia in Japan for the Edo era. When the young directors backstage at NHK (Nippon Hōsō Kyōkai, "Japan broadcasting corpo-

The White Heron Castle, Himeji, completed in 1609

ration") wish to project a traditional background for one of their soap operas, they faithfully reconstruct a scene from the Edo era (1603–1867). The narrow alleys, the carefully reproduced architecture of the houses, the gorgeous costumes of the protagonists—all these are drawn from Japan under the Tokugawa hegemony. It is true that much of what is generally regarded as the uniquely Japanese contribution to world culture did emerge during the relative disorder of the Muromachi. But its character had evolved from an earlier feudal era, the Kamakura, and both owed a large debt to an earlier classical period, the Heian. Yet the strands of all three ages came together in the long and peaceful Edo era.

As far as art was concerned, the relative isolation of Japan for two centuries had the effect of turning the country in on itself. Edo became the most completely Japanese era in the cultural history of the nation. There appeared, of course, the great public buildings in Chinese style, monuments symbolic of Tokugawa power. Classical schools, however, declined in originality and strength. At the same time, popular art and folk crafts flowered as never before. Nor was good taste confined to the metropoli—ancient Kyōto, brash Ōsaka, and colorful Edo—the period of peace and prosperity saw a widespread diffusion of arts and letters. Each of dozens of castle towns became a center for arts and crafts; each of a number of daimyō, a patron for artists and craftsmen. In the towns, subsidy was available from samurai and wealthy merchants. Of all the periods of art, Edo is the best represented in Western museums.

Tokugawa Treasures

There were two tributaries to the mainstream of Edo culture. The lifestyle of the samurai class (which included the Tokugawa, related families, and the various daimyō) tended to carry forward without much change the aristocratic emphasis inherited from the Muromachi and Azuchi-Momoyama eras. In sharp contrast to these upper-class arts were the marvelous inventions of the townsmen and the folk crafts of the agrarian villages.

In public architecture, too, representations ranged from the elaborately ornate to the elegantly simple. The Momoyama tradition of decoration was best reproduced in the mausoleum dedicated to the founder of the Tokugawa, Ieyasu, and located at Nikkō (1617). The various structures at Nikkō were to many Japanese "splendid" (*kekkō*), to others, gaudy in the extreme, even fussy. For example, the Gate of Sunlight (Yōmeimon) has, despite its embellishment, been regarded as a masterpiece of Japanese architecture.

The elaborate mode of Momoyama painting went in several different directions. The Kanō school, for example, divided and evolved into the decorative art of Kōrin (1658–1715). The *Yamato-e* tradition of Japanese painting also produced new schools, the forerunners of those that would produce the woodblock prints of the next century.

Built about the same time as the Nikkō mausoleum but at the other extreme were the small, residential rather than public, detached palaces of Shūgakuin and Katsura outside of Kyōto. The latter particularly has received worldwide acclaim. The principles of design renowned architects have cherished in Katsura are extreme simplicity, economy in materials, functionalism in form, and beauty in design. Floors were covered with squares made of thick dyed rush (*tatami*). Once again, the colors of nature were preferred over paint. An old twisted tree trunk, rather than finished lumber, ran alongside the alcove (*tokonoma*) and held up the latticed ceiling. Just outside, worn stones served as steps, pebbles as a bed for cool flowing water, deep eaves as protection against changes in weather.

Katsura, of course, must be considered a public work of architecture. But often overlooked is the fact that the ordinary farmhouse had a similar simple unity. Now rare in either original or reconstructed form, the classic Japanese farmhouse is a prime contribution to the world of architecture. The key features include the use of plain wood, rush matting, and the alcove, already described; also the sliding weather partitions (*amado*), thick sliding panels (*fusuma*), light rice-paper partitions (*shōji*), and the separate bath (*ofuro*) and facilities (*benjo*). Many of these features have been picked up and reproduced in modern houses throughout the world; if not, they appear in almost every interior decorating magazine as models of the best in modern design.

Mention should also be made of the host of fittings and accoutrements that lay between the samurai and commoner traditions. Many have withstood the impact of the West and modernization and remain today as heirlooms from Japan's rich past. There were nests of boxes (called *inrō*); miniature carvings (*netsuke*) that served as toggles; sword guards (*tsuba*) of iron inlaid with gold and silver; folded paper figures (*origami*); and, of course, utensils for the tea ceremony and flower arranging.

Chōnin *Culture*

Japanese like to think back to a traditional rustic lifestyle, which they often identify with the agrarian village rooted in rice culture. Nevertheless, much of the traditional heritage came not from the countryside where most Japanese indeed lived but from the network of castle towns built up under Tokugawa hegemony. In these towns lived most of the samurai, but the tone of the settlements was sounded by townsmen (*chōnin*). *Chōnin* culture was shared by samurai and nonsamurai alike.

During the Edo era, there were actually at least three urban traditions. Kyōto retained an aristocratic outlook, but the nobility had little functional significance. Thus Kyōto became a center for elegant crafts. Ōsaka came to represent the mercantile outlook, even though in Neo-Confucian theory merchants had low status. Edo, originally settled by the warrior class, came to represent a synthesized mass culture.

Popular culture emerged from an urban environment dominated by the entertainment district, the pleasure quarter, and the way of the courtesan. The mood was captured by racy novels, popular drama, and garish woodblock prints. For a long time after, Japanese regarded all these phenomena as common, even vulgar, since they dealt with an ephemeral world in sharp contrast to the universe of Confucian learning and Buddhist meditation.

The most brilliant flowering of *chōnin* culture appeared in the Genroku era (1688–1704), the style of which was essentially Japanese in origin and little affected by outside influence. A key to understanding the culture lay in a Japanese term, *ukiyo.* Originally, as Donald Keene instructs us, the word meant simply "the sad world"; by means of a pun, it came to mean "the floating world."

First to capture this atmosphere was Ihara Saikaku (1642–93), the first important novelist in five centuries since the Lady Murasaki. Indeed, some of his work was deliberately modeled on the classic Genji tale. Saikaku was at once poetic in the classic style, flavorful in the aesthetic style, and colloquial in Genroku style. On the surface, his work was highly secular, dealing with the latest fads, fashions, and foibles of his society. At the depths, his writing reflected the traditional Japanese view of the evanescence of everyday life. He worked with three types of stories: tales of love, stories about merchants, and samurai legends. Of these, the first were perhaps the most successful and the last, the most popular. His masterpiece, *Five Women Who Loved Love,*[6] was filled with an intimate knowledge of Genroku society.

Saikaku the realist has been contrasted with Japan's Shakespeare, Chikamatsu Monzaemon (1653–1725) the romanticist. To appreciate Chikamatsu we must recall how urban culture had produced new dramatic forms with far greater popular appeal than that of the aristocratic Nō. The seventeenth and eighteenth centuries witnessed development of a puppet-theater, in which the action was carried by amazingly realistic doll-figures about half life-size. Dialogue and commentary by chorus, something like the Greek version, were intoned from a platform on the side by reciters (*jōruri*). Originally a kind of storytelling in the street, *jōruri* was elevated by Chikamatsu to become *bunraku* puppet-theater, which then worked a powerful influence on the more popular Kabuki dance-drama.

Bunraku Puppet Theater and the late Yoshida Bungoro, manipulator

Chikamatsu preferred, in fact, to write for the puppets, because he was suspicious of Kabuki actors who often added their own variations to his plays and used them as vehicles to promote their own success. Often, parts of the plays were adaptations of classical Nō drama, but the language was entirely colloquial. Chikamatsu wrote two kinds of plays: domestic tragedies, like modern soap operas, and historical pieces, which were fictionalized versions of famous events and personages.[7] Although the former were superior as literature, the latter were more beloved by audiences. In the historical dramas, the playwright, like Shakespeare, demonstrated clearly that persons of status also felt ordinary human emotions and suffered tragedy. In the domestic pieces, characters displayed noble human feelings (*ninjō*) in conflict with duty (*giri*). And thus were the lifestyles of samurai and commoners fused.

After Chikamatsu, Kabuki was performed by live actors (always men), with huge and revolving stages, runways through the audiences, elaborate but stylized scenery, and brilliant costumes. Often the jerky halting movements of the ancestor-puppets were faithfully reproduced on the Kabuki stage. Contents of the drama never quite matched the skills demonstrated by Chikamatsu, but the Kabuki as theater became more and more splendid, what Joshua Logan has called among the best in the world. Doubtless the most popular play in grand Kabuki has been the *Treasury of Loyal Retainers*.[8]

This floating life, this novel urban style, this popular drama, all were captured in one of the most remarkable and famous Japanese contributions to the world of art: the woodblock print depicting the floating world (*ukiyo-e*). Tapping an ancestry of aesthetic expression in the *Yamato-e* style of Japanese painting and the black-and-white impressionism of Zen art, the new genre was deeply folkish in inspiration. It was used to illustrate novels, to advertise dramatic productions, and to commemorate famous places of natural beauty along well-traveled routes. When in the nineteenth century these prints reached the West, their impact was great on a number of artists: Dégas, Manet, Monet, and Whistler among the impressionists; Toulouse-Lautrec, Van Gogh, and Gauguin among the postimpressionists. The *ukiyo-e* woodblock print has enjoyed more popularity in the West than any other form of Japanese art. Indeed, the influence was so strong that a word was coined in French, *Japonisme*, to describe the effects on Western art.[9]

The first to turn the energy in Japanese painting to woodblocks was Hishikawa Moronobu (died about 1694), who realized that prints could enjoy wider circulation than individually painted scrolls. He and Torii Kiyonobu (1664–1729) specialized in portraits of Kabuki actors. At first the prints were in monochrome, but later color was added. Then the work became a collaboration among the publisher, who commissioned and financed the enterprise; the artist, who prepared the overall design; the skilled engraver, who transferred the design to block; and the printer, who painstakingly fitted each block of different color to the paper and turned out the finished print.

Harunobu (1724–70) was primarily responsible for developing the color print; he concentrated on the human figure, either the courtesan or the Kabuki actor. Utamaro (1754–1806) was most famous for his "picture girls," well-known courtesans and geisha. Meanwhile, entire families were devoting themselves to subjects drawn from Kabuki. Possibly theatrical prints were carried to their heights by Sharaku (eighteenth and nineteenth centuries), whose portraits of actors often also constituted biting caricatures of public figures. Of all these artists, Sharaku constitutes the enigma, for he worked only a brief ten months (1794–95) and disappeared again into obscurity.

By the end of the eighteenth century, *ukiyo-e* printmakers had turned from figures to landscapes. The artists pictured everyday life and well-known scenes. Increasingly under the subtle influence of Western prints, they began to use perspective in their designs. Two men were not only famous during their own lives but are celebrated for their designs to this day: Hokusai (1760–1849) and Hiroshige (1797–1858).

Hokusai is said to have produced some 35,000 original prints; Hiroshige, more than 5,000 original prints. (Commonly, artisans later pulled duplicate prints from the blocks originally prepared by the artists.) Hokusai has become best known for his thirty-six views of Mt. Fuji. Hiroshige is remembered for his delightfully graphic sketches of the fifty-three stages on the old Tōkaidō road.

Even those who on the surface seemed to reject the superficiality of the floating world were nonetheless profoundly affected by it. Bashō Matsuo (1644–94), for example, seemed to work apart from the Genroku bustle, like a Zen hermit of an earlier age. His striking poetry symbolized the fact that the older (*tanka*) verse forms had fallen out of favor, replaced by an even briefer and more indirect style of expression (the *haiku*, a 3-line, 17-syllable poem with a structure of 5-7-5). In the hands of Bashō, the shorter *haiku* poem probably reached its zenith. By convention, the lines were associated with seasons of the year. They also painted in sparse words a picture of nature while evoking a deep human emotion. In one sense, the detachment of Bashō has been exaggerated—his poetry, like that of Saikaku, plumbed the Edo style.

During the Tokugawa period, Japan was admittedly a storehouse of Chinese art. In some cases, the original had disappeared on the continent. Was the Japanese artistic impulse, then, mere imitation? The answer is no, since in any Japanese collection of Edo art the very finest items have unmistakable Japanese qualities. With an acknowledged debt to continental—as well as indigenous—origins, the *Yamato-e* painting, the Bizen pot, the carved Nō mask, and the woodblock print nonetheless stand out as unique contributions to the world of art. At the very least, Japan's art and literature document a long tradition, a tradition with which Japanese faced the impact of the West and the process of modernization.

There has been some speculation whether, after the uniquely Japanese quality of Edo style, the art of Japan has been able to withstand the typhoon of Western influence. An answer to that question must be postponed until Chapter 4, after an examination of the process of modernization as a whole has been made.

4

Kindaika (*Modernization*)

In 1968 members of conservative groups, which had dominated Japanese politics for two decades, prepared to celebrate the 100th anniversary of the "Meiji Restoration." The year also marked a century of cultural exchange with the West. Those who opposed the conservative majority tried to ignore the anniversary. They denounced the appearance of "absolutism" in Meiji Japan and decried the "inevitable results": militarism, "fascism," aggression, imperialism, war, and a disastrous defeat.

For over a century many Japanese have been fascinated by the term "modernization" (*kindaika*) and intrigued by the process. Ordinary Japanese have used the word as a generalization. Scholars and critics have tried to sharpen the concept to describe a complicated historical process, usually without success. Many Japanese and some foreign observers, too, have argued that modernization has been proceeding over the last *two* centuries. Others have replied that, whatever has happened since the 1860s, the net result has *not* been a completely modern Japan. It must also be noted that whatever the process and the results, only a minority among the Japanese have taken a look at the various modern masks and then deliberately decided to lay them aside in favor of purely traditional visages.

In pursuit of the modern muse, thoughtful Japanese and outsiders have reflected on the ancestry of nineteenth- and twentieth-century Japan. Modern Japanese society, most have agreed, has been a descendant of Tokugawa feudalism. Here again, ordinary Japanese have used the word *feudalism* as a generalization. Usually usage has implied a pejorative judgment. Some observers have, however, tried to confine use of the term to a neutral and comparative description of a complicated set of values and institutions.

73

THE TOKUGAWA PERIOD

The Tokugawa era (also known as the Edo period, 1603–1867) has had, in the words of John Hall, a "bad press."[1] A second generation of foreigners in Japan (the first were the Catholic missionaries, who arrived in the sixteenth century) were particularly harsh in their denunciation of "isolated, feudal Japan" under the Tokugawa family. These were the foreign employees of the new Meiji government.

For example, William Elliot Griffis, probably the most widely read of the old Japan hands, spoke of Tokugawa society as a "chamber of horrors." Men like Griffis simply could not wait until Japan became a Western, Christian outpost in Asia.

In their first forays into comparative history, some Japanese scholars used the newly coined term for feudalism (*hōken*) in an objective, historical context, rather than in a pejorative sense. They followed in the footsteps of European historians like Federic William Maitland. In the 1920s, other Japanese began to adopt a Marxist definition of the term and thereby revealed a growing ambivalence. On the one hand, they regarded feudalism as a stage in development on the way to a bourgeois, modern age. For them, feudalism in Japan had been a welcome—even an inevitable—development in the evolution dictated by the laws of history. On the other hand, they denounced the persistence of "feudal characteristics" in what should have been a middle-class Meiji revolution. Marxist coloration has been found in much twentieth-century Japanese thinking (including non-Marxist analysis) and accounts for the generally negative view in which feudalism has been held.

In contemporary Japan, popular attitudes toward Tokugawa feudalism have also been somewhat ambivalent. The average Japanese, voracious reader and avid television viewer, has displayed a distinct nostalgia for the swashbuckling samurai, the resourceful merchants, and the attendant ladies of the Edo era vividly pictured in best-seller historical novels and in TV daytime drama. This is in part a yearning for times past, similar to the American fascination with a frontier long closed. There is a vague feeling that Japan is not what it was. And in Japan, as in the United States, the cynic responds that the old society never resembled the fictional re-creation of it.

Serious scholarship, in which both Japanese and foreign observers have engaged, has produced a sharp division of attitude toward prenineteenth century feudal Japan. Nor is the debate strictly of academic interest, for the assumptions have profoundly affected contemporary policy considerations. Without plunging into the controversy in detail, one can usefully note what the various views have in common.

Preconditions

In most of the descriptions of the transition from Tokugawa to Meiji Japan, it has been assumed that a traditional system, which was fully developed during the Edo era, served to color the society in the later period.These traditional norms were analyzed in Chapter 2 and may be summarized here. Values were achieved and reinforced in family-style groups, within a patriarchal hierarchical structure. Individuals received blessings and, in return, assumed social obligations. There were no universal norms; rather there was a situational ethic. For some observers, such values made up feudal residues that blocked any hope for a truly modern society; for others, they provided the firm foundation for modern society, Japanese-style.

Feudal Residues. In the first view, Tokugawa feudalism provided the background out of which Meiji and the later Japan emerged. Feudal society had been dominated by a narrow privileged ruling class, with an oppressed peasantry living on the threshold of starvation. Famine had been endemic. In place of political movements had been intrigues, plots, treachery, assassinations, and arbitrary justice. Moreover, under Tokugawa hegemony, particularly in the late stages marked by contradictions, feudalism had lived on beyond its natural life and left "hideous wounds upon the minds and spirits of both rulers and ruled."[2]

Much of the Japanese writing on this theme (as well as some foreign scholarship that picked it up and translated it abroad) attempted to explain a paradox. True, Tokugawa contradictions had indeed led to the overthrow of the feudal regime. The result was *not*, however, a full-fledged bourgeois revolt. Instead, according to this line of thought, the long-suffering Japanese had been burdened with Meiji militarist absolutism. And thus the laws of history must have been amended in the Japanese case. The feudal residues guaranteed an abortive modernization and eventual disaster. This interpretation of history appeared most prominently in the 1930s and 1940s in a vain rearguard action against the steady advance of militarism in Japan.[3] After a respite, an echo of the same interpretation reappeared in postwar Japan. The feudal residues have been further reinforced, the argument has continued, by the alliance between Japan and the U.S. military-industrial complex. Thus, the inevitable revolution into modernity has been postponed.

The Revisionist View. In the other view as well, Tokugawa feudalism provided the background from which modern Japan emerged. Those who have expressed the revisionist view of history, however, have attributed to the Edo era an entirely different set of qualities. Less committed to a specific evolutionary theory, they have argued that traditional norms and practices were not necessarily displaced by modern institutions. Indeed, the Japanese case has dramatically illustrated the capacity for the two to coexist. The Japanese scholars and the foreign observers who have expressed this revisionist view have arbitrarily been dubbed the modernizationists.[4]

In modernization theory, traditional Japanese emphases on group loyalty, group coherence, group decisionmaking by consensus, and obligations to the group (as compared with an emphasis on individual rights) made possible the smooth and swift transition from Tokugawa to Meiji Japan. Thus, the Edo era had been more accurately a postfeudal or a national feudal society.

John Hall, Marius Jansen, and other Western scholars have tried to refute the dogma that held that the Tokugawa system constituted a "refeudalization" in order to rebut the view that Japan in the Edo era "stagnated in isolation." Empirical evidence amassed by postwar Japanese researchers and by outsiders has shown that the country, although it was indeed relatively isolated, experienced a considerable amount of growth. Chapter 2 summarized conclusions drawn from such empirical findings.

In brief recapitulation, under the Tokugawa system truly feudal forms came to be replaced by a highly developed (if, until Meiji, classbound) bureaucracy staffed by samurai. While lip service was paid to the principle that administration remained in the hands of a military (feudal) elite, the samurai increasingly underwent a significant transformation into a civil bureaucracy.

Thus the delicate check-and-balance system built up by the Tokugawa marched straight into modern Meiji Japan. A new central government in the hands of an oligarchy of former samurai displaced the feudal military headquarters (*bakufu*). Modern prefectures replaced the old domains. Members of a new civil service stepped into the slots that had been assigned to feudal bureaucrats. Thus by the mid-nineteenth century, during the Edo era, the prototype of the modern Meiji government of the late nineteenth century had already been established.

The modernizationists have admitted that Tokugawa authorities preserved a feudal outlook, almost to the very end. They have insisted, however, that Edo society at large developed well beyond the boundaries of a feudal system. The castle towns laid down the foundations for Japan's modern urban structure. There was rapid economic growth (without foreign trade) in a commercial society. Literacy levels were very high.

Tokugawa thinkers even contributed to a theory of social organicism. The shōgun-to-daimyō relationship was one of the whole to the parts. Administrators stood at the top not for the sake of personal or family interests, but for the benefit of the country. One may doubt the true motivation, of course, but at least the rationale for power seemed almost modern. Policy was made in the name of the national interest and for public welfare. Despite the continuing formal division into domains, an inchoate sense of nationalism never entirely disappeared. Respect for the imperial tradition was sustained so that Japan had, in the late nineteenth century, a clear alternative form of rule. After the Meiji Restoration, Confucianism provided the matrix of loyalty to the emperor and to the state.

In these various ways, according to the modernizationists, Tokugawa Japan was in fact a protomodern society. Feudal values, although they were definitely not the product of cultural contact with Europe, made it possible for Japan to accommodate to the West and allowed the nation to begin the process of modernization quickly. This thesis has been described as a grand design drawn by Edwin O. Reischauer, the father-figure of the modernizationists; it has also been denounced as the "Reischauer line."

It should be noted that both the modernizationists and their critics recognize the existence of a link between the Tokugawa period and what followed. From one vantage, feudalism of the Edo era was a dynamic hinge on the door opening to modernity. From the other, Japan's was an abortive modernization that never escaped the effects of feudalism.

Both views imply that the second coming of the West (most dramatically symbolized by Commodore Matthew C. Perry's black ships) was the important occasion for, but not the basic cause of, the transition. The Western impact was a catalyst that precipitated modern elements out of a traditional solution. As a catalyst it was historically significant.

Impact of the West

Many Westerners came to be interested in the maturation of Japan in the nineteenth century. They became intrigued by things Japanese displayed in the world's fairs. They also read descriptions of exotic Japan written by the early foreign employees, the "Japan helpers." Meanwhile, Japanese came to be fascinated by outsiders' views of Japan. As a result, the role of the West has been exaggerated (by Westerners and on occasion by Japanese) to such an extent that often nineteenth-century change was discussed in terms of the "Westernization" of Japan. Fortunately, this culture-bound term has been cast out of the literature, giving way to the more neutral "impact of the West."

Having abandoned Westernization as an inaccurate description, however, one encounters different kinds of difficulties with its replacement. "Impact of the West," too, is a generalization. But the critical questions are: Impact on whom and under what circumstances? Does the impact consist of certain people, of specific goods, and of identifiable ideas? Again, it must be recognized that many Japanese thought that what they were engaged in was something like the Westernization of Japan.

In any case, goods, people, and ideas reaching Japan in the 1850s and 1860s were not the first to come from the West. The Japanese had experienced what has been called the "Christian century" (1549–1638) before Tokugawa seclusion. After the Franciscans and Jesuits were expelled or crucified, Dutch traders remained in Nagasaki. Dutch studies subtly influenced Tokugawa thought. After Perry but before the flood of foreign influence, Dutch missions advised the Japanese on naval affairs (in 1854–59). The British came to Nagasaki in 1858. The French, however, did the most for the fading shogunate. They established the Yokusuka naval complex in 1865.

Although it has been the convention to say that the United States opened Japan, between 1853 and 1868 U.S. advisers were few. Samuel R. Brown and Dr. J. C. Hepburn in Yokohama and Guido Verbeck in Nagasaki were missionary educators. Since the teaching of Christianity was still banned, they worked as teachers of English and as students of Japanese. Then there was William Elliot Griffis of Rutgers College, who made the long foray to Japan while the Tokugawa still administered the country. He observed and described the collapse of the feudal domain in Fukui and the establishment of the new prefecture. He also taught science in Tokyo at an institution that later became Tokyo Imperial University. Griffis's most important role, however, was to be the unofficial historian of the early foreign employees.

THE MEIJI MODERNIZERS

When, in 1868, the Tokugawa banners were folded away and the flag of the new regime unfurled, as usual it was in the name of the emperor that the transition was effected. In theory it was a restoration of power to the Meiji emperor and, therefore, it has been called the Meiji Restoration. In practice, the drama was stage-managed by re-markably young samurai from the outer domains, men who formed an oligarchy. Japanese scholars have drawn a distinction between the early stages (1868–69) of *restoration* and the later (1870–90) of *renovation*. In the latter period, the fad for foreign goods, the prominent role played by foreign advisers, and the wide currency of translated foreign ideas were important enough to lead observers (including Japanese) to conclude that the country was indeed undergoing Westernization.

More Western Impact

After the restoration, the new government followed in the footsteps of the shogunate in seeking knowledge throughout the world. The search was now enshrined in policy, however, in a famous imperial promise, usually referred to as the Charter Oath (April 1868). Specifically, the fourth and fifth articles pledged a breaking off of "evil customs of the past" and a search for knowledge "to strengthen the national polity."

If the outcome did not constitute the Westernization of Japan, the Meiji experiment certainly did represent one of the first instances of development with foreign aid. The least desirable form such aid could take as far as the modernizers were concerned was the borrowing of capital from abroad. However, the Meiji leaders did, like their predecessors (the Tokugawa authorities), advocate collecting information through translation (technical assistance, so to speak, through books), but there were obvious limitations in this method. They also sent young Japanese overseas to study and to become, as Robert Schwantes phrased it, nineteenth-century "counterparts." Trained Japanese were intended to replace, and soon did supplant, employed foreigners and alien advisers.

In all of these processes of absorption of foreign techniques, the Japanese made all the critical decisions. They adapted the Western models of development to their own needs. They mobilized domestic resources and paid most of the costs. Above all, as many of the prominent foreign employees soon discovered, the Meiji modernizers adhered firmly to the policy of *Japanese* control and management of the transition.

In an historical context, then, it is fair to say that in the nineteenth century Japan was no more a second-class Western state than it was a minor Chinese state between the seventh and ninth centuries, a Christian outpost as a result of missionary activity in the sixteenth century, or a U.S. society in the late 1940s.

Nonetheless, Japanese have not forgotten the contributions of the foreign employees. They have been impressed by the statistics compiled by Hazel Jones, who estimates that the advisers provided the Meiji government with more than 9,500 man-years of service. The peak years of foreign advice were 1877 and 1888. After the 1890s the number of foreign employees declined sharply as the Japanese took firm control of further growth.[5]

The Modern Society

The various steps taken by the modernizers, aided by foreign advisers, may be quickly reviewed. In 1868 the new regime maneuvered the former feudal lords into returning their land registers to the emperor. In many cases the daimyō became the Emperor's new governors in 1871 when the modern prefectures were carved out of the former domains. In 1872 came the fundamental law of education (followed in 1890 by the more formal Imperial Rescript on Education). Thereafter each son and daughter was socialized into becoming a subject of the emperor. In 1873 the Imperial Rescript on Conscription gave each male, whether of samurai descent or not, the privilege of laying down his life for the modern state. Also in 1873 each person gained the modern privilege of being taxed directly by the central government.

Meanwhile, the central administration was streamlined, on the surface modeled after European-style cabinets, court complexes, and civil and criminal legal systems. In fact, all the institutions were firmly in the hands of an oligarchy, which, in typical Japanese fashion, eventually made up a body of elder statesmen (*genrō*). The capstone was the promulgation, after meticulous preparation (and little discussion), of a new constitution in 1889.[6]

In the economic realm, banks were established, railways built, lighthouses erected, and port facilities improved. Government-sponsored strategic industries were nurtured and privately financed small-scale businesses encouraged. The Japanese began (and continue to perform, with one major interruption in the 1940s) the modern miracle.

Meiji Japan thus provided the historical data on which to speculate about the whole process of development, although it was not until later that systematic analysis was labeled "modernization" in a technical sense. Most often outsiders, familiar with the Western model, referred to Japan's emergence as "late development." Now, a century later, it has become apparent that the Japanese transition occurred quite early in the game, as compared with presently developing areas in Southeast Asia, Africa, Latin America, and the Middle East. Whether Japan's success at modernization provides a model for other societies is a moot issue. Conditions in today's world and in developing nations are quite different from those in the nineteenth century and in Japan. Nonetheless, the Japan experience has contributed to development theory.

Viewed historically, Meiji Japan faced what has been called a crisis of security in the form of perceived pressure from the Western powers. Nonetheless, the new regime enjoyed sufficient stability to withstand external and internal threats and to allow for planning and action. A strong sense of nationalism was widely shared, even during the Tokugawa period, at least by members of the ruling elite. The capacity of traditional attitudes to coexist with, and contribute to, modern institutions has been noted. The best illustration was the use made, for better or worse, of the venerable imperial tradition.

Restoration and the subsequent renovation were rationalized in the name of the emperor. This was not an instance of "divine right" in the European sense. Rather, to borrow the felicitous phrase of a constitution drafted much later on, it was a matter of the emperor serving as the symbol of the state and of the unity of the people.

Politics. The politics of Meiji Japan, like the politics of any modern society, witnessed the enormous expansion of the sphere of governmental action. Individual Japanese became involved, some for the very first time, in the political process, just as human beings in modern societies elsewhere became subjects, citizens, comrades, or cadres. Heretofore many Japanese had been socialized in relatively self-sufficient, extralegal agrarian villages. Many had been touched only indirectly by government. Most Japanese had been members of the great mass of East Asian peasants (like the Chinese, who were to be described by Sun Yat-sen several generation after the Meiji transition as "a loose sheet of sand").

The modern state is marked by a high degree of integration, and politics tends to become what the social scientists call a "system." The system usually leads to the egalitarian involvement of the masses. They in turn may be administered under the iron law of oligarchy or they may move toward participatory democracy. The case of Meiji Japan offers a strong argument to the effect that if swift and efficient modernization is desired, then firm direction should be provided by an oligarchy. During at least the first decades of the transition, the Meiji modernizers were not interrupted in their tasks by rapidly rising expectations on the part of the disciplined and obedient Japanese.

Edwin Reischauer has cited the fact that the men of Meiji set a course under which Japan would, in a generation or two, alter its society from one in which position was primarily determined by heredity to one in which status depended largely on the education and achievements of the individual. Again, tradition and modernity were mixed: The result was not a society in which all were to be considered equal, but a meritocracy in which sophisticated screening processes were used to select the elite.

Other features of the Meiji system have contributed to our understanding of the modern state. Government evolved a system of allocating and terminating roles of political leadership that on balance took into account achievement rather than ascriptive status. Thus emerged the celebrated Japanese civil service. Recruitment for this service was (and remains) largely the function of the great public institutions of higher education, most particularly Tokyo Imperial University (now the University of Tokyo). Max Weber would have quickly identified in the civil service a certain functional differentiation that reflected the growing division of labor and specificity of roles in a modern society. Henceforth public administration was marked by an increasingly secular, impersonal, and—it was claimed in Japan as elsewhere—rational system of decisionmaking.

Psychology. The psychology of modern humans is marked by a difference in attitude. In contrast to their ancestors, who were born, grew up, and died without ever expecting to see anything but a traditional society, leaders of the modern state (and gradually most individuals, too) come to regard change as desirable, necessary, and even inevitable. This has something to do with the scientific revolution. Doubtless increasing human control over the forces of nature prepare people for the systematic, purposeful application of their energies to a more rational control of the social environment.

Changes in the thrust of Japanese slogans clearly illustrate the shifts in attitude on the part of transition leaders. Faced with the sea-coming aliens, spokesmen first called for "reverence to the emperor, expulsion of the barbarians." Instrumental politics then articulated "unity of the military and court." The planners moved on to specific steps "to strengthen the army, to enrich the country." Human goals, even ideals, were expressed in "civilization and enlightenment" (*bummei kaika*).

Even in the late Tokugawa era, certainly by the Meiji period, horizons were steadily expanded. Eyes lifted from the tiny villages to the already established towns, from the towns to the industrial cities, and from the port cities to the world. Men (more quickly than women) acquired geographic and psychic mobility.

Economy. The economy of the modern society moves toward diversification. The term is more precise than "industrialization" because, particularly with those called "late developers," the economy leapfrogs through the familiar industrial revolution into more advanced stages of technology. First agriculture and then industrial know-how provide the surplus for further development; diversification guarantees a sustained increase per capita in gross national product (GNP). People *think* they are better off.

Although the Meiji modernizers were often impatient with what they regarded as slow economic progress, between 1885 and the end of the century the GNP doubled; the rate of increase rose from 1.2 percent per annum in 1885 to 4.1 percent in 1898; per capita GNP increased from about 65 yen in 1885 to 115 yen in 1898.[7] Indeed, Japan was launched on a voyage of unequalled growth, which was to last, with some interruptions, for a century. This achievement is documented in more detail in Chapters 6 and 7.

Society. A modern society tends to be much more complex than its traditional forebear. Borrowing a metaphor from natural science, it has moved from a simple single-cell organism through differentiation into a complicated multicell organism. Beyond the figure of speech are some interesting historical issues.

Many descriptions of the modern condition include the degree of urbanization as a measure of cultural change. Most members of non-modernized societies live in a rural rather than in an urban environment. Traditional East Asia, dominated by rice culture, is still largely rural.

The majority of the Japanese population, during the Edo era, also lived in agrarian villages. And yet, as has been observed, many of the traditional values of premodern Japan were formulated by the minority who lived in lively cities. The problem can be partially solved by using a concept designed by Gideon Sjoberg, who has identified the preindustrial city. This is to say, there have been rural areas that have displayed at least protomodern features; there have been urban areas that have carried tradition to the city.

Moreover, if urban settlement has indeed been the mode of modern man, then it cannot be identified simply with industry. Cities appeared before the industrial revolution. Well after the Meiji transformation, Japanese cities encompassed much more than industry, moving into the high-tech function (see Chapter 7).

There is also the question whether urbanization constitutes the cause of change, the effect, or both. In the late Tokugawa and in early Meiji, the implanted treaty ports caused a temporary Westernization of surrounding Japanese life. After the Meiji Restoration and during the renovation stage, directed political reform resulted in a grand expansion of a new *Japanese* bureaucracy, whereby city-based administrative organs became integrative agencies affecting the nation.

Ernest Weisman, in his United Nations study of the Japanese city, reported that urban development was effected under the leadership of, first, the Tokugawa feudal, and then, the modern Meiji, oligarchy.[8] The situation was not unique: Development of cities in other parts of the world illustrates the critical role of the *political* apparatus as the key independent variable.

Finally, urbanization, the growth of a commercial society, and the political style in the Japanese transition reveal the difficulty in generalizing from experience in Western Europe and in the United States. In Japan the government usually made the critical choices and played the leading role in entrepreneurship. As one Japanese put it, "political businessmen" were "samurai-in-spirit" and "merchants-in-talent." Historians have searched the Meiji record in vain for middle-class entrepreneurs expected to lead Japan out of feudalism into a society managed by the bourgeoisie.

To repeat, the Meiji oligarchy made political choices. Their decisions were, of course, affected but not determined by economic factors. The process was an example of the autonomy of choice, the very core of policymaking in a modern society.

THE DILEMMAS OF GROWTH

In the above analysis, one may immediately detect several problems in the use of the term *modernization.* Inevitably, the concept implies a built-in bias related to time. Primitive and traditional societies are often viewed as "backward." The modern society is usually regarded as looking "forward." Modernization is not only change; it appears to constitute progress. The appearance is what Charles Frankel has referred to as "changes in the concept of social time."

Although most Japanese have thought that by modernizing they were progressing, in fact it is impossible to find an empirical link between the two processes. As the economist points out with statistical evidence, it is true that in modern times the GNP per capita in Japan has increased. Moreover, Japanese have come to boast of a record longevity rate in the world. It is also true, however, that contemporary Japanese (like their North American counterparts) have discovered that satisfaction of material wants and a long life only bring them to the more subtle problem of affluence. Japanese live longer perhaps only to survive in "the lonely crowd." Modernization made possible electric lights and streetcars on the Ginza. But it brought Zero aircraft and bombs to Pearl Harbor and B29s over Japan itself. In any objective evaluation, modernization holds both promise and peril.

Taishō Transition

Meiji was an imperial era name (*nengō*)[9] for the period beginning in 1868. By allusion to Chinese classics Meiji meant "enlightened government." It also was the posthumous name for the emperor who died in 1912. His successor was known as the Taishō emperor. This classical term referred to "great rectification, adjustment."

This contrast is not meant to imply that the year 1912 marked a clean break between the Meiji (1868–1912) and the Taishō (1912–25) eras. Nonetheless, the periodization does provide an interesting milestone. In his masterpiece, *Kokoro* (The heart, 1914), the novelist Natsume Sōseki brings his main character to suicide on hearing of the death of Emperor Meiji and of the sympathetic suicides of the national hero General Nogi and his wife. In fiction at least, it was the end of an era.

Change from the steady diet of faith in progress had, of course, begun even in the late Meiji period. Culture, heretofore identified solely with the nation-family and the state, came to have a life of its own. Industrialization had continued apace, with concomitant shifts in class lines. By 1907 universal education (for six years) was achieved; higher education (mostly for males) expanded rapidly. New, white-collar, middle-class Japanese came to resemble those found in urban Western countries. They stood between the elite leaders of the Meiji period and the mass citizenry of Taishō. Intellectuals, spawned by the new middle class, in turn began to express doubts about consumerism, mass culture, and the new secularism. Contemporary literature clearly revealed a tension between new values and the older statist claims of imperial Japan. The historian Harry D. Harootunian has hinted that Taishō may have marked the height of individualism as we know it in the West. Writers of the time denounced "individualistic dissipation," "preoccupation with carnal desire," and "the general celebration of luxury."[10]

Nor did political institutions move precisely along the path laid out by the Meiji modernizers. In 1890 they had established, in response to forces released from the Pandora's box, a bicameral Diet (legislature). A lower House of Representatives was to be elected by male voters who were upper-level taxpayers (less than 1 percent of the population). An upper House of Peers (modeled on England's House of Lords) represented a select peerage. Despite these restrictions, this Japanese legislature turned out to be the first successful parliamentary experiment outside the West.

During the first decade of the government's experience with the Diet, control of even limited representative government proved far more difficult than had been anticipated. The legislature vigorously opposed the founding fathers until the latter adopted an officially backed party system of their own. In the decade after the turn of the century, the old oligarchs retired in favor of their bureaucratic followers. Until 1912–13, protégés of the elder statesmen alternated in holding the position of prime minister. By 1918, in response to press and public pressure for "normal constitutional government," the bureaucrats had retreated before cabinets dominated by political parties.

Meanwhile, the number of voters was steadily expanding as tax requirements were lowered in 1900 and again in 1919. By 1920, the original limited electorate (1890: 450,872) had grown sevenfold (1924: 3,288,368). Finally, in 1925 the vote was given to all adult males, and by the next election the electorate had quadrupled (1928: 12,405,056). Cabinets were being chosen from one of two parties with similar ideological stances. Indeed, Japan between Taishō and the early Shōwa period (1926–1931) seemed to combine modernization and democracy. This span of time has been referred to as Taishō democracy.

Critics and Crisis

Beneath the giddy changes that produced a jazz age in Tokyo—the appearance of outspoken modern boys (*mobo*), flappers (*moga*) with Clara Bow haircuts, and a wave of liberal, even radical, ideas—there were intractable problems. Diet control over cabinets was only a superficial convenience, for men like the last *genrō*, Saionji Kimmochi, exercised the imperial prerogative by screening the leadership of Japan. A gap between traditional-rural and modern-urban sectors of society was matched by a dual structure in the economy. There was a difference in productivity as well as in wages between new industry with advanced technology and what Freda Utley called "Japan's feet of clay" in agriculture. Growth, encouraged in Japan by the absence of competition on the part of European powers (who were engaged in World War I), gave way to stagnation, which was born of resumed competition and matured in deep depression.

Soon the force of tradition, so useful in the early modernization process, was channeled into reaction. Purists began to link behavior in the Diet, corruption in the corporations, the selfishness of individualism, and a liberal lifestyle. Often these were blamed on Western influence. Such trends were reflected in the culture of "Taishō adjustment."

In the literature of late Meiji and early Taishō, some poetry flirted very tentatively with Western form, but most of it continued to be written in classical Japanese. Perhaps understandably, the theater never really broke with colorful Kabuki in favor of the introspective dialogues of Western drama. Modern literature, to the majority of the Japanese, has meant the novel or short story. And in subtle style Japanese writers have served as the social critics—the skeptical observers of modernization and its effects.

At first, however, novelists embraced modernity and even imitated Western style. The primary effect was seen in the abandonment of classical Japanese and the adoption of colloquial language. Soon Japanese writers found their own idioms for new views. For example, *Ukigumo* (The drifting cloud, 1889)[11] by Futabatei Shimei was probably Japan's first modern novel (according to Donald Keene), because it adopted everyday language and depicted the dilemma of a young man, a product of strict samurai training, facing the values of a new and strange society.

Many of the modern Japanese novelists gradually unfolded a pattern of development. The first step was to experience strong influence from the West. The second was to borrow writing methods from abroad and use them in novels. The third was to turn their backs on Western influence, to peer back into traditional sources, and to develop individual Japanese styles. Three examples have been offered by Donald Keene.[12]

Mori Ōgai (1862–1922) studied medicine in Germany in the late 1880s, and his early stories carry a romantic German tenor. After his return home, his writing began to picture life in Meiji Japan.[13] He, too, was apparently deeply moved by the Emperor Meiji's death and the Nogi suicides and in his later writing confined himself to true tales about samurai.

Although his novels are quite different, the experiences of Natsume Sōseki (1867–1916) were similar to those of Mori. His early writing clearly reveals the influence of his residence in England in the early 1900s. Later back home, he became obsessed with the disastrous effects of individualism (what he called "egoism"). His best-known work (already cited) is Kokoro,[14] in which a man is tortured by the memory of his betrayal of a friend.

The third representative writer is one who lived through the late Meiji and early Taishō periods and the war and on into another period of puzzling change after defeat. Tanizaki Junichirō (1886–1965) early on also moved from unashamed admiration for the West to a position of doubt. In his novel Tade kuu mushi (Some prefer nettles, 1928),[15] he skillfully spun certain contrasts between backgrounds and, with psychological insight, wove conflicts among the main characters. Thus Tokyo is a city of horns and headlights, movies and modern youth, beauty parlors and hot baths. Ōsaka, with its bunraku puppet theater, represents Japan's past. In this tale the young mistress of the hero's father-in-law, a doll-like, dream-like figure, represents tradition; the Eurasian prostitute, a lingering interest in the West; the hero's stylish young wife, the inevitably modern. The latter's father sums it all up in a phrase: "Misako's education has been half old and half new, and all this modernness of hers is a pretty thin veneer."

Japanese art, as might be expected, responded to the epochal changes in a bewildering variety of ways. In the late sixteenth century, artists working in a genre painting style produced outlandish scenes, screens showing the arrival of alien ships, and portraits of tall foreigners with huge noses. This was called "southern barbarian" (namban) art (which has since achieved museum quality rarity). There was some Western influence on the Edo woodblock prints. Although the quality of the so-called Yokohama prints declined during the early Meiji era, their subjects have proved to be of great historic interest. They picture the European style of the Yokohama treaty port, the Napoleonic costumes of the first Diet, and the artifacts of the modern technology, such as

Japanese pottery and the late master potter Hamada Shōji

the telegraph pole, the railroad, and Victorian-era architecture. The majority of artists, however, continued to work in Japanese style. Meanwhile an American, Ernest Fenollosa, and his disciple, Okakura Kazukō, were responsible for impressing on Japanese consciousness the greatness of the Japanese artistic tradition.

In some areas, for example in architecture, modernization did equal Westernization. The Ginza area, today Tokyo's most famous shopping district, was developed artificially and quite consciously to be a Western showplace for Japan. The marvelous old red brick structures (of which only one or two survive) represented one of the deliberate attempts to convince the Western powers of Japan's emergence as a modern state and of the need to relinquish the unequal treaties inherited by the Meiji regime. Similarly, in public sculpture Western-trained artists erected imposing statues of the cultural heroes of the Meiji transformation.

Meanwhile, in a strange shift caused by cultural contact, Japanese decorative arts became extremely popular in the West, and native pieces were eagerly collected by foreigners. In the provinces at least, fine folk art continued to express Japan's traditions through the Meiji period. By the early Shōwa era, Japanese artists (particularly those working in woodblock style) were beginning to adapt Western techniques to their needs. Their creations were traditional and ultramodern at the same time.

MILITARISM AND MODERNIZATION

Until 1945 Japan could be accurately described as another of those thoroughly modern nations in which military considerations gradually served as dynamic shapers of policy and even culture, rather than instruments in the hands of civilian policymakers concerned with defense. To some, as has been noted, this was an inheritance from feudalism. To others, the phenomenon was a much more complicated aspect of the modern condition.

Origins

Some of the elements of Tokugawa feudalism (for example, the samurai dominance of early Meiji government) inherited by Japanese in the modern era did encourage militarism. The Japanese quickly moved toward mass conscription, however, and a conscript army certainly cannot be described as feudal. It was more in tune with modern mobilization for war.

One of the factors that permitted the rise of militarism in Japan was a fatal flaw in the modern Meiji Constitution, which allowed army and navy leaders direct access to the imperial symbol. In addition to this provision was the practice of appointing only officers on active duty to the cabinet as service ministers (men who were therefore under armed forces orders). The privileged position of the military, however, did not forestall the rise of Taishō democracy or, for a time at least, civilian control of the military. Nonetheless, social critics like Katō Shūichi have argued that there was a connection linking rapid Meiji modernization, Taishō liberalism (with very shallow roots), and aggressive Shōwa militarism.

At first Japan's wars in the modern period were fought in remote areas, with only a modicum of discomfort at home. Japan's domestic reforms and successful war against China (1894–95) led England, as well as the other powers, to relinquish extraterritorial rights in 1899. Victory over Russia (1904–05) set Japan on the then modern road of empire, to the control of Taiwan, adding the acquisition of southern

Manchuria, the lower half of Sakhalin, and, eventually, Korea. After World War I, Japan joined the Allied victors and in the process sought concessions from China, seized German concessions in Shantung, and took control of former German islands in the Pacific. The latter were colonies thinly disguised as mandates. No wonder that many Japanese came to believe that war solved internal problems and paid direct dividends.

In the late 1920s and in the 1930s, the imperial army clearly revealed the commingling of traditional and modern forces and, at the same time, the torque of change. On the one hand, the army was the legitimate heir to the samurai code; its officers and sergeants were highly paternalistic to the conscripts. On the other hand, leadership was increasingly drawn from the underprivileged peasant classes (not from samurai families); ordinary soldiers, too, came from the countryside. Japan provides one more illustration of the historical role of the armed forces—often among the most important agencies for modernization and among the most significant groups for socialization. As Lucian Pye has suggested, the soldier has been to some extent a modernized man.

Young military officers were among the most strident critics of Taishō democracy, of what they considered corrupting influences at home, and of what they regarded as Western ideas. One of their solutions was to propose a "Shōwa restoration" of power to the emperor, whom they would then manipulate for their own purposes. The young militarists of the early Shōwa period were not altogether reactionary, however; they were often puzzlingly radical and even revolutionary. Drawn from depressed rural areas and returned as veterans to poverty-stricken villages, peasant conscripts of the imperial army in the 1930s and early 1940s stood against landlords, against capitalists and their "bourgeois" political parties, and against Communists.

Mobilization: Successes and Failures

After 1937 Japan was in almost uninterrupted conflict for almost eight years. Before and during the great Pacific conflict, every single Japanese was to feel the effects of modern war.

The chief differences of opinion within Japan among civilian politicians, financial leaders, generals, admirals, and imperial advisers were over the issue of how far Japan should become involved on the Asian mainland. There was no disagreement over the assumption that Japan was destined to become the leader of East Asia. Differences were related to methods, not basic aims.

China's unification and growing resistance to Japan's assumed role of policemen dedicated to the extermination of Communism in Asia appeared to many Japanese to be aided by the Western powers, which had special privileges in conflict with those of Japan. It was claimed that Japan's very survival was being threatened by encirclement on the part of the United States, Great Britain, and the Soviet Union.

Various shifts in government, from the first Konoye cabinet (1937) to the selection of General Tōjō (1941), were indicative of the struggle over methods, not over basic aims, between extremists and moderates. Army extremists became convinced that direct action was necessary to solve "the China problem," regardless of consequences. Moderates, who were drawn from the imperial navy, the diplomatic corps, and business circles, hoped that Japan's objectives in East Asia could be achieved without friction with the Western powers. By June 1941, naval circles, alarmed over oil shortages and a U.S. embargo, joined those in favor of direct action and advocated moving into Southeast Asia even at the risk of war. As Japan's aims became more clearly identified with those of the Axis powers, moderates retreated and acquiesced to the plans of those who advocated forceful action. Although the Japanese did not make a single identifiable decision to go to war, they narrowed the options to such an extent that all policy issues had to be settled in a military manner.

Perhaps the closest historical parallel (although there are marked differences) was the (at first almost unconscious) involvement of the United States in Vietnam in the 1960s. Somehow the need to police Southeast Asia came to be perceived as a U.S. "national interest" (just as the new order in East Asia had been the "imperial will" for Japan in the 1940s). Slowly and inexorably, all other options were eroded so that Americans came to handle all policy matters in Vietnam in military fashion (just as Japanese had allowed militarism to rise in the 1940s). After 1965 Japanese were quick to point out that Americans were repeating some of the earlier Japanese mistakes (including a bankrupt military campaign on the continent of Asia).

To return to the case of Japan in the 1940s, the plunge into world war held a paradox. Various Japanese leaders did develop plans for industrial mobilization and even a blueprint for a monolithic state dedicated to total war. Sketches included designs for the grandiose Imperial Rule Assistance Association, which was to become a transcendent national party. The old political parties were subjected to "voluntary dissolution," but the association was never a success. In similar fashion, plans were drawn for a new authoritarian political structure (the *shintaisei*). Characteristic Japanese indirection in politics and diffusion of power in government, however, led to a surprising failure to mobilize completely and eventually contributed to defeat. There was no corporate state (as in Fascist Italy), no dictator with popular support (as in Nazi Germany), and, indeed, no formal change in the old Meiji political structure. The experience has cast doubt on the wisdom of applying to Japan's wartime society the modern term *fascism,* although Japanese opponents to the rise of the militarists freely used the word. Even in the postwar period, those who opposed what they saw as a revival of militarism applied the pejorative word *fascist* to their opponents.

Another interesting contradiction—thus far hidden in the multifold effects of modern militarism and the impact of war—was the experience of women, the largest social group in the country. Thomas H. R. Havens has informed us that during the war years the state continued to encourage male precedence, to sanctify motherhood, and to applaud supportive home-front activities by women's groups. Large numbers of unmarried women were mustered late in the disaster, but the government to the very end avoided fully mobilizing them in the war effort. Yet modern war further eroded the patriarchal ideology in a manner difficult to describe.[16] Due attention has been paid to the Occupation reforms, which laid out a new path for Japanese women after the war. These and the modern miracle of further economic development, which so affected the entire family as a unit, have been emphasized (and will be summarized below). The modernizing effect of the war on women has thus far been neglected.

Certainly women of the postwar period took the lead in the pacifist, antimilitarist stance of the Japanese.[17] They were joined by the surviving veterans of the war, youth, and intellectuals and reinforced by the writings of social critics and novelists. Ōoka Shōhei (born 1909), for example, produced the most vivid denunciation of militarism and war in his *Nobi* (Fires on the plain, 1952).[18] Writing from a home for the mentally ill on the outskirts of Tokyo, the author's chief character, former Private Tamura, reflects on the horrors of the Philippine campaign. Then he adds:

> The reports in the newspapers, which reach me morning and evening even in this secluded spot, seem to be trying to force me into the thing that I want least of all, namely, another war. Wars may be advantageous to the small group of gentlemen who direct them, and I therefore leave these people aside; what baffles me is all the other men and women who now once again seem so anxious to be deluded by these gentlemen. Perhaps they will not understand until they have gone through experiences like those I had in the Philippine mountains; then their eyes will be opened.

Toward the end of the Pacific war and after the defeat of Japan, the Allied victors (led by the United States) explicitly proclaimed and implicitly revealed through Occupation policies that "irresponsible militarism," in the words of the Potsdam Proclamation, had misled the Japanese people, brought disaster to their nation, and delivered woe to the world. The defeated Japanese, then thoroughly disillusioned with the leadership of the early Shōwa era, agreed. Therein lies the origin of modern antimilitarist sentiment in Japan.

One irony in postwar demilitarization policy was the fact that distaste for German and Japanese military adventures had led Americans to make militarism a vital factor in U.S. life during World War II. After a brief respite, Americans plunged into the cold war, fought in Korea, and became involved in the disastrous Vietnam War. Viewed historically, militarism—rather like the phenomena of feudalism and modernization—defies generalization and requires precise definition.

5

Postwar Politics

Since the mid-nineteenth century, change in Japan has involved choice, the essence of politics. In the 1860s and 1870s, the Japanese chose to open their nation, to seek knowledge throughout the world, and to modernize the traditional society. In the 1920s they chose a path of development that included an incipient, parliamentary democracy. In the 1930s they followed a path that led to militarism, aggression, war, and defeat.

Some Japanese revisionist historians have argued that Japan has had little or no choice in modern times. Their country was dragged out into the Western nation-state system ("opened" was the euphemism); Japan was steadily hemmed in by Western imperialism, so that the nation responded in kind; in 1941 a metronome was set to ticking, so that the country was forced into what the Japanese hoped would be a limited war. The result was total disaster.

And yet surprisingly, defeat, the downfall of militarism in Japan, and the nature and timing of the surrender also allowed choice. In other words, there were significant conditions in the so-called unconditional surrender. There was, to put it inelegantly, a shotgun wedding between the chief avenging victor and the vanquished. The result was a marriage that would last at least into the 1950s. In 1952, when Japan regained the exercise of free will—and could have had a divorce—the bond continued to hold for four decades, in sickness and in health, for better or for worse.

Again, revisionists have argued that Japan was punished improperly for "aggression," which was never committed. There was no choice in 1945; little or no choice during the Occupation (1945–52); and narrow choices, if any at all, in the majority peace treaty, the security arrangement with the United States, and Japanese foreign policy that followed the end of alien administration. Japan (the figure of speech was altered) became an adopted child, orphaned in the cold war.

95

MORE MODERNIZATION

A great deal of the literature on Japan's defeat has held, with the remarkable vision available to hindsight, that the island nation lost because it entered a protracted struggle that it could not win. The fact remains that, for millions on either side, the conflict was in doubt until almost the very end. Furthermore, such ex post facto reasoning has tended to overlook the political occasion for surrender in favor of a somewhat futile attempt to weigh the causes of defeat.

The Politics of Surrender

It is understandable that Americans, with a growing guilt complex, would continue to emphasize the first atomic bombs as punctuation marks that closed the conflict. Use of the new and awesome weapon was only one among a number of factors, however, in the timing of the surrender. It was probably not the chief reason why underground peace movements in Tokyo were able eventually to maneuver toward the decision for peace.

First, appalling though the effects were, especially for the future of all mankind, the immediate psychological impact was remarkably localized to Hiroshima and Nagasaki. The bombs affected Japan's shaken leadership more than they did ordinary civilians, for it was only after the war that the cloud of secrecy enshrouding the bombings disappeared. This is not to say, of course, that the long-range political effects of the use of atomic weapons were insignificant. On August 5, 1989, in Hiroshima, Mayor Araki Takeshi opened the second World Conference of Mayors for Peace through Intercity Solidarity. Some 230 representatives from 120 cities in 26 countries took part in the five-day conference, marking the forty-fourth anniversary of the world's first use of an atomic weapon.

In 1945, however, Japanese had given every indication that they could and would continue to resist. The will to defend Japan had continued after the equally devastating, if less efficient, fire raids on some ninety Japanese cities (with twenty more than half destroyed). Stubborn defenders in the living hell of Iwo and catastrophic casualties in the so-called iron typhoon of Okinawa demonstrated to invaders that the Japanese would defend the main islands to the death, with bamboo spears if necessary.

Japan's leadership and the man in the street were much more deeply affected by Russia's entry into the Pacific war on August 8, 1945, just six days before the actual surrender. In this last-minute violation of a neutrality pact, the Soviet Union had signaled to the Japanese what a joint occupation would be like. The Japanese were almost desperate to seek a formula by which they could surrender to the United States.

The key factors determining the occasion for surrender—short of total annihilation—involved political decisions on both sides. The bankrupt formula of unconditional surrender, cavalierly enunciated by Franklin D. Roosevelt at Casablanca (1942), had to be modified at Potsdam (1945). The Allied proclamation at Potsdam hinted at the conditions, so to speak, in unconditional surrender. In good negotiating style, the terms were deliberately left vague. Until that stage, the peace party in Japan had no leverage; die-hard militarists could incite the people to greater sacrifices because of unconditional surrender. Beyond that point, those who sought peace could use the oldest political symbol known to Japan. On August 15, 1945, the emperor's voice, recorded for broadcast for the first time, read an imperial rescript accepting the Potsdam Proclamation. Speaking for the Japanese people, Emperor Hirohito decided "to endure the unendurable." The political significance of all these developments, without trying to rank them in order, may be quickly summarized:

- Japan, like the other Axis powers, was militarily occupied, but the country was not divided into zones as was Germany.
- Japan was occupied in theory by the Allied powers, but in fact was controlled by Americans.
- Japan was not placed under the direct military government of alien forces, as was Italy.
- Japan maintained a government intact and was soon administering the Occupation under, of course, strict supervision.[1]

The Occupation: Directed Change

Almost as soon as the Allied, overwhelmingly U.S., Occupation got under way, another significant political effect of the developments that led to surrender became apparent. Perhaps because they were Americans, the occupationnaires launched a program of what Robert Ward has called "planned political change."[2]

The Occupation was of historical significance, then, because it constituted more modernization. Like the tightly controlled experiment carried out in the nineteenth century by Meiji modernizers, the effort was applied from the top down on obedient and disciplined Japanese subjects. Like the leaders of Meiji, who were drawn from a military (samurai) class, the majority of the occupationnaires were drawn from a military command, Allied soldiers who had fought their way out of Corregidor to Australia and thence across the southwest Pacific to the Philippines and eventually to Japan.

Unlike the Meiji modernizers, the directors of the Occupation were dedicated democrats. Although the elements need not be linked in an objective definition, to the occupationnaires modernization meant "democratization" (whatever the latter might have meant). Although militarism had often been part and parcel of modernization in Japan, at first the Americans found it impossible to visualize democracy without demilitarization of the Japanese. It was one of the small ironies of history that directed change, planned development, democratization, and demilitarization were implemented in occupied Japan by professional military personnel. Unlike the Meiji modernizers, the Taishō democrats, and the early Shōwa militarists, the U.S. occupationnaires were, of course, aliens.

The people attached to the Supreme Commander for the Allied Powers (SCAP) liked to refer to the process of change as "induced revolution." This process was monitored (sometimes criticized as going too far, at other times encouraged to go further) by means of belated Allied policy guidance drafted by an international body in Washington (the Far Eastern Commission) and, on the spot, by a small watchdog group (the Allied Council for Japan, in Tokyo).

Initial policy guidelines set forth goals of demilitarization, collection of reparations, and the restitution of all overseas territories. The Japanese government was to remove all obstacles to the "revival of democratic tendencies among the Japanese people." The reference, which surprised many hard-line critics of Japan, was, of course, to the promising development of Taishō democracy prior to the rise of the military. One other provision seemed to offer an intriguing dilemma: Occupation forces would be withdrawn when Allied objectives were reached and when a responsible government was established "in accordance with the freely expressed will of the Japanese people." During the Occupation the Japanese, on the one hand, were under the strict control of a military headquarters and, on the other, were to eventually determine the form of their political institutions by their own expressed wishes.

In any case, the first steps in directed change were punitive. They included a three-stage purge of wartime leaders (almost 9,000 Diet members, local government officers, and financial and industrial leaders) as well as indictments upheld between 1946 and 1948 of twenty-five major war criminals (seven were hanged).

The occupationnaires then turned to positive political planning. Once again, as at the time of surrender, the stance designed for Emperor Hirohito was crucial. A majority of the Japanese elite had decided not to resist (but to delay) revision of Japan's earlier organic law, the Meiji Constitution (1889). Indeed, they had very shrewdly arranged for the Shōwa emperor to write his own preamble to a new constitution. In the Imperial Rescript of January 1, 1946, he clearly explained that his status as a symbol did not depend on myths about his divinity. Rather, his importance lay in representing the state and the unity of the people. The very manner of the presentation may have confirmed in the minds of many Japanese the very special invulnerability of the throne to simply verbal or strictly legal definition. Moreover the rescript was another step that probably helped avoid removal of the emperor as a war criminal. Other moves preparatory to constitution making had to do with the emperor's subjects.

By October 1945, a virtual bill of rights had been adopted following SCAP fiat. Political prisoners were ordered released, and drastic modifications were made in police organization. Restrictions on political and religious liberty were forbidden. SCAP made quite clear that they wanted state and religious sects (the prime examples, Shintō shrines) separated.

Because the occupationnaires were Americans, they thought of social engineering primarily in legal terms. In their view, modernization meant democratization and democracy was assured by constitutionalism. Just as the Meiji Constitution was the capstone of the structure built by the nineteenth-century modernizers, so the MacArthur Constitution (1947) was the cornerstone in the foundation laid down by the twentieth-century modernizers.

By February 1946, Supreme Commander Douglas MacArthur and his advisers had become impatient with the Japanese penchant for delaying revision of the Meiji Constitution. On February 4 behind closed doors, General Courtney Whitney, chief of the Government Section of SCAP, briefed his assembled personnel. They were, in effect, to become a constitutional convention for Japan! Full discussion was to be allowed within the section in preparing the so-called Whitney draft, except for three points, which General MacArthur explicitly required. First, the emperor's powers were to be exercised constitutionally and then only according to the basic will of the people. Second, Japan was to renounce war forever. Third, the Japanese would abolish "all vestiges of feudalism." The general added a cryptic word of advice: "Pattern budget after British system."

In brief summary, thereafter the Japanese leaders writhed, delayed, discussed, and (for reasons that will be detailed) eventually gave in. The Whitney draft with modifications became a SCAP version of an organic law; the SCAP draft with minor changes became a cabinet version; the government draft finally became the new Constitution of Japan. It was promulgated by the emperor on November 3, 1946, the anniversary of the birth of his ancestor, the Meiji emperor. It went into effect on May 3, 1947 (which to the present has been celebrated as Constitution Day). Product of persuasion, some cajolery, and more than a hint of threat, the new constitution articulated at the time more a U.S. outlook than a Japanese viewpoint.

Because the so-called MacArthur Constitution has remained intact, unamended, and in effect to date, its characteristics will be described by implication in the section on the contemporary governance of Japan. For the moment, we may single out five features of the cabinet draft (none of which was fundamentally altered by subsequent formal or informal revision).

First, the draft redefined the emperor's powers. The sovereign was to become "the symbol of the State and of the unity of the people." His position derived "from the will of the people with whom resides sovereign power" (Chap. I, Art. 1).

Second, the Diet was to become "the highest organ of state power" (Chap. IV, Art. 41). Executive power was to be vested in the Cabinet (Chap. V, Art. 65), which would be collectively responsible to the Diet. In one of the few Japanese contributions to the structure, the draft provided for an upper House of Councillors (replacing the old House of Peers) as well as a lower House of Representatives (Chap. IV, Art. 42).

Third, the draft offered (after firm U.S. prodding) an extensive bill of rights (Chap. III). They were not subject to law as were rights under the Meiji Constitution. This was the section that most clearly reflected U.S. thinking. Rights included the familiar life, liberty, and the pursuit of happiness and provided for equality before the law. In addition, they went beyond U.S. constitutional definitions by guaranteeing academic freedom, the right to select residence, collective bargaining, and full employment. Indeed, as one wag described them at the time, the rights were so extensive that it is doubtful they would have passed the U.S. Senate.

Fourth, the Japanese government agreed to press for a complete renovation of the judicial system. The "whole judicial power" was to be vested in a supreme court and attendant inferior benches (Chap. VI, Art. 76). Again, the U.S. hallmark was left on the judicial process by a provision that the supreme court was to determine the constitutionality of any law, order, regulation, or official act (Chap. VI, Art. 81). The Japanese, not by tradition committed to defining social norms by adversary law, have since warily skirted this thoroughly American procedure of judicial review.

Fifth, and doubtless most significant, was the provision that has lent the organic law another of its names, the Peace Constitution. The idealistic ideas about security thrust upon the Japanese originally included abolition of war as a sovereign right; Japan's renunciation of war as a means for settling disputes "and even for preserving its own security";

and refusal to authorize any army, navy, or air forces or to confer "rights of belligerency" upon any Japanese force.[3] Fortunately (from the point of view of later SCAP officials whose views of security were to differ from the first draft) and unfortunately (from the point of view of Japanese opposed to any rearmament), the final operational section of the constitution was reworded, reportedly by Japanese lawyers, to leave an ambiguity about security.

In the version eventually accepted by the Japanese, a wordy preamble of the constitution prepared the way by stating, "We the Japanese people, desire peace for all time." Leaving aside the imported Jeffersonian overtones, the statement accurately reflected Japanese sentiment at the time (and in the 1980s probably still articulates the desire of the majority). Considerably less realistic, in light of later developments, was the implied faith in the new United Nations, with the Japanese people "trusting in the justice and faith of the peace-loving peoples of the world."

The final operational provision (Chap. II, Art. 9) reiterated sincere Japanese aspiration "to an international peace based on justice and order." Using the old and ill-defined formula from the Kellogg-Briand Pact (1927), Japan forever renounced war "as a sovereign right of the nation and the threat or use of force as a means of settling international disputes. *In order to accomplish the aim of the preceding paragraph*, land, sea, and air forces, as well as other war potential, will never be maintained. The right of belligerency of the state will not be recognized" (italics added). Thus, in the celebrated Article 9, the stage was set for some lively postwar politics.

Strict constructionists have since argued that Japanese armed forces, by whatever name, and even the presence of U.S. forces in Japan, have been unconstitutional. Conservative realists have countered that, by the peace provision, Japan did indeed forswear the use of armed forces as instruments of national policy *on offense*; the nation did not and could not, however, waive the inherent right of *self-defense*. In either case, the Peace Constitution of Japan has set a very interesting precedent for the vanquished, for the victors, and for the world.

Other Occupation Reforms

In general, SCAP headquarters vigorously followed the path of reform of Japanese institutions for about one-half the Occupation period, that is, until 1948. SCAP directed countless changes during this phase, but three major economic programs may be singled out.

Personnel in SCAP, in the best U.S. trustbusting tradition, identified the giant *zaibatsu* (literally, "finance cliques") and ordered them dissolved. As a result, the huge combines—firms like Mitsui, Mitsubishi, and

Sumitomo—were broken up. Eventually this antitrust drive stalled because the United States became more interested in reconstruction than in reform after 1948. The *zaibatsu* as a *method* of doing business (in contrast with *zaibatsu* as specific prewar firms) was familiar and useful to the Japanese, and they revived the form.

A second directed change had to do with labor. Occupationnaires encouraged creation of the Labor Ministry (1947), which sponsored legislation designed to nurture modern labor-management relations. Under SCAP encouragement, trade union membership, which at the surrender totaled only 5,300, rose to a total of almost 7 million by 1949. The sudden rise of labor as a postwar political force proved to be an embarrassment to SCAP, which on January 31, 1947, had to step in and forbid a general strike that would have disrupted further directed change.

The third significant area invaded by SCAP was the agrarian countryside—the directed change, a breathtaking land reform. In brief, the program, which was administered by the Japanese, allowed 3 million cultivators to acquire 5 million acres of land. In the short run, redistribution of land by parcelization probably reduced agricultural production. In the long run, land reform greatly strengthened the rural economy, increased production, and provided a stable foundation on which to build post-Occupation government.

All the formal statistical data on land reform was not as impressive as informal direct observation of the effects on rural Japan in the 1950s. The little hamlets looked the same, except for the fact that technology was steadily upgrading production. Lives revolved around the same crop cycle. Age and status were still used to choose village leadership. However, one link in the chain of tradition had been broken. In many fine old houses, the collection of documents signifying descent from lower samurai or magistrate status through village chief to modern elective post was still intact. The distinctive signature in cursive script of a prominent Meiji or Taishō political boss still dominated the largest room of the house. His political stakes, driven into the Japanese soil, had once marked out a circle of political influence subtly mixed with the pattern of landholding. The documents and sometimes even the political memento were for sale, though, because the land surrounding the fine house now belonged to the neighbors.[4]

The Occupation: In Retrospect

The balance sheet of assets and liabilities during the Occupation can only be filled in after an analysis of development since the end of this truly unique experiment. We shall hazard a few comments here on this stage of modernization and on directed change.

Anyone interested in the social sciences should find of interest the classic dilemma that the occupationnaires faced when they held almost absolute control of Japan. Should they let history take its course, allowing the Japanese, after a disastrous defeat, to slowly find their own way to reorganize society? Or should the men of SCAP erect a scaffolding for renovated institutions and expect that everyday habits would eventually develop to fill in the frame, something like the Japanese manner of constructing a building from the top down?

Because the Americans were impatient modernizers and because they so firmly believed in the desirability of change, they chose the latter method. Shortly thereafter in the United States, there was a parallel between the experiment in Japan and a quietly revolutionary step taken at home. In *Brown* vs. *Board of Education* (1954) the Supreme Court clearly enunciated the principle of unconstitutionality in segregation. The Court hoped that social habits would eventually fill in the framework of law.

Many Americans and some Japanese believe that the true modernization of Japan got under way only with the defeat and the Occupation. Those with a longer historical perspective have adopted the description of that momentous period chosen by Kazuo Kawai, who referred to the 1945–52 era as "Japan's American interlude."

The phrase was not designed to denigrate the U.S. effort: Rather, it quite appropriately served to emphasize what had gone on before and to highlight what happened after the Occupation. If one has faith in growth, the experiment was a success story. If, on the contrary, one has grave doubts about some of the results of growth and marches to the beat of a different drum in the development parade, the Occupation accounted in part for what was wrong with Japan in the late 1950s and 1960s.

THE POLITICS OF PEACE

When one considers the tremendous changes that have occurred in Japan over the past century, the relative stability of formal political institutions is astounding. When the peace treaty finally took effect on April 28, 1952, observers were gloomy in their estimates of the staying power of largely alien, imposed institutions. The MacArthur Constitution had been practically thrust on the Japanese; furthermore, in parts its language sounded strange to their ears. All the more surprising, the organic law has yet to be formally amended (and informed Japan watchers see little or no chance for revision in the immediate future). There were a number of reasons for the unexpected stability.

The Reverse Course Stalls

As we have seen, the Occupation operated with relative efficiency, finished its work rather swiftly, and left a reservoir of goodwill. Prime Minister Yoshida Shigeru later wrote, "Judged by results, it can be frankly admitted that Allied (of course, predominantly American) occupation policy was a success." After 1948, the Americans turned (some would say, because of the cold war) from reform to reconstruction. In the latter half of the Occupation, the aims of SCAP came closer and closer to the desires of a majority of the Japanese. In other words, postwar politics meant that increasingly *the Japanese* were making choices on the path toward further development.

Moreover, the framework imposed by the occupationnaires was never so inflexible as to rule out adjustment. For example, like its counterpart, the Constitution of the United States, the Japanese organic law allowed informal revision in practice. Thus (despite Art. 9) a covert and then a limited rearmament for defense was not only permitted but was also encouraged by the Americans, particularly after the outbreak of the Korean conflict. In practice, Japanese increasingly cast aside unworkable principles of federalism (encompassed in Chap. VIII), which had been imported by SCAP. Informal change saw the reestablishment of centralized control of fiscal policy, education, and, to some extent, police standards.

Finally, beginning in the early 1950s, socioeconomic changes within Japan picked up speed at a geometric rate, making an accepted organic law out of what appeared to be a highly idealistic constitution and turning alien institutions into viable organizations. Some of the credit can be attributed to the afterglow of the Occupation; the major impetus, however, was thenceforth in the hands of the new Japanese modernizers. They in turn were identifiable descendants of the men of Meiji.

Establishment Gambles

It should come as no surprise to learn that because of the nature of the surrender, the evolution of the Occupation, and the results of developments after 1950, political institutions in Japan have been dominated (into the 1980s at least) by conservatives. Once again, the Japanese plunged into change—into further modernization, if you will—paradoxically in order to conserve their unique lifestyle. They thus followed in the footsteps of Meiji leaders, who were the first to engage in conservative modernization. Since 1952, conservatives (often called the Establishment) have deliberately engaged in two gambles.

First, post-treaty conservative governments (enduring the unendurable, to paraphrase previous surrender terms) purchased the return of the exercise of sovereignty and their security by placing Japan under a U.S. defensive umbrella. There were advantages and disadvantages to the decision. For at least two decades Japanese enjoyed a security greater than the one they had when they poured out their wealth to maintain the imperial army and navy. Since the war, Japan has spent less on defense from its per capita GNP than any comparable power in the world. As long as Americans continued to make this commitment to defend Japan—and as long as the commitment was credible to all parties concerned—this system offered protection.

On the other hand, security arrangements with the United States necessitated a continuing U.S. presence (in the form of military bases). Like most people in such circumstances, the Japanese never have become enamored of involuntary tourists. Even after the military presence was limited in Okinawa and restricted (to air and naval units) on the main islands, for a proud people the security was at the cost of dependence. Japan's security policy was literally set at Pearl Harbor.

When Americans acted in an unpredictable and even irresponsible manner (and on occasion they did, in the opinion of the Japanese), the protected Japanese were concerned. For example, the late stages of the Korean conflict, which threatened to become a major war against China, and the Vietnam engagement, which also could have become a thrust against China, worried the Japanese. Finally, if and when the United States wavered or if its commitment no longer seemed credible, then odds on the gamble of the U.S. defensive umbrella would lengthen. One day Japanese were bound to ask, if the United States were at the same time threatened by a major power, would Americans still spring to the defense of Japan?

The second gamble taken by the conservative governments was parlayed out of the first. After 1952, having seen to security, Japan's leaders pledged to achieve an unprecedented growth rate, which would double the GNP in a decade. The results were beyond the wildest dreams of even the most sanguine conservative politician, for Japan at one point doubled its GNP in only seven years.

Growth as a Religion

From about the time of the peace treaty (1952) until the stormy days of renewal of the U.S. security arrangement (1960), and even until about the 1973 "oil shock," much noise was made by opposition politicians and conservative leaders about foreign affairs and security. Every cabinet after 1960 and into the 1970s, however, paid primary attention on a day-by-day basis to questions of growth, full employment, the balance of payments, and exports.

Japan thus entered upon a second stage "takeoff"; urbanization accelerated at an exponential rate; a new white-collar, middle class of wage earners (the celebrated *sarari man*) emerged; and Japanese plunged into what at times seemed to be an obsession with durable consumer goods. Citizens watched GNP statistics like they followed basefall scores or the results of *sumō* wrestling matches. Japan was in second or third place; the country had arrived in the heavyweight class. To shift the image, growth became the only religion besides ritual in which Japanese had faith.

Economic growth, as measured in the economist's precise terms of *increasing GNP per capita,* constitutes one of the cleanest, least complicated definitions of modernization. It is a neutral definition and has the advantage of being measurable. The effects of what Lawrence Olson has called "economism" in 1960s Japan run well beyond the economic, however. GNP per capita is only an average: it does not take into account the tangible and hidden costs of growth; it says little or nothing about the distribution of the fruits of growth; and, in terms of net national welfare, it ignores any goal other than the acquisition of goods and services.

In any case, growth has left its stamp on Japan's post-1952 politics. For several decades, it provided the platform on which conservative leaders mounted perennial political victories. Opposition politicians were voices crying in the wilderness.

THE GOVERNANCE OF JAPAN

In the period of growth of the 1960s (as in the 1860s) the Japanese were able to carry with them a large measure of traditional behavior to protect themselves against the shocks of change. The real tests of stability and instability were to come later. Meanwhile, conservative leaders supported by a majority of the public built as efficient a machine to encourage further growth as the world has ever seen.

The Conservative Establishment

Despite the often revolutionary nature of the directed change sponsored by SCAP, the overall effect of the Occupation was conservative, as was that of the New Deal (a ghost of which lived on in Japan long after its demise in the United States). Large numbers of Japanese acquired a stake in stability, even in the new democracy.

Post-1952 politics in Japan came to be dominated by groups that deliberately conserved Japanese values while they were engaged in further modernization. They included the majority Liberal Democratic Party (LDP—Jiyūminshutō), supportive business organizations, and the

National Indoor Gymnasium, Tokyo, designed by Tange Kenzo for 1964
Olympiad

closely allied elite of the civil service bureaucracy. Some observers
borrowed British terminology to describe this intricate structure as the
Establishment. One Japanese political scientist called it a "pluralistic
hegemony." Those who tried to compete with Japanese businesses came
to use a pejorative term, "Japan, Inc."[5]

There are, however, several reasons why any conspiracy thesis
misses the mark in trying to describe conservative hegemony. First, the
opposition (made up of the so-called progressive parties) has always
held just enough power to act as monitor and to guard against any

reactionary course. Second, conservative forces were never organized in any monolithic fashion. Various parts of the Establishment, particularly the several factions of the LDP, have engaged in intense competition and have served to aggregate various interests in the society. In this sense, they have played the roles of political parties. Finally, in the 1970s the absolute majority enjoyed by the LDP gave way to coalition by compromise; and in the late 1980s, to a direct challenge to the party's rule.

A somewhat unflattering view might see the majority of the Japanese in the 1960s armed with relative security provided by someone else; enjoying increasing levels of prosperity discounted by inflation; and content to drift in the mainstream of what they thought was progress. In the 1960s and 1970s, the LDP rolled up impressive (if in each election slightly decreasing) majorities (see Table 5.1). As Akuto Hiroshi has pointed out, since the merger of conservative parties in 1955, party preferences have changed very little. The LDP started by enjoying just below 50 percent of respondents, dipped sharply in 1960 (the year of the riots attendant upon renewal of the U.S. security treaty), rose to above 50 percent in 1964, and then hovered around 45 percent through the election of 1976 (see Figure 5.1). Akuto expressed his belief "that supporters of the Liberal Democratic Party have a 'mainstream' life style and are simple-minded, good at taking a broad view of things and, relatively speaking, realistic."[6]

The LDP. Post-1952 conservative political parties had as their ancestors embryonic political organizations established as early as 1874. By 1900 the oligarch Itō Hirobumi had seen the necessity of seeking party support and founded the Constitutional Political Friends Association (Rikken Seiyūkai). In 1918 this party provided the platform for the first full-fledged party cabinet under Prime Minister Hara Takashi. From 1924 until 1932 parties followed a pattern of alternate rule, power being shared between the Constitutional People's Party (Rikken Minseitō) and the Political Friends (Seiyūkai). By the 1940s even conservative groups had dissolved themselves in favor of transcendental or military-led cabinets.

Immediately after the surrender, a number of new parties made their appearance either as new entities or as resurrected forms of the old conservative parties. The 1946 election constituted an important milestone in that it enfranchised all adults, both male and female. As a result of this election, the president of the Liberal Party (Jiyūtō), Yoshida Shigeru, was selected to be prime minister and began perfecting the politics of Occupation patronage. There was a brief interruption of conservative rule with Japan's first (and only) socialist government (1947), followed by a coalition cabinet (1948). Yoshida then returned to form

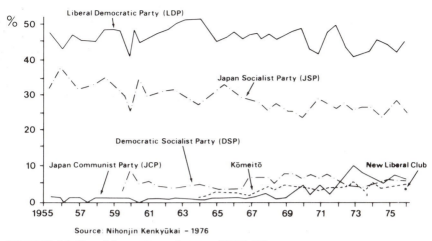

Source: Nihonjin Kenkyūkai - 1976

FIGURE 5.1 Trend in party preference (1955–76)

his second (October 1948) and third (1949) cabinets. During this period, he laid down firm foundations for continuing conservative rule by preparing protégés like Ikeda Hayato and Satō Eisaku (who later became prime ministers). Yoshida's fourth (1952–53) and fifth (1953–54) cabinets saw consummation of the peace treaty.

Following the treaty there was a brief reaction to Occupation reforms, under the leadership of the formerly purged Hatoyama Ichirō. His second (1955) and third (1955–56) cabinets were marked by the reappearance of other purged politicians such as Miki Bukichi, Ishibashi Tanzan, Kishi Nobusuke, and Kōno Ichirō. Hatoyama's most important contribution, however, was the merger of the Liberal and Democratic (Minshutō) parties in November 1955. Thus was formed the Liberal Democratic Party (LDP; Jiyūminshutō), which one witticism described as neither liberal nor democratic, nor even a party.

The LDP gave the prime minister's mantle to Kishi Nobusuke twice (1956–58 and 1958–60). He lost it in 1960 during the riots protesting the renewal of the U.S. security treaty. His retirement then opened the way for what has been called the Yoshida School, premiers who implemented the U.S.-Japanese security arrangement in subdued style and emphasized economic growth. Ikeda Hayato was particularly famous for his low-key approach. He was succeeded by Satō Eisaku, who held the post of prime minister for a record seven years and eight months (1964–72).

A second glance at election results (again, see Table 5.1) will reveal the reason for another unflattering interpretation of LDP strength during this long period of conservative dominance. It is quite clear that the

TABLE 5.1   Votes and Seats in Lower House (Elections 1967–90)

Party	Year	Popular Vote	Percent	Seats*	Percent
Liberal	1967	22,447,834	48.8	277	57.0
Democrats	1969	22,381,566	47.6	288	59.2
(LDP)	1972	24,563,078	46.8	271	55.2
	1976	25,653,624	41.8	249	48.7
	1979	24,084,127	44.6	248	48.5
	1980	28,262,411	47.9	284	56.0
	1983	25,982,781	45.8	250	49.0
	1986	29,875,501	49.4	300	59.0
	1990	30,315,410	46.1	275	54.0
Socialists	1967	12,826,099	27.9	140	28.8
(JSP)	1969	10,074,099	21.4	90	18.5
	1972	11,478,600	21.9	118	24.0
	1976	11,713,005	20.7	123	24.1
	1979	10,643,448	19.1	107	20.9
	1980	11,400,747	19.3	107	20.9
	1983	11,065,080	19.5	112	22.0
	1986	10,412,584	17.2	85	17.0
	1990	13,025,468	23.4	136	27.0
Democratic	1967	3,404,462	7.4	30	6.2
Socialists	1969	3,636,590	7.8	31	6.4
(DSP)	1972	3,659,922	7.7	19	3.9
	1976	3,554,075	6.3	29	5.7
	1979	3,663,691	6.8	35	6.8
	1980	3,896,728	6.6	32	6.0
	1983	4,129,907	7.3	38	7.0
	1986	3,895,858	6.4	26	5.0
	1990	3,178,949	4.8	14	3.0
Clean Govern-	1967	2,472,371	5.4	25	5.1
ment Party	1969	5,124,666	10.9	47	9.7
(Kōmeito)	1972	4,436,631	8.5	29	5.9
	1976	6,177,300	10.9	55	10.0
	1979	5,282,682	9.8	57	11.2
	1980	5,329,924	9.0	33	6.0
	1983	5,745,750	10.1	58	11.0
	1986	5,701,277	9.4	56	11.0
	1990	5,242,674	8.0	45	9.0

(continues)

TABLE 5.1 (*continued*)

Party	Year	Popular Vote	Percent	Seats*	Percent
Communists	1967	2,190,563	4.8	5	1.0
(JCP)	1969	3,199,031	6.8	14	2.9
	1972	5,496,827	10.5	38	7.7
	1976	5,878,192	10.4	17	3.3
	1979	5,625,526	10.4	39	7.6
	1980	5,803,613	9.8	29	6.0
	1983	5,302,485	9.3	26	5.0
	1986	5,313,246	8.8	26	5.0
	1990	5,226,985	8.0	16	3.0
Minor Parties	1967	2,635,232	5.8	9	1.9
and	1969	2,573,932	5.5	16	3.3
Independents	1972	2,788,549	5.3	16	3.0
	1976	3,272,575	5.8	11	2.0
	1979	4,710,637	9.0	25	5.0
	1980	4,335,363	7.0	26	5.0
	1983	4,553,687	8.1	27	5.0
	1986	5,250,140	9.0	19	4.0
	1990	4,866,054	7.4	26	5.0
Totals	1967	45,996,561	100.0	486	100.0
	1969	46,989,884	100.0	486	100.0
	1972	52,423,477	100.0	491	100.0
	1976	56,612,755	100.0	511	100.0
	1979	54,010,108	100.0	511	100.0
	1980	59,028,834	100.0	511	100.0
	1983	56,779,690	100.0	511	100.0
	1986	60,448,609	100.0	512	100.0
	1990	65,704,290	100.0	512	100.0

Note: In most of the elections listed, candidates who ran as independents later (after the election) were drafted by, or joined, the LDP. Thus the LDP could boast of a larger number of seats in the Diet than were represented as a result of the election. In this fashion the LDP added to its Diet strength by from two to nine seats (the independent totals reduced accordingly).

percentage of LDP seats in the Diet was regularly higher than the LDP percentage of the popular vote. This was a result of gerrymandering, whereby LDP bastions in the conservative countryside were overrepresented as compared with opposition parties' bases in the cities (that is, it took more voters in urban areas to return a progressive member than was required in rural areas to return a LDP Diet member).

Having been successful in every postwar election save one, the LDP could have and (as the opposition charged) often did exercise a dictatorship of the majority. According to one description, up through the 1960s Japan had not a two-party but a one-and-a-half-party system. Nonetheless, even in the heady days of growth and successive LDP victories, the party was constitutionally, politically, and morally accountable to the public; through elections the LDP was indirectly monitored by the citizens; the majority was subject to minority pressure in the Diet; and always the LDP cabinets were audited by the media.

Despite the checks and balances, the LDP enjoyed hegemony until the 1970s. There were several reasons for the party's success. First, the top LDP chieftains, high-level bureaucrats, and powerful business leaders built a closely knit network. Business provided the bulk of LDP funds through the Federation of Economic Organizations (Keidanren). Bureaucrats cooperated until they reached the retirement age of fifty-five, when they entered upon a second career in business, in the public corporations, or even in the Diet under LDP sponsorship.

A second reason for LDP success is also often listed as one of the weaknesses of the party. The LDP has been a conglomeration of readily identifiable factions (*habatsu*) whose interminable maneuvers have frequently been criticized even by LDP leaders themselves. Thus, although the prime minister is technically elected by the Diet, it has been apparent from alert newspaper commentary that the selection has actually been made by factional groups within the perennial majority party. Even more obvious has been the need to balance these forces in order to form a cabinet. And yet the factions have represented a marriage of convenience between the traditional and the modern. A patron-client relationship—something like the traditional parent-child (*oyabun-kobun*) relationship—fills a traditional need and has at the same time come to serve a new function. Patrons within the Establishment serve their clients in constituencies, and thus reflect various interests.

This brings us to a third, related reason for LDP success. The party engaged in what Watanuki Jōji has called "organizational clientelism," the provision of public benefits to constituencies through personal sponsoring associations (*kōenkai*) that back LDP candidates. Indeed, the party was very responsive in meeting personal, regional, and occupational demands.

Finally, the LDP made it a rule to tolerate extremely prolonged Diet sessions (one notable exception caused the 1960 riots); so on the surface it appeared to be the most tolerant majority party in the world. In fact, discussion was often limited to procedural matters, with little or no time for substance, with the result that policy was decided within

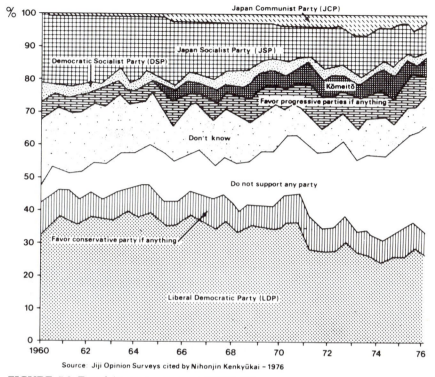

FIGURE 5.2 Trend in party support (1960–76)

the LDP. Even so, LDP members have had opportunities to express a wide range of views within the majority. Some (the doves) supported and engaged in travel to the People's Republic of China before recognition; some (the hawks) held onto ties with Taiwan. Some made contact with the Democratic People's Republic of (North) Korea; others enjoyed close business ties with the Republic of (South) Korea.

The patron-client structure also has supplied one of the LDP's most serious problems. It has been said that in the 1960s the minimum expenditure per month by a Diet member in a nonelection period was 3 million yen (then about U.S. $10,000). LDP factions spent five times more than the total spent by all four opposition parties. This did not take into account the "hidden" money.

Indeed, so serious were the earlier "black mist" scandals and so alarming the money politics of Prime Minister Tanaka Kakuei that they contributed to the decline of LDP support. The statistics on "party support" as compared with those on "party preference" demonstrate a subtle difference and reveal a critical party problem (see Figure 5.2). Through the late 1960s and into the 1970s the number of disenchanted was growing. They were voters who would name the LDP when asked for party preference but who would not feel obliged to go to the polls voluntarily in party support. For Japanese politics at large, it was significant that apparently many voters were leaving the LDP to support no party.

Yet another criticism of the LDP has been that the party has been marked by bureaucratic leadership. Certainly many of the cabinets—for example, those of Ikeda Hayato, Satō Eisaku, Tanaka Kakuei, Fukuda Takeo, and Ōhira Masayoshi—have been indelibly marked by failings of bureaucrats: passivity, inability to search out new approaches, and conservative equilibrium (among factions). It has also been charged that the LDP has converted the entire administrative state into a bureaucracy dedicated to keeping the LDP in power.

The Bureaucracy. It is safe to say that the bureaucracy has been as little changed by war, defeat, Occupation reform, and post-treaty change as any sector of Japanese government. In the long process of change since the nineteenth century the "ministers of modernization" were the first to emerge. Second, the demands of modern war and mobilization strengthened the strategic position of the bureaucrats. Third, even under the Occupation, directed change demanded clean lines of authority from SCAP to the Japanese government. Finally, the Establishment effectively used the power of government to launch the society into an orbit of advanced economic development. In modern times, each and every step has served what may be called the administrative state.

These are also the reasons why the cabinet has been more important in its role as head of the civil service than as a leadership drawn from the Diet. In recent times the cabinet has consisted of twelve ministries: Finance (inner citadel of the bureaucracy), the powerful Ministry of International Trade and Industry (MITI), Foreign Affairs, Education, Health and Welfare, Agriculture, Forestry and Fisheries, Transport, Posts and Telecommunications, Labor, Construction, Justice, and Home Affairs. Spokesman for, and a member of, the cabinet has been the Chief Cabinet Secretary. Also members have been the heads of seven agencies: Defense; Management and Coordination; Hokkaidō and Okinawa Development; Economic Planning; Science and Technology; Environment; and National Land. Equal to all these ministries combined in the exercise of power has been the Office of the Prime Minister (something like the U.S. Executive Office of the President). The specific roles of several of these ministries, particularly of Finance and of MITI, will be discussed in Chapter 6.

Business and Politics. Perhaps the most popular criticism of the LDP denounces the organization as "only a party of business." That the LDP has had close ties with big business and, indeed, has received much of its funding from large corporations, there can be no doubt.

Once again, however, this oversimplification obscures the fact that business circles, like the Establishment, represent plural interests. Business organizations differ sharply in their approaches to problems. They range from the staid Federation of Economic Organizations (Keidanren, something like the National Association of Manufacturers in the United States) to the more liberal Japan Association of Corporate Executives (Keizai Dōyūkai). Far more conservative are the Japan Chamber of Commerce and Industry (Nisshō), representing small- and medium-sized industries; and the Japan Federation of Employers' Associations (Nikkeiren).

It must always be remembered that big business was not the sole clientele of the LDP. A persistent stronghold of traditional values and political support for conservative parties, the old urban middle class figured as well. The LDP has also enjoyed support from what increasingly must be called the sector of agribusiness. Through their agricultural cooperative associations, farmers have held a tight rein on local conservative candidates. The fact is that through the 1960s the LDP garnered support from every sector of the society.

Establishment Policies and Policymaking

In the period when Japanese attention was riveted on growth, the LDP very effectively tapped majority desires in its policies. Thus, although further industrialization and urbanization were expected to alter traditional values, the emerging middle class remained essentially conservative.

LDP campaign pamphlets emphasized, for example, the need for "A Bright Japan, An Abundant Life." In 1960 the party's so-called Philosophers' Group aired a policy statement that, in typical political party style, tried to strike a happy medium between neoconservatism and neonationalism on the one hand and the necessity for "revised capitalism"—provision of full employment and rectification of imbalances of wealth—on the other. The LDP repudiated prewar racism, adhered to parliamentarism, rejected the dogmas of Marxism, and promised to eliminate undemocratic activities. In short, the LDP emphasized "that the entire [Japanese] people should be turned into the middle class and owners of capital."

The foreign policy platforms of the LDP revealed one paradox. Although successive conservative governments supported continuation of the Tokyo-Washington security axis, the LDP also favored an eventual revision of the U.S.-inspired constitution. Party leaders wanted to remove constitutional barriers (in Art. 9) to a further buildup of defense forces, and wanted to effect a gradual withdrawal of U.S. forces from bases in Japan. No conservative leader ever dared advocate the acquisition of nuclear arms, even for defense.

Although the LDP regularly supported close political and economic (as well as security) ties with the United States, the party bowed to the force of nationalism and favored the return of Okinawa to Japanese control. (Reversion of administrative control over the Ryūkyū chain to Japan was the price the United States paid in 1970 in order to obtain renewal of the U.S.-Japanese security arrangement.)

Under the new constitution (Chap. IV, Art. 41), the Diet was supposed to become the highest organ of state power and the sole lawmaking body of the state. The ultimate weapon of the Diet was its control over all appropriations and expenditures, over taxation and revenue, and over monitoring the budget. In theory, then, the legislature was expected to enjoy constitutional supremacy over the executive.

In fact, planning of the legislative program in postwar Japan (as has been the case in all so-called administrative states) has increasingly been a function of the cabinet. The cabinet, actually a coalition of LDP faction leaders, has in turn relied almost totally on the bureaucracy. Drafting of legislative bills has almost always begun in administrative departments. Individual members' bills have been scarce in Japan.

After careful scrutiny by a legislative bureau of the cabinet, a draft bill has then been submitted to rigorous study by various vice-ministers, permanent members of the higher civil service. At this cabinet level, too, there has been an increasing use of public commissions to study, modify, and even to implement legislative objectives. Once approved, the bill has been introduced in the legislature, where, particularly in the committees, the government has been questioned and opposition testimony aired. Until the mid-1970s, however, the LDP controlled all committees, so it was entirely up to the majority party whether or not a bill would be amended. On very serious issues, the opposition had only one alternative course: to boycott the Diet and committee sessions, particularly those dealing with the annual budget.

Empirical research on public policymaking, even in the 1960s when the LDP was dominant, has clearly revealed that the process was nonetheless heterogeneous. To repeat, the Establishment was a pluralistic hegemony. In some cases (as in the United States) conservative policymaking was closed (even to business circles) and out of public view. For example, Prime Minister Tanaka's visit to Beijing in 1972 and the decision to normalize relations with the People's Republic of China were carried out despite substantial opposition within the LDP, among a minority of the bureaucracy, and in some segments of big business.

In other cases, a large number of individuals and pressure groups in the public at large were critical to the outcome of policymaking. To cite two issues—the pollution problem in the early 1970s grew to dominate a session of the legislature (the Pollution Diet of 1970), despite the reluctance of the LDP and business to face the dilemma; and the perennial procedure for setting of rice price support has involved a complicated network of segments of the LDP and of the various ministries.

In some other instances, such as the issues involving investment liberalization and university enrollment expansion, the presumably all-powerful Establishment remained inactive until forced to respond to pressure applied by foreign governments or by broad sectors of the domestic society. At other times (to cite again Tokyo's decision to normalize relations with Beijing), the government was capable of major innovation.[7]

The Emperor

Where does the emperor fit into this conservative scheme of government? By tradition, the emperor has usually been a ceremonial symbol rather than an operating executive. The new constitution assigned no power to him and made the imperial institution subject to the authority of the cabinet.

Nonetheless, the opposition to the Establishment has argued, not without reason, that the imperial symbol has been used to bolster LDP rule. Even progressive parties have been very reluctant to tamper with what is also essentially a symbol of stability amid rapid change. The death of Emperor Hirohito on January 7, 1989, however, reopened the delicate question of the imperial system.

Hirohito's demise ended the 62-year reign of Japan's 124th ruler, whose incumbency was the longest in Japanese history. During the turbulent era, the emperor had in fact served as a symbol of several Japans. To recapitulate, as early as 1921 when Hirohito as prince regent acted for the ailing Taishō Emperor, the nation was in the twilight of progressive change, known to Japanese as "Taishō democracy." In the 1930s and 1940s, the emperor in uniform and astride his white horse, Shiroyuki, came to represent militarism, aggression, and, eventually, a disastrous defeat. After November 1946, Hirohito eased into the more comfortable role of "symbol of the State and the unity of the people." In the 1950s, he symbolically presided over the remarkable reconstruction of Japan and in the 1960s and 1970s, over the nation's rise to become a major world economic power.

Some measure of Japan's international status was indicated by the fact that representatives of some 163 countries and 26 international organizations were represented at Hirohito's funeral, held in Tokyo on February 24, 1989.

On the domestic front, as the emperor's condition worsened after September 1988, reactions were sharply divided. Tōkai University historian Sato Isao probably captured the majority mood when he stated that most Japanese felt affection for the emperor "as someone who had shared wartime and postwar hardships with them." According to a Kyōdō News Service poll conducted in January 1989, two of every three Japanese entertained warm feelings for the late emperor. On the other hand, about half the younger generation said that they were indifferent to the transition marked by Hirohito's death. With youth, apathy seemed to outweigh antipathy to the imperial system.

Two traditional taboos marked Hirohito's long illness. A media vigil, involving over one thousand journalists, produced mere scraps of information about the emperor's medical condition. Only with his death was it reported that he had suffered from duodenal cancer. It was as though the Imperial Household bureaucracy felt that an emperor should not be subject to ordinary human frailty. Doubtless Hirohito, the quite accomplished marine biologist, would have protested this violation of scientific canon, had he been in control of information.

The other taboo, which was also dissolved by the death, was the reluctance of many Japanese to discuss openly the historic role of Hirohito in Shōwa, the era of "enlightened peace" (1926–1988). With the passing of the emperor, the floodgates of opinion were opened.

Millions of citizens continued to appear before the Imperial Palace and at prefectural stations to record condolences, as they had registered concern during the long illness. Many public events were cancelled. A sizable number of Japanese, however, protested that media obsession with Hirohito's illness and death had tended to revive and to glorify the traditional imperial system which, they argued, had been supplanted by modern constitutional democracy. And in Japan, where business had become the nation's business, on the first working day after the imperial death, investors on the Tokyo Stock Exchange sent the Nikkei average up to an all-time high of 30,678.39.

Extreme rightist groups, who are wont to cruise Tokyo streets with vans decked out in military and ultranationalist symbols (augmented by deafening recordings), argued that it was time to abandon the "imposed constitution" and to reestablish the emperor as formal head of state. They were furious when the mayor of Nagasaki restated a simple and apparent truth.

In December 1988, Mayor Motoshima Hitoshi had released a short statement (just 13 characters in Japanese), which said, "I think the emperor cannot be free from responsibility for the war." Indeed, the statement simply reflected one made by Hirohito himself, in an interview with General MacArthur on September 26, 1945, to the effect that the emperor "bore sole responsibility for every political and military decision made and taken by my people in the conduct of the war."

Nevertheless, the reaction was immediate and furious. The mass media seized on the issue; the mayor was relieved of a post with the Nagasaki chapter of the LDP; and his office received thousands of letters, postcards, and telephone calls in protest. He and his family received death threats from rightists. On January 18, 1990, at Nagasaki City Hall, a member of an extremist group shot the 67-hear-old mayor in the back at point-blank range. Soon thereafter, doctors described Motoshima's condition as "stable."

According to careful outside observers, the illness and death of the emperor did usefully serve to air out the significant issue of an imperial system in modern Japanese democracy. Events forced a thorough review of the tumultuous Shōwa era, in general, and of Hirohito's role in particular. The weight of evidence seems to show that Hirohito was neither emotionally nor politically trained to be a decisionmaker. He never, with one or two exceptions, directly expressed policy opinions. True, in 1941 he gave oblique consent to the opening of hostilities, with the naive proviso that the United States should be given advance notice. In an imperial conference held on August 10, 1945, at the invitation of the prime minister, he did break a deadlock in advisory opinion. Then, and in the famous broadcast of August 15, 1945, announcing the surrender, the emperor stated that Japan must "bear the unbearable." As in most other cases, he doubtless came to this conclusion with advice. His responsibility for the war, shared with all the Japanese people, was beyond challenge. "War guilt"—for the Emperor or for all Japanese— was another complicated matter.[8]

The immediate succession went off smoothly, as Akihito ascended the Chrysanthemum Throne as Japan's 125th ruler. Authorities gave him the reign name Heisei ("peace and concord"). On January 9, 1989, in his first audience ceremony, Akihito pledged to be "ever with the people" and to "observe the constitution of Japan . . . and discharge my duties in accordance with it." In several subtle ways, the emperor and his poised wife, Empress Michiko, demonstrated their wordly experience gained from extensive travel abroad. Moreover, they have shown a tendency to mix more with ordinary Japanese citizens than was common in previous imperial custom.

Controversy was not, however, dead. After the traditional year of mourning, the Imperial Household Agency announced that the formal enthronement of Akihito would take place in Tokyo (rather than at the traditional site, Kyōto) on November 12, 1990. The first, a public ceremony (*Soku-no-rei*) would be attended by 2,500, including 500 foreign guests. In contrast, a private ceremony (*Daijōsai*) with religious overtones, was to be conducted November 22–23 for the Imperial Household. The government decision to use public Imperial Court funds, even for the private religious ritual, promptly inspired protests on constitutional grounds.

The Progressive Opposition

Although growth was the religion in the decade of the 1960s and conservative politicians the high priests of ritual GNP, it should also come as no surprise that growth created problems. Inflation (which discounted growth), environmental disruption, a severe housing shortage, and the rise of traffic accidents—all of these offered the opposition forces salient issues on which, by and large, they failed to capitalize at least until the 1970s.

Successful conservatives and the frustrated opposition were coming up against an increased political apathy spreading among the new urban middle class. It was a studied indifference, the result of high levels of literacy, expanded technology, scientism, and pragmatism. This new middle class also suffered from being in the center of a polarized struggle between the Establishment and those in opposition, who chose to call themselves the "progressives." On the one hand, the Establishment and a large majority of the middle class were distrustful of Marxism and anyone whom they regarded as a Marxist. The LDP stressed authority and order, often at the expense of minority rights. On the other hand, some of the thinking and much of the rhetoric of the organized opposition have been framed in Marxist categories. The progressives proved to be quite inept, often inappropriately carrying irrelevant nineteenth-century dogma into the political fray.

Dissenting Interests. Trade unions have provided the bulk of support for the progressives, specifically for the socialist parties. A criticism of labor (and of its support for socialist parties) has been that unions represent only a fraction of the total population (indeed, only about 29 percent of all employed personnel).

Splits in the labor movement have paralleled those among socialists. Formed in July 1950, the General Council of Trade Unions (Sōhyō) embraced about 4.5 million members (about 12 percent of the labor force, but some 55 percent of organized labor). The All-Japan Congress of Trade Unions (Dōmei Kaigi), formed in 1964, had a membership of something over 2 million (or 6 percent of the labor force). Sōhyō was critical of the communists but backed leftist socialists. In 1951, it refused to join the International Federation of Free Trade Unions abroad. Dōmei lent support to moderate socialists. Other labor voices were heard from employees of public corporations, like railway workers; there was also the smaller but vocal dissent articulated by the Japan Teachers Union (Nikkyōso). Later developments in the reorganization of union labor are described below, in Chapter 7.

Labor has often been joined, and socialists supported, by the *interi* (the intellectuals, meaning almost anyone who has gone to college). Although some intellectuals have lent counsel to conservative governments as advisers, there has been a wide gulf between most of the *interi* and the Establishment. Older intellectuals have been profoundly affected by the rhetoric, if not the thought, of Marxism. There has also been an opinion gap between intellectuals and the middle class, although the latter pay exaggerated respect to television commentators and social critics drawn from the *interi*.

Although opposition forces have never fully capitalized on the tensions generated by rapid growth, they have gained support from those who have not enjoyed a proportionate share of the rising GNP: some technical and professional personnel, some white-collar workers (including low-level bureaucrats), small merchants, some youth, and many women. The latter have been particularly concerned with the effects of inflation on household accounts. Other dissenters have joined opposition parties in criticizing "money politics" within the LDP, in uncovering scandals related to aircraft purchases (in the 1970s), in opposing covert rearmament, and in defending the Peace Constitution. The urban proletariat, called by one observer the "industrial peasants," on occasion have shown a penchant for real radicalism.

The Socialists. The major and most effective opposition to the LDP has been provided by various kinds of socialist parties. The Japan Socialist Party (JSP, Nihon Shakaitō) was established in November 1945. "Japan" was used to distinguish these socialists from brethren abroad. The term "social democratic," a frequent translation, was meant to link Japanese socialists to similarly named groups abroad. In fact, the JSP inherited diverse elements—Christian reformism, proletarian radicalism, evolutionary and scientific socialism—from prewar social democracy. In June 1947, Katayama Tetsu became prime minister and formed the first socialist party government in Japan's history. Unfortunately, this experiment in responsible opposition failed in the face of severe economic recession and a series of scandals. Since then the JSP, never tempered by the probability of achieving power nor moderated by the responsibility of governance, has perceived politics as a struggle and has looked at political issues in doctrinaire terms.

Right- and left-wing socialists first parted company over the issue of the majority peace treaty in 1952. The threat of the merger of the conservative parties and a complicated compromise brought the two wings back together in October 1955. Although the JSP could count on relatively large blocks of voters at election time, the party had only a small regular membership (of about 60,000) and represented only one large pressure group (labor).

JSP platforms, as contrasted with those of the LDP, have been more wordy, more theoretical, and more intensely concentrated on issues of foreign policy and security. On the domestic front, the dominant left faction defines the JSP as a class-mass party. Although the aim of the party ought to continue to be socialist revolution, under existing conditions this revolution should be effected only by peaceful means. The objectives of the JSP have been to increase the income of the "working public," to reform the economic system in order to limit the profits of large enterprises, to nurture backward industries (agriculture, fisheries, and forestry), and to perfect a social security system.

The foreign policy platform of the JSP, like that of the LDP, has revealed an interesting paradox. Whereas the party fought against any revision of the U.S.-imposed Peace Constitution, it also advocated termination of the U.S.-Japan security treaty. In its place a collective security arrangement among Japan, the Soviet Union, the People's Republic of China, and the United States was urged. Asia was to become a nuclear-free area. Early on, the JSP advocated recognition of the People's Republic as the official government of China.

In 1959 a long-simmering rebellion against domination of the JSP by left-wing factions and against the party's preoccupation with class struggle led Nishio Suehiro to bolt from the JSP with forty members of his faction. In 1960 rebels established the Democratic Socialist Party (DSP). Steadily improving economic conditions offered a promise that the DSP would grow, as had moderate social democrats in Western Europe and the Labour Party in Britain. By the 1972 election, however, support for the DSP had declined to a point where the party commanded only 7 percent of the popular vote (the JSP, 22 percent).

The Communists. In the case of the Japan Communist Party (JCP, Nihon Kyōsantō), as in the JSP, "Japan" was used to distinguish the Communists from other such parties abroad. Often the JCP proved to be far less radical than the JSP. Severely hampered by splits among the pro-Soviet, the pro-Chinese, and neutral wings, the JCP nevertheless became one of the most pragmatic of the progressive parties at the local level. In the 1972 election, it grew to become the third-ranking opposition party (10.5 percent of the popular vote and thirty-eight lower house seats). Moreover, all along, the JCP could boast of a larger rank-and-file membership than that of all the other progressive parties combined.

The Kōmeitō. Far more strident than the Communists—and for a time their bitter enemies—were the voices representing one of the new religions (see Chapter 8). They in turn produced another opposition party. The social programs of the so-called Value Creation Society (Sōka Gakkai), founded in 1946, appealed to *rōnin* students, the sick, the dispossessed, and some women, particularly those who resented the discrimination inherent in a male society. Sōka Gakkai soon grew in membership to 15 million. Its political arm was the Kōmeitō (Clean government party), but later under pressure there was a formal division between the religious and the political organizations.

The Kōmeitō entered the lists slowly and deliberately, and in 1967 finally put up candidates in the election for the lower house. By 1969 it commanded 10 percent of the total popular vote and had increased the number of its seats to forty-seven in the House of Representatives. By 1972, however, its share of the popular vote had fallen sharply (to 8.5 percent), so it held only twenty-nine seats in the lower house. The Kōmeitō has in effect represented both tradition (in its emphasis on discipline, its familial face-to-face cells, its appeal to nationalism, and its exclusivist religious beliefs) and modern style (in its efficient organization, its great material wealth, and its establishment of an overseas network).

Forces opposed to the LDP did not leave a clear stamp on Japan's politics until the decade of the 1970s. By then, Japan had entered a new period, the era of the postindustrial society, and Japanese faced new issues. Even in the growth cycle of the 1960s, however, charts showing support for all parties (see Figure 5.2) clearly reveal an apolitical trend, as the individual became less confident of the effectiveness of any government, party, or leader. "Japanese people became decreasingly political."[9]

Those Japanese who supported no party rose from about 10 percent in 1960 to more than 30 percent by 1975. Elections were obviously in a state of flux as the floating middle-class vote moved from established parties into this or that coalition. Another way of describing this development is to state that affluence practically eliminated the possibility of revolution, so sharp ideological differences played a more and more minor role. Somewhat paradoxically, but easily understandable in terms of traditional style, the image of a party and of a candidate became more vital than ideology in Japanese thinking.

6

The Postwar Economy

At times it seems that only the skilled economist, the mathematically trained econometrician, or the knowledgeable businessperson can understand the intricate Japanese economy and its complicated relationships with the outer world. Lay people can attempt, however, to grasp the manner in which the basic culture of Japan has directly affected Japanese economic behavior. They can understand, in reverse, how economic development has profoundly affected—but not determined—the lifestyle of the Japanese.

THE BACKGROUND

The modern economy, including particularly the industrial structure of a nation, is conditioned by the variety of people's wants expressed in the market (demand); by the level of development (specifically the technological capacity) and the relative accessibility of raw materials (supply); and finally, by the underlying value system, the social structure, and individual behavior. These last factors constitute what is called the culture or national character.

Social Foundations

In the postwar—and even more in the post-treaty—period, Japan has been controlled in the political realm by those whom we have called the Establishment. Both in politics and in the economy, according to some foreigners, "Japan, Inc.," has offered restraint of trade and unfair competition. The Japanese economist Kitamura Hiroshi has used more moderate terms—in French, *dirigisme*, and in English, sponsored or controlled capitalism.[1] Others have argued that the Japanese economy has constituted an unrestrained free enterprise system. The first task is to try to reconcile these apparent contradictions.

A review of the evolution of Japanese traditions has clearly revealed the persistence of one value that the process of modernization has not eroded. In Japan there has always been a highly developed sense of loyalty and dedication to some kind of close-knit group (the family, the clan, the domain, or the firm). In the nineteenth century, modern technology and a kind of capitalism were transplanted to Japan, but a certain degree of collectivism remained in the highly homogeneous society. One can detect indelible marks of group loyalty on, for example, the management of individual enterprises.

Japan's unique modern system began to take shape in the mid-1930s. For a Japanese high school graduate (at age eighteen to nineteen) or college graduate (at age twenty-two to twenty-three), the most important consideration in choosing a position has been the balance sheet of benefits to be gained at the time of retirement (at age fifty-five to sixty). Before the 1970s the Japanese employee seldom changed employers; the employer seldom discharged employees before their retirement. The worker always knew that the pay scale depended, at least in the first half of his career, on seniority (only later would merit and achievement produce differential rewards).

During a period of rapid growth (for example, in the 1960s), businesses grew steadily and took on more employees each year. Thus the younger came to outnumber the older employees; so the labor force was in the form of a pyramid. The elders could be paid well if the younger workers would accept lower wages for a time; the latter could look forward to accelerating pay raises and promotions, providing business remained on the upswing. During a short slump it did not pay to discharge employees who had made a lifetime commitment to the firm. It did make sense to sell below cost in a recession, particularly if the transaction involved exports; so unemployment figures were not a good indication of the status of the economy. Of course, during a slump there might be an increase in the number of underemployed employees, called sunshine boys if they were elders of not high merit, men who were marking time until retirement. In times of prolonged recession (and perhaps when the economy entered a later stage, as we shall observe), the pyramid of workers had a tendency to convert into a diamond. The ratio of highly paid, slightly older employees increased in proportion to newly recruited workers.

Such practices have imparted very special characteristics to Japanese economic behavior. First, there is the frequently observed strong corporate loyalty: employees have come to believe that they share the same fate as their employers. Second, there is the very low level of unemployment, even during recession. These are the more traditional collectivist features.

On the other hand, government decisions in the economic sphere were (and still are) arrived at through a long and complicated process of consultation, compromise, and consensus between bureaucrats and party faction leaders, between the bureaucracy and business, and among all groups in the Establishment. Despite appearances, there has been a surprising degree of decentralization in the decisionmaking process. This is a mixed (traditional and modern) feature of the system.

There has also been a lively, even fierce, competition among the units of the oligopolistic economic structure. This is the modern, capitalist feature of the economy. Here is yet another illustration of the coexistence of traditional values and modern techniques in Japan. After a brief review of the history of the development of this unique system,[2] it will be possible to assay some of its accomplishments, its shortcomings, and (in Chapter 7) its most recent transformation.

Modern Development

Japan's economy has often been included among those called late developers, and it has usually been regarded as one of the most rapid developers. As a matter of fact, the emergence of the modern economy in Japan was relatively early as compared with nations now regarded as part of the developing world. Indeed, few realize that Japan's growth predates both the post–World War II era and the pre-war Taishō and Meiji periods. The "economic miracle" reflects development reaching back to the late nineteenth century.

By comparing the economic growth rates of ten nations (including the United States) to Japan's, Columbia University economist James Nakamura found that Japanese performance has been unmatched in the last century. Except for the "take-off" period (1870–1913) and the era of the Pacific war (1938–53), progress was more rapid for Japan than the other nations. Even the proto-modern Edo era (1603–1867) saw a substantial rise in per capita production. Between 1885 and the end the century, Japan's GNP doubled.[3]

At first the Japanese fully exploited traditional crafts, for example, silk, and the earliest industries to modernize were in the field of textiles, for example, cotton. In the last two decades of the nineteenth century, the Japanese were content to upgrade traditional light industries (paper, sugar refining, food processing) before establishing new production (cement, glass, beer) on a sound footing. Already by the late 1880s, agricultural and light industries had provided the necessary surplus for further expansion.

The era of World War I saw remarkable growth. The nation's military contribution to the Allied victory was minimal; Japan was left to concentrate on industrial growth at home and on the expansion of overseas markets. Export surpluses meant that the country no longer had to rely on agriculture and traditional crafts to provide for further investment.

This decade (1910–20) nonetheless saw modern industry grafted onto a labor-rich semideveloped economy. This was not a phenomenon confined to Japan, but its effects were striking there because of the size of the contrasting sectors. On the one hand were the traditional labor-intensive pursuits (agriculture, fishing); on the other were the modern Western-sytle projects requiring capital equipment and therefore investment.

In the 1920s, the widening gap between the two sectors, together with worldwide depression, contributed to political tension in Japan. The modern sector continued to expand as weaker firms were weeded out and the celebrated *zaibatsu* (finance cliques) reached their peak of power. By the early 1920s the output of workers in the modernized sector was at least four times higher and wages were at least three times higher than comparable figures for workers in the traditional sector. Continued growth hid the serious social and economic problems in the more traditional parts of society. This double standard or "dual structure," as economists identified it, was destined to play an important role in the postwar economy.

In the late 1930s and early 1940s mobilization for war and engagement in conflict served to widen the gap in the dual structure. Exploitation of colonies and military ventures on the mainland also added to the number of corporations in the modern sector. New firms competed with the older *zaibatsu* in Manchuria and in the outer empire. In one decade (1930–40), the output of modern industries doubled, and even production contributed by agriculture increased by about 25 percent. Since most of the surplus was plowed back into heavy industry and military expenditures, the standard of living of ordinary Japanese did not go up proportionately. Moreover, total war destroyed half of Japan's industrial capacity. Yet the skeleton of modern infrastructure survived and a skilled labor force soon reappeared after the surrender.

Occupation policies had a long-lasting effect on Japan's economy. In the beginning (just after the defeat of Japan), the results of SCAP's actions were quite mixed. On the positive side, massive economic assistance, particularly the provison of food, staved off complete collapse and chaos. Initial encouragement of, and then later limitatations on, activities of the labor movement had both positive and negative effects. In terms of immediate agricultural production, the impact of land reform was probably negative; but in the long run, this quiet revolution laid down a firm foundation in the countryside for solid growth. Many of the older corporations, stripped of the *zaibatsu* form and operating procedure, found it difficult to carry on.

U.S. contributions to Japan's revival were nevertheless impressive. The United States lent technical assistance and in 1949 helped "rationalize" the economy under the Dodge Plan. Technology flowed to Japan under favorable licensing agreements and the Japanese mastered U.S. managerial techniques. With help, the successful Japan Productivity Center was established. In the early 1950s, the Korean conflict brought Japan a windfall: some $3 billion in offshore procurement for the U.S. and UN war effort.

THE ERA OF GROWTH

Historians may come to characterize the middle decades of the twentieth century as an era of rather unexpected economic development. After all, for a century critics had accurately identified inherent weaknesses in capitalist economies, pointing to contradictions in their uneven development, and had made dire predictions about the future of the capitalist world. Although it could empirically be demonstrated that economic factors were not the sole, or even the major, ones that caused conflict or even imperialism, it can be said that war was a terrible waste of economic resources. Two world wars had indeed threatened to wreck free economies.

One difficulty with the gloomy forecasts was that so-called pure capitalist systems, if they ever existed at all, lasted for only brief periods of time. Reforms came as much from the conservative center and right as from the reformist left; as a result, in the 1950s there was a remarkable resurgence of what were called free enterprise economies. In fact, these were already mixed capitalist economies for the most part, with varying degrees of public control melded with private sectors (as in the U.S. economy). Until the 1970s and the stage to which the Club of Rome referred as the "limits of growth," such economies registered outstanding performances and thus reinforced foundations for further expansion. The difficult problem of equitable distribution of proceeds of production may have remained, but by and large there was enough diffusion of the fruits of growth to make for a sharp reduction in the ideological component of politics and to permit the majority to believe that revolution was not probable.

In this pattern of postwar development, Japan's economy is unique both in its record of aggregate growth and in its relatively low expenditure on social overhead. The Japanese economy is also an exception in that, in the process of growth, the share of national income accruing to labor tended to decline during the two decades (1950–70).

The Post-Treaty Miracle

It will be recalled that, in one of the Establishment's deliberate gambles, leaders promised that with limited defense expenditures, the nation would achieve an unprecedented growth rate. The GNP was to double in a decade. The results in the 1960s were beyond the conservatives' fondest hopes.

Because the annual average rate of real growth needed to double the GNP was 7.2 percent, public attention focused on that benchmark. The Japanese followed GNP statistics just as they watched baseball scores. GNP (in current prices) almost doubled in *five* years (1955–60), more than doubled again in the following *five* years (1960–65), and then rose another 2.5 times in *five* years (1965–70). In this last period, the *annual* rate of growth (in real terms) averaged 12.1 percent.

By 1971 the GNP was estimated at more than $250 billion, with income per capita about $2,000. Of course, such figures were in gross amounts and even per capita income was only an average. Neither spoke to the equally basic issue of equitable distribution; high GNP levels said less about the quality of life. Such problems were to be seriously raised in the 1970s.

Nonetheless, growth had become the dominant Japanese religion. Between 1962 and 1971 Japan rose to have the world's third largest GNP after the U.S. and the USSR (see Table 6.1). Indeed, the Japanese were the originators of supply-side economics. They supported a record savings rate to provide investment funds; they plunged into advanced technology and plant modernization. Above all, firms invested in human resources, training for a skilled labor force. Lifetime employment (in big corporations), seniority, and a bonus system based on profits made workers powerful constituencies.

Factors in this performance included, first, conscious planning. The Economic Planning Agency (EPA; Keizai Kikaku Chō) made the projections. The strategy was to expand the modern industrial sector enough to draw more personnel into high-productivity, higher-wage employment from the agricultural sector and thus to dissolve the dual structure.

In another sense, no plan was able to keep up with growth. Except in years when there were fiscally induced recessions (1958, 1962, 1965, 1971), growth rates consistently outstripped projections. The result was a kind of rolling plan, meaning that projections were revised after each year according to annual production results.

Hugh Patrick has expressed the opinion that the powerful Ministry of Finance actually provided planning problems. Members of this ministry, as noneconomists and products of the public law faculty of the University of Tokyo, were legalistic in outlook—prior to the 1970s they worshipped balanced or surplus budgets. Finance bureaucrats saw themselves as leaders, certain that they made the best prime ministers.[4]

A second factor in the success story (if growth is considered success) had to do with the extraordinary capacity of the Japanese public to

TABLE 6.1   Selected Economic Indicators: An International Comparison

Economic Indicators	Japan	USA	UK	West Germany	France	Italy
1971 GNP ($ billions)	256.6	1050.4	146.7	235.5	176.7	108.2
1971 National income per capita	1991.0	4133.0	2026.0	3056.0	2646.0	1635.0
1972 Steel production (metric tons)	96.9	120.7	25.3	43.7	23.9	19.7
1972 Exports ($ billions)	28.6	49.1	24.3	46.2	26.0	18.5
1969 Passenger cars (per 1000 pop.)	67.0	526.0	250.0	244.0	270.0	185.0
1970 TV sets (per 1000 pop.)	215.0	412.0	293.0	272.0	201.0	181.0
1966 Social Security ($ per capita)	52.3	258.9	229.1	332.7	321.2	177.9

Sources: Bank of Japan; EPA; adapted from Kitamura, *Choices for the Japanese Economy.*

save. Statistics showing international comparisons of GNP and savings (in Table 6.2) revealed Japanese willingness to forgo consumption. Surpluses were plowed back into capital goods and more growth. In any case, a majority consistently approved the path chosen.

A third factor involved technology. Because a large proportion of the nation's industrial plant was destroyed or scrapped during and after the war, by the 1960s the bulk of Japanese equipment was new (ten years old or less). Japanese also made breakthroughs of their own, for example, in the adaptation and use of transistors. There was a new emphasis on quality in Japanese products.

The technological phase of growth was directed by a newly created agency, the Ministry of International Trade and Industry (MITI). Widely known by the initials of its English name, even in Japan, MITI was somewhat similar in function to the U.S. Department of Commerce. Its mission reflected the strong commitment of the government to make Japanese industries competitive in the world. MITI assembled information,

TABLE 6.2   Gross Savings: An International Comparison,
1961–70 Average

Country	Ratio: gross savings to GNP	Share of GNP household savings	Ratio: personal savings to disposable income
Japan	37.5%	12.6%	18.6%
W. Germany	27.4	9.8	14.7
France	25.7	8.7	12.5
Italy	23.6	11.6	15.4
UK	18.7	3.8	5.6
USA	15.4	4.4	6.4

Note: Order of countries in accordance with gross savings-GNP ratios; derived
from Bank of Japan
Source: Kitamura, *Choices*

maintained its own research institutes, and made critical judgments on
markets in which Japanese business was apt to be competitive. It also
financed the Japan External Trade Organization (known in the United
States as JETRO).[5]

Aspects of the Growth Economy

In the postwar growth period, the Japanese economy benefited
from what has been called "a very accommodating labor force." This
meant that a large proportion of workers entered industry directly from
rural areas, where traditional norms such as discipline, obedience, and
respect for superiors were highly valued. So long as it could be tapped,
the pool of workers provided by the traditional side of the dual structure
also played a significant role in the supply of mobile and relatively
cheap labor.

In the process of growth the modern sector not only absorbed all
the increases in size of labor force (and there were some resulting from
the postwar baby boom), but also drew substantial numbers of workers
out of agriculture and small-scale industry. In the decade 1955–65 the
proportion of the labor force in primary pursuits fell from 40 percent
to 25 percent. By the early 1970s it fell below 20 percent (in 1975 it
was 14 percent). Meanwhile, the proportion of the labor force engaged
in secondary pursuits rose from 23 percent to 32 percent between 1955
and 1965. By 1975 it was at 34 percent (see Table 7.1[6]). The share of
GNP produced by the primary sector fell from 23 percent in 1955 to
less than 10 percent in 1965. Over the same period the share produced
by manufacturing increased from 23 percent to over 33 percent. As a
result, the end of the dual structure was in sight. Whereas in the 1950s

employment as a *sarari man* in a large company with good wages and generous benefits was seen as the privilege of a minority, by the mid-1960s it seemed to be within the reach of everyone.

These are some of the reasons why Kitamura recognized the dual structure as an important factor in postwar growth. Japan was following along the path of developed Western countries, but with a lag of three or more decades. In the early 1960s, for the first time in Japan's history, the demand for and supply of labor came into equilibrium. The other striking characteristic in a kind of telescoped development (to be examined in Chapter 7), was the dramatic increase in the relative importance of the tertiary sector (services, finance, software, information retrieval, and such).

Japan's growth economy of the 1960s was characterized by the relatively high proportion of GNP devoted to fixed capital formation, as compared with other developed nations. Within the manufacturing sector, the biggest gains were made by heavy industy—for example, engineering, automobiles, chemicals, and electronics.

Emphasis on status in the GNP race illustrates the strong Japanese sense of hierarchy. Close attention to status in the outside world also has tended to distort the condition of various parts of the economy inside Japan. Thus, although Japan may have forged ahead in various industries in the advanced sector (like steel, shipbuilding, and auto-mobiles), as late as the 1970s, overall per capita productivity in Japan was still only half that of the United States and about two-thirds that of West Germany. Moreover, the obsession with fixed capital formation produced the counterpart of a steadily declining relative share of national product devoted to personal consumption and benefits. Public outlays on collective services (social security, education, and care of the aged) remained relatively modest in relation to capacity.

One aspect of this dynamic, booming economy proved to be politically significant in the early 1970s. The location of capital-intensive industry on a relatively small, rather fragile land area resulted in an incredibly high-density economy. Overcrowding in cities and environ-mental disruption were not unrelated to the system of management under which growth took place. The conservative governments eventually would have to come to grips with this shadow on growth.

Throughout the period of growth there were also imbalances and shortages that acted as fiscal brakes consciously applied to achieve slowdowns, to spot scarcities of labor, and to manage rising living costs. The last deprived the Japanese of full participation in the gains, and some referred to the income doubling plan as "the price doubling plan." A medium-term economic plan (1965) tried to ease some of the pressures by paying more attention to the improvement of efficiency in agriculture

Quality control in assembling computers

Kanagawa Science Park (near Yokohama)

and in small business. It also aimed to reduce strains in public housing, transportation, and social services. These developments represented the beginning of the Japanese consciousness that growth itself had limits.

As far as distributive shares of manufacturing output were concerned, Japan displayed a relatively low consumption ratio. Put the other way around, there was a bias in favor of the maximization of private capital. The standard of living of the ordinary Japanese, although it must have risen quite substantially over the growth of years, nonetheless was never tuned to the high pitch of economic growth.

Foreigners, too, played significant roles in Japan's growth. They provided needed technical assistance (as they had in the nineteenth century) and supplied direct and indirect investment in order to fill the gap between necessary imports and available exports. The steady growth of exports fulfilled two needs: It matched an increasing demand for raw materials (coal, iron ore, and particularly petroleum products), and it met the increasing Japanese demand for imports of consumer goods. Between 1962 and 1971, exports rose more than fourfold in current prices.

In the early 1970s, before the oil shock of 1973, Japan's remarkable growth, rising levels of exports, and balance-of-payments surplus began to arouse hostility abroad. After the 1973 energy crisis, of course, Japan's problem appeared to be in the realm of deficits rather than surpluses. The shortfalls, however, were soon displaced by towering surpluses. Japan came to represent to nations abroad the need to accommodate to a shift in world economic relationships.

Toward the end of the 1960s, Japan, like any other advanced country, began to come to grips with problems inherent in rapid growth. From the early 1970s on, it became apparent that the postwar international financial system, based on the overwhelming superiority of the United States, was heading for transition, if not breakdown. Especially in the United States, there was a temptation to link the Japanese accumulation of foreign exchange with U.S. balance-of-payments problems and the decline in the value of the dollar. Therefore, the Japanese were subjected for the first time to intensive pressures: antidumping measures, special quotas, and what were euphemistically called "voluntary export restraints." This was the period, too, in which negative images were called forth: the Tokyo management as "Japan, Inc.," the businesspeople overseas as "ugly Japanese," and all of the people of the island nation as "transistor salesmen," "economic animals," and "workaholics."

7
The Postindustrial Society

The industrial revolution, which was in fact a prolonged evolution, unleashed a set of forces that vastly changed traditional societies. Economies became diversified, populations migrated to the cities, new classes and political alignments appeared, and standards of living rose. It has most often been identified as happening in England from the middle of the eighteenth century to the late nineteenth century, a little later in France and Germany, after the Civil War in the United States, and in the early twentieth century in Japan.

SOME MODELS OF CHANGE

The literature of the nineteenth and early twentieth centuries was full of attempts to analyze the changes arising out of the industrial revolution, to evolve a science of history, and to prescribe how societies should respond to change. Then the establishment of new states after World War II stimulated countless new attempts to understand the patterns of change, now frequently termed modernization or development. Japan has presented a special case in this literature as the first non-Western state to become a major industrial power; this section will therefore briefly survey the development of moderniztaion theory and its application to Japan.

Older Models

Sliding over for the sake of simplicity certain preliminary stages of history (marked by gathering, primitive, and slave societies), earlier explanations of development usually began with the *traditional society*, which was defined as having rested on a subsistence, usually agrarian, economy. In traditional societies, legal authority and political rule were legitimized by religious or secular tradition. Traditional political culture was parochial. It demonstrated a low level of countrywide mass involvement. Social integration was tenuous, except in primary groups. There was also a low level of education.

After World War II, theorists attempted to apply concepts of change to so-called non-Western societies and to further unravel the tapestry of development. Thus emerged an image of the *transitional society*. Transition was marked by instability of the traditional order and, according to Samuel Huntington, "political decay" of traditional authority. Lucian Pye spoke of the "crisis of identity" in transitional, non-Western states.

In older models, the *modern society* was assumed to be a highly industrialized and integrated unit. Descriptions ran well beyond the economic and technological, however, and embraced social organization, the role of the individual, and the ethos. Subjects or citizens were actively involved in a political system through integrating interest groups, political parties, and (particularly in democracies) elections. The decisionmaking process was described as being rational and sometimes accountable to the people. The role of the individual, previously rooted in familial status, was now increasingly determined by achievement. Often these criteria of merit conflicted with a growing egalitarian spirit bred in a modern society. All members of the modern society did not engage equally in making decisions. Stability was sustained, however, by the appearance of the characteristics listed above, in what Gabriel Almond and Sidney Verba called a "civic political culture."[1]

An impertinent question, one which challenges the surety of the ideas in modernization theory, inevitably emerges from older models of change. To paraphrase Marx, will the modern condition be historically marked by the withering away of change? Newer theories, to which the Japanese experience contributes valuable data, seem to indicate that development can move beyond modern (if by that term is meant the familiar industrial society of the twentieth century).

Contemporary Models

Contemporary models of development pick up certain constructs from the older models. Thus at the *primary stage* of development the mode of production is usually rooted in agriculture, forestry, and fishing (the primary pursuits). The *secondary stage* is a product of the industrial revolution. (The labor force is engaged mainly in industry, that is, in secondary pursuits.) The *tertiary stage* is characterized by the emergence of employment in highly skilled tasks and services (tertiary pursuits).

According to contemporary theory, when the surplus from industrial output rises to a certain point, there is a parallel shift of demand and supply into services. The service sector, including trade, transportation, finance, insurance, management (both government and business), real estate brokerage, information retrieval, and the like, accounts for most of the output and absorbs most of the labor force. Thus, the primary stage is distinguished by a labor-intensive economy; the secondary, by a capital-intensive economy; and the tertiary, by a knowledge-intensive economy. The third stage has also been labeled the postindustrial era, but there are many other indicative names.[2]

The term *postindustrial* was coined rather casually in the late 1950s in an analysis of work and leisure. Reshaped and applied by Daniel Bell, it came to be used in a much wider context. Whereas much of development theory had hitherto been drawn from, and applied to, Western experience, the ideas of Bell and of some who followed his theoretical path extended the analysis to the unusual case of Japan.

Zbigniew Brzezinski, for example, described in essence three Americas: first, the preindustrial society of Appalachia, the sharecropper and the migrant worker; second, the industrial world of the highly unionized blue-collar worker; and third, the "technetronic" realm of the technician, information retrieval specialist, and scientist. As a member of the Trilateral Commission (which included delegates from Japan, the United States, and Western Europe), Brzezinski devoted six months to study and writing in Tokyo. He found the "sudden blossoming" of Japan—its emergence from traditional, graduation from modern industrial, and entry into postindustrial society—to be something very "fragile."[3] By this he meant to indicate that the Japanese were uncertain as to their mission at this stage of development.

Most studies of advanced industrial societies, however, still concentrated on the experience of Western Europe and North America. The old industrial societies had undergone sufficient alteration to justify viewing the results as something qualitatively different. The range of terminology reflected the search for clues. Some observers spoke of "the technological society." Others called it "the temporary society." Still others referred to "the postmodern, active society." With a little more specificity, John K. Galbraith (who has been widely published, read, and heard in Japan) wrote about "the new industrial state." Samuel Huntington was the first to describe "postindustrial politics." Two authors carried the concept one step further by analyzing policy in a "postwelfare state." Martin Heisler studied political structures and processes in Western Europe, concentrating on "postindustrial democracies."

Japan: Traditional, Modern, and Postmodern

Meanwhile, in the period just before Japan's explosive growth in the 1960s, there was a certain ambivalence to be found in descriptions of Japanese society. It was recognized as highly industrialized and affluent, something like its advanced Western counterparts. Japan's economy still suffered from a dual structure, however, marked by the coexistence of traditional and modern elements. The Japanese political process remained somewhat unfamiliar and therefore not quite modern (as presumably Western political processes were). A distinguished Japanese sociologist referred to Japan as "an open society made up of closed components." It was, moreover, a highly stratified society with mutually exclusive group loyalties—a social structure that had been little changed by industrialization and Western imports.[4]

There was other evidence of the effective admixture of traditional and modern, of the coexistence of preindustrial mores and behavior and the values of an industrial age. These characteristics will be summarized in Chapter 8 after first examining here the results of the variable forces of Japanese tradition, of modernization, and of contemporary transition.

In the 1960s, when Japan plunged into high-speed growth, social critics pointed out that the nation boasted a highly literate citizenry and a skilled labor force, both attuned to the media. They described Japan as an information society (*jōhō shakai*).

In his evaluation of Japanese politics, Taketsugu Tsurutani used three cross-national criteria to determine a society's status as postindustrial.[5] First, in the postindustrial society a majority of the labor force is engaged in the tertiary or service sector. Since 1973 Japan has had over half of its workers involved in trade, finance, management and the provision of software (see Table 7.1). Statistics for 1989 revealed that 57 percent of the work force was in services.

Second, in the postindustrial society the service sector generates a larger proportion of the GNP than do the primary and secondary sectors combined. In Japan, the service sector has generated over 50 percent of GNP since the early 1960s.

Third, the postindustrial society builds on the gains furnished by the earlier industrial revolution: high levels of production, national income per capita, savings, and investment. Japan's astounding records on these fronts is fairly well known and has been noted as a "miracle," but the figures continue to be startling.

By 1965 Japan had joined members of the world's exclusive club of nations with GNPs greater than $100 billion, and by 1971 its GNP was estimated at more than $250 billion. In 1974, however, the GNP growth rate actually dipped below zero. Inflation, a stringent fiscal policy, loss of purchasing power, and a deficit in balance of payments all contributed to a negative attitude toward growth. Two OPEC-driven "oil shocks" (1973–74 and 1979–80) introduced notes of sobriety. Private consumption as a proportion of GNP was running at only 52 percent (as compared with 60 percent in the United States and in Europe).

TABLE 7.1 Japan, 1934–89: Population, Employment by Sector, and Consumer Price Index

	1934–36 (annual)	1950	1960	1970	1980	1985	1989
1. Population (millions)	69.3	84.1	94.3	104.7	117.1	121.0	123.1
2. Employed population (millions)	31.2	35.6	43.7	52.6	55.3	58.1	62.4
3. Employed by sector (millions):							
Primary (agriculture, forestry, fisheries)	14.5	17.2	14.2	10.1	7.4	5.1	5.3
Secondary (manufacturing, mining, construction industry)	6.8	7.8	12.8	17.7	19.3	19.9	21.0
Tertiary (wholesale, retail, finance, transport, services, government)	9.9	10.6	16.7	24.3	30.2	32.8	35.8
4. Consumer price index							
(1934–36=100)	100.0	219.9	308.0	988.8	—	—	—
(1980=100)	—	—	—	—	100.0	114.4	119.2

Source: Statistics drawn from *Nihon tōkei geppō* (Monthly Statistics of Japan) (Tokyo: Management & Coordination Agency, Statistics Bureau, various dates).

Moreover, in the mid-1970s Japanese felt a sense of growing crisis because of problems in the environment (an issue discussed later in this chapter). The *Asahi Shimbun* captured the public's mood when it coined the slogan "To hell with GNP!"

Nonetheless, in 1975 the GNP passed the next milestone, $300 billion. Another turning point came in 1983, a decade after the first oil shock. Oil shortages gave way to glut. There was a resurgence of technological progress. Recovery was led by growth in the electronics industry, followed by production of new ceramics and discoveries in biotechnology. Just as important were developments on the demand side, as both labor unions and consumers welcomed technological innovation.

New service industries arose in child care, education, sports, visual media, music, personnel placement, and data processing. Deregulation and privatization, particularly under the administration of Prime Minister

Nakasone Yasuhiro (1982–87), eased public deficits and invigorated the private sector. Decentralization spurred the spread of industry from the Tōkaidō zone to outlying areas. The Ina Basin in mountainous Nagano Prefecture, for example, attracted such a vast electronics complex that local residents began referring to the area as the "Silicon Valley of Japan."

Up until two or three centuries ago, the annual income per capita in most societies was below $200. The Japanese had a per capita income of $200 (in 1965 dollars) in 1920. A generation ago, before the impact of inflation was felt worldwide, it was thought that the highest stage of industrial society would be attained when the annual per capita income reached $2,000, which was marked as the threshold of the mass-consumption phase. It was predicted (about 1965) that the postindustrial stage would be entered when the annual per capita income passed $4,000.

Japan did not reach the high industrial level until 1967, about twenty-five years later than the United States. Within five years, however, the country was hurled into the mass-consumption stage. In March 1978, the annual per capita income of the Japanese stood at $7,167 (as compared with $9,150 for Americans). This level provided the potential for greater expenditure (as compared with the past) and relatively high levels of savings (as compared with those of the other advanced industrial nations). Between 1961 and 1970, Japanese gross savings out of GNP averaged 37.5 percent (more than twice the U.S. rate; see Table 6.2).

Meanwhile, in 1970, Herman Kahn had predicted that Japan would reach the postindustrial stage late in the decade. Japanese were very excited over Kahn's prediction that their nation would attain superstate status before the end of the century.

These, then, are the basic characteristics of the new society: a majority of the labor force in the tertiary sector; services contributing a larger proportion of GNP than do activities in the primary and secondary sectors combined; and high levels of production, income per capita, expenditures, and savings. These characteristics are, however, both a cause of the new society and an effect of it. Moreover, a society's economy and its technological development are but subsystems in a wide matrix of social forces. Cultural and political affairs continue to exercise a striking degree of autonomy.[6]

ASPECTS OF, AND NEW ISSUES IN, THE POSTINDUSTRIAL SOCIETY

Perhaps the most important aspect of Japan's emergence as a postindustrial society was that its celebrated insularity was being steadily eroded in the 1970s, well before the Japanese fad of "internationalization." On the domestic front, the postindustrial society seemed to be "bigger": more places, more people, more things. What was called psychic mobility in the modernization process was named "the eclipse of distance" by Daniel Bell.

Members of the new society were—or expected soon to be—richer in material goods. The ideal became the way of the suburban, white-collar, salaried man. The postindustrial society appeared to be easier to live in because of achievements of advanced technology and the expansion of leisure time. By the end of the 1980s, Japanese were on the threshold of a five-day work week; traditionally socialized to hard work, they fretted over whether they could effectively use the leisure time! Nevertheless, although most Japanese (as will be noted in Chapter 8) came to think of themselves as members of the middle class, few thought they were affluent. According to a 1987 survey by the Office of the Prime Minister, only 38.6 percent of respondents felt they were "well off."

Bridling the Economy

Japan's growth in the 1960s and early 1970s was still immeasurably aided by the dual structure of the economy. Fully 90 percent of entrants into the labor force came directly out of junior high school; many of these had migrated from rural areas. In Japan, as in most postindustrial societies, labor-management conflict was not eliminated, but it was better institutionalized (although it might be argued that Japanese collective bargaining rights had been won by fiat in the MacArthur Constitution rather than on the picket line). The advent of advanced technology increased productivity. Although wages increased at a faster pace than did labor's output, the share of national income accruing to workers did not increase as rapidly as did Japan's GNP.

By the late 1970s, some 95 percent of Japanese youth was going on to senior high school (a rate about the same as that in the United States; the Japanese dropout rate was, however, lower than the American). Moreover, the proportion of Japanese young people entering college or university was rising steadily (to between 35 and 40 percent in 1978). Such changes had an effect on the mobility of youth. The future of the lifetime employment system (in larger industries) came to be openly discussed. And the labor force was making new demands: rewards commensurate with performance (rather than with seniority); avoidance of overtime (unless there was extra pay); and longer periods of leisure. Workers came to expect more money, more consumer durables, more luxuries, and more holidays. It began to dawn on many Japanese that they had been working hard simply to increase exports—in other words, to fulfill outsiders' demands.

Japanese labor nevertheless remained relatively accommodating. The minor economic setbacks of the mid-1970s and Japan's only prolonged recessions in the post-treaty period (1978–79 and 1981–83) moderated the stridency of labor's voice. Moreover, workers finally were (or soon hoped to be) property owners with a family-centered life, with longer paid vacations, and even some travel. They were thus less motivated by or interested in militant political activity that might challenge the system.

As was the case with advanced industrial nations of the West, by the late 1970s the Japanese economy was feeling the effects of rising relative labor costs and a high energy bill. The result was a shift in the very structure of the economy. For example, Japan suffered a loss of comparative advantage, particularly in labor-intensive industries, where Japanese wages were high as compared with those of developing countries. Certain industries that had been competitive at one time were no longer so: nonferrous mineral smelting, refining, and processing as well as petrochemicals, textiles, and consumer electronics.

In 1971, a semipublic body, the Council on Industrial Structure, drafted a scheme whereby resource-hungry and pollution-prone heavy industry would give way to knowledge-intensive projects that could economize on energy. Again, in 1973, a private group of economists projected a drastic industrial reorganization to be carried out through 1985. They proposed that steel production be limited to meeting domestic demand and that the construction of oil refineries and petrochemical plants be slowed. In 1974, the Council on Industrial Structure again prepared a long-term vision of industrial structure within a revised growth plan. The report expected the economy to grow at an annual average rate not exceeding 6 percent.[7]

Such proposed adjustments created conflicts within the celebrated but by no means monolithic Establishment. On the one hand, Ministry of International Trade and Industry (MITI) bureaucrats were among the first to visualize a medium-growth economy, one to be complemented by those of developing neighbors in Southeast Asia, Korea, and Taiwan. Such areas had come to enjoy a comparative labor advantage. MITI's response to the oil shocks of the 1970s, for example, was to try to reduce energy-intensive industries in favor of knowledge-intensive in-dustries.

On the other hand, leaders of large and entrenched industries organized in Keidanren at first were opposed to such plans. Well represented in the LDP, such interests had kept Japan in largely protected industries where effective trade barriers were quite high—for example, smelting and refining raw materials, textiles, and food processing (the most heavily protected of all). It was exports from precisely these industries that generated international friction.

Ideas of adjustment were soon reflected in Japan's official plan for a mixed (public-private) economy for the period 1973–77. At least the rhetoric of policy objectives was altered to reflect social change: to create a "rich and balanced environment," to guarantee "affluent and stabilized living conditions," to curb inflation, and to contribute to "international harmony." Granted, these were highly subjective concepts; nonetheless, planners were trying to reflect a consensus that had shifted from obsession with GNP growth to concern for what was called net national welfare (NNW).

Divergent views continued to surface even among conservatives and, on occasion, agencies reversed positions. After the recessions of the late 1970s, MITI officials and some big-business federations favored continued high growth. Such a course, they argued, would bring full employment, providing the Japanese were willing to put up with inflation. Representatives of large industries in Keidanren and MITI were delighted with the U.S. pressure on Japan (on the eve of the 1979 Tokyo summit) to maintain a higher (7 percent) growth rate.

The Ministry of Finance and the Bank of Japan, however, countered with the idea that the era of high growth was over. They argued that a 4- to 5-percent rate of expansion would be quite respectable and would meet the threat of inflation. These finance bureaucrats opposed stimulative policies because, in their opinion, the country was undergoing a significant change in demographic structure (the graying of Japan). Eventually the elderly would constitute a drain because of transfer payments (in the form of social security benefits), and any tax reduction would make later increases in tax rates difficult. Finance planners were also concerned over the steady rise in internal debt. In 1978 the total had reached about ¥20 trillion (general-accounts deficit, ¥11 trillion; prefectural and local deficits, ¥5.6 trillion; national-railway and public-corporation deficits, ¥3 trillion) or about $85 billion at 1990 exchange rates (compared with the U.S. federal deficit of about $60 billion and the states' and municipalities' deficits of about $30 billion).

Enough has been said about levels of income to make clear that the religion of postindustrial societies has been growth, most often matched by consumerism. Japan of the late 1960s and early 1970s shared the faith, but in an unusual manner. Many Japanese still worshipped from afar by continuing their high rate of savings and dedicating disposable funds to consumer durables. (In those days, unlike Americans many Japanese avoided installment debts except—at a mature age—for expenditure on a house.) By the 1970s, about one-third of all households

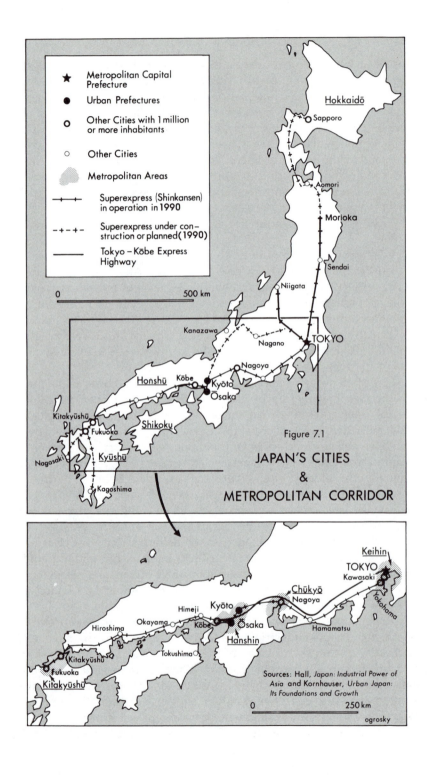

Legend:

★ Metropolitan Capital Prefecture

● Urban Prefectures

◉ Other Cities with 1 million or more inhabitants

○ Other Cities

▓ Metropolitan Areas

—+—+— Superexpress (Shinkansen) in operation in 1990

-+-+- Superexpress under construction or planned (1990)

——— Tokyo – Kōbe Express Highway

0 ————— 500 km

Figure 7.1

JAPAN'S CITIES
&
METROPOLITAN CORRIDOR

Hokkaidō
Sapporo
Aomori
Morioka
Sendai
Niigata
Kanazawa
Nagano
TOKYO
Nagoya
Honshū
Kōbe
Kyōto
Ōsaka
Kitakyūshū
Fukuoka
Shikoku
Nagasaki
Kyūshū
Kagoshima

Keihin
TOKYO
Kawasaki
Chūkyō
Nagoya
Yokohama
Kyōto
Himeji
Hiroshima
Okayama
Kōbe
Ōsaka
Hamamatsu
Hanshin
Tokushima
Kitakyūshū
Fukuoka
Kitakyūshū

Sources: Hall, *Japan: Industrial Power of Asia* and Kornhauser, *Urban Japan: Its Foundations and Growth*

0 ————— 250 km

ogrosky

Bullet trains (*Shinkansen*), which have a maximum speed of 125 miles per hour

had acquired automobiles. Virtually every house had come to have a refrigerator, a color television set, and an electric washing machine (the "three imperial treasures" of the 1960s).

Contrary to predictions, the Japanese savings rate remained high throughout the 1970s, but the purposes of thrift were different. Whereas in the 1960s savings helped finance purchases of consumer durables, in the next decade public opinion surveys revealed marked concern for income in old age. Faced with greater difficulty in getting a job after retirement and with as yet low levels of social security benefits, Japanese continued to save—for survival.

As the per capita income level approached those of the most advanced countries, the per capita stock of social overhead capital in Japan—parks, recreation facilities, hospitals, and the like—lagged behind private investment in additional production. The per capita stock of housing remained even lower. In May 1979, the sensitive Japanese were stung by a description emanating from a European Community report, which called them "workaholics living in rabbit hutches." They ruefully admitted that they existed in confined living quarters called *apāto* ("apartment," the general term, or "mansion," an almost ludicrous euphemism).

The most serious challenge to the new society came, however, from abroad. In the 1970s, the ratio of current-account surplus to GNP had remained modest. It moved from 1 percent (1971) to 2.2 percent (1973); as the yen's value climbed, the ratio tapered off to 1.7 percent (1978). The oil shocks proved effective in reducing Japan's external surplus, which actually moved temporarily into the red.

The new wave of growth in the 1980s saw the ratio of current-account surplus to GNP rise from 1.8 percent (1983) to 2.8 percent (1984) and then to 3.6 percent (1985). By 1984, Japan's excess of exports over imports amounted to nearly $34 billion, as automobiles, steel, and electronic goods poured across the Pacific. Because of an overvalued dollar, Japan became the second-largest trading partner (after Canada) of the United States. Towering trade and current-account surpluses were bringing Japan into disfavor abroad.

The fiscal year which ended March 30, 1986 saw Japan's GNP reach $1,626 billion, making the nation the second-largest economic power in the world. By the end of 1985, Japan had become the world's largest creditor nation, with assets exceeding liabilities by $120 billion.

Trade surpluses provided domestic byproducts. It began to occur to some Japanese—particularly youth—that they should demand some of the rewards for diligence *now*, in place of export-driven growth and further investment in industry toward returns *later*.

The conflicts over growth policies were settled in the arena of politics. By 1978 the combined opposition outnumbered the LDP, which had been identified with high growth. In the House of Representatives Budget Committee, the progressive parties blocked all legislation until the LDP administration agreed to a tax cut. Some commentators linked the poor showing of the LDP in the October 1979 lower house elections to the party's support of national tax increases. Prime Minister Ōhira Masayoshi withdrew the proposal for a tax rise. After the summit of 1979, it was apparent that Japan would have difficulty maintaining the previously pledged 7 percent growth rate.

In 1986, a committee chaired by the late Governor Maekawa Haruo of the Bank of Japan unveiled a report on the future of the economy. The Maekawa report noted Japan's important position in the world and the parallel problem of the current-account surplus. It stated that Japan "should make effective use of the surplus funds now available to promote qualitative improvements in national living standards and harmonious growth of the world's economy." The report set five specific goals: (1) expansion of domestic demand, (2) conversion of industrial structure, (3) improvement of market access for imports, (4) stabilization of exchange rates, and (5) promotion of international financial cooperation. Among

Japanese, it was agreed that the economy no longer would be export led; growth was to come from expansion of domestic demand.

Despite the ongoing effects of restructuring the economy, Japan's growth continued in the fiscal year ending March 31, 1989. The nominal GNP reached ¥371.3 trillion (about $2,800 billion), with per capita GNP estimated at $19,642 (compared with $18,403 in the United States). The rise, however, was generated by about a 5 percent increase in domestic demand, with a drop of 0.7 percent in external demand.

As a result, the critical ratio of current-account surplus to GNP decreased from 4.5 percent (fiscal 1986) to 3.5 percent (fiscal 1987). Nonetheless, by May 1988 Japan's foreign exchange reserves reached an all-time high of $87.24 billion (2.5 percent of world reserves).

Among the advanced industrial nations, Japan has enjoyed a modest inflation rate. In June 1989, the consumer price index stood at 104.2 (May 1985 = 100). Unemployment also has remained low, the seasonally adjusted rate running between 2.3 percent (November 1988) and 2.4 percent (May 1989).[8]

Governing the Citizens

High levels of production and relatively wide distribution of income results in a society with a low level of status polarization and, often, a large middle class. In such a society, a network of active secondary associations makes its appearance. This pluralism, according to one stream of Western political thought, provides the foundation for a stable democracy. The individual, however, becomes skeptical of strong attachment to any one cause or any one organization. (Incidentally, this postindustrial society can lean toward a capitalist, a mixed, or a socialist economy.)

Certainly Japan after the 1960s illustrated many of the political characteristics listed. The Establishment had laid out plans for a mixed economy. The LDP was the reconnaissance party, which constantly pierced new frontiers of growth. The prospect of affluence caused even the progressive opposition to alter its platforms toward pragmatic, moderate, centrist programs.

That postindustrial growth has reinforced a stable democracy in Japan does not necessarily mean that all Japanese have participated equally in decisionmaking. Rather, the majority has come to feel that all have been treated equally. Many Japanese were convinced that they were creating the world's most egalitarian society. They did, however, put their own twist on familiar political ideas. Ever since the war's end, labels like "democracy" have evoked favorable responses among Japanese. "Liberalism" has not become a pejorative term. What has often puzzled visiting Americans, however, is that "socialism" has also been viewed with favor, whereas "capitalism" has often been held in low esteem. And into the 1970s, "individualism" continued to be regarded as self-ishness.

Historically viewed, the salient issues of the preindustrial society were the choice of elites and relevant questions of political justice. The preindustrial society was vertically fragmented.

At the industrial stage, the salient issues were management of the economy and questions of economic justice. Thinkers wrestled with problems of production (capitalism and laissez-faire) and the ideas of Adam Smith and Social Darwinism or with problems of distribution (socialism and communism) and the ideas of Marx, Lenin, and Mao. The industrial society tended to be horizontally fragmented.[9]

In the transition between late-industrial and early-postindustrial society in Japan of the 1960s, a certain residual behavior was apparent. In a kind of cultural lag, many politicians (for example, in the Establishment) continued to emphasize production and growth. Others (in the progressive opposition) continued to hammer on issues of equitable distribution. All groups became both vertically and horizontally fragmented. Perhaps most important, salient political issues transcended management, labor, class and even political party.

In public opinion surveys there was evidence of a change from traditional submissiveness to authority to demands for participation in politics, even to active protest. In 1953, a majority of Japanese over twenty years of age were still prepared to leave the settlement of issues to "competent politicians." By 1973, a majority explicitly disagreed with this traditional attitude (51 percent disagreed; only 23 percent agreed; and 15 percent sidestepped the issue by urging a case-by-case approach). Moreover, two decades later (1984) a majority (54 percent) of respondents in a poll conducted by the Prime Minister's Office expressed the view that the people's wishes were not reflected in national policy.[10]

The shift in opinion was caused by a significant mixture of factors, including increased knowledge, a product of the information explosion, and increased participatory motivation, a product of the new politics. At the same time, the shift indicated decreased trust in institutional channels (political parties and elections), a byproduct of sophistication and skepticism.

Such political trends were matched with a change in attitude toward growth. Previously ignored costs of expansion had become politicized. The effects of heavy industrialization—inflation, pollution, traffic accidents, neglect of housing and such—were felt in egalitarian fashion. This was because both income and costs were more equitably distributed and because information about the effects was readily available. Uneasiness with growth, combined with unwillingness to give up its fruits, led to myriad demands that simply could not be met. For example, the average Japanese was provided with one of the world's finest public transportation systems, but like his American counterpart insisted on the inalienable right to drive an automobile to work, while denouncing the level of photochemical smog in Tokyo.

Postindustrial transition profoundly affected components of Japan's political system, if not the overall shape of political leadership. Such effects can be illustrated by a summary of the evolving status of political parties and their allies.

The LDP. The Liberal Democrats have continued (with one exception, the upper house election of July 1989) to win national elections. With a majority in the lower house, the LDP has continued to select the prime minister and to form cabinets. To be more accurate, factions within the LDP have continued to select the chief political leaders. Persistent rule, however, has camouflaged a wide range of political changes, including alteration of the LDP itself and even of its factions. The traditional pattern of strong patron-client relations has given way to a far more complicated matrix of power. The party's structure, its voter base, its role in policymaking, and its choice of policies have shifted with time.[11]

In July 1976 the LDP suffered its only split. A handful of LDP Diet members defected and formed the New Liberal Club (NLC), which for a time seemed to promise the appearance of a new conservative party relying on young, middle-class urbanites. The "NLC boom" was not sustained, however, as the LDP was able to adjust its priorities to the needs of city dwellers.

Although the bureaucracy remained powerful, of course, the increase in participatory politics came to establish political parties (particularly the LDP) as the holders of the levers of power. They controlled the prime ministership and the cabinet. They steadily coopted the bureaucrats (one in four LDP Diet members were graduates from the bureaucracy) and forced them to collaborate with the new breed of politicians.

By the 1970s, policy debates had begun to displace ideological confrontation. This trend both enhanced the status of the Diet and also led to complications for the LDP. Legislative rules required a party to have more than a simple majority in order to control legislation in committees. When the LDP was challenged, in the 1970s, the party was forced to become expert at "coalition politics" out of the need to coordinate policies with the opposition. The challenge also led the LDP to create social programs to meet the needs of the new postindustrial society.

Such demands led the LDP to become something other than simply a coalition of factions. Put another way, factions have evolved from being a "family" into a "village." By the late 1980s there were only four major factions (led by Nakasone, Takeshita Noboru, Miyazawa Kiichi, and Abe Shintarō) and one minor group (led by Kōmoto Toshio). Tradition joined transition as strict seniority rules were applied to recruit leadership of the party. Senior status rested less on age than on the date of election to the Diet and the number of times elected. Despite the seniority rules, interfactional competition became so complex as to require power sharing in distribution of top posts.

The main positions in the LDP became the president (concurrently prime minister), the secretary-general, the chairperson of the Executive Council, and the chairperson of the Policy Affairs Research Council (PARC). The press usually listed the "big four," alongside members of the cabinet, as the chief wielders of power.

Recruitment of the prime minister became more contentious as back-room maneuvers among many factions gave way to, on some occasions, a kind of party primary to select among four or more candidates; or to, on other occasions, consensus building among fewer and bigger factions and candidates. The era (1956–1972) when Japan had only three prime ministers (Kishi, Ikeda, Satō) gave way to a period (1972–1989) when no less than nine LDP leaders held the post (Tanaka, Miki Takeo, Fukuda, Ōhira, Suzuki Zenko, Nakasone, Takeshita, Uno Sousuke, and Kaifu Toshiki). Increased activity in the rank and file of the LDP led to a growth in size of the party from 455,000 members (in 1977) to over 3 million members (in 1979).

In 1990, factions continue to determine the LDP leadership, but Diet members increasingly influence which policies the party will support. Diet members have steadily increased their skills and participation in policymaking in response to the complicated challenges of the postindustrial society. The result has been not so much a fading of the power of the bureaucracy, but an increase in the influence of middle-level, particularly LDP, Diet members.

The link between the Diet and the bureaucracy is, in typical Japanese fashion, a flexible, almost informal grouping regularly referred to as a "tribe" (*zoku*). Majority party members eventually find a niche in the PARC, specifically in its divisions, which correspond to standing committees in the Diet and to ministries under the cabinet. Knowledgeable members must rely on the bureaucracy, which continues to mark up cabinet bills. Such a network offers the ambitious party member expertise on current policy issues. It helps the legislator to represent constituents in local support groups (the *kōenkai*). To the senior LDP Diet member, the tribe offers experience, which will be useful in the cabinet and in party leadership. In the rapid transition of the society, there are tribes that specialize in agriculture (a traditional LDP "vineyard"), in taxes (a potential pitfall for any leader), in construction (here Tanaka mastered the secrets of success), and in several other areas.

Perhaps the most important links are those between the networks, in Tokyo, and interests in prefectures, urban settlements, towns, and villages, at the local level. On this front too, the LDP has seen change. Tradition is served in a way by the remarkable reservoir of second-generation candidates, sons or adopted sons-in-law of previous Diet members. Overlapping this pool is an increasing number of local politicians who are LDP Diet candidates.

Meanwhile, the voter base of the LDP has also reflected the effects of societal change. Mention has been made of the rise in number of voters who support *no* party (see Figure 5.1 in Chapter 5, above). According to an *Asahi Shimbun* poll of July 1986, some 39.6 percent of the electorate was in this category. Nevertheless, when asked whether they would vote in the next election, only 10 percent of respondents listed no party. Gerald Curtis has described the "new independents" as a Japanese variant of the "dealignment phenomenon" also evident in Western Europe and in the United States. Almost always the independents made up a loss to the progressive parties, while a hard core continued traditional support for the LDP. This pattern allowed prediction of a close election (for the lower house, 1990), but a surprisingly strong showing by the LDP, as independents at the last moment gravitated to the majority party.

The continued success of the LDP rested, however, mainly on its relative adroitness in shifting policy to meet the changing demands of the new society. In the low-growth economy of the 1970s, the party was driven to remain in power: It forsook austerity budgets and embraced programs to stimulate domestic demand. These programs included extension of public works projects into rural Japan and expansion of expenditures for health care and social security, a surprising platform for a welfare state under the aegis of the conservative LDP. Nowhere was the impact more strongly felt than in the countryside.

After redistricting in 1986, rural areas remained overrepresented but nevertheless accounted for only one in four LDP supporters. The farmer-voter, however, had also changed. Most (70 percent) had come to derive only part of their income from farming; this part-time farming was dubbed "*san-chan* agriculture" (for grandpa, *ojii-chan*; grandma, *obā-chan*; and mother, *okā-chan*). The bulk of their income was derived from increased employment in semirural towns and in prefectural-urban centers. The bureaucracy cooperated with the LDP in promoting policies designed to cushion the postindustrial transition in rural areas (with farm, particularly rice, subsidies) and to spread growth to less prosperous regions (with construction subsidies).

No one better represented the policy shift from austerity to welfare than did a relatively young upstart politician. Tanaka Kakuei (prime minister, 1972–74) had made his fortune in construction and had become a master of political patronage. His programs produced a new generation of wealthy businessmen spread around the prefectures and, naturally, they became a new support base for the LDP.

When critics began to complain about the problems caused by rapid growth, Prime Minister Tanaka responded by writing *Building a New Japan*,[12] which soon became a best-seller. His plan for remodeling the Japanese archipelago was scuttled by speculation, inflation in land prices, and later, his own involvement in a corruption trial. Nevertheless, the 1974 budget marked the high-water mark for expenditure on social welfare, which exceeded spending on public works. Support for the LDP among urban voters increased because of the welfare programs. The party's gains were almost all matched by losses suffered by opposition parties.

The oil shock of 1973, with attendant inflation and recession, led all Japanese seriously to reconsider expansionary policies identified with the LDP. The Lockheed scandal of February 1976 (an affair that directly implicated former Prime Minister Tanaka for bribes connected with aircraft procurement) convinced many Japanese that rapid growth was also accompanied by structural political weaknesses. The selection of a quiet reformer, Miki Takeo (prime minister, 1974–76), represented an attempt to stave off the decline of the LDP majority. In 1976 the party, however, suffered an election debacle in the lower house, maintaining a working majority only with the aid of independents.

Prime Minister Fukuda Takeo (1976–78) was the first to try to maneuver conservative policies through a Diet in which LDP and opposition strengths, at least in the committees, were evenly matched. His policy was a compromise, an advocacy of a pattern of "permanent residential zones" interspersed with industrial areas to ease the housing shortage.

The first Ōhira cabinet (1978–79) marked a major milestone as Japan played host at the Tokyo summit meeting of advanced industrial powers in June 1979 (described in Chapter 9). At home, Ōhira's "pastoral cities" plan to create medium-sized cities in sparsely settled regions had to be abandoned by his second cabinet (1979–80), which was formed with a narrow margin of support. After Prime Minister Ōhira's death in June 1980, the party used a revived majority to form a cabinet around Suzuki Zenko (prime minister, 1980–82), a sixty-two-year-old veteran compromiser among LDP factions.

In October 1982, Prime Minister Suzuki, grappling with a stubborn recession, falling revenues, and (unusual for Japan at that date) growing budget deficits, announced that he would not seek reelection as LDP president. Despite efforts of party leaders to settle the succession issue behind the screen, a party primary had to be held and Nakasone Yasuhiro captured 58 percent of the votes of rank-and-file party members. In the final voting for party president, Nakasone won a comfortable majority among LDP Diet members and was then selected by the Diet to be prime minister.

Nakasone (prime minister, 1982–87) had long been touted as a party president. In fact, he was an unusual LDP leader, direct, outspoken, and therefore, often mistrusted in factional circles. Beyond doubt he became the leader of Japan best known abroad and thus admired, and as often suspect, at home. Perhaps most important, he became identified with the LDP's successful change to a policy of austerity.

A number of factors had helped build a consensus in favor of austerity: fear of inflation, a temporary public mood in favor of frugality, and a broad opposition to tax increases. Once public opinion shifted toward the importance of deficit-cutting budgets, there was no party that could outbid the LDP in its commitments to cut government waste, to reorganize public administration, and to privatize public corporations. In 1980 a government commission chaired by Dokō Toshio (former president of Keidanren) had underscored the need for administrative reform. Nakasone had joined the austerity bandwagon even before becoming prime minister, when, in a sensitive post as head of the Administrative Management Agency (AMA, later renamed the Management and Coordination Agency), had taken a political gamble and tackled the task of reform.

As prime minister, Nakasone racked up several successes by fulfilling the promise to deregulate giant public corporations. Nippon Telephone and Telegraph Corporation (NTT) went private in 1985. Bonds remaining in the hands of the government helped offset budget deficits. In 1987, the government disbanded the debt-ridden Japanese National Railways (JNR), in favor of seven privatized regional Japanese Railways (JR). Japan Air Lines (JAL) was also deregulated late in 1987. During the Nakasone regime and that of his successor, Takeshita, administrative reform resulted in a 4 percent reduction in the number of central government employees and a 20 percent cut in the number of employees of public corporations.

After Prime Minister Nakasone led the LDP to a sweeping victory in the dual elections for half the upper house and the lower house in July 1986, he was able to break two precedents. The party extended his second term as president (and thus, as prime minister) to October 1987, but refused to grant him a third term. That month Nakasone became only the second leader in postwar history to name his successor as party president.

Takeshita Noboru (prime minister, 1987–89) inherited two problems from his predecessor. In 1988 he pushed a package of tax reforms (originally proposed by Nakasone) through the lower house. This included an unpopular value-added tax of 3 percent. Unfortunately, the levy coincided with the unmasking of the latest spectre to haunt the LDP.

In the early years of growth, the world of big business and finance (*zaikai*) had provided the major source of funding for the Establishment, specifically for the LDP and its factions. In postindustrial transition, the relative importance of the *zaikai* tapered off as nontraditional firms and regional entrepreneurs gained in strength. Earlier "black mist" scandals and the Lockheed affair in the 1970s had shaken the LDP and led to cosmetic party reform. Strict *campaign* limitations (in a 1975 revision of election laws) left untouched unreported funds paid to factions and *individual* contributions derived from private fund-raising parties. It was reported that many senior Diet members regularly sold several thousand tickets for these so-called parties at ¥30,000 each (about $200). Corporations often bought blocks of tickets and listed them as deductible business expenses, not as political contributions. In the new high-tech age, stock also passed from hand to hand. The problem of "money politics" (literally, money-power politics, *kinken seiji*) came to a head in the so-called Recruit scandal.

This affair involved the private transfer of stock, before it was publicly listed, by an aggressive new firm, the Recruit Cosmos Company. In June 1989, after a 260-day inquiry, the Justice Ministry made a final report to the Diet. By that time former Prime Minister Nakasone had been publicly questioned and his staff implicated in "Recruit-gate," although he was not indicted. Takeshita had been forced to resign and three cabinet ministers had left their posts. Some sixteen businessmen and politicians (two Diet members) were indicted; fourteen were arrested; and almost all the major LDP leaders were included in a list of "gray" figures tainted by the probe.

As a result, public support for the LDP eroded rapidly. When Takeshita had come to power in 1987, his cabinet enjoyed a high level of approval (58.6 percent in *Kyōdō News* surveys). In March 1989 support fell to an all-time low (12.6 percent). One apparent consequence was the party's loss of the upper house election held in July 1989.

Indeed, Takeshita's successor, Uno Sousuke (prime minister, June–August 1989) inherited Recruit damage and served only a few months. He was also hampered by rumors of a personal involvement with a part-time geisha.

In August 1989, the LDP settled on Kaifu Toshiki to complete Uno's party presidential term (until October 30). On August 9 he was elected prime minister by the lower house. With some trepidation, since Kaifu had a weak base in the party, the LDP continued his presidency, pending the outcome of the lower house election to be held in February 1990. As has been indicated, in that election disaffected supporters of the LDP returned in droves to vote for the party, which captured over 46 percent of the popular vote and over 275 seats in the Diet. Most observers agreed that these results were not so much a ringing endorsement of the perennial winner as they were evidence of the public's lack of faith in any alternative.

The Opposition. The regular success of the conservatives might leave the impression that there was no dissent in postindustrial Japan. Such was not the case. At the local level first, an impressive array of urban protest movements and of local government leaders appeared. Although they were often backed by the progressive (*kyūshin-teki*) political parties, they were given a distinctive name, the reformist band (*kakushin-ha*), denoting a more activist posture. For a time the reformists were able to recruit voters who had not regularly supported any party.

The floodtide of a floating vote came with the rise of citizens' movements, which in turn produced what were called the "urban communes." The best example was the reelection of Minobe Ryōkichi to a second term as governor of the Tokyo metropolis in 1971. He was backed by a coalition of progressives and independents and rolled to a landslide victory, garnering the largest number of votes cast for any single candidate in the election history of Japan. (In April 1979, at age seventy-five and after three terms, Minobe chose not to run again.) Into the late 1970s the progressive parties, particularly the Socialists (JSP), made steady gains. The 1979 lower house election, however, saw re-establishment of LDP control over local government.

Women had lent strong support to Governor Minobe and other reformist leaders. Housewives fought pitched battles over new issues: inflation, which affected household accounts; quality control, which touched on consumer goods at higher prices; traffic accidents, which impinged on childrens' welfare; housing shortages, which caused cramped family lifestyles; and environmental disruption (*kōgai*), which lowered the quality of life in favor of growth. The media so widely advertised the problem of environment that leaders found it necessary to devote parliamentary attention to the issue. One legislative session (1970) became known as the "Pollution Diet."[13]

Also among opposition forces were extremists, those who seized upon the disarray of transition and engaged in political action for the sake of expression, rather than for the purpose of influencing decision.

At one pole was the infamous Red Army, which favored a kind of romanticism and nostalgia for the class struggle. At the other were ultra-nationalist groups, like the Society of the Shield (formed by Mishima Yukio), which were seized with nostalgia for traditional "basic Japan." In a dramatic but pointless gesture, Mishima committed ritual suicide in front of derisive members of Japan's Self-Defense Forces. Scattered between the poles were small coalitions, which sporadically but forcefully opposed expansion of the Narita International Airport.

In November 1985, on the eve of privatization of the national railways, a "middle-core" band of radical students sabotaged rail and signal systems to disrupt the morning rush-hour commute. In May 1986, others lobbed homemade shells toward public sites in an attempt to interrupt the second summit meeting held in Tokyo. Symbolic protest, however, remained on the outer fringes of the new society.

Meanwhile, the major problem faced by the opposition was the steady decline of the JSP, the largest dissenting party. The decline was compounded by the inheritance of ideological infighting from prewar socialist movements, by the party's inability to adapt to change, and by conservatives' successes. As LDP factions declined in numbers but grew in size, JSP factions practically disappeared.

In the 1970s, recruitment to the JSP was almost entirely from ranks of the General Council of Trade Unions (Sōhyō). Similarly, when Nishio bolted from the JSP somewhat earlier and formed the Democratic Socialist Party (DSP), he turned to the All-Japan Congress of Trade Unions (Dōmei Kaigi) for support. In both cases, the socialist groups became one-interest parties. Sōhyō, in particular, encouraged the JSP to stick with outdated, dogmatic, leftist slogans dealing with "class struggle," and to support the federation in its annual "spring offensive" for higher wages.

Meanwhile, both socialist parties were wounded by the flight of workers from organized unions. In the late 1980s, the proportion of organized workers in the total labor force dropped to 29 percent. Moreover, support of the JSP by labor declined rapidly. In the 1960s, more than half the blue-collar workers had supported the party; by the mid-1970s, only about one-third lent support to the JSP. (In 1986, the LDP garnered the support of 46 percent of blue-collar workers.)

In fact, JSP platforms had little to do with union support or with local bases of power. In the Diet, socialists came to be among the forces most opposed to change. Successful socialist candidates tended their own factions and their own constituencies. At times they engaged in disruption in the form of boycotts of legislative, specifically budget, proceedings; at times they quietly agreed to lift boycotts in exchange for LDP support of minor bills. In short, the JSP became a professional, permanent opposition. The DSP, in contrast, became more ready to form a coalition with the LDP if the latter were to stagger seriously.

It made no difference whether the country was in a period of rapid growth or in an era of austerity—the JSP's policy choices were not adroit. In the early 1970s, progressive governors and reformist mayors led the JSP in denouncing the Establishment for emphasizing growth and neglecting welfare. Around 1975, the same forces belatedly joined in the cry for reduced deficits. The JSP regularly and appropriately offered a list of problems on the welfare front (social security for the aged, the need for health insurance, and neglect of the infrastructure). In the late 1980s, however, the party futilely led the barricade defense against the LDP's consumption tax, but offered no clear alternative to cover demanded welfare expenditures.

Policy designed for the shrinking agricultural sector offered another illustration of the failure of the JSP and the Communists (JCP) to capitalize on social issues and to challenge conservative dominance. During the drive to austerity in the 1970s, as the LDP began to shave rice subsidies, opposition parties had hopes of taking on the majority party in one of their citadels of power. The JSP became even more protectionist than the LDP had been and supported *higher* producer prices for rice. At the same time, the party demanded *lower* consumer prices. The confusion did little to attract urban voters, who should have become the natural allies of the opposition.

In 1977, during the revolt of the urban communes, the popular mayor of Yokohama, Asukata Ichio, was chosen to be chairman of the JSP and to bring compromise among warring party groups. Asukata's membership drive was designed to broaden the base of the party and to lessen the dependence on Sōhyō. The chairman was successful in increasing the local membership from 40,000 to 65,000.

In 1983, Chairman Ishibashi Masashi proclaimed inauguration of a "post-Marxist new JSP." In the LDP victory and JSP debacle of the 1986 dual elections, however, JSP representation fell to 83 seats in the lower house, the lowest level since 1949. Ishibashi took responsibility for the defeat and resigned. He was succeeded by Doi Takako, a professor of international law, Diet member for seven terms, and the first woman to head a major political party in Japan.

Women and labor union members—potentially the strongest allies of the progressive parties—have since tried valiantly to revive the JSP. Chairwoman Doi immediately launched a campaign to lead the party into pragmatic politics. Under her guidance, the JSP moved toward establishment of European-style democratic socialism in Japan. She strove to create a new image for the party, one that would appeal especially to women and to younger voters.

Throughout 1988 and into 1989, Doi led the JSP and other opposition parties in a call for reform of money politics and a redesign of the economy to provide affluence to ordinary Japanese. In April 1989, she organized a boycott of Diet proceedings, demanding a legislative inquiry into the Recruit scandal. With executives of the DSP and of Kōmeitō, in cooperation with the labor federations, the JSP organized a forum to discuss formation of a shadow coalition, ready to succeed the LDP.

In the campaign for the upper house election of July 1989, the Socialists demanded repeal of the 3 percent consumption tax and democratic reform of political parties. Although the JSP made startling gains in the election (in fact, a defeat for the LDP), public opinion polls revealed that support for the party grew from 11 percent to only 18 percent of the voters (only a little more than the percentage of seats held in the lower house).

The 1989 election probably marked the high point of what the press dubbed "the Madonna boom" (after the popular singer). In the subsequent election of a leader by the Diet, Doi Takako became the first female to be nominated for the post of prime minister. (The LDP majority in the dominant lower house, however, led to the selection of Kaifu Toshiki).

As of October 1989, women held 33 seats (13 percent) of the total of 252 in the upper house. JSP women numbered 15. In the lower house, only 7 (2 JSP, 5 JCP) women were seated. The impact of Doi's charismatic style was, nonetheless, felt in the majority LDP too. In August Prime Minister Kaifu had selected two women to serve in his cabinet, including Moriyama Mayumi as chief cabinet secretary, the first woman to hold the sensitive post. (After the limited LDP victory in the February 1990 election for the lower house Kaifu returned to an all-male cabinet.)

Once again, failure of the opposition to unseat the LDP was due to the inability of the progressive forces to mount an effective coalition. On November 21, 1989, the thirty-nine-year-old General Council (Sōhyō) officially went out of existence, in favor of the new Japanese Trade Union Confederation (Shin Rengō no Kai). In November 1987, when the new confederation was first organized, the Japan Federation (Dōmei) also disbanded. Rengō, to use the short name, thus commanded 8 million union members (65 percent of organized workers) and became the noncommunist world's third largest labor federation.

In the 1989 election (for half the upper house), Rengō sponsored its own candidates and took 11 seats (added to one held). The confederation insisted, however, that its Diet members were expected to serve only as bridges among opposition members. "Labor has carried out an historic reconciliation," Chairman Yamagishi Akira announced. "The opposition parties should follow suit." He urged the noncommunist opposition—

the JSP, DSP, Kōmeitō, and smaller socialist groups—to cooperate toward victory in the next (the lower house) election. Traditional infighting over ideology, and foreign and security policies, however, won out.

True, the JSP also won a victory of sorts in the 1990 election. The party increased the number of seats held in the lower house by 53 (from 83 to 136). Nevertheless, although the LDP lost 20 seats (from 295 to 275), it still won a majority in the house. One reason was that JSP gains were almost all at the expense of other opposition parties. Noteworthy was the fact that Doi was not able to persuade her "party of labor unions, old-school liberals, and of men" (as it has been described) to place women candidates in many slots occupied by male incumbents. The JSP elected only seven women to the lower house.

There has been little doubt in the minds of objective observers that the Japanese people have deserved a more effective opposition. The long monopoly of power, which has on occasion bred arrogance and corruption in LDP circles; the persistence of trade-balance issues, which have alienated allies abroad; and the neglect of social overhead in the new society—all have cried out for alternative solutions.

Coping in the New Society

As of June 1989, Japan's population had practically stabilized at 123 million. The annual rate of increase was less than 0.5 percent, the lowest rise in three decades. Obviously, Japan's demographic problems were matters of quality, not quantity. Significant were changes in location of, and subtle structural shifts in, population.

Perhaps the most important feature of postindustrial society is the continuing migration of people to urban centers. Thus, whereas in the 1960s planners thought in terms of *industrial deconcentration,* in the 1970s and 1980s they have had to deal with *postindustrial concentration* because of neotechnical growth. Skilled users of presses, photocopiers, computers, fax machines and software still find it convenient to be close to one another. Now, however, they can be in suburban or in relatively remote conurbations that stretch, with a few rural interruptions, between Tokyo and Kitakyūshū (see Figure 7.1). A pattern of megalopolitan settlement is woven from strands of high-speed communication and transportation. This may mean, for example, a net decrease in the population of the ward (-*ku*) areas of Tokyo; a startling increase in the population of the major metropolitan areas (Tokyo, Osaka, and Nagoya, which contained almost 43 percent of the total population in 1985); and a net decrease in rural population.

Although population losses suffered by rural areas slowed from 3.4 percent annually in 1971 to only 0.6 percent in 1978, a white paper issued by the National Land Agency reported (1978) that there were 1,093 villages, towns and even cities (34 percent of all communities) that could be categorized as "depopulated areas." More specifically, the report indicated a drop of 20.7 percent in the number of children up to fourteen years of age in the depopulated areas (the overall national figure climbed 8.3 percent), and a drop of 6.1 percent in the number of those age fifteen to sixty-four (as against a 5.1 percent rise nationally). Meanwhile, in depopulated areas the number of persons sixty-five years of age or older increased by 9.8 percent. One could not avoid the image of a rural society of elders deserted by the city-bound younger generation.

Mass urban society has most often been pictured as leading inevitably to alienation of individuals and eventual collapse of the social structure. There has been a nostalgia for the old unified traditional society. The Japanese case certainly supports the observation that although urbanization gives rise to geographic and social mobility, it has had concomitant disintegrating effects on isolated rural (and in that sense, traditional) society. Just as Oscar Lewis has found a marvelously intricate new pattern of social ties in the urban settlements of San Juan and of Spanish Harlem in New York City, however, so observers of the Japanese urban scene have found complex admixtures of traditional and modern in the wards of Tokyo.[14]

The end of the Pacific war was a distinct demarcation point between age groups of Japanese. As of March 1977, slightly over 50 percent of all Japanese had been born since 1945. (By 1988, some 64 percent has been born since the war.) Built into the affluent society was a generation of young people who have grown up with almost no exposure to war, scarcity, or want. Youth had become a "new breed" (described in Chapter 8).

The Graying of Japan

By the end of the 1970s, one aspect of postindustrial society warranted special attention in the media. Japanese referred to the phenomenon as "the graying of Japan," or sometimes as "the twilight of Japan" (*Nihon no yugure*). The elders involved were called "the silver society."

Although Japan is not yet the oldest nation in the world (in 1985, 10.3 percent of its population was over sixty-five, compared with 11.7 percent in the United States), it is the fastest aging society in a rapidly graying world. Of greater interest are projections made by a statistics bureau of the Prime Minister's Office that predicts that the total number of elderly could reach 25 million in the second decade of the twenty-first century (see Figure 7.2). Other data showed that in 2025, the old-

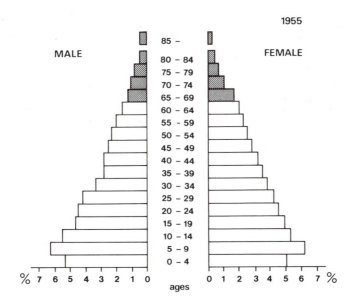

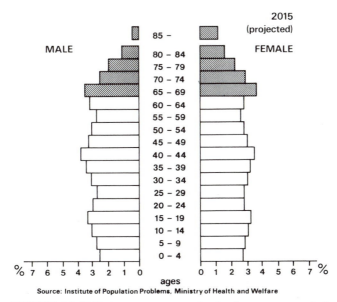

Source: Institute of Population Problems, Ministry of Health and Welfare

FIGURE 7.2 Shift in age structure in the Japanese population

age group could approach 24 percent of the population, a rate that would probably surpass those of other advanced industrial nations.

Experts listed three factors in accounting for this demographic development. First was the postwar baby boom, which provided the original population base. Second was the subsequent sudden drop in the birthrate, a decrease that naturally raised the percentage of elderly in the total population. Third was the extension of average life expectancy in Japan. In 1987 life expectancy at birth reached the highest levels in the world: 75.61 years for males and 81.39 years for females.

Demographers and economists have agreed that the graying of Japan has come at the very worst time, when the nation is adjusting to medium or slow growth. Management has preferred to await attrition rather than to engage in arbitrary layoffs. As a result, according to Kuroda Toshio of Nippon University, one of every five productive-age workers (ages fifteen to sixty-four) is now over fifty-five years of age. Nevertheless, unemployment and underemployment are overwhelmingly concentrated in the older age group. Aging of the work force has compelled companies to review their personnel policies (for example, the tradition of lifetime employment, and the treatment of the older staff) and to cope with increases in personnel costs, clogged channels for promotion, and sagging morale among younger workers.

At both extremes, among youth and in the "silver society," coping with rapidly changing social conditions has been a major feature of postindustrial life in Japan. Perhaps this is the reason so many Japanese have been preoccupied with the query, who *are* we Japanese?

8

Kokuminsei: Who Are We Japanese?

There are those who argue strenuously against any further use of the term *national character*. Critics of the concept claim that Japanese have used this vague idea to exaggerate the unique in their lifestyle. Many Japanese have gone on to imply that foreigners can never really understand their way of life.

Toward the end of the Pacific war and immediately thereafter, national character studies of the Japanese belonged in the field of pathology. In good Freudian and post-Freudian style, foreign analysts took (secondhand) clinical findings related to Japanese individuals and lifted them to the status of national averages in order to account for puzzling traits (of the enemy). They then arrived at policy recommendations for victory, occupation, and so forth. Thus, to cite extreme cases, it was said that the Japanese adult personality was warped by earlier strict toilet training (just as Russian character was once said to emanate from the practice of swaddling the infant in the fields, and the American personality was said to be indelibly marked by timed breast-feeding in babyhood). Yet the formulas thus derived and applied to Japanese perhaps made more sense than the assumptions behind the policy of unconditional surrender, in which the Japanese were treated as subhumans.

Unfortunately, national averages always obscure individual differences. Whether national character studies belong in the soft or the hard sciences, they are certainly persistent and subjectively important. The literature has, however, often confused such subtly different aspects of national style as modal personality, culture, ideology, attitudes, and behavior.

SOME GUIDELINES

Interestingly enough, in light of the flurry of activity in the behavioral sciences in the postwar period, the link between attitude and behavior

Food preparation on television

is at best tenuous. There is no invariate connection between orientation, ideology, or national character and action. Often, of course, national character has been described in terms of certain striking behavior emanating from acculturation. The connection, however, has not been empirically demonstrated and what we think we know so far is based on instinct.

Of attitudes and behavior, attitudes are more measurable. They are revealed in opinion surveys and the emergent patterns are often called *kokuminsei* (national character) by the most skillful Japanese pollsters.

Attitudes

> Kawabata Yasunari, Japan's Nobel Prize–winning novelist, once said that the Japanese communicate through quiet understanding, a kind of telepathy, since for them the truth lies in the implicit rather than in the stated. They have a word for it: *ishin-denshin,* "communication by the heart."[1]

To ascertain "attitudes, or ways of thinking, of a people," observers now enjoy a wealth of data thanks to, among others, Japan's Institute of Statistical Mathematics of the Ministry of Education. Every five years since 1953 the Institute has conducted a survey in depth. Respondents have been a random sample of 5,400 persons, on average, chosen from several hundred cities, towns, and villages throughout the country.

The institute has pointed out that revealed attitudes lend themselves to various interpretations. For example, one can identify those points of view that Japanese share with citizens of other countries. Such an approach might strengthen what some sociologists call the "convergence thesis," namely, that because of universal historical experiences, stages of economic growth, and the impact of advanced technology, Japanese will soon resemble their counterparts in other developed nations. More often, however, the tendency is to concentrate on differences between Japanese and non-Japanese. This is called the *cross-societal dimension*.

Beginning in the 1970s, the institute added a control—data drawn from surveys of Japanese-Americans (*Nikkeijin*) in Hawaii. One of the striking features uncovered has been the breadth of Japanese-American attitudes (indicating widely individual outlooks), compared with the similarity of Japanese attitudes (indicating the importance of the group even in this period of national transition). These different configurations led one institute member to conclude, "Japanese-Americans are not Japanese," but "we Japanese are Japanese!"

Another approach looks at opinion distribution among the Japanese. This produces a map of the topology of national character known as the *intrasocietal dimension*.

Yet another view results from a look at opinions and attitudes that a majority of the Japanese hold. This is called the *modality dimension*. In the surveys conducted by the institute, a "majority" consists of not only two-thirds of the total sample of respondents, but also two-thirds across-the-board regardless of distributions according to sex, age, or level of education. The institute has identified a few key attitudes held by such a majority.

Such findings suggest one last approach. The institute has been interested in the degree and direction of change in attitudes (including no change) over time. For the purpose of identifying national character, obviously opinions that remain relatively persistent are significant. This aspect the institute labels the *constancy dimension*.

The institute's approach to identifying national character suggests that it essentially agrees with Alex Inkeles, who has stated that national character "refers to relatively enduring personality characteristics and patterns that are modal among adult members of a society." By way of comparison, the institute more precisely concluded: "Therefore, it is reasonable to define the study of the (Japanese) national character as the search for those attitudes which satisfy constancy, homogeneity, modality, and cross-societal difference."[2]

Behavior

> . . . In spite of substantial changes common to industrialization, an examination of the real personal relations that prevail upon individuals in a social context reveals a surprisingly marked degree of variation from society to society, which shows that the traditional aspects of each society have been retained.[3]

As to the attitudes-behavior link, anthropologists have argued that a social organization (and social behavior), in order to be viable in the long run, must complement—not clash with—basic systems of belief shared by the society. Some social scientists have gone one step further: National character, it has been argued, forms the substratum that nurtures, shapes, and supports the political regime of a nation. From there, doctrine has on occasion wandered into the swamp of generalization. By way of illustration, it was once fashionable to say that relatively open, individualistic, egalitarian, and autonomous personality types supported stable democracies (like the United Kingdom and the United States) and that closed, rigid, aggressive, and hierarchically oriented personality types supported authoritarian regimes (like Nazi Germany). Such a quantum leap from individual personality clinically observed to national characterization is too great.

Middle-level linkage, short of broad generalization, is far safer. For example, a good query by U.S. Occupation officials might have been, how will Japanese attitudes affect the development of post-treaty Japan? If some attitudes are relatively open and egalitarian, but not individualistic and if others reveal a more closed, hierarchically oriented but not aggressive stance, will the resultant behavior be stable and democratic or rigid and authoritarian? In the case of Japan, we can connect certain persistent attitudes with certain specific practices. The former seem to nurture the latter.

SOME PATTERNS

The most important findings in surveys conducted by the Institute of Statistical Mathematics and by other polling agencies provide evidence of a clear reassertion of traditional elements in Japanese national character. Understandably, the turmoil of total war, devastating defeat, and "reorientation" by the Occupation (even though it was benign) at first served to throw into doubt the validity of prewar and wartime values. Later polls demonstrated, however, that the Japanese continued to regain self-respect and to view their own culture without the antipathy that so often marked immediate postwar attitudes.

Although the prewar dogma of the superiority of the Yamato people has subsided due to media exposure and scientific enlightenment, an implicit racism survives (as it does in America) in the belief that racial differences do exist and are socially important. Nonetheless, contemporary Japanese nationalism is less virulent than the prewar and wartime varieties. It is still spun out of Japan's insular location, formidable languarge barriers, and a cultural gulf between Japanese and outsiders, particularly Westerners. Continued use of the notorious term *gaijin* (foreigners) stems not so much from a myth of superiority as from ignorance. Insular feelings are enhanced by the tightly knit homogeneity of the Japanese. (Minorities, as will be noted, are "invisible.") If there is any group of people about whom generalization in terms of national character can be made, it is the Japanese.

The Group

In traditional Japan, the group or collectivity received the primary consideration. The concept of group duty, rather than individual right or privileges, determined the basis for action and behavior.[4]

As one might expect, attachment to the group has remained an attitude shared by the majority in surveys conducted by the institute. Ezra Vogel, the Harvard sociologist very well known in Japan, has referred to this phenomenon as "a continuing value consensus." Even in the earliest survey taken after the Occupation (1953), some 68 percent of respondents found contentment to be contingent on the eventual success of the group writ large, the family-nation of Japan.

Japanese group consciousness, however, has not been all that simple to describe, in part because of the apparently paradoxical emphasis on individual merit and achievement as measures of success. Japanese resolve this paradox by often focusing on achievement generated through groups (the family, the firm, or the ministry), which most often remain objects of devotion and determine rewards.

In earlier institute surveys, respondents were given hypothetical situations: The president of a firm is seeking to hire a new employee and on the qualifying examination, the highest grade is won by an unknown; the second highest, by a relative of the president. It is made clear that either candidate will perform satisfactorily in the position. Whom should the president hire? Almost three-quarters of those polled advised hiring the candidate with the higher grade.

Respondents were given another choice: You may have a department chief who sticks to the work rules, who seldom demands overtime, but who does nothing in matters not connected with the job; *or* you may

Japanese wedding: Exchange of nuptial cups (*sansankudo*)

have a chief who is much more flexible about rules, who occasionally demands overtime work, but who also looks after you personally in matters not connected with the job. Over 80 percent preferred the latter kind of department head.

Put in a slightly different fashion, a survey conducted for the Prime Minister's Office tested the idea, "mixing with superiors outside work is desirable," among youth (eighteen to twenty-four years old) in eleven different countries. In the first poll (1972), the ratio of workers who supported the proposition was highest in Japan (73.4 percent), ranking ahead of developing countries (South Korea, the Philippines, Yugoslavia, and Brazil). In the third poll (1983), the ratio in Japan declined (to 63.3 percent) but remained relatively high compared with developed Western countries (the United States, France, United Kingdom, West Germany, and Switzerland).[5]

This is not to say that Japanese never feel a certain malaise because of membership in the group. Indeed, they often do, and they recognize that discontent arises from the pressures of conformity and from the tight web of obligations to the group. In a Prime Minister's Office poll (1983), although some 80 percent of respondents said they "had a worthwhile purpose in life," only 46 percent replied that it was because of their work, either "definitely" or "to some extent."

Conformism

Stories embracing Japanese tradition—Kabuki theater and even contemporary literature and some films—are sprinkled with descriptions of the group-centered society in which human feelings (*ninjō*) and bonds of moral obligation (*giri*) are central concerns. It is widely believed even in Japan that in the postwar period such traditional norms, along with filial piety (*oyakoko*), suffered notable' declines among attitudes shared by Japanese.

It is often pointed out that there has been a shift from elaborate relationship-oriented pronouns of address to the more neutral "you" and even "I." Results of surveys made by the Institute of Statistical Mathematics in the early post-treaty period (1953–58), however, showed that although a majority (60 percent) used new language, many still preferred traditional modes of address.

The 1970s saw a remarkable resurgence of traditional values. In the 1973 survey, given a multiple choice, respondents most often singled out filial piety (on 63 percent of replies) and the need to repay a moral debt (*ongaeshi*, which appeared on 43 percent of the replies). In the 1978 survey, these precepts again ranked first and second.

The trend can be explained in yet another way. In the earlier surveys, those who believed that it is best to conform to society's established mores accounted for no more than about one-third of the respondents. In the sixth survey (1978) the total rose to 42 percent. In parallel, the immediately postwar norm of respect for individual rights and freedom remained relatively strong (from 40 to 45 percent of replies) through the early 1970s. Such respect ran far higher (60 percent) among younger respondents (the twenty- to twenty-four-year-old group). The prevalance of such opinions inspired fear among elders that the younger generation was widely infected by privatism (called *mai homism*, "my homeism," by the Japanese).

In 1978, to the surprise of the pollsters, respect for individual rights among youth plummeted (to 46 percent) and remained at a level below that of some of the older Japanese (the thirty-five- to thirty-nine-year-old group). Such results signified neither the acceptance of "individualism" (in, for example, American style) nor antipathy to "individual rights" (in Japanese style).

Indeed, a government White Paper on Youth (1986) reflected continuing adult ambivalence with respect to the lifestyle of young Japanese people. The paper surveyed youth: how they lived, what they owned, what they thought and felt, and what they believed should be accomplished for their future. There were signs, the report stated, that both adults and youth had become "tired" of certain trends: permissiveness (including overfeeding children) and the "examination hell" and the coupled desire to have youngsters enter the "best" high schools and universities. "The value of individualism that was supposedly obtained with postwar democracy has ever since been more a negative than a positive thing," the paper concluded.[6]

Lifetime Commitment and Hierarchy

Traditional and transition, orientation and behavior come together in pledges of lifelong loyalty to the group in Japan. Cultural historians trace contemporary concern for the group back to the rural villages and Tokugawa towns.

The traditional household (*ie*) and its preindustrial urban equivalent (the *iemoto*) tended to blend ascriptive factors such as place of birth, kinship, and status with newer criteria such as ability, perseverance, and achievement. Other groups, pseudo-kinship bands, were dedicated to traditional pursuits, such as tea ceremony, Kabuki, *sumō* wrestling, flower arranging, *koto* playing, garden design, cooking and (today) photography, architecture, and certain academic disciplines. One key to the guildlike structure of these latter groups was the master-disciple relationship, wherein an experienced elder (*sensei*) offered training, support, and protection in return for followers' permanent obligation (*on*).

The group in contemporary Japan is, of course, a far more differentiated structure. The Japanese anthropologist Nakane Chie has identified the process whereby individuals with different "attributes" become members of a group, which exerts lasting pressure on the members to conform to its "frame." This insight is best understood by means of an everyday illustration.

At the Stonier Graduate School of Banking at Rutgers University, American in-service trainees tended to respond to introductions with name followed by profession. An American was a "banker," and it was only later that one learned that he was a product of the Harvard Business School, a former intern at the Bank of America, and then an employee of CitiBank. In summer institutes for Japanese bankers, who were at Rutgers to learn about American business and to upgrade their English conversation, the trainee usually responded to introductions with family name followed by affiliation with a firm. Yamamoto was a "Sumitomo salary man," a product of Tokyo University, and only later did one surmise that he was a cashier, a loan officer, or an investment counsel. The Americans were from a society mainly marked by attributes, whereas the Japanese came from one dominated by frames.

Attitudes and actions merge in a system of lifetime employment (at least in large corporations). One must be careful discussing this system because recent studies have shown that the practice may not have been as widespread as thought and that it is under erosion. It is true, however, that many Japanese still think in terms of loyalty to one firm for a lifetime and the security provided thereby.

In large firms, institutional practices underline the ideal: Employees are recruited directly from schools, with an emphasis on their adaptability and willingness to conform. They receive a variety of group benefits: bonuses paid twice a year equal to salaries for three to five months (these are the sources of much of household savings), company housing in the early years, and family benefits (for education, marriages, child care, illness, and death).

It perhaps follows logically that Japanese groups are characterized by the seniority system (*nenkō joretsu*). Seniority, like group consciousness, however, is not easily described. Seniority does not rest simply on age. More often a career is measured from the date of graduation from a university; in a firm, from date of employment; and, as has been noted, in the Diet by the date of first election to the legislature. In the first decade or so, advancement is strictly according to seniority; in the middle years, merit and achievement may affect status.

There are numerous examples of vertical group orientation in organizations other than the firm. In general, seniority rules in the party faction (*habatsu*), finance circle (*zaibatsu*), and bureaucracy (*kambatsu*), particularly in the higher civil service. One can also detect horizontal layers (according to seniority), like school cliques (*gakubatsu*), running across the vertical frames. In the ministries above the rank of bureau chief, over 80 percent of positions are filled by Tōdai (University of Tokyo) graduates. Tōdai graduates also preside over some 300 large corporations.

The Rule of Consensus

A fourth pattern of behavior, linked to basic Japanese orientation, has received rather widespread attention. The distinctively Japanese method of making decisions involves circulation and consensus, which are rooted in the principle of collective responsibility.

In the Japanese system, the respected elder leader is where he is because he has, over time, come to know and trust subordinates and they, him. Leaders (the "legitimizers") obtain all the views of subordinates before announcing a final decision. The objective of this arrangement is to retain harmony within the group. The style of the system is of the patron-client relationship in a vertical chain, with mutual reliance ("interpenetration") on horizontal levels. It has been recognized that the middle level in such a system is in a critical position, strategically located to accommodate different views from the lower half, to transmit them to the upper half, and then to implement policy from top to bottom.

To offer an illustration drawn from politics, in the postwar era the Japanese prime minister, though an important figure, has probably been less powerful than a determined and energetic U.S. president. The prime minister must seek consensus among powerful elites (the legislators, various LDP factions, and components of the world of finance—the *zaikai*) and at the same time avoid too frequent and outright confrontation with the opposition. The ideal is a public consensus, not a decision resting on a slim majority, say of 51 to 49.

Thus Japanese view with amazement the frequent and public squabbles in Washington between department heads and the president, among cabinet members, and between the majority and minority parties. The diplomatic historian Hosoya Chihiro has concluded, "The Japanese Prime Minister seems to perceive his role in the accommodating of different views of the various leaders and agencies concerned, so as to secure their maximum support, rather than asserting his own priorities in order to lead his nation in a certain direction."[7]

On occasion Japanese have argued that their system is more democratic than its U.S. counterpart in that the views of a variety of members of the society are taken into account before a decision. On the other hand, Japanese consensual decisionmaking has been called less democratic than the American system in that it allows less interaction and feedback between the makers of decisions on the one side and the legislature, the interested public, and various interest groups on the other.

Shortcomings in the Japanese system have been identified with an elaborate documentary process called *ringisei*. The exercise includes a system (*-sei*) rooted in a practice called "reverential inquiry" (*rinshin*) into the intentions of a superior. The byproduct is a memorandum (*ringisho*), which is painstakingly disseminated from a desk up to a bureau chief, then down to another desk and up to another bureau chief, until all the bureaus have been covered. The document then proceeds to division level, across-and-up, across-and-up to departments, and eventually to the government minister (or the firm president). Any amendment, of course, requires a filtering through the whole process again. At best, the system guarantees an eventual consensus and guards against any nasty surprises.

Japanese have roundly denounced the process as dominated by a "feudal lord system" (*tonosama hōshiki*) at its worst and, when a horizontal veto is absolute, as marred by the interference of "compost emperors" (*hiryō tennō*). An equally colorful figure of speech has been applied by Maruyama Masao to the academic world, describing its extreme compartmentalization according to specialties and relative lack of collegial ties. He has written that this helps produce "our octopus-pot culture."

Somewhat more seriously, the above examples should explain why a meeting between Japanese and, let us say, Americans will, as often as not, produce a minor crisis rather than an agreement. The American side will be somewhat blunt, impatient, openly adversarial, and expressive of the wish to reach a decision quickly so as to catch the United Airlines flight back to New York. The Japanese will be circumspect (often called "devious" by Americans), expressive of the wish to reach a lasting compromise, polite in offering another cup of tea to break the silence, and quite willing to prolong the exchange until a sound human relationship (if not a decision) is reached.

SOME EFFECTS

Although some traditional norms have remained firm or have been revived in Japanese minds, there have also been changes of attitude detected by opinion surveys. Because of the rising level of material well-being in the 1960s, however, Japanese have become more conservative. Their conservatism does not apply just to the political arena and, as was indicated in Chapters 5 and 7, does not necessarily imply support for a specific party.

The Family

. . . The decrease in members per household and the simplified realationships in families are symptoms that the whole family structure has been undergoing considerable change in recent years; the change is expressed by the term . . . *kaku kazoku* (meaning "change to the nuclear family") in Japanese.[8]

Observers agree that there is no society in the world in which family life has been regarded with such satisfaction as in Japan. At the height of transition into the postindustrial era in the 1970s, Institute surveys revealed that the level of satisfaction with the family rose steadily from 80 to 86 percent of respondents. The level satisfaction with Japanese society at large was much lower.

There was, of course, a significant impact on the family from structural change in the whole society. The graying of Japan resulted in a sharp increase in the number of nuclear families consisting of only the elderly. By 1977, according to the White Paper on National Welfare, such units numbered almost 2 million (5.6 percent of the nation's households). Nevertheless, about three-quarters of Japan's over-sixty-five population were still sharing homes with their married children (as compared with about one-third in either Europe or the United States).

Japanese experts, however, have referred to "quasi-household sharing," where the older live with the younger family members in the same house but with strictly separated living space and household accounts; and to "quasi-household separation," where the elders live separately but at a distance short enough to allow daily contact. In one case, an elderly woman reflected happily, "I live upstairs and my son and his family live downstairs, close enough to carry soup back and forth without its getting cold." In another case, a young wife commented, "The privacy of each family is well guarded. I'm free from psychological warfare with my mother-in-law."[9]

The family, by the way, has reached far beyond the modern nuclear unit, even beyond the traditional extended group of relatives, to affect all of Japanese society. Thus, in the nineteenth century leaders spoke of the "family-state," in order to encourage national unity. For a long time in Japan, one has been able to refer to the "familistic company" or to "familistic management." And until recently, the family has served as the main substitute for state-provided social security.

The Salary Man

The *sarari man* is awakened by an alarm-radio at 6:00 A.M. Getting into a suit, he shaves with an electric razor. . . . Glancing at his digital watch from time to time, he washes down his ham-and-egg breakfast with coffee. Putting his shoes on at the entrance, he dashes out to the nearest bus stop. Transferring to a subway, he reaches his office downtown in a ferro-concrete building and sits at his desk all day, carrying on his assigned tasks. For lunch he goes down to a basement restaurant for a hamburger steak. On his way back from work he might stop by a favorite bar with his colleagues for a shot of whiskey (Suntory or, if celebrating, Johnny Walker). On the way home, he reads an evening newspaper, printed with the latest computer technology. He checks on his favorite baseball team. For dinner at home he has a mini steak from Australia and a beer (Kirin). After dinner, he enjoys a Beethoven symphony played on a stereo, while drinking black tea. Promptly at 11:00 P.M. he watches NHK news and the weather forecast on television. Before retiring, he goes to the family Buddhist altar and thanks an Indian-derived deity that he is one hundred percent Japanese.[10]

The snapshot presented above is not, of course, a portrait of the "average" Japanese male head-of-household. It is, rather, an impression of the suburban, white-collar salary man. Nor is it a caricature, for most Japanese men regard the *sarari man* as the ideal to which they aspire.

In 1962, according to surveys conducted by the Prime Minister's Office, 76 percent of all respondents had come to consider themselves to be members of the middle class; after 1979, over 90 percent of all Japanese claimed that status. Three of four who did, however, differentiated the ideal from reality. Further evidence that middle-class living did not match the ideal was provided by a life insurance survey of the late 1970s that revealed that respondents defined middle-class status as an individual family in a separate house with a small garden and an annual income of ¥5 million (about $22,725). By the late 1980s, although per capita GNP approached the equivalent of $20,000, as has been noted, many Japanese were still living in "rabbit hutches."

The rosy outlook generated by the era of high growth nevertheless was not entirely eroded. Despite sporadic recessions, as of 1978 almost half the Japanese thought the future looked bright (compared with 38 percent in 1973).

Paradoxes found in data on attitudes in the postindustrial society are to be expected. Thus a preference for acquiring wealth as life's goal has never been high (about 17 percent of respondents from 1958 to 1969) and actually declined (to 14 percent in 1978).

"To live according to one's own taste"—this would seem to be a form of individualism to the outsider—became the top goal (about 40 percent of respondents). "Acquiring fame" has ranked low in Japan (only about 2 percent of respondents). Perhaps this attitude reflects the firm belief among Japanese that real power always resides behind the screen.

Women

> The two things that have become stronger since the end of the war are women and nylon stockings.
>
> (*postwar saying*)

Even Japanese men would readily admit that the stability of the family depends in large part on the attitudes and behavior of women. Tradition dictates that the Japanese female be socialized to be content and that she act to accommodate the wishes of males. It is also known that the status of women in traditional-rural and in preindustrial-urban Japan was often higher than was apparent. Today, of course, women play a quite different role in the family and in society.

A glance at some of the data will lay to rest many of the myths inherited from tradition and may correct outsiders' misconceptions. First, it is apparent that women have come to be vital parts in the service-centered, postindustrial society. In surveys conducted by the Prime Minister's Office in 1984, about 55 percent of all female respondents (almost 78 percent of unmarried women) held a job. Women in their forties accounted for 70 percent of those employed. The usual pattern has been for women to work prior to and immediately after marriage; to remain at home to rear the child(ren); and to return to work after the children have entered school.

Second, in the tertiary sector many jobs, particularly those held by women, have been temporary. During a recession (January 1987) when unemployment reached a record 3 percent, female workers were the first to lose their jobs, and a larger proportion of the female than the male labor force was out of work.

Third, in a recent opinion survey, over 70 percent of all respondents expressed general satisfaction with Japanese family life; even more significant, almost 76 percent of the women were quite happy with their lives.[11]

Nonetheless, the lot of the woman in Japan is not enviable in the opinion of a resident expert, Iwao Sumiko of Keiō University. She believes that of the women in four cultures, Swedish women have the highest status, followed by American women, then French, and finally Japanese. And yet, to the despair of outside feminist observers, Japanese women's attitudes seem to indicate a steady shift to "neo-traditional" standards.

Responses have been sought to the query, "If you could be born again, would you rather be a man or a woman?" In 1958, only 27 percent of female respondents wanted to remain women. This proportion rose steadily to 36 percent in 1963, to 48 percent in 1968, to 51 percent in 1973, and to 52 percent in 1984. (It would, of course, be equally accurate to note that almost one-half of women in 1984 did not want to remain female.)

Professor Iwao has offered three possible reasons that Japanese women would choose to be reborn the same sex despite their low status. Women are wary of the pressures outside the household to which Japanese men are subjected. They share the satisfaction felt by most Japanese in family affairs. They truly enjoy the childrearing, if not the housework.[12]

Finally, to cite a readily observed factor, in the crucial realm of making decisions in the family, the Japanese wife is in a paramount position. In almost 74 percent of all Japanese families, the woman of the house determines the flow of daily living expenses (as compared with about 20 percent of American families). In short, *okusan* manages—perhaps she has always managed—the household budget. Financial firms in Tokyo are well aware of this fact of life and target housewives in order to win savings and investment accounts.

In the extreme division of labor within the family, the wife regards herself as the center of the household enterprise. It should be added that the Japanese woman, once having satisfied herself in the care of her children, is increasingly interested in an outside career. Nor is the nuclear family an ideal over the whole lifespan: a majority of women expect to live with or close to their children and eventually to become the central figure in a semi-extended family.

Youth and Education

> Japanese society places a high value on interpersonal relations and ability to cooperate with others. The Japanese believe that being a member of a well-organized and tightly knit group that works hard toward common goals is a natural and pleasurable experience. Schools reflect this cultural priority.[13]

To speak of youth in Japan is to talk first about education. As in other countries, a specific pattern of values unique to the particular society is transmitted through enculturation in schools. As Nobuo Shimahara has pointed out in his study of Japanese education, such value orientations "are tenacious, persistent, suborganic principles that resist pressures for change brought about by the institutional transformation of such a society."[14]

College Entrance Examinations. In January each year, over 300,000 high school seniors and graduates take uniform college entrance examinations, which are held in over 200 universities and testing sites throughout Japan. The first battery of tests covers Japanese language and general science; the second, English, mathematics, and social studies. In March, successful candidates then sit for supplemental examinations at universities of their choices. It would be difficult for outsiders to grasp how much these examinations dominate the lives of youngsters (and their mothers) from nursery-school days onward.

The examinations and the education geared toward them have efficiently incorporated both certain traditional, particularistic (Japanese) norms and certain modern, universalistic (worldwide) orientations essential for success in the postindustrial era. The system has above all been instrumental in the recruitment of the Japanese elite. Ezra Vogel's study of city life in the early stages of the growth era (Tokyo in the 1960s) identified arduous preparation for the examinations as a kind of rite de passage through which a young person (more often a male) proved that he had the ability and stamina to become a salary man.[15]

The impact of the entrance tests has been felt not only by the select elite, but by *all* Japanese, who are educated in schools geared to the examinations. The system has identified early not only the future white-collar leadership, but also the necessary "gray-collar" technicians, accountants, supervisors, blue-collar laborers, and unskilled help. Japanese education cultivates a relatively unaltered pattern of values among all Japanese and can therefore be called conservative.

Entrance tests, denounced as the "infernal examination hell," do, however, serve as key devices for socialization into the new, postmodern society as well. Emphasis during the adolescent years on universalistic achievement orientations—mastery of mathematics, acquaintance with modern science, familiarity with languages—gives way in adulthood to a focus on particularistic norms—loyalty to the group, a sense of hierarchy, the desire to conform. The great divide is the college entrance examination.

Education Reform. The Japanese educational system, not well known abroad but widely admired for its achievements, does, of course, have some problems. The demanding curriculum is difficult for slow learners, and three factors compound this problem. First, Japanese teachers believe that effort alone can compensate for differences in ability. Second, there are few provisions for diagnosis of learning disabilities and for remedial assistance. Third, schools use automatic promotion, which increases the pressure on those who fall behind.

There is some student violence in Japan, but much less than in the United States. It includes confrontation with teachers, violence against other students, and vandalism. The unique problem of bullying (*ijime*)— intimidation and tormenting of individuals, especially by bands of students—is probably the darker side of group action. Bullied students usually have some characteristics which set them apart. In the first half of 1985, some 1,000 students were involved in bullying incidents which required police interventions. A Tokyo hotline received over 1,300 calls in six months, mostly from lower secondary-school students.

These and other problems led to a national consensus that Japan's famous education system needed drastic reform. In one of the summit meetings between President Reagan and Prime Minister Nakasone in 1983, the leaders agreed to mount a cooperative study of each other's educational practices.[16] Nakasone selected education reform to be a major plank in his political platform.

In April 1987, the National Council on Education Reform submitted recommendations designed to help the educational system cope with a new information-oriented society and to facilitate internationalization of Japan's education in order to meet the demands of the twenty-first century. The council called for simplified textbook screening and a new district system, which would give parents and children greater freedom in choosing schools. Earlier, the Council has presented proposals stressing the need for establishment of Japanese identity among children and, at the same time, for respect of their individuality.

The New Breed. It should be noted that the council's recommendations did not signify that "individualism" (in, for example, an American style) had suddenly become socially acceptable in Japan. In fact, a paradox shows up in an expression, "the new human breed" (*shinjinrui*), applied to youth. First used as a derogatory description of an affluent, apathetic generation, the term took on the connotation that such youth were cool, self-possessed, and sure winners. Having grown up without first-hand knowledge of depression or war, young Japanese came to reflect diversity, if not individuality. With first-hand experience with consumerism and even affluence, they became members of the worldwide "now" generation.

An illustration of youth's "very own" principles was the response by some 80 percent that they were increasingly sensitive to friends' demands for privacy. In traditional Japan this luxury scarcely existed, whereas youngsters in the 1980s demanded privacy as an essential. Although the new breed were the children of change, they were also conservative to the core and wished to protect the status quo.

Religion

The literature on modernization is rich in generalizations on how societies shift to an emphasis on achievement as opposed to ascriptive status and on rationality (science) as opposed to religion (or myth). Even in the Tokugawa period, the Japanese began to move toward a more secular, highly pragmatic outlook.

Modern, and particularly postwar, Japanese attitudes shifted steadily toward pragmatism and scientism. Yet, an interesting aspect of all postindustrial societies (including Japan) is the continued search for transcendent truth to offset the torque of change. Many "new religions" appeared in postwar Japan. Among them was the Sōka Gakkai, which (as has been noted) gave rise to a political party, the Kōmeitō.

Nevertheless, when confronted with the blunt question "Do you have any personal religious faith?" three out of four Japanese responded (1973) that they did not. In the same poll, however, almost 80 percent replied that a "religious attitude" is important! Data have indicated that the Japanese, like other people, grow more religious as they grow older. On the other hand, in the twenty- to twenty-four-year-old group (1973 survey) only 8 percent of the respondents were religious-minded; later (1978) the portion doubled to 17 percent. Of course, many otherwise secular Japanese continue to mark changes in the life cycle with Shintō ceremony or with Buddhist ritual.

"Japaneseness"

From perusing results of attitude surveys and descriptions of Japanese behavior, it is possible to surmise that the Japanese entered the 1990s not unalterably conmitted to any ideology. Contemporary culture has maintained substantial elements of traditional beliefs: emphasis on the family, the group, and the nation-family, rather than on the individual; stress on discipline and duty, rather than on personal freedom; reliance on distinctions of status (measured by merit as the Japanese define it), rather than on social equality.

To assume that interest in Japanese national character is confined to outsiders—that residents of Japan take it for granted—would be an error. There has been, according to one Japanese editor, an "introspection boom" in the nation in recent years. A variety of terms such as "theory of Japanese culture," "theory of Japanese society," and "theory of national character" has appeared in media. Singled out in the next chapter is, in loose translation, "theory of Japaneseness" (*Nihonron*).

There are two reasons for the immense popularity of *Nihonron* in Japan. First, it articulates the need for self-identification amidst the tsunami of change that for a century has washed over Japan. Second, even in its mildest form the theory assumes Japanese uniqueness. To become truly "internationalized," as former Prime Minister Nakasone put it, the nation must explore its own roots. Thus the venerable uniqueness argument can serve to offset any lingering perception of inferiority. Because *Nihonron* is one side of the yen coin, it is best to discuss it further, along with the other side known as internationalization (*kokusaika*), in the next, the last, chapter.

9

Kokusaika:
The Internationalization
of Japan

Perhaps no country has offered the fascinating variety of international experience that Japan has enjoyed and suffered in modern times. Rapid, often urgent, shifts in international status have left deep impressions on Japanese images of the outer world and outsiders' images of Japan.

THE VARIOUS JAPANS

Reared as a kind of distant younger cousin of the great Middle Kingdom of China within the venerable Confucian family of nations, Japan was the first member of that family to modernize and to seek status in the Western nation-state system on its own terms. Thereafter the Japanese did not completely forget the sinic tradition. Even before the 1978 normalization of relations with mainland China, it seemed inevitable that Japan would seek to return to some of its main roots, which were Chinese.

In the early 1940s, the Greater East Asia Coprosperity Sphere was, in its more idealist aspects, an attempt to reject the Western nation-state system, to rebuild a hierarchical family of nations, and to construct a new order for Asia, in which Japan would play the self-appointed role of elder brother. Some have said that what the Japanese had previously failed to accomplish by force, they brought about in the postwar period by economic clout.

In the late 1940s Japan was prepared, with U.S. aid, to return to the peaceful life. Since the Japanese were among those with the most vivid experiences of the horrors of war, they had come to believe conclusively in the idea—before it was incorporated in their new constitution—that they should renounce force as an instrument of national policy. In the 1950s, 1960s, and 1970s, largely with the help of the United States, they were reintroduced to the harsh realities of diplomacy and force. A Japanese observer sadly expressed the idea that his people had moved from a prewar fear of dependence to a postwar fear of independence.

In the era after the peace treaty, Japan's diplomacy has been denounced for being pragmatic. It could equally be praised for pursuing what Japanese call an omnidirectional course, unencumbered by the need to press for any dogma or ideology. Japan has been a pioneer in an experiment to try to conduct diplomacy within a multistate system without large components of force. This effort will doubtless fail, unless the world system is changed radically, but the ideal can scarcely be faulted.

Meanwhile Japan, although it was still *in* Asia, began to feel less and less *of* Asia. The island nation had been the first to emerge from the old Confucian tradition, the first from East Asia to pass through the threshold of modernity, and among the first in the world to enter the postindustrial stage. (In recent years, however, Japan's modernization has been matched by the explosive growth of the four "little tigers," South Korea, Taiwan, Hong Kong, and Singapore.)

Consider for a moment the claim often made by Japanese that their nation remains an Asian power. Certainly this is the broadest kind of generalization. More specifically, Japan is the first member of the Confucian family to test the process of modernization with tools inherited from the old orders in East Asia. Even more specifically, today Japan is an island nation anchored in Northeast Asia. As such it is a regional power with political, economic, and security concerns directed to the continent and to the Korean peninsula. Japan is located on the western edge of the Pacific basin (a link to the West) and on the northern edge of Southeast Asia (the nearest link to the developing world). Tokyo is now a world city. Its connections are aerial routes, mapped on great circles (the fastest links to the developed world). Japan is a U.S. ally— sometimes regarded as reliable, sometimes as reluctant—with bonds forged from lively educational and cultural exchange, limited security links, reciprocal investment, and commerce. Regardless of attitudes on either side of the Pacific bridge, Japan and the United States enjoy the largest cross-ocean trade in the history of the world.

THE ECONOMIC SUPERPOWER

As a postindustrial power, Japan has grown economically closer to the advanced industrial democracies of the West than to the newly industrializing countries (NICs). In June 1979, Tokyo hosted the first summit conference of advanced industrial nations to be held in Asia. (In May 1986, Japan welcomed a summit conference in Tokyo for the second time.) Japan had joined those nations that accounted for about 55 percent of world economic activity.

To the Tokyo Summit

Beginning in 1974, efforts had been made in the Organization for Economic Cooperation and Development (OECD) and the International Monetary Fund (IMF) to encourage a greater awareness of the linkages among the economic policies of the industrialized democracies. The first steps were taken at the Bonn summit in July 1978. All the countries with a balance-of-payments surplus—Japan, Germany, France, Italy, and the United Kingdom—pledged to achieve somewhat higher growth rates. Japan was to attempt, at the same time, to reduce its large global trade surplus. The United States, the major deficit country, was to give priority to reducing inflation, to promoting exports, and to developing a domestic energy policy that would reduce its demand for imported oil. All these powers failed to reach their goals during 1978.

Despite the fact that in June 1978 OECD members renewed their 1974 pledge to avoid trade restrictions, measures to restrict or to cartelize trade proliferated. Protectionist measures and import quotas adopted by the United States were of particular concern to Japan. The European Community also increased restrictions on the import of steel and electronic products. Japan undertook no new import restrictions but continued to be severely criticized for failure to pursue more vigorously liberalization of trade. By the end of 1978, the amount of world trade under some sort of restriction was higher than the proportion the year before. It had reached a total of 3 to 5 percent of the total trade (estimated at from $30 to $40 billion).[1]

For the Japanese, the conference of leading industrial democracies, which met June 28–29, 1979, was like a semicolon in a long sentence. The first clause described how the Japanese, over the course of a century, had struggled to modernize and, over three postwar decades, had worked hard, sacrificed, and saved in order to become members of an exclusive group. Japan had attended such conferences before, but the 1979 meeting was dubbed the Tokyo summit. It was the first held outside the Western world. The second clause, beyond the punctuation, began to describe the dangers of becoming an economic superpower. Certainly for Japan, the Tokyo summit was an historic turning point marking a shift from one set of problems—those of development and modernization—to another set of problems—those of the postindustrial era.

It was immediately apparent that Japan did indeed belong to the elite group of nations. Nevertheless, the Japanese delegation, representing a nation with an economy larger than that of West Germany, for example, was less confident than the Germans. A closer look at Japan's status may explain the ambivalence.

On the eve of the Tokyo summit, Japan enjoyed the highest rate of growth among the advanced democracies. (The rate of 9.1 percent shown in Figure 9.1 was an estimate, annualized in nominal terms.) This growth was dependent on a high rate of exports. The Japanese feared trouble from the Western powers on this account and felt quite vulnerable.

Among the summit powers, Japan was the second-lowest consumer of energy (measured by per capita use of oil) (Fig. 9.1). Japanese were using less than half the energy consumed by the profligate Americans and yet felt far more vulnerable. About 99 percent of Japan's supplies of petroleum products had to be purchased abroad, the total bill in 1978 accounting for 30 percent of Japan's imports (or about $23.4 billion).

Japan's was the lowest inflation rate among those of the seven summit nations. Like the West Germans, the Japanese preferred to avoid overheating their economy. This strategy brought criticism from the Americans, who wanted a higher growth rate in Japan. Although the strength of the yen abroad reflected the truly remarkable growth of productive capacity, its weakness at home demonstrated that the real purchasing power of Japanese currency continued to decline because the consumption standard was kept low.

The lowest unemployment percentage also belonged to Japan, but because many firms chose not to lay off workers when production dropped, this rate was a poor indicator of the level of business activity. At the time of the summit, a stubborn recession continued and under-employment was widespread.

By far the most sensitive issue at the Tokyo summit—in fact, at each subsequent conference—was the reaction of Western partners to Japan's towering trade surplus. Although, as has been noted, Japan's current-account surplus was only 1.7 percent of GNP in 1978, Japan came to the summit with a surplus estimated for 1979 at $10.5 billion (as compared with an estimated U.S. deficit of $11 billion). The effects were volatile, for they indicated the beginning of a profound shift in the structure of the international economy.

The Japanese carefully prepared for the important, but routine, conference scheduled for June 1979. The planned agenda included issues of growth, inflation, employment, trade, relations between the North (developed nations) and South (developing nations), and currency equilibrium. Then on June 26, the very eve of the summit, the OPEC powers

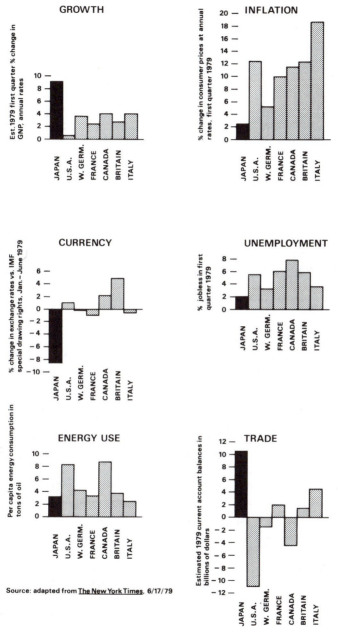

FIGURE 9.1 Major powers at the Tokyo Summit (1979)

met in Geneva and announced higher oil prices. The effect, particularly on Japan, was electric. In a flash the Tokyo conference became the energy summit.

Beyond the Summit

Meanwhile, within Japan it became apparent that Japan had already entered a lower-growth era. In June 1979, immediately after the summit, Prime Minister Ōhira predicted that, despite the OPEC oil price increases, Japan would achieve the revised goal of a 6.3 percent growth rate. In August 1979, the Economic Council offered a new seven-year plan (1979–86) calling for an inflation-adjusted growth rate of 5.7 percent per annum.

In fact, in 1979 Japan's growth rate fell to 5.2 percent, in 1980 to 4.2 percent, and in 1981 to 2.9 percent. In the fourth quarter of 1981, exports softened, causing the first actual drop in real GNP since the aftermath of the 1973 oil shock.

Although in many ways the pattern in both the United States and Japan was similar, there was one glaring difference. For the year 1981, Japan, despite lower growth, had a 54 percent increase in exports; the United States had a drop of 14 percent. It was the beginning of a new phase in the relations between the two Pacific powers.

THE TOKYO-WASHINGTON AXIS

Since the peace treaty, which took effect in 1952, there have been two major issues that have dominated the relations between Japan and the United States. The first, security policy, was particularly important early in the postwar period. In the 1950s and 1960s, Japan's domestic politics resonated with shrill cries about the U.S.-Japan security treaty of 1951. Renewal of the treaty amidst riots, in the summer of 1960, threatened to tear up the delicate roots of democracy transplanted from prewar Japan. In 1970, however, automatic renewal of the treaty was something of an anticlimax, owing in part to the U.S. decision to return Okinawa to Japanese administration. The antagonism toward the treaty is evidence that the Japanese have slowly moved to overcome their "postwar fear of independence." As of 1990, however, despite continued friction with the USSR in the North Pacific, relaxation of East-West tension in Europe has relegated the security issue to second place.

The other issue, trade, has also been significantly affected by Japan's escape from dependence as well as its subsequent emergence as a superpower in the economic realm. Ever since the Tokyo summit of 1979, the U.S.-Japan relationship has revolved mainly around persistent disputes over the trade balance between the two states. A subtle tie between one issue, trade and investment, and the other, the security problem, has developed. By the opening of the 1990s, although the United States continued by treaty to provide a one-way security guarantee to Japan, the latter was in a real sense financing not only this bilateral arrangement but also underwriting much of the U.S. economy. This anomaly, as Kenneth Pyle has pointed out, has transformed the U.S. troops in Japan into a mercenary force defending Japan in return for Japanese capital.[2] On both issues (trade and security), Japanese and U.S. interests call for negotiation of a new Tokyo-Washington agreement.

Trade and Investment

The Japanese might well be forgiven for confusion over the various U.S. positions with regard to Japan's economic performance. In the rapid-growth era of the 1960s, Washington complained to Tokyo about excessive growth rates and aggressive export drives. In the slower growth era of the 1970s, Washington urged Japan to maintain a higher growth rate in order to stimulate Japanese imports from the United States. In the late 1970s, although Japan's balance-of-trade surplus had begun to decline, Americans remained irked by Japan's trade policies.

A senior business counselor, former Undersecretary of State George Ball, described U.S. attitudes in a wry fashion. He noted that Japan not only enjoyed a trade surplus with the United States, but that the type of goods it imported made the U.S. appear as an underdeveloped country. Japan was taking large quantities of American food and raw materials, while manufactured goods accounted for only two-fifths of its imports from the United States.[3]

In such a climate, specific differences tended to exacerbate the general trade problem. In the early 1970s, the focus was on textiles (to Japanese, a politically sensitive issue). Then it was the U.S. steel trigger price mechanism (a thinly disguised trade barrier). Just before the Tokyo summit, the Japanese allowed a dispute to escalate by their slow, deliberate negotiating style, thus endangering the success of President Carter's visit and the summit itself. The issue had to do with procurement: specifically, purchases made by Nippon Telephone and Telegraph (today's semi-public NTT) and reciprocal access to the Japanese market by U.S. companies like American Telephone and Telegraph (AT&T). At the eleventh hour, negotiators in Tokyo hammered out a compromise based on principles of reciprocity.

During fiscal 1978–79, just prior to the Tokyo summit, the U.S. economy recovered relatively rapidly from the recession of the mid-1970s, and generated a demand for imports that Japan's exporters stood ready to satisfy. Because of lagging recovery in Japan, however, a reciprocal demand for American goods did not develop.

Various reasons have been cited for the relatively poor U.S. performance in Japan: subtle barriers to imports, difficulties of access to the Japanese market, a traditionally weak U.S. effort to promote exports and, in some cases, the inferior quality of U.S. products (as judged by consumers in Japan).

In fact, few Americans understood the trade statistics. Faced with rising unemployment, automobile workers in Detroit took to bashing in the hoods and windshields of fuel-efficient Japanese cars. Some managers refused workers the right to park their Toyotas, Nissans, and Hondas in plant parking lots. (Later, a member of the U.S. Congress smashed a Japanese-made television set on the steps of the Capitol.)

It was significant that automobiles provided the new Reagan administration its first experience in trade negotiations with Japan. In May 1981 in Tokyo, a "non-agreement" emerged from the "non-negotiations" between the U.S. Trade Representative, William Brock, and Japanese officials. They agreed to hold Japan's auto shipments to the United States to 1.68 million units in the first year and to allow that total to increase in the following years by only 16.5 percent of whatever growth occured in the U.S. market. Quota legislation in the U.S. Congress was then tabled, although proposals for restrictions continued to surface.

By the late 1980s, the voluntary limit on Japan's auto exports stood at 2.3 million units, but the curb had become academic. Japanese car exports to the United States were running about 10 percent below the ceiling as automakers increasingly assembled their cars in that country. For example, in 1989 the Honda Accord became the best-selling model in the United States. Honda's Marysville, Ohio, plant assembled 363,668 Accords and Civics during that year and by then, most engines and parts, previously imported, were subcontracted from U.S. firms. A two-door Honda was actually being exported to Japan. U.S. Trade Representative Carla Hills warned European Community nations not to adopt import curbs on these cars: Japanese cars built in the United States, she stated, were to be treated as American cars!

Meanwhile, one after another point of friction marked the two partners' trade relations. In 1982, Washington singled out for attack twenty-two Japanese import quotas on fishery and agricultural products. In March 1983, Tokyo adopted a package of import-promotion measures, including simplification of standards and certification (regarded by Americans as invisible barriers). In August 1984, the United States and Japan initialed an agreement to increase Japanese imports of beef and citrus products. This step was bitterly opposed by the Central Union of Agricultural Cooperatives (Zenchū). Earlier that year, then–Vice President George Bush was in Tokyo, urging Japan to liberalize its financial markets. In April 1985, Prime Minister Nakasone in Tokyo unveiled a "substantial" package of measures designed to reduce the trade balance. The United States countered with demands for negotiations on imports of pharmaceuticals, telecommunications facilities, and electronic products. At the end of July 1986, Washington obtained a semiconductor pact assuring a "fair share" of the Japanese market. In March 1987, however, Tokyo was disturbed by Washington's continued investigations of Japanese barriers to imports of superconductors. In April, President Reagan reluctantly imposed 100 percent penalty duties on the $300 million in Japanese computers, television sets and power tools shipped to the United States annually.

Despite the long and arduous negotiations, Japan's exports to the United States (about one-third of the Japanese total) continued to be a problem. In 1984, those exports soared 40 percent and the trade surplus reached $60 billion. In 1985, the surplus declined but still totaled $49.7 billion (about one-third the total U.S. trade deficit). In the fiscal year ending March 31, 1987, the trade surplus rose again to $52 billion. Although it declined over the next three fiscal years, it was still at $42.7 billion (March 31, 1990).

In May 1989, on the recommendation of the trade representative, Carla Hills, President Bush named Japan an "unfair trader" under provision 301 of the 1988 Omnibus Trade Act. Washington sought liberalization of restrictions on imports of superconductors, lumber products and intellectual properties (patents, copyrights and trademarks). Tokyo refused to negotiate under the "unilateral" threat of 301 and appealed under the General Agreement on Tariffs and Trade (GATT). During Prime Minister Kaifu's first visit to the United States in September 1989, he promised Bush that Japan would nevertheless give priority to the trade surplus.

Troublesome as these trade negotiations were to both the Japanese and to the Americans, their impact was not nearly so dramatic as that of a byproduct of the surplus. As has been noted, by the end of 1985 Japan had become the world's largest creditor nation. Profits from trade were being converted into investments abroad. In 1986, Japan's holdings of foreign securities reached a total of $145.8 billion (86 percent of which were in the U.S. market). During the year 1989, according to a Washington-based nonprofit group, Japanese investment in the United States (including direct and portfolio holdings) rose 11 percent to reach a total of $1.98 trillion. Investment in American real estate, however, fell about 11 percent. The Japanese had begun to turn to Western, and even Eastern, Europe for real estate opportunities.

Again, the statistics were abstractions to the average American, who was much more impressed with high-profile, symbolic purchases by the Japanese. In September 1989, the Sony Corporation bought

Columbia Pictures, described as the "soul" of American motion picture
and television culture. And in October, the Mitusbishi Estate Company
bought a 51 percent stake in the Rockefeller Group, which operated
fourteen buildings in New York City including world-famous Rockefeller
Center. The Japanese explained that "these were not hostile takeovers."
Nevertheless, late in October the Japanese government warned business
leaders to avoid "conspicuous" purchasees of property in the United
States. At the same time, the Finance Minstry indicated that it would
give favorable tax treatment to Japanese companies which made phil-
anthropic grants to their host communities. As for the Americans, it
made no difference that the Canadians, the British, and the Dutch were
leading purchasers of securities and real estate, especially in New York.
Japan was in the process of "buying up America."

At this point the two powers entered a novel—and a very dangerous—phase of trade negotiations. In the summer of 1989, Washington
opened talks with Tokyo on what Americans called the Structural
Impediments Initiative. Even the Japanese media dubbed these negotiations "SII." Few Americans were, however, aware that the formal title
in Japanese was Japan-U.S. Structure Conference (*Nichibei kōzō kyōgi*),
with no references to "impediments."

In brief, the U.S. negotiators argued that Japanese should increase
their spending (thus giving a boost to imports) and decrease their savings.
(In other words, to paraphrase a song in *My Fair Lady*, "Why can't the
Japanese be more like us Americans?") They also called for an increase
in public expenditures, especially on the neglected Japanese infrastructure,
from the current 7 percent to 12 percent of GNP. The U.S. side boldly
demanded repeal of Japan's Large-Scale Retail Stores Law, which made
it difficult for Japanese chains (and, of course, foreign retailers) to enter
neighborhoods dominated by Japanese mon-and-pop establishments.
Moreover, the U.S. team argued that Japan's Federal Trade Commission
should raise penalties for violations of antimonopoly regulations, especially bidrigging in the construction business. Finally, U.S. negotiators
sought changes in Japan's price mechaniam (which caused the gap
between domestic goods and imports) and called attention to skyrocketing
land prices.

The Japanese responded in kind. The prime reason for the U.S.
trade deficit, they argued, was the U.S. budget deficit and higher interest
rates. Americans should increase their savings, decrease their spending
and, especially, cut down on the number of credit cards carried by each
individual. (In other words, "Why can't the Americans be more like us
Japanese?") The United States should design an industrial policy, with
particular attention to the generation and promotion of exports. American

businesses are bound by concern for short-term profits and quarterly balance sheets, and a lack of long-term staying power.

Tokyo pointed to the comparatively poor education record in the United States and singled out inadequate worker training. In identifying American structural barriers, as Fred Bergsten (of the Institute for International Economics) explained, the Japanese were applying "reciprocal *gaiatsu*" (external pressure). Beyond the formal Japanese government position, there were oblique hints made by critics that Tokyo could, if need be, sell its supercomputers to the USSR. A more serious threat, informally aired, was that Japanese could reduce or even cut off investment in U.S. Treasury securities, thus withdrawing much of the underwriting of the U.S. budget deficit.

As usual the Japanese were dilatory, this time in recognizing the importance of their trade balances, and they interminably dragged out negotiations. Their pinpointing of American spending habits struck at the very psyche of the United States. Repeated mention of the Washington budget deficit highlighted the severe paralysis in U.S. policy implementation. The Americans, on the other side, in what Japanese regarded as an arrogant stance, called for a minor revolution in Japan's society. Objective observers expressed doubt that success could result from such negotiations. Forgetting the lessons of history, Washington ignored the difficulties Tokyo always faces when confronted with a deadline and a metronome ticking.

In February 1990, when Prime Minister Kaifu had registered his modest victory, President Bush telephoned Tokyo to offer an invitation (Japanese media reported that Kaifu was "summoned") to meet in California to discuss the trade balance. In order to enhance the possibility of success in SII negotiations, the president removed Japan from the list of nations subject to provision 301. In May, a Japanese interim report and a joint report were prepared toward a final statement due June 25–26, on the eve of the Houston summit conference scheduled for July.

As usual the Americans were quite specific: they asked Japan to increase public expenditures to 9 percent of GNP; they requested quarterly reports on antimonopoly violations; and they demanded copies of all of Tokyo's administrative guidance directives. As usual the Japanese responded in very general terms: The Finance Ministry cited an IMF report that recommended Japan's maintenance of a surplus in order to help pay for reconstruction in Eastern Europe and for aid to the developing nations.

Interestingly enough, a survey of fifty vernacular newspapers in Japan revealed that although the press expressed resentment of U.S. pressure, a majority of the papers saw a net gain from reform inspired by SII. The *Chūnichi-Tokyo Shimbun* noted that Japanese decisions by traditional consensus made dramatic reform difficult. *Kōbe Shimbun* commented that SII was actually a domestic problem, which had to be solved immediately. Japan's Wall Street Journal, the *Nihon Keizai Shimbun*, urged the LDP and the JSP to be aware that SII could support the interests of Japanese consumers, particularly those in large cities. *Mainichi* called SII "an entirely new two-way experiment" but warned that "sweat and toil" were required on the U.S. side as well. *Yomiuri* predicted that structural reform "will produce a stronger Japan." One editorial slyly remarked that SII would make of Japan a more formidable competitor for the United States.[4]

To the social scientist, whether Japanese or American, the impediments dispute was of significance beyond the perennial details of trade negotiation. In posture and style, both sides revealed divergent traits of national character.

A New American Sport: Japan-Bashing

During the Pacific war, of course, anti-Japanese and anti-American sentiments were common currency in the two nations. For several decades after the Occupation, however, extreme views were relegated to demagogues on both sides of the Pacific. U.S. citizens became quite knowledgeable about the "new Japan" and generally held the Japanese people in high regard. As late as 1978, favorable views were sustained. In April that year, a special survey of attitudes of Americans toward Japan was commissioned by Potomac Associates, of Washington, D.C. Some 22 percent of respondents chose "very favorable" to describe their view of Japan, and another 50 percent selected "somewhat favorable," for a total of 72 percent positively disposed. (Only 17 percent list "somewhat unfavorable" or "very unfavorable" and 11 percent were undecided.)

By 1990, it was apparent that trade friction had begun to erode the favorable attitude. According to a survey conducted in the United States for Japan's Foreign Ministry in the period January to March, 44 percent of U.S. respondents still regarded Japan as a "dependable ally," but this figure was down 6 percent from a previous poll. Moreover, some 40 percent considered Japan to be an "unreliable ally."

Indeed, former U.S. Ambassador James Hodgson saw a distinct shift in U.S. attitudes. "Today," he wrote, "we see the demagogues reflecting public opinion rather than attempting to lead it." The ambassador was most alarmed by one survey that revealed that 63 percent of the Americans polled believed that the United States had more to fear from Japan than from the Soviet Union.[5] Obviously, the Washington-Moscow thaw was being matched by icy conditions in Washington-Tokyo trade relations.

The Japanese in turn became alarmed over an orchestrated attack mounted by a band of critics known, both in Washington and in Tokyo, as the "revisionists." These foreign commentators covered a wide range of opinion. In brief summary, the majority expressed the view that Japan is unique and peculiar, a riddle difficult to solve by ordinary (that is, "Western") reasoning. To these writers the Japanese are players who follow completely different rules. In contrast, a minority argued that Japan actually has a familiar power structure, but the wielders of influence carefully camouflage the system, hiding it from Japanese and foreigners alike. In either case, they charged Japan is in the world but not of it.

The first volley in the revisionists' assult was fired by a well-known journalist, the late Theodore White, in his 1985 article, "The Danger From Japan," in the *New York Times Magazine*. White's anti-Japanese and anti–Nationalist Chinese feelings had first been aired in his best seller, *Thunder Out of China* (1946). Japanese were shocked by his argument that, although the United States had won the military conflict, Japan was on the way to victory in the "economic war."

In the late 1980s, the voices of revisionists became more strident. Another journalist, James Fallows, Washington editor of the *Atlantic*, carried forward the military terminology with essays on "Containing Japan" and "How to Conquer Japan." In a separate book, he elaborated on what he called the polar opposites of the Japanese and U.S. societies. He stressed the importance of the United States' "Being Abnormal," in contrast with other nations' "Being Normal." In a strange way, he thus matched the Japanese penchant for claiming uniqueness in their society. And in his doctrine of containment, he borrowed in bizarre fashion the terminology of an almost forgotten scheme designed by George Kennan and applied to the Soviet Union.

Other revisionists included former American trade negotiators like Clyde Prestowitz, who had understandably become frustrated by lengthy negotiations with the Japanese. He denounced Tokyo's tendency to view the world in terms of "us versus them," but succeeded in reinforcing the trend in his book. The well-known economist Peter Drucker joined the chorus by charging Japan with practicing "adversarial trade." Even Chalmers Johnson, an old Japan hand who had impressed Americans and Japanese with his analysis of *MITI and the Japanese Miracle*, came to argue that Americans must either imitate Japanese thinking and behavior or defend themselves against expansion of Japanese influence. The Japanese were perhaps most shaken by the position taken by Michael Armacost, who succeeded the fatherly Mike Mansfield as U.S. ambassador. In a speech delivered in July 1989, the official U.S. government representative echoed the revisionist line: Americans, he said, were growing more conscious of Japan's economic challenge as the Soviet threat receded.

The managing director of Keidanren in Tokyo, Nukuzawa Kazuo, pointed out that old friends who had cooperated with Japan and enjoyed seeing its reconstruction and growth had been nominated by critics to be members of the "chrysanthemum club." Nominees included "Japanologists" (a favorite target of revisionists), Washington consultants with Japanese clients, and former U.S. diplomats who had served in Tokyo.

Two examples may be cited to show that criticism of Japan was not the monopoly of Americans. A British journalist, Bill Emmot, wrote *The Sun Also Sets*, which attracted attention in Japan. Karl van Wolferen, a Dutch reporter with considerable experience in Japan, set out to defend the "revisionists" (he always put the word in quotation marks). The author placed himself among the "realists." In a widely reviewed book, the author referred to Japan as a "stateless state," something short of a sovereign nation. He argued that Japan has displayed all the trappings of a parliamentary regime but has not developed a real democracy, and that power has been diffused among groups and rests in a "system." Many of his readers concluded that Japan was even more an inscrutable "enigma" than they had thought.[6]

The Japanese Response

Many Japanese, in letters to editors and in television comment, reponded to U.S. pressure in the SII negotiations and to foreign criticism. The reaction was, however, marked not so much by anger as by a characteristic sense of vulnerability. The government in Tokyo was, of course, defensive but circumspect.

The most notable countercharges emanated from two prominent Japanese. Morita Akio, the influential chairman of the Sony Corporation and widely known in the United States through television interviews, mentioned possible friction in an informal autobiography published in English.[7] In this book and in numerous stump speeches in Japan and in the United States, he reflected his own international experience, especially his residence in the United States. He did not, however, hesitate in singling out the United States' structural weaknesses. He sounded familar themes: Americans should save more, borrow less, think more in long-range business terms, and study harder. The United States should complain less about Japan and start emulating it.

The other leader was Ishihara Shintaro, novelist and member of the Diet, who had mounted a token candidacy for the position of LDP president when Kaifu was selected. He was little known in the United States until he coauthored a book with Sony's Morita. The title was *A Japan That Can Say No* (*No to ieru Nippon*). The Foreign Ministry and even the LDP were concerned when they learned that a pirated English translation was being circulated in Washington and in Silicon Valley. Nonetheless, Ishihara vetoed several offers to publish an authorized English-language version and Morita concurred, stating that the book was "intended for a Japanese readership."

What sparked controversy was Ishihara's free use of the term *minzoku* (race)—even the erroneous idea of a "pure Japanese race"—to explain his nation's success. Morita later publicly disassociated himself from this concept. Although Ishihara's bite seemed less than his bark, when he visited Washington in January 1990, his opinions stirred up old arguments. He refused to back down on the notion that U.S. use of atomic bombs, at the close of the war, was racially motivated. As a result, he claimed, Japanese have been forced to "pay" for an "illusion," the U.S.-Japan security treaty. Racism, he added, has also been the root cause of trade friction.

It is interesting to note that Ishihara was willing to accept the results of SII negotiations that would aid Japanese consumers. In an interview in Tokyo in April 1990, Ishihara stated: "We are witnessing the advent of an era of individualism in which the business structure which has served the interests of business, politicians and bureaucrats at the expense of those of consumers must be abolished even by taking advantage of external pressure." He did argue, however, that Japan should not give in to U.S. "blackmail" under provision 301.

The U.S.-Japan Security Treaty

Fortunately, as has been noted, the security arrangement between Tokyo and Washington came to play a less controversial role in the 1980s and in 1990 than it did in previous decades. It offered fewer points of friction than did the trade disputes. On June 22, 1990, President Bush talked by telephone with Prime Minister Kaifu in a formal celebration of the thirtieth anniversary of the security pact that was renegotiated in 1960 and renewed in 1970 and in 1980. There were several reasons why the commemoration was only a matter of pleasant protocol.

At the time of the anniversary, Washington continued to maintain 50,000 military personnel in Japan (out of a total of 120,000 in the Asia-Pacific region). Faced with the fact that peace was breaking out all over Eastern Europe, however, the Americans were talking of reducing their forces worldwide, including those in Japan and Korea. That budgetary considerations were supplanting the strategies of the cold war was clearly illustrated by U.S. Defense Secretary Dick Cheney's visit to Tokyo in February 1990. Cheney asked for continued qualitative improvement of Japanese forces and assumption by Tokyo of all (rather than the current 40 percent) of the costs of maintaining U.S. forces in Japan. Ishikawa

Yozo, the director general of Japan's Defense Agency (JDA), gave a tentatively favorable response.

Otherwise, Japanese opinion was mixed. Matsunaga Nobuo, Japan's respected former ambassador to the United States, cautioned that it was premature to revise the security treaty. East-West relations in Europe, he stated, should not be confused with the U.S.-Japan alliance. A former foreign minister, Okita Saburo, agreed and argued that communism in Asia is basically different—more heterogeneous—than in Europe. Asai Motofumi, however, blamed the treaty for inhibiting an independent Japanese foreign policy. (Asai, a veteran of twenty-five years in the Foreign Ministry and then a professor at the University of Tokyo, circulated his views in a book on foreign policy.)[8] In any case, Japanese security policy—once rooted in the Tokyo-Washington alliance—increasingly concerned the establishment of an independent defense posture in the world and, as such, is better described below.

JAPAN AND THE WORLD

Over a century ago, reponding to foreign pressure, the Meiji modernizers initiated a program of "civilization and enlightenment" in order to produce a "strong defense, a rich country." After the Pacific war, U.S. modernizers, in cooperation with the Japanese, set about reconstruction, economic reform, and "democratization." In the late 1980s, a conservative coalition of Japanese leaders was sponsoring additional change, more modernization, and "restructuring" of the economy. Although restructuring appeared to be a domestic issue, in fact it was drafted under extreme foreign pressure. Additional steps taken with the Structural Impediments Initiative promised once again to alter the face of Japan.

As Japan entered the last decade of the twentieth century, there were no black ships off its shores. (There were still gray ships flying the stars and stripes moving in and out of Tokyo Bay, it is true.) There were no "men of high purpose" (*shishi*), armed with two swords and striding the streets in search of "foreign barbarians." (There were modern samurai–*sarari man* types, armed with brief cases and laptop computers.) The friendly but firm American occupationnaires, who had come to "reorient" Japan, had long since gone home. (A few members of the "chrysanthemum club" returned now and then, as diplomats, consultants, or observers; and there were quite a few blue-eyed, long-nose business types striding the streets of Marunouchi.) And yet it was—to borrow the immortal phrase of that U.S. cultural hero and baseball star, Yogi Berra (a familiar name in Japan)—"*déjà vu* all over again." In newfound summit status, in trade disputes with an ally, in the search for an

All Nippon Airways Hotel, Roppongi, Tokyo

appropriate defense policy, and in the search for identity, Japan was once again facing a profound transition.

The Search for a Defense Policy

Despite some claims to the contrary, up through the 1980s most evidence pointed to a Japanese determination to build only *defensive* capability. The constitutional and legal limitations on Japan's armaments have been discussed (in Chapter 5, above). Suffice to say that no government in Tokyo has been willing to move openly beyond the assumptions that the Self-Defense Forces (SDF) are exclusively for defense and that only after an attack would arms be used. Japan's stance has excluded any policy of containment and has permitted no "strategic offense" for purposes of defense.

The SDF is equipped with only conventional weapons: no long-range missiles, no long-distance bombers, and no attack aircraft carriers are in its arsenal. In addition, every postwar government and leader has been explicit in upholding the three nonnuclear principles: not to produce nuclear weapons, not to possess them, and not to allow such weapons to be brought into Japan. (There is every reason to believe, however, that the United States has occasionally and surreptitiously ignored Japan's sensitivity in these matters by bringing nuclear weapons into or carrying them through Japan without formal consultation.)

As a matter of fact, as of the late 1980s the demands placed on an independent Japanese defense policy had been modest. Tokyo had met Washington's request for burden sharing, particularly when both felt a stronger Soviet threat. When the United States had taken action not solely in response to that threat and had acted in an unpredictable manner, for example, in the last stages of the Korean conflict, the Japanese had become uneasy. Similarly, the U.S. engagement in Vietnam, which could have also become a thrust against China, had been worrisome.

After the Korean conflict reached stalemate, Japan normalized relations with South Korea in 1965. Through the 1980s, Japan had no formal diplomatic contact with North Korea. Indeed, while the Korean peninsula continued to provide a flashpoint between Seoul and Pyong-yang, Tokyo continued to provide a base—and even strategic cooper-ation—to the United States and South Korea. The Team Spirit '83 and Team Spirit '84 maneuvers were among the largest exercises held in East Asia since World War II. Recently, with détente between Washington and Moscow, tension even in the Korean peninsula has decreased. There were widespread predictions that Japan, China, the USSR and the United States would move to encourage the two Koreas to reconcile their differences. Tokyo's potential role was enhanced when, in May 1990,

South Korea's President Roh Tae Woo visited Tokyo and received from Emperor Akihito an unambiguous apology for Japanese treatment of Koreans before 1945.

After the Vietnam conflict wound down, Japan welcomed the U.S. rapprochement with Peking, and Tokyo normalized relations with China in 1978. Although the Japanese, like the Americans, were disturbed over the Tiananmen Square repression in June 1989, Tokyo was very cautious in its complaints and Japanese businesses moved back into China more quickly than American ones did.

Paradoxically, the U.S. withdrawal from Vietnam was a relief to Japan in one sense and a concern in another. By the late 1970s, Japanese media openly predicted the steady reduction of the U.S. military presence in East Asia. The impression was reinforced by announcement of the Nixon Doctrine in Guam, whereby Asians were henceforth expected to defend Asians. To this was added President Carter's campaign pledge to eventually withdraw troops from Korea. (After the Tokyo summit of 1979, Carter visited Seoul and indefinitely suspended the pledge.)

In 1981, at the National Press Club in Washington, Prime Minister Suzuki Zenkō unveiled a plan whereby Japan would provide "defense coverage" along sea lanes up to 1,000 nautical miles from the home islands. Ono Seiichiro, executive director of the Research Institute for Peace and Security, expressed the opinion that Japan "must cooperate to maintain world peace, especially to promote the stability of the Asia and Pacific region."

In May 1990, Director General Ishikawa of Japan's Defense Agency (JDA) went on a tour of Thailand, Malaysia, and Australia. It was the first visit by a Japanese military official since the war. A specific Japanese interest was identified by J. N. Mak, an analyst with the Institute of Strategic International Studies in Malaysia. An increased Japanese role in Southeast Asia, he stated, specifically to protect the Malacca Strait, was "inevitable." The passage has been, of course, crucial to Japan for the import of energy from the Middle East.

There was one other point of friction that continued, even at the beginning of the 1990s, to encourage Japan's defense planning. Beginning in 1976 and through the mid-1980s, Japan had by cabinet decision limited defense spending to 1 percent of GNP. Meanwhile, as the JDA director general noted, the USSR was then spending 12–14 percent of GNP on defense. The nearest evidence of Soviet expenditure was to be found in what Japanese called the "northern territories." The term referred to a cluster of islands in the southern Kuriles, islets claimed by Japan but occupied by the USSR since 1945. Tokyo normalized relations with Moscow in 1956, but no peace treaty has been concluded because of this territorial dispute. In the 1980s, Tokyo became alarmed over Soviet deployment of division-strength forces in the islands, of aircraft carriers in the northern Pacific, and of Backfire bombers and SS-20 missiles in Soviet Asia.

By 1990, there was widespread hope in Tokyo that a long-planned visit by President Mikhail Gorbachev would begin to reduce, if not eliminate, the northern territories dispute. Director General Ishikawa of the JDA admitted that the Soviets were in the process of reducing forces, even in Asia. He argued, however, that Japan should continue to improve the quality of its defense forces.

Some time before, a group of international relations specialists, headed by Inoki Masamichi, had drafted a report for Prime Minister Ōhira Masayoshi. The fundamental change of the 1970s, they wrote, was the end of U.S. superiority, both military and economic. This new group of military realists called themselves "internationalists" (*kokusai-ha*, compared with older established realists called "domesticists," *kokunai-ha*). They saw as an alternative to fading U.S. power not unarmed neutrality but increased defense effort.

By the early 1980s, a sharp shift in public attitudes toward defense was also apparent. The consensus held that Japan, in the so-called Peace Constitution, had never relinquished its right of self-defense. Those who agreed with the proposition, "the SDF is necessary," rose to a record 82.6 percent of respondents (only 7.5 percent disagreed; 9.9 percent had no opinion). There were, however, contradictions in survey results. Two-thirds of those polled remained opposed to constitutional revision to clarify the status of the SDF; an equal proportion was opposed to using maritime forces (MSDF) to protect sea lanes. By the mid-1980s, on the other hand, there was a clear majority in support of increased force levels.[9] In short, at that time general threat perceptions and security consciousness had grown, but a rather powerful undercurrent of pacifism remained and limited planning for defense.

Tokyo, like other world capitals, was shocked by Iraq's invasion of Kuwait on August 2, 1990, when the expected cold-war peace was shattered. Stung by criticism that Japan's contribution to the U.S.-led international effort to contain the Gulf crisis had been "too little, too late," on September 14 the Kaifu government doubled the $1 billion in aid promised to multinational forces in Saudi Arabia. In addition, Japan pledged an additional $2 billion in loans to Egypt, Jordan, and Turkey.

A more significant development was the introduction of legislation designed to facilitate sending Self-Defense Forces personnel to the Middle East. The bill would create a provisional U.N. Peace Cooperation Corps consisting of civilian experts, medical officers, and members of the SDF. It was quite clear that Prime Minister Kaifu was not prepared to seek revision of the Self-Defense Forces Law, in effect to test the constitutional prohibition of use of SDF personnel overseas for combat.

In the early 1990s, forces were pulling Japan in different directions. Relaxation of East-West tension affected Japanese thinking, easing the pressure on Tokyo to rearm more rapidly. The West was, however, expecting Japan to exercise influence equal to the nation's economic status. An increasingly independent Japanese security policy was to be expected, if it was not inevitable. The search for a viable policy was only a metaphor illustrating Japan's general search for identity.

The Search for Identity

Early in 1990, a new movie pitted Japan's perennial favorite "Godzilla" the monster, against "Biollante" in a story of international intrigue. Director Omori Kazuki changed the traditional story line by including foreigners—Arabs, Britons, and Americans—as well as English-speaking Japanese in the plot. It marked, one reviewer noted, the internationalization of Godzilla and was unusual in that both foreigners and Japanese turned out to be "the bad guys."

It would be easy to dismiss the film as being, in effect, a comic strip in a different form. In Japan, however, such a characterization would not obscure the film's importance because comic books (*manga*) play a much wider role there than in other countries. Japanese *manga* have been used to teach languages, to instruct citizens in the intricate workings of government, to simplify arcane economic data about the trade balance, and to spin out rather racy soap operas. Godzilla is now giving a bow to a serious effort (or the latest fad), internationalization (*kokusaika*).

Internationalization was the topic of public comment earlier, but probably turned into a "boom" with Prime Minister Nakasone Yasuhiro (and later with Takeshita and Kaifu). In a ceremony held in Tokyo in February 1987 marking National Foundation Day, Nakasone referred to the need to transform Japan into an "international state" (*kokusai kokka*). As an internationalist, the prime minister said, "It is necessary to understand the diversified principles concerning life and different cultures of various foreign countries and to promote mutual cooperation." As a confirmed nationalist, however, he doubled back to stress the necessity "for Japanese, in order that they might move to a more cosmopolitan view, to know themselves and Japan." Thus, as has been noted above (in Chapter 8), Nakasone returned to a discussion of "Japaneseness" (*Nihonron*).

Regardless of the debate over "Japaneseness" or the fad favoring internationalization, the facts reveal a profound change in Japan's noted inward-looking nature. Space allows mentioning only a few indicators. For example, Japanese are now going abroad in droves as tourists, as scholars, and even teachers, and as business personnel forced to manage in a foreign culture.

In 1986 the Transport Ministry presented an overseas travel promotion campaign aimed at doubling the number of travelers abroad in the 1987–1991 period. In that year, some 5.5 million Japanese traveled abroad. With the yen's dramatic appreciation, the number increased to 6.8 million in 1987, to 8.4 million in 1988, and to over 10 million in 1989. Popular destinations included the United States, South Korea, Taiwan, France, Singapore, Switzerland, Australia, and Canada. Total travel expenditures, according to the Japan Travel Bureau (JTB), were expected to exceed a record ¥16.4 trillion in 1990. According to the Finance Ministry, increased travel abroad has resulted in a $2–3 billion annual reduction in Japan's current account surplus.

The effects of travel, however, run far beyond current-account statistics. When they return, tourists obviously shed some of the traditional Japanese insularity. Nor are the effects confined to tourists. It has been estimated that at any one time there are some 35,000 Japanese nationals working and residing in the metropolitan New York region, many of whom live there for several years. There are as many, or more, in California. More than one commuter train out of northern New Jersey has been dubbed "the Orient express."

Travel gains have not flowed one way, the JTB noted. A record 3.35 million foreign tourists were expected to visit Japan in 1990. The Osaka Flower Exhibition—the latest in a series of spectacular world's fairs mounted in Japan—was planned to host 20 million guests, Japanese and foreign, during the year.

Changes effected by the Finance Ministry between 1985 and 1988 allowed foreign security firms to engage in brokerage and in the trust market (which taps lucrative pension funds) in Tokyo. Widows in longevity investment clubs soon saw unfamiliar names popping up in listings by the Tokyo Stock Exchange (TSE): Barclay's, Chase Manhattan, CitiCorp, Bank of America, Deutsche Bank, Smith Barney, Salomon Brothers, and First Boston. Despite the backbreaking costs of maintaining a branch office in Tokyo, many foreign firms assigned hundreds of personnel to Japan.

In 1988, the TSE became the world's largest in terms of market capitalization of listed shares. The TSE lost the rank temporarily in the period January–March 1990, when the 225-issue Nikkei stock average plunged to unfamiliar lows. The sharp drop itself revealed clearly the restructuring of the economy, deregulation of financial markets, and internationalization of business. Tokyo portfolio managers, like their counterparts in New York, fretted over the invasion of "program trading." Meanwhile, the public saw plainly that heretofore all-powerful bureaucrats exercised considerably less leverage in controlling the liberalized financial markets.

One other illustration may be offered to demonstrate the progress of internationalization. In the 1980s, as a result of intense trade friction, Japan steadily increased its funding of official development assistance (ODA). In 1989, Japan surpassed the United States to become the world's largest donor of foreign aid, providing nearly $9 billion in ODA, according to the Foreign Ministry. The total was down almost 2 percent from the previous year in dollar terms but increased 5.6 percent in yen terms.

For two decades, Japan's ODA has been supplemented by a small but significant program called the Japan Overseas Cooperation Volunteers (JOVC). This "peace corps" has dispatched nearly 9,000 volunteers between twenty and thirty-five years of age to developing countries in Africa, Asia, and Latin America. Forty-four nations have received or are receiving JOVC personnel and some thirty private Japanese companies have institutionalized temporary leave for JOVC volunteers.

Japan as a Case Study

Prior to Japan's emergence as a superpower in the postwar era, the nation's development was often measured against historical principles derived from experience outside Japan. For several centuries prior to the nineteenth, Japanese themselves believed that the clearest expression of most universal truths was to be found in China's history and in the Confucian classics. If Japanese society on occasion strayed from the Chinese exemplar, it was because domestic arrangements were, if not distorted, unique.

In the nineteenth century, when Japan emerged from the old orders of the Confucian family of nations, the society was again often measured against standards derived from non-Japanese experience. An assumption, persistent among foreigners and often made by the Japanese themselves, was that what happened after 1868 was the Westernization of Japan. On some occasions, sophisticated Japanese promoted the idea (through the ballroom societies of Tokyo and the red brick–clad Ginza) in order to enable their nation to escape from the Western-imposed unequal treaties. Thus, although some appeared to adopt the West as a model, they were fiercely protecting the essentials of Japanese lifestyle.

Modernization theory, as postulated by foreigners and by Japanese, was certainly enriched by the inclusion of data from the experience of Japan. Strive as they did to use "non-Western" development to universalize their disciplines, however, the modernizationists too followed a kind of convergence thesis. Once again the construct and much of the terminology were derived from a model historically and culturally determined by Western experience. Merit and accomplishment, differentiation of roles and specialization, the unseen hand of the market, the civic culture and participatory democracy—all these were propositions inherited from Western political sociology. In the contemporary policy arena as well, American assumptions behind the Structural Impediments Initiative were that Japan must follow the American model.

Even before Japan's explosive growth of the 1960s, for the first time it was hinted that Japanese experience too might provide a guide for development. In a May 1979 speech to a United Nations conference in Manila, Prime Minister Ōhira sounded the theme again, suggesting that Japan's modern history might offer a model to developing nations, mainly those in Southeast Asia.

Current conditions in developing nations have, however, been quite different from those found in feudal (protomodern) Japan. Many of these countries have suffered from a lack of national identity, have been burdened with politicized religious feuds, have been handicapped by shortages of skilled business leaders and of trained labor, and have plodded along with low levels of education. Of those which apparently succeeded—for example, the four "little tigers" (South Korea, Taiwan, Hong Kong, and Singapore)—it is an interesting but moot point whether they have emerged because they shared the Confucian tradition or because they have displayed their own talents, different from Japan's.

Even more surprising to Japanese, it was suggested that the nation's experience might prove useful to economically advanced countries as well. The way was paved by futurologists who, writing in the early 1970s, predicted the emergence of the Japanese superstate by the 1980s. Some even claimed that the twenty-first would be Japan's century. In the late 1970s, Japanese became very excited over the appearance of a book by a Harvard University professor and skilled observer of Japan, Ezra Vogel, who nominated the nation as one of the world's most effective industrial democracies. With a title, *Japan as Number One*, that intrigued Japanese readers and led them to overlook the subtitle (*Lessons for America*), the translation quickly became a best-seller in Japan.[10]

Specifically, Vogel expressed the belief that Americans could learn a lot from Japan's success if they were only willing to pay attention. He offered a masterful analysis of Japan's continuing modernization, the country's efficient organization, its skill in adapting technology, its patience in marketing, and its well-educated and disciplined labor force. A good deal more dubious was his assumption that Japan might serve as a model for other postindustrial powers.

At the close of the 1980s, Japanese and outsiders alike were struck by the publication of an essay in a U.S. neoconservative newsletter; no doubt the Japanese were intrigued by the fact that it was written by a U.S. government official, Francis Fukuyama, who was obviously an American of Japanese ancestry. He had just completed service as deputy director of the State Department's policy planning staff. The article's title was "The End of History?"[11]

In brief summary, Fukuyama borrowed the Hegelian concept that history is a dialectical process with a beginning, a middle, and an end. It is propelled by overcoming contradictions between thesis and antithesis. In effect, the author also followed Hegel in believing that history in the narrower sense is the history of ideology. This includes the history of thought about "first principles," including those related to social and political organization. "The end of history, then," Fukuyama wrote, means not the end of worldly events but the end of evolution of thought about such first principles." That evolution came to rest in liberal-democratic states descended from the French and American revolutions, he concluded.

Fukuyama stated that there have been two major challenges to liberalism: fascism and communism. Fascism was destroyed as a living ideology in World War II. Communism's challenge has been more severe. Surely, however, the class issue has been resolved in the West, he continued. "In Japan, the fact that the essential elements of political and economic liberalism have been so successfully grafted onto unique national traditions and institutions, guarantees their survival in the long run." Maoism, far from being the pattern of Asia's future, he argued, has been discredited and has become an anachronism.

The Fukuyama thesis was promptly publicized in Japan and was dismissed as being too simplistic. As a reviewer pointed out, the end of history (including conflict) is scarcely in sight so long as the world continues to be racked with ethnic strains.[12] The Soviet Union, with or without Marxism, has yet to solve such problems. Over a century after its civil war, the United States still wrestles with problems of race relations.

The Japanese have turned this issue of race relations upside down. The majority, still confortable in the myth of monoethnic nation-family, has yet to be completely tested in the ability to handle "outsiders." Japanese treatment of minorities living *in* Japan, for example, some one million Koreans or persons of Korean ancestry, is not exemplary. Two-thirds of this minority are second or third generation and were born, reared, and educated in Japan; yet they are still regarded as "non-Japanese." Until recently, they have been fingerprinted as "aliens." It is no wonder that even friends of the Japanese wonder if they are really ready for internationalization.

It is necessary to add only one or two points on Fukuyama's ideas. The author of "The End of History?" would have done better, even on his own terms, to have borrowed the felicitous phrase of Daniel Bell, "the end of ideology." To the extent that the Japanese are well educated, well informed by vigorous media, scientifically and technologically sophisticated, live in an open society, and are, therefore, highly pragmatic and suspicious of dogma, they are perhaps witness to the end of ideology. It is quite another point whether Western "liberalism" has been "grafted" onto Japanese tradition or tradition has been pruned to produce the shape of the postmodern trees of democracy and capitalism.

In either case, across the East China Sea the Tiananmen Square incident and aftermath demonstrate clearly that even the age of ideology is not at an end quite yet. Nor is conflict completely subdued—witness the Gulf crisis in the Middle East.

The end-of-history idea is, in fact, a disguised version of the convergence thesis. It parallels the belief by many business leaders in the West (and some Japanese tycoons) that "the market" throughout the world will eventually displace ideology and even politics. It is a contemporary form of economic determinism, which comes close to matching Marxism, Leninism, and Maoism in naivete. Its main fault is, however, that in declaring the end of history, it neglects history and culture. These *are*, of course, what make Japan unique. The Japanese, in turn, must learn that China is unique, so is the United States, and so are developed and developing nations. Even tentative solutions to the perils and promise of the postindustrial age worked out by the Japanese should be of comparative interest, regardless of whether they were once imported in part or are immediately transferable abroad.

Indeed, a world without history—with all nations essentially alike— is unlikely. Fortunately, too, for such a world would be a dull planet.

Notes

CHAPTER 1

1. Komatsu Sakyō; *Nihon chimbotsu* [Japan sinks] (New York: Harper & Row, 1976).

2. Kawabata Yasunari, *Yukiguni* [Snow country] (New York: Knopf, 1957).

3. Using the entry "Disasters—Natural" in the *Britannica Book of the Year* as a sample for a six-year span, one can test the incidence of typhoons, deluges, and earthquakes: 1983—May, Northern Honshū: earthquake, tsunami, 58 killed. 1984—June, Kumamoto: deluge, landslide, 14 killed. 1985—February, Omi: snowy landslide, 10 killed; July, Numazu: typhoon Irma, 19 killed; August, Kyūshū: typhoon Pat, 26 killed or missing. 1986—January, Niigata: snowy avalanche, 13 killed; March, Tokyo: snowstorm, 13 killed. 1987—July, S. Korea and Kyūshū: typhoon Thelma, subsequent storms, severe flooding. 1988—none. *Encyclopedia Britannica, Book of the Year* 1984, 1985, 1986, 1987, 1988, 1989 (Chicago: Encyclopedia Britannica).

4. For one of the first uses of the term, see Jean Gottman, *Megalopolis: The Urbanized Northeastern Seaboard of the United States* (New York: The Twentieth Century Fund, 1961). With a Greek root, the term received further currency through the systematic research at the Athens Center of Ekistics, under the world-renowned planner C. A. Doxiadis. See also C. Nagashima, "Megalopolis in Japan," *Ekistics* 24, no. 140 (July 1967), pp. 6–14.

CHAPTER 2

1. Langdon Warner, *The Enduring Art of Japan* (New York: Grove Press, 1952), Chap. 2, "Shinto, Nurse of the Arts."

2. In the 1980s a verbal fad—called a "boom" in Japan—centered on the "theory of Japan" (*Nihonron*), that is, speculation on the nature of the Japanese. Such speculation is described more fully below (Chapter 8).

3. George Sansom, *A History of Japan to 1334* (Stanford: Stanford University Press, 1958), p. 62.

4. John Whitney Hall, *Japan, from Prehistory to Modern Times* (New York: Dell, 1970), Chap. 6, "The Aristocratic Age."

5. The *Heike monogatari* [Tales of the Heike] in a modern retelling appeared as a serialized novel written by the Japanese author of best-sellers Yoshikawa Eiji; the original was translated as *The Heike Story* by F. W. Uramatsu (New York: Knopf, 1956).

6. John Hall described the relationship in *Government and Local Power in Japan* (Princeton: Princeton University Press, 1966), pp. 255 ff.

7. In June 1989 a University of Michigan anthropologist, C. Loring Brace, unveiled a controversial theory on the origin of the samurai. Having examined skeletal remains from the battle of Kamakura (1333), he concluded that the lowly Ainu, not "ethnic Japanese," were the true descendants of Jōmon; that samurai were descended from the Ainu; and finally, that most modern Japanese are descended from Yayoi. There was irony in the theory because the Ainu, so long discriminated against, might thus have had a genetic effect on the traditional ruling classes of Japan. Many historians, particularly those in Japan, held strong reservations about the idea.

There was also irony in the fact that in 1986, in the Diet, former Prime Minister Nakasone Yasuhiro defended the government against charges of discrimination toward Ainu. They have, he argued, already intermingled with the Japanese people. He himself, he added, probably carried a rich infusion of Ainu blood, witness "my heavy beard and thick eyebrows."

8. The definition is adapted from K. Asakawa, "Some Aspects of Japanese Feudal Institutions," *Transactions of the Asiatic Society of Japan* 46, pt. 1 (1918), cited by John Whitney Hall, "Feudalism in Japan—A Reassessment," *Comparative Studies in Society and History* 5, no. 1 (October 1962):30.

CHAPTER 3

1. John Whitney Hall, "The Visual Arts and Japanese Culture," in John W. Hall and Richard K. Beardsley, *Twelve Doors to Japan* (New York: McGraw Hill, 1965), p. 273.

2. *The* Manyōshū; *One Thousand Poems Selected and Translated from the Japanese* (Tokyo: Iwanami Shoten, 1940; distr. in U.S. by the University of Chicago Press, 1941), vol. 4, p. 600; vol. 3, p. 351.

3. From the *Genji monogatari* [Tale of Genji], we are fortunate in having two translations, both fine works of literature in themselves. The long-standing, classic rendition was accomplished by Arthur Waley, trans., *The Tale of Genji* (Boston: Houghton Mifflin, 1925–33), 6 vols. Now we have, as well, the version by Edward G. Seidensticker, trans., *The Tale of Genji* (New York: Knopf, 1976), 2 vols.

4. There are masterly translations of both works: Minoru Shinoda, *The Founding of the Kamakura Shogunate* (with selected translations of the *Azuma Kagami*) (New York: Columbia University Press, 1960); *The Taiheiki: A Chronicle of Medieval Japan*, trans. with introduction and notes by Helen Craig McCullough (New York: Columbia University Press, 1959).

5. In our day, we have a complicated psychological novel about the burning of the Kinkakuji by a deranged youth, a work written by a controversial author,

Mishima Yukio: *Kinkakuji* [The temple of the golden pavilion] (New York: Knopf, 1958), trans. by Ivan Morris.

6. *Kōshoku gonin onna* (1686), trans. by William Theodore DeBary (Rutland, Vermont: Tuttle, 1965). Often, as explained in the preface, a writer or artist was known by his given (pen or artistic) name: thus, Saikaku.

7. Again we are fortunate in having a faithful translation: *Chikamatsu Monzaemon: Major Plays*, trans. by Donald Keene (New York: Columbia University Press, 1961).

8. *Chūshingura* (1748) was a drama about the forty-seven *rōnin* (masterless samurai). The kabuki play was all the more interesting because, *in fact*, some such incident *did occur*. In brief, when a minor daimyō drew sword against an antagonist within the shogunal headquarters, he was ordered to commit suicide. In the fictional version, his retainers led dissolute lives (in Kyōto) to allay suspicion, then took the head of their late lord's enemy and, on order of the authorities, committed *seppuku*. Even today, their graves in the quiet courtyard of Sengakuji in Tokyo are visited by Japanese tourists. Professor Henry D. Smith II of Columbia University has addressed the puzzle that the story still enjoys—in numerous publications, movies, and TV dramas—widespread popularity in contemporary Japan.

9. A major activity of the new International Center for Japonisme of the Jane Voorhees Zimmerli Art Museum, Rutgers University, was the organization of an international conference (May 13–14, 1988), chaired by Dr. Edwin O. Reischauer. "Perspectives on Japonisme: The Japanese Influence on America" was matched by an exhibit of *ukiyo-e*, originally collected by Matsukata Kōjirō. See *The Matsukata Collection of Ukuiyo-e Prints; Masterpieces from the Tokyo National Museum*, catalogue by Julia Meech (New Brunswick: Rutgers, the State University of New Jersey, 1988). Proceedings of the conference were later published (1989).

CHAPTER 4

1. John Whitney Hall, "The New Look of Tokugawa History," in J. W. Hall and Marius B. Jansen, eds., *Studies in the Institutional History of Early Modern Japan* (Princeton: Princeton University Press, 1968), p. 55.

2. The words are chosen from the writing of an influential Canadian historian, E. Herbert Norman, in a previously unpublished manuscript, "Feudal Background of Japanese Politics." The analysis in detail is to be found in Norman's seminal work, *Japan's Emergence as a Modern State; Political and Economic Problems of the Meiji Period* (New York: Institute of Pacific Relations, 1940). Pantheon Asian Library has reprinted this work in a paperback edition, which includes a long introduction by the editor, the unpublished essay "Feudal Background" (1944), and other materials. See John W. Dower, ed., *Origins of the Modern Japanese State: Selected Writings of E. H. Norman* (New York: Pantheon Books, 1975). Some who have revived Norman have claimed that the modernizationists have deliberately suppressed Norman's pioneering work.

3. John Dower states that Norman, writing in the 1940s, tried to identify the feudal origins of, and to denounce, Japanese militarism. Dower, *Origins*, p.

86. Writing in the context of later scholarship, critics of Norman and opponents of the Norman revivalists have suspected that, for the revivalists, the Canadian's interpretations of feudalism served (in the 1960s) a contemporary social need; that is, they drew a parallel between Japanese militarism of the 1940s and U.S. policy in the 1960s, as illustrated in the Vietnam disaster.

4. The modernization literature related to Japan is remarkably rich. Witness the series of six volumes *Studies in the Modernization of Japan,* published by the Conference on Modern Japan of the Association for Asian Studies: Marius B. Jansen, ed., *Changing Japanese Attitudes Toward Modernization* (1965); R. P. Dore, ed., *Aspects of Social Change in Modern Japan* (1968); Robert E. Ward, ed., *Political Development in Modern Japan* (1968); William W. Lockwood, ed., *The State and Economic Enterprise in Japan* (1965); Donald Shively, ed., *Tradition and Modernization in Japanese Culture* (1971); and James William Morley, ed., *Dilemmas of Growth in Prewar Japan* (1971) (all volumes, Princeton: Princeton University Press). Lest the reader think I am hiding behind scholarship, I am affiliated with the "modernization" school.

5. Ardath W. Burks, ed., *The Modernizers: Overseas Students, Foreign Employees, and Meiji Japan* (Boulder, Colo.: Westview Press, 1985); Shimada Tadashi, ed., *Za yatoi: Oyatoi gaikokujin no sogoteki kenkyū* [The yatoi: A comprehensive study of hired foreigners] (Kyōto: 1987).

6. Understandably, little was made of the century anniversary in 1989 of the Meiji Constitution. Instead, attention was centered on the new Constitution of Japan (1947): For example, a symposium, "The Constitution of Japan—The Fifth Decade," was held at the Duke University School of Law, September 14–16, 1989. In the panels, numerous comparisons were drawn between the first (the Meiji) and the new (the "MacArthur") constitutions.

7. The yen figures are for internal comparison. They are figured roughly on 1934–36—that is, prewar—prices (¥1 = $0.33).

8. Kagaku Gijutsu-chō Shigenkyoku (Science and technology agency, resources bureau), *Kokuren ohosadan oboegaki* [Report on a U.N. mission] (Tokyo, 1963).

9. In June 1979 Diet legislation formally reinstituted the imperial era name system. Opponents argued that the bill had turned the clock back, served to deify the emperor, and violated the new constitution, which declared that sovereignty rests with the people. *Japan Times Weekly* (International Edition) 19, no. 24 (June 16, 1979).

10. H. D. Harootunian, "Introduction: A Sense of an Ending and the Problem of Taishō," in Bernard S. Silberman and H. D. Harootunian, eds., *Japan in Crisis: Essays on Taishō Democracy* (Princeton: Princeton University Press, 1974), especially p. 20. This volume should be read with the modernization series listed in n.4 above. Also see E. O. Reischauer, "What Went Wrong," in Morley, ed., *Dilemmas of Growth.*

11. *Ukigumo,* trans. by Marleigh G. Ryan as *Japan's First Modern Novel* (New York: Columbia University Press, 1967).

12. In this section I have leaned heavily on the succinct and witty survey prepared by my colleague and coauthor Donald Keene, entitled "Literature," in

Arthur E. Tiedemann, ed., *An Introduction to Japanese Civilization* (New York: Columbia University Press, 1974), vol. 12, pp. 376–421. I have also been enlightened by the work of my colleague Janet A. Walker, *The Japanese Novel of the Meiji Period and the Ideal of Individualism* (Princeton: Princeton University Press, 1979). This author examines modern selfhood (*kindaiteki jiga*) in literature.

13. *Gan* [The wild geese, 1913], trans. by Kingo Ochiai and Sanford Goldstein (Tokyo: Tuttle, 1959).

14. *Kokoro*, trans. by Edwin McClellan (Chicago: Regnery, 1957).

15. *Tade kuu mushi*, trans. by Edward G. Seidensticker (New York: Knopf, 1955). The quotation later in this paragraph is from p. 189.

16. Thomas H. R. Havens, "Women and War in Japan, 1937–45," *American Historical Review* 80, no. 4 (October 1975):913–34.

17. See Dorothy Robins-Mowry, *The Hidden Sun: Women of Modern Japan* (Boulder, Colo.: Westview Press, 1983). Dr. Robins-Mowry, an officer in the U.S. Information Agency, served as a specialist on women's affairs during the ambassadorship of Edwin O. Reischauer.

18. *Nobi* [Fires on the plain], trans. by Ivan Morris (New York: Knopf, 1969; Penguin Books, 1969). Quotation is from the paperback edition, p. 232.

CHAPTER 5

1. Some explanatory comments need to be added. First, there were no zones *except* for three areas: (1) part of the Ogasawara (Bonin) chain held outright (returned by the United States on November 15, 1967); (2) most of the Ryūkyū chain (including Okinawa) was held under direct U.S. Army administration (Okinawa reverted to Japanese administration on May 15, 1972); and (3) what the Japanese call the "northern territories" (the southern Kurile chain claimed by the Japanese), which have remained under the Soviet Union's control. Second, until 1948, about 5,000 British Commonwealth Occupation Forces (BCOF) were based in southern Honshū and in Shikoku. Third, the Eighth Army and each of the various corps had a military government section, but as far as SCAP headquarters in Tokyo was concerned, military government was simply one staff activity at tactical level. Finally, supervision was provided by the Supreme Commander for the Allied Powers (SCAP), a post occupied by U.S. General of the Army Douglas MacArthur (until April 11, 1951, when President Harry Truman replaced MacArthur with General Matthew B. Ridgway).

2. Robert E. Ward, "Reflections on the Allied Occupation and Planned Political Change in Japan," in Robert E. Ward, ed., *Political Development in Modern Japan* (Princeton: Princeton University Press, 1968).

3. There can be little doubt that these earlier extreme ideas originated with General MacArthur. They were taken from his notes, February 3, 1946, in *Political Reorientation of Japan, September 1945 to September 1948*, Report of Government Section, Supreme Commander for the Allied Powers (Washington: U.S. Government Printing Office, 1949), vol. 1, p. 102.

4. I made these observations of rural Japan in the 1950s; see Part 2, "The Government and Politics of Japan," in Paul M. A. Linebarger (ed.), Djang Chu,

and Ardath W. Burks, *Far Eastern Governments and Politics* (Princeton: D. Van Nostrand Company, rev. ed. 1956), especially p. 539.

5. The label "Japan, Inc.," apparently originated from a cover story of *Time*, May 10, 1971.

6. Akuto made an interesting connection between political preference (support of the LDP) and consumer behavior. The supporters of the LDP usually choose top brands, he reported, for they will simply think that the best-known products must have high quality to enjoy top status and that it is realistic and safe to choose them. See Akuto Hiroshi, "Changing Political Culture in Japan," in *Text of the Seminar on "Changing Values in Modern Japan,"* organized by Nihonjin Kenkyūkai in association with Japan Society (Tokyo: Nihonjin Kenkyūkai, n.d. [ca. 1977]), p. 92.

7. Such empirical research has been incorporated in T. J. Pempel, ed., *Policymaking in Contemporary Japan* (Ithaca: Cornell University Press, 1977); see conclusion (Chap. 9).

8. An earlier work in English, challenging the standard portrait of the Shōwa emperor, by David Bergamini suffered from shoddy scholarship and advocacy of a conspiratorial thesis. A more recent biography by Edward Behr offered a more credible account of a man with limited power who operated at some distance from the powerless figure pictured in official works. See *Hirohito* by Edward Behr (New York: Villard Books, 1989).

9. Remarks by Professor Akuto Hiroshi, seminar with members of the Nihonjin Kenkyūkai, Japan House, New York, March 14, 1977 (see citation to text, n.6 above).

CHAPTER 6

1. Though not an economist, I had the privilege of reviewing an excellent and succinct analysis: Hiroshi Kitamura, *Choices for the Japanese Economy* (London: Royal Institute of International Affairs, 1976). The study was an extremely helpful guide in preparing this chapter.

2. Once again, I am grateful to have had the survey of a colleague and coauthor: E. S. Crawcour, "The Modern Economy," in Arthur E. Tiedemann, ed., *An Introduction to Japanese Civilization* (New York: Columbia University Press, 1974), pp. 487–513.

3. This thesis was developed in James Nakamura (with Hiroshi Shimbo), "The Market, Human Capital Formation, and Economic Development in Tokugawa Japan" (unpublished paper for University Seminar on Economic History, Columbia University, March 3, 1988).

4. See the proceedings of a conference hosted by the East Asian Institute of Columbia University, held March 24, 1978: Edward J. Lincoln, ed., *Japan's Changing Political Economy* (Washington: United States–Japan Trade Council, 1978).

5. For a detailed analysis, see Chalmers Johnson, *MITI and the Japanese Miracle: The Growth of Industrial Policy, 1925–1975* (Stanford: Stanford University Press, 1982).

6. Please note that the figures given in Table 7.1 are for numbers of persons (out of the total gainfully employed). Thus there were 16.1 million in primary pursuits (out of a labor force of 39.3 million) in 1955; only 11.7 million (out of 47.6 million) in 1965; and 7.4 million (out of 53.1 million) in 1975. By way of comparison, the total number in secondary pursuits increased from 9.2 million (1955) to 15.2 million (1965) and to 18.1 million (1975) out of the totals for the labor force in respective years. Also note that by 1975, the total number in tertiary (service) pursuits (27.6 million) exceeded the totals for primary and secondary combined (25.5 million).

CHAPTER 7

1. Gabriel Almond and Sidney Verba, eds. *The Civic Culture: Political Attitudes and Democracy in Five Nations* (Boston: Little, Brown, 1965).

2. Many economists have been loathe to accept the term "postindustrial." They have argued that laws of supply and demand still govern economic activity; their analyses have always taken into account, on the supply side, production of goods *and services*. What is implied in "postindustrial"—an economy with relatively greater emphasis on services—is only a matter of *degree*, these critics state. Other economists and some social scientists have, however, become interested in "extra-market" forces (the realm of "all other factors being equal"), specifically in a change of degree so significant as to become a matter of *kind*. A postindustrial society, then, is of interest because the transition to such a society involves deep and wide social and political alterations, which interact with economic change.

3. Zbigniew Brzezinski, *The Fragile Blossom: Crisis and Change in Japan* (New York: Harper and Row, 1972), Introduction, p. xii.

4. Chie Nakane, *Japanese Society* (Berkeley and Los Angeles: University of California Press, 1970), p. 149.

5. Taketsugu Tsurutani, *Political Change in Japan; Response to Postindustrial Challenge* (New York: McKay, 1977), especially Chapter 1.

6. This point is clearly made in Chapter 1 of James William Morley, ed. *Prologue to the Future; the United States and Japan in the Postindustrial Age* (New York: D.C. Heath for Japan Society, Inc., 1974).

7. *Sangyō kōzō no chōki bijion* [Long-term vision of industrial structure] (Tokyo: Council on Industrial Structure, September 1974).

8. Although I have not cited it directly here, I have leaned heavily on data collected for a special report on Japan, "The Economic Superpower," *The Americana Annual 1990* (Danbury, CT: Grolier, 1990).

9. See Tsurutani, *Political Change*, pp. 33–34.

10. Watanuki Jōji, *Politics in Postwar Japanese Society* (Tokyo: University of Tokyo Press, 1977), pp. 32–33; also Prime Minister's Office, "Public Opinion Survey on Society and the State" (Tokyo: Foreign Press Center, S-84-6, June 1984).

11. The best treatment of change within a seemingly unchanging LDP is to be found in Gerald L. Curtis, *The Japanese Way of Politics* (New York: Columbia University Press, 1988), especially Chapter 3.

12. First published in Japan as *Nippon rettō kaizō-ron* (1972), the volume appeared in an English-language edition the following year: Tanaka Kakuei, *Building a New Japan; A Plan for Remodeling the Japanese Archipelago* (Tokyo: Simul Press, 1973).

13. In the summer of 1970, I participated in a series of miniseminars on *kōgai*, held with metropolitan desk editors at sites ranging from Sapporo in the north to Fukuoka in the south. In June 1972, I was an informal observer of the Japanese delegation to the UN Conference on Human Environment, held in Stockholm.

14. See, for example, the excellent study of the emergent politico-religious group by James W. White, *The Sokagakkai and Mass Society* (Stanford: Stanford University Press, 1970); also Theodore C. Bestor, *Neighborhood Tokyo* (Stanford: Stanford University Press, 1989).

CHAPTER 8

1. John K. Emmerson, *Japanese and Americans in a New World in a New Age* (New York: Japan Information Service, Consulate General of Japan, 1974).

2. This analysis of the surveys draws heavily on the notes I took at an invitational seminar with members of the Nihonjin Kenkyūkai, held at Japan House in New York, on March 14, 1977. Notes were buttressed by the *Text of the Seminar on "Changing Values in Modern Japan"* (Tokyo: Nihonjin Kenkyūkai (NK), n.d. [ca. 1977]). The quotation is from the lead article by Hayashi Chikio, "Changes in Japanese Thought During the Past Twenty Years," Appendix I (Kokuminsei Survey), p. 36.

3. Nakane Chie, *Human Relations in Japan*, Summary translation of *Tateshakai no ningen kankei* [Personal relations in a vertical society] (Tokyo: Ministry of Foreign Affairs, 1972), p. 3.

4. Yoshiharu Scott Matsumoto, "Contemporary Japan: The Individual and the Group," *Transactions* of the American Philosophical Society, new series, 50.1 (1960), p. 7.

5. Prime Minister's Office, "Outline of the Third International Survey of Youth Attitudes" (Tokyo: Foreign Press Center, S-84-4, April 1984); also "Public Opinion Survey on Attitude Towards Work" (Tokyo: Foreign Press Center, S-83-8, August 1983).

6. For an analysis of the white paper, see editorial, "Our 'New Human Breed'," *The Japan Times Weekly*, 27.1 (January 3, 1987), p. 12.

7. Hosoya Chihiro, "Characteristics of the Foreign Policy Decision-Making System in Japan," paper prepared for delivery at the 1972 meeting of the International Studies Association, Dallas, Texas, March 14–18, 1972.

8. Tadashi Fukutake, *Japanese Society Today* 2nd ed. (Tokyo: University of Tokyo Press, 1981), p. 33.

9. In the late 1980s, Anne Stanaway, a director for Public Broadcasting System television, received a grant to study Japan as an aging society. See her article, "The Happiness and Longevity Club," *Japan Society Newsletter*, XXXIV.7 (April 1987).

10. Adapted, with acknowledgment to Harumi Befu, "Civilization and Culture: Japan in Search of Identity," *Japanese Civilization in the Modern World* (Osaka: Senri Ethnological Studies No. 16, National Museum of Ethnology, 1984), p. 59.

11. Prime Minister's Office, "Public Opinion on the Life of the Nation" (Tokyo: Foreign Press Center, S-85-4, September 1985); "Public Opinion Survey on Women" and "Public Opinion Survey on Women (II)" (Tokyo: Foreign Press Center, S-84-15 and S-85-1, November 1984 and April 1985).

12. Iwao Sumiko, "A Full Life for Modern Japanese Women," Nihonjin Kenkyūkai *"Changing Values,"* cited; also Prime Minister's Office, "International Comparison on Youth and Family" (Tokyo: Foreign Press Center, S-82-6, June 1982).

13. U.S. Department of Education, *Japanese Education Today* (Washington, D.C.: U.S. Government Printing Office, 1987), p. 3.

14. Nobuo K. Shimahara, *Adaptation and Education in Japan* (New York: Praeger, 1979), p. 2.

15. Ezra Vogel, *Japan's New Middle Class* (Berkeley: University of California Press, 1965), p. 40; see also Shimahara's comment, *Adaptation,* p. 82.

16. Two colleagues were members of the respective commissions: on the Japanese side, Kaneko Tadashi, my former graduate assistant (now with the Kokuritsu Kyōiku Kenkyūjo [National Institute for Educational Research]); on the U.S. side, Nobuo Shimahara (cited above), Associate Dean, Graduate School of Education, Rutgers. Dean Shimahara kept me up to date on the U.S. commission's work, *Japanese Education Today* (cited above).

CHAPTER 9

1. Although the United States–Japan Trade Council in Washington, D.C. acts openly as a registered agent of the Japanese government, it has also objectively provided the interested public a valuable range of data and analysis on Japan's international trade position. For example, the council has published a series of reference books, including the first edition of the *Yearbook of U.S.– Japan Economic Relations 1978* (Washington: United States–Japan Trade Council, March 1979). Much of the review of Japan's status in 1978, as the nation approached the summit, is based on the *Yearbook.*

2. Kenneth B. Pyle, "The Burden of Japanese History and the Politics of Burden-Sharing," in John H. Makin and Donald C. Hellmann, eds. *Sharing World Leadership* (Washington: American Enterprise Institute, 1989). In this context Pyle cited Robert Gilpin, *The Political Economy of International Relations* (Princeton: Princeton University Press, 1987).

3. *Japan, the United States, and the World; A Conversation with George W. Ball, Robert C. Christopher, and Fuji Kamiya* (New York: Japan Information Center, June 1979), p. 8.

4. "Shimen tembō" [Views on all sides] from *Shimbun kyōkai-hō,* weekly publication of the Newspaper Publishers and Editors Association of Japan, April 1990.

5. James D. Hodgson, "U.S. Public Attitude Toward Japan Is the Main Threat" (Opinion), *The Japan Times Weekly International Edition*, April 9–15, 1990.

6. See a lecture by Karl van Wolferen, "The Enigma of Japanese Power: A Response to Misunderstandings," given at the International House of Japan, Tokyo, Jan. 9, 1990, and an excellent capsule survey of revisionist thought, "The American Revisionists' Hostile View of Japan" by Homma Nagayo, Professor, The Tokyo Woman's Christian University, in *IHJ Bulletin*, quarterly publication of the International House of Japan, Tokyo, 10.2 (Spring 1990). Jon Woronoff, *Politics the Japanese Way* (New York: St. Martin's Press, 1988) is a strange treatment more like a Japanese cartoon (*manga*) than the "scientific approach" it purports to be.

7. Akio Morita, *Made in Japan; Akio Morita and SONY* (New York: E.F. Dutton, 1986).

8. Asai Motofumi, *Nihon gaikō—hansei to nenkan* [Japanese foreign policy—reconsideration and turnabout] (Tokyo: Iwanami, 1989).

9. See, for example, Japan's Defense Agency, "Summary of 'Defense of Japan'," (Tokyo: Foreign Press Center, W-84-11, September 1984).

10. Herman Kahn, *The Emerging Japanese Superstate: Challenge and Response* (Englewood Cliffs, N.J.: Prentice-Hall, 1970); Ezra Vogel, *Japan as Number One: Lessons for America* (Cambridge: Harvard University Press, 1979).

11. The original essay appeared in a narrowly circulated newsletter in 1989. Fukuyama's ideas were given wider public attention in his article "Entering Post-History," *New Perspectives Quarterly*, 6:49–52 (Fall 1989). It was given additional attention in James Atlas, "What Is Fukuyama Saying? And to Whom Is He Saying It?" *New York Times Magazine*, October 22, 1989.

12. Afrin Bey, "Third World and 'History'," *The Japan Times*, Weekly Overseas Edition, December 9, 1989.

Annotated Bibliography

In the postwar era and especially in the last decade, there has been a veritable explosion of excellent books—written in, or translated into, English—about Japan. The following list, with entries annotated, contains standard and recent studies that have proved useful in the development of the topics indicated.

AN INTRODUCTION: JAPAN AND THE JAPANESE

Hall, John W., and Beardsley, Richard K., eds. *Twelve Doors to Japan*. New York: McGraw Hill, 1965.
  Essays designed for upper division and graduate students.

Reischauer, Edwin O. *The Japanese*. Cambridge, Mass.: Belnap–Harvard University Press, 1978.
  An authoritative analysis, arranged by topics, by the former U.S. ambassador to Japan.

Tiedemann, Arthur E., ed. *An Introduction to Japanese Civilization*. New York and London: Columbia University Press, 1974.
  A series of eighteen essays written by scholars and designed for lower division students; nine devoted to the chronological history of Japan and another nine to various aspects of Japan's culture.

Varley, H. Paul. *Japanese Culture, A Short History*, rev. ed. New York: Praeger, 1977.
  An introduction, with emphasis on Japan's cultural contributions to the world.

LANDSCAPE AND SETTLEMENT

International Society for Educational Information, ed. *Atlas of Japan; Physical, Economic and Social*. Tokyo: International Society for Educational Information, 1970.

A series of seventy-five maps of comparable scales, reproduced in color; with accompanying text (English, French, Spanish).

Isida, Ryuziro. *Geography of Japan*. Tokyo: Kokusai Bunka Shinkōkai (distributed by the East West Center Press, University of Hawaii), 1961.
A topical survey of geography, extensively illustrated with maps and photographs.

Kornhauser, David. *Urban Japan; Its Foundation and Growth*. London and New York: Longman, 1976.
An introduction to Japan's geography with emphasis on contemporary urban settlement, but with ample attention as well to the historic and rural background.

HISTORY: TRADITION, TRANSITION, AND MODERNIZATION

Dore, R. P. *Education in Tokugawa Japan*. Berkeley and Los Angeles: University of California Press, 1965.
A British scholar analyzes the preconditions for, and one of the major factors in, Japan's rapid modernization.

Dower, John W., ed. *Origins of the Modern Japanese State: Selected Writings of E. H. Norman*. New York: Pantheon Books, 1975.
An annotated reproduction of the work of a pioneering Canadian historian; the editor's notes provide a kind of rebuttal to the "modernization" approach.

Hall, John Whitney. *Japan from Prehistory to Modern Times*. New York: Dell Publishing Co., 1970.
With special emphasis on ancient and early modern (Tokugawa) history.

Hall, John Whitney, and Jansen, Marius B., eds. *Studies in the Institutional History of Early Modern Japan*. Princeton: Princeton University Press, 1968.
Emphasizes the dynamic rather than the dormant aspects of the Tokugawa era; see especially Hall's analysis of feudalism and the "new look" of Tokugawa history.

Jansen, Marius B., ed. *Changing Japanese Attitudes Toward Modernization*. Princeton: Princeton University Press, 1965.
The first volume (and the history volume) in the six-volume series, "Studies in the Modernization of Japan," published by the Conference on Modern Japan of the Association for Asian Studies.

Morley, James W., ed. *Dilemmas of Growth in Prewar Japan*. Princeton: Princeton University Press, 1971.
One of the six-volume "Modernization" series; devoted to the resultant tensions of growth.

Sansom, G. B. *Japan, A Short Cultural History*, rev. ed. New York: Appleton-Century-Crofts, 1943.
 The classic one-volume survey of Japanese history and culture.

Silberman, Bernard S., and Harootunian, H. D., eds. *Japan in Crisis: Essays on Taishō Democracy.* Princeton: Princeton University Press, 1974.
 A companion to, and in a sense a criticism of, the "Modernization" series; U.S. and Japanese scholars explain how the earlier drive to development changed between 1900 and 1945.

Titus, David. *Palace and Politics in Prewar Japan.* New York: Columbia University Press, 1974.
 A study of the role of the palace in Japanese government between 1868 and 1945; especially from 1921 to 1945 (including the career of Kido Kōichi).

CULTURE: ART, LITERATURE, AND THOUGHT

Seidensticker, Edward G., trans. *The Tale of Genji.* New York: Alfred Knopf, Inc., 1976.
 A modern rendition of one of the world's earliest novels, written by Lady Murasaki.

Shively, Donald, ed. *Tradition and Modernization in Japanese Culture.* Princeton: Princeton University Press, 1971.
 One of the six-volume "Modernization" series; the volume devoted to thought and culture.

Tsunoda, Ryusaku, deBary, W. Theodore, and Keene, Donald, eds. *Sources of Japanese Tradition.* New York: Columbia University Press, 1958.
 Original Japanese sources translated into English; selections tell what the Japanese thought of themselves and of their culture across history.

Waley, Arthur, trans. *The Tale of Genji.* Boston: Houghton Mifflin, 1952–53.
 The older, classic rendition of the novel by Lady Murasaki.

Warner, Langdon. *The Enduring Art of Japan.* New York: Grove Press, 1952.
 Japan's history as seen through its art.

POSTWAR POLITICS

Burks, Ardath W. "The Government and Politics of Japan." In *Far Eastern Governments and Politics,* ed. by Paul M. A. Linebarger with coauthor, Djang Chu. Princeton: D. Van Nostrand Co. (rev. ed.), 1956.
 Modern Japanese political institutions within a setting of political history and compared with those of China.

Curtis, Gerald L. *The Japanese Way of Politics.* New York: Columbia University Press, 1988.
> The first book written in English that deals primarily with the evolution of Japanese politics over three decades of Liberal Democratic Party (LDP) rule.

Hrebenar, Ronald J., ed. *The Japanese Party System.* Boulder: Westview Press, 1988.
> A series of essays by recognized experts on Japanese politics and political parties.

Ishida, Takeshi. *Japanese Political Culture.* New Brunswick, N.J.: Transaction Books, 1983.
> A critical analysis of the Japanese value system as it is translated into political institutions.

Pempel, T. J. *Policy and Politics in Japan: Creative Conservatism.* Philadelphia: Temple University Press, 1982.
> Policymaking processes examined through six case studies: economic policy, labor-management relations, social welfare, higher education, environmental protection, and administrative reform.

Pharr, Susan J. *Political Women in Japan.* Berkeley: The University of California Press, 1981.
> Based on interviews with 100 young Japanese women who are active in voluntary political groups.

Steiner, Kurt, Krauss, Ellis S., and Flanagan, Scott C., eds. *Political Opposition and Local Politics in Japan.* Princeton: Princeton University Press, 1980.
> A survey, including contributions by six authors in addition to the editors, of progressive opposition forces at the local level.

Ward, Robert E., ed. *Political Development in Modern Japan.* Princeton: Princeton University Press, 1968.
> One of the six-volume "Modernization" series; the political science volume.

POSTWAR ECONOMY

East Asian Institute of Columbia University. *Japan's Changing Political Economy 1978,* ed. by Edward Lincoln. Washington, D.C.: United States–Japan Trade Council, 1978.
> Proceedings of a conference chaired by Robert S. Ingersoll, former U.S. ambassador to Japan.

Hollerman, Leon, ed. *Japan and the United States: Economic and Political Adversaries.* Boulder: Westview Press, 1979.

Japanese and U.S. contributors discuss economic and political controversies between the two countries.

Lockwood, William W., ed. *The State and Economic Enterprise in Modern Japan.* Princeton: Princeton University Press, 1965.
One of the six-volume "Modernization" series; the economics volume.

Patrick, Hugh, and Rosovsky, Henry, eds. *Asia's New Giant: How the Japanese Economy Works.* Washington: Brookings Institution, 1976.
The monumental 1000-page compendium prepared by American economists, assisted by Japanese, and based on approximately six months of field work in Japan (1973–74).

United States–Japan Trade Council. *Yearbook of U.S.–Japan Economic Relations.* Washington, D.C.: U.S.–Japan Trade Council, March 1979–(annual).
A survey of U.S.–Japan trade problems, with up-to-date statistics on Japan's economy.

POSTINDUSTRIAL SOCIETY AND NATIONAL CHARACTER

Almond, Gabriel A., and Verba, Sidney, eds. *The Civic Culture: Political Attitudes and Democracy in Five Nations.* Boston: Little, Brown, 1965.
A comparative study of modern industrialized integrated societies and their political life.

Beauchamp, Edward R., ed. *Learning to Be Japanese: Selected Readings on Japanese Society and Education.* Hamden, Conn.: Linnet Books, 1978.
A collection of journal articles and book chapters on the roots of modern education in Japan.

Bell, Daniel. *The Coming of Post-Industrial Society: A Venture on Social Forecasting.* New York: Basic Books, 1973.
An early identification of the postindustrial phenomenon.

Brzezinski, Zbigniew. *The Fragile Blossom: Crisis and Changes in Japan.* New York: Harper and Row, 1972.
Theory of the "technetronic" society applied to Japan.

Gibney, Frank. *Japan, the Fragile Superpower.* New York: Norton, 1975.
A critical, at times acerbic, analysis, mainly of Japanese business and management.

Kahn, Herman. *The Emerging Japanese Superstate: Challenge and Response.* Englewood Cliffs, N.J.: Prentice-Hall, 1970.
An early prediction, by the Director of the Hudson Institute, that the twenty-first would be the Japanese century.

Morley, James Williams, ed. *Prologue to the Future: The United States and Japan in the Postindustrial Age.* Lexington, Mass.: D.C. Heath, 1974.
Papers by a panel of Japanese and American social scientists, under the auspices of the Japan Society and the Johnson Foundation.

Nihonjin Kenkyūkai (NK). *Text of the Seminar on "Changing Values in Modern Japan."* Tokyo: Nihonjin Kenkyūkai, in association with the Japan Society, New York, n.d. (ca. 1976).
Summary and interpretation of data collected over a period of two decades (1953–73) by the Institute of Statistical Mathematics, Tokyo; including the essay by Iwao Sumiko, "A Full Life for Modern Japanese Women."

Nakane, Chie. *Japanese Society.* Berkeley and Los Angeles: University of California Press, 1970.
Authoritative analysis of the "frame" social structure of Japan.

Shimahara, Nobuo K. *Adaptation and Education in Japan.* New York: Praeger, 1979.
A field study of Japanese education and particularly of the college entrance examinations; a case study of the socialization of youth.

Vogel, Ezra. *Japan's New Middle Class.* Berkeley and Los Angeles: University of California Press, 1965.
The salary man and his family in a Tokyo suburb.

White, James W. *The Sōkagakkai and Mass Society.* Stanford: Stanford University Press, 1970.
Urban life, alienation, and new religion as a buffer.

JAPAN: SOME COMPARISONS

Craig, Albert M., ed. *Japan, A Comparative View.* Princeton: Princeton University Press, 1979.
Japan and Japanese culture as equally valid standards for comparisons; a series of essays by Western and Japanese scholars.

Vogel, Ezra. *Japan as Number One: Lessons for America.* Cambridge: Harvard University Press, 1979.
Japan as an organizational object lesson; the Japanese translation was a best-seller in Japan.

About the Book and Author

Japan has been among the first of the handful of countries to move "beyond modern," and in this third edition of a much-praised book, Ardath Burks brings the blur of the nation's rapid change into focus. In his newly revised and updated *Japan*, Professor Burks also traces the history of the Japanese, exploring their traditions, their continuity, and their cultural heritage. He devotes a chapter to the remarkable "introspection boom" (*Nihonron*): the Japanese asking, "Who are we Japanese?" In discussing the country's swift modernization, the author looks not only at the initial transition from primary agriculture to an industrial economy but also at the current evolution into a service-centered society.

On both domestic and international levels, the book evaluates the maturing of Japanese industry and its growing investment abroad, as well as the global tensions fueled by Japan's enormous trade surpluses. In response to the intense trade pressure it feels, the country is beginning to shift from export-driven growth to a consumer-oriented economy, a shift that will demand the building of a heretofore neglected, yet essential, infrastructure of housing and transportation.

The author analyzes domestic political developments including the regime of Nakasone Yasuhiro and the fall of Takeshita Noboru and Uno Sousuke, precipitated by financial scandal within the majority Liberal Democratic Party (LDP). Burks assesses the formidable tasks facing the revamped ruling LDP as its new generation of younger leaders grapples with an evolving economy, an expanding regional role, and the dissatisfaction of women and young people who have begun to rebel against the growth ethic and their marginalized role in society.

In his well-drawn, lucid portrait of this complex country, Professor Burks reflects on Japan as a nation in historical transition, envisioning a postindustrial future filled with friction and promise. As he writes in his introduction, "Americans and Japanese too often look at each other through opposite ends of a telescope. This brief study is an attempt to bring Japan and its people into sharper focus for the English-speaking world."

Ardath W. Burks is professor emeritus of Asian studies at Rutgers, the State University of New Jersey. In retirement he continues to consult, preparing political and investment risk surveys for Wall Street firms seeking to better understand Japan and its role in the world economy.

Index

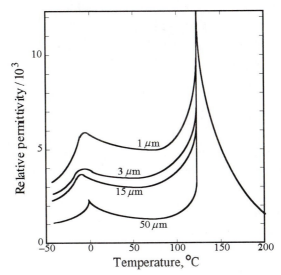

Figure 15.17

Effect of grain size on permittivity of $BaTO_3$.

Sintering conditions can also have an important effect on the permittivity. The replacement of various aliovalent cations such as La^{3+} and Nb^{5+} in $BaTiO_3$ has also been shown to inhibit grain growth which, as seen in Fig. 15.17, has the effect of increasing the permittivity below T_C. Finally, lower-valency substitutions such as Mn^{3+} on Ti^{4+} sites act as acceptors and enable high-resistivity dielectrics to be sintered in low-partial-pressure atmospheres.

Examples of a number of ceramic ferroelectric crystals and some of their properties are listed in Table 15.6.

EXPERIMENTAL DETAILS

The signature of a ferroelectric material is the hysteresis loop. This loop can be measured in a variety of ways, one of which is by making use of the electric circuit shown schematically in Fig. 15.18. A circuit voltage across the ferroelectric crystal is applied to the horizontal plates of an oscilloscope. The vertical plates are attached to a linear capacitor in series with the ferroelectric crystal. Since the voltage generated across the linear capacitor is proportional to the polarization of the ferroelectric, the oscilloscope will display the hysteresis loop.

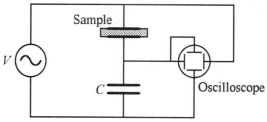

Figure 15.18
Circuit used to measure ferroelectric hysteresis.

15.7.3 Antiferroelectric Ceramics

In some perovskite ceramics, the instability that occurs at the Curie temperature is not ferroelectric but rather antiferroelectric. In antiferroelectric crystals, the neighboring lines of ions are displaced in opposite senses which creates two alternating dipole sublattices of equivalent but opposite polarization. Consequently, the net polarization is zero, and the dielectric constant does change at the transition temperature. Examples of antiferroelectric crystals are WO_3, $NaNbO_3$, $PbZrO_3$, and $PbHfO_3$.

In general, the difference in energies between the ferroelectric and antiferroelectric states is quite small (a few joules per mole); consequently, phase transitions between the two states occur readily and can be brought about by slight variations in composition or the application of strong electric fields.

15.7.4 Piezoelectric Ceramics

Piezoelectric materials are solids that are capable of converting mechanical energy to electrical energy and vice versa. This is shown schematically in Fig. 15.19a, where the application of a stress changes the polarization. When an external force is applied to produce a compressive or tensile strain in the ceramic, a change is generated in the dipole moment, and a voltage is developed across the ceramic (Fig. 15.19a). The opposite is also true; application of an electric field will result in a change in the dimensions of the crystal (Fig. 15.19b).

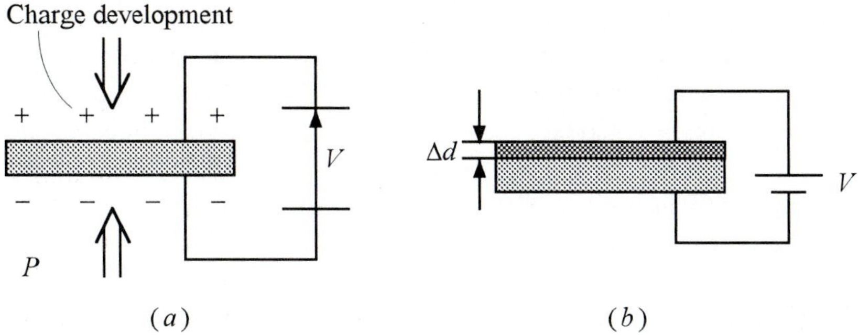

Charge development

Figure 15.19

(*a*) The direct piezoelectric effect is that polarization charges are created by stress.

(*b*) The inverse effect is that a strain is produced as a result of the applied stress.

The main uses of piezoceramics are in the generation of charge at high voltages, detection of mechanical vibrations, control of frequency, and generation of acoustic and ultrasonic vibrations. Most, if not all, commercial piezoelectric materials are based on ferroelectric crystals. The first commercially developed piezoelectric material was $BaTiO_3$. One of the most widely exploited piezoelectric materials today, however, is based on the $Pb(Ti, Zr)O_3$ or PZT solid solution system.

To produce a useful piezoelectric material, a permanent dipole has to be frozen in the piezoelectric. This is usually done by applying an electric field as the specimen is cooled through the Curie temperature. This process is known as **poling** and results in the alignment of the dipoles, and an electrostatic permanent dipole results.

15.8
SUMMARY

1. The presence of uncompensated or unpaired electron spins and their revolution around themselves (spin magnetic moment) and around their nuclei (orbital magnetic moment) endow the atoms or ions with a net magnetic moment. The net magnetic moment of an ion is the sum of the individual contributions from all unpaired electrons.

2. These magnetic moments can
   (i) not interact with one another, in which case the solid is a paramagnet and obeys Curie's law where the susceptibility is inversely proportional to temperature. Thermal randomization at higher temperature reduces the susceptibility.
   (ii) interact in such a way that adjacent moments tend to align themselves in the same direction as the applied field intensity, in which case the solid is a ferromagnet and will spontaneously magnetize below a certain critical temperature T_C. The solid will also obey the Curie-Weiss law (above T_C, it will behave paramagnetically). To lower the energy of the system the magnetization will not occur uniformly, but will occur in domains. It is the movement of these domains, which are separated by domain walls, that is responsible for the hysteresis loops typical of ferromagnetic materials.
   (iii) interact is such a way that the adjacent moments align themselves in opposite directions. If the adjacent moments are exactly equal they cancel each other out; and the solid is an *antiferromagnet*. However, if the adjacent moments are unequal they will *not* cancel, and the solid will possess a net magnetic moment. Such materials are known as *ferrimagnets* and constitute all known magnetic ceramics. Phenomenologically ferrimagnets are indistinguishable from ferromagnets.
3. Magnetic ceramics are ferrielectric and are classified according to their crystal structure into spinels, hexagonal ferrites, and garnets.
4. The interaction between, and the alignment of, adjacent dipoles in a solid give rise to ferroelectricity.
5. The interaction between adjacent moments (magnetic as well as dipolar) causes the solid to exhibit a critical temperature below which spontaneous magnetization or polarization sets in, where all the moments are aligned parallel to one another in small microscopic domains. The rotation and growth of these domains in externally applied fields give rise to hysteresis loops and remnant magnetization or polarization, whichever the case may be.

APPENDIX 15A

Orbital Magnetic Quantum Number

In Eq. (15.25), **L** is the summation of m_l, rather than l. The reason is shown schematically in Fig. 15.20, for the case where $l = 3$; in the presence of a magnetizing field H, the electron's orbital angular momentum is quantized along the direction of H. The physical significance of m_l and why it is referred to as the orbital magnetic quantum number should now be more transparent — it is the projection of the orbital angular momentum l along the direction of the applied magnetic field intensity.

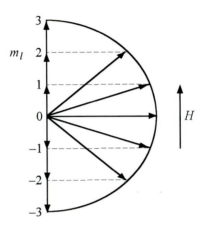

Figure 15.20
The total number of quantized allowed projections of the orbital angular momentum for $l = 3$ in the direction of the applied magnetic field is determined by m_l. Note that for $l = 3$ there are seven allowed projections: 3, 2, 1, 0, –1, –2, and –3.

PROBLEMS

15.1. In the Bohr model for the hydrogen atom,[287] the electron travels in a circular orbit of radius 5.3×10^{-11} m. Calculate the current and magnetic moment associated with this electron. How does the moment calculated compare to μ_B? *Hint*: Calculate the potential energy of the electron and equate it to the vibration energy.
Answer: $i \approx 1.05$ mA, $\mu_{\text{orb}} \approx 9.25 \times 10^{-24}$ A·m²

---

[287] It is important to note that, this problem notwithstanding, s electrons do not have an orbital moment. Quantum mechanics predicts that their angular momentum is zero, since $l = 0$.

15.2. (a) Derive Eq. (15.29).

(b) A solid with electron spins is placed in a magnetizing field H of 1.6×10^6 A/m. The number of spins parallel to the field was 3 times as large as the number of antiparallel spins. What was the temperature of the system? State all assumptions.

Answer: $T \approx 2.45$ K

(c) At what temperature would you expect all spins be aligned to the field? At what temperature would the number of spins upward exactly equal the number of spins downward?

15.3. (a) Show that if the assumption $\mu_{ion}B \ll kT$ is not made, then the susceptibility will be given by

$$M = (N_1 - N_2)\mu_{ion} = N\mu_{ion} \tanh\left(\frac{\mu_{ion}B}{kT}\right)$$

Plot this function as a function of $\mu_{ion}B/kT$. What conclusions can you reach about the behavior of the solid at high fields or very low temperatures?

(b) For a solid placed in a field of 2 T, calculate the error in using this equation as opposed to Eq. (15.29) at 300 K. You may assume $\mu_{ion} = \mu_B$.

Answer: ≈ 0.001 percent

(c) Repeat part (b) at 10 K.

Answer: ≈ 0.6 percent

15.4. A beam of electrons enters a uniform magnetic field of 1.2 T. What is the energy difference between the electrons whose spins are parallel to and those whose spins are antiparallel to the field? State all assumptions.

Answer: 1.4×10^{-4} eV

15.5. (a) For a ferromagnetic solid, derive the following expression:

$$T_C = \frac{\lambda\mu_{ion}^2\mu_0 N}{k} = 3\lambda C$$

(b) For ferromagnetic iron, $C \approx 1$ and T_C is 1043 K. Calculate the values of μ_{ion} and λ.

Answer: $\mu_{ion} = 2.13\mu_B$; $\lambda \approx 350$

15.6. The susceptibility of a Gd^{3+} containing salt was measured as a function of temperature. The data are shown below. Are these results consistent with Curie's law? If yes, then calculate the Curie constant (graphically) and the effective Bohr magnetons per ion.

T, K	100	142	200	300
χ, cm^3/mol	6.9×10^{-5}	5×10^{-5}	3.5×10^{-5}	2.4×10^{-5}

Information you may find useful: Molecular weight of salt $= 851$ g/mol and its density is 3 g/cm^3.

15.7. ZnO·Fe$_2$O$_3$ is antiferromagnetic. If it is known that this compound is a regular spinel, suggest a model that would explain the antiferromagnetism.

15.8. Sketch the magnetization versus temperature curve for Fe$_3$O$_4$.

15.9. Each ion in an iron crystal contributes on average $2.22\mu_B$. In Fe$_3$O$_4$, however, each Fe ion contributes an average of $4.08\mu_B$. How can you rationalize this result?

15.10. Consider Fe(Ni$_x$Fe$_{2-x}$)O$_4$. What value of x would result in a net moment per formula unit of exactly $2\mu_B$. *Hint:* The Ni^{2+} occupies the octahedral sites.
Answer: $x = 1$

15.11. (*a*) Calculate the spin-only magnetic moments of Ni^{2+}, Zn^{2+}, and Fe^{3+}.
(*b*) Nickel ferrite (NiO·Fe$_2$O$_3$) and zinc ferrite (ZnO·Fe$_2$O$_3$) have inverse and normal spinel structures, respectively. The two compounds form mixed ferrites. Assuming that the coupling between the ions is the same as in magnetite and that the orbital momenta are quenched, calculate the magnetic moment per formula unit for (Zn$_{0.25}$Ni$_{0.75}$O)·Fe$_2$O$_3$.

15.12. Figure 15.21 presents magnetization curves for a magnetic ceramic fired at two different temperatures, shown on the plot.
(*a*) Explain the general reasons for the shape of these curves. In other words, what processes determine the shape of the curves at low, intermediate, and large values of the applied magnetic field?
(*b*) What changes in the material might explain the change in magnetic behavior with increased firing temperatures?

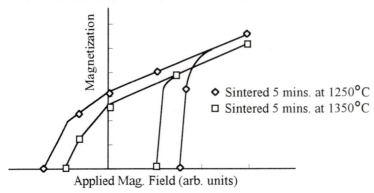

Figure 15.21
Upper half of the hysteresis loop for a barium ferrite as a function of processing time and temperature.

15.13. Using the data given in Table 15.6, estimate the value of the interaction factor β in $BaTiO_3$.
   Answer: 2.3×10^{-3}

15.14. (*a*) Calculate the dc capacitance of a $BaTiO_3$ capacitor 1 μm thick and 1 cm^2 in area that is operating near T_C.
   Answer: $1.4 \, \mu F$
   (*b*) Calculate the total dipole moment for a 2-mm thick diameter disk of $BaTiO_3$.
   (*c*) Calculate the unit cell geometry of a unit cell of $BaTiO_3$ that is subjected to an electrical field applied in such a way as to increase the polarization of the unit cell to $0.18 \, C/m^2$.

15.15. Estimate the density of Ar at which its polarization would go to infinity. State all assumptions. The atomic weight of Ar is 39.94 g/mol.
   Answer: $9.57 \, g/cm^3$

ADDITIONAL READING

1. D. Jiles, *Introduction to Magnetism and Magnetic Materials*, Chapman & Hall, London, 1991.
2. L. L. Hench and J. K. West, *Principles of Electronic Ceramics*, Wiley, New York, 1990.
3. B. Jaffe, W. R. Cook, and H. Jaffe, *Piezoelectric Ceramics*, Academic Press, New York, 1971.

4. E. Fatuzzo and W. J. Merz, *Ferroelectricity,* North-Holland, Amsterdam, 1967.

5. H. Frohlich, *Theory of Dielectrics*, 2d ed., Oxford Science Publications, 1958.

6. M. E. Lines and A. M. Glass, *Ferroelectrics and Related Materials*, Oxford Science Publications, 1977.

7. N. Ashcroft and N. Mermin, *Solid State Physics*, Holt-Saunders International Ed., 1976.

8. C. Kittel, *Introduction to Solid State Physics*, 6th ed, Wiley, New York, 1988.

9. K. J. Standley, *Oxide Magnetic Materials*, 2d ed., Oxford Science Publications, 1992.

10. A. H. Morrish, *The Physical Principles of Magnetism*, Wiley, New York, 1965.

11. A. J. Moulson and J. M. Herbert, *Electroceramics*, Chapman & Hall, London, 1990.

12. D. W. Richerson, *Modern Ceramic Engineering*, 2d ed., Marcel Dekker, New York, 1992.

13. J. M. Herbert, *Ceramic Dielectrics and Capacitors*, Gordon and Breach, London, 1985.

14. J. M. Herbert, *Ferroelectric Transducers and Sensors*, Gordon and Breach, London, 1982

15. J. C. Burfoot & G. W. Taylor, *Polar Dielectrics and Their Applications*, Macmillan, London, 1979.

16. R. C. Buchanan, ed., *Ceramic Materials for Electronics*, Marcel Dekker, New York, 1986.

17. L. M. Levinson, ed., *Electronic Ceramics*, Marcel Dekker, New York, 1988.

18. J. C. Burfoot, *Ferroelectrics, an Introduction to the Physical Principles*, Nostrand, London, 1967.

CHAPTER 16

Optical Properties

White sunlight Newton *saw, is not so pure;*
A Spectrum bared the Rainbow to his view.
Each Element absorbs its signature:
Go add a negative Electron to
Potassium Chloride; it turns deep blue,
As Chromium incarnadines Sapphire.
Wavelengths, absorbed are reemitted through
Fluorescence, Phosphorescence, and the higher
Intensities that deadly Laser Beams *require.*

John Updike, *Dance of the Solids*[†]

16.1
INTRODUCTION

Since the dawn of civilization, gems and glasses have been prized for their transparency, brilliance, and colors. The allure was mainly aesthetic, fueled by the rarity of some of these gems. With the advent of optical communications and computing, the optical properties of glasses and ceramics have become even more important, an importance which cannot be overemphasized. For example, in its 150th anniversary issue devoted to the key technologies for the 21st century, *Scientific American*[288] devoted an article to all-optical networks. Today commercial fiber-optic networks are based on the ability of very thin, cylindrical conduits of glass to transmit information at tens of gigabits[289] of information per second. This multigigabit transport of information is fast enough to move an edition

---

[†]   J. Updike, **Midpoint and other Poems**, A. Knopf, Inc., New York, New York, 1969. Reprinted with permission.

[288] V. Chan, *Scientific American*, September 1995, p. 72.

[289] A gigabit is 1 billion bits; a terabit is 1 trillion bits.

of the *Encyclopedia Britannica* from coast to coast in 1 s! In theory, a fiber can transport 25 terabits of information, an amount sufficient to carry *simultaneously* all the telephone calls in the United States on Mother's Day (one of the busiest days of the year).

Alternating currents, infrared radiation, microwaves, visible light, X-rays, ultraviolet light, etc., all produce oscillating electromagnetic fields differing only in their frequencies. And although sometimes they are thought of as being distinct, they do constitute a continuum known as the *electromagnetic* (EM) spectrum that spans 24 orders of magnitude of frequencies or wavelengths. (See Fig. 16.1.) Within that spectrum, the visible range occupies a small window from 0.4 to 0.7 μm or 1.65 to 3.0 eV.[290]

All EM radiation will interact with solids in some fashion or other. Understanding the nature of this interaction has been and remains invaluable in deciphering and unlocking the mysteries of matter. For instance, it is arguable, and with good justification, that one of the most important techniques to study the solid state has been X-ray diffraction. Other spectroscopic techniques are as varied as radiation sources and what is being monitored, i.e., reflected, refracted, absorbed rays, etc.

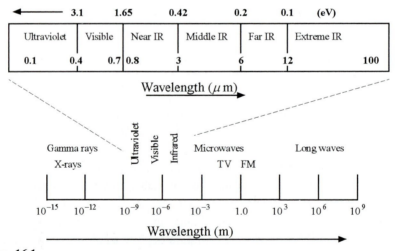

Figure 16.1

Electromagnetic spectrum. The visible spectrum constitutes a small window between 0.4 and 0.7 μm or 1.7 and 3.1 eV.

---

290 Remember, in vacuum the wavelength λ and frequency ν are related by $\nu = c/\lambda$, where c is the velocity of light. The energy, on the other hand, is given by $E = h\nu$, where h is Planck's constant.

In what follows, the various interactions between electromagnetic radiation and ceramics will be discussed. However, the phenomena described in this chapter relate mostly to the "optical" region of the electromagnetic spectrum, which includes wavelengths from 50 nm to 100 μm (25 to 0.1 eV). In other words, the discussion will be limited to the part of the spectrum shown at the top of Fig. 16.1. Furthermore, only insulating ceramics will be dealt with here — the cases where the concentration of free electrons is large will not be considered.

16.2
BASIC PRINCIPLES

When a beam of light or electromagnetic radiation impinges on a solid (Fig. 16.2), that radiation can be:

- *Transmitted* through the sample
- *Absorbed* by the sample
- *Scattered* at various angles

The scattered waves can be coherent or incoherent (see App. 16A for more details). When the scattered waves constructively interfere with one another, the scattering is termed *coherent*. Light scattered in the opposite direction of the incident beam leads to reflection. Light scattered in the same direction as the incident beam and recombining with it gives rise to refraction. The recombination of the scattered beams can also give rise to diffraction, where the intensity of the diffracted beams depends on the relative positions of the atoms and is thus used to determine the position of atoms in a solid (e.g., X-ray diffraction, see Chap. 3). Incoherent interference, on the other hand, gives rise to other forms of scattering, such as Rayleigh scattering.

For a total incident flux of photons I_0 energy conservation requires that

$$I_0 = I_T + I_R + I_A$$

where I_T, I_R, and I_A represent the transmitted, reflected, and absorbed intensities, respectively. The intensity I is the energy flux per unit area and has units of $J/(m^2 \cdot s)$. Dividing both sides of this equation by I_0 yields

$$1 = T + R + A \tag{16.1}$$

where T, R, and A represent, respectively, the fraction of light transmitted, reflected, and absorbed.

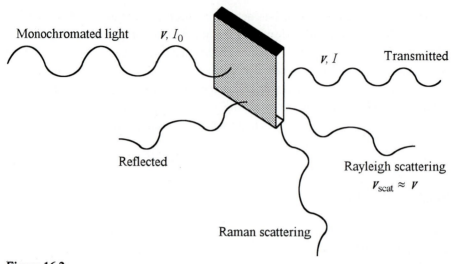

Figure 16.2

Various interactions between radiation and solids. The monochromatic ray with frequency ν and intensity I_0 can be transmitted (with the same frequency but a reduced intensity), scattered (Rayleigh and/or Raman), absorbed, or reflected.

In the following sections and throughout this chapter the relationship between the makeup of a solid and its optical properties is discussed. The optical properties of greatest interest here are the refractive index n, which for low-loss materials determines the reflectivity and transmissivity, and the various processes responsible for absorption and/or scattering.

Refraction

A common example of refraction is the apparent bending of light rays as they pass from one medium to another, e.g., a rod immersed in a fluid will appear bent. The extent of this effect is characterized by a fundamental property of all materials, namely, the refractive index n. When light encounters a boundary between two materials with different refractive indices, for reasons that are touched upon below, its velocity and direction will change abruptly, a phenomenon called *refraction*. As discussed below, the physics behind what gives rise to n is intimately related to the *electronic polarizability* of the atoms or ions in a solid. To understand the physical origin of the refractive index, it is useful to make the following two simplifying assumptions: First, the frequency of the applied field is much greater than ω_{ion} but smaller than ω_0, the natural frequency of vibration of the electronic cloud. Second, $k_e'' = 0$; in other words, the electronic charges are all oscillating in phase with the applied field. As these charged particles oscillate, they in turn reradiate an electromagnetic wave of the same frequency, creating their own electric field, which interacts with and slows down the incident field.[291]

As noted, the major effect of the interaction of the incident and reradiated waves is to make the velocity of the transmitted light appear to have traveled through the solid (v_{sol}) more slowly than through vacuum (v_{vac}), which leads to perhaps the simplest definition of n, namely,

$$n = \frac{v_{vac}}{v_{sol}} \tag{16.2}$$

Refer to Fig. 16.3. Another equivalent definition is

$$n = \frac{\sin i}{\sin r} \tag{16.3}$$

Typical values of n are listed in Table 16.1, from which it is obvious that for most ceramics n lies between 1.2 and 2.6. Note that all the values are greater than 1, indicating that the velocity of light in a medium is always less than that in vacuum.

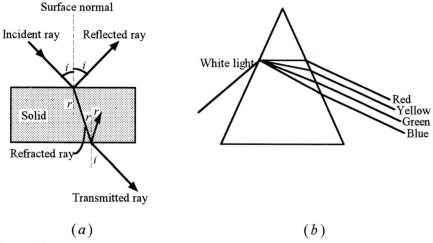

(a)   (b)

Figure 16.3

(a) Light refraction and reflection. Each interface reflects and refracts a portion of the incident beam. (b) Light refraction by a prism.

---

291 The exact details of how this occurs are beyond the scope of this book but are excellently described in R. P. Feynman, R. B. Leighton, and M. Sands, *The Feynman Lectures on Physics*, vol. 1, Chap. 31, Addison-Wesley, Reading, Massachusetts, 1963.

TABLE 16.1
Refractive Indices of Selected Ceramic Materials

Material	n	Material	n
		Halides and Sulfides	
CaF_2	1.430	NaF	1.330
BaF_2	1.480	NaI	1.770
KBr	1.560	PbF_2	1.780
KCl	1.510	PbS	3.910
LiF	1.390	TlBr	2.370
NaCl	1.550	ZnS	2.200
		Oxides	
Al_2O_3 (Sapphire)	1.760	PbO	2.610
$3Al_2O_3 \cdot 2SiO_2$	1.640	TiO_2	2.710
$BaTiO_3$	2.400	SrO	1.810
BaO	1.980	$SrTiO_3$	2.490
BeO	1.720	Y_2O_3	1.920
$CaCO_3$	1.658, 1.486	ZnO	2.000
$MgAl_2O_4$	1.720	$ZrSiO_4$	1.950
MgO	1.740	ZrO_2	2.190
		Covalent Ceramics	
C (diamond)	2.424	$\alpha - SiO_2$ (quartz)	1.544, 1.553
$\alpha - SiC$	2.680		
		Glasses	
Pb-silicate glasses	2.500	Soda-lime-silicate glass	1.510
Fused quartz	2.126	$Na_2O - CaO - SiO_2$[†]	1.458
Pyrex®[‡]	1.470	Vycor®[§]	1.458

[†]   Dense optical flint
[‡]   borosilicate
[§]   96% silica

In the more general case where k_e'' cannot be neglected, n has to be complex, i.e.,[292]

$$\hat{n} = n + i\kappa \qquad (16.4)$$

where κ is called the **extinction coefficient** or **absorption index** and is a measure of the absorbing capability of a material. Kappa (κ) should not be confused with the k_e' or k_e'', although they are related (see below).

As discussed in Chap. 14, at frequencies greater than about 10^{15} s^{-1}, only electrons can follow the field and all other polarization mechanisms including ionic

---

[292] It is instructive to note the similarity between this equation and Eq. (14.20).

polarization drop out. In this situation, it can be shown (not too easily) that the following relationships between the electronic polarizability parameters k'_e and k''_e, on one hand, and n and κ, on the other, hold:

$$k'_e = n^2 - \kappa^2 = 1 + \frac{e^2 N}{\varepsilon_0 m_e} \frac{\omega_0^2 - \omega^2}{\left(\omega_0^2 - \omega^2\right)^2 + f^2 \omega^2} \tag{16.5}$$

and

$$k''_e = 2n\kappa = \frac{e^2 N}{\varepsilon_0 m_e} \frac{\omega f}{\left(\omega_0^2 - \omega^2\right)^2 + f^2 \omega^2} \tag{16.6}$$

Furthermore, from these two equations it can be easily shown that (see Prob. 16.1)

$$n = \frac{1}{\sqrt{2}} \sqrt{\left(k'^2_e + k''^2_e\right)^{1/2} + k'_e} \tag{16.7}$$

and

$$\kappa = \frac{1}{\sqrt{2}} \sqrt{\left(k'^2_e + k''^2_e\right)^{1/2} - k'_e} \tag{16.8}$$

Equations (16.5) to (16.8), are important for the following reasons:

1. They clearly demonstrate the one-to-one correspondence between electronic polarization and n. Typically, for ceramics k''_e is on the order of 0.01 to 0.0001 (see Table 14.1); consequently, without much loss in accuracy, k''^2_e can be neglected with respect to k'_e in Eq. (16.7), in which case

$$\boxed{n = \sqrt{k'_e}} \tag{16.9}$$

2. Since k'_e is a function of frequency, it follows that the refractive index also has to be a function of frequency (see Worked Example 16.1). This change in refractive index with frequency or wavelength is called **dispersion**. Typical dispersion curves for a number of ceramics are shown in Fig. 16.4, where it is clear that the refractive index increases as the frequency of light increases.

3. Even at the high end of the range of k''_e (that is, 0.01) and assuming the lowest value of k'_e possible, that is, 1, the κ calculated from Eq. (16.8) is on the order of 0.005 and thus can be ignored for most applications. Note that this conclusion is valid only as long as the system is far from resonance.

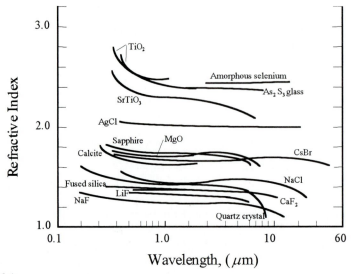

Figure 16.4
Change in refractive index with frequency for select glasses and crystals.

WORKED EXAMPLE 16.1. Based on Eq. (16.5), explain the phenomenon of light refraction by a prism shown in Fig. 16.3b. What does that say about the nature of white light?

Answer

According to Eq. (16.5), for $\omega < \omega_0$, k'_e increases with increasing frequency of the incident light. Consequently, according to Eq. (16.9), n should also increase with increasing frequency. In other words, a higher-frequency light (e.g., blue) will be refracted or deflected by a larger angle than a lower-frequency light (e.g., red light), as observed in Fig. 16.3b. This simple experiment makes it clear that "white" light is composed of a spectrum of frequencies.

Reflectivity

Not all light that is incident on a surface is refracted; as shown in Fig. 16.3, a portion of it can be reflected. It can be shown, again not too easily, that the reflectivity of a perfectly smooth solid surface at *normal* incidence is given by

$$R = \frac{(n-1)^2 + \kappa^2}{(n+1)^2 + \kappa^2} \tag{16.10}$$

which is known as **Fresnel's** formula.

Again, given that for most ceramics and glasses $\kappa \ll 1$, it follows that reflectivity is simply related to n. For example, lead-silicate glasses, also known as crystal glasses,[293] with refractive indices of about 2.6, would reflect about 20 percent of the incident light (which explains why crystal glass sparkles). By contrast, a typical soda-lime silicate glass with an n of about 1.5 only reflects about 4 percent.

Interestingly enough, near resonance, n will increase dramatically, and so, according to Eq. (16.10), will reflectivity. This occurs because the various secondary waves from the atoms in the surface will cooperate to produce a reflected wavefront traveling at an angle equal to the angle of incidence. *Selective* reflection is thus a phenomenon of resonance and occurs strongly near those wavelengths corresponding to natural frequencies of bound charges in the substance, i.e., near resonance. The substance will *not* transmit light of these wavelengths; instead, it reflects strongly. True absorption (see below), where the light is converted to heat (i.e., processes associated with k_e'' of κ), also occurs at these frequencies to a greater or lesser extent because of the large amplitudes of vibrating charges involved. If true absorption were entirely absent, however, the reflecting power would be 100 percent at the wavelengths in question.

Absorbance and transmittance

The transmittance T through a transparent medium is proportional to the amount of light that is neither reflected nor absorbed. For low-loss (low-absorbing) materials, the absorption A in Eq. 16.1, can be neglected, and $T = 1 - R$. In other words, the fraction of light not reflected is transmitted.

In general, however, as light passes through a medium, it is attenuated or lost by one of two mechanisms: Either it is absorbed, i.e., the light is transformed to heat; or it is scattered, i.e., a portion of the beam is deflected.[294]

Intrinsic absorption. In Chap. 14 the power dissipation per unit volume in a dielectric was shown to be [Eq. (14.25)]

$$P_V = \frac{1}{2}\sigma_{ac}E_0^2 = \frac{1}{2}(\sigma_{dc} + \omega k_e''\varepsilon_0)E_0^2 \tag{16.11}$$

where E_0 is the applied electric field. Energy conservation dictates that in the absence of any other energy-dissipating mechanisms, this loss will result in a

---

[293] This is an unfortunate nomenclature. There is nothing crystalline about this glass, or any other glass for that matter.

[294] A good example of scattering is the way that rays of sunlight from a window are made visible by very fine dust particles suspended in air.

decrease in the intensity I of the light, that is, $P_v = -dI/dx$, passing through a material of thickness dx. Furthermore, it can be shown[295] that the intensity of light in a medium of refractive index n is given by

$$I = \frac{n\varepsilon_0 c E_0^2}{2} \tag{16.12}$$

where c is the velocity of light. Ignoring σ_{dc} in Eq. (16.11), which for most insulators and optical materials is an excellent assumption (see Prob. 16.1), noting that $k_e'' = 2n\kappa$, and combining Eqs. (16.11) and (16.12), one obtains

$$\frac{dI}{dx} = -\frac{2I\omega\kappa}{c} = -\alpha_a I \tag{16.13}$$

Integrating from the initial intensity I_0 to the final or transmitted intensity I_T gives

$$\boxed{\frac{I_T}{I_0} = \exp(-\alpha_a x)} \tag{16.14}$$

where x is the optical path length and α_a is the **absorption constant**, given by $2\omega\kappa/c$. Here α_a is measured in m^{-1} and is clearly a function of frequency.

Note that α_a is proportional to k_e'', which in turn reflects the fact that the oscillating charges *not* in phase with the applied EM field are the ones responsible for the absorption. For an ideal dielectric, k_e'', κ, and α_a all vanish, and no energy will be absorbed (see Worked Example 16.2). Finally, note that when the frequency of the incident radiation approaches the resonance frequency of either the bonding electrons or the ions, then strong absorptions occur, absorptions that, as discussed below, are ultimately responsible for delineating the frequency range over which a material is transparent.

Absorbance by impurity ions. As discussed in greater detail below, impurity ions in a material can *selectively* absorb light at specific wavelengths. Such a chemical species is called a **chromophore**. In such a case, attenuation is proportional to the path traveled dx and the concentration of absorbing centers c_i, as described by the **Beer-Lambert law**:

$$-\frac{dI}{dx} = \varepsilon_{BL} c_i I \tag{16.15}$$

where ε_{BL} is a constant that depends on the impurities and the medium in which they reside; ε_{BL} is referred to sometimes as the **linear absorption coefficient** and

---

[295] See, e.g., R. P. Feynman, R. B. Leighton, and M. Sands, *The Feynman Lectures on Physics*, vol. 1, pp. 31-110, Addison-Wesley, Reading, Massachusetts, 1963.

sometimes as the **extinction coefficient.** Once again, integrating this expression yields

$$\boxed{\frac{I_T}{I_0} = \exp(-\varepsilon_{BL} c_i x)} \tag{16.16}$$

This is an important result because it predicts that the reading of a radiation detector (which measures the rate of flow of energy per unit area and unit time) will decrease exponentially with the thickness of the medium and the concentration of absorbing centers.

An implicit assumption made in deriving Eqs. (16.14) and (16.16) was that scattering could be neglected. In general, however, the loss coefficient must account for all losses, hence

$$\alpha_{tot} = \alpha_a + \varepsilon_{BL} c_i + \alpha_s \tag{16.17}$$

where α_s is the absorption coefficient due to scattering (Sec. 16.4). It follows that in the most general case

$$\boxed{I_T = I_0 \exp(-\alpha_{tot} x)} \tag{16.18}$$

In many cases either of these mechanisms may be negligible with respect to the other, but it is important to realize the existence of these processes and the fact that in many cases more than one mechanism may be operating.[296]

EXPERIMENTAL DETAILS: MEASURING OPTICAL PROPERTIES

Clearly two important optical properties are n and κ. Described briefly below is a method by which these parameters can be measured.[297]

---

[296] It is not possible to distinguish between absorption and scattering losses from a simple measurement of the attenuation; both phenomena cause attenuation. One method to differentiate between the two, however, is to measure the light intensity at all angles. If the measurements show that all the light taken away from the original beam reappears as scattered light, the conclusion is that scattering — not absorption — is responsible for the attenuation. If the energy is absorbed, it will not disappear, but reappears at a different frequency, i.e., as heat.

[297] Implicit in this discussion is that the material is fully dense and pore-free, with a grain size that is either much smaller than or much greater than the wavelength of the incident radiation. If that were not the case, as discussed in greater detail below, scattering would have to be taken into account. Furthermore, it is assumed that the material is pure enough that the $\varepsilon_{BL} c_i$ term can be neglected.

The basic idea is to measure both the transmission and the reflectivity of a thin slab of material, which is usually carried out in a device known as a *spectrophotometer.* Figure 16.5 schematically illustrates the four major components of such a device: a source of radiation, a monochromator, the sample, and a number of detectors. In a typical experiment, both T and R are measured, preferably simultaneously. Because of multiple reflections at the various planes of the crystal, T is not given by Eq. (16.14), but rather by[298]

$$T = \frac{I_{out}}{I_0} = \frac{(1-R)^2 e^{-\alpha_a x}}{1-R^2 e^{-2\alpha_a x}} \tag{16.19}$$

where R is given by Eq. (16.10) and x is the thickness of the sample. Note that this expression is only valid for normal incidence and is identical to Eq. (16.14) when $R = 0$.

In principle, the observed reflectivity R_{obs} can be measured by providing a second detector position, as shown in Fig. 16.5. The observed reflectivity is related to R by

$$R_{obs} \approx R\left(1 + \frac{I_T}{I_0} e^{-\alpha_a x}\right) \tag{16.20}$$

Thus by measuring both the reflectivity and the transmittivity of a sample, n and κ can be calculated from Eqs. (16.10), (16.19), and (16.20) (see Worked Example 16.2).

An alternate approach is to measure the transmission of two different samples of different thicknesses with identical reflectivities.[299]

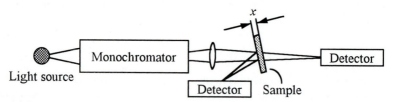

Figure 16.5
Schematic of measuring the optical constants of a solid.

[298] J. C. Slater, *Electromagnetic Theory,* McGraw-Hill, New York, 1941.

[299] Interestingly enough (and it is left as an exercise to the reader to show that) strongly absorbing samples are also quite reflective, and vice versa.

WORKED EXAMPLE 16.2. If 80 percent of an Na lamp light incident on a 1-mm glass panel is transmitted and 4 percent is reflected, determine the n and κ for this glass. The wavelength of Na light is 0.59 μm or 5.1×10^{14} s^{-1}.

Answer

Given that $R = 0.04$, applying Eq. (16.10) and ignoring κ (see below) yield $n = 1.5$. To calculate α_a use is made of Eq. (16.19). However, given that the loss is small, the second term in the denominator can be neglected, and Eq. (16.19) simplifies to

$$T = (1 - R)^2 e^{-\alpha_a x}$$
$$0.8 = (1 - 0.04)^2 e^{-\alpha_a (0.001)}$$

Solving for α_a yields 141 m^{-1}. Furthermore, given that $\alpha_a = 2\omega\kappa/c$, it follows that

$$\kappa = \frac{\alpha_a c}{4\pi\nu} = \frac{141(3 \times 10^8)}{(4 \times 3.14)(5.1 \times 10^{14})} = 6.6 \times 10^{-6}$$

Note that the error in ignoring κ in Eq. (16.10) is fully justified.

WORKED EXAMPLE 16.3. Carefully describe the changes that occur to an EM ray when it impinges on a solid for which (*a*) $n = 2$ and $\alpha_{tot} = \alpha_a = 0$ and (*b*) $n = 2$ and $\alpha_{tot} = \alpha_a = 0.4$.

Answer

(*a*) When a wave propagating in vacuum impinges normally on a solid for which $n = 2$ and $\alpha_a = 0$, the outcome is shown schematically in Fig. 16.6a. Since $n = 2$, the velocity of the wave is halved, which implies that the wavelength also is halved inside the solid. Note that because $\alpha_a = 0$, the intensity of the transmitted light and its *frequency* remain constant throughout.

(*b*) When, α_a is nonzero, the intensity of the transmitted wave is reduced as a result of absorption by the solid (Fig. 16.6b). For simplicity in both these cases, reflection was neglected.

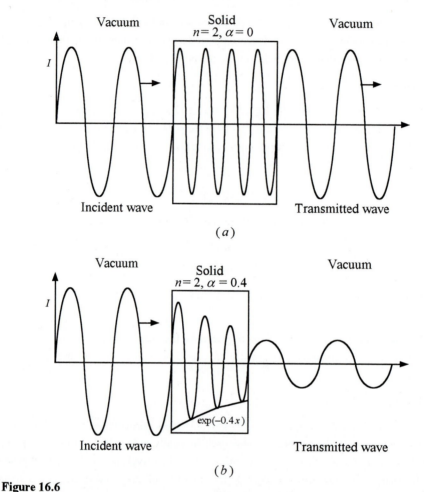

Figure 16.6

Schematic of changes that occur to an electromagnetic wave transmitted through a solid with (a) $n = 2$ and $\alpha_a = 0$. Note the halving in wavelength. Since $\alpha_a = 0$, there is no loss in intensity or energy of the wave as it passes through the solid. (b) Here $n = 2$ and $\alpha_a = 0.4$. The sample absorbs a portion of the energy of the incident wave, and the intensity of the transmitted wave is thereby reduced. Since $n = 2$, the wavelength in the solid is again half that in vacuum.

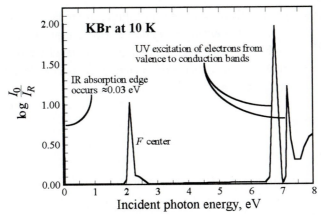

Figure 16.7
Spectral reflectance of KBr over a wide energy range of incident radiation.[300] Note that at resonance, the solid becomes very reflective, whereas away from resonance, most of the light is transmitted.

16.3
ABSORPTION AND TRANSMISSION

In the previous section, the relationships between transmittance and absorbance were described, with few details given. In this section, some of the specifics are elucidated. Scattering is dealt with separately in Sec. 16.4.

The complexity of the situation is depicted in Fig. 16.7, where the reflectance[301] of KBr at 10 K is plotted as a function of incident photon energy over a wide range from the IR to the UV. The salient features are an IR absorption edge at 0.03 eV, an absorption peak at about 2 eV, and a number of absorption peaks in the UV part of the spectrum around 7 eV.

From the previous discussion it is clear that the requirements for a material to be transparent are the absence of strong absorption and/or scattering in the visible range. The range over which a solid is transparent is called the **transmission range** and is bounded on the high-frequency (low-wavelength) side by UV absorption phenomena and on the low-frequency side by IR absorption. The spectral transmission ranges of a number of ceramics are compared in Figs. 16.8

---

300 It is worth noting here that the information embedded in this graph, between the IR absorption edge and the UV spectrum, is the same information that appears in Fig. 14.13*b* over the same range.

301 As noted earlier, near resonance, crystals become considerably more reflective.

and 16.9, from which it is obvious that most ceramics are indeed transparent over a wide range of frequencies. For example, window glass transmits light from 1×10^{15} to $\approx 7.5 \times 10^{13}$ s^{-1}, which is why, not surprisingly, it is used for windows. It is interesting to note that typical semiconductor materials such as Si and GaAs are only transparent in the IR range.

16.3.1 UV Range

Electronic resonance

This was discussed in Sec. 16.2. The factors that affect the frequency at which resonance occurs were discussed in Chap. 14 and will not be repeated here, except to point out that in glasses the formation of nonbridging oxygens tends to decrease the frequencies (increase the wavelength) at which resonance will occur. This is clear from Fig. 16.4 — quartz is transparent to higher frequencies than window glass that contains nonbridging oxygens. It was also noted in Chap. 14 (see Fig. 14.6) that S^{2-}, Te^{2-}, and Se^{2-} are some of the more polarizable ions. It is thus not surprising that ceramics containing these ions tend to be opaque in the visible spectrum and have absorption edges that are shifted into the IR range; CdS (Fig. 16.8) is a good example.

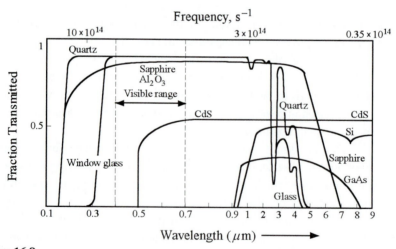

Figure 16.8
Spectral transmission of a number of materials.

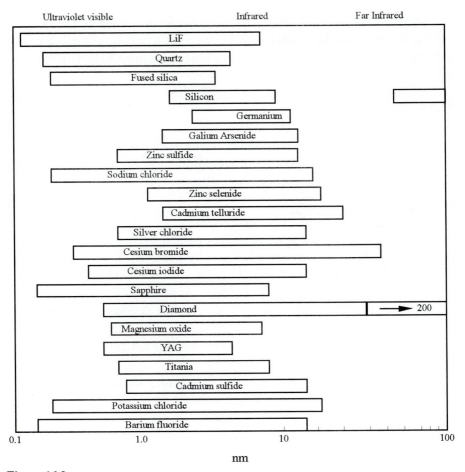

Figure 16.9
Useful transmission ranges for optical materials.

Photoelectric effect

As discussed in Chap. 2, insulating crystalline materials exhibit an energy gap E_g between their valence and conduction bands. When the incident photon with energies is greater than E_g, that is,

$$h\nu > E_g \tag{16.21}$$

it will be absorbed by promoting an electron from the valence band to the conduction band. This is known as the **photoelectric effect**, and in addition to increasing the conductivity of the solid, it results in the absorption of the incident wave. It is important to note that if the light has an energy less than E_g, absorption will not occur. Hence a well-established technique for measuring the band gap of a material is to measure the conduction of the sample as a function of the frequency of incident light — the frequency at which the onset of photoconductivity is related to E_g through Eq. (16.21).

16.3.2 Visible Range

For appreciable absorption to occur in the visible range, electronic transitions must occur. The nature of these transitions can result from various sources, as described now.

Transition-metal cations

It is well known that the colors of crystals and minerals are a strong function of the type of dopant or impurity atoms, especially transition-metal cations, present. For example, rubies are red and some sapphires are blue, yet both are essentially Al_2O_3. It is only by doping Al_2O_3 with parts per million of Cr ions, e.g., that the magnificent red color develops. Similarly, the sapphire develops its blue color as a result of Ni doping. Since rubies are red and pure alumina is transparent, it follows that the Cr ions must absorb blue light and transmit the red light, which is what is registered by the eyes.

To explain this phenomenon the **ligand field theory** has been proposed which successfully accounts for the coloring and magnetic properties of many transition-metal-containing ceramics. The basic idea is that the transition metal interacts directionally with the ligands surrounding it in such a fashion that its energy levels are split. A **ligand** is a negatively charged, nonspherical environment surrounding the centrally located transition-metal ion and is partially covalently bonded with it. A free transition-metal ion will have five d orbitals (see Fig. 2.2b) that are degenerate in energy and in which the electrons can be found with equal probability (Fig. 16.11a). However, if that ion is placed in a field where the ligands surround it octahedrally, as shown in Fig. 16.10a and b, symmetry dictates that the d_{z^2} and $d_{x^2-y^2}$ orbitals (Fig. 16.10a) will be repelled more strongly than any of the d_{xy}, d_{xz}, or d_{zx} orbitals. This in turn will result in a splitting of the orbital energies such that the repulsion of the surrounding ligands increases the energy of the d_{z^2} and $d_{x^2-y^2}$ orbitals relative to the d_{xy}, d_{xz}, and d_{zx} orbitals. The resulting energy scheme is shown in Fig. 16.11c.

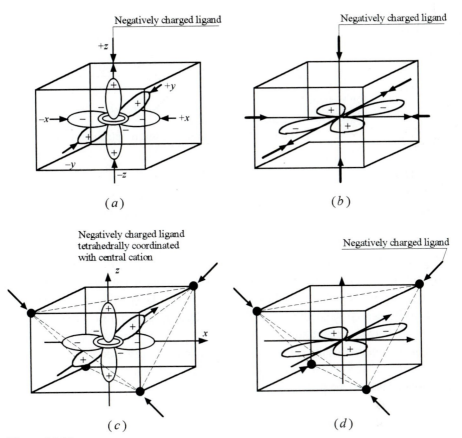

Figure 16.10
Interaction of octahedral and tetrahedral ligands with d orbitals. (a) Interaction of d_{z^2} and $d_{x^2-y^2}$ orbitals with six ligands in octahedral field. (b) Interaction of d_{xy}, d_{xz}, and d_{zx} orbitals with six ligands in octahedral field. (c) Interaction of d_{z^2} and $d_{x^2-y^2}$ orbitals with four ligands in tetrahedral field. (d) Interaction of d_{z^2} and $d_{x^2-y^2}$ orbitals with four ligands in tetrahedral field.

If the surrounding ligands are tetrahedrally coordinated (Fig. 16.10c and d), the opposite occurs — the d_{xy}, d_{xz}, and d_{zx} orbitals are now the ones that are repelled more strongly by the ligand field and consequently have the higher energies. The corresponding energy diagram in this case is shown in Fig. 16.11d.

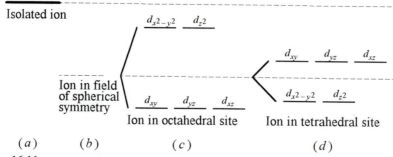

Figure 16.11

Energy of d orbitals of transition-metal cations (a) in a free or isolated ion, (b) in a field of spherical symmetry, (c) with ion in octahedral site, (d) with ion in tetrahedral site.

It is this energy split that gives rise to the various colors observed in ceramics and glasses. If the energy of the incoming photon is close to the energy difference between the d orbitals, then absorption of that incident wave will occur and the electrons will be promoted from the lower to the higher level. The magnitude of the energy split, and consequently the resulting color, depends on the strength of the interaction between the transition ion and the host crystal as well as on the coordination number of the central ions. This can be clearly seen in Table 16.2, in which various absorption bands for transition-metal ions in soda-lime silicate glasses are summarized.

TABLE 16.2
Absorption maxima and colors of transition-metal ions in soda-lime silicate glasses

Ion	Number of d electrons	Absorption maximum, μm	Coordination number with oxygen	Color
Cr^{3+}	1	0.660	6	Green
V^{3+}	2	0.645	6	Green
Fe^{2+}	6	1.100	4 or 6?	Blue
Mn^{3+}	4	0.500	6	Purple
Mn^{2+}	5	0.435	4 or 6?	Brown
Ni^{2+}	8	1.330	6	Purple
Cu^{2+}	9	0.790	6	Blue

It is worth noting here that the probability of the transition decreases as the energy of the incident light differs significantly from the energy split between the d orbitals. In other words, the maximum probability for transition occurs when the energy of the incident light is the same as the energy split between the levels. Furthermore, objects remain colored as they are observed, because the electrons that are excited rapidly lose their energy to their immediate surroundings as heat (i.e., at a different frequency) and hence the number of ions available for excitation remains approximately constant even though absorption occurs.

Absorption by color centers produced by radiation or reduction

It is often observed that ceramics, especially oxides, will turn black when they are heavily reduced or exposed to strong radiation for extended periods. In either case, the formation of **color centers** is responsible for the observed phenomena. A color center is an impurity or a defect onto which an electron or a hole is locally bound. For example, how the reduction of an oxide can result in the formation of both $V_O^{\bullet}$ and V_O^{x} defects was discussed in some detail in Chap. 6 (see Fig. 6.4a and b). In the context of this chapter, both are considered color centers. If E_d for these defects is the energy needed to liberate the electron into the conduction band (see Fig. 7.12c), it follows that light of that frequency will be absorbed. Note here that *all* incident wavelengths with energies equal to or greater than E_d, and not just the ones that are $\approx E_d$, will be absorbed because electrons can be promoted into any level in the conduction band which has a finite width.

The formation of color centers in the alkali halides, especially silver, has been studied extensively and in great detail, in an attempt to understand the photographic process. At least half a dozen color centers have been identified in these materials, of which the most widely studied is probably the F *center*, defined as an electron trapped at an anion vacancy. The name comes from the German word for color *Farbe*. In the case of KBr, the F center (Fig. 16.7) is believed to be an electron trapped at a bromine vacancy. The F center can be modeled by assuming the electron is trapped in a box of side d, which scales with the lattice parameter of the alkali halide. The F center transition is believed to be between the ground and first excited state of this particle in a box. This model, while crude, qualitatively explains the data for many of the alkali halide F center spectra.

Absorption by microscopic second phases

Small metallic particles dispersed in glasses scatter light and can create striking colors. This phenomenon is essentially a scattering effect and is discussed in greater detail in the next section.

WORKED EXAMPLE 16.4. Given the band structure shown in Fig. 16.12a, schematically sketch the optical absorption spectrum as a function of incident photon energy. Also discuss the expected photoconductivity. The arrows in Fig. 16.12a denote the allowable transitions. All others are not allowed.

Answer

The absorption spectrum is shown in Fig. 16.12b and is characterized by the following: Below E_1 there is no absorption. The first absorption peak centered on E_1 corresponds to the excitation of an electron from the ground state to the excited state of the imperfection. This transition does not affect the photoconductivity, however, because the electron is still localized. The next absorption centered on E_2 is due to the excitation of an electron from the ground state of the imperfection to the conduction band. This will give rise to a current. Finally, the last absorption centered on E_g is due to the intrinsic transitions across the band gap.

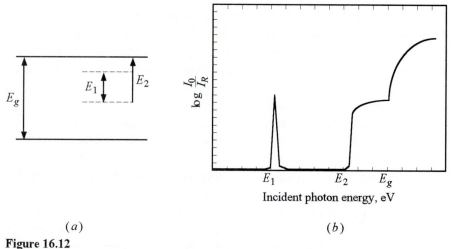

(a) (b)

Figure 16.12
(a) Energy-level diagram. (b) Corresponding optical response.

16.3.3 IR range

Ionic polarization

This phenomenon was dealt with in detail in Sec. 14.4.2. As the frequency of the incident light approaches the natural frequency of vibration of the ions in a solid ω_{ion}, resonance occurs and energy is transferred from the incident light to the solid. In other words, the incident wave is adsorbed. The frequency at which this occurs is called the **IR absorption edge**; as discussed in Sec. 14.4.2, it depends on the strength of the ionic bond as reflected by the natural frequency of vibration of the ions ω_{ion}, their charges, and their mass. These factors and their effect on the IR absorption edge are clearly demonstrated in Fig. 16.13, where the IR absorption edge is plotted as a function of frequency of incident light for a number of ceramic crystals.

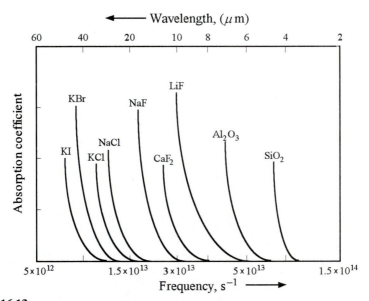

Figure 16.13

Infrared absorption edges of select ceramic crystals. Note the correlation between the IR edge and the melting points of the solids.

16.4
SCATTERING AND OPACITY

Given that most ceramics have band gaps in excess of 1 eV (see Table 2.6), and based on the foregoing discussion, the inevitable conclusion is that most ceramics are intrinsically transparent. However, everyday experience indicates that with the notable exception of glasses, most ceramics are *not* transparent but rather are opaque. As noted above and elaborated on in this section, the reason for this state of affairs is not related to any absorption mechanisms per se, but is due to scattering of incident light by pores and/or grain boundaries present within the ceramic. It is important to note here, however, that dense single crystals of most ceramics are indeed transparent, with gems being excellent examples.

Systems that are optically heterogeneous scatter light such as transparent media containing small particles or pores. Scattering is probably most easily described as being reflections from internal surfaces. Figure 16.14 schematically illustrates how a light beam is scattered by an isolated spherical void. Note that the emerging rays are no longer parallel.

By neglecting multiple and intrinsic scattering and absorption due to impurities, Eq. (16.18) simplifies to

$$\frac{I_T}{I_0} = \exp(-\alpha_s x) \tag{16.22}$$

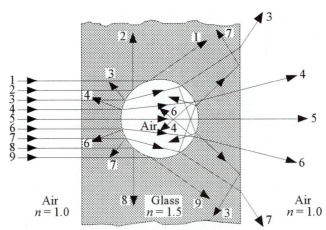

Figure 16.14

Light scattering by a spherical pore in an otherwise homogeneous medium.

where α_s was defined earlier as the scattering coefficient, sometimes referred to as the **turbidity** or **extinction coefficient**.

Assuming there are N_s scatterers per unit volume, each with a radius r_s, it follows that the intensity scattered per unit volume in any given direction is simply proportional to N_s times the intensity scattered by one particle. In other words

$$\alpha_s = Q_s N_s \pi r_s^2 \tag{16.23}$$

where Q_s is a dimensionless constant that depends on the angle between the incident and scattered light as well as the relative size of the particles to the wavelength of the incident light.

At this point it is useful to consider two limiting cases. First are the particles whose dimensions are small with respect to the wavelength of the light, that is, $r_s \ll \lambda$. In this case, the scattering in the forward direction is equal to the scattering in the backward direction, and it can be shown that[302]

$$Q_s = (\text{const}) \left(\frac{r_s}{\lambda} \right)^4 \left(n_{\text{matrix}}^2 - n_{\text{scatter}}^2 \right)^2 \tag{16.24}$$

where the n_i represent the refractive indices of the matrix and scattering particles. This type of scattering is known as **Rayleigh scattering** and pertains to single scattering by independent spheres of identical size. In other words, this is under experimental conditions in which the particles are so far from one another that each is subjected to a parallel beam of light and has sufficient room to form its own scattering pattern, undisturbed by the presence of other particles (see App. 16B for more details).

Second are particles that are very large compared to the wavelength of light, or $r_s \gg \lambda$. Here it can be shown that the total energy scattered is simply twice the amount it can intercept, or

$$Q_s = 2 \tag{16.25}$$

In other words the total light scattered by a particle of radius r_s is simply twice the cross-sectional area of that particle.[303]

Finally, note that if the volume fraction of the scattering phase is f_p, then

---

[302] See, e.g., H. C. van de Hulst, *Light Scattering by Small Particles*, Dover, New York, 1981.

[303] That a particle of area A removes *twice* the energy it can intercept is known as the *extinction paradox*. After all, common experience tells us that the shadow of an object is usually equal to the object — not twice as large! The paradox is removed when the assumptions made to derive Eq. (16.25) are taken into account, namely, that (1) all scattered light including that at small angles is removed and (2) the observation is made at a very great distance, i.e., far beyond the zone where a shadow can be distinguished.

$$f_p = \frac{4}{3}\pi r_s^3 N_s \qquad (16.26)$$

Based on the preceding discussion, the following salient points are noteworthy:

1. Scattering of small particles is a very strong function of the wavelength of incident radiation. Consequently, blue light is scattered much more strongly than red light. This phenomenon is responsible for blue skies and red sunsets. At sunset, the sun is observed directly, and it appears red because the blue light has been selectively scattered away from the direct beams. During the day, the molecules and dust particles in the atmosphere scatter the blue light through various angles, rendering the sky blue.

2. Scattering by small particles occurs only to the extent that there is a difference between the refractive indices of the matrix and of the scatterers. In ceramics, pores, with $n = 1$, are very potent scatterers. It is for the same reason that TiO_2 is added to latex to create white paint.

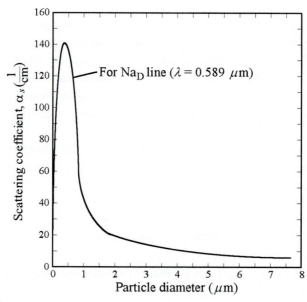

Figure 16.15

Effect of particle size on the scattering coefficient of a fixed volume of particles. The light used was monochromatic, with a wavelength of 0.589 μm.

3. Scattering is a strong function of particle size. By assuming a fixed volume of particles and combining Eqs. (16.23), (16.24), and (16.26), it is not difficult to show that for very small particles, α_s is proportional to r_s^3. Conversely, by combining Eqs. (16.23), (16.25), and (16.26), the result that α_s scales as $1/r_s$ is readily obtainable. The effect of particle size on the scattering coefficient is illustrated in Fig. 16.15. The maximum in scattering occurs when the particle size is about equal to the wavelength of incident radiation.

EXPERIMENTAL DETAILS: MEASURING LIGHT SCATTERING

A typical arrangement for the study of light scattering is shown in Fig. 16.16. The detector is mounted so that it can measure the angular dependence θ of the intensity of the scattered light from the direction of the incident beam. The scattering coefficient α_s is determined by the integration of the scattered intensity at all angles to the incident beam. Note that this arrangement is needed to differentiate between scattering and absorption.

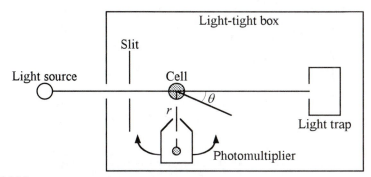

Figure 16.16
Basic construction of light-scattering apparatus.

16.5
FIBER OPTICS AND OPTICAL COMMUNICATION

A fiber-optic waveguide is a thin device composed of a high-refractive-index substance which is completely surrounded by a lower-refractive-index one. The situation is depicted in Fig. 16.17a, where according to **Snell's law**,

$$n \sin \phi = n' \sin \phi' \qquad (16.27)$$

If the angle of incidence is greater than a critical angle ϕ_c the total internal reflection will occur rather than refraction, as shown in Fig. 16.17b. This angle is given by Snell's law when $\phi' = 90°$, or

$$\sin \phi_c = \frac{n'}{n} \qquad (16.28)$$

Thus in an optical waveguide, some of the light which is launched into the high-index core is carried along that region by reflecting off the interface with the low-index cladding, as shown in Fig. 16.17c.

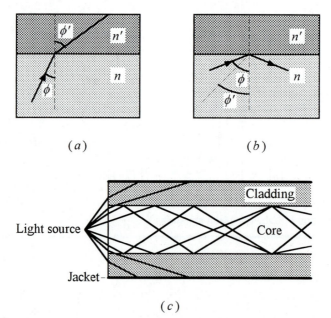

(a) (b)

(c)

Figure 16.17
(a) Snell's law of refraction. (b) Total internal reflection. (c) Light rays from point source enter at many angles. Rays that impinge at an angle that is less than the critical angle are guided down the optical waveguide by total internal reflection.

The process of optical telecommunications consists of four parts:

1. The electric signal is coded digitally and converted to an optical signal.
2. The optical signal consisting of high-frequency laser pulses is sent along the waveguide, which is a silica fiber with a core and a cladding. The core carries the light, and the cladding guides the light through the core.
3. As the light travels along the fibers, it broadens and weakens; hence, the signals have to reamplified periodically.
4. The signal is received and decoded by converting the light pulses back to electric signals in a form that a telephone or computer can interpret.

What is of interest in this section is the transmission medium and what limits the length over which the light pulses that represent the digital 0s or 1s can be transmitted without distortion or attenuation.

For short distances of roughly 1 km or less, polymer waveguides can be used. For longer distances, however, the losses are unacceptable and inorganic glasses have to be used. To date, the material of choice is extremely pure silica glass fibers. Ideally an optical fiber should be loss-free, in which case the signal would not be attenuated. The attenuation is usually expressed in decibels (dB) as

$$dB = 10\log\left(\frac{\text{power output}}{\text{power input}}\right) = 10\log\frac{I_T}{I_0} \qquad (16.29)$$

where I_T/I_0 is the ratio of the intensity at the detector to that at the source.[304]

A number of phenomena contribute to the scattering, absorption, and overall deterioration of an optical signal as it travels down a waveguide. These are discussed in some detail in the following sections. Typical absorption or loss data for a silica optical fiber are shown in Fig. 16.18, where the following salient points are noteworthy:

1. Above 5 μm, absorptions of the Si–O–Si bond network, i.e., ionic polarizations, become important.
2. Trace amounts of impurities, particularly transition-metal oxides, can have a profound effect on absorption due to electronic transitions alluded to earlier. Furthermore, the presence of Si–OH in the glass can cause significant absorption due to overtones of the O–H bond vibrations. It is important to control these impurities since they lie in the useful transmission window. For example, it has been estimated that one part per billion of these impurities could lead to 1 dB/km loss in silicate glass (see Prob. 16.25).

---

[304] The conversion from absorbance to dB/km can be made by using $1/cm = 4.3 \times 10^5$ dB/km (se Prob. 16.23).

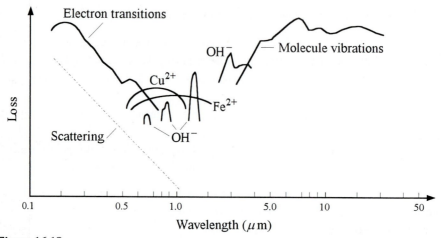

Figure 16.18
Sources of optical losses in fused silica.[305]

3. Electronic transitions of the glass become important at wavelengths shorter than 0.5 μm.

4. Density and composition fluctuations are inherent in glasses and lead to scattering. For example, Rayleigh scattering in fused silica amounts to about 0.7 dB/km at 1 μm. Since scattering scales as $1/\lambda^4$, it becomes most important at shorter wavelengths.

In addition to these mechanisms, scattering due to defects such as pores, inclusions, or dust particles introduced in the fiber during processing have to be eliminated. Another source of scattering is irregularities in fiber diameter; this is especially important if the diameter fluctuations are regularly and closely spaced (<1 mm apart).

Another limitation to the information-carrying capacity of a fiber waveguide is how close together the light pulses can be transmitted without overlapping. This is usually determined by pulse broadening. One reason for this is called *differential delay*, which results from light traveling different paths through the fiber. To avoid the problem, waveguides are often constructed with a graded index, with the composition at the center being silica-doped with germania and the amount of germania decreasing radially outward. Since germania has a higher n than silica, n will decrease with increasing distance from the center of the fiber. Given that light travels faster in low-index media, the light waves that travel off center travel at a

305 Adapted from W. G. French, *Journal of Material Education*, Penn State University Press, College Park, Pennsylvania, 1979, p. 341.

faster rate than those transmitted down the center of the fiber, which tends to minimize the undesirable broadening.

The transmission capacity, defined as the highest bit rate times the maximum transmission length, has increased by roughly an order of magnitude every 4 years since 1975. By 1978, 1 billion bits (1 Gbit) could be transmitted each second through a system 10 km long. The transmission capacity was thus 10 gigabit-kilometers per second. During the next 3 years, improved technology increased capacity to 100 Gbit·km/s. This was done by reducing the size of the core to create "single-mode" fibers, which forced the light to travel at nearly uniform velocity, which greatly reduced dispersion. The second advance was in developing transmitters and receivers that could handle light at 1.3 μm, a wavelength in which silica is more transparent (Fig. 16.18). In 1982, the third generation began to appear as researchers developed processing techniques that increased the purity of the silica fibers in the 1.2 to 1.6 μm range. This improvement raised the transmission capacity to hundreds of gigabites.

The development of erbium-doped silica glasses in the late 1980s ushered in a new generation of light wave communications systems with transmission capacities on the order of thousands of gigabit-kilometers per second. The Er ions embedded in the glass amplify the signal as they absorb infrared radiation produced by a laser diode chip at a wavelength of 1.48 or 0.98 μm. The light is absorbed by the Er atoms by pumping them to a higher energy level. When a weakened signal enters the Er-doped fiber, the excited Er atoms transfer their energy to the weakened optical signal, which in turn is regenerated. This was a major breakthrough for several reasons. The first is that it eliminated the need for signal regenerators or repeaters. The repeaters convert the light to an electric current, amplify the current, and transform it back to light. The Er-doped fibers do not interrupt the path of the light as it propagates through the fiber.

16.6
SUMMARY

When electromagnetic radiation proceeds from one medium to another, some of it is reflected, some is absorbed, and some is transmitted.

1. Electronic polarization results in the retardation of EM radiation, which is directly responsible for refraction. The index of refraction n quantifies the degree of bending or retardation. n is directly related to electronic polarization, which in turn is determined by the polarizability of the atoms or ions in the solid. The more polarizable the ions or atoms, the larger the index of refraction.

2. The reflectivity of a surface depends on its index of refraction as well. Insulators with high indices of refraction are more reflective than ones with low n's.
3. The processes by which light is absorbed by solids are several and include:

    *   The photoelectric effect where electrons absorb the incident light and are promoted into the conduction band. For this process to occur the energy of the incident light has to be greater than the band gap of the material. For ceramics this energy is typically in the UV range.
    *   The split in energy of transition-metal ion d and f orbitals, as a result of their interaction with their local environment, gives rise to selective absorption in the visible range. It is this absorption that is responsible for the striking colors exhibited by glasses and gems.
    *   Reduction and radiation can give rise to color centers defined as an impurity or defect onto which an electron or hole is locally bound. The localization of the electron and its promotion to higher energy levels give rise to absorption.

4. In the IR range, absorption is usually associated with ionic polarization in which the ionic lattice as a whole absorbs the radiation and starts vibrating in resonance with the applied field. The most important factor is the strength of the ionic bond; stronger bonds result in higher resonance frequencies.
5. In addition to absorption, light can be scattered in different directions. Scattering is distinguishable from absorption in that the energy of the incident light is not absorbed by the sample but simply scattered in various directions. Scattering is a complex function of the density of scatterers, their relative size with respect to the wavelength of the incident light, and the relative values of the refractive indices of the scatterers and the medium in which they reside. In general scattering is a maximum when the size of the scatterers is of the order of the wavelength of the incident light.
6. Optical communication depends on the ability of very thin silica fibers to transmit light signals over large distances with little attenuation. The glass fiber is designed such that its outside surface has a lower refractive index than its center, which results in the total internal reflection of the optical signal within the fiber. In other words, the light signal is confined within the fiber with little loss and essentially acts as an optical waveguide.

APPENDIX 16A

Coherence

A requirement for refraction, reflection, and diffraction is that the light beams be coherent.

Typically, light from common sources such as the sun or incandescent lamp filaments is incoherent because the emitting atoms in such sources act independently rather than cooperatively. Coherent and incoherent light are treated differently. For completely coherent light, the amplitudes of the waves are added vectorially, and the resultant amplitude is squared to obtain a quantity proportional to the resultant intensity. For completely incoherent light beams, first the amplitudes of the light are squared to obtain a quantity proportional to the intensities and then the intensities are added to obtain the resultant intensity. This procedure is consistent with the fact that for completely independent light sources, the intensity at every point is greater than the intensity due to either of the light sources acting alone.

APPENDIX 16B

Assumptions Made in Deriving Eq. (16.24)

Four assumptions are made in deriving Eq. (16.24):

1. The scattered light has the same frequency as the incident light, which in turn is monochromatic, i.e., confined to one frequency.
2. The scatterers are assumed to be independent; i.e., there are no cooperative effects between scatterers, hence there is no systematic relation between the phases of the scattered beams. To ensure independent scattering, it is estimated that the distance between scatterers should be about 3 times the radius of the particles. This assumption allows for the intensities scattered by the various particles to be simply added without regard to phase. In other words, intensities rather than amplitudes are added, as noted above.
3. Multiple scattering is neglected. In other words, it is assumed that each particle is exposed to the light of the original beam. Scattering where a particle is exposed to light scattered by other particles is termed *multiple scattering* and is neglected. To ensure that this condition is met, the sample has to be thin or

dilute. This implies that if there are N_s scattering centers, the intensity of the scattered beam is simply N_s times that removed by a simple particle.
4. The scattering centers are isotropic and the same size.

PROBLEMS

16.1. Typical values for k_e'' for ceramics range from 0.01 to 0.0001. Estimate the value of σ_{dc} below which it can be safely neglected when one is dealing with optical properties. State all assumptions.
Answer: $\approx 0.01 \; 1/(\Omega \cdot m)$

16.2. (*a*) Refer to Table 16.1. Identify the materials with the highest and lowest values of n. Explain the differences in terms of what you know about polarizabilities of the constituent ions.
(*b*) What differences in the indices of refraction and dispersion would you expect between LiF and PbS? Explain.
(*c*) Which oxide would you expect to have the higher index of refraction, MgO or BaO? Explain.

16.3. If a highly reflective surface is required, should one use a material with a high or low index of refraction? Explain.

16.4. One way to tell whether a glass plate is made of pure silica or soda-lime silica glass is to view it on edge. The silica plate is clear, whereas the window glass is green. Explain.

16.5. It was noted in Experimental Details that it is possible to measure α_a by measuring the transmission of two different samples of different thicknesses that have identical reflectivities. Describe the experimental setup you would use to carry out the measurements, what you would measure, and how you would extract α from the results you obtain. Why is it important that the two samples have the same reflectivity?

16.6. For an ion for which $\varepsilon_{BL} = 10 \; m^{-1} \, \%^{-1}$, answer the following questions:
(*a*) If the concentration of the light-absorbing ions in a solution is tripled, how does the transmission change if the thickness is 1 cm?
Answer: $I_1 = 1.22 I_2$
(*b*) How must the thickness of the sample be altered to keep the transmission invariant through the two solutions?
Answer: $x_1 = 3x_2$

16.7. A 40-cm glass rod has an absorption coefficient α_a of 0.429 m^{-1}. If 50 percent of the light entering one end of the rod is transmitted, determine

(a) The scattering coefficient α_s

Answer: 1.304 m^{-1}

(b) The total coefficient α_{tot}

Answer: 1.733 m^{-1}

16.8. (a) Experimentally in IR absorption, two absorption bands are measured at 3000 and 750 cm^{-1}. One is suspected to be due to a C–H stretching vibration, while the other is suspected to be due to a C–Cl stretching vibration. Assign each absorption band to its appropriate bond. Explain your answer.

(b) Repeat part (a) for C–O and C=O; the absorption bands measured were at 1000 and 1700 cm^{-1}. Which band corresponds to which bond? Explain your answer.

16.9. (a) Weaker bonds and heavier ions are preferable for extended IR transmission. Is this statement true or false? Explain, using examples from Fig. 16.9.

(b) Which of the following three materials will transmit IR radiation to the longest wavelength, MgO, SrO, or BaO? Explain.

16.10. Which of the following materials do you anticipate to be transparent to visible light? Explain.

Material	Diamond	ZnS	CdS	PbTe
Band gap, eV	5.4	3.54	2.42	0.25

16.11. (a) What material would you use for a prism for infrared investigations?

(b) Which material would you use in making lenses for an ultraviolet spectrograph?

16.12. The transmitted light through a 5-mm sample of CdS which has a band gap of 2.4 eV is observed. Under these conditions, what is

(a) The color of the sample?

(b) Cu can dissolve in CdS as an impurity and has an energy level that is normally, in the dark, electron-occupied and lies 1.0 eV above the valence band of CdS. What color changes, if any, would you expect as the Cu concentration increases from 1 to 1000 ppm?

(c) The band gap of CdS increases with temperature according to $E_g = 2.56 - 5.2 \times 10^{-4}$ T. What color changes do you expect in the transmitted light as CdS is heated from 0 to 1000 K?

16.13. Crystals of NaCl show strong absorption of electromagnetic radiation at a wavelength of about 0.6 μm. Assume this is due to the vibration of individual atoms.

(a) Calculate the frequency of vibrations.

Answer: 5×10^{12} Hz

(b) Calculate the potential energy of a sodium ion as a function of distance r from its equilibrium position, assuming the vibration to be simple harmonic.

Answer: $1.89 \times 10^{-4} r^2$ J, where r is in angstroms

16.14. Calculate the ratio of molecules in a typical excited rotational, vibrational, and electronic energy level to that in the lowest energy state at 25 and 1000°C, taking the levels to be 30, 1000, and 40,000 cm^{-1}, respectively, above the lowest energy state.

16.15. The experimental values for absorption energies in electron volts of F centers in alkali halides are listed below. Plot these values as a function of lattice parameter, and develop a qualitative model to explain the results. *Hint:* Think of a particle in a box.

LiCl   3.1 eV   NaCl   2.7 eV   KCl   2.2 eV   RbCl   2.0 eV

16.16. Using sketches, explain why s and p orbitals are unaffected (i.e., do not split) by ligands in octahedral fields.

16.17. Rayleigh scattering is a strong function of particle size. Plot the functional dependence of the scattering coefficient as a function of r for a given wavelength of light and volume fraction of scattering particles.

16.18. Typically, TiO_2 particles are added to latex, a polymeric base with a refractive index of 1.5, to make white paint.

(a) Discuss why TiO_2 is a good candidate for this application.

(b) On the market you find three TiO_2 particle sizes with narrow distributions and an average particle size of 0.2, 2.0, and 20 μm. Which would you use to make white paint, and why?

16.19. A 40-cm glass rod absorbs 15 percent of the light entering at one end. When it is subjected to intense radiation, tiny particles are produced in it that give rise to Rayleigh scattering. After radiation the rod transmits 55 percent of the light. Calculate

(a) The absorption coefficient α_a

Answer: 0.406 m^{-1}

(b) The scattering coefficient α_s

Answer: 1.09 m^{-1}

16.20. The surface of a glass plate is rough on the scale of the incident light wavelength. Use a sketch to show what happens when the beam of light strikes the surface at a glancing incidence. Show what happens when the surface is wet with a liquid of equal refractive index.

16.21. Why do car headlights appear brighter when the road is wet?

16.22. (a) Explain why optical waveguides often have a refractive index gradient.

(b) Show how n can be calculated given knowledge of the critical angle.

16.23. Show that 1 cm$^{-1} = 4.3 \times 10^5$ dB/km.

16.24. (a) In an optical communications network, the ratio of the light intensity at the source to that at the detector is 10^{-6}. What is the loss in decibels in this system?

Answer: −60 dB

(b) The attenuation of ordinary soda-lime silicate glass is about −3000 dB/km. What fraction of the light signal will be lost in 1 meter?

Answer: One-half

16.25. A certain glass containing 500 ppm of Cr^{3+} ions absorbs 10 percent of the incident light in 10 cm. Assume the Cr^{3+} ions are responsible for the absorbance.

(a) What is the absorbance loss in dB/km of the original glass?

Answer: 45.7 dB/km

(b) What must the concentration of Cr^{3+} be so that the absorbance is 10 percent in 100 m?

Answer: 0.5 ppm

(c) Calculate the loss (dB/km) for the 100 m sample.

Answer: 4.6 dB/km

16.26. (a) What is the critical angle for total internal reflection for an optical fiber with a core refractive index of 1.52 and a cladding of 1.46?

Answer: 74°

(*b*) Repeat part (*a*) for a system for which the core refractive index is 1.46 and that for the cladding is 1.46.

16.27. (*a*) A ceramic body containing 0.25 vol % spherical pores transmits 50 percent of the incident light and scatters 50 percent in 1-mm thickness. Estimate the average diameter of the pores. State all necessary assumptions.

Answer: 10.8 μm

(*b*) Calculate the fraction of light transmitted if the average diameter of the pores is 1 μm. What does this result imply about the requirements for obtaining polycrystalline transparent ceramics?

Answer: 0.05 percent

16.28. (*a*) Assuming the constant in Eq. (16.24) is 30, calculate the fraction of light transmitted through 5 cm of a solution with a concentration of 10^{25} m^{-3} of scatterers for which the diameter is 1.2 nm. You can assume the incident light is monochromatic with 0.6 μm wavelength. You can further assume that the relative dielectric constant of the solution at this wavelength is 2.25, while that of the particles is near 1.

Answer: 99.997 percent

(*b*) Repeat part (*a*) for particles with a diameter of 12 nm.

Answer: $\approx 3 \times 10^{-10}$ percent

16.29. (*a*) Why do you think that low-fat milk is more translucent than regular milk? Explain.

(*b*) Why do you think fog headlights are yellow? Explain.

16.30. ZnS has a band gap of 3.64 eV. When doped with Cu^{2+}, it emits radiation at 670 nm. When zinc vacancies are produced by the incorporation of Cl^- ions, the radiation is centered on 440 nm.

(*a*) Write the incorporation reaction that results in the formation of the zinc vacancies.

(*b*) Using a sketch, locate the impurity levels in the band gap in relation to the valence band.

16.31. You are asked to compare the values of elastic modulus, thermal conductivity, and thermal expansion coefficients in the temperature range of 50 to 800°C, of optical-quality polycrystalline MgF_2 from different sources ranging in impurity levels from 300 to 5000 ppm impurity content. Do you expect the results to be almost the same or markedly different for these properties and samples? Explain. List other physical property measurements that you expect to be (*a*) more variable, (*b*) less variable between samples than those measured. Explain.

What property can you think would be (*c*) most variable and (*d*) least variable between samples? Explain. Include in your list magnetic, electrical, thermal, mechanical, and optical properties.

ADDITIONAL READING

1. J. N. Hodgson, *Optical Absorption and Dispersion in Solids*, Chapman & Hall, London, 1976.
2. F. Wooten, *Optical Properties of Solids*, Academic Press, New York, 1972.
3. L. L. Hench and J. K. West, *Principles of Electronic Ceramics*, Wiley-Interscience, New York, 1990.
4. A. J. Moulson and J. H. Herbert, *Electroceramics*, Chapman & Hall, London, 1990.
5. R. P. Feynman, R. B. Leighton, and M. Sands, *The Feynman Lectures on Physics*, vols. 1 and 3, Addison-Wesley, Reading, Massachusetts, 1963.
6. A. Javan, "*The Optical Properties of Materials*," *Scientific American*, 1967.
7. W. D. Kingery, H. K. Bowen, and D. R. Uhlmann, *Introduction to Ceramics*, 2d ed.,Wiley, New York, 1976.
8. J. S. Cook, "Communications by Optical Fibers," *Scientific American*, **229**:28–35, Nov. 1973.
9. M. E. Lines, "The Search for Very Low Loss Fiber-Optics Materials," *Science*, **226**:663, 1984.
10. E. Dusurvire, "Lightwave Communications: The Fifth Generation," *Scientific American*, **266:(1)** 114, 1993.
11. B. E. A. Saleh and M. C. Teich, *Fundamentals of Photonics*, Wiley, New York, 1991.
12. F. A. Jenkins and H. E. White, *Fundamentals of Optics*, 4th ed., McGraw-Hill, New York, 1976.
13. H. C. van de Hulst, *Light Scattering by Small Particles*, Dover, New York, 1981.
14. B. A. Leyland, *Introduction to Laser Physics*, Wiley, New York, 1966.

Index

Physical Constants

Gas Constant	R	8.31467 J/K·mol
Boltzmann's Constant	k	1.381×10^{-23} J/mol K $= 8.62 \times 10^{-5}$ eV/atom K
Plank's Constant	h	$6.625 \times 10^{-34} (J \cdot s)$
Electronic charge	e	1.6×10^{-19} C
Velocity of light	c	2.998×10^8 m/s
Permittivity of free space	ε_0	8.85×10^{-12} J/C²m
Rest mass of electron	m_e	9.11×10^{-31} (kg)
Avogadro's Constant	N_A	6.022×10^{23} particles/mole
Gravitational Acceleration	g	9.81 m²/s
Faraday's Constant	F	$96,487$ C/equivalent

Conversions

Length

$1\,m = 10^{10}\,Å$	$1\,Å = 10^{-10}\,m$
$1\,m = 10^9\,nm$	$1\,nm = 10^{-9}\,m$
$1\,m = 10^6\,\mu$	$1\,\mu = 10^{-6}\,m$
$1\,m = 10^3\,mm$	$1\,mm = 10^{-3}\,m$
$1\,m = 10^2\,cm$	$1\,cm = 10^{-2}\,m$
$1\,mm = 0.0394\,in$	$1\,in = 25.4\,mm$
$1\,cm = 0.394\,in$	$1\,in = 2.54\,cm$
$1\,m = 3.28\,ft$	$1\,ft = 0.3048\,m$

Mass

$1\,Mg = 10^3\,kg$	$1\,kg = 10^{-3}\,Mg$
$1\,kg = 10^3\,g$	$1\,g = 10^{-3}\,kg$
$1\,kg = 2.205\,lb_m$	$1\,lb_m = 0.4536\,kg$
$1\,g = 2.205 \times 10^{-3}\,lb_m$	$1\,lb_m = 453.6\,g$

Area

$1\,m^2 = 10^4\,cm^2$	$1\,cm^2 = 10^{-4}\,m^2$
$1\,mm^2 = 10^{-2}\,cm^2$	$1\,cm^2 = 10^2\,mm^2$
$1\,m^2 = 10.76\,ft^2$	$1\,ft^2 = 0.093\,m^2$
$1\,cm^2 = 0.1550\,in^2$	$1\,in^2 = 6.452\,cm^2$

Volume

$1\,m^3 = 10^6\,cm^3$	$1\,cm^3 = 10^{-6}\,m^3$
$1\,mm^3 = 10^{-3}\,cm^3$	$1\,cm^3 = 10^3\,mm^3$
$1\,m^3 = 35.32\,ft^3$	$1\,ft^3 = 0.0283\,m^3$
$1\,cm^3 = 0.0610\,in^3$	$1\,in^3 = 16.39\,cm^3$
$1\,L = 10^3\,cm^3$	$1\,cm^3 = 10^{-3}\,L$
$1\,gal(US) = 3.785\,L$	$1\,L = 0.264\,gal$

Density

$1\,kg/m^3 = 10^{-3}\,g/cm^3$	$1\,g/cm^3 = 10^3\,kg/m^3$
$1\,Mg/m^3 = 1\,g/m^3$	$1\,g/m^3 = 1\,Mg/m^3$
$1\,kg/cm^3 = 0.0624\,lb_m/ft^3$	$1\,lb_m/ft^3 = 16.02\,kg/m^3$
$1\,g/cm^3 = 62.4\,lb_m/ft^3$	$1\,lb_m/ft^3 = 1.602 \times 10^{-2}\,g/m^3$
$1\,g/cm^3 = 0.0361\,lb_m/in^3$	$1\,lb_m/in^3 = 27.7\,g/cm^3$

Force

$1\,N = CV/m = J/m$
$1\,N = 10^5$ dynes
$1\,N = 0.2248\,lb_f$
$1\,dyne = 10^{-5}\,N$
$1\,lb_f = 4.448\,N$

Energy

$1\,J = 6.24 \times 10^{18}$ eV	$1\,J = 0.239$ cal
$1\,eV = 3.83 \times 10^{-26}$ cal	$1\,Btu = 252.0$ cal
$1\,J = 1\,N \cdot m = 1\,W \cdot s$	

$1\,eV = 1.602 \times 10^{11}$ J	$1\,J = 10^{-7}$ ergs
$1\,cal = 2.61 \times 10^{10}$ eV	$1\,cal = 4.184$ J
$1\,eV/particle = 96,500$ J/mole	

Photon energy: $E = 1.24$ eV at $\lambda = 1\,\mu m$

Thermal energy (@300 K) $kT = 0.0258$ eV